Contents

The National Garden Scheme

A company limited by guarantee. Registered in England & Wales. Charity No. 1112664. Company N

Registered & Head Office: Hatchlands Park, East Clandon, Guildford, Surrey, GU4 7RT.
01483 211535 www.ngs.org.uk

Published by Constable, an imprint of Little, Brown Book Group, Carmelite House, 50 V
London EC4Y 0DZ. An Hachette UK Company www.hachette.co.uk www.littlebrown.c

Foreword

We opened our garden in support of the National Garden Scheme for more than 20 years, so I know the joy of sharing the garden I love, with visitors. It gives such pleasure to young and old whilst raising money for this exceptional charity. The great outdoors is the place to breathe and live life and gives so much pleasure to many. At the same time, for 95 years it has proved a superbly effective form of fundraising, as shown by the results achieved in 2022 when more than £3.1 million (all raised at gardens) was donated to the National Garden Scheme's nursing and health beneficiaries, taking the total donated to more than £67 million.

In 2022 we were almost back to the same level of fundraising as pre Covid-19. The year also witnessed a return to the glorious variety of gardens, from country houses to groups of allotments, from tiny town patches and group village gardens to hospice and community gardens. This variety is increasingly the hallmark of the National Garden Scheme and demonstrates a diversity that we all applaud.

Some of the gardens have opened for decades, none more distinguished than Sandringham, the Royal Family's Norfolk home. Her Majesty Queen Elizabeth II gave her lifelong support for the National Garden Scheme. Her Majesty opened the gardens of three of her homes; Sandringham, Frogmore House in Windsor Great Park and Windlesham Moor in Surrey, the first country home of Princess Elizabeth after her marriage to The Duke of Edinburgh. Our charity is indeed privileged to have had such loyal royal support.

2023 will be a glorious year for the National Garden Scheme and I would like to say thank you in advance to all of you: visitors; garden owners; and volunteers who will make this happen.

Mary Berry

Dame Mary Berry

President

Who's who

Chairman's message

I must start by thanking our President, Dame Mary Berry, for her warm and uplifting Foreword to this year's Garden Visitor's Handbook. She expresses the enjoyment that we all get from opening our gardens and welcoming visitors, something that enables us to start every year with a sense of expectation and the ambition to nurture our gardens to new heights.

But I am also aware of the hard work and commitment that goes into opening your garden and so I want to echo Dame Mary's thanks to the thousands of you whose gardens are listed in this year's book. If you are not currently someone who opens your garden, I hope that browsing the pages and then planning some visits might inspire you to consider doing it. We are always looking for new gardens and on page 17 you will find an article giving you more details about what is involved in opening to support the National Garden Scheme.

I know from experience that the rewards of opening your garden are uniquely varied. They are a combination of pride and fulfilment, the enjoyment of welcoming strangers and sharing a common pleasure, and contributing to the National Garden Scheme's remarkable fundraising efforts and its support for some of our best-loved charities. You will find the details of our 2022 donations – all from funds raised at gardens during the year – on page 5.

One of the National Garden Scheme's assets that deserves to be better known, is the large group of about 1,000 gardens that open on a by arrangement basis. This offers you the flexibility of choosing a date that suits you and booking a private visit that often produces a memorably personal experience. You will find more about our by arrangement gardens on page 24.

The National Garden Scheme has always been about local communities so I am delighted that we have expanded our donations portfolio to include our Community Garden Grants scheme. This has now supported hundreds of wonderful projects and you can read more about it on page 22.

Rupert Tyler

Rupert Tyler

Discovering the nation's best gardens

The National Garden Scheme gives visitors unique access to exceptional private gardens and raises impressive amounts of money for nursing and health charities through admissions, teas and cake.

Originally established to raise funds for district nurses, we are now the most significant charitable funder of nursing in the UK and our beneficiaries include Macmillan Cancer Support, Marie Curie, Hospice UK, Parkinson's UK, The Queen's Nursing Institute and Carers Trust. Year by year, the cumulative impact of our donations grows incrementally, helping our beneficiaries make their own important contribution to the nation's health and care and strengthening the partnerships that we have with them all.

In 2022, despite the long summer drought, garden visiting returned to near normal levels following the pandemic and in November, thanks to the generosity of our garden owners, volunteers and supporters we were able to make our largest ever annual donation to our beneficiary charities of £3.11 million.

The National Garden Scheme doesn't just open beautiful gardens for charity – we are passionate about the physical and mental health benefits of gardens too. We fund projects which promote gardens and gardening as therapy, support charities doing amazing work in gardens and health and give grants for community gardening projects. Our funding also supports the training of gardeners and offers respite to horticultural workers who have fallen on difficult times.

With over 3,500 gardens opening across England, Wales, Northern Ireland and the Channel Islands in 2023 there are plenty for you to explore. We hope you enjoy revisiting old favourites and discovering new horticultural gems. Most gardens open on one or more specific dates, but many also open by arrangement, offering the facility for you to book a private visit.

You can also nominate your favourite National Garden Scheme Garden for the *Nation's Favourite Gardens* competition run in association with *The English Garden* magazine. Visit www.theenglishgarden.co.uk/ngs to find out more.

YOUR GARDEN VISITS HELP CHANGE LIVES

In 2022 the National Garden Scheme donated £3.11 million to our beneficiaries, providing critical support to nursing and health charities.

Marie Curie
£450,000

Macmillan Cancer Support
£450,000

Hospice UK
£450,000

Carers Trust
£350,000

The Queen's Nursing Institute
£400,000

Parkinson's UK
£350,000

SUPPORT FOR GARDENERS

- English Heritage £125,000
- Perennial £100,000
- Working for Gardeners Association £65,000
- National Botanic Garden, Wales £20,000
- Professional Gardeners' Trust £20,000
- Garden Museum £10,000

GARDENS AND HEALTH CHARITIES

- Maggie's £100,000
- Horatio's Garden £90,000
- ABF The Soldiers' Charity £80,000
- Greenfingers Charity £50,000

Thank you!

To find out more visit **ngs.org.uk/beneficiaries**

DONATE

Charity number: 1112664

OUT OF THE ORDINARY

Tips on using your Handbook

This book lists all the gardens opening for the National Garden Scheme between January 2023 and early 2024. It is arranged alphabetically in county sections within each nation, each including a map, calendar of opening dates and details of each garden with illustrations of some.

Symbols explained

NEW Gardens opening for the first time this year or re-opening after a long break.

◆ Garden also opens on non-National Garden Scheme days. (Gardens which carry this symbol contribute to the National Garden Scheme either by opening on a specific day(s) and/or by giving a guaranteed contribution.)

Wheelchair access to at least the main features of the garden.

Dogs on short leads welcome.

Plants usually for sale.

Card payments accepted.

NPC Plant Heritage National Plant Collection.

Gardens that offer accommodation.

Refreshments are available, normally at a charge.

Picnics welcome.

D Garden designed by a Fellow, Member, Pre-registered Member, or Student of The Society of Garden Designers.

Garden accessible to coaches. Coach sizes vary so please contact the garden owner or County Organiser in advance to check details.

Group Visits Group Organisers may contact the County Organiser or a garden owner direct to organise a group visit to a particular county or garden. Otherwise contact the National Garden Scheme office on 01483 211535.

Funds raised In most cases all funds raised at our open gardens comes to the National Garden Scheme. However, there are some instances where income from teas or a percentage of admissions is given to another charity.

Children must be accompanied by an adult.

Photography is at the discretion of the garden owner; please check first. Photographs must not be used for sale or reproduction without prior permission of the owner.

Toilet facilities are not guaranteed at all gardens.

If you cannot find the information you require from a garden owner or County Organiser, call the National Garden Scheme office on 01483 211535.

There are many ways to book and enjoy a visit to our gardens...

Cashless payments and online booking

Many gardens accept card payments. Look for the cashless symbol in the garden listing on our website.

Pre-booking available

For over 3,000 gardens you can purchase tickets online in advance, as well as purchasing them on the gate on the day. Look for the Pre-booking available symbol in the garden listing.

Pre-booking essential

The **Pre-booking essential** symbol in the garden listing means you need to pre-book your visit. You will also see either:

Book now

A book now button which allows you to book your tickets through the National Garden Scheme

OR

Owner Info

If there is no book now button, you will need to book your tickets direct with the garden owner. Click on the Owner Info tab to find their contact details

Special Events

Some gardens offer Special Events which may include a tour, an exclusive visit or a meal. These must be pre-booked, have restricted availability and sell out quickly. To book your ticket and sign up to receive the latest updates visit **ngs.org.uk/special-garden-events**.

Open by Arrangement

Over 1,100 gardens invite you to visit By Arrangement – so that you can visit on a date to suit you. Contact the garden owner to discuss availability and book your visit, for more details see page 24.

The impact of our donations to nursing

Founded in 1927 to support community nurses, the National Garden Scheme has raised over £67 million at our garden gates for our beneficiary charities. In 2022, the lion's share of our £3.11 million donation went to some of the UK's best-loved nursing and health charities including Carers Trust, Hospice UK, Macmillan Cancer Support, Marie Curie, Parkinson's UK and The Queen's Nursing Institute.

Since the pandemic, these indispensable charities have continued to provide vital services to patients and communities, and support for the NHS in the face of the wider healthcare challenges and cost of living crisis. The long-term nature of funding from the National Garden Scheme allows these charities to continue to plan and provide critical specialist community nursing services, end-of-life care and respite for families and carers.

Below: Parkinson's nurse with patient

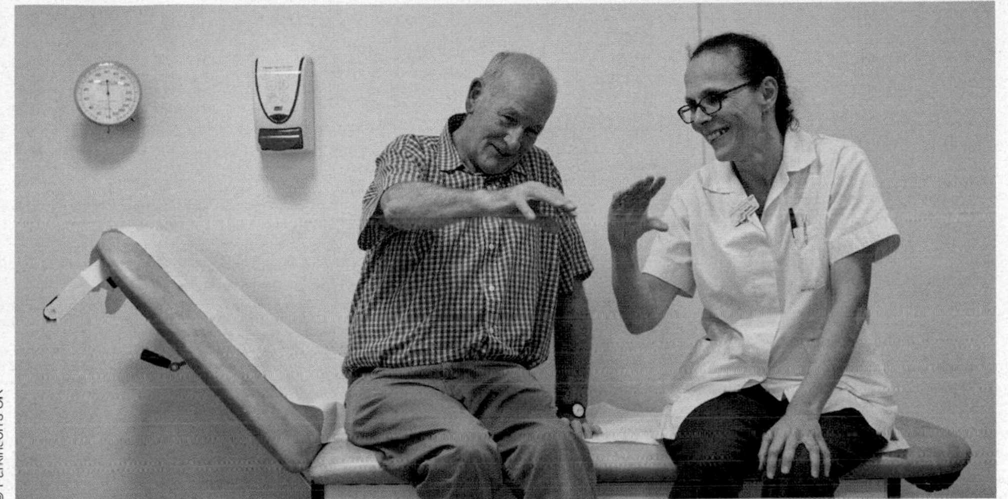

© Parkinson's UK

© Marie Curie

In 2022, we celebrated the tenth anniversary of our support for Parkinson's UK. In that time, we have donated over £1.7 million to bring better care, treatments and quality of life for those living with Parkinson's, and it is now estimated that around 7,000 patients currently benefit from the support of the seventeen nursing posts that we have funded.

Our support in 2022 has also helped:

- More than 1,900 people at the Y Bwthyn NGS Macmillan Specialist Palliative Care Unit in Glamorgan, to receive treatment.
- Provide the equivalent of 2,652 night shifts by Marie Curie nurses.
- Contribute to more than 300,000 adults and children – along with their loved ones and carers – receiving support across the hospice network.
- Fund the Queen's Nurse Programme, a national network of 2,000 highly skilled and qualified nurses working in the community in England, Wales, Northern Ireland, the Channel Islands and the Isle of Man.

National Garden Scheme Chairman Rupert Tyler said: "Despite the worst of the pandemic having passed, our beneficiaries continue to support those in often dire need and who have now been confronted with the challenge of the new cost of living crisis. This continues to place unbearable pressure on many aspects of their work, and we are delighted to be able to continue our support in such a meaningful way."

Above: Marie Curie nurse for 19 years, Caty Hollis works at Bradford Hospice
Opposite: QNI nurse Sandra working in the homeless health service in South London

Gardens and Health: people and communities

Our Gardens and Health programme raises awareness of the physical and mental health benefits of gardens and gardening for everyone. We work to promote gardens and health throughout the year, linking service users from our beneficiaries with free garden visits and funding Gardens and Health projects. We also fund a number of specific Gardens and Health charities which create gardens within health care settings or provide horticultural therapy or training to those in need.

We have a long-term commitment to Horatio's Garden, which builds beautiful outdoor spaces for garden therapy and helps promote wellbeing for patients, visitors and staff at NHS spinal units. We have been providing funding since 2015, and in that time, we have helped support the creation of four new gardens.

We are also committed to funding gardens at Maggie's centres. These uplifting places with professional staff on hand to offer the support people with cancer and their families need, combine iconic architecture with gorgeous gardens. Maggie's Chief Executive, Dame Laura Lee says: "Gardens are so important to the work that Maggie's centres offer. Our doors are open to everyone with cancer and our gardens are there to uplift, to calm, to be a place of reflection. Beauty has the capacity to move the spirit and change the soul and to give a gift back to people that life is important in the here and now.

"We're so grateful to the National Garden Scheme for their continued support and funding of our green spaces which offer that extra special gift at a very distressing time in people's lives."

All of the gardens we support at Horatio's Garden and Maggie's have been enjoyed by thousands of service users during 2022.

The National Garden Scheme also funds horticultural therapy for serving soldiers and veterans through ABF The Soldiers' Charity. In 2022 our funding provided life-changing education, training and employment opportunities in the outdoor sector, including horticulture and gardening, for ten individuals and reached over 300 more through charity partners providing horticultural related activities and health projects for larger groups.

Below: Tomatoes and flowers grown at Maggie's are all part of the patient experience

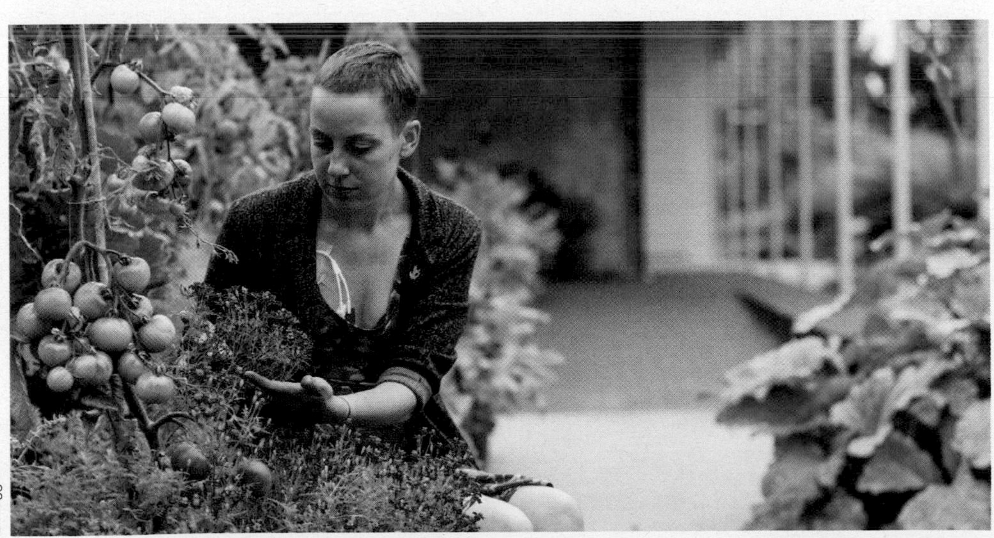

© Maggie's

A Big Garden Garden Get-together

Once again, our annual Great British Garden Party fundraiser, run in partnership with event sponsor Talking Tables, saw supporters up and down the country, hosting events and parties in support of some of the UK's best-loved nursing and health charities. From picnics in the park to afternoon tea in the garden, plant sales and raffles to treasure hunts and book swaps, our supporters were hugely creative, and between them raised an amazing £30,000.

"I didn't know much about the National Garden Scheme but I spotted a Facebook post promoting their fundraiser, the Great British Garden Party and immediately thought 'I can do that!' So many of us have been helped at some point in our lives by the charities which the National Garden Scheme support – in my case Macmillan had provided my family and I with invaluable support. To share tea and cake and raise money for these fantastic causes – what is not to love!" said Wendy who organised a community event in her small hamlet in Northumberland.

Above: Milly hosted an afternoon tea for a group of friends

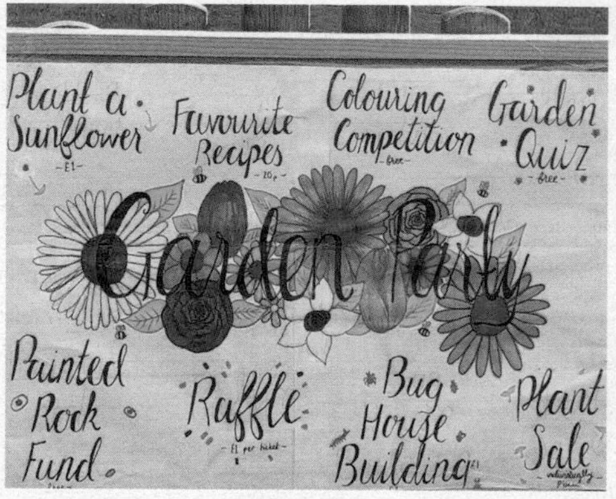

Above: A myriad of fundraising ideas at Wendy's community event

You don't need to have lots of money or space to get involved. If your space is tiny, you can invite two people at a time for tea and cake and still help raise money for a good cause. And if you are doing something bigger you don't have to do it alone. It's amazing how many people will get involved if you only ask, and sharing ideas and skillsets make it all so much fun.

You can do it too! Sign up to host your own garden fundraiser in 2023: ngs.org.uk/gardenparty

Left: Talking Tables held a staff picnic in the park to raise funds

Opening your garden can bring huge rewards

Thanks to the generosity of garden owners, volunteers and visitors we have donated over £67 million to nursing and health charities since 1927, and in 2022 we donated £3.11 million. We can only donate such life-changing amounts by having wonderful gardens to share.

We are constantly seeking out new and beautiful gardens to open, whether that be rolling country acres, cottage gardens, town gardens, wildlife-friendly spaces, community allotments or plantsman's havens. They all offer a unique and very personal experience to our garden-loving visitors and help raise money for great causes.

Opening your garden to the public may seem daunting at first, but if you're passionate about your garden and your friends and family tell you how lovely it is, then it's very likely other people will want to visit your garden too. Some of our garden owners have shared with us just why they find opening their garden such a rewarding experience.

Paddock Allotments, London

Thirty one allotments – representing over 2,400 plot holders – opened for the National Garden Scheme in 2022. These amazing communities of enthusiastic growers offered the visitor so much; from tips on growing coriander, artichokes, asparagus and blueberries to dahlias and damsons to a deep sense of solace and wellbeing. "Opening the allotments for the National Garden Scheme is a wonderful day. Plot holders love to be asked about what they're growing, and visitors really enjoy the tea and cakes and ploughman's lunches," says plot holder Cynthia

Below: National Garden Scheme open day at Paddock Allotments. See page 373 for details of the 2023 opening

Above: Garden owners Stewart and Louise Woskett. See page 203 for details of their 2023 open dates

© Joe Wainwright

Above: John tries to have something flowering every month of the year and has seen wildlife flock to his modest size plot. Open in 2023, see page 471 for details

Laurel cottage – our dream garden

In the first few months of purchasing their new home in Bishop's Stortford, which had fallen into disrepair, owners Stewart and Louise began to wonder whether they'd made the right decision. But they had a vision for the house and for the garden and were determined to see it through. Today, Laurel Cottage and its dream garden are still a work in progress. "It still feels young to me," admits Stewart who is lucky to visit lots of gardens as part of his job, borrowing ideas and being given cuttings and plants from clients who know about the project. "Each plant is special and has personal meaning, so the garden is being created by lots of people, lots of generosity and kindness."

Approaching the National Garden Scheme to see whether the garden was up to scratch was a nervous moment for Stewart but he need not have worried. "I can't believe how warmly we were welcomed into the National Garden Scheme family or the level of support we've received," says Stewart. "And although heading towards our first opening was really nerve-wracking, it's one of the most exciting things we've ever done. It's always been an ambition to open for the National Garden Scheme and it didn't disappoint."

153 Willoughbridge: A festival of flowers and wildlife

Only the postman uses the front door at John Butcher's rural home near Market Drayton, Shropshire. Plants have blocked access to all but the letterbox. "Visitors have to come down the side of the house," he says. "It's a fragrant jungle down there, buzzing with insect life."

John's perennial borders are packed with plants, and this means that they provide ample shelter for amphibians and small mammals, too. His bug hotels attract solitary bees and wasps, and birds flock to his feeders and nest in the garden. There's always the chance of finding something new in the garden: "One year we had a hummingbird hawk moth for the first time and while filling a gap in the bin store's green roof, I discovered a bumblebee nest. It makes gardening like this feel so worthwhile."

John has been buoyed by the positive response to his garden at his open garden days. "Visitors love all the ideas to promote biodiversity. It goes to show what's possible in even the smallest garden."

Open your garden with the National Garden Scheme

You'll join a community of individuals passionate about gardens and help raise money for many of the UK's most loved nursing and health charities. Whatever its size or style, if your garden has quality, character and interest we'd love to hear from you.

Call us on 01483 211535 or email hello@ngs.org.uk

Highfield Farm – a garden defined by its plants

Highfield farm in Monmouthshire has over 1400 cultivars, many of them rarities, densely planted over three acres to generate an exuberant display across the seasons. "On open garden days, when I am asked about why we open, I explain my admiration for the Macmillan nurses who looked after my mother, and this often develops into moving conversations with our visitors about their personal experiences and their gratitude to the nurses for their support. This not only provides a meaningful context for the garden opening and the visitors' donations, but also a real sense of fulfilment for our participation in the National Garden Scheme," says garden owner Jenny Lloyd.

If you would like to get involved and open your garden with the National Garden Scheme, call us on 01483 211535 or email us at hello@ngs.org.uk.

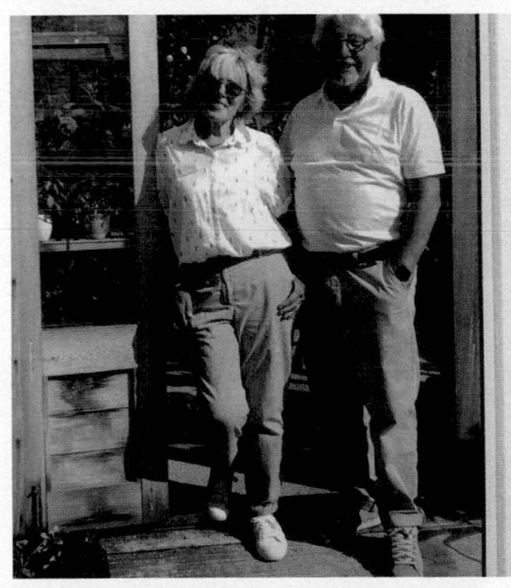

Above: Garden owners Jenny and Roger Lloyd. See page 680 for details of their 2023 openings

Our Community Garden Grants

In addition to our annual donations to nursing and health charities, the National Garden Scheme awards grants to help community gardening projects, celebrating with the presentation of a special plaque on completion of the work. Postponed due to Covid restrictions, the grants resumed in 2022 funding 76 projects with a total of £187,000.

"We want to see this funding going to the heart of community projects, helping to invigorate the people they support and introducing new audiences to the huge benefits that gardens and gardening bring to their health, wellbeing and to the surrounding environment."
Danny Clarke, National Garden Scheme Ambassador, gardener and TV presenter

Above: Families enjoying the allotments, learning and wildlife at the CRC project in Nottingham. © Chronicle and Echo

Below: Jubilee Corner Extension, Par Bay Community Trust, Cornwall

Projects from Northumberland to Cornwall across 33 counties benefitted from a Community Garden Grant in 2022, including:

The Jubilee Corner Extension in Cornwall that extends an existing community garden to serve the community of St Blazey, Par and Tywardreath. Sonia Clyne, who had the initial vision for the project, says: "It's wonderful to see the transformative effects the garden has on people, relieving anxiety, building confidence and self-esteem, and bringing such joy."

St Michael's Hospice, in Herefordshire received a grant to help improve access to the garden. "The garden provides a place that everyone can visit to relax, spend time with family and friends and simply enjoy the peace and quiet, enhancing both physical and mental health and wellbeing. The pathways have been widened and the bends softened, and new gravel and resin laid, all to allow for easier and safer movement for wheelchair users and their families," says Tony Larkin, Grants and Trusts Executive at the hospice.

Home-Start Slough received a grant for a women's allotment project which helps empower women through building up friendships and connecting with nature to improve their emotional and physical wellbeing. Gardening has been proven to boost mental and physical health and when a lack of local perinatal support for women living with anxiety and low mood was identified in Slough, it led a midwifery team to create an allotment project to encourage pregnant and new mums to grow and cultivate vegetables.

One woman referred to the project when she was pregnant said: "I love spending time outdoors but I've no nice space where I live. Going to the allotment group with other pregnant women became a sanctuary for me. I love planting and it felt so good to be outside in the fresh air. Everything I put my hands on bloomed. Mentally, emotionally, physically, and socially – the allotment group changed my life."

Grant applications open each autumn. Find out more at: https://ngs.org.uk/who-we-are/community-garden-grants/

Gardens open by arrangement – a personal visit for you and your friends

The National Garden Scheme opens over 3,500 private gardens to the public each year. Most visitors are familiar with the gardens that open their gates to everyone on set days, but in 2023, 718 of them also open By Arrangement.

The gorgeous Llanllyr in Lampeter, Ceredigion (pictured opposite) for example, is open to groups By Arrangement from April – September outside of its published public opening day of 18th June. A further 293 open only By Arrangement in 2023 like Linden in County Down and Dip on the Hill in Suffolk (pictured below).

These wonderful, often hidden gems are really worth exploring, especially if you are keen to return to a garden you love, perhaps missed a public open day or simply want to explore somewhere completely unchartered. By Arrangement gardens cater for different group sizes from as few as one to 20+, it all depends on the size and accessibility of the garden, some have fixed prices for entry others are agreed in discussion with the garden owner.

Below: Dip-on-the-Hill, Suffolk (See page 511 for more information)

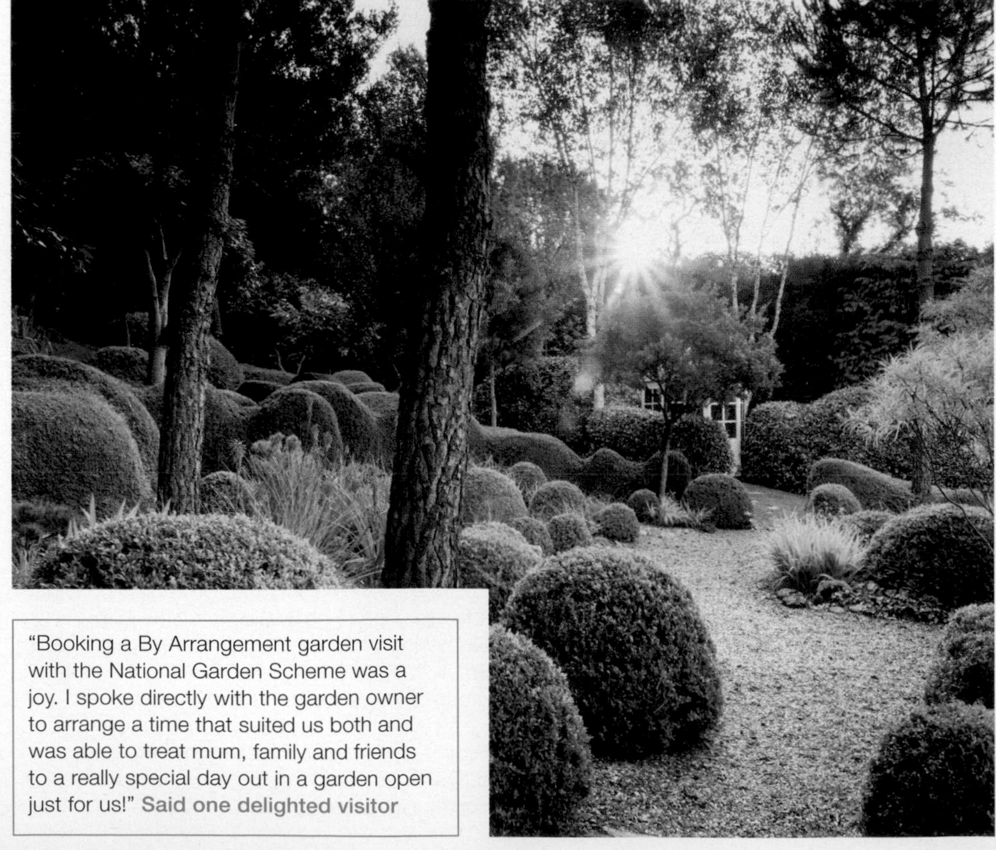

"Booking a By Arrangement garden visit with the National Garden Scheme was a joy. I spoke directly with the garden owner to arrange a time that suited us both and was able to treat mum, family and friends to a really special day out in a garden open just for us!" **Said one delighted visitor**

Below: Llanllyr in Lampeter, Ceredigion (See page 664 for more information)

© Helen Harrison

Your questions answered...

What is a By Arrangement garden?
A garden open By Arrangement accepts visitors to their garden for a more personalised visit, on a pre-agreed date. Usually, the garden will advertise a date range when they are willing to open so that visitors can decide on the most suitable option for them. This can often be a wide date range, for example March-October, so there are plenty of opportunities to find a date to suit you and your group or family and friends.

What are the benefits of booking a visit in advance?
Booking a By Arrangement garden gives you direct contact with the garden owner, allowing you to liaise on everything from convenient times and dates, to what you'll see and do once you're there.

Why choose By Arrangement?
Did you miss an open day or perhaps you would like to bring your family back for a special occasion? Contact any garden offering By Arrangement visits to discuss options.

Booking your By Arrangement visit
All our gardens opening in 2023 are available to view online at https://findagarden.ngs.org.uk/ From the garden information page you will find the garden owner's contact details. They are waiting to hear from you and will be delighted to help you create the perfect visit.

The garden owner will confirm all the details and whether they require a deposit/full payment ahead of the visit.

As with all garden visits organised through the National Garden Scheme, By Arrangement visits support our nursing and health beneficiaries.

BEDFORDSHIRE

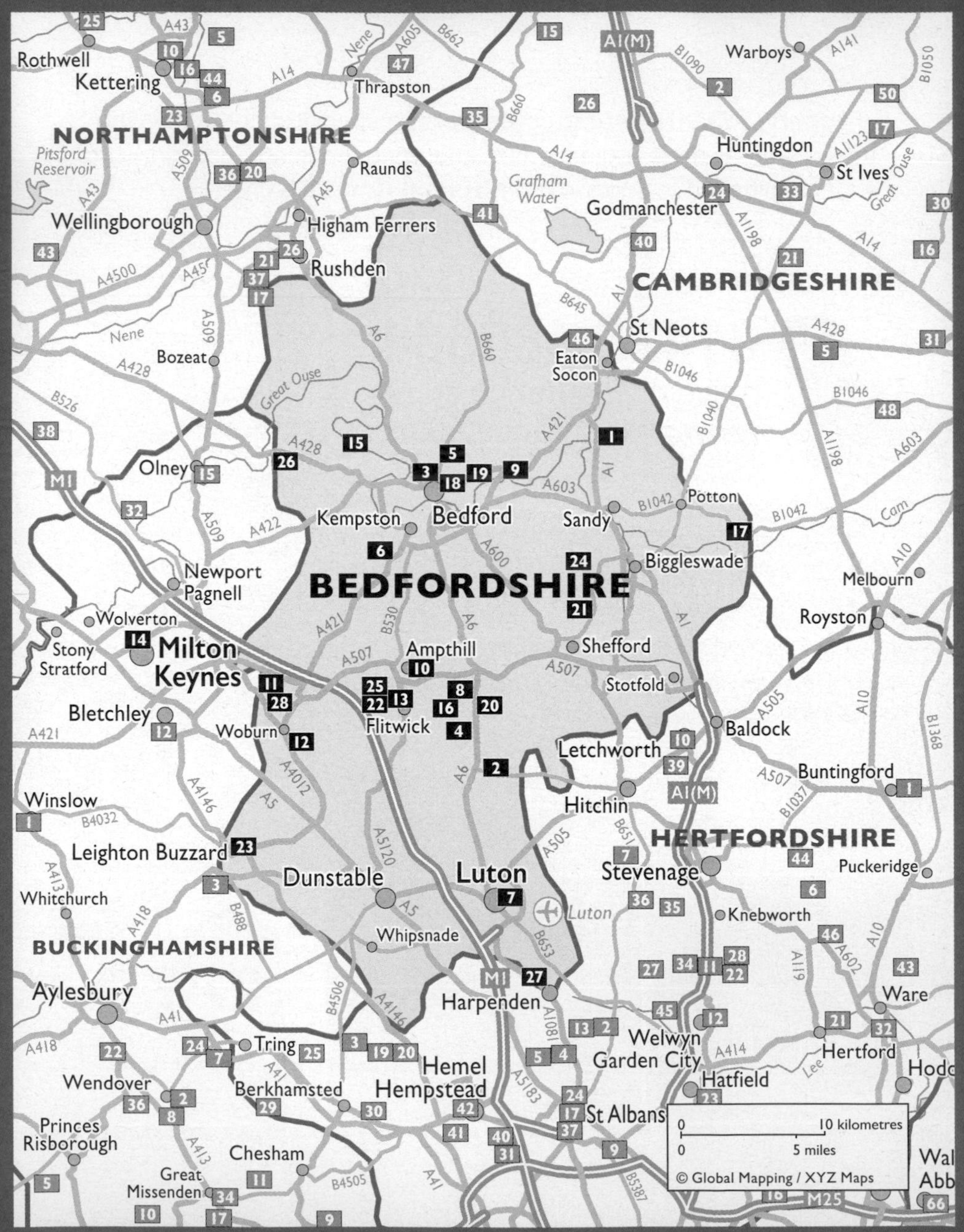

Dear Visitor, We are delighted to welcome you to the National Garden Scheme in Bedfordshire. Your visits make a huge difference, and thanks to your generosity and that of our garden owners and volunteers, we raised £44K in 2022 for our nursing and health, and gardening charities.

This year, our open garden season begins on Sunday 22nd January at 127 Stoke Road which houses one of the National Collections of Snowdrops. You can enjoy beautiful planting and inspirational designs at our gardens from early spring to late autumn. Our garden owners are always delighted to share their passion with you. There are themed gardens such as the serene Japanese garden at 22 Elmsdale Road and the meandering Texan dry riverbed garden at Bedford Heights. The walled garden at Howbury Hall is a working garden run as a cut flower business and 80 West Hill is a winner of 2 National Awards for Design and Construction. A duo of 18th century gardens designed by Capability Brown – Southill Park and The Walled Garden at Luton Hoo are also opening for the National Garden Scheme.

Several of our gardens are also open by arrangement throughout the year for group visits. As the season draws to a close, we will conclude the year with an evening of lights at Townsend Farmhouse and autumn colours at King's Arms Garden and at the Swiss Garden.

Thank you for your support to the National Garden Scheme, and we look forward to welcoming you to our gardens in Bedfordshire.

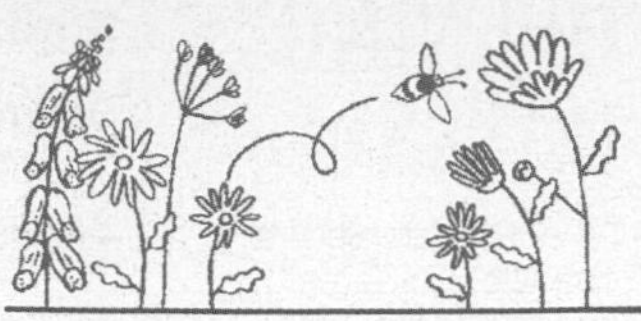

Volunteers

County Organiser
Indi Jackson
01525 713798
indi.jackson@ngs.org.uk

County Treasurer
Colin Davies
01525 712721
colin.davies@ngs.org.uk

Press Officer
Lucy Debenham
lucy.debenham@ngs.org.uk

Facebook & Twitter
Lucy Debenham
(as above)

Booklet Co-ordinators
Indi Jackson
(as above)

Alex Ballance
alexballance@yahoo.co.uk

Photography
Venetia Barrington
venetiajanesgarden@gmail.com

Talks Co-ordinator
Indi Jackson
(as above)

Assistant County Organisers
Geoff & Davina Barrett
geoffanddean@gmail.com

Alex Ballance
(as above)

@bedfordshire.ngs
@NGSBeds
@ngsbedfordshire

Left: The Manor House, Stevington

OPENING DATES

All entries subject to change. For latest information check **www.ngs.org.uk**

Map locator numbers are shown to the right of each garden name.

January

Sunday 22nd
127 Stoke Road 23

February

Snowdrop Festival

Sunday 26th
◆ King's Arms Garden 10

March

Thursday 2nd
The Swiss Garden 24

Sunday 5th
127 Stoke Road 23

Sunday 26th
NEW 204 Hart Lane 7

April

Saturday 22nd
22 Elmsdale Road 6

Sunday 23rd
22 Elmsdale Road 6

Sunday 30th
Townsend Farmhouse 25

May

Monday 1st
Townsend Farmhouse 25

Sunday 7th
The Old Rectory, Wrestlingworth 17

Sunday 14th
NEW 204 Hart Lane 7
Lake End House 11

Saturday 20th
Church Farm 4

Sunday 21st
Church Farm 4
◆ The Manor House, Stevington 15
NEW 80 West Hill 28

Sunday 28th
The Old Rectory, Wrestlingworth 17
Steppingley Village Gardens 22

Monday 29th
Steppingley Village Gardens 22

June

Saturday 3rd
NEW Mill Lane Gardens 16

Sunday 4th
Barton le Clay Gardens 2
NEW Mill Lane Gardens 16
Southill Park 21

Sunday 11th
Lake End House 11

Sunday 18th
Turvey Village Gardens 26

Saturday 24th
Hollington Farm 8

Sunday 25th
Hollington Farm 8

July

Sunday 9th
Lindy Lea 13

Saturday 15th
Bedford Heights 3

Sunday 16th
Lake End House 11

Sunday 23rd
NEW 204 Hart Lane 7

August

Friday 4th
◆ The Walled Garden 27

Saturday 12th
4 St Andrews Road 18

Sunday 13th
Lake End House 11

Sunday 20th
NEW Silsoe Village Gardens 20

Saturday 26th
4 St Andrews Road 18
1a St Augustine's Road 19

Monday 28th
15 Douglas Road 5

September

Sunday 10th
Howbury Hall Garden 9

Saturday 16th
40 Leighton Street 12

Sunday 17th
Lake End House 11

Sunday 24th
NEW 204 Hart Lane 7

October

Saturday 28th
Townsend Farmhouse 25

Sunday 29th
◆ King's Arms Garden 10

November

Thursday 2nd
The Swiss Garden 24

By Arrangement

Arrange a personalised garden visit with your club, or group of friends, on a date to suit you. See individual garden entries for full details.

10 Alder Wynd, Silsoe Village Gardens 20
1c Bakers Lane 1
Church Farm 4
Lake End House 11
1 Linton Close 14
The Old Rectory, Wrestlingworth 17
The Round Cottage, Silsoe Village Gardens 20
4 St Andrews Road 18
1a St Augustine's Road 19

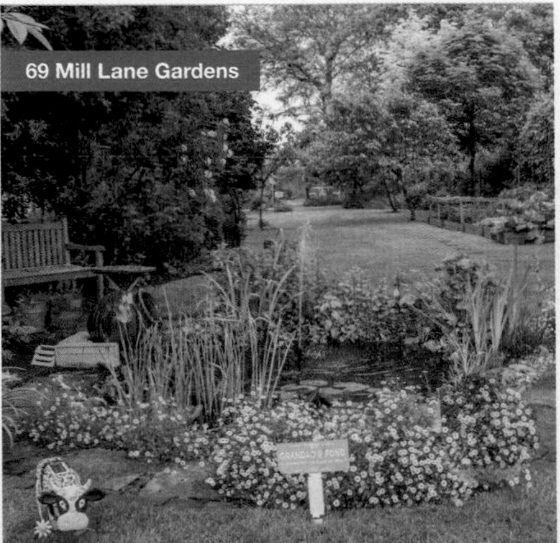
69 Mill Lane Gardens

In 2022 the National Garden Scheme donated £3.11 million to our nursing and health beneficiaries

THE GARDENS

1 1C BAKERS LANE

Tempsford, Sandy, SG19 2BJ. Juliet & David Pennington, 01767 640482, juliet.pennington01@gmail.com. *1m N of Sandy on A1 & Station Rd is on the E side. Bakers Ln is 300 yds down Station Rd on L.* **Visits by arrangement Feb to Oct for groups of up to 20. Adm £6, chd free.**

This is a garden for all seasons with interesting and unusual plants. It comes to life with a winter border planted with colourful cornus and evergreen shrubs, underplanted here with snowdrops and hellebores. Gravel areas are carpeted with miniature cyclamen. It develops through the seasons culminating in early autumn with a spectacular show of sun-loving herbaceous plants. Plant sale: exotic and unusual plants. No refreshments but visitors are welcome to bring a picnic to enjoy in the garden. Sorry no dogs.

GROUP OPENING

2 BARTON LE CLAY GARDENS

Manor Road, Barton-Le-Clay, Bedford, MK45 4NR. *Between Luton and Bedford. Old A6 through Barton-le-Clay Village, Manor Rd is off Bedford Rd. Parking in paddock.* **Sun 4 June (2-5). Combined adm £6, chd free. Light refreshments.**

THE MANOR HOUSE
David Pilcher.

WAYSIDE COTTAGE
Nigel Barrett.

2 gardens in the attractive village of Barton-le-Clay surrounded by beautiful Bedfordshire countryside. The gardens are beautifully landscaped with picturesque stream and bridge over a natural river. At the Manor House, colourful stream side planting inc an abundance of arum lilies and a sunken garden with lily pond. A well-stocked pond with a fountain and waterfalls and a variety of outbuildings nestled within the old walled garden create a tranquil scene at Wayside Cottage. Partial wheelchair access, 2ft wide bridges.

3 BEDFORD HEIGHTS

Brickhill Drive, Bedford, MK41 7PH. Graham A Pavey, www.grahamapavey.co.uk. *Park in main car park & not at Travelodge.* **Sat 15 July (10-2). Adm £5, chd free. Light refreshments.**

Originally built by Texas Instruments, a Texan theme runs throughout the building and gardens. The entrance simulates an arroyo or dry riverbed, and is filled with succulents and cacti, plus a mixture of grasses and herbaceous plants, many of which are Texan natives. There are also 3 courtyard gardens, where less hardy plants thrive in the almost frost free environment. No wheelchair access to 2 of the courtyards but can be viewed from a platform.

4 CHURCH FARM

Church Road, Pulloxhill, Bedford, MK45 5HD. Sue & Keith Miles, 07941 593152. *Parking by kind permission of The Cross Keys Pub, Pulloxhill High St, MK45 5HB (approx 500yds from garden). Yellow signs from the Cross Keys to Church Rd & past parish church.* **Sat 20, Sun 21 May (2-5). Adm £6, chd free. Visits also by arrangement Apr to Aug for groups of 10 to 20.**

Mixed colourful planting to the rear of the house leading to a long sunny border of drought tolerant plants and large topiary subjects. Beyond the farmyard a wildlife pond, small kitchen garden and wildflower orchard with a rose walk. A cutting garden and countryside views. Livestock may be present in adjoining fields so sorry, no dogs.

5 15 DOUGLAS ROAD

Bedford, MK41 7YF. Peter & Penny Berrington. *B660 Kimbolton Rd, turn into Avon Dr, R into Tyne Cres, & Douglas Rd on R.* **Mon 28 Aug (2-5). Adm £5, chd free. Home-made teas.**

This medium sized town garden has a series of outdoor rooms, separated by hedges, fences and arches. There is seating for visitors to relax and enjoy the peace and seclusion. A circular cottage garden with a central decorative fire pit, separated by plum and apple trees from vegetables and herbs grown in raised beds. The pond is home to newts, snails and 'Brickhill' frogs. Wheelchair access to most parts of the garden.

The Round Cottage

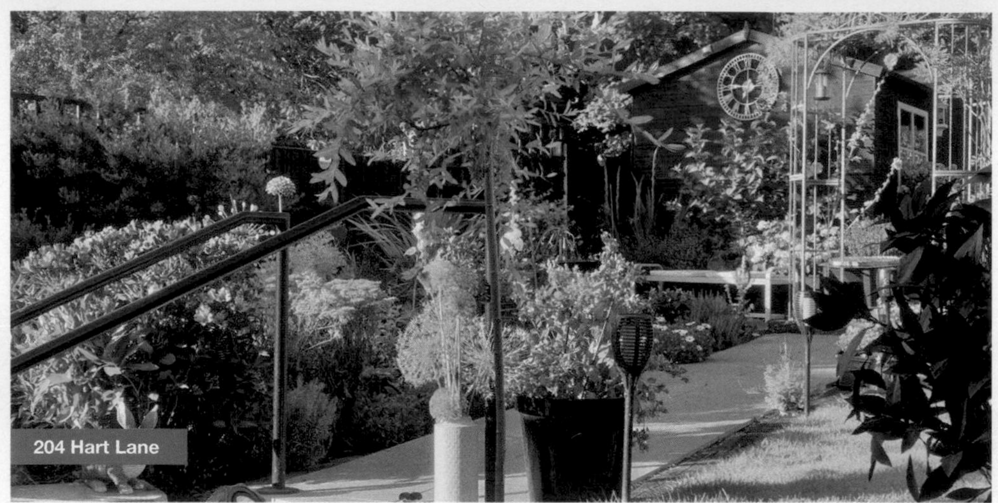
204 Hart Lane

6 22 ELMSDALE ROAD
Wootton, Bedford, MK43 9JN. Roy & Dianne Richards. *Follow signs to Wootton, turn R at The Cock Pub, follow to Elmsdale Rd.* **Sat 22, Sun 23 Apr (1.30-5). Adm £5, chd free. Home-made teas.**
Topiary garden greets visitors before they enter a genuine Japanese Feng Shui garden inc bonsai. Large collection of Japanese plants, Koi pond, lily pond and a Japanese Tea House. The garden was created from scratch by the owners about 20 yrs ago and has many interesting features inc Japanese lanterns and the Kneeling Archer terracotta soldier from China.

7 NEW 204 HART LANE
Luton, LU2 0JH. Alison Bilcock. *Street parking only.* **Sun 26 Mar, Sun 14 May, Sun 23 July, Sun 24 Sept (2-5). Adm £5, chd free. Home-made teas.**
A small gem, with year round seasonal plants, mature shrubs, a relaxing water feature and different seating areas to enjoy the garden with a cream tea. There are hidden animals for children to find, something for all the family. Home-made jams and preserves for sale.

8 HOLLINGTON FARM
Flitton Hill, Maulden, Bedford, MK45 2BE. Susan & John Rickatson. *Off A507 between Clophill & Ampthill- take Silsoe/ Flitton then bear R & follow yellow signs.* **Sat 24, Sun 25 June (1-5.30). Adm £6, chd free. Cream teas.**
2 acre country garden developed around an old farmhouse over a 25 year period. Trees and shrubs are now mature impressive specimens. Semi formal areas near the house inc a small parterre, pergola, pond and borders. In outer parts there is a small woodland planted in the 1990's and a farm wildflower meadow on Flitton Hill with views over the River Flit valley. Play areas for children. Wheelchair access, some steps and slopes.

9 HOWBURY HALL GARDEN
Howbury Hall Estate, Renhold, Bedford, MK41 0JB. Julian Polhill & Lucy Copeman, www.howburyfarmflowers.co.uk. *Leave A421 at A428/Gt Barford exit towards Bedford. Entrance to house & gardens ½m on R. Parking in field.* **Sun 10 Sept (2-5). Adm £6, chd free. Home-made teas.**
A late Victorian garden designed with mature trees, sweeping lawns and herbaceous borders. The large walled garden is a working garden, where 1 half is dedicated to growing a large variety of vegetables whilst the other is run as a cut flower business. In the woodland area, walking towards the large pond, the outside of a disused ice house can be seen. Gravel paths and lawns may be difficult for smaller wheelchairs.

10 ◆ KING'S ARMS GARDEN
Brinsmade Road, Ampthill, Bedford, MK45 2PP. Ampthill Town Council, www.ampthill-tc.gov.uk/amenities/kings-arm-garden. *Free parking in town centre. Entrance opp Old Market Place, down King's Arms Yard.* **For NGS: Sun 26 Feb, Sun 29 Oct (2-4). Adm £4, chd free. Light refreshments. For other opening times and information, please visit garden website.**
Small woodland garden of about 1½ acres created by plantsman, the late William Nourish. Trees, shrubs, bulbs and many interesting collections throughout the year. Since 1987, the garden has been maintained by 'The Friends of the Garden' on behalf of Ampthill Town Council. Charming woodland garden with mass plantings of snowdrops and early spring bulbs. Beautiful autumn colours in October. Wheelchair friendly path running around most of the garden.

11 LAKE END HOUSE
Mill Lane, Woburn Sands, Milton Keynes, MK17 8SP. Mr & Mrs G Barrett, 07831 110959, geoffanddean@gmail.com. *From A5130, turn into Weathercock Ln, to Burrows Close & to Mill Ln.* **Sun 14 May, Sun 11 June, Sun 16 July, Sun 13 Aug, Sun 17 Sept (2-5). Adm £6, chd free. Light refreshments. Visits also by arrangement May to Sept for groups of 10 to 40.**
A large 3 acre lake created on a brown field site, provides a wonderful

watery centrepiece for this extensive modern garden. The modern house (not open) overlooks the lake, while on the other side, a Japanese themed garden with a tea house, red wooden bridge, small pond and cloud-pruned trees. The planting is naturalistic around the lake with willows, alders, cornuses providing sanctuary for waterfowl. Suitable for wheelchair users.

12 40 LEIGHTON STREET
Woburn, MK17 9PH. Ron & Rita Chidley. *500yds from centre of village L side of rd. Very limited on rd parking outside house.* **Sat 16 Sept (2-5). Adm £5, chd free. Home-made teas.**
Large cottage style garden in 3 parts. There are several 'rooms' with interesting features to explore and many quiet seating areas. There are perennials, climbers, vegetables, shrubs, trees and 2 ponds with fish. In late summer and autumn there is an abundance of colour from dahlias, fuchsias, grasses and late flowering roses. Many pots of unusual and tender plants.

13 LINDY LEA
Ampthill Road, Steppingley, Bedford, MK45 1AB. Roy & Linda Collins. *Next to Steppingley Hospital. Park in hospital car park.* **Sun 9 July (2-5). Adm £5, chd free. Home-made teas.**
Set within an acre, this garden is a haven for wildlife with 2 water features, cottage garden style perennial plantings, variety of shrubs and mature trees. There is also a vegetable garden and a sunny terrace furnished with pots and climbers. Wheelchair access: some paths may be difficult to negotiate.

14 1 LINTON CLOSE
Heelands, Milton Keynes, MK13 7NR. Pat & John Partridge, 01908 220571, pip1941@icloud.com. *From H4 Dansteed Way, take Stainton Dr, Linton Close is 3rd R.* **Visits by arrangement 2 Apr to 24 Sept for groups of up to 6. Adm £10. Light refreshments.**
Small town garden developed over 30 years from heavy clay and builder's rubble. Gravel garden with sedum and dwarf conifer collection at the front. The back garden is planted with hardy evergreen shrubs for structure and perennials providing year-round interest and colour. We are attempting to garden more sustainably.

15 ◆ THE MANOR HOUSE, STEVINGTON
Church Road, Stevington, Bedford, MK43 7QB. Kathy Brown, 01234 822064, info@kathybrownsgarden.com, www.kathybrownsgarden.com. *5 miles NW of Bedford. Off A428 through Bromham. If using SatNav please enter entire address, not just postcode.* **For NGS: Sun 21 May (10.30-4.30). Adm £8.50, chd £4. Pre-booking essential, please visit www.ngs.org.uk for information & booking. Home-made teas.**
For other opening times and information, please phone, email or visit garden website.
The Manor House Garden has many different rooms, inc 6 art inspired gardens. Early roses and wisteria along with several different clematis montana festoon the walls and pergolas; foxgloves, poppies and peonies provide colour down below. Elsewhere, avenues of white stemmed birches, gingko, eucalyptus, and metasequoia leads the eye to further parts of the garden. This garden will be closed between 1 - 2pm. 85% of the garden accessible for wheelchair users. Disabled WC.

GROUP OPENING

16 NEW MILL LANE GARDENS
Mill Lane, Greenfield, Bedford, MK45 5DG. *In the centre of Greenfield off High St. From Flitwick, Mill Ln is on L in the centre of the village. Parking at the village hall further on the R. Room for disabled parking only at each house.* **Sat 3, Sun 4 June (1-5). Combined adm £6, chd free. Home-made teas at 69 Mill Lane.**

69 MILL LANE
Pat Rishton.

70 MILL LANE
Lesley Arthur.

NEW 71 MILL LANE
Carol Mcclurg.

Greenfield is an attractive and varied Bedfordshire village close to Flitwick and Ampthill. The village has houses of many different characters inc several thatches and a recently refurbished village pub. Mill Lane was previously in the centre of a fruit growing area and was famous for its strawberry fields. The 3 gardens are interesting and varied inc cottage garden style perennials, courtyards, fruit gardens and ponds, both ornamental and for wildlife. River Flit runs through 70 Mill Lane. Most of the gardens are accessible to wheelchairs but there are some steps, banks and gravel paths. Plenty of seating provided. 69 & 71 Mill Lane fully accessible. Sorry no dogs.

17 THE OLD RECTORY, WRESTLINGWORTH
Church Lane, Wrestlingworth, Sandy, SG19 2EU.
Josephine Hoy, 01767 631204, hoyjosephine@hotmail.co.uk. *The Old Rectory is at the top of Church Ln, which is well signed, behind the church.* **Sun 7, Sun 28 May (2-5). Adm £7, chd free. Visits also by arrangement 1 Apr to 11 June for groups of up to 30. Teas will be served for groups.**
4 acre garden full of colour and interest. The owner has a free style of gardening sensitive to wildlife. Beds overflowing with tulips, alliums, bearded iris, peonies, poppies, geraniums and much more. Beautiful mature trees and many more planted in the last 30 yrs inc a large selection of betulas. Gravel gardens, box hedging and clipped balls, woodland garden and wildflower meadows. Wheelchair access may be restricted to grass paths.

The National Garden Scheme searches the length and breadth of England, Wales, Northern Ireland and the Channel Islands for the very best private gardens

18 4 ST ANDREWS ROAD

Bedford, MK40 2LJ. Amanda Dimmock, 07710273942, amandawdimmock@aol.com. *Off Park Ave.* **Sat 12 Aug (10-4). Adm £3.50, chd free. Sat 26 Aug (12-4). Combined adm with 1a St Augustine's Road £6, chd free. Light refreshments at 1A St Augustine's Road on Saturday 26th August. Visits also by arrangement 1 May to 17 Sept for groups of 6 to 8. Visit duration max of 1 hr.**

An elegant, multi-functional urban garden (30ft x 70ft) set within a semi-formal layout featuring clipped beech hedges softened by colourful shrubs and perennials and a rose pergola. Focal points have been created to be enjoyed from both within the garden and from the house.

19 1A ST AUGUSTINE'S ROAD

Bedford, MK40 2NB. Chris Bamforth Damp, 01234 353730/01234 353465, chrisdamp@mac.com. *St Augustine's Rd is on L off Kimbolton Rd as you leave the centre of Bedford.* **Sat 26 Aug (12-4.30). Combined adm with 4 St Andrews Road £6, chd free. Home-made teas. Admission to 1A St Augustine's Road only £5.00/chd free. Visits also by arrangement 1 June to 1 Oct for groups of up to 30.**

A colourful town garden with herbaceous borders, climbers, a greenhouse and pond. Planted in cottage garden style with traditional flowers, the borders overflow with late summer annuals and perennials inc salvias and rudbeckia. The pretty terrace next to the house is lined with ferns and hostas. The owners also make home-made chutneys and preserves which can be purchased on the day. The garden is wheelchair accessible.

GROUP OPENING

20 NEW SILSOE VILLAGE GARDENS

Alder Wynd, Silsoe, Bedford, MK45 4GQ. *Access to the village is via the A6 or the A507. Once in the village follow the yellow signs.* **Sun 20 Aug (2-5). Combined adm £7, chd free. Home-made teas at Round Cottage.**

10 ALDER WYND
David & Frances Hampson, 01525 861356, mail@davidhampson.com.
Visits also by arrangement 1 July to 2 Sept for groups of up to 20.

NEW NEWBURY PLANTS
Bruce Liddle.

THE ROUND COTTAGE
Peter & Jane Gregory, 01525 862343, peter.gregory44@hotmail.co.uk.
Visits also by arrangement 1 June to 1 Sept for groups of 5 to 20.

The historic village of Silsoe dates back to the Viking era, but is best known today as the home of Wrest Park. 10 Alder Wynd is a small modern garden, constructed 9 years ago and is the runner up in 2 national gardening awards. A wide range of seasonal perennials, trees and shrubs give this garden year round appeal. Notable for its collection of giant hostas, ferns, exotic perennials and cannas. Round Cottage is a pretty cottage garden surrounding an 1820's thatched cottage. Traditional style flower beds where perennials mix with exotics from around the world plus a sprinkle of annuals. Large Koi pond with waterfall, Victorian style greenhouse. There is a deep lit well. The Arboretum at Newbury Farm is a private 4 acre arboretum, started over 30 years ago by Maurice and Carol Clarke. Enjoy a stroll through the collection of over 90 trees, planted against a backdrop of rolling Bedfordshire farmland, as well as through a recently constructed gravel garden. The 3 gardens are short drive away from each other.

21 SOUTHILL PARK

Southill, nr Biggleswade, SG18 9LL. Mr & Mrs Charles Whitbread. *In the village of Southill, 5m from the A1.* **Sun 4 June (2-5). Adm £6, chd free. Cream teas.**

Southill Park 1st opened its gates to NGS visitors in 1927. A large garden with mature trees and flowering shrubs, herbaceous borders, a formal rose garden, sunken garden, ponds and kitchen garden. It is on the south side of the 1795 Palladian house. The parkland was designed by Lancelot 'Capability' Brown. A large conservatory houses the tropical collection.

GROUP OPENING

22 STEPPINGLEY VILLAGE GARDENS

Steppingley, Bedford, MK45 5AT. *Follow signs to Steppingley, pick up yellow signs from village centre.* **Sun 28, Mon 29 May (2-5). Combined adm £7, chd free. Home-made teas at Townsend Farmhouse.**

MIDDLE BARN
John & Sally Eilbeck.

37 RECTORY ROAD
Bill & Julie Neilson.

TOP BARN
Tim & Nicky Kemp.

TOWNSEND FARMHOUSE
Hugh & Indi Jackson.
(See separate entry)

Steppingley is a picturesque Bedfordshire village on the Greensand Ridge, close to Ampthill, Flitwick and Woburn. Although a few older buildings survive, most of Steppingley was built by the 7th Duke of Bedford between 1840 and 1872. 4 gardens in the village offer an interesting mix of planting styles and design to inc pretty courtyards, cottage garden style perennial borders, ponds, a Victorian well, glasshouses, an orchard, vegetable gardens, a herb garden, wildlife havens and country views. Livestock inc chickens, ducks and fish.

23 127 STOKE ROAD

Linslade, Leighton Buzzard, LU7 2SR. Steve Owen. *Opp the turn for Rothschild Rd.* **Sun 22 Jan, Sun 5 Mar (11-4). Adm £5, chd free.**

The garden houses the National Collection of Galanthus (snowdrop), with nearly 2,000 different varieties. Most are grown in the garden setting but some specialised ones can be seen on display racks. There is a Japanese themed garden and a raised alpine bed. Other plants grown inc daphnes and a collection of over 100 maple trees. The alpine house contains a collection of Primula allionii.

NPC

80 West Hill

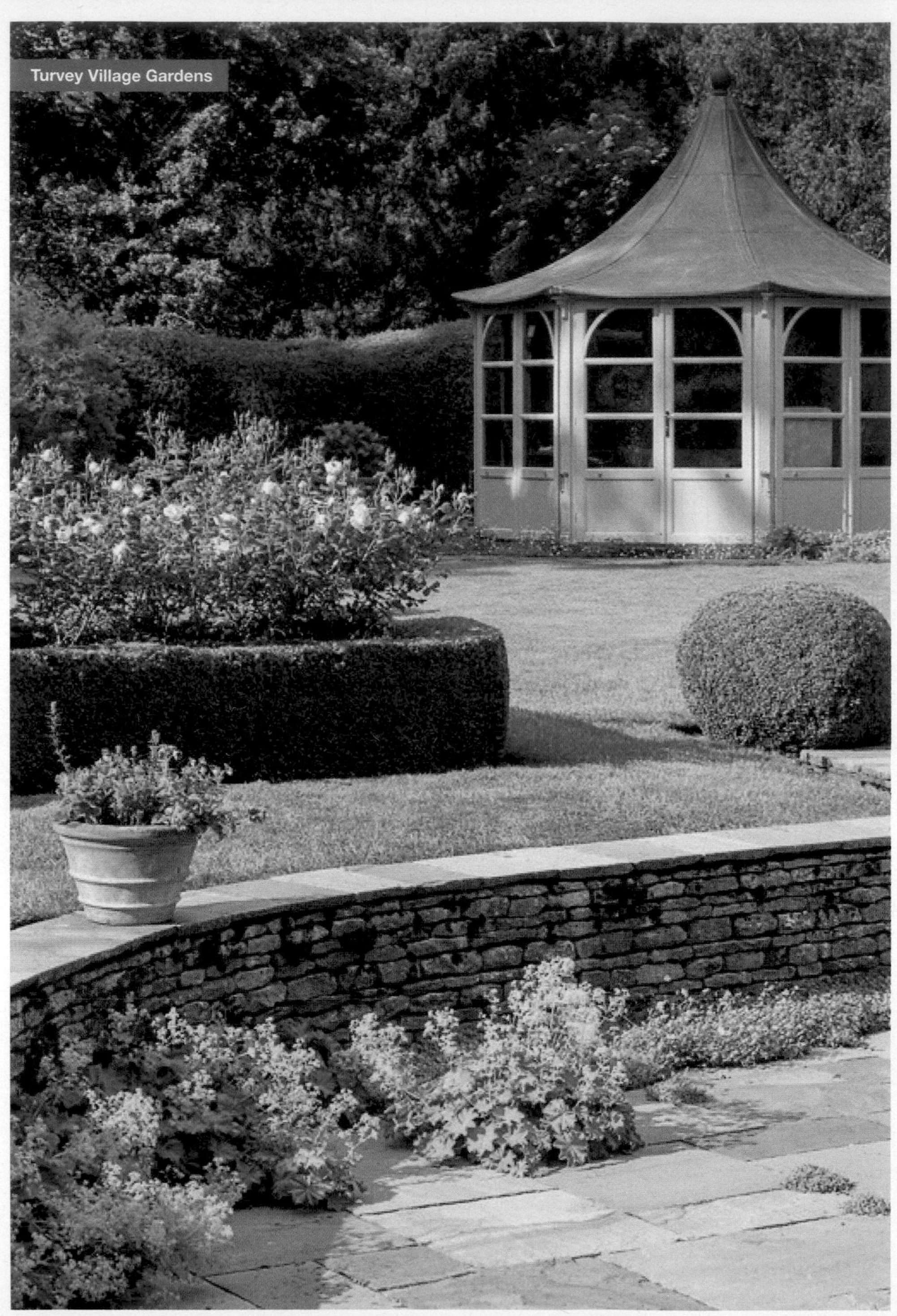

Turvey Village Gardens

24 THE SWISS GARDEN
Alder Drive, Biggleswade, SG18 9ER. The Shuttleworth Trust, www.shuttleworth.org/explore/swiss-garden/. *2m W of Biggleswade. Signed from A1 & A600. New entrance. Use postcode SG18 9DT & follow directional signs to Alder Dr.* **Thur 2 Mar (10-5); Thur 2 Nov (10-4). Adm £10, chd free. Light refreshments at the Runway Cafe.**
This enchanting garden was created in the 'Swiss Picturesque' style for the 3rd Lord Ongley in the early C19 and reopened in July 2014 after a major HLF-funded restoration. Serpentine paths lead to cleverly contrived vistas, many of which focus on the thatched Swiss Cottage. Beautiful wrought-iron bridges, ponds, sweeping lawns and the magnificent pulhamite-lined Grotto Fernery are a few highlights of this lovely romantic landscape. Evergreen trees dominate the landscape and the garden contains several county champions inc a rare group of weeping hollies and an imposing cedar of lebanon, the tallest in Bedfordshire. Admission inc the Swiss Garden, collection (vintage aircraft and vehicles), the woodland sculpture trail, the shrubbery walk up to the house and parkland adjacent to the estate lake. The Swiss Garden contains 13 listed buildings and artefacts in total. The pathways in the Swiss Garden are firm and even, with minimal gradients, and most are suitable for access by wheelchair users.

25 TOWNSEND FARMHOUSE
Rectory Road, Steppingley, Bedford, MK45 5AT. Hugh & Indi Jackson. *Follow directions to Steppingley village and pick up yellow signs from village centre.* **Sun 30 Apr, Mon 1 May (2-5). Adm £6, chd free. Home-made teas. Evening opening Sat 28 Oct (5.30-8.30). Adm £10, chd £2. Pre-booking essential, please visit www.ngs.org.uk for information & booking. Light refreshments. Opening with Steppingley Village Gardens on Sun 28, Mon 29 May.**
Country garden with tree lined driveway and many perennial borders. Pretty cobbled courtyard with a glasshouse and a Victorian well 30 metres deep, viewed through a glass top. Colourful display of tulips and spring bulbs in early May and a spectacular evening of lanterns, diya lamps and flower rangoli in late October.

GROUP OPENING

26 TURVEY VILLAGE GARDENS
High Street, Turvey, Bedford, MK43 8EP. *Follow arrow signs from village centre. Parking in field behind Chantry House.* **Sun 18 June (2-5). Combined adm £7, chd free. Home-made teas in medieval church of All Saints.**

CHANTRY HOUSE
Sheila & Anthony Ormerod.

NEW **GABLE END**
Chris & Wendy Knell.

7 THE GREEN
Paul & Rosemary Gentry.

NEW **PEPPERS**
Liz & Mark Upex.

The historic village of Turvey lies beside the River Ouse and is recorded in Domesday Book of 1086 as a parish in the Hundred of Willey. 4 gardens in the village offer a varied mix of design and interest. Chantry House is a 1½ acre garden, approached by a drive bounded by mature yew and box cloud hedges where the medieval church of All Saints overlooks the garden at this point. A south facing lawn is flanked by high rose covered walls and interspersed by herbaceous beds. 7 The Green is an informal village garden. A wide terrace with comfortable seating and a myriad of planted pots overlooks a central lawn and flowerbeds which slope down to flowering cherries and a view of pasture and mature trees. Gable End is a medium-sized garden consisting of four distinct areas. A range of planting inc fruit and vegetables are enjoyed here. Peppers is a small, informal village garden with mixed planting aimed at attracting wildlife and achieving year-round interest. Wheelchair access via gravel paths and few shallow distanced steps.

27 ◆ THE WALLED GARDEN
Luton, LU1 4LF. Luton Hoo Estate, www.lutonhooestate.co.uk. *From A1081 turn at West Hyde Rd. After 100 metres turn L through black gates, follow signs to Walled Garden.* **For NGS: Fri 4 Aug (11-3). Adm £6, chd free. Light refreshments at Woodyard Coffee Shop.**
For other opening times and information, please visit garden website.
The 5 acre Luton Hoo Estate Walled Garden was designed by Capability Brown and established by noted botanist and former Prime Minister, Lord Bute, in the late 1760s. The Walled Garden is now the focus of an incredible volunteer project and continues to be researched, restored, repaired and re-imagined for the enjoyment of all. Unique service buildings inc a vinery, fernery and propagation houses. Exhibition of Victorian tools. Uneven surfaces reflecting the age and history of the site.

28 NEW 80 WEST HILL
Aspley Guise, Milton Keynes, MK17 8DX. Mr James & Mrs Helen Scott, www.thegardenco.co.uk/residential/modern-sanctuary/. *6m SE of Milton Keynes; close to Jn13 of the M1. Leave junc 13 of M1. 2m W towards Woburn Sands. On West Hill, turn R onto Ln marked 80-92. No parking ex for disabled visitors, please park at Cricket Club (next R on West Hill).* **Sun 21 May (12-4.30). Adm £5, chd free. Home-made teas.**
Contemporary front and rear gardens to complement a modern house within a walled garden plot. The rear garden is based on strong geometry with naturalistic planting inc several bespoke features. The front garden feels more organic, with various uses of natural stone and with planting to suit some dry sunny spots as well as woodland shade. Winner of 2 National Awards for design and construction. Features inc fire pit area, water feature, glasshouse, secluded seating area. There is careful use of Yorkshire stone illustrating the journey from boulder to slab.

9,500 patients and their families are being supported across four Horatio's gardens thanks to our annual donations

BERKSHIRE

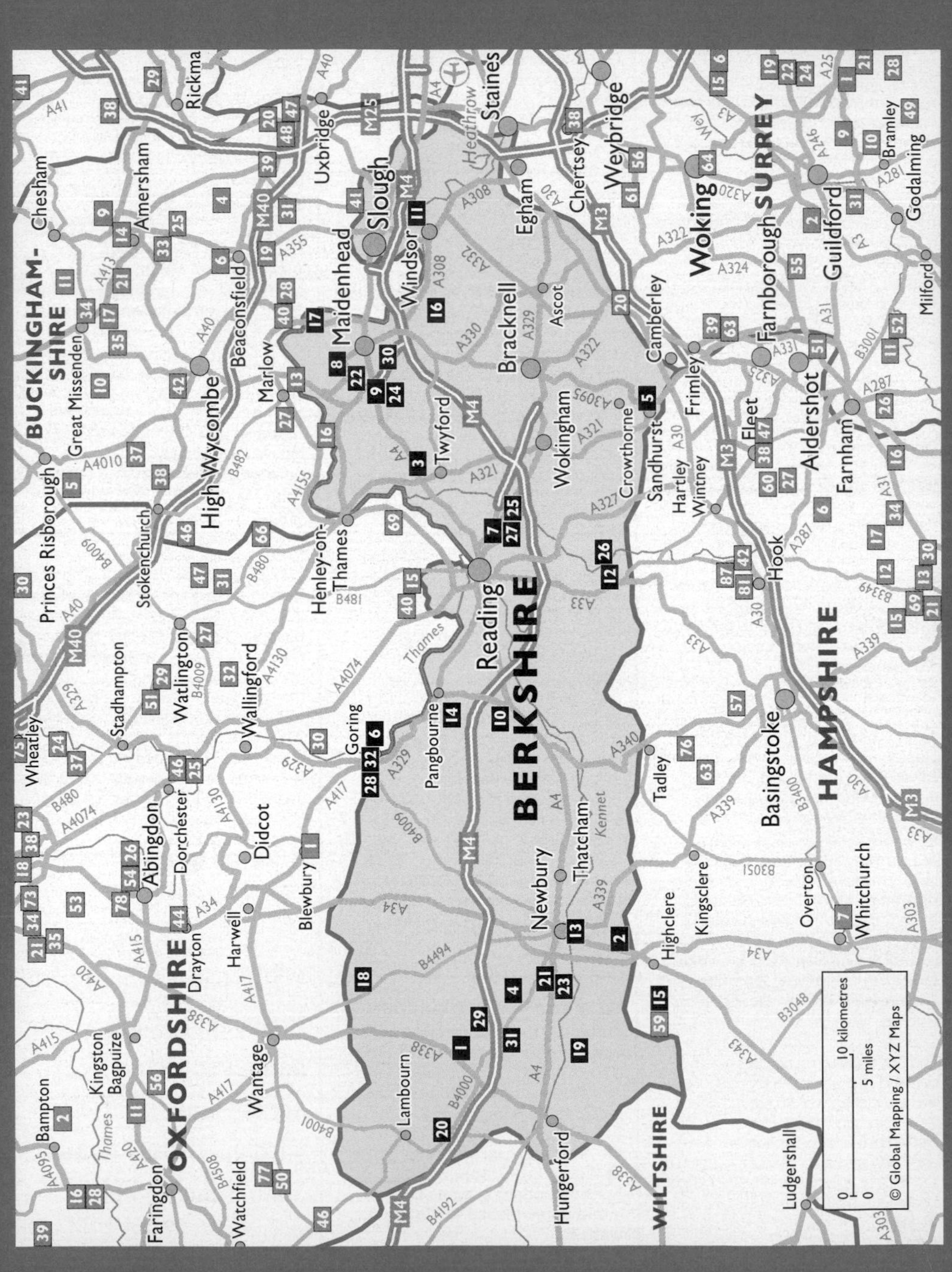
BERKSHIRE
BUCKINGHAM-SHIRE
OXFORDSHIRE
WILTSHIRE
HAMPSHIRE
SURREY
Reading
Slough
Maidenhead
Windsor
Bracknell
Wokingham
Twyford
Newbury
Thatcham
Hungerford
Lambourn
Pangbourne
Goring
Henley-on-Thames
Marlow
High Wycombe
Beaconsfield
Amersham
Chesham
Great Missenden
Princes Risborough
Stokenchurch
Watlington
Wallingford
Stadhampton
Wheatley
Abingdon
Dorchester
Didcot
Blewbury
Harwell
Drayton
Wantage
Kingston Bagpuize
Bampton
Faringdon
Watchfield
Uxbridge
Rickma
Staines
Egham
Chertsey
Weybridge
Woking
Ascot
Camberley
Frimley
Farnborough
Guildford
Godalming
Bramley
Milford
Aldershot
Farnham
Fleet
Crowthorne
Sandhurst
Hartley Wintney
Hook
Basingstoke
Tadley
Kingsclere
Highclere
Overton
Whitchurch
Ludgershall
Heathrow
Thames
Kennet
Wey
M4
M40
M25
M3
10 kilometres
5 miles
© Global Mapping / XYZ Maps

The Royal County of Berkshire offers a wonderful mix of natural beauty and historic landmarks that are reflected in the portfolio of gardens opening for the National Garden Scheme.

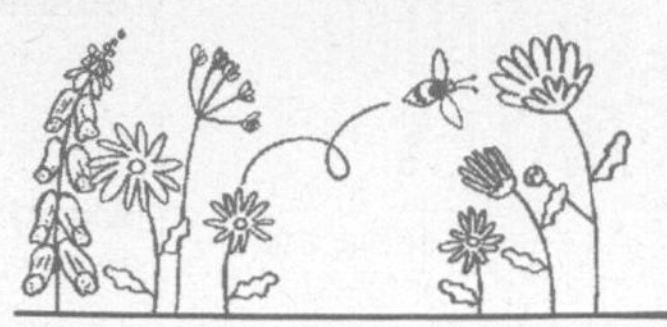

Our gardens come in all shapes, sizes and styles, from urban oases and community plots, to village gems and country house landscapes. We are delighted to welcome seven gardens opening for the first time this year. From east to west across the county they include 13 Broom Acres in Sandhurst, Middle Barton in Fifield, 6 Beverley Gardens In Wargrave , Brierley House and Suvikuja in Swallowfield Village, Belvedere near Newbury and Ashbrooke in Great Shefford. It is also a pleasure to welcome back Wickham House.

In addition to our open days, fifteen gardens offer opportunities for private visits 'By Arrangement' and you can contact the garden openers directly to organise them. You will find more information and photographs of all our gardens on the website.

So while some gardens may capture your interest due to their designers or their historic setting, most have evolved thanks to the efforts of their enthusiastic owners. We hope that all National Garden Scheme gardens offer moments of inspiration and pleasure, and look forward to welcoming you soon.

© Sussie Bell

Volunteers

County Organiser
Heather Skinner
01189 737197
heather.skinner@ngs.org.uk

County Treasurer
Hugh Priestley
01189 744349 Fri – Mon
hughpriestley@aol.com

Booklet Co-ordinator
Heather Skinner
(as above)

Assistant County Organisers
Claire Fletcher
claire.fletcher@ngs.org.uk

Carolyn Foster
07768 566482
candrfoster@btinternet.com

Rebecca Thomas
01491 628302
rebecca.thomas@ngs.org.uk

Bob Weston
01635 550240
bob.weston@ngs.org.uk

@BerksNationalGardenScheme
@BerksNGS

Left: The Old Rectory, Farnborough

OPENING DATES

All entries subject to change. For latest information check **www.ngs.org.uk**

Map locator numbers are shown to the right of each garden name.

Extended openings are shown at the beginning of the month.

February

Snowdrop Festival

Wednesday 8th
◆ Welford Park 29

March

Saturday 11th
Stubbings House 24

Sunday 12th
Stubbings House 24

Thursday 16th
St Timothee 22

April

Wednesday 12th
Malverleys 15

Sunday 16th
The Old Rectory, Farnborough 18
Rookwood Farm House 21

Wednesday 26th
Rooksnest 20

Thursday 27th
Odney Club 17

May

Daily from Tuesday 9th to Saturday 13th
61 Sutcliffe Avenue 25

Saturday 13th
Stubbings House 24

Sunday 14th
The Old Rectory, Farnborough 18
Stubbings House 24

Monday 29th
Rookwood Farm House 21

June

Every Tuesday from Tuesday 13th
68 Church Road 7

Daily from Tuesday 6th to Saturday 10th
61 Sutcliffe Avenue 25

Sunday 4th
Swallowfield Village Gardens 26

Saturday 10th
NEW 13 Broom Acres 5
7 The Knapp 27
Thurle Grange 28
Wynders 32

Sunday 11th
St Timothee 22
7 The Knapp 27
Thurle Grange 28
Wynders 32

Saturday 17th
Chaddleworth 6
7 The Knapp 27

Sunday 18th
NEW 6 Beverley Gardens 3
7 The Knapp 27
Wickham House 31

Saturday 24th
NEW 13 Broom Acres 5

Sunday 25th
NEW 6 Beverley Gardens 3

Tuesday 27th
Wembury 30

Wednesday 28th
The Old Rectory, Farnborough 18
Rooksnest 20

Friday 30th
Wembury 30

July

Every Tuesday
68 Church Road 7

Sunday 2nd
Deepwood Stud Farm 9

Tuesday 4th
Eton College Gardens 11

Thursday 6th
NEW Ashbrooke 1
St Timothee 22

Saturday 8th
NEW 13 Broom Acres 5

Sunday 9th
Island Cottage 13

Wednesday 12th
Malverleys 15

Saturday 15th
NEW Belvedere 2

Sunday 16th
NEW Middle Barton 16

Saturday 22nd
NEW 13 Broom Acres 5

Sunday 23rd
NEW Ashbrooke 1

August

Every Tuesday
68 Church Road 7

Daily from Tuesday 22nd to Saturday 26th
61 Sutcliffe Avenue 25

Tuesday 1st
Wembury 30

Thursday 3rd
NEW 6 Beverley Gardens 3

Saturday 5th
NEW 13 Broom Acres 5

Wednesday 9th
The Old Rectory, Farnborough 18

Sunday 27th
Stockcross House 23

September

Saturday 2nd
NEW Belvedere 2

Tuesday 12th
Wembury 30

November

Thursday 2nd
St Timothee 22

By Arrangement

Arrange a personalised garden visit with your club, or group of friends, on a date to suit you. See individual garden entries for full details.

NEW Ashbrooke 1
NEW Belvedere 2
Boxford House 4
NEW 13 Broom Acres 5
68 Church Road 7
Compton Elms 8
Deepwood Stud Farm 9
Handpost 12
Island Cottage 13
Lower Bowden Manor 14
The Old Rectory Inkpen 19
Rooksnest 20
St Timothee 22
Stockcross House 23
7 The Knapp 27
Wynders 32

National Garden Scheme gardens are identified by their yellow road signs and posters. You can expect a garden of quality, character and interest, a warm welcome and plenty of home-made cakes!

THE GARDENS

1 NEW ASHBROOKE

Wantage Road, Great Shefford, Hungerford, RG17 7DA. Jackie & Roger Frith, 07584 598189, jackiefrith2468@gmail.com. *5m NE of Hungerford. From M4 (J14) follow A338 towards Wantage for 2m into Great Shefford village. Ashbrook is on R opp Station Rd. From Wantage on A338, Ashbrook is on L. Parking in Station Rd.* **Thur 6, Sun 23 July (10.30-4). Adm £4, chd free. Pre-booking essential, please visit www.ngs.org.uk for information & booking. Light refreshments. Two hour timed slots at 10.30am or 2pm. Visits also by arrangement 1 July to 20 Aug for groups of 10 to 30.**

A small narrow garden, divided into three areas running down to the River Lambourn. Front with roses, shrubs, dahlias, tubs and baskets. The middle is a very wet garden down to the river with mixed herbaceous borders and planting to suit such conditions. Lower garden has a small pond, two greenhouses, vegetable plots and exhibition chrysanthemums on banks of the river.

2 NEW BELVEDERE

Garden Close Lane, Newbury, RG14 6PP. Noushin Garrett. *7m S of M4 J13. Take A34 exit Highclere/ Wash Common then A343 towards Newbury. Parking at St George's Church, Wash Common RG14 6NU with 600 metres walk to garden.* **Sat 15 July, Sat 2 Sept (10.30-4.30). Adm £5, chd free. Pre-booking essential, please visit www.ngs.org.uk for information & booking. Two hour timed slots at 10.30am or 2.30pm. Visits also by arrangement June to Aug for groups of 10 to 30. Visits in June & Aug only.**

A garden of 2½ acres consisting of herbaceous borders, rose garden, fruit trees with woodland, bamboo and rhododendron walks. A formal parterre garden with dahlias. Sorry no parking, WC or teas at garden. Limited wheelchair access.

3 NEW 6 BEVERLEY GARDENS

Wargrave, Reading, RG10 8ED. Patricia Vella & Jon Black. *A4 from Maidenhead take B477 to Wargrave, turn R into Silverdale Rd. A4 from Reading take A321 into Wargrave, turn R at lights, ¼ m turn L into Silverdale Rd, Beverley Gardens is on R.* **Sun 18, Sun 25 June, Thur 3 Aug (1-5.30). Adm £4, chd free. Pre-booking essential, please visit www.ngs.org.uk for information & booking. Home-made teas. Two hour timed slots at 1pm or 3.30pm.**

The main garden has primarily exotic and Mediterranean planting with small paths through some of the beds, inviting you to take a closer look. This leads into a kitchen garden with greenhouse, chickens and ducks. There is a small woodland garden with stumpery and fire pit. The garden has plenty of places to sit and relax. Wheelchair access over gravel drive leading to a level garden with some bark paths. No access through main beds.

4 BOXFORD HOUSE

Boxford, Newbury, RG20 8DP. Tammy Darvell, Head Gardener, 07802 883084, tammydarvell@hotmail.com. *4m NW of Newbury. Directions will be provided on booking.* **Visits by arrangement May to Aug for groups of 20 to 50. Adm £11, chd free. Guided tour & refreshments inc.**

Beautiful large family garden extensively developed over the past 10 yrs. Emphasis on roses and scent throughout the 5 acre main garden. Old and new orchards, laburnum tunnel, formal and colourful herbaceous borders. Handsome formal terraces, pond, water features and garden woodland areas. Inviting cottage garden and productive vegetable gardens.

5 NEW 13 BROOM ACRES

Sandhurst, GU47 8PN. Sunil Patel, 07974 403077, garden@sunilpatel.co.uk, www.sunilpatel.co.uk. *Between Sandhurst & Crowthorne. From Crowthorne/Sandhurst road turn E into Greenways then 1st R into Broom Acres. Street parking.* **Sat 10, Sat 24 June, Sat 8, Sat 22 July, Sat 5 Aug (1-5). Adm £5, chd free. Pre-booking essential, please visit www.ngs.org.uk for information & booking. Home-made teas. Visits also by arrangement June to Sept for groups of 10 to 30.**

A spellbinding ¼ acre suburban romantic style garden with shrubbery, exotics, specimen trees, scented climbers and mixed herbaceous borders. Densely planted and decadently flowering, a feast for all the senses. Come for the garden, stay for the magic and discover why romance remains compelling, timeless and enduring.

6 CHADDLEWORTH

Streatley, Reading, RG8 9PR. Susan & Simon Carter, 07711 420586, www.chaddleworthbedandbreakfast.com. *Just off A329 leaving Streatley towards Wallingford. From T-lights in Streatley take A329 towards Wallingford. Pass L turn for A417. Take next R signed Cleeve Court, then Chaddleworth is 1st on R along lane.* **Sat 17 June (11-4). Adm £5, chd free. Home-made teas.**

Chaddleworth's gardens surround the house on all four sides with a variety of garden styles. Several lawned areas with different borders, small herbaceous, ornamental grasses, wildflower area, wild orchids, vegetable garden, rose path, small cutting garden. Mostly wheelchair accessible with some gravel.

7 68 CHURCH ROAD

Earley, Reading, RG6 1HU. Pat Burton, 07809 613850, patsi777@virginmedia.com. *E of Reading. Off A4 at Shepherd's House Hill, turn into Pitts Ln, into Church Rd, across r'about, 4th property on L.* **Every Tue 13 June to 29 Aug (2-5). Adm £5, chd free. Pre-booking essential, please visit www.ngs.org.uk for information & booking. Home-made teas. Visits also by arrangement 19 June to 31 Aug for groups of 5 to 25.**

A fascinating urban garden with changing elements throughout the seasons. Different areas showcase a variety of interesting plants, pergola, outdoor dining room and small summerhouse. The working greenhouse is home to alpines grown year-round and seen as specimen plants in the garden. A living roof covers the pergola. Tiered areas of hostas and an abundance of summer bedding.

13 Broom Acres

8 COMPTON ELMS

Marlow Road, Pinkneys Green, Maidenhead, SL6 6NR. Alison Kellett, kellettaj@gmail.com. *Situated at end of gravel road located opp & in between The Arbour & The Golden Ball Pub & Kitchen on the A308.* **Visits by arrangement 21 Mar to 6 Apr for groups of 15 to 40. Adm £10, chd free. Light refreshments inc.**

A delightful spring garden set in a sunken woodland, lovingly recovered from clay pit workings. The atmospheric garden is filled with snowdrops, primroses, hellebores and fritillaria, interspersed with anemone and narcissi under a canopy of ash and beech.

9 DEEPWOOD STUD FARM

Henley Road, Stubbings, nr Maidenhead, SL6 6QW. Mr & Mrs E Goodwin, 01628 822684, ed.goodwin@deepwood.co. *2m W of Maidenhead. M4 J8/9 take A404M N. 2nd exit for A4 to Maidenhead. L at 1st r'about on A4130 Henley, approx 1m on R.* **Sun 2 July (2-5). Adm £5, chd free. Home-made teas. Visits also by arrangement Mar to Oct for groups of 10+.**

4 acres of formal and informal gardens within a stud farm, so great roses! Small lake with Monet style bridge and three further water features. Several neo-classical follies and statues. Walled garden with windows cut in to admire the views and horses. Woodland walk and enough hanging baskets to decorate a pub! Partial wheelchair access.

10 ◆ ENGLEFIELD HOUSE GARDEN

Englefield, Theale, Reading, RG7 5EN. Lord & Lady Benyon, 01189 302504, peter.carson@englefield.co.uk, www.englefieldestate.co.uk. *6m W of Reading. M4 J12. Take A4 towards Theale. 2nd r'about take A340 to Pangbourne. After ⅙m entrance on the L.* **For opening times and information, please phone, email or visit garden website.**

The 12 acre garden descends dramatically from the hill above the historic house through woodland where mature native trees mix with rhododendrons and camellias. Drifts of spring and summer planting are followed by striking autumn colour. Stone balustrades enclose the lower terrace with lawns, roses and mixed borders. A stream meanders through the woodland. Open every Monday from Apr-Sept (10am-6pm) and Oct-Mar (10am-4pm). Please check the Englefield Estate website for any changes before travelling. Wheelchair access to some parts of the garden.

11 ETON COLLEGE GARDENS

Eton, nr Windsor, SL4 6DB. Eton College. *½m N of Windsor. Parking signed off B3022, Slough Rd. Walk from car park across playing fields to entry. Follow signs for tickets & maps sold at gazebo nr entrance.* **Tue 4 July (12-4.30). Adm £6, chd free. Home-made teas.**

A rare chance to visit a group of central College gardens surrounded by historic school buildings, inc Luxmoore's garden on a small island in the Thames reached across two attractive bridges. Also, an opportunity to explore the fascinating Eton College Natural History Museum, the Museum of Eton Life and a small group of other private gardens. Sorry, the gardens are not suitable for wheelchairs due to gravel, steps and uneven ground.

12 HANDPOST

Basingstoke Road, Swallowfield, Reading, RG7 1PU. Faith Ramsay, 07801 239937, faith@mycountrygarden.co.uk, www.mycountrygarden.co.uk. *From M4 J11, take A33 S. At 1st T-lights turn L on B3349 Basingstoke Rd. Follow road for 2¾m, garden on L, opp Barge Ln. Parking for 12 cars max.* **Visits by arrangement 1 May to 1 Sept for groups of 10 to 35. Adm £10. Home-made teas inc. Donation to Thrive.**

4 acre designer's garden with many areas of interest. Features inc four lovely long herbaceous borders attractively and densely planted in different colour sections, a formal rose garden, an old orchard with a grass meadow, pretty pond and peaceful wooded area. Large variety of plants, trees and a productive fruit and vegetable patch. Largely accessible with some gravel areas.

13 ISLAND COTTAGE

West Mills, Newbury, RG14 5HT. Karen Swaffield, karen@lockisland.com. *7m from M4 J13. Park in town centre car parks. Walk past St Nicholas Church or between Côte Restaurant & Holland & Barrett to canal. 200yds to swing bridge & follow signs. Limited side road parking.* **Sun 9 July (2-5). Adm £4, chd free. Visits also by arrangement 28 May to 30 Sept for groups of up to 22.**

A pretty, small waterside garden set between the River Kennet and the Kennet and Avon Canal. Interesting combinations of colour and texture to look at rather than walk through, although visitors are welcome to do that too. A deck overlooks a sluiceway towards a lawn and border. Started from scratch in 2005 and mostly again after the floods of 2014. Being simplified for manageability! Home-made teas unless very wet weather.

14 LOWER BOWDEN MANOR

Bowden Green, Pangbourne, RG8 8JL. Juliette & Robert Cox-Nicol, 07552 217872, robert.cox-nicol@orange.fr. *1½m W of Pangbourne. Directions provided on booking.* **Visits by arrangement for groups of up to 50. Adm £7, chd free. Home-made teas by prior request.**

A 5 acre designer's garden with stunning views and where structure predominates. Specimen trees show contrasting bark and foliage. A marble 'Pan' plays to a pond with boulders and boulder-shaped evergreens. Versailles planters with standard topiaries line a rill. A stumpery leads to the orchard's carpet of daffodils and later a wave of white hydrangeas. Steps and gravel, but most areas accessible. Dogs welcome on leads.

15 MALVERLEYS

Fullers Lane, East End, Newbury, RG20 0AA. *A34 S of Newbury, exit signed for Highclere. The gate will open 30 mins before each tour. Directions will be emailed prior to opening.* **Wed 12 Apr, Wed 12 July (10.30-4.30). Adm £17.50, chd free. Pre-booking essential, please visit www.ngs.org.uk for information & booking. Visits start promptly at 10.30am, 1pm or 3pm with guided tour by Head Gardener. Tea or coffee inc.**

10 acres of dynamic gardens which have been developed over the last 11 yrs to inc magnificent mixed borders and a series of contrasting yew hedged rooms, hosting flame borders, a cool garden, a pond garden and new stumpery. A vegetable garden with striking fruit cages sit within a walled garden, also encompassing a white garden. Meadows open out to views over the parkland. Sorry, due to steps and uneven paths, the garden is not suitable for wheelchairs.

16 NEW MIDDLE BARTON

Coningsby Lane, Fifield, Maidenhead, SL6 2PF. Sam Perkins & Sophia Gerth, www.instagram.com/the_gherkins_garden. *Between Windsor & Maidenhead. From A308 Windsor Rd take Fifield Rd S for ⅛m, then R into Coningsby Ln & follow NGS signs.* **Sun 16 July (10-4). Adm £5, chd free. Pre-booking essential, please visit www.ngs.org.uk for information & booking. Home-made teas. Two hour timed slots at 10am, 12pm or 2pm.**

A stunning and compact garden with Dutch-influenced, deep herbaceous perennial borders, a small wildflower meadow and a contemporary Corten steel vegetable patch. Designed and created since 2020 by a novice yet enthusiastic gardening couple. We have achieved a naturalistic, wildlife friendly garden that holds interest in every month of the yr to thrive alongside a busy family life.

17 ODNEY CLUB

Odney Lane, Cookham, SL6 9SR. John Lewis Partnership. *3m N of Maidenhead. Off A4094 S of Cookham Bridge. Signs to car park in grounds.* **Thur 27 Apr (12-4). Adm £5, chd free. Light refreshments.**

This 120 acre site is beside the Thames with lovely riverside walks. A favourite with Stanley Spencer who featured our magnolia in his work. Lovely wisteria, specimen trees, side gardens, spring bedding and ornamental lake. The John Lewis Partnership Heritage Centre will be open, showcasing the textile archive and items illustrating the history of John Lewis and Waitrose. Dogs on leads welcome (Guide Dogs only indoors). Wheelchair access with some gravel paths.

18 THE OLD RECTORY, FARNBOROUGH

nr Wantage, OX12 8NX. Mr & Mrs Michael Todhunter. *4m SE of Wantage. Take B4494 Wantage-Newbury road, after 4m turn E at sign for Farnborough. Approx 1m to village, Old Rectory on L.* **Sun 16 Apr, Sun 14 May (2-5); Wed 28 June, Wed 9 Aug (11-4). Adm £5, chd free. Home-made teas. Donation to Farnborough PCC.**

In a series of immaculately tended garden rooms, inc herbaceous borders, arboretum, secret garden, roses, vegetables and bog garden, there is an explosion of rare and interesting plants, beautifully combined for colour and texture. With stunning views across the countryside, it is the perfect setting for the 1749 rectory (not open), once home of John Betjeman, in memory of whom John Piper created a window in the local church.

19 THE OLD RECTORY INKPEN
Lower Green, Inkpen, RG17 9DS. Mrs C McKeon, 07850 678239, oldrectoryinkpen@gmail.com. *4m SE of Hungerford. From centre of Kintbury at the Xrds, take Inkpen Rd. After ½m turn R, then go approx 3m (passing Crown & Garter Pub, then Inkpen Village Hall on L). Next to St Michael's Church.* **Visits by arrangement June to Sept for groups of up to 30. Adm £10, chd free. Light refreshments inc.**
On a gentle hillside with views of the Inkpen Hill, the Old Rectory offers a peaceful setting for this pretty 5⅛ acre garden. Originally an historic 1½ hectare formal garden to the south of the C17 Rectory, it is on a gentle slope surrounded by later additions of a walled kitchen garden, herbaceous borders, pleached lime walks and a recently cultivated wildflower meadow.

20 ROOKSNEST
Ermin Street, Lambourn Woodlands, RG17 7SB. Rooksnest Estate, 07787 085565, gardens@rooksnest.net. *2m S of Lambourn on B4000. From M4 J14, take A338 Wantage Rd, turn 1st L onto B4000 (Ermin St). Rooksnest signed after 3m.* **Wed 26 Apr, Wed 28 June (11-4). Adm £6, chd free. Light refreshments. Visits also by arrangement 5 Apr to 28 June for groups of 15 to 20.**
Approx 10 acre, exceptionally fine traditional English garden. Rose garden, herbaceous garden, pond garden, herb garden, fruit, vegetable and cutting garden and glasshouses. Many specimen trees and fine shrubs, orchard and terraces. Garden mostly designed by Arabella Lennox-Boyd. Limited WC facilities. Most areas have step free wheelchair access, although surface consists of gravel and mowed grass.

21 ROOKWOOD FARM HOUSE
Stockcross, Newbury, RG20 8JX. The Hon Rupert & Charlotte Digby, 01488 608676, charlotte@rookwoodhouse.co.uk, www.rookwoodhouse.co.uk. *3m W of Newbury. M4 J13, A34(S). After 3m exit for A4(W) to Hungerford. At 2nd r'about take B4000 towards Stockcross, after approx ¾m turn R into Rookwood.* **Sun 16 Apr, Mon 29 May (11-5). Adm £5, chd free. Home-made teas & light refreshments.**
This exciting valley garden, a work in progress, has elements all visitors can enjoy. A rose covered pergola, fabulous tulips, giant alliums, and a recently developed jungle garden with cannas, bananas and echiums. A kitchen garden features a parterre of raised beds, which along with a bog garden and colour themed herbaceous planting, all make Rookwood well worth a visit. WC available.

22 ST TIMOTHEE
Darlings Lane, Pinkneys Green, Maidenhead, SL6 6PA. Sarah & Sal Pajwani, 07976 892667, pajwanisarah@gmail.com. *1m N of Maidenhead. M4 J8/9 to A404M. 3rd exit onto A4 to Maidenhead. L at 1st r'about to A4130 Henley Rd. After ½m turn R onto Pinkneys Drive. At Pinkneys Arms Pub, turn L into Lee Ln, follow NGS signs.* **Sun 11 June (11-4). Adm £7, chd free. Talk & Walk events on Thur 16 Mar (10.30-12.30); Thur 6 July (2-4); Thur 2 Nov (10.30-12.30). Adm £16, chd free. Pre-booking essential, please visit www.ngs.org.uk for information & booking. Home-made teas. Visits also by arrangement 17 Jan to 9 Nov for groups of 10 to 50. Introductory talks on seasonal topics also offered.**
A 2 acre country garden planted for year-round interest with a variety of different colour themed borders each featuring a wide range of hardy perennials, shrubs and ornamental grasses. Also inc a box parterre, wildlife pond and rose terrace together with areas of long grass and beautiful mature trees, all set against the backdrop of a 1930s house. Small group 'Talk & Walk' events inc tea and cake are held on a variety of seasonal topics. Pre-booking essential. 16 March: The joy of Spring Bulbs. 6 July: Successional Planting. 2 Nov: Ornamental Grasses and the Autumn Garden.

23 STOCKCROSS HOUSE
Church Road, Stockcross, Newbury, RG20 8LP. Susan & Edward Vandyk, 01488 608810, Info@stockcrosshousegarden.co.uk. *3m W of Newbury. M4 J13, A34(S). After 3m exit A4(W) to Hungerford. At 2nd r'about take B4000, 1m to Stockcross, 2nd L into Church Rd. Coach parking available.* **Sun 27 Aug (12-5). Adm £5, chd free. Home-made teas. Visits also by arrangement May to Sept for groups of 10 to 50.**
Romantic 2 acre garden set around a former rectory (not open). Deep mixed borders with emphasis on strong, complementary colour combinations. Large orangery, wisteria and clematis clad pergola, folly with pond reflecting the church tower. Croquet lawn and pavilion. Naturalistic planting and pond on lower level. Small stumpery, fernery and kitchen garden. Plants from garden for sale. Partial wheelchair access with some gravelled areas.

24 STUBBINGS HOUSE
Stubbings Lane, Henley Road, Maidenhead, SL6 6QL. Mr & Mrs D Good, www.stubbingsnursery.co.uk. *From A404(M) W of Maidenhead, exit at A4 r'about & follow signs to Maidenhead. At the small r'about turn L towards Stubbings. Take the next L onto Stubbings Ln.* **Sat 11, Sun 12 Mar, Sat 13, Sun 14 May (10-4). Adm £4, chd free. Light refreshments.**
Parkland garden accessed via adjacent retail nursery. Set around C18 Grade II listed house (not open), home to Queen Wilhelmina of Netherlands in WW2. Large lawn with ha-ha and woodland walks. Notable trees inc historic cedars and araucaria. March brings an abundance of daffodils and in May a 60 metre wall of wisteria. Attractions inc a C18 icehouse and access to adjacent NT woodland. On site licenced café offering breakfast, snacks, cream teas and seasonal lunches. Wheelchair access to a level site with firm gravel paths.

25 61 SUTCLIFFE AVENUE
Earley, Reading, RG6 7JN. Sue & Dave Wilder. *3m SE of Reading. On Wokingham Rd (A329), from E, pass Showcase Cinema, then L at Co-op (Meadow Rd), then R into Sutcliffe Ave. If coming from W, turn R at Co-op.* **Daily Tue 9 May to Sat 13 May, Tue 6 June to Sat 10 June, Tue 22 Aug to Sat 26 Aug (10-4). Adm £3.50, chd free. Pre-booking essential, please visit www.ngs.org.uk for information & booking. Home-made teas. Two hour timed slots at 10am or 2pm.**

A characterful urban wildlife friendly garden, offering the relaxing sense of a walk in the countryside. An inviting path winds past wildflower islands, through a rose covered arch, to a cutting garden and small pond. Features inc tree trunks creatively recycled and a chicken run.

♿ ✿ ☕

GROUP OPENING

26 SWALLOWFIELD VILLAGE GARDENS

The Street, Swallowfield, RG7 1QY. *5m S of Reading. From M4 J11 take A33 S. At 1st T-lights turn L on B3349 signed Swallowfield. In the village follow signs for parking. Buy tickets & map in Swallowfield Medical Practice car park, opp Crown Pub.* **Sun 4 June (2-5.30). Combined adm £8, chd free. Home-made teas at Brambles.**

BIRD IN HAND HOUSE
Margaret McDonald.

BRAMBLES
Sarah & Martyn Dadds.

NEW **BRIERLEY HOUSE**
Jean & Richard Trinder.

THE FIRS
Harmi Kandohla & Mark Binns.

9 FOXBOROUGH
Jeremy Bayliss.

LAMBS FARMHOUSE
Eva Koskuba.

NORKETT COTTAGE
Jenny Spencer.

NEW **SUVIKUJA**
Catherine Glover & Richard Hoyle.

This year Swallowfield is offering eight gardens to visit. A number are in the village itself or just outside, so there is a mix of walking to some, with those in different directions needing a car or bicycle to reach them comfortably. Whilst each provides its own character and interest, they all nestle in countryside by the Blackwater and Loddon rivers with an abundance of wildlife and lovely views. The garden owners, many of whom are members of the local Horticultural Society are always happy to chat and share their enthusiasm and experience. See website for info about individual gardens. Plants for sale. Wheelchair access to many gardens, however some have gravel drives, slopes and uneven ground.

♿ ✿ ☕ 🔊

Wickham House

Middle Barton

27 7 THE KNAPP

Earley, Reading, RG6 7DD. Ann & Alex McKie, 07881 451708, annmckie@hotmail.co.uk. *A329 from Wokingham to Reading. Turn L into Aldbourne Ave just before junction with B3350 & 1st L into The Knapp.* **Sat 10, Sun 11, Sat 17, Sun 18 June (10.30-3.30). Adm £4, chd free. Pre-booking essential, please visit www.ngs.org.uk for information & booking. Home-made teas. Two hour timed slots at 10.30am or 1.30pm. Visits also by arrangement 22 May to 8 July for groups of 5 to 20.**

A mature suburban garden, surrounded by trees and divided into rooms with lawns, island beds, borders, gravel garden, pond and an area with wild flowers. The highlights are the roses, particularly rambling roses over gazebos and arches and the wildflower areas.

28 THURLE GRANGE

Rectory Road, Streatley, Reading, RG8 9QH. David Juster. *From Streatley, N on A329 Wallingford Rd. Fork L onto A417 Wantage Rd, then fork L again along Rectory Rd. Past golf course, downhill, past stables. Thurle Grange is on R. Parking will be signed.* **Sat 10, Sun 11 June (10-4). Combined adm with Wynders £6, chd free.**

Recently developed 1 acre garden set around attractive country house (c1900, not open) with lovely views across peaceful valley. Centred upon a splendid catalpa tree with a mix of formal and informal successional planting for year-round interest. Features inc rose filled parterre, wildflower area, yew avenue and richly planted herbaceous borders. Home-made teas at Wynders.

29 ◆ WELFORD PARK

Welford, Newbury, RG20 8HU. Mrs J H Puxley, 01488 608691, snowdrops@welfordpark.co.uk, www.welfordpark.co.uk. *6m NW of Newbury. M4 J13, A34(S). After 3m exit for A4(W) to Hungerford. At 2nd r'about take B4000, after 4m turn R signed Welford. Entrance on Newbury-Lambourn road.* **For NGS: Wed 8 Feb (11-4). Adm £12, chd £4.50. Light refreshments. For other opening times and information, please phone, email or visit garden website.**

One of the finest natural snowdrop woodlands in the country and a wonderful display of hellebores, galanthus cultivars, winter flowering shrubs throughout the extensive gardens. This is an NGS 1927 pioneer garden on the River Lambourn set around Queen Anne House (not open). Also, the stunning setting for Great British Bake Off 2014 - 2019/22. Refreshments in large tent serving light lunches, teas and cakes. Also, hot drinks and snacks at Shepherds Hut in garden. Children friendly and dogs welcome on leads. Coach parties please book in advance. Paths are as wheelchair friendly as possible, please watch for signs to avoid damp areas.

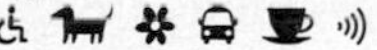

30 WEMBURY

Altwood Close, Maidenhead, SL6 4PP. Carolyn Foster. *W side of Maidenhead, S of A4. M4 J8/9, then A404M J9B, 3rd exit to A4 to Maidenhead, R at 1st r'about, 2nd L onto Altwood Rd, then 5th R, follow NGS signs.* **Tue 27, Fri 30 June, Tue 1 Aug, Tue 12 Sept (10.30-4). Adm £4, chd free. Pre-booking essential, please visit www.ngs.org.uk for information & booking. Home-made teas. Two hour timed slots at 10.30am or 2pm.**

A wildlife friendly, plant lover's cottage garden with borders generously planted with bulbs, perennials, grasses and shrubs for successional interest for every season. Productive vegetable garden and greenhouses. Many pots and baskets with seasonal annuals and tender perennials especially salvias.

31 WICKHAM HOUSE

Wickham, Newbury, RG20 8HD. Mr & Mrs James D'Arcy. *7m NW of Newbury or 6m NE of Hungerford. From M4 J14, take A338(N) signed Wantage. Approx ¾m turn R onto B4000. Drive through Wickham. Entrance 100yd uphill on R. From Newbury take B4000, house on L 50yds past the Wickham speed limit sign.* **Sun 18 June (11.30-3.30). Adm £5, chd free. Home-made teas & gammon rolls.**

In a beautiful country house setting, this exceptional ½ acre walled garden was created from scratch in 2008. Designed by Robin Templar-Williams, the different rooms have distinct themes and colour schemes. Delightful arched clematis and rose walkway. Wide variety of trees, planting, pots brimming with colour and places to sit and enjoy the views. Separate cutting and vegetable garden. Wheelchair access over gravel paths.

32 WYNDERS

Rectory Road, Streatley, Reading, RG8 9QA. Marcus & Emma Francis, 07920 712571, marcus.francis@fsp-law.com, www.instagram.com/marcusfrancis21. *From Streatley, N on A329 Wallingford Rd. Fork L onto A417 Wantage Rd, then fork L again along Rectory Rd. After golf course, downhill & parking will be signed. Wynders is on L after stables.* **Sat 10, Sun 11 June (10-4). Combined adm with Thurle Grange £6, chd free. Home-made teas. Visits also by arrangement 10 Feb to 31 Oct for groups of 10 to 50. Tea, cake & tour with Head Gardener inc.**

Densely planted ¾ acre garden surrounded by stunning countryside. Over 60 roses, 150ft long border, formal garden, gravel garden, cutting beds, entertainment for children. Free roaming wildfowl and waterfowl. Classic and modern sports cars on display. Something for everyone. Outside of opening weekend in June, fabulous autumn colours, winter garden and spring bulb displays. Partial wheelchair access over gravel drive, please call in advance for reserved disabled parking.

BUCKINGHAMSHIRE

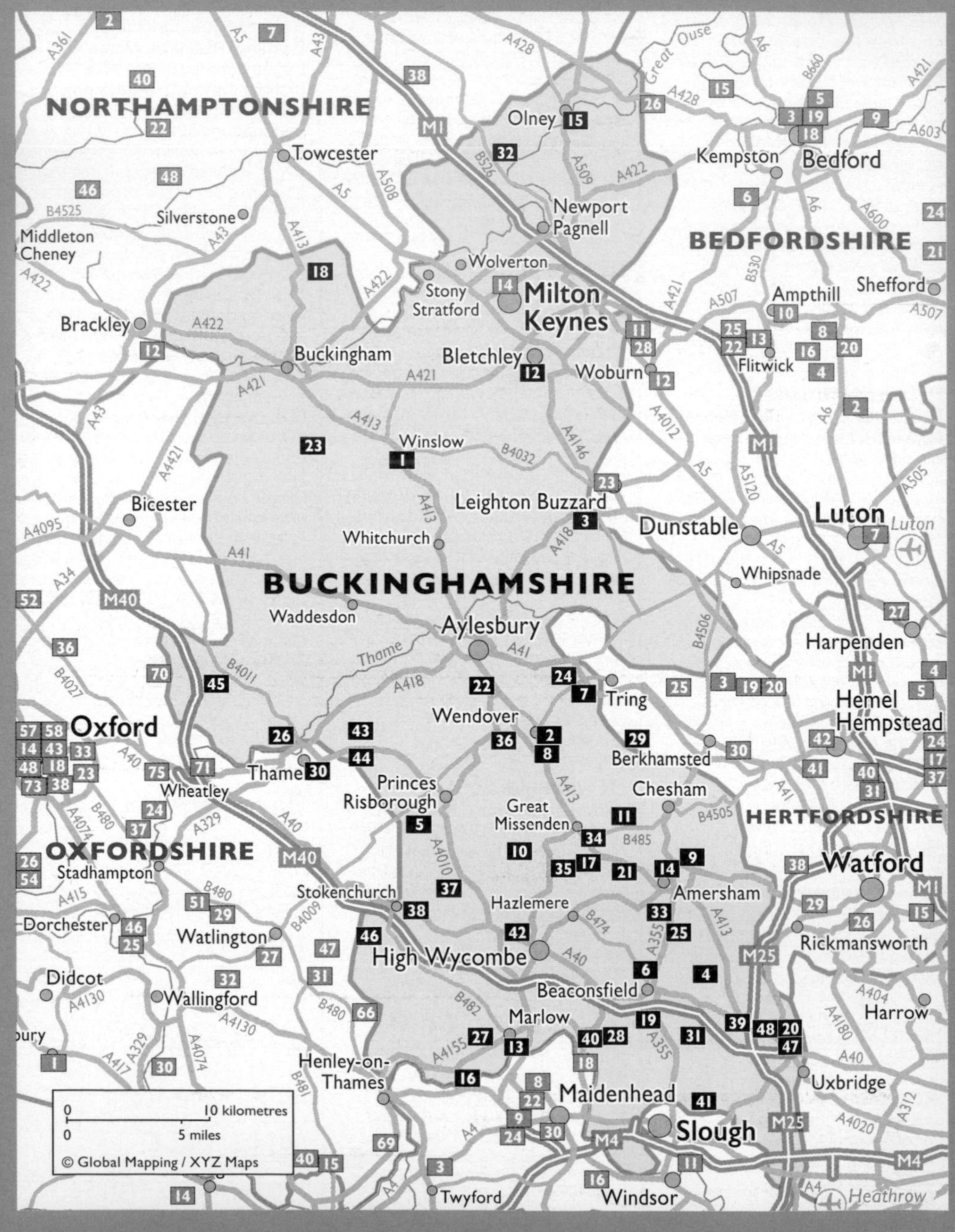

Buckinghamshire has a beautiful and varied landscape; edged by the River Thames to the south, crossed by the Chiltern Hills, and with the Vale of Aylesbury stretching to the north.

This year Buckinghamshire will hold three group openings, many of which can be found in villages of thatched or brick and flint cottages.

Many Buckinghamshire gardens have been used as locations for films and television, with the Pinewood Studios nearby and excellent proximity to London.

We also boast historical gardens including Ascott, Cowper and Newton Museum Gardens, Hall Barn, and Stoke Poges Memorial Gardens (Grade I listed).

Most of our gardens offer homemade tea and cakes to round off a lovely afternoon, visitors can leave knowing they have enjoyed a wonderful visit and helped raise money for nursing and health charities at the same time.

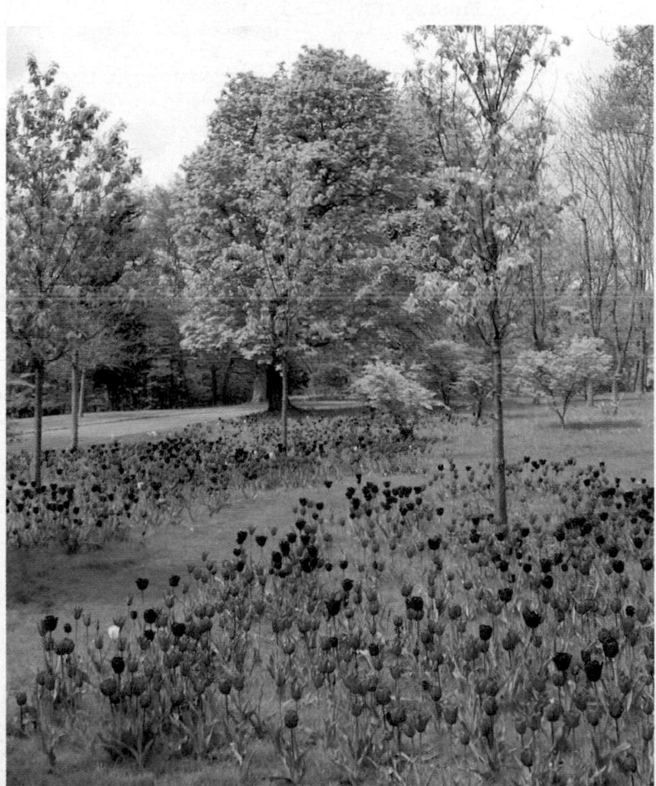

Volunteers

County Organiser
Maggie Bateson
01494 866265
maggiebateson@gmail.com

County Treasurer
Tim Hart
01494 837328
timgc.hart@btinternet.com

Publicity
Sandra Wetherall
01494 862264
sandracwetherall@gmail.com

Social Media
Stella Vaines
07711 420621
stella@bakersclose.com

Talks
Janice Cross
01494 728291
janice.cross@ngs.org.uk

Booklet Co-ordinator
Maggie Bateson
(as above)

Assistant County Organisers
Janice Cross
(as above)

Judy Hart
01494 837328
judy.elgood@gmail.com

Mhairi Sharpley
01494 782870
mhairisharpley@btinternet.com

Stella Vaines
(as above)

@BucksNGS

@BucksNGS

@national_garden_scheme_bucks

Left: Orchard House

OPENING DATES

All entries subject to change. For latest information check **www.ngs.org.uk**

Map locator numbers are shown to the right of each garden name.

February

Snowdrop Festival

Sunday 19th
Hollydyke House 21

March

Wednesday 15th
Montana 29

Sunday 26th
Wind in the Willows 48

April

Sunday 9th
Overstroud Cottage 34

Sunday 16th
Long Crendon Gardens 26

Sunday 23rd
Abbots House 1
The Lantern Cottage 24
Orchard House 33

Sunday 30th
Beech House 4

May

Monday 1st
Turn End 43

Sunday 14th
Overstroud Cottage 34
◆ Stoke Poges Memorial Gardens 41

Tuesday 16th
Red Kites 37

Wednesday 17th
Montana 29

Sunday 21st
The Plough 36

Sunday 28th
18 Brownswood Road 6
Higher Denham Gardens 20
◆ Nether Winchendon House 30

Monday 29th
Glebe Farm 18

June

Sunday 4th
Abbots House 1
Overstroud Cottage 34
The Plough 36
Tythrop Park 44
The White House 47

Saturday 10th
Acer Corner 2
◆ Cowper & Newton Museum Gardens 15

Sunday 11th
Acer Corner 2
Bledlow Manor 5
Canal Cottage 7
126 Church Green Road 12
◆ Cowper & Newton Museum Gardens 15
The Lantern Cottage 24
Long Crendon Gardens 26

Tuesday 13th
126 Church Green Road 12

Wednesday 14th
The Walled Garden, Wormsley 46

Thursday 15th
Lords Wood 27

Saturday 17th
St Michaels Convent 39

Saturday 24th
NEW Chiltern Haven 11
Horatio's Garden 22
Old Park Barn 32
Robin Hill 38

Sunday 25th
Old Park Barn 32

July

Saturday 1st
◆ Ascott 3
Little Missenden Gardens 25

Sunday 2nd
Chiltern Forage Farm 10
Little Missenden Gardens 25
Wadzana 45

Sunday 9th
NEW 18 Copperkins Lane 14
Fressingwood 17

Tuesday 11th
Red Kites 37

Wednesday 12th
8 Claremont Road 13

Thursday 13th
8 Claremont Road 13

Saturday 15th
8 Claremont Road 13

Sunday 16th
8 Claremont Road 13

Wednesday 19th
Montana 29

Sunday 23rd
Old Keepers 31

August

Wednesday 2nd
Danesfield House 16

Sunday 27th
◆ Nether Winchendon House 30

September

Saturday 16th
Acer Corner 2

Sunday 17th
Acer Corner 2

By Arrangement

Arrange a personalised garden visit with your club, or group of friends, on a date to suit you. See individual garden entries for full details.

Abbots House 1
Acer Corner 2
Beech House 4
Cedar House 8
Chesham Bois House 9
Glebe Farm 18
Hall Barn 19
Hollydyke House 21
Kingsbridge Farm 23
Magnolia House 28
Montana 29
Old Keepers 31
Old Park Barn 32
Overstroud Cottage 34
Peterley Corner Cottage 35
Red Kites 37
Robin Hill 38
The Shades 40
1 Talbot Avenue 42
Wind in the Willows 48

Our donation of £2.5 million over five years has helped fund the new Y Bwthyn NGS Macmillan Specialist Palliative Care Unit in Wales. 1,900 inpatients have been supported since its opening

THE GARDENS

1 ABBOTS HOUSE

10 Church Street, Winslow, MK18 3AN. Mrs Jane Rennie, 01296 712326, jane@renniemail.com. *9m N of Aylesbury. A413 into Winslow. From town centre take Horn St & R into Church St, L fork at top. Entrance through door in wall, Church Walk. Parking in town centre & adjacent streets.* **Sun 23 Apr, Sun 4 June (1-5). Adm £4.50, chd free. Home-made teas. Visits also by arrangement 24 Apr to 28 July for groups of 5 to 15.**

Behind red brick walls a 3/4 acre garden on four different levels, each with unique planting and atmosphere. Lower lawn with white wisteria arbour and pond, upper lawn with rose pergola and woodland, pool area with grasses, Victorian kitchen garden and wild meadow. Spring bulbs in wild areas and woodland, many tulips in Apr, remaining areas peak in June/July. Water feature, sculptures and many pots. Rennie's Winslow Cider and Cider Lush available. Partial wheelchair access due to steps to garden levels. Guide dogs and medical-aid dogs only.

2 ACER CORNER

10 Manor Road, Wendover, HP22 6HQ. Jo Naiman, 07958 319234, jo@acercorner.com, www.acercorner.com. *3m S of Aylesbury. Follow A413 into Wendover. L at clock tower r'about into Aylesbury Rd. R at next r'about into Wharf Rd, continue past schools on L, garden on R.* **Sat 10, Sun 11 June, Sat 16, Sun 17 Sept (2-5). Adm £4, chd free. Visits also by arrangement 1 Apr to 24 Sept for groups of up to 20.**

Garden designer's garden with Japanese influence and large collection of Japanese maples. The enclosed front garden is Japanese in style. Back garden is divided into three areas; patio area recently redesigned in the Japanese style, densely planted area with many acers and roses, also a corner which inc a productive greenhouse and interesting planting.

3 ◆ ASCOTT

Ascott Estate, Wing, Leighton Buzzard, LU7 0PP. The National Trust, info@ascottestate.co.uk, www.ascottestate.co.uk. *2m SW of Leighton Buzzard, 8m NE of Aylesbury. Via A418. Buses: 150 Aylesbury - Milton Keynes, 250 Aylesbury & Milton Keynes. Access is via the visitors entrance, use postcode LU7 0ND.* **For NGS: Sat 1 July (12-5). Adm £7, chd £3. Pre-booking essential, please visit national-trust-ascott.checkfront.com/reserve for information & booking. Light refreshments. NT members are required to pay to enter the gardens on NGS days. For other opening times and information, please email or visit garden website.**

Combining Victorian formality with early C20 natural style and recent plantings to lead it into the C21. A completed garden designed by Jacques and Peter Wirtz and also a Richard Long sculpture. Terraced lawns with specimen and ornamental trees, panoramic views to the Chilterns. Naturalised bulbs, mirror image herbaceous borders and impressive topiary inc box and yew sundial. Ascott House is closed on NGS Days The Ascott tearoom will be open until 15 mins before closing, serving a range of hot and cold refreshments. Please contact the Estate Office to reserve a wheelchair in advance of your visit.

4 BEECH HOUSE

Long Wood Drive, Jordans, Beaconsfield, HP9 2SS. Sue & Ray Edwards, 01494 875580, raychessmad@hotmail.com. *From A40, L to Seer Green & Jordans for approx 1m, turn into Jordans Way on R, Long Wood Dr 1st L. From A413, turn into Chalfont St Giles, straight ahead until L signed Jordans, 1st L Jordans Way.* **Sun 30 Apr (2-5). Adm £4, chd free. Biscuits & cordials. Visits also by arrangement 12 Mar to 31 Aug.**

2 acre plantsman's garden built up over the last 35 yrs with a wide range of flowering and foliage plants in a variety of habitats providing a show all yr. Many bulbs, perennials, shrubs, roses and grasses provide continuous interest. Numerous trees planted for their flowers, foliage, ornamental bark and autumn display. Two attractive meadows are always a popular feature. Wheelchair access dependent upon weather conditions.

5 BLEDLOW MANOR

Off Perry Lane, Bledlow, nr Princes Risborough, HP27 9PA. Lord & Lady Carrington, www.carington.co.uk/gardens. *9m NW of High Wycombe, 3m SW of Princes Risborough. 1/2m off B4009 in middle of Bledlow village. For SatNav use postcode HP27 9PA.* **Sun 11 June (2-5). Adm £7, chd free. Home-made teas.**

The present garden covering around 12 acres inc the walled kitchen garden crisscrossed by paths with vegetables, fruit, herbs and flowers. The sculpture garden, the replanted Granary garden with fountain and borders, as well as the individual paved gardens and parterres divided by yew hedges and more. The Lyde water garden formed out of old cress beds fed by numerous springs. Partial wheelchair access via steps or sloped grass to enter gardens.

6 18 BROWNSWOOD ROAD

Beaconsfield, HP9 2NU. John & Bernadette Thompson. *From New Town turn R into Ledborough Ln, L into Sandleswood Rd, 2nd R into Brownswood Rd.* **Sun 28 May (2-5.30). Adm £4, chd free. Home-made teas.**

A plant filled garden designed by Barbara Hunt. A harmonious arrangement of arcs and circles introduces a rhythm that leads through the garden. Sweeping box curves, gravel beds, brick edging and lush planting. A restrained use of purples and reds dazzle against a grey and green background. There has been considerable replanning and replanting during the winter.

In 2022 we donated £450,000 to Marie Curie which equates to 23,871 hours of community nursing care

7 CANAL COTTAGE
11 Wharf Row, Buckland Road, Buckland, Aylesbury, HP22 5LJ. Angela Hale. *4m E of Aylesbury. On the A41 exit at Aston Clinton/ Wendover. Down Lower Icknield Way onto Buckland Rd. Park along road & walk over white bridge. Garden is signed.* **Sun 11 June (2-5). Adm £3, chd free. Open nearby The Lantern Cottage.**
A peaceful garden celebrates what can be achieved in a long narrow strip behind a cottage (not open); landscaped to create a variety of rooms. Open deep borders of perennials welcome the visitor. Shaped lawns lead you through a shady rose arbour undergrown with a variety of ferns and hostas. Final room is a pond, seating area with Mediterranean terracing and many pots.

8 CEDAR HOUSE
Bacombe Lane, Wendover, HP22 6EQ. Sarah Nicholson, 01296 622131, sarahhnicholson@btinternet.com. *5m SE Aylesbury. From Gt Missenden take A413 into Wendover. Take 1st L before row of cottages, house at top of lane. Parking for 10 cars only.* **Visits by arrangement 11 Feb to 17 Sept for groups of 10 to 30. Adm £5, chd free. Light refreshments.**
A plantsman's garden in the Chiltern Hills with a great variety of trees, shrubs and plants. A sloping lawn leads to a natural swimming pond with wild flowers inc native orchids. A lodge greenhouse and a good collection of half-hardy plants in pots. Local artist's sculptures can be viewed. Picnics welcome on prior request. Wheelchair access over gentle sloping lawn.

9 CHESHAM BOIS HOUSE
85 Bois Lane, Chesham Bois, Amersham, HP6 6DF. Julia Plaistowe, 01494 726476, plaistowejulia@gmail.com, cheshamboishouse.co.uk. *1m N of Amersham-on-the-Hill. Follow Sycamore Rd (main shopping centre road of Amersham), which becomes Bois Ln. Do not use SatNav once in lane as you will be led astray.* **Visits by arrangement Mar to Aug. Adm £5, chd free. Home-made teas.**
3 acre beautiful garden with primroses, daffodils and hellebores in early spring. Interesting for most of the yr with lovely herbaceous borders, rill with small ornamental canal, walled garden, old orchard with wildlife pond, and handsome trees some of which are topiary. It is a peaceful oasis. Wheelchair access via gravel at front of house.

10 CHILTERN FORAGE FARM
Spring Coppice Lane, Speen, Princes Risborough, HP27 0SU. Emma Plunket, www.chilternforagefarm. *Follow signs to gate at top of the hill & car park in field.* **Sun 2 July (2.30-5). Adm £5, chd free. Light refreshments.**
Tours at 2.30pm and 4pm of this project under development in attractive hillside setting. Pasture being restored to hay meadows and planted with fruit trees and soft fruit. Tips for foraging, encouraging wild flowers and creation of wildlife habitats. On calm sunny day great for spotting butterflies and moths.

11 NEW CHILTERN HAVEN
Pednor, Chesham, HP5 2SX. Mr Jonathan & Mrs Clare Waters. *2m W of Chesham, 3m E of Great Missenden. Chesham/Great Missenden turn off B485 to Little Hundridge Ln. 1/8m R at T-junction, R & parking on R. From Chesham Chartridge Ln, turn L to Westdean Ln, then R to Pednor Bottom, after 1½m parking on R.* **Sat 24 June (2-5). Adm £5, chd free. Home-made teas.**
A country garden over 1½ acres with a wide range of flowering and foliage plants. Bulbs, perennials, roses, shrubs and trees have been chosen to provide continuous interest, with colour concentrated in purple, pink, blue and lemon. The lawns around the house have deep, curving borders with sunny and shady aspects. There is also a wildlife pond, old orchard, small vegetable garden and greenhouse.

12 126 CHURCH GREEN ROAD
Bletchley, Milton Keynes, MK3 6DD. Janice & David Hale. *13m E of Buckingham, 11m N of Leighton Buzzard. Turn R off B4034 into Church Green Rd, take L turn at mini r'about.* **Sun 11 June (2-6); Tue 13 June (2-5.30). Adm £4, chd free. Light refreshments.**
A gentle sloping mature garden of ½ acre is a plant lover's delight, which inc a small formal garden, shady areas and mixed borders of shrubs, perennials and roses. Features inc a thatched Wendy house, pergolas, formal pond, wildlife pond, productive fruit and vegetable garden, two greenhouses and patio.

13 8 CLAREMONT ROAD
Marlow, SL7 1BW. Andi Gallagher. *No parking at garden, but short walk from all town car parks. From town centre walk S on High St, L on Institute Rd, L on Beaufort Gardens & R on Claremont Rd.* **Wed 12, Thur 13, Sat 15, Sun 16 July (2-5). Adm £5, chd free.**
Paintings, prints and pots to see and buy in this small town garden owned by an artist gardener. The unusual house was built in 2015. Gravel paths divide the rectangular beds filled with herbaceous perennials, grasses and ferns. A cow trough water feature and owner's ceramics add surprise. A gate leads to a deliberate wild area with fruit trees and art studio with garden related paintings.

14 NEW 18 COPPERKINS LANE
Amersham, HP6 5QF. Chris Ludlam. *1m N of Amersham. 15 min walk from Amersham Tube/Train Stn. Off Amersham/Chesham road A416. Parking on road only, please park with consideration to neighbours.* **Sun 9 July (12-5). Adm £4.50, chd free. Pre-booking essential, please visit www.ngs.org.uk for information & booking. Home-made teas.**
A fabulous compact ½ acre garden with lawn, herbaceous borders, wildflower meadows, orchard, raised vegetable beds, treehouse, summerhouse, and pool. Originally designed as a white garden, it now inc calming purple, blue and pink colours. Plenty of roses, grasses, salvia, nepeta, peony, delphinium, lupin, geranium, foxglove, allium, lavender, astrantia, pittosporum, and ornamental trees. Wheelchair access over deep gravel path, plus two sloped tiled paths on entry.

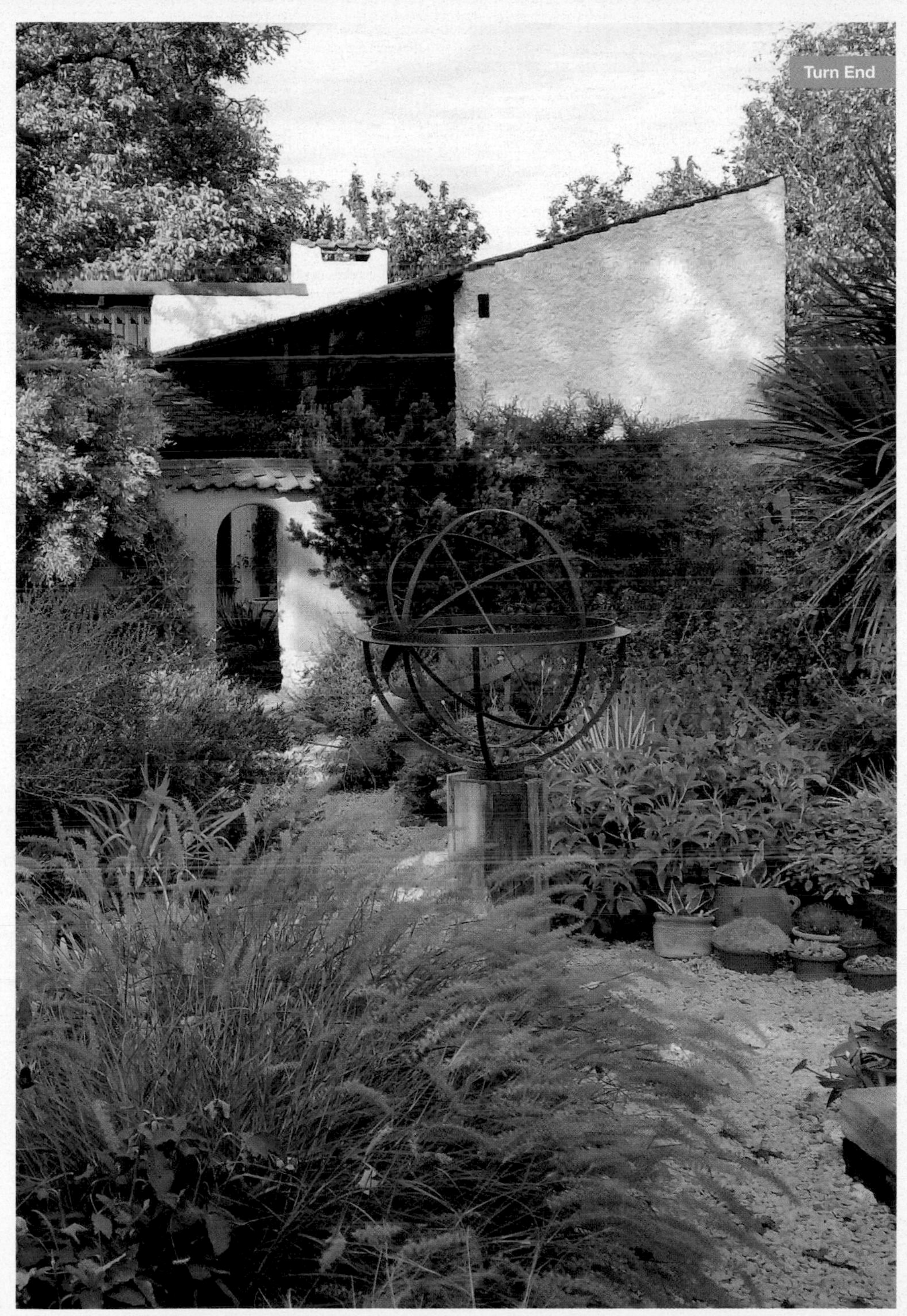

Turn End

15 ◆ COWPER & NEWTON MUSEUM GARDENS
Orchard Side, Market Place, Olney, MK46 4AJ. Cowper and Newton Museum, 01234 711516, house-manager@cowperandnewtonmuseum.org.uk, www.cowperandnewtonmuseum.org.uk. *5m N of Newport Pagnell. 12m S of Wellingborough. On A509. Please park in public car park in East St.* **For NGS: Sat 10, Sun 11 June (11-4.30). Adm £4, chd free. Home-made teas. For other opening times and information, please phone, email or visit garden website.**
The tranquil Flower Garden of C18 poet William Cowper, who said 'Gardening was of all employments, that in which I succeeded best', has plants introduced prior to his death in 1800, many mentioned in his writings. The Summer House Garden with Cowper's 'verse manufactory', now a Victorian Kitchen Garden, has new and heritage vegetables organically grown, also a herb border and medicinal plant bed. Features inc lacemaking demonstrations and local artists painting live art on both days. Georgian dancers on Sun 11 June. Wheelchair access on mostly hard paths.

16 DANESFIELD HOUSE
Henley Road, Marlow, SL7 2EY. Danesfield House Hotel, 01628 891010, amoorin@danesfieldhouse.co.uk, www.danesfieldhouse.co.uk. *3m from Marlow. On the A4155 between Marlow & Henley-on-Thames. Signed on the LHS Danesfield House Hotel & Spa.* **Wed 2 Aug (10-4). Adm £5, chd free.**
The gardens at Danesfield were completed in 1901 by Robert Hudson, the Sunlight Soap magnate who built the house. Since the house opened as a hotel in 1991, the gardens have been admired by several thousand guests each yr. However, in 2009 it was discovered that the gardens contained outstanding examples of pulhamite in both the formal gardens and the waterfall areas. The 100 yr old topiary is also outstanding. Part of the grounds inc an Iron Age fort. Tour with Head Gardener, Dan Lawrence at 10.30am or 1.30pm only (max 30 per tour). Tours must be pre-booked and lunch and afternoon tea must be pre-booked by contacting Alex Moorin on amoorin@danesfieldhouse.co.uk or 01628 891010. Wheelchair access on gravel paths through parts of the garden.

17 FRESSINGWOOD
Hare Lane, Little Kingshill, Great Missenden, HP16 0EF. John & Maggie Bateson. *1m S of Great Missenden, 4m W of Amersham. From the A413 at Chiltern Hospital, turn L signed Gt & Lt Kingshill. Take 1st L into Nags Head Ln. Turn R under railway bridge, then L into New Rd & continue to Hare Ln.* **Sun 9 July (2-5.30). Adm £4.50, chd free. Home-made cake & elderflower cordial.**
Thoughtfully designed and structured garden with year-round colour and many interesting features inc herbaceous borders, a shrubbery with ferns, hostas, grasses and hellebores. Small formal garden, pergolas with roses and clematis. A variety of topiary and small garden rooms. A central feature incorporating water with grasses. Large bonsai collection.

Bledlow Manor

© Val Corbett

18 GLEBE FARM
Lillingstone Lovell, Buckingham, MK18 5BB. Mr David Hilliard, 01280 860384, thehilliards@talk21.com, sites. google.com/site/glebefarmbarn. *Off A413, 5m N of Buckingham & 2m S of Whittlebury. From A5 at Potterspury, turn off A5 & follow signs to Lillingstone Lovell.* **Mon 29 May (1.30-5.30). Adm £4, chd free. Home-made teas. Visits also by arrangement in June for groups of 5 to 10.**
A large cottage garden with an exuberance of colourful planting and winding gravel paths, amongst lawns and herbaceous borders on two levels. Ponds, a wishing well, vegetable beds, a knot garden, a small walled garden and an old tractor feature. Everything combines to make a beautiful garden full of surprises.

19 HALL BARN
Windsor End, Beaconsfield, HP9 2SG. Mrs Farncombe, 01494 677788, garden@thefarncombes.com. *½m S of Beaconsfield. Lodge gate 300yds S of St Mary & All Saints' Church in Old Town centre. Please do not use SatNav.* **Visits by arrangement. Adm £5, chd free. Home-made teas provided for groups of 10+.**
Historical landscaped garden laid out between 1680-1730 for the poet Edmund Waller and his descendants. Features 300 yr old cloud formation yew hedges, formal lake and vistas ending with classical buildings and statues. Wooded walks around the grove offer respite from the heat on sunny days. One of the original NGS garden openings of 1927.

GROUP OPENING

20 HIGHER DENHAM GARDENS
Higher Denham, UB9 5EA. *6m E of Beaconsfield. Turn off A412, approx ½m N of junction with A40 into Old Rectory Ln. After 1m enter Higher Denham straight ahead. Tickets for all gardens at community hall, 70yds into the village.* **Sun 28 May (2-5). Combined adm £6, chd free. Home-made teas in village hall. Donation to Higher Denham Community CIO (Garden Upkeep Fund).**

NEW **11 ASHCROFT DRIVE**
Robert Brooks.

33B LOWER ROAD
Mr & Mrs J Hughes.

5 SIDE ROAD
Jane Blyth.

WIND IN THE WILLOWS
Ron James.
(See separate entry)

At least four gardens will open in the delightful Misbourne chalk stream valley, and Wind in the Willows also opens separately in March. Wind in the Willows has over 350 shrubs and trees, informal woodland and wild gardens inc riverside and bog plantings and a collection of 80 hostas and 12 striped roses in 3 acres. A new water lily pond has been created within the river. 'Really different' was a typical visitor comment. The garden at 5 Side Road is medium sized with lawns, borders and shrubs, and many features which children will love. A big conifer has been removed and the area replanted for greater interest. 33B Lower Road is an informal wildlife friendly garden with a variety of shrubs, perennials, ferns and vegetables. 11 Ashcroft Drive a unique and not to be missed garden laid out in a Persian style stocked with typical Persian plants. This garden is a short walk or drive away from the other three. In May the owner of Wind in the Willows will lead optional guided tours of the garden starting at 2.30pm and 4pm. Tours generally last approx 1 hour. Partial wheelchair access.

21 HOLLYDYKE HOUSE
Little Missenden, HP7 0RD. Bob & Sandra Wetherall, 01494 862264, sandracwetherall@gmail.com. *Off A413 between Great Missenden & Amersham. Parking, please use field adj to house, weather permitting.* **Sun 19 Feb (11-4). Adm £4, chd free. Light refreshments. Opening with Little Missenden Gardens on Sat 1, Sun 2 July. Visits also by arrangement for groups of 10+.**
3 acre garden surrounds Hollydyke House (not open) with year-round interest. During Feb we hope to see hellebores, crocus and carpets of snowdrops. You will be able to appreciate the structure in winter with the trees, colourful barks, seed heads and grasses. Walk down the road to see more snowdrops surrounding C10 Saxon Church, St John The Baptist. Join a tour to learn about its fascinating history. Wheelchair access on gravel paths.

22 HORATIO'S GARDEN
National Spinal Injuries Centre (NSIC), Stoke Mandeville Hospital, Mandeville Road, Stoke Mandeville, Aylesbury, HP21 8AL. Amy Moffett, www.horatiosgarden.org.uk. *The closest car park to Horatio's Garden is at Stoke Mandeville Hospital, Car Park B, opp Asda. Free parking on open day.* **Sat 24 June (2-5). Adm £5, chd free. Home-made teas in the garden room.**
Opened in Sept 2018, Horatio's Garden at the National Spinal Injuries Centre, Stoke Mandeville Hospital is designed by Joe Swift. The fully accessible garden for patients with spinal injuries has been part funded by the NGS. The beautiful space is cleverly designed to bring the sights, sounds and scents of nature into the heart of the NHS. Everything is high quality and carefully designed to bring benefit to patients who often have lengthy stays in hospital. The garden features a contemporary garden room, designed by Andrew Wells as well as a stunning Griffin Glasshouse. We also have a wonderful wildflower meadow. Please come along and meet the Head Gardener and volunteer team and taste our delicious tea and home-made cake! The garden is fully accessible, having been designed specifically for patients in wheelchairs or hospital beds.

23 KINGSBRIDGE FARM
Steeple Claydon, MK18 2EJ. Mr & Mrs Tom Aldous, 01296 730224. *3m S of Buckingham. Halfway between Padbury & Steeple Claydon. Xrds signed to Kingsbridge Only.* **Visits by arrangement Mar to July for groups of 6+. Adm £6, chd free. Home-made teas in our cosy, converted barn.**
Stunning and exceptional 6 acre garden, ever evolving! Main lawn is enclosed by softly curving, colour themed herbaceous borders with gazebo, topiary, clipped yews, pleached hornbeams leading to ha-ha and countryside beyond. A natural stream with bog plants and nesting kingfishers, meanders serenely through shrub and woodland gardens with many walks. A garden always evolving, to visit again and again.

24 THE LANTERN COTTAGE

26 Green End Street, Aston Clinton, Aylesbury, HP22 5JE. Jacki Connell. *3m E of Aylesbury. From Aylesbury take A41E. At large r'about, continue straight (signed Aston Clinton). At next r'about continue straight onto London Rd. L at The Bell Pub, parking on Green End St.* **Sun 23 Apr, Sun 11 June (2-5). Adm £3, chd free. Home-made teas. Open nearby Canal Cottage on 11 June only.**

At The Lantern Cottage, spring hellebores and an abundance of tulips give way to an early summer display of roses, peonies, bearded iris, alliums and climbers inc various clematis, wisteria and akebia. A wide selection of salvias and herbaceous perennials, mostly raised from seed and cuttings. Pelargonioums provide year-round colour in the conservatory and the greenhouse is always full!

GROUP OPENING

25 LITTLE MISSENDEN GARDENS

Amersham, HP7 0RD. *On A413 between Great Missenden & Old Amersham. Please park in signed car parks.* **Sat 1, Sun 2 July (2-6). Combined adm £8, chd free. Home-made teas at Hollydyke House & the church.**

BOURN'S MEADOW
Roger & Sandra Connor.

NEW **GRANARY BARN**
Jane Hill.

HOLLYDYKE HOUSE
Bob & Sandra Wetherall.
(See separate entry)

KINGS BARN
Mr & Mrs Playle.

LITTLE MISSENDEN CE INFANT SCHOOL
Little Missenden School.

MANOR FARM HOUSE
Evan Bazzard.

NEW **MILLARD'S BARN**
Rosemary Higgs.

MISSENDEN HOUSE
Sara Jane Ambrose.

MISSENDEN LODGE
Mr & Mrs R Kimber.

ORCHARD COTTAGE
Peter & Jeannie Macewan.

TOWN FARM COTTAGE
Mr & Mrs Tim Garnham.

TUDOR COTTAGE
H Graham.

THE WHITE HOUSE
Mr & Mrs Harris.

A variety of gardens set in this attractive Chiltern village in an AONB. You can start off at one end of the village and wander through stopping off halfway for tea at the beautiful Anglo-Saxon church built in 975. The church has recently received a lottery grant for restoration and many medieval wall paintings have been found. Tours will be given of these discoveries. The gardens reflect different style houses inc several old cottages, converted barns, Elizabethan and Georgian houses (houses not open). There are herbaceous borders, shrubs, trees, old fashioned roses, hostas, topiary, koi and lily ponds, kitchen gardens, play areas for children and the River Misbourne runs through a few. Some gardens are highly colourful and others just green and peaceful. Beekeeper at Hollydyke House. Village used many times for filming inc the newly released Disney film Cruella. Partial wheelchair access to some gardens due to gravel paths and steps.

GROUP OPENING

26 LONG CRENDON GARDENS

Long Crendon, HP18 9AN. *2m N of Thame. Park in the village as there is no parking at individual gardens.* **Sun 16 Apr, Sun 11 June (2-6). Combined adm £6, chd free. Home-made teas in Church House, High St (Apr) & St Mary's Church (June).**

BAKER'S CLOSE
Mr & Mrs Peter Vaines.
Open on Sun 16 Apr

BARRY'S CLOSE
Mr & Mrs Richard Salmon.
Open on Sun 16 Apr

BRINDLES
Sarah Chapman.
Open on Sun 11 June

NEW **11 CARTERS LANE**
Mr D & Mrs C Chadwick.
Open on Sun 11 June

COP CLOSE
Sandra & Tony Phipkin.
Open on Sun 11 June

25 ELM TREES
Carol & Mike Price.
Open on Sun 16 Apr

43 HIGH STREET
James & Laura Solyom.
Open on Sun 11 June

MANOR FARM BUNGALOW
Tracy Russell & David Newman.
Open on Sun 11 June

MANOR HOUSE
Mr & Mrs West.
Open on Sun 16 Apr

TOMPSONS FARM
Mr & Mrs T Moynihan.
Open on Sun 11 June

Four gardens will open on Sun 16 Apr. Baker's Close with1000s of daffodils, tulips and narcissi, shrubs and wild area. Barry's Close with spring flowering trees, borders and water garden. 25 Elm Trees a cottage style organic garden planted to encourage wildlife. Manor House a large garden with views to the Chilterns, two ornamental lakes, a variety of spring bulbs and shrubs. Six gardens will open on Sun 11 June. Brindles is a totally organic garden with natural swimming pool, beehives, roses and vegetable garden. Cop Close with mixed borders, vegetable garden, wildflower bank and damp garden. 11 Carters Lane is new this yr with mixed borders containing roses, tender perennials and annuals as well as a vegetable garden. Manor Farm Bungalow has been designed for wildlife with a newly created pond. The Smithy, 43 High Street is a recently planted cottage front garden designed by a local garden designer. Tompsons Farm, a large woodland garden with large lake and herbaceous borders. Partial wheelchair access to the following gardens: April: Barry's Close, Manor House June: Tompson's Farm, Manor Farm Bungalow.

In 2022, 20,000 people were supported by Perennial caseworkers and prevent teams thanks to our funding

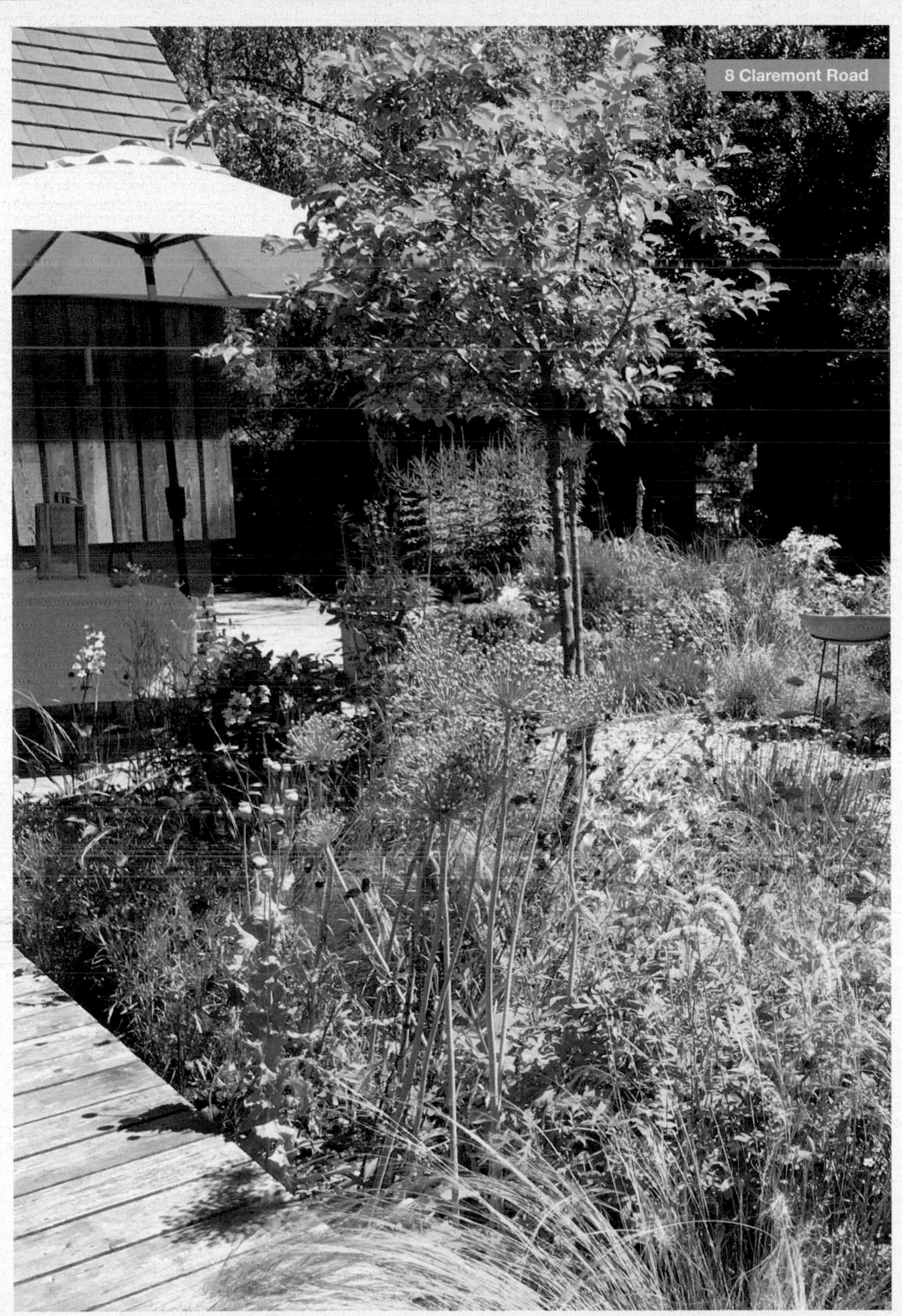

8 Claremont Road

27 LORDS WOOD
Frieth Road, Marlow Common, SL7 2QS. Mr & Mrs Messum. *1½m NW Marlow. From Marlow turn off the A4155 at Platts Garage into Oxford Rd & Chalkpit Ln towards Frieth for 1½m, 100yds past the Marlow Common Rd turn L down a made-up bridlepath & follow parking signs.* **Thur 15 June (11-4.30). Adm £5, chd free. Home-made teas.**
'An outpost of Old Bloomsbury in Marlow Woods' was how diarist Frances Partridge described Lords Wood. James and Alex Strachey entertained many of the Bloomsbury Group inc Lytton Strachey and Dora Carrington. The 5 acres surrounding the house (not open) showcase sculpture, water features, extensive mature borders, flower and herb gardens, orchard, and woodland walks with spectacular views over the Chilterns. Partial wheelchair access; gravel paths, steep slopes and open water.

28 MAGNOLIA HOUSE
Grange Drive, Wooburn Green, Wooburn, HP10 0QD. Elaine & Alan Ford, 01628 525818, lanforddesigns@gmail.com, sites.google.com/site/lanforddesigns. *On A4094 2m SW of A40 between Bourne End & Wooburn. From Wooburn Church, direction Maidenhead, Grange Drive is on L before r'about. From Bourne End, L at 2 mini r'abouts, then 1st R.* **Visits by arrangement 25 Feb to 23 July for groups of up to 25. Adm £6, chd free. Home-made teas. Visits can be combined with The Shades next door.**
½ acre garden with mature trees inc large magnolia. Cacti, fernery, stream, two ponds, greenhouses, aviary, 10,000 snowdrops, hellebores, bluebells and over 60 varieties of hosta. Child friendly. Constantly being changed and updated. Partial wheelchair access.

29 MONTANA
Shire Lane, Cholesbury, HP23 6NA. Diana Garner *3m NW of Chesham. From Wigginton turn R after Champneys, 2nd R onto Shire Ln. From Cholesbury turn on to Cholesbury Rd by cricket club, 1st L is Shire Ln. Montana ½m down on LHS by footpath sign.* **Wed 15 Mar, Wed 17 May, Wed 19 July (11-2). Adm £4, chd free. Pre-booking essential, please phone 01494 758347 or email montana@cholesbury.net for information & booking. Home-made teas. Visits also by arrangement Mar to July for groups of up to 35.**
A peaceful large garden surrounded by trees, and planted with rare and unusual flowering trees, shrubs, perennials underplanted with bulbs. Shade loving plants, kitchen garden and small meadow with apiary. A gate leads to a 3 acre mixed deciduous wood with level paths, a large fernery planted in an old clay pit and an avenue of thousands of daffodils and acers. Lots of seats to enjoy the quiet. An unmanicured garden high in the Chiltern Hills. Surrounding fields have been permanent pasture for more than 100 yrs. Covered open barn for teas in wet weather. The majority of paths in the garden and wood are wheelchair friendly.

30 ◆ NETHER WINCHENDON HOUSE
Nether Winchendon, nr Aylesbury, HP18 0DY. Mr Robert Spencer Bernard, 01844 290101, Contactus@netherwinchendonhouse.com, www.nwhouse.co.uk. *6m SW of Aylesbury, 6m from Thame. Approx 4m from Thame on A418, turn 1st L to Cuddington, turn L at Xrds, downhill turn R & R again to parking by house.* **For NGS: Sun 28 May, Sun 27 Aug (2-5.30). Adm £4, chd free. Pre-booking preferred. For other opening times and information, please phone, email or visit garden website.**
Nether Winchendon House has fine, and rare trees set in an inspiring and stunning landscape with parkland. The south lawn runs down to the River Thame. A Founder NGS Member (1927). Enchanting and romantic Mediaeval and Tudor House, one of the most romantic of the historic houses of England and listed Grade I. Dogs on leads welcome. Picturesque small village with an interesting church.

31 OLD KEEPERS
Village Lane, Hedgerley, SL2 3UY. Rob Cooper, Robcooper1612@gmail.com. *Parking in village hall on Kiln Ln, garden on private lane opp.* **Sun 23 July (11-4). Adm £4.50, chd free. Pre-booking essential, please visit www.ngs.org.uk for information & booking. Home-made teas. Visits also by arrangement 1 June to 15 Oct for groups of 15 to 30.**
Recently established, 1½ acre garden combining extensive perennial borders, an orchard, meadow and woodland edge gardens, surrounding a Grade II listed former brickmaker's cottage (not open). Large selection of roses and an interesting use of rusted metalwork made by British craftsmen throughout the garden. Beautiful views across the garden and village, with strategically positioned benches and perching spots to enjoy the view.

32 OLD PARK BARN
Dag Lane, Stoke Goldington, MK16 8NY. Emily & James Chua, 01908 551092, emilychua51@yahoo.com. *4m N of Newport Pagnell on B526. Park on High St. A short walk up Dag Ln. Disabled parking for four cars near garden via Orchard Way. Please contact us for coach parking details.* **Sat 24, Sun 25 June (1.30-5). Adm £5, chd free. Home-made teas. Visits also by arrangement 7 June to 14 July for groups of 10 to 50.**
A garden of almost 3 acres made from a rough field over 25 yrs ago. Near the house (not open) a series of terraces cut into the sloping site create the formal garden with long and cross vistas, lawns and deep borders. The aim is to provide interest throughout the yr with naturalistic planting and views borrowed from the surrounding countryside. Beyond is a wildlife pond, meadow and woodland garden. Partial wheelchair access.

33 ORCHARD HOUSE
Tower Road, Coleshill, Amersham, HP7 0LB. Mr & Mrs Douglas Livesey. *From Amersham Old Town take the A355 to Beaconsfield. Appox ¾m along this road at top of hill, take 1st R into Tower Rd. Parking in cricket club grounds.* **Sun 23 Apr (2-5). Adm £5, chd free. Home-made teas in the barn.**
The 5 acre garden inc several wooded areas with eco bug hotels for wildlife. Two ponds with wild flower planting, large avenues of silver birches, a bog garden with board walk and a wildflower meadow. There is a cut flower garden and a dramatic collection of spring bulbs set amongst an acer glade. Wheelchair access with sloping lawn in rear garden.

34 OVERSTROUD COTTAGE
The Dell, Frith Hill, Great Missenden, HP16 9QE. Mr & Mrs Jonathan Brooke, 01494 862701, susanmbrooke@outlook.com. *½m E Great Missenden. Turn E off A413 at Great Missenden onto B485 Frith Hill to Chesham Rd. White Gothic cottage set back in lay-by 100yds uphill on L. Parking on R at church.* **Sun 9 Apr, Sun 14 May, Sun 4 June (2-5). Adm £4, chd free. Cream teas at parish church. Visits also by arrangement 4 Apr to 29 June for groups of 15 to 20.**
Artistic chalk garden on two levels. Collection of C17/C18 plants inc auriculas, hellebores, bulbs, pulmonarias, peonies, geraniums, dahlias, herbs and succulents. Many antique species and rambling roses. Potager and lily pond. Blue and white ribbon border. Cottage was once C17 fever house for Missenden Abbey. Features inc a garden studio with painting exhibition (share of flower painting proceeds to NGS).

35 PETERLEY CORNER COTTAGE
Perks Lane, Prestwood, Great Missenden, HP16 0JH. Dawn Philipps, 01494 862198, dawn.philipps@googlemail.com. *Turn into Perks Ln from Wycombe Rd (A4128), Peterley Corner Cottage is 3rd house on the L.* **Visits by arrangement 8 May to 30 July for groups of 10 to 40. Adm £5, chd free. Light refreshments.**
A 3 acre mature garden inc an acre of wild flowers and indigenous trees. Surrounded by tall hedges and wood, the garden has evolved over the last 30 yrs. There are many specimen trees and mature roses inc a Paul's Himalaya Musk. A large herbaceous border runs alongside the formal lawns with other borders like heathers and shrubs. A more recent addition is an enclosed potager.

36 THE PLOUGH
Chalkshire Road, Terrick, Aylesbury, HP17 0TJ. John & Sue Stewart. *2m W of Wendover. Entrance to garden & car park signed off B4009 Nash Lee Rd. 200yds E of Terrick r'about. Access to garden from field car park.* **Sun 21 May, Sun 4 June (1-5). Adm £5, chd free. Home-made teas.**
Formal organic garden with open views to the Chiltern countryside. Designed as a series of outdoor rooms around a listed former C18 inn (not open), inc border, parterre, vegetable and fruit gardens, and a newly planted orchard. Delicious home-made teas in our barn and adjacent entrance courtyard. Jams and apple juice for sale, made with fruits from the garden. Cash only please.

37 RED KITES
46 Haw Lane, Bledlow Ridge, HP14 4JJ. Mag & Les Terry, 01494 481474, lesterry747@gmail.com. *4m S of Princes Risborough. Off A4010 halfway between Princes Risborough & West Wycombe. At Hearing Dogs sign in Saunderton turn into Haw Ln, then ¾m on L up the hill.* **Tue 16 May, Tue 11 July (2-5). Adm £5, chd free. Home-made teas. Visits also by arrangement 15 May to 9 Sept for groups of 20+.**
This Chiltern hillside garden of 1½ acres is planted for year-round interest and has superb views. Lovingly and beautifully maintained, there are several different areas to relax in, each with their own character. Wildflower orchard, mixed borders, pond, vegetables, woodland area and a lovely hidden garden. Many climbers used in the garden which changes significantly through the seasons.

38 ROBIN HILL
Water End, Stokenchurch, High Wycombe, HP14 3XQ. Caroline Renshaw & Stuart Yates, 07957 394134, Info@cazrenshawdesigns.co.uk. *2m from M40 J5 Stokenchurch. Turn off A40 just S of Stokenchurch towards Radnage, then 1st R to Waterend & then follow signs.* **Sat 24 June (12-4). Adm £4.50, chd free. Home-made teas. Visits also by arrangement 15 May to 24 Sept for groups of 15 to 30.**
Relaxed designer's garden at the start of The Chilterns. Informal cottage garden/prairie style with a variety of trees, woodland edge plants, shade borders, shrubs, perennials and grass borders blending into pastureland, now restored to beautiful long grass meadow. Mature cherry orchard, chickens, vegetable garden. Planting constantly evolving to better suit our changing climate, and benefit wildlife. Wheelchair access over mainly flat and lawned garden, but no paths.

39 ST MICHAELS CONVENT
Vicarage Way, Gerrards Cross, SL9 8AT. Sisters of the Church. *15 min walk from Gerrards Cross Stn. 10 mins from East Common buses. Parking in nearby street, limited parking at St Michael's.* **Sat 17 June (2-4.30). Adm by donation. Home-made teas.**
Recently acquired garden, having been neglected for many yrs, is now being developed by the community as a place for quiet, reflection and to gaze upon beauty. Inc a walled garden with vegetables, labyrinth and pond, a shady woodland dell and a recently built chapel. Colourful borders, beds and mature majestic trees. Come and see how the garden is developing and growing!

40 THE SHADES
High Wycombe, HP10 0QD. Pauline & Maurice Kirkpatrick, 01628 522540. *On A4094 2m SW of A40 between Bourne End & Wooburn. From Wooburn Church, direction Maidenhead, Grange Drive is on L before r'about. From Bourne End, L at 2 mini r'abouts, then 1st R.* **Visits by arrangement 25 Feb to 23 July for groups of up to 25. Combined visit with Magnolia House.**
The Shades drive is approached through mature trees, areas of shade loving plants, beds of shrubs, 60 various roses and herbaceous plants. The rear garden with natural well surrounded by plants, shrubs and acers. A green slate water feature and scree garden with alpine plants completes the garden. Light refreshments at Magnolia House. Partial wheelchair access.

Our 2022 donation supports 1,700 Queen's Nurses working in the community in England, Wales, Northern Ireland, the Channel Islands and the Isle of Man

41 ◆ STOKE POGES MEMORIAL GARDENS
Church Lane, Stoke Poges, Slough, SL2 4NZ. Buckinghamshire Council, 01753 523744, memorial.gardens@buckinghamshire.gov.uk, www.buckinghamshire.gov.uk.
1m N of Slough, 4m S of Gerrards Cross. Follow signs to Stoke Poges & from there to the Memorial Gardens. Car park opp main entrance, disabled visitor parking in the gardens. Weekend disabled access through churchyard. **For NGS: Sun 14 May (1-4.30). Adm £5, chd free. Home-made teas. For other opening times and information, please phone, email or visit garden website.**
Unique 22 acre Grade I registered garden constructed 1934-9 with a contemporary garden extension. Rock and water gardens, sunken colonnade, rose garden, 500 individual gated gardens, beautiful mature trees and newly landscaped areas. Guided tours every hour. Guide dogs only.

42 1 TALBOT AVENUE
Downley, High Wycombe, HP13 5HZ. Mr Alan Mayes, 01494 451044, alan.mayes2@btopenworld.com.
From Downley T-lights off West Wycombe Rd, take Plomer Hill turn off, then 2nd L into Westover Rd, then 2nd L into Talbot Ave. **Visits by arrangement May to Sept for groups of up to 20. Adm £4, chd free. Tea.**
A Japanese garden, shielded from the upper garden level by Shoji screens. A winding path leads you over a traditional Japanese bridge by a pond and waterfall and invites you through a moongate to reveal a purpose built tea house, all surrounded by traditional Japanese planting inc maples, cherry blossom trees, azaleas and rhododendrons. Ornamental grasses and bamboo complement the hard landscaping with feature cloud tree and checkerboard garden path.

Chiltern Haven

43 TURN END
Townside, Haddenham, Aylesbury, HP17 8BG. Margaret & Peter Aldington, turnendgarden@gmail.com, www.turnend.org.uk. *3m NE of Thame, 5m SW of Aylesbury. Exit A418 to Haddenham. Turn at Rising Sun Pub into Townside. See Turn End website for parking info for this event. Street parking very limited. Please park with consideration for residents.* **Mon 1 May (2-5). Adm £5, chd free. Pre-booking essential, please visit www.ngs.org.uk for information & booking. Home-made teas.**
Grade II registered. Series of garden rooms, each with a different planting style enveloping architect's own Grade II* listed house (not open). Dry garden, formal box garden, sunken gardens, mixed borders around curving lawn, all framed by ancient walls and mature trees. Bulbs, irises, wisteria, roses, ferns and climbers. Courtyards with pools, pergolas, secluded seating and Victorian coach house. Open artist's studio.

44 TYTHROP PARK
Kingsey, HP17 8LT. Nick & Chrissie Wheeler. *2m E of Thame, 4m NW of Princes Risborough. Via A4129, at T-junction in Kingsey turn towards Haddenham, take L turn on bend. Parking in field on L.* **Sun 4 June (2-5.30). Adm £7, chd free. Home-made teas.**
10 acres of garden surrounds a C17 Grade I listed manor house (not open). This large and varied garden blends traditional and contemporary styles, featuring pool borders rich in grasses with a green and white theme, walled kitchen and cutting garden with large greenhouse at its heart, box parterre, deep mixed borders, water feature, rose garden, wildflower meadow and many old trees and shrubs.

45 WADZANA
8 Lynnens View, Oakley, HP18 9LQ. Wendy & Peter Hopcroft. *Within 8m of Oxford, Bicester & Thame. 6m from M40. Offroad parking NE of Headington past crematorium through Horton-cum-Studley. SE of Bicester & NW from Thame on B4011. In Oakley follow signs for Oxford. Turn into Lynnens View. SatNav use HP18 9RD.* **Sun 2 July (1-5). Adm £4.50, chd free. Home-made teas.**
A young but well developed garden landscaped into individually themed areas inc herbaceous, silver birch copse, formal, herb, meadow and pergola leading to stumpery. Views over neighbouring fields gives sense of space. Extensive vegetable garden and greenhouse. Seating for tea and cake near fountain and water feature. Virtual exhibition of Wendy's paintings with opportunity to purchase. Wheelchair access to part of garden and water feature.

SPECIAL EVENT

46 THE WALLED GARDEN, WORMSLEY
Wormsley, Stokenchurch, High Wycombe, HP14 3YE. Wormsley Estate. *Leave M40 at J5. Turn towards Ibstone. Entrance to estate is ¼m on R. NB: 20mph speed limit on estate. Please do not drive on grass verges.* **Wed 14 June (10-3). Adm £7, chd free. Pre-booking essential, please visit www.ngs.org.uk for information & booking. Home-made teas.**
The Walled Garden at Wormsley Estate is a 2 acre garden providing flowers, vegetables and tranquil contemplative space for the family. For many yrs the garden was neglected until Sir Paul Getty purchased the estate in the mid 1980s. In 1991 the garden was redesigned and has changed over the yrs, but remains true to the original brief. Wheelchair access to grounds, but no WC facilities.

47 THE WHITE HOUSE
Village Road, Denham Village, UB9 5BE. Mr & Mrs P G Courtenay-Luck. *3m NW of Uxbridge, 7m E of Beaconsfield. Signed from A40 or A412. Parking in village road. The White House is in centre of village, opp St Mary's Church. Chiltern Line Train Stn, Denham.* **Sun 4 June (2-5). Adm £6, chd free. Cream teas.**
Well established 6 acre formal garden in picturesque setting. Mature trees and hedges with River Misbourne meandering through lawns. Shrubberies, flower beds, rockery, rose garden and orchard. Large walled garden. Herb garden, vegetable plot and Victorian greenhouses. Wheelchair access with gravel entrance and path to gardens.

48 WIND IN THE WILLOWS
Moorhouse Farm Lane, Off Lower Road, Higher Denham, UB9 5EN. Ron James, 07740 177038, r.james@company-doc.co.uk. *6m E of Beaconsfield. Turn off A412, approx ½m N of junction with A40 into Old Rectory Ln. After 1m enter Higher Denham straight ahead. Take lane next to the community centre & Wind in the Willows is the 1st house on L.* **Sun 26 Mar (2-5). Adm £5, chd free. Light refreshments in the community hall. Opening with Higher Denham Gardens on Sun 28 May. Visits also by arrangement Mar to Sept for groups of 10+. Donation to Higher Denham Community CIO (Garden Upkeep Fund).**
3 acre wildlife friendly, year-round garden, comprising informal, woodland and wild gardens, separated by streams lined by iris, primulas and astilbe. Over 350 shrubs and trees, many variegated or uncommon, marginal and bog plantings inc 80 hostas and a bed of stripy roses. Stunning was the word most often used by visitors. 'Best private garden I have visited in 20 yrs of NGS visits' said another. Although unlikely to be seen on busy open days, 65 species of bird and 13 species of butterfly have been seen in and over the garden, which is also home to the now endangered water vole (Water Rat in the book Wind in the Willows), frogs and toads. Wheelchair access over gravel paths and spongy lawns.

Our 2022 donation to ABF The Soldier's Charity meant 372 soldiers and veterans were helped through horticultural therapy

CAMBRIDGESHIRE

Historic cathedral cities, wide open skies and unique fenland landscapes combine to make Cambridgeshire special.

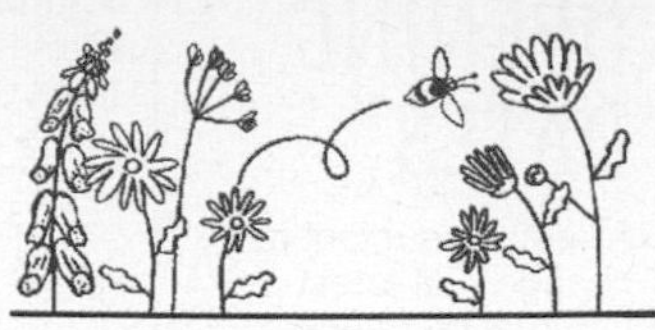

Our generous gardeners invite you to come and take a closer look at gardens to delight and surprise. From the splendid city gardens to parkland magnificence and rural idylls in isolated hamlets the county has something to suit all tastes.

Stroll around our group gardens and be inspired by their diversity and interest. Discover community gardens, some the outcome of pandemic experiences and examples of triumph over adversity. Explore spaces planned for the dry East Anglian climate and others on rich fenland soil. See the contemporary and the historic; small town courtyards, large country gardens and those maintained for wildlife and the environment. Be inspired by innovative and creative ideas and talk to our enthusiastic and knowledgeable gardeners.

Some are open by arrangement where your host will happily share plant knowledge, anecdotes and plant passions with you - so long as you have time to listen!

Begin the year with a visit to a snowdrop garden in February. Later enjoy a summer afternoon with friends and family. Relax and unwind with good tea and cake in the surroundings of a beautiful garden. End with the spectacular autumn tree colour changes.

Whenever you visit, you can be sure that you will receive a warm welcome and a memorable day out.

Left: King's College Fellow's Garden and Provosts Garden

Volunteers

County Organiser
Jenny Marks 07956 049257
jenny.marks@ngs.org.uk

Deputy County Organiser
Pam Bullivant 01353 667355
pam.bullivant@ngs.org.uk

County Treasurer
Jackie Hutchinson 01954 710950
jackie.hutchinson@ngs.org.uk

Booklet Coordinator
Jenny Marks (As above)

Publicity
Penny Miles 07771 516448
penny.miles@ngs.org.uk

Social Media
Hetty Dean
hetty.dean@ngs.org.uk

Assistant County Organisers
Jacqui Latten-Quinn 07941 279571
jacqui.quinn@ngs.org.uk

Penny Miles (As above)

Jane Pearson 07890 080303
jane.pearson@ngs.org.uk

Barbara Stalker 07800 575100
barbara.stalker@ngs.org.uk

Annette White 01638 730876
annette.white@ngs.org.uk

Claire Winfrey 01733 380216
claire.winfrey@ngs.org.uk

Julia Rigby 07742 432726
julia.rigby@ngs,org.uk

National Garden Scheme Cambs @GardenCambs cambsngs

OPENING DATES

All entries subject to change. For latest information check **www.ngs.org.uk**

Extended openings are shown at the beginning of the month.

Map locator numbers are shown to the right of each garden name.

January

Every Monday to Friday from Tuesday 3rd
Robinson College 43

Every Saturday and Sunday from Saturday 7th
Robinson College 43

February

Snowdrop Festival

Every Monday to Friday
Robinson College 43

Every Saturday and Sunday
Robinson College 43

Sunday 19th
Clover Cottage 8

Sunday 26th
Clover Cottage 8

March

Every Monday to Friday
Robinson College 43

Every Saturday and Sunday
Robinson College 43

Sunday 5th
Clover Cottage 8

Sunday 26th
Kirtling Tower 28
Trinity College Fellows' Garden 49

April

Every Monday to Friday to Friday 14th
Robinson College 43

Every Saturday and Sunday to Sunday 9th
Robinson College 43

Saturday 1st
NEW Heath Fruit Farm 17

Sunday 2nd
Netherhall Manor 37

Saturday 15th
NEW Heath Fruit Farm 17

Sunday 30th
Chaucer Road Gardens 7

May

Every Friday, Saturday and Sunday
23A Perry Road 40

Monday 1st
Chaucer Road Gardens 7

Sunday 7th
Netherhall Manor 37

Sunday 21st
◆ Ferrar House 15
NEW Khususi 26
Madingley Hall 31
Molesworth House 35

Saturday 27th
NEW Coveney & Wardy Hill Gardens 10
High Bank Cottage 19

Sunday 28th
Cambourne Gardens 5
Cottage Garden 9
NEW Coveney & Wardy Hill Gardens 10
Elm House 13
High Bank Cottage 19
Island Hall 24
Sutton Gardens 47

June

Every Friday, Saturday and Sunday
23A Perry Road 40

Saturday 3rd
Staploe Gardens 46
Twin Tarns 50

Sunday 4th
Clover Cottage 8
Ely Open Gardens 14
Staploe Gardens 46
Twin Tarns 50

Monday 5th
Abbots Ripton Hall 2

Saturday 10th
Abbots Barn 1
Beaver Lodge 3

Sunday 11th
Abbots Barn 1
Beaver Lodge 3
Clover Cottage 8
Duxford Gardens 11
NEW 5 Nelsons Lane 36
The Old Rectory 39

Tuesday 13th
Abbots Ripton Hall 2

Saturday 17th
NEW Khususi 26
Little Oak 30

Sunday 18th
Burwell Village Gardens 4
Green End Farm 16
Isaacson's 23
Linton Gardens 29
Little Oak 30
Toft Gardens 48

Monday 19th
Abbots Ripton Hall 2

Saturday 24th
Little Oak 30
Marmalade Lane, Cohousing Community Garden 34
Preachers Passing 41

Sunday 25th
Kirtling Tower 28
Little Oak 30
Preachers Passing 41
NEW Westley Waterless Gardens 51

Tuesday 27th
Abbots Ripton Hall 2

July

Every Friday, Saturday and Sunday
23A Perry Road 40

Every Saturday and Sunday from Saturday 8th
Robinson College 43

Every Monday to Friday from Wednesday 5th
Robinson College 43

Saturday 1st
Castor House 6
38 Norfolk Terrace Garden 38

Sunday 2nd
Castor House 6
Highsett Cambridge 20
NEW Hilton Gardens 21
38 Norfolk Terrace Garden 38
Sawston Gardens 44

Monday 3rd
Abbots Ripton Hall 2

Sunday 9th
King's College Fellows' Garden and Provost's Garden 27

Sunday 23rd
NEW Stapleford Village Gardens 45

August

Every Friday, Saturday and Sunday
23A Perry Road 40

Every Saturday and Sunday to Sunday 6th
Robinson College 43

Every Monday to Friday until Friday 11th. Then from Wednesday 30th
Robinson College 43

Sunday 6th
Netherhall Manor 37

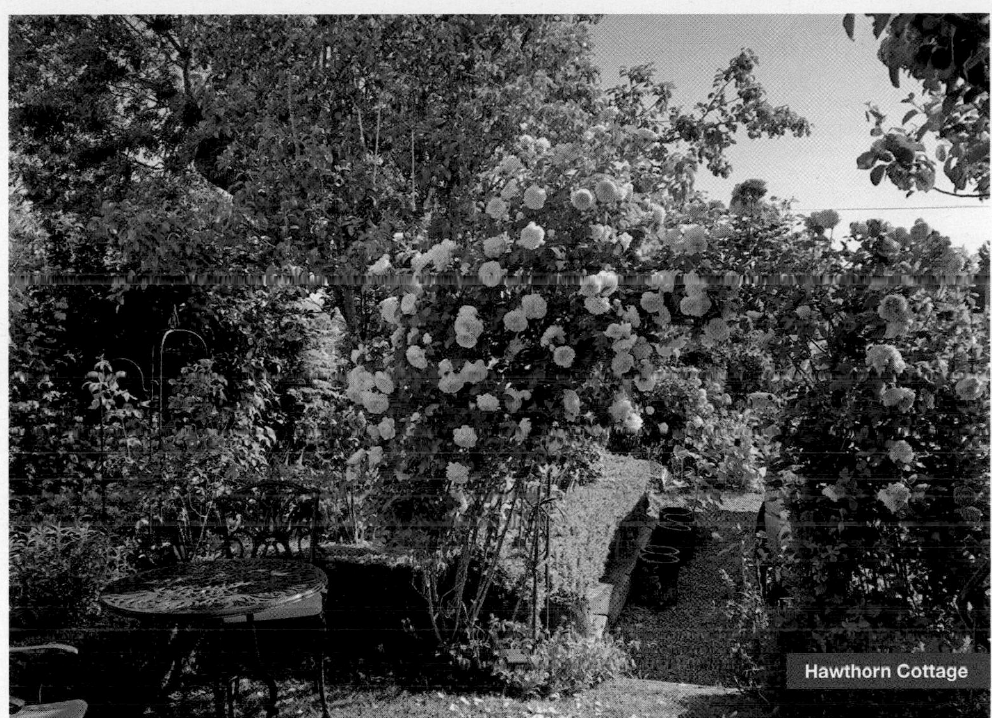
Hawthorn Cottage

Saturday 12th
Beaver Lodge 3

Sunday 13th
Beaver Lodge 3
Netherhall Manor 37

Thursday 17th
◆ Elgood's Brewery Gardens 12

September

Every Monday to Friday
Robinson College 43

Every Saturday and Sunday
Robinson College 43

Saturday 9th
Linton Gardens 29

Sunday 10th
◆ The Manor, Hemingford Grey 33

Saturday 16th
NEW Ramsey Walled Garden 42

October

Every Monday to Friday
Robinson College 43

Every Saturday and Sunday
Robinson College 43

November

Every Monday to Friday
Robinson College 43

Every Saturday and Sunday
Robinson College 43

December

Every Monday to Friday to Friday 22nd
Robinson College 43

Every Saturday and Sunday to Sunday 17th
Robinson College 43

By Arrangement

Arrange a personalised garden visit with your club, or group of friends, on a date to suit you. See individual garden entries for full details.

Beaver Lodge 3
Castor House 6
38 Chapel Street, Ely Open Gardens 14
Church Lane House, Westley Waterless Gardens 51
NEW Church View, Westley Waterless Gardens 51
Clover Cottage 8
Cottage Garden 9
Drovers, Coveney & Wardy Hill Gardens 10
6 Hemingford Crescent 18
NEW The Hey Chapel, Coveney & Wardy Hill Gardens 10
14 High Street, Linton Gardens 29
Horseshoe Farm 22
Isaacson's 23
Juniper House 25
67 Main Street, Yaxley 32
Molesworth House 35
Netherhall Manor 37
NEW Newnham Farmhouse, Burwell Village Gardens 4
The Old Baptist Chapel, Sutton Gardens 47
The Old Rectory 39
23A Perry Road 40
Staploe Gardens 46
Twin Tarns 50
Upwater Lodge, Chaucer Road Gardens 7
The Windmill 52

THE GARDENS

1 ABBOTS BARN

Southorpe, Stamford, PE9 3BX. Carl & Vanessa Brown. *5m SE of Stamford, 8m W of Peterborough. On entering Southorpe Village from A47, 1st house on L.* **Sat 10, Sun 11 June (1-5.30). Adm £6, chd free. Home-made teas. Vegans and gluten free catered for.**
2 acre garden surrounding a converted Georgian barn; a large family garden with herbaceous borders and wildflower meadow. An allotment-sized vegetable plot, orchard, Mediterranean gravel garden and large glasshouse on site. We also have a 25 year-old woodland with mown grass paths, beach garden and wildlife pond. A large courtyard garden with croquet lawn and cutting flower garden completes Abbots Barn. We maintain the garden using organic methods, providing a rich wildlife habitat. Free range bantams in the garden.

2 ABBOTS RIPTON HALL

Abbots Ripton, Huntingdon, PE28 2PQ. The Lord & Lady De Ramsey, www.abbotsriptonhall.co.uk. *2m N of Huntingdon. On B1090.* **Mon 5, Tue 13, Mon 19, Tue 27 June, Mon 3 July (2-4). Adm £15.**
8½ acres of quintessential English garden with 5 acre lake developed since 1937 with much work done recently by the current owners. 150yd 'cottage' double border, mixed borders, old shrub roses garden, Victorian rose arch walk. Tender Mediterranean grey border. Bog garden, fine collection of trees, inc 60 different oaks. Many follies, each with its own character. Over 1600 plant species. Wheelchair access to some gravel paths. Cashless payments only. Last entry 3pm.Teas available at Village Store & lunches at The Elm.

3 BEAVER LODGE

Henson Road, March, PE15 8BA. Mr & Mrs Nielson-Bom, 01354 656185/07946 891866, beaverbom@gmail.com. *A141 to Wisbech Rd into March, L into Westwood Ave, at end, R into Henson Rd, property on R opp school playground.* **Sat 10, Sun 11 June, Sat 12, Sun 13 Aug (12-4). Adm £3, chd free. Home-made teas. Visits also by arrangement May to Nov for groups of 5 to 30.**
An impeccable oriental garden with more than 120 large and small bonsai trees, acers, pagodas, oriental statues, water features and pond with koi carp, creating a peaceful and relaxing atmosphere. The garden is divided into different rooms, 1 with a Mediterranean feel inc tree ferns, lemon trees, bougainvilleas and a great variety of plants and water fountain. Plenty of sitting areas.

GROUP OPENING

4 BURWELL VILLAGE GARDENS

Burwell, CB25 0BB. *10m NE of Cambridge, 4m NW of Newmarket via the B1102/ B1103.* **Sun 18 June (12-5). Combined adm with Isaacson's £6, chd free. Home-made teas at Isaacson's.**

9 THE BRIARS
Ms Marianne Hall and Mr Tony Hollyer.

NEW **NEWNHAM FARMHOUSE**
Mr and Mrs Quentin Cooke, 07802 857150, q.cooke@ntlworld.com.
Visits also by arrangement 1 Mar to 1 Nov for groups of 5+. Refreshments can be organised in advance of visit.

3 ROMAN CLOSE
Mrs Dove.

SPRING VIEW
Colin Smith.
NPC

Ramsey Walled Garden

3 Roman Close is a small, newly established garden. Cottage garden planting, pretty and full borders with roses and many other perennials and annuals. Pots and hanging baskets. Spring View houses the National Collection of Yucca, a huge diversity of forms and sizes, demonstrating drought tolerant planting with other species as well, a specialist plantsman's garden and eco house. 9 The Briars is an environmentally friendly garden with pond and variety of flowers, fruit and vegetables and featuring a living wall. Newnham Farmhouse has a walled garden surrounding a C16 thatched cottage. The garden has over 50 Japanese acers. The borders are packed, and a lawn that is filled with wildflowers rather than just grass. There are ponds front and back and a few interesting sculptures. See also Isaacsons' separate listing; a plantsman's garden of striking diversity in a series of different zones and incorporating medieval plants, in the grounds of a clunch stone former lodging range dating from 1340.

GROUP OPENING

5 CAMBOURNE GARDENS

Great Cambourne, CB23 6AH. *8m W of Cambridge on A428. From A428: take Cambourne junc into Great Cambourne. From B1198, enter village at Lower Cambourne & drive through to Great Cambourne. Follow yellow signs via either route to start at any garden.* **Sun 28 May (11-5). Combined adm £6, chd free. Home-made teas at 13 Fenbridge. Coffee & biscotti at 43 Monkfield Lane.**

13 FENBRIDGE
Lucinda & Tony Williams.

88 GREENHAZE LANE
Darren & Irette Murray.

128 JEAVONS LANE
Mr Tim Smith.

8 LANGATE GREEN
Steve & Julie Friend.

5 MAYFIELD WAY
Debbie & Mike Perry.

43 MONKFIELD LANE
Tony & Penny Miles.

A unique and inspiring modern group, all created from new build in just a few years. This selection of 6 demonstrates how imagination and gardening skill can be combined in a short time to create great effects from unpromising and awkward beginnings. The grouping inc foliage gardens with collections of carnivorous pitcher plants and hosta; a suntrap garden for play, socialising and colour; gardens with ponds, a vegetable plot and many other beautiful borders showing their owners' creativity and love of growing fine plants well. Cambourne is one of Cambridgeshire's newest communities, and this grouping showcases the happy, vibrant place our village has become. Superb examples of beauty and creativity in a newly built environment.

6 CASTOR HOUSE

Peterborough Rd, Castor, Peterborough, PE5 7AX. Mr & Mrs Ian Winfrey, 01733 380216, ian@winfrey.co.uk, www.castorhousegardens.co.uk. *4m W of Peterborough. House on main Peterborough Rd in Castor. Parking in paddock off Water Ln.* **Sat 1, Sun 2 July (2-5). Adm £7.50, chd free. Home-made teas. Visits also by arrangement 26 June to 28 July for groups of 20 to 50.**
Main gardens redesigned 2009 within well established 12 acre garden. Multiple themed garden rooms, spring fed ponds and stream gardens, 60 + different roses, peony and prunus walk, colour themed borders, potager, stumpery, Loggia and summerhouses, hydrangea walk, woodland plant beds, mature specimen trees and woodland walks to large wildlife pond with views to enjoy.

GROUP OPENING

7 CHAUCER ROAD GARDENS

Cambridge, CB2 7EB. *1m S of Cambridge. Off Trumpington Rd (A1309), nr Brooklands Ave junc. Parking available at MRC Psychology Dept on Chaucer Rd.* **Sun 30 Apr, Mon 1 May (2-5). Combined adm £8, chd free. Home-made teas at Upwater Lodge.**

11 CHAUCER ROAD
Mark & Jigs Hill.

12 CHAUCER ROAD
Mr & Mrs Bradley.

16 CHAUCER ROAD
Mrs V Albutt.

UPWATER LODGE
Mr & Mrs George Pearson, 07890 080303, jmp@pearson.co.uk.
Visits also by arrangement 3 Apr to 30 July for groups of 15 to 40.

Chaucer Road Gardens are fine examples of large Edwardian gardens tucked away on the south side of Cambridge. They are all very different in size and character, some with glimpses of a bygone age. There are magnificent old trees, fine lawns, ponds and a watermeadow with river frontage and rare breed sheep. Planting is varied, sometimes unusual, and in constant flux. Cakes made with garden fruit where possible. Swings and climbing ropes. Plant stall possible but please email to check. Wheelchair access some gravel areas and grassy paths with fairly gentle slopes.

8 CLOVER COTTAGE

50 Streetly End, West Wickham, CB21 4RP. Mr Paul & Mrs Shirley Shadford, 01223 893122, shirleyshadford@live.co.uk. *3m from Linton, 3m from Haverhill & 2m from Balsham. From Horseheath turn L, from Balsham turn R, thatched cottage opp triangle of grass next to old windmill.* **Sun 19, Sun 26 Feb, Sun 5 Mar (2-4). Adm £3.50, chd free. Light refreshments. Sun 4, Sun 11 June (12-4.30). Adm £4, chd free. Home-made teas. Visits also by arrangement 19 Feb to 30 June.**
In winter find a flowering cherry tree, borders of snowdrops, aconites, iris reticulata, hellebores and miniature narcissus throughout the packed small garden which has inspiring ideas on use of space. Pond and arbour, raised beds of fruit and vegetables. In summer arches of roses and clematis, hardy geraniums, delightful borders of English roses and herbaceous plants. Snowdrops, hellebores and spring flowering bulbs for sale for the snowdrop festival, plants also for sale in June.

9 COTTAGE GARDEN
79 Sedgwick Street, Cambridge, CB1 3AL. Rosie Wilson, 07805 443818, liccycat@icloud.com. *Cambridge City. Off Mill Rd, S of the railway bridge.* **Sun 28 May (2-6). Adm £4, chd free. Home-made teas. Visits also by arrangement 29 May to 31 July for groups of up to 20.**
Small, long, narrow and planted in the cottage garden style with over 40 roses, some on arches and growing through trees. Particularly planned to encourage wildlife with small pond, mature trees and shrubs. Perennials and some unusual plants interspersed with sculptures.

GROUP OPENING

10 NEW COVENEY & WARDY HILL GARDENS
Coveney, Ely, CB6 2DN. *From A142 or A10 Ely bypass, follow signs for Coveney. For Wardy Hill from A142 sp Witcham.* **Sat 27, Sun 28 May (1-5). Combined adm £6, chd free. Home-made teas at some gardens.**

DROVERS
Mr David Guyer, 07859 917947, guyer100@yahoo.co.uk.
Visits also by arrangement May to Sept. Jointly open with The Hey Chapel.

NEW **HAWTHORN COTTAGE**
Lesley & Philip Wenn.

NEW **THE HEY CHAPEL**
Mr Peter Ross, 07813798945, peter_ross@dai.com.
Visits also by arrangement May to Sept Jointly open with Drovers.

NEW **2 SCHOOL LANE**
Mrs Jane McCartney.

The fenland parish of Coveney overlooks Ely cathedral and welcomes you to 4 gardens: 2 School Lane, a plantsman's garden, compact, but planned for year-round interest. Some unusual plants, patio, rock garden and veg plot. Nearby, Hawthorn Cottage is a true cottage garden of roses and clematis with bee friendly herbaceous borders. A flourishing wildlife pond leads to a shaded patio with bonsai. Take a moment to visit the churchyard with natural planting. In Wardy Hill, the Hey Chapel shows its character the moment you walk up the winding path to a view of the cathedral and a magnificent veg garden then back down through a series of garden 'rooms', seating area, rose arbours and pond. Drovers is based on Chinese pleasure garden philosophy but with English naturalist planting that wanders by a wildlife pond, beneath mature trees to a rose garden and potted terrace.

GROUP OPENING

11 DUXFORD GARDENS
Duxford, CB22 4RP. *Most gardens are close to the centre of the village of Duxford. S of the A505 between M11 J10 & Sawston. Parking at Temple Farmhouse, CB22 4PT* **Sun 11 June (2-6). Combined adm £7, chd free. Home-made teas.**

BUSTLERS COTTAGE
John & Jenny Marks.

DUXFORD MILL
Mrs Frankie Bridgwood.

2 GREEN STREET
Mr Bruce Crockford.

16 ICKLETON ROAD
Claire James.

31 ST PETER'S STREET
Mr David Baker.

Several village gardens of different sizes and characters. Duxford Mill gardens are always enchanting, with the mill pond, the mill race and the river defining the way the 11 acres are gardened. The tiny, charming 2 Green Street is always full of colour and interest. 31 St Peter's Street, on a steep slope, is filled with colour and surprises, and also has a dry rockery in front of the house. Bustlers Cottage has an acre of cottage garden inc vegetable garden and recently planted hedges. Stop here and try a home-made cream tea and buy some plants. Finally at 16 Ickleton Road, the garden is packed full of unusual plants, some for sale. Wheelchair access, some gravel paths and a few steps, mostly avoidable.

In 2022 the National Garden Scheme donated £3.11 million to our nursing and health beneficiaries

12 ◆ ELGOOD'S BREWERY GARDENS
North Brink, Wisbech, PE13 1LW. Elgood & Sons Ltd, 01945 583160, info@elgoods-brewery.co.uk, www.elgoods-brewery.co.uk. *1m W of town centre. Leave A47 towards Wisbech Centre. Cross river to North Brink. Follow river & brown signs to brewery & car park beyond.* **For NGS: Thur 17 Aug (11-4). Adm £5, chd free. Light refreshments. For other opening times and information, please phone, email or visit garden website.**
Approx 4 acres of peaceful garden featuring 250 yr old specimen trees providing a framework to lawns, lake, rockery, herb garden and maze. Wheelchair access to Visitor Centre and most areas of the garden.

13 ELM HOUSE
Main Road, Elm, Wisbech, PE14 0AB. Mrs Diana Bullard. *2½m SW of Wisbech. From A1101 take B1101, signed Elm, Friday Bridge. Elm House is ⅓m on L, well signed.* **Sun 28 May (1-5). Adm £5, chd free. Light refreshments.**
Elm House has a lovely walled garden with arboretum, many rare trees and shrubs, mixed perennials and a 3 acre flower meadow. There will be live music at the event. Children and guide dogs welcome. Most paths suitable- uneven surfaces in flower meadow.

GROUP OPENING

14 ELY OPEN GARDENS
Ely, CB7 4DL. www.elyopengardens.com. *14m N of Cambridge. Parking at Barton Rd, The Grange Council Offices, St.Mary's St, Brays Ln and Newnham St. All in easy reach of most gardens. Map on website or from elyopengardens.com.* **Sun 4 June (12-6). Combined adm £10, chd free. Light refreshments. Cold drinks & ice creams available in Bishop of Huntingdon's garden. Refreshments also at 1 Merlin Drive.**

BISHOP OF HUNTINGDON'S GARDEN
Dagmar, Bishop of Huntingdon.

THE BISHOP'S HOUSE
The Bishop of Ely.

12 CHAPEL STREET
Ken & Linda Ellis,
elyopengardens.com.

38 CHAPEL STREET
Peter & Julia Williams,
01353 659161,
peterrcwilliams@btinternet.com.
Visits also by arrangement Apr to Oct for groups of up to 35. Parking available on forecourt for up to 4 cars.

5B DOWNHAM ROAD
Mr Christopher Cain.

1 MERLIN DRIVE
Dr Chris Wood.

1 ROBINS CLOSE
Mr Brian & Mrs Ann Mitchell.

A delightful and varied group of gardens in an historic cathedral city: The Bishop's House, adjoining Ely Cathedral, has mixed planting and formal rose garden bordered by wisteria. The Bishop of Huntingdon's garden is a family garden, with herbaceous borders, pond, orchard and more, with view to cathedral. 12 Chapel Street, a small town garden, shows varied gardening interests, from alpines to herbaceous plants and shrubs and a railway! 38 Chapel Street has year-round interest and is bursting with unusual and interesting planting. 1 Merlin Drive and 1 Robins Close are modern estate gardens. The former has a collection of subtropical plants, and ferns, the latter has raised beds, vegetable and a small dry garden. 5b Downham Road has 2 small distinct walled gardens. One is cool and shady, the other shows how careful planting can create a tranquil garden in a challenging plot. Wheelchair access to areas of most gardens.

15 ◆ FERRAR HOUSE
Little Gidding, Huntingdon, PE28 5RJ. Mrs Susan Capp, 01832 293383, info@ferrarhouse.co.uk, www.ferrarhouse.co.uk. *Take Mill Rd from Great Gidding (turn at Fox & Hounds) then after 1m turn R down single track lane. Car Park at Ferrar House.* **For NGS: Sun 21 May (10.30-4.30). Adm £5, chd free. Home-made teas in tent on lawn. Open nearby Khususi. For other opening times and information, please phone, email or visit garden website.**
The peaceful garden of a Retreat House with beautiful countryside views. History talks in adjacent Church of St John's. Inspiration for T.S. Eliot's 'Little Gidding' quartet. Tranquil seating set amongst the walled bee friendly flower beds, knowledgeable gardeners on hand to discuss plantings, used books for sale, lawn with family games. Annual meadow area. Walled veg garden being redeveloped. WC accessible at Ferrar House.

16 GREEN END FARM
Over Road, Longstanton, Cambridge, CB24 3DW. Sylvia Newman, www.sngardendesign.co.uk. *From A14 take direction of Longstanton, at r'about take 2nd exit, at next r'about turn L (this shows a dead end on the sign) Garden 200 metres on L.* **Sun 18 June (11-4). Adm £6, chd free. Home-made teas.**
A developing garden that's beginning to blend well with the farm. An interesting combination of new and established spaces executed with a design eye. An established orchard with beehives; 2 wildlife ponds. An outside kitchen and social space inc pool, productive kitchen and cutting garden. Doves, chickens, bees and sheep complete the picture!

17 NEW HEATH FRUIT FARM
The Heath, Bluntisham, Huntingdon, PE28 3LQ. Rob and Mary Bousfield, www.heathfruitfarm.co.uk. *4 m N of St Ives Cambs. Drive out of Bluntisham at Wood End; we are on the R after water towers. If you come from St Ives to Wheatsheaf Xroads take The Heath on your R. We are on the L after a few bends.* **Sat 1, Sat 15 Apr (10-2). Adm £4, chd free. Home-made teas.**
A traditional orchard of 25 acres, boasting a range of apples, pears, apricots, plums, gages and cherries. Visit for a tour in the spring to behold a magnificent array of blossom and buzzing bees, while brown hares gambol around in springtime madness. Guided tours on the hour focusing on fruit growing, wildlife and the history of a Cambridgeshire orchard. Dogs allowed but must be kept on a lead.

Abbots Barn

18 6 HEMINGFORD CRESCENT
Stanground, Peterborough, PE2 8LL. Michael & Nick Mitchell, 07880 871763, michaelandnick64@gmail.com. *S side of city centre. 10 mins drive from A1M. Take Whittlesey Rd B1092. Coming out of town look for Apple Green petrol stn on L. Turn into Coneygree Rd. Follow rd round, Hemingford Cres is 6th turning on L.* **Visits by arrangement June to Aug for groups of 5 to 30. Adm £4, chd free.**
Medium sized city garden. Inspired by our travels to Morocco, Egypt and India. Mixture of exotic planting, shrubs and perennials. Various seating areas inc an outside dining area and an enclosed Moroccan/ Egyptian room. Raised beds and pond area. Michael is an artist and his work is available for viewing.

19 HIGH BANK COTTAGE
Kirkgate, Tydd St Giles, Wisbech, PE13 5NE. Mrs F Savill. *Cambridgeshire/ Lincolnshire border. Off the A1101.* **Sat 27, Sun 28 May (10.30-4). Adm £5, chd £1. Light refreshments. Home-made cakes and biscuits with tea, coffee, and cold drinks. Golf course opp has restaurant and cafe.**
The garden is a tranquil oasis from the hurry of life. A cottage garden, mainly, but it has many mature trees and shrubs. It is separated into different areas with seating so that the views of the garden can be enjoyed. There are 2 ponds, one of which has a few fish, after the otter took the rest, and a river bank with areas for wildlife, and also an allotment. Plants and garden related craft items for sale.

GROUP OPENING

20 HIGHSETT CAMBRIDGE
Cambridge, CB2 1NZ. Alice Fleet. *Centre of Cambridge. Via Station Rd, Tenison Rd, 1st L Tenison Ave, entrance ahead. SatNav CB2 1NZ. Regret no parking available in Highsett. Cash payments only.* **Sun 2 July (2-5). Combined adm £6, chd free. Home-made teas at 82 Highsett.**
Several delightful town gardens will be open to view in this well known housing complex within central Cambridge. Set in large communal grounds with fine specimen trees, many 70+ years old, and lawns. A haven for children and wildlife. Architect Eric Lyons planned the whole estate in the late 1950's with a mixture of flats, small houses and large town houses, the very ethos of tranquil living space for all generations. Several of the open gardens have been skilfully modernised by garden designers. Most paths accessible into the gardens.

GROUP OPENING

21 NEW HILTON GARDENS
The Green, Hilton, Huntingdon, PE28 9NB. *Parking is in the village hall car park and in residential streets.* **Sun 2 July (12-5). Combined adm £7, chd free. Light refreshments at the village hall. Artisan ice cream from Monach Farm will be available.**

NEW **GARDEN HOUSE**
Angela Potton.

NEW **1 GROVE END**
Malcolm Lynn.

NEW **HILTON COMMUNITY GARDEN**
Hilton Community.

NEW **OAKTREE FARM**
Dawn and Christian Hill.

NEW **16 SCOTTS CLOSE**
Rosemary & Peter Blake.

NEW **27 SCOTTS CRESCENT**
Mrs Sandy Monk.

NEW **WISTOW**
Claire & Nick Sarkies.

The Manor, Hemingford Grey

© Howard Rice

6 private gardens will open for the National Garden Scheme for the 1st time in 2023. Hilton village features a turf maze which was originally cut in 1660 and a parish church dating back to the C12. Both of these will be open to visit. The magnificent village green of 27 acres will be walked across, via the Platinum Jubilee "Queen's Green Canopy" to reach the Community Garden which was established in February 2021. At the Community Garden are raised beds for vegetables, a fruit cage, trugs for herbs and salads, a wildflower garden native hedging and shrubs. This garden has transformed the ground and provides a calm and tranquil setting for villagers to spend time and relax. The 6 private gardens offer a range of landscaped areas with some unusual perennials, over 40 different clematis in one garden, vegetables, trees and shrubs. Access to the community garden is on a rutted farmers track and some of the gardens have a gravel path.

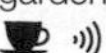

22 HORSESHOE FARM

Chatteris Road, Somersham, Huntingdon, PE28 3DR. Neil & Claire Callan, 01354 693546, nccallan@yahoo.co.uk. *9m NE of St Ives, Cambs. Easy access from the A14. Situated on E side of B1050, 4m N of Somersham Village. Parking for 8 cars on the drive.* **Visits by arrangement 1 May to 2 July for groups of up to 20. Adm £5, chd free. Home-made teas.**

This ¾ acre plant-lovers' garden has a large pond with summerhouse and decking, bog garden, alpine troughs, mixed rainbow island beds with over 40 varieties of bearded irises, water features, a small hazel woodland area, wildlife meadow, secret corners and a lookout tower for wide Fenland views and bird watching.

23 ISAACSON'S

6 High Street, Burwell, CB25 0HB. Dr Richard & Dr Caroline Dyer, 01638 601742, richard@familydyer.com. *10m NE of Cambridge, 4m NW of Newmarket. Behind a tall yew hedge with topiary, & grass triangle, at the S end of the village where Isaacson Rd turns off the High St. Approx 400 yds S of the church.* **Sun 18 June (12-5). Combined adm with Burwell Village Gardens £6, chd free. Home-made teas. Visits also by arrangement 2 May to 16 July for groups of 10 to 24. Entrance fee incl refreshments.**

Warmed and sheltered by the medieval walls of a C14 house (the oldest in Burwell) the garden is richly diverse in format and planting. There are 'theatres' (auricula, pinks), hints of Snowdonia, the Mediterranean, and a French potager, Anglesey Abbey in the snowdrop season and much else. Throughout there are many interesting, unusual and rare plants. Around each corner a new vista surprises.

24 ISLAND HALL

Godmanchester, PE29 2BA. Grace Vane Percy, 01480 459676, enquire@islandhall.com, www.islandhall.com. *1mile S of Huntingdon (A1). 15 m NW of Cambridge (A14). In centre of Godmanchester next to free Mill Yard car park.* **Sun 28 May (10.30-4.30). Adm £5, chd free. Home-made teas.**

3 acre grounds. Tranquil riverside setting with mature trees. Chinese bridge over Saxon mill race to embowered island with wild flowers. Garden restored in 1983 to mid C18 formal design, with box hedging, clipped hornbeams, parterres, topiary, good vistas over borrowed landscape and C18 wrought iron and stone urns. The ornamental island has been replanted with Princeton elms (ulmus americana). Mid C18 mansion (not open).

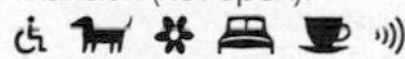

25 JUNIPER HOUSE

Cross Drove Coates, Whittlesey, Peterborough, PE7 2HJ. Mrs Jeni Cairns, 07541 229447, jenicairns@btinternet.com, www.juniperhouseemporium.com. *Stay on A605 until you reach Coates village green and turn up S green. R off S green & L onto lane & follow signs. Please be aware of uneven rd surfaces.* **Visits by arrangement June & July for groups of 5 to 30. Adm £5, chd free.**

Artist and designer Jeni Cairns has been creating a garden for enjoyment and nature over the past 13 yrs, It was previously the home of her grandparents for 60 yrs. Jeni is passionate about plants and learning as much as possible through experimentation and creativity. She takes inspiration from her garden to create her sculptures and decorative metal work.

26 NEW KHUSUSI

Church Road, Buckworth, Huntingdon, PE28 5AL. Mrs Justine Harding. *10 min drive from the A1/A14 interchange (Brampton Hut). At Alconbury Weston turn onto Buckworth Rd. Stay on the rd for 2.2 m until you see a sign for L turn to Buckworth. Follow this for ½ m into Buckworth.* **Sun 21 May (12-5). Open nearby Ferrar House. Sat 17 June (12-5). Adm £5, chd free. Light refreshments.**

The ½ acre garden enjoys open rural views and relaxed wildlife-friendly planting that is sympathetic to its setting. It features a variety of mature trees, ferns, hostas and euphorbias, and a well. Natural ornamentation inc a water-feature created from local stones, and tree stumps. The hot, dry aspect supports a mature olive tree and succulent planters. The garden is accessed via a gravel drive. The main section is open lawn which does have gentle slopes to different areas.

27 KING'S COLLEGE FELLOWS' GARDEN AND PROVOST'S GARDEN

Queen's Road, Cambridge, CB2 1ST. Provost & Scholars of King's College, tinyurl.com/kingscol. *In Cambridge, the Backs. Entry by gate at junc of Queen's Rd & West Rd or at King's Parade. Parking at Lion Yard, short walk, or some pay & display places in West Rd & Queen's Rd.* **Sun 9 July (12-4). Adm £6, chd free. Home-made teas.**

Fine example of a Victorian garden with rare specimen trees. Rond Pont entrance leads to a small woodland walk, herbaceous and sub-tropical borders, rose pergola, kitchen/ allotment garden and orchard. The Provost's garden opens with kind permission, offering a rare glimpse of an Arts and Crafts design. Chance to view the new wildflower meadow created on the former Great Lawn. Wheelchair access, gravel paths.

In 2022 the National Garden Scheme donated £3.11 million to our nursing and health beneficiaries

28 KIRTLING TOWER

Newmarket Road, Kirtling, CB8 9PA. The Lord & Lady Fairhaven. *6m SE of Newmarket. From Newmarket head towards village of Saxon St, through village to Kirtling. Turn L at the war memorial; the rd is signed to Upend & the entrance is signed on the L.* **Sun 26 Mar, Sun 25 June (11-4). Adm £6, chd free. Light refreshments.**

Surrounded by a moat, formal gardens and acres of parkland. In the spring there are swathes of daffodils, narcissi, crocus, muscari, chionodoxa and tulips. Closer to the house are vast lawn areas. Large Secret Garden and Cutting and Vegetable Garden. In the summer the Walled Garden has superb herbaceous borders with anthemis, hemerocallis, geraniums and delphiniums. The Victorian Garden is filled with peonies. Views of surrounding countryside. A Classic car display will be in attendance. Many of the paths and routes around the garden are grass - they are accessible by wheelchairs, but can be hard work if wet.

GROUP OPENING

29 LINTON GARDENS

Linton, CB21 4JB. Henry & Sarah Bennett. *Linton Village College Car Park accessed from A1307 between A11 and Haverhill, then short walk following yellow signs. The postcode for the car park is CB21 4JB.* **Sun 18 June (2-6). Combined adm £8, chd free. Sat 9 Sept (2-6). Combined adm £5, chd free. Light refreshments in the Maltings Courtyard at 94 High Street.**

14 HIGH STREET
Chris Tyler-Smith and Yali Xue, ylxcts@gmail.com.
Open on all dates
Visits also by arrangement 2 Jan to 15 Dec for groups of 2 to 6.

94 HIGH STREET
Rosemary Wellings.
Open on Sun 18 June

LINTON HOUSE
Steve Meeks.
Open on Sun 18 June

NEW **20 MILL LANE**
Tom and Katy Meeks.
Open on Sat 9 Sept

MILLICENT HOUSE
Sue Anderson.
Open on Sun 18 June

QUEENS HOUSE
Michael & Alison Wilcockson.
Open on Sun 18 June

SUMMERFIELD HOUSE
Ray & Bridget Linsey.
Open on Sun 18 June

Several gardens in the middle of Linton. Opening in June, the largest, Linton House, has the river Granta flowing through, and a pretty bridge. A rose walk, huge trees, and interesting planting, as well as a mulberry tree older than the house. Summerfield House is an immaculate garden with a plantsman's collection of herbaceous perennials and shrubs, sympathetic hard landscaping and a pleached lime wall. Millicent House which has a long-established garden again with some unusual planting. 94 High Street is the smallest garden, a secret garden, full of herbs and other edible plants, as well as surprising artefacts. It is found behind the owner's art gallery. Queens House has a hedged formal garden with statuary which reflects the age and feel of the Grade II* listed house. Next to it is 14 High Street, where most of the planting is white. The entrance belies the size of the garden, much of which is designed with wildlife in mind. Opens twice, in Sept with 20 Mill Lane, whose cottage garden has beautiful showing of dahlias and salvias as well as abundant edible produce. Limited parking in the village. Linton Village College has kindly agreed to make their carpark available on the day. The car park will be locked at 6.30pm after which time it will not be possible to remove any car.

30 LITTLE OAK

66 Station Road, Willingham, Cambridge, CB24 5HG. Mr & Mrs Eileen Hughes, www.littleoak.org.uk/garden. *4m N of A14 J25 nr Cambridge. Easy to find on the main rd in the village. Driveway parking for disabled use only.* **Sat 17, Sun 18, Sat 24, Sun 25 June (1-5). Adm £5, chd free. Home-made teas. Home-made cakes also available. Picnics welcome.**

Michael and Eileen welcome you to their 1 acre garden. Featuring a 50' Laburnum Walk with perennial borders, ponds, cottage garden, fruit cage and kitchen garden. There are also greenhouses, growing tunnels, an orchard, Mediterranean, rose garden, coppice and chickens. Shepherds hut, wood turning and model railways - weather permitting. Main garden is wheelchair accessible.

31 MADINGLEY HALL

Cambridge, CB23 8AQ. University of Cambridge, 01223 746222, reservations@madingleyhall.co.uk, www.madingleyhall.co.uk. *4m W of Cambridge. 1m from M11, J13. Located in the centre of Madingley village. Entrance adjacent to mini r'about.* **Sun 21 May (2.30-5.30). Adm £6, chd free. Home-made teas at St Mary Magdalene Church, in the grounds of Madingley Hall.**

C16 Hall (not open) set in 8 acres of attractive grounds landscaped by Capability Brown. Features inc landscaped walled garden with hazel walk. Historic meadow, topiary, mature trees and wide variety of hardy plants.

32 67 MAIN STREET, YAXLEY

Peterborough, PE7 3LZ. Mrs Karen Woods, 07787 864426, karenandstevewoods@yahoo.co.uk. *S of Peterborough, approx 3m from J16 on A1. On entering Yaxley, turn R into Dovecote Ln. At the bottom, turn L onto Main St. Find us approx 1m on R.* **Visits by arrangement 1 May to 17 Sept for groups of up to 25. Adm £5, chd free. Home-made teas.**

1 acre working organic cottage garden providing something for the kitchen and vase year-round. Seasonal cut flowers grown in raised beds and traditional borders. Allotment style vegetable garden. Paddock with fruit trees and views over the fen to the rear. Chickens for eggs. Wreath workshops and flower demonstrations available, please email for more details.

33 ◆ THE MANOR, HEMINGFORD GREY

Hemingford Grey, PE28 9BN. Mrs D S Boston, 01480 463134, diana_boston@hotmail.com, www.greenknowe.co.uk. *4m E of Huntingdon. Off A1307. Parking for NGS opening day only, in field off double bends between Hemingford Grey & Hemingford Abbots. This will be signed. Entrance to garden*

via small gate off river towpath. **For NGS: Sun 10 Sept (11-5). Adm £6, chd free. Home-made teas. For other opening times and information, please phone, email or visit garden website.**
Garden designed by author Lucy Boston, surrounds C12 manor house on which Green Knowe books based (house open by appt). 3 acre 'cottage' garden with topiary, snowdrops, old roses, extensive collection of irises inc Dykes Medal winners and Cedric Morris varieties, herbaceous borders with mainly scented plants. Meadow with mown paths. Enclosed by river, moat and wilderness. Gravel paths, wheelchairs are encouraged to use the lawns.

34 MARMALADE LANE, COHOUSING COMMUNITY GARDEN

09 Marmalade Lane, Orchard Park, Cambridge, CB4 2ZE. Cambridge Cohousing Ltd, www.facebook.com/CambridgeK1CoHousingProject. *Please arrive at the Common House & take a seat outside. A guide will conduct you. If possible arrive by public transport. Parking on site for disabled visitors. Location close to Orchard Park East Bus Stop & Mere Rd bus stop. Cambridge North Stn is a short taxi or bus ride away.* **Sat 24 June (10-5). Adm £5, chd free. Pre-booking essential, please visit www.ngs.org.uk for information & booking. Home-made teas.**
Community Garden owned, created and looked after by residents. Primarily a wildflower meadow with allotment, small copse, patio, lawn and play space. Chickens and a small hedgehog sanctuary were introduced in 2021, with 3 tiny ponds to attract frogs being a particular success. A new lawn and play area were also created in lockdown. A path circumnavigates the main garden.

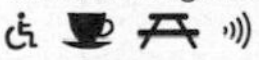

35 MOLESWORTH HOUSE

Molesworth, PE28 0QD. John & Gilly Prentis, 07771 918250, gillyprentis@gmail.com. *10 m W of Huntingdon off the A14. Next to the church in Molesworth.* **Sun 21 May (2-6). Adm £5, chd free. Home-made teas. Visits also by arrangement.**
Molesworth House is an old 3 acre rectory garden with everything that you'd both expect and hope for, given its Victorian past. There are surprising corners to this traditional take on a happy and relaxed garden - come and see for yourself. Accessible to wheelchairs although there is some gravel.

36 NEW 5 NELSONS LANE

Haddenham, Ely, CB6 3UH. Mr Peter Wilson, haddenhamgardener.tumblr.com. *10 m N of Cambridge and 7 m from Ely. From A10 N of Cambridge, take A1123 at Stretham.Follow for 2½m to Haddenham, bearing L just before water tower as you enter the village. L at bottom of hill. No parking at all in Nelsons Ln.* **Sun 11 June (12-5). Adm £4, chd free. Light refreshments at Haddenham Arts Centre.**
Extending to half an acre the garden has been created from scratch over 16 years and inc numerous areas from the central herbaceous borders surrounded by yew hedges to a hidden slate garden, vegetable and fruit beds and wild areas to support nature. A long range of traditional buildings houses summerhouse and potting sheds for tender edibles such as peppers and chilies.

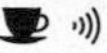

37 NETHERHALL MANOR

Tanners Lane, Soham, CB7 5AB. Timothy Clark, 01353 720269, timothy.r.clark@btinternet.com. *6m Ely, 6m Newmarket. Enter Soham from Newmarket, Tanners Ln 2nd R 100yds after cemetery. Enter Soham from Ely, Tanners Ln 2nd L after War Memorial.* **Sun 2 Apr, Sun 7 May, Sun 6, Sun 13 Aug (2-5). Adm £3, chd free. Home-made teas. Visits also by arrangement for groups of 8 to 15.**
Elegant garden 'touched with antiquity'. Good Gardens Guide 2000. Unusual garden appealing to those with historical interest in individual collections of plant groups: March-old primroses, daffodils, Victorian double flowered hyacinths and first garden hellebore hybrids. May-old English tulips, Crown Imperials. Aug-Victorian pelargonium, heliotrope, calceolaria, Red list for Plant Heritage. Author-Margery Fish's Country Gardening and Mary McMurtrie's Country Garden Flowers Historic Plants 1500-1900. Author's books for sale. Featured on Gardeners' World 3 times. Wheelchair access flat garden.

38 38 NORFOLK TERRACE GARDEN

Cambridge, CB1 2NG. John Tordoff & Maurice Reeve. *Central Cambridge. A603 East Rd turn R into St Matthews St to Norfolk St, L into Blossom St & Norfolk Terrace is at the end.* **Sat 1, Sun 2 July (11-6). Adm £3, chd free. Light refreshments.**
A small, paved courtyard garden in Moroccan style. Masses of colour, backed by oriental arches. An ornamental pool done in patterned tiles.The garden won third prize in the 2018 Gardener's World national competition. It is also inc in the book 'The Secret Gardens of East Anglia'.The owners' previous, London garden, was named by BBC Gardeners' World as 'Best Small Garden in Britain'. There will be a display of recent paintings by John Tordoff and handmade books by Maurice Reeve.

39 THE OLD RECTORY

312 Main Road, Parson Drove, Wisbech, PE13 4LF. Helen Roberts, 07818 070641, yogahelen@talk21.com. *SW of Wisbech. From Peterborough on A47 follow signs to Parson Drove, L after Thorney Toll. From Wisbech follow the B1166 through Leverington Common.* **Sun 11 June (11-4). Adm £5, chd free. Home-made teas. Visits also by arrangement 14 May to 30 July for groups of 10+. Adm inc refreshments.**
This garden represents the personality of this Georgian house (not open), classical and formal at the front and friendly and a bit crazy at the back. A walled cottage style of about an acre with an abundance of plants and roses leading onto wildflower meadows with walking paths. There are also 3 ponds each one also with a different personality.

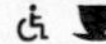

In 2022, our donations to Carers Trust meant that 456,089 unpaid carers were supported across the UK

40 23A PERRY ROAD

Buckden, St Neots, PE19 5XG. David & Valerie Bunnage, 01480 810553, d.bunnage@btinternet.com. *5m S of Huntingdon on A1. From A1 Buckden r'about take B661, Perry Rd approx 300yds on L.* **Every Fri, Sat and Sun 1 May to 31 Aug (2-4). Adm £5, chd free. Pre-booking essential, please visit www.ngs.org.uk for information & booking. Visits also by arrangement. Coaches welcome.**

Approx 1 acre garden of many garden designs inc Japanese interlinked by gravel paths. Plantsmans garden, 155 acers and unusual shrubs. Quirky garden with interesting features some narrow paths not suitable for wheelchairs. Regret no dogs. WC available.

41 PREACHERS PASSING

55 Station Road, Tilbrook, Huntingdon, PE28 0JT. Keith & Rosamund Nancekievill. *Tilbrook 4½m S of J16 on A14. Station Rd in Tilbrook can be accessed from the B645 or B660. Preachers Passing faces the small bridge over the River Til at a sharp bend in Station Rd with All Saints church behind it.* **Sat 24, Sun 25 June (10.30-5). Adm £5, chd free. Light refreshments. Home made cakes, ice cream, teas & soft drinks.**

¾ acre garden fits into its pastoral setting. Near the house, parterre, courtyard and terrace offer formality; but beyond, prairie planting leads to a wildlife pond, rock gardens, meadow, stumpery, rose garden and copse. Enjoy different views from arbour, honeysuckle-covered swing or scattered benches. Deciduous trees and perennials give changing colour. Here are open spaces and hidden places.

42 NEW RAMSEY WALLED GARDEN

Wood Lane, Ramsey, Huntingdon, PE26 2XD. Jane Sills, www.ramseywalledgarden.org. *Just N of Ramsey Town Centre. Follow B1096 out of Ramsey. Just after last house on R, turn R opp cemetery. Take track to Ramsey Rural Museum. Park under trees. See map on website.* **Sat 16 Sept (2-5). Adm £5, chd free. Home-made teas at Ramsey Rural Museum.**

A Victorian Walled Garden, located within the grounds of Ramsey Abbey, has been restored by volunteers and is an enchanting secret in the heart of Ramsey. Dating back to 1840, this one acre garden is dedicated to growing fruit, vegetables and flowers. Features inc a magnificent 33m glasshouse, an apple tunnel planted with Cambridgeshire varieties, a dahlia border and a collection of salvias. The garden is wheelchair accessible but is 450m from the car parking area. Motorised buggies can access the garden.

43 ROBINSON COLLEGE

Grange Road, Cambridge, CB3 9AN. Warden and Fellows, www.robinson.cam.ac.uk/about-robinson/gardens/national-gardens-scheme. *Garden at main Robinson College site, report to Porters' Lodge. There is only on-street parking.* **Daily weekdays 10-4, weekends 2-4, from Tue 3 Jan. Closed Sat 15 April-Tues 4 July, Sat 12 Aug - Tue 29 Aug, Sat 23-Sun 31 Dec. Adm £5, chd free. Prebooking essential, please visit www.ngs.org.uk for information and booking.**

10 original Edwardian gardens are linked to central wild woodland water garden focusing on Bin Brook with small lake at heart of site. This gives a feeling of park and informal woodland, while at the same time keeping the sense of older more formal gardens beyond. Central area has a wide lawn running down to the lake framed by many mature stately trees with much of the original planting intact. More recent planting inc herbaceous borders and commemorative trees. Tickets must be booked on line in advance. Entry inc guidebook, please collect from Porters' Lodge. No picnics. Children must be accompanied at all times. NB from time to time some parts, or occasionally all, of Robinson College gardens may be closed for safety reasons involving work by contractors and our maintenance staff. Ask at Porters' Lodge for wheelchair access.

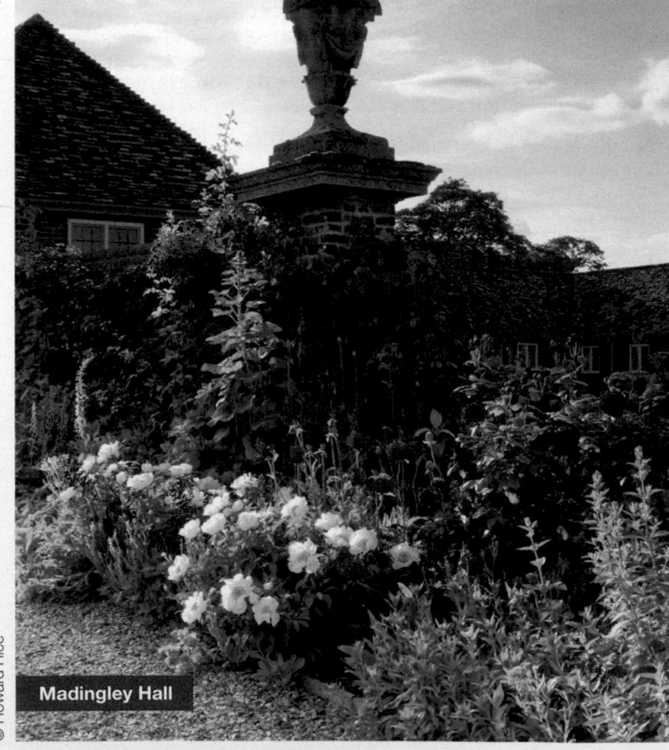

Madingley Hall

GROUP OPENING

44 SAWSTON GARDENS

Sawston, CB22 3HY. *5m SE of Cambridge. Midway between Saffron Walden & Cambridge on A1301.* **Sun 2 July (1-5). Combined adm £6, chd free. Home-made teas in Mary Challis Garden, 68 High Street. Refreshments served from 1.30-4.00pm.**

34 CAMBRIDGE ROAD
Mrs Mary Hollyhead.

◆ MARY CHALLIS GARDEN
A M Challis Trust Ltd, 01223 560816, chair@challistrust.org.uk, www.challistrust.org.uk.

11 MILL LANE
Tim & Rosie Phillips.

10 WYNEMARES
Mr Lee Kirby.

4 gardens in this large South Cambs. village. 11 Mill Lane is an attractive C16/C19 house with an impressive semi-circular front lawn edged with roses and mixed borders, and a secluded sun-dappled garden at the rear. 34 Cambridge Road features a long rear garden, with meticulously maintained lawns, colourful borders, mature trees and shrubs, small pond and fruit and vegetable plots. The 2 acre Mary Challis Garden has a tranquil setting in the heart of the village, with extensive lawns, wildflower meadow, herbaceous borders, orchard, vegetable beds, vinehouse and beehives. 10 Wynemares presents colourful and creative plantings in a small developing garden. Nearly all parts of the gardens are accessible by wheelchair.

GROUP OPENING

45 NEW STAPLEFORD VILLAGE GARDENS

Stapleford, Cambridge, CB22 5BL. *4m S of Cambridge on A1301.* **Sun 23, Sun 23 July (1-6). Combined adm £5, chd free.**

3A DUKES MEADOW
Jan & Lee Gruncell.
Open on Sat 22, Sun 23 July

5 PRIAMS WAY
Tony Smith.
Open on Sat 22, Sun 23 July

So much interest in two contrasting gardens. At 5 Priams Way there is lots of colour, unusual plants and playful design. A large productive grape vine and several pots of variously coloured camellias lead to the landscaped garden, with a pond and a lush lawn surrounded by wide curves of herbaceous borders, restocked following last year's drought. Seating in the family barbecue area allows visitors to stop for a few minutes and admire the acer shrubs and trees, and an unusual oak leaf hydrangea. New in the last few years is an expanding collection of hibiscus plants in several different colours. Metalwork sculptures, mostly made locally, is used to enhance some planting or as stand alone features. 3A Dukes Meadow has some unusual cultivars and plants, many propogated by the owners. This is a garden with surprises at each turn, lots of sun, interesting ways of using stumps and some plants in unusual places. Colourful and engaging everywhere you look.

GROUP OPENING

46 STAPLOE GARDENS

Staploe, St Neots, PE19 5JA. 07702 707880, caroline-falling@hotmail.co.uk. *Great North Rd in W part of St Neots. At r'about just N of the Co-op store, exit westwards on Duloe Rd. Follow this under the A1, through the village of Duloe & on to Staploe.* **Sat 3, Sun 4 June (1-5). Combined adm £6, chd free. Home-made teas at Falling Water House. Visits also by arrangement 1 Apr to 1 Nov for groups of 10+.**

FALLING WATER HOUSE
Caroline Kent.

OLD FARM COTTAGE
Sir Graham & Lady Fry.

Old Farm Cottage: Formal and terraced beds with roses, and lawns, surrounding thatched house (not open), leading to 3 acres of orchard, wildflower meadow, woodland and wetland maintained for wildlife. Ginkgo, loquat and manuka trees grown from seed. Falling Water House: a mature woodland garden, constructed around several century old trees inc 3 wellingtonia. Kitchen garden potager, and herbaceous borders, planted to attract bees and wildlife, through which meandering paths have created hidden vistas. Wheelchair access, Old Farm Cottage has uneven ground and 1 steep slope.

GROUP OPENING

47 SUTTON GARDENS

Sutton, Ely, CB6 2QQ. *6m W of Ely. From A142 turn L at r'about on to B1381 to Earith. Parking at Brooklands Centre 1 km on R.* **Sun 28 May (2-6). Combined adm £7, chd free. Home-made teas at The Burystead.**

THE BURYSTEAD
Sarah Cleverdon & Stephen Tebboth.

61 HIGH STREET
Ms Kate Travers & Mr Jon Megginson, 01353 778427, jonmegginson1@gmail.com.

THE OLD BAPTIST CHAPEL
Janet Porter & Steve Newton, 01353 777493, janet.porter@cantab.net.
Visits also by arrangement May to July inc access to house and information on its history and conversion.

NEW **87 THE ROW**
Mrs Alison Beale.

89 THE ROW
Andrew Thompson.

The Old Baptist Chapel is a new garden created around an C18 building converted to a family home. It inc the adjacent burial ground. The emphasis is on year-round interest and minimal maintenance. 61 The High Street is a tiny shaded garden with a host of features: interesting trees, herb and vegetable plots, water features, a greenhouse, boxes for birds, bats and bugs, chickens and a large collection of acers. 87 The Row is a spacious family garden with pergola, herbaceous beds, shrub border, mature native trees, vegetable beds and fruit trees. 89 The Row is an informal garden on a slight slope with a variety of planted areas inc mature trees, shrubs, fernery and herbaceous. There is also a small vegetable plot inc a greenhouse, and a wide variety of fruit trees. Scented plants are important throughout the garden. The Burystead is a ½ acre walled courtyard garden of formal design within a large country garden against the backdrop of a restored C16 thatched barn.

GROUP OPENING

48 TOFT GARDENS

Toft, Cambridge, CB23 2RY. *8m W of Cambridge. Park on main rd or near church & follow yellow signs.* **Sun 18 June (12-5). Combined adm £5, chd free. Home-made teas in Toft People's Hall.**

21 COMBERTON ROAD
Mr Sheppard.

THE GIG HOUSE
Jane Tebbit.

MANOR COTTAGE
Mrs Mary Paxman.

The Gig House: cottage garden with roses, clematis, perennials, cut flower beds and vegetables. Also, plants for natural dyes, dry beds, wildflowers, a small wildlife pond, tiny orchard and plenty of places to sit. Manor Cottage: cottage garden extended and developed by the current owner over 30 yrs. Herbaceous borders, vegetable garden, greenhouse and orchard covering an acre, as well as a paddock and 2 wooded areas. Borders extended and replanted in 2020 when various grasses, salvias and alliums were added. 21 Comberton Road: the basic concept was conceived before moving in, calling on a wide range of sources and experiences. 4 zones in 0.18 acres, 123 taxa, 450 plants; 1 afternoon to view this newly planted garden.

49 TRINITY COLLEGE FELLOWS' GARDEN

Queen's Road, Cambridge, CB3 9AQ. Master and Fellows of Trinity College, www.trin.cam.ac.uk/about/gardens. *Short walk from city centre. At the Northampton St/Madingley Rd end of Queen's Road, close to Garret Hostel Ln.* **Sun 26 Mar (1-4). Adm £4, chd free. Light refreshments. Special dietary requirements are catered for.**

Interesting historic garden of about 8 acres with impressive specimen trees, mixed borders, drifts of spring bulbs and informal lawns with notable influences throughout from Fellows over the years. Across the gently flowing Bin Brook to Burrell's Field, you will find some modern planting styles and plants nestled amongst the accommodation blocks in smaller intimate gardens. Members of the Gardens Department will be on hand to answer any questions and serve homemade cakes and drinks. Plant sales are also available. Wheelchair access - some gravel paths.

50 TWIN TARNS

6 Pinfold Lane, Somersham, PE28 3EQ. Michael & Frances Robinson, 07938 174536, mikerobinson987@btinternet.com. *Easy access from the A14. 4m NE of St Ives. Turn onto Church St. Pinfold Ln is next to the church. Please park on Church Street as access is narrow & limited.* **Sat 3, Sun 4 June (1-6). Adm £5, chd free. Cream teas. Visits also by arrangement May to Aug.**

One acre wildlife garden with formal borders, kitchen garden and ponds, large rockery, mini woodland, wildflower meadow (June/July). Topiary, rose walk, greenhouses. Character oak bridge, veranda and treehouse. Adjacent to C13 village church.

GROUP OPENING

51 NEW WESTLEY WATERLESS GARDENS

Westley Waterless, Newmarket, CB8 0RL. *10m E of Cambridge, 5m S of Newmarket. Both gardens are near the entrance to Church Ln. Parking along the Main St and part way up Church Ln.* **Sun 25 June (2-5). Combined adm £5, chd free. Home-made teas at Church Lane House.**

CHURCH LANE HOUSE
Dr Lucy Crosby, lucycrosby.vet@gmail.com.
Visits also by arrangement 20 May to 30 Sept.

NEW **CHURCH VIEW**
Diana Hall, 07974 667298, boohall@icloud.com.
Visits also by arrangement Apr to Sept for groups of up to 20.

Church Lane House is just under an acre in size, comprising a wildlife pond, herbaceous borders, cut flowers, dye plants, orchard wildflower meadow, Victorian style greenhouse and kitchen garden, plus chickens, guinea pigs and bees. This is a family friendly garden with plenty of seating areas to take in the views. Church View is a country garden which inc a small flower farm. There are many old fashioned scented roses, both shrub and climbing plus herbaceous borders, raised growing beds and 2 polytunnels. The garden inc a wildlife pond and flowering plant-filled pots. The farm growing area changes season by season. Access is mostly on grass and some paving areas.

52 THE WINDMILL

10 Cambridge Road, Impington, CB24 9NU. Pippa & Steve Temple, 07775 446443, mill.impington@ntlworld.com, www.impingtonmill.org. *2½m N of Cambridge. Off A14 at J32, B1049 to Histon, L into Cambridge Rd at T-lights, follow Cambridge Rd round to R, the Windmill is approx 400yds on L.* **Visits by arrangement 1 Apr to 30 Aug. Adm £5, chd free. Light refreshments inc coffee, tea, wine and nibbles by prior agreement.**

A previously romantic wilderness of 1½ acres surrounding windmill, now filled with bulbs, perennial beds, pergolas, bog gardens, grass bed and herb bank. Secret paths and wild areas with thuggish roses maintain the romance. Millstone seating area in smouldering borders contrasts with the pastel colours of the remainder of the garden. Also 'Pond Life' seat, 'Tree God' and amazing compost area! The Windmill - an C18 smock on C19 tower on C17 base on C16 foundations - is being restored. Guide dogs only.

Our 2022 donation to Hospice UK meant that 911 NHS and Social Care staff accessed individual counselling support

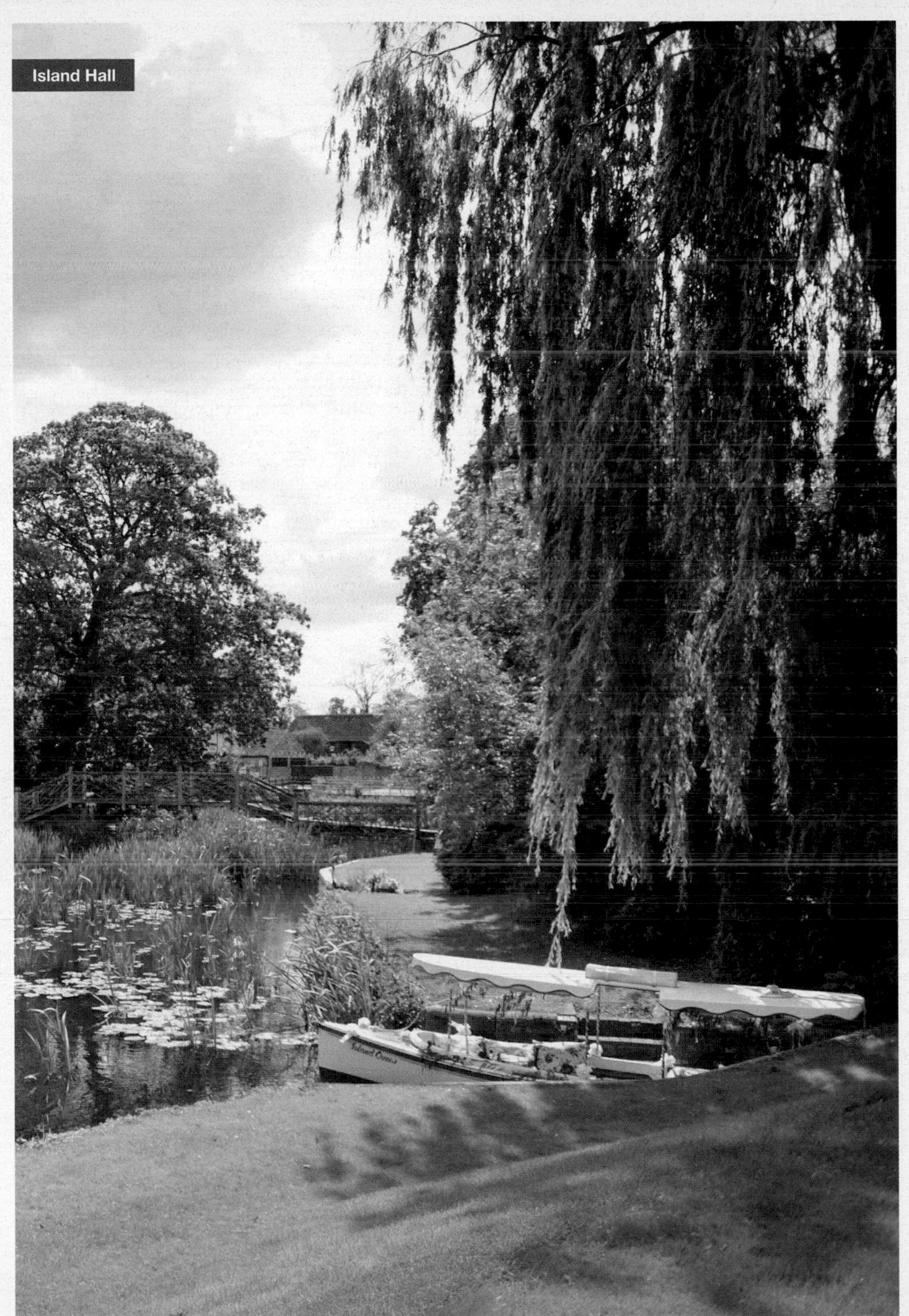

Island Hall

CHESHIRE & WIRRAL

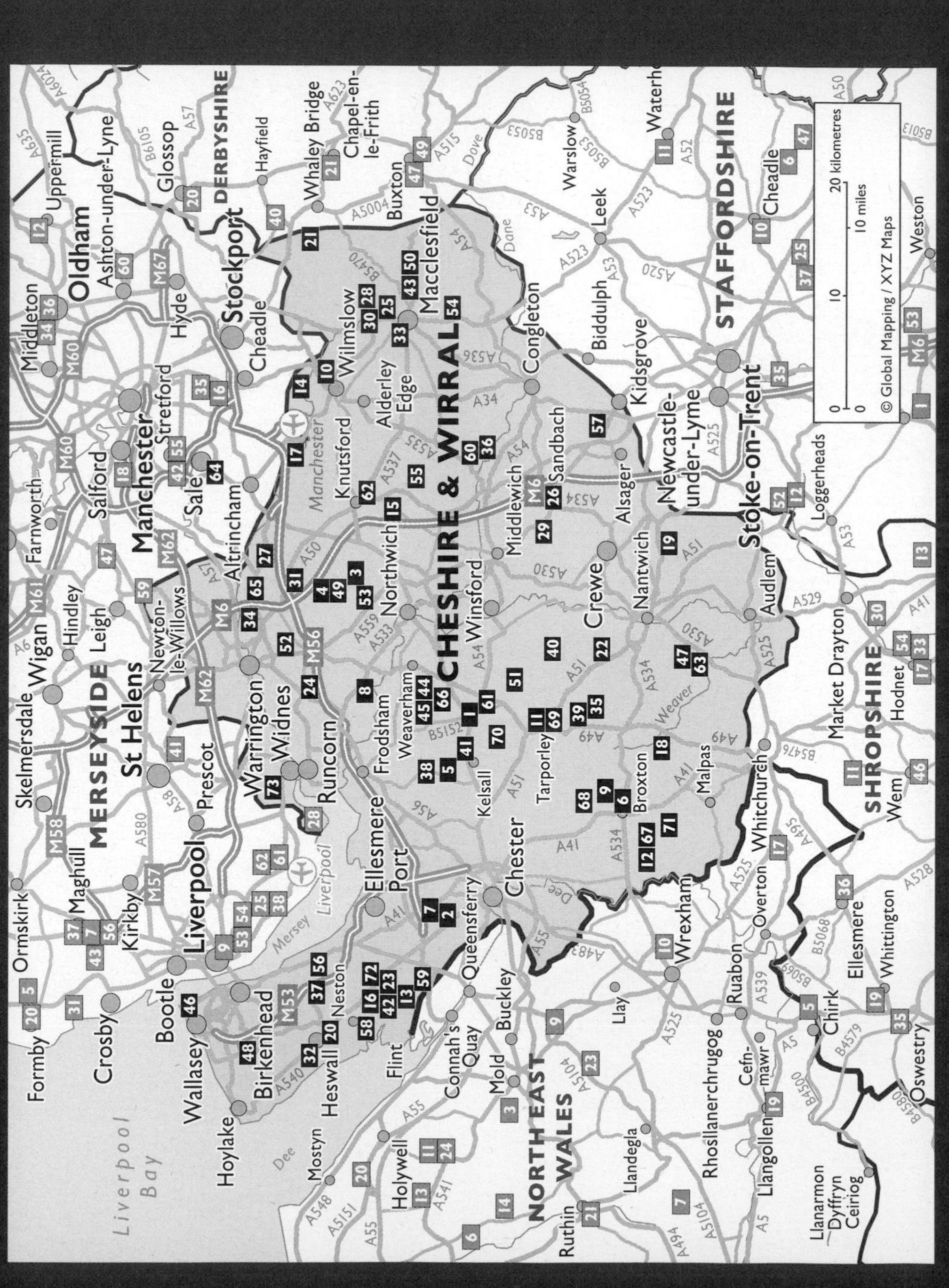

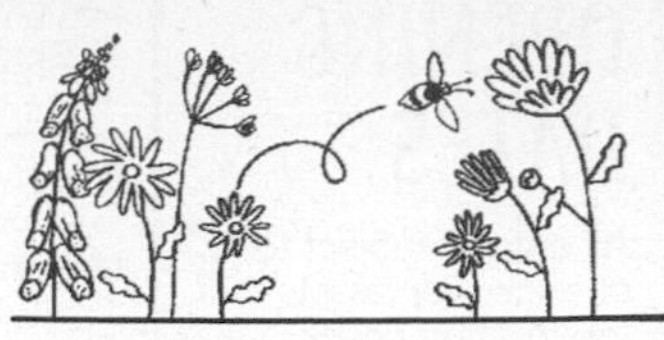

The area of Cheshire and Wirral comprises what are now the four administrative regions of West Cheshire and Chester, East Cheshire, Warrington and Wirral, together with gardens in the south of Greater Manchester, Trafford and Stockport.

The perception of the area is that of a fertile county dominated by the Cheshire Plain, but to the extreme west it enjoys a mild maritime climate, with gardens often sitting on sandstone and sandy soils and enjoying mildly acidic conditions.

A large sandstone ridge also rises out of the landscape, running some 30-odd miles from north to south. Many gardens grow ericaceous-loving plants, although in some areas, the slightly acidic soil is quite clayey. But the soil is rarely too extreme to prevent the growing of a wide range of plants, both woody and herbaceous.

As one travels east and the region rises up the foothills of the Pennine range, the seasons become somewhat harsher, with spring starting a few weeks later than in the coastal region.

As well as being home to one of the RHS's major shows, the region's gardens include two National Garden Scheme 'founder' gardens in Arley Hall and Peover Hall, as well as the University of Liverpool Botanic Garden at Ness.

Below: Maggie's, Wirral

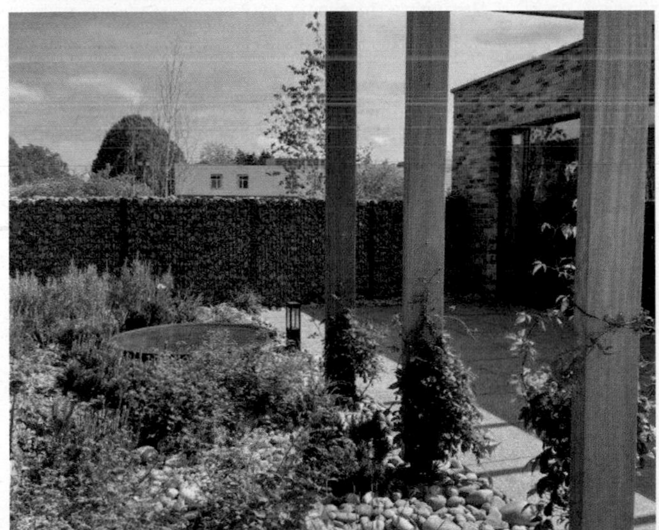

@National Garden Scheme Cheshire & Wirral   CheshireWirrNGS

Volunteers

County Organiser
Janet Bashforth
07809 030525
jan.bashforth@ngs.org.uk

County Treasurer
John Hinde
0151 3530032
johnhinde059@gmail.com

Booklet Co-ordinator
John Hinde (see above)

Booklet Advertising
Richard Goodyear
07775 924929
richard.goodyear@ngs.org.uk

Social Media & Publicity
Jacquie Denyer
jacquie.denyer@ngs.org.uk

Photographer
Liz Mitchell 01260 291409
liz.mitchell@ngs.org.uk

Talks
Mike Porter 01925 753488
porters@mikeandgailporter.co.uk

Assistant County Organisers
Sue Bryant 0161 928 3819
suewestlakebryant@btinternet.com

Jean Davies 01606 892383
mrsjeandavies@gmail.com

Linda Enderby 07949 496747
linda.enderby@ngs.org.uk

Sandra Fairclough 0151 342 4645
sandrafairclough51@gmail.com

Richard Goodyear (see above)

Juliet Hill 01829 732804
t.hill573@btinternet.com

Romy Holmes 01829 732053
romy@holmes-email.co.uk

Mike Porter (see above)

OPENING DATES

All entries subject to change. For latest information check **www.ngs.org.uk**

Map locator numbers are shown to the right of each garden name.

February

Snowdrop Festival

Every Friday
Briarfield 13

Sunday 26th
Bucklow Farm 15

March

Sunday 5th
NEW Oulton Park House 51

April

Sunday 2nd
Parm Place 53

Saturday 8th
Adswood 2
◆ Poulton Hall 56

Sunday 9th
Adswood 2
◆ Poulton Hall 56

Sunday 16th
Briarfield 13
Long Acre 35

Tuesday 18th
◆ Arley Hall & Gardens 4

Saturday 22nd
NEW Parvey Lodge 54

Sunday 23rd
Hill Farm 29
NEW Parvey Lodge 54
The Wonky Garden 73

Saturday 29th
Adswood 2

Sunday 30th
Adswood 2
Laskey Farm 34
NEW Norley Court 45
Somerford 62
10 Statham Avenue 65

May

Monday 1st
Framley 23
Laskey Farm 34
10 Statham Avenue 65

Thursday 4th
◆ Cholmondeley Castle Gardens 18

Saturday 6th
Brooke Cottage 14

Sunday 7th
Brooke Cottage 14

Saturday 13th
64 Carr Wood 17
Hathaway 25

Sunday 14th
◆ Abbeywood Gardens 1
Hathaway 25
NEW Highbank 26
Hill Farm 29
◆ Stonyford Cottage 66
Tiresford 69

Saturday 20th
The Old Parsonage 49

Sunday 21st
Bankhead 6
Manley Knoll 38
Mayfield House 39
◆ Mount Pleasant 41
The Old Parsonage 49
34 Stanley Mount 64
Tirley Garth Gardens 70

Saturday 27th
Laskey Farm 34
10 Statham Avenue 65

Sunday 28th
Laskey Farm 34
Rowley House 60
10 Statham Avenue 65

June

Thursday 1st
15 Park Crescent 52

Saturday 3rd
NEW Brandon House 12

Sunday 4th
NEW Brandon House 12
Tattenhall Hall 68

Saturday 10th
Drake Carr 21
◆ Peover Hall Gardens 55

Sunday 11th
Bucklow Farm 15
NEW 18 Dee Park Road 20
Drake Carr 21
Hill Farm 29
The Homestead 31
60 Kennedy Avenue 33
◆ Peover Hall Gardens 55

Wednesday 14th
NEW 18 Dee Park Road 20
10 Statham Avenue 65

Friday 16th
Somerford 62

Saturday 17th
NEW New Inn Farm 43
One House Walled Garden 50
Willaston Grange 72

Sunday 18th
NEW 18 Dee Park Road 20
Hall Lane Farm 24
NEW Maggie's Wirral 37
NEW New Inn Farm 43
24 Old Greasby Road 48
One House Walled Garden 50

Saturday 24th
All Fours Farm 3
18 Highfield Road 28

Sunday 25th
All Fours Farm 3
Burton Village Gardens 16
18 Highfield Road 28
15 Park Crescent 52
Parm Place 53
Sandymere 61

July

Saturday 1st
◆ Bluebell Cottage Gardens 8

Sunday 2nd
Ashton Grange 5
◆ Bluebell Cottage Gardens 8
The Wonky Garden 73

Wednesday 5th
10 Statham Avenue 65

Thursday 6th
5 Cobbs Lane 19

Saturday 8th
5 Cobbs Lane 19

Sunday 9th
The Homestead 31
Rowley House 60

Saturday 15th
Laskey Farm 34
Milford House Farm 40
The Old Byre 47
NEW Sound Manor 63

Sunday 16th
Laskey Farm 34
The Old Byre 47
NEW Sound Manor 63

Saturday 22nd
The Birches 7
Stretton Old Hall 67

Sunday 23rd
The Birches 7
Norley Bank Farm 44
Rose Brae 58
Stretton Old Hall 67

Sunday 30th
The Firs 22
Norley Bank Farm 44
Somerford 62

August

Sunday 6th
The Firs 22

Monday 7th
Norley Bank Farm 44

Sunday 27th
Laskey Farm 34

Monday 28th
Laskey Farm 34

September

Sunday 3rd
◆ Mount Pleasant 41

Sunday 10th
The Wonky Garden 73

Tuesday 12th
◆ Ness Botanic Gardens 42

Sunday 17th
◆ Abbeywood Gardens 1

October

Sunday 8th
◆ The Lovell Quinta Arboretum 36

Saturday 14th
NEW Parvey Lodge 54

Sunday 15th
NEW Parvey Lodge 54

February 2024

Sunday 25th
Bucklow Farm 15

By Arrangement

Arrange a personalised garden visit with your club, or group of friends, on a date to suit you. See individual garden entries for full details.

Adswood 2
All Fours Farm 3
Ashton Grange 5
Bankhead 6
Bolesworth Castle 9
Bollin House 10
Bowmere Cottage 11
Briarfield 13
NEW 18 Dee Park Road 20
The Firs 22
213 Higher Lane 27
Hill Farm 29
Hilltop 30
The Homestead 31
Inglewood 32
Laskey Farm 34
Long Acre 35
Mayfield House 39
Milford House Farm 40
Norley Bank Farm 44
NEW 6 Oakland Vale 46
15 Park Crescent 52
Parm Place 53
NEW Parvey Lodge 54
Rosewood 59
Sandymere 61
10 Statham Avenue 65
Stretton Old Hall 67
Tattenhall Hall 68
Trustwood, Burton Village Gardens 16
The Well House 71
The Wonky Garden 73

THE GARDENS

1 ◆ ABBEYWOOD GARDENS

Chester Road, Delamere, Northwich, CW8 2HS. The Rowlinson Family, 01606 889477, info@abbeywoodestate.co.uk, www.abbeywoodestate.co.uk. *11m E of Chester. On the A556 facing Delamere Church.* **For NGS: Sun 14 May, Sun 17 Sept (9-5). Adm £6, chd free. Light refreshments. Restaurant in Garden. For other opening times and information, please phone, email or visit garden website.**

Superb setting near Delamere Forest. Total area 45 acres inc mature woodland, new woodland and new arboretum all with connecting pathways. Approx 4½ acres of gardens surrounding large Edwardian house. Vegetable garden, exotic garden, chapel garden, pool garden, woodland garden, lawned area with beds.

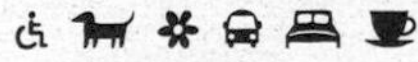

2 ADSWOOD

Townfield Lane, Mollington, CH1 6LB. Ken & Helen Black, 01244 851327, keneblack@outlook.com, www.kenblackclematis.com. *3m N of Chester. From Wirral take A540 towards Chester. Cross A55 at r'about, past Wheatsheaf pub, turn L into Overwood Lane. At T junction turn R into Townfield Lane. NGS sign 100 yds on the L.* **Sat 8, Sun 9, Sat 29, Sun 30 Apr (10-4). Adm £5, chd free. Pre-booking essential, please visit www.ngs.org.uk for information & booking. Light refreshments. Pls note the garden will be closed between 12 - 2 pm each day. Visits also by arrangement 1 May to 30 July for groups of 8 to 25.**

A cottage garden with a wide range of perennials, climbers and English roses. Some borders revamped for 2023. The garden has over 150 varieties of clematis, providing colour throughout the year. There are several seating areas inc a pavilion and new summerhouse. Book on April dates for tours of early spring flowering clematis - as featured on Gardeners' World. Packs of young spring clematis to start your own collection can be purchased. There are no steps but the front drive and some of the garden paths are gravelled and the side access to the rear garden is quite narrow.

3 ALL FOURS FARM

Colliers Lane, Aston by Budworth, Northwich, CW9 6NF. Mr & Mrs Evans, 01565 733286. *M6 J19, take A556 towards Northwich. Turn immed R, past The Windmill pub. Turn R after approx 1m, follow rd, garden on L after approx 2m. Direct access available for drop off & collection for those with limited mobility.* **Sat 24, Sun 25 June (10-4). Adm £5, chd free. Light refreshments. Visits also by arrangement 17 June to 2 July for groups of 25+.**

A traditional and well established country garden with a wide range of roses, hardy shrubs, bulbs, perennials and annuals. You will also find a small vegetable garden, pond and greenhouse as well as vintage machinery and original features from its days as a working farm. The garden is adjacent to the family's traditional rose nursery. The majority of the garden is accessible by wheelchair.

4 ◆ ARLEY HALL & GARDENS

Arley, Northwich, CW9 6NA. Viscount Ashbrook, 01565 777353, enquiries@arleyhallandgardens.com, www.arleyhallandgardens.com. *10m from Warrington. Signed from J9 & 10 (M56) & J19 & 20 (M6) (20 min from Tatton Park, 40 min to Manchester). Please follow the brown tourist signs.* **For NGS: Tue 18 Apr (10-5). Adm £12, chd £5. Light refreshments in The Gardener's Kitchen Cafe. For other opening times and information, please phone, email or visit garden website.**

Within Arley's 8 acres of formal garden there are many different areas, each with its own distinctive character. Beyond the Chapel is The Grove, a well established arboretum and a woodland walk of about another 6 or 7 acres. Gardens are wheelchair accessible, however parts of the estate have cobbles which can prove a little difficult for manual wheelchairs.

5 ASHTON GRANGE

Grange Road, Ashton Hayes, Chester, CH3 8AE. Martin & Kate Slack, 01829 759172, kateslack1@icloud.com. *8m E of Chester. Grange Rd is a single track road off B5393, by the village sign at N end of Ashton Hayes.* **Sun 2 July (12.30-4.30). Adm £6, chd free. Home-made teas. Visits also by arrangement May to Aug for groups of 10 to 30.**

Ashton Grange has a traditional country house garden with sweeping lawns, herbaceous borders, rhododendrons, an impressive kitchen garden and unusual trees. You can wander through mature woodlands, see tree carvings, visit the wildflower meadow with lovely views and sit by a large wildlife pond. The garden, meadows and woodlands open to visitors extend to approximately 12 acres. Picnics welcome. Wheelchair access to most parts of the garden but woodland paths could be difficult. Limited parking for wheelchair users.

6 BANKHEAD

Old Coach Road, Barnhill, Chester, CH3 9JL. Simon & Sian Preston, 07970 794456, Sian@simonpreston.org. *10 m S of Chester. End of Old Coach Rd at the junction with A534.* **Sun 21 May (1-5). Adm £7, chd free. Home-made teas. Visits also by arrangement 22 Apr to 25 June for groups of 10 to 25. Light refreshments can be made available.**

Two acres of terraced gardens developed from Victorian times with spectacular views south and west over the Dee valley towards the Welsh hills. Rose garden, herbaceous borders, large pond with Japanese style garden, rhododendrons and azaleas. Small orchard and vegetable garden. Shetland ponies and chickens. Some steps to the main terrace but can reach most areas on paths with slight inclines. Parking close to the garden for disabled access.

7 THE BIRCHES

Grove Road, Mollington, Chester, CH1 6LG. Martin Bentley & Colin Williams. *3m N of Chester. From the A540 Parkgate Rd turn onto Coal Pit Lane & follow around until it becomes Grove Rd. Pass riding stables on R, gardens approx ¼m further on R.* **Sat 22, Sun 23 July (12-5). Adm £5, chd free. Home-made teas.**

½ acre of gardens comprising of front night garden with pale/white herbaceous borders. Back garden split into 5 distinct areas: Koi and wildlife pond with two large herbaceous borders, grasses and tropical border with fernery, orchard with hens and bees, vegetable garden and a wildlife stream with bog plants and sunken garden. Level garden with lawn or pathways to most areas.

8 ◆ BLUEBELL COTTAGE GARDENS

Lodge Lane, Dutton, WA4 4HP. Sue Beesley, 01928 713718, info@bluebellcottage.co.uk, www.bluebellcottage.co.uk. *5m NW of Northwich. From M56 (J10) take A49 to Whitchurch. After 3m turn R at T-lights towards Runcorn/ Dutton on A533. Then 1st L. Signed with brown tourism signs from A533.* **For NGS: Sat 1, Sun 2 July (10-5). Adm £5, chd free. Home-made teas. For other opening times and information, please phone, email or visit garden website.**

South facing country garden wrapped around a cottage on a quiet rural lane in the heart of Cheshire. Packed with thousands of rare and familiar hardy herbaceous perennials, shrubs and trees. Unusual plants available at adjacent nursery. New walled garden with greenhouse and raised vegetable beds. The opening dates coincide with the peak of flowering in the herbaceous borders. Picnics welcome in the meadow area adjacent to the car park or at the benches in the nursery. The garden is on a gentle slope with wide lawn paths. All areas are accessible. We have a wider accessible WC.

9 BOLESWORTH CASTLE

Tattenhall, CH3 9HQ. Mrs Anthony Barbour, 01829 782210, dcb@bolesworth.com, www.bolesworth.com. *8m S of Chester on A41. Enter through Broxton gate on Bolesworth Hill Rd using postcode CH3 9HN.* **Visits by arrangement 17 Apr to 26 May for groups of 10+. Adm £6, chd free.**

The Rock Walk at Bolesworth was established in 1986 by Anthony Barbour. Planted up with rhododendrons, azaleas and camellias it is a fine woodland walk, worth the fairly steep climb from the Castle. The 2022 drought brought some losses, but by April/May 2023 it will be restored and feature a Camellia Walk as well as large leaf rhododendrons and evergreen and deciduous azaleas.

10 BOLLIN HOUSE

Hollies Lane, Wilmslow, SK9 2BW. Angela Ferguson & Gerry Lemon, 07828 207492, fergusonang@doctors.org.uk. *From Wilmslow past stn & proceed to T-junction. Turn L onto Adlington Rd. Proceed for ½m, then turn R into Hollies Lane (just after One Oak Lane). Drive to the end of Hollies Lane & follow yellow signage. Park on Browns Lane (other side Adlington Rd) or Hollies Lane.* **Visits by arrangement 15 May to 15 July for groups of 8 to 30. Parking for up to 5 cars available. Tea & cake or scones, cold drinks also available.**

Our garden has established deep borders full of perennials, a wildflower meadow and a new formal area with parterres and central water feature. The meadow, with its annual and perennial wildflower areas (with mown pathways and benches), attracts lots of bees, dragonflies and other insects and butterflies. Bollin House is in an idyllic location with the garden, orchard and meadow dropping into the Bollin valley. The River Bollin flows along this valley and on the opp side the fields lead up to views of Alderley Edge. Ramps to gravel lined paths to most of the garden. Some narrow paths through borders. Mown pathways in the meadow.

11 BOWMERE COTTAGE

Bowmere Road, Tarporley, CW6 0BS. Romy & Tom Holmes, 01829 732053, romy@holmes-email.co.uk. *10m E of Chester. From Tarporley High St (old A49) take Eaton Rd signed Eaton. After 100 metres take R fork into Bowmere Rd, Garden 100 metres on L.* **Visits by arrangement 8 June to 31 July for groups of 5 to 30. Adm £5, chd free. Home-made teas.**

A colourful and relaxing 1 acre country style garden around a Grade II listed house. The lawns are surrounded by well-stocked herbaceous and shrub borders and rose covered pergolas. There are 2 plant filled courtyard gardens and

Stretton Old Hall

© Liz Mitchell

a small vegetable garden. Shrubs and rambling roses, clematis, hardy geraniums and a wide range of mostly hardy plants make this a very traditional English garden.

12 NEW BRANDON HOUSE

Stretton, Tilston, Malpas, SY14 7JA. Simon & Christine Cottrell. *12m S of Chester, 3m N of Malpas. Travelling S along A41 from Chester.Turn R onto A534 towards Wrexham.Turn L at Cock O Barton pub & property is approx 1m on L with thatched roof.* **Sat 3, Sun 4 June (11-4.30). Adm £5, chd free. Home-made teas.**

With views over the Cheshire countryside a formal courtyard opens up to a large stunning wisteria. The garden covers 1½ acres where formal planting combines with cottage garden flowers. David Austen roses abound. Fed from a small pond a rill runs through frothy Alchemilla. A woodland copse opens to a newly created wildflower meadow and pond.

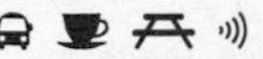

13 BRIARFIELD

The Rake, Burton, Neston, CH64 5TL. Liz Carter, 07711 813732, carter.burton@btinternet.com, *9m NW of Chester. Turn off A540 at Willaston-Burton Xrds T-lights & follow rd for 1m to Burton village centre.* **Every Fri 3 Feb to 24 Feb (10-4). Adm £4, chd free. Sun 16 Apr (1-5). Adm £5, chd free. Home-made teas in the church just along the lane from the garden on 16 April only. Opening with Burton Village Gardens on Sun 25 June. Visits also by arrangement 27 Mar to 30 Sept.**

Tucked under the south facing side of Burton Wood the garden is home to many specialist and unusual plants, some available in plant sale. This 2 acre garden is on two sites, a short walk along an unmade lane. Trees, shrubs, colourful herbaceous, bulbs, alpines and water features compete for attention. Deliberately left untidy through the winter for wildlife, the snowdrops love it. Erythronium are a feature of the garden in April and May. Always changing, Liz can't resist a new plant! Rare and unusual plants sold (70% to NGS) from the drive each Friday from 9am to 5pm, Feb to Oct. Check updates on Briarfield Gardens Facebook.

14 BROOKE COTTAGE

Church Road, Handforth, SK9 3LT. Barry & Melanie Davy. *1m N of Wilmslow. Centre of Handforth, behind Health Centre. Turn off Wilmslow Rd at St Chads, follow Church Rd round to R. Garden last on L. Parking in Health Centre car park.* **Sat 6, Sun 7 May (12-5). Adm £4.50, chd free. Home-made teas.**

This garden is divided into 3 distinct areas. A shady woodland area of ferns, azaleas, rhododendrons, camellias, magnolia, erythroniums, trilliums, arisaema and foxgloves. Patio with many hostas, daylilies, large-leaved plants and pond. Borders with grasses, perennials, euphorbia, alliums and tulips. Anthriscus, aquilegia, persicaria and astrantia create meadow effect popular with insects.

15 BUCKLOW FARM

Pinfold Lane, Plumley, Knutsford, WA16 9RP. Dawn & Peter Freeman. *2m S of Knutsford. M6 J19, A556 Chester. L at 2nd set of T-lights. In 1¼ m, L at concealed Xrds. 1st R. From Knutsford A5033, L at Sudlow Lane, becomes Pinfold Lane.* **Sun 26 Feb (12.30-4). Light refreshments. Sun 11 June (1-5). Cream teas. Adm £4, chd free. Mulled wine in Feb. Cream teas in June. 2024: Sun 25 Feb. Donation to Knutsford Methodist Church.**

Country garden with shrubs, perennial borders, rambling roses, herb garden, vegetable patch, meadow, wildlife pond/water feature and alpines. Landscaped and planted over the last 30 yrs with recorded changes. Free range hens. Carpet of snowdrops and spring bulbs. Leaf, stem and berries to show colour in autumn and winter. Cobbled yard from car park, but wheelchairs can be dropped off near gate.

GROUP OPENING

16 BURTON VILLAGE GARDENS

Burton, Neston, CH64 5SJ. *9m NW of Chester. Turn off A540 at Willaston-Burton Xrds T-lights & follow road for 1m to Burton. Maps given to visitors. Buy your ticket at 1st garden & keep it for entry to other gardens.* **Sun 25 June (11-5). Combined adm £6, chd free. Home-made teas in the Sport and Social Club behind the village hall.**

BRIARFIELD
Liz Carter.
(See separate entry)

◆ BURTON MANOR WALLED GARDEN
Friends of Burton Manor Gardens CIO, 0151 336 6154, www.burtonmanorgardens.org.uk.

THE COACH HOUSE
Jane & Mike Davies, 0151 353 0074, burtontower@googlemail.com.

TRUSTWOOD
Peter & Lin Friend, 0151 336 7118, lin@trustwoodbnb.uk, www.trustwoodbnb.uk.
Visits also by arrangement May to July for groups of up to 10.

With our light sandy soil, often on a south facing slope, we are focussing on drought tolerant planting and copious mulching. The Coach House, has stunning views of the Welsh hills. The garden is full of colour with herbaceous borders, a cut flower area plus a productive fruit and vegetable patch The wooded hillside is work in progress. Briarfield's sheltered site is home to many unusual plants, some available in the plant sale at the house. The 1½ acre main garden invites exploration not only for its variety of plants but also for the imaginative use of ceramic sculptures. Period planting with a splendid vegetable garden surrounds the restored Edwardian glasshouse in Burton Manor's walled garden. Paths lead past a sunken garden and terraces to views across the Cheshire countryside. Trustwood is a relaxed country garden, a haven for wildlife with emphasis on British native trees planted along the drive and the use of insect friendly plants. Chickens roam in the woodland. Briarfield & The Coach House are too hilly for wheelchairs, with steep slopes and steps.

17 64 CARR WOOD

Hale Barns, Altrincham, WA15 0EP. Mr David Booth. *10m S of Manchester city centre. 2m from J6 M56: Take A538 to Hale Barns. L at 'triangle' by church into Wicker Lane & L at mini r'about into Chapel Lane & 1st R into Carr Wood.* **Sat 13 May (1-5). Adm £5, chd free. Home-made teas.**

⅔ acre landscaped, south facing garden overlooking Bollin valley laid out in 1959 by Clibrans of Altrincham. Gently sloping lawn, woodland walk, seating areas and terrace, extensive mixed shrub and plant borders. Ample parking on Carr Wood. Wheelchair access to terrace overlooking main garden.

18 ◆ CHOLMONDELEY CASTLE GARDENS

Cholmondeley, Malpas, SY14 8AH. The Cholmondeley Gardens Trust, 01829 720203, eo@chol-estates.co.uk, www.cholmondeleycastle.com. *4m NE of Malpas SatNav SY14 8ET. Signed from A41 Chester-Whitchurch rd & A49 Whitchurch-Tarporley rd.* **For NGS: Thur 4 May (10-5). Adm £8.50, chd £4. For other opening times and information, please phone, email or visit garden website.**

70 acres of romantically landscaped gardens with fine views and eye-catching water features, which still manages to retain its intimacy. Beautiful mature trees form a background to millions of spring bulbs and superb plant collections inc magnolias, rhododendrons and camellias. Large display of daffodils. Magnificent magnolias. One of the finest features of the gardens are its trees, many of which are rare and unusual, Cholmondeley Gardens is home to over 40 county champion trees. 100m long double mixed herbaceous border and Rose Garden with 250 rose varieties. Fresh coffee, lunches and cakes daily with locally sourced products located in the heart of the Gardens. Partial wheelchair access.

19 5 COBBS LANE

Hough, Crewe, CW2 5JN. David & Linda Race. *4m S of Crewe. M6 J16 r'about take A500 (Nantwich). Next r'about 1st exit Keele. Next r'about straight (Hough). After 1m L into Cobbs Ln. From A51 Nantwich bypass to A500 r'about 3rd exit (Shavington), after 3m (past White Hart), R into Cobbs Ln. Park at village hall (300 metres).* **Thur 6, Sat 8 July (11-5). Adm £5, chd free. Home-made teas at village hall 300 metres up Cobbs Lane. Toilet facilities are available.**

A plant person's ⅔ acre garden with island beds, wide cottage style herbaceous borders with bark paths. A large variety of hardy and some unusual perennials. Interesting features, shrubs, grasses and trees, with places to sit and enjoy the surroundings. A water feature runs to a small pond, a wildlife friendly garden containing a woodland area bordered by a small stream.

20 NEW 18 DEE PARK ROAD

Dee Park Road, Wirral, CH60 3RQ. Mrs Shay & Mr Les Whitehead, 07778 309671, shaywhitehead@gmail.com. *Heswall. J4 off M53. Follow signs for Brimstage, Heswall, follow to end. Turn L at r'about, take 1st L onto A540 to Chester, 1st R down Gayton Lane. Dee Park is on the bottom L.*

Hathaway

Sun 11, Wed 14, Sun 18 June (11-5). Adm £4, chd free. Home-made teas. Wine also available. Visits also by arrangement 8 June to 23 July.
A long suburban garden backing onto private woodland. Mature garden laid to lawns, with mixed borders, leading on to numerous seating areas. The garden features a very unusual triple hexagonal greenhouse, which was originally purchased from the Liverpool Garden Festival site, c1984, before taking you through to a shaded wooded area where there are numerous different types of hosta on display. Wheelchair accessible with one step to negotiate.

21 DRAKE CARR

Mudhurst Lane, Higher Disley, SK12 2AN. Alan & Joan Morris.
8m SE of Stockport, 12m NE of Macclesfield. From A6 in Disley centre turn into Buxton Old Rd, go up hill 1m & turn R into Mudhurst Lane. After ⅓m park on layby or grass verge. No parking at garden. Approx 150 metre walk.

Sat 10, Sun 11 June (11-5). Adm £5, chd free. Home-made teas and gluten-free cakes.
½ acre cottage garden in beautiful rural setting with natural stream running into wildlife pond with many native species. Surrounding C17 stone cottage, the garden, containing borders, shrubs and vegetable plot, is on several levels divided by grassed areas, slopes and steps. This blends into a boarded walk through bog garden, mature wooded area and stream-side walk into a wildflower meadow.

22 THE FIRS

Old Chester Road, Barbridge, Nantwich, CW5 6AY. Richard & Valerie Goodyear, 07775 924929, richard.goodyear@ngs.org.uk. *3m N of Nantwich on A51. After entering Barbridge turn R at Xrds after 100 metres. The Firs is 2nd house on L.* **Sun 30 July, Sun 6 Aug (11-4). Adm £4.50, chd free. Home-made teas. Visits also by arrangement 30 Apr to 13 Aug for groups of 10 to 40.**

Canalside garden set idyllically by a wide section of the Shropshire Union Canal with long frontage. Garden alongside canal with varied trees, shrubs and herbaceous beds, with some wild areas. All leading down to an observatory and Japanese torii gate at the far end of the garden with views across fields. Usually have nesting friendly swans with cygnets May to September.

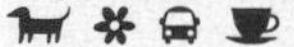

23 FRAMLEY

Hadlow Road, Willaston, Neston, CH64 2US. Mrs Sally Reader. *½m S of Willaston village centre. From Willaston Green, proceed along Hadlow Rd, crossing the Wirral Way. Framley is the next house on R. From Chester High Rd, proceed along Hadlow Rd toward Willaston. Framley is third house on L.* **Mon 1 May (10-4). Adm £5, chd free. Home-made teas.**

This 5 acre garden holds many hidden gems. Comprising extensive mature wooded areas, underplanted with a variety of interesting and unusual woodland plants - all at their very best in spring. A selection of deep seasonal borders surround a mystical sunken garden, planted to suit its challenging conditions. Wide lawns and sandstone paths invite you to discover what lies around every corner. Please phone ahead for parking instructions for wheelchair users - access around much of the garden although the woodland paths may be challenging.

24 HALL LANE FARM

Hall Lane, Daresbury, Warrington, WA4 4AF. Sir Michael & Lady Beverley Bibby. *1m from J11 M56. Leave M56 at J11, head towards Warrington take 2nd R turn into Daresbury village go around sharp bend then take L into Daresbury Lane. Entrance on L after 100 yds.* **Sun 18 June (12.30-4.30). Adm £5, chd free. Home-made teas.**

The 2 acres of private formal garden originally designed by Arabella Lennox-Boyd are arranged in a 'gardens within gardens' style to create a series of enclosed spaces each with their own character and style. The gardens inc a vegetable garden, orchard, Koi pond, as well as lawns and a treehouse.

25 HATHAWAY

1 Pool End Road, Tytherington, Macclesfield, SK10 2LB. Mr & Mrs Cordingley. *2m N of Macclesfield, ½m from The Tytherington Club. From Stockport follow A523 past the Butley Ash pub, at r'about turn R on A538 for Tytherington. From Knutsford follow A537 at A538, L for Tytherington. From Leek follow A523 at A537 L & 1st R A538.* **Sat 13, Sun 14 May (10.30-4). Adm £4, chd free. Light refreshments. Teas/coffee/cordial & cakes.**

SW facing garden of approx ⅕ acre, laid out in two parts. Lawn area surrounded by mature, colourful perennial borders. Rose arbour.

Framley

Our small pond has now become a large one, located in the lawn, with goldfish. Larger patio. Small mature wooded area with winding paths on a lower level. Front, laid to lawn with 2 main borders separated by a small grass area. Good wheelchair access to most of garden except wood area.

26 NEW HIGHBANK

110 Bradwall Road, Sandbach, CW11 1AW. Peter & Coral Hulland. *Approx 1m N of Sandbach town centre. The garden runs between Bradwall Rd and Twemlow Ave, but access & limited parking are on Twemlow Ave only (CW11 1GL).* **Sun 14 May (10.30-4.30). Adm £4, chd free. Light refreshments.**

Small town garden originally a builders yard extending to a ⅓ acre. Grounds to rear and front inc a small wildlife pond, water features, greenhouse and a number of seating areas. Laid mainly to lawns with well-stocked borders inc trees, unusual shrubs and many perennials, some of which are allowed to naturalise throughout the garden.

27 213 HIGHER LANE

Lymm, WA13 0RN. Mark Stevenson, 07470 715007, mjs.stevenson@btinternet.com. *From M56 J7 follow the signs for Lymm A56. The garden is approx 300 metres from the Jolly Thresher pub towards Lymm on R.* **Visits by arrangement Apr to Oct Small donation required for refreshments. Adm £4, chd free.**

Established on an acre of grounds, segregated into various planting schemes. Herbaceous borders, specimen rhododendrons, hydrangeas, ferns and grasses, gingers and subtropical plants, along with so much more. A pond complete with fish and waterfalls surrounded by Japanese acers with far reaching views across fields and woodland. Many rare and unusual plants and shrubs with wooded pathways. 70% of the garden is accessible to wheelchair users.

28 18 HIGHFIELD ROAD

Bollington, Macclesfield, SK10 5LR. Mrs Melita Turner. *3m N of Macclesfield. A523 to Stockport. Turn R at B5090 r'about signed Bollington. Pass under viaduct. Take next R (by Library) up Hurst Lane. Turn R into Highfield Rd. Property on L. Park on wider road just beyond.* **Sat 24, Sun 25 June (10-4.30). Adm £4.50, chd free. Home-made teas.**

This small terraced garden packed with plants was designed by Melita and has evolved over the past 15 yrs. This plantswoman is a plantaholic and RHS Certificate holder. An attempt has been made to combine formality through structural planting with a more casual look influenced by the style of Christopher Lloyd. Some minor changes for 2023.

29 HILL FARM

Mill Lane, Moston, Sandbach, CW11 3PS. Mrs Chris & Mr Richard House, 01270 526264, housecr2002@yahoo.co.uk. *2m NW of Sandbach. From Sandbach town centre take A533 towards Middlewich. After Fox pub take next L (Mill Lane) to a canal bridge & turn L. Hill Farm 400m on R just before post box.* **Sun 23 Apr, Sun 14 May, Sun 11 June (11-5). Adm £5, chd free. Light refreshments. Visits also by arrangement 1 May to 1 Oct for groups of up to 25.**

The garden extends to approx ½ acre and is made up of a series of gardens inc a formal courtyard with a pond and vegetable garden with south facing wall. An orchard and wildflower meadow were established about 4 yrs ago. A principal feature is a woodland garden which supports a rich variety of woodland plants. This was extended in April 2020 to inc a pond and grass/herbaceous borders. A level garden, the majority of which can be accessed by wheelchair.

30 HILLTOP

Flash Lane, Prestbury, SK10 4ED. Martin Gardner, 07768 337525, hughmartingardner@gmail.com, www.yourhilltopwedding.com. *2m N of Macclesfield. A523 to Stockport. Turn R at B5090 r'about signed Bollington, after ½m turn L at Cock & Pheasant pub into Flash Lane. At bottom of lane turn R into cul-de-sac. Hilltop Country House signed on R.* **Visits by arrangement 1 May to 30 Aug for groups of 8 to 20. Adm £6, chd free. Light refreshments. Tea/Coffee & a cake £2 per person.**

Interesting country garden of approx 4 acres. Woodland walk, parterre, herb garden, herbaceous borders, dry stone walled terracing, lily ponds with waterfall. Wisteria clad 1693 house (not open), ancient trees, orchard with magnificent views to Pennines and to the west. Partial wheelchair access, disabled WC, easy parking.

31 THE HOMESTEAD

2 Fanners Lane, High Legh, Knutsford, WA16 0RZ. Janet Bashforth, 07809 030525, janbash43@sky.com. *J20 M6/J9 M56 at Lymm interchange take A50 for Knutsford, after 1m turn R into Heath Lane then 1st R into Fanners Lane. Follow parking signs.* **Sun 11 June, Sun 9 July (11-4.30). Adm £4.50, chd free. Home-made teas. Visits also by arrangement May to July for groups of 10 to 40. Refreshments available for group visits.**

This compact gem of a garden has been created over the last 7yrs by a keen gardener and plantswoman. Enter past groups of liquidambar and white stemmed birch. Colour themed areas with many perennials, shrubs and trees. Past topiary and obelisks covered with many varieties of clematis and roses, enjoy the colours of the hot border, a decorative greenhouse and pond complete the picture. Further on there is a small pond with water lilies and iris. Many types of roses and clematis adorn the fencing along the paths and into the trees.

32 INGLEWOOD

4 Birchmere, Heswall, CH60 6TN. Colin & Sandra Fairclough, 07715 546406, sandrafairclough51@gmail.com. *6m S of Birkenhead. From A540 Devon Doorway/Clegg Arms r'about go through Heswall. ¼m after Tesco, R into Quarry Rd East, 2nd L into Tower Rd North & L into Birchmere.* **Visits by arrangement 3 Apr to 14 July for groups of 12 to 25. Home-made teas avail at extra cost. Adm £5, chd free.**

Beautiful ½ acre garden with stream, large koi pond, 'beach' with grasses, wildlife pond and bog area. Brimming with shrubs, bulbs, acers, conifers, rhododendrons, herbaceous plants and hosta border. Interesting features inc hand cart, wood carvings, bug hotel and Indian dog gates leading to a secret garden. Lots of seating to enjoy refreshments.

33 60 KENNEDY AVENUE

Macclesfield, SK10 3DE. Bill North. *5min NW of Macclesfield Town Centre. Take A537 Cumberland St, 3rd r'bout (West Park) take B5087 Prestbury Rd. Take 5th turn on L into Kennedy Ave, last house on L before Brampton Ave opp Belong Care Home.* **Sun 11 June (12-5). Adm £5, chd free. Home-made teas.**

Small suburban garden which featured in 2017 August edition of Amateur Gardening magazine. The garden is reopening after a year's absence following a redesign of a number of features. For a small urban garden it certainly packs a punch. Design inc many features and is well worth a visit. Wheelchair access will require a carer to guide up the drive and will only provide access to part of the garden.

34 LASKEY FARM

Laskey Lane, Thelwall, Warrington, WA4 2TF. Howard & Wendy Platt, 07785 262478, howardplatt@lockergroup.com, www.laskeyfarm.com. *2m from M6/M56. From M56/M6 follow directions to Lymm. At T-junction turn L onto the A56 in Warrington direction. Turn R onto Lymm Rd. Turn R onto Laskey Lane.* **Sun 30 Apr, Mon 1, Sat 27, Sun 28 May (11-4), open nearby 10 Statham Avenue. Sat 15, Sun 16 July, Sun 27, Mon 28 Aug (11-4). Adm £6, chd £1. Home-made teas. Visits also by arrangement 1 May to 25 Aug for groups of 12 to 50. Joint visits with nearby 10 Statham Ave may be arranged.**

1½ acre garden inc herbaceous and rose borders, vegetable area, a greenhouse, parterre and a maze showcasing grasses and prairie style planting. Interconnected pools for wildlife, specimen koi and terrapins form an unusual water garden which features a swimming pond. There is a treehouse plus a number of birds and animals. Family friendly, we offer a treasure hunt, and a mini menagerie consisting of chickens, guinea fowl and guinea pigs. Most areas of the garden may be accessed by wheelchair.

35 LONG ACRE

Wyche Lane, Bunbury, CW6 9PS. Margaret & Michael Bourne, 01829 260944, mjbourne249@tiscali.co.uk. *3½ m SE of Tarporley. In Bunbury village turn into Wyche Lane by Nags Head pub car park, garden 400yds on L.* **Sun 16 Apr (2-5). Adm £5, chd free. Home-made teas. Visits also by arrangement 11 Apr to 24 June for groups of 10+. Refreshments avail on request. Donation to St Boniface Church and Bunbury Village Hall.**

Plantswoman's garden of approx 1 acre with unusual and rare plants and trees inc Kentucky Coffee Tree, Sciadopitys, Kalopanax picta and others. Pool garden, exotic conservatory with bananas, anthuriums and medinilla, herbaceous, greenhouses with Clivia in spring and Disa orchids in summer. Spring garden with camellias, magnolias, bulbs; roses and lilies in summer.

36 ◆ THE LOVELL QUINTA ARBORETUM

Swettenham, CW12 2LF. Tatton Garden Society, 01565 831981, admin@tattongardensociety.org.uk, www.lovellquintaarboretum.co.uk. *4m NW of Congleton. Turn off A54 N 2m W of Congleton or turn E off A535 at Twemlow Green, NE of Holmes Chapel. Follow signs to Swettenham. Park at Swettenham Arms. What3words app - twins. sheds.keepers.* **For NGS: Sun 8 Oct (1.30-4). Adm £5, chd free. For other opening times and information, please phone, email or visit garden website.**

This 28 acre arboretum has been established since the1960s and contains around 2,500 trees and shrubs, some very rare. National Collections of Quercus, Pinus and Fraxinus. A large selection of oak, a private collection of hebes plus autumn flowering, fruiting and colourful trees and shrubs. Waymarked walks. Autumn colour, winter walk and spring bulbs. Refreshments at the Swettenham Arms during licensed hours or by arrangement. Care required but wheelchairs can access much of the arboretum on the mown paths.

NPC

37 NEW MAGGIE'S WIRRAL

Clatterbridge Road, Bebington, Wirral, CH63 4JY. Kathy Wright Centre Manager. *Clatterbridge Hospital, Wirral. Leave M53 at Junction 4. Follow signs for Clatterbridge Hospital. Follow signs for Maggie's through hospital site to end of road, if barrier is operating tell security you are going to Maggie's.* **Sun 18 June (11-4). Adm £5, chd free. Light refreshments.**

Maggies Wirral invites us in via a winding path through sun loving planting sheltered by wavy stone wall. Centre is set within varied colourful gardens. Peaceful woodland area attracts wildlife, Mediterranean garden has a shallow pool and sensory plants, paved areas for quiet contemplation and more active therapeutic activities. Greenhouse and growing area encourages involvement. Smooth surfaces throughout garden allow full accessibility.

38 MANLEY KNOLL

Manley Road, Manley, WA6 9DX. Mr & Mrs James Timpson, www.manleyknoll.com. *3m N of Tarvin. On B5393, via Ashton & Mouldsworth. 3m S of Frodsham, via Alvanley.* **Sun 21 May (12-5). Adm £5, chd free. Home-made teas.**

Arts & Crafts garden created early 1900s. Covering 6 acres, divided into different rooms encompassing parterres, clipped yew hedging, ornamental ponds and herbaceous borders. Banks of rhododendron and azaleas frame a far-reaching view of the Cheshire Plain. Also a magical quarry/folly garden with waterfall and woodland walks.

39 MAYFIELD HOUSE

Moss Lane, Bunbury Heath, Tarporley, CW6 9SY. Mr & Mrs J France Hayhurst, jeanniefh@me.com. *Off A49 on Tarporley/Whitchurch Rd. Moss Lane is opp School Lane which leads to Bunbury village. There's a yellow speed camera (30 mph) across the road from the house.* **Sun 21 May (12.30-4.30). Adm £5, chd free. Home-made teas. Visits also by arrangement 24 Apr to 13 Oct for groups of 12 to 28.**

A thoroughly English garden with a wealth of colour and variety throughout the year. A background of fine trees, defined areas bordered by mixed hedging, masses of rhododendrons, azaleas, camellias,hydrangeas and colourful shrubs. Clematis and wisteria festoon the walls in early summer. Easy access and random seating areas. Small lake with an island and broad lawns with glades of foxgloves. Garden statuary, large pond with fish, swimming pool. Gravel drive. Very

60 Kennedy Avenue

wide wrought iron gates lead to the garden and it is entirely navigable.

40 MILFORD HOUSE FARM

Long Lane, Wettenhall, Winsford, CW7 4DN. Chris & Heather Pope, 07887 760930, hclp@btinternet.com. *3m E of Tarporley. From A51 at Alpraham turn R into Long Lane. Proceed for 2½m to St David's Church on L. Milford House Farm is 150 metres further on L. Park at church or those with limited mobility can park at house.* **Sat 15 July (12-5). Adm £4, chd free. Home-made teas in marquee at garden. Gluten free, dairy free options available. Visits also by arrangement June & July for groups of 5 to 26. Coaches by arrangement only. Lunches avail on request.**

A large country garden created over 20 yrs with lawn, mixed borders, hot, white and pastel areas. Modern walled garden for vegetables, fruit, greenhouse, potting shed, dahlias, wall shrubs and display of exotics, succulents, tender plants, water feature. Orchard for fruit, native, ornamental trees. Wildlife pond with toads and newts. Gardened on organic lines to increase biodiversity and attract wildlife. Picnics in orchard. Visits by arrangement can inc woodland walks (weather dependent) Children's activities. Walled garden has gravel and flags. Accessible WC at church.

41 ◆ MOUNT PLEASANT

Yeld Lane, Kelsall, CW6 0TB. Dave Darlington & Louise Worthington, 01829 751592, louisedarlington@btinternet.com, www.mountpleasantgardens.co.uk. *8m E of Chester. Off A54 at T-lights into Kelsall. Turn into Yeld Lane opp Farmers Arms pub, 200yds on L. Do not follow SatNav directions.* **For NGS: Sun 21 May, Sun 3 Sept (11-4). Adm £7, chd £5. Tea. For other opening times and information, please phone, email or visit garden website.**

10 acres of landscaped garden and woodland started in 1994 with impressive views over the Cheshire countryside. Steeply terraced in places. Specimen trees, rhododendrons, azaleas, conifers, mixed and herbaceous borders; 4 ponds, formal and wildlife. Vegetable garden, stumpery with tree ferns, sculptures, wildflower meadow and Japanese garden. Bog garden, tropical garden. Sculpture trail and exhibition.

The National Garden Scheme searches the length and breadth of England, Wales, Northern Ireland and the Channel Islands for the very best private gardens

42 ◆ NESS BOTANIC GARDENS
Neston Road, Ness, Neston, CH64 4AY. The University of Liverpool, 0151 795 6300, nessgdns@liverpool.ac.uk, www.liverpool.ac.uk/ness-gardens. *10m NW of Chester. Off A540. M53 J4, follow signs M56 & A5117 (signed N Wales). Turn onto A540 follow signs for Hoylake. Ness Gardens is signed locally. 487 bus takes visitors travelling from Liverpool, Birkenhead etc.* **For NGS: Tue 12 Sept (10-4). Adm £7.50, chd £3.50. Light refreshments in our Café, based in our Visitor Centre. For other opening times and information, please phone, email or visit garden website.**
Looking out over the dramatic views of the Dee estuary from a lofty perch on the Wirral peninsula, Ness Botanic Gardens boasts 64 acres of landscaped and natural gardens overflowing with horticultural treasures. With a delightfully peaceful atmosphere, a wide array of events taking place, plus a café and gorgeous open spaces it is a great fun-filled day out for all the family. National Collections of Sorbus and Betula. Herbaceous borders, Rock Garden, Mediterranean Bank, Potager and conservation area. Wheelchairs are available free to hire (donations gratefully accepted) but advance booking is highly recommended.
NPC

43 NEW NEW INN FARM
Buxton New Road, Rainow, Macclesfield, SK11 0AD. Mark & Louise Simms. *2½m NE Macclesfield on A537 Macclesfield to Buxton rd. Car parking at One House Walled Garden, 500m walk to New Inn Farm. Drop off possible, disabled parking for 2-3 cars.* **Sat 17, Sun 18 June (10-5). Adm £4, chd free. Home-made teas.**
Beautiful cottage style garden constructed on hillside with spectacular views to the Welsh Hills and Cheshire plain, ⅓ acre garden leading from stone flagged patio, with natural stone trough to lawned area. Established herbaceous borders inc rhododendrons, azaleas and mature trees. Paths leading to wildlife pond, greenhouse and newly planted wildlife garden inc folly and cairn.

44 NORLEY BANK FARM
Cow Lane, Norley, Frodsham, WA6 8PJ. Margaret & Neil Holding, 07828 913961, neil.holding@hotmail.com. *Near Delamere Forest. From the Tigers Head pub in the centre of Norley village keep the pub on L, carry straight on through the village for approx 300 metres. Cow Lane is on the R.* **Sun 23, Sun 30 July, Mon 7 Aug (11-5). Adm £5, chd free. Home-made teas. Some allergy free refreshments available. Visits also by arrangement 10 July to 28 Aug for groups of 10+.**
A garden well-stocked with perennials, annuals and shrubs that wraps around this traditional Cheshire farmhouse. There is an enclosed cut flower and nursery garden surrounding a greenhouse and just beyond the house a vegetable garden. Wander past this to a large wildlife pond with a backdrop of pollinating flowers and a a recently formed stumpery. All set in about 2 acres. Free range hens, donkeys and Coloured Ryeland sheep.

45 NEW NORLEY COURT
Marsh Lane, Norley, Frodsham, WA6 8NY. Clare Albinson. *20 mins E of Chester. Close to Delamere Forest. A556 - Stoneyford Ln/ Cheese Hill/Cow Ln, then turn L. A49 Acton Ln/Station Rd L at church next R into Marsh Ln. Norley Ct is at top of hill. Parking at Tigers Head pub 200 yds on R from Marsh Ln.* **Sun 30 Apr (2-5). Adm £6, chd free. Home-made teas.**
Norley Court is a large spring garden with rhododendrons, azaleas, pieris, a bluebell wood and embothyrium flowering against a backdrop of North Cheshire views. A variety of trees are planted along the banks, inc cornus, a handkerchief tree, Judas trees, mountain ash trees and a strawberry tree. There is slope access to most areas of the garden, though some of these are quite steep and grassed.

46 NEW 6 OAKLAND VALE
Wallasey, CH45 1LQ. Ian Butler & Charles Stringer, 07803 937715, ianbut@btinternet.com. *On Magazines Promenade, New Brighton. J1 M53 , signs to New Brighton attractions on A554. R at Morrisons r'about, up hill on A554. L at Vaughan Rd, down to river. Park before river. No cars on prom. Oakland Vale on L facing river.* **Visits by arrangement in Aug for groups of up to 12. Refreshments or home-made teas by prior arrangement. Adm £4, chd free.**
A very small walled town garden densely filled with tropical and exotic style plants. Bananas, tree ferns, cordylines and bamboo along with tetrapanax and paulownia form the backdrop to more unusual 'exotic' plants and a pond. The terraced front garden overlooking the Mersey is filled with grasses, ferns, phormiums and echiums along with drought tolerant plants. Located on Magazines Promenade offering interesting river/ beach walks with dockland and Liverpool views.

47 THE OLD BYRE
Sound Lane, Sound, Nantwich, CW5 8BE. Gill Farrington. *11m from J16 from Nantwich signs for A530 Whitchurch. At Sound School take 1st R Wrenbury Heath Rd. Parking at garden for visitors with limited mobility. Additional parking at Sound Manor a short walk away.* **Sat 15, Sun 16 July (10-4). Adm £4, chd free. Tea. Open nearby Sound Manor.**
5 acre working smallholding. 4 acres used for grazing and beehives, the rest is garden with a large cider apple orchard, wildlife pond, vegetable and cutting garden patch, herbaceous borders poultry and sheep. We aim to be wildlife friendly and re-use, and recycle everything we can hence some weedy and untidy areas where items are stored ready to be re-used or incorporated into a new project. We produce up to 2 tons of cider apples a year which are used by a local cider maker just 3 miles away. Homemade crafts using recycled materials will be on sale inc yarn made from the fleece of our Ouessant sheep.

48 24 OLD GREASBY ROAD
Upton, Wirral, CH49 6LT. Lesley Whorton & Jon Price. *Approx 1m from J2A M53 (Upton bypass). M53 J2 follow Upton sign. At r'about (J2A) straight on to Upton bypass. At 2nd r'about, turn L by Upton Cricket Club. 24 Old Greasby Rd is on L.* **Sun 18 June (11-4). Adm £4.50, chd free. Home-made teas.**
A multi-interest and surprising suburban garden. Both front and rear gardens incorporate innovative features designed for climbing and rambling roses, clematis, underplanted with cottage garden plants with a very productive kitchen garden. Garden transformed from a boggy, heavily-clayed and flood-damaged suburban garden to a peaceful and productive location.

The Birches

49 THE OLD PARSONAGE
Back Lane, Arley Green, Northwich, CW9 6LZ. The Hon Rowland & Mrs Flower, www.arleyhallandgardens.com. *5m NNE of Northwich. 3m NNE of Great Budworth. M6 J19 & 20 & M56 J10. Follow signs to Arley Hall & Gardens. From Arley Hall notices to Old Parsonage which lies across park at Arley Green (approx 1m).* **Sat 20, Sun 21 May (2-5). Adm £6, chd free. Home-made teas.**
2 acre garden in attractive and secretive rural setting in secluded part of Arley Estate, with ancient yew hedges, herbaceous and mixed borders, shrub roses, climbers, leading to woodland garden and unfenced pond with gunnera and water plants. Rhododendrons, azaleas, meconopsis, cardiocrinum, some interesting and unusual trees. Wheelchair access over mown grass, some slopes and bumps and rougher grass further away from the house.

50 ONE HOUSE WALLED GARDEN
off Buxton New Road, Rainow, SK11 0AD. Louise Baylis. *2½m NE of Macclesfield. Just off A537 Macclesfield to Buxton rd. 2½m from Macclesfield Station.* **Sat 17, Sun 18 June (10-5). Adm £5, chd free. Home-made teas.**
An historic early C18 walled kitchen garden, hidden for 60 yrs and restored by volunteers. This romantic and atmospheric garden has a wide range of vegetables, flowers and old tools. There is an orchard with friendly pigs, a wildlife area and pond, woodland walk with foxgloves and views and a traditional greenhouse with ornamental and edible crops.

51 NEW OULTON PARK HOUSE
Oulton Park, Oulton, Tarporley, CW6 9BL. Bill & Rosie Spiegelberg. *Little Budworth. Take Coach Rd Off A49 opp Tarporley Garden Centre down to "T" junction L & then R down private drive. Slow over 3 sleeping policemen then yellow signs.* **Sun 5 Mar (1.30-4.30). Adm £5, chd free. Home-made teas.**
This 4 acre garden boasts rhododendrons, azaleas, snowdrops, and carpets of crocus in early spring. There are mature trees, a woodland walk with Daphnes and a herbaceous border. A small water feature and orangery complete the picture.

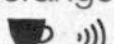

52 15 PARK CRESCENT
Appleton, Warrington, WA4 5JJ. Linda & Mark Enderby, 07949 496747, lmaenderby@outlook.com. *2½m S of Warrington. From M56 J10 take A49 towards Warrington for 1½m. At 2nd set of lights turn R into Lyons Lane, then 1st R into Park Crescent. No15 is last house on R.* **Thur 1, Sun 25 June (11.30-4.30). Adm £5, chd free. Light refreshments. Prosecco. Visits also by arrangement 12 June to 11 Aug for groups of 15 to 50.**
An abundant garden containing many unusual plants, trees and a mini orchard. A cascade, ponds and planting encourage wildlife. There are raised beds, many roses in various forms and herbaceous borders. The garden has been split into distinct areas on different levels each with their own vista drawing one through the garden. Some unusual plants and a good example of planting under a lime tree in clay soil and shade.

53 PARM PLACE
High Street, Great Budworth, CW9 6HF. Jane Fairclough, 01606 891131, janefair@btinternet.com. *3m N of*

The Homestead

Northwich. Great Budworth on E side of A559 between Northwich & Warrington, 4m from J10 M56, also 4m from J19 M6. Parm Place is W of village on S side of High St. **Sun 2 Apr, Sun 25 June (1-5). Adm £5, chd free. Visits also by arrangement Apr to Aug for groups of 10 to 30. Donation to Great Ormond Street Hospital.**
Well-stocked ½ acre plantswoman's garden with stunning views towards S Cheshire. Curving lawns, parterre, shrubs, colour coordinated herbaceous borders, roses, water features, rockery, gravel bed with some grasses. Fruit and vegetable plots. In spring large collection of bulbs and flowers, camellias, hellebores and blossom.

54 NEW PARVEY LODGE

Parvey Lane, Sutton, Macclesfield, SK11 0HX. Mrs Tanya Walker, 07789 528093, tanya.v.walker@gmail.com. *Located on Parvey Lane in the heart of the village of Sutton. Close proximity to Fairways Garden Centre, Sutton Hall & Sutton PO.* **Sat 22, Sun 23 Apr, Sat 14, Sun 15 Oct (10-5). Adm £5, chd free. Cream teas. Light refreshments. Visits also by arrangement 22 Apr to 15 Oct for groups of 10+. Please book at least a month in advance.**
A beautiful privately owned 3 acre garden with different areas to explore. Formal garden, Himalayan Cedar majestically situated at the front of the house, plenty of acer, fruit trees, shaped box hedge, tennis court lawn, lots of bulbs, snowdrops, daffodils. rhododendron and camellia, birds and wildlife (deer). Best to visit in spring and late autumn. Most of the garden is wheelchair friendly.

55 ◆ PEOVER HALL GARDENS

Over Peover, Knutsford, WA16 9HW. Mr & Mrs Brooks, 01565 654107, bookings@peoverhall.com, www.peoverhall.com. *4m S of Knutsford. Do not rely on SatNav. From A50/Holmes Chapel Rd at Whipping Stocks pub turn onto Stocks Lane. Follow R onto Grotto Ln ¼m turn R onto Goostrey Ln. Main entrance on R on bend through white gates.* **For NGS: Sat 10, Sun 11 June (2-5). Adm £7, chd free. Cream teas in the Park House Tea Room.** For other opening times and information, please phone, email or visit garden website.
The extensive formal gardens to Peover Hall feature a series of 'garden rooms' filled with clipped box, water garden, Romanesque loggia, warm brick walls, unusual doors, secret passageways, beautiful topiary work and walled gardens, rockery, rhododendrons and pleached limes. Peover Hall, a Grade II* listed Elizabethan family house dating from 1585, provides a fine backdrop.
The Grade I listed Carolean Stables which are of significant architectural importance will be open to view. Tours of Peover Hall will also be available over the weekend with Mr & Mrs Brooks. Partial wheelchair access to garden - wheelchair users please ask the car-park attendant for parking on hard standing rather than on the grass.

56 ◆ POULTON HALL

Poulton Lancelyn, Bebington, Wirral, CH63 9LN. The Poulton Hall Estate Trust & Poulton Hall Walled Garden Charitable Trust, 07836 590875, info@poultonhall.co.uk, www.poultonhall.co.uk. *2m S of Bebington. From M53, J4 towards Bebington; at T-lights R along Poulton Rd; house 1m on R.* **For NGS: Sat 8, Sun 9 Apr (2-5). Adm £6, chd free. Cream teas may be booked in advance via our website. For other opening times and information, please phone, email or visit garden website.**
3 acres, lawns fronting house, wildflower meadow. Surprise approach to walled garden, with reminders of Roger Lancelyn Green's retellings, Excalibur, Robin Hood and Jabberwocky. Scented sundial garden for the visually impaired. Memorial sculpture for Richard Lancelyn Green by Sue Sharples. Rose, nursery rhyme, witch, herb and oriental gardens and new Memories Reading room. Level gravel paths. Separate wheelchair access (not across parking field). Disabled WC.

57 ◆ RODE HALL

Church Lane, Scholar Green, ST7 3QP. Randle & Amanda Baker Wilbraham, 01270 873237, enquiries@rodehall.co.uk, www.rodehall.co.uk. *5m SW of Congleton. Between Scholar Green (A34) & Rode Heath (A50).* **For opening times and information, please phone, email or visit garden website.**
Nesfield's terrace and rose garden with stunning view over Humphry Repton's landscape is a feature of Rode, as is the woodland garden with terraced rock garden and grotto. Other attractions inc the walk to the lake with a view of Birthday Island complete with heronry, restored ice house, working 2 acre walled kitchen garden and Italian garden. Fine display of snowdrops in Feb and bluebells in May. Snowdrop walks February 4th - Mar 4th 11-4pm, Wed - Sun. Bluebell walks in May. Summer: Weds and Bank Hol Mons until end of Sep, 11-5. Courtyard Kitchen offering wide variety of refreshments.

58 ROSE BRAE

Earle Drive, Parkgate, Neston, CH64 6RY. Joe & Carole Rae. *11m NW of Chester. From A540 take B5134 at the Hinderton Arms towards Neston. Turn R at the T- lights in Neston & L at the Cross onto Parkgate Rd, B5135. Earle Drive is ½m on R.* **Sun 23 July (1-4.30). Adm £5, chd free. Home-made teas.**
An all season, half acre garden comprising bulbs, trees, shrubs, perennials and climbers. In July roses, clematis, hydrangeas, phlox and agapanthus predominate together with ferns, grasses, flowering trees, topiary, a bulb lawn and a formal pool with water lilies.

59 ROSEWOOD

Old Hall Lane, Puddington, Neston, CH64 5SP. Mr & Mrs C E J Brabin, 0151 353 1193, angela.brabin@btinternet.com. *8m N of Chester. From A540 turn down Puddington Lane, 1½m. Park by village green. Walk 30yds to Old Hall Lane, turn L through archway into garden.* **Visits by arrangement for groups of up to 40. Adm £4, chd free. Tea.**
Year-round garden; thousands of snowdrops in Feb, camellias in autumn, winter and spring. Rhododendrons in April/May and unusual flowering trees from March to June. Autumn cyclamen in quantity from Aug to Nov. Perhaps the greatest delight to owners are 2 large *Cornus capitata*, flowering in June. Bees kept in the garden. Honey sometimes available.

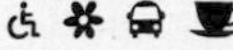

60 ROWLEY HOUSE

Forty Acre Lane, Kermincham, Holmes Chapel, CW4 8DX. Tim & Juliet Foden. *3m ENE from Holmes Chapel. J18 M6 to Holmes Chapel, then take A535 (Macclesfield). Take R turn in Twemlow (Swettenham) at Yellow Broom restaurant. Rowley House ½m on L.* **Sun 28 May, Sun 9 July (11-4.30). Adm £5, chd free. Home-made teas.**

Our aim is to give nature a home and create a place of beauty. There is a formal courtyard garden and informal gardens featuring rare trees, and herbaceous borders, a pond with swamp cypress and woodland walk with maples, rhododendrons, ferns and shade-loving plants. Beyond the garden there are wildflower meadows, natural ponds and a wood with ancient oaks. Also wood sculptures by Andy Burgess. Many unusual plants and trees, also many areas dedicated to wildlife. Leaflet provided with map and details of garden.

61 SANDYMERE

Middlewich Road, Cotebrook, CW6 9EH. Sir John Timpson, rme2000@aol.com. *5m N of Tarporley. On A54 approx 300yds W of T-lights at Xrds of A49/A54.* **Sun 25 June (12-4). Adm £7, chd free. Home-made teas. Visits also by arrangement 1 May to 28 July for groups of 5 to 20.**

16 landscaped acres of beautiful Cheshire countryside with terraces, walled garden, extensive woodland walks and an amazing hosta garden. Turn each corner and you find another gem with lots of different water features inc a rill built in 2014, which links the main lawn to the hostas. Look out for our new Japanese themed garden. Partial wheelchair access.

62 SOMERFORD

19 Leycester Road, Knutsford, WA16 8QR. Emma Dearman & Joe Morris. *1m from centre of Knutsford. Somerford is opp Leycester Close. Park on Leycester Rd or Legh Rd with consideration to local residents.* **Sun 30 Apr (11-4). Adm £4, chd free. Home-made teas. Evening opening Fri 16 June (5-8). Adm £6. Wine. Sun 30 July (11-5). Adm £4, chd free. Home-made teas.**

Majestic trees surround this 1½ acre garden. Hard landscaping and sculptures complement lush herbaceous and perennial borders. Lawns are separated by a magnificent oak pergola. Cube-headed hornbeams lead into the snail-trail walk and informal lawn and fernery. Most of the garden is accessible by wheelchair. Phone 07984 311352 for further information.

63 NEW SOUND MANOR

Wrenbury Heath Road, Sound, Nantwich, CW5 8BT. Gill & Richard Ratcliffe. *5 minutes drive from the A530 Nantwich to Whitchurch road. Our drive is next to a line of bollards.* **Sat 15, Sun 16 July (10-4). Adm £5, chd free. Open nearby The Old Byre.**

Our 0.4 acre country garden is within our 10 acre equestrian property. The garden wraps around an old farmhouse festooned with climbers. The part cobbled yard is lined with pots of summer bedding and tender perennials. There is a raised alpine bed, a large ornamental pond and a bog garden. The overflowing mixed borders have specimen trees, shrubs, perennials and magnificent *Stipa gigantea* grasses.

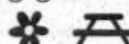

64 34 STANLEY MOUNT

Sale, Manchester, M33 4AE. Debbie & Steve Bedford. *1m from J7 M60 heading S on A56 (Washway Rd) towards Altrincham. From A56 Altrincham N towards Sale (Washway Rd) Turning for Stanley Mount off Washway Rd, No 34 on R.* **Sun 21 May (1-5). Adm £4, chd free. Home-made teas.**

Informal small suburban garden designed and created from scratch. Mature trees, hedges, shrubs and climbers provide the framework. Planted areas feature herbaceous perennials, grasses, bulbs, clematis, roses,hostas, ferns and groundcover. Seated areas, creative pots, productive greenhouse, cacti, succulents and water feature.

65 10 STATHAM AVENUE

Lymm, WA13 9NH. Mike & Gail Porter, 01925 753488, porters@mikeandgailporter.co.uk, youtu.be/53zmVWfZa-s. *Approx 1m from J20 M6 /M56 interchange. From M/way follow B5158 to Lymm. Take A56 Booth's Hill Rd, L towards Warrington, turn R on to Barsbank Lane, pass under Bridgewater canal, after 50 metres turn R onto Statham Ave. No 10 is 100 metres on R.* **Sun 30 Apr, Mon 1, Sat 27, Sun 28 May (11-4), open nearby Laskey Farm. Wed 14 June, Wed 5 July (11-4). Adm £5, chd free. Home-made teas. Enjoy home-made cakes or try Gail's famous meringues with fresh fruit and cream. Visits also by arrangement 24 Apr to 31 July for groups of 10 to 30. Joint visits with nearby Laskey Farm can be arranged.**

Beautifully structured ¼ acre south facing terraced garden rising to the Bridgewater towpath. Hazel arch opens to clay paved courtyard with peach trees and kitchen herb bed. Rose pillars lead to lush herbaceous beds and quiet shaded areas, bordered by azaleas and rhododendrons in spring, peaceful, pastel shades in early summer and hydrangeas and fuchsias in late summer. Interesting garden buildings. A treasure hunt/quiz to keep the children occupied. Delicious refreshments to satisfy the grown ups.

66 ◆ STONYFORD COTTAGE

Stonyford Lane, Oakmere, CW8 2TF. Janet & Tony Overland, 01606 888970, info@stonyfordcottagegardens.co.uk, www.stonyfordcottagegardens.co.uk. *5m SW of Northwich. From Northwich take A556 towards Chester. ¾m past A49 junction turn R into Stonyford Lane. Entrance ½m on L.* **For NGS: Sun 14 May (11-4). Adm £5, chd free. Cream teas in our Tea Room. Light lunches & home-made cakes available. For other opening times and information, please phone, email or visit garden website.**

Set around a tranquil pool, this Monet style landscape has a wealth of moisture loving plants, inc iris and candelabra primulas. Drier areas feature unusual perennials and rarer trees and shrubs. Woodland paths meander through shade and bog plantings, along boarded walks, across wild natural areas with views over the pool to the cottage gardens. Unusual plants available at the adjacent nursery. Open Tues - Fri, Apr - Oct 10-5pm. Some gravel paths.

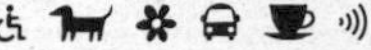

67 STRETTON OLD HALL

Stretton, Tilston, Malpas, SY14 7JA. Stephen Gore Head Gardener, Strettonoldhallgardens@hotmail.com. *5m N of Malpas. From Chester follow A41, Broxton r'about follow signs for Stretton Water Mill, turn L at Cock o Barton. 2m on L.* **Sat 22, Sun 23 July (11-5). Adm £8, chd free. Home-made teas. Visits also by arrangement 1 June to 24 Sept.**
5 acre Cheshire countryside garden with a planting style best described as controlled exuberance with a definite emphasis upon perennials, colour, form and scale. Divided into several discrete and individual gardens inc stunning herbaceous borders, scree garden, walled kitchen garden and glasshouse. Wildflower meadows, wildlife walk around the lake with breathtaking vistas in every direction. Regional Winner, The English Garden's The Nation's Favourite Gardens 2021.

68 TATTENHALL HALL

High Street, Tattenhall, Chester, CH3 9PX. Jen & Nick Benefield, Chris Evered & Jannie Hollins, 01829 770654, janniehollins@gmail.com. *8m S of Chester on A41. Turn L to Tattenhall, through village, turn R at Letters pub, past War Memorial on L through sandstone pillared gates. Park on rd or in village car park.* **Sun 4 June (2-5.30). Adm £5, chd free. Home-made teas. Visits also by arrangement May to July.**
Plant enthusiasts' garden around Jacobean house (not open). 4½ acres, wildflower meadows, interesting trees, large pond, stream, walled garden, colour themed borders, succession planting, spinney walk with shade plants, yew terrace overlooking meadow, views to hills. Glasshouse and vegetable garden. Wildlife friendly, sometimes untidy garden, interest year-round, continuing to develop. Partial wheelchair access due to gravel paths, cobbles and some steps.

69 TIRESFORD

Tarporley, CW6 9LY. Susanna Posnett, 07989 306425. *Tiresford is on A49 Tarporley bypass. It is adjacent to the farm on R leaving Tarporley in the direction of Four Lane Ends T-lights.* **Sun 14 May (2-5). Adm £5, chd free. Home-made teas. Ice-Cream.**
Established 1930s garden undergoing a major restoration project to reinstate it to its former glory. Fabulous views of both Beeston and Peckforton Castles offer a wonderful backdrop in which to relax and enjoy a delicious tea. This year we hope to realise our long term plan to recreate the kitchen garden and to repair the fountain amongst the new hot borders of the sunken garden. The house has been transformed into a stylish 6 bedroom B&B. There is parking for wheelchair users next to the house and access to a disabled WC in the house.

70 TIRLEY GARTH GARDENS

Mallows Way, Willington, Tarporley, CW6 0RQ. *2m N of Tarporley. 2m S of Kelsall. Entrance 500yds from village of Utkinton. At N of Tarporley take Utkinton rd.* **Sun 21 May (1-5). Adm £5, chd free. Home-made teas.**
40 acre garden, terraced & landscaped, designed by Thomas Mawson who is considered the leading exponent of garden design in early C20. It is the only Grade II* Arts & Crafts garden in Cheshire that remains complete and in excellent condition. The gardens are an important example of an early C20 garden laid out in both formal and informal styles. By early May the garden is bursting into flower with almost 3000 rhododendron and azalea, many 100 yrs old. Art Exhibition by local Artists.

71 THE WELL HOUSE

Wet Lane, Tilston, Malpas, SY14 7DP. Mrs S H French-Greenslade, 01829 250332. *3m NW of Malpas. On A41, 1st R after Broxton r'about, L on Malpas Rd through Tilston. House on L.* **Visits by arrangement 13 Feb to 29 Sept for groups of 10 to 25. Pre-booked refreshments for small groups only.**
1 acre cottage garden, bridge over natural stream, spring bulbs, perennials, herbs and shrubs. Triple ponds. Adjoining ¾ acre field made into wildflower meadow; first seeding late 2003. Large bog area of kingcups and ragged robin. February for snowdrop walk. Victorian parlour and collector's items on show. Dogs on leads only.

72 WILLASTON GRANGE

Hadlow Road, Willaston, Neston, CH64 2UN. Mr & Mrs M Mitchell. *On A540 (Chester High Rd) take the B5151 into Willaston. From the village green in the centre of Willaston turn onto Hadlow Rd, Willaston Grange is ½m on L.* **Sat 17 June (12-5). Adm £5, chd free. Home-made teas.**
Willaston Grange is an Arts & Craft property. 13 yrs ago the house and gardens had been derelict for 3 yrs. After months of detailed planning, the significant restoration work began. The gardens extend to 6 acres with a small lake, wide range of mature and rare tree species, herbaceous border, woodland and vegetable gardens, orchard and magical treehouse. The fully restored Arts & Crafts house provides the perfect backdrop for a summer visit, along with afternoon tea and live music. Most areas accessible by wheelchair.

73 THE WONKY GARDEN

Ditton Community Centre, Dundalk Road, Widnes, WA8 8DF. Mrs Angela Hayler, 07976 373979, thewonkygarden@gmail.com, en-gb.facebook.com/thewonkygarden. *From the Widnes exit of the A533 turn L (Lowerhouse La) then L at the r'about. The Community Centre is ½m up on the R. The garden is behind the Centre.* **Sun 23 Apr, Sun 2 July, Sun 10 Sept (11-4). Adm £5, chd free. Light refreshments. Visits also by arrangement May to Sept.**
The flower garden is our show garden, the focus for horticultural therapy/nature based activities. It has large herbaceous borders, trees and shrubs, planting focussing on the senses and wildlife. We grow masses of edibles and cut flowers in the allotment garden, 80% is donated to support food poverty. The children's 'explorify' garden nestles between the two and is used to support young families. The garden is designed and managed by a wonderful group of volunteers. We support many community groups and individuals of all ages and abilities inc schools, colleges and work experience. Our focus is on supporting physical and mental health, isolation and loneliness. An accessible path (1½ meters wide) extends from the car park through the herbaceous and children's nature garden and into the allotment.

CORNWALL

ISLES OF SCILLY

Tresco
St Martin's
Bryher
Hugh Town
St Mary's
St Agnes

The Isles of Scilly lie about 28 miles or 45 kilometres south west of Land's End

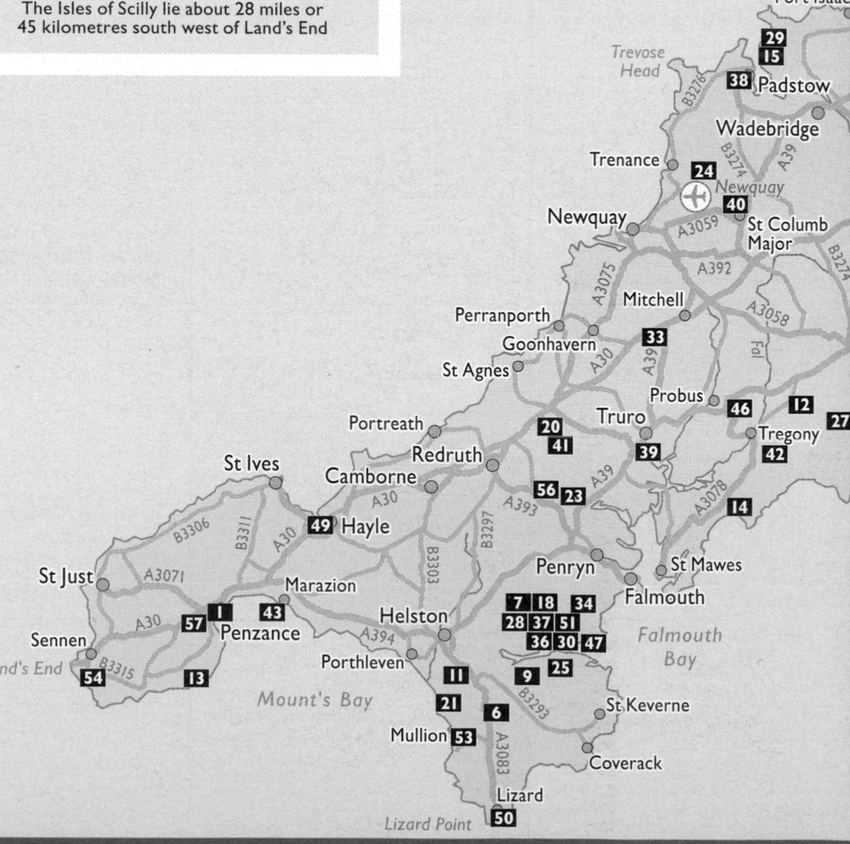

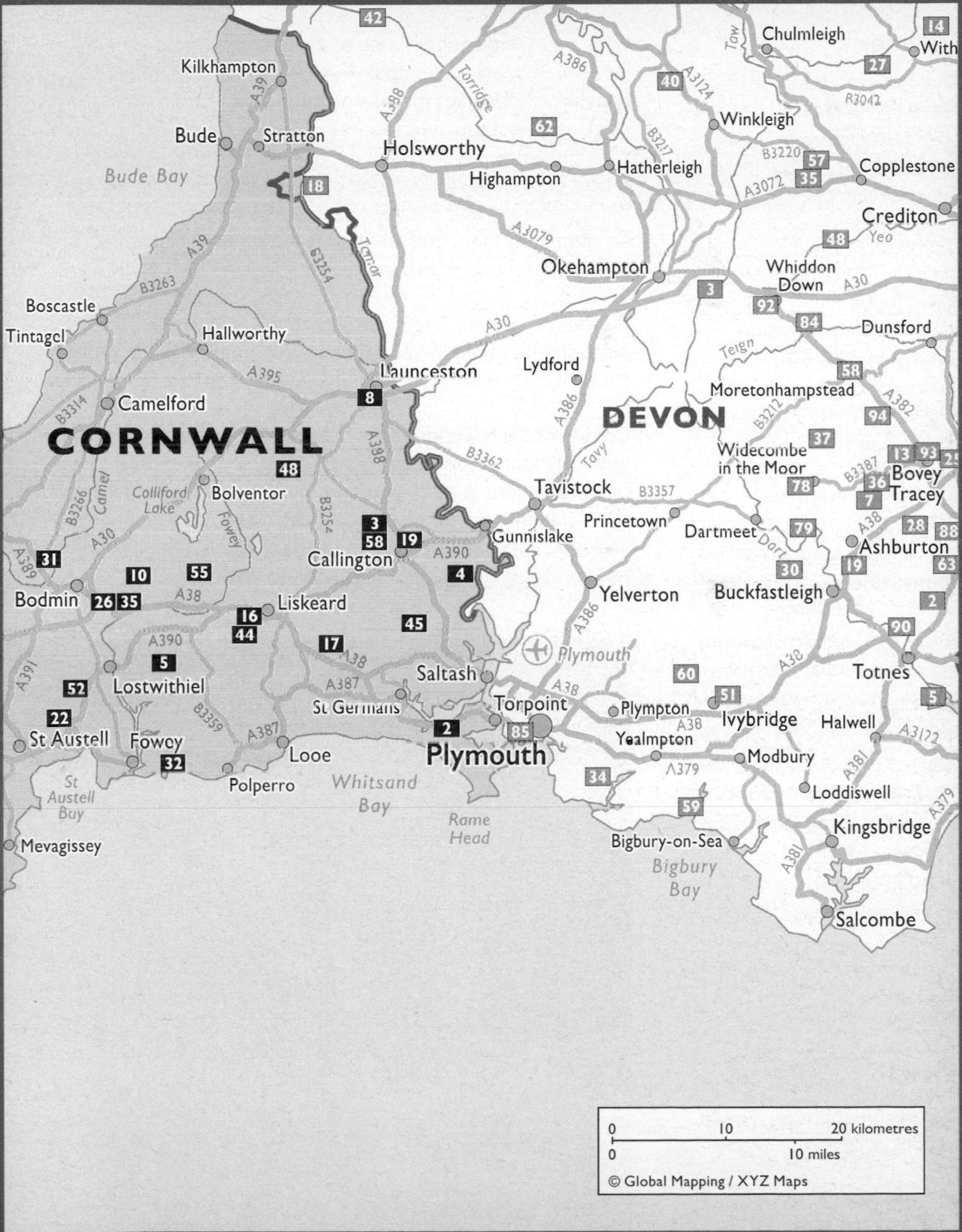

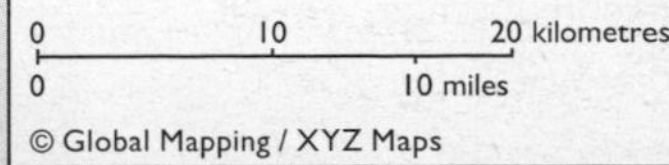
0 10 20 kilometres
0 10 miles
© Global Mapping / XYZ Maps
CORNWALL
DEVON
Kilkhampton
Bude
Stratton
Bude Bay
Holsworthy
Highampton
Hatherleigh
Chulmleigh
Winkleigh
Copplestone
Crediton
Okehampton
Whiddon Down
Dunsford
Boscastle
Tintagel
Hallworthy
Launceston
Lydford
Moretonhampstead
Camelford
Bolventor
Colliford Lake
Widecombe in the Moor
Bovey Tracey
Tavistock
Princetown
Dartmeet
Gunnislake
Callington
Ashburton
Bodmin
Liskeard
Yelverton
Buckfastleigh
Lostwithiel
Saltash
Totnes
St Germans
Torpoint
Plympton
Ivybridge
Halwell
St Austell
Fowey
Looe
Plymouth
Yealmpton
Modbury
Polperro
Whitsand Bay
Loddiswell
St Austell Bay
Rame Head
Kingsbridge
Mevagissey
Bigbury-on-Sea
Bigbury Bay
Salcombe

Volunteers

County Organiser
Claire Woodbine
07483 244318
claire.woodbine@ngs.org.uk

County Treasurer
Marie Tolhurst
marie.tolhurst@ngs.org.uk

Publicity
Laura Tucker
laura.tucker@ngs.org.uk

Booklet Co-ordinator
Position vacant Please contact **Claire Woodbine** (see above)

Photographer
Keith Tucker
keith.tucker@ngs.org.uk

Assistant County Organisers
Christopher Harvey Clark
01872 530165
suffree2012@gmail.com

Ginnie Clotworthy 01208 872612
ginnie.clotworthy@ngs.org.uk

Sara Gadd 07814 885141
sara.gadd@ngs.org.uk

Alison O'Connor 01726 882460
tregoose@tregoose.co.uk

Libby Pidcock 01208 821303
libby.pidcock@ngs.org.uk

Sue Newton 07786 367610
sue.newton@ngs.org.uk

@CornwallNGS
@cornwall.ngs

Cornwall has some of the most beautiful natural landscapes to be found anywhere in the world.

Here, you will discover some of the country's most extraordinary gardens, a spectacular coastline, internationally famous surfing beaches, windswept moors and countless historic sites. Cornish gardens reflect this huge variety of environments particularly well.

A host of National Collections of magnolias, camellias, rhododendrons and azaleas, as well as exotic Mediterranean semitropical plants and an abundance of other plants flourish in our acid soils and mild climate.

Surrounded by the warm currents of the Gulf Stream, with our warm damp air in summer and mild moist winters, growth flourishes throughout most of the year.

Cornwall boasts an impressive variety of beautiful places open to the National Garden Scheme. These range from beautifully designed gardens centred on ancient manor houses, to moorland water gardens, gardens with epic sea views, a serene and secluded Japanese garden and, returning this year, the wonderful Antony Woodland Garden & Woodland Walk.

Right: Antony Woodland Garden and Woodland Walk

OPENING DATES

All entries subject to change. For latest information check **www.ngs.org.uk**

Map locator numbers are shown to the right of each garden name.

March

Friday 3rd
NEW ◆ Antony Woodland Garden & Woodland Walk 2

Saturday 25th
The Lodge 26
Pinsla Garden 35

Sunday 26th
The Lodge 26
Pinsla Garden 35
◆ Prideaux Place 38

April

Saturday 1st
◆ Trewidden Garden 57

Monday 10th
Riverside Cottage 39

Tuesday 11th
Riverside Cottage 39

Sunday 16th
Carwinnick 12

Monday 17th
◆ Pencarrow 31

Tuesday 18th
Riverside Cottage 39

Wednesday 19th
Pinsla Garden 35
NEW Trevina House 55

Thursday 20th
NEW Trevina House 55

Saturday 22nd
◆ Chygurno 13
NEW Rose Morran 40

Sunday 23rd
Anvil Cottage 3
◆ Chygurno 13
Ethnevas Cottage 18
NEW Rose Morran 40
Windmills 58

Tuesday 25th
Riverside Cottage 39

Friday 28th
NEW ◆ Antony Woodland Garden & Woodland Walk 2

Sunday 30th
Cardinham Gardens 10
NEW Rosevallon Barn 42

May

Tuesday 2nd
Riverside Cottage 39

Sunday 7th
◆ Boconnoc 5
East Down Barn 17
Navas Hill House 30
South Lea 45
Trebartha Estate Garden and Country Garden at Lemarne 48
◆ Trevince Estate Gardens 56

Tuesday 9th
Pinsla Garden 35
Riverside Cottage 39

Wednesday 10th
NEW Trevina House 55

Thursday 11th
NEW Trevina House 55

Saturday 13th
Greenwith Meadow 23

Sunday 14th
Greenwith Meadow 23
◆ The Japanese Garden 24
Lower Penbothidnow 28

Monday 15th
◆ The Japanese Garden 24

Tuesday 16th
Pinsla Garden 35
Riverside Cottage 39

Thursday 18th
South Bosent 44

Tuesday 23rd
Pinsla Garden 35

Wednesday 31st
◆ Kestle Barton 25

June

Friday 2nd
Bucks Head House Garden 7

Saturday 3rd
NEW Trevesco 53

Sunday 4th
NEW Trevesco 53

Wednesday 7th
Gardens Cottage 22
◆ Kestle Barton 25

Thursday 8th
Gardens Cottage 22

Friday 9th
Bucks Head House Garden 7

Sunday 11th
South Lea 45
Suffree Farm 46

Tuesday 13th
Pinsla Garden 35

Wednesday 14th
◆ Kestle Barton 25
Pinsla Garden 35

Friday 16th
Bucks Head House Garden 7
Dobwalls Gardens 16

Saturday 17th
Dobwalls Gardens 16
NEW Fox Hollow 20

Sunday 18th
Alverton Cottage 1
Caervallack 9
NEW Firste Park 19
NEW Fox Hollow 20

Tuesday 20th
Pinsla Garden 35

Wednesday 21st
◆ Kestle Barton 25
Pinsla Garden 35
NEW Trevina House 55

Thursday 22nd
NEW Trevina House 55

Friday 23rd
Bucks Head House Garden 7

Saturday 24th
NEW Trelan 49

Wednesday 28th
◆ Kestle Barton 25

Friday 30th
Bucks Head House Garden 7
Dobwalls Gardens 16

July

Saturday 1st
Dobwalls Gardens 16
◆ Roseland House 41

Sunday 2nd
◆ Roseland House 41

Wednesday 5th
Gardens Cottage 22
◆ Kestle Barton 25

Thursday 6th
Gardens Cottage 22

Friday 7th
NEW ◆ Antony Woodland Garden & Woodland Walk 2
Bucks Head House Garden 7

Saturday 8th
Trevilley 54

Sunday 9th
Cardinham Gardens 10

Wednesday 12th
◆ Kestle Barton 25

Friday 14th
Bucks Head House Garden 7

Saturday 15th
NEW Firste Park 19
NEW Pennycomequick 33

Sunday 16th
NEW Firste Park 19
NEW Pennycomequick 33

Wednesday 19th
◆ Kestle Barton 25

Friday 21st
Bucks Head House Garden 7

Saturday 22nd
◆ Chygurno 13
Dobwalls Gardens 16
NEW Trelan 49

Sunday 23rd
◆ Chygurno 13
Dobwalls Gardens 16

Wednesday 26th
◆ Kestle Barton 25

Friday 28th
Bucks Head House Garden 7

Sunday 30th
Byeways 8

August

Wednesday 2nd
◆ Kestle Barton 25

Friday 4th
Bucks Head House Garden 7

Sunday 6th
Crugsillick Manor 14
Lower Penbothidnow 28

Wednesday 9th
◆ Kestle Barton 25

Friday 11th
Bucks Head House Garden 7

Saturday 12th
Dobwalls Gardens 16

Sunday 13th
Anvil Cottage 3
Dobwalls Gardens 16
Windmills 58

Tuesday 15th
◆ Bonython Manor 6

Wednesday 16th
◆ Kestle Barton 25

Friday 18th
Bucks Head House Garden 7

Sunday 20th
NEW Firste Park 19

Wednesday 23rd
Gardens Cottage 22
◆ Kestle Barton 25

Thursday 24th
Gardens Cottage 22

Friday 25th
Bucks Head House Garden 7

Wednesday 30th
◆ Kestle Barton 25

September

Saturday 2nd
NEW Daymer House 15
NEW Rose Morran 40

Sunday 3rd
NEW Rose Morran 40

Wednesday 6th
◆ Kestle Barton 25

Saturday 9th
NEW Firste Park 19

Sunday 10th
Towan House 47

Wednesday 13th
◆ Kestle Barton 25
South Bosent 44
NEW Trevina House 55

Thursday 14th
South Bosent 44
NEW Trevina House 55

Saturday 16th
Ash Barn 4

Sunday 17th
Ash Barn 4
NEW Daymer House 15

Wednesday 20th
◆ Kestle Barton 25

Wednesday 27th
◆ Kestle Barton 25

October

Sunday 1st
Trebartha Estate Garden and Country Garden at Lemarne 48

Thursday 5th
NEW Trevina House 55

Friday 6th
NEW ◆ Antony Woodland Garden & Woodland Walk 2
NEW Trevina House 55

By Arrangement

Arrange a personalised garden visit with your club, or group of friends, on a date to suit you. See individual garden entries for full details.

Anvil Cottage 3
Bucks Head House Garden 7
Caervallack 9
Carminowe Valley Garden 11
Crugsillick Manor 14
Dove Cottage, Dobwalls Gardens 16
East Down Barn 17
Ethnevas Cottage 18
Garden Cottage 21
Gardens Cottage 22
9 Higman Close, Dobwalls Gardens 16
Lower Penbothidnow 28
Malibu 29
Pendower House 32
Penwarne 34
Pinsla Garden 35
Port Navas Chapel 36
NEW Rose Morran 40
NEW Rosevallon Barn 42
South Bosent 44
South Lea 45
Towan House 47
NEW Trelan 49
NEW Tremichele 50
Trenarth 51
Trethew 52
Trevilley 54
Windmills 58

Caervallack

THE GARDENS

1 ALVERTON COTTAGE

Alverton Road, Penzance, TR18 4TG. David & Lizzie Puddifoot. *Next door to YMCA. About 600 metres from Penlee car park travelling towards A30. Morrab Gardens are also close to car park.* **Sun 18 June (2-5.30). Adm £4, chd free. Home-made teas.**

Alverton Cottage is a Grade II listed Regency house. The garden is modest in size though large for a Penzance garden, south facing and sheltered by mature trees inc elms. Very large monkey-puzzle tree and holm oak. Laid out in the 1860s, we have added a succulent area, fernery and have cleared and replanted since 2019. In nearby Morrab Sub tropical Gardens the 'Friends' will also welcome visitors. Off road disabled parking but wheelchair to garden terrace only.

2 NEW ◆ ANTONY WOODLAND GARDEN & WOODLAND WALK

Ferry Lane, Torpoint, PL11 3AB. Sir Richard Carew-Pole, 01752 815303, woodlandgarden@antonyestate.com, www.antonywoodlandgarden.com. *2m NW of Torpoint. From the main gate off A374, proceed along Ferry Lane past Antony House. Shortly after passing Antony House, you will see Broomhill Tearoom ahead, with car park on R, beneath the trees.* **For NGS: Fri 3 Mar, Fri 28 Apr, Fri 7 July, Fri 6 Oct (10.30-5). Adm £8, chd free. Light refreshments at Broomhill Tearoom. Cream teas, home-made cakes, coffee, light lunches. For other opening times and information, please phone, email or visit garden website.**

Antony Woodland Garden is one of the most beautiful gardens in Cornwall. It is a haven of serenity and peace, the perfect place for a pleasant walk, a picnic, for children to explore and for those interested in gardening, to enjoy a magnificent variety of plants. As an "International Camellia Garden of Excellence" it holds the National Collection of *Camellia japonica*. It has the most stunning array of magnolias, beautiful walks along the river's edge, contemporary sculpture and carpets of wild flowers, waiting to be discovered.

NPC

Riverside Cottage

3 ANVIL COTTAGE

South Hill, PL17 7LP. Geoff & Barbara Clemerson, 01579 362623, gcclemerson@gmail.com. *3m NW of Callington. Head N on A388 from Callington centre. After ½m L onto South Hill Rd (signed South Hill), straight on for 3m. Gardens on R just before St Sampson's Church.* **Sun 23 Apr, Sun 13 Aug (1-5). Combined adm with Windmills £6, chd free. Home-made teas. Gluten free refreshments available. Visits also by arrangement 18 Apr to 6 Sept for groups of up to 30. Combined with Windmills next door.**

Essentially, this is a plantsman's garden. Winding paths lead through a series of themed rooms with familiar, rare and unusual plants. Steps lead up to a raised viewpoint looking west towards Caradon Hill and Bodmin Moor. Other paths take you on a circular route through a rose garden, hot beds, a tropical area and a secret garden.

4 ASH BARN

Callington, PL17 8BP. Mary Martin. *3m E of Callington. Leave Callington on A390 (Tavistock direction) take 1st R signed Harrowbarrow. Turn R at T junction in village, 1m down hill to Glamorgan Mill. Parking on L, 200 metres walk on road to signed orchard.* **Sat 16, Sun 17 Sept (10-5). Adm £5, chd free. Tea in orchard.**

Enjoy autumnal colour and fruitfulness in our tranquil 6 acre orchard. For 40 yrs James and Mary have lovingly grown and researched these rare cultivars of apples, pears and cherries collected from all over Cornwall and Devon. Lily pond and varied sward cutting regime for maximum wildlife habitat. Mixed woodland underplanted with Tamar Valley narcissi, bluebells and primroses.

5 ◆ BOCONNOC

Lostwithiel, PL22 0RG. Fortescue Family, 01208 872507, events@boconnoc.com, www.boconnoc.com. *Off A390 between Liskeard & Lostwithiel. From East Taphouse follow signs to Boconnoc. (SatNav does not work well in this area).* **For NGS: Sun 7 May (1-5). Adm £6, chd free. For other opening times and information, please phone, email or visit garden website.**

20 acres surrounded by parkland and woods with magnificent trees, flowering shrubs and stunning views. The gardens are set amongst mature trees which provide the backcloth for exotic spring flowering shrubs, woodland plants, with newly planted magnolias and a fine collection of hydrangeas. Bathhouse built in 1804, obelisk built in 1771, house dating from Domesday, deer park and C15 church.

6 ◆ BONYTHON MANOR
Cury Cross Lanes, Helston, TR12 7BA. Mr & Mrs Richard Nathan, 01326 240550, sbonython@gmail.com, www.bonythonmanor.co.uk. *5m S of Helston. On main A3083 Helston to Lizard Rd. Turn L at Cury Cross Lanes (Wheel Inn). Entrance 300yds on R.* **For NGS: Tue 15 Aug (2-4). Adm £11, chd £2. Tea, coffee, fruit juices and home-made cakes. For other opening times and information, please phone, email or visit garden website.**
Magnificent 20 acre colour garden inc sweeping hydrangea drive to Georgian manor (not open). Herbaceous walled garden, potager with vegetables and picking flowers; 3 lakes in valley planted with ornamental grasses, perennials and South African flowers. A 'must see' for all seasons colour.

7 BUCKS HEAD HOUSE GARDEN
Trengove Cross, Constantine, TR11 5QR. Deborah Baker, 07801 444916, deborah.fwbaker@gmail.com, deborahfwbaker.com/home.html. *5m SW of Falmouth. A394 towards Helston, L at Edgcumbe towards Gweek/Constantine. Proceed for 0.8 m then L towards Constantine. Further 0.8m, garden on L at Trengove Cross.* **Every Fri 2 June to 25 Aug (2-5). Adm £6, chd free. Tea. Visits also by arrangement 2 June to 25 Aug for groups of up to 20.**
Enchanting cottage and woodland gardens of native and rare trees, shrubs and perennials, encouraging biodiversity. The site of 1½ acres is on a south facing Cornish hillside with panoramic views. Protected by essential windbreak hedging, the inspiring collection of plants has been chosen to create intrigue and increase diversity. During the summer months art works will be on display. Wheelchair access to lower garden and to woodland.

8 BYEWAYS
Dunheved Road, Launceston, PL15 9JE. Tony Reddicliffe. *Launceston town centre. 100yds from multi-storey car park past offices of Cornish & Devon Post into Dunheved Rd, 3rd bungalow on R.* **Sun 30 July (1-5). Adm £5, chd free. Home-made teas.**
Small town garden developed over 11 yrs by enthusiastic amateur gardeners. Herbaceous borders, rockery. Tropicals inc bananas, gingers and senecio. Stream and water features. Roof garden. Japanese inspired tea house and courtyard with bridge. Fig tree and Pawlonia flank area giving secluded seating. Living pergola. New area of ponds and water features with removal of sheds and overgrown areas.

9 CAERVALLACK
St Martin, Helston, TR12 6DF. Matt Robinson & Louise McClary, 01326 221130, mat@build-art.co.uk. *5m SE of Helston. Go through Mawgan village, over 2 bridges, past Gear Farm shop; go past turning on L, garden next farmhouse on L. Parking is in field opp entrance.* **Sun 18 June (1.30-4). Adm £6, chd free. Tea & home-made cakes; scones, clotted cream & jam. Visits also by arrangement 1 Apr to 1 Oct for groups of up to 30. A few days notice required.**
Romantic garden arranged into rooms, the collaboration between an artist and an architect. Colour and form of plants structured by topiary, hedging and architectural experiments in cob, brick, slate & concrete. Grade II listed farmhouse and ancient orchard. Roses and wisteria a speciality. 28yrs in the making. Arts & Crafts cob walls and paving; heroic 54ft pedestrian footbridge, 5 sided meditation studio, cast concrete ponds and amphitheatre; coppice and wildflower meadow; mature orchard and vegetable plot. Topiary throughout.

GROUP OPENING

10 CARDINHAM GARDENS
Cardinham, Bodmin, PL30 4BN. *5m NE of Bodmin. From A30 or Bodmin take A38 towards Plymouth, 1st L at r'about, past Crematorium & Cardinham Woods turning. Signs & garden map at Cardinham Parish Hall, opp Cardinham Church.* **Sun 30 Apr, Sun 9 July (11-5). Combined adm £6, chd free. Home-made teas at Deviock Farm on April 30th & in Cardinham Parish Hall on July 9th. Teas from 2pm to 5pm.**

THE BEECHES
PL30 4BJ. Rosemary Rowe. Open on Sun 9 July

NEW **DEVIOCK FARM**
PL30 4DA. Bernie & Sue Muir. Open on Sun 30 Apr

PINSLA GARDEN
PL30 4AY. Mark & Claire Woodbine. Open on all dates
(See separate entry)

NEW **TRESLEA**
PL30 4DL. Loraine & Nigel Pollard. Open on Sun 9 July

The rural parish of Cardinham on the SW side of Bodmin Moor has a wide range of ecosystems as the moorland merges into woodland and ancient downs and meadows. April 30th group opening has 2 gardens: Pinsla Garden with naturalistic planting, stone circle and vegetable garden in peaceful woodland and Deviock Farm, a 2 acre south facing wildlife friendly garden with many mature trees and shrubs developed over past five yrs. Enclosed farmhouse garden, pond, damp areas, vegetables and a disused waterwheel pit feature. On July 9th, 3 gardens are open: The Beeches, a small rural idyll, adapted for easier maintenance, with shrubs, colourful containers, roses and climbers. Treslea has a separate good-sized orchard/vegetable plot. The garden has been constructed over 20 yrs by the present owners providing a haven for wildlife with herbaceous borders, roses, magnolias and wisteria. Picturesque valley on the edge of Bodmin moor. Pinsla Garden will show continuous colour with seasonal planting.

11 CARMINOWE VALLEY GARDEN
Tangies, Gunwalloe, TR12 7PU. Mr & Mrs Peter Stanley, 01326 565868, stanley.m2@sky.com. *3m SW of Helston. A3083 Helston-Lizard rd. R opp main gate to Culdrose. 1m downhill, garden on R.* **Visits by arrangement May & June for groups of 5 to 20. Adm £6, chd free.**
Overlooking the beautiful Carminowe valley towards Loe Pool, this abundant garden combines native oak woodland, babbling brook and large natural pond with more formal areas. Wildflowers, mown pathways, shrubberies, orchard. Enclosed

cottage garden, spring colours and roses in early summer provide great variety and contrast. Gravel paths, slopes.

12 CARWINNICK

Grampound, Truro, TR2 4RJ. Mr Bryan Coode. *Between Truro & St Austell, off A390 at Hewas Water/Sticker. Signs displayed on A390 showing turn-off point down B3287 towards Tregony. Garden 1m on R. Access is down single track lane (one way system), so should only be approached from this direction.* **Sun 16 Apr (2-5). Adm £5, chd free. Home-made teas.**

Once a Medieval farm, over the last 50 yrs a home to 2 families. There has been much tree planting and development of the 5 acres which has borrowed the landscapes and vistas, so the woods at Heligan to E and turbines at Gorran can both be seen. Spring garden with nearly 50 camellias, daffodils and other early shrubs and plants. Mainly semi-tropical plants around tennis court. A fine Magnolia Lanarth may be in flower if we are lucky. Level ground across gravel and lawns.

13 ◆ CHYGURNO

Lamorna, TR19 6XH. Dr & Mrs Robert Moule, 01736 732153, rmoule010@btinternet.com. *4m S of Penzance. Off B3315. Follow signs for The Lamorna Cove Hotel. Garden is at top of hill, past Hotel on L.* **For NGS: Sat 22, Sun 23 Apr, Sat 22, Sun 23 July (2-5). Adm £5, chd free. For other opening times and information, please phone or email.**

Beautiful, unique, 3 acre cliffside garden overlooking Lamorna Cove. Planting started in 1998, mainly southern hemisphere shrubs and exotics with hydrangeas, camellias and rhododendrons. Woodland area with tree ferns set against large granite outcrops. Garden terraced with steep steps and paths. Plenty of benches so you can take a rest and enjoy the wonderful views.

14 CRUGSILLICK MANOR

Ruan High Lanes, Truro, TR2 5LJ. Dr Alison Agnew & Mr Brian Yule, 07538 218201, alisonagnew@icloud.com. *On Roseland Peninsula. Turn off A390 Truro-St Austell rd onto A3078 towards St Mawes. Approx 5m after Tregony turn 1st L after Ruan High Lanes towards Veryan, garden is 200yds on R. Limited parking.* **Sun 6 Aug (11-6). Adm £5, chd free. Tea, coffee, soft drinks, cakes & light lunches. Visits also by arrangement 15 Apr to 29 Oct for groups of 10 to 40. As parking is limited pls enquire.**

2 acre garden, substantially re-landscaped and planted, mostly over last 10 yrs. To the side of the C17/C18 house, a wooded bank drops down to a walled kitchen garden and hot garden. In front, sweeping yew hedges and paths define oval lawns and broad mixed borders. On a lower terrace, the focus is a large pond and the planting is predominantly exotic flowering trees and shrubs. Wheelchair access to the central level of the garden, the house and cafe. Garden is on several levels connected by fairly steep sloping gravel paths.

15 NEW DAYMER HOUSE

Daymer Bay, Trebetherick, Wadebridge, PL27 6SA. Linda Burrows, 01208 862639, burrows_linda@hotmail.com. *Next to the beach between Rock & Polzeath. Follow road down towards Daymer Bay beach. Please park in beach car park, only space for a few cars at house. Garden is down unmade lane to St Enodoc Church, last turning on L before car park.* **Sat 2, Sun 17 Sept (1-6). Adm £5, chd free. Home-made teas.**

2 acres of wild coastal gardens in the grounds of an Arts & Crafts house. The garden is of Edwardian style enjoying a certain amount of formalism, with 'outside rooms' created using dry stone walls, rustic paths and stone flagged terraces that merge into woodland glades and wild areas. There is an interesting mix of lawns, mature trees, woodland, herbaceous borders, pond and vegetable garden. Paths are uneven but should be wide enough to access. Most of the garden is uneven lawn with no paths.

GROUP OPENING

16 DOBWALLS GARDENS

Dobwalls, Liskeard, PL14 4LR. *Do not use postcode for SatNav. In Dobwalls at double mini r'about take road to Duloe. Yellow signs will then direct. Limited field parking at Dove Cottage, limited street parking at 9 Higman Close.* **Fri 16, Sat 17, Fri 30 June, Sat 1, Sat 22, Sun 23 July, Sat 12, Sun 13 Aug (1-5). Combined adm £8, chd free. Home-made teas at South Bosent.**

DOVE COTTAGE

PL14 4LR. Becky Martin, 07871 368227, beckymartin1@hotmail.com. **Visits also by arrangement 16 June to 31 Aug for groups of up to 24. Combined with 9 Higman Close.**

9 HIGMAN CLOSE

PL14 4LW. Jim Stephens & Sue Martin, 01579 321074, J.l.stephens@btinternet.com. **Visits also by arrangement 16 June to 31 Aug for groups of up to 24. Combined with Dove Cottage.**

SOUTH BOSENT

PL14 4LX. Adrienne Lloyd & Trish Wilson.
(See separate entry)

Three very different gardens within a few minutes drive from each other. The garden at Dove Cottage was replanted from scratch in 2017 showing what can be achieved in 6 yrs. Separate areas with different types of planting. Emphasis on colour. Narrow winding paths, shady pergola, deck with tropical planting, tiny sunroom, lush foliage area. 9 Higman Close, a restless, constantly changing, unapologetically busy garden, full of good plants with colour, scent and surprises. Glasshouse overflowing with cacti and succulents, some over 35 yrs old. The garden at South Bosent is an example of work in progress, currently being developed from farmland. The aim is to create a combination of interesting plants coupled with habitat for wildlife over a total of 9 ½ acres. Several garden areas, woodland gardens, a meadow, ponds of varying sizes, inc new rill and waterfall. In spring, the bluebell wood trail runs alongside the stream.

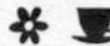

17 EAST DOWN BARN

Menheniot, Liskeard, PL14 3QU. David & Shelley Lockett, 07803 159662. *S side of village nr cricket ground. Turn off A38 at Hayloft restaurant/railway stn junction & head towards Menheniot village. Follow NGS signs from sharp LH bend as you enter village.* **Sun 7 May (1-4). Adm £4, chd free. Home-made teas. Visits also by arrangement 27 Mar to 2 June for groups of 15 to 30.**

Garden laid down between 1986-1991 with the conversion of the barn into a home and covers almost ½ acre of east sloping land with stream running north-south acting as the easterly boundary. 3 terraces before garden starts to level out at the stream. Garden won awards in the early years under the stewardship of the original owners. Ducks are in residence in the stream so regret no dogs.

18 ETHNEVAS COTTAGE

Constantine, Falmouth, TR11 5PY. Lyn Watson & Ray Chun, 01326 340076, ethnevas@outlook.com. *6m SW of Falmouth. Nearest main rds A39, A394. Follow signs for Constantine. At lower village sign, at bottom of winding hill, turn off on private lane. Garden ¾m up hill.* **Sun 23 Apr (1-5). Adm £5, chd free. Home-made teas. Visits also by arrangement Apr to June for groups of up to 20. Couples are welcome.**

Isolated granite cottage in 2 acres. Intimate flower and vegetable garden. Bridge over stream to large pond and primrose path through semi-wild bog area. Hillside with grass paths among native and exotic trees. Many camellias and rhododendrons. Mixed shrubs and herbaceous beds, wildflower glade, spring bulbs. A garden of discovery with hidden delights.

19 NEW FIRSTE PARK

Winsor Lane, Kelly Bray, Callington, PL17 8HD. Mrs Tina Monahan. *From Callington towards Kelly Bray, just before Swingletree pub, turn R into Station Rd, about ¼m Firste Park is on the L top of Winsor Lane. Parking signposted in adjoining field.* **Sun 18 June, Sat 15, Sun 16 July, Sun 20 Aug, Sat 9 Sept (11-4). Adm £5, chd free. Light refreshments.**

1950s house with mature trees, flower gardens established about 5 yrs ago with just over 1 acre incorporating a waterfall, pond and lawned areas. Packed with many plants and shrubs inc over 100 named roses, several varieties of hydrangeas and perfumed plants in abundance. There is an outside kitchen area, pergolas, a fruit and vegetable garden with cut flowers which we also use for dried flowers. Most of the garden is accessible but some gravel pathways.

20 NEW FOX HOLLOW

Station Road, Carnhot, Chacewater, Truro, TR4 8PA. Adele & Gary Peters-Float. *Enter Blackwater, take lane next to Citroen garage signposted Chacewater, Station Rd, 1m on R just before train viaduct. Ignore SatNav once on this lane. Park in lane, few extra spaces opp.* **Sat 17, Sun 18 June (11-4). Adm £5, chd free. Home-made teas.**

Approx 1 acre garden rediscovered over 7 yrs by the owners. Lovingly redesigned so the 4 generations living here can enjoy. Planted with herbaceous beds, traditional cottage flowers, wildflower areas, vegetable garden and orchard. A garden of discovery, meandering paths lead through honeysuckle arches to seating areas with far reaching views. Greenhouses with many plants. Most areas accessible to wheelchairs.

21 GARDEN COTTAGE

Gunwalloe, Helston, TR12 7QB. Dan & Beth Tarling, 01326 241906, beth@gunwalloe.com, www.gunwalloecottages.co.uk. *Just beyond Halzephron Inn at Gunwalloe. Cream cottage with green windows.* **Visits by arrangement Apr to Sept. Adm £5, chd £5. Cream teas.**

Coastal cottage garden. Small garden with traditional cottage flowers, vegetable garden, greenhouse and meadow with far reaching views. Instagram: seaview_gunwalloe. Garden featured in Country Living magazine.

22 GARDENS COTTAGE

Prideaux, St Blazey, PL24 2SS. Sue & Roger Paine, 07786 367610, sue.newton@btinternet.com, en-gb.facebook.com/gardenscottageprideaux. *1m from railway Xing on A390 in St Blazey. Turn into Prideaux Rd opp Gulf petrol station in St Blazey (signed Luxulyan). Proceed ½m. Turn R (signed Luxulyan Valley and Prideaux) & follow signs.* **Wed 7, Thur 8 June, Wed 5, Thur 6 July, Wed 23, Thur 24 Aug (2-5). Adm £5, chd free. Light refreshments. Visits also by arrangement 29 May to 1 Sept for groups of 12 to 30.**

Set in a tranquil location on the edge of Luxulyan valley, work commenced on creating a 1½ acre garden from scratch in winter 2015. Now in its 9th year the garden has matured in a way that's sympathetic to its surrounding landscape, has year-round interest with lots of colour, is productive and simply feels good to be in. Check our Facebook page for special events.

23 GREENWITH MEADOW

Greenwith, Perranwell Station, Truro, TR3 7LS. Kathy Atkins. *Off A39 between Truro & Falmouth. In Perranwell village go up hill signed Greenwith. At Xrds go straight on & garden is 2nd on R. Parking further up on road, well tucked in or in Silverhill Rd on R.* **Sat 13, Sun 14 May (12-6). Adm £5, chd free.**

Smallish garden made in the last 6 yrs by artist owner. Flower borders, roses, gravel garden, small meadow areas, and wildlife pond. Many sitting areas and lots of plants in containers. Artists studio open with paintings to see or buy (20% to NGS).

24 ◆ THE JAPANESE GARDEN

St Mawgan, TR8 4ET. Natalie Hore & Stuart Ellison, 01637 860116, info@japanesegarden.co.uk, www.japanesegarden.co.uk. *6m E of Newquay. St Mawgan village is directly below Newquay Airport. Follow brown & white road signs on A3059 & B3276.* **For NGS: Sun 14, Mon 15 May (10-6). Adm £5, chd £2.50. For other opening times and information, please phone, email or visit garden website.**

Discover an oasis of tranquillity in a Japanese-style Cornish garden, set in approx 1 acre. Spectacular Japanese maples and azaleas, symbolic teahouse, koi pond, bamboo grove, stroll woodland, zen and moss gardens. A place created for contemplation and meditation. Adm free to gift shop, bonsai and plant areas. Featured in BBC2 Big Dreams Small Spaces. Refreshments available in village a short walk from garden. 90% wheelchair accessible, with gravel paths.

25 ◆ KESTLE BARTON

Manaccan, Helston, TR12 6HU. Karen Townsend, 01326 231811, info@kestlebarton.co.uk, www.kestlebarton.co.uk. *10m S of Helston. Leave Helston on A3083 towards Lizard. At the R'about take 1st exit onto B3293 & follow signs towards St Keverne; after Trelowarren turn L and follow the brown signs for appprox 4m.* **For NGS: Every Wed 31 May to 27 Sept (10.30-5). Adm by donation. Modest Tea Room in garden. Cash only honesty box. Tea/ coffee, cakes, ice creams, apple juice from our own orchards. For other opening times and information, please phone, email or visit garden website.**

A delightful garden near Frenchmans Creek, on the Lizard, which is the setting for Kestle Barton Gallery; wildflower meadow, Cornish orchard with named varieties and a formal garden with prairie planting in blocks by James Alexander Sinclair. It is a riot of colour in summer and continues to delight well into late summer. Art Gallery. Good wheelchair access and reasonably accessible WC. Dogs on leads welcome.

26 THE LODGE

Fletchersbridge, Bodmin, PL30 4AN. Mr Tony Ryde. *2m E of Bodmin. From A38 at Glynn Crematorium r'about take rd towards Cardinham & continue down to hamlet of Fletchersbridge. Park at Stable Art on R. Short walk to garden 1st R over river bridge.* **Sat 25, Sun 26 Mar (12.30-5.30). Adm £5, chd free. Home-made teas. Open nearby Pinsla Garden.**

3 acre riverside garden created since 1998, specialising in trees and shrubs chosen for their flowers, foliage and form, and embracing a Gothic lodge remodelled in 2016, once part of the Glynn estate. Water garden with ponds, waterfalls and abstract sculptures. Magnolias, azaleas, rhododendrons, camellias, prunus and spring bulbs star in late March. Wheelchair access to gravelled areas around house and along left side of garden.

27 ◆ THE LOST GARDENS OF HELIGAN

Pentewan, St Austell, PL26 6EN. Heligan Gardens Ltd, 01726 845100, heligan.reception@heligan.com, www.heligan.com. *5m S of St Austell. From St Austell take B3273 signed Mevagissey, follow signs.* **For opening times and information, please phone, email or visit garden website.**

Lose yourself in the mysterious world of The Lost Gardens where an exotic sub-tropical jungle, atmospheric Victorian pleasure grounds, an interactive wildlife project and the finest productive gardens in Britain all await your discovery. Wheelchair access to Northern gardens. Armchair tour shows video of unreachable areas. Wheelchairs available at reception free of charge.

28 LOWER PENBOTHIDNOW

Penbothidno Lane, Constantine, Falmouth, TR11 5AU. Dorothy Livingston, 07785 254975, Dorothy.livingston@me.com. *Signs point downhill on Fore St ⅓ m & top of lane, opp Save the Children Shop at bottom of Fore St - 200 metres at end of lane. Park by church or on street. Disabled parking only at property.* **Sun 14 May, Sun 6 Aug (2-5.30). Adm £6, chd free. Home-made teas. Visits also by arrangement Mar to Oct for groups of up to 20. Access not suitable for coaches.**

Hillside garden of about 2 acres with different areas planted in different styles and plant selections, ranging from formal to wildflower meadow. Surrounded by field and woods. Colour year-round with high point for shrubs and trees in late spring and with herbaceous border, hydrangeas and flowering shrubs in Jul/Aug. Ample outside seating. Plenty to interest a plantsperson. Spring and summer shrubs and flowering trees. Formal garden with clipped hedges and wisteria. Tree fern grove. Southern hemisphere plants. Summer herbaceous border, lawns and sculpture. WC in Studio available. Wheelchair can access drive and main lawns and see over or into most areas of the garden inc the formal garden.

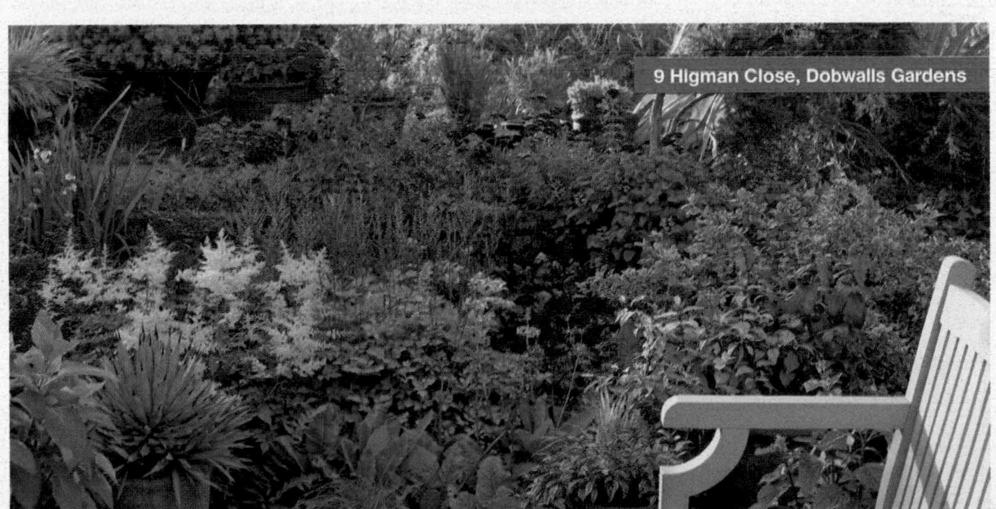

9 Higman Close, Dobwalls Gardens

29 MALIBU

Tristram Cliff, Polzeath, Wadebridge, PL27 6TP. Nick Pickles, 07944 414006, nickdpickles@gmail.com. *Travel via Pityme/Rock/Trebetherick past Oystercatcher pub on L. Take 2nd L signposted Tristram Caravan Park/ Cracking Crab Restaurant. Then L signposted Tristram Cliff, up to end house. Free parking for coast & beach.* **Visits by arrangement for groups of up to 20. Adm £4, chd free. Cream teas, coffee, cakes & light lunches avail by prior agreement.**

Just yards away from the Coast Path with stunning views over the beach and Pentire headland, Malibu consists of a sheltered compact garden to the rear and a general interest front garden. An interesting mix of features, different rockeries and strong focus on succulents, ferns and herbaceous plants. Greenhouse with many plants make this a rounded garden experience. We welcome dogs on leads. Wheelchairs can access the front garden with ease. Rear garden can only be accessed with assistance. Ground floor WC available.

30 NAVAS HILL HOUSE

Bosanath Valley, Mawnan Smith, Falmouth, TR11 5LL. Aline & Richard Turner, 01326 251233, alineturner@btinternet.com. *1½m from Trebah & Glendurgan Gdns. Head for Mawnan Smith, pass Trebah & Glendurgan Gdns then follow yellow signs. Don't follow SatNav which suggests you turn R before Mawnan Smith - congestion alert!* **Sun 7 May (2-5.30). Adm £5, chd free. Cream teas on the veranda/terrace. 'All you can eat' £5.00.**

8½ acre elevated valley garden with paddocks, woodland, kitchen garden and ornamental areas. The ornamental garden consists of 2 plantsman areas with specialist trees and shrubs, walled rose garden, water features and rockery. Young and established wooded areas with bluebells, camellia walks and young large leafed rhododendrons. Seating areas with views across wooded valley. There is plenty of free parking at the property. The Trengilly Singers will entertain us about 3.30pm. Partial wheelchair access, some gravel and grass paths.

31 ◆ PENCARROW

Washaway, Bodmin, PL30 3AG. Molesworth-St Aubyn Family, 01208 841369, info@pencarrow.co.uk, www.pencarrow.co.uk. *4m NW of Bodmin. Signed off A389 & B3266. Free parking.* **For NGS: Mon 17 Apr (10-5). Adm £8, chd free. Light refreshments. For other opening times and information, please phone, email or visit garden website.**

50 acres of tranquil, family-owned Grade II* listed gardens. Superb specimen conifers, azaleas, magnolias and camellias galore. Many varieties of rhododendron give a blaze of spring colour; blue hydrangeas line the mile-long carriage drive throughout the summer. Discover the Iron Age hill fort, lake, Italian gardens and granite rock garden. Dogs welcome, café and children's play area. Gravel paths, some steep slopes.

32 PENDOWER HOUSE

Lanteglos-by-Fowey, PL23 1NJ. Mr Roger Lamb, 01726 870884, rl@rogerlamb.com. *Near Polruan off B3359 towards Bodinnick. 2m from Fowey using the Bodinnick ferry. Please ask for directions when arranging to visit the garden. Do not use SatNav.* **Visits by arrangement 14 June to 14 July for groups of up to 30. Adm £5, chd free. Home-made teas. Other refreshments available by prior arrangement.**

Set in the heart of Daphne du Maurier country in its own valley this established garden, surrounding a Georgian rectory, is now undergoing a revival having been wild and neglected for some years. It has formal herbaceous terraces, a cottage garden, orchard, ponds, streams and a C19 shrub garden with a fine collection of azaleas, camellias and rhododendrons plus rare mature specimen trees.

33 NEW PENNYCOMEQUICK

Zelah, Truro, TR4 9JD. Andrew & Pat Harvey. *8m N of Truro. From Truro head N on A39 to the r'about at Carland Cross. Join the A30 and head W for 1½m then turn L signed St Allen. The 1st entrance on R is Pennycomequick drive.* **Sat 15, Sun 16 July (2-5). Adm £5, chd free. Home-made teas.**

A 2 acre garden, 40 yrs in the making, transformed from bleak windswept fields to a haven of trees, shrubs, perennials, wildflowers and an area growing flowers for cutting and drying. Stroll beside the colourful riot of the new herbaceous border to the calm tranquillity of the Hydrangea walk.

34 PENWARNE

Mawnan Smith, Falmouth, TR11 5PH. Mrs R Sawyer, penwarnegarden@gmail.com. *1m outside Mawnan Smith towards Mabe Burnthouse & Falmouth.* **Visits by arrangement Feb to May Adm £10.00 with Guided Tour, £7.50 without tour, chd free.**

Originally planted in the late C19, this 12 acre garden inc extensive plantings of camellias, rhododendrons and azaleas. Special features inc large magnolias and a number of fine mature trees inc copper beech, handkerchief tree and Himalayan cedar. The walled garden, believed to be the site of a medieval chapel, houses herbaceous planting, climbing roses and fruit trees. Historic house and gardens with many mature specimens.

35 PINSLA GARDEN

Glynn, nr Cardinham, Bodmin, PL30 4AY. Mark & Claire Woodbine, 07483 244318, claire.woodbine@ngs.org.uk, www.pinslagarden.wordpress.com. *3½m E of Bodmin. From A30 or Bodmin take A38 towards Plymouth, 1st L at r'about, past Crematorium & Cardinham woods turning. Go towards Cardinham village, up steep hill, 2m on R.* **Sat 25, Sun 26 Mar (12.30-5.30), open nearby The Lodge. Wed 19 Apr (9-5), open nearby Trevina House. Tue 9, Tue 16, Tue 23 May, Tue 13, Wed 14, Tue 20 June (9-5). Wed 21 June (9-5), open nearby Trevina House. Adm £5, chd free. Opening with Cardinham Gardens on Sun 30 Apr, Sun 9 July. Visits also by arrangement 20 Apr to 16 July for groups of up to 20.**

Surround yourself with deep nature. Pinsla is a tranquil cottage garden buzzing with insects enjoying the sheltered sunny edge of a wild wood. Lose yourself in a colourful tapestry of naturalistic planting. There are lots of unusual planting combinations, cloud pruning, intricate paths, garden art and a stone circle. Sorry, no teas but you are welcome to bring your own thermos. Partial wheelchair access as some paths are narrow and bumpy.

36 PORT NAVAS CHAPEL

Port Navas, Constantine, Falmouth, TR11 5RQ. Keith Wilkins & Linda World, 01326 341206, keithwilkins47@gmail.com. *2m E of Constantine, nr Falmouth. From the direction of Constantine, we are the 2nd driveway on R after the 'Port Navas' sign.* **Visits by arrangement for groups of up to 20. Adm £5, chd free.**

Japanese style garden set in 3/4 acre of woodland, next to the old Methodist Chapel, with ornamental ponds and waterfalls, rock formations, Tsukubai water feature, granite lanterns, woodland stream, geodesic dome and Zen garden. Partial wheelchair access to driveway and sloping grass areas.

37 ◆ POTAGER GARDEN

High Cross, Constantine, Falmouth, TR11 5RF. Mr Mark Harris, 01326 341258, enquiries@potagergarden.org, www.potagergarden.org. *5m SW of Falmouth. From Falmouth, follow signs to Constantine. From Helston, drive through Constantine and continue towards Falmouth.* **For opening times and information, please phone, email or visit garden website.**

Potager has emerged from the bramble choked wilderness of an abandoned plant nursery. With mature trees which were once nursery stock and lush herbaceous planting interspersed with fruit and vegetables Potager Garden aims to demonstrate the beauty of productive organic gardening. There are games to play, hammocks to laze in and boules and badminton to enjoy. Potager café serving vegetarian food all day.

38 ◆ PRIDEAUX PLACE

Padstow, PL28 8RP. Mr & Mrs Peter Prideaux-Brune, 01841 532411, office@prideauxplace.co.uk, www.prideauxplace.co.uk. *1/2m W of Padstow centre. Off A389 Wadebridge to Padstow Rd. Follow brown historic house signs.* **For NGS: Sun 26 Mar (10.30-5). Adm £8, chd free. Fully licensed cafe offering light lunches, cream teas etc. For other opening times and information, please phone, email or visit garden website.**

A garden of vistas, overlooking its ancient deer park and Camel Estuary to Bodmin Moor and beyond. The gardens suffered badly after the Second World War, but today's exciting and extensive restoration programme is well underway, bringing the gardens back to their former glory.

39 RIVERSIDE COTTAGE

St. Clement, Truro, TR1 1SZ. Billa & Nick Jeans. *1½m SE of Truro. From Trafalgar r'about on A39 in Truro, follow signs for St Clement, up St Clement Hill. R at top of hill, continue to car park by river; Riverside Cottage is first cottage on L.* **Mon 10 Apr (10.30-4.30). Every Tue 11 Apr to 16 May (10.30-4.30). Adm by donation. Tea, coffee & cakes.**

Small wildlife friendly cottage garden overlooking beautiful Tresillian River. Wildflower areas, soft fruit and raised bed vegetable patch. Old apple trees, sloping grassy paths and steps and plenty of seats affording places to rest to enjoy views down river. Sadly not suitable for wheelchairs.

40 NEW ROSE MORRAN

Talskiddy, St Columb, TR9 6EB. Peter & Jenny Brandreth, 07711 517159, jennybrandreth@yahoo.com. *2 m N of St Columb Major. From A39 take Talskiddy/St Eval turn. Then take 2nd R signed Talskiddy & follow yellow NGS signs.* **Sat 22, Sun 23 Apr, Sat 2, Sun 3 Sept (11-5). Adm £5, chd free. Light refreshments. Visits also by arrangement 8 Apr to 29 Oct for groups of up to 20.**

Colour, shape and texture define this North Cornwall garden. The property and gardens extend to about an acre, with grassland, wild areas and Cornish hedges completing the 2 acre plot. Planted over the past 10 yrs there is a mix of shrubs and trees interplanted with perennials to ensure year-round display. Vegetable beds and a polytunnel with tender plants make this a rounded garden visit. There are level areas to view and the whole garden is accessible via grass slopes but some are quite steep. There is one step to the inside WC.

41 ◆ ROSELAND HOUSE

Chacewater, TR4 8QB. Mr & Mrs Pridham, 01872 560451, charlie@roselandhouse.co.uk, www.roselandhouse.co.uk. *4m W of Truro. Park in village car park (100yds) or on surrounding rds.* **For NGS: Sat 1, Sun 2 July (1-5). Adm £5, chd free. Home-made teas. For other opening times and information, please phone, email or visit garden website.**

The 1 acre garden is a mass of summer colour when the National Collection of *Clematis viticella* is at its peak in July, other climbing plants abound lending both foliage and scent. The Victorian conservatory and greenhouses are also full of unusual and interesting plants. The garden features 2 ponds. Some slopes.

NPC

42 NEW ROSEVALLON BARN

Tregony, Truro, TR2 5TS. Jenny Ralph, 07971 436344, rosevallon@icloud.com. *From main road in Tregony turn into Cuby Close on opp side of the road from St Cuby. Follow lane for just over 1m, pass Pencoose, then turn R down lane onto farm track.* **Sun 30 Apr (1.30-5). Adm £5, chd free. Home-made teas. Visits also by arrangement 30 Apr to 7 May.**

Rural setting surrounded by farmland. Converted barn with 3 acres of garden planted by present owners. Small stream and pond fed from spring water, herbaceous borders, shrubs, coppice with snowdrops, bluebells, primroses, fritillary. Vegetables and fruit, wildflower meadow, gravel garden, year-round interest. Bricked drive and accessible grass areas.

43 ◆ ST MICHAEL'S MOUNT

Marazion, TR17 0HS. James & Mary St Levan, www.stmichaelsmount.co.uk. *2½m E of Penzance. ½m from shore at Marazion by Causeway.* **For opening times and information, please visit garden website, or www.ngs.org.uk for details of pop up openings.**

Infuse your senses with colour and scent in the unique sub-tropical gardens basking in the mild climate and salty breeze. Clinging to granite slopes the terraced beds tier steeply to the ocean's edge, boasting tender exotics from places such as Mexico, the Canary Islands and South Africa. Laundry lawn, mackerel bank, pill box, gun emplacement, tiered terraces, well, tortoise lawn. Walled gardens, seagull seat.

44 SOUTH BOSENT

Liskeard, PL14 4LX. Adrienne Lloyd & Trish Wilson, 07539 256855, lloydadj@btinternet.com, www.southbosentgardens.co.uk. *2½ m W of Liskeard. From r'about at junction of A390 & A38 take turning to Dobwalls. At mini-r'about R to Duloe, after 1¼ m at Xrds turn R. Garden on L after ¼ m.* **Thur 18 May, Wed 13, Thur 14 Sept (2-5). Adm £5, chd free. Home-made teas. Opening with Dobwalls Gardens on Fri 16, Sat 17, Fri 30 June, Sat 1, Sat 22, Sun 23 July, Sat 12, Sun 13 Aug. Visits also by arrangement 18 May to 17 Sept.**

This garden is an example of work in progress currently being developed from farmland. The aim is to create a combination of interesting plants coupled with habitat for wildlife over a total of 9½ acres. There are several garden areas, woodland gardens, a meadow, ponds of varying sizes, inc a rill and waterfall. In spring, the bluebell wood trail runs alongside the stream. Main garden area accessible. No wheelchair access to bluebell wood due to steps.

45 SOUTH LEA

Pillaton, Saltash, PL12 6QS. Viv & Tony Laurillard, 01579 350629, tonyandviv.laurillard@gmail.com. *Opp Weary Friar pub. 4m S of Callington. Signed from r'abouts on A388 at St Mellion & Hatt, & on A38 at Landrake. Roadside parking. Please do not park in pub car park.* **Sun 7 May, Sun 11 June (1-5). Adm £6, chd free. Home-made teas. Visits also by arrangement May & June for groups of 10 to 30.**

In the front a path winds through interesting landscaping with a small pond and unusual planting. Tropical beds by front door with palms, cannas, etc. The back garden, with views over the valley, is a pretty picture in May with spring bulbs and clematis, with attractive mixed borders in June. Lawns are separated by a fair sized fish pond and the small woodland area is enchanting in spring. Plenty of seating. Due to steps, wheelchair access only to front dry garden and rear terrace, from which much of garden can be viewed.

46 SUFFREE FARM

Probus, Truro, TR2 4HL. Chris & Wendy Harvey Clark. *1m S of centre of Probus. Take A3078 (Tregony/St Mawes) turning off the A390 Probus bypass. After 600 yds layby on R of road. Through layby down gravel track. Parking in field round bend.* **Sun 11 June (12-5).**

Pennycomequick

Adm £5, chd free. Home-made teas.
Cornish farmhouse surrounded by 2 acre garden overlooking beautiful wooded valley. Borders, Mediterranean garden, wildflower areas, woodland plantation, Shetland ponies and pigs. Beautiful views across a wooded valley. A gentle downhill path leads to the gate. Thereafter the garden is easy to negotiate with a wheelchair.

47 TOWAN HOUSE

Old Church Road, Mawnan Smith, Falmouth, TR11 5HX. Mr Dave & Mrs Tessa Thomson, 01326 250378, tandd.thomson@btinternet.com. *On entering Mawnan Smith, fork L at the Red Lion pub. When you see the gate to Nansidwell, turn R into Old Church Rd. Follow NGS signs.* **Sun 10 Sept (2-5). Adm £6, chd free. Home-made teas. Visits also by arrangement 11 June to 31 Aug for groups of 5 to 15.**
Towan House is a coastal garden with some unusual plants and views to St Mawes, St Anthony Head lighthouse and beyond with easy access to the SW coast path overlooking Falmouth Bay. It is approximately ¼ acre and divided into 2 gardens, one exposed to the north and east winds, the other sheltered allowing tropical plants to flourish such as 'Hedychium', Tetrapanax and cannas.

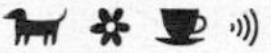

48 TREBARTHA ESTATE GARDEN AND COUNTRY GARDEN AT LEMARNE

Trebartha, nr Launceston, PL15 7PD. The Latham Family. *6m SW of Launceston. North Hill, SW of Launceston nr junction of B3254 & B3257. No coaches.* **Sun 7 May, Sun 1 Oct (2-5). Adm £8, chd free. Home-made teas.**
Historic landscape gardens featuring streams, cascades, rocks and woodlands, inc fine trees, bluebells, ornamental walled garden and country garden at Lemarne. Ongoing development of C19 American garden and C18 fish ponds. Allow at least 1 hour for a circular walk. Some steep and rough paths, which can be slippery when wet. Stout footwear advised. October opening for autumn colour.

49 NEW TRELAN

Wharf Road, Lelant, St Ives, TR26 3DU. Nick Williams, 07817 425157, nick@nickgwilliams.com. *Close to St Uny Church, Lelant. Trelan Gardens are off Wharf Rd which leads to a car park on Dynamite Quay. Car park signed at the end of Wharf Rd and the gardens are 200m down on R.* **Sat 24 June, Sat 22 July (2-6). Adm £6, chd free. Light refreshments. Visits also by arrangement 24 June to 22 July for groups of up to 8.(Fridays only).**
Trelan has a distinctly tropical feel. There is a swimming pond surrounded by lush vegetation, an Italianate sunken garden, a fern garden, numerous young trees and countless *Echium pininana*. It is a feast for the eyes.

50 NEW TREMICHELE

Housel Bay, The Lizard, Helston, TR12 7PG. Mike & Helen Painton, 01326 291370. *On the Lizard peninsular approx 10m from Helston. Follow directions to Lizard village. Park on the village green, parking is by voluntary contribution. Walk along Beacon Terrace & follow signs for Housel Bay Hotel.* **Visits by arrangement May to July for groups of 6 to 15. Adm £5, chd free. Tea, coffee, cold drinks, home-made cake & cookies.**
A newly created (2016) coastal garden of approx 1 acre overlooking the Atlantic and the Lizard Lighthouse. It is a steep sloping garden divided into 3 terraces accessed by slopes or steps. A productive kitchen garden, with raised beds and fruit trees. Greenhouse and cold frames. A wildflower meadow with access to the coast path. Banks have a mix of planting inc hardy exotics.

51 TRENARTH

High Cross, Constantine, Falmouth, TR11 5JN. Lucie Nottingham, 01326 340444, lmnottingham@btinternet.com, www.trenarthgardens.com. *6m SW of Falmouth. Main rd A39/A394 Truro to Helston, follow Constantine signs. High X garage turn L for Mawnan, 30yds on R down dead end lane, Trenarth is ½m at end of lane.* **Visits by arrangement Feb to Nov for groups of up to 30. Garden tours & light refreshments available on request. Adm £5, chd free.**
4 acres round C17 farmhouse in peaceful pastoral setting. Year-round and varied. Emphasis on tender, unusual plants, structure and form. C16 courtyard, listed garden walls, holm oak avenue, yew rooms, vegetable garden, traditional potting shed, orchard, palm and gravel area with close planting inc agapanthus, agave and dierama, woodland area with children's interest. Circular walk down ancient green lane via animal pond to Trenarth Bridge, returning through woods. Abundant wildlife. Bees in tree bole, lesser horseshoe bat colony, swallows, wildflowers and butterflies. Family friendly, children's play area, the Wolery, and plenty of room to run, jump and climb.

52 TRETHEW

Lanlivery, nr Bodmin, PL30 5BZ. Ginnie & Giles Clotworthy, 01208 872612, ginnieclotworthy@hotmail.co.uk. *3 m W of Lostwithiel. On rd between Lanlivery and Luxulyan. 1m from Lanlivery. Ask for directions when booking. Do not use SatNav.* **Visits by arrangement 15 May to 30 Sept for groups of up to 20. (excluding July & Aug) Individuals also welcome. Adm £5, chd free. Teas available on request.**
Series of profusely planted and colourful areas surrounding an ancient Cornish farmhouse. Features inc terracing with wisteria and rose covered pergola, gazebo and herbaceous borders within yew and beech hedges, all overlooking orchard with old fashioned roses and pond beyond. Late summer colour with salvias, dahlias and Michaelmas daisies. Magnificent views. This is a garden on a hill with flat areas around the house. Uneven drive with gravel.

9,500 patients and their families are being supported across four Horatio's gardens thanks to our annual donations

53 NEW TREVESCO
7 Commons Close, Mullion, Helston, TR12 7HY. Colin Read. little.trout@yahoo.com. *Situated on the W side of Mullion on the road to Poldhu Cove & Cury. ¼ m from village centre car parks.* **Sat 3, Sun 4 June (11-4). Adm £5, chd free. Home-made teas.**
This ¼ acre garden has been established over 14 yrs in a sub-tropical style with rarer unusual plants inc puya, furcraea, tetrapanax, pseudopanax and several larger feature plants e.g. olive trees, *Phoenix canariensis* and *Ceanothus arboreus*. Many tender plants have been established without winter protection very successfully due to local climate. There is also a pond.

54 TREVILLEY
Sennen, Penzance, TR19 7AH. Patrick Gale & Aidan Hicks, 01736 871808, trevilley@btinternet.com. *For walkers, Trevilley lies on the footpath from Trevescan to Polgigga & Nanjizal. On the Open Day follow the signs into our parking field. If visiting by arrangement, Trevilley is up an unmade track outside Trevescan. If you get to a white house you need to reverse a little!* **Sat 8 July (1-5). Adm £5, chd free. Visits also by arrangement 15 June to 15 July for groups of 6 to 30. By arrangement visitors are welcome to bring a picnic lunch or tea.**
Eccentric, romantic and constantly evolving garden, as befits the intense creativity of its owners, carved out of an expanse of concrete farmyard over 20 yrs. Inc elaborate network of decorative cobbling, pools, container garden, vegetable garden, shade garden, the largely subtropical mowhay garden and both owners' studios but arguably its glory is the westernmost walled rose garden in England. Dogs are welcome on leads. The garden visit can form the climax of enjoying the circular coast path walk from Land's End to Nanjizal and back across the fields. Cream teas are available in the Apple Tree Cafe in Trevescan. Visitors may leave their cars with us while they walk there across the fields. The rose garden is accessible by wheelchair though much of its surface is uneven due to decorative cobbling. Beyond that lie narrow paths and steps.

55 NEW TREVINA HOUSE
St. Neot, Liskeard, PL14 6NR. Kevin & Sue Wright, www.facebook.com/TrevinaInCornwall. *1m N of St Neot. From A38 turn off at Halfwayhouse pub. Head up hill to Goonzion Downs. Bear L & keep straight on towards Colliford Lake. From A30 take Colliford Lake/ St Neot turning. Head towards St Neot.* **Wed 19 Apr (10-5), open nearby Pinsla Garden. Thur 20 Apr, Wed 10, Thur 11 May (10-5). Wed 21 June (10-5), open nearby Pinsla Garden. Thur 22 June, Wed 13, Thur 14 Sept, Thur 5, Fri 6 Oct (10-5). Adm £5. Light refreshments.**
The gardens at Trevina are both old and new. From a Victorian cottage garden with original cobbled paths to rewilded woodland. A trout pond occupies the site of a medieval fish pond and the site of a Cornish Round. There is a traditional kitchen garden and organic orchard. The gardens have colour and interest year-round, from the first snowdrops, bluebells, apple blossom and ripe apples in October.

56 ◆ TREVINCE ESTATE GARDENS
Gwennap, Redruth, TR16 6BA. Richard & Trish Stone, 01209 822725, info@trevince.co.uk, www.trevince.co.uk. *From A30 turn off at Scorrier A3047, then take B3298 past St Day & through Carharrack. After 1m turn L into Gwennap.* **For NGS: Sun 7 May (10-4). Adm £7, chd free. Home-made teas. For other opening times and information, please phone, email or visit garden website.**
A haven for garden lovers and curious souls. An old family estate garden, reimagined for modern times. A no-dig productive walled garden is at the heart of the space. A woodland walk with a collection of large leaved rhododendrons among old oaks. Inc several 30 yr old conifers planted as a a living conservation seed bank. Wheelchairs can access the walled garden and Pond Garden. All terrain wheelchairs can also access the Shrubbery and Oriental Glade.

57 ◆ TREWIDDEN GARDEN
Buryas Bridge, Penzance, TR20 8TT. Mr Alverne Bolitho - Richard Morton, Head Gardener, 01736 364275/363021, contact@trewiddengarden.co.uk, www.trewiddengarden.co.uk. *2m W of Penzance. Entry on A30 just before Buryas Bridge. Postcode for SatNav is TR19 6AU.* **For NGS: Sat 1 Apr (10.30-5.30). Adm £8.50, chd free. For other opening times and information, please phone, email or visit garden website.**
Historic Victorian garden with magnolias, camellias and magnificent tree ferns planted within ancient tin workings. Tender, rare and unusual exotic plantings create a riot of colour thoughout the season. Water features, specimen trees and artefacts from Cornwall's tin industry provide a wide range of interest for all. Last entry is 4:30pm.

NPC

58 WINDMILLS
South Hill, Callington, PL17 7LP. Sue & Peter Tunnicliffe, 01579 363981, tunnicliffesue@gmail.com. *3m NW of Callington. Head N from Callington A388, after about ½ m turn L onto South Hill Rd (signed South Hill). Straight on for 3m, gardens on R just before church.* **Sun 23 Apr, Sun 13 Aug (1-5). Combined adm with Anvil Cottage £6, chd free. Home-made teas. Gluten free cakes available. Visits also by arrangement May to Aug for groups of 10 to 30. Combined with Anvil Cottage next door.**
Next to medieval church and on the site of an old rectory there are still signs in places of that long gone building. A garden full of surprises, formal paths and steps lead up from the flower beds to extensive vegetable and soft fruit area. More paths lead to a pond, past a pergola, and down into large lawns with trees and shrubs and chickens.

Our annual donations to Parkinson's UK meant 7,000 patients are currently benefiting from support of a Parkinson's Nurse

Chygurno

CUMBRIA

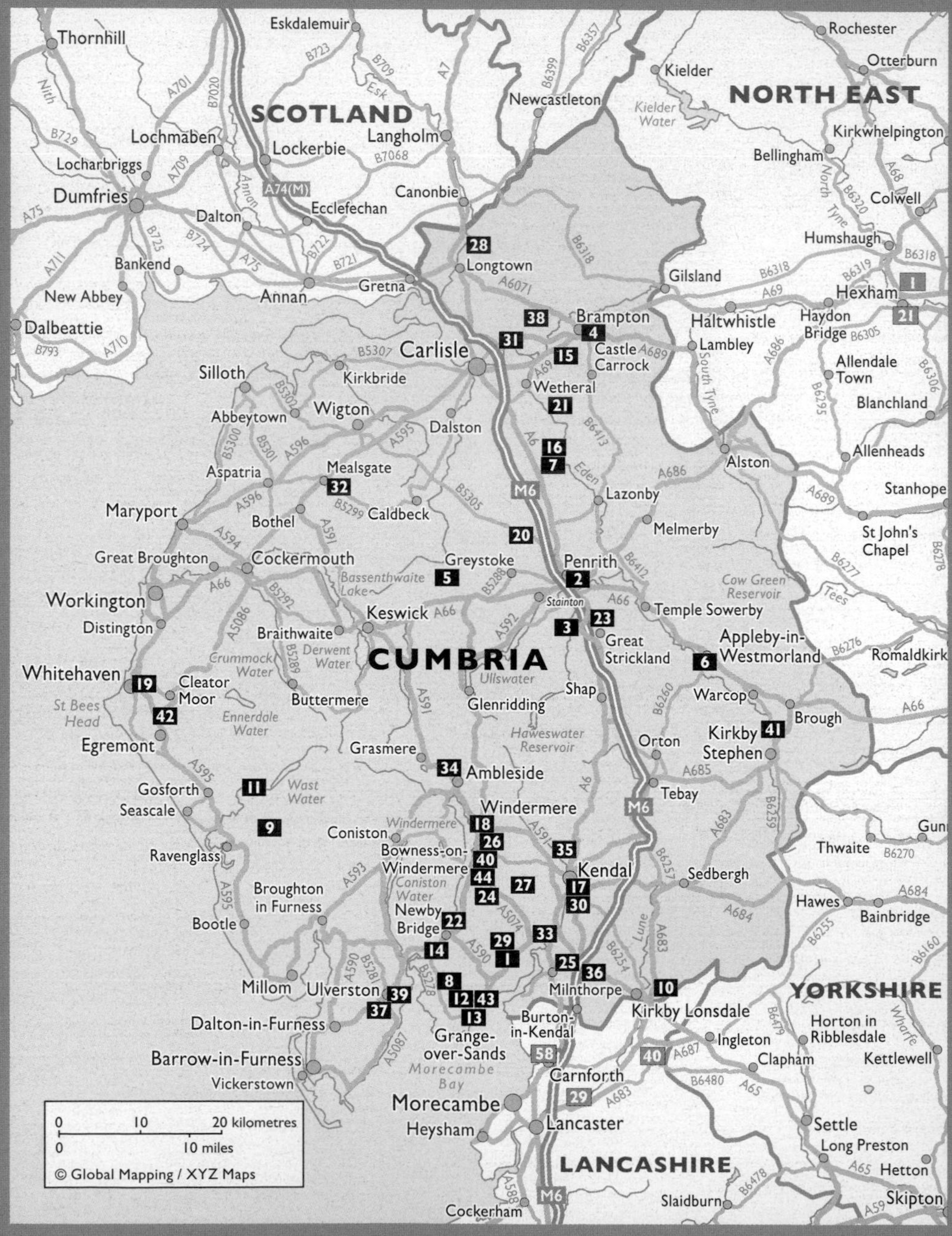
SCOTLAND
NORTH EAST
CUMBRIA
YORKSHIRE
LANCASHIRE
Thornhill
Eskdalemuir
Rochester
Otterburn
Kielder
Kielder Water
Newcastleton
Lochmaben
Lockerbie
Langholm
Locharbriggs
Dumfries
Dalton
Ecclefechan
Canonbie
Kirkwhelpington
Bellingham
Colwell
Humshaugh
Bankend
New Abbey
Dalbeattie
Annan
Gretna
Longtown
Gilsland
Hexham
Haltwhistle
Haydon Bridge
Lambley
Brampton
Castle Carrock
Carlisle
Kirkbride
Silloth
Wetheral
Allendale Town
Blanchland
Abbeytown
Wigton
Dalston
Allenheads
Alston
Aspatria
Mealsgate
Lazonby
Stanhope
Maryport
Bothel
Caldbeck
Melmerby
St John's Chapel
Great Broughton
Cockermouth
Greystoke
Penrith
Bassenthwaite Lake
Stainton
Workington
Keswick
Temple Sowerby
Cow Green Reservoir
Distington
Braithwaite
Great Strickland
Appleby-in-Westmorland
Romaldkirk
Whitehaven
Cleator Moor
Crummock Water
Derwent Water
Ullswater
St Bees Head
Buttermere
Glenridding
Shap
Warcop
Brough
Ennerdale Water
Haweswater Reservoir
Egremont
Grasmere
Orton
Kirkby Stephen
Ambleside
Gosforth
Wast Water
Tebay
Seascale
Windermere
Coniston
Ravenglass
Bowness-on-Windermere
Kendal
Thwaite
Sedbergh
Broughton in Furness
Coniston Water
Hawes
Bainbridge
Newby Bridge
Bootle
Millom
Ulverston
Milnthorpe
Kirkby Lonsdale
Dalton-in-Furness
Burton-in-Kendal
Horton in Ribblesdale
Grange-over-Sands
Ingleton
Barrow-in-Furness
Clapham
Kettlewell
Vickerstown
Morecambe Bay
Carnforth
Morecambe
Settle
Heysham
Lancaster
Long Preston
Hetton
Slaidburn
Skipton
Cockerham
Nith
Esk
Annan
North Tyne
South Tyne
Eden
Tees
Lune
Wharfe
A74(M)
M6
A66
A6
A591
A595
A590
A683
A684
A686
A689
A69
0 10 20 kilometres
0 10 miles
© Global Mapping / XYZ Maps

CUMBRIA is where Wordsworth `wandered lonely as a cloud' before finding his `host of golden daffodils'. A county where part has been a UNESCO World Heritage Site since 2017.

It offers mountains, fells, lakes, tarns, waterfalls and a rugged coastline softened by estuaries and rolling open country spaces through which rivers, streams and becks wander and hurry on their way to the sea.

Amongst all of this natural beauty are our gardens, offering havens of tranquillity, panoramic views of distant lakes and mountains - family gardens flourishing alongside stately homes set in their vast estates.

It is a county of contrasts in scale and interest, where a visit to some of our gardens will offer colour, relaxation, evidence of love and care and perhaps an idea or two to take back home. Perhaps even a plant or two to create a corner of our county in your own patch.

Below: Esk Bank

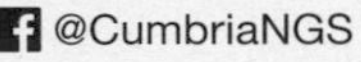 @CumbriaNGS

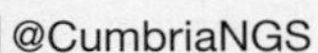 @CumbriaNGS

Volunteers

County Organisers

North
Alannah Rylands
01697 320413
alannah.rylands@ngs.org.uk

South
Linda Greening
01524 781624
greening@ngs.org.uk

County Treasurer
Cate Bowman
01228 573903
cate.bowman@ngs.org.uk

Publicity – Publications & Special Interest
Carole Berryman
01539 443649
carole.berryman@outlook.com

Publicity – Social Media
Gráinne Jakobson
01946 813017
grainne.jakobson@ngs.org.uk

Booklet Co-ordinator
Cate Bowman
01228 573903
cate.bowman@ngs.org.uk

Assistant County Organisers
Carole Berryman (as above)

Bruno Gouillon
01539 532317
brunog45@hotmail.com

Sarah Byrne
015395 34405
sarah.byrne@ngs.org.uk

Gráinne Jakobson (as above)

Christine Davidson
07966 524302
christine.davidson@ngs.org.uk

Belinda Quigley
07738 005388
belinda.quigley@ngs.org.uk

OPENING DATES

All entries subject to change. For latest information check **www.ngs.org.uk**

Extended openings are shown at the beginning of the month.

Map locator numbers are shown to the right of each garden name.

February

Snowdrop Festival

Every Saturday from Saturday 11th
Summerdale House 36

Sunday 19th
Summerdale House 36

March

Every Saturday
Summerdale House 36

Sunday 19th
◆ Holehird Gardens 18
◆ Rydal Hall 34

April

Every Saturday
Summerdale House 36

Saturday 8th
Hazel Cottage 16

Sunday 9th
Hazel Cottage 16

Sunday 16th
◆ Rydal Hall 34

Sunday 23rd
Low Fell West 24
Summerdale House 36

May

Every Saturday
Summerdale House 36

Sunday 7th
Windy Hall 40

Sunday 14th
◆ Netherby Hall 28

Sunday 21st
Crumble Cottages 8
Matson Ground 26

Saturday 27th
Coombe Eden 7
Galesyke 11
Hazel Cottage 16

Sunday 28th
Coombe Eden 7
Galesyke 11
Grange over Sands Hidden Gardens 13
Hazel Cottage 16

June

Friday 2nd
◆ Swarthmoor Hall 37

Saturday 3rd
◆ Swarthmoor Hall 37

Sunday 4th
Abi and Tom's Garden Plants 1
NEW Brampton East Gardens 4
The Old Vicarage 29
Quarry Hill House 32
Windy Hall 40
Yewbarrow House 43

Saturday 10th
Sprint Mill 35

Sunday 11th
◆ Hutton-In-The-Forest 20
8 Oxenholme Road 30
NEW 5 Primrose Bank 31
Summerdale House 36

Saturday 17th
Coombe Eden 7
Hazel Cottage 16
Middle Blakebank 27

Sunday 18th
Askham Hall 3
Coombe Eden 7
Hazel Cottage 16
Sprint Mill 35
Tithe Barn 38
Yews 44

Saturday 24th
NEW Eskbank 9

Sunday 25th
NEW Eskbank 9
Grange over Sands Hidden Gardens 13
Ivy House 21

July

Saturday 1st
Woodend House 42

Sunday 2nd
Ulverston Town & Country Gardens 39
Woodend House 42
Yewbarrow House 43

Thursday 6th
◆ Holehird Gardens 18
Larch Cottage Nurseries 23

Sunday 9th
Haverthwaite Lodge 14
Hayton Village Gardens 15
Lakeside Hotel 22
Winton Park 41

Saturday 22nd
NEW Heron Hill Primary School 17

Sunday 23rd
Crumble Cottages 8
◆ Rydal Hall 34

Sunday 30th
Grange Fell Allotments 12

August

Saturday 5th
NEW Holly House 19

Sunday 6th
NEW Holly House 19
Yewbarrow House 43

Thursday 10th
Larch Cottage Nurseries 23

Saturday 12th
Galesyke 11

Sunday 13th
Galesyke 11
Ivy House 21

Sunday 20th
Fell Yeat 10
◆ Rydal Hall 34

September

Saturday 2nd
Middle Blakebank 27

Sunday 3rd
Yewbarrow House 43

Thursday 7th
Larch Cottage Nurseries 23

October

Sunday 15th
Low Fell West 24

By Arrangement

Arrange a personalised garden visit with your club, or group of friends, on a date to suit you. See individual garden entries for full details.

NEW Aimshaugh 2
Chapelside 5
Church View 6
Coombe Eden 7
Crumble Cottages 8
Galesyke 11
Grange Fell Allotments 12
Grange over Sands Hidden Gardens 13
Ivy House 21
Low Fell West 24
Lower Rowell Farm & Cottage 25
Matson Ground 26
Middle Blakebank 27
8 Oxenholme Road 30
NEW 5 Primrose Bank 31
Rose Croft 33
Sprint Mill 35
Stonechats, Hayton Village Gardens 15
Summerdale House 36
Woodend House 42
Yewbarrow House 43

In 2022 the National Garden Scheme donated £3.11 million to our nursing and health beneficiaries

THE GARDENS

1 ABI AND TOM'S GARDEN PLANTS

Halecat, Witherslack, Grange-Over-Sands, LA11 6RT. Abi & Tom Attwood, www.abiandtom.co.uk. *20 mins from Kendal. From A590 turn N to Witherslack. Follow brown tourist signs to Halecat. Rail Grange-over-sands 5m. Bus X6 2m. NCR 70.* **Sun 4 June (10-5). Combined adm with The Old Vicarage £6, chd free. Home-made teas.**

The 1 acre nursery garden is a fusion of traditional horticultural values with modern approaches to the display, growing and use of plant material. Our full range of perennials can be seen growing alongside one another in themed borders be they shady damp corners or south facing hot spots. The propagating areas, stock beds and family garden, normally closed to visitors, will be open on the NGS day. More than 1,000 different herbaceous perennials are grown in the nursery, many that are excellent for wildlife. For other opening times and information please visit our website. Sloping site that has no steps but steep inclines in places.

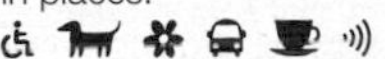

2 NEW AIMSHAUGH

Barco Avenue, Penrith, CA11 8LZ. Mrs Anne Frazer, 01768 606138, annefrazer@outlook.com. *Turn into town at the A66/A6 r'about. Through 2 sets of lights, then turn R onto Old London Rd. At the top of the hill turn R onto Folly Lane. L onto Barco Ave, R side.* **Visits by arrangement June & July for groups of up to 20. Adm £4, chd free. Light refreshments. Home-made cakes. Refreshments to be ordered on booking.**

A plantsman's garden still under development on this ⅔ acre plot with more than 1100 varieties of plants so far. Almost all plants are hardy for easy maintenance and planting is of restricted height to preserve the panoramic views of the Lake District. Materials that otherwise would go to landfill are reused in some features. Various methods used to reduce water usage on this sandy soil.

3 ASKHAM HALL

Askham, Penrith, CA10 2PF. Charles Lowther, 01931 712350, enquiries@askhamhall.co.uk, www.askhamhall.co.uk. *5m S of Penrith. Turn off A6 for Lowther & Askham.* **Sun 18 June (11.30-7). Adm £6, chd free. Light refreshments. Donation to Askham & Lowther Churches.**

Askham Hall is a Pele Tower incorporating C14, C16 and early C18 elements in a courtyard plan. Opened in 2013 with luxury accommodation, a restaurant, outdoor heated pool and wedding barn. Splendid formal garden with terraces of herbaceous borders and topiary, dating back to C17. Meadow area with trees and pond, kitchen gardens and animal trails. Café serving wood-fired pizzas, cake, hot & cold drinks with indoor & outdoor seating.

GROUP OPENING

4 NEW BRAMPTON EAST GARDENS

Brampton, CA8 1EX. *Gardens located at the eastern side of Brampton, just off the A69 main rd. Nr the Sands, Brampton.* **Sun 4 June (1-5). Combined adm £5, chd free. Home-made teas.**

NEW **HUFFNPUFF**
CA8 1EX. Bridget Barling.

NEW **IVY COTTAGE**
CA8 1UB. Mrs Sally Nelson.

NEW **LOVER'S LANE COMMUNITY GARDEN**
CA8 1TN. Julia Turnton.

NEW **31 MOAT STREET**
CA8 1UJ. Kathy Arther.

NEW **RANDELSON HOUSE**
CA8 1UJ. Heather Tipler.

A selection of gardens inc a community garden producing organic produce, a back-lane garden, a small courtyard garden, a medium-sized diverse garden and a new-build garden on a steep slope called HuffnPuff. The house was built of straw with a wooden frame in 2019, using many eco ideas. The garden is south-facing on a steep slope and is terraced and organically grown, with seating and summerhouses. Ivy Cottage is a medium sized garden with several different areas on a sloping site. Mecanopsis is grown here. Lovers Lane Community garden has around 40 members and aims to be inclusive and organic, experimenting with new ways of gardening in trial beds. They encourage flowering 'weeds' and other flowers to attract pollinators, producing fruit, vegetables, herbs and their own compost. 31 Moat St is a small garden which is situated behind old, sandstone and terraced houses. Art will also be on display. Randelson House has a small courtyard garden, with shady and sunny areas.

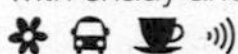

5 CHAPELSIDE

Mungrisdale, Penrith, CA11 0XR. Tricia & Robin Acland, 01768 779672, rtacland@gmail.com. *12m W of Penrith. On A66 take minor rd N signed Mungrisdale. After 2m, sharp bends, garden on L immed after tiny church on R. Use church car park at foot of our short drive. On C2C Reivers 71, 10 cycle routes.* **Visits by arrangement 11 May to 13 July Refreshments by arrangement for larger groups. Adm £5, chd free.**

1 acre windy garden below fell, round C18 farmhouse and outbuildings. Fine views. Tiny stream, large pond. Herbaceous, gravel, alpine and shade areas, bulbs in grass. Wide range of plants, many unusual. Relaxed planting regime. Run on organic lines. Art constructions in and out, local stone used creatively.

6 CHURCH VIEW

Bongate, Appleby-in-Westmorland, CA16 6UN. Mrs H Holmes, 01768 351397, engcougars@outlook.com. *0.4m SE of Appleby town centre. A66 W take B6542 for 2m St Michael's Church on L garden opp. A66 E take B6542 & continue to Royal Oak Inn, garden next door, opp church.* **Visits by arrangement May to Sept for groups of up to 30. Adm £5, chd free.**

A series of narrow pathways weave through gardens overflowing with lush, immersive, cottage-garden style planting. Graceful ornamental grasses, a wealth of diverse herbaceous perennials, along with bulbs, roses, and some shrubs; thoughtfully combined to provide layers of texture, colour and scent, with a very long season of interest lasting well into late summer.

7 COOMBE EDEN
Armathwaite, Carlisle, CA4 9PQ. Belinda & Mike, 07738 005388. *8m SE of Carlisle. Turn off A6 just S of High Hesket signed Armathwaite. Continue to bottom of hill where garden can be found on R turn for Lazonby.* **Sat 27, Sun 28 May, Sat 17, Sun 18 June (12-5). Combined adm with Hazel Cottage £8, chd free. Home-made teas at Hazel Cottage. Cream teas available at Coombe Eden for group visits only & must be pre booked. Visits also by arrangement 29 May to 30 July for groups of up to 20.May be poss to combine with Hazel Cottage if available.**
A garden of traditional and contemporary beds. Steep banks down to stream with pretty Japanese style bridge. Large rhododendrons give a breathtaking show in May and June. New formal garden and vegetable areas added in 2020.

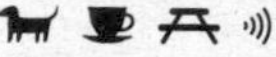

8 CRUMBLE COTTAGES
Beckside, Cartmel, LA11 7SP. Sarah Byrne & Stewart Cowe, 015395 34405, sarah@crumblecottages.co.uk, www.crumblecottages.co.uk. *1m from Cartmel up past the racecourse. Please note car SatNav brings you directly here however mobile phone SatNav take you up the wrong lane.* **Sun 21 May, Sun 23 July (10.30-4). Adm £5, chd £1. Home-made teas. Ice cream & home-made raspberry sauce. Visits also by arrangement Apr to Oct for groups of 20+. Please contact Sarah.**
The gardens inc wildflower meadows, 1$\frac{1}{2}$ acre water gardens built just over 7 yrs ago to improve the biodiversity of the area, walled kitchen gardens which have a Grade II listed wall with 7 bee boles. Cut flower and butterfly borders. Ornamental planting around the house divided into a number of planting styles. All areas designed to encourage wildlife. We now have hedgehogs! Wheelchair access to ornamental areas by house and outlying areas of garden but only by a wheelchair that can cope with uneven ground.

9 NEW ESK BANK
Eskdale, Holmrook, CA19 1UE. Pooja & Adrian Norton. *A595 to Holmrook, then to Eskdale Green, park at the Green station. Walk over line and down drive beech hedges. By train to Ravenglass for Eskdale, then La'al Ratty steam train to the Green station.* **Sat 24 June (11-5); Sun 25 June (11-4). Adm £5, chd free. Cream teas. Donation to Akshaya Patra Foundation UK.**
Nestled in Eskdale, the $\frac{3}{4}$ acre cottage-style garden is on a south-facing slope with 360 panoramic views of the valley. There is a mix of perennial beds and herbaceous borders with many roses and shrubs inc rhododendrons, magnolia and azaleas. There is a summerhouse and a large kitchen garden with a greenhouse. The sloping lawns have an orchard, wildflower borders and a small pond. Public WC available in Eskdale village next to shop.

10 FELL YEAT
Casterton, Kirkby Lonsdale, LA6 2JW. Mrs A E Benson. *1m E of Casterton village. On the rd to Bull Pot. Leave A65 at Devils Bridge, follow A683 for 1m, take the R fork to High Casterton at golf course, straight across at 2 sets of Xrds, house on L, $\frac{1}{4}$m from no-through-rd sign.* **Sun 20 Aug (12-5). Adm £4, chd free.**
1 acre garden with many unusual trees and shrubs. A developing wood is becoming the dominant feature, along with a small meadow, which has orchids. As the garden has matured the emphasis has been increasingly on wildlife habitat, creating a wilder feel and home to birds of prey. Explore the fernery and maturing stumpery and new grotto house with rocks and ferns. Large collection of hydrangeas. Adjoining nursery specialising in ferns, hostas, hydrangeas and many unusual plants.

11 GALESYKE
Wasdale, CA20 1ET. Christine McKinley, 01946 726267, mckinley2112@sky.com. *In Wasdale valley, between Nether Wasdale village and the lake. From Gosforth, follow signs to Nether Wasdale & then to lake, approx 5m. From Santon Bridge follow signs to Wasdale then to lake, approx 2$\frac{1}{4}$m.* **Sat 27, Sun 28 May, Sat 12, Sun 13 Aug (10-5). Adm £4, chd free. Cream teas. Visits also by arrangement May to Aug.**
4 acre garden combining formal areas, shrubberies and woodland, dissected by the River Irt which can be crossed by a suspension bridge. In spring the garden is vibrant with azaleas and rhododendrons, in summer by an impressive collection of colourful hydrangeas. Set in the heart of the Wasdale valley the garden has an unforgettable backdrop of the Screes and the high Lakeland fells. There is a magnificent Victorian sandstone sundial in the garden.

12 GRANGE FELL ALLOTMENTS
Fell Road, Grange-Over-Sands, LA11 6HB. Mr Bruno Gouillon, 01539 532317, brunog45@hotmail.com. *Opp Grange Fell Golf Club. Rail 1.3m, Bus 1m X6, NCR 70.* **Sun 30 July (11-4.30). Adm £4, chd free. Light refreshments. Visits also by arrangement 30 Apr to 2 Oct for groups of 10 to 30.**
The allotments are managed by Grange Town Council. Opened in 2010, 30 plots are now rented out and offer a wide selection of gardening styles and techniques. The majority of plots grow a mixture of vegetables, fruit trees and flowers. There are a few communal areas where local fruit tree varieties have been donated by plot holders with herbaceous borders and annuals.

GROUP OPENING

13 GRANGE OVER SANDS HIDDEN GARDENS
Grange-Over-Sands, LA11 7AF. Bruno Gouillon, 01539 532317, brunog45@hotmail.com. *Off Kents Bank Rd, 3 gardens on Cart Lane then last garden up Carter Rd for Shrublands. Rail 1.4m; Bus X6; NCR 70.* **Sun 28 May, Sun 25 June (11-4). Combined adm £5, chd free. Light refreshments. Visits also by arrangement Apr to Sept for groups of 10 to 30. Donation to St Mary's Hopsice.**

21 CART LANE
Veronica Cameron.

ELDER COTTAGE
Bruno Gouillon & Andrew Fairey.

HAWTHORNE COTTAGE
Mrs Carroll & Mr John Ashton.

SHRUBLANDS
Jon & Avril Trevorrow.

4 very different gardens hidden down narrow lanes off the road south out of Grange. Off Kents Bank Road, 3 gardens on Cart Lane all back onto the railway embankment, providing shelter from the wind but also creating a frost pocket. The garden at 21 Cart Lane is a series of rooms designed to create an element of surprise with fruit and vegetables in raised beds. Elder Cottage is an organised riot of fruit trees, vegetables, shrubby perennials and herbaceous plants. Productive and peaceful. Hawthorne Cottage has been redesigned and replanted over the last 2 yrs to create a garden with colour and interest. Up the hill on Carter Rd, Shrublands is a ¾ acre garden situated on a hillside overlooking Morecambe Bay. Visitors with mobility issues can access Shrublands & 21 Cart Ln, but can only view Elder Cottage from the roadside & Hawthorne Cottage from the gate.

14 HAVERTHWAITE LODGE

Haverthwaite, LA12 8AJ. David Snowdon. *100yds off A590 at Haverthwaite. Turn S off A590 opp Haverthwaite train stn. Bus 6, NCR 70. Parking on the roadside at the top of the drive. No vehicle access on the drive except for those with mobility aids.* **Sun 9 July (11-4). Adm £5, chd free. Open nearby Lakeside Hotel.**
Traditional Lake District garden that has been redesigned and replanted. A series of terraces lead down to the River Leven and wooded area. Inc a rose garden, cutting garden, dell area, rock terrace, herbaceous borders and many interesting mature shrubs. Although unsuitable for wheelchairs or those using mobility aids, visitors may view the garden from the terrace. In a stunning setting the garden is surrounded by oak woodland and was once a place of C18 and C19 industry. The Lakeside Hotel, 2½ m east, is also open on the 9th July and will be offering refreshments.

In 2022 we donated £450,000 to Marie Curie which equates to 23,871 hours of community nursing care

Windy Hall

© Caole Drake

Summerdale House

GROUP OPENING

15 HAYTON VILLAGE GARDENS

Hayton, Brampton, CA8 9HR. Facebook.HaytonOpenGardens. *5m E of M6 J43 at Carlisle, ½m S of A69. 3m W of Brampton. Signed to Hayton off A69. Narrow roads: Please park 1 side only & if poss park less centrally leaving space for less able near pub. Tickets sold in E of village (Townhead), W & middle with map. Map also available on Facebook.* **Sun 9 July (12-5). Combined adm £5, chd free. Home-made teas at Hayton Village Primary School, often with live music. Also cold refreshments in one of the gardens outside if suitable. Donation to Hayton Village Primary School.**

ARNWOOD
Joanne Reeves-Brown.

ASH TREE FARM
Mr & Mrs J Dowling.

BECK COTTAGE
Fiona Cox.

CHESTNUT COTTAGE
Barry Brian.

THE GARTH
Frank O'Connor & Linda Mages.

HAYTON C OF E PRIMARY SCHOOL
Hayton C of E Primary School, www.hayton.cumbria.sch.uk.

HEMPGARTH
Sheila & David Heslop.

HOPE COTTAGE
Adam & Samantha Grant.

KINRARA
Tim & Alison Brown, Facebook.KinraraOpenGardens.

MILLBROOK
M & J Carruthers.

THE PADDOCK
Phil & Louise Jones.

STONECHATS
Anna & Mic Mayhew, 07909 920384.
Visits also by arrangement.

TOWNFOOT HOUSE
Alison Springall.

TOWNHEAD COTTAGE
Chris & Pam Haynes.

A valley of gardens of very varied size and styles mostly around old stone cottages. Smaller and larger cottage gardens, courtyards and containers, woodland walks, exuberant borders, frogs, pools, colour and texture throughout. Home-made teas and often live music at the school where the children create gardens annually within the main school garden (RHS award). Gardens additional to those listed also generally open or visible. Kinrara is designed by an artist and an architect with multiple garden design experience. Accessibility ranges from full to none. Note that gardens are spread out and some walking needed to see all. Come early or late for easier parking.

16 HAZEL COTTAGE

Armathwaite, Carlisle, CA4 9PG. Mr D Ryland & Mr J Thexton. *8m SE of Carlisle. Turn off A6 just S of High Hesket signed Armathwaite, after 2m house facing you at T-junction. 1¼m walk from Armathwaite train stn.* **Sat 8, Sun 9 Apr (12-5). Adm £5, chd free. Sat 27, Sun 28 May, Sat 17, Sun 18 June (12-5). Combined adm with Coombe Eden £8, chd free. Home-made teas.**

Flower arrangers and plantsman's garden. Extending to approx 5 acres. Inc mature herbaceous borders, pergola, ponds and planting of disused railway siding providing home to wildlife. Many variegated and unusual plants. Varied areas, planted for all seasons, south facing, some gentle slopes, ever changing. April's openings coincide with local Daffodil Walk. Only partial access, small steps to WC area. Main garden planted on gentle slope.

17 NEW HERON HILL PRIMARY SCHOOL

Hayfell Avenue, Kendal, LA9 7JH. Karen Harper. *8m from J36 M6. The school is on the southern edge of Kendal. Turn off Burton Rd at the T-lights into Heron Hill which leads onto Hayfell Ave.* **Sat 22 July (10-4.30). Adm £5, chd free. Home-made cakes, tea & coffee.**

The extensive grounds at Heron Hill Primary School offer a variety of wonderful environments for all to enjoy inc a shady quiet garden, wildflower meadows, woodland, beetle bank, pollinator planting schemes, raised beds and polytunnel for vegetables. We provide a diverse habitat for a wide range of wildlife as well as looking after our own honeybees.

18 ◆ HOLEHIRD GARDENS

Patterdale Road, Windermere, LA23 1NP. Lakeland Horticultural Society, 015394 46008, enquiries@holehirdgardens.org.uk, www.holehirdgardens.org.uk. *1m N of Windermere. On A592, Windermere to Patterdale rd.* **For NGS: Sun 19 Mar (10-4); Thur 6 July (10-5). Adm £5, chd free. Self-service hot drinks available for July opening only, not March. For other opening times and information, please phone, email or visit garden website. Donation to Plant Heritage.**

Run by volunteers of the Lakeland Horticultural Society to promote knowledge of gardening in Lakeland conditions. On fellside overlooking Windermere, the 12 acres provide interest year-round. 4 National Collections (astilbe, daboecia, polystichum, meconopsis). Lakeland Collection of hydrangeas. Walled garden has colourful mixed borders and island beds. Alpine beds and display houses. Wheelchair access limited to walled garden and beds accessible from drive.

NPC

19 NEW HOLLY HOUSE

37 Holly Terrace, Hensingham, Whitehaven, CA28 8RF. Dave & Lynne Todhunter. *1m from Whitehaven Town Centre. A595 from N turn L at T-lights signposted Hensingham follow B5295 to mini-r'about in Hensingham Sq take 1st exit, Holly House on L across rd from Globe Pub. Car park in Square/Main St.* **Sat 5, Sun 6 Aug (12-5). Adm £5, chd free. Home-made teas.**

Tucked behind a large Grade II listed Georgian house, the gardens of Holly House are a hidden gem. Split into 3 sections they comprise the top garden with terrace, lawn and herbaceous borders. The middle garden has a hydrangea walk, formal rose garden, wildflower/orchard area, fruit and vegetable beds and large greenhouse. The woodland garden has some steep paths which may be muddy/slippy at times. Wheelchair access to most of the garden, the woodland garden is not suitable for wheelchairs. 'Keep Clear' section at the front for DROP OFF only.

20 ◆ HUTTON-IN-THE-FOREST

Penrith, CA11 9TH. Lord & Lady Inglewood, 01768 484449, info@hutton-in-the-forest.co.uk, www.hutton-in-the-forest.co.uk. *6m NW of Penrith. On B5305, 2m from exit 41 of M6 towards Wigton.* **For NGS: Sun 11 June (10-5). Adm £8, chd £1. Light refreshments. For other opening times and information, please phone, email or visit garden website.**

Hutton-in-the-Forest is surrounded on two sides by distinctive yew topiary and grass terraces - which to the south lead to C18 lake and cascade. 1730s walled garden is full of old fruit trees, tulips in spring, roses and an extensive collection of herbaceous plants in summer. Our gravel paths can make it difficult to push a wheelchair in places.

21 IVY HOUSE

Cumwhitton, CA8 9EX. Martin Johns & Ian Forrest, 01228 561851, martinjohns193@btinternet.com. *6m E of Carlisle. At the bridge at Warwick Bridge on A69 take turning to Great Corby & Cumwhitton. Through Great Corby & woodland until you reach a T-junction. Turn R.* **Sun 25 June, Sun 13 Aug (1-5). Adm £5, chd free. Light refreshments. Visits also by arrangement 1 Apr to 17 Sept.**

Approx 2 acres of sloping fellside garden with meandering paths leading to a series of 'rooms': inc pond, fern garden, mixed beds and vegetable and herb gardens. Copse with meadow leading down to beck. Trees, shrubs, ferns, grasses, bamboos, evergreens and perennials planted with emphasis on variety of texture and colour. Sadly unsuitable for wheelchairs. WC available.

22 LAKESIDE HOTEL

Lake Windermere, Newby Bridge, LA12 8AT. Classic Lodges, 015395 30001, reservations@lakesidehotel.co.uk. *1m N of Newby Bridge. Turn N off A590 across R Leven at Newby Bridge along W side of Windermere.* **Sun 9 July (11-4.30). Adm £5, chd free. Light refreshments. Open nearby Haverthwaite Lodge. Donation to Cumbria Wildlife Trust.**

On the shores of Lake Windermere. Created for year-round interest, with choice plants. Herbaceous borders and foliage shrubs, scented and winter interest plants, seasonal bedding. Roof garden with developing fragrant garden. Espaliered local heritage apple varieties. Annual wildflower verge. Currently registering for the National Collection of Scented Pelargoniums. Some slopes & gravel paths.

In 2022, our donations to Carers Trust meant that 456,089 unpaid carers were supported across the UK

23 LARCH COTTAGE NURSERIES

Melkinthorpe, Penrith, CA10 2DR. Peter Stott, www.larchcottage.co.uk. *From N leave M6 J40 take A6 S. From S leave M6 J39 take A6 N signed off A6.* **Thur 6 July, Thur 10 Aug, Thur 7 Sept (1-4). Adm £5, chd free.**

A unique nursery and garden designed and built over the past 35 yrs, the gardens inc lawns, flowing perennial borders, rare and unusual shrubs, trees, small orchard and a kitchen garden. A natural stream runs into a small lake - a haven for wildlife and birds. At the head of the lake stands a small frescoed chapel designed and built by the owner for family use. Japanese Dry garden, ponds and Italianesque columned garden specifically for shade plants, the Italianesque tumbled down walls are draped in greenery acting as a backdrop for the borders filled with stock plants. Newly designed and constructed lower gardens and chapel. Accessible to wheelchair users although the paths are rocky in places.

24 LOW FELL WEST

Crosthwaite, Kendal, LA8 8JG. Barbie & John Handley, 015395 68297, barbie@handleyfamily.co.uk. *4½m S of Bowness. Off A5074, turn W just S of Damson Dene Hotel. Follow lane for ½m.* **Sun 23 Apr (12-5); Sun 15 Oct (11-4). Adm £5, chd free. Light refreshments. Visits also by arrangement for groups of up to 30.Nearest coach access in layby A5074. Steep ½m walk to garden.**

This 2 acre woodland garden in the tranquil Winster valley has extensive views to the Pennines. The four season garden, restored since 2003, inc expanses of rock planted sympathetically with grasses, unusual trees and shrubs, climaxing for autumn colour. There are native hedges and areas of plant rich meadows. A woodland area houses a gypsy caravan and there is direct access to Cumbria Wildlife Trust's Barkbooth Reserve of Oak woodland, bluebells and open fellside. Wheelchair access to much of the garden, but some rough paths and steep slopes.

25 LOWER ROWELL FARM & COTTAGE

Milnthorpe, LA7 7LU. John & Mavis Robinson & Julie & Andy Welton, 015395 62270. *Approx 2m from Milnthorpe, 2m from Crooklands. Signed to Rowell off B6385, Milnthorpe to Crooklands Rd. Garden ½m up lane on L.* **Visits by arrangement 7 May to 9 July for groups of 10 to 30.Refreshments by arrangement. Adm £5, chd free.**

Approx 1¼ acre garden with views to Farleton Knott and Lakeland hills. Unusual trees and shrubs, plus perennial borders. Architectural pruning, greenhouse, polytunnel with tropical plants, cottage gravel garden and vegetable plot. Fabulous display of snowdrops in spring followed by other spring flowers, with colour most of the year. Wildlife ponds and 2 friendly pet hens.

26 MATSON GROUND

Windermere, LA23 2NH. Matson Ground Estate Co Ltd, 07831 831918, sam@matsonground.co.uk. *⅔m E of Bowness. Turn N off B5284 signed Heathwaite. From E 100yds after Windermere Golf Club, from W 400yds after Windy Hall Rd. Rail 2½m; Bus 1m, 6, 599, 755, 800; NCR 6 (1m).* **Sun 21 May (1-5). Adm £6.50, chd free. Home-made teas. Visits also by arrangement 2 Jan to 22 Dec for groups of 5 to 30. Tour of garden with Head Gardener. No refreshments available for group visits.**

2 acres of mature, south facing gardens. A good mix of formal and informal planting inc topiary features, herbaceous and shrub borders, wildflower areas, stream leading to a large pond and developing arboretum. Rose garden, rockery, topiary terrace borders, ha-ha. Productive, walled kitchen garden c 1862, a wide assortment of fruit, vegetables, cut flowers, cobnuts and herbs. Greenhouse.

27 MIDDLE BLAKEBANK

Underbarrow, Kendal, LA8 8HP. Mrs Hilary Crowe, 07713 608963, hfcmbb@aol.com. *Lyth valley between Underbarrow & Crosthwaite. E off A5074 to Crosthwaite. The garden is on Broom Lane, a turning between Crosthwaite & Underbarrow signed Red Scar & Broom Farm.* **Sat 17 June, Sat 2 Sept (11-4). Adm £5, chd free. Home-made teas. Visits also by arrangement 17 Apr to 7 Sept for groups of 10 to 25.**

The garden extends to 4½ acres and overlooks the Lyth valley with extensive views south to Morecambe Bay and east to the Howgills. We have orchards, a wildflower meadow and more formal garden with a range of outbuildings. Over the last 8 yrs the garden has been developed with plantings that provide varying colour and texture year-round. Some parts of the garden are accessed by steps. We enjoy providing home-made cakes and sandwiches - under cover in large barn if necessary!

28 ◆ NETHERBY HALL

Longtown, Carlisle, CA6 5PR. Mr Gerald & Mrs Margo Smith, 01228 792732, info@netherbyhall.co.uk, netherbyhall.co.uk. *Take Junction 44 from the M6 & follow the A7 to Longtown. Take Netherby rd & follow it for about 2m. Netherby Hall is on L.* **For NGS: Sun 14 May (10-4). Adm £6, chd free. The tea room will be open in the Orangery serving tea, coffee, cakes & light refreshments. For other opening times and information, please phone, email or visit garden website.**

Netherby Hall Garden is 36 acres, consisting of a 1½ acre walled kitchen garden which produces fruit, vegetables and herbs for the Pentonbridge Inn restaurant, with herbaceous borders, a Victorian pleasure ground, woodlands with many fine specimen trees, lawns with azaleas and rhododendrons and a wildflower meadow, all set in more than 200 acres of designed landscape bordering the River Esk. Ample parking. Carriage paths allow access around the Victorian pleasure grounds & walled garden.

29 THE OLD VICARAGE

Church Road, Witherslack, Grange-Over-Sands, LA11 6RS. Mr Alaisdhair & Mrs Mary MacPhie. *On Church Rd between Dean Barwick School & the church green. Follow brown signs to Halecat until final L turn sign where instead of turning L carry on past the school. First building on L 100 yds after school. Enter through gates on left.*

Sun 4 June (11-4.30). Combined adm with Abi and Tom's Garden Plants £6, chd free. Home-made teas.
1 acre garden on edge of ancient woodland with old and newer specimen trees inc magnificent *Cedrus libani* from the original Georgian planting. Beech and walnut surrounding well mown lawns and a bog garden with stepping stones. Shrubbed, flowering and herbaceous beds. Woodland walk along limestone pavement to attractive fernery. Close to Halecat Garden and Nursery Centre approx 800 yds. Access level from driveway and then over lawned areas and gravel paths.

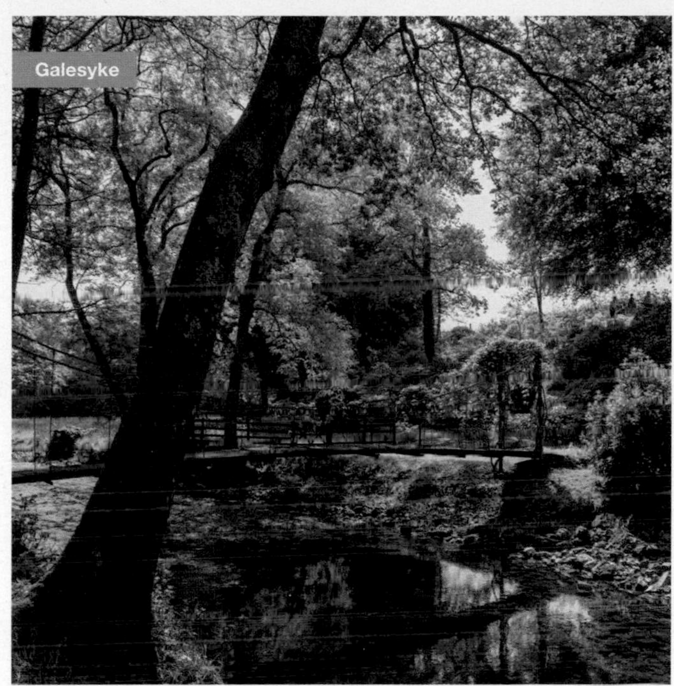
Galesyke

30 8 OXENHOLME ROAD
Kendal, LA9 7NJ. Mr John & Mrs Frances Davenport, 01539 720934, frandav8@btinternet.com. *SE Kendal. From A65 (Burton Rd, Kendal/Kirkby Lonsdale) take B6254 (Oxenholme Rd). No.8 is 1st house on L beyond red post box.* **Sun 11 June (10-5). Adm £4.50, chd free. Home-made teas. Visits also by arrangement 16 Apr to 15 Oct for groups of up to 50.**
Artist and potter's garden of approx ½ acre of mixed planting designed for year-round interest, inc two linked small ponds. Roses, grasses and colour themed borders surround the house, with a gravel garden at the front, as well as a number of woodland plant areas, some vegetables, fruit and orchard areas and lots of sitting spaces. Designed to attract birds and insects. John is a ceramic artist and Frances is a painter. Paintings and pots are a feature of the garden display. We offer tea,coffee, biscuits andcakes to eat at seating around garden. Garden essentially level, but access to WC is up steps.

31 NEW 5 PRIMROSE BANK
Crosby-On-Eden, Carlisle, CA6 4QT. Mark & Cate Bowman, 07773 471298, catebowman@icloud.com. *3m NE of Carlisle off A689. From M6 J44 take A689 towards Hexham. At the next r'about take the 1st turn towards Brampton/Hexham. Take the 2nd turn on R to Crosby on Eden. Park at the village hall.* **Sun 11 June (12-4). Adm £4, chd free. Home-made teas. Visits also by arrangement 19 June to 23 June for groups of up to 50.**
A surprise garden in a modern housing development, making use of reclaimed scrubland next to the Hadrian's Wall Path and close to the River Eden. Mixed productive and flower garden, featuring extensive vegetable garden with polytunnel, small orchard, rose garden, wildlife pond, wide mixed borders. Rose clad pergola. Gravel paths unsuitable for wheelchairs.

32 QUARRY HILL HOUSE
Boltongate, Mealsgate, Wigton, CA7 1AE. Mr Charles Woodhouse & Mrs Philippa Irving. *⅓m W of Boltongate, 6m SSW of Wigton, 13m NW of Keswick, 10m E Cockermouth. M6 J41, direction Wigton, through Hesket Newmarket, Caldbeck & Boltongate on B5299, entrance gates Quarry Hill House drive on R. On A595 Carlisle to Cockermouth turn L for Boltongate, Ireby.* **Sun 4 June (1.30-6). Adm £5, chd free. Home-made teas. Donation to Cumbria Community Foundation: Quarry Hill Grassroots Fund.**
3 acre parkland setting, country house (not open), woodland garden with marvellous views of Skiddaw, Binsey and the Northern Fells and also the Solway and Scotland. Outstanding trees, some ancient and rare and many specimen, shrubs, herbaceous borders, potager vegetable garden.

33 ROSE CROFT
Levens, Kendal, LA8 8PH. Enid Fraser, 07976 977018, enidfraser123@btinternet.com. *Approx 4m from J36 on M6. From J36 take A590 toward Barrow. R turn Levens, L at pub, follow rd to garden on L. From A6, into Levens, past shop, bear L, over Xrds, downhill. L turn signed 'PV Dobson'. Garden on R after Dobsons.* **Visits by arrangement June to Sept for groups of up to 25. Adm £5, chd free. Tea, soft drinks, biscuits.**
This 'secret' garden's richness in herbaceous plants, grasses, shrubs and trees belies the twin challenges of thin topsoil and an acutely sloping site. With grand views across the Lyth valley, steps and pathways carry down to lower-level lawns, streamside summerhouse and wildflowers. Long season of interest with designed elements providing bountiful displays well into late summer.

34 ◆ **RYDAL HALL**
Ambleside, LA22 9LX. Diocese of Carlisle, 01539 432050, gardens@rydalhall.org, www.rydalhall.org. *2m N of Ambleside. E from A591 at Rydal signed Rydal Hall. Bus 555, 599, X8, X55; NCR 6.* **For NGS: Sun 19 Mar, Sun 16 Apr, Sun 23 July, Sun 20 Aug (9-4). Adm by donation. Light refreshments in tea shop on site. For other opening times and information, please phone, email or visit garden website.**
Forty acres of park, woodland and gardens to explore. The formal Thomas Mawson garden has fine examples of herbaceous planting, seasonal displays and magnificent views of the Lakeland Fells. Enjoy the peaceful atmosphere created in the Quiet Garden with informal planting around the pond and stunning views of the waterfalls from The Grot, the UK's first viewing station. Partial wheelchair access, top terrace only.

35 **SPRINT MILL**
Burneside, Kendal, LA8 9AQ. Edward & Romola Acland, 01539 725168/07806 065602, mail@sprintmill.uk. *2m N of Kendal. From Burneside follow signs towards Skelsmergh for ½m, then L into drive of Sprint Mill. Or from A6, about 1m N of Kendal follow signs towards Burneside, then R into drive.* **Sat 10, Sun 18 June (10.30-5). Adm £4, chd free. Light refreshments. Visits also by arrangement Mar to Sept.**
Unorthodox organically run garden,the wild and natural alongside provision of owners' fruit, vegetables and firewood. Idyllic riverside setting, 5 acres to explore inc wooded riverbank with hand-crafted seats. Large vegetable and soft fruit area, following no dig and permaculture principles. Hand tools prevail. Historic water mill with original turbine. The 3 storey building houses owner's art studio and personal museum, inc collection of old hand tools associated with rural crafts. Goats, hens, ducks, rope swing, very family friendly. Short walk to our flower rich hay meadows. Access for wheelchairs to some parts of both garden and mill.

36 **SUMMERDALE HOUSE**
Cow Brow, Nook, Lupton, LA6 1PE. David & Gail Sheals, sheals@btinternet.com, www.summerdalegardenplants.co.uk. *7m S of Kendal, 5m W of Kirkby Lonsdale. From J36 M6 take A65 towards Kirkby Lonsdale, at Nook take R turn Farleton. Location not signed on highway. Detailed directions available on our website.* **Every Sat 11 Feb to 27 May (11-4.30). Sun 19 Feb, Sun 23 Apr, Sun 11 June (11-4.30). Home-made teas. Adm £5, chd free. Home-made teas on Sundays only & home-made soup on 19th Feb. Visits also by arrangement 9 Feb to 9 June for groups of 10+. No coaches.**
1½ acre part-walled country garden set around C18 former vicarage. Several defined areas have been created by hedges, each with its own theme and linked by intricate cobbled pathways. Relaxed natural planting in a formal structure. Rural setting with fine views across to Farleton Fell. Large collections of auricula, primulas and snowdrops. Adjoining RHS Gold Medal winning nursery. Auricula display during season.
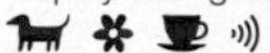

37 ◆ **SWARTHMOOR HALL**
Swarthmoor Hall Lane, Ulverston, LA12 0JQ. T Miller, 01229 583204, info@swarthmoorhall.co.uk, www.swarthmoorhall.co.uk. *1½m SW of Ulverston. A590 to Ulverston. Turn off to Ulverston railway stn. Follow Brown tourist signs to Hall. Rail 0.9m. NCR70 & 700 (1m).* **For NGS: Fri 2, Sat 3 June (10.30-3.30). Adm £3, chd free. Cream teas. For other opening times and information, please phone, email or visit garden website.**
Traditional English country garden surrounding a Grade II* C17 hall. Cottage style borders, woodland garden. Quiet Garden dedicated to peaceful contemplation, ornamental vegetable garden and hay meadow. Please check the website for opening times for the historic hall.

38 **TITHE BARN**
Laversdale, Irthington, Carlisle, CA6 4PJ. Mr Gordon & Mrs Christine Davidson, 01228 573090, christinedavidson7@sky.com. *8m N E of Carlisle. ½m from Carlisle Lake District Airport. From 6071 turn for Laversdale, from M6 J44 follow A689 Hexham/Brampton/Airport. Follow NGS signs.* **Sun 18 June (1-5). Adm £4, chd free. Home-made teas.**
Set on a slight incline, the thatched property has stunning views of the Lake District, Pennines and Scottish Border hills. Planting follows the cottage garden style. The surrounding walls, arches, grottos and quirky features have all been designed and created by the owners. There is a peaceful sitting glade beside a rill and pond. This property also offers self-catering accommodation (sleeps 2) within the grounds of the garden.

GROUP OPENING

39 **ULVERSTON TOWN & COUNTRY GARDENS**
Oubas Hill, Ulverston, LA12 7LA. *5 gardens within 2m of the centre of Ulverston. Ulverston is approached on the A590. Although public transport is available to Ulverston (Rail & buses 6 & X6) the gardens are spread beyond normal walking range. Maps will be avail at all gardens* **Sun 2 July (10-5). Combined adm £5, chd free. Light refreshments at Hamilton Grove.**

104 BIRCHWOOD DRIVE
LA12 9NY. Jane Parker.

HAMILTON GROVE
LA12 7LB. Helen & Martin Cooper.

11 OUBAS HILL
LA12 7LA. Janet & David Parratt.

14 OUBAS HILL
LA12 7LA. Pat Bentley.

NEW **QUAKER LODGE**
LA12 0JS. Mel Elliston.

5 gardens in and around the historic market town of Ulverston, with its canal and lighthouse monument to Sir John Barrow. All will surprise. 3 of the gardens deal successfully with steep slopes, one is relatively large with a balance between formal and country styles, one majors on bio-diversity, two cram great variety into modest sized gardens. A trip around all 5 is a distance of 6 miles.

40 WINDY HALL
Crook Road, Windermere, LA23 3JA. Luke Evans & Sophie Bryde. *½m S of Bowness-on-Windermere. On western end of B5284, pink house up Linthwaite House Hotel driveway. Rail 2.6m; Bus 1m, 6, 599, 755, 800; NCR 6 (1m).* **Sun 7 May, Sun 4 June (10-5). Adm £5, chd free. Light refreshments. Teas and home-made cake.**
Crafted and nurtured by David and Diane Kinsman over 40 years, the 6 acre gardens are eclectic and unique within the lakeland landscape. Carved out of the fellside this garden is steeped in character and charm. Sophie and Luke are now caring for the garden and continuing David and Diane's legacy. Mosses and ferns, rare Hebridean sheep, exotic waterfowl.

41 WINTON PARK
Appleby Road, Kirkby Stephen, CA17 4PG. Mr Anthony Kilvington, www.wintonparkgardens.co.uk. *2m N of Kirkby Stephen. On A685 turn L signed Gt Musgrave/Warcop (B6259). After approx 1m turn L as signed.* **Sun 9 July (11-4). Adm £6, chd free. Light refreshments.**
5 acre country garden bordered by the banks of the River Eden with stunning views. Many fine conifers, acers and rhododendrons, herbaceous borders, hostas, ferns, grasses, heathers and several hundred roses. Four formal ponds plus rock pool. Partial wheelchair access.

42 WOODEND HOUSE
Woodend, Egremont, CA22 2TA. Grainne & Richard Jakobson, 019468 13017, gmjakobson22@gmail.com. *2m S of Whitehaven. Take the A595 from Whitehaven towards Egremont. On leaving Bigrigg take 1st turn L. Go down hill, garden at bottom on R opp Woodend Farm.* **Sat 1, Sun 2 July (11-5.30). Adm £4, chd free. Home-made teas. Visits also by arrangement May to Sept for groups of up to 25.**
A beautiful garden tucked away in a small hamlet. Meandering gravel paths lead around the garden with imaginative, colourful planting and quirky features. Take a look around a productive, organic potager using no-dig methods, wildlife pond, mini spring and summer meadows and sit in the pretty summerhouse. Designed to be beautiful year-round and wildlife friendly. Children's trail with prizes. The gravel drive and paths are difficult for wheelchairs but more mobile visitors can access the main seating areas in the rear garden.

43 YEWBARROW HOUSE
Hampsfell Road, Grange-over-Sands, LA11 6BE. Jonathan & Margaret Denby, 07733 322394, jonathan@bestlakesbreaks.co.uk, www.yewbarrowhouse.co.uk. *¼m from town centre. Proceed along Hampsfell Rd passing a house called Yewbarrow to brow of hill then turn L onto a lane signed 'Charney Wood/ Yewbarrow Wood' & sharp L again. Rail 0.7m, Bus X6, NCR 70.* **Sun 4 June, Sun 2 July, Sun 6 Aug, Sun 3 Sept (11-4). Adm £5, chd free. Light refreshments. Visits also by arrangement May to Sept for groups of 10 to 50.**
More Cornwall than Cumbria' according to Country Life, a colourful 4 acre garden filled with exotic and rare plants, with dramatic views over Morecambe Bay. Outstanding features inc the Orangery, the Japanese garden with infinity pool, the Italian terraces and the restored Victorian kitchen garden. Dahlias, cannas and colourful exotica are a speciality. Find us on YouTube.

44 YEWS
Storrs Park, Bowness-on-Windermere, LA23 3JR. Sir Christopher & Lady Scott, www.yewsestate.com. *1.4m S of Bowness-on-Windermere. On A5074 about 0.3m N of Blackwell, The Arts & Crafts House. Rail Windermere 3m, What3words app - ballots.speak. fractions.* **Sun 18 June (11-5). Adm £6, chd free. Home-made teas.**
7 acre garden overlooking Windermere has been undergoing extensive redevelopment since 2018. Formal gardens designed by Mawson and Avray Tipping inc sunken borders, rose garden and croquet lawn. Newly planted woodland walk, bog garden and renovated Yew maze. Naturalistic planting, areas left for wildlife and many fine trees. New kitchen garden built 2022 and original Messenger glasshouse. There is some level access to formal areas of the garden but only via gravel paths.

Askham Hall

DERBYSHIRE

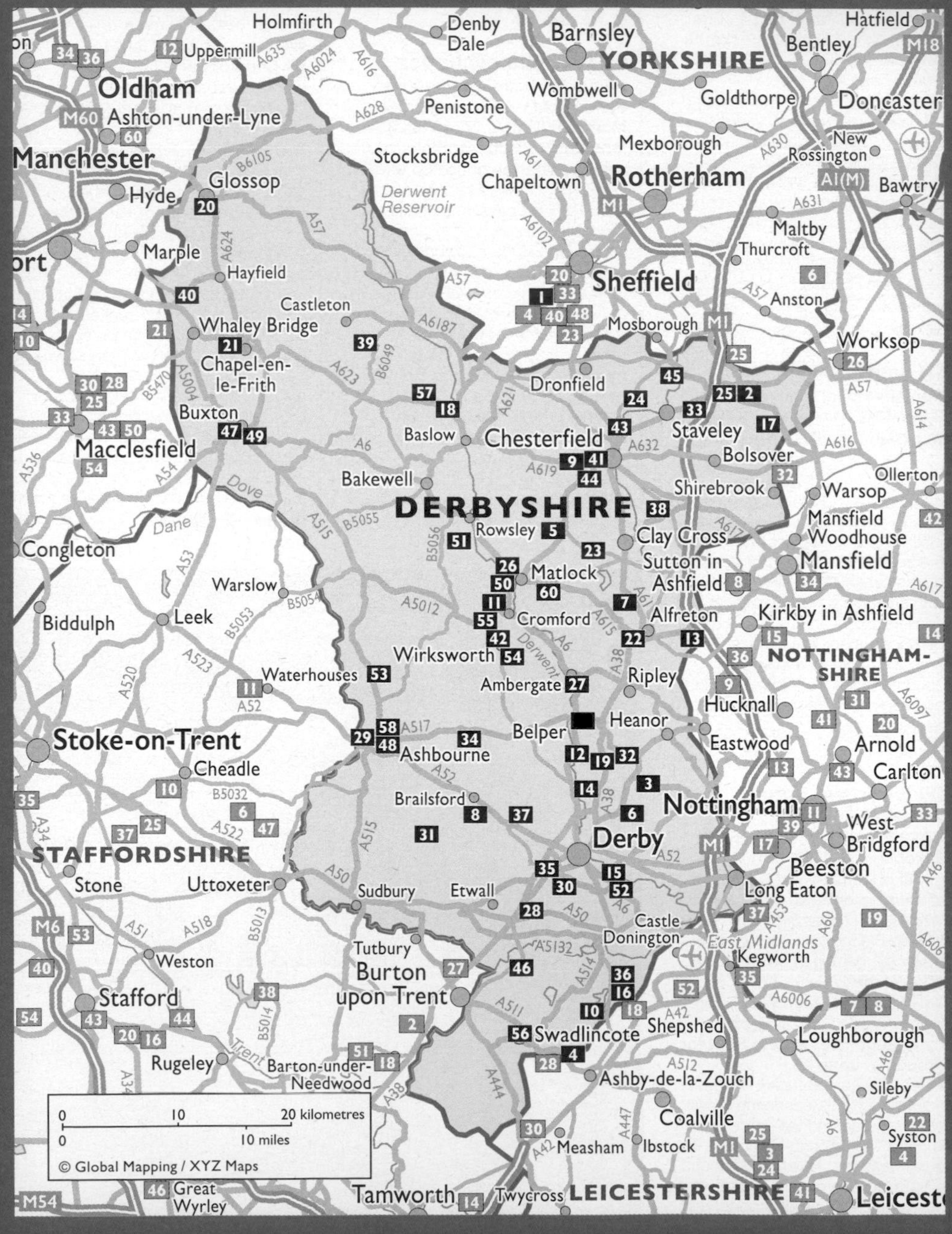
Holmfirth
Denby Dale
Barnsley
YORKSHIRE
Hatfield
Bentley
Uppermill
Oldham
Wombwell
Goldthorpe
Doncaster
Penistone
Ashton-under-Lyne
Manchester
Mexborough
New Rossington
Stocksbridge
Chapeltown
Rotherham
Bawtry
Glossop
Hyde
Derwent Reservoir
Maltby
Marple
Thurcroft
Hayfield
Sheffield
Castleton
Anston
Whaley Bridge
Mosborough
Worksop
Chapel-en-le-Frith
Dronfield
Buxton
Staveley
Baslow
Chesterfield
Macclesfield
Bolsover
Bakewell
Shirebrook
Ollerton
Warsop
DERBYSHIRE
Mansfield Woodhouse
Rowsley
Clay Cross
Congleton
Sutton in Ashfield
Mansfield
Matlock
Warslow
Biddulph
Leek
Cromford
Alfreton
Kirkby in Ashfield
NOTTINGHAM-SHIRE
Wirksworth
Waterhouses
Ambergate
Ripley
Hucknall
Heanor
Belper
Eastwood
Stoke-on-Trent
Ashbourne
Arnold
Cheadle
Carlton
Brailsford
Nottingham
West Bridgford
Derby
STAFFORDSHIRE
Beeston
Stone
Uttoxeter
Sudbury
Etwall
Long Eaton
Castle Donington
East Midlands
Tutbury
Kegworth
Weston
Burton upon Trent
Stafford
Shepshed
Swadlincote
Loughborough
Rugeley
Barton-under-Needwood
Ashby-de-la-Zouch
Sileby
Coalville
Syston
Measham
Ibstock
Great Wyrley
Tamworth
Twycross
LEICESTERSHIRE
Leicest
Dove
Dane
Derwent
Trent
0 10 20 kilometres
0 10 miles
© Global Mapping / XYZ Maps

Derbyshire is the county where the Midlands meets the North, and visitors are attracted to the rugged hills of the High Peak, the high moorlands near Sheffield and the unspoilt countryside of the Dales.

There are many stately homes in the county with world famous gardens, delightful private country gardens, and interesting small cottage and town gardens.

Some of the northern gardens have spectacular views across the Peak District; their planting reflecting the rigours of the climate and long, cold winters. In the Derbyshire Dales, stone walls give way to hedges and the countryside is hilly with many trees, good agricultural land and very pretty villages.

South of Derby the land is much flatter, the architecture has a Midlands look with red brick replacing stone and softer planting in the gardens.

The east side of Derbyshire is different again, reflecting the recent past with small pit villages, and looking towards the rolling countryside of Nottinghamshire. There are fast road links with other parts of the country via the M1 and M6, making a day trip to a Derbyshire garden a very easy choice.

Below: Stanton In Peak Gardens

Volunteers

County Organiser
Hildegard Wiesehofer 07809 883393
hildegard@ngs.org.uk

County Treasurer
Ann Wilkinson 07831 598396
anne.wilkinson@ngs.org.uk

Social Media
Tracy & Bill Reid
07932 977314
billandtracyreid@ngs.org.uk

Booklet Co-ordinator
Malcolm & Wendy Fisher
0115 9664 322
wendy.fisher111@btinternet.com

Dave & Valerie Booth 01283 221167
valerie.booth1955@gmail.com

Group Visit Co-ordinator
Pauline Little 01283 702267
plittle@hotmail.co.uk

Assistant County Organisers
Gill & Colin Hancock 01159 301061
gillandcolinhancock@gmail.com

Jane Lennox 07939 012634
jane@lennoxonline.net

Pauline Little (See above)

Christine & Vernon Sanderson
01246 570830
christine.r.sanderson@uwclub.net

Kate & Peter Spencer 01629 822499
pandkspencer@gmail.com

Paul & Kathy Harvey 01629 822218
pandk.harvey@ngs.org.uk

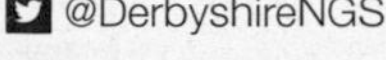

@ngsderbyshire
@DerbyshireNGS
@derbyshirengs

OPENING DATES

All entries subject to change. For latest information check **www.ngs.org.uk**

Map locator numbers are shown to the right of each garden name.

February

Snowdrop Festival

Saturday 11th
The Dower House 16

Sunday 12th
The Dower House 16

Sunday 26th
The Old Vicarage 42

March

Saturday 18th
◆ Cascades Gardens 11

Saturday 25th
Chevin Brae 12

April

Monday 10th
◆ The Burrows Gardens 8

Wednesday 12th
334 Belper Road 3

Sunday 16th
NEW Yew Tree Bungalow 60

Saturday 22nd
◆ Cascades Gardens 11

Sunday 23rd
The Paddock 43

Wednesday 26th
NEW Greenacres 19

Sunday 30th
12 Ansell Road 1
Barlborough Gardens 2
◆ Meynell Langley Trials Garden 37

May

Monday 1st
12 Ansell Road 1
The Limes 29

Wednesday 3rd
◆ Renishaw Hall & Gardens 45

Sunday 7th
NEW Snitterton Hall 50
26 Stiles Road 52

Tuesday 9th
Brierley Farm 5

Sunday 14th
Littleover Lane Allotments 30
27 Wash Green 54

Friday 19th
◆ Cascades Gardens 11

Saturday 20th
Longford Hall Farm 31

Sunday 21st
334 Belper Road 3
Fir Croft 18
Moorfields 38
26 Stiles Road 52
15 Windmill Lane 58

Sunday 28th
12 Ansell Road 1
Broomfield Hall 6
88 Church Street West 13
◆ Meynell Langley Trials Garden 37

Monday 29th
12 Ansell Road 1
◆ The Burrows Gardens 8
Repton NGS Village Gardens 46
◆ Tissington Hall 53

June

Saturday 3rd
The Holly Tree 26
Longford Hall Farm 31

Sunday 4th
Fir Croft 18
Highfield House 22
The Holly Tree 26

Saturday 10th
The Dower House 16
The Smithy 49

Sunday 11th
The Dower House 16
The Smithy 49

Wednesday 14th
◆ Bluebell Arboretum and Nursery 4

Friday 16th
◆ Cascades Gardens 11

Saturday 17th
Holmlea 27
◆ Melbourne Hall Gardens 36
Wild in the Country 57

Sunday 18th
Fir Croft 18
Holmlea 27
◆ Melbourne Hall Gardens 36
13 Westfield Road 56
NEW Yew Tree Bungalow 60

Friday 23rd
330 Old Road 41

Saturday 24th
12 Ansell Road 1
◆ Calke Abbey 10
88 Church Street West 13
Elmton Gardens 17
330 Old Road 41

Sunday 25th
12 Ansell Road 1
◆ The Burrows Gardens 8
Elmton Gardens 17
High Roost 20
Littleover Lane Allotments 30
25 Melbourne Close 35
◆ Meynell Langley Trials Garden 37

July

Saturday 1st
Barlborough Gardens 2

Sunday 2nd
Barlborough Gardens 2
Hill Cottage 23

Wednesday 5th
◆ Renishaw Hall & Gardens 45

Saturday 8th
12 Ansell Road 1
New Mills School 40
The Old Vicarage 42

Sunday 9th
12 Ansell Road 1
8 Curzon Lane 15
New Mills School 40
The Old Vicarage 42
The Paddock 43

Wednesday 12th
◆ Bluebell Arboretum and Nursery 4

Saturday 15th
Longford Hall Farm 31
Repton NGS Village Gardens 46

Sunday 16th
72 Burnside Avenue 7

Saturday 22nd
Byways 9
◆ Cascades Gardens 11
Stanton in Peak Gardens 51

Sunday 23rd
◆ The Burrows Gardens 8
Byways 9
8 Curzon Lane 15
◆ Meynell Langley Trials Garden 37
Stanton in Peak Gardens 51
26 Windmill Rise 59

Saturday 29th
NEW Mastin Moor Gardens and Allotments 33

Sunday 30th
NEW Mastin Moor Gardens and Allotments 33

August

Sunday 6th
Barlborough Gardens 2
8 Curzon Lane 15
9 Main Street 32
13 Westfield Road 56

Wednesday 9th
NEW Greenacres 19

Thursday 10th
72 Burnside Avenue 7

Sunday 13th
Littleover Lane Allotments 30

Monday 14th
◆ Tissington Hall 53

Wednesday 16th
◆ Bluebell Arboretum and Nursery 4

Saturday 19th
◆ Cascades Gardens 11
Chevin Brae 12

Sunday 20th
72 Burnside Avenue 7

Monday 21st
◆ Tissington Hall 53

Sunday 27th
◆ Meynell Langley Trials Garden 37
27 Wash Green 54
12 Water Lane 55

Monday 28th
◆ The Burrows Gardens 8
◆ Tissington Hall 53
12 Water Lane 55

September

Sunday 3rd
The Hollies 25

Saturday 9th
NEW Landown Farm 28

Sunday 10th
Broomfield Hall 6
Coxbench Hall 14
NEW Landown Farm 28
NEW Yew Tree Bungalow 60

Saturday 16th
Holmlea 27

Sunday 17th
Holmlea 27
◆ Meynell Langley Trials Garden 37

Saturday 23rd
◆ Calke Abbey 10

By Arrangement

Arrange a personalised garden visit with your club, or group of friends, on a date to suit you. See individual garden entries for full details.

12 Ansell Road 1
Barlborough Gardens 2
334 Belper Road 3
Brindley Farm 5
72 Burnside Avenue 7
Byways 9
10 Chestnut Way, Repton NGS Village Gardens 46
Chevin Brae 12
88 Church Street West 13
Coxbench Hall 14
8 Curzon Lane 15
The Dower House 16
NEW Greenacres 19
High Roost 20
Higher Crossings 21
Highfield House 22
Hill Cottage 23
Hillside 24
Littleover Lane Allotments 30
9 Main Street 32
Meadow Cottage 34
25 Melbourne Close 35
Nether Moor House 39
New Mills School 40
The Old Vicarage 42
The Paddock 43
Park Hall 44
Riversvale Lodge 47
40 St Oswald's Crescent 48
The Smithy 49
27 Wash Green 54
NEW Yew Tree Bungalow 60

Yew Tree Bungalow

THE GARDENS

1 12 ANSELL ROAD

Ecclesall, Sheffield, S11 7PE. Dave Darwent, 01142 665881, dave@poptasticdave.co.uk. *Approx 3m SW of City Centre. Travel to Ringinglow Rd (88 bus), then Edale Rd (opp Ecclesall C of E Primary School). 3rd R - Ansell Rd. No 12 on L ¾ way down.* **Sun 30 Apr, Mon 1, Sun 28, Mon 29 May, Sat 24, Sun 25 June, Sat 8, Sun 9 July (11-6). Adm £3.50, chd free. Light refreshments. Vegan & gluten free options available. Visits also by arrangement 10 Apr to 4 Aug for groups of up to 15. By arrangement adm inc refreshments.**

Now in its 95th year since being created by my grandparents, this is a suburban mixed productive and flower garden retaining many original plants and features as well as the original layout. A book documenting the history of the garden has been published and is on sale to raise further funds for charity. Original rustic pergola with 90+ year old roses. Then and now pictures of the garden in 1929 and 1950's vs present. Map of landmarks up to 55m away which can be seen from garden. Seven water features. Wide variety of unique-recipe home-made cakes available. Small winter garden.

GROUP OPENING

2 BARLBOROUGH GARDENS

Clowne Road, Barlborough, Chesterfield, S43 4EH. Christine Sanderson, 07956 203184, christine.r.sanderson@uwclub.net, www.facebook.com/barlboroughgardens. *7m NE of Chesterfield. Off A619 midway between Chesterfield & Worksop. ½m E M1, J30. Follow signs for Barlborough then yellow NGS signs. Parking in village.* **Sun 30 Apr (11.30-4.30). Combined adm £5, chd free. Sat 1, Sun 2 July (11.30-4.30). Combined adm £6, chd free. Sun 6 Aug (11.30-4.30). Combined adm £5, chd free. Light refreshments at The Hollies (all dates) & at Raisewells House (selected dates). Visits also by arrangement May to Aug for groups of 10 to 30. By arrangement adm price inc private tour of two gardens & refreshments.**

CLARENDON
Neil & Lorraine Jones.
Open on Sun 30 Apr, Sat 1, Sun 2 July

THE HOLLIES
Vernon & Christine Sanderson.
Open on all dates
(See separate entry)

LINDWAY
Thomas & Margaret Pettinger.
Open on Sat 1, Sun 2 July

RAISWELLS HOUSE
Mr Andrew & Mrs Rosie Dale.
Open on Sat 1, Sun 2 July, Sun 6 Aug

Barlborough is an attractive historic village and a range of interesting buildings can be seen all around the village centre. The village is situated close to Renishaw Hall for possible combined visit. Map detailing location of all gardens is issued with adm ticket. For more information visit our Facebook page - see details above. Barlborough Gardens celebrated 10 years of opening for National Garden Scheme in 2021. The Hollies inc an area influenced by the Majorelle Garden in Marrakech. This makeover was achieved using upcycled items together with appropriate planting. It was featured in an episode of TVs "Love Your Garden". Partial wheelchair access at The Hollies.

In 2022, 20,000 people were supported by Perennial caseworkers and prevent teams thanks to our funding

13 Westfield Road

3 334 BELPER ROAD

Stanley Common, DE7 6FY. Gill & Colin Hancock, 01159 301061, gillandcolinhancock@gmail.com. *7m N of Derby. 3m W of Ilkeston. On A609, ¾m from Rose & Crown Xrds (A608). Please park in field up farm drive.* **Wed 12 Apr, Sun 21 May (11.30-4.30). Adm £4, chd free. Home-made teas. Dairy & gluten free cakes available. Visits also by arrangement 26 Feb to 31 July for groups of 5 to 10.**

Beautiful country garden with many attractive features inc a laburnum tunnel, rose and wisteria domes, old workmen's hut, wildlife pond. Take a walk through the 10 acres of woodland and glades to a ½ acre lake. Organic vegetable garden. See snowdrops and hellebores in February, cowslips in April and wild orchids and rose meadow in June. Plenty of seating to enjoy home made cakes. Children welcome with plenty of activities to keep them entertained. Abundant wildlife. Paths round wood lake. Some of the garden not suitable for wheelchairs and walking aids.

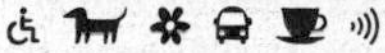

4 ◆ BLUEBELL ARBORETUM AND NURSERY

Annwell Lane, Smisby, Ashby de la Zouch, LE65 2TA. Robert & Suzette Vernon, 01530 413700, sales@bluebellnursery.com, www.bluebellnursery.com. *1m NW of Ashby-de-la-Zouch. Arboretum is clearly signed in Annwell Lane (follow brown signs), ¼m S, through village of Smisby off B5006, between Ticknall & Ashby-de-la Zouch. Free parking.* **For NGS: Wed 14 June, Wed 12 July, Wed 16 Aug (9-4). Adm £5, chd free. For other opening times and information, please phone, email or visit garden website.**

Beautiful 9 acre woodland garden with a large collection of rare trees and shrubs. Interest throughout the yr with spring flowers, cool leafy areas in summer and sensational autumn colour. Many information posters describing the more obscure plants. Adjacent specialist tree and shrub nursery. Please be aware this is not a wood full of bluebells, despite the name. The woodland garden is fully labelled and the staff can answer questions or talk at length about any of the trees or shrubs on display. Please wear sturdy, waterproof footwear during or after wet weather.

5 BRIERLEY FARM

Mill Lane, Brockhurst, Ashover, Chesterfield, S45 0HS. Anne & David Wilkinson, 07831 598396, annelwilkinson@btinternet.com. *6m SW of Chesterfield, 4m NE of Matlock, 1½m NW of Ashover. From the A632 turn W into Alicehead Rd, L into Swinger Ln, 2nd L into Brockhurst Ln & R into Mill Ln. Brierley Farm is the only house on the L.* **Tue 9 May (12-4). Adm £6. Home-made teas. Visits also by arrangement 18 Apr to 23 June for groups of 15 to 30. Afternoons only.**

This 5 acre site features a 2 acre garden and lawns and 3 acre woodland with paths. The sloping garden leads down to a bridge over the stream into the woodland. The high water table feeds three linked ponds and areas of bog garden. Raised stone-walled beds feature bulbs, hellebore, geum and rhododendron in the spring followed by seasonal perennial and annual flowering plants into the autumn.

6 BROOMFIELD HALL

Morley, Ilkeston, DE7 6DN. Derby College, www.facebook.com/BroomfieldPlantCentre. *4m N of Derby. 6m S of Heanor on A608.* **Sun 28 May, Sun 10 Sept (10-4). Adm £5, chd free. Home-made teas in a pop-up volunteer run cafe.**

25 acres of constantly developing educational Victorian gardens/ woodlands maintained by volunteers and students. Herbaceous borders, walled garden, themed gardens, rose garden, potager, prairie plantings, Japanese garden, tropical garden, winter garden, plant centre. Light refreshments, entertainment, cacti, carnivorous, bonsai, fuchsia specialists and craft stalls. A gem of Derbyshire. Most of garden is accessible to wheelchair users and we are all on hand to help.

7 72 BURNSIDE AVENUE

Shirland, Alfreton, DE55 6AE. A & S Wilson-Hunt, 07856 900022, sandrawh16@gmail.com. *2½m from Alfreton. 3½m from Clay Cross on A61. From the A61 turn onto Hallfieldgate Lane, onto Byron St. R onto Burnside Ave. Follow the road & garden is last bungalow on L.* **Sun 16 July, Thur 10, Sun 20 Aug (12.30-5). Adm £3, chd free. Pre-booking essential, please visit www.ngs.org.uk for information & booking. Cream teas. Visits also by arrangement 9 July to 10 Sept for groups of up to 6.**

Delightful small garden packed with herbaceous borders and some tropical planting. Dahlias, Cannas and Phlox give colour with hanging baskets, tree ferns and sedum roof offer plenty of interest. Level garden with paved seating area offering a peaceful haven to enjoy refreshments. Wide variety of plants to see. Animal safari around the garden. Can you find them all? Ask about hot composting. Evening visits and cream teas are possible for groups when booking a 'by arrangement' visit.

8 ◆ THE BURROWS GARDENS

Burrows Lane, Brailsford, Ashbourne, DE6 3BU. Mrs N M Dalton, 01335 360745, enquiries@burrowsgardens.com, www.burrowsgardens.com. *5m SE of Ashbourne; 5m NW of Derby. A52 from Derby: turn L opp sign for Wild Park Leisure 1m before village of Brailsford. ¼m & at grass triangle head straight over through wrought iron gates.* **For NGS: Mon 10 Apr, Mon 29 May, Sun 25 June, Sun 23 July, Mon 28 Aug (11-4). Adm £5, chd free. Home-made teas. For other opening times and information, please phone, email or visit garden website.**

Immaculate lawns show off exotic rare plants and trees, mixing with old favourites in this outstanding garden. A wide variety of styles from temple to Cornish, Italian to English, gloriously designed and displayed. This is a must see garden. Most of garden accessible to wheelchairs.

Our 2022 donation supports 1,700 Queen's Nurses working in the community in England, Wales, Northern Ireland, the Channel Islands and the Isle of Man

9 BYWAYS
7A Brookfield Avenue, Brookside, Chesterfield, S40 3NX. Terry & Eileen Kelly, 07414 827813, telkel1@aol.com. *1½m W of Chesterfield. Follow A619 from Chesterfield towards Baslow. Brookfield Ave is 2nd R after Brookfield Sch. Please park on Chatsworth Rd (A619).* **Sat 22, Sun 23 July (11.30-4.30). Adm £3.50, chd free. Light refreshments inc gluten free option. Visits also by arrangement 8 July to 12 Aug for groups of 10 to 25. Donation to Ashgate Hospice.**
Three times winners of the Best Back Garden over 80 sqm, and also twice winners of Best Front Garden, Best Container Garden and Best Hanging Basket in Chesterfield in Bloom. Well established perennial borders inc helenium, monardas, phlox, grasses, acers. Rockery and many planters containing acers, pelargoniums, ferns and hostas. Greenhouse with a display of pelargoniums. Large pergola.

10 ◆ CALKE ABBEY
Ticknall, DE73 7LE. National Trust, 01332 863822, calkeabbey@nationaltrust.org.uk, www.nationaltrust.org.uk/calke. *10m S of Derby. On A514 at Ticknall between Swadlincote & Melbourne. For SatNav use DE73 7JF.* **For NGS: Sat 24 June, Sat 23 Sept (9.30-4.30). Adm £7, chd £3.50. Light refreshments. Adm likely to increase in 2023. For other opening times and information, please phone, email or visit garden website.**
With peeling paintwork and overgrown courtyards, Calke Abbey tells the story of the dramatic decline of a country-house estate. The productive kitchen garden, impressive collection of glasshouses, garden buildings and tunnels hint at the work of past gardeners, while today the flower garden, herbaceous borders and unique auricula theatre providing stunning displays all year. Restaurant at main visitor facilities for light refreshments and locally sourced food. House is also open. Buggy available for those with limited mobility. Trampers also available to book for 3 hours. Please call 01332 863822 to book.

11 ◆ CASCADES GARDENS
Clatterway, Bonsall, Matlock, DE4 2AH. Alan & Alesia Clements, 07967 337404, alan.clements@cascadesgardens.com, www.cascadesgardens.com. *5m SW of Matlock. From Cromford A6 T-lights turn towards Wirksworth. Turn R along Via Gellia, signed Buxton & Bonsall. After 1m turn R up hill towards Bonsall. Garden entrance at top of hill. Park in village car park.* **For NGS: Sat 18 Mar, Sat 22 Apr, Fri 19 May, Fri 16 June, Sat 22 July, Sat 19 Aug (12-4). Adm £8, chd £4. Light refreshments. For other opening times and information, please phone, email or visit garden website.**
The Meditation Garden and Bonsai centre: Fascinating 4 acre peaceful garden in spectacular natural surroundings with woodland, cliffs, stream, pond and an old limestone quarry. Inspired by Japanese gardens and Buddhist philosophy, secluded garden rooms for relaxation and reflection. Beautiful landscape with a wide collection of unusual perennials, conifers, shrubs and trees. Plants for sale. Hellebore month in March. Mostly wheelchair accessible. Gravel paths, some steep slopes.

12 CHEVIN BRAE
Chevin Road, Milford, Belper, DE56 0QH. Dr David Moreton, 07778 004374, davidmoretonchevinbrae@gmail.com. *1½m S of Belper. Coming from S on A6 turn L at Strutt Arms & cont up Chevin Rd. Park on Chevin Rd. After 300 yds follow arrow to L up Morrells Lane. After 300 yds Chevin Brae on L with silver garage.* **Sat 25 Mar, Sat 19 Aug (1-5). Adm £3, chd free. Home-made teas. Visits also by arrangement 1 Feb to 27 Oct for groups of up to 16.**
A large garden, with swathes of daffodils in the orchard a spring feature. Extensive wildflower planting along edge of wood features aconites, snowdrops, wood anemones, fritillaries and dog tooth violets. Other parts of garden will have hellebores and early camelias. In the summer, the large flower borders and rose trellises give much colour, and the vegetable and fruit gardens are at their peak. A steep drive and two flights of steps need to be climbed to access the garden Tea and home-made cakes, pastries and biscuits, many of which feature fruit and jam from the garden.

13 88 CHURCH STREET WEST
Pinxton, Nottingham, NG16 6PU. Rosemary Ahmed, 07842 141210. *Pinxton is approx 1m from J28 of the M1. At motorway island take the B6019 towards Alfreton & take the first turning on L Pinxton In which turns into Alfreton rd, the garden is signposted from there.* **Sun 28 May, Sat 24 June (1-5). Adm £3.50, chd free. Light refreshments. Visits also by arrangement 22 Apr to 24 Sept for groups of up to 15.**
Plantwomens garden,developed over 25 yrs to inc collections of hardy geraniums, agaves, ferns, persicaria and grasses.The garden has a quirky mix of salvage items.A wildflower lawn is a work in progress, a small rockery, upcycled garden room, mixed borders and a large patio with pot displays.

14 COXBENCH HALL
Alfreton Road, Coxbench, Derby, DE21 5BB. Mr Brian Ballin, 01332 880200, office@coxbench-hall.co.uk, www.coxbench-hall.co.uk. *4m N of Derby close to A38. After passing thru Little Eaton, turn L onto Alfreton Rd for 1m, Coxbench Hall is on L next to Fox & Hounds pub between Little Eaton & Holbrook. From A38, take Kilburn turn & go towards Little Eaton.* **Sun 10 Sept (2.30-4.30). Adm £3, chd free. Light refreshments inc home-made diabetic & gluten free cakes. Visits also by arrangement for groups of up to 15.**
Former ancestral Georgian Home of the Meynell family, the gardens focus on sustainability, organic and wildlife friendly themes. There is a tropical style garden, ponds, hosta's, vegetables, C18 potting shed, veteran yew tree, kitchen garden and hillside rhododendron garden with a stoned path by the upper woodland overlooking the gardens below. We grow vegetables next to the C18 potting shed, have a veteran (500-800 year old) yew tree, tropical style garden, hosta's, rhododendrons and woodland walks. Our wheelchair accessible gardens are developed to inspire the senses of our residents via different colours, textures and fragrances of plants. Most of garden is lawned or block paved inc a block paved path around the edges of the main lawn. Regret no wheelchair access to woodland area.

15 8 CURZON LANE

Alvaston, Derby, DE24 8QS. John & Marian Gray, 01332 601596, maz@curzongarden.com, www.curzongarden.com. *2m SE of Derby city centre. From city centre take A6 (London Rd) towards Alvaston. Curzon Lane on L, approx ½m before Alvaston shops.* **Sun 9, Sun 23 July, Sun 6 Aug (12-5). Adm £3.50, chd free. Light refreshments. Visits also by arrangement July & Aug for groups of 10 to 30.**

Mature garden with lawns, borders packed full with perennials, shrubs and small trees, tropical planting and hot border. Ornamental and wildlife ponds, greenhouse with different varieties of tomato, cucumber, peppers and chillies. Well-stocked vegetable plot. Gravel area and large patio with container planting.

16 THE DOWER HOUSE

Church Square, Melbourne, DE73 8JH. William & Griselda Kerr, 07799 883777, griseldakerr@btinternet.com. *6m S of Derby. 5m W of exit 23A M1. 4m N of exit 13 M42. When in Church Sq, turn R just before the church by a blue sign giving service times. Gates are 50 yds ahead.* **Sat 11, Sun 12 Feb (10-4); Sat 10 June (10-5). Light refreshments. Sun 11 June (10-5). Adm £5, chd free. Visits also by arrangement 22 Jan to 1 Nov for groups of 10+ excl weekends. Tea sometimes possible for small groups.**

Beautiful view of Melbourne Pool from balustraded terrace running length of 1829 house. Garden drops steeply by paths and steps to lawn with herbaceous borders and bank of some 60 shrubs. Numerous paths lead to different areas of the garden, providing varied planting opportunities inc a bog garden, glade, shrubbery, grasses, herb and vegetable garden, rose tunnel, orchard and small woodland. Hidden paths and different areas for children to explore and various animals such as a bronze crocodile and stone dragon to find. They should be supervised by an adult at all times due to the proximity of water. Wheelchair access to top half of the garden only. Shoes with a good grip are highly recommended as slopes are steep. No parking within 50 yards.

GROUP OPENING

17 ELMTON GARDENS

Elmton, Worksop, S80 4LS. *2m from Creswell, 3m from Clowne, 5m from J30, M1. From M1 J30 take A616 to Newark. Follow approx 4m. Turn R at Elmton signpost. At junction turn R, the village centre is in ½m.* **Sat 24, Sun 25 June (1-5). Combined adm £5, chd free. Cream teas at the Old Schoolroom next to the church. Food also available all day at the Elm Tree Inn.**

THE BARN
Mrs Anne Merrick.

THE COTTAGE
Judith Maughan & Roy Reeves.

ELM TREE COTTAGE
Mark & Linda Hopkinson.

ELM TREE FARM
Angie & Tim Caulton.

PEAR TREE COTTAGE
Geoff & Janet Cutts.

PINFOLD
Nikki Kirsop & Barry Davies.

Elmton is a lovely little village situated on a stretch of rare unimproved Magnesian limestone grassland with quaking grass, bee orchids and harebells all set in the middle of attractive, rolling farm land. It has a pub, which serves food all day, a church and a village green with award winning wildlife conservation area and newly restored Pinfold. Cream teas are served in the old schoolroom. Garden opening normally coincides with Elmton Well Dressing, with three well dressings on display. There is a brass band and a local history exhibition. The very colourful but different open gardens have wonderful views. They show a range of gardening styles, themed beds and have a commitment to fruit and vegetable growing. Elmton received a gold award and was voted best small village for the 7th time and best wildlife and conservation area in the East Midlands in Bloom competition in 2019.

18 FIR CROFT

Froggatt Road, Calver, S32 3ZD. Dr S B Furness, www.alpineplantcentre.co.uk. *4m N of Bakewell. At junction of B6001 with A625 (formerly B6054), adjacent to Froggat Edge Garage.* **Sun 21 May, Sun 4, Sun 18 June (2-5). Adm by donation.**

Massive scree with many varieties. Plantsman's garden; rockeries; water garden and nursery; extensive collection (over 3000 varieties) of alpines; conifers; over 800 sempervivums, 500 saxifrages and 350 primulas. Many new varieties not seen anywhere else in the UK. Huge new tufa wall planted with many rare alpines and sempervivums. Wheelchair access to part of garden only.

19 NEW GREENACRES

Makeney Road, Holbrook, Belper, DE56 0TF. Veronica Holtom, 07749 277927, veronica.cooke1@icloud.com. *5m N of Derby. Approach via Makeney Rd, using private drive for Holbrook Hall Residential Care Home. Approach from Portway is not recommended, SatNav may suggest it - Ignore.* **Wed 26 Apr, Wed 9 Aug (2-5.30). Adm £5, chd free. Home-made teas. Visits also by arrangement 10 Apr to 30 Sept for groups of 10 to 25.**

1½ acres of garden in a beautiful rural Derbyshire setting with far reaching countryside views. Many birch and pine trees, a very large selection of camellias, hellebores and bulbs in spring and herbaceous and mixed beds and borders all year. Lovely pond and rock garden with cascades, water lilies and alpines. Lawns, paths, paved areas and walkways with strategically sited seating. 90% of the site is accessible. There are few steps and the majority of the site is level.

Our 2022 donation to ABF The Soldier's Charity meant 372 soldiers and veterans were helped through horticultural therapy

20 HIGH ROOST

27 Storthmeadow Road, Simmondley, Glossop, SK13 6UZ. Peter & Christina Harris, 01457 863888, harrispeter448@gmail.com. *¾m SW of Glossop. From Glossop A57 to M/CL at 2nd r'about, up Simmondley Ln nr top R turn. From Marple A626 to Glossop, in Chworth R up Town Ln past Hare & Hound pub 2nd L.* **Sun 25 June (12-4). Adm £3, chd free. Light refreshments. Visits also by arrangement 15 May to 15 July for groups of up to 40. Donation to Donkey Sanctuary.**

Garden on terraced slopes, views over fields and hills. Winding paths, archways and steps explore different garden rooms packed with plants, designed to attract wildlife. Alpine bed, gravel gardens; vegetable garden, water features, statuary, troughs and planters. A garden which needs exploring to discover its secrets tucked away in hidden corners. Craft stall, children's garden quiz and lucky dip.

21 HIGHER CROSSINGS

Crossings Road, Chapel-en-le-Frith, High Peak, SK23 9RX. Malcolm & Christine Hoskins, 01298 812970, malcolm275@btinternet.com. *Chapel-en-le-frith. Turn off B5470 N from Chapel-en-le-Frith on Crossings Rd signed Whitehough/Chinley. Higher Crossings is 2nd house on R beyond 1st xrds. Park best before xrds on Crossings Rd or L on Eccles Rd.* **Visits by arrangement 1 May to 23 Sept for groups of 10 to 20. Adm £4, chd free. Light refreshments.**

Nearly 2 acres of formal terraced country garden, sweeping lawns and magnificent Peak District views. Rhododendrons, acers, azaleas, hostas, herbaceous borders, Japanese garden. Mature specimen trees and shrubs leading through a dell. Beautiful stone terrace and sitting areas. Garden gate leading into meadow. Tea and biscuits or home-made cakes available (not inc in admission price). Wheelchair access around the house and upper terrace and gravel garden.

22 HIGHFIELD HOUSE

Wingfield Road, Oakerthorpe, Alfreton, DE55 7AP. Paul & Ruth Peat & Janet & Brian Costall, 01773 521342, highfieldhouseopengardens@hotmail.co.uk, www.highfieldhouse.weebly.com. *Rear of Alfreton Golf Club. A615 Alfreton-Matlock Rd.* **Sun 4 June (10.30-5.30). Adm £3, chd free. Home-made teas. Visits also by arrangement 13 Feb to 28 Feb for groups of 15+. £10 adm inc soup, roll, tea & cake February, June & July.**

Lovely country garden of approx 1 acre, incorporating a shady garden, woodland, pond, laburnum tunnel, orchard, herbaceous borders and vegetable garden. Fabulous AGA baked cakes and lunches. Groups welcome by appointment inc 14th-28th February for snowdrops with afternoon tea or lunch inside by the fire. Lovely walk to Derbyshire Wildlife Trust nature reserve to see orchids in June. Some steps, slopes and gravel areas.

The Smithy

23 HILL COTTAGE
Ashover Road, Littlemoor, Ashover, nr Chesterfield, S45 0BL. Jane Tomlinson & Tim Walls, 07946 388185, lavenderhen@aol.com. *Littlemoor. 1.8m from Ashover village, 6.3m from Chesterfield & 6.1m from Matlock. Hill Cottage is on Ashover Rd (also known as Stubben Edge Ln). Opp the end of Eastwood Ln.* **Sun 2 July (10.30-4.30). Adm £4, chd free. Light refreshments. Visits also by arrangement 3 July to 31 Aug for groups of up to 15. £6.50 for adm, tea & cake. £4 for adm only.**
Hill Cottage is a lovely example of an English country cottage garden. Whilst small, the garden has full, colourful and fragrant mixed borders with hostas and roses in pots along with a heart shaped lawn. A greenhouse full of chillies and scented pelargoniums and a small vegetable patch. Views over a pastoral landscape to Ogston reservoir. A wide variety of perennials and annuals grown in pots and containers.

24 HILLSIDE
286 Handley Road, New Whittington, Chesterfield, S43 2ET. Mr E J Lee, 01246 454960, eric.lee5@btinternet.com. *3m N of Chesterfield. Between B6056 & B6052.* **Visits by arrangement for groups of up to 50. Adm £4, chd free. Light refreshments.**
There is a Japanese feature showcasing Japanese plants and a tree trail with 60 named trees, both small and large. There is a Himalayan bed with 40 species grown from wild collected seeds and an Asian area with bamboos, acers and a Chinese border. Other attractions are pools, streams and bog gardens.

25 THE HOLLIES
87 Clowne Road, Barlborough, Chesterfield, S43 4EH. Vernon & Christine Sanderson, www.facebook.com/barlboroughgardens. *7m NE of Chesterfield. Off A619 midway between Chesterfield & Worksop. ½m E M1, J30. Follow signs for Barlborough then yellow NGS signs. Parking along Clowne Rd.* **Sun 3 Sept (11.30-4.30). Adm £3.50, chd free. Light refreshments. The 'Cool Café' cake menu will inc a number of autumnal home-made cakes. Opening with Barlborough Gardens on Sun 30 Apr, Sat 1, Sun 2 July, Sun 6 Aug.**
The Hollies maximises the unusual garden layout and inc shade area, patio garden with Moroccan corner, cottage border plus fruit trees. A wide selection of home-made cakes on offer inc gluten free, sugar free and vegan options. Home-made preserves and jam jar posies for sale. Extensive view across arable farmland. An area of the garden, influenced by the Majorelle Garden in Marrakech, was achieved using upcycled items together with appropriate planting. This garden project was featured on an episode of "Love Your Garden". Partial wheelchair access.

26 THE HOLLY TREE
21 Hackney Road, Hackney, Matlock, DE4 2PX. Carl Hodgkinson. *½m NW of Matlock, off A6. Take A6 NW past bus stn & 1st R up Dimple Rd. At T-junction, turn R & immed L, for Farley & Hackney. Take 1st L onto Hackney Rd. Continue ¾m.* **Sat 3, Sun 4 June (11-4.30). Adm £4, chd free. Home-made teas.**
In excess of 1½ acres and set on a steeply sloping south facing site, sheltering behind a high retaining wall. It contains a bog garden, herbaceous borders, pond, vegetable plot, fruit trees and shrubs, honeybees and chickens. Extensively terraced with many paths and steps. Spectacular views across the Derwent valley.

27 HOLMLEA
Derby Road, Ambergate, Belper, DE56 2EJ. Bill & Tracy Reid. *On the A6 in Ambergate - between Belper & Matlock. 9m N of Derby, 6m S of Matlock. Easy access from M1 J28. On A6 next to Bridge House Cafe. Additional Parking at Anila Restaurant opp Hurt Arms.* **Sat 17, Sun 18 June, Sat 16, Sun 17 Sept (12-5). Adm £4, chd free. Home-made teas.**
Discover our 1½ acre garden. You will find the conventional and unusual - some might say eccentric! Formal garden, kitchen garden, gravel garden, boules court, 'Disused Railway', canal lock water feature, riverside walk and more. New for 2023, teas in 'The Pavilion'. Family friendly with activities for kids, dogs on leads welcome - beware of the free ranging poultry. Adjacent to Shining Cliff Woods, SSI. Find us on Facebook at 'Holmlea Gardens'. Wheelchair route around main features of the garden. Unfortunately the Riverside Walk is not suitable for wheelchairs.

28 NEW LANDOWN FARM
Bakeacre Lane, Findern, Derby, DE65 6BH. Mrs Laina Froom, www.facebook.com/FroomsDahlias. *Park in the car park & enjoy the garden with its dahlia borders before heading 200 metres up the track where you will find Frooms Dahlias.* **Sat 9, Sun 10 Sept (10.30-3.30). Adm £4, chd free. Light refreshments.**
The garden is located in an equestrian setting and is mainly laid to lawn, with raised beds and borders brimming with 250 dahlia varieties in the spectacular showcase beds. There is a pond and seating areas. We offer a 'make your own eco vase' activity and a wide selection of dahlia cuttings/plants. A horse trailer cafe sets the scene enabling you to enjoy refreshments watching the horses. Mature farm garden surrounded by horses grazing in the paddocks. Quirky features to enjoy. Accessible for wheelchair users, who can drive right up to the garden and the showbeds.

29 THE LIMES
Belle Vue Road, Ashbourne, DE6 1AT. Adam & Karen Noble, limevue@peakbreaks.co.uk. *From Market place take Union Street , then L into Belle Vue Rd. The Limes is 2nd house on L behind a curved wall. Enter via the pedestrian gate. Parking in town car parks or further along road.* **Mon 1 May (1-4). Adm £3.50, chd free. Home-made teas.**
A secret garden overlooking the roof tops of Ashbourne town centre. Designed as a series of rooms cascading down multiple levels. Sun deck. Scent garden, down past the hidden loungers to the fish pond, home to large friendly koi and onward to the main borders with cottage style planting, shade planting and a large prairie border.

30 LITTLEOVER LANE ALLOTMENTS

19 Littleover Lane, Normanton, Derby, DE23 6JF. Mr David Kenyon, 07745 227230, davidkenyon@tinyworld.co.uk, littleoverlaneallotments.org.uk. *On Littleover Ln, opp the junction with Foremark Avenue. Just off the Derby Outer Ring Road (A5111). At the Normanton Park r'about turn into Stenson Rd then R into Littleover Ln. The Main Gates are on the L as you travel down the rd.* **Sun 14 May, Sun 25 June, Sun 13 Aug (11-3.30). Adm £5, chd free. Light refreshments. Visits also by arrangement 1 Apr to 17 Sept for groups of 5 to 30.**

A quiet oasis just off Derby's Outer Ring Road, hidden away in a residential area. A private site of nearly 12 acres, established in 1920, cultivated in a variety of ways, inc organic, no dig and potager style. Come and chat with plot holders about their edibles and ornamentals. Many exotic and Heritage varieties cultivated. Later in the year produce can be available to sample. On site disabled parking available. All avenues have been stoned but site is on a slope and extensive so some areas not accessible.

31 LONGFORD HALL FARM

Longford, Ashbourne, DE6 3DS. Liz Wolfenden, Longfordhallfarmholidaycottages.com. *From A52 take turn to Hollington. At T junction (Long Ln) turn R & follow signs. Use drive with Longford Hall Farm sign at the entrance.* **Sat 20 May, Sat 3 June, Sat 15 July (1.30-5). Adm £5, chd free. Home-made teas.**

The large front garden is mainly shrubs hydrangeas, roses, ferns, grasses and Hostas beds. These surround a modern pond and fountain. There are on going projects in this garden. The walled garden has large traditional herbaceous perennial borders, landscaped rill and small orchard. Most of the garden is on one level.

32 9 MAIN STREET

Horsley Woodhouse, Ilkeston, DE7 6AU. Ms Alison Napier, 01332 881629, ibhillib@btinternet.com. *3m SW of Heanor. 6m N of Derby. Turn off A608 Derby to Heanor Rd at Smalley, towards Belper, (A609). Garden on A609, 1m from Smalley turning.* **Sun 6 Aug (1.30-4.30). Adm £4, chd free. Cream teas. Visits also by arrangement 5 Apr to 13 Sept for groups of 5+.**

⅓ acre hilltop garden overlooking lovely farmland view. Terracing, borders, lawns and pergola create space for an informal layout with planting for colour effect. Features inc large wildlife pond with water lilies, bog garden and small formal pool. Emphasis on carefully selected herbaceous perennials mixed with shrubs and old fashioned roses. Gravel garden for sun loving plants and scree garden, both developed from former drive. Please ensure you bring your own carrier bags for the plant stall. Wide collection of homegrown plants for sale. All parts of the garden accessible to wheelchairs. Wheelchair adapted WC.

33 NEW MASTIN MOOR GARDENS AND ALLOTMENTS

Worksop Road, Mastin Moor, Chesterfield, S43 3DN. Mastin Moor Gardens & Allotments CIO, mastinmoorgardensandallotments.com. *Immed below the Bolsover Rd Xrds in Mastin Moor. Access is easiest using the layby on Bolsover Rd approx. 100 metres from the Xrd junction with the A619 Worksop Rd.* **Sat 29, Sun 30 July (10-4). Adm £4, chd free. Light refreshments.**

This public access (open 24/7) three acre site in Mastin Moor Derbyshire has 'open plan' allotments and gardens. Established in the early C21 the gardens inc an arboretum, pond, bog garden, orchard, picnic area and sensory garden in addition to some 25 allotments and memorials. There are over a mile of paths around the site. A composting toilet is available. A wide variety of trees, shrubs, wildflowers and fruiting trees as well as flower and herb displays in addition to open allotments. The site is fully available for wheelchair and motor scooter access although some paths may be subject to the effects of wet weather at times.

34 MEADOW COTTAGE

1 Russell Square, Hulland Ward, Ashbourne, DE6 3EA. Michael & Sally Halls, 01335 372064, mvaehalls@gmail.com. *5m E of Ashbourne. There is a lane between & on the same side as the 2 garages in Hulland Ward with a finger post to the Medical Centre. Meadow Cottage is just down the lane to the R.* **Visits by arrangement 8 May to 26 Aug. Adm £6, chd free. Light refreshments.**

Created from what was a field in 1995, this garden faces south and is on a sloping site. There is a wooded area, a wildflower orchard, herbaceous and vegetable areas. The soil is heavy clay so roses do well here. At 750' above sea level, growing seasons are later and shorter. The lawns are all original meadow. There are bees and hens too. The access to the garden is over a gravel area and the slope is significant down the garden but can be mastered with care.

35 25 MELBOURNE CLOSE

Mickleover, Derby, DE3 9LG. Marie Wilton, marie-mmt@hotmail.com. *3m from Derby city centre. W from Royal Derby Hospital. At 2nd set of T-lights past hosp take R filter lane onto Western Rd, 2nd R onto Brisbane Rd. Melbourne Close is 3rd R but no parking. Park on nearby roads.* **Sun 25 June (1-5). Adm £3.50, chd free. Home-made teas. Visits also by arrangement June & July for groups of 5 to 15.**

Suburban cottage style garden developed over the past 4 yrs. Mixed colour themed borders and beds with an emphasis on structure and texture, together with a small wildlife pond and a gravel area with olive trees, lavender and herbs. Seasonal planting in containers and hanging baskets together with climbing plants over archways. There are various types of bamboo in pots providing screening.

36 ◆ MELBOURNE HALL GARDENS

Church Square, Melbourne, Derby, DE73 8EN. Melbourne Gardens Charity, 01332 862502, info@melbournehall.com, www.melbournehall.com. *6m S of Derby. At Melbourne Market Place turn into Church St, go down to Church Sq. Garden entrance across visitor centre next to Melbourne Hall tea room.* **For NGS: Sat 17, Sun 18 June (1-5). Adm £10, chd £5. Sitooterie - Coffee/Ice Cream Takeaway - Melbourne Hall Courtyard. For other opening times and information, please phone, email or visit garden website.**

A 17 acre historic garden with an abundance of rare trees and shrubs. Woodland and waterside planting with extensive herbaceous borders. Meconopsis, candelabra primulas, various Styrax and Cornus kousa. Other garden features inc Bakewells wrought iron arbour, a yew tunnel and fine C18 statuary and water features. 300yr old trees, waterside planting, feature hedges and herbaceous borders. Fine statuary and stonework. Don't forget to visit our pigs, alpacas, goats and various other animals in their garden enclosures. Gravel paths, uneven surface in places, some steep slopes.

37 ◆ MEYNELL LANGLEY TRIALS GARDEN

Lodge Lane (off Flagshaw Lane), Kirk Langley, Ashbourne, DE6 4NT. Robert & Karen Walker, 01332 824358, enquiries@meynellgardens.com, www.meynell-langley-gardens.co.uk. *4m W of Derby, nr Kedleston Hall. Head W out of Derby on A52. At Kirk Langley turn R onto Flagshaw Ln (signed to Kedleston Hall) then R onto Lodge Ln. Follow Meynell Langley Gardens signs.* **For NGS: Sun 30 Apr, Sun 28 May, Sun 25 June, Sun 23 July, Sun 27 Aug, Sun 17 Sept (10-4). Adm £5, chd free. For other opening times and information, please phone, email or visit garden website.**

Completely re-designed during 2020 with new glasshouse and patio area incorporating water rills and small ponds. New wildlife and fish ponds also added. Displays and trials of new and existing varieties of bedding plants, herbaceous perennials and vegetable plants grown at the adjacent nursery. Over 180 hanging baskets and floral displays. Adjacent tea rooms serving lunches and refreshments daily. Plant sales from adjacent nursery. Level ground and firm grass and some hard paths.

38 MOORFIELDS

261 Chesterfield Road, Temple Normanton, Chesterfield, S42 5DE. Peter, Janet & Stephen Wright. *4m SE of Chesterfield. From Chesterfield take A617 for 2m, turn on to B6039 through Temple Normanton, taking R fork signed Tibshelf, B6039. Garden ¼m on R. Limited parking.* **Sun 21 May (1-5). Adm £4, chd free. Light refreshments.**

Two adjoining gardens, each planted for seasonal colour, which during the late spring and early summer feature perennials inc alliums, lupins, camassias and bearded irises. The larger garden has mature, mixed island beds and borders, a gravel garden to the front, a small wildflower area, large wildlife pond, orchard, soft fruit beds and vegetable garden. Seasonal colour. Open aspect with extensive views to mid Derbyshire.

39 NETHER MOOR HOUSE

Granby Road, Bradwell, Hope Valley, S33 9HU. Anna & Rod Smallwood, 07973 738846, annamsmallwood@gmail.com, nethermoorgarden.uk. *15m W of Sheffield on the B6049 off the Hope Valley (A6187). From N via A6187 & B6049 R at playing field (Gore Ln), up Smalldale. From S via B6049 L at playing field, (Town Ln) up Smalldale. Turn L onto Granby Rd. Park on L side of rd, well tucked in.* **Visits by arrangement 1 Mar to 9 July for groups of 6 to 20. Adm £5, chd free. Tea & biscuits available for group visits.**

A country garden with views of Hope Valley. Mature trees. Front terraces with colourful planting and box hedges, below lawn and mixed borders, unusual plants. Japanese patio garden: small pond, pools of planting, massive arches, mix of Japanese and traditional plants. At the back, a long border, sloping grass, sculptures, wild flowers, fruit and vegetables. Sit and take in the views. Snowdrops, hellebores, daffodils, peonies, Japanese Tai Haku, ginkgo and handkerchief trees, rambling and shrub roses.

D

40 NEW MILLS SCHOOL

Church Lane, New Mills, High Peak, SK22 4NR. Mr Craig Pickering, 07774 567635, cpickering@newmillsschool.co.uk, www.newmillsschool.co.uk. *12m NNW of Buxton. From A6 take A6105 signed New Mills, Hayfield. At C of E Church turn L onto Church Lane. Sch on L. Parking on site.* **Sat 8, Sun 9 July (2-5). Adm £4, chd free. Light refreshments in School Library. Visits also by arrangement 10 July to 7 Aug for groups of 15 to 20.**

Mixed herbaceous perennials/shrub borders, with mature trees and lawns and gravel border situated in the semi rural setting of the High Peak inc a Grade II listed building with four themed quads. The school was awarded highly commended in the School Garden 2019 RHS Tatton Show and won the Best High School Garden and the People's Choice Award. Hot and cold beverages and cakes are available. Ramps allow wheelchair access to most of outside, flower beds and into Grade II listed building and library.

41 330 OLD ROAD

Brampton, Chesterfield, S40 3QH. Christine Stubbs & Julia Stubbs. *Approx 1½ m from town centre. 50 yds from junc with Storrs Rd. 1st house next to grazing field; on-road parking available adjacent to tree-lined roadside stone wall.* **Fri 23, Sat 24 June (10.30-5). Adm £4, chd free. Home-made teas.**

Deceptive ⅓ acre plot of mature trees, landscaped lawns, orchard and cottage style planting. Unusual perennials, species groups such as astrantia, lychnis, thalictrum, heuchera and 40+ clematis. Acers, actea, hosta, ferns and acanthus lie within this interesting garden. Through a hidden gate, another smaller plot of similar planting, with delphinium, helenium, Echinacea, acers and hosta. Winner Chesterfield in Bloom Best Large Garden.

42 THE OLD VICARAGE

The Fields, Middleton by Wirksworth, Matlock, DE4 4NH. Jane Irwing, 01629 825010, irwingjane@gmail.com. *Garden is behind church on Main St & nr school. Travelling N on A6 from Derby turn L at Cromford, at top of hill turn R onto Porter Ln, at T-lights, turn R onto Main St. Park on Main St near DWT. Walk through churchyard. No parking at house.* **Sun 26 Feb (11-4); Sat 8, Sun 9 July (11-5). Adm £4, chd free. Home-made teas at the house, limited covered seating if raining. Visits also by arrangement 4 Mar to 16 Sept for groups of 5 to 25.**

Glorious garden with mixed flowering borders and mature trees in gentle valley with fantastic views to Black Rocks. All-season interest. Acid loving plants such as camellias and rhododendrons are grown in pots in courtyard garden. Tender ferns and exotic plants grown in fernery. Beyond is the orchard, fruit garden, vegetable patch and greenhouse, the home of honey bees, doves and hens. Spectacular Rambling Rector rose over front of house late June to early July.

43 THE PADDOCK
12 Manknell Rd, Whittington Moor, Chesterfield, S41 8LZ. Mel & Wendy Taylor, 01246 451001, debijt9276@gmail.com. *2m N of Chesterfield. Whittington Moor just off A61 between Sheffield & Chesterfield. Parking at Victoria Working Mens Club, garden signed from here.* **Sun 23 Apr, Sun 9 July (11-5). Adm £3.50, chd free. Cream teas. Visits also by arrangement 9 Apr to 30 Sept.**
½ acre garden incorporating small formal garden, stream and koi filled pond. Stone path over bridge, up some steps, past small copse, across the stream at the top and back down again. Past herbaceous border towards a pergola where cream teas can be enjoyed.

44 PARK HALL
Walton Back Lane, Walton, Chesterfield, S42 7LT. Kim Staniforth, 07785 784439, kim.staniforth@btinternet.com. *2m SW of Chesterfield centre. From town on A619 L into Somersall Lane. On A632 R into Acorn Ridge. Park on field side only of Walton Back Lane.* **Visits by arrangement Apr to July for groups of 10+. Minimum group charge £50. Adm £6, chd free. Light refreshments.**
Romantic 2 acre plantsmans garden, in a stunningly beautiful setting surrounding C17 house (not open) four main rooms, terraced garden, parkland area with forest trees, croquet lawn, sunken garden with arbours, pergolas, pleached hedge, topiary, statuary, roses, rhododendrons, camellias, several water features. Runner-up in Daily Telegraph Great British Gardens Competition 2018. Two steps down to gain access to garden.

45 ◆ RENISHAW HALL & GARDENS
Renishaw Park, Sheffield, S21 3WB. Alexandra Hayward, 01246 432310, enquiries@renishaw-hall.co.uk, www.renishaw-hall.co.uk. *10m from Sheffield city centre. By car: Renishaw Hall only 3m from J30 on M1, well signed from junction r'about.* **For NGS: Wed 3 May, Wed 5 July (10.30-4.30). Adm £9, chd £4.50. Home-made teas at the Cafe within the courtyard. For other opening times and information, please phone, email or visit garden website.**
Renishaw Hall and Gardens boasts 7 acres of stunning gardens created by Sir George Sitwell in 1885. The Italianate gardens feature various rooms with extravagant herbaceous borders. Rose gardens, rare trees and shrubs, National Collection of Yuccas, sculptures, woodland walks and lakes create a magical and engaging garden experience. The National Garden Scheme openings coincide with the bluebells being in flower on Wednesday 3rd May and the roses blooming on Wednesday 5th July. For Café bookings please phone the office on 01246 432310. Wheelchair route around garden, map will be given out with full instruction.

NPC

GROUP OPENING

46 REPTON NGS VILLAGE GARDENS
Repton, Derby, DE65 6FQ. *6m S of Derby. From A38/A50, S of Derby, follow signs to Willington, then Repton, then R at r'about towards Newton Solney to reach 10 Chestnut Way, other gardens signposted from the r'about.* **Mon 29 May, Sat 15 July (1-5.30). Combined adm £6, chd free. Home-made teas at 10 Chestnut Way.**

10 CHESTNUT WAY
DE65 6FQ. Pauline Little, 01283 702267, plittle@hotmail.co.uk.
Open on all dates
Visits also by arrangement Feb to Oct for groups of 5 to 25. £8.00 inc refreshments.

22 PINFOLD CLOSE
DE65 6FR. Mr & Mrs O Jowett.
Open on all dates

REPTON ALLOTMENTS
DE65 6FX. Mr A Topping.
Open on Sat 15 July

Repton is a thriving village dating back to Anglo Saxon times. The village gardens are all very different, ranging from the very small to very large, several of them have new features for 2023. 10 Chestnut Way is a plantaholic's garden much changed over winter - be prepared to be surprised. 22 Pinfold Close is a small garden but is packed full with a special interest in tropical plants and has a new orchid house. Repton Allotments is a small set of allotments currently undergoing a revival with community area, new polytunnel and attractive views across Derbyshire. All gardens have plenty of seats. Some gardens have grass or gravel paths but most areas wheelchair accessible.

47 RIVERSVALE LODGE
Riversvale, Buxton, SK17 6UZ. Dena Lewis, 01298 70504, lewis.riversvale@btinternet.com. *½m from Buxton town centre just off the A53 to Leek. Turn R off St John's Rd, down Gadley Ln. This becomes single track so cars need to park at the top of the lane. Short walk down to the garden, following the NGS signs.* **Visits by arrangement Apr to Sept for groups of up to 15. Adm £5, chd free. Adm inc tea/ coffee & cake.**
My garden is small but full of plants collected over nearly 30 yrs. The aim is for interest throughout the seasons, based on foliage, texture and unusual plants which will flourish at over 1000ft. A wide range of insects has resulted. The site provides a clay loam soil and a range of aspects so that many plant groups thrive.

48 40 ST OSWALD'S CRESCENT
Ashbourne, DE6 1FS. Anne McSkimming, 01335 342235, anneofashbourne@hotmail.com. *From the centre car park (Shaw Croft) in Ashbourne, go along Park Rd. Turn L into Park Ave then 2nd R into St Oswald's Crescent. Follow the road til you see signs. Park on road.* **Visits by arrangement in June for groups of up to 8. Adm £4, chd free. Light refreshments.**
Small enclosed town garden, developed by a plantswoman, from scratch 11yrs ago. Trees, lots of shrubs, perennials, climbers. Some unusual specimens. Wide and narrow beds. Plenty of colour and interest throughout the season.

49 THE SMITHY
Church Street, Buxton, SK17 6HD. Roddie & Kate MacLean, 07753 896848, director@creative-heritage.net. *200 metres S of Buxton Market Place, in Higher Buxton. Located on access-only Church St, which cuts corner between B5059 & A515. Walk*

Chevin Brae

S from Buxton Market Place car park. Take slight R off A515, between Scriveners Bookshop & The Swan Inn. **Sat 10, Sun 11 June (11-5). Adm £3.50, chd free. Visits also by arrangement 1 May to 20 Aug for groups of 10 to 20.**
Small oasis of calm in the town centre, designed by its architect-owners. Pretty colour-themed borders and dappled tree cover. Many visitors comment about how we've optimised the use of the space without appearing to cram it all in. Herbaceous borders, wildlife pond, octagonal greenhouse, raised vegetable beds and several different seating areas for eating outdoors.

50 NEW SNITTERTON HALL

Snitterton, Matlock, DE4 2JG. Simon Haslam & Kate Alcock. *1m nw of Matlock. Follow signs from Matlock to Winster, heading up the hill past Sainsbury's. Ignore SatNav if it sends you to Matlock Meadows Ice creams & stay on the road for ½m until you see our signs.* **Sun 7 May (10.30-4.30). Adm £8, chd free. Home-made teas.**
Snitterton Hall is a Grade 1 listed late Elizabethan manor house, and its gardens extend to over 4 acres. While little of the original C16 and C17 gardens remain, there are areas of formal planting, a productive vegetable garden and more naturally planted areas extending into the surrounding countryside - with a delightful collection of sculptures. The gardens are a real haven. Lots of garden sculptures - particularly driftwood sculptures by James Doran-Webb and bronzes by Helen Sinclair.

In 2022 the National Garden Scheme donated £3.11 million to our nursing and health beneficiaries

GROUP OPENING

51 STANTON IN PEAK GARDENS

Stanton-In-The-Peak, Matlock, DE4 2LR. *At the top of the hill in Stanton in Peak, on the rd to Birchover. Stanton in Peak is 5m S of Bakewell. Turn off the A6 at Rowsley, or at the B5056 & follow signs up the hill.* **Sat 22, Sun 23 July (1-5). Combined adm £5, chd free. Home-made teas at 2 Haddon View. At Woodend a pop-up pub serves real ale from the barrel.**

2 HADDON VIEW
Steve Tompkins.

WOODEND COTTAGE
Will Chandler.

Stanton in Peak is a hillside, stone village with glorious views, and is a Conservation Area in the Peak District National Park. Two gardens are open; Steve's garden at 2 Haddon View is at the top of the village and is one tenth acre crammed with plants. Follow the winding path up the garden with a few steps. There are cacti flowering in the greenhouse, lots of pots, herbaceous borders, three wildlife ponds with red and pink water lilies, a koi pond, rhododendrons and a cosy summerhouse. Just down the hill is Woodend where Will has constructed charming roadside stone follies on a strip of raised land along the road. The hidden rear garden has diverse planting, and an extended vegetable plot. All with breathtaking views. A pop-up 'pub' will have draught beer from a local brewery.

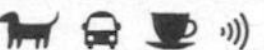

52 26 STILES ROAD

Alvaston, Derby, DE24 0PG. Mr Colin Summerfield & Karen Wild. *Off A6. 2m SE of Derby city centre. From city take A6 (London Rd) towards Alvaston. At Blue Peter Island take L into Beech Ave 1st R into Kelmoor Rd & 1st L into Stiles Rd.* **Sun 7, Sun 21 May (1-5). Adm £3, chd free. Home-made teas.**
Our small town garden is full of surprises. Pass through our new wisteria tunnel and fernery to find 2 ponds, a stream, mini woodland walk, seating areas, an eclectic mix of plants throughout the garden, veg plot and mature fruit trees, over 30 clematis,10 acers, 3 wisteria. Front garden features an aubrietia wall, azaleas and clematis plus much more. Close to Elvaston Castle. Half of garden is wheelchair accessible, most can be seen from patio area.

53 ◆ TISSINGTON HALL

Tissington, Ashbourne, DE6 1RA. Sir Richard & Lady FitzHerbert, 07836 782439, sirrichard@tissingtonhall.co.uk, www.tissingtonhall.co.uk. *4m N of Ashbourne. E of A515 on Ashbourne to Buxton Rd in centre of the beautiful Estate Village of Tissington.* **For NGS: Mon 29 May, Mon 14, Mon 21, Mon 28 Aug (12-3). Adm £7, chd £3.50. Tea at Award winning Herbert's Fine English Tearooms. Tel 01335 350501. For other opening times and information, please phone, email or visit garden website.**
Large garden celebrating over 85yrs in the National Garden Scheme, with stunning rose garden on west terrace, herbaceous borders and 5 acres of grounds. Herbert's Tearooms Andrew Holmes Butchers and Onawick Candle workshop also open in the village. Wheelchair access advice from ticket seller. Please seek staff and we shall park you nearer the gardens.

Park Hall

54 27 WASH GREEN
Wirksworth, Matlock, DE4 4FD. Mr Paul & Mrs Kathy Harvey, 01629 822218, pandkharvey@btinternet.com. *⅓m E of Wirksworth centre. From Wirksworth centre, follow B5035 towards Whatstandwell. Cauldwell St leads over railway bridge to Wash Green, 200 metres up steep hill on L. Park in town or uphill from garden entry.* **Sun 14 May, Sun 27 Aug (11-4). Adm £4, chd free. Home-made teas inc gluten free options. Visits also by arrangement 1 Apr to 29 Oct for groups of up to 30.**
Secluded 1 acre garden, with outstanding 360 deg views. Inner enclosed area has topiary, pergola and lawn surrounded by mixed borders, with paved seating area. The larger part of the garden has sweeping lawns, with large borders, beds, wildlife pond with summer house, bog garden, areas of woodland and specimen trees. A quarter of the plot is a productive fruit and vegetable garden with polytunnel. Drop off at property entry for wheelchair access. The inner garden has flat paths with good views of whole garden.

55 12 WATER LANE
Middleton, Matlock, DE4 4LY. Hildegard Wiesehofer, 07809 883393, wiesehofer@btinternet.com. *Approx 2½m SW of Matlock. 1½m NW of Wirksworth, 8m from Ashbourne. From Derby: At A6 & B5023 intersection take Wirksworth Rd. Follow NGS signs. From Ashbourne: take Matlock Rd to Middleton. Follow NGS signs. Park on main road. Limited parking in Water Lane.* **Sun 27, Mon 28 Aug (11-4). Adm £4, chd free. Home-made teas inc gluten free options.**
Small, eclectic hillside cottage garden on different levels, created as a series of rooms. Each room has been designed to capture the stunning views over Derbyshire and Nottinghamshire. Mini woodland walk, ponds, eastern and 'infinity' garden. Emphasis is on holistic, sustainable and organic principles. Glorious views and short distance from High Peak Trail, Middleton Top and Engine House.

56 13 WESTFIELD ROAD
Swadlincote, DE11 0BG. Val & Dave Booth. *5m E of Burton-on-Trent, off A511. Take A511 from Burton-on-Trent. Follow signs for Swadlincote. Turn R into Springfield Rd, take 3rd R into Westfield Rd.* **Sun 18 June, Sun 6 Aug (1-5). Adm £3.50, chd free. Home-made teas.**
A deceptive country-style garden in Swadlincote, a real gem. The garden is on two levels of approx ½ acre. Packed herbaceous borders designed for colour. Shrubs, baskets and tubs. Lots of roses. Greenhouses, raised-bed vegetable area, fruit trees and two ponds. Free range chicken area. Plenty of seating to relax and take in the wonderful planting from two passionate gardeners.

57 WILD IN THE COUNTRY
Hawkhill Road, Eyam, Hope Valley, S32 5QQ. Mrs Gill Bagshawe, www.wildinthecountryflowers.co.uk. *In Eyam, follow signs to public car park. Located next to Eyam Museum & opp public car park on Hawkhill Rd.* **Sat 17 June (11-4). Adm £3, chd free.**
A rectangular plot devoted totally to growing flowers and foliage for cutting. Sweet pea, rose, larkspur, cornflower, nigella, ammi. All the florist's favourites can be found here. There is a tea room, a village pub and several cafes in the village to enjoy refreshments.

58 15 WINDMILL LANE
Ashbourne, DE6 1EY. Jean Ross & Chris Duncan. *Take A515 (Buxton Rd) from the market place & at the top of the hill turn R into Windmill Lane; house by 4th tree on L.* **Sun 21 May (1.30-5). Adm £3, chd free.**
Densely planted town garden providing ideas others may wish to develop. Using a limited palette (mainly white, pink, mauve and burgundy) and emphasising leaf shape and colour, the key aims in establishing this garden were all yr interest; low maintenance; attraction of pollinators; and growing some soft fruit and veg. Garden now contains many drought-resistant plants. Extensive views.

59 26 WINDMILL RISE
Belper, DE56 1GQ. Kathy Fairweather. *From Belper Market Place take Chesterfield Rd towards Heage. Top of hill, 1st R Marsh Lane, 1st R Windmill Lane, 1st R Windmill Rise - limited parking on Windmill Rise - disabled mainly.* **Sun 23 July (11.30-4). Adm £3, chd free. Light refreshments inc gluten free & vegan options.**
Behind a deceptively ordinary façade, lies a real surprise. A lush oasis, much larger than expected, with an amazing collection of rare and unusual plants. A truly plant lovers' organic garden divided into sections: woodland, Japanese, secret garden, cottage, edible, ponds and small stream. Many seating areas, inc a new summerhouse in which to enjoy a variety of refreshments. Home made delicious cakes and light lunches available.

60 NEW YEW TREE BUNGALOW
Thatchers Lane, Tansley, Matlock, DE4 5FD. Jayne Conquest, 07745 093177, jayneconquest@btinternet.com. *2m E of Matlock. On A615, 2nd R after Tavern at Tansley. Parking in Charles Gregory & Sons Timber yard car park opp Tavern Inn.* **Sun 16 Apr, Sun 18 June, Sun 10 Sept (11-4.30). Adm £5, chd free. Home-made teas. Visits also by arrangement 1 Mar to 9 Sept for groups of 6 to 30.**
Plants women's ½ acre cottage style garden on a slope created by the present owners over 30 years. Many mixed borders planted with many choice and unusual trees, shrubs, herbaceous perennial, bulbs and annuals to provide year-round interest. Several seats to view different aspects of the garden. Shady area planted with a wide range of ferns. Vegetable and fruit garden. The garden is sloping but most of garden can be seen from patio.

Our 2022 donation to Hospice UK meant that 911 NHS and Social Care staff accessed individual counselling support

DEVON

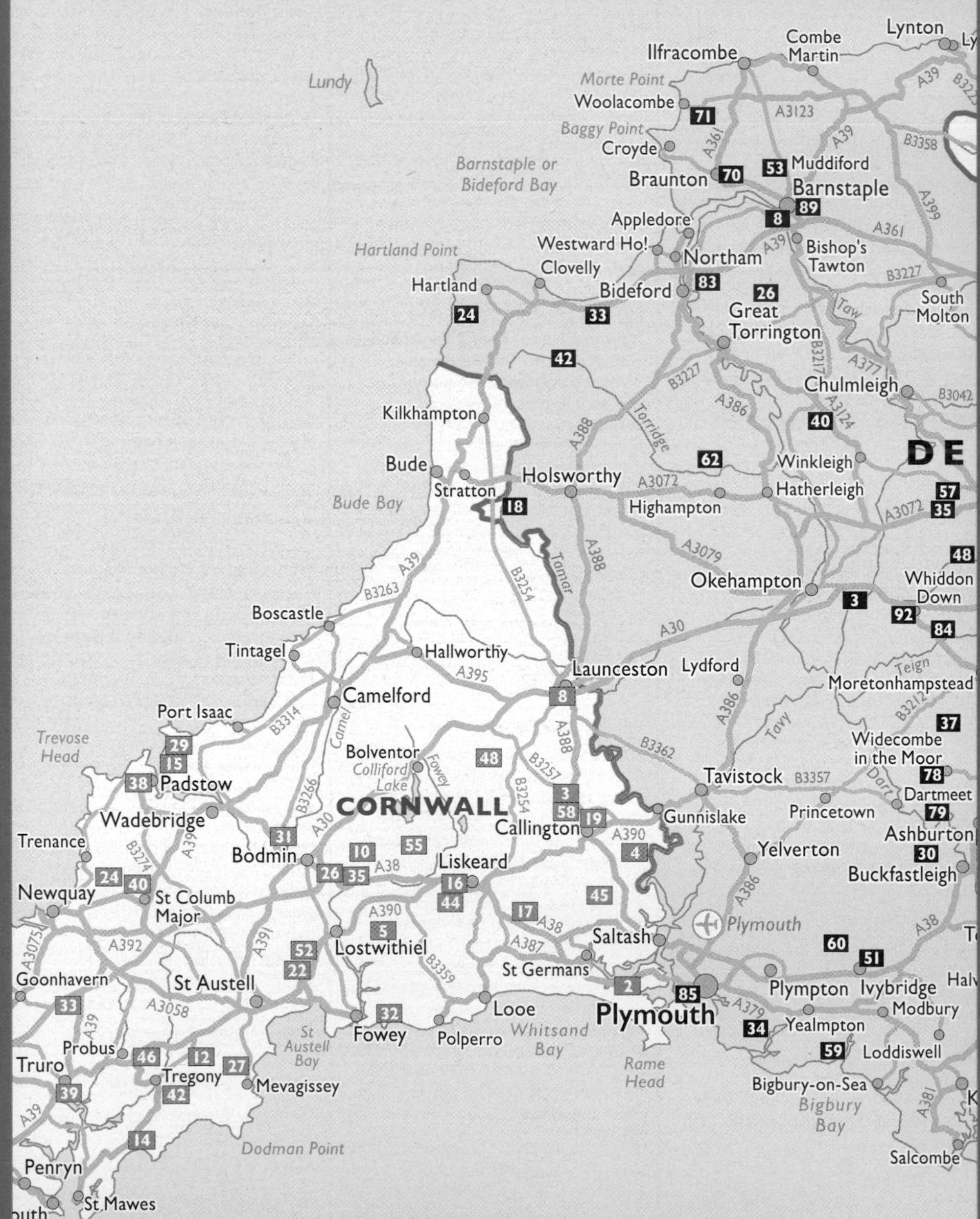
Lundy
Ilfracombe
Combe Martin
Lynton
Morte Point
Woolacombe
71
A3123
A39
Baggy Point
Croyde
A361
B3358
Barnstaple or Bideford Bay
Braunton
70
53
Muddiford
Barnstaple
89
8
A399
Appledore
A361
Hartland Point
Westward Ho!
Northam
A39
Bishop's Tawton
Clovelly
B3227
Hartland
Bideford
83
26
South Molton
24
33
Great Torrington
Taw
42
B3217
A377
B3227
Chulmleigh
B3042
Kilkhampton
A386
A3124
Torridge
40
A388
DE
Bude
62
Winkleigh
Holsworthy
A3072
Hatherleigh
57
Stratton
Highampton
A3072
35
Bude Bay
18
A3079
48
A388
A39
B3263
B3254
Tamar
Okehampton
Whiddon Down
3
92
84
Boscastle
A30
Tintagel
Hallworthy
Launceston
Lydford
Teign
A395
Moretonhampstead
Camelford
8
Port Isaac
B3314
A388
A386
B3212
Trevose Head
29
15
Camel
Tavy
37
Bolventor
48
B3362
Widecombe in the Moor
Colliford Lake
Fowey
B3257
78
38
Padstow
B3266
B3254
3
Tavistock
B3357
Dart
Dartmeet
CORNWALL
58
19
Gunnislake
Princetown
79
Wadebridge
A30
Callington
Ashburton
A39
31
A390
Trenance
B3274
Bodmin
10
55
Liskeard
4
Yelverton
30
A38
Buckfastleigh
24
40
26
35
16
Newquay
St Columb Major
44
45
A386
A390
17
A38
Plymouth
A391
5
A38
A392
52
Lostwithiel
A387
Saltash
60
A3075
22
B3359
51
St Germans
Goonhavern
St Austell
2
Plympton
Ivybridge
A3058
85
A379
Modbury
33
Looe
Plymouth
Fowey
32
Whitsand Bay
34
Yealmpton
A39
Polperro
St Austell Bay
59
Loddiswell
Probus
46
12
Rame Head
Truro
27
Tregony
Mevagissey
Bigbury-on-Sea
39
42
Bigbury Bay
A39
A381
14
Dodman Point
Salcombe
Penryn
St Mawes

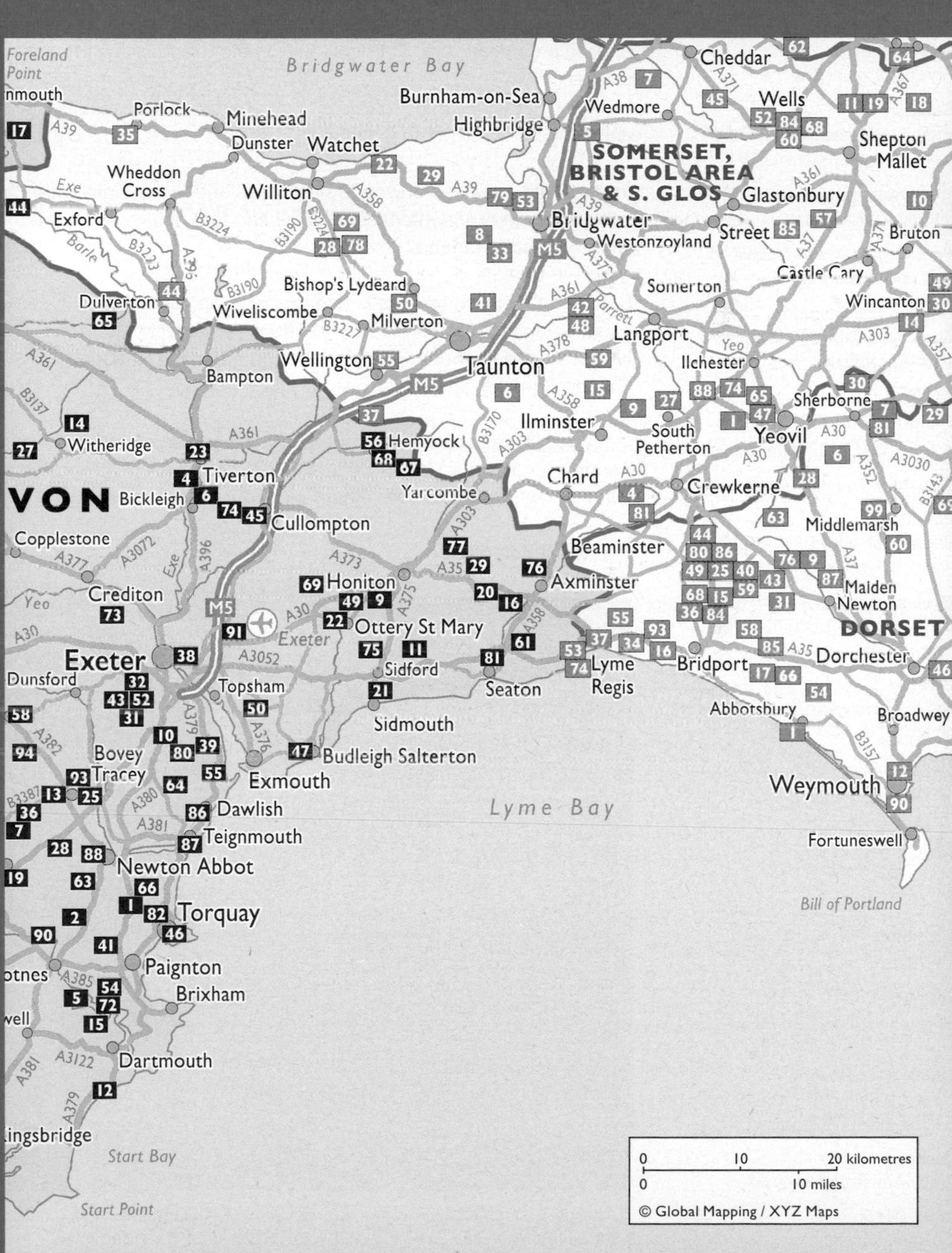
Foreland Point
Bridgwater Bay
Cheddar
Burnham-on-Sea
Wedmore
Wells
Porlock
Minehead
Highbridge
Shepton Mallet
Dunster
Watchet
SOMERSET, BRISTOL AREA & S. GLOS
Wheddon Cross
Williton
Glastonbury
Exe
Exford
Bridgwater
Street
Bruton
Barle
Westonzoyland
Castle Cary
Dulverton
Bishop's Lydeard
Somerton
Wiveliscombe
Milverton
Wincanton
Langport
Parrett
Wellington
Taunton
Ilchester
Yeo
Bampton
Sherborne
Ilminster
South Petherton
Yeovil
Witheridge
Hemyock
Tiverton
Chard
Crewkerne
VON
Bickleigh
Yarcombe
Cullompton
Middlemarsh
Copplestone
Beaminster
Honiton
Axminster
Maiden Newton
Crediton
Exeter
Ottery St Mary
DORSET
Exeter
Dorchester
Lyme Regis
Bridport
Dunsford
Sidford
Seaton
Topsham
Abbotsbury
Broadwey
Sidmouth
Bovey Tracey
Budleigh Salterton
Exmouth
Weymouth
Dawlish
Lyme Bay
Teignmouth
Fortuneswell
Newton Abbot
Bill of Portland
Torquay
Paignton
Brixham
Dartmouth
Kingsbridge
Start Bay
Start Point
0 10 20 kilometres
0 10 miles
© Global Mapping / XYZ Maps

Volunteers

County Organiser & Central Devon
Miranda Allhusen 01647 440296
Miranda@allhusen.co.uk

County Treasurer
Nigel Hall 01884 38812
nigel.hall@ngs.org.uk

Publicity
Brian Mackness 01626 356004
brianmackness@clara.co.uk

Tracy Armstrong 07729 225549
tracygorsty@aol.com

Paul Vincent 01803 722227
paul.vincent@ngs.org.uk

Booklet Co-ordinator
Anne Sercombe 01626 923170
anne.sercombe@ngs.org.uk

Talks
Julia Tremlett 01392 832671
jandjtremlett@hotmail.com

Assistant County Organisers

East Devon
Penny Walmsley 01404 831375
walyp_uk@yahoo.co.uk

Exeter
Jenny Phillips 01392 254076
jennypips25@hotmail.co.uk

Exmoor
Angela Percival 01598 741343
angela.percival@ngs.org.uk

North Devon
Jo Hynes
hynesjo@gmail.com

North East Devon
Jill Hall 01884 38812
jill22hall@gmail.com

Plymouth
Position vacant

South Hams
Sally Vincent 01803 722227
sallyvincent14@gmail.com

South West Devon
Naomi Hindley 01364 654902
naomi.hindley@ngs.org.uk

Torbay
Tracy Armstrong (see above)

West Devon
Position Vacant

Devon is a county of great contrasts in geography and climate, and therefore also in gardening.

The rugged north coast has terraces clinging precariously to hillsides so steep that the faint-hearted would never contemplate making a garden there. But here, and on the rolling hills and deep valleys of Exmoor, despite a constant battle with the elements, National Garden Scheme gardeners create remarkable results by choosing hardy plants that withstand the high winds and salty air.

In the south, in peaceful wooded estuaries and tucked into warm valleys, gardens grow bananas, palms and fruit usually associated with the Mediterranean.

Between these two terrains is a third: Dartmoor, 365 square miles of rugged moorland rising to 2000 feet, presents its own horticultural demands. Typically, here too are many National Garden Scheme gardens.

In idyllic villages scattered throughout this very large county, in gardens large and small, in single manors and in village groups within thriving communities – gardeners pursue their passion.

Below: Sutton Mead

@Devon NGS | @DevonNGS | @ngsdevon

OPENING DATES

All entries subject to change. For latest information check **www.ngs.org.uk**

Extended openings are shown at the beginning of the month.

Map locator numbers are shown to the right of each garden name.

February

Snowdrop Festival

Friday 3rd
Higher Cherubeer 40

Friday 10th
Higher Cherubeer 40

Saturday 11th
The Mount, Delamore 60

Sunday 12th
Ashley Court 4
The Mount, Delamore 60

Saturday 18th
Higher Cherubeer 40

Sunday 26th
East Worlington House 27

March

Every Saturday and Sunday from Saturday 11th
Haldon Grange 31

Sunday 5th
East Worlington House 27

Saturday 18th
Samlingstead 71

Sunday 19th
Bickham House 10
Chevithorne Barton 23

Wednesday 22nd
Haldon Grange 31

Saturday 25th
Idestone Barton 43

Sunday 26th
Heathercombe 37

Wednesday 29th
Haldon Grange 31

April

Every Saturday and Sunday
Haldon Grange 31

Saturday 1st
Monkscroft 57

Sunday 2nd
Bickham House 10
High Garden 39
Monkscroft 57
NEW Sherwood 73
Upper Gorwell House 89

Saturday 8th
Ashley Court 4
Byes Reach 21

Sunday 9th
Andrew's Corner 3
Byes Reach 21
Kia-Ora Farm & Gardens 45

Monday 10th
Andrew's Corner 3
Byes Reach 21
Haldon Grange 31
Kia-Ora Farm & Gardens 45

Friday 14th
Sidbury Manor 75

Sunday 16th
Andrew's Corner 3
Sidbury Manor 75

Wednesday 19th
Haldon Grange 31

Friday 21st
Musselbrook Cottage Garden 62
Pounds 67

Saturday 22nd
Musselbrook Cottage Garden 62
Pounds 67
Regency House 68

Sunday 23rd
Bickham House 10
Kia-Ora Farm & Gardens 45
Musselbrook Cottage Garden 62
Regency House 68
St Merryn 70
Whitstone Farm 93

Saturday 29th
Greatcombe 30
South Wood Farm 77
Torview 88

Sunday 30th
Andrew's Corner 3
Greatcombe 30
Kia Ora Farm & Gardens 45
South Wood Farm 77
Torview 88
Whitstone Farm 93

May

Every Saturday and Sunday
Haldon Grange 31

Monday 1st
Andrew's Corner 3
Greatcombe 30
Haldon Grange 31
Kia-Ora Farm & Gardens 45
Torview 88

Saturday 6th
NEW Bagtor Mill Gardens 7

Sunday 7th
NEW Bagtor Mill Gardens 7
Bickham House 10
Heathercombe 37
Middle Well 54
◆ Mothecombe House 59
Spring Lodge 80
Upper Gorwell House 89

Monday 8th
Haldon Grange 31

Wednesday 10th
Avenue Cottage 5

Thursday 11th
Avenue Cottage 5

Friday 12th
Musselbrook Cottage Garden 62

Saturday 13th
Musselbrook Cottage Garden 62
Spitchwick Manor 79
NEW Stonehouse Lawn Tennis Club 85

Sunday 14th
Goodwill Gardens 28
Heathercombe 37
Hole Farm 42
Kia-Ora Farm & Gardens 45
Musselbrook Cottage Garden 62
St Merryn 70
NEW Sherwood 73
Spitchwick Manor 79
NEW Stonehouse Lawn Tennis Club 85

Wednesday 17th
Haldon Grange 31

Friday 19th
NEW Beech Walk House 9
NEW Haytor Gardens 36
Moretonhampstead Gardens 58

Saturday 20th
Bradford Tracey House 14
NEW Haytor Gardens 36
Heathercombe 37
Moretonhampstead Gardens 58
The Old Vicarage 65

Sunday 21st
NEW Beech Walk House 9
Bickham House 10
Bradford Tracey House 14
Chevithorne Barton 23
NEW Haytor Gardens 36
Heathercombe 37
Moretonhampstead Gardens 58
The Old Vicarage 65

Tuesday 23rd
Heathercombe 37

Wednesday 24th
Heathercombe 37

Thursday 25th
Heathercombe 37

Friday 26th
Heathercombe 37

Saturday 27th
Breach 16
Brendon Gardens 17
◆ Goren Farm 29
Greatcombe 30
Heathercombe 37

Languard Place 46
Springfield House 81
Stone Farm 83

Sunday 28th
Andrew's Corner 3
Breach 16
Brendon Gardens 17
The Bridge Mill 18
◆ Cadhay 22
Goodwill Gardens 28
◆ Goren Farm 29
Greatcombe 30
Heathercombe 37
Higher Cherubeer 40
Kia-Ora Farm & Gardens 45
Languard Place 46
Stone Farm 83

Monday 29th
Andrew's Corner 3
◆ Cadhay 22
Goodwill Gardens 28
Greatcombe 30
Haldon Grange 31
Kia-Ora Farm & Gardens 45

June

Every Saturday and Sunday to Sunday 11th
Haldon Grange 31

Friday 2nd
Idestone Barton 43
Southcombe Barn 78

Saturday 3rd
Abbotskerswell Gardens 1
Dunley House 25
Higher Orchard Cottage 41
Idestone Barton 43
Littlefield 50
Whiddon Goyle 92
NEW Willowrey Farm 94

Sunday 4th
Abbotskerswell Gardens 1
Bickham House 10
Dunley House 25
Hayne 35
High Garden 39
Higher Orchard Cottage 41
Littlefield 50
Whiddon Goyle 92
NEW Willowrey Farm 94

Monday 5th
Southcombe Barn 78

Friday 9th
Bramble Torre 15
Marshall Farm 52
Mill House 56
Pounds 67
Regency House 68
Southcombe Barn 78

Sandridge Park

Saturday 10th
Bramble Torre 15
East Woodlands Farmhouse 26
Heathercombe 37
Mill House 56
NEW Ogwell Gardens 63
Pounds 67
Regency House 68
NEW Sandridge Park 72

Sunday 11th
Bramble Torre 15
◆ Docton Mill 24
East Woodlands Farmhouse 26
Heathercombe 37
Kia-Ora Farm & Gardens 45
Marshall Farm 52
Mill House 56
NEW Ogwell Gardens 63
Regency House 68
St Merryn 70
NEW Sandridge Park 72
NEW Sherwood 73
Upper Gorwell House 89

Monday 12th
Southcombe Barn 78

Tuesday 13th
Heathercombe 37

Wednesday 14th
Heathercombe 37

Thursday 15th
Heathercombe 37

Friday 16th
Greatcombe 30
Heathercombe 37
Musselbrook Cottage Garden 62

Saturday 17th
Ashley Court 4
Bovey Tracey Gardens 13
Greatcombe 30
Harbour Lights 33
Heathercombe 37
Musselbrook Cottage Garden 62
Regency House 68
Russet House 69
Shutelake 74
Teignmouth Gardens 87
West Clyst Barnyard Gardens 91

Sunday 18th
Bovey Tracey Gardens 13
Greatcombe 30
Harbour Lights 33
Heathercombe 37
Musselbrook Cottage Garden 62
Regency House 68
Russet House 69
Shutelake 74
Teignmouth Gardens 87
West Clyst Barnyard Gardens 91

Tuesday 20th
Heathercombe 37

Wednesday 21st
Heathercombe 37

Thursday 22nd
NEW ◆ Blackbury Honey Farm 11
Heathercombe 37

Friday 23rd
Heathercombe 37

Saturday 24th
Heathercombe 37
Lewis Cottage 48
Musbury Barton 61
Regency House 68

Sunday 25th
Bickham House 10
Heathercombe 37
Kia-Ora Farm & Gardens 45
Lewis Cottage 48
Musbury Barton 61
Regency House 68

Tuesday 27th
Heathercombe 37

Wednesday 28th
Heathercombe 37

Thursday 29th
Heathercombe 37

Friday 30th
Greatcombe 30
Heathercombe 37
Socks Orchard 76

July

Saturday 1st
Backswood Farm 6
Greatcombe 30
Heathercombe 37
NEW Heavitree's Unexpected Gardens 38
NEW 2 Jubilee Villas 44
Socks Orchard 76

Stone Farm 83

Sunday 2nd
Backswood Farm 6
Greatcombe 30
Heathercombe 37
NEW Heavitree's Unexpected Gardens 38
High Garden 39
NEW 2 Jubilee Villas 44
Socks Orchard 76
Stone Farm 83

Friday 7th
◆ Marwood Hill Garden 53

Saturday 8th
Springfield House 81

Sunday 9th
Barnstaple World Gardens 8
Bickham House 10
Kia-Ora Farm & Gardens 45
Upper Gorwell House 89

Thursday 13th
NEW ◆ Blackbury Honey Farm 11

Friday 14th
Greatcombe 30
NEW The Walled Garden 90

Saturday 15th
Greatcombe 30
Samlingstead 71

Sunday 16th
Greatcombe 30
NEW The Walled Garden 90

Friday 21st
Musselbrook Cottage Garden 62

Saturday 22nd
Am Brook Meadow 2
NEW 2 Middlewood 55
Musselbrook Cottage Garden 62

Sunday 23rd
Am Brook Meadow 2
Bickham House 10
Kia-Ora Farm & Gardens 45
NEW 2 Middlewood 55
Musselbrook Cottage Garden 62

Friday 28th
Greatcombe 30

Saturday 29th
Greatcombe 30
Lewis Cottage 48
The Old School House 64
Squirrels 82

Sunday 30th
Avenue Cottage 5
Chevithorne Barton 23
Greatcombe 30
Lewis Cottage 48
The Old School House 64
Squirrels 82
Whitstone Farm 93

August

Saturday 5th
Brendon Gardens 17
Byes Reach 21

Sunday 6th
Bickham House 10
Brendon Gardens 17
Byes Reach 21
Kia-Ora Farm & Gardens 45

Saturday 12th
The Old Vicarage 65

Sunday 13th
Little Ash Bungalow 49
The Old Vicarage 65

Sunday 20th
Bickham House 10
Kia-Ora Farm & Gardens 45

Saturday 26th
Lewis Cottage 48
Stone Farm 83

Sunday 27th
◆ Cadhay 22
Kia-Ora Farm & Gardens 45
Lewis Cottage 48
Stone Farm 83

Monday 28th
Ashley Court 4
◆ Cadhay 22
Kia-Ora Farm & Gardens 45

September

Saturday 2nd
Moretonhampstead Gardens 58

Sunday 3rd
Bickham House 10
High Garden 39
Moretonhampstead Gardens 58

Saturday 9th
Brocton Cottage 19

Sunday 10th
Brocton Cottage 19
Hole Farm 42
Kia-Ora Farm & Gardens 45

Friday 15th
Pounds 67

Saturday 16th
Pounds 67
South Wood Farm 77

Sunday 17th
South Wood Farm 77

Sunday 24th
Kia-Ora Farm & Gardens 45
Upper Gorwell House 89

October

Friday 6th
◆ Marwood Hill Garden 53

Saturday 7th
◆ Stone Lane Gardens 84

Saturday 14th
Dunley House 25
◆ Stone Lane Gardens 84

Sunday 15th
Dunley House 25
NEW Sherwood 73

Saturday 21st
Regency House 68

Sunday 22nd
Chevithorne Barton 23
Regency House 68

February 2024

Friday 2nd
Higher Cherubeer 40

Friday 9th
Higher Cherubeer 40

Saturday 17th
Higher Cherubeer 40

By Arrangement

Arrange a personalised garden visit with your club, or group of friends, on a date to suit you. See individual garden entries for full details.

Andrew's Corner 3
Avenue Cottage 5
Backswood Farm 6
Bickham House 10
Breach 16
Brocton Cottage 19
Byes Reach 21
East Woodlands Farmhouse 26
Haldon Grange 31
Hall Farm, Brendon Gardens 17
Halscombe Farm 32
Harbour Lights 33
The Haven 34
Heathercombe 37
Higher Cherubeer 40
Higher Tippacott Farm, Brendon Gardens 17
Kia-Ora Farm & Gardens 45
Lee Ford 47
Lewis Cottage 48
Little Ash Bungalow 49
Middle Well 54
NEW 2 Middlewood 55
The Mount, Delamore 60
Musselbrook Cottage Garden 62
The Old Vicarage 65
Regency House 68
St Merryn 70
Samlingstead 71
NEW Sherwood 73
Shutelake 74
South Wood Farm 77
Springfield House 81
Squirrels 82
Stone Farm 83
Stonelands House 86
Sutton Mead, Moretonhampstead Gardens 58
Whiddon Goyle 92
Whitstone Farm 93
NEW Willowrey Farm 94

THE GARDENS

GROUP OPENING

1 ABBOTSKERSWELL GARDENS

Abbotskerswell, TQ12 5PN. *2m SW of Newton Abbot town centre. A381 Newton Abbot/Totnes rd. Sharp L turn from Newton Abbot, R from Totnes. Field parking at Fairfield. Maps available at all gardens & at Church House.* **Sat 3, Sun 4 June (1-5). Combined adm £7, chd free. Home-made teas at Church House in the village. Teas available from 2pm. Maps and tickets from 1pm.**

ABBOTSFORD
Wendy & Phil Grierson.

ABBOTSKERSWELL ALLOTMENTS
Tasha Mundy.

1 ABBOTSWELL COTTAGES
Jane Taylor.

BRIAR COTTAGE
Peggy & David Munden.

FAIRFIELD
Brian Mackness.

1 FORDE CLOSE
Ann Allen.

THE POTTERY
Mr D & Mrs B Dubash.

7 WILTON WAY
Mr Vernon & Mrs Cindy Stunt.

16 WILTON WAY
Katy & Chris Yates.

For 2023 Abbotskerswell offers eight gardens plus the village allotments; ranging from very small to large, the gardens offer a wide range of planting styles and innovative landscaping. Cottage gardens, terracing, wildflower areas, specialist plants and an arboretum. Ideas for every type and size of garden. Visitors are welcome to picnic in the field or arboretum at Fairfield. Sales of plants, garden produce, jams and chutneys. Disabled access to three gardens. Partial access to most others.

2 AM BROOK MEADOW

Torbryan, Ipplepen, Newton Abbot, TQ12 5UP. Jennie & Jethro Marles. *5m from Newton Abbot on A381. Leaving A381 at Causeway Cross go through Ipplepen village, heading towards Broadhempston. Stay on Orley Rd for ¾m. At Poole Cross turn L signed Totnes, then turn 2nd L into field parking.* **Sat 22, Sun 23 July (2-6.30). Adm £6, chd free.**
Country garden developed over past 17 yrs to encourage wildlife. Perennial native wildflower meadows, large ponds with ducks and swans, streams and wild areas covering 10 acres are accessible by gravel and grass pathways. Formal courtyard garden with water features and herbaceous borders and prairie-style planting together with poultry and bees close by. Wheelchair access to most gravel path areas is good, but grass pathways in larger wildflower meadow are weather dependent.

3 ANDREW'S CORNER

Skaigh Lane, Belstone, EX20 1RD. Robin & Edwina Hill, 01837 840332, edwinarobinhill@outlook.com, www.andrewscorner.garden. *3m E of Okehampton. Signed to Belstone from A30. In village turn L, signed Skaigh. Follow NGS signs. Garden approx ½m on R. Visitors may be dropped off at house, parking in nearby field.* **Sun 9, Mon 10, Sun 16, Sun 30 Apr, Mon 1, Sun 28, Mon 29 May (2-5). Adm £5, chd free. Home-made teas. Visits also by arrangement Feb to June.**
Join us as we enter our 52nd year of opening and take a walk on the wild side in this tranquil moorland garden. April openings highlight magnolias, trillium and the lovely erythroniums. Early May the maples, rhododendrons and unusual shrubs provide interest, late May brings the flowering davidia, cornus, embothrium and the spectacular blue poppies. Wheelchair access difficult when wet.

4 ASHLEY COURT

Ashley, Tiverton, EX16 5PD. Tara Fraser & Nigel Jones, 07768 878015, hello@ashleycourtdevon.co.uk, www.ashleycourtdevon.co.uk. *1m S of Tiverton on the Bickleigh road (A396). Turn off the A396 to Ashley & then immed L. Take the drive to the L of Ashley Court Lodge Cottage (don't go up Ashley Back Lane where the SatNav will direct you).* **Sun 12 Feb, Sat 8 Apr, Sat 17 June, Mon 28 Aug (12.30-4.30). Adm £6, chd free. Speciality teas inc many delicious vegan recipes & cakes containing fruits & vegetables from the walled kitchen garden.**
Ashley Court is a small Regency country house with an historically interesting walled kitchen garden currently undergoing restoration. It is unusually situated in a deep valley and has a frost window and the remains of several glasshouses and cold frames. View the apple loft, root stores, stable buildings, woodland walk and lawns, borders and beautiful mature trees. A walk through discovery and history. The walled garden probably pre-dates the 1805 house as there was previously an older Ashley Court on the other side of the garden. Peculiar features such as a curved garden wall and a frost window make it unusual.

5 AVENUE COTTAGE

Ashprington, Totnes, TQ9 7UT. Mr Richard Pitts & Mr David Sykes, 01803 732769, richard.pitts@btinternet.com, www.avenuecottage.com. *3m SE of Totnes. A381 Totnes to Kingsbridge for 1m; L for Ashprington, into village then L by pub. Garden ¼m on R after Sharpham Estate sign.* **Wed 10, Thur 11 May, Sun 30 July (10.30-4). Adm £5, chd free. Pre-booking essential, please visit www.ngs.org.uk for information & booking. Home-made teas. Visits also by arrangement Mar to Sept for groups of up to 25.**
11 acre woodland valley garden, forming part of the Grade II* Sharpham House Landscape. The garden contains many rare and unusual trees and shrubs with views over the River Dart AONB and Sharpham. Azaleas, hydrangeas and magnolias are a speciality. A spring rises in the garden under a spectacularly large rhododendron feeding two ponds and a small stream running through a meadow with mown paths.

6 BACKSWOOD FARM

Bickleigh, Tiverton, EX16 8RA. Andrew Hughes, 01884 855005, andrew@backswood.co.uk. *2m SE of Tiverton. Take A396 off the A361. Turn L to Butterleigh opp Tesco, L at mini r'about, after 150 yds turn R up Exeter Hill for 2m then 1st R to Bickleigh. After 500 yds turn 1st*

R into farm entrance. **Sat 1, Sun 2 July (2-5). Adm £6, chd free. Visits also by arrangement Apr to Aug for groups of 15 to 30. Teas for group visits can be arranged on request.**
Created with a real passion, this newly created 2 acre nature garden provides many uniquely designed homes for wildlife. Wander through the flower meadow visiting individually designed areas, structures, water features and ponds. Seating areas afford stunning views towards Exmoor and Dartmoor. Both native and herbaceous plants have been chosen to benefit insect and bird life. Evolving every year. Dogs on leads. WC available. Not suitable for pushed wheelchairs.

GROUP OPENING

7 NEW BAGTOR MILL GARDENS
Ilsington, Newton Abbot, TQ13 9RT. Mark Wills. *Coming from Haytor at Smokey Cross take the road to Bickington & at Birchanger Cross turn R to Bagtor. Coming from Liverton/Ilsington carry on towards Haytor until Smokey Cross & turn L.* **Sat 6, Sun 7 May (10-4). Combined adm £5, chd free. Cream teas.**

NEW BAGTOR MILL
Mark Wills.

NEW LEMON COTTAGE
Phil & Jamie Richards.

Bagtor Mill is a C17 corn mill beside the River Lemon surrounded by a wildlife garden with rhododendrons, camellias, ponds, bluebells and massive gunnera. The 21 acres of ancient woodland that joins the garden in a steep valley setting is a SSSI following an old packhorse track made by charcoal burners. Opposite is Lemon Cottage with a spring-fed pond, a Japanese-inspired garden and 14 acres of woodland with views across Dartmoor. While the main paths are accessible, not all paths in both gardens are.

GROUP OPENING

8 BARNSTAPLE WORLD GARDENS
Anne Crescent, Little Elche, Barnstaple, EX31 3AF. *NW of Barnstaple centre. 31 Anne Cres L off Old Torrington Rd into Phillips Ave then follow signs: 21 Becklake Close from A3125 to Roundswell on Westermoor Way: 25 Elmfield Rd from B3233 L at Bickington PO.* **Sun 9 July (11-4). Combined adm £5, chd free. Light refreshments. Japanese snacks, teas and cakes.**

31 ANNE CRESCENT
Mr Gavin Hendry.

21 BECKLAKE CLOSE
Karen & Steve Moss.

25 ELMFIELD ROAD
Nigel & Carol Oates.

Explore 3 very different spaces, one focusing on everything Japanese, one on tropical plants, and one to interest the plantsman. 31 Anne Crescent: N Devon's Little Elche. Small urban L-shaped tropical garden complete with wide variety of palms, agaves, bananas, cacti, tree ferns and delightful pond loaded with a wide variety of colourful koi. 21 Becklake Close: All things Japanese. 25 Elmfield Rd: Front: mixed borders and gravel area containing hardy palms and Mount Etna Broom. Rear: L shaped patio leading to lawn edged by mixed borders of unusual plants. The overall impression is of a colourful but calm place, enhanced by the sound of the stream trickling past the end of the garden. A garden to attract those interested in plants with a difference.

The Walled Garden

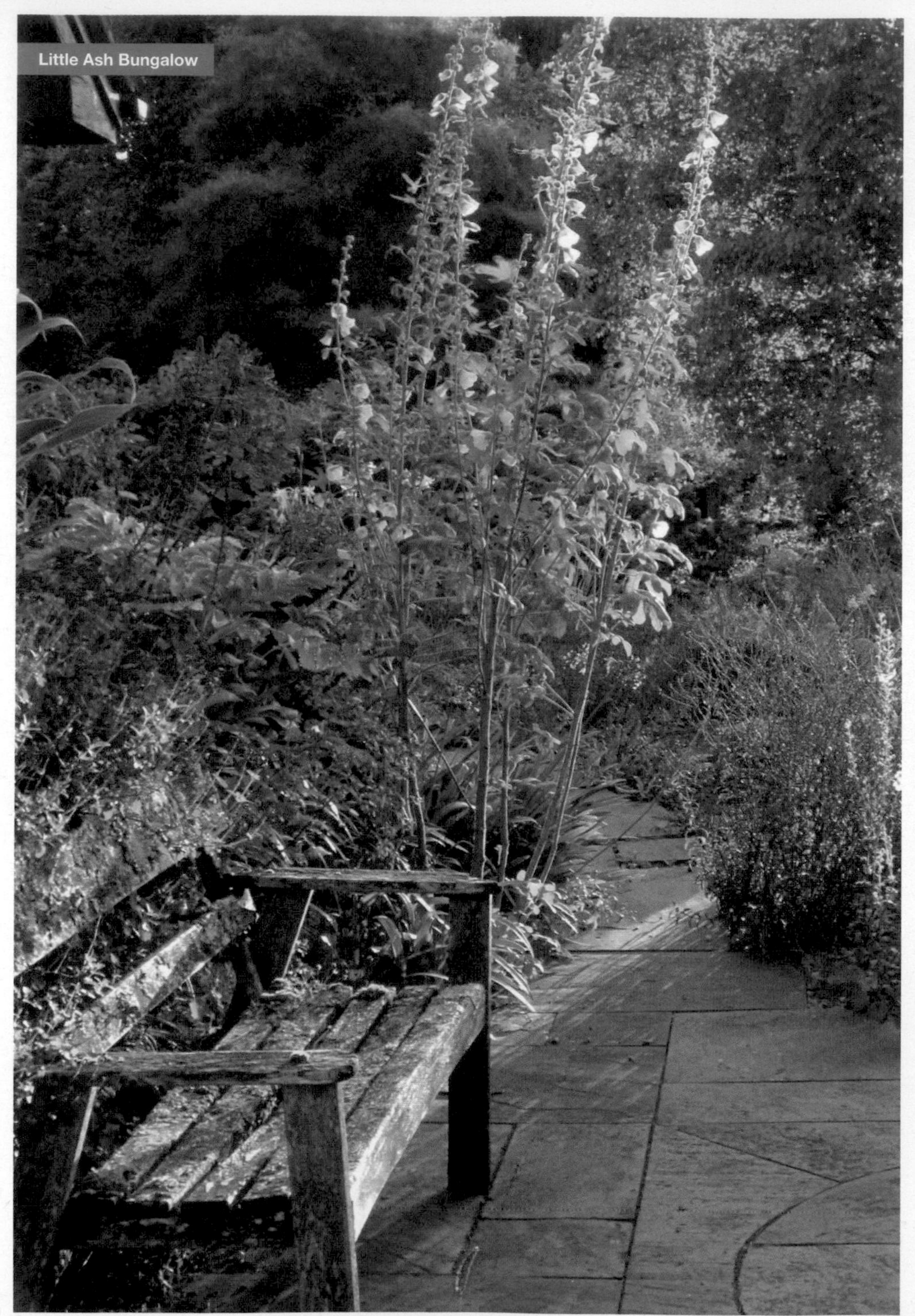
Little Ash Bungalow

9 NEW **BEECH WALK HOUSE**
Gittisham, Honiton, EX14 3AB. Nelly & Richard Marker. *2m SW of Honiton. From Honiton follow signs for Sidmouth onto Sidmouth rd. At the sign for Gittisham, turn R onto Beech Walk. Continue for ½m to reach the entrance of the drive.* **Fri 19, Sun 21 May (1-5). Adm £6, chd free. Home-made teas.**
7 acre garden in an AONB with views across the Devon countryside. A well established/mature collection of rhododendron, camellias, magnolia, wisterias and other shrubs. The lower half of the garden contains many original features of the original Grade I listed Combe House. Other features inc mixed borders, meadow, stumpery, vegetable garden, rockery and bluebell woods.

10 BICKHAM HOUSE
Kenn, Exeter, EX6 7XL. Julia Tremlett, 01392 832671, jandjtremlett@hotmail.com. *6m S of Exeter, 1m off A38. Leave A38 at Kennford Services, follow signs to Kenn, 1st R in village, follow lane for ¾m to end of no through rd. Only use SatNav once you are in the village of Kenn.* **Sun 19 Mar, Sun 2, Sun 23 Apr, Sun 7, Sun 21 May, Sun 4, Sun 25 June, Sun 9, Sun 23 July, Sun 6, Sun 20 Aug, Sun 3 Sept (1.30-5). Adm £6, chd free. Home-made teas. Visits also by arrangement 20 Mar to 31 Aug.**
6 acres with lawns, borders, mature trees. Formal parterre with lily pond. Walled garden with colourful profusion of vegetables and flowers. Palm tree avenue leading to millennium summerhouse. Late summer colour with dahlias, crocosmia, agapanthus etc. Cactus and succulent greenhouse. Pelargonium collection. Lakeside walk. WC, disabled access.

11 NEW **◆ BLACKBURY HONEY FARM**
Southleigh, Colyton, EX24 6JF. Maureen Basterfield, 01404 871600, maureen@basterfield.com, www.blackburyfarm.co.uk. *6m S of Honiton. From Honiton take the Sidmouth rd, A375. Turn L at Hare & Hounds. Straight ahead then turn L after brown sign. From Exeter direction, take A3952 & follow brown signs from Branscombe Cross.* **For NGS: Thur 22 June, Thur 13 July (10-4). Adm £5, chd free. Cream teas. Light lunches. Vegetarian & gluten free options usually available. For other opening times and information, please phone, email or visit garden website.**
On a working Honey Farm, a garden for all pollinators, alive with buzzing and fluttering from spring to autumn. Large beds and borders with a wide range of colourful perennials and annuals providing pollen and nectar. Shrubs and trees underplanted with bulbs. Herb garden, polytunnel growing salad and vegetables for the tearoom. 20 acres of wildflower meadows, large apple orchard, plums, damsons, apricots and peaches. Courses and demonstrations available. Apple pressing in the autumn.

12 ◆ BLACKPOOL GARDENS
Dartmouth, TQ6 0RG. Sir Geoffrey Newman, 01803 771801, beach@blackpoolsands.co.uk, www.blackpoolsands.co.uk. *3m SW of Dartmouth. From Dartmouth follow brown signs to Blackpool Sands on A379. Entry tickets, parking, WCs & refreshments available at Blackpool Sands. Sorry, no dogs permitted.* **For opening times and information, please phone, email or visit garden website.**
Carefully restored C19 subtropical plantsman's garden with collection of mature and newly planted tender and unusual trees, shrubs and carpet of spring flowers. Paths and steps lead gradually uphill and above the Captain's Seat offering fine coastal views. Recent plantings follow the southern hemisphere theme with callistemons, pittosporums, acacias and buddlejas.

GROUP OPENING

13 BOVEY TRACEY GARDENS
Bovey Tracey, TQ13 9NA. *6m N of Newton Abbot. Take A382 to Bovey Tracey. Car parking in central car park for Parke View and on local roads for other gardens. 2 miles between furthest two gardens.* **Sat 17, Sun 18 June (1.30-5.30). Combined adm £7, chd free. Home-made teas at Gleam Tor.**

5 BRIDGE COTTAGES
TQ13 9DR. Cath Valentine.

GLEAM TOR
TQ13 9DH. Gillian & Colin Liddy.

GREEN HEDGES
TQ13 9LZ. Alan & Linda Jackson.

PARKE VIEW
TQ13 9AD. Peter & Judy Hall.

2 REDWOODS
TQ13 9YG. Mrs Julia Mooney.

11 ST PETER'S CLOSE
TQ13 9ES. Pauline & Keith Gregory.

Nestling in the Dartmoor foothills, Bovey Tracey offers a wide range of gardens. Colourful 11 St Peter's Close has a mini railway and a watercolours exhibition, while Green Hedges is packed full of interest: colourful borders, shade plants, organic vegetables, greenhouses and a small stream. In the centre lies Parke View, a romantic 1 acre garden with meandering paths and old stone walls between separate areas, many unusual plants, trees, shrubs and colour themed herbaceous borders. There's lots to enjoy at Gleam Tor - long herbaceous border, white garden, wildflower meadow, prairie planting, wildlife pond and interesting 'memory patio' (plus Colin's legendary cakes for tea). 2 Redwoods has mature trees, a Dartmoor leat, a fernery, a sunny gravel garden and acid loving shrubs, while 5 Bridge Cottages is a productive, imaginatively designed and quirky cottage garden within an historical pottery site. Please note that cashless payment for tickets or plants is available only at Parke View and Gleam Tor (for teas here too). The other gardens are cash only. Partial wheelchair access, none at St Peter's Close or Green Hedges.

National Garden Scheme gardens are identified by their yellow road signs and posters. You can expect a garden of quality, character and interest, a warm welcome and plenty of home-made cakes!

14 BRADFORD TRACEY HOUSE

Witheridge, Tiverton, EX16 8QG. Elizabeth Wilkinson. *20 mins from Tiverton. Postcode will get you to thatched lodge at bottom of drive which has 2 bouncing hares on the top. From Witheridge NW via Bendley Hill. From A361 take the Rackenford turn.* **Sat 20, Sun 21 May (1.30-5). Adm £6, chd free. Tea & cake served takeaway style to enjoy in the garden or inside.**

A pleasure garden set around a Regency hunting lodge combining flowers with grasses, shrubs and huge trees in a natural and joyful space. It is planted for productivity and sustainability giving harvests of wonderful flowers, fruits, herbs and vegetables (and weeds!). There are beautiful views over the lake, forest walks, deep blowsy borders, an ancient wisteria and an oriental treehouse garden.

15 BRAMBLE TORRE

Dittisham, nr Dartmouth, TQ6 0HZ. Paul & Sally Vincent, www.rainingsideways.com. *¾m from Dittisham. Leave A3122 at Sportsman's Arms. Drop down into village, at Red Lion turn L to Cornworthy. Continue ¾m, Bramble Torre straight ahead. Follow signs to car park.* **Fri 9, Sat 10, Sun 11 June (2-5). Adm £5, chd free. Cream teas.**

Set in 30 acres of farmland in the beautiful South Hams, the 3 acre garden follows a rambling stream through a steep valley: lily pond, herbaceous borders, roses, camellias, lawns and shrubs, a formal herb and vegetable garden. All are dominated by a huge embothrium glowing scarlet in late spring against a sometimes blue sky! Well behaved dogs on leads welcome. Partial wheelchair access, parts of garden very steep and uneven. Tea area with wheelchair access and excellent garden view.

16 BREACH

Shute Road, Kilmington, Axminster, EX13 7ST. Judith Chapman & BJ Lewis, 01297 35159, jachapman16@btinternet.com. *1½m W of Axminster off A35. Turn L off the A35 at the War Memorial, continue up Shute Rd, past farm on R & after 150m, Breach is a short walk along a byway to the L. Parking is on Shute Rd.* **Sat 27, Sun 28 May (1.30-5). Adm £5, chd free. Home-made teas. Visits also by arrangement 6 May to 29 Oct for groups of 8 to 30.**

3+ acres with many mature trees as well as unusual trees planted recently, e.g. Hoheria, *Cornus mas*, Nyssa. Herbaceous borders, shrubberies, vegetable/fruit area. Bog garden using a natural spring and 2 ponds attract dragon flies. Disused shale tennis court is becoming a wildflower area with orchids. Soil is mainly acidic and a band of specimen rhododendrons was planted 12 yrs ago. Development continues. Partial wheelchair access.

GROUP OPENING

17 BRENDON GARDENS

Brendon, Lynton, EX35 6PU. *1m S of A39 N Devon coast rd between Porlock and Lynton.* **Sat 27, Sun 28 May, Sat 5, Sun 6 Aug (12-5). Combined adm £6, chd free. Light refreshments at Higher Tippacott Farm. Light lunches, home-made cakes, inc gluten free & cream teas. WC.**

NEW **BARN FARM**
Andrew and Debra Hodges.

HALL FARM
Karen Wall, 01598 741343, lalindevon@yahoo.co.uk.
Visits also by arrangement 10 Apr to 8 Sept.

HIGHER TIPPACOTT FARM
Angela & Malcolm Percival, 01598 741343, lalindevon@yahoo.co.uk.
Visits also by arrangement 10 Apr to 8 Sept.

NEW **MEADPOOL HOUSE**
Gillian Reynolds.

Stunning part of Exmoor National Park. Excellent walking along river. Barn Farm: Stylish courtyard garden on two levels incorporating walls and old stone buildings. Roses, shrubs and other cottage garden flowers and vegetables. Adjoining paddock with copse and young orchard of Devon varieties. Hall Farm: C16 longhouse set in 2 acres of tranquil mature gardens, with lake and wild area beyond. Idyllic setting with views. Rare Whitebred Shorthorn cattle. Higher Tippacott Farm: 950ft alt. on moor, facing south overlooking its own valley with stream and pond. Interesting planting on many interconnecting levels with lawns. Young fruit trees in meadow. Vegetable patch with sea glimpse. Lovely views along valley and up to high moorland. Meadpool House: Spacious and restful riverside garden with flat terrain. Interesting paved areas, borders, trees, feature walls and a productive gated kitchen garden. Plants, books and bric a brac for sale at Higher Tippacott Farm. Display of vintage telephones and toys.

18 THE BRIDGE MILL

Mill Rd, Bridgerule, Holsworthy, EX22 7EL. Rosie & Alan Beat, www.thebridgemill.org.uk. *In Bridgerule village on R Tamar between Bude & Holsworthy. Between the chapel by river bridge & church at top of hill. Garden is at bottom of hill opp Short & Abbott agricultural engineers. See website for detailed directions.* **Sun 28 May (11-5). Adm £5, chd free. Home-made teas in the garden with friendly ducks if fine, or in the stable if wet. Plenty of dry seating.**

One acre organic gardens around mill house and restored working water mill. Cottage garden style planting; herb garden with medicinal and dye plants; productive fruit and vegetable garden, and wild woodland and water garden by mill. 16 acre smallholding: lake and riverside walks, wildflower meadows, friendly livestock, sheep, ducks and hens. The historic water mill was restored to working order in April 2012 and in 2017 Alan was awarded a plaque by the Society for the Protection of Ancient Buildings for his sympathetic restoration of the buildings. Exhibition of garden embroideries by Linda Chilton. Wheelchair access to some of gardens plus accessible WC.

19 BROCTON COTTAGE

Pear Tree, Ashburton, Newton Abbot, TQ13 7QZ. Mrs Naomi Hindley, 01364 654902, hindley1@clara.co.uk. *¼m from A38. Leave A38 at Princetown turning and turn R at end of slip road. Parking is restricted so if possible park by Shell garage & walk up road to thatched house.* **Sat 9, Sun 10 Sept (1-5). Adm £5, chd free. Home-made teas. Visits also by arrangement for groups of 5+.**

1.3 acres recovered from neglect, starting to look wonderful. Orchard, woodland, ponds and productive areas linked to established and newly planted herbaceous borders and shrubberies. A new bed has been developed with autumn planting. Wonderful views over Devon countryside. Dogs on leads only. Dahlias grown from seed and grass bed with *Stipa gigantica* are features in September. Lunches also available at The Dartmoor Lodge Motel, Furzleigh Mill Hotel and various outlets in Ashburton town centre. There is electric wheelchair access to much of the garden. Manual wheelchairs more limited due to gradient but can access terrace for teas and views.

20 ◆ BURROW FARM GARDENS
Dalwood, Axminster, EX13 7ET. Mary & John Benger, www.burrowfarmgardens.co.uk. *3½ m W of Axminster. From A35 turn N at Taunton Xrds then follow brown signs.* **For opening times and information, please visit garden website.**

This beautiful 13 acre garden has unusual trees, shrubs and herbaceous plants. Traditional summerhouse looks towards lake and ancient oak woodland with rhododendrons and azaleas. Early spring interest and superb autumn colour. The more formal Millennium garden features a rill. Anniversary garden featuring late summer perennials and grasses. Winner of the Nations Favourite Garden 2021. Café, nursery and gift shop with range of garden ironwork. Various events inc spring and summer plant fair and open air theatre held at garden each yr. Visit events page on website for more details.

21 BYES REACH
26 Coulsdon Road, Sidmouth, EX10 9JP. Lynette Talbot & Peter Endersby, 07767 773374, latalbot01@gmail.com. *From Exeter on A3052 turn R at lights onto Sidford Rd A375. Turn ½ m L into Coulsdon Rd.* **Sat 8, Sun 9, Mon 10 Apr, Sat 5, Sun 6 Aug (1.30-5). Adm £3, chd free. Home-made teas. Gluten & lactose free options for cakes. Visits also by arrangement 1 May to 1 Sept for groups of 5 to 20.**

¼ acre garden with colour themed hot border at front. At rear of house: potager style vegetable garden with raised beds, seasonal colour-themed beds, studio and greenhouse, 20 metre fruit arched walkway, rill, ferns, rockery, collection of hostas, pond. Seating in secluded niches with views to garden. Easy access.

22 ◆ CADHAY
Ottery St Mary, EX11 1QT. Rupert Thistlethwayte, 01404 813511, jayne@cadhay.org.uk, www.cadhay.org.uk. *1m NW of Ottery St Mary. On B3176 between Ottery St Mary & Fairmile. From E exit A30 at Iron Bridge. From W exit A30 at Patteson's Cross, follow brown signs for Cadhay.* **For NGS: Sun 28, Mon 29 May, Sun 27, Mon 28 Aug (2-5). Adm £5, chd £1. Cream teas. For other opening times and information, please phone, email or visit garden website.**
Tranquil 2 acre setting for Tudor manor house. 2 medieval fish ponds surrounded by rhododendrons, gunnera, hostas and flag iris. Roses, clematis, lilies and hellebores surround walled water garden. 120ft herbaceous border walk informally planted. Magnificent display of dahlias throughout. Walled kitchen gardens have been turned into allotments and old apple store is now a tearoom. Gravel paths.

23 CHEVITHORNE BARTON
Chevithorne, Tiverton, EX16 7QB. Head Gardener, chevithornebarton.co.uk. *3m NE of Tiverton. Follow yellow signs from A361, A396 or Sampford Peverell.* **Sun 19 Mar, Sun 21 May, Sun 30 July, Sun 22 Oct (2-4.30). Adm £6, chd free.**
Walled garden, summer borders and Robinsonian inspired woodland of rare trees and shrubs. In spring, garden features a large collection of magnolias, camellias, and rhododendrons with grass paths meandering through a sea of bluebells. Home to National Collection of Quercus (Oaks). Lots of autumn colour.

NPC

24 ◆ DOCTON MILL
Lymebridge, Hartland, EX39 6EA. Lana & John Borrett, 01237 441369, docton.mill@btconnect.com, www.doctonmill.co.uk. *8m W of Clovelly. Follow brown tourist signs on A39 nr Clovelly.* **For NGS: Sun 11 June (11-4.30). Adm £5, chd free. Cream teas and light lunches available all day. For other opening times and information, please phone, email or visit garden website.**
Situated in a stunning valley location. The garden surrounds the original mill pond and the microclimate created within the wooded valley enables tender species to flourish. Recent planting of herbaceous, stream and summer garden give variety through the season.

25 DUNLEY HOUSE
Bovey Tracey, TQ13 9PW. Mr & Mrs F Gilbert. *2m E of Bovey Tracey on rd to Hennock. From A38 going W turn off slip rd R towards Chudleigh Knighton on B3344, in village follow yellow signs to Dunley House. From A38 eastwards turn off on Chudleigh K slip rd L and follow signs.* **Sat 3, Sun 4 June, Sat 14, Sun 15 Oct (2-5). Adm £6, chd free. Home-made teas.**
9 acre garden set among mature oaks, sequoiadendrons and a huge liquidambar started from a wilderness in mid eighties. Rhododendrons, camellias and over 40 different magnolias. Arboretum, walled garden with borders and fruit and vegetables, rose garden and new enclosed garden with lily pond. Large pond renovated in 2016 with new plantings. Woodland walk around perimeter of property.

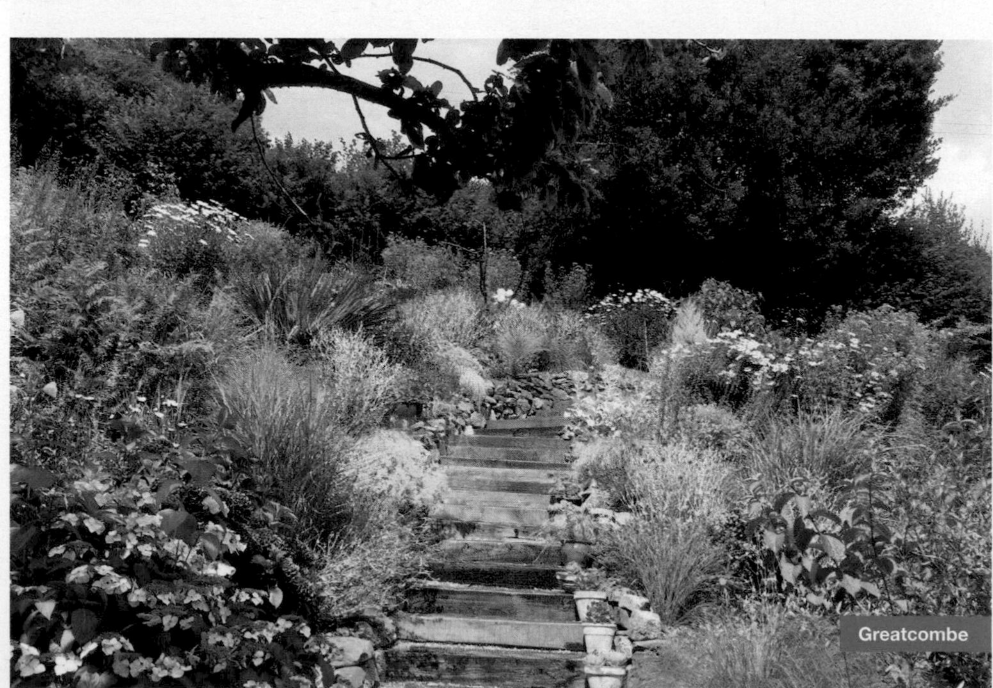
Greatcombe

26 EAST WOODLANDS FARMHOUSE
Alverdiscott, Newton Tracey, Barnstaple, EX31 3PP. Ed & Heather Holt, heatherholtexmoor@gmail.com. *5m NE of Great Torrington, 5m S of Barnstaple, off B3232. From Grt Torrington turn R into single track rd before Alverdiscott; & from Barnstaple turn L after Alverdiscott. 1m down rd R fork at Y-junction.* **Sat 10, Sun 11 June (2-5). Adm £5, chd free. Home-made teas. Gluten free cakes available. Visits also by arrangement 3 June to 5 Aug for groups of 10 to 25.**
East Woodlands is a beautiful RHS inspired and designed garden full of rooms packed with plants, shrubs and trees. Enjoy the spectacular bamboos, flowing grasses, colourful roses, Mediterranean, Japanese, cottage and bog gardens (unfenced pond), all set in an acre looking out over N Devon countryside. Occasional live music. Lots of seating areas. Partial wheelchair access.

27 EAST WORLINGTON HOUSE
East Worlington, Witheridge, Crediton, EX17 4TS. Barnabas & Campie Hurst-Bannister. *In centre of E Worlington, 2m W of Witheridge. From Witheridge Square R to E Worlington. After 1½m R at T-junction in Drayford, then L to Worlington. After ½m L at T-junction. 200 yds on L. Parking nearby, disabled parking at house.* **Sun 26 Feb, Sun 5 Mar (1.30-5). Adm £5, chd free. Cream teas in thatched parish hall next to house.**
Thousands of purple crocuses feature in this 2 acre garden, set in a lovely position with views down the valley to Little Dart river. These spectacular crocuses have spread over many years through the garden and into the neighbouring churchyard. Walks from the garden across the river and into the woods. Dogs on leads please.

28 GOODWILL GARDENS
Bickington, Newton Abbot, TQ12 6JY. Michael Hext. *Part of Granny Pats Tiny Farmshop & Nursery on old A38 Liverton to Bickington rd. Past Welcome Stranger pub next to layby on R before Bickington look for blue shed.* **Sun 14, Sun 28, Mon 29 May (1-4.30). Adm £5, chd free. Cream teas on covered decking area.**
A small nursery and smallholding with a large wildflower area, also shrubs, herbaceous perennial beds, flowers full of insects and birds, no-dig cut flower area, sweet peas, dahlias and lots more. Soft fruit and vegetable beds, pet pigs and hens and polytunnels. The garden, which was started in 1990 on a steep site, is now laid out in terraces. New this yr is the start of a community allotment site.

29 ♦ GOREN FARM
Broadhayes, Stockland, Honiton, EX14 9EN. Julian Pady, 01404 881335, gorenfarm@hotmail.com, www.goren.co.uk/pages/open-days. *6m E of Honiton, 6m W of Axminster. Go to Stockland television mast. Head 100 metres N signed from Ridge Cross, head E towards ridge.* **For NGS: Sat 27 May (11-6); Sun 28 May (10-6). Adm £5, chd £2. Cream teas. Groups of more than 4 are asked to book their cream teas in advance. The farm has a licence to serve cider. For other opening times and information, please phone, email or visit garden website.**
Wander through 50 acres of natural species rich wildflower meadows. Easy access footpaths. Dozens of varieties of wild flowers and grasses. Thousands of orchids from early June and butterflies in July. Stunning views of Blackdown Hills. Georgian house and walled gardens. In August it is late season and the meadows are setting seed, ready to be mown. Butterflies, beetles and many other insects are in abundance on the grasses and wildflower seed heads. Café, pick your own and farm shop. Nature trail with species information signs and picnic tables. Visit the cider museum, walled gardens and greenhouses. Various music events are held throughout the summer. Partial wheelchair access to meadows. Dogs welcome on a lead only, please clean up after your pet.

30 GREATCOMBE
Holne, Newton Abbot, TQ13 7SP. Robbie & Sarah Richardson, 07725 314887, www.facebook.com/greatcombegardens *4m NW Ashburton via Holne Bridge & Holne village. 4m NE Buckfastleigh via Scorriton. Narrow lanes. Large car park adjacent to garden.* **Sat 29, Sun 30 Apr, Mon 1, Sat 27, Sun 28, Mon 29 May, Fri 16, Sat 17, Sun 18, Fri 30 June, Sat 1, Sun 2, Fri 14, Sat 15, Sun 16, Fri 28, Sat 29, Sun 30 July (1-5). Adm £5, chd free. Home-made teas.**
An oasis of a garden hidden in a Dartmoor valley known for its unusual planting and colour schemes (and delicious home-made teas) set in 2 acres and evoking a deep sense of tranquillity. Moorland stream, spring flowering shrubs and bulbs, summer climbers, perennial borders, wide variety of hydrangeas and grasses, a haven for pollinators. Plant nursery. Artist's studio. Metal plant supports in all sizes.

31 HALDON GRANGE
Dunchideock, Exeter, EX6 7YE. Ted Phythian, 01392 832349, judithphythian@yahoo.com, youtu.be/G0gluYoWncA. *5m SW of Exeter. From A30 through Ide village to Dunchideock 5m. L to Lord Haldon, Haldon Grange is next L. From A38 (S) turn L on top of Haldon Hill follow Dunchideock signs, R at village centre to Lord Haldon.* **Every Sat and Sun 11 Mar to 11 June (1-5). Wed 22, Wed 29 Mar, Mon 10, Wed 19 Apr, Mon 1, Mon 8, Wed 17, Mon 29 May (1-5). Adm £6, chd free. Home-made teas. Our visitors are welcome to bring a picnic. Visits also by arrangement 11 Mar to 11 June for groups of 10+. Access for small coach size only. Regret no dogs.**
Peaceful, well established 13 acre garden, parts dating back to 1770s. This hidden gem boasts one of the largest collections of rhododendrons, azaleas, magnolias and camellias. Interspersed with mature and rare trees and complemented by a lake and cascading ponds. 6 acre arboretum, large lilac circle, wisteria pergola with views over Exeter and Woodbury completes this family run treasure. Plants are on sale, also teas and home bakes (cash only). On site car parking. WC available, alcohol wipes and hand gel provided. Strictly no dogs. Wheelchair access to main parts of garden.

32 HALSCOMBE FARM

Halscombe Lane, Ide, Exeter, EX2 9TQ. Prof J Rawlings, jgshayward@tiscali.co.uk. *From Exeter go through Ide to mini r'about take 2nd exit and continue to L turn into Halscombe Lane.* **Visits by arrangement May to Sept.**

Farmhouse garden created over last 10 yrs. Large collection of old roses and peonies, long and colourful mixed borders, productive fruit cage and vegetable garden all set within a wonderful borrowed landscape.

33 HARBOUR LIGHTS

Horns Cross, Bideford, EX39 5DW. Brian & Faith Butler, 01237 451627, brian.nfu@gmail.com, harbourlightsgarden.org. *8m W of Bideford, 3m E of Clovelly. On main A39 between Bideford and Clovelly, halfway between Hoops Inn & Bucks Cross. There will be a union jack flag & yellow arrow signs at the entrance.* **Sat 17, Sun 18 June (11-5). Adm £4.50, chd free. Cream teas, cakes, light lunches, wine all available. Visits also by arrangement 8 June to 12 Aug for groups of 12+.**

½ acre colourful garden with Lundy views. A garden of wit, humour, unusual ideas, installation art, puzzles, volcano and many surprises. Water features, shrubs, foliage area, grasses in an unusual setting, fernery, bonsai and polytunnel, plus masses of colourful plants. You will never have seen a garden like this! Free leaflet. We like our visitors to leave with a smile! Child friendly. A 'must visit' unique and interactive garden with intriguing artwork of various kinds, original plantings and ideas.

34 THE HAVEN

Wembury Road, Hollacombe, Wembury, South Hams, PL9 0DQ. Mrs S Norton & Mr J Norton, 01752 862149, suenorton1@hotmail.co.uk. *20mins from Plymouth city centre. Use A379 Plymouth to Kingsbridge Rd. At Elburton r'about follow signs to Wembury. Parking on roadside. Bus stop nearby on Wembury Rd. Route 48 from Plymouth.* **Visits by arrangement 28 Feb to 31 May for groups of up to 20. Adm £5, chd free. Cream teas.**

½ acre sloping plantsman's garden in South Hams AONB. Tearoom and seating areas. 2 ponds. Substantial collection of large flowering Asiatic and hybrid tree magnolias. Large collection of camellias inc *Camellia reticulata*. Rare dwarf, weeping and slow growing conifers. Daphnes, early azaleas and rhododendrons, spring bulbs and hellebores. Wheelchair access to top part of garden.

35 HAYNE

Zeal Monachorum, Crediton, EX17 6DE. Tim & Milla Herniman, www.haynedevon.co.uk. *½m S of Zeal Monachorum. From Zeal Monachorum, keeping church on L, drive through village. Continue on this road for ⅓m, garden drive is 1st entrance on R.* **Sun 4 June (2-6). Adm £5, chd free. Home-made teas.**

In the magical walled garden exciting new planting blends with the beautiful mature wisteria, purple and white, and rambling wild roses in combination with a more modern Piet Oudolf style perennial planting surrounding the recently renovated grade II* farm buildings. Extensive cut flower planting adds to the scene. Magic,

Lower Coombe Cottage, Teignmouth Gardens

mystery and soul by the spadeful! Live jazz band. Disabled WC. Wheelchair access to walled garden through orchard.

GROUP OPENING

36 NEW HAYTOR GARDENS

Haytor, Newton Abbot, TQ13 9XT. Judy Gordon Jones. *300m from Haytor on road to Ilsington. From Bovey Tracey, turn L at red phone box. After approx. 300m Brambly Wood signposted on R. Garden 50m. Haytor House next to this turning. Parking 100m further down the road in field on L.* **Fri 19, Sat 20, Sun 21 May (11-5). Combined adm £6, chd free. Home-made teas at Haytor House.**

NEW **BRAMBLY WOOD**
Lindsay & Laurie Davidson.

NEW **HAYTOR HOUSE**
Judy Gordon Jones & Hilary Townsend.

Haytor Gardens offers 2 very different gardens on the edge of Dartmoor with views across to Torbay and the Teign estuary. The 1 acre garden at Brambly Wood has been developed over the past 40 yrs with packed herbaceous borders, a pond and extensive planting of rhododendrons, azaleas and camellias looking their best in spring with tranquil seating areas throughout the garden. Visitors are also welcome to view the owners' studios showing artisan pottery and textile work. The ¼ acre garden at Haytor House was designed along more classical lines as a project for horticultural students in the 1930s and provides distinctive planting within partitioned areas. An inner walled garden with a folly leads to 3 further distinctive garden 'rooms' with a lily pond, raised beds, mixed herbaceous borders and large shrubs and acers among its many features.

37 HEATHERCOMBE

Manaton, nr Bovey Tracey, TQ13 9XE. Claude & Margaret Pike Woodlands Trust, 01626 354404, gardens@pike.me.uk, www.heathercombe.com. *7m NW of Bovey Tracey. From Bovey Tracey take scenic B3387 to Haytor/ Widecombe. 1.7m past Haytor Rocks (before Widecombe Hill) turn R to Hound Tor & Manaton. 1.4m past Hound Tor turn L at Heatree Cross to Heathercombe.* **Sun 26 Mar, Sun 7, Sun 14 May (1.30-5.30). Every Tue to Sun 20 May to 28 May (1.30-5.30). Every Tue to Sun 10 June to 2 July (11.30-5.30). Adm £6, chd free. Self service tea & coffee. Do bring a picnic. Please pay for your tickets in advance where possible via www.ngs.org.uk. Visits also by arrangement Apr to Oct. Donation to Rowcroft Hospice.**

Tranquil secluded valley with streams running through woods, ponds and lake. 30 acres of spring/summer interest - daffodils, extensive bluebells, large displays of rhododendrons, many unusual specimen trees, cottage garden, orchard, wildflower meadow, bog/ fern/woodland gardens, woodland walks and sculptures. 2 miles of mainly level sandy paths with many benches and new lakeside shelter. Disabled reserved parking close to tea room & WC.

GROUP OPENING

38 NEW HEAVITREE'S UNEXPECTED GARDENS

North Street, Heavitree, Exeter, EX1 2RH. orlando@orlandomurrin.com. *Blue Jay Way (5a North St) is accessed from the exit of the Co-op supermarket. Map to all the gardens will be available from Blue Jay Way.* **Sat 1, Sun 2 July (1-5). Combined adm £5, chd free. Light refreshments.**

NEW **BLUE JAY WAY**
Orlando Murrin.

NEW **ELYSIUM LODGE**
Nigel Bagnall & Dana Littlepage Smith.

NEW **31 REGENTS PARK**
Katie Sellek.

NEW **14 THIRD AVENUE**
Emma Newbery.

A quartet of dramatically different city gardens on a 2km circular walk. Blue Jay Way is a plantsman's guerrilla garden, boldly dominating the exit to the Co-op Car Park. Many themes, including blue and white flowers, jungle leaves and scented plants. 31 Regents Park is an insect-friendly garden, inspired by a visit to a Greek olive grove, and currently being future-proofed for climate change. A small back garden featuring hot colours. 14 Third Avenue is a steep terraced garden on challenging clay soil, with dramatic landscaping and carefully curated colour palette. Architectural features inc gabion wall, Indian sandstone, timber and pebbles. Generously planted front garden is a highlight in the local street scene. Elysium Lodge, on the edge of Higher Cemetery, is one of the most secret gardens in Exeter. Thread your way through a sequence of exciting and surprising garden rooms, with naturalistic planting, tree echiums, pond, roses and strategic seats and benches from which to sit and marvel. Start in a supermarket exit, and finish in a cemetery.

39 HIGH GARDEN

Chiverstone Lane, Kenton, EX6 8NJ. Chris & Sharon Britton, www.highgardennurserykenton.wordpress.com. *5m S of Exeter on A379 Dawlish Rd. Leaving Kenton towards Exeter, L into Chiverstone Lane, 50yds along lane.* **Sun 2 Apr, Sun 4 June, Sun 2 July, Sun 3 Sept (12-5). Adm £5, chd free. Home-made teas.**

Stunning garden of over 4 acres. Huge range of trees, shrubs, perennials, grasses, climbers and exotics planted over past 15yrs Great use of foliage gives texture and substance as well as offsetting the floral display. 70 metre summer herbaceous border. Over 40 individual mixed beds surrounded by meandering grass walkways. Exciting new formal plantings. Large range of plants from adjoining plantsman's nursery. Slightly sloping site but the few steps can be avoided.

Our donation of £2.5 million over five years has helped fund the new Y Bwthyn NGS Macmillan Specialist Palliative Care Unit in Wales. 1,900 inpatients have been supported since its opening

40 HIGHER CHERUBEER
Dolton, Winkleigh, EX19 8PP. Jo & Tom Hynes, hynesjo@gmail.com. *2m E of Dolton. From A3124 turn S towards Stafford Moor Fisheries, take 1st R, garden 500m on L.* **Fri 3, Fri 10, Sat 18 Feb (2-4.30); Sun 28 May (2.30-5.30). Adm £5, chd free. Home-made teas. 2024: Fri 2, Fri 9, Sat 17 Feb. Visits also by arrangement 1 Feb to 2 July for groups of 10+. Sep & Oct visits for autumn flowering Cyclamen also available.**
1¾ acre country garden with gravelled courtyard and paths, raised beds, alpine house, lawns, herbaceous borders, woodland beds with naturalised cyclamen and snowdrops, kitchen garden with large greenhouse and orchard. Winter openings for National Collection of Cyclamen species, hellebores and over 400 snowdrop varieties.
NPC

41 HIGHER ORCHARD COTTAGE
Aptor, Marldon, Paignton, TQ3 1SQ. Mrs Jenny Saunders, http://www.facebook.com/HigherOrchardCottage. *1m SW of Marldon. A380 Torquay to Paignton. At Churscombe Cross r'about R for Marldon, L towards Berry Pomeroy, take 2nd R into Farthing Lane. Follow for exactly 1m. Turn R at NGS sign & follow signs for parking.* **Sat 3, Sun 4 June (11-5). Adm £5, chd free. Home-made teas. Refreshments from 2pm provided by Marldon Garden Club if weather permits. Outside seating only.**
Two acre garden with generous colourful herbaceous borders, wildlife pond, specimen trees and shrubs, productive vegetable beds and grass path walks through wildflower meadows in lovely countryside.
Each year we host sculpture and art installations by local artists and crafters.

42 HOLE FARM
Woolsery, Bideford, EX39 5RF. Heather Alford. *11m SW of Bideford. Follow directions for Woolfardisworthy, signed from A39 at Bucks Cross. From village follow NGS signs from school for approx 2m.* **Sun 14 May, Sun 10 Sept (2-6). Adm £5, chd free. Home-made teas in converted barn.**
3 acres of exciting gardens with established waterfall, ponds, vegetable and bog garden. Terraces and features, inc round house, have all been created using natural stone from original farm quarry. Peaceful walks through Culm grassland and water meadows border the River Torridge and host a range of wildlife. Home to a herd of pedigree native Devon cattle.

43 IDESTONE BARTON
Dunchideock, Exeter, EX2 9UE. Mr & Mrs James Studholme. *From Ide take rd to Dunchideock. After 500 metres fork R (signed Idestone). Take 1st L after 1m. Follow over Xrds and down steep sided S bend. Car park signed on R.* **Sat 25 Mar, Fri 2, Sat 3 June (12-5). Adm £6, chd free. Teas available June dates only.**
Romantic 6 acre country garden in unspoilt countryside only 3m from Exeter. Built on 5 different levels, with several distinctive rooms, garden features yew-hedged kitchen garden, croquet lawn, rose terrace, orchard, swimming pond and arboretum. No teas available for March opening.

44 NEW 2 JUBILEE VILLAS
Simonsbath, Minehead, TA24 7SH. Mark & Emma Hawkins. *5m W of Exford. On entering Simonsbath from Exford, pass the church on R & take the next R turn to Ashcombe car park.* **Sat 1, Sun 2 July (11-4). Adm £4, chd free.**
A ½ acre garden for plant lovers, situated at an altitude of 1100ft on Exmoor. There are deep herbaceous borders with some rare and unusual plants, a gravel garden, a shady garden and a wildlife pond. The mature trees and hedges give the garden structure beyond its age. There is a greenhouse area from which the owners propagate plants for their own garden and for sale. Parking at Ashcombe car park where there are WC facilities. Pub and tearoom close by.

45 KIA-ORA FARM & GARDENS
Knowle Lane, Cullompton, EX15 1PZ. Mrs M B Disney, 01884 32347, rosie@kia-orafarm.co.uk, www.kia-orafarm.co.uk. *On W side of Cullompton & 6m SE of Tiverton. M5 J28, through town centre to r'about, 3rd exit R, top of Swallow Way turn L into Knowle Lane, garden beside Cullompton Rugby Club.* **Sun 9, Mon 10, Sun 23, Sun 30 Apr, Mon 1, Sun 14, Sun 28, Mon 29 May, Sun 11, Sun 25 June, Sun 9, Sun 23 July, Sun 6, Sun 20, Sun 27, Mon 28 Aug, Sun 10, Sun 24 Sept (2-5.30). Adm £4, chd free. Home-made teas. Teas, plants & sales not for NGS. Visits also by arrangement Apr to Sept Please state NGS when booking.**
Charming, peaceful 10 acre garden with lawns, lakes and ponds. Water features with swans, ducks and other wildlife. Mature trees, shrubs, rhododendrons, azaleas, heathers, roses, herbaceous borders and rockeries. Nursery avenue, novelty crazy golf. Stroll leisurely around and finish by sitting back, enjoying a traditional home-made Devonshire cream tea or choose from the wide selection of cakes!

46 LANGUARD PLACE
Middle Warberry Road, Torquay, TQ1 1RS. Alison & Steve Dockray. *From Babbacombe Rd opp St Matthias Church, turn into Higher Warberry Rd, then L into Middle Warberry Rd. Approx 800yds up hill to junction on R. Languard Place is on the corner.* **Sat 27, Sun 28 May (1-5.30). Adm £5, chd free. Light refreshments. Teas, coffees and cakes available in the garden.**
This small level south facing cottage style organic garden is a plantsman's delight. Enjoy the herbaceous borders packed with roses, clematis and many unusual perennials. Archways and paths lead to a separate Japanese style area with natural wildlife pond and gazebo, acers and bamboos. There's a productive greenhouse, vegetable and fruit areas and a wisteria arch. Limited wheelchair access.

47 LEE FORD
Knowle Village, Budleigh Salterton, EX9 7AJ. Mr & Mrs N Lindsay-Fynn, 01395 445894, crescent@leeford.co.uk, www.leeford.co.uk. *3½m E of Exmouth. For SatNav use postcode EX9 6AL.* **Visits by arrangement 3 Apr to 29 Sept for groups of 10 to 20. Adm £7, chd free. Tea and cake served in conservatory. Donation to Lindsay-Fynn Trust.**
Extensive, formal and woodland garden, largely developed in 1950s, but recently much extended with mass displays of camellias, rhododendrons and azaleas, inc many rare varieties. Traditional walled garden filled with

fruit and vegetables, herb garden, bog garden, rose garden, hydrangea collection, greenhouses. Ornamental conservatory with collection of pot plants. Direct access to Pedestrian route and National Cycle Network route 2 which follows old railway linking Exmouth to Budleigh Salterton. Garden ideal destination for cycle clubs or rambling groups. Formal gardens are lawn with gravel paths. Moderately steep slope to woodland garden on tarmac with gravel paths in woodland.

48 LEWIS COTTAGE

Spreyton, nr Crediton, EX17 5AA. Mr & Mrs M Pell and Mr R Orton, 07773 785939, rworton@mac.com, www.lewiscottageplants.co.uk. *5m NE of Spreyton, 8m W of Crediton. From Hillerton X keep stone X to your R. Drive approx 1½m, Lewis Cottage on L, drive down farm track. From Crediton follow A377 N. Turn L at Barnstaple X junction, then 2m from Colebrooke Church.* **Sat 24, Sun 25 June, Sat 29, Sun 30 July, Sat 26, Sun 27 Aug (12-5). Adm £5, chd free. Light refreshments. Visits also by arrangement 15 May to 31 Aug for groups of 12 to 24.**

4 acre garden located on SW facing slope in rural mid Devon. Evolved primarily over last 30 yrs, harnessing and working with the natural landscape. Using informal planting and natural formal structures to create a garden that reflects the souls of those who garden in it, it is an incredibly personal space that is a joy to share. Spring camassia cricket pitch, rose garden, large natural dew pond, woodland walks, bog garden, hornbeam rondel, winter garden, hot and cool herbaceous borders, fruit and vegetable garden, outdoor poetry reading room and plant nursery selling plants mostly propagated from the garden.

49 LITTLE ASH BUNGALOW

Fenny Bridges, Honiton, EX14 3BL. Helen & Brian Brown, 01404 850941, helenlittleash@hotmail.com, www.facebook.com/littleashgarden. *3m W of Honiton. Leave A30 at 1st turn off from Honiton 1m, Patteson's Cross from Exeter ½m & follow NGS signs.* **Sun 13 Aug (1-5). Adm £5, chd free. Light refreshments. Visits also by arrangement 1 Aug to 21 Oct for groups of 10+. Easy access and parking for coaches.**

Country garden of 1½ acres, packed with different and unusual herbaceous perennials, trees, shrubs and bamboos. Designed for year-round interest, wildlife and owners' pleasure. Naturalistic planting in colour coordinated mixed borders, highlighted by metal sculptures, providing foreground to the view. Natural stream, pond and damp woodland area, mini wildlife meadows and raised gravel/alpine garden. Regional Winner, The English Garden's The Nation's Favourite Gardens in 2021. Grass paths.

50 LITTLEFIELD

Parsonage Way, Woodbury, Exeter, EX5 1HY. Caryn Vanstone & Bruno Dalbiez. *Please park cars in the village and follow signs to a narrow driveway shared with Summer Lodge. Access to driveway from Parsonage Way, opp the stone cross on junction with Pound Lane.* **Sat 3, Sun 4 June (11-4). Adm £4, chd £2. Cream teas. Scones, home-made biscuits & jam, Devon clotted cream.**

½ acre eco-garden. Rescued from derelict land in 2009/10, only 13m wide, 150m long, divided into herbaceous, shrub and tree planting, with large vegetable area, fruit and orchard with chickens and beehive. Plantsperson's garden stocked with large range of varieties, in colourful combinations. Sculptures, unusual ironwork, wildlife pond all add charm and interest. Managed using permaculture techniques. Large displays showing photos, and telling the story of the garden's development since 2009. Other displays introduce the visitor to the key ideas of permaculture and high-sustainability gardening, with wildlife and healthy ecosystems at its heart. The entire garden can be accessed in a wheelchair, but be aware that most paths are gravel and can be soft.

51 ◆ LUKESLAND

Harford, Ivybridge, PL21 0JF. Mrs R Howell and Mr & Mrs J Howell, 01752 691749, lorna.lukesland@gmail.com, www.lukesland.co.uk. *10m E of Plymouth. Turn off A38 at Ivybridge. 1½m N on Harford rd, E side of Erme valley. Beware of using SatNavs as these can be very misleading.* **For opening times and information, please phone, email or visit garden website.**

25 acres of flowering shrubs, wild flowers and rare trees with pinetum in Dartmoor National Park. Beautiful setting of small valley around Addicombe Brook with lakes, numerous waterfalls and pools. Extensive and impressive collections of camellias, rhododendrons, azaleas and acers; also spectacular *Magnolia campbellii* and huge *Davidia involucrata*. Superb spring and autumn colour. Open Suns, Weds and BH (11-5) 12 March - 11 June and 11 Oct - 15 November. Adm £7.50, under 16s free. Group discount for parties of 20+. Group tours available by appointment. Children's trail.

52 MARSHALL FARM

Ide, nr Exeter, EX2 9TN. Jenny Tuckett. *Between Ide and Dunchideock. Drive through Ide to top of village r'about, straight on for 1½m. Turn R onto concrete drive, parking in farmyard at rear of property.* **Fri 9, Sun 11 June (1-5). Adm £4, chd free. Home-made teas. Donation to Devon Air Ambulance.**

Garden approached along lane lined with homegrown lime, oak and chestnut trees. A country garden created approx 1967. One acre featuring wildflower gardens, gravel beds, pond, parterre garden and a vegetable and cutting garden. Wildlife ponds set in old orchard. Stunning views of Woodbury, Sidmouth gap and Haldon.

The National Garden Scheme searches the length and breadth of England, Wales, Northern Ireland and the Channel Islands for the very best private gardens

53 ◆ MARWOOD HILL GARDEN
Marwood, nr Guineaford, Barnstaple, EX31 4EB. Dr J A Snowdon, 01271 342528, info@marwoodhillgarden.co.uk, www.marwoodhillgarden.co.uk. *4m N of Barnstaple. Signed from A361 & B3230. Look out for brown signs. See website for map & directions. Coach & car park.* **For NGS: Fri 7 July, Fri 6 Oct (10-4.30). Adm £9.50, chd £5. Garden Tea Room offers selection of light refreshments throughout the day, all home-made or locally sourced. For other opening times and information, please phone, email or visit garden website.**
Marwood Hill is a very special private garden covering an area of 20 acres with lakes and set in a valley tucked away in N Devon. From early spring snowdrops through to late autumn there is always a colourful surprise around every turn. National Collections of astilbe, *Iris ensata* and tulbaghia, large collections of camellia, rhododendron and magnolia. Partial wheelchair access.

54 MIDDLE WELL
Waddeton Road, Stoke Gabriel, Totnes, TQ9 6RL. Neil & Pamela Millward, 01803 782981, neilandpamela@talktalk.net. *A385 Totnes towards Paignton. Turn off A385 at Parkers Arms, Collaton St. Mary. After 1m, turn L at Four Cross.* **Sun 7 May (11-5). Adm £6, chd free. Home-made teas. Morning tea/coffee. Visits also by arrangement Apr to Oct Small coaches disembark at gate, large coaches need to offload 300m away.**
Tranquil 2 acre garden plus woodland and streams contain a wealth of interesting plants chosen for colour, form and long season of interest. Many seats from which to enjoy the sound of water. Interesting structural features (rill, summerhouse, pergola, cobbling, slate bridge). Heady mix of exciting perennials, shrubs, bulbs, climbers and specimen trees. Vegetable garden. Child friendly. Display by The Totnes Potters. Architectural features offset by striking planting. Regional Finalist in The English Garden magazine's The Nation's Favourite Gardens Competition 2021. Mostly accessible by wheelchair.

55 NEW 2 MIDDLEWOOD
Cockwood, Exeter, EX6 8RN. Cliff & Chris Curd, 07881 442031, cliffswheelbarrow@gmail.com. *1m S of Starcross. A379 from Dawlish, R at Cofton Cross, passing Cofton Holidays. Middlewood ½m on R. A379 from Exeter, turn L at Cockwood Harbour. Turn R, pass The Ship. Park on Church Rd, not in Middlewood.* **Sat 22 July (11-5); Sun 23 July (1-5). Adm £5, chd £2. Light refreshments. Visits also by arrangement 5 Aug to 31 Aug for groups of up to 10. Children welcome but should be supervised due to garden terrain.**
Entered via a courtyard, a former market garden on a north facing slope, with steep, uneven paths. View over the Exe. A food production and wildlife garden with a self-sufficiency ethos; greenhouse, polytunnel, fruit cage, raised vegetable beds, wild area, ponds. Productive fruit bushes and trees inc exotics. Not accessible for wheelchair users.

56 MILL HOUSE
Whitehall, Hemyock, Cullompton, EX15 3UQ. Vanessa Worrall. *1m from Hemyock centre. Travel towards Culmstock from Hemyock turn 1st R, over 2 bridges. Garden is 1st on the R.* **Fri 9, Sat 10, Sun 11 June (11-4). Adm £6, chd free. Home-made teas. Open nearby Regency House.**
A delightful abundant cottage garden, packed full of cottage favourites. Lupins, oriental poppies, geraniums all nestle with pops of alliums, roses, hollyhocks and peonies. The leat garden has a wonderful perennial border alongside an orchard of apple and pear. The Mill leat is bordered with hydrangeas, specimen trees and rhododendrons, a delight to sit watch and experience the sound of water. The garden is owned by artist Vanessa Worrall who will be opening her studio to coincide with the garden opening.

57 MONKSCROFT
Zeal Monachorum, Crediton, EX17 6DG. Mr Ken & Mrs Jane Hogg. *Lane opp church. Parking in farmyard.* **Sat 1, Sun 2 Apr (12-4.30). Adm £5, chd free. Home-made teas.**
Pretty, medium sized garden of oldest cottage in village. Packed with spring colours, primroses, daffodils, tulips, magnolias and camellias. Views to far hills. New exotic garden. Also tranquil fishing lake with daffodils and wildflowers in beautiful setting, home to resident kingfisher. Steep walk to lake approx 20mins, or 5mins by car. Dogs on leads welcome. WC at lake.

GROUP OPENING

58 MORETONHAMPSTEAD GARDENS
Moretonhampstead, TQ13 8PW. *12m W of Exeter, 12m N of Newton Abbot. Signs from the Xrd of A382 and B3212. On E slopes of Dartmoor National Park. Parking at both gardens.* **Fri 19, Sat 20, Sun 21 May, Sat 2, Sun 3 Sept (1-5). Combined adm £7, chd free. Home-made teas at both gardens. Hot soup & savoury scones at Sutton Mead.**

MARDON
Graham & Mary Wilson.

SUTTON MEAD
Edward & Miranda Allhusen, 01647 440296, miranda@allhusen.co.uk.
Visits also by arrangement Apr to Sept. Guided tour with owner included.

2 large gardens on edge of moorland town. One in a wooded valley, the other higher up with magnificent views of Dartmoor. Both have mature orchards and year-round vegetable gardens. Substantial rhododendron, azalea and tree planting, croquet lawns, summer colour and woodland walks through hydrangeas and acers. Mardon: 4 acres based on its original Edwardian design. Long herbaceous border and formal granite terraces, stunning grasses. Fernery and colourful bog garden beside stream fed pond with its thatched boathouse. Arboretum. Sutton Mead: also 4 acres, shrub lined drive. Lawns surrounding granite lined pond with seat at water's edge. Unusual planting, dahlias, grasses, bog garden, rill fed round pond, secluded seating and gothic concrete greenhouse. Sedum roofed summerhouse. Enjoy the views as you wander through the woods. Dogs on leads welcome. Teas are a must.

South Wood Farm

Hayne

59 ◆ MOTHECOMBE HOUSE

Mothecombe, Holbeton, Plymouth, PL8 1LA. Mr & Mrs J Mildmay-White, jmw@flete.co.uk, www.flete.co.uk/gardens. *12m E of Plymouth. From A379 between Yealmpton & Modbury turn S for Holbeton. Continue 2m to Mothecombe.* **For NGS: Sun 7 May (11-5). Adm £6, chd free. Home-made teas.**

Queen Anne House (not open) with Lutyens additions and terraces set in private estate hamlet. Walled gardens, orchard with spring bulbs and magnolias, camellia walk leading through bluebell woods to streams and pond. Mothecombe Garden is managed for wildlife and pollinators. Bee garden with 250 lavenders in 16 varieties. New project to manage adjacent 6 acre meadow as a traditional wildflower pasture. Sandy beach at bottom of garden, unusual shaped large *Liriodendron tulipifera*. Tea, coffee and cake in the garden. Lunches at The Schoolhouse, Mothecombe village. Gravel paths, two slopes.

60 THE MOUNT, DELAMORE

Cornwood, Ivybridge, PL21 9QP. Mr & Mrs Gavin Dollard, 01752 837605, nicdelamore@gmail.com. *Please park in car park for Delamore Park Offices not in village. From Ivybridge turn L at Xrds in Cornwood village keep pub on L, follow wall on R to sharp R bend, turn R.* **Sat 11, Sun 12 Feb (10.30-3.30). Adm £5, chd free. Visits also by arrangement 6 Feb to 10 Mar.**

Welcome one of the first signs of spring by wandering through swathes of thousands of snowdrops in this lovely wood. Closer to the village than to Delamore Gardens, paths meander through a sea of these lovely plants, some of which are unique to Delamore and which were sold as posies to Covent Garden market as late as 2002. The Cornwood Inn in the village is now community owned and open, serving excellent food. Main house and garden open for sculpture exhibition every day in May.

61 MUSBURY BARTON

Musbury, Axminster, EX13 8BB. Lt Col Anthony Drake. *3m S of Axminster off A358. Turn E into village, follow yellow signs. Garden next to church, parking for 12 cars, otherwise park on road in village.* **Sat 24, Sun 25 June (1.30-5). Adm £5, chd free.**

This 6 acre garden surrounds a traditional Devon longhouse. It is planted with much imagination. There are many rare and unusual things to see, inc 2000 roses in beds, some of them edged with local stone. A stream in a steep valley tumbles through the garden. Visit the avenue of horse chestnuts and the arboretum with interesting trees. Enjoy the many vantage points in this unusual garden with a surprise round every corner.

62 MUSSELBROOK COTTAGE GARDEN

Sheepwash, Beaworthy, EX21 5PE. Richard Coward, 01409 231677, coward.richard@sky.com. *1.3m N of Sheepwash. SatNav may be misleading. A3072 to Highampton. Through Sheepwash. L on track signed Lake Farm. A386 S of Merton take rd to Petrockstow. Up hill opp, eventually L. After 350 yds turn R down track signed Lake Farm.* **Fri 21, Sat 22, Sun 23 Apr, Fri 12, Sat 13, Sun 14 May, Fri 16, Sat 17, Sun 18 June, Fri 21, Sat 22, Sun 23 July (11-4.30). Adm £5, chd free. Visits also by arrangement 26 Apr to 18 Oct for groups of 6 to 16. Teas served for group visits. All visitors can bring their own picnics.**

1 acre naturalistic/wildlife/plantsman's garden for year-round interest. Rare/unusual plants on sloping site. 13 ponds (koi, lilies), stream, Japanese garden, Mediterranean garden, wildflower meadow, massed bulbs. Hundreds of ericaceous plants inc camellias, magnolias, rhododendrons, acers, and hydrangeas. Dierama, crocosmia and many grasses. Superb autumn colour. Aquatic nursery inc water lilies. Planting extremely labour intensive as ground is full of rocks. A mattock soon became my indispensable tool, even for planting bulbs. The aquatic plant nursery can be visited by separate arrangement.

GROUP OPENING

63 NEW OGWELL GARDENS
Ogwell, TQ12 6AH. *1½ m SW of Newton Abbot. Turn R at the r'about on the Totnes rd. Go straight on up the hill to the village green. Turn R between the 2 parts of the green down to the Memorial Hall for tickets and a map.* **Sat 10, Sun 11 June (11-5). Combined adm £7, chd free. Home-made teas at Ogwell Memorial Hall, Ogwell, Newton Abbot. TQ12 6AJ.**

NEW BUTTERCOMBE COTTAGE
Helen Clarke.

NEW OGWELL MILL
Jamie & Caroline Colville.

NEW 1 ORCHARD CLOSE
Kevin & Louise Sessions.

NEW PADDOCKS
Gwyn Hughes.

NEW ROSELEA
Andy Harper.

NEW RYDON BRAKE
Paul & Pauline Wynter.

NEW 4 SUNNY HOLLOW
Paul & Anthea Martin.

NEW WILLOW TREE HOUSE
Dan & Katy Farrell-Wright.

NEW WINDYRIDGE
Rebecca & Tim Flower.

In its first year of opening the Ogwell group offers nine gardens of varying styles, inc small cottage gardens, those focused on encouraging wildlife, one with a swimming pond and striking planting, and large gardens with ponds, a bonsai collection and other attractions.

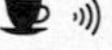

64 THE OLD SCHOOL HOUSE
Ashcombe, Dawlish, EX7 0QB. Vanessa Hurley. *From A38 at Kennford take A380 towards Torquay. Turn off onto B3192 signed to Teignmouth, take 1st exit off r'about follow windy road for 1½ m 1st R after church and follow signs.* **Sat 29 July (1-5); Sun 30 July (1.30-5). Adm £5, chd free. Light refreshments. Home-made cakes for sale.**
The Old School House is part of Ashcombe Estate. Created by Queen's Nurse Vanessa and husband Chaz. It is an artistic palette of colour with a hint of the theatrical. Set in over ½ acre there is a variety of plants, shrubs, vines, bananas and trees attracting insects and wildlife plus an interesting array of quirky recycled materials. Dawlish water stream circles around the pretty garden. There are gravelled areas to entrances and some uneven ground. Some disabled parking outside house.

65 THE OLD VICARAGE
West Anstey, South Molton, EX36 3PE. Tuck & Juliet Moss, 01398 341604, julietm@onetel.com. *9m E of South Molton. From S Molton go E on B3227 to Jubilee Inn. From Tiverton r'about take A396 7m to B3227 then L to Jubilee Inn. Follow NGS signs to garden.* **Sat 20, Sun 21 May, Sat 12, Sun 13 Aug (12.30-5). Adm £5, chd free. Cream teas. Visits also by arrangement for groups of up to 30.**
Croquet lawn leads to multi-level garden overlooking 3 large ponds with winding paths, climbing roses and overviews. Brook with waterfall flows through garden past fascinating summerhouse built by owner. Benched deck overhangs first pond. Features rhododendrons, azaleas and primulas in spring and large collection of wonderful hydrangeas in August. A wall fountain is mounted on handsome, traditional dry wall above house. Access by path through kitchen garden. A number of smaller standing stones echoing local Devon tradition have been added as entertainment.

66 ◆ PLANT WORLD
St Marychurch Road, Newton Abbot, TQ12 4SE. Ray Brown, 01803 872939, info@plant-world-seeds.com, www.plant-world-gardens.co.uk. *2m SE of Newton Abbot. 1½ m from Penn Inn turn-off on A380. Follow brown tourist signs at end of A380 dual carriageway from Exeter.* **For opening times and information, please phone, email or visit garden website.**
The four acres of landscape gardens with fabulous views have been called Devon's 'Little Outdoor Eden'. Representing five continents, they offer an extensive collection of rare and exotic plants from around the world. Superb mature cottage garden and Mediterranean garden will delight the visitor. Attractive viewpoint café, picnic area and shop.

67 POUNDS
Hemyock, Cullompton, EX15 3QS. Diana Elliott, 01823 680802, shillingscottage@yahoo.co.uk, www.poundsfarm.co.uk. *8m N of Honiton. M5 J26. From ornate village pump, nr pub & church, turn up rd signed Dunkeswell Abbey. Entrance ½ m on R. Park in field. Short walk up to garden on R.* **Fri 21 Apr (1-4). Sat 22 Apr, Fri 9, Sat 10 June (1-4), open nearby Regency House. Fri 15, Sat 16 Sept (1-4). Adm £5, chd free.**
Cottage garden of lawns, colourful borders and roses, set within low flint walls with distant views. Slate paths lead through an acer grove to a swimming pool, amid scented borders. Beyond lies a traditional ridge and furrow orchard, with a rose hedge, where apple, pear, plum and cherries grow among ornamental trees. Further on, an area of raised beds combine vegetables with flowers for cutting. Some steps, but most of the garden accessible via sloping grass, concrete, slate or gravel paths.

68 REGENCY HOUSE
Hemyock, EX15 3RQ. Mrs Jenny Parsons, 07772 998982, jenny.parsons@btinternet.com, www.regencyhousehemyock.co.uk. *8m N of Honiton. M5 J26/27. From Catherine Wheel pub & church in Hemyock take Dunkeswell-Honiton Rd. Entrance ½ m on R. Please do not drive on the long grass alongside the drive. Disabled parking (only) at house.* **Sat 22 Apr (2-5), open nearby Pounds. Sun 23 Apr (2-5). Fri 9, Sat 10 June (2-5), open nearby Pounds. Sun 11, Sat 17, Sun 18, Sat 24, Sun 25 June, Sat 21, Sun 22 Oct (2-5). Adm £6, chd free. Home-made teas on terrace. Visits also by arrangement 22 Apr to 16 Oct.**
5 acre plantsman's garden approached across a little ford. Many interesting and unusual trees and shrubs. Visitors can try their hand at identifying plants with the very comprehensive and amusing plant list. Plenty of space to eat your own picnic. Walled vegetable and fruit garden, lake, ponds, bog plantings and sweeping lawns. Horses, Dexter cattle and Jacob sheep. A tranquil space. Gently sloping gravel paths give wheelchair access to the walled garden, lawns, borders and terrace, where teas are served.

69 RUSSET HOUSE

Talaton, Exeter, EX5 2RL. Liz & Alan Franklin, www.russethouse.uk. *16m NE Exeter. Exit A30 onto B3177 signed Fenny Bridges & follow brown signs for Escot. At Escot, straight on 1½m to Talaton. Last house on L. Park in road if space otherwise church car park.* **Sat 17, Sun 18 June (2-6). Adm £4, chd free. Home-made teas.**

¾ acre garden with stunning far reaching views. Mature orchard with 70 varieties. Dense planting of trees, shrubs, roses, clematis and perennials in cottage informal style. Walled back garden has a meandering path past the pond to a productive vegetable and soft fruit area with a greenhouse. Patio has seating areas and a large number of pots. More seating elsewhere allowing enjoyment of lovely views. All the trees in the orchard are labelled as are most trees, shrubs and perennials elsewhere in the garden. Whilst small the vegetable area is highly productive with the use of raised beds and container planting.

70 ST MERRYN

Higher Park Road, Braunton, EX33 2LG. Dr W & Mrs Ros Bradford, 01271 813805, ros@st-merryn.co.uk. *5m W of Barnstaple. A361, R at 30mph sign, then at mini r'about, R into Lower Park Rd, then L into Seven Acre Lane, top of lane R into Higher Park Rd. Pink house 200 yds on R.* **Sun 23 Apr, Sun 14 May, Sun 11 June (2-5.30). Adm £5, chd free. Cream teas. Visits also by arrangement 17 Apr to 13 Aug for groups of up to 20.**

1 acre garden which has been renovated and developed over the last 47 yrs with foliage, flowers, scent, colour-themed borders. Sheltered seating. Artist-owner hopes to lead the visitor around the garden using the winding paths to the thatched summerhouse, ponds, sculptural bonfire, hens, and swimming pool. Earth removed in the construction of the Art Gallery has been used to construct a grassy knoll.

71 SAMLINGSTEAD

Near Roadway Corner, Woolacombe, EX34 7HL. Roland & Marion Grzybek, 01271 870886, roland135@msn.com. *1m outside Woolacombe. Stay on A361 all the way to Woolacombe. Passing through town head up Chalacombe Hill, L at T-junction, garden 150metres on L.* **Sat 18 Mar, Sat 15 July (10.30-3). Adm £5, chd free. Cream teas in 'The Swallows' a purpose built out-building. Hot sausage rolls and cakes, tea, coffee & soft drinks will also be available. Visits also by arrangement Mar to Sept.**

Garden is within 2 mins of N Devon coastline and Woolacombe AONB. 6 distinct areas; cottage garden at front, patio garden to one side, swallows garden at rear, meadow garden, orchard and field (500m walk with newly planted hedgerow). Slightly sloping ground so whilst wheelchair access is available to most parts of garden certain areas may require assistance.

72 NEW SANDRIDGE PARK

Stoke Gabriel, Totnes, TQ9 6RL. Mark & Rosemary Yallop. *Stoke Gabriel is 4½m from Totnes on the A385 via Aish. Sandridge Park is 1½m from Stoke Gabriel on Waddeton Rd.* **Sat 10, Sun 11 June (11-5). Adm £6, chd free. Home-made teas.**

High above the River Dart the Sandridge Park Estate has a Nash villa of 1805 at its heart. Close to the house are formal gardens and ponds, a newly-built walled kitchen garden and orchard, and lawns overlooking the river. There is an extensive collection of roses and specimen trees, a fern garden and woodland walks along the banks of the Dart. Gravel paths around the house and some steps in places. Woodland walks not suitable for wheelchairs.

73 NEW SHERWOOD

Newton St Cyres, Exeter, EX5 5BT. Nicola Chambers, 07702 895435, nikkinew2012@yahoo.com. *2m SE of Crediton. Off A377 Exeter to Barnstaple rd, ¾m Crediton side of Newton St Cyres, signed Sherwood, entrance to drive in 1¾m. Do not follow SatNav once on lane, don't fork R, just drive straight & keep going.* **Sun 2 Apr, Sun 14 May, Sun 11 June, Sun 15 Oct (10-5). Adm £7.50, chd £3. Pre-booking essential, please visit www.ngs.org.uk for information & booking. Home-made teas. Visits also by arrangement 2 Apr to 15 Oct.**

23 acres, 2 steep wooded valleys. Wild flowers, spring bulbs, especially daffodils; extensive collections of magnolias, camellias, rhododendrons, azaleas, berberis, heathers, maples, cotoneasters, buddleias, hydrangeas, cornus and epimedium. National Collections of Magnolias, Knaphill azaleas and berberis. Regret no dogs. Suitable footwear required due to areas of steep terrain.

74 SHUTELAKE

Butterleigh, Cullompton, EX15 1PG. Jill & Nigel Hall, 01884 38812, jill22hall@gmail.com. *3m W of Cullompton; 3m S of Tiverton. Follow signs for Silverton from Butterleigh village. Take L fork 100yds after entrance to Pound Farm. Car park sign on L after 150yds.* **Sat 17, Sun 18 June (11-5). Adm £6, chd free. Home-made teas in the studio barn if inclement. Snack lunches & cakes inc GF/Veg. Visits also by arrangement June to Sept for groups of 10 to 20.**

Shutelake is a working farm. The garden melds subtly with surrounding landscape, in the way its terraces complement the tumbling Devon countryside. Formal close to the house, with a Mediterranean feel. Terraced borders full of warm perennials in summer. Beyond a natural pond teeming with wildlife feeds the Burn stream. Woodland paths and sculptures. Check NGS website for ad hoc pop-up openings.

75 SIDBURY MANOR

Sidmouth, EX10 0QE. Lady Cave, www.sidburymanor.com. *1m NW of Sidbury. Sidbury village is on A375, S of Honiton, N of Sidmouth.* **Fri 14, Sun 16 Apr (2-5). Adm £6, chd free. Cream teas.**

Built in 1870s this Victorian manor house built by owner's family and set within E Devon AONB comes complete with 20 acres of garden inc substantial walled gardens, extensive arboretum containing many fine trees and shrubs, a number of champion trees, and areas devoted to magnolias, rhododendrons and camellias. Partial wheelchair access.

76 SOCKS ORCHARD

Smallridge, Axminster, EX13 7JN. Michael & Hilary Pritchard. *2m from Axminster. From Axminster on A358 L at Weycroft Mill T-lights. Pass Ridgeway Hotel on L. Continue on*

lane for ½m. Park in field opp. **Fri 30 June, Sat 1, Sun 2 July (1-5). Adm £5, chd free. Home-made teas.**
1 acre plus plantaholic's garden designed for year-round interest and colour. Many specimen trees, large collection of herbaceous plants, over 200 roses, woodland shrubs, dahlias, gravel and grass borders, small orchard, vegetable patch. Steep bank with trees, shrubs and wild flowers. Viewpoint at top. Alpine troughs and alpine house. Scented leaf pelargonium and succulents collections. Chickens. Featured in Good Housekeeping magazine. Wheelchair access to most of garden.

77 SOUTH WOOD FARM

Cotleigh, Honiton, EX14 9HU. Professor Clive Potter, Southwoodfarmgarden@gmail.com. *3m NE of Honiton. From Honiton head N on A30, take 1st R past Otter Valley Field Kitchen layby. Follow for 1m. Go straight over Xrds and take first L. Entrance after 1m on R.* **Sat 29, Sun 30 Apr, Sat 16, Sun 17 Sept (2-5). Adm £6, chd free. Home-made teas. Visits also by arrangement 2 May to 1 Sept for groups of 15 to 30.**
Designed by renowned Arne Maynard around C17 thatched farmhouse, country garden exemplifying how contemporary design can be integrated into a traditional setting. Herbaceous borders, roses, yew topiary, knot garden, wildflower meadows, orchards, lean-to greenhouses and a mouthwatering kitchen garden create an unforgettable sense of place. Rare opportunity to visit spring garden in all its glory with 5000 bulbs inc 2500 tulips and hundreds of camassias in flower meadow. Regional Finalist, The English Garden's The Nation's Favourite Gardens 2019. Gravel pathways, cobbles and steps.

78 SOUTHCOMBE BARN

Widecombe-in-the-Moor, Newton Abbot, TQ13 7TU. Tom Dixon & Vashti Cassinelli, info@southcombebarn.com, www.southcombebarn.com. *6m W of Bovey Tracey. B3387 from Bovey Tracey after village church take rd SW for 400yds then sharp R signed Southcombe, after 200yds pass C17 farmhouse and park on L.* **Fri 2, Mon 5, Fri 9, Mon 12 June (11.30-3.30). Adm £5, chd free.**
Southcombe Barn, idyllically located on the edge of Dartmoor, has a natural look with 3 acres of vibrant wildflower meadows and flowering trees with mown grass paths running alongside a gently babbling stream. Art Gallery featuring local artists will be open and there are tearooms in the village.

79 SPITCHWICK MANOR

Poundsgate, Newton Abbot, TQ13 7PB. Mr & Mrs P Simpson. *4m NW of Ashburton. Princetown rd from Ashburton through Poundsgate, 1st R at Lodge. From Princetown L at Poundsgate sign. Past Lodge. Park after 300yds at Xrds.* **Sat 13, Sun 14 May (11-4.30). Adm £6, chd free. Cream teas.**
6½ acre garden with extensive beautiful views. Mature garden undergoing refreshment. A variety of different areas; lower ancient walled garden with glasshouses, formal rose garden with fountain, camellia walk with small leat and secret garden with Lady Ashburton's plunge pool built 1773. 2.6 acre vegetable garden sheltered by high granite walls housing 9 allotments and lily pond. Mostly wheelchair access.

80 SPRING LODGE

Kenton, Exeter, EX6 8EY. David & Ann Blandford. *Between Lyson & Oxton. Head for Lyson then Oxton.* **Sun 7 May (10-5). Adm £4, chd free. Pre-booking essential, please visit www.ngs.org.uk for information & booking. Tea.**
Originally part of the Georgian pleasure gardens to Oxton House, the garden at Spring Lodge boasts a hermit's cave, dramatic cliffs and a stream which runs under the house. Built in the quarry of the big house this ½ acre garden is on many levels, with picturesque vistas and lush planting. Pre booking essential due to restricted parking.

81 SPRINGFIELD HOUSE

Seaton Road, Colyford, EX24 6QW. Wendy Pountney, 01297 552481. *Starting on A3052 coast rd, at Colyford PO take Seaton Rd. House 500m on L. Ample parking in field.* **Sat 27 May, Sat 8 July (10.30-6). Adm £5, chd free. Home-made teas. Visits also by arrangement Apr to Oct for groups of up to 30. Refreshments by arrangement.**
1 acre garden with mainly new planting. Numerous beds, majority of plants from cuttings and seed keeping cost to minimum, full of colour spring to autumn. Vegetable garden, fruit cage and orchard with ducks and chickens. Large formal pond. Wonderful views over River Axe and path to bird sanctuary, which is well worth a visit, leads from the garden.

82 SQUIRRELS

98 Barton Road, Torquay, TQ2 7NS. Graham & Carol Starkie, 01803 329241, calgra@talktalk.net. *5m S of Newton Abbot. From Newton Abbot take A380 to Torquay. After ASDA store on L, turn L at T-lights up Old Woods Hill. 1st L into Barton Rd. Bungalow 200yds on L. Also could turn by B&Q. Parking nearby.* **Sat 29, Sun 30 July (2-5). Adm £5, chd free. Home-made teas. Visits also by arrangement 1 Aug to 6 Aug.**
Plantsman's small town environmental garden, landscaped with small ponds and 7ft waterfall. Interlinked through abutilons to Japanese, Italianate, Spanish, tropical areas. Specialising in fruit inc peaches, figs, kiwi. Tender plants inc bananas, tree fern, brugmansia, lantanas, oleanders. Collection of fuchsia, dahlias, abutilons, bougainvillea. Environmental and Superclass Gold Medal Winners. 27 hidden rainwater storage containers. Advice on free electric from solar panels and solar hot water heating and fruit pruning. Three sculptures. Many topiary birds and balls. Huge 20ft Torbay palm. 9ft geranium. 15ft abutilons. New Moroccan and Spanish courtyard with tender succulents etc.

9,500 patients and their families are being supported across four Horatio's gardens thanks to our annual donations

83 STONE FARM

Alverdiscott Rd, Bideford, EX39 4PN. Mr & Mrs Ray Auvray, 01237 421420, rayauvray@icloud.com. *1½m from Bideford towards Alverdiscott. From Bideford cross river using Old Bridge and turn L onto Barnstaple Rd. 2nd R onto Manteo Way and 1st L at mini r'about.* **Sat 27, Sun 28 May, Sat 1, Sun 2 July, Sat 26, Sun 27 Aug (2-5). Adm £4.50, chd free. Home-made teas. Visits also by arrangement 20 May to 28 Aug for groups of 10 to 20.**

1 acre country garden with striking herbaceous borders, dry stone wall terracing, white garden, hot garden and dahlia beds. Extensive fully organic vegetable gardens with polytunnels and ¼ acre walled garden, together with an orchard with traditional apples and pears. Some gravel paths but wheelchair access to whole garden with some help.

84 ◆ STONE LANE GARDENS

Stone Farm, Stone Lane, Chagford, TQ13 8JU. Stone Lane Gardens Charitable Trust, 01647 231311, admin@stonelanegardens.com, www.stonelanegardens.com. *Halfway between Chagford & Whiddon Down, close to A382. 2.3m from Chagford, 1½m from Castle Drogo, 2½m from A30 Whiddon Down via Long Lane.* **For NGS: Sat 7, Sat 14 Oct (10-6). Adm £7, chd £3.50. Light refreshments in small tea room and tea garden. For other opening times and information, please phone, email or visit garden website. Donation to Plant Heritage.**

Outstanding and unusual 5 acre arboretum and water garden on edge of Dartmoor National Park. Beautiful National Collection of Betula species with lovely colourful peeling bark, from dark brown, reds, orange, pink and white. Interesting underplanting. Meadow walks and lovely views. Summer sculpture exhibition situated in the garden. Tripadvisor award winner. Near to NT Castle Drogo and garden. RHS Partner Garden. Partial wheelchair access to the gardens.

NPC

85 NEW STONEHOUSE LAWN TENNIS CLUB

Durnford Street, Devil's Point, Plymouth, PL1 3QR. *Car parks at Devil's Point and Cremyll St. Follow signs to Royal William Yard but just after St Paul's Church carry straight on do not turn L to the Yard.* **Sat 13, Sun 14 May (10.30-4.30). Adm £5, chd free. Cream teas.**

In grounds leased from The Edgcumbe Estate the Tennis Club was established about 100 yrs ago and over time the gardens have been improved and extended. The major features are the outstanding view across Plymouth Sound to the breakwater and the proximity to the sea means frosts are rare and tender plants can be grown. Drop off point at main gate and gentle slope to the gardens. Regret no dogs.

86 STONELANDS HOUSE

Stonelands Bridge, Dawlish, EX7 9BL. Mr Kerim Derhalli (Owner) Mr Saul Walker (Head Gardener), 07815 807832, saulwalkerstonelands@outlook.com. *Outskirts of NW Dawlish. From A380 take junction for B3192 & follow signs for Teignmouth, after 2m L at crossroads onto Luscombe Hill, further 2m, main gate on L.* **Visits by arrangement Apr to June for groups of 10 to 20. Adm £7, chd free. Home-made teas. Please book refreshments through Head Gardener, separate payment on the day.**

Beautiful 12 acre pleasure garden surrounding early C19 property designed by John Nash. Many mature specimen trees, shrubs and rhododendrons, large formal lawn, recently landscaped herbaceous beds, vegetable garden, woodland garden, orchard with wildflower meadow and riverside walk.

An atmospheric and delightful horticultural secret. Wheelchair access to lower area of gardens. Paths through woodland, meadow and riverside walk may be unsuitable.

D

GROUP OPENING

87 TEIGNMOUTH GARDENS

Cliff Road, Teignmouth, TQ14 8TW. *½m from Teignmouth town centre. 5m E of Newton Abbot. 11m S of Exeter. Purchase ticket for all gardens at first garden visited, a map will be provided showing location of gardens and parking.* **Sat 17, Sun 18 June (1-5). Combined adm £7, chd free. Home-made teas at Bitton House, next to The Orangery.**

7 COOMBE AVENUE
Stewart & Pat Henchie.

21 GORWAY
Mrs Christine Richman.

26 HAZELDOWN ROAD
Mrs Ann Sadler.

LOWER COOMBE COTTAGE
Tim & Tracy Armstrong.

LOWER HOLCOMBE HOUSE
Joanne Sparks.

THE ORANGERY
Teignmouth Town Council.

SEA VISTA
Shirley & Tim Williams.

65 TEIGNMOUTH ROAD
Mr Terry Rogers.

Seven gardens, plus the beautifully restored Orangery, are opening in the picturesque coastal town of Teignmouth. A wide range of garden styles and sizes can be explored this year from very small but inspiring manicured gardens to large wildlife havens. Features inc Mediterranean courtyards, Japanese garden, greenhouses, pollinator friendly planting, exotic plants, streams and ponds, fruit and vegetable beds and stunning sea views from some of the gardens. Home-made teas can be enjoyed at Bitton House next to The Orangery. Tickets are valid for both days. Partial wheelchair access at some gardens.

88 TORVIEW

44 Highweek Village, Newton Abbot, TQ12 1QQ. Ms Penny Hammond. *On N of Newton Abbot accessed via A38. From Plymouth: A38 to Goodstone, A383 past Hele Park & L onto Mile End Rd. From Exeter: A38 to Drumbridges then A382 past Forches Cross & take next R. Please park in nearby streets.* **Sat 29, Sun 30 Apr, Mon 1 May (12-5). Adm £6, chd free. Home-made teas.**

Owned by two semi-retired horticulturists: Mediterranean formal front garden with wisteria clad Georgian house and small alpine house. Rear courtyard with tree ferns, pots/troughs, lean-to 7m conservatory with tender plants and climbers. Steps to 30x20m walled garden - flowers, vegetables and trained fruit. Shade tunnel of woodland plants. Many rare and unusual plants.

89 UPPER GORWELL HOUSE
Goodleigh Rd, Barnstaple, EX32 7JP. Dr J A Marston, www.gorwellhousegarden.co.uk. *3/4m E of Barnstaple centre on Bratton Fleming rd. Drive entrance between 2 lodges on L coming uphill (Bear St) approx 3/4m from Barnstaple centre. Take R fork at end of long drive. New garden entrance to R of house up steep slope.* **Sun 2 Apr, Sun 7 May, Sun 11 June, Sun 9 July, Sun 24 Sept (2-6). Adm £6, chd free. Cream teas. Visitors are welcome to bring their own picnics.**
Created mostly since 1979, this 4 acre garden overlooking the Taw estuary has a benign microclimate which allows many rare and tender plants to grow and thrive, both in the open and in the walled garden. Several strategically placed follies complement the enclosures and vistas within the garden. Mostly wheelchair access but some very steep slopes at first to get into garden.

90 NEW THE WALLED GARDEN
Nelson Coach House, Staverton, Totnes, TQ9 6PA. Chris & Carmen Timpson, www.instagram.com/thewalled.garden. *8mins from Totnes. From Totnes/Ashburton A384 direction 1st property on R after the Staverton village sign. Parking around village or at pub if you are having lunch there.* **Fri 14, Sun 16 July (11-5). Adm £5, chd free.**
A 1 acre historic walled garden and adjacent orchard undergoing re-imagination since 2018. Productive kitchen garden with peaches and nectarines thriving on the hot wall, dry garden with mature olive trees and drought resistant planting, water rill and paradise pond, orchard and small scale wildflower meadow. Lunch/refreshments at Sea Trout pub. Top of garden not accessible but dry garden, kitchen garden and orchard are via gravel paths with some slopes.

GROUP OPENING

91 WEST CLYST BARNYARD GARDENS
Westclyst, Exeter, EX1 3TR. *From Pinhoe take B3181 towards Broadclyst. At Westclyst T-lights continue straight past speed camera 1st R onto Private Road over M5 bridge, R into West Clyst Barnyard.* **Sat 17, Sun 18 June (1-5). Combined adm £5, chd free. Home-made teas.**

6 WEST CLYST BARNYARD
Alan & Toni Coulson.

7 WEST CLYST BARNYARD
Malcolm & Ethel Hillier.

These gardens have been planted in farmland around a converted barnyard of a medieval farm. There are 2 gardens with a wildflower meadow, wildlife ponds, bog garden, David Austin roses, magnolias and many trees and shrubs. Cars may be driven to the gate of No 7 for disabled access to the gardens but then please park in the car park.

92 WHIDDON GOYLE
Whiddon Down, Okehampton, EX20 2QJ. Mr & Mrs Lethbridge, 07561 440319, cillethbridge@aol.com. *From Whiddon Down, follow signs for Okehampton, take 2nd exit at r'about. Whiddon Goyle is on L, signposted.* **Sat 3, Sun 4 June (11-4). Adm £6, chd free. Home-made teas. Visits also by arrangement 4 June to 29 Sept.**
Whiddon Goyle enjoys stunning views over Dartmoor and sits 1000 ft above sea level. Built in 1930s it is cleverly designed to protect its 2 acre garden against the Dartmoor weather. It enjoys many features inc a rockery, croquet lawn, rose garden, herbaceous borders, ponds, a newly developed cut flower garden which supplies seasonal blooms, along with a pair of majestic monkey trees. Access is via a gravelled driveway on a slope.

93 WHITSTONE FARM
Whitstone Lane, Bovey Tracey, TQ13 9NA. Katie & Alan Bunn, 01626 832258, klbbovey@gmail.com. *1/2m N of Bovey Tracey. From A382 turn L opp golf club, after 1/3m L at swinging sign 'Private road leading to Whitstone'. Follow NGS signs.* **Sun 23, Sun 30 Apr, Sun 30 July (1.30-4.30). Adm £6, chd free. Pre-booking essential, please visit www.ngs.org.uk for information & booking. Tea and home-made cakes, gluten free option. Visits also by arrangement 5 Apr to 6 Sept for groups of up to 25. Garden groups welcome.**
Nearly 4 acres of steep hillside garden with stunning views of Haytor and Dartmoor. Snowdrops start in January followed by bluebells throughout the garden. Arboretum planted 45 yrs ago, over 200 trees from all over the world inc magnolias, camellias, acers, alders, betula, davidias and sorbus. Always colour in the garden and wonderful tree bark. Major plantings of rhododendrons and cornus. Late flowering eucryphia (National Collection) and hydrangeas. Display of architectural and metal sculptures and ornaments. Partial access to lower terraces for wheelchair users.

NPC

94 NEW WILLOWRey FARM
Moretonhampstead Road, Lustleigh, Newton Abbot, TQ13 9SN. Mr & Mrs Anthony Walters, 01647 277307, fiona@druidhouse.co.uk. *1½m N of Lustleigh & S of Moretonhampstead on A382.* **Sat 3, Sun 4 June (1-5). Adm £6, chd £3. Home-made teas. Visits also by arrangement June to Oct.**
Amidst 80 acres of woodland and pasture an ancient granite cross passage house is surrounded by formal gardens. Granite walled garden with an abundance of herbaceous plants, intorlinking ponds, ericaceous shrubs and trees, riverside woodland walk over an old granite bridge, orchards, kitchen garden, fruit cages and Cluckington Palace! Dogs welcome on leads.

Our annual donations to Parkinson's UK meant 7,000 patients are currently benefiting from support of a Parkinson's Nurse

DORSET

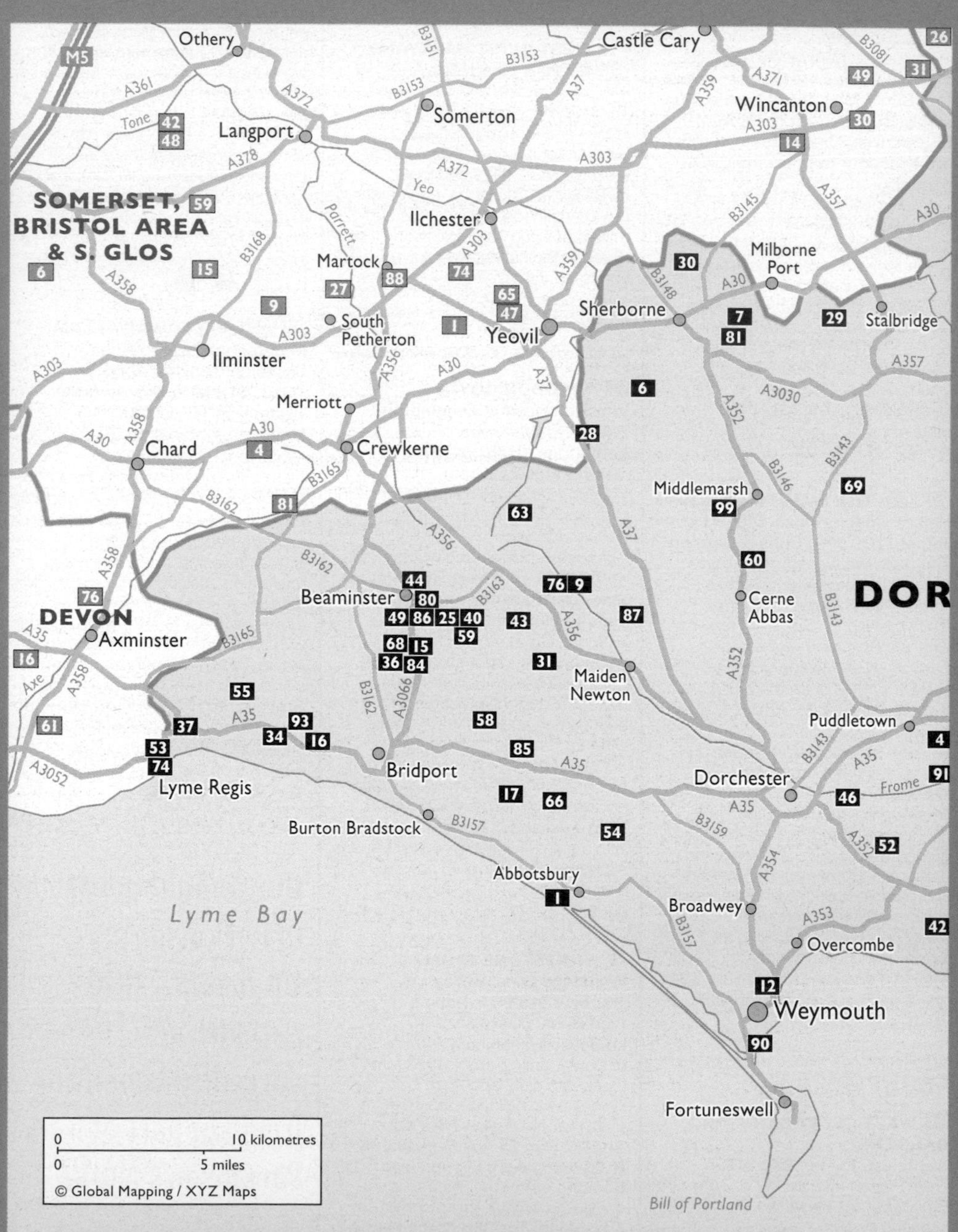

Othery
Castle Cary
M5
A361
Tone
42
48
Langport
A372
B3151
B3153
Somerton
A37
A359
A371
B3081
26
49
31
Wincanton
30
A303
14
A378
A372
Yeo
Parrett
Ilchester
A357
A30
SOMERSET, BRISTOL AREA & S. GLOS
59
6
15
B3168
Martock
88
27
74
A303
A359
65
47
30
B3145
Milborne Port
B3148
A358
9
South Petherton
1
Yeovil
Sherborne
A30
7
81
29
Stalbridge
Ilminster
A303
A356
A30
A37
A357
A3030
6
Merriott
A352
Chard
A30
4
Crewkerne
28
B3143
A358
B3165
Middlemarsh
B3146
69
B3162
81
63
99
A37
A356
60
B3162
76
9
A358
76
DEVON
Beaminster
44
80
B3163
87
Cerne Abbas
DOR
A35
Axminster
49
86
25
40
59
43
A356
B3143
B3165
16
68
15
36
84
31
Maiden Newton
A352
Axe
A358
55
B3162
A3066
61
37
A35
93
34
16
58
Puddletown
53
74
A3052
Lyme Regis
Bridport
85
A35
B3143
4
A35
91
Dorchester
17
66
46
Frome
A35
Burton Bradstock
B3157
B3159
54
A352
52
A354
Abbotsbury
1
Lyme Bay
Broadwey
B3157
A353
42
Overcombe
12
Weymouth
90
Fortuneswell
0
10 kilometres
0
5 miles
© Global Mapping / XYZ Maps
Bill of Portland

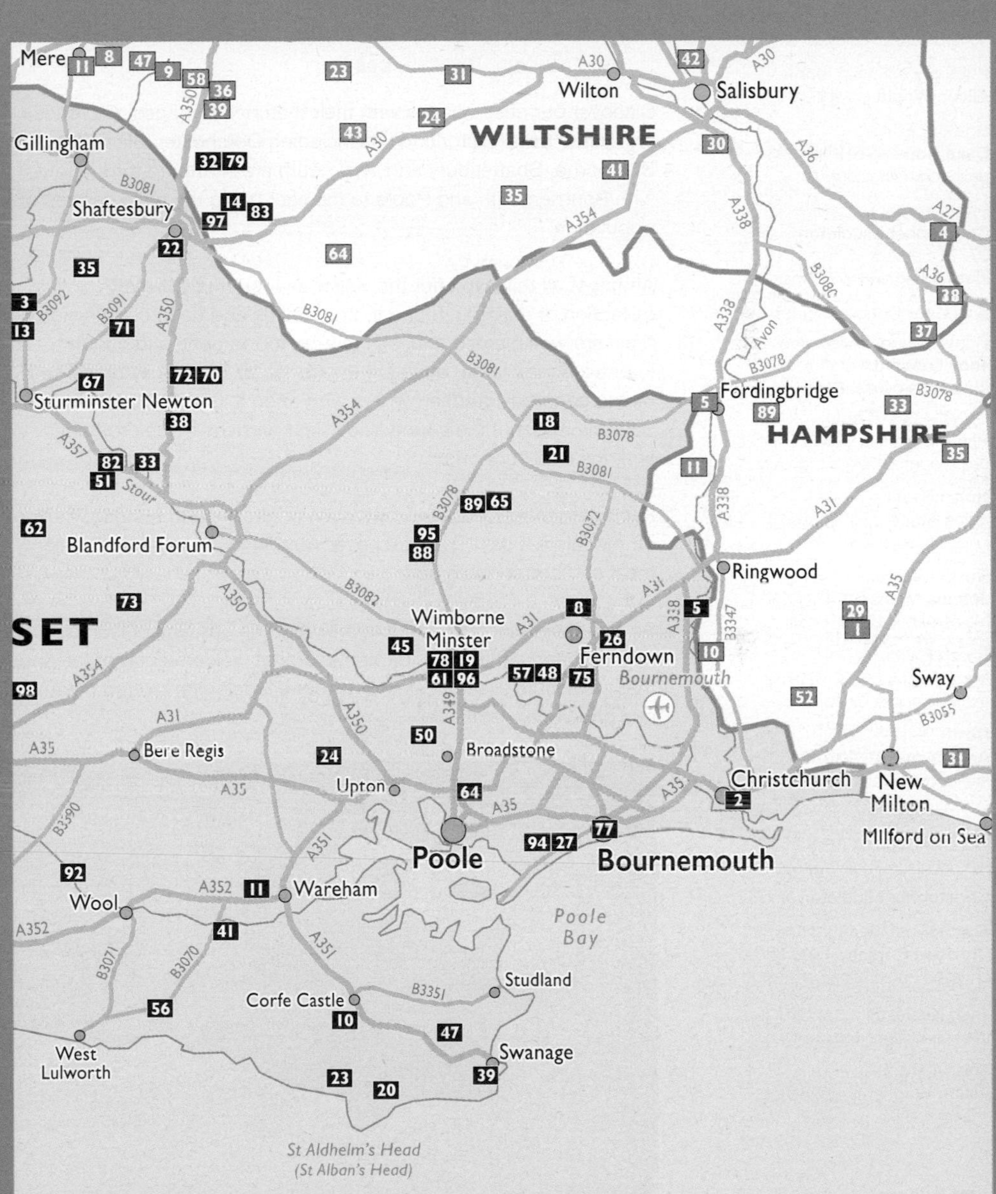
Mere
Gillingham
Shaftesbury
Sturminster Newton
Blandford Forum
SET
Wilton
Salisbury
WILTSHIRE
Fordingbridge
HAMPSHIRE
Ringwood
Wimborne Minster
Ferndown
Bournemouth
Sway
Bere Regis
Broadstone
Upton
Christchurch
New Milton
Milford on Sea
Poole
Bournemouth
Wool
Wareham
Poole Bay
Studland
Corfe Castle
Swanage
West Lulworth
St Aldhelm's Head
(St Alban's Head)
Stour
Avon
A30
A350
A354
A36
A338
A27
B3081
B3092
B3091
A357
B3078
B3072
B3082
A31
B3347
A35
B3055
A349
B3390
A351
A352
B3071
B3070
B3351
B3308C

Volunteers

County Organiser
Alison Wright 01935 83652
alison.wright@ngs.org.uk

County Treasurer
Richard Smedley 01202 528286
richard@carter-coley.co.uk

Publicity & Social Media
Alison Wright (as above)

Booklet Editor
Dave Rorke 07513 384998
dave.rorke@ngs.org.uk

Photographer
Christopher Middleton
07771 596458
christophermiddleton@mac.com

Assistant County Organisers

Central East/ Wimborne
Tony Leonard 07711 643445
tony.leonard@ngs.org.uk

Central East/Poole
Phil Broomfield 07810 646123
phil.broomfield@ngs.org.uk

North East
Jules Attlee 07837 289964
julesattlee@icloud.com

North Central
Joanna Mains 01747 839831
mainsmanor@tiscali.co.uk

South East
Mary Angus 01202 872789
mary@gladestock.co.uk

North West
Annie Dove 01300 345450
anniedove1@btinternet.com

South Central
Pip Davidson 07765 404248
pip.davidson@ngs.org.uk

Christopher Middleton (see above)

South West/Beaminster
Christine Corson 01308 863923
christine.corson@ngs.org.uk

Trish Neale 01308 863790
trish.neale@ngs.org.uk

West Central
Alison Wright (as above)

@NGS.Dorset
@DorsetNGS
@dorset_national_garden_scheme

Dorset is not on the way to anywhere. We have no cathedral and no motorways. The county has been inhabited forever and the constantly varying landscape is dotted with prehistoric earthworks and ancient monuments, bordered to the south by the magnificent Jurassic Coast.

Discover our cosy villages with their thatched cottages, churches and pubs. Small historic towns including Dorchester, Blandford, Sherborne, Shaftesbury and Weymouth are scattered throughout, with Bournemouth and Poole to the east being the main centres of population.

Amongst all this, we offer the visitor a wonderfully diverse collection of gardens, found in both towns and deep countryside. They are well planted and vary in size, topography and content. In between the larger ones are the tiniest, all beautifully presented by the generous garden owners who open for the National Garden Scheme. Most of the county's loveliest gardens in their romantic settings also support us.

Each garden rewards the visitor with originality and brings joy, even on the rainiest day! They are never very far away from an excellent meal and comfortable bed.

So do come, discover and explore what the gardens of Dorset have to offer with the added bonus of that welcome cup of tea and that irresistible slice of cake, or a scone laden with clotted cream and strawberry jam!

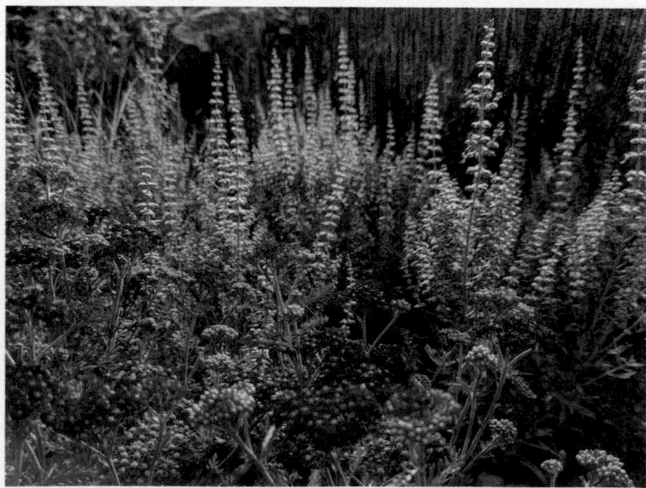

Above: Glenholme Herbs

OPENING DATES

All entries subject to change. For latest information check **www.ngs.org.uk**

Extended openings are shown at the beginning of the month.

Map locator numbers are shown to the right of each garden name.

February

Snowdrop Festival

Every day from Friday 17th to Tuesday 21st
Lawsbrook 51

Friday 10th
The Old Vicarage 71

Sunday 12th
The Old Vicarage 71

Saturday 25th
Manor Farm, Hampreston 57

Sunday 26th
Manor Farm, Hampreston 57

March

Sunday 12th
Frankham Farm 28

Saturday 18th
Chideock Manor 16

Sunday 19th
Chideock Manor 16
The Old Vicarage 71

April

Saturday 1st
◆ Cranborne Manor Garden 18

Wednesday 5th
◆ Edmondsham House 21
Knitson Old Farmhouse 47

Monday 10th
◆ Edmondsham House 21
The Old Rectory, Pulham 69

Wednesday 12th
◆ Edmondsham House 21
Knitson Old Farmhouse 47

Thursday 13th
The Old Rectory, Pulham 69

Sunday 16th
The Old Vicarage 71

Wednesday 19th
◆ Edmondsham House 21
Horn Park 44

Friday 21st
Slape Manor 84

Saturday 22nd
Slape Manor 84

Sunday 23rd
Broomhill 9
Frankham Farm 28
Slape Manor 84

Monday 24th
NEW Grays Farmhouse 31

Wednesday 26th
◆ Edmondsham House 21

Saturday 29th
NEW 27A Forest Road 27
NEW Grays Farmhouse 31

Sunday 30th
Annalal's Gallery 2
NEW 27A Forest Road 27
Holworth Farmhouse 42
Western Gardens 94

May

Monday 1st
Holworth Farmhouse 42

Wednesday 3rd
Knitson Old Farmhouse 47

Sunday 7th
The Old Rectory, Litton Cheney 66
The Pines 75

Wednesday 10th
Knitson Old Farmhouse 47
NEW Oakdale Library Gardens 64
The Old Rectory, Litton Cheney 66

Friday 12th
24 Carlton Road North 12

Saturday 13th
24 Carlton Road North 12
Falconers 24
Harcombe House 34
The Pines 75
Well Cottage 93

Sunday 14th
22 Avon Avenue 5
24 Carlton Road North 12
Falconers 24
Harcombe House 34
The Old Vicarage 71
Well Cottage 93
Wincombe Park 97

Monday 15th
24 Carlton Road North 12

Wednesday 17th
The Old Rectory, Netherbury 68
Wincombe Park 97

Saturday 20th
Edwardstowe 22
NEW 18 Ford Close 26
NEW 22 Lancaster Drive 50

Sunday 21st
Edwardstowe 22
NEW 18 Ford Close 26
NEW 22 Lancaster Drive 50
The Secret Garden and Serles House 78

Monday 22nd
NEW Oakdale Library Gardens 64

Saturday 27th
NEW Myrtle Cottage 62
White House 95

Sunday 28th
Annalal's Gallery 2
◆ Careys Secret Garden 11
Holworth Farmhouse 42
NEW Myrtle Cottage 62
The Old Rectory, Manston 67
Staddlestones 88
White House 95

Monday 29th
Holworth Farmhouse 42
Staddlestones 88

Wednesday 31st
The Old Rectory, Manston 67

June

Thursday 1st
Stanbridge Mill 89

Friday 2nd
Lower Abbotts Wootton Farm 55
Stanbridge Mill 89

Saturday 3rd
NEW 27A Forest Road 27
Old Down House 65
NEW Swallows Rest 90

Sunday 4th
Broomhill 9
NEW Bucknowle Coach House 10
Carraway Barn 13
NEW 18 Ford Close 26
NEW 27A Forest Road 27
Frankham Farm 28
Manor Farm, Hampreston 57
Old Down House 65
NEW Swallows Rest 90

Tuesday 6th
Eastington Farm 20
Encombe House 23

Wednesday 7th
Horn Park 44
Knitson Old Farmhouse 47
Old Down House 65

Thursday 8th
Carraway Barn 13

Friday 9th
Mappercombe Manor 58

Saturday 10th
Philipston House 73

Sunday 11th
NEW Ardhurst 3
Bembury Farm 6
Carraway Barn 13

Mappercombe Manor 58
The Secret Garden and Serles House 78
NEW Shillingstone Gardens 82
Utopia 91
NEW Yew Tree House 99

Tuesday 13th
◆ Holme for Gardens 41

Wednesday 14th
Bembury Farm 6
Farrs 25
Knitson Old Farmhouse 47

Friday 16th
NEW Chilcombe House 17

Saturday 17th
Chideock Manor 16
The Hollow, Blandford Forum 38
Hooke Farm 43
Stable Court 87

Sunday 18th
Black Shed 7
Chideock Manor 16
The Hollow, Blandford Forum 38
Hooke Farm 43
Lewell Lodge 52
The Old School House 70
NEW Penmead Farm 72
Stable Court 87
Staddlestones 88
Western Gardens 94

Tuesday 20th
◆ Littlebredy Walled Gardens 54

Wednesday 21st
Deans Court 19
The Hollow, Blandford Forum 38

Thursday 22nd
Little Cliff 53
The Old School House 70
NEW Penmead Farm 72

Saturday 24th
◆ Hogchester Farm 37

Sunday 25th
Annalal's Gallery 2
NEW Ardhurst 3
22 Avon Avenue 5
NEW The Chantry 15
Hanford School 33
NEW Hingsdon 36
Little Cliff 53

Tuesday 27th
◆ Littlebredy Walled Gardens 54

Thursday 29th
Yardes Cottage 98

July

Every Wednesday
The Hollow, Swanage 39

Saturday 1st
Semley Grange 79
Slape Manor 84
White House 95
Yardes Cottage 98

Sunday 2nd
NEW Bucknowle Coach House 10
Holworth Farmhouse 42
The Old Rectory, Litton Cheney 66
Slape Manor 84
White House 95

Thursday 6th
Lulworth Castle House 56

Saturday 8th
NEW Glenholme Herbs 30

Sunday 9th
NEW Glenholme Herbs 30

Saturday 15th
◆ Hogchester Farm 37

Sunday 16th
Broomhill 9
Hilltop 35
Manor Farm, Hampreston 57

Wednesday 19th
Deans Court 19

Thursday 20th
Little Cliff 53

Saturday 22nd
NEW 27A Forest Road 27

Sunday 23rd
NEW 27A Forest Road 27
Hilltop 35
Little Cliff 53

Saturday 29th
NEW Hollymoor Farm 40

Sunday 30th
Annalal's Gallery 2
22 Avon Avenue 5
Black Shed 7
Hilltop 35

August

Every Wednesday
The Hollow, Swanage 39

Sunday 6th
Annalal's Gallery 2
Manor Farm, Hampreston 57
The Old Rectory, Pulham 69

Thursday 10th
Broomhill 9
The Old Rectory, Pulham 69

Sunday 13th
Hilltop 35

Saturday 19th
Brook View Care Home 8

Sunday 20th
Brook View Care Home 8
Castle Rings 14
Hilltop 35

Wednesday 23rd
Farrs 25

Sunday 27th
Annalal's Gallery 2
22 Avon Avenue 5
Staddlestones 88

Monday 28th
Staddlestones 88

September

Sunday 3rd
NEW Ardhurst 3

Saturday 9th
NEW 27A Forest Road 27
1 Pine Walk 74

Sunday 10th
NEW 27A Forest Road 27
1 Pine Walk 74

Tuesday 12th
◆ Holme for Gardens 41

Sunday 17th
Annalal's Gallery 2
NEW Ardhurst 3

Friday 22nd
◆ Knoll Gardens 48

October

Wednesday 4th
◆ Edmondsham House 21

Wednesday 11th
◆ Edmondsham House 21

Sunday 15th
Frankham Farm 28

Wednesday 18th
◆ Edmondsham House 21

Wednesday 25th
◆ Edmondsham House 21

December

Sunday 3rd
Annalal's Gallery 2

Sunday 17th
Annalal's Gallery 2

By Arrangement

Arrange a personalised garden visit with your club, or group of friends, on a date to suit you. See individual garden entries for full details.

Bembury Farm 6
Broomhill 9
Carraway Barn 13
NEW 27A Forest Road 27
Frith House 29
Grove House 32
The Hollow, Blandford Forum 38
The Hollow, Swanage 39
Holworth Farmhouse 42
Knitson Old Farmhouse 47
Knowle Cottage 49
Lawsbrook 51
Manor Farm, Hampreston 57
NEW Myrtle Cottage 62
Norwood House 63
Old Down House 65
The Old Rectory, Manston 67

The Chantry

THE GARDENS

1 ◆ ABBOTSBURY GARDENS
Abbotsbury, Weymouth, DT3 4LA. Ilchester Estates, 01305 871387, info@abbotsbury-tourism.co.uk, www.abbotsburygardens.co.uk. *8m W of Weymouth. From B3157 Weymouth-Bridport, 200yds W of Abbotsbury village.* **For opening times and information, please phone, email or visit garden website.**
30 acres, started in 1760 and considerably extended in C19. Much recent replanting. The maritime microclimate enables this Mediterranean and southern hemisphere garden to grow rare and tender plants. National Collection of Hoherias (flowering Aug in NZ garden). Woodland valley with ponds, stream and hillside walk to view the Jurassic Coast. Open all year except for Christmas week. Partial wheelchair access, some very steep paths and rolled gravel but we have a selected wheelchair route with sections of tarmac hard surface.

   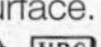 NPC 

2 ANNALAL'S GALLERY
25 Millhams Street, Christchurch, BH23 1DN. Anna & Lal Sims, www.annasims.co.uk. *Town centre. Park in Saxon Square PCP - exit to Millhams St via alley at side of church.* **Sun 30 Apr, Sun 28 May, Sun 25 June, Sun 30 July, Sun 6, Sun 27 Aug, Sun 17 Sept, Sun 3, Sun 17 Dec (2-4). Adm £3, chd free.**
Enchanting 180 yr old cottage, home of two Royal Academy artists. 32ft x 12½ ft garden on 3 patio levels. Pencil gate leads to colourful scented Victorian walled garden. Sculptures and paintings hide among the flowers and shrubs. Unusual studio and garden room. Mural of a life-size greyhound makes the cottage easy to find and adds a smile to people's faces.

3 NEW ARDHURST
Nash Lane, Marnhull, Sturminster Newton, DT10 1JZ. Mr & Mrs Ed Highnam. *From A30 East Stour, B3092 3m to Marnhull, Crown Inn on R. Turn R down Church Hill, follow NGS yellow signs. From Sturminster Newton, 3m into Marnhull, turn L down Church Hill, follow yellow signs.* **Sun 11 June (12-3.30). Home-made teas. Open nearby Carraway Barn. Sun 25 June (1-4.30). Light refreshments. Sun 3 Sept (1-4.30). Home-made teas. Sun 17 Sept (1-4.30). Light refreshments. Adm £7, chd free. Wine offered in September.**
The garden has been 6yrs in the making. Plantsman's content and borders in reconstruction, large netted vegetable area, water harvesting, and all work in progress. All plants for pollination and year-round interest. Parking in field indicated. Parking in driveway for wheelchair access only. Driveway is slightly rough, and rest of the garden is laid to lawn with patio to the rear for shade.

National Garden Scheme gardens are identified by their yellow road signs and posters. You can expect a garden of quality, character and interest, a warm welcome and plenty of home-made cakes!

4 ◆ ATHELHAMPTON HOUSE & GARDENS
Athelhampton, Dorchester, DT2 7LG. Giles Keating, 01305 848363, enquiries@athelhampton.co.uk, www.athelhampton.co.uk. *5m E of Dorchester signed off A35 at Puddletown.* **For opening times and information, please phone, email or visit garden website.**
The gardens date from 1891 and inc the Great Court with 12 giant yew topiary pyramids overlooked by two terraced pavilions. This glorious Grade I architectural garden is full of vistas and surprises with spectacular fountains and the River Piddle flowing through it. Wheelchair map to guide you around the gardens. There are accessible toilets in Visitor Centre. Please see our Accessibility Guide on our website.
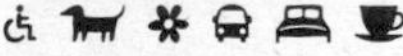

5 22 AVON AVENUE
Avon Castle, Ringwood, BH24 2BH. Terry & Dawn Heaver, dawnandterry@yahoo.com. *Past Ringwood from E A31 turn L after garage, L again into Matchams Ln, Avon Castle 1m on L. A31 from W turn R into Boundary Ln, then L into Matchams Ln, Avon Ave ½m on R.* **Sun 14 May, Sun 25 June, Sun 30 July, Sun 27 Aug (12-5). Adm £5, chd free. Home-made teas.**
Japanese themed water garden featuring granite sculptures, ponds, waterfalls, azaleas, rhododendrons cloud topiary and a collection of goldfish and water lilies. Children must be under parental supervision due to large, deep water pond. No dogs please.

6 BEMBURY FARM
Bembury Lane, Thornford, Sherborne, DT9 6QF. Sir John & Lady Garnier, 01935 873551, dodie.garnier32@gmail.com. *Bottom of Bembury Lane, N of Thornford village. 6m E of Yeovil, 3m W of Sherborne on Yetminster road. Follow signs in village. Parking in field.* **Sun 11, Wed 14 June (2-6). Adm £8, chd free. Home-made teas. Visits also by arrangement Apr to Sept for groups of 10+.**
Created and developed since 1996 this peaceful garden has lawns and large herbaceous borders informally planted with interesting perennials around unusual trees, shrubs and roses. Large collection of clematis; also a pretty woodland walk, wildflower corner, lily pond, oak circle, yew hedges with peacock, clipped hornbeam round kitchen garden and plenty of seating to sit and reflect.

7 BLACK SHED
Blackmarsh Farm, Dodds Cross, Sherborne, DT9 4JX. Paul & Helen Stickland, www.blackshed.flowers/blog. *From Sherborne, follow A30 towards Shaftesbury. Black Shed approx 1m E at Blackmarsh Farm, on L, next to The Toy Barn. Large car park shared with The Toy Barn.* **Sun 18 June, Sun 30 July (1-5). Adm £5, chd free. Home-made teas.**
Over 200 colourful and productive flower beds growing a sophisticated selection of cut flowers and foliage to supply florists and the public for weddings, events and occasions throughout the seasons. Traditional garden favourites, delphiniums, larkspur, foxgloves, scabious and dahlias alongside more unusual perennials, foliage plants and grasses, creating a stunning and unique display. A warm welcome and generous advice on creating your own cut flower garden is offered. Easy access from gravel car park. Wide grass pathways enabling access for wheelchairs. Gently sloping site.

8 BROOK VIEW CARE HOME
Riverside Road, West Moors, Ferndown, BH22 0LQ. Charles Hubberstey, www.brookviewcare.co.uk. *Past village shops, L into Riverside Rd, Brook View Care Home is on R after 100 metres. Parking onsite or nearby roads.* **Sat 19, Sun 20 Aug (11-5). Adm £3.50, chd free. Home-made teas.**
Our colourful and vibrant garden is spread over two main areas, one warm and sunny, the other cooler and shadier. A delightful pond area, games lawn and mixed borders, then walking past our greenhouse leads to the fruit and vegetable gardens. Produce is eagerly used by the kitchen, and residents will help out with the production of the bedding plants, all expertly managed by our gardener.
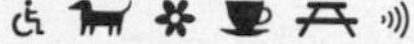

9 BROOMHILL
Rampisham, Dorchester, DT2 0PT. Mr & Mrs D Parry, 07775 806 875, carol.parry2@btopenworld.com. *11m NW of Dorchester. From Dorchester A37 Yeovil, 9m L Evershot. From Yeovil A37 Dorchester, 7m R Evershot. Follow signs. From Crewkerne A356, 1½m after Rampisham Garage L Rampisham. Follow signs.* **Sun 23 Apr, Sun 4 June, Sun 16 July, Thur 10 Aug (2-5). Adm £5, chd free. Home-made teas. Visits also by arrangement 1 May to 14 Aug for groups of 8 to 45. Refreshments on request.**
A former farmyard transformed into a delightful, tranquil garden set in 2 acres. Clipped box, island beds and borders planted with shrubs, roses, grasses, masses of unusual perennials and choice annuals to give vibrancy and colour into the autumn. Lawns and paths lead to a less formal area with large wildlife pond, meadow, shaded areas, bog garden, late summer border. Orchard and vegetable garden. Gravel entrance, the rest is grass, some gentle slopes.

10 NEW BUCKNOWLE COACH HOUSE
Bucknowle, Wareham, BH20 5PQ. Johnathan Ford. *1m W of Corfe Castle. W from Corfe Castle on the Church Knowle/Kimmeridge Rd then L down track then R past the entrance to Bucknowle House.* **Sun 4 June, Sun 2 July (12-4). Adm £5, chd free.**
Veteran Yew and Holm Oak provide the backdrop for extensive new plantings of tall grasses and nectar rich flowers. Raised vegetable and cut flower beds lead though to an orchard with a modern Japanese inspired studio. Purbeck stone terraces provide many vantage points from which to survey stunning vistas accompanied by the sound of flowing water and cooling pools.

11 ◆ CAREYS SECRET GARDEN
Wareham, BH20 7PG. Simon Constantine, hello@careyssecretgarden.co.uk, www.careyssecretgarden.co.uk. *We send the exact location once you have booked a ticket. The garden is within a 3m radius of Wareham, (5 min drive from the train stn). 11m from Poole & 19m*

from Dorchester. **For NGS: Sun 28 May (10.30-3.30). Adm £7.50, chd £2.75. Pre-booking essential, please visit https://bookings.careyssecretgarden.co.uk/ for information & booking. Light refreshments in the walled garden. Please feel free to bring along your own re-usable cup to help us reduce waste. For other opening times and information, please email or visit garden website.**

Behind a 150 yr old wall, situated just outside of Wareham, sits 3 ½ acres in the midst of transformation. Left untouched for more than 40 yrs, this garden is about to flourish again, with a focus on permaculture and rewilding. Awarded Gold in Dorset Tourism's 'Business of the Year' 2021/2022 and Silver in South West Tourism Awards 'New Business of the Year' 2021/2022. Those with mobility issues are welcome to contact us in advance and we can tailor your visit to your needs accordingly.

12 24 CARLTON ROAD NORTH

Weymouth, DT4 7PY. Anne & Rob Tracey. *8m S of Dorchester. A354 from Dorchester, almost opp Rembrandt Hotel R into Carlton Rd North. From Town Centre follow esplanade towards A354 Dorchester, L into Carlton Rd North.* **Fri 12, Sat 13, Sun 14, Mon 15 May (2-5). Adm £4, chd free. Home-made teas.**

Town garden near the sea. Long garden on several levels. Steps and narrow sloping paths lead to beds and borders filled with trees, shrubs and herbaceous plants inc many unusual varieties. A garden which continues to evolve and reflect an interest in texture, shape and colour. Wildlife is encouraged. Raised beds in front garden create a space for vegetable growing.

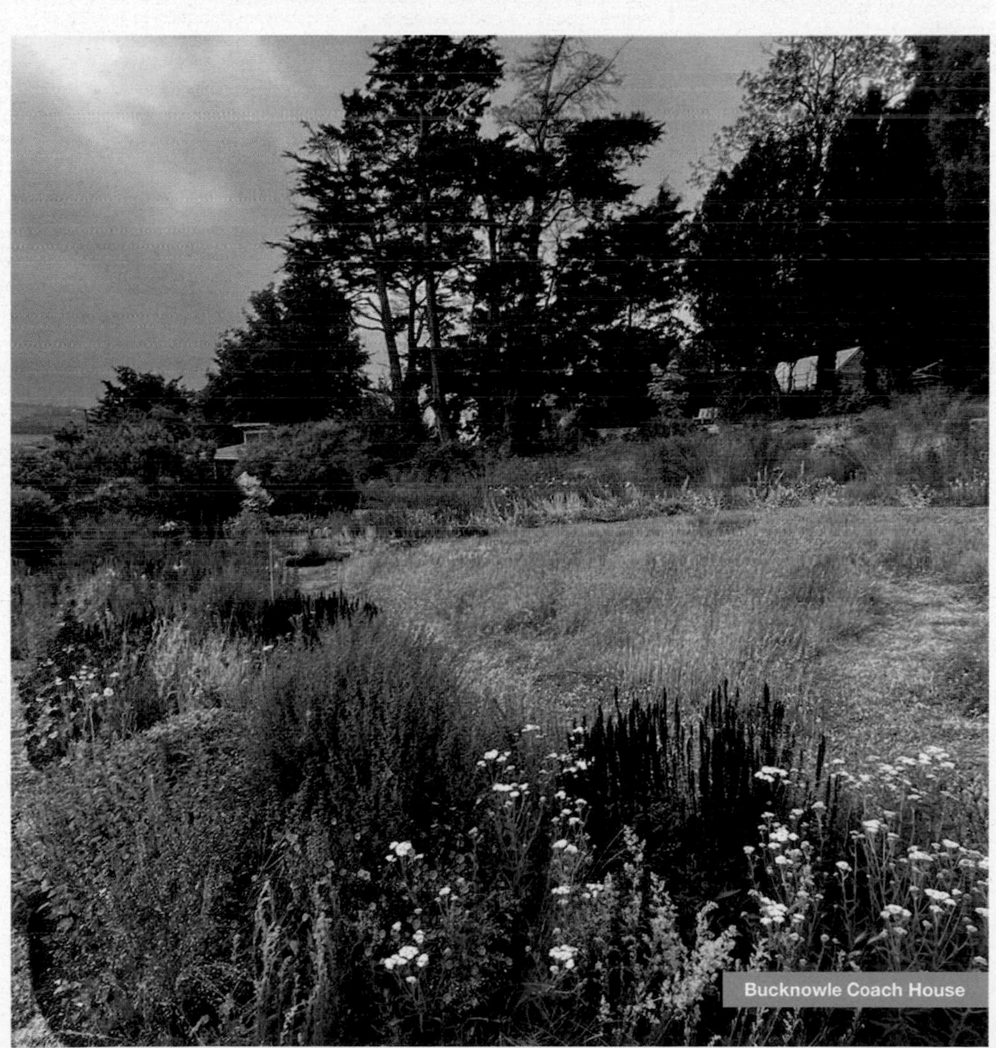

Bucknowle Coach House

13 CARRAWAY BARN

Carraway Lane, Marnhull, Sturminster Newton, DT10 1NJ. Catherine & Mark Turner, 07905 960281, Carrawaybarn.ngs@gmail.com. *From Shaftesbury A30 & B3092. ½m after The Crown turn R into Carraway Lane. From Sturminster Newton B3092 2.8m turn L into Carraway Lane. Bear R behind 1st house, until in large courtyard.* **Sun 4, Thur 8 June (2-5). Sun 11 June (12-4.30), open nearby Ardhurst. Adm £6, chd free. Pre-booking essential, please visit www.ngs.org.uk for information & booking. Home-made teas. Visits also by arrangement in June for groups of up to 40.**

Set in 2 acres, around C19 former barn. In recent years, a natural swimming pond, large shrub border, waterfall and wildflower area (with beehives) have been created. Shady woodland walks, a white border of hydrangeas, hostas and ferns lead to the beautiful established walled garden, where deep borders are planted with roses, peonies, alliums, geraniums and topiary, encircling a water lily pond. Partial wheelchair access; gravelled courtyard and walled garden, gently sloping lawns.

14 CASTLE RINGS

Donhead St. Mary, Shaftesbury, SP7 9BZ. Michael Thomas. *2 m N of Shaftesbury. A350 Shaftesbury/ Warminster road, past Wincombe Business Park. 1st R signed Wincombe & Donhead St Mary.* **Sun 20 Aug (2-5). Adm £5, chd free.**

Small, long garden in two parts laid out beside an Iron Age hill fort with spectacular views. A formal area with colourful planting and pots is followed by a paved area with more pots and steps leading through informal planting. Topiary, roses and clematis on tripods.

15 NEW THE CHANTRY

Chantry Street, Netherbury, Bridport, DT6 5NB. Peter Higginson. *1m S of Beaminster, follow signs for Netherbury. Garden is in centre near church.* **Sun 25 June (2-6). Combined adm with Hingsdon £8, chd free. Light refreshments at Hingsdon.**

Set in the middle of the village and within stone walls and hedges, a 1 acre established traditional garden of lawns, trees, shrubs and colourful mixed borders on a gently sloping site, with a formal pond and some new planting.

16 CHIDEOCK MANOR

Chideock, Bridport, DT6 6LF. Mr & Mrs Howard Coates. *2m W of Bridport on A35. In centre of village turn N at church. The Manor is ¼m along this rd on R.* **Sat 18, Sun 19 Mar, Sat 17, Sun 18 June (2-5). Adm £10, chd free. Home-made teas.**

6/7 acres of formal and informal gardens. Bog garden beside stream and series of ponds. Yew hedges and mature trees. Lime and crab apple walks, herbaceous borders, colourful rose and clematis arches, fernery and nuttery. Walled vegetable garden and orchard. Woodland and lakeside walks. Fine views. Partial wheelchair access.

17 NEW CHILCOMBE HOUSE

Chilcombe, Bridport, DT6 4PN. Kate Hubbard. *Take sharp turn off dual carriageway on A35 4m E of Bridport, 9m W of Dorchester.* **Fri 16 June (1-5). Adm £6, chd free.**

Hillside garden of 2 acres with sweeping views across the valley. Created by the late American-born artist John Hubbard (1931-2017), it is a magical and much-admired garden. Densely planted, it includes courtyards, wild areas and a walled garden divided into a series of rooms, created by box balls and pinnacles, beech hedging and interspersed with soft floral planting. The borders inc foxgloves, germaniums, herbs, old English roses and tender perennials.

18 ◆ CRANBORNE MANOR GARDEN

Cranborne, BH21 5PP. Viscount Cranborne, 01725 517289, info@cranborne.co.uk, www.cranborne.co.uk. *10m N of Wimborne on B3078. Enter garden via Cranborne Garden Centre, on L as you enter top of village.* **For NGS: Sat 1 Apr (9.30-5). Adm £6.50, chd £1. Light refreshments. For other opening times and information, please phone, email or visit garden website. Donation to other charities.**

Beautiful and historic garden laid out in C17 by John Tradescant and enlarged in C20, featuring several gardens surrounded by walls and yew hedges: blue and white garden, cottage style and mount gardens, water and wild garden. Many interesting plants, with fine trees and avenues. Mostly wheelchair access.

19 DEANS COURT

Deans Court Lane, Wimborne Minster, BH21 1EE. Sir William Hanham, 01202 849314, info@deanscourt.org, www.deanscourt.org. *Pedestrian entrance (no parking) is on Deans Ct Lane. Vehicle entrance via Poole Road (BH21 1QF).* **Wed 21 June, Wed 19 July (11-5). Adm £7, chd free. Donation to Friends of Victoria Hospital, Wimborne.**

13 acres of peaceful, partly wild gardens in ancient monastic setting with mature specimen trees, Saxon fish pond, herb garden, orchard and apiary beside River Allen close to town centre. First Soil Association accredited garden, within C18 serpentine walls. The Permaculture system has been introduced here with chemical free produce supplying the Deans Court Café (open) nearby. For disabled access, please contact us in advance of visiting to help us understand your specific needs regarding accessibility, and work out the best plan. Follow signs within grounds for disabled parking closer to gardens. Deeper gravel on some paths. See website extended description re access.

SPECIAL EVENT

20 EASTINGTON FARM

Worth Matravers, Swanage, BH19 3LF. Rachel James. *In Corfe Castle village take the R turn signed Kingston. Take the 2nd R signed Worth Matravers. Eastington Farm is 650 metres on L, long gravel drive & parking at the end of the field.* **Tue 6 June (10.30-5). Combined adm with Encombe House £90, chd free. Pre-booking essential, please visit www.ngs.org.uk for information & booking. Please arrive promptly at 10.30am for tea, coffee & light refreshments.**

The garden has been created over the last 25 yrs and sits within dry Purbeck stone walls with views to the sea. It is divided into different garden rooms with planting themes, all surrounding the C16/17 house. These inc more formal yew hedge pyramids, lonicera balls and cloud topiary alongside soft floral planting, with an orchard of wild flowers, and a working vegetable garden. A buffet lunch with wine will be provided at nearby Encombe House. Please arrive there between 12.30 and 1pm.

21 ◆ EDMONDSHAM HOUSE
Edmondsham, Wimborne, BH21 5RE. Mrs Julia Smith, 01725 517207, julia.edmondsham@homeuser.net. *9m NE of Wimborne. 9m W of Ringwood. Between Cranborne & Verwood. Edmondsham off B3081. Wheelchair access and disabled parking at West front door.* **For NGS: Every Wed 5 Apr to 26 Apr (2-5). Mon 10 Apr (2-5). Every Wed 4 Oct to 25 Oct (2-5). Adm £4, chd £1. Tea, coffee, cake & soft drinks 3.30 to 4.00 pm in Edmondsham House, Weds only. For other opening times and information, please phone or email. Donation to Prama Care.**
6 acres of mature gardens, grounds, views, trees, rare shrubs, spring bulbs and shaped hedges surrounding C16/C18 house, giving much to explore inc C12 church adjacent to garden. Large Victorian walled garden is productive and managed organically (since 1984) using 'no dig' vegetable beds. Wide herbaceous borders planted for seasonal colour. Traditional potting shed, cob wall, sunken greenhouse. Coaches by appointment only. Some grass and gravel paths.

22 EDWARDSTOWE
50-52 Bimport, Shaftesbury, SP7 8BA. Mike & Louise Madgwick. *Park in town's main car park. Walk along Bimport (B3091) 500 metres, Edwardstowe last house on L.* **Sat 20, Sun 21 May (11-4.30). Adm £4, chd free.**
Parts of the garden were extensively remodelled during 2018, a new greenhouse and potting shed have been added, along with changing pathways and the vegetable garden layout. Long borders have been replanted in places with trees managed letting in more light.

SPECIAL EVENT

23 ENCOMBE HOUSE
Kingston, Corfe Castle, Wareham, BH20 5LW. James & Arabella Gaggero. *Drive through Corfe turn R to Kingston. Turn R at The Scott Arms. Drive past church, 300yds 2nd turning L. Turn onto tarmac driveway signed Encombe. Follow signs for parking.* **Tue 6 June (10.30-5). Combined adm with Eastington Farm £90, chd free. Pre-booking essential, please visit www.ngs.org.uk for information & booking. Light buffet lunch with wines.**
The historic Encombe House and Estate is nestled within a unique stunning valley in the Purbeck hills. The garden has been extensively redeveloped since 2009, with a modern, sympathetic design for the garden by Tom Stuart-Smith. The main garden to the south of the house inc large sweeping borders filled with grasses and perennials, alongside extensive lawns, lake and deep herbaceous beds. A limited number of tickets have been made available for this special one day event and private tour of the gardens and grounds. Following your visit to Eastington Farm pls make your way to Encombe where there will be a light buffet lunch, hosted by the owners, James & Arabella Gaggero. After lunch an introductory talk will be given and then there will be the opportunity to explore the gardens accompanied by the owners and gardening team.

24 FALCONERS
89 High Street, Lytchett Matravers, Poole, BH16 6BJ. Hazel & David Dent. *6m W from Poole. Garden is past village hall at end of High St, on L. No parking at the garden, please park considerately in the village.* **Sat 13, Sun 14 May (10.30-4.30). Adm £3.50, chd free. Home-made teas.**
Behind the gate of 150 yr old Falconers Cottage lies a ¼ acre mature garden with some interesting plants. The characterful cottage and garden have been cared for and enhanced by the current custodians. As you wander round the many aspects, inc pond, herbaceous bed, climbers and vegetable plot, you will find restful areas to sit and enjoy this garden in spring.

SPECIAL EVENT

25 FARRS
3, Whitcombe Rd, Beaminster, DT8 3NB. Mr & Mrs John Makepeace, 01308 862204, info@johnmakepeacefurniture.com, www.johnmakepeacefurniture.com. *Southern edge of Beaminster. On B3163. Car parking in the Square or Yarn Barton Car Park, or side streets of Beaminster.* **Wed 14 June, Wed 23 Aug (2.30-4.30). Adm £40. Pre-booking essential, please visit www.ngs.org.uk for information & booking. Cream teas in the house or garden, weather dependent.**
Enjoy several distinctive walled gardens, rolling lawns, sculpture and giant topiary around one of Beaminster's historic town houses. John's inspirational grasses garden, Jennie's riotous potager with an oak fruit cage. Glasshouse, straw bale studio, geese in orchard. Remarkable trees, planked and seasoning in open sided barn for future furniture commissions. A limited number of tickets have been made available for these two special afternoon openings, hosted by John and Jennie Makepeace. There will be a warm welcome from John at 2.30pm in the main rooms of the house, with a talk on his furniture design and recent commissions. This will be followed by the opportunity to wander through the beautiful walled gardens. At 3.30pm, Jennie will give a talk on plants and this will be followed by a cream tea served in the house. Some gravel paths, alternative wheelchair route through orchard.

26 NEW 18 FORD CLOSE
Ferndown, BH22 8AA. Carolyn Best. *1m outside of Ferndown town centre toward Ringwood. Just off the A348 heading S toward Ferndown. 1st turning after the Smugglers Haunt pub. The road is a dead end. Parking available along the length of Ford Close.* **Sat 20, Sun 21 May, Sun 4 June (1-5). Adm £4, chd free. Home-made teas.**
Small evolving garden. 60ft back garden across 2 levels. Lower level contains decking and planted pots. Upper level a combination of perennial borders, fruit and vegetable beds and grapevines, from which we produce wine.

Eastington Farm

27 NEW **27A FOREST ROAD**
Poole, BH13 6DQ.
Paul James Allen,
paul@unusualplantsuk.co.uk,
unusualplantsuk.co.uk. *Located off the Avenue between Branksome Chine & Westbourne. Turn into Tower Rd W (L from Branksome end, R from Westbourne end) then 1st R into Forest Rd.* **Sat 29, Sun 30 Apr (10-4). Adm £4, chd free. Sat 3, Sun 4 June, Sat 22, Sun 23 July, Sat 9, Sun 10 Sept (10-4). Adm £4. Teas, coffee and cakes. Visits also by arrangement 3 June to 10 Sept.**
An exciting garden comprising a courtyard style front garden and jungle themed back packed full with vibrant unusual plants, many rare and exotic. Over 1800 different types inc over 100 Salvias and 50 Abutilons alone! A plant lover's paradise.

28 **FRANKHAM FARM**
Ryme Intrinseca, Sherborne, DT9 6JT. Susan Ross MBE, 07594 427365, neilandsusanross@gmail.com.
3m S of Yeovil. Just off A37 - turn next to Hamish's farm shop signed to Ryme Intrinseca, go over small bridge and up hill, drive is on L. **Sun 12 Mar, Sun 23 Apr, Sun 4 June, Sun 15 Oct (12-5). Adm £6, chd free. Light refreshments in our newly converted barn (no steps). BBQ with our own farm produced beef, lamb & pork, vegetarian soup, home-made cakes by village bakers.**
3½ acre garden, created since 1960 by the late Jo Earle for year-round interest. This large and lovely garden is filled with a wide variety of well grown plants, roses, unusual labelled shrubs and trees. Productive vegetable garden. Clematis and other climbers. Spring bulbs through to autumn colour, particularly oaks. Sorry, no dogs. Ramp available for the two steps to the garden. Modern WCs inc disabled.

29 **FRITH HOUSE**
Stalbridge, DT10 2SD. Mr & Mrs Patrick Sclater, 01963 250809, rosalynsclater@btinternet.com.
5m E of Sherborne. Between Milborne Port & Stalbridge. From A30 1m, follow sign to Stalbridge. From Stalbridge 2m and turn W by PO. **Visits by arrangement May & June for groups of 15 to 50. Mondays to Fridays inclusive. Adm £10, chd free. Home-made teas included in admission.**
Approached down a long drive with fine views, 4 acres of garden around Edwardian house and self-contained hamlet. Range of mature trees, lakes and flower borders. House terrace edged by rose border and featuring Lutyensesque wall fountain and game larder. Well-stocked kitchen gardens. Woodland walks with masses of bluebells in spring. Garden with pretty walks set amidst working farm. Accessible gravel paths.

30 NEW **GLENHOLME HERBS**
Penmore Road, Sandford Orcas, Sherborne, DT9 4SE.
Maxine & Rob Kellaway,
www.glenholmeherbs.co.uk. *3m N of Sherborne. Please see directions on our website & type Glenholme Herbs into Google maps to find us as the postcode will take you to the wrong location.* **Sat 8, Sun 9 July**

(2-5). Adm £4, chd free. Home-made teas.
Paths meander through large, colourful beds, inspired by Piet Oudolf, featuring a wide selection of herbs and salvias along with grasses, verbena and echinacea. Planted with wildlife in mind and alive with pollinators. The garden also features a beautiful natural swimming pond. A mixture of grass and firm gravel paths.

31 NEW GRAYS FARMHOUSE
Clift Lane, Toller Porcorum, Dorchester, DT2 0EJ. Rosie and Roger Britton, www.rosiebritton.com. *2m W of Toller Porcorum. A35 from r'about E of Bridport approx 3m, then L, signed Askerswell. Continue past Spyways over Eggardon Hill, under bridge by Powerstock Common. ½m, L down long farm track to end.* **Mon 24, Sat 29 Apr (12.30-5). Adm £5, chd free. Home-made teas.**
A succession of spring bulbs and wildflowers adds colour to this country garden overlooking fields and adjoining ancient bluebell woods. The 3½ acres inc a wildlife pond, stream, and lightly wooded glade as well as more formal planting near the house. Rosie's studio has cards and paintings for sale, many inspired by flowers and gardens. The interesting plant stall offers varied/unusual plants. Also on display in Rosie's studio will be a number of sketchbooks and journals featuring the garden over the past decade. Some gravel but mostly lawn with very gentle gradient. Parking next to house for the disabled.

32 GROVE HOUSE
Semley, Shaftesbury, SP7 9AP. Judy & Peter Williamson, 07976 391723, judy@judywilliamson.net. *A350 turn to Semley. Grove House approx ½m on L just before railway bridge. From Semley village leave pub on R go under railway bridge (approx ⅓m) Grove House 2nd on R.* **Visits by arrangement June & July. Adm by donation.**
Classic English garden divided into rooms. Yew hedges and topiary. Rose garden. Large herbaceous border. Late summer hot garden with grasses. Lawns. Mown walks through specimen trees. Several good places to eat nearby.

33 HANFORD SCHOOL
Child Okeford, Blandford Forum, DT11 8HN. Rory & Georgina Johnston. *From Blandford take A350 to Shaftesbury; 2m after Stourpaine turn L for Hanford. From Shaftesbury take A350 to Poole; after Iwerne Courtney turn R to Hanford. NGS signage from A350 & A357.* **Sun 25 June (2-4.30). Adm £5, chd free. Home-made teas.**
Perhaps the only school in England with a working kitchen garden growing quantities of seasonal vegetables, fruit and flowers for the table. The rolling lawns host sports matches, gymnastics, dance and plays while ancient cedars look on. The stable clock chimes the hour and the chapel presides over it all. Teas in Great Hall (think Hogwarts). What a place to go to school or visit. Several steps/ramp to main house. No wheelchair access to WC.

34 HARCOMBE HOUSE
Pitman's Lane, Morcombelake, Bridport, DT6 6EB. John & Rachael Willmott. *5m W of Bridport. Take A35 from Bridport, turn R to Whitchurch Canonicorum just past The Artwave Gallery into Tizzards Knapp. Then immed R into Pitmans Lane & follow for approx 400m. Park in paddock on L.* **Sat 13, Sun 14 May (1-6). Combined adm with Well Cottage £9, chd free. Home-made teas.**
Landscaped into hillside 500ft above the Char valley with spectacular views across Charmouth and Lyme Bay. Steeply sloping site comprises a formal garden, wild garden, and woodland area. The beautiful formal garden has been rediscovered, restored and replanted in a natural and relaxed style over last 15 yrs with an abundance of shrubs and perennials, many unusual and visually stunning. Majestic rhododendrons, azaleas and hostas complement spring bulbs and bluebells in a blaze of colour in May/June. Challenging for the less mobile visitor and definitely unsuitable for wheelchairs and buggies.

35 HILLTOP
Woodville, Stour Provost, Gillingham, SP8 5LY. Josse & Brian Emerson, www.hilltopgarden.co.uk. *7m N of Sturminster Newton, 5m W of Shaftesbury. On B3092 turn E at Stour Provost Xrds, signed Woodville. After 1¼m thatched cottage on R. On A30, 4m W of Shaftesbury, turn S opp Kings Arms. 2nd turning on R signed Woodville, 100 yds on L.* **Sun 16, Sun 23, Sun 30 July, Sun 13, Sun 20 Aug (2-6). Adm £4, chd free. Home-made teas.**
Summer at Hilltop is a gorgeous riot of colour and scent, the old thatched cottage barely visible amongst the flowers. Unusual annuals and perennials grow alongside the traditional and familiar, boldly combining to make a spectacular display, which attracts an abundance of wildlife. Always something new, the unique, gothic garden loo is a great success.

36 NEW HINGSDON
Netherbury, Bridport, DT6 5NQ. Anne Peck. *Between Bridport and Beaminster. 1½m from A3066 or 1m from B3162 signed to Netherbury.* **Sun 25 June (2-6). Combined adm with The Chantry £8, chd free. Home-made teas.**
Hingsdon is a hilltop garden of about 2 acres with spectacular panoramic views. There are many unusual shrubs, mostly planted within the last 15 yrs, a mixed border of two halves (cool and hot), rose garden and a large kitchen garden. There is also a small arboretum with an idiosyncratic collection of over 90 trees. The main garden is accessible, although sloping, but the arboretum is too steep for wheelchair access. The kitchen garden has steps.

In 2022 we donated £450,000 to Marie Curie which equates to 23,871 hours of community nursing care

37 ◆ HOGCHESTER FARM
Axminster Road, Charmouth, Bridport, DT6 6BY. Mr Rob Powell, 07714 291846, rob@hogchester.com. *A35 Charmouth Rd, follow dual carriageway and take turning at signs for Hogchester Farm.* **For NGS: Sat 24 June, Sat 15 July (9-6). Adm £4, chd £1. Café on hillside offering tea, coffee, soft drinks, cakes & light refreshments. For other opening times and information, please phone or email.**
Hogchester Farm is a collaboration between those seeking connection with nature and themselves through conservation therapy and the arts. The 75 acre old dairy farm has been largely gifted to nature which has helped to preserve the overflowing abundance of natural life. Having worked closely with the Dorset Wildlife Trust, Hogchester Farm has been able to preserve wild meadows and wilding areas which are filled with local flora and fauna inc wild orchids, foxgloves and primroses. The farm offers something for everyone, making a great family day out with a treasure hunt for children in the wild meadow area, many animals including rare breed sheep, goats, pigs and horses, as well as horticulture based therapy for general mental health and well-being. Talks given on the transformation of the dairy farm into a nature reserve.

38 THE HOLLOW, BLANDFORD FORUM
Tower Hill, Iwerne Minster, Blandford Forum, DT11 8NJ. Sue Le Prevost, 01747 812173, sue.leprevost@hotmail.co.uk. *Between Blandford & Shaftesbury. Follow signs on A350 to Iwerne Minster. Turn off at Talbot Inn, continue straight to The Chalk, bear R along Watery Lane for parking in Parish Field on R. 5 min uphill walk to house.* **Sat 17, Sun 18, Wed 21 June (2-5). Adm £4, chd free. Home-made cakes & gluten-free available. Visits also by arrangement 6 Mar to 3 May for groups of up to 20.**
Hillside cottage garden built on chalk, about ⅓ acre with an interesting variety of plants in borders that line the numerous sloping pathways. Water features for wildlife and well placed seating areas to sit back and enjoy the views. Productive fruit and vegetable garden in converted paddock with raised beds and greenhouses. A high maintenance garden which is constantly evolving. Use of different methods to plant steep banks.

39 THE HOLLOW, SWANAGE
25 Newton Road, Swanage, BH19 2EA. Suzanne Nutbeem, 01929 423662, gdnsuzanne@gmail.com. *½m S of Swanage town centre. From town follow signs to Durlston Country Park. At top of hill turn R at red postbox into Bon Accord Rd. 4th turn R into Newton Rd.* **Every Wed 5 July to 30 Aug (2-5.30). Adm £4, chd free. Visits also by arrangement 5 July to 30 Aug.**
Wander in a dramatic sunken former stone quarry, a surprising garden at the top of a hill above the seaside town of Swanage. Stone terraces with many unusual shrubs and grasses form a beautiful pattern of colour and foliage attracting happy butterflies and bees. Pieces of mediaeval London Bridge lurk in the walls. Steps have elegant handrails. WC available. Exceptionally wide range of plants inc cacti and airplants.

40 NEW HOLLYMOOR FARM
Hollymoor Lane, Beaminster, DT8 3NL. Jacqui Fleet. *From B3163 Whitcombe Rd turn into East St. Continue to join Hollymoor Lane. Go past the farmhouse, parking is in field on R after rd narrows (SatNav may direct you down lane on L).* **Sat 29 July (11-5). Adm £5, chd free. Home-made teas.**
The main garden has been established over 25 yrs. It offers a formal backdrop of pollarded trees and pleached hornbeam trees, which provide elegant structure alongside soft floral planting. The garden has a central circular bed surrounded by box hedging, from which stem a number of pathways lined with lavender and box balls. Lattice and arches provide a framework, all of which are covered with climbing clematis and roses. In addition to this, there are a number of vegetable beds which have been softened with wildlife friendly planting and flowers. Beyond the garden there is a large paddock with areas left to go wild, a mature orchard and greenhouses. Within the paddock there is a large pen with Burford Brown chickens and the greenhouses are utilised for seedlings and propagation. Level site, some steps but most areas accessible. Some gravel & short grass areas to navigate.

41 ◆ HOLME FOR GARDENS
West Holme Farm, Wareham, BH20 6AQ. Simon Goldsack, 01929 554716, simon@holmefg.co.uk, www.holmefg.co.uk. *Easy to find on the B3070 road to Lulworth 2m out of Wareham.* **For NGS: Tue 13 June (9.30-5.30); Tue 12 Sept (9.30-5). Adm £6, chd free. Light refreshments in The Orchard Café. For other opening times and information, please phone, email or visit garden website.**
Extensive formal and informal gardens strongly influenced by Hidcote Manor and The Laskett. The garden is made up of distinct rooms separated by hedges and taller planting. Extensive collection of trees, shrubs, perennials and annuals sourced from across the UK. Spectacular wildflower meadows. Grass amphitheatre, Holme henge garden, lavender avenue, cutting garden, pear tunnel, hot borders, white borders, ornamental grasses, unusual trees and shrubs. Grass paths are kept in good order and soil is well drained so wheelchair access is reasonable except immediately after heavy rain.
NPC

42 HOLWORTH FARMHOUSE
Holworth, Dorchester, DT2 8NH. Anthony & Philippa Bush, 01305 852242, bushinarcadia@yahoo.co.uk. *7m E of Dorchester. 1m S of A352. Follow signs to Holworth. Through farmyard with duckpond on R. 1st L after 200yds of rough track. Ignore No Access signs.* **Sun 30 Apr, Mon 1, Sun 28, Mon 29 May, Sun 2 July (2-5). Adm £5, chd free. Home-made teas. Visits also by arrangement 7 May to 30 Sept.**
Escape briefly from the cares and pressures of the world. Visit this unusual garden tucked away in an area of extraordinary peace and tranquillity chosen by the monks of Milton Abbey. The garden was created 40 years ago and is constantly evolving. Many mature and unusual trees and shrubs, numerous borders with seats to ponder and reflect. Vegetables, water and wild spaces. Beautiful unspoilt views. Many birds and butterflies.

Norwood House

43 HOOKE FARM

Hooke, Beaminster, DT8 3NZ. Julia Hailes MBE & Jamie Macdonald, www.juliahailes.com. *Coming from Crewkerne, Yeovil or Dorchester, turn R at church & R again after 500 yds, just before pond. From Beaminster or Bridport turn L at Hooke woods, past Hooke Court & L after pond.* **Sat 17, Sun 18 June (10.30-5.30). Adm £8, chd £4. Locally sourced lunch, cream teas & refreshments throughout the day.**

Hooke Farm has been transformed into a wildlife haven with bird boxes, bat caves, butterflies and bee-friendly wildflower meadows. The landscaping project inc a series of ponds in a wetland area, orchard trees, woodland planting all interlinked with mown paths through swaying grass. There are standing stones, a stilted henhouse, a giant throne and driftwood stags. There will be talks and guided tours on both afternoons, covering different aspects of wilding and environmentally friendly gardening. This will include butterflies, moths, reptiles, weed control, tree maintenance, grassland management and how to attract insects and birds. We are peat-free. There are areas of the garden that cannot be accessed by wheelchairs but there's still plenty to see if you can manage slopes.

44 HORN PARK

Tunnel Rd, Beaminster, DT8 3HB. Mr & Mrs David Ashcroft. *1½m N of Beaminster. On A3066 from Beaminster, L before tunnel (see signs).* **Wed 19 Apr, Wed 7 June (2.30-4.30). Adm £5, chd free. Home-made teas.**

Large plantsman's garden with magnificent views over Dorset countryside towards the sea. Many rare and mature plants and shrubs in terrraced, herbaceous, rock and water gardens. Woodland garden and walks in bluebell woods. Good amount of spring interest with magnolia, rhododendron and bulbs which are followed by roses and herbaceous planting, wildflower meadow with 164 varieties inc orchids.

Staddlestones

45 ◆ KINGSTON LACY
Wimborne Minster, BH21 4EA. National Trust, 01202 883402, kingstonlacy@nationaltrust.org.uk, www.nationaltrust.org.uk/kingston-lacy. *2½m W of Wimborne Minster. On Wimborne-Blandford rd B3082.* **For opening times and information, please phone, email or visit garden website.**
35 acres of formal garden, incorporating parterre and sunk garden planted with Edwardian schemes during spring and summer. 5 acre kitchen garden and allotments. Victorian fernery containing over 35 varieties. Rose garden, mixed herbaceous borders, vast formal lawns and Japanese garden restored to Henrietta Bankes' creation of 1910. National Collection of Convallaria and *Anemone nemorosa*. Snowdrops, blossom, bluebells, autumn colour and Christmas light display. Deep gravel on some paths but lawns suitable for wheelchairs. Slope to visitor reception and South lawn. Dogs allowed in some areas of woodland.
NPC

46 ◆ KINGSTON MAURWARD GARDENS AND ANIMAL PARK
Kingston Maurward, Dorchester, DT2 8PY. Kingston Maurward College, 01305 215003, events@kmc.ac.uk, www.morekmc.com. *1m E of Dorchester off A35.. Follow brown Tourist Information signs.* **For opening times and information, please phone, email or visit garden website.**
Stepping into the grounds you will be greeted with 35 impressive acres of formal gardens. During the late spring and summer months, our National Collection of penstemons and salvias display a lustrous rainbow of purples, pinks, blues and whites, leading you on through the ample hedges and stonework balustrades. An added treat is the Elizabethan walled garden, offering a new vision of enchantment. Open early Jan to mid Dec. Hours will vary in winter depending on conditions - check garden website or call before visiting. Partial wheelchair access only, gravel paths, steps and steep slopes. Map provided at entry, highlighting the most suitable routes.
NPC

47 KNITSON OLD FARMHOUSE
Corfe Castle, Wareham, BH20 5JB. Rachel Helfer, 01929 421681, rjehelfer@gmail.com. *Purbeck. Between Corfe Castle & Swanage. Follow the A351 3m E from Corfe Castle. Turn L signed Knitson. After 1m fork R. We are on L after ¼m.* **Wed 5, Wed 12 Apr, Wed 3, Wed 10 May, Wed 7, Wed 14 June (12-4.30). Adm £4, chd free. Home-made teas. Visits also by arrangement Feb to Nov for groups of 8 to 20. Home-made teas on request.**
Mature cottage garden nestled at base of chalk downland. Herbaceous borders, rockeries, climbers and shrubs, evolved and designed over 60 yrs for year-round colour and interest. Large wildlife friendly kitchen garden for self sufficiency. Historical stone artefacts are used in the garden design. Ancient trees and shrubs are retained as integral to garden design. Over 20 varieties of fruits and many vegetables year-round are the basis for a sustainable lifestyle. Garden is on a slope, main lawn and tea area are level but there are uneven, sloping paths.

48 ◆ KNOLL GARDENS
Hampreston, Wimborne, BH21 7ND. Mr Neil Lucas, 01202 873931, enquiries@knollgardens.co.uk, www.knollgardens.co.uk. *2½m W of Ferndown. Brown tourist signs from all directions, including from A31. Car park on site.* **For NGS: Fri 22 Sept (10-5). Adm £7.95, chd £5.95. Light refreshments. For other opening times and information, please phone, email or visit garden website.**
A unique, naturalistic and calming garden, renowned for its whispering ornamental grasses, it also surprises and delights with an abundance of show-stopping flowering perennials. A stunning backdrop of trees and shrubs add drama to this wildlife and environmentally-friendly garden. On site nursery selling quality plants, with expert advice readily available. Some slopes. Various surfaces inc gravel, paving, grass and bark.

49 KNOWLE COTTAGE
No.1, Shorts Lane, Beaminster, DT8 3BD. Claire & Guy Fender, 01308 863923, christinecorson@gmail.com. *Near St Mary's Church. Park in main square or in main town car park & walk down Church Lane and R onto single track Shorts Lane. Entry is through blue gates after 1st cottage on L.* **Visits by arrangement 30 Apr to 9 July for groups of 10 to 15. Sundays only.**
Combined with 21A The Square and Shadrack House, Knowle Cottage is a large 1½ acre garden with 35 metre long south facing herbaceous border with year-round colour. Formal rose garden within a circular floral planting is bound on 3 sides by lavender. Small orchard and vegetables in raised beds in adjacent walled area, with whole garden leading to small stream, and bridge to pasture. Beds accessed from level grass, slope not suitable for wheelchairs. Limited outside seating available. Dogs welcome on leads.

50 NEW 22 LANCASTER DRIVE
Broadstone, BH18 9EL. Karen Wiltshire. *2 mins from Broadstone village centre. From B3074 Higher Blandford Rd, take first L onto Springdale Rd, 2nd R onto Springdale Ave, then head straight on to Lancaster Dr, the house will be on the R. Park on Lancaster Dr.* **Sat 20, Sun 21 May (12-5). Adm £5, chd free. Light refreshments.**
360ft natural woodland garden that is hidden in the heart of Broadstone. Mature oaks and a natural spring make up a magical woodland setting that's left for wildlife to live without disturbance. New acers have been planted in the woodland to add seasonal colour. A new gravel garden closer to the house with water features fed from the natural spring.

51 LAWSBROOK
Brodham Way, Shillingstone, DT11 0TE. Clive Nelson, 07771 658846, cne70bl@aol.com, www.facebook.com/Lawsbrook. *5m NW of Blandford. To Shillingstone on A357. Turn up Gunn Lane (past Wessex Ave on L & Everetts Lane on R) then turn R at next opportunity as road bends to L. Lawsbrook 250 metres on R.* **Daily Fri 17 Feb to Tue 21 Feb (10-4). Adm £4, chd free. Light refreshments. Home-made cakes, tea, coffee, juices. Visits also by arrangement 14 Feb to 31 May.**
The garden has over 130 different tree species spread over 6 acres. Native species and many unusual specimens inc Dawn redwood, Damyio oak and Wollemi pine. Extensive snowdrop display in Feb. You can be assured of a relaxed and friendly day out, with guided tour inc if desired. Children's activities, all day teas/cakes and dogs are very welcome. Gravel path at entrance, grass paths over whole garden.

52 LEWELL LODGE
West Knighton, Dorchester, DT2 8RP. Rose & Charles Joly. *3m SE of Dorchester. Turn off A35 onto A352 towards Wareham, take West Stafford bypass, turn R for West Knighton at T junction. ¾m turn R up drive after village sign.* **Sun 18 June (2-5). Adm £6, chd free. Home-made teas.**
Elegant 2 acre classic English garden designed by present owners over last 25 yrs, surrounding Gothic Revival house. Double herbaceous borders enclosed by yew hedges with old fashioned roses. Shrub beds edged with box and large pyramided hornbeam hedge. Crab apple tunnel, box parterre and pleached hornbeam avenue. Large walled garden, many mature trees and woodland walk. Garden is level but parking is on gravel and there are gravel pathways.

53 LITTLE CLIFF
Sidmouth Road, Lyme Regis, DT7 3EQ. Mrs Debbie Bell. *Edge of Lyme Regis. Turn off A35 onto B3165 to Lyme Regis. Through Uplyme to mini r'about by Travis Perkins. 3rd exit on R, up to fork & L down Sidmouth Rd following NGS arrows from mini r'about. Garden on R.* **Thur 22, Sun 25 June, Thur 20, Sun 23 July (2-5). Adm £4.50, chd free. Home-made teas.**
South facing seaward, Little Cliff looks out over spectacular views of Lyme Bay. Spacious garden sloping down hillside through series of garden rooms where visual treats unfold. Vibrant herbaceous borders inc hot garden and bog garden intermingled with mature specimen trees, shrubs and wall climbers. A new large greenhouse with unusual subtropical plants and succulents has been added this year. Jungle walkway at bottom of garden, Indian influenced pavilion leading to white borders and vegetable garden. New hidden garden amongst the palms and rich colours of the hot garden, and new Mediterranean terrace garden at top by house.

54 ◆ LITTLEBREDY WALLED GARDENS
Littlebredy, DT2 9HL. The Walled Garden Workshop, 01305 898055, secretary@wgw.org.uk, www.littlebredy.com. *8m W of Dorchester. 10m E of Bridport. 1½m S of A35. NGS days: park on village green then walk 300yds. For the less mobile (and on normal open days) use gardens car park.* **For NGS: Tue 20, Tue 27 June (2-6). Adm £6, chd free. Home-made teas. For other opening times and information, please phone, email or visit garden website.**
1 acre walled garden on south facing slopes of Bride River valley. Herbaceous borders, riverside rose walk, lavender beds and potager vegetable and cut flower gardens. Partial wheelchair access, some steep grass slopes. For disabled parking please follow signs to main entrance.

55 LOWER ABBOTTS WOOTTON FARM

Whitchurch Canonicorum, Bridport, DT6 6NL. Clare Trenchard. *6m W of Bridport. Well signed from A35 at Morecombe Lake (2m) and Bottle Inn at Marshwood on B3165 (1½m). Some disabled off-road parking.* **Evening opening Fri 2 June (5.30-8). Adm £6, chd free. Light savoury snacks with a glass of wine.**

Sculptor owner reflects her creative flair in garden form, shape and colour. Open gravel garden contrasts with main garden consisting of lawns, borders and garden rooms, making a perfect setting for sculptures. The naturally edged pond provides a tranquil moment of calm and tranquillity, but beware of being led down the garden path by the running hares!

SPECIAL EVENT

56 LULWORTH CASTLE HOUSE

Lulworth Park, East Lulworth, Wareham, BH20 5QS. James & Sara Weld. *In the village of E Lulworth enter through the main entrance to Lulworth Castle and Park & follow the signs.* **Thur 6 July (10.30-1). Adm £30, chd free. Pre-booking essential, please visit www.ngs.org.uk for information & booking.**

Large coastal garden next to Lulworth Castle with views to the sea. Pleasure grounds surround a walled garden filled with roses and perennials, behind which sits a working kitchen garden with ornate fruit cage and greenhouses. In front of the house the scented walk leads to a wildflower meadow, lavender labyrinth and Islamic garden with rills and fountains. A limited number of tickets have been made available for this special one-day event, kindly hosted by the owners, James & Sara Weld. Meet at the front the house on the main drive for an introductory talk on the gardens, followed by a guided tour accompanied by Sara and her gardening team. After visiting the garden a set menu lunch will be made available at the nearby Weld Arms from 1pm (not included in admission price). To make a booking pls visit: https://theweldarms.co.uk or telephone: 01929 400211.

57 MANOR FARM, HAMPRESTON

Wimborne, BH21 7LX. Guy & Anne Trehane, 01202 574223, anne.trehane@live.co.uk. *2½m E of Wimborne, 2½m W of Ferndown. From Canford Bottom r'about on A31, take exit B3073 Ham Lane. ½m turn R at Hampreston Xrds. House at bottom of village.* **Sat 25 Feb (10-1). Light refreshments. Sun 26 Feb (1-4); Sun 4 June, Sun 16 July, Sun 6 Aug (1-5). Home-made teas. Adm £5, chd free. Visits also by arrangement 30 May to 4 Aug for groups of 15 to 30.**

Traditional farmhouse garden designed and cared for by 3 generations of the Trehane family through over 100 yrs of farming and gardening at Hampreston. Garden is noted for its herbaceous borders and rose beds within box and yew hedges. Mature shrubbery, water and bog garden. Open for hellebores in Feb. Excellent plants as usual for sale at openings. Hellebores for sale in Feb.

58 MAPPERCOMBE MANOR

Nettlecombe, Bridport, DT6 3SS. Arthur Crutchley. *4m NE of Bridport. From A3066 turn E signed W Milton & Powerstock. After 3m leave Powerstock on your L, bear R at Marquis of Lorne pub, entrance drive 150yds ahead.* **Fri 9, Sun 11 June (1-5). Adm £6, chd free. Home-made teas.**

Elegant Manor House that was originally a monks' rest house with stew pond and dovecote. South facing gardens on 4 levels with ancient monastic route and sweeping views. Approx 4 acres of gardens with well-planted borders, inc echiums, salvias and euphorbia. Apart from stonework and mature trees, garden mostly replanted in last 25 yrs. Many old roses and a haven for bees and butterflies. Dogs on leads. Partial wheelchair access, gravel and stone paths, steps. 150 yd walk from car park, limited parking by house.

59 ◆ MAPPERTON GARDENS

Mapperton, Beaminster, DT8 3NR. The Earl & Countess of Sandwich, 01308 862645, office@mapperton.com, www.mapperton.com. *6m N of Bridport. Off A356/A3066. 2m SE of Beaminster off B3163.* **For opening times and information, please phone, email or visit garden website.**

Terraced valley gardens surrounding Tudor/Jacobean manor house. On upper levels, walled croquet lawn, orangery and Italianate formal garden with fountains, topiary and grottoes. Below, C17 summerhouse and fishponds. Lower garden with shrubs and rare trees, leading to woodland and spring gardens. Partial wheelchair access (lawn and upper levels).

60 ◆ MINTERNE GARDEN

Minterne House, Minterne Magna, Dorchester, DT2 7AU. Lord Digby, 01300 341370, enquiries@minterne.co.uk, www.minterne.co.uk. *2m N of Cerne Abbas. On A352 Dorchester-Sherborne road.* **For opening times and information, please phone, email or visit garden website.**

As seen on BBC Gardeners' World and voted one of the 10 prettiest gardens in England by The Times. Famed for their display of historic rhododendrons, azaleas, Japanese cherries and magnolias in April/May when the garden is at its peak. Small lakes, streams and cascades offer new vistas at each turn around the 1m horseshoe shaped gardens covering 23 acres. The season ends with spectacular autumn colour. Snowdrops in Feb. Spring bulbs, blossom and bluebells in April. Easter trails for children. Over 200 acers provide spectacular autumn colour in Sept/Oct, inc Halloween trails for children. Fireworks at end of October.

61 ◆ MUSEUM OF EAST DORSET

23-29 High Street, Wimborne Minster, BH21 1HR. Museum of East Dorset, 01202 882533, info@museumofeastdorset.co.uk, www.museumofeastdorset.co.uk. *Wimborne is just off A31. From W take B3078, from E take B3073 towards town centre. From Poole and Bournemouth enter town from S on A341.* **For opening times and information, please phone, email or visit garden website.**

Tranquil walled garden tucked away in the centre of Wimborne. Colourful herbaceous borders and heritage orchard trees line the path which stretches 100m down to the mill stream. This year for the first time we will be running a series of evening garden lectures from 6pm to 8.30pm.

Planned dates are 7 July, 4 Aug and 1 Sep 2023. Please see www.ngs.org.uk for more details. Wheelchair access throughout the site.

62 NEW MYRTLE COTTAGE

Woolland, Blandford Forum, DT11 0ES. Brian & Lynn Baker, 01258 817432, brian.baker15@btinternet.com. *7m W of Blandford Forum. Situated at the base of Bulbarrow Hill, pass the church on your L, pass the turning to Ibberton on the R. Myrtle Cottage is on the R after the Elwood Centre.* **Sat 27, Sun 28 May (10-4). Adm £4, chd free. Home-made teas. Visits also by arrangement Mar to Sept Also by prior arrangement most weekdays throughout the year.**

A small to medium size segmented garden, sympathetic to wildlife with a small wildflower meadow and pond, part flower part fruit and vegetable, a mix for everyone. Interesting hosta pots and numerous chilli plant varieties grown from seed. Gravel paths, otherwise some areas not accessible.

63 NORWOOD HOUSE

Corscombe, Dorchester, DT2 0PD. Mr & Mrs Jonathan Lewis, 07836 600185, jonathan.lewis@livegroup.co.uk. *Equidistant between Halstock & Corscombe. 1m from Fox Inn towards Halstock. 1m from national speed limit sign leaving Halstock towards Corscombe. Parking in field off the main drive.* **Visits by arrangement May to Sept for groups of 10+.**

Hidden away in a stunning West Dorset valley, the gardens of Norwood House are full of subtle floral colour, with glorious purple, pink and white borders surrounding the house and lawns. The planting is elegant and modern, with box topiary providing formal structure, interspersed with soft planting of verbena, grasses, *Alchemilla mollis* and crocosmia. The borders also inc many varieties of geranium and other perennials and shrubs. In front of the house is a large wildflower meadow and a lake, which offers the most spectacular view. We can assist wheelchair users to access the gardens only some parts of which are inaccessible.

64 NEW OAKDALE LIBRARY GARDENS

Wimborne Road, Poole, BH15 3EF. Oakdale Library Gardens Association, www.facebook.com/Oakdalelibrarygardens. *Corner of Wimborne Rd & Dorchester Rd. Number 5/6 bus towards Canford Heath. Bus stop directly adjacent to Library. Free parking on site.* **Wed 10 May (2-5); Mon 22 May (10-1). Adm by donation. Light refreshments on NGS open days only.**

Award winning gardens comprising of the 'Bookerie' Reading and Rhyme time garden where wildlife is welcomed with bee friendly planting, an Insect mansion and pond. Also a Commemorative Garden, a nautical themed garden, herb garden and children's adventure trail. The gardens have been designed and maintained by volunteers. Featured in '111 places in Poole that you shouldn't miss ' by Katherine Bebo. The Bookerie is open during Library opening hours. Other gardens open at all times. Plant sales on NGS open days. Full wheelchair access in all the gardens except the children's adventure trail.

65 OLD DOWN HOUSE

Horton, Wimborne, BH21 7HL. Dr & Mrs Colin Davidson, 07765 404248, pipdavidson59@gmail.com. *7½m N of Wimborne. Horton Inn at junction of B3078 with Horton Rd, pick up yellow signs leading up through North Farm. No garden access from Matterley Drove. 5min walk to garden down farm track.* **Sat 3, Sun 4, Wed 7 June (2-5). Adm £4.50, chd free. Home-made teas in garden with some undercover areas available. Visits also by arrangement 5 June to 30 June.**

Nestled down a farm track this ¾ acre garden on chalk surrounds a C18 farmhouse. Stunning views over Horton Tower and farmland. Cottage garden planting with formal elements. Climbing roses clothe pergola and house walls along with stunning *Wisteria sinensis* and banksian rose. Mainly walled potager, well-stocked. Chickens. Interesting and unusual locally grown plants for sale.

66 THE OLD RECTORY, LITTON CHENEY

Litton Cheney, Dorchester, DT2 9AH. Richard & Emily Cave. *9m W of Dorchester. 1m S of A35, 6m E of Bridport. Park in village and follow signs.* **Sun 7, Wed 10 May, Sun 2 July (11-5). Adm £7, chd free. Home-made teas.**

Steep paths lead to beguiling 4 acres of natural woodland with many springs, streams, 2 pools one a natural swimming pool planted with native plants. Formal front garden, designed by Arne Maynard, with pleached crabtree border, topiary and soft planting including tulips, peonies, roses and verbascums. Walled garden with informal planting, kitchen garden, orchard and 350 rose bushes for a cut flower business.

67 THE OLD RECTORY, MANSTON

Manston, Sturminster Newton, DT10 1EX. Andrew & Judith Hussey, 01258 474673, judithhussey@hotmail.com. *6m S of Shaftesbury, 2½m N of Sturminster Newton. From Shaftesbury, take B3091. On reaching Manston, past Plough Inn, L for Child Okeford on R-hand bend. Old Rectory last house on L.* **Sun 28, Wed 31 May (2-5.30). Adm £7, chd free. Home-made teas. Visits also by arrangement 1 Apr to 15 Sept.**

Beautifully restored 5 acre garden. South facing wall with 120ft herbaceous border edged by old brick path. Enclosed yew hedge flower garden. Wildflower meadow marked with mown paths and young plantation of mixed hardwoods. Well maintained walled Victorian kitchen garden with new picking flower section. Large new greenhouse also installed. Knot garden now well established.

Our 2022 donation supports 1,700 Queen's Nurses working in the community in England, Wales, Northern Ireland, the Channel Islands and the Isle of Man

The Old Rectory, Manston

SPECIAL EVENT

68 THE OLD RECTORY, NETHERBURY

Beaminster, DT6 5NB. Simon & Amanda Mehigan, oldrectorynetherbury.tumblr.com. *2m SW of Beaminster. Full directions will be sent prior to the event.* **Wed 17 May (11-1). Adm £25, chd free. Pre-booking essential, please visit www.ngs.org.uk for information & booking. Light refreshments.**

5 acre garden designed and maintained by present owners over last 28 yrs. Formal areas with topiary near house, large drifts of naturalistic planting elsewhere. Spring bulbs inc species tulips and erythroniums are a speciality. Extensive bog garden with pond and stream, large collection of candelabra primulas and other moisture lovers. Hornbeam walk. Wildflower areas and orchards. Decorative kitchen/cutting garden. A limited number of tickets have been made available for this special morning opening. Admission price inc an introductory talk about the development of the garden from the garden owners, Simon & Amanda. A garden plan will be provided, as well as light refreshments.

69 THE OLD RECTORY, PULHAM

Dorchester, DT2 7EA. Mr & Mrs N Elliott, 01258 817595, gilly.elliott@hotmail.com, www.instagram.com/theoldrectory_pulham. *13m N of Dorchester. 8m SE of Sherborne. On B3143 turn E at Xrds in Pulham. Signed Cannings Court.* **Mon 10, Thur 13 Apr, Sun 6, Thur 10 Aug (2-5). Adm £8, chd free. Home-made teas. Visits also by arrangement 1 Apr to 1 Sept for groups of 10 to 50.**

4 acres of formal and informal gardens surrounding C18 rectory with splendid views. Yew pyramid allées and hedges, circular herbaceous borders with late summer colour. Exuberantly planted terrace, purple and white beds. Box parterres, mature trees, pond, sheets of daffodils, tulips, glorious churchyard. Ha-ha, pleached hornbeam circle. Enchanting bog garden with stream and islands.10 acres woodland walks. Mostly wheelchair accessible.

70 THE OLD SCHOOL HOUSE

The Street, Sutton Waldron, Blandford Forum, DT11 8NZ. David Milanes. *Turn into Sutton Waldron from A350, continue for 300 yds, 1st house on L in The Street. Entrance past house through gates*

in wall. **Sun 18, Thur 22 June (1-6). Adm £5, chd free. Open nearby Penmead Farm.**
Village garden(.8 acre) laid out in last 9 yrs with planting of hedges into rooms inc orchard, secret garden and pergola walkway. Recent addition of raised bed for growing vegetables. Strong framework of existing large trees, beds are mostly planted with roses and herbaceous plants. Pleached hornbeam screen. A designer's garden with interesting semi-tender plants close to house. Refreshments available at Penmead Farm in the village open on same days. Level lawns.

71 THE OLD VICARAGE

East Orchard, Shaftesbury, SP7 0BA. Miss Tina Wright, 01747 811744, tina_lon@msn.com. *4½m S of Shaftesbury, 3½m N of Sturminster Newton. On B3091 between 90 degree bend & layby with defibrillator red phone box. Parking is on the opp corner towards Hartgrove.* **Fri 10, Sun 12 Feb, Sun 19 Mar, Sun 16 Apr, Sun 14 May (2-5). Adm £6, chd free. Home-made teas in garden but inside if wet in winter. Visits also by arrangement 2 Jan to 30 Nov By arrangement groups with tea & cake inc £10 per head.**
1.7 acre wildlife garden with hundreds of different snowdrops, crocus and other winter flowering shrubs. Followed by other bulbs inc large swathes of narcissi and hundreds of tulips, camassias and alliums. A wonderful stream meanders down to a pond and there are lovely reflections from the swimming pond, the first to be built in Dorset. Grotto, old Victorian man pushing his lawn mower which his owner purchased brand new in 1866. Pond dipping, swing and other children's attractions. New acre being gradually planted. Cakes inc gluten free, and vegans are also catered for. Not suitable for wheelchairs if very wet.

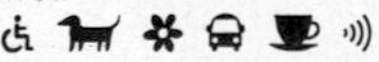

72 NEW PENMEAD FARM

The Street, Sutton Waldron, Blandford Forum, DT11 8PF. Matthew & Claire Cripps. *5½m S of Shaftesbury. From A350 turn into the village of Sutton Waldron. Drive through the village and entrance is approx ¼m on R immed after the road bridge which straddles the stream.* **Sun 18 June (2-6.30); Thur 22 June (3-8). Adm £5, chd free. Home-made teas. Open nearby The Old School House. Wine & light refreshments will be available from 6.30 pm on Thursday opening.**
Situated on the site of an old brickworks the property is bordered by over 30 mature oak trees. The garden comprises a woodland and stream (Fontmell Brook) walk, meadows, substantial vegetable garden, orchard, spring fed pond and small semi walled garden. Being on clay soil roses thrive. Views to Pen Hill and Fontmell Down. Some sloping paths and can be wet under foot. Accessible bar steep slope down to stream and some gravel paths.

73 PHILIPSTON HOUSE

Winterborne Clenston, Blandford Forum, DT11 0NR. Mark & Ana Hudson. *2 km N of Winterborne Whitechurch & 1 km S of Crown Inn in Winterborne Stickland. Park in signed track/field off road, nr Bourne Farm Cottage. Enter garden from field.* **Sat 10 June (2-5.30). Adm £5, chd free. Cream teas cost £5 per person - pls bring cash.**
Charming 2 acre garden with lovely views in Winterborne valley. Many unusual trees, rambling roses, wisteria, mixed borders and shrubs. Sculptures. Rose parterre, walled garden, swimming pool garden, vegetable garden. Stream with bridge over to wooded shady area with cedarwood pavilion. Orchard with mown paths planted with spring bulbs. Sorry, no WC available. Dogs welcome on leads, pls clear up after them. Wheelchair access is good providing it is dry.

74 1 PINE WALK

Lyme Regis, DT7 3LA. Mrs Erika Savory, 07802 884794, erika.savory@btconnect.com. *Pls park in Holmbush Car Park at top of Cobb Rd.* **Sat 9, Sun 10 Sept (11-4). Adm £7.50, chd free. Home-made teas. Visits also by arrangement in Sept for groups of up to 20.**
Unconventional ½ acre, multi level garden above Lyme Bay, adjoining NT's Ware Cliffs. Abundantly planted with an exotic range of shrubs, cannas, gingers and magnificent ferns. Apart from a rose and hydrangea collection, planting reflects owner's love of Southern Africa inc staggering succulents and late summer colour explosion featuring drifts of salvias, dahlias, asters, grasses and rudbeckia.

75 THE PINES

15 Longacre Drive, Ferndown, BH22 9EE. Ian Gallimore. *½m from centre of Ferndown. R off A348 towards Poole, Longacre Drive is almost opp M&S Foodhall.* **Sun 7, Sat 13 May (2-6). Adm £4, chd free. Light refreshments.**
Suburban garden. Heavily planted with mostly perennials, trees and exotic type plants to give a secluded and private feel. 2 ponds, seating areas and outdoor kitchen/BBQ area.

76 PUGIN HALL

Rampisham, nr Dorchester, DT2 0PR. Mr & Mrs Tim Wright, 01935 83652, wright.alison68@yahoo.com. *Near centre of village. NW of Dorchester. From Dorchester A37 Yeovil, 9m L Evershot, follow signs. From Crewkerne A356, take 1st L to Rampisham, Pugin Hall is on L after ½m.* **Visits by arrangement Apr to Sept for groups of 10+. Includes talk on the history of Pugin Hall & garden. Home-made teas.**
Pugin Hall was once Rampisham Rectory, designed in 1847 by Augustus Pugin, who also helped to design the interior of the Houses of Parliament. A Grade I listed building, it is surrounded by 4½ acres of garden, inc a large front lawn with rhododendrons and perennial borders, a walled cut flower and fruit garden, orchard and beyond the River Frome a woodland walk. The walled garden is planted with shrubs, roses, clematis, masses of unusual perennials, salvias and Japanese anemones against a backdrop of espalier fruit trees, olive trees, bay trees and box hedging with spirals. Pugin Hall is the only intact Pugin designed building currently in private ownership and is considered to be the most complete example of domestic architecture designed by him. The plan of the house encompasses Pugin's characteristic pinwheel design: an arrangement of rooms whose axis rotate about a central hall and lends itself well to the varying effects of light and shade within.

77 RUSSELL-COTES ART GALLERY & MUSEUM
East Cliff, Bournemouth, BH1 3AA. Phil Broomfield, 07810 646123, russellcotes@bcpcouncil.gov.uk, www.russellcotes.com. *On Bournemouth's East Cliff Promenade. Situated next to the Royal Bath Hotel, 2 mins walk from Bournemouth Pier. The closest car park is Bath Road Sth. Parking also available on the cliff top.* **Visits by arrangement 3 Apr to 2 Oct for groups of 10+. Weekdays or weekends, excluding Mondays.**
Enjoy a private garden tour of this sub-tropical garden sited on the cliff top, overlooking the sea, full of a wide variety of plants from around the globe. East Cliff Hall was the home of Sir Merton and Lady Annie Russell-Cotes. The garden was restored allowing for modern access with areas retaining original 1901 design conceived by the founders such as the ivy clad grotto and Japanese influence. The Russell-Cotes Café serves a delicious range of light lunches, teas, coffees, and cakes. Some gravel paths and no wheelchair access to terrace.

78 THE SECRET GARDEN AND SERLES HOUSE
47 Victoria Road, Wimborne, BH21 1EN. Chris & Bridget Ryan. *Centre of Wimborne. On B3082 W of town, very near hospital, Westfield car park 300yds. Off road parking close by.* **Sun 21 May, Sun 11 June (1-5). Adm £4.50, chd free. Home-made teas. Donation to Wimborne Civic Society and The Arts' Society.**
The former home of the late Ian Willis, who lived here for just under 40 yrs. Alan Titchmarsh described this amusingly creative garden as 'one of the best 10 private gardens in Britain'. The ingenious use of unusual plants complements the imaginative treasure trove of garden objects d'art. Wheelchair access to garden only. Narrow steps may prohibit wide wheelchairs.

79 SEMLEY GRANGE
Semley, Shaftesbury, SP7 9AP. Mr & Mrs Reid Scott. *From A350 take turning to Semley, continue along road & under railway bridge. Take 1st R up Sem Hill. Semley Grange is 1st on L - parking on green.* **Sat 1 July (12-5). Adm £5, chd free.**
Large garden recreated in last 10 yrs. Herbaceous border, lawn and wildflower meadow intersected by paths and planted with numerous bulbs. The garden has been greatly expanded by introducing many standard weeping roses, new mixed borders, pergolas and raised beds for dahlias, underplanted with alliums. Numerous fruit and ornamental trees introduced over last 10 yrs.

80 SHADRACK HOUSE
Shadrack Street, Beaminster, DT8 3BE. Mr Hugh Lindsay, 01308 863923, christine.corson@ngs.org.uk. *Centre of Beaminster. Within 5 mins walk of 21A The Square.* **Visits by arrangement 30 Apr to 9 July for groups of 10 to 15. Open on Sundays only.**
Combined with 21A The Square and Knowle Cottage, Shadrack is a delightful small mature garden, hidden behind high walls, with an abundance of roses, clematis, shrubs, and climbers galore. Entrance through garden gates below the house, with views over to the church. Beaminster Festival end of June, beginning of July.

81 ◆ SHERBORNE CASTLE
New Rd, Sherborne, DT9 5NR. Mr E Wingfield Digby, www.sherbornecastle.com. *½m E of Sherborne. On New Rd B3145. Follow brown signs from A30 & A352.* **For opening times and information, please visit garden website.**
40+ acres. Grade I Capability Brown garden with magnificent vistas across surrounding landscape, inc lake and views to ruined castle. Herbaceous planting, notable trees, mixed ornamental planting and managed wilderness are linked together with lawn and pathways. Short and long walks available. Partial wheelchair access, gravel paths, steep slopes, steps.

GROUP OPENING

82 NEW SHILLINGSTONE GARDENS
Shillingstone, DT11 0SL. Margaret Kennard, 01258 863771, margaretkennarduk@gmail.com. *4m W of Blandford Forum. The village lies on the A357 between Blandford Forum and Sturminster Newton.* **Sun 11 June (11-5). Combined adm £8, chd free. Home-made teas at Shillingstone Station Project.**

NEW CHERRY COTTAGE
Lal & Gloria Ratnayake.
NEW HAMBLEDON RISE
Tim & Margaret Kennard.
NEW MANOR HOUSE
Tom & Claire Downes.
NEW SHILLINGSTONE HOUSE
Michael & Caroline Salt.

All four gardens are in the centre of this pretty village which lies close to the River Stour with Hambledon Hill as a backdrop. The two larger gardens have some magnificent trees, traditional borders, old brick walls supporting multiple rambling roses. One has an old fashioned walled kitchen garden mixing vegetables, fruit and flowers and the other has ornamental fruit cages, raised beds and an unusual but successful way of growing strawberries. Nearby a cottage garden packed with exotic plants, Bonsai, pond and herbs. The last garden is a 'work in progress', a new bed with roses and wisteria this year. Behind the house a fine Magnolia, wildlife pond and mixed borders.

Our 2022 donation to ABF The Soldier's Charity meant 372 soldiers and veterans were helped through horticultural therapy

27A Forest Road

84 SLAPE MANOR

Netherbury, Bridport, DT6 5LH. Paul Mulholland & Tarsha Finney, 07534 676148, info@slapemanor.com. *1m S of Beaminster. Turn W off A3066 to Netherbury. House ½m S of Netherbury on back rd to Bridport signed Waytown.* **Fri 21, Sat 22, Sun 23 Apr, Sat 1, Sun 2 July (11-5). Adm £6, chd free. Home-made teas. Visits also by arrangement Jan to Nov for groups of 10+. Pls discuss refreshment options when booking your visit.**

River valley garden in a process of transformation. Spacious lawns, wildflower meadows, and primula fringed streams leading down to a lake. Walk over the stream with magnificent hostas, gunneras and horizontal *Cryptomeria japonica* 'Elegans' around the lake. Admire the mature wellingtonias, ancient wisterias, rhododendrons and planting around the house. Kitchen garden renovation underway. Slape Manor is one of the inspirations behind Chelsea 2022 Gold medal winning and Best in Show Garden designed by Urquhart and Hunt with Rewilding Britain. Mostly flat with some sloping paths and steps. Ground is often wet and boggy.

85 SOUTH EGGARDON HOUSE

Askerswell, Dorchester, DT2 9EP. Buffy Sacher, 07920 520280, buffysacher@gmail.com. *From Dorchester on A35 turn R for Askerswell. Through village at T junction. Go straight over. From Bridport on A35 take 1st turning to Askerswell. Pass Spyway Inn on R. Turn next L.* **Visits by arrangement May to Sept for groups of 5+.**

5 acres of formal and informal gardens designed around a 2000 yr old yew tree and lake. Water garden with streams pond and lake. Woodland walk, orchard, wild garden, large herbaceous borders filled with roses and perennials. Ornamental kitchen garden and vegetable garden.

86 21A THE SQUARE

Beaminster, DT8 3AU. Christine Corson, 01308 863923, christine.corson@ngs.org.uk. *6m N of Bridport. 6m S of Crewkerne on B3162. 5 mins walk from public car park.* **Visits by arrangement 30 Apr to 9 July for groups of 10 to 15. Open on Sundays only.**

Combined with Knowle Cottage and Shadrack House, all within walking distance of each other, the garden at 21A The Square is now 7 yrs old. It is a generous square walled garden, with a lovely view over the town and to the hills. A flower arrangers delight, everything that can be put in a vase, fresh or dried. Shrubs, borders and 2 huge silver cedars, about 100 yrs old at the very bottom. Lots of pots. Awarded 1st prize in Medium Gardens category in the 2021 Melplash Show. Very wheelchair friendly, house and garden flat, and ramp from house into garden.

87 STABLE COURT

Chalmington, Dorchester, DT2 0HB. Jenny & James Shanahan, 01300 321345, jennyshanahan8@gmail.com. *From A37 travel towards Cattistock & Chalmington for 1½m. R at triangle to Chalmington. ½m house on R, red letter box at gate. From Cattistock take 1st turn R then next L at triangle.* **Sat 17, Sun 18 June (2-5.30). Adm £6, chd free. Home-made teas. Visits also by arrangement 1 May to 30 May for groups of up to 20.**

This exuberant garden was begun in 2010. Extending to about 1½ acres, it is naturalistic in style with a shrubbery, gravel garden, wild garden and pond where many trees have been planted and which are maturing wonderfully. Overflowing with roses scrambling up trees, over hedges and walls. More formal garden closer to house with lawns and herbaceous borders. Lovely views over Dorset countryside. Exhibition of paintings in studio. Gravel path partway around garden, otherwise the paths are grass, not suitable for wheelchairs in wet weather.

88 STADDLESTONES

14 Witchampton Mill, Witchampton, Wimborne, BH21 5DE. Annette & Richard Lockwood, 01258 841405, richardglockwood@yahoo.co.uk. *5m N of Wimborne off B3078. Follow signs through village and park in sports field, 7 min walk to garden, limited disabled parking near garden.* **Sun 28, Mon 29 May, Sun 18 June, Sun 27, Mon 28 Aug (2-5). Adm £5, chd free. Cream teas. Visits also by arrangement 16 Apr to 1 Oct.**

A beautiful setting for a cottage garden with colour themed borders, pleached limes and hidden gems, leading over chalk stream to shady area which has some unusual plants. Plenty of areas just to sit and enjoy the wildlife. Wire bird sculptures by local artist. Wheelchair access to first half of garden.

89 STANBRIDGE MILL

Gussage All Saints, BH21 5EP. Lord & Lady Phillimore. *7m N of Wimborne. On B3078 to Cranborne 150yds from Horton Inn on Shaftesbury road.* **Thur 1, Fri 2 June (10-5.30). Adm £7.50, chd free. Home-made teas.**

Hidden garden created in 1990s around C18 water mill (not open) on River Allen. Series of linked formal gardens featuring herbaceous and iris borders, pleached limes, white walk and new laburnum walk. 20 acre nature reserve with reed beds and established shelter belts. Grazing meadows with wild flowers Some areas around river not suitable for wheelchairs.

90 NEW SWALLOWS REST

41 South Road, Wyke Regis, Weymouth, DT4 9NR. Keith & Jane Smith, 07747 753656, info@swallowsrestselfcatering.co.uk. *Weymouth A354 towards Portland, R at lights on Boot Hill into Wyke Rd, drive over min r'about, past church, turn L into Wyke Sq. Follow Westhill Rd to bottom & turn R into Swallows Rest B&B.* **Sat 3, Sun 4 June (2-6). Adm £6, chd free. Home-made teas.**

This unusual seaside garden emerged from the year 2000, a gradual process with beds being excavated from a field on this former dairy farm. Limey topsoil was brought in to neutralise the heavy clay. All the trees and hedges were planted as small saplings. Due to the extremely high winds and salty sea air, it was necessary to form wind breaks. The garden is now packed with a vast array of trees, shrubs, roses and perennials. The garden's charm is enhanced with scores of free-range ducks, chickens, bantams and geese. Stunning views of Portland, Chesil Beach and the Fleet complete this delightful and surprising garden. Ponds, gazebo, hot tub, sculptures and summerhouse. Monet style bridge.

91 UTOPIA

Tincleton, Dorchester, DT2 8QP. Nick & Sharon Spiller. *Take signs to Tincleton from Dorchester, Puddletown. Pick up garden signs in the village.* **Sun 11 June (1-5). Adm £5, chd £4. Light refreshments.**

Approximately ½ acre of secluded, peaceful garden made up of several rooms inspired by different themes. Inspiration is taken from Mediterranean and Italian gardens, woodland space, water gardens and the traditional country vegetable garden. Seating is scattered throughout to enable you to sit and enjoy the different spaces and take advantage of both sun and shade.

92 ◆ THE WALLED GARDEN

Moreton, Dorchester, DT2 8RH. Kelsi Dean-Bluck, 01929 405685, info@thewalledgardenmoreton.co.uk, walledgardenmoreton.co.uk. *In the village of Moreton, near Crossways. Look for the brown signs out on the main road.* **For opening times and information, please phone, email or visit garden website.**

The Walled Garden is a beautiful 5 acre landscaped formal garden situated in Moreton, Dorset. The village is close to the historic market town of Dorchester and situated on the River Frome. A wide variety of perennial plants sit in the borders, which have been styled in original Georgian and Victorian designs. Sculpture from various local artists, family area and play park, animal area, and plant shop, plus on site cafe.

93 WELL COTTAGE

Ryall, Bridport, DT6 6EJ. John & Heather Coley, 01297 489066, jfrcoley@btinternet.com. *Less than 1m N of A35 from Morcombelake. From E: R by farm shop in Morcombelake. Garden 0.9m on R. From W: L entering Morcombelake, immed R by village hall and L on Pitmans Lane to T junc. Turn R, Well Cott on L. Parking on site and nearby.* **Sat 13, Sun 14 May (1-6). Combined adm with Harcombe House £9, chd free. Home-made teas at Harcombe House. Visits also by arrangement 8 Apr to 31 Aug for groups of 10+.**

1+acre garden brought back to life since 2012. There is now much more light after some trees were taken down and new areas have been cultivated. A number of distinct areas, some quite surprising but most still enjoy wonderful views over Marshwood Vale. Heather's textile art studio will be open to view. Runner up in best large garden category at Melplash Show.

94 WESTERN GARDENS

24A Western Ave, Branksome Park, Poole, BH13 7AN. Mr Peter Jackson, 01202 708388, pjbranpark@gmail.com. *3m W of Bournemouth. From S end Wessex Way (A338) at gyratory take The*

Avenue, 2nd exit. At T-lights turn R into Western Rd then at bottom of hill L. At church turn R into Western Ave. **Sun 30 Apr, Sun 18 June (2-5). Adm £6, chd free. Home-made teas. Visits also by arrangement 17 Apr to 22 Sept for groups of 20+.**
'This secluded and magical 1acre garden captures the spirit of warmer climes and begs for repeated visits' (Gardening Which?). Created over 40 yrs it offers enormous variety with rose, Mediterranean courtyard and woodland gardens, herbaceous borders and cherry tree and camellia walk. Lush foliage and vibrant flowers give year-round colour and interest enhanced by sculpture and topiary. Plants and home-made jams and chutneys for sale. Wheelchair access to 3/4 garden.

95 WHITE HOUSE

Newtown, Witchampton, Wimborne, BH21 5AU. Mr Tim Read, 01258 840438, tim@witchampton.org. *5m N of Wimborne off B3078. Travel through village of Witchampton towards Newtown for 800m. Pass Crichel House's castellated gates on L. White House is a modern house sitting back from road on L after further 300m.* **Sat 27, Sun 28 May, Sat 1, Sun 2 July (1-5). Adm £4.50, chd free. Light refreshments. Visits also by arrangement Apr to Sept for groups of up to 15.**
1½ acre garden set on different levels, with a Mediterranean feel, planted to encourage wildlife and pollinators. Wildflower border, pond surrounded by moisture loving plants, prairie planting of grasses and perennials, orchard. Chainsaw sculptures of birds of prey. Reasonable wheelchair access to all but the top level of the garden.

96 ◆ WIMBORNE MODEL TOWN & GARDENS

16 King Street, Wimborne, BH21 1DY. Wimborne Minster Model Town Ltd Registered Charity No 298116, 01202 881924, info@wimborne-modeltown.com, www.wimborne-modeltown.com. *2 mins walk from Wimborne town centre & the Minster Church. Follow Wimborne signs from A31; from Poole/Bournemouth follow Wimborne signs on A341; from N follow Wimborne signs B3082/B3078. Public parking opp in King St car park.* **For opening times and information, please phone, email or visit garden website.**
Attractive garden surrounding intriguing model town buildings, inc original 1950s miniature buildings of Wimborne. Herbaceous borders, rockery, perennials, shrubs and rare trees. Miniature river system inc bog garden and other water features. Sensory area inc vegetable garden, grasses, a seasonally fragrant and colourful border, wind and water features. Plentiful seating. Open 1 April - 29 October (10am - 5pm). Seniors discount; groups welcome. Tearoom, shop, crazy golf, miniature dolls' house collection, 00 gauge model railway, Wendy Street play area, and a 1/10th scale model of the model town.

97 WINCOMBE PARK

Shaftesbury, SP7 9AB. John & Phoebe Fortescue, 01747 852161, pacfortescue@gmail.com, www.wincombepark.com. *2m N of Shaftesbury. A350 Shaftesbury to Warminster, past Wincombe Business Park, 1st R signed Wincombe & Donhead St Mary. 3/4m on R.* **Sun 14, Wed 17 May (2-5). Adm £8, chd free. Home-made teas. Dairy & gluten free options available. Visits also by arrangement 1 May to 11 June for groups of 10 to 40. Private tour & refreshments can be provided.**
Extensive mature garden with sweeping panoramic views from lawn over parkland to lake and enchanting woods through which you can wander amongst bluebells. Garden is a riot of colour in spring with azaleas, camellias and rhododendrons in flower amongst shrubs and unusual trees. Beautiful walled kitchen garden. Partial wheelchair access only, slopes and gravel paths.

98 YARDES COTTAGE

Dewlish, Dorchester, DT2 7LT. Christine & Ross Robertson, 07774 855152, christine.m.robertson@hotmail.com. *9m NE of Dorchester. From A35 Puddletown/A354 junction, take B3142 & immed turn R onto Long Lane. Follow road around bend to R, then turn R through Dewlish to garden. Disabled parking only.* **Thur 29 June, Sat 1 July (1-5). Adm £5, chd free. Home-made teas. Visits also by arrangement 29 June to 21 July for groups of up to 10.**
Country cottage garden of 1½ acres bordering the Devil's Brook, having a wealth of different planting areas inc woodland, stream, small lake, extensive lawns fringed with formal herbaceous borders, vegetable and soft fruit areas. Much of the planting encourages insect life and supports our bees and hens. Yardes Cottage honey for sale.

99 NEW YEW TREE HOUSE

Hermitage Lane, Hermitage, Dorchester, DT2 7BB. Anna Vines, annavinesgardens.co.uk. *7m S of Sherborne. Turn towards Holnest off A352, Dorchester-Sherborne road. Continue 2m to Hermitage. In Hermitage, new build house on N side at the western end of village green.* **Sun 11 June (12-5). Adm £5, chd free. Home-made teas.**
A recently landscaped and planted ½ acre plot, surrounding a new-build eco house. The house and garden were designed to relate to each other and the surrounding rural landscape. The garden comprises a small kitchen garden, orchard, Mediterranean garden, boules court and herbaceous borders, each providing their own ambience, while informal perennial borders enclose terraces around the house. Photographic display in the barn within the garden. Gravel driveway.

Our donation of £2.5 million over five years has helped fund the new Y Bwthyn NGS Macmillan Specialist Palliative Care Unit in Wales. 1,900 inpatients have been supported since its opening

ESSEX

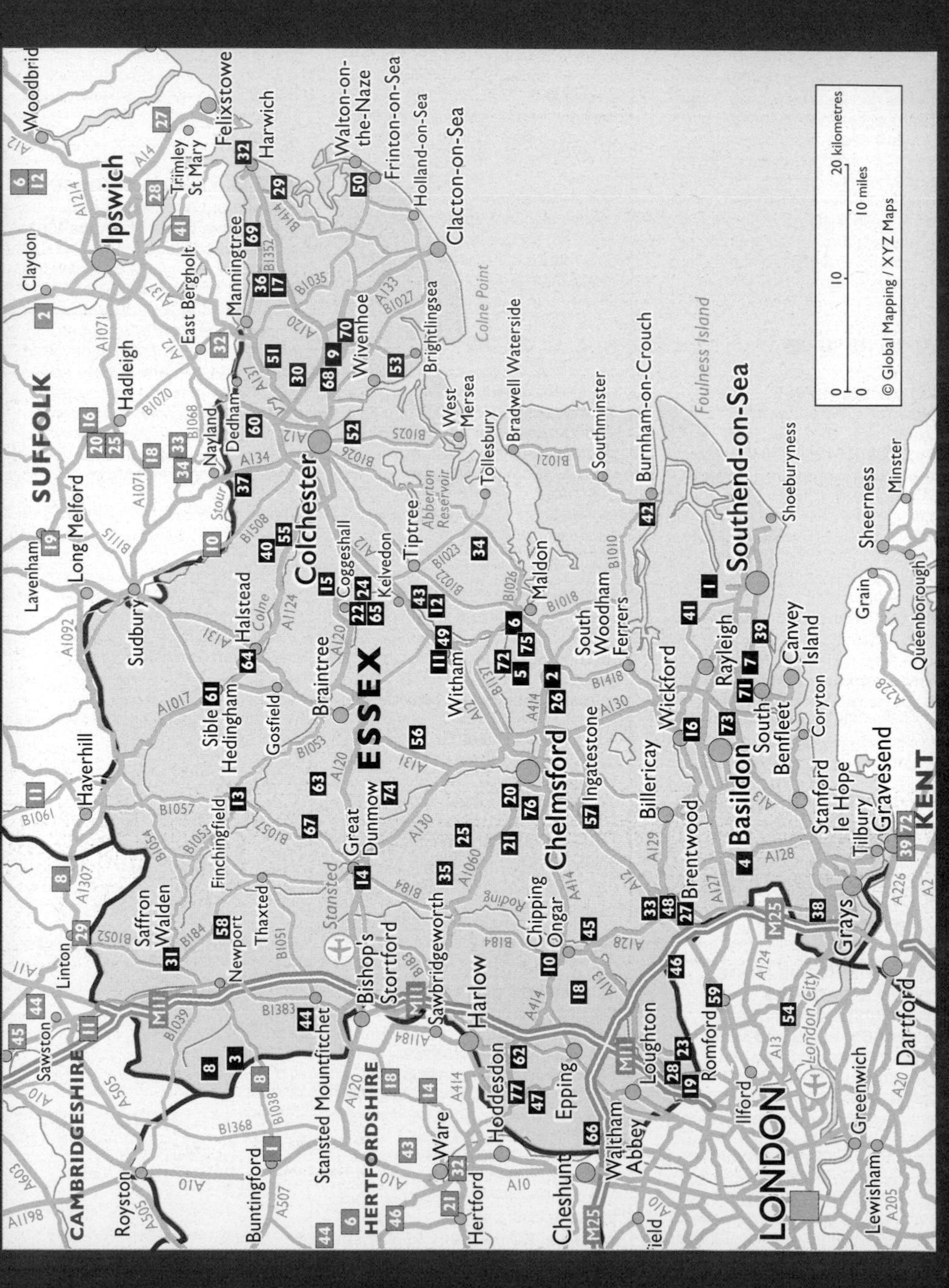
SUFFOLK
ESSEX
CAMBRIDGESHIRE
HERTFORDSHIRE
KENT
LONDON
Woodbridge
Felixstowe
Ipswich
Claydon
Trimley St Mary
Harwich
Walton-on-the-Naze
Frinton-on-Sea
Holland-on-Sea
Clacton-on-Sea
Manningtree
East Bergholt
Hadleigh
Dedham
Nayland
Lavenham
Long Melford
Sudbury
Colchester
Wivenhoe
Brightlingsea
Colne Point
West Mersea
Tollesbury
Bradwell Waterside
Southminster
Burnham-on-Crouch
Foulness Island
Southend-on-Sea
Shoeburyness
Tiptree
Abberton Reservoir
Coggeshall
Kelvedon
Halstead
Braintree
Witham
Maldon
South Woodham Ferrers
Rayleigh
Canvey Island
Wickford
Basildon
South Benfleet
Coryton
Chelmsford
Ingatestone
Billericay
Brentwood
Stanford le Hope
Tilbury
Gravesend
Grays
Dartford
Sheerness
Minster
Grain
Queenborough
Haverhill
Sible Hedingham
Gosfield
Finchingfield
Great Dunmow
Thaxted
Saffron Walden
Newport
Linton
Sawston
Stansted Mountfitchet
Bishop's Stortford
Stansted
Sawbridgeworth
Harlow
Chipping Ongar
Epping
Loughton
Romford
Ilford
Waltham Abbey
Cheshunt
Hoddesdon
Ware
Hertford
Royston
Buntingford
Greenwich
Lewisham
London City
Stour
Colne
Roding
20 kilometres
10 miles
© Global Mapping / XYZ Maps

With its mix of villages, small towns, rivers and coast line, Essex provides some of the most varied garden styles in the country.

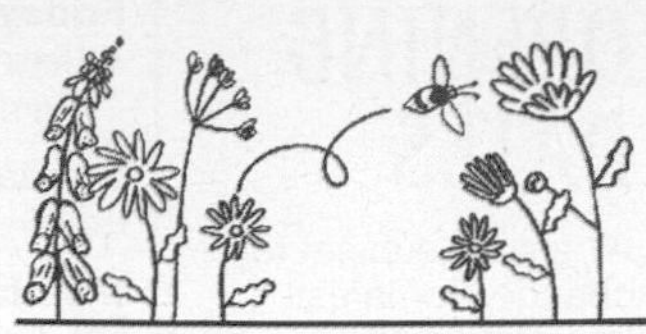

Despite its proximity to London, Essex is a largely rural county, its rich, fertile soil and long hours of sunshine make it possible to grow a wide variety of plants and our garden owners make the most of this opportunity. We have some grand country estates, small town gardens with much in between, as well as community gardens and allotments.

Our year starts with snowdrop gardens in February, runs through tulip gardens in the spring and makes the most of the high summer months with gardens full of roses and colourful perennials. Our climate allows the season to extend through and beyond late summer when many gardens feature more exotic and tropical planting schemes making this a good time to explore.

Many of our gardens are happy to arrange private group visits, often from gardening clubs and increasingly from groups of friends or family. But whether you visit on one of the open days listed in the following pages or By Arrangement you can be sure of a warm welcome. Our garden owners are usually on hand to talk about their garden to ensure you leave with a little more knowledge and lots of inspiration.

We look forward to sharing our gardens with you.

Below: Beth Chatto's Plants and Gardens

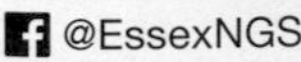 @EssexNGS  @EssexNGS 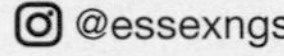 @essexngs

Volunteers

County Organiser
Susan Copeland
07534 006179
susan.copeland@ngs.org.uk

County Treasurer
Richard Steers
07392 426490
steers123@aol.com

Publicity & Social Media Co-ordinator
Debbie Thomson
07759 226579
debbie.thomson@ngs.org.uk

Publicity & Social Media Assistant
Alan Gamblin 07720 446797
alan.gamblin@ngs.org.uk

Booklet Co-ordinator and Publicity Assistant
Doug Copeland 07483 839387
doug.copeland@ngs.org.uk

Assistant County Organisers
Tricia Brett 01255 870415
tricia.brett@ngs.org.uk

Avril & Roger Cole-Jones
01245 225726
randacj@gmail.com

Lesley Gamblin 07801 445299
lesley.gamblin@ngs.org.uk

Linda & Frank Jewson
01992 714047
linda.jewson@ngs.org.uk

Victoria Kennedy 01702 420373
victoria.kennedy@ngs.org.uk

Frances Vincent 07766 707379
frances.vincent@ngs.org.uk

Talks
Ed Fairey 07780 685634
ed@faireyassociates.co.uk

County Photographer
Caroline Cassell 07973 551196
caroline.cassell@ngs.org.uk

OPENING DATES

All entries subject to change. For latest information check **www.ngs.org.uk**

Extended openings are shown at the beginning of the month.

Map locator numbers are shown to the right of each garden name.

January

Sunday 22nd
◆ Green Island 30

February

Snowdrop Festival

Wednesday 8th
Dragons 21

Sunday 12th
Grove Lodge 31

Sunday 19th
Grove Lodge 31
Longyard Cottage 47

Saturday 25th
Horkesley Hall 37

March

Saturday 11th
◆ Beth Chatto's Plants & Gardens 9

Saturday 18th
Pear Trees 58

Sunday 19th
Pear Trees 58

Sunday 26th
NEW 1 Bridge Barn 14

April

Every Thursday
Barnards Farm 4

Every Thursday and Friday
Feeringbury Manor 24

Friday 14th
◆ Beeleigh Abbey Gardens 6

Sunday 16th
Heyrons 35

Monday 17th
NEW 2 Cedar Avenue 16

Sunday 23rd
NEW 9 Beresford Gardens 7
Ulting Wick 72
Writtle University College 76

Wednesday 26th
NEW Pollyfield 60

Thursday 27th
Grange Farm Community Garden 28

Friday 28th
Ulting Wick 72

Saturday 29th
Loxley House 48

Sunday 30th
Bucklers Farmhouse 15

May

Every Thursday
Barnards Farm 4

Every Thursday and Friday
Feeringbury Manor 24

Every Thursday
Grange Farm Community Garden 28

Monday 1st
Furzelea 26

Sunday 7th
Bassetts 5
◆ Green Island 30

Friday 12th
Scrips House 65

Sunday 14th
1 Whitehouse Cottages 75

Sunday 21st
April Cottage 3
Bessie's End 8
May Cottage 50
The Old Rectory 56

Wednesday 24th
Oak Farm 55

Friday 26th
Scrips House 65

Saturday 27th
Isabella's Garden 39

Sunday 28th
Chippins 17
The Gates 27
Horkesley Hall 37
Isabella's Garden 39

Monday 29th
Fairwinds 23
Laurel Cottage 44
Rookwoods 61

June

Every Thursday
Barnards Farm 4

Every Thursday and Friday
Feeringbury Manor 24

Every Thursday
Grange Farm Community Garden 28

Saturday 3rd
NEW Pollyfield 60

Sunday 4th
NEW 9 Beresford Gardens 7
Furzelea 26
May Cottage 50
Peacocks 57
Silver Birches 66
Walnut Tree Cottage 74

Tuesday 6th
Braxted Park Estate 12
8 Dene Court 20

Friday 9th
Scrips House 65

Saturday 10th
NEW Allways 1
Moverons 53

Sunday 11th
Blake Hall 10
Bucklers Farmhouse 15
The Delves 19
Jankes House 40
Moverons 53
Oak Farm 55
NEW 291 Thundersley Park Road 71

Monday 12th
NEW 291 Thundersley Park Road 71

Saturday 17th
9 Malyon Road 49
Pear Trees 58
NEW Thatched Cottage 70

Sunday 18th
NEW 2 Cedar Avenue 16
Clunes House 18
NEW 1 & 2 Grapevine Cottages 29
9 Malyon Road 49
Pear Trees 58
NEW Thatched Cottage 70

Tuesday 20th
8 Dene Court 20

Friday 23rd
Scrips House 65

Saturday 24th
18 Pettits Boulevard, RM1 59

Sunday 25th
Fudlers Hall 25
Havendell 34
61 Humber Avenue 38
18 Pettits Boulevard, RM1 59
NEW 291 Thundersley Park Road 71

Monday 26th
NEW 291 Thundersley Park Road 71

Wednesday 28th
Keeway 42

July

Every Thursday
Barnards Farm 4

Every Thursday and Friday
Feeringbury Manor 24

Every Thursday
Grange Farm Community Garden 28

Saturday 1st
Brick House 13
Grange Farm Community Garden 28
Keeway 42
9 Malyon Road 49

Sunday 2nd
Barnards Farm 4
Chippins 17

262 Hatch Road 33
2 Hope Cottages 36
9 Malyon Road 49
Peacocks 57
1 Whitehouse Cottages 75
Wychwood 77

Wednesday 5th
8 Dene Court 20
Little Myles 45
Long House Plants 46
Oak Farm 55

Friday 7th
NEW 59 East Street 22

Saturday 8th
NEW 59 East Street 22
69 Rundells - The Secret Garden 62

Sunday 9th
Furzelea 26

Tuesday 11th
Ulting Wick 72

Wednesday 12th
Little Myles 45

Friday 14th
8 Dene Court 20

Sunday 16th
Harwich Gardens 32
262 Hatch Road 33

Sunday 23rd
The Delves 19
NEW 207 Mersea Road 52
The Old Rectory 56

Saturday 29th
Isabella's Garden 39

Sunday 30th
Isabella's Garden 39
Laurel Cottage 44
NEW 207 Mersea Road 52

August

Every Thursday
Barnards Farm 4

Every Thursday
Grange Farm Community Garden 28

Tuesday 1st
8 Dene Court 20

Wednesday 2nd
Long House Plants 46

Sunday 6th
Kamala 41

Sunday 13th
NEW 207 Mersea Road 52

Wednesday 16th
8 Dene Court 20

Sunday 20th
2 Hope Cottages 36

Sunday 27th
NEW 207 Mersea Road 52

Monday 28th
Ulting Wick 72

Tuesday 29th
8 Dene Court 20

September

Every Thursday and Friday
Feeringbury Manor 24

Friday 1st
Ulting Wick 72

Saturday 2nd
18 Pettits Boulevard, RM1 59

Sunday 3rd
Barnards Farm 4
Kamala 41
18 Pettits Boulevard, RM1 59

Wednesday 6th
Long House Plants 46

Sunday 10th
Oak Farm 55
Sandy Lodge 64

Sunday 17th
Furzelea 26

October

Every Thursday and Friday to Saturday 7th
Feeringbury Manor 24

Wednesday 4th
◆ Beth Chatto's Plants & Gardens 9

Sunday 8th
◆ Green Island 30

Saturday 21st
Saling Hall 63

December

Every day from Thursday 14th to Tuesday 19th
14 Una Road 73

January 2024

Sunday 28th
◆ Green Island 30

By Arrangement

Arrange a personalised garden visit with your club, or group of friends, on a date to suit you. See individual garden entries for full details.

Appledore, 46 Hopping Jacks Lane 2
Bassetts 5
NEW Blunts Hall 11
Chippins 17
The Delves 19
8 Dene Court 20
Dragons 21
NEW 59 East Street 22
Fairwinds 23
Feeringbury Manor 24
Fudlers Hall 25
Furzelea 26
Grange Farm Community Garden 28
Grove Lodge 31
Heyrons 35
Horkesley Hall 37
61 Humber Avenue 38
Isabella's Garden 39
Keeway 42
Kelvedon Hall 43
May Cottage 50
Mayfield Farm 51
Moverons 53
Peacocks 57
Pear Trees 58
Rookwoods 61
69 Rundells - The Secret Garden 62
8 St Helens Green, Harwich Gardens 32
Sandy Lodge 64
Silver Birches 66
Snares Hill Cottage 67
Spring Cottage 68
Strandlands 69
Ulting Wick 72
Walnut Tree Cottage 74

In 2022 the National Garden Scheme donated £3.11 million to our nursing and health beneficiaries

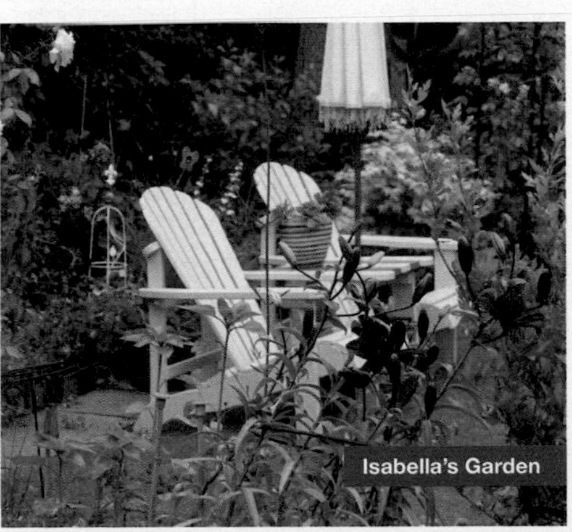
Isabella's Garden

THE GARDENS

1 NEW ALLWAYS

14 St Andrews Road, Rochford, SS4 1NP. Ms Liz Grant. *Traverse down St Andrews Rd. The blue house can be found 100m on the R.* **Sat 10 June (10-4). Adm £5, chd free. Light refreshments. Teas, coffees, cold drinks & home made cakes.**

A cottage style garden of a listed 1930's Arts & Crafts house. A mature wisteria greets visitors on the front of the house. The rear garden has an area of lawn flanked by perennial borders with a viburnum arch leading to an attractive pond with Monet style bridge, pebbled beach area and a mature Indian Bean tree. A new wildlife pond and bog garden has been created in 2022 in the woodland area.

2 APPLEDORE, 46 HOPPING JACKS LANE

Danbury, CM3 4PJ. Gerry & Lynne Collins, 01245 221341, lynnemorley@ymail.com, www.lynnemorleygardendesign.co.uk. *Danbury, Essex. From Eves Corner small public car park, take the narrow one way rd into Hopping Jacks Ln. Garden is approx 400 metres on the R opp Simmons Way. RHS Hyde Hall is within 4m of garden.* **Visits by arrangement Apr to June for groups of 10 to 18. Light refreshments. Tea, coffee and cakes.**

Tranquil village garden featuring mature trees, shady paths, sunny lawn and borders. Paths circulate around the garden leading to a moon gate framing our secret garden, where you can sit and relax a while. Planting inc a growing collection of over 40 Japanese Acers with Ferns and Hostas providing the understorey. Dahlias in the borders and Acers in flaming foliage provide autumn interest.

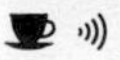

3 APRIL COTTAGE

Sheepcote Green, Clavering, Saffron Walden, CB11 4SJ. Anne & Neil Harris. *6 m W of Saffron Walden. At Newport take B1038. After 3m turn R in Clavering, signed Langley Upper Green. 1m further Sheepcote Green is midway between the villages of Clavering & Langley. Bessie's End is 5min drive away.* **Sun 21 May (12-5). Combined adm with Bessie's End £6, chd free. Home-made teas in the Community Hall on Langley Upper Green.**

A charming thatched cottage which boasts a well-established, colourful cottage garden, filmed in 2020 with Alan Titchmarsh for the 'Love Your Cottage Garden Special.' Packed with many unusual plants; old fashioned roses, clematis, herbaceous perennials galore, wildlife and ornamental ponds, hosta collection and all to be discovered along winding paths. A viewing platform offers wonderful views.

4 BARNARDS FARM

Brentwood Road, West Horndon, CM13 3FY. Bernard & Sylvia Holmes & The Christabella Charitable Trust, 07504 210405, vanessa@barnardsfarm.eu, www.barnardsfarm.eu. *What3words app - tulip.folds.statue. 5m S of Brentwood. On A128 1½ m S of A127 Halfway House flyover. From junc continue S on A128 under the railway bridge. Garden on R just past bridge. Please use postcode CM13 3FY.* **Every Thur 6 Apr to 31 Aug (11-4.30). Adm £10, chd free. Light refreshments. Sun 2 July, Sun 3 Sept (1-5.30). Adm £14, chd free. Pre-booking essential, please visit www.ngs.org.uk for information & booking.**

So much to explore! Climb the Belvedere for the wider view and take the train for a woodland adventure. Spring bulbs and blossom, summer beds and borders, ponds, lakes and streams, walled vegetable plot. 'Japanese garden', sculptures grand and quirky enhance and delight. See the new Talia-May Avenue and Angel sculptures. Sorry, no dogs except guide dogs. Barnards Miniature Railway rides (BMR): Separate charges apply. Sunday extras: Bernard's Sculpture tour 2.30pm Car collection. 1920s Cycle shop. Archery. Model T Ford Rides. Collect loyalty points on Thur visits and earn a free Sun or Thur entry. Season Tickets available Aviators welcome (PPO). Picnics welcome. Wheelchair accessible WC Golf Buggy tours available.

5 BASSETTS

Bassetts Lane, Little Baddow, Chelmsford, CM3 4BZ. Mrs Margaret Chalmers, 079401 79572, magschalmers@btinternet.com. *1m down Tofts Chase becomes Bassetts Ln. Big yellow house on L, wooden gates, red brick wall. From Spring Elms Ln go down Bassetts Ln, L at the bottom of the hill. continue for quarter of a mile.* **Sun 7 May (10-5). Adm £5, chd free. Visits also by arrangement 14 Apr to 12 Oct.**

2 acre garden, tennis court, swimming pool surrounding an early C17 house (not open)with plants for year-round interest set on gently sloping ground with lovely distant views of the Essex countryside. Shrub borders and mature ornamental trees, an orchard and two natural ponds. Many places to sit and relax. No refreshments but bring a picnic and enjoy the views. Please check wheelchair access with garden owner 01245 226768.

6 ◆ BEELEIGH ABBEY GARDENS

Abbey Turning, Beeleigh, Maldon, CM9 6LL. Catherine Foyle, 07506 867122, www.visitmaldon.co.uk/beeleigh-abbey. *1m NW of Maldon. Leaving Maldon via London Rd take 1st R after Cemetery into Abbey Turning.* **For NGS: Fri 14 Apr (10.30-4.30). Adm £7, chd £2.50. Home-made teas. For other opening times and information, please phone or visit garden website.**

3 acres of secluded gardens in rural historic setting. Mature trees surround variety of planting and water features, woodland walks under planted with bulbs leading to tidal river, cottage garden, kitchen garden, orchard, wildflower meadow, rose garden, wisteria walk, magnolia trees, lawn with 85yd long herbaceous border. Scenic backdrop of remains of C12 abbey incorporated into private house (not open). Refreshments inc hot and cold drinks, cakes and rolls. Wheelchair access, gravel paths, some gentle slopes and some steps. Large WC with ramp and handlebars.

7 NEW **9 BERESFORD GARDENS**

Benfleet, SS7 2SB. Geoff and Jackie Blackledge. *W of Hadleigh town centre. Please avoid parking on 'even' numbered side of Beresford Gardens. From Victoria House r'about on A13 take A129 Rayleigh Rd. Beresford 1st rd on R. From A127 Rayleigh Weir take A129 to Hadleigh. Beresford Gdns last turning on L. Parking at Hadleigh Baptist Church (SS7 2SE).* **Sun 23 Apr, Sun 4 June (1-5). Adm £5, chd free. Home-made teas. Coffees & home-made cakes also available.**

A suburban garden adjoining ancient woodland rich with local wildlife. A pond, lawns and borders with mixed planting provide colour and interest through the seasons. Greenhouses, succulents. Seating around the garden to relax and enjoy refreshments. Level access to and around the garden with no steps except for access to a small deck area.

In 2022, our donations to Carers Trust meant that 456,089 unpaid carers were supported across the UK

Saling Hall

8 BESSIE'S END

Langley Upper Green, Saffron Walden, CB11 4RY. Susan & Doug Copeland. *7m W of Saffron Walden. At Newport take B1038 After 3m turn R at Clavering, signed Langley. Upper Green 3m further. At village green turn R. Garden 200yds on R. Parking at community hall. April Cottage is 5min drive away.* **Sun 21 May (12-5). Combined adm with April Cottage £6, chd free. Home-made teas in the Community Hall on the village green close to the garden.**

Formerly part of the gardens of Wickets, Bessie's End is an acre of wild flower landscaped meadow with far reaching rural views, an orchard and many specimen trees. The garden also boasts a Mediterranean style courtyard and a parterre enclosed by espalier apple trees. A delightful shepherd's hut completes the scene. Lots of seating to relax and enjoy the peace 'far from the madding crowd'. Terracotta pots for sale. Gravel drive.

9 ◆ BETH CHATTO'S PLANTS & GARDENS

Elmstead Market, Colchester, CO7 7DB. Beth Chatto's Plants & Gardens, 01206 822007, customer@bethchatto.co.uk, www.bethchatto.co.uk. *¼m E of Elmstead Market. On A133 Colchester to Clacton Rd in village of Elmstead Market.* **For NGS: Sat 11 Mar, Wed 4 Oct (10-5). Adm £13.50, chd £4.50. Inc sandwiches, paninis, soup, salads, cakes, hot & cold beverages. For other opening times and information, please phone, email or visit garden website.**

Internationally famous gardens, inc dry, damp, shade, reservoir and woodland areas. The result of over 60 yrs of hard work and application of the huge body of plant knowledge possessed by Beth Chatto and her husband Andrew. Visitors cannot fail to be affected by the peace and beauty of the garden. Beth is renowned internationally for her books, her gardens and her influence on the world of gardening and plants. Please visit website for up to date visiting details and pre-booking. Picnic area available in the adjacent field. Disabled WC & parking. Wheelchair access around all of the gardens - on gravel or grass (concrete in Nursery, Welcome Centre & Gardener's shop areas).

10 BLAKE HALL

Bobbingworth, CM5 0DG. Mr & Mrs H Capel Cure, www.blakehall.co.uk. *10m W of Chelmsford. Just off A414 between Four Wantz r'about in Ongar & Talbot r'about in N Weald. Signed on A414.* **Sun 11 June (10.30-4). Adm £6, chd free. Home-made teas in C17 barn.**

25 acres of mature gardens within the historic setting of Blake Hall (not open). Arboretum with broad variety of specimen and spectacular ancient trees. Lawns overlooking countryside. With its a beautiful rose garden, herbaceous border and many spectacular rambling roses. June is a wonderful time of the year to visit the gardens at Blake Hall. Some gravel paths.

11 NEW BLUNTS HALL

Blunts Hall Drive, Witham, CM8 1LX. Alan & Lesley Gamblin, 07720 446797, alan.gamblin@ngs.org.uk. *Blunts Hall Drive is off of Blunts Hall Rd.* **Visits by arrangement 13 Feb to 18 Aug for groups of 10 to 30. No visits during April. Weekdays only. Limited parking. Adm £10, chd free. Home-made teas. Please contact us to discuss any**

Peacocks

specific requirements.
Restored 3 acre Victorian garden. Courtyard Garden, Terrace leading down to lawns with re-instated Parterre surrounded by herbaceous borders. Orchard with old and new fruit trees and vegetable plot. Listed Ancient Monument. Woodland walk around spring-fed pond. Fernery. Steps lead down to front lawns recently planted with a yew avenue. Snowdrops and aconites in early spring. Specimen trees.

12 BRAXTED PARK ESTATE

Braxted Park Road, Great Braxted, Witham, CM8 3EN. Mr Duncan & Mrs Nicky Clark, www.braxtedpark.com. *A12 N, by-pass Witham. Turn L to Rivenhall & Silver End by pub called the Fox (now closed) At T junc turn R to Gt Braxted & Witham. Follow brown sign to Braxted Pk (NOT Braxted Golf Course). Down drive to Car Park.* **Tue 6 June (10-4). Adm £7, chd free. Light refreshments in the Walled Garden Pavilion. Light lunches and afternoon tea available.**

Idyllic Braxted Park, a prestigious events venue, welcomes gardeners to enjoy its tranquil surroundings. Extensive borders of perennials, shrubs and roses abound. Walled Garden features themed gardens radiating from the central fountain and parasol mulberry trees. Themes inc a Black and White Garden, Italian Garden and English Garden, each containing a wealth of planting inspiration. A native wildflower meadow extends from the house (not open) towards the lake. Head Gardener, Andrea Cooper, will be on hand to answer questions and guided tours for a further donation of £5pp payable upon the day. An extensive new native meadow has been created between house and lakes. A walk around the lake is not to be missed. Guide dogs only.

13 BRICK HOUSE

The Green, Finchingfield, Braintree, CM7 4JS. Mr Graham & Mrs Susan Tobbell. *9m NW of Braintree. Brick House in the centre of the picturesque village of Finchingfield, at the Xrds of the B1053 & B1057 in rural NW Essex.* **Sat 1 July (2-6). Adm £6, chd free. Home-made teas. Coffee, soft drinks, bottled water & a variety of cakes also available.**

This recently renovated period property and garden covers 1½ acres, with a brook running through the middle. The garden features contemporary sculptures, and several distinct planting styles; most notably a hidden scented cottage-style garden, a crinkle-crankle walled area with an oriental feel and some rather glorious high summer herbaceous borders inspired by the New Perennial Movement. Modern marble sculptures by Paul Vanstone add drama and contrast to the soft landscaping. Garden is professionally designed to be a calming space.

14 NEW 1 BRIDGE BARN

Stortford Road, Dunmow, CM6 1SG. Mark, Ruma and Eve Lacey. *Outskirts of Gt Dunmow. In Dunmow take B1256 to Tesco. R at Woodside Way r'bout, for street parking. Pedestrian access from Tesco to Stortford Rd. Cross rd & turn L at r'about. Private rd for pedestrians only to gdn.* **Sun 26 Mar (10-5). Adm £4, chd free. Light refreshments.**

A small garden full of a wide variety of rare and common plants under the canopy of the catalpa bignonioides and a metal sculpture designed by Mark Lacey. Colour and scents abound. The scent of Viburnum Elwesii is present as you enter the garden late March/early April. Pachyphragma Macrophylla, Hacquetia Epipactis and Daphne Odora flower during Spring. Limited parking for disabled on request. Please call 07531725326 to reserve a space. Gravel driveway. Brick & stone paths in the garden.

15 BUCKLERS FARMHOUSE

Buckleys Lane, Coggeshall, Colchester, CO6 1SB. Ann Bartleet. *1½m E of Coggeshall. 1'm N of A120 via Salmons Ln.* **Sun 30 Apr, Sun 11 June (1.30-4.30). Adm £6, chd free. Cream teas. Refreshments can be taken outside or in my traditional Essex Barn.**

Traditional country garden, with blowsy borders, yew hedges, topiary and a little knot garden. A very tranquil, magical place, with two ponds and a pool - is it just for the reflections or can you swim in it? In the spring we have about 1,000 tulips. Later, there is a terrace full of succulents, almost no room for sitting! This 2 acre garden has been created by the owners over the last 50 yrs. There is gravel, but about 50% of the garden is accessible with a wheelchair.

16 NEW 2 CEDAR AVENUE

Wickford, SS12 9DT. Mr Chris Cheswright. *Off Nevendon Rd, 5 mins from Wickford High St. From Basildon on A132 turn 1st R onto Nevendon Rd at BP r'bout, R onto Park Dr, 2nd R Cedar Ave.* **Mon 17 Apr, Sun 18 June (11-4). Adm £6, chd free. Home-made teas.**

A large town garden with a dry front garden, pots arranged on a large patio with bulbs followed by tropical style planting. Collections of hostas, succulents, carnivorous plants, alpines, hardy bromeliads and coleus are on display in appropriate seasons outdoors and in greenhouses. Planting of trees, perennials, grasses with spring bulbs in the lawn. Areas for wildlife and a pond. Access via sideway to house. The majority of the garden is accessible via a flat lawn area.

17 CHIPPINS

Heath Road, Bradfield, CO11 2UZ. Kit & Ceri Leese, 01255 870730, ceriandkit@gmail.com. *3m E of Manningtree. Take A137 from Manningtree Stn, turn L opp garage. Take 1st R towards Clacton. At Radio Mast turn L into Bradfield continue through village. Bungalow is opp primary school.* **Sun 28 May (11-4). Home-made teas. Sun 2 July (11-4). Adm £4, chd free. Combined entry with 2 Hope Cottages £6. Visits also by arrangement 21 May to 30 July for groups of 5 to 35.**

Artist's and plantaholics' paradise packed with interest. Spring heralds hostas, aquilegia and irises with meandering stream, wildlife pond and Horace the Huge! Vast range of colourful summer perennials in large mixed borders, featuring daylilies. Over 100 pots and hanging baskets! Exotic border with cannas and dahlias. Studio in conservatory with paintings and etching press. Kit is a landscape artist and printmaker, pictures always on display. Afternoon tea with delicious homemade cakes is also available for small parties (min of 5) on specific days if booked in advance.

18 CLUNES HOUSE

Mill Lane, Toot Hill, Ongar, CM5 9SF. Dr Hugh & Mrs Elaine Taylor. *4m from J7 M11 & 4m from the towns of Ongar & Epping. The garden is at the end of Mill Ln in Toot Hill.* **Sun 18 June (12-4). Adm £6, chd free. Home-made teas.**

This is a beautiful, and interesting traditional country garden with wonderful views over the Essex countryside. A pergola with roses, clematis and wisteria leads into a woodland walk and pond surrounded by herbaceous borders and sculpture. A path through the wildlife meadow takes you into the orchard and a small holding with pigs and sheep, ending at the kitchen garden and cut flower beds. Original World War II air raid shelter. The garden is adjacent to the 'Essex Way' long distance walking trail.

19 THE DELVES

37 Turpins Lane, Chigwell, IG8 8AZ. Fabrice Aru & Martin Thurston, 0208 5050 739, martin.thurston@talktalk.net. *Between Woodford & Epping. Chigwell, 2m from N Circular Rd at Woodford, follow the signs for Chigwell (A113) through Woodford Bridge into Manor Rd & turn L, Bus 275 & W14.* **Sun 11 June, Sun 23 July (11-6.30). Adm £4, chd free. Visits also by arrangement 14 May to 29 Oct for groups of up to 10.**

An unexpected hidden, magical, part-walled garden showing how much can be achieved in a small space. An oasis of calm with densely planted rich, lush foliage, tree ferns, hostas, topiary and an abundance of well maintained shrubs complemented by a small pond and 3 water features designed for year-round interest. Featured on BBC Gardeners' World and ITV Good Morning Britain.

20 8 DENE COURT

Chignall Road, Chelmsford, CM1 2JQ. Mrs Sheila Chapman, 01245 266156. *W of Chelmsford (Parkway). Take A1060 Roxwell Rd for 1m. Turn R at T-lights into Chignall Rd. Dene Court 3rd exit on R. Parking in Chignall Rd.* **Tue 6, Tue 20 June, Wed 5, Fri 14 July, Tue 1, Wed 16, Tue 29 Aug (12-4). Adm £3.50, chd free. Visits also by arrangement 1 June to 6 Sept for groups of 10 to 40.**

Beautifully maintained and designed compact garden (250sq yds). Owner is well-known RHS gold medal-winning exhibitor (now retired). Circular lawn, long pergola and walls festooned with roses and climbers. Large selection of unusual clematis. Densely-planted colour coordinated perennials add interest from June to Sept in this immaculate garden.

21 DRAGONS

Boyton Cross, Chelmsford, CM1 4LS. Mrs Margot Grice, 01245 248651, margot@snowdragons.co.uk. *3m W of Chelmsford. On A1060. ½m W of The Hare Pub.* **Wed 8 Feb (11.30-3). Adm £5, chd free. Light refreshments. Visits also by arrangement 12 Jan to 31 Oct for groups of 10+.**

Galanthus are a passion. A plantswoman's ¾ acre garden, planted to encourage wildlife. Sumptuous colour-themed borders with striking plant combinations, featuring specimen plants, fernery, clematis and grasses. Meandering paths lead to ponds, patio, scree garden and small vegetable garden. 2 summerhouses, one overlooking stream and farmland.

22 NEW 59 EAST STREET

Coggeshall, Colchester, CO6 1SJ. Sara Impey and Tom Fenton, 07740 928369, tjhfenton@gmail.com,www.instagram.com/thegardenat59eaststreet/. *12 m W of Colchester, 16m NE of Chelmsford, 2m N of the A12, just off the A120. There is a public car park, free for 2 hrs, entered from Stoneham St. 5 mins walk. Car park also at Co-op.* **Fri 7 July (2-7). Tea. Sat 8 July (11-4). Light refreshments. Adm £5, chd free. Visits also by arrangement 12 Sept to 30 Sept. Please give at least 2 days notice.**

Part of an old walled garden, mostly replanted in the last 3 years, and still developing. A lavender walk lined with espaliered apples and pears leads to a fountain. Peach cage. Greenhouse. Box parterre. Grotto. Small Med garden. Exotics. Wide range of perennials, shrubs, roses, grasses and bamboo with unusual plants, grapes, veg, soft fruit, under an ancient and massive copper beech. Please contact us in advance if you want onsite parking. The paths are all wheelchair accessible and there is an accessible WC.

23 FAIRWINDS

Chapel Lane, Chigwell Row, IG7 6JJ. Sue & David Coates, 07731 796467, scoates@forest.org.uk. *2m SE of Chigwell, Grange Hill Tube turn R at exit, 10 mins walk uphill. Nr M25 J26. From N Circular Waterworks r'about, signed Chigwell. Fork R for Manor/Lambourne Rd. Park in Lodge Close Car Park. 2 disabled cars can park by the house. Please display blue badge.* **Mon 29 May (2-5). Adm £5, chd free. Home-made teas. Visits also by arrangement Mar to Sept for groups of 8 to 20.**

Gravelled front garden and 3 differently styled back garden spaces with planting changes every year. Established wildflower meadow in main lawn. Meander, sit, relax and enjoy. Beyond the rustic fence, lies the wildlife pond and vegetable plot. Planting influenced by Beth Chatto, Penelope Hobhouse, Christopher Lloyd and Piet Oudolf. Happy insects in bee house, bug houses, log piles, bantam hens, meadow and sampling the spring pollen. Newts in pond. There be dragons a plenty!! Wheelchair access: wood chip paths in woodland area may require assistance.

24 FEERINGBURY MANOR

Coggeshall Road, Feering, CO5 9RB. Mr & Mrs Giles Coode-Adams, 01376 561946, seca@btinternet.com. *Between Feering & Coggeshall on Coggeshall Rd, 1m from Feering village.* **Every Thur and Fri 6 Apr to 28 July (10-4). Every Thur and Fri 1 Sept to 7 Oct (10-4). Adm £6, chd free. Visits also by arrangement 1 Jan to 29 Dec. Donation to Feering Church.**

There is always plenty to see in this peaceful 10 acre garden with 2 ponds and river Blackwater. Jewelled lawn in early April then spectacular tulips and blossom lead on to a huge number of different and colourful plants, many unusual, culminating in a purple explosion of michaelmas daisies in late Sept. No wheelchair access to arboretum, steep slope.

25 FUDLERS HALL

Fox Road, Mashbury, CM1 4TJ. Mr & Mrs A J Meacock, 01245 231335. *7m NW of Chelmsford. Chelmsford take A1060, R into Chignal Rd. ½m L to Chignal St James approx 5m, 2nd R into Fox Rd signed Gt Waltham. Fudlers From Gt Waltham. Take Barrack Ln for 3m.* **Sun 25 June (2-5). Adm £6, chd free. Home-made teas. Visits also by arrangement 26 June to 14 July for groups of 15 to 50.**

An award winning, romantic 2 acre garden surrounding C17 farmhouse with lovely pastoral views, across the Chelmer Valley. Old walls divide garden into many rooms, each having a different character, featuring long herbaceous borders, ropes and pergolas festooned with rambling old fashioned roses. Enjoy the vibrant hot border in late summer. Yew hedged kitchen garden. Ample seating. Wonderful views across Chelmer Valley. 2 new flower beds, many roses. 500yr old yew tree. Wheelchair access: gravel farmyard and 30ft path to gardens, all of which is level lawn.

26 FURZELEA

Bicknacre Road, Danbury, CM3 4JR. Avril & Roger Cole-Jones, 01245 225726, randacj@gmail.com. *4m E of Chelmsford, 4m W of Maldon A414 to Danbury. At village centre S into Mayes Ln. 1st R. Past Cricketers Pub, L on to Bicknacre Rd. Use NT car park on L. Garden 50 m on R or parking 200 m past on L. Use NT Common paths from back of car park.* **Mon 1 May, Sun 4 June, Sun 9 July, Sun 17 Sept (11-5). Adm £6, chd free. Home-made teas. Visits also by arrangement 16 Apr to 1 Oct for groups of 15 to 50. £10 per person incl refreshments.**

Country garden of nearly 1 acre that has evolved overtime to showcase the seasons by titillating the senses with colour, scent, texture and form. Paths, steps and archways lead you onwards through intimate spaces and flowing lawns. Clipped box hedges edge flower beds of tulips in spring, roses, hemerocallis and other herbaceous plants in summer, dahlias, grasses, salvias and exotics in autumn. Black and White Garden plus exotics and many unusual plants add to the visitors interest. International garden tour visitors voted us the best garden in their recent tours of East Anglia. Short walk to Danbury Country Park and Lakes and short drive to RHS Hyde Hall.

27 THE GATES

London Road Cemetery, London Road, Brentwood, CM14 4QW. Ms Mary Yiannoullou, www.frontlinepartnership.org. *At the rear of the cemetery. Enter the Cemetery (opp Tesco Express). Drive to the rear of the cemetery, our site on L. Limited parking within but plenty in the surrounding rds.* **Sun 28 May (11.30-3.30). Adm £4, chd free. Home-made teas in the Tea House. A selection of home-made cakes available.**

A horticultural project that offers local citizens, inc vulnerable adults, an opportunity to develop new skills within a horticultural setting. Greenhouses and allotments for raising bedding plants, educational workshops and growing fruit and vegetables. Walks through the Woodland Dell and Sensory area. Large variety of plants and produce for sale. Enjoy the mosaic garden, visit the apiary and relax in the various seating areas around the site. Areas accessed via slopes and ramps.

28 GRANGE FARM COMMUNITY GARDEN

High Road, Chigwell, IG7 6DP. Mrs Sally Panrucker, 07809 151020, spanrucker@vaef.org.uk. *Turn off High Rd onto Grange Farm Ln, the Grange Farm Centre is approx. 500 m on the L (plenty of free parking).* **Every Thur 27 Apr to 31 Aug (11-1). Sat 1 July (11-3.30). Home-made teas. Adm £4, chd free. All cakes made by Ace Cookery Members. Only hot drinks available on Thursday openings. Visits also by arrangement 1 June to 27 July.**

An allotment project called Ace Gardening, maintained by a group of adults with learning disabilities set within the beautiful grounds of The Grange Farm Centre, adjoining Chigwell Meadows. All staff and volunteers are trained using the THRIVE therapeutic approach to gardening. We encourage members to be fully involved in all aspects of gardening: choosing plants and long-term planning. Range of goods for sale, inc hanging baskets, planters, home-made jams and pickles. Meet Ace Members and Staff from the Essex Wildlife Trust.

29 NEW 1 & 2 GRAPEVINE COTTAGES

Harwich Road, Little Oakley, Harwich, CO12 5JD. Mrs Ann Hebden. *SW of Harwich. A120 from Colchester signposted Harwich, take 3rd exit at Ramsey r'about. Turn R onto Mayes Ln, opp church. R at T junc onto B1414, location ½m on R.* **Sun 18 June (11-4). Adm £4, chd free. Home-made teas.**

Hidden behind an C18 cottage, a beautiful, tranquil cottage garden brimming with interesting plants and shrubs. Pergolas and arches are bedecked with roses, clematis and climbers inc a grapevine. The garden features a wildlife pond. The ivy-clad well displays one of the many garden sculptures to delight the visitor. Grasses and succulents capture the eye.

30 ◆ GREEN ISLAND

Park Road, Ardleigh, CO7 7SP. Fiona Edmond, 01206 230455, greenislandgardens@gmail.com, www.greenislandgardens.co.uk. *3m NE of Colchester. From Ardleigh village centre, take B1029 towards Great Bromley. Park Rd is 2nd on R after level Xing. Garden is last on L.* **For NGS: Sun 22 Jan, Sun 7 May, Sun 8 Oct (10-4). Adm £9, chd £2.50. Light refreshments. Light Lunches, cream teas, cakes, ice creams all available in the tearoom. Picnics welcome in the car park area only. 2024: Sun 28 Jan. For other opening times and information, please phone, email or visit garden website. Donation to Plant Heritage.**

A garden for all seasons, highlights inc bluebells, azaleas, autumn colour, winter hamamelis and snowdrops. A plantsman's paradise. Carved within 20 acre mature woodland are huge island beds, Japanese garden, terrace, gravel, seaside and water gardens, all packed with rare and unusual plants. Bluebells Bazaar weekend 13/14 May, Autumn colour weekend 14/15 Oct, Snowdrops 28 Jan 2024. National collections of Hamamelis cvs and Camellias (autumn and winter flowering). Flat and easy walking /pushing wheelchairs. Ramps at entrance and tearoom. Disabled parking and WC.

NPC

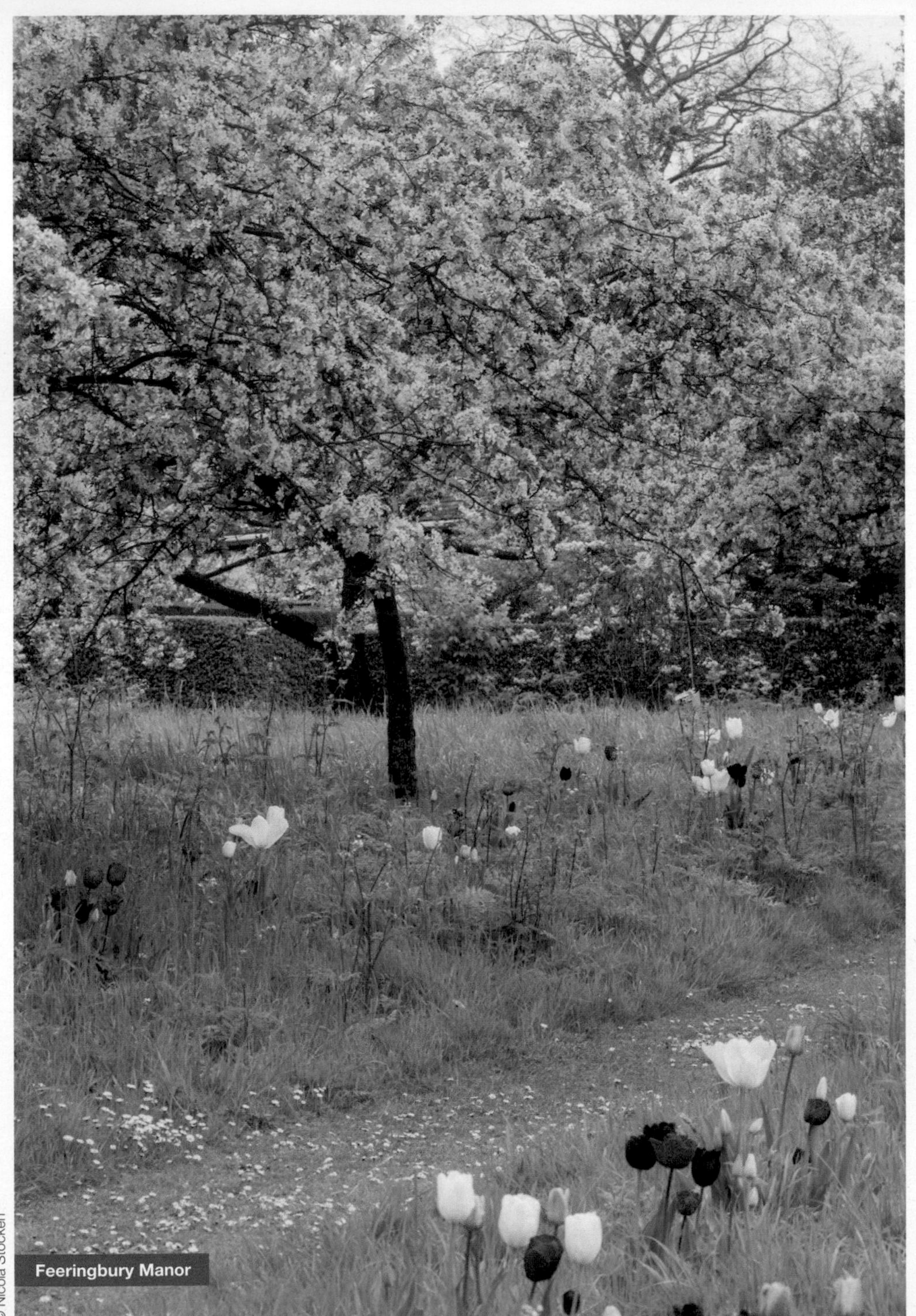

Feeringbury Manor

31 GROVE LODGE
3 Chater's Hill, Saffron Walden, CB10 2AB. Chris Shennan, 01799 522271. *Approx 10 mins walk from town centre. Facing The Common on E side, about 100 yds from the turf maze. Note: Chater's Hill is one way.* **Sun 12, Sun 19 Feb (2-5). Adm £5, chd free. Home-made teas. Visits also by arrangement 2 Jan to 20 Dec. Refreshments will be tailored to suit the size of group.**
A large walled garden close to the town centre with unusually high biodiversity (e.g. 17 species of butterfly recorded) close to the turf maze and Norman castle. Semi-woodland on light tree-draining chalk soil allows bulbs, hellebores and winter-flowering shrubs to thrive. 2 ponds, topiary, orchard and small vegetable garden blend some formality with informal areas where wildlife thrives. A profusion of winter aconites, snowdrops, other bulbs and spring blossom. Lots of seating from which to enjoy the garden. Garden appearing in 2023 Feb edition of Garden Answers magazine. Wheelchair access via fairly steep drive leading to terrace from which the garden may be viewed.

GROUP OPENING

32 HARWICH GARDENS
Harwich, CO12 3NH. *Centre of Old Harwich. Car park on Wellington Rd (CO12 3DT) within 50 m of St Helens Green. Street parking available.* **Sun 16 July (11-4). Combined adm £5, chd free. Refreshments all day at 8 St. Helens Green.**

NEW **17 CHURCH STREET**
Kay Hughes.

63 CHURCH STREET
Sue & Richard Watts.

42 KINGS QUAY STREET
Mrs Elizabeth Crame.

QUAYSIDE COURT
Dave Burton.

8 ST HELENS GREEN
Frances Vincent, 07766707379, francesvincent@icloud.com.
Visits also by arrangement 17 July to 31 Aug for groups of 10 to 14. Afternoon tea incl sandwiches, savoury, cake and prosecco £12.

The gardens all within walking distance in the historical town of Harwich. 8 St Helens Green, just 100m from the sea. A small town garden, with dahlias, hydrangeas mixed with perennials, small fruit trees and pots. 63 Church Street is a long, narrow walled courtyard with shrubs, climbers, perennials and a small vegetable garden packed into the garden. Architectural features inc an Elizabethan window with original glass. Quayside Court, a community garden, unusually boasts a sunken garden hidden from view at the end of the car park which features a pond, vegetable patch, roses and climbers. 42 Kings Quay Street courtyard garden was only landscaped last year. Raised brick beds incorporate established plants such as a grape vine, David Austin roses, tree peony and a combination of edible and ornamental plants.17 Church Street is a newly developed walled courtyard garden at the back of the former post office. Raised beds with a variety of planting inc veg. There will be an art display at 8 St. Helens Green with commission to the National Garden Scheme on any purchases.

33 262 HATCH ROAD
Pilgrims Hatch, Brentwood, CM15 9QR. Mike & Liz Thomas. *2m N of Brentwood town centre. On A128 N toward Ongar turn R onto Doddinghurst Rd at mini r'about (to Brentwood Centre) After the Centre turn next L into Hatch Rd. Garden 4th on R.* **Sun 2, Sun 16 July (11.30-4.30). Adm £5, chd free. Home-made teas.**
A formal frontage with lavender. An eclectic rear garden of around an acre divided into 'rooms' with themed borders, several ponds, three greenhouses, fruit and vegetable plots and oriental garden. There is also a secret white garden, spring and summer wildflower meadows,Yin and Yang borders, a folly and an exotic area. There is plenty of seating to enjoy the views and a cup of tea and cake.

34 HAVENDELL
Beckingham Street, Tolleshunt Major, CM9 8LJ. Malcolm & Val. *5m E of Maldon 3m W of Tiptree. From B1022 take Loamy Hill Rd. At the Xrds L into Witham Rd. Follow NGS signs.* **Sun 25 June (11-4.30). Adm £5, chd free. Home-made teas.**
This beautiful and tranquil garden set in ⅓ acre, will amaze and inspire you as you walk among the herbaceous and shrub borders. The well established trees give height and shade, whilst the the wisteria (now over 20 years old) has a huge winding trunk. Meandering paths and seating areas entice you to take time out in a garden that surrounds you with nature. WC available.

35 HEYRONS
High Easter, Chelmsford, CM1 4QN. Mr Richard Wollaston, 01245 231428, richard.wollaston@gmail.com. *½m outside High Easter towards Good Easter. Via High Easter: through village, turn L downhill & ¼m on R. Via Good Easter: After 2m on L ½ way up hill before the village. Via Leaden Roding: After 2m turn L over bridge. ¼ m on L.* **Sun 16 Apr (11-4). Adm £6, chd free. Home-made teas. Refreshments cost £3 for tea or coffee & cake. This is in addition to the entry charge. Visits also by arrangement Mar to Oct for groups of 10 to 30.**
A garden of 4 parts originally created for family enjoyment within and around an ancient, restored farmyard. An Essex barn and brick farm buildings surround an intensely planted walled garden. A terraced area with mature trees, herbaceous and shady beds leads up to a rose garden. Beyond is an open area with grass beds, tennis court, big skies, a small meadow and fine views over Essex countryside. Wheelchair on gravel driveway can be difficult, but drop off access can be arranged on request.

Our 2022 donation to Hospice UK meant that 911 NHS and Social Care staff accessed individual counselling support

36 2 HOPE COTTAGES

Heath Road, Bradfield, Manningtree, CO11 2UZ.
Martin Ford. *3m E of Manningtree. 9m W of Harwich. Take A137 from Manningtree Station. Turn L opp garage. Take 1st R towards Clacton. At Radio Mast turn L into Bradfield continue through village. Garden is opp just past the Primary School.* **Sun 2 July (11-4), open nearby Chippins. Combined entry with Chippins £6. Sun 20 Aug (11-4). Adm £4, chd free.**
This is a tranquil, country style garden overlooking the serene open countryside. The beautifully planted garden reflects the owner's love of trees. In early summer the rich variety of roses and peonies, give the garden a soft, romantic feel. Followed later with splashes of vibrant colour from dahlias, salvias, agapanthus and sweet peas. Tea, coffee and biscuits available.

37 HORKESLEY HALL

Vinesse Road, Little Horkesley, Colchester, CO6 4DB. Mr & Mrs Johnny Eddis, 07808 599290, horkesleyhall@hotmail.com, www.airbnb.co.uk/rooms/10354093. *6m N of Colchester City Centre. 2m W of A134, 10 mins from A12. On Vinesse Rd, look for the grass triangle with tree in middle, turn into Little Horkesley Church car park. Go to very far end - access is via low double black gates.* **Sat 25 Feb, Sun 28 May (12.30-4.30). Adm £6, chd free. Home-made teas in St Peter & St Paul's Church for Feb opening. Soups also available. Refreshments in May will be in the garden: home-made cakes and more. Visits also by arrangement 10 Feb to 30 Sept for groups of 10+.**
Magical parkland, traditional English garden setting with 8 acres of romantic garden surrounding classical house overlooking lake. Major 20ft balancing stones sculpture. Exceptional trees. A developing Snowdrop collection and winter walk. Over 50 varieties of iris from disbanded National Collection. Walled garden. Charming enclosed swimming pool garden to relax with teas. Excellent Plant Stall. We hope to have a talk on snowdrops in the church followed by an open garden and teas to remember Brooke, who was a special friend of the garden. For more information please email garden owner or view our social media on Facebook & Instagram. Partial wheelchair access to some areas, gravel paths and slopes but easy access to tea area with lovely views over lake and garden.

38 61 HUMBER AVENUE

South Ockendon, RM15 5JW. Mr Greg & Mrs Kasia Purton-Dmowski, 07711 721629, kasia.purtondmowski@gmail.com. *M25- J31 to Thurrock Services, or Grays from A13, at r'about take exit onto Ship Ln to Aveley. Turn R then 2nd exit at r'about onto Stifford Rd/B1335. Take Foyle Dr to Humber Ave.* **Sun 25 June (12-5). Adm £4, chd free. Home-made teas. Gluten Free refreshments available. Visits also by arrangement May to Sept for groups of 10 to 20.**
A suburban garden close to Nature Reserve, (30m x 14m) featuring an old cherry tree from the orchard of the famous Belhus Mansion. The garden features abundant large borders planted heavily with herbaceous plants, small trees, roses, lilies, and ornamental grasses with oriental senses mixed with a traditional English feel with shady areas with hostas, tree ferns and Japanese style plants. The owners are interested in design and are members of Sogetsu International School, original arrangements will be on display during open day.

39 ISABELLA'S GARDEN

42 Theobalds Road, Leigh-On-Sea, SS9 2NE. Mrs Elizabeth Isabella Ling-Locke, 01702 714424, ling_locke@yahoo.co.uk. *Take A13 towards Southend on Sea. As you pass 'Welcome to Leigh-on-Sea' sign, turn R at T- lights onto Thames Dr then L onto Western Rd, carry on 0.6m then R onto Theobalds Rd.* **Sat 27, Sun 28 May, Sat 29, Sun 30 July (12-4.30). Adm £5, chd free. Home-made teas. Home-made cakes inc gluten free and vegan options. Visits also by arrangement 3 June to 30 July for groups of 10 to 25.**
This enchanting garden is bursting with a profusion of colour from early spring through to the autumn months. Roses, clematis, agapanthus, herbaceous plants, alpines, pots with unusual succulents and cacti fill every corner of this garden. There is a wildlife pond. Lilies and water features as well as other garden ornaments which are to be found hiding within the shrubbery and throughout the garden. This garden is situated in Leigh on Sea, and just a 5 min walk from the cockle sheds of Old Leigh and Leigh railway station. Wheelchair access to the majority of the garden is accessible, there are steps near to the house.

40 JANKES HOUSE

Jankes Green, Wakes Colne, Colchester, CO6 2AT.
Bridget Marshall. *8m NW of Colchester. From A1124 Halstead Rd turn R at Wakes Colne opp village shop to Station Rd. Take 2nd R after station entrance into Jankes Ln. House signed. Parking on grass verge.* **Sun 11 June (12-4). Adm £5, chd free. Home-made teas. Combined entry with Oak Farm £8.**
½ acre Traditional English country garden with amazing rural views. Garden has been artistically created by owner. Beautiful mixed borders featuring roses, clematis and architectural shrubs. Camassia in wild grass area. Many young specimen trees. Dry garden. Delightful pond and patio area. Vegetable garden and small orchard. Intimate places to sit and relax. Garden Owner's artistic cards and home-made preserves for sale. Wheelchair access to whole garden.

41 KAMALA

262 Main Road, Hawkwell, Hockley, SS5 4NW. Karen Mann. *3m NE of Rayleigh. From A127 at Rayleigh Weir take B1013 towards Hockley. Garden on L after White Hart Pub & village green.* **Sun 6 Aug, Sun 3 Sept (12-5.30). Adm £6, chd free. Home-made teas.**
Spectacular herbaceous borders which sing with colour as displays of salvia are surpassed by Dahlia drifts. Gingers, Brugmansia, various bananas, bamboos and canna add an exotic note. Grasses sway above the blooms, giving movement. Rest in the rose clad pergola while listening to the 2 Amazon Parrots in the aviary. The garden features an RHS accredited Dahlia named "Jake Mann". Trees and shrubs inc Acers, Catalpa aurea, Cercis 'Forest Pansy' and a large unusual Sinocalycanthus (Chinese

Allspice) a stunning, rare plant with fantastic flowers.

42 KEEWAY

Ferry Road, Creeksea, nr Burnham-on-Crouch, CM0 8PL. John & Sue Ketteley, 01621 782083, sueketteley@hotmail.com. *2m W of Burnham-on-Crouch. B1010 to Burnham on Crouch. At town sign take 1st R into Ferry Rd signed Creeksea & Burnham Golf Club & follow NGS signs.* **Wed 28 June, Sat 1 July (2-5). Adm £5, chd free. Home-made teas. Visits also by arrangement 12 June to 31 July for groups of 10+.**

Large, mature country garden with stunning views over the River Crouch. Formal terraces surround the house with steps leading to sweeping lawns, mixed borders packed full of bulbs and perennials, formal rose and herb garden with interesting water feature. Further afield there are wilder areas, paddocks and lake. A productive greenhouse, vegetable and cutting gardens complete the picture.

43 KELVEDON HALL

Kelvedon, Colchester, CO5 9BN. Mr & Mrs Jack Inglis, 07973 795955, v_inglis@btinternet.com. *Near Colchester. Take Maldon Rd direction Great Braxted from Kelvedon High St. Go over R Blackwater bridge & bridge over A12 At T-junc turn R onto Kelvedon Rd. Take 1st L, single gravel road, oak tree on corner.* **Visits by arrangement 3 Apr to 30 June for groups of 20 to 40. Adm £8, chd free. Home-made teas in the Courtyard Garden or in the Pool Walled Garden - weather permitting.**

Varied 6 acre garden surrounding a gorgeous C18 house. A blend of formal and informal spaces interspersed with modern sculpture. Pleached hornbeam and yew and box topiary provide structure. A courtyard walled garden juxtaposes a modern walled pool garden, both providing season long displays. Herbaceous borders offset an abundance of roses around the house. Lily covered ponds with a wet garden.

44 LAUREL COTTAGE

88 The Street, Manuden, Bishop's Stortford, CM23 1DS. Stewart and Louise Woskett. *Approx 3m N of Bishop's Stortford. From B/Stort along Rye St. At the Mountbatten Restaurant r'about continue straight on to Hazel End Rd. In ½m turn L to stay on Hazelend Rd. Continue past Yew Tree pub. Gdn is last Cottage on R.* **Mon 29 May, Sun 30 July (1-5). Adm £5. Light refreshments.**

Romantic, quintessential thatched cottage garden. Features intimate pathways meandering through borders with an abundance of cottage garden favourites and 1or 2 surprises. Full of rustic charm and an ancient Yew tree. Tranquil seating areas to sit and enjoy fragrant, colourful borders. Hidden treasures enhance this characterful and quirky garden. Created in just 4 years from a very neglected plot. Parking at Community Centre car park adjacent to cottage and playing field.

45 LITTLE MYLES

Ongar Road, Stondon Massey, CM15 0LD. Judy & Adrian Cowan. *1½m SE of Chipping Ongar. Off A128 at Stag Pub, Marden Ash, towards Stondon Massey. Over bridge, 1st house on R after 'S' bend. 400yds Ongar side of Stondon Church.* **Wed 5, Wed 12 July (1.30-5). Adm £5, chd free. Home-made teas.**

A 3 acre romantic garden that has shifted its emphasis to providing an abundance of nectar-rich flowers for struggling bees and butterflies. The Herb garden full of vipers bugloss and roses. Meandering paths past full borders, themed hidden gardens, hornbeam pergola, sculptures and tranquil benches. Crafts and handmade herbal cosmetics for sale. Explorers sheet and map for children. New expanded flower meadow full of colourful nectar-rich flowers for bees and butterflies. Gravel paths. No disabled WC available.

46 LONG HOUSE PLANTS

Church Road, Noak Hill, Romford, RM4 1LD. Tim Carter, www.longhouse-plants.co.uk. *3½m NW of J28, M25. Take A1023 Brentwood. At 1st T-lights, turn L to South Weald after 0.8m turn L at T-junc. After 1.6m turn L, over M25. ½m turn R into Church Rd, nursery opp church. Disabled Car Parking.* **Wed 5 July, Wed 2 Aug, Wed 6 Sept (11-4). Adm £6, chd £3. Home-made teas.**

A beautiful garden - yes, but one with a purpose. Long House Plants has been producing homegrown plants for more than 10 years and here is a chance to see where it all begins! With wide paths and plenty of seats carefully placed to enjoy the plants and views. It has been thoughtfully designed so that the collections of plants look great together through all seasons. Paths are suitable for wheelchairs and mobility scooters. Disabled WC in nursery. 2 cobbled areas not suitable.

47 LONGYARD COTTAGE

Betts Lane, Nazeing, Waltham Abbey, EN9 2DA. Jackie & John Copping. *Nr Harlow, Essex. The garden is. 3½m S of Harlow & is midway between M25 junc Waltham Abbey and M11 junc (15mins approx by car from either to Betts Lane). Opp red telephone box.* **Sun 19 Feb (11-4). Adm £4, chd free. Tea.**

Longyard Cottage is an interesting ¾ of an acre garden situated within yards of an SSSI site, a C11 Church, a myriad of footpaths and a fine display of snowdrops and spring bulbs. This conceptual garden is based on a 'journey' and depicted through the use of paths which take you through 3 distinct areas, with its own characteristics. It's a tactile garden with which you can engage or simply sit and relax.

48 LOXLEY HOUSE

49 Robin Hood Road, Brentwood, CM15 9EL. Robert & Helen Smith. *1m N of Brentwood town centre. On A128 N towards Ongar turn R onto Doddinghurst Rd at mini r'about. Take the 1st rd on L into Robin Hood Rd. 2 houses before the bend on L.* **Sat 29 Apr (11-3). Adm £4, chd free. Home-made teas.**

On entering the rear garden you will be surprised and delighted by this town garden. A colourful patio with pots and containers. Steps up onto a lawn with circular beds surrounded by hedges, herbaceous borders, trees and climbers. 2 water features, 1 a Japanese theme and another with ferns in a quiet seating area. The garden is planted to offer colour throughout the seasons.

49 9 MALYON ROAD

Witham, CM8 1DF. Maureen & Stephen Hicks. *Car park at bottom of High St opp Swan Pub, 5 min walk from car park cross High St by pedestrian crossing onto River Walk, follow path, take 1st R turn into Luard Way, Malyon Rd straight ahead.* **Sat 17, Sun 18 June, Sat 1, Sun 2 July (10.30-4). Adm £4, chd free. Light refreshments. We provide tea, coffee, cakes, sausage rolls and soft drinks.**

Large town garden with mature trees and shrubs made up of a series of garden rooms. Flower beds, pond, summerhouse and greenhouse giving all year interest. Plenty of places to sit and relax with paths that take you on a tour of the garden. A quiet hidden place not expected in a busy town. Our garden is a hidden oasis, where you can escape from the hustle and bustle of busy life. On the edge of town, close to the Witham's rambling Town River Walk.

50 MAY COTTAGE

19 Walton Road, Kirby-Le-Soken, Frinton-On-Sea, CO13 0DU. Julie Abbey, 07885 875822, jools.abbey@hotmail.com. *Kirby-Le-Soken, Essex, CO13 0DU. On the B1034, 2 m before Walton on Naze.* **Sun 21 May, Sun 4 June (11-4). Adm £5, chd free. Light refreshments. Visits also by arrangement 21 May to 18 June.**

A quintessential English country garden divided into 3 compartments with 3 ponds, a stream and a small waterfall. A variety of planting often grown from cuttings. Follow path down to Bakers Oven, a building over 100 years old. A picket gate leads to a secret garden with a winding path, and an area only planted in 2020 leads to a summerhouse. The garden is full of interesting artefacts.

51 MAYFIELD FARM

Hungerdown Lane, Ardleigh, CO7 7LZ. Ed Fairey & Jennifer Hughes, 07736679887, jen@faireyassociates.co.uk. *3½ m NE of Colchester. From Ardleigh village centre continue on the A137 to Manningtree. 3rd R, Tile Barn Ln. R again Hungerdown Ln. 500yds garden on R.* **Visits by arrangement 1 Mar to 15 Oct for groups of 10+. Adm £10, chd free. Home-made teas. Adm inc tea & cake. Guided tour with the Garden Owner.**

Mayfield Farm is a 7 acre garden which only a few years ago was largely a field, with a huge glasshouse and polytunnels. Now filled with many beautiful beds and borders, we have also created a secret garden and planted yew hedging for topiary. Over a 100,000 bulbs have been planted in the garden during the past few years to create a wonderful spring display. Always a new and exciting project being undertaken! Mostly flat but areas of gravel, grass and paddock for parking.

52 NEW 207 MERSEA ROAD

Colchester, CO2 8PN. Kay and Rod. *S part of Colchester. From the town centre drive 2 m up Mersea Rd. Garden on the L side of the rd past the cemetery & mini r'about on the B1025.* **Sun 23, Sun 30 July, Sun 13, Sun 27 Aug (11-4). Adm £4, chd free. Home-made teas. Also home-made cake & soft drinks.**

The garden is 185 ft long divided in to sections. There is a decking area with pots leading to a unique fountain made from waste slates. A circular lawn is surrounded by herbaceous perennials, annuals, trees and shrubs with cabin. Behind the cabin are ferns and more perennials. Beyond is a patio with pergola, patio, pond, long raised beds with subtropical planting, veg beds and hens. Most of the garden is accessible (grass paths). No large motorised chairs.

53 MOVERONS

Brightlingsea, CO7 0SB. Lesley & Payne Gunfield, lesleyorrock@me.com, www.moverons.co.uk. *7m SE of Colchester. At old church turn R signed Moverons Farm. Follow ln & garden signs for approx 1m. Beware some SatNavs take you the wrong side of the river. Open garden with Sculpture Exhibition.* **Sat 10, Sun 11 June (10.30-5). Adm £5, chd free. Home-made teas. Visits also by arrangement 1 June to 15 Sept for groups of 10+.**

Tranquil 4 acre garden in touch with its surroundings and stunning estuary views. A wide variety of planting in mixed borders to suit different growing conditions. Large natural ponds, plenty of seating areas, sculptures and barn for rainy day teas! Our Reflection pool garden and courtyard have been completely redeveloped. Magnificent trees some over 300yrs old give this garden real presence. Most of the garden is accessible by wheelchair via grass and gravel paths. There are some steps and bark paths.

55 OAK FARM

Vernons Road, Wakes Colne, Colchester, CO6 2AH. Ann & Peter Chillingworth. *6 m NW of Colchester. Vernons Rd off A1124 between Ford St & Chappel. From Colchester, 3rd R after Ford St; Oak Farm is 200m up ln on R. From Halstead, 2nd L after viaduct in Chappel.* **Wed 24 May (2-5). Sun 11 June (2-5), open nearby Jankes House. Wed 5 July, Sun 10 Sept (2-5). Adm £5, chd free. Home-made teas. 11 June combined entrance with Jankes House £8.**

Informal farmhouse garden of about 1 acre on an exposed site with extensive views south and west across the Colne Valley. Shrubberies, lawned borders and shady spots where people can enjoy refreshments. Look out for the secret garden. On-going projects inc the rose avenue, Mediterranean bed and prairie garden. Wheelchair access although there are some steps in places, access can be gained to most of the garden.

National Garden Scheme gardens are identified by their yellow road signs and posters. You can expect a garden of quality, character and interest, a warm welcome and plenty of home-made cakes!

Horkesley Hall

© Marcus Harpur

56 THE OLD RECTORY

Boreham Road, Great Leighs, Chelmsford, CM3 1PP. Pauline & Neil Leigh-Collyer. *Approx ½m outside the village of Great Leighs. From Boreham village (J19 off A12) turn into Waltham Rd & travel about 5m, Garden on L. From Great Leighs travel on Boreham Rd for ¾m, garden on R. Garden is found via the post code.* **Sun 21 May, Sun 23 July (12-4.30). Adm £7, chd free. Home-made teas.**

Wander around 4 acres of mature gardens surrounded by open countryside. Aspects inc herbaceous borders, a delightful courtyard, lake, fountain, sweeping lawns, arched walkways and many specimen trees. Beautiful wisteria clothing house. Many seating areas to enjoy alternative vistas. Extensive car parking available on grass off the main road. Picnics allowed. Main areas accessible for wheelchairs. Some areas limited by gravel paths and steps.

57 PEACOCKS

Main Road, Margaretting, CM4 9HY. Phil Torr, 07802 472382, phil.torr@btinternet.com. *Margaretting Village Centre. From village x-roads go 100 yds in the direction of Ingatestone. Entrance gates found on L set back 50ft from rd.* **Sun 4 June, Sun 2 July (11-4). Adm £5. Visits also by arrangement May & June for groups of 15+. Donation to St Francis Hospice.**

10 acre varied garden ideal for picnics with several seating areas. Mature native and specimen trees. Restored horticultural buildings. A series of garden rooms inc Walled Paradise Gardens, Garden of Reconciliation, Alhambra fusion. Long herbaceous/ mixed border. Sunken dell and waterfall. Temple of Antheia on the banks of a lily lake. Large areas for wildlife inc woodland walk, nuttery and orchard. Traditionally managed large wildflower meadow. Picnics are encouraged. Display of old Margaretting postcards. Garden sculpture. Most of garden wheelchair accessible.

58 PEAR TREES

Elder Street, Wimbish, Saffron Walden, CB10 2XA. Mrs Julie Bayliss, 07929802373, peartrees.jb@gmail.com. *Between Saffron Walden & Thaxted. Elder St is a hamlet of Wimbish, also signed Carver Barracks, Turn R off B184 from Saffron Walden or L from Thaxted. Pear Trees is opp entrance to Carver Barracks.* **Sat 18, Sun 19 Mar, Sat 17, Sun 18 June (1-5). Adm £5, chd free. Light refreshments. Home-made cakes & cheese scones. Also hot soup in March. Visits also by arrangement 11 Mar to 25 June for groups of 5 to 15.**

Pear Trees is a south facing garden nearly ½ acre with open views across farmland to Rowney Woods. Front garden has mixed fruit trees underplanted with spring bulbs. Rear garden features mature oak tree and many other trees. A garden allotment boasts a wide range of soft fruit and seasonal vegetables. There are herbaceous borders, raised beds. Patio and seating to enjoy a hot drink and home-made cake. Pear Trees is a garden committed to benefit insects and birds. From the gravel drive there is a concrete path to rear garden. No steps.

1 & 2 Grapevine Cottages

59 18 PETTITS BOULEVARD, RM1
Rise Park, Romford, RM1 4PL. Peter & Lynn Nutley. *From M25 take A12 towards London, turn R at Pettits Ln junc then R again into Pettits Blvd or Romford Stn. 103 or 499 bus to Romford Fire Stn & follow yellow signs.* **Sat 24, Sun 25 June, Sat 2, Sun 3 Sept (1-5). Adm £4, chd free. Home-made teas.**
The garden is 80ft x 23ft on three levels with an ornamental pond, patio area with shrubs and perennials, many in pots. Large eucalyptus tree leads to a woodland themed area with many ferns and hostas. There are agricultural implements and garden ornaments giving a unique and quirky feel to the garden. Tranquil seating areas situated throughout.

60 NEW POLLYFIELD
Chapel Road, Langham, Colchester, CO4 5NY. Laura Benns. *3m N of Colchester, a short distance from A12. 2nd to last house on L after turning into Chapel Rd from Moor Rd.* **Wed 26 Apr, Sat 3 June (11-4). Adm £4, chd free. Home-made teas.**
The garden is mature in nature and 1½ acres in size. It consists of several areas, inc a formal lawn, wild meadow woodland, mixed and herbaceous beds, a pond and patio area. There is an extensive vegetable garden, inc a variety of soft fruits and mature fruit trees. Cottage garden planting and a large bearded iris bed add colour and interest in early summer. Disabled parking only on drive. Garden mainly flat although some uneven ground in woodland area.

61 ROOKWOODS
Yeldham Road, Sible Hedingham, CO9 3QG. Peter & Sandra Robinson, 07770 957111, sandy1989@btinternet.com, www.rookwoodsgarden.com. *8m NW of Halstead. Entering Sible Hedingham from direction of Haverhill on A1017 take 1st R just after 30mph sign. Coming through SH from the Braintree direction turn L just before the 40mph leaving the village.* **Mon 29 May (10.30-4.30). Adm £6, chd free. Home-made teas. Visits also by arrangement 22 May to 29 Sept for groups of 8 to 30.**
Rookwoods is a tranquil garden where you can enjoy a variety of mature trees, herbaceous borders, a shrubbery, pleached hornbeam rooms and a walk through an ancient oak wood. Discover our work-in-progress wildflower bed, along with a buttercup meadow. There is no need to walk far. You can come and linger over tea, under a dreamy wisteria canopy while enjoying views across the garden. Wheelchair access gravel drive.

62 69 RUNDELLS - THE SECRET GARDEN
Harlow, CM18 7HD. Mr & Mrs K Naunton, 01279 303471, k_naunton@hotmail.com. *3m from J7 M11. A414 exit T-lights take L exit Southern Way, mini r'about 1st exit Trotters Rd leading into Commonside Rd, after shops on left 3rd L into Rundells.* **Sat 8 July (2-5). Adm £3, chd free. Home-made teas. Visits also by arrangement May to Sept.**
Featured on Alan Titchmarsh's first 'Love Your Garden' series ('The Secret Garden') 69, Rundells is a very colourful, small town garden packed with a wide variety of shrubs, perennials, herbaceous and bedding plants in over 200 assorted containers. Hard landscaping on different levels has a summerhouse, various seating areas and water features. Steep steps. Access to adjacent allotment open to view. Various small secluded seating areas. A small fairy garden has been added to give interest for younger visitors. The garden is next to a large allotment and this is open to view with lots of interesting features inc a bee apiary. Honey and other produce for sale (conditions permitting).

63 SALING HALL
The Street, Great Saling, Braintree, CM7 5DT. Matthew & Jennifer O'Connell. *Parking at The Salings Millenium Hall (CM7 5DW). 1st R turn on entry into Great Saling from the S, with a short walk N through Great Saling village to Saling Hall.* **Sat 21 Oct (10-4). Adm £7, chd free. Tea & cakes available.**

Opened for NGS by Hugh & Judy Johnson until 2011, and now undergoing an ambitious programme of works by new owners. This special 12-acre Grade II listed garden combines a historic ornamental walled garden with a large surrounding park, predominantly an arboretum full of rare trees, with kitchen garden, orchard, wildflower meadow and water garden; Autumn colour is a particular treat here.

64 SANDY LODGE

Howe Drive, Hedingham Road, Halstead, CO9 2QL. Emma & Rick Rengasamy, 07899 920002, emmarengasamy@gmail.com. *8m NE of Braintree. Turn off Hedingham Rd into Ashlong Grove. Howe Drive is on L. Please park in Ashlong Grove.* **Sun 10 Sept (11-5). Adm £4.50, chd free. Home-made teas. Visits also by arrangement May to Sept for groups of 15 to 30.**

A stunning garden with views over Halstead; there is something to see every day. A ¾ acre garden created over 10 years. Enter our gravel Bee Border, and wander through the Winter Wedding border to the Woodlands Walk and on to the meadow. Stroll across to double borders and long border with flowing prairie planting up towards cabin and BBQ. Lots of seating. Winners of the Ideal Home magazine Best Readers Garden Award 2021 with full article published in Ideal Home August 2022.

65 SCRIPS HOUSE

Cut Hedge Lane, Coggeshall, CO6 1RL. Mr James & Mrs Sophie Bardrick. *From A12 N/S take Kelvedon/Feering exit. Turn L/R passed Kelvedon stn. Continue for 2m turn L into Scrips Rd. From A120 take Coggeshall exit. Drive through village towards Kelvedon. Turn R up Scrips Rd.* **Fri 12, Fri 26 May, Fri 9, Fri 23 June (10-3). Adm £7, chd free.**

A large garden with further paddocks. Inc a mature woodland with a memorial avenue of fastigiate oaks. The garden is divided into small areas inc a walled pool garden and spring garden. A pleached lime walk leads to a white garden flanked by two 30m herbaceous borders. There is an ornamental pond with ducks, an ample vegetable garden and fruit cage, an orchard and a chicken run. Wheelchair access to most areas. Cash only, no change given. Picnics welcome.

66 SILVER BIRCHES

Quendon Drive, Waltham Abbey, EN9 1LG. Frank & Linda Jewson, 01992 714047, linda.jewson@ngs.org. *M25, J26 to Waltham Abbey. At T-lights by McD turn R to r'about. Take 2nd exit to next r'about. Take 3rd exit (A112) to T-lights. L to Monkswood Ave follow yellow signage.* **Sun 4 June (11.30-5.30). Adm £5, chd free. Home-made teas. Visits also by arrangement 20 May to 16 Sept.**

The garden boasts 3 lawns on the 2 levels. This surprisingly secluded garden has many mixed borders packed with colour. Mature shrubs and trees create a short woodland walk. Crystal clear water flows through a shady area of the garden. Chimney Pots for sale. There are areas which would not be suitable for wheelchairs.

67 SNARES HILL COTTAGE

Duck End, Stebbing, CM6 3RY. Pete & Liz Stabler, 01371 856565, petestabler@gmail.com. *Between Dunmow & Bardfield. On B1057 from Great Dunmow to Great Bardfield, ½m after Bran End on L.* **Visits by arrangement Apr to Sept for groups of 8+. Home-made teas.**

A 'quintessential English Garden' - Gardeners' World. Our quirky 1½ acre garden has surprises round every corner and many interesting sculptures. A natural swimming pool is bordered by romantic flower beds, herb garden and Victorian folly. A bog garden borders woods and leads to silver birch copse, beach garden and 'Roman' temple. Classic cars, shepherds hut, natural swimming pond, numerous water features and sculptures.

68 SPRING COTTAGE

Chapel Lane, Elmstead Market, CO7 7AG. Mr Roger & Mrs Sharon Sciachettano, sharons4852@gmail.com. *3m from Colchester. Overlooking village green N of A133 through Elmstead Market. Parking limited adjacent to cottage, village car park nearby on S side of A133.* **Visits by arrangement July & Aug for groups of 10 to 20. Home-made teas. We provide a gluten free chocolate cake.**

From Acteas to Zauschenerias and Aressima to Zebra grass we hope our large variety of plants will please. Our award winning garden features a range of styles and habitats e.g. woodland dell, stumpery, Mediterranean area, perennial borders and pond. Our C17 thatched cottage and garden show case a number of plants found at the world famous Beth Chatto gardens ½m down the road. In late summer we can promise colour and scent to delight the senses and a range of unusual species to please the plant enthusiast. We also have a productive vegetable patch.

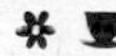

69 STRANDLANDS

off Rectory Road, Wrabness, Manningtree, CO11 2TX. Jenny Edmunds, 01255 886260, strandlands@outlook.com. *1km along farm track from the corner of Rectory Rd. If using a SatNav, the post code will leave you at the corner of Rectory Rd. Turn onto a farm track, signed to Woodcutters Cottage & Strandlands, & continue for 1km.* **Visits by arrangement 29 Apr to 2 July for groups of 12 to 25. Adm £6, chd free. Light refreshments. Tea, coffee, soft drinks and home-made cake: £4 per person.**

Cottage surrounded by 4 acres of land bordering beautiful and unspoilt Stour Estuary. 1 acre of decorative garden: formal courtyard with yew, and box hedges, lily pond, summerhouse; 2 large island beds, secret 'moon garden', 3 acres of wildlife meadows with groups of native trees, and large wildlife pond. Panoramic views of the Stour Estuary from the bottom of the garden. Grayson Perry's 'A House for Essex' can be seen just one field away from Strandlands. Wheelchair access mostly accessible and flat although parking area is gravelled.

In 2022 we donated £450,000 to Marie Curie which equates to 23,871 hours of community nursing care

70 NEW THATCHED COTTAGE
Main Road, Frating, Colchester, CO7 7DJ. Elaine Craig-Bennett and Robin Buck. *6 m E of Colchester. Located on the A133 between The King's Arms Public House and The Village Hall (parking) opp Haggar's Ln.* **Sat 17, Sun 18 June (12-4). Adm £4, chd free. Home-made teas at Rosemary Cottage garden owned by Sue Critchley.**
A thatched cottage garden with many varieties of old and David Austin roses interplanted with a wide range of perennials. Several pathways through arches, weave through the garden taking you past a gazebo, a summerhouse and a bridged pond. Raised brick beds host a variety of common and unusual plants. There are many planted pots and hanging baskets. A plant sale. WC facilities and parking available at the Village Hall . Other stalls in the garden of Rosemary Cottage.

71 NEW 291 THUNDERSLEY PARK ROAD
Benfleet, SS7 1AH. Mr John & Mrs Janet Fenner. *Top of Bread & Cheese Hill, Benfleet, Essex. R turn at the top of Bread & Cheese Hill.* **Sun 11, Mon 12, Sun 25, Mon 26 June (10-5). Adm £6, chd free. Home-made teas.**
This interesting garden set on a steep Essex hill is full of hidden treasures. The owner, Janet Fenner has created different garden rooms, incl a Chinese garden, a dry Mediterranean garden, a

Dragons

'Ravello' bed, a sunken garden and an English garden. A variety of large patios have added interest with large topiary plants, hanging baskets and beautiful annuals that add bursts of colour.

72 ULTING WICK

Crouchmans Farm Road, Ulting, Maldon, CM9 6QX. Mr & Mrs B Burrough, 07984614947, philippa.burrough@btinternet.com, www.ultingwickgarden.co.uk. *3m NW of Maldon. Take R turning to Ulting off B1019 as you exit Hatfield Peverel by a green. Garden on R after 2 M.* **Sun 23, Fri 28 Apr, Tue 11 July, Mon 28 Aug, Fri 1 Sept (2-5). Adm £7.50, chd free. Visits also by arrangement 1 Feb to 1 Oct for groups of 15 to 60. Donation to All Saints Ulting Church.**

Listed black barns provide backdrop for vibrant and exuberant planting in 8 acres. Thousands of snowdrops, tulips, flowing innovative spring planting, herbaceous borders, pond, mature weeping willows, kitchen garden, dramatic late summer beds with zingy, tender, exotic plant combinations. Drought tolerant perennial and mini annual wildflower meadows. Woodland. Many plants propagated in-house. Unusual plants for sale. Beautiful dog walks along the River Chelmer from the garden. Regional Winner, The English Garden's The Nation's Favourite Gardens 2021. Some gravel around the house but main areas of interest are accessible for wheelchairs.

73 14 UNA ROAD

Bowers Gifford, Basildon, SS13 2HU. Mr John & Mrs Barbara Spooner. *1m E of Basildon on B1464 between Pitsea & Saddlers Farm r'about (A130/A13.) From Pitsea turn L into Pound Ln. From Southend (A127) X A130/A1245 junc. 1m turn L & follow yellow signs.* **Evening openings Thur 14 Dec to Tue 19 Dec (4.30-7.30). Adm £4, chd free. Light refreshments.**

This garden is opening to display a wonderland of Christmas lights decorations throughout the half acre garden. The displays will inc a Gnome Christmas Village to delight the children and displays of penguins and polar bears and much more. Refreshments will be available inc hot chocolate, mulled wine and mince pies.

74 WALNUT TREE COTTAGE

Cobblers Green, Felsted, Dunmow, CM6 3LX. Mrs Susan Monk, susan_monk@hotmail.com. *B1417 brings you into Felsted from both ends of the village then turn into Causeway End Rd, ½m down the road you will see a yellow sign on your L to turn R into the paddock.* **Sun 4 June (11-4). Adm £6, chd free. Home-made teas. Visits also by arrangement 3 Apr to 30 Sept for groups of 10 to 15.**

A traditional cottage garden framed by wonderful views over the Essex countryside and planted with well-stocked flower borders and lawns that slop gently down to a natural pond. A productive garden borders the well maintained paddock where chickens and guinea fowl roam. Large patio and other seating areas provide a place to relax and appreciate the garden. Wheelchair access, there is a path that makes most of the garden accessible.

75 1 WHITEHOUSE COTTAGES

Blue Mill Lane, Woodham Walter, CM9 6LR. Mrs Shelley Rand. *In between Maldon & Danbury, short drive from A12. From A414 Danbury, turn L at The Anchor & continue into the village. Directly after the white village gates at the far end of the village, turn R into Blue Mill Ln.* **Sun 14 May, Sun 2 July (10-3). Adm £5, chd free. Home-made teas.**

Nestled betwixt farmland in rural Essex, is our small secret garden, that has a wonderful charm and serenity to it. Set in 3½ acres, mostly paddocks, a little plot of loveliness wraps around our Victorian cottage, and roses smother the porch in June. A meandering lawn takes you through beds and borders softly planted with a cottage feel, a haven for wildlife and people alike. Dean Harris, a local artist blacksmith, will have a pop-up forge on site making and selling metal plant accessories. Parking available a short walk up the lane near The Cats pub, and for those who might struggle to walk, there's parking at the property in the paddock.

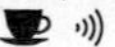

76 WRITTLE UNIVERSITY COLLEGE

Writtle, CM1 3RR. Writtle University College, www.writtle.ac.uk. *4m W of Chelmsford. On A414, nr Writtle village. Parking available on the main campus.* **Sun 23 Apr (10-4). Adm £5, chd free. Light refreshments in The Garden Room(main campus) & Lordship tea room (Lordship campus).**

Writtle University college has 15 acres of informal lawns with naturalized bulbs and wildflowers. Large tree collection, mixed shrubs, herbaceous border, dry/ Mediterranean borders, seasonal bedding and landscaped glasshouses. All gardens have been designed and built by our students studying a wide range of horticultural courses. Wheelchair access: some gravel, however majority of areas accessible to all.

77 WYCHWOOD

Epping Road, Roydon, CM19 5DW. Mrs Madeleine Paine. *At Tylers Cross r'about head in the direction of Roydon. Garden on R approx 400m from the r'about. Parking is available in Redrick's nursery next to garden.* **Sun 2 July (12-5.30). Adm £5, chd free. Home-made teas.**

A garden approx. ¾ acre with a large pond, attracting much wildlife, as well as the owners resident ducks. Free ranging chickens roam in the shrubbery and budgerigar aviary. There are numerous features inc vegetable and fruit plot, mixed shrub and herbaceous borders, 1920's summerhouse and Scandinavian cabin. Old fashioned style roses are a particular feature of the garden.

In 2022 the National Garden Scheme donated £3.11 million to our nursing and health beneficiaries

GLOUCESTERSHIRE

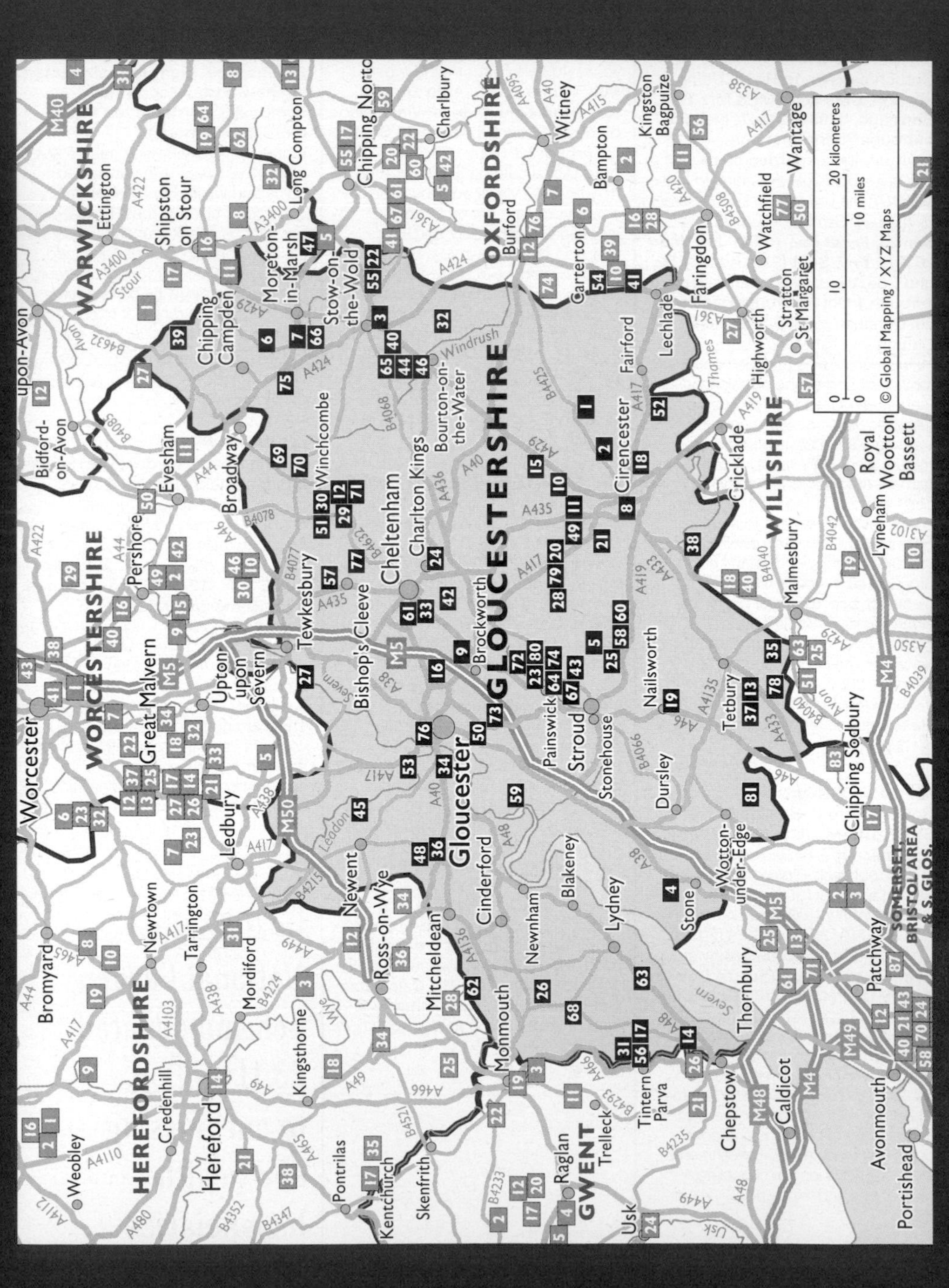

Gloucestershire is one of the most beautiful counties in England, spanning as it does a large part of the area known as the Cotswolds as well as the Forest of Dean and Wye and Severn Valleys.

The Cotswolds is an expanse of gently sloping green hills, wooded valleys and ancient, picturesque towns and villages; it is designated as an area of Outstanding Natural Beauty, and its quintessentially English charm attracts many visitors.

Like the county itself many of the gardens that open for the National Garden Scheme are simply quite outstanding. There are significant gardens which open for the public as well, such as Kiftsgate and Bourton House. There are also some large private gardens which only open for us, such as Daylesford and Charlton Down House.

There are however many more modest private gardens whose doors only open on the National Garden Scheme open day, such as Bowling Green Road in Cirencester with over 300 varieties of Hemerocallis. This tiny garden has now opened for over 40 years. The National collection of Rambling Roses is held at Moor Wood and that of Juglans and Pterocarya at Upton Wold.

Several very attractive Cotswold villages also open their gardens and a wonderful day can be had strolling from cottage to house marvelling at both the standard of the gardens and the beauty of the wonderful buildings, only to pause for the obligatory tea and cake!

© Mark Bolton

Above: Clouds Rest

@gloucestershirengs

@Glosngs

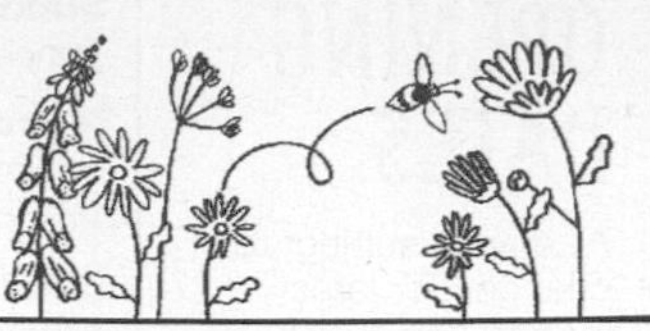

Volunteers

County Organiser
Vanessa Berridge
01242 609535
vanessa.berridge@ngs.org.uk

County Treasurer
Pam Sissons
01242 573942
pam.sissons@ngs.org.uk

Social Media
Mandy Bradshaw
01242 512491
mandy.bradshaw@ngs.org.uk

Publicity
Ruth Chivers
01452 542493
ruth.chivers@ngs.org.uk

Booklet Coordinator
Vanessa Graham
07595 880261
vanessa.graham@ngs.org.uk

By Arrangement Coordinator
Simone Seward
01242 573733
simone.seward@ngs.org.uk

Assistant County Organisers
Yvonne Bennetts
yvonne.bennetts@ngs.org.uk

Jackie Healy 07747 186302
jackie.healy@ngs.org.uk

Ali James 01869 350654
ali.james@ngs.org.uk

Valerie Kent 01993 823294
valerie.kent@ngs.org.uk

Immy Lee 07801 816340
immy.lee@ngs.org.uk

Sally Oates 01285 841320
sally.oates@ngs.org.uk

Rose Parrott 07853 164924
rosemary.parrott@ngs.org.uk

Liz Ramsay 01242 672676
liz.ramsay@ngs.org.uk

OPENING DATES

All entries subject to change. For latest information check **www.ngs.org.uk**

Map locator numbers are shown to the right of each garden name.

January

Sunday 29th
Home Farm 36

February

Snowdrop Festival

Monday 6th
Downton House 23

Tuesday 7th
Downton House 23

Sunday 12th
Home Farm 36
Trench Hill 72

Saturday 18th
Cotswold Farm 20

Sunday 19th
Cotswold Farm 20
Trench Hill 72

March

Sunday 5th
Home Farm 36

Sunday 19th
Trench Hill 72

Sunday 26th
Sheepehouse Cottage 67

April

Saturday 1st
South Lodge 68

Sunday 2nd
◆ Highnam Court 34
Home Farm 36

Wednesday 5th
Brockworth Court 9
Kemble House 38

Sunday 9th
Trench Hill 72

Monday 10th
◆ Kiftsgate Court 39
Trench Hill 72

Sunday 16th
◆ Upton Wold 75

Wednesday 19th
Barnsley House 2
Daylesford House 22

Saturday 22nd
The Gate 29

Sunday 23rd
Blockley Gardens 6
Charlton Down House 13
◆ The Coach House Garden 18

Tuesday 25th
Wortley House 81

Wednesday 26th
Lords of the Manor Hotel 44
Westaway 77

Sunday 30th
Eastcombe and Bussage Gardens 25
Home Farm 36

May

Monday 1st
Eastcombe and Bussage Gardens 25

Saturday 6th
NEW 1 Cobden Villas 19

Sunday 7th
NEW 1 Cobden Villas 19
◆ Highnam Court 34
Ramblers 63
South Lodge 68
Trench Hill 72

Thursday 11th
Richmond Painswick Retirement Village 64

Saturday 13th
Richmond Painswick Retirement Village 64

Sunday 14th
NEW Penny Leaze 58
NEW Pontings Farm 60

Monday 15th
NEW Penny Leaze 58
NEW Pontings Farm 60

Friday 19th
◆ The Garden at Miserden 28
The Old Vicarage 53

Saturday 20th
Charingworth Court 12
The Old Vicarage 53

Sunday 21st
Charingworth Court 12
NEW 11 Hatherley Court Road 33
The Old Vicarage 53
◆ Stanway Fountain & Water Garden 70

Thursday 25th
Richmond Painswick Retirement Village 64

Saturday 27th
Hookshouse Pottery 37
Little Orchard 43
Richmond Painswick Retirement Village 64

Sunday 28th
NEW Beechwood Park 3
Greenfields, Brockweir Common 31
Hookshouse Pottery 37
Little Orchard 43
Pasture Farm 55
Tuffley Gardens 73

Monday 29th
Hookshouse Pottery 37
Pasture Farm 55

Tuesday 30th
Hookshouse Pottery 37

Wednesday 31st
Hookshouse Pottery 37
Rockcliffe 65

June

Saturday 3rd
NEW 1 Cobden Villas 19
Cotswold Farm 20
Forthampton Court 27

Sunday 4th
◆ The Coach House Garden 18
NEW 1 Cobden Villas 19
Cotswold Farm 20
Forthampton Court 27
◆ Highnam Court 34
Langford Downs Farm 41
◆ Oxleaze Farm 54

Monday 5th
Berkeley Castle 4

Wednesday 7th
Trench Hill 72

Thursday 8th
Richmond Painswick Retirement Village 64

Friday 9th
NEW Winstone Glebe 79

Saturday 10th
NEW Chase End 14
NEW Madams 45
Richmond Painswick Retirement Village 64
South Lodge 68
NEW 11 Upper Washwell 74

Sunday 11th
Blockley Gardens 6
Charlton Down House 13
Hodges Barn 35
NEW Madams 45
Stanton Village Gardens 69
Weir Reach 76

Monday 12th
Hodges Barn 35

Wednesday 14th
Oak House 51
Trench Hill 72
Weir Reach 76

Friday 16th
Hookshouse Pottery 37

Saturday 17th
Oak House 51

Sunday 18th
Bisley Gardens 5

Tuesday 20th
Wortley House 81

Wednesday 21st
Trench Hill 72

Thursday 22nd
Richmond Painswick Retirement Village 64

Saturday 24th
Chedworth Gardens 15
Richmond Painswick Retirement Village 64

Sunday 25th
NEW Churchdown Village Junior School 16
Moor Wood 49

Wednesday 28th
Lords of the Manor Hotel 44
NEW The Manor 46
Rockcliffe 65
Trench Hill 72

Friday 30th
Hookshouse Pottery 37

July

Saturday 1st
Perrywood House 59

Sunday 2nd
◆ Highnam Court 34
Kirkham Farm 40
Leckhampton Court Hospice 42
Perrywood House 59

Wednesday 5th
Awkward Hill Cottage 1

Thursday 6th
Charlton Down House 13

Saturday 8th
◆ Cerney House Gardens 11

Sunday 9th
Greenfields, Little Rissington 32
◆ Sezincote 66
NEW Woodchester Park House 80

Thursday 13th
Charlton Down House 13

Sunday 16th
Trench Hill 72
◆ Westonbirt School Gardens 78

Thursday 27th
Charlton Down House 13

August

Thursday 3rd
Charlton Down House 13

Sunday 6th
◆ Highnam Court 34

Saturday 12th
NEW 39 Neven Place 50

Sunday 13th
NEW 39 Neven Place 50
◆ Stanway Fountain & Water Garden 70

Monday 14th
◆ Kiftsgate Court 39

Thursday 17th
Charlton Down House 13

Sunday 20th
◆ Bourton House Garden 7
The Manor 47

Thursday 24th
Charlton Down House 13

Sunday 27th
East Court 24

Monday 28th
NEW Calmsden Manor 10
East Court 24

September

Saturday 2nd
Cotswold Farm 20

Sunday 3rd
NEW Beechwood Park 3
Cotswold Farm 20
◆ Highnam Court 34
The Old Rectory, Quenington 52
Sheepehouse Cottage 67

Wednesday 6th
Lords of the Manor Hotel 44
NEW The Manor 46

Sunday 10th
Clouds Rest 17
The Patch 56
Trench Hill 72

Sunday 17th
NEW Penny Leaze 58
NEW Pontings Farm 60
NEW 37 Queen's Road 61

Monday 18th
NEW Penny Leaze 58
NEW Pontings Farm 60

Friday 22nd
◆ Sudeley Castle & Gardens 71

Wednesday 27th
Brockworth Court 9

January 2024

Sunday 28th
Home Farm 36

February 2024

Sunday 11th
Home Farm 36
Trench Hill 72

Saturday 17th
Cotswold Farm 20

Sunday 18th
Cotswold Farm 20
Trench Hill 72

By Arrangement

Arrange a personalised garden visit with your club, or group of friends, on a date to suit you. See individual garden entries for full details.

Awkward Hill Cottage 1
Berkeley Castle 4
25 Bowling Green Road 8
Brockworth Court 9
Charingworth Court 12
Charlton Down House 13
NEW Chase End 14
Daglingworth House 21
Downton House 23
20 Forsdene Walk 26
The Gate 29
Green Bough 30
Greenfields, Brockweir Common 31
NEW 11 Hatherley Court Road 33
Home Farm 36
Keens Cottage, Chedworth Gardens 15
Kemble House 38
Little Orchard 43
Monks Spout Cottage 48
Moor Wood 49
Oak House 51
Pasture Farm 55
The Patch 56
Pear Tree Cottage 57
Radnors 62
Sheepehouse Cottage 67
South Lodge 68
Trench Hill 72
Weir Reach 76
Westaway 77
Wortley House 81

Greenfields, Brockweir Common

THE GARDENS

1 AWKWARD HILL COTTAGE

Awkward Hill, Bibury, GL7 5NH. Mrs Victoria Summerley, 07718 384269, v.summerley@hotmail.com, www.awkwardhill.co.uk. *No parking at property, or on Hawkers Hill or Awkward Hill. Best to park in village & walk past Arlington Row up Awkward Hill, or up Hawkers Hill from Catherine Wheel pub.* **Wed 5 July (2-6). Adm £5, chd free. Home-made teas. Visits also by arrangement 5 June to 31 Aug for groups of 10 to 30.Regret no children.**

An ever-evolving country garden in one of the most beautiful villages in the Cotswolds, designed to reflect the local landscape and encourage wildlife. Planting is both formal and informal, contributing year-round interest with lots of colour and texture. Pond and waterfall, chickens Wonderful views over neighbouring meadow and woodland, small pond with jetty, 2 sunny terraces and plenty of places to sit and relax, both in sun and shade.

2 BARNSLEY HOUSE

Barnsley, Cirencester, GL7 5EE. Calcot Health & Leisure Ltd, 01285 740000, reception@barnsleyhouse.com, www.barnsleyhouse.com. *4m NE of Cirencester. From Cirencester, take B4425 to Barnsley. House entrance on R as you enter village.* **Wed 19 Apr (10-3). Adm £5, chd free. Tea.**

The beautiful garden at Barnsley House, created by Rosemary Verey, is one of England's finest and most famous gardens inc knot garden, potager garden and mixed borders in her successional planting style. The house also has an extensive kitchen garden which will be open with plants and vegetables available for purchase. Narrow paths mean restricted wheelchair access but happy to provide assistance.

3 NEW BEECHWOOD PARK

Fosseway, Stow On The Wold, Cheltenham, GL54 1FP. Brio Retirement Living. *Head out of Stow towards Bourton on the Water. Pass a BP garage on the R & a cemetery on the L. Beechwood Park will be just past this on the L.* **Sun 28 May, Sun 3 Sept (10-3). Adm £5, chd free. Tea, coffee, cake & full bar.**

Five beautifully designed, modern courtyard gardens set in a semi-urban community in Stow-on-the-Wold. The gardens and main lawn area with feature pod are set against a contemporary design of Cotswold apartment communities. Full wheelchair access plus disabled toilet facilities.

4 BERKELEY CASTLE

Berkeley, GL13 9PJ. Charles Berkeley, 01453 810303, info@berkeley-castle.com, www.berkeley-castle.com. *Halfway between Bristol & Gloucester, 10mins from J13 &14 of M5. Follow signs to Berkeley from A38 & B4066. Visitor entrance L off Canonbury St, just before town centre.* **Mon 5 June (11-5). Adm £6, chd £3. Light refreshments in the Castle. Visits also by arrangement 2 Apr to 31 Oct for groups of 20 to 50.**

Unique historic garden of a keen plantsman, with far-reaching views across the River Severn. Gardens contain many rare plants which thrive in the warm microclimate against the stone walls of this medieval castle. Woodland, historic trees and stunning terraced borders. The admission price does not inc entrance into the castle.

Paulmead, Bisley Gardens

GROUP OPENING

5 BISLEY GARDENS

Wells Road, Bisley, Stroud, GL6 7AG. *Gardens & car park well signed in Bisley village. Gardens on S edge of village at head of Toadsmoor Valley, N of A419 Stroud to Cirencester road.* **Sun 18 June (2-6). Combined adm £5, chd free.**

PAULMEAD
Judy & Philip Howard and Tom & Emma Howard.

WELLS COTTAGE
Mr & Mrs Michael Flint.

Two beautiful gardens with differing styles. Paulmead: 1 acre landscaped garden constructed in stages over last 25 yrs. Terraced on 3 main levels. Natural stream garden, herbaceous and shrub borders, formal vegetable garden, summerhouse overlooking pond. Unusual treehouse. Development of new garden around hen house, Inc ha-ha. Wells Cottage: Just under 1 acre, terraced on several levels with formal topiary and beautiful views over the valley. Much informal planting of trees and shrubs to give colour and texture. Lawns and herbaceous borders. Collection of grasses. Formal pond area. Rambling roses on chain pergola. Vegetable garden with raised beds.

GROUP OPENING

6 BLOCKLEY GARDENS

Blockley, Moreton in Marsh, GL56 9DB. *3m NW of Moreton-in-Marsh. Just off the A44 Morton-in-Marsh to Evesham road.* **Sun 23 Apr, Sun 11 June (2-6). Combined adm £8, chd free. Home-made teas at St George's Hall.**

BELL BANK
Mr & Mrs C D Walters.
Open on Sun 11 June

BLOCKLEY ALLOTMENTS
Blockley & District Allotment Ass, www.blockleyallotments.wixsite.com/blockleyallotments.
Open on Sun 11 June

CHURCH GATES
Mrs Brenda Salmon.
Open on all dates

COLEBROOK HOUSE
Mr & Mrs G Apsion, www.instagram.com/colebrook_house_gardens.
Open on all dates

ELM HOUSE
Chris & Val Scragg.
Open on Sun 23 Apr

NEW **GARDEN HOUSE**
Mr Nick & Mrs Ginny Williams-Ellis.
Open on Sun 11 June

THE MANOR HOUSE
Zoe Thompson.
Open on Sun 11 June

PORCH HOUSE
Mr & Mrs Johnson.
Open on Sun 23 Apr

SNUGBOROUGH MILL
Rupert & Mandy Williams-Ellis, 01386 701310, rupert.williams-ellis@talk21.com.
Open on Sun 11 June

WOODRUFF
Paul & Maggie Adams.
Open on all dates

This popular historic hillside village has a great variety of high quality, well-stocked gardens - large and small, old and new. Blockley Brook, an attractive stream, which flows right through the village, graces some of the gardens; these inc gardens of former water mills, with millponds attached. From some gardens there are wonderful rural views. Children welcome, but close supervision is essential. Regional Finalist, The English Garden's The Nation's Favourite Gardens 2021.

7 ♦ BOURTON HOUSE GARDEN

Bourton-on-the-Hill, GL56 9AE. Mr & Mrs R Quintus, 01386 700754, info@bourtonhouse.com, www.bourtonhouse.com. *2m W of Moreton-in-Marsh. On A44.* **For NGS: Sun 20 Aug (10-5). Adm £9, chd free. Light refreshments & home-made cakes in Grade I Listed C16 tithe barn. For other opening times and information, please phone, email or visit garden website.**

Award-winning, 3 acre garden featuring imaginative topiary, wide herbaceous borders with many rare, unusual and exotic plants, water features, unique shade house and many creatively planted pots. Fabulous at any time of year, but magnificent in summer months and early autumn. Walk around a 7 acre pasture, with free printed guide to specimen trees available. 70% access for wheelchairs. Disabled WC.

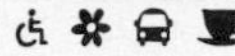

8 25 BOWLING GREEN ROAD

Cirencester, GL7 2HD. Mrs Sue Beck, 01285 653778, sjb@beck-hems.org.uk. *On NW edge of Cirencester. Take A435 to Spitalgate/Whiteway T-lights, turn into The Whiteway (Chedworth turn), then 1st L into Bowling Green Rd, garden in bend in rd between Nos 23 & 27.* **Visits by arrangement 2 July to 10 Sept for groups of up to 20. Adm £4.50, chd free. Tea/coffee/juice/biscuits can be provided for small groups or individuals if requested in advance.**

Welcome to this naturalistic garden, where plants are allowed to take centre stage and do the talking. Wander at will in a mini-jungle of curvaceous clematis, gorgeous grasses, romantic roses, heavenly hemerocallis and plentiful perennials, glimpsing a graceful giraffe, friendly frogs and even a unicorn. Rated by visitors as an amazing hidden gem with a unique atmosphere. View The Chatty Gardener's blog, https://thechattygardener.com/plants-and-more/how-to-grow-hemerocallis/ on growing Hemerocallis and the garden owner having a daylily cultivar registered by UK Hybridizer to mark her 40th Anniversary of opening for NGS.

9 BROCKWORTH COURT

Court Road, Brockworth, GL3 4QU. Tim & Bridget Wiltshire, 01452 862938, timwiltshire@hotmail.co.uk. *6m E of Gloucester. 6m W of Cheltenham. Adj St Georges Church on Court Rd. From A46 turn into Mill Lane, turn R, L, R at T junctions. From Ermin St, turn into Ermin Park, then R at r'about then L at next r'about.* **Wed 5 Apr, Wed 27 Sept (2-6). Adm £6, chd free. Home-made teas. Visits also by arrangement Apr to Oct for groups of 5 to 30.**

This intense yet informal tapestry style garden beautifully complements the period manor house which it surrounds. Organic, naturalistic, with informal cottage-style planting areas that seamlessly blend together. Natural fish pond, with Monet bridge leading to small island with thatched Fiji house. Kitchen garden once cultivated by monks. Views to Crickley and Coopers Hill. Adjacent Norman church (open). Historic tithe barn, manor house visited by Henry VIII and Anne Boleyn in 1535. Partial wheelchair access.

10 NEW **CALMSDEN MANOR**
Calmsden, Cirencester, GL7 5ET. Mr M & Mrs J Tufnell. *5m N of Cirencester. Turn to Calmsden off A429 at The Stump pub. On entering the village Manor gates are straight ahead of you.* **Mon 28 Aug (2.30-6.30). Adm £10, chd free. Home-made teas.**
With borders originally designed by Mary Keen, Calmsden Manor is a well loved garden. It is a treat for plant lovers, design enthusiasts, vegetable growers and those simply looking for a lovely day out in the Cotswolds. Inc herbaceous borders, a small arboretum, wild grass and flower meadow, woodland and walled garden for fruit, vegetables and cut flowers. There is much to enjoy.

11 ◆ **CERNEY HOUSE GARDENS**
North Cerney, Cirencester, GL7 7BX. Mr N W Angus & Dr J Angus, 01285 831300, janet@cerneygardens.com, www.cerneygardens.com. *4m NW of Cirencester. On A435 Cheltenham rd turn L opp Bathurst Arms, follow rd past church up hill, then go straight towards pillared gates on R (signed Cerney House).* **For NGS: Sat 8 July (10-7). Adm £6, chd £1. For other opening times and information, please phone, email or visit garden website.**
A romantic English garden for all seasons. There is a secluded Victorian walled garden featuring herbaceous borders overflowing with colour. In the summer the magnificent display of rambling romantic roses come to life. Enjoy our woodland walk, extended nature trail and new medicinal herb garden. Dogs welcome. Walled garden accessible for electric wheelchairs. Gravel paths and inclines may not suit manual wheelchairs.

12 **CHARINGWORTH COURT**
Broadway Road, Winchcombe, GL54 5JN. Susan & Richard Wakeford, 07791 353779, susanwakeford@gmail.com, www.charingworthcourt cotswoldsgarden.com. *8m NE of Cheltenham. 400 metres N of Winchcombe town centre on B4632. Limited parking along Broadway Rd. Town car parks in Bull Lane, Chandos St (short stay) and all day parking (£1) in Back Lane. Map on our website.* **Sat 20, Sun 21 May (11-5.30). Adm £5, chd free. Light refreshments. Visits also by arrangement 2 May to 15 July for groups of 10 to 30. Light refreshments can be arranged.**
Artistically and lovingly created 1½ acre garden surrounding restored Georgian/Tudor house (not open). Relaxed country style with Japanese influences, large pond, sculpture and walled vegetable/flower garden, created over 25 yrs from a blank canvas. Mature copper beech trees, Cedar of Lebanon and Wellingtonia; and younger trees replacing an earlier excess of *Cupressus leylandii*. Partial access due to gravel paths which can be challenging but several areas accessible without steps. Some disabled parking next to house.

13 **CHARLTON DOWN HOUSE**
Charlton Down, Tetbury, GL8 8TZ. Neil & Julie Record, cdh.groupbookings@gmail.com. *From Tetbury, take A433 towards Bath for 1½ m; turn R (north) just before the Hare & Hounds, then R again after 200yds into Hookshouse Lane. Charlton Down House is 600yds on R.* **Sun 23 Apr, Sun 11 June (11-5); Thur 6, Thur 13, Thur 27 July, Thur 3, Thur 17, Thur 24 Aug (1-5). Adm £7, chd free. Home-made teas. Six by arrangement visits also available 24 Apr to 31 Aug for groups of 20 to 35.**
Extensive country house gardens in 180 acre equestrian estate. Formal terraces, perennial borders, walled topiary garden, enclosed cut flower garden and large glasshouse. Newly planted copse. Rescue animals. Ample parking. Largely flat terrain; most garden areas accessible.

14 NEW **CHASE END**
Tidenham Chase, Chepstow, NP16 7JN. Tim & Penny Wright, pennyjwright63@gmail.com. *On the B4228 between Chepstow & St Briavels. First house on R after Boughspring Lane, travelling on the B4228 towards St. Briavels. Parking 2nd entrance on R, through gate into top field. No parking on main road.* **Sat 10 June (11-5). Adm £5, chd free. Home-made teas. Visits also by arrangement 11 June to 18 June for groups of up to 30.**
A 4 acre, south facing, sloping site overlooking the Severn, where soil and wildlife are nurtured with organic and no-dig methods. There are 3 acres of mature wildflower meadows, with orchards and a vineyard, and 1 acre of more formal garden with mature herbaceous borders, trees and shrubs (notably roses and echiums), 2 ponds with bog gardens, an area of prairie planting and a greenhouse.

GROUP OPENING

15 **CHEDWORTH GARDENS**
Chedworth, Cheltenham, GL54 4AN. *7m NE of Cirencester. Off Fosseway, A429 between Stow-on-the-Wold (12m) & Cirencester. Park & Ride in field by village hall (signed). Tickets available from village hall.* **Sat 24 June (10-5). Combined adm £10, chd free. Light refreshments at Chedworth Village Hall. Donation to Chedworth & District Horticultural Society.**

ABSOLAM'S ORCHARD
Richard & Bettina Abraham.

ADAMS POOL
Bee Brockman.

NEW **CHEDWORTH HOUSE**
Chedworth House.

COBBLERS COTTAGE
Ceri Powell & Ajay Shah.

KEENS COTTAGE
Sue & Steve Bradbury, 07831 464956, bradburydesigns@aol.com, www.bradburydesigns.co.uk.
Visits also by arrangement 13 May to 22 July for groups of 5 to 20. Please contact Sue Bradbury for details on private garden tours.

THE OXBYRE
Chedworth Open Gardens.

NEW **WHITES BARN**
Lindsey & Marie Haslett.

WINDSOR COTTAGE
Peter & Annette Seymour.

Varied collection of country gardens, nestling throughout the mile long Chedworth Valley with tributary of River Coln running below. Stunning views. Featuring lots of ideas to inspire. Afternoon teas and light lunches served in village hall throughout the day. WC available plus plant sales and stalls and visit our Horticultural Show marquee open in the afternoon.

16 NEW **CHURCHDOWN VILLAGE JUNIOR SCHOOL**
Station Road, Churchdown, Gloucester, GL3 2JX. www.churchdownvillage-jun.gloucs.sch.uk. *4.5m NE of Gloucester. The school is between St Andrew's Church & Churchdown Community Centre. Pls follow signage.* **Sun 25 June (2-5). Adm £5, chd free. Tea, coffee & soft drinks. Home-made cakes.**
RHS School Gardening Award winning site featuring a working allotment with polytunnel, a walk-through forest school with willow tunnel, shelter and wildlife pond, and a large playing field. The site is a learning environment for all our children, with the allotment and polytunnel run by an enthusiastic after school Garden Club. Live music will celebrate our first NGS opening.

17 CLOUDS REST
Brockweir, Chepstow, NP16 7NW. Mrs Jan Bastord. *In the Wye Valley, 6.7m N of Chepstow & 10.6m S of Monmouth, off A466, across Brockweir Bridge.* **Sun 10 Sept (12-5). Combined adm with The Patch £8, chd free. Tea & cake served at a marquee in our woodland area.**
The garden at Clouds Rest was started in 2012 from a south-westerly facing stony paddock, with views across the Wye Valley. Its many gravel pathways meander through herbaceous beds with a mixture of roses, then a wide selection of Michaelmas daisies in September. Easy parking in our paddock. New additions inc woodland area and orchard with lily pond. Bog garden. Lovely orchard.

18 ◆ THE COACH HOUSE GARDEN
Church Lane, Ampney Crucis, Cirencester, GL7 5RY. Mr & Mrs Nicholas Tanner, 01285 850256, mel@thecoachhousegarden.co.uk, www.thecoachhousegarden.co.uk. *3m E of Cirencester. Turn into village from A417, immed before Crown of Ampney Brook Inn. Over hump-back bridge, parking to R on cricket field (weather permitting) or signed nearby field. Disabled parking nr house.* **For NGS: Sun 23 Apr, Sun 4 June (2-5.30). Adm £6, chd free. Home-made teas. For other opening times and information, please phone, email or visit garden website.**
Approx 1½ acres, full of structure and design. Garden is divided into rooms inc rill garden, gravel garden, rose garden, herbaceous borders, pleached lime allee and potager. Created over last 34+ yrs by present owners and constantly evolving. New potting shed and greenhouse and wildlife pond added since 2018. Visitors welcome from April - June (groups of 15+), please see website.

19 NEW **1 COBDEN VILLAS**
Walkley Wood, Nailsworth, Stroud, GL6 0RT. Sue Ratcliffe. *Parking in Nailswoth town centre with a 10 min walk, L at the Britannia pub onto Horsley Rd; R onto Pike Lane. There is limited on road parking in Pike Lane/Meadow Bank.* **Sat 6, Sun 7 May, Sat 3, Sun 4 June (10-4). Adm £5, chd free. Tea, coffee & cakes.**
An organic and wildlife friendly garden, divided into distinct areas. A small mixed woodland area provides separation from the adjacent lane with underplanting of shade loving native and unusual plants. Terraced beds and borders adopt cottage garden planting ethos with extensive ground cover and shrubs. A rill and small pond provides a home for frogs and dragonflies.

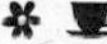

20 COTSWOLD FARM
Duntisbourne Abbots, Cirencester, GL7 7JS. John & Sarah Birchall, www.cotswoldfarmgardens.org.uk. *5m NW of Cirencester off old A417. From Cirencester L signed Duntisbourne Abbots Services, R and R underpass. Drive ahead. From Gloucester L signed Duntisbourne Abbots Services. Pass Services. Drive L.* **Sat 18, Sun 19 Feb (11-3). Light refreshments. Sat 3, Sun 4 June, Sat 2, Sun 3 Sept (2-5). Home-made teas. Adm £7.50, chd free. 2024: Sat 17, Sun 18 Feb. Donation to A Rocha.**
This beautiful Arts & Crafts garden overlooks a quiet valley on descending levels with terraces designed by Norman Jewson in the1930s. Enclosed by Cotswold stone walls and yew hedges, the garden has year-round interest inc a snowdrop collection with over 80 varieties. The terraces, shrub garden, herbaceous borders and bog garden are full of scent and colour from spring to autumn. Rare orchid walks. Picnics welcome. Wheelchair access to main terrace only (no wheelchair access to WCs).

21 DAGLINGWORTH HOUSE
Daglingworth, Cirencester, GL7 7AG. David & Henrietta Howard, 07970 122122, ettajhoward@gmail.com. *3m N of Cirencester off A417/419. House with blue gate beside church in centre of Daglingworth, at end of No Through Road.* **Visits by arrangement 24 Apr to 30 Aug for groups of 5 to 20. Adm £8, chd free. Light refreshments for groups only.**
Walled garden, temple, grotto, and pools. Classical garden of 2½ acres, with humorous contemporary twist. Hedges, topiary shapes, herbaceous borders. Pergolas, grass garden, wildflower meadow, woodland, cascade and mirror canal. Sunken garden. Pretty Cotswold village setting beside church. Visitor comment: 'It breaks every rule of gardening - but it's wonderful!'.

22 DAYLESFORD HOUSE
Daylesford, GL56 0YG. Lord & Lady Bamford. *5m West of Chipping Norton. Off A436 between Stow-on-the-Wold & Chipping Norton.* **Wed 19 Apr (1-5). Adm £6, chd free. Home-made teas.**
Magnificent C18 landscape grounds created in 1790 for Warren Hastings, greatly restored and enhanced by present owners under organic regime. Lakeside and woodland walks within natural wildflower meadows. Large formal walled garden, centred around orchid, peach and working glasshouses. Trellised rose garden. Collection of citrus within period orangery. Secret garden, pavilion, formal pools. Very large garden with substantial distances. The owners of Daylesford House have specifically requested that photographs are NOT taken in their garden or grounds. No dogs allowed except guide dogs. Teas cash only. Partial wheelchair access.

23 DOWNTON HOUSE
Gloucester St, Painswick, GL6 6QN. Ms Jane Kilpatrick, 01452 813861, info@janekilpatrick.co.uk. *4m N of Stroud. Entry to garden is via Hollyhock Lane only. Please note: NO cars in Lane. Park in Stamages Lane village car park off A46 below church, or in Churchill Way, 1st L off Gloucester St B4073.* **Mon 6, Tue 7 Feb (10.30-3). Adm £5, chd free. Light refreshments. Coffee & home-made cake from Lucy and her Horsebox. Visits also by arrangement 1 Feb to 15 Sept for groups of 8 to 20. Weekdays only.**
The owner is co-author of 'The Galanthophiles : 160 Years of Snowdrop Devotees' (2018, Orphans Publishing). The walled ⅓ acre garden in the centre of Painswick provides the perfect setting for her collection of Galanthus cultivars. In addition, naturalised drifts of *Galanthus nivalis* and *Galanthus atkinsii* flourish amongst winter-flowering shrubs, cyclamen, *Iris reticulata* and Polystichum ferns.

24 EAST COURT
East End Road, Charlton Kings, Cheltenham, GL53 8QN. Ben White. *The garden will be signed off the A40/ East End Road junction. Parking in the Balcarras School car park, opp the garden, will be signposted.* **Sun 27, Mon 28 Aug (10-6). Adm £7, chd free. Home-made teas.**
A garden on 2 levels. The upper around the 1806 house (not open) is traditionally styled with formal beds and lawns and 3 majestic purple

Woodchester Park House

beech trees some 200 yrs old. The in house designed 2½ acre lower garden – now 7 yrs old – is in complete contrast. Winding brick paths, swathes of herbaceous plantings towered over by arching Datisca, Miscanthus and Paulownia. An experimental south facing walled area enjoys exciting exotic and unusual tender perennials and annuals. The new pond has attracted a wide range of aquatic wildlife. The garden is always evolving, so visit us and share our excitement. Mostly accessible to wheelchairs – please ask for help where needed. Toilets are up a single step.

GROUP OPENING

25 EASTCOMBE AND BUSSAGE GARDENS

Eastcombe, Stroud, GL6 7EB. *3m E of Stroud. Maps available on the day from Eastcombe Village Hall GL6 7EB & Redwood GL6 8AZ. On street parking only. Tickets valid Sunday & Monday due to length of trail (2 miles).* **Sun 30 Apr, Mon 1 May (1.30-5.30). Combined adm £8, chd free. Light refreshments at Eastcombe Village Hall. Ice creams at Hawkley Cottage.**

CADSONBURY
Natalie & Glen Beswetherick.

17 FARMCOTE CLOSE
John & Sheila Coyle.

20 FARMCOTE CLOSE
Ian & Dawn Sim.

21 FARMCOTE CLOSE
Mr & Mrs Robert Bryant.

HAWKLEY COTTAGE
Helen & Gerwin Westendorp.

12 HIDCOTE CLOSE
Mr K Walker.

HIGHLANDS
Helen & Bob Watkinson.

1 JASMINE COTTAGE
Mrs June Gardiner.

1 THE LAURELS
Andrew & Ruth Fraser.

REDWOOD
Heather Collins.

NEW **WHITE HOUSE**
Jane Gandy.

WOODVIEW COTTAGE
Julian & Eileen Horn-Smith.

YEW TREE COTTAGE
Andy & Sue Green.

Medium and small gardens in a variety of styles and settings within this picturesque hilltop village location, with its spectacular views of the Toadsmoor Valley. In addition, one large garden is located in the bottom of the valley, approachable only on foot as are some of the other gardens. Full descriptions of each garden and those with wheelchair access can be found on the National Garden Scheme website. Please show any pre-booked tickets at village hall or Redwood. WC at village hall and Hawkley Cottage. Plants for sale at village hall and some gardens.

26 20 FORSDENE WALK

Coalway, Coleford, GL16 7JZ. Pamela Buckland, 01594 837179. *From Coleford take Lydney/ Chepstow Rd at T-lights. L after police station ½m up hill turn L at Xrds then 2nd R (Old Road) straight on at minor Xrds then L into Forsdene Walk.* **Visits by arrangement May to Sept for groups of up to 15. Adm £3.50. Home-made teas.**
Corner garden filled with interest and design ideas to maximise smaller spaces. A series of interlinking colour themed rooms, some on different levels. Packed with perennials, grasses and ferns. A pergola, small man-made stream, fruit and vegetables and pots in abundance on gravelled areas.

27 FORTHAMPTON COURT

Forthampton, Tewkesbury, GL19 4RD. Alan & Anabel Mackinnon. *W of Tewkesbury. From Tewkesbury A438 to Ledbury. After 2m turn L to Forthampton. At Xrds go L towards Chaceley. Go 1m turn L at Xrds.* **Sat 3, Sun 4 June (12.30-4.30). Adm £7, chd free. Home-made teas.**
Charming and varied garden surrounding N Gloucestershire medieval manor house (not open) within sight of Tewkesbury Abbey. Inc borders, lawns, roses and magnificent Victorian vegetable garden. Disabled drop off at entrance, some gravel paths and uneven areas.

28 ◆ THE GARDEN AT MISERDEN

Miserden, nr Stroud, GL6 7JA. Mr Nicholas Wills, 01285 821303, hello@miserden.org, www.miserden.org. *6m NW of Cirencester. Leave A417 for Birdlip, drive through Whiteway & follow signs for Miserden.* **For NGS: Fri 19 May (10-5). Adm £9, chd free. Light refreshments at The Glasshouse Café. For other opening times and information, please phone, email or visit garden website.**
A timeless walled garden designed in the C17 with a wonderful sense of peace and tranquillity. Known for its magnificent mixed borders and Lutyens' Yew Walk and quaint grass steps; there is also an ancient mulberry tree, enchanting arboretum and stunning views across the Golden Valley. Potting Shed Shop, Glasshouse Café, a variety of workshops, pop-up restaurant. Afternoon Tea experiences in the garden and woodland. Routes around the garden are gravel or grass, there are alternative routes to those that have steps. Disabled WC at the café.

29 THE GATE

80 North Street, Winchcombe, GL54 5PS. Vanessa Berridge & Chris Evans, 01242 609535, vanessa.berridge@sky.com. *Winchcombe is on B4632 midway between Cheltenham & Broadway. Parking behind Library in Back Lane, 50 yds from The Gate. Entry via Cowl Lane.* **Sat 22 Apr (2-6). Adm £4, chd free. Home-made teas. Visits also by arrangement Apr to Sept for groups of 5 to 25. Adm for By Arrangement visits £9 inc home-made tea.**
Compact cottage-style garden planted with bulbs in spring, and with summer perennials, annuals, climbers and herbs in the walled courtyard of C17 former coaching inn. Also a separate, productive, walled kitchen garden with espaliers and other fruit trees. Wheelchair access to most areas of the garden; other areas are partially visible from negotiable paths.

30 GREEN BOUGH

Market Lane, Greet, Winchcombe, GL54 5BL. Mary & Barry Roberts, 07966 528646, barryandmary@gmail.com. *1¼m N of Winchcombe. From Winchcombe take B4078. After railway bridge, R into Becketts Lane, immed L into Market Lane. Garden on R at Mill Lane junction.* **Visits by arrangement Apr to Sept for groups of up to 10. Parking available on driveway for 3 cars & roadside parking in Market Lane. Adm £4, chd free. Home-made teas £4 per person.**

Small country garden, informally planted for all seasons starting in early spring with massed bulbs, flowers, shrubs and trees surrounding the house. Many of the plants are either grown from seed or propagated from cuttings by the owner, to give generous drifts of colour in wide mixed borders and in pots and planters on the terraces.

31 GREENFIELDS, BROCKWEIR COMMON

Brockweir, Chepstow, NP16 7NU. Jackie Healy, 07747 186302, jackie@greenfields.garden, www.greenfields.garden. *Located between Chepstow & Monmouth. A446 from M'mouth. Go thru Llandogo, L to Brockweir. From Chepstow, thru Tintern, R to B'weir. Go over bridge, pass pub up hill, 1st L, follow lane to fork, L at fork. 1st property on R. No coaches.* **Sun 28 May (1-5). Adm £5, chd free. Home-made teas. Visits also by arrangement 1 June to 28 July for groups of up to 40.**

Greenfields is a 1½ acre plantsperson's gem of a garden set in the beautiful Wye Valley AONB. The garden features many mature trees and numerous rare/unusual plants and shrubs planted as discrete gardens. Greenfields is the passion and work of 'Head Gardener' Jackie, who has a long interest in the propagation of plants. Greenfields has been featured in Garden Answers magazine. Some areas of the garden are not accessible by wheelchair, and the WC is not wheelchair accessible.

32 GREENFIELDS, LITTLE RISSINGTON

Cheltenham, GL54 2NA. Mrs Diana MacKenzie-Charrington. *On Rissington Rd between Bourton-on-the-Water & Little Rissington, opp turn to Grt Rissington (Leasow Lane). SatNav using postcode does not take you to house.* **Sun 9 July (2-5). Adm £6, chd free. Home-made teas.**

The honey coloured Georgian Cotswold stone house sits in 2 acres of garden, created by current owners over last 22 yrs. Lawns are edged with borders full of flowers and flowering bulbs. A small pond and stream overlook fields. Bantams roam freely. Mature apple trees in wild garden, greenhouse in working vegetable garden. Regret no dogs. Partial wheelchair access.

33 NEW 11 HATHERLEY COURT ROAD

Cheltenham, GL51 3AQ. Mike & Rosemary Richards, 07884 493823, gaiatraditionalgardeners@gmail.com. *Hatherley. Next to Hatherley Court Road entrance to Hatherley Park.* **Sun 21 May (2-4). Adm £5, chd free. Pre-booking essential, please visit www.ngs.org.uk for information & booking. Visits also by arrangement 1 May to 18 May for groups of up to 20.**

A large urban garden developed over the last 25 yrs by the current owners. Their vision is to create an informal nature lovers' garden where there is tranquillity away from the town and a time for mental refreshment. The space consists of paths, herbaceous border, pond, wildflower spread, silver birch copse and lush green vegetation of the dry river bed. Refreshments are available at 'The Butterfly Box,' Hatherley Park 9-5pm everyday. Although the far area of the garden is inaccessible for wheelchairs, the raised decking area gives an overall view of the garden.

34 ◆ HIGHNAM COURT

Highnam, Gloucester, GL2 8DP. Mr R J Head, 01684 292875 (Mike Bennett), mike.highnamcourt@gmail.com, www.HighnamCourt.co.uk. *2m W of Gloucester. On A40/A48 junction from Gloucester to Ross or Chepstow. At this r'about take exit at 3 o'clock if coming from Gloucester direction. Do NOT go into Highnam village.* **For NGS: Sun 2 Apr, Sun 7 May, Sun 4 June, Sun 2 July, Sun 6 Aug, Sun 3 Sept (11-4.30). Adm £5, chd free. Light refreshments. For other opening times and information, please phone, email or visit garden website. Donation to other charities.**

40 acres of Victorian landscaped gardens surrounding magnificent Grade I house (not open), set out by artist Thomas Gambier Parry. Lakes, shrubberies and listed Pulhamite water gardens with grottos and fernery. Exciting ornamental lakes, and woodland areas. Extensive 1 acre rose garden and many features, inc numerous wood carvings. Group visits for 25 or more on weekdays by prior arrangement.

35 HODGES BARN

Shipton Moyne, Tetbury, GL8 8PR. Mr & Mrs N Hornby, www.hodgesbarn.com. *3m S of Tetbury. On Malmesbury side of village.* **Sun 11, Mon 12 June (2-6). Adm £7, chd free. Home-made teas at the Pool House.**

Very unusual C15 dovecote converted into family home. Cotswold stone walls host climbing and rambling roses, clematis, vines, hydrangeas and together with yew, rose and tapestry hedges create formality around house. Mixed shrub and herbaceous borders, shrub roses, water garden, woodland garden planted with cherries and magnolias. Vegetable and picking flower garden.

36 HOME FARM

Newent Lane, Huntley, GL19 3HQ. Mrs T Freeman, 01452 830210, torillfreeman@gmail.com. *4m S of Newent. On B4216 ½m off A40 in Huntley travelling towards Newent.* **Sun 29 Jan, Sun 12 Feb, Sun 5 Mar, Sun 2, Sun 30 Apr (11-4). Adm £4, chd free. 2024: Sun 28 Jan, Sun 11 Feb. Visits also by arrangement Jan to Apr.**

Set in elevated position with exceptional views. 1m walk through woods and fields to show carpets of spring flowers. Enclosed garden with fern border, sundial and heather bed. White and mixed shrub borders. Stout footwear advisable in winter. Two delightful cafés within a mile.

37 HOOKSHOUSE POTTERY

Hookshouse Lane, Tetbury, GL8 8TZ. Lise & Christopher White, www.hookshousepottery.co.uk. *2½m SW of Tetbury. Follow signs from A433 at Hare & Hounds Hotel, Westonbirt. Alternatively take A4135 out of Tetbury towards Dursley & follow signs after ½m on L.* **Daily Sat**

27 May to Wed 31 May (11-5.30). Fri 16, Fri 30 June (1.30-5.30). Adm £5, chd free. Home-made teas.
Garden offers a combination of long perspectives and intimate corners. Planting inc wide variety of perennials, with emphasis on colour interest throughout the seasons. Herbaceous borders, new woodland garden and flower meadow, water garden containing treatment ponds (unfenced) and flowform cascades. Sculptural features. Kitchen garden with raised beds, orchard. Run on organic principles. Pottery showroom with hand thrown woodfired pots inc frost proof garden pots. Art and craft exhibition (May 27th to May 31st.) inc garden furniture and sculptures. Garden games and treehouse. Mostly wheelchair accessible.

38 KEMBLE HOUSE

Kemble, Cirencester, GL7 6AD. Jill Kingston, 07798 830287, kingsjill50@gmail.com. *Approaching Kemble on the A429 from Cirencester, turn L onto School Rd then R onto Church Rd, pass Kemble Church on L. Kemble House is the next house on L.* **Wed 5 Apr (2-5). Adm £5, chd free. Visits also by arrangement 5 Apr to 23 June.**
A landscaped garden with many tall lime trees. Herbaceous borders line the lawns. The main one in front of the house is a grass tennis court that was laid in the 1880s. There is a walled garden with many fruit trees and two rose gardens. Two paddocks surround the property with Hebridean sheep. Wheelchair access possible, gravel pathways and some steps.

39 ◆ KIFTSGATE COURT

Chipping Campden, GL55 6LN. Mr & Mrs J G Chambers, 01386 438777, info@kiftsgate.co.uk, www.kiftsgate.co.uk. *4m NE of Chipping Campden. Opp Hidcote NT Garden.* **For NGS: Mon 10 Apr (2-6). Home-made teas. Mon 14 Aug (12-6). Cream teas. Adm £11, chd £3.50. For other opening times and information, please phone, email or visit garden website.**
Magnificent situation and views, many unusual plants and shrubs, tree peonies, hydrangeas, abutilons, species and old-fashioned roses inc largest rose in England, *Rosa filipes* 'Kiftsgate'.

40 KIRKHAM FARM

Upper Slaughter, Cheltenham, GL54 2JS. Mr & Mrs John Wills. *On road between Lower Slaughter & Lower Swell. Opp farm buildings on roadside. 1½ m W of Fosseway A429 & SW of Stow on the Wold.* **Sun 2 July (11-5). Adm £6, chd free. Tea. Delicious home-made cakes.**
A country garden overlooking lovely views with several mixed borders that always have a succession of colour. Gravel gardens, trees and shrubs and a hidden pool garden, raised beds, cutting beds and a developing wildflower bank give plenty of interest. Teas are served in our beautifully restored stone barn. The majority of the garden is accessible.

41 LANGFORD DOWNS FARM

Langford Downs Farm, nr Lechlade, GL7 3QL. Mr & Mrs Gavin MacEchern. *Access from A361 via layby behind copse - 6m from Burford towards Lechlade on the R OR 2m from Lechlade towards Burford on the L (NOT in Langford).* **Sun 4 June (2-6). Adm £7, chd free. Home-made teas.**
Cotswold house built in 2009 and new garden created from blank canvas. Extensive tree planting. Good sized garden comprising mixed tree, shrub and herbaceous borders. Traditional hedges comprising arches and windows and crab apple espalier. Extensive cut flower and vegetable garden, interesting courtyard with raised nursery beds. Cotswold pond with natural spring, bug hotels, magic garden. All areas wheelchair friendly.

42 LECKHAMPTON COURT HOSPICE

Church Road, Leckhampton, Cheltenham, GL53 0QJ. www.sueryder.org/care-centres/hospices/leckhampton-court-hospice. *2m SW of Cheltenham. From Church Rd take driveway by church signed Sue Ryder Leckhampton Court Hospice & follow parking signs.* **Sun 2 July (11-4). Adm £5, chd free. Light refreshments. Visitors welcome to picnic on our back lawn (bring your own chairs & picnic blankets).**
Set within this Grade II* listed medieval estate, the informal gardens at Leckhampton Court Hospice surround the buildings combining lawns and planted beds. Feature garden designed by Peter Dowle, RHS Chelsea gold medal winner. Highlights inc woodland walk around lake and into woodland, numerous protected mature trees and a terrace with magnificent views across Cheltenham towards Malvern. A golden maple tree planted by His Majesty King Charles III to commemorate his 70th birthday is situated along the Woodland Walk. Wheelchair access in some areas: from back of reception to Sir Charles Irving terrace; woodland walk; main courtyard.

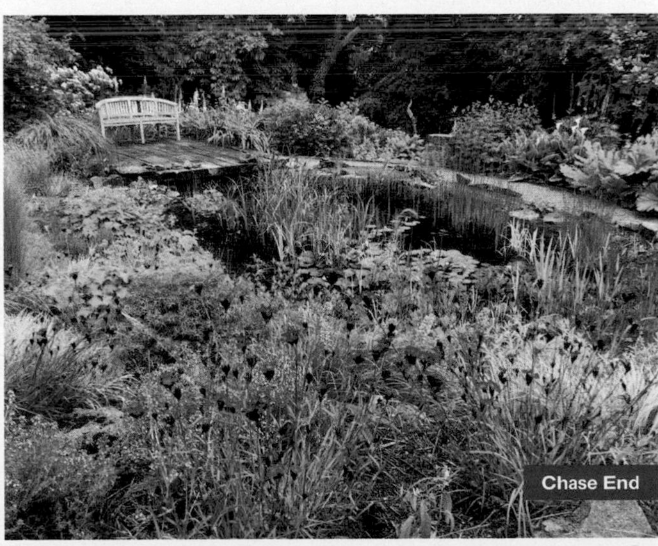

Chase End

43 LITTLE ORCHARD
Slad, Stroud, GL6 7QD. Rod & Terry Clifford, 01452 813944, terryclifford.tlc@gmail.com. *2m from Stroud, 10 m from Cheltenham. Last property on L in Slad village before leaving 30mph speed limit travelling from Stroud to Birdlip on B4070. SatNav may not bring you directly to property. Parking on verge opp.* **Sat 27, Sun 28 May (10.30-5). Adm £5, chd free. Home-made teas. Cider tasting available. Visits also by arrangement May to Sept for groups of 5 to 30. Talk on evolution of the house, garden & Slad Valley Cider.**
One acre garden created from wilderness on a challenging, steeply sloping site using many reclaimed materials, stonework and statuary. Multiple terraces and garden 'rooms' enhanced with different planting styles and water features to complement the natural surroundings. Many seating areas with stunning views. Children's Trail. Cidery offering tastings. Exhibition and sale of local artists' work. Wheelchair access possible but challenging due to the severity of slopes and steps. Please phone for further details.

44 LORDS OF THE MANOR HOTEL
Upper Slaughter, Cheltenham, GL54 2JD. Mike Dron (Head Gardener), 01451 820243, reservations@lordsofthemanor.com, www.lordsofthemanor.com. *2m W of Bourton on the Water. From Fosse way follow signs for the Slaughters from close to Bourton-on-the-Water (toward Stow). From B4077 (Stanway Hill road) coming from Tewkesbury direction, follow signs 2m after Ford village.* **Wed 26 Apr (10-3). Adm £7, chd £5. Wed 28 June, Wed 6 Sept (10-3). Combined adm with The Manor £10, chd £5. Refreshments are available from the hotel, from light snacks to afternoon tea. Pre booking advised.**
Classic Cotswold country garden with a very English blend of semi-formal and informal, merging beautifully with the surrounding landscape. Beautiful walled garden, established wildflower meadow, the River Eye. The herb garden and stunning bog garden were originally designed by Julie Toll around 2012. Wildlife garden, croquet lawn, courtyard and Victorian skating pond. Steps in walled garden & bog garden. Regret that due to the age & layout of the building there are no wheelchair-friendly facilities within the hotel.

45 NEW MADAMS
Eden's Hill, Upleadon, Newent, GL18 1EE. Phil & Pauline Cooke, www.upleadon-village.co.uk/madams-ngs. *3m ENE of Newent, 10m NNW of Gloucester. From Newent Xrds take B4216 Dymock, 1st R Tewkesbury Rd for 2.6m. At Xrds L to Eden's Hill. From Gloucester A40 to Ross-on-Wye, R B4215 to Newent, 3.4m R onto Gloucester Rd. 2.8m to Xrds.* **Sat 10, Sun 11 June (1-6). Adm £5, chd free. Home-made teas.**
A 2 acre country woodland garden with far reaching views. The garden is still developing with a variety of herbaceous and mixed borders and a small wildflower area with an abundance of common spotted orchids, small formal pond and vegetable patch. Many mature trees and areas left wild for nature. Partial wheelchair access possible with gravel on drive, grass slopes and some steps. Limited disabled parking at the house.

46 NEW THE MANOR
Upper Slaughter, Cheltenham, GL54 2JG. James Bodenstedt. *From B4068 follow signs for the Slaughters then onto Becky Hill, last property on L. From A429 go through Lower Slaughter over the River Eye onto Becky Hill, 1st property on R.* **Wed 28 June, Wed 6 Sept (10-5). Combined adm with Lords of the Manor Hotel £10, chd £5. Light refreshments.**
The 8 acre garden is part of a C17 manor house consisting of formal and informal gardens inc a rose garden, herbaceous and shrub borders, pergola walk, an arboretum with a large pond, orchard and kitchen and herb garden. The various gardens are on different levels and are currently going through a complete restoration. There are a mixture of gravel and stone paths, steps and lawns. Wheelchair access can be difficult in places due to step access only. However, what cannot be accessed directly can be viewed from the main lawn.

47 THE MANOR
Little Compton, Moreton-In-Marsh, GL56 0RZ. Reed Foundation (Charity), www.reedbusinessschool.co.uk. *Next to church in Little Compton. ½m from A44 or 2m from A3400. Follow signs to Little Compton, then yellow NGS signs.* **Sun 20 Aug (1-5). Adm £6, chd free. Home-made teas in churchyard next door.**
Extensive Arts & Crafts garden with meadow and arboreta. Footpaths around our fields, playground between car park and garden. Croquet, and tennis free to play. Children and dogs welcome.

48 MONKS SPOUT COTTAGE
Glasshouse Hill, May Hill, Longhope, GL17 0NN. Nigel & Jane Jackson, 07767 858295, monksspout@icloud.com. *1m from NT May Hill, in the hamlet of Glasshouse. Access via lane (which is also a public footpath called the Wysis Way) immed adjacent to Glasshouse Inn.* **Visits by arrangement Feb to July for groups of up to 40. Adm £7.50, chd free. Hot drinks & soup/roll in the winter. Hot drinks & cake in the spring/summer.**
The ⅔ acre garden is adjacent to Castle Hill Wood which is the backdrop for a mix of herbaceous borders, lawns and ponds, with spring snowdrops, greenhouse, stream and large display of insectivorous plants, mostly planted outside but some in the greenhouse. With several mature trees and recently planted acers, there is a mix of shade and sun creating both damp and dry planting opportunities. Garden sculptures. Adjacent public footpaths through the wood (not part of the garden) where visitors can see the remains of a ringwork castle dating from C12.

49 MOOR WOOD
Woodmancote, GL7 7EB. Mr & Mrs Henry Robinson, 01285 831397, henry@moorwoodhouse.co.uk, www.moorwoodroses.co.uk. *3½m NW of Cirencester. Turn L off A435 to Cheltenham at N Cerney, signed Woodmancote 1¼m; entrance in village on L beside lodge with white gates.* **Sun 25 June (2-6). Adm £5, chd free. Home-made teas. Visits also by arrangement 15 June to 2 July. Coaches cannot access**

drive, so will mean a 150 metre walk.
2 acres of shrub, orchard and wildflower gardens in beautiful isolated valley setting. Holder of National Collection of Rambler Roses. June 20th to 30th is usually the best time for the roses.

50 NEW 39 NEVEN PLACE

Gloucester, GL1 5NF. Mr Chris & Mrs Jenny Brooker. *Approx 3m N of J12 M5. Follow A38 to Bristol Rd, then into Tuffley Ave, Tuffley Cres, Manu Marble Way to Neven Place. On street parking.* **Sat 12, Sun 13 Aug (10.30-4). Adm £3, chd free. Home-made teas.**
A jungle/exotic small scale garden on a relatively new development, featuring tree ferns, palms, bamboos, bananas, a rill, raised beds and other interesting features.

51 OAK HOUSE

Greenway Lane, Gretton, Cheltenham, GL54 5ER. Paul & Sue Hughes, 01242 603990, ppphug@gmail.com. *In centre of Gretton. Signed Gretton from B4077, approx 3m from A46 Teddington Hands r'about. Greenway Lane 300m R past railway bridge. Parking on main road with short walk to garden.* **Wed 14, Sat 17 June (11-4.30). Adm £5, chd free. Home-made teas. Visits also by arrangement Apr to Oct for groups of up to 20.**
A one acre secret garden divided into rooms. Gradually developed over the last 30 yrs. Many places to sit and enjoy the scent of honeysuckle, philadelphus and over 50 varieties of roses. Wildflower meadow, gazebo and summerhouse. Formal lily pond and small wildlife pond. Some quirky features. You may even see a fairy.

52 THE OLD RECTORY, QUENINGTON

Church Rd, Quenington, Cirencester, GL7 5BN. Mr & Mrs David Abel Smith, www.queningtonoldrectory.com. *8m NE of Cirencester. Opp St Swithins Church at bottom of village. Garden well signed once in village.* **Sun 3 Sept (2-5). Adm £8, chd free. Home-made teas.**
On the banks of the mill race and the River Coln, this is an organic garden of great variety. Mature trees, large vegetable garden, herbaceous borders, shade garden, pool and bog gardens. Homegrown plants for sale. Permanent sculpture collection in the gardens. The majority of the garden is accessible by wheelchair.

53 THE OLD VICARAGE

Murrells End, Hartpury, GL19 3DF. Mrs Carol Huckvale. *5m NW of Gloucester. From Over r'about on A40 N Gloucester bypass, take A417 NW to Hartpury (Ledbury Rd). After Maisemore, turn L at signs for Hartpury College. House 1m on R. Follow signs for parking.* **Fri 19, Sat 20, Sun 21 May (1-4.30). Adm £5, chd free. Home-made teas.**
Tranquil garden of about 2 acres, with yew oval, mature trees, steps to croquet lawn, mixed borders around main lawn, potager, fruit trees. Work in progress developing wildflower meadow area and dry, shady woodland walk. The neighbouring vegetable garden may also be open to visitors during the 3 days of opening.

54 ◆ OXLEAZE FARM

Between Eastleach & Filkins, Lechlade, GL7 3RB. Mr & Mrs Charles Mann, 07786 918502, chipps@oxleaze.co.uk, www.oxleaze.co.uk/oxleaze-garden. *5m S of Burford, 3m N of Lechlade off A361 to W (signed Barringtons). Take 2nd L then follow signs.* **For NGS: Sun 4 June (2-6). Adm £6, chd free. Coffee/tea & home-made cakes available for NGS opening & visiting groups. For other opening times and information, please phone, email or visit garden website.**
Set among beautiful traditional farm buildings, plantsperson's good size garden created by owners over 35 yrs. Year-round interest; mixed borders, vegetable potager, decorative fruit cages, pond/bog garden, bees, potting shed, meadow, and topiary for structure when the flowers fade. Garden rooms off central lawn with corners in which to enjoy this organic Cotswold garden. Groups welcome by arrangement. Mostly wheelchair access.

55 PASTURE FARM

Upper Oddington, GL56 0XG. Mr & Mrs John LLoyd, 07850 154095/01451 830203, ljmlloyd@yahoo.com. *3m W of Stow-on-the-Wold. Just off A436, midway between Upper & Lower Oddington.* **Sun 28, Mon 29 May (11-6). Adm £6, chd free. Home-made teas. Visits also by arrangement 28 May to 29 May for groups of up to 20.**
Informal country garden developed over 30 yrs by current owners. Mixed borders, topiary, orchard and many species of trees. Gravel garden and rambling roses in 'the ruins'. A concrete garden and wildflower area leads to vegetable patch. Large spring-fed pond with ducks. Also bantams, chickens, black Welsh sheep and Kunekune pigs. A special and very large plant sale in aid of Kate's Home Nursing. Public footpath across 2 small fields arrives at C11 church, St Nicholas, with doom paintings, set in ancient woodlands. Truly worth a visit (See Simon Jenkins' Book of Churches). Mostly wheelchair accessible.

56 THE PATCH

Hollywell Lane, Brockweir, Chepstow, NP16 7PJ. Mrs Immy Lee, 07801 816340, thepatch.brockweir@gmail.com, www.thepatchbrockweir.com. *6.7m N of Chepstow & 10.6m S of Monmouth, off A466, across Brockweir Bridge. From Monmouth & Chepstow direction follow satnav to Brockweir Bridge, then yellow signs. From Coleford/Gloucester direction follow SatNav to Hewelsfield Xrds, then yellow signs.* **Sun 10 Sept (12-5). Combined adm with Clouds Rest £8, chd free. Home-made teas. Visits also by arrangement 30 May to 30 July for groups of up to 40. Possibility of combining visits with Greenfields, Brockweir.**
Rural ¼ acre garden with far reaching views across the Wye Valley, containing 60+ repeat flowering roses and a variety of shrubs and perennials, providing interest year-round. The garden open day is in combination with Clouds Rest. The gardens are linked by an easy drive or a 15 min stony walk along the Offa's Dyke path. Ample parking at both gardens. Partial wheelchair access.

57 PEAR TREE COTTAGE

58 Malleson Road, Gotherington, nr Cheltenham, GL52 9EX. Mr & Mrs E Manders-Trett, 01242 674592, mmanderstrett@gmail.com. *4m N of Cheltenham. From A435, travelling N, turn R into Gotherington at garage 1m after Bishop's Cleeve bypass. Garden on L 100yds past Shutter Inn.* **Visits by arrangement 12 Mar to 2 July for groups of up to 30. Light refreshments. Tea/coffee & cake available.**

Mainly informal country garden of approx ½ acre with pond and gravel garden. Herbaceous borders, trees and shrubs surround lawns and seating areas. Wild garden and orchard lead to greenhouses, vegetable garden and beehives. Spring bulbs, early summer perennials and shrubs particularly colourful. Gravel drive and several shallow steps can be overcome for wheelchair users with prior notice.

58 NEW PENNY LEAZE

Sturmyes Road, France Lynch, Stroud, GL6 8LU. Pam Meecham. *Between Stroud and Cirencester. From Old Neighbourhood turn into Abnash, then Burrcombe Rd, follow thru to Highfield Way, R into Lynch Rd, past King's Head Pub, take L fork into Sturmyes Rd, follow for ¼ m. Limited parking.* **Sun 14, Mon 15 May, Sun 17, Mon 18 Sept (12-4). Combined adm with Pontings Farm £8, chd free. Home-made teas.**

Seasonal planting for pollinators and wildlife in mind. Nestled into a south-east facing slope, where we have attempted to balance wildness, formal gardening and tiny orchard, while blending in with the natural shape and character of the land. The meandering paths lead to tranquil, shady, seating areas.

59 PERRYWOOD HOUSE

Longney, Gloucester, GL2 3SN. Gill & Mike Farmer. *7m SW of Gloucester, 4m W of Quedgeley. From N: R off B4008 at Tesco r'about. R at 2nd r'about then straight over 2 r'abouts then signed. From S: L off A38 in Moreton Valence to Epney/Longney, over canal bridge, R at T junction then signed.* **Sat 1, Sun 2 July (11-5). Adm £5, chd free. Home-made teas.**

One acre plant lover's garden in the Severn Vale surrounded by open farmland. Established over 20 yrs ago an informal country garden with mature trees and shrubs, inc colourful herbaceous borders, a small pond, and planted containers. There are plenty of places to sit and enjoy the garden. Lots of interesting plants for sale. All areas accessible with level lawns and gravel drives. Disabled parking available.

60 NEW PONTINGS FARM

France Lynch, Stroud, GL6 8LX. Mr & Mrs Randall. *Between Stroud and Cirencester. From Old Neighbourhood turn into Abnash, then Burrcombe Rd, follow thru to Highfield Way, R at T junction signed Avenis Green, follow Hillside lane ½ m to village, Pontings Farm on L before post box.* **Sun 14, Mon 15 May, Sun 17, Mon 18 Sept (12-4). Combined adm with Penny Leaze £8, chd free. Cream teas.**

One acre accessible hillside garden with views of the Golden Valley and beyond. Bee and wildlife friendly with wildflower meadow, pond, and orchard. Raised beds for vegetables and flowers. Our prairie style and gravel gardens thrive in our changing climate of hotter drier summers. The woodland is an ongoing project planted with birch trees, magnolias, Scots pines and hawthorn. There is a short walk of 10 mins between Pontings Farm and Penny Leaze. Directions will be signed through this pretty Cotswold village. Penny Leaze has very limited parking, and the lanes are narrow. Sat Nav can become inaccurate. Parking for both gardens in field, disabled parking at the garden. Majority of the garden is wheelchair/small scooter accessible (not Penny Leaze).

61 NEW 37 QUEEN'S ROAD

Queens Rd, Cheltenham, GL50 2LX. Geraldine & Richard Pinch. *4 mins walk from Cheltenham Spa train stn. Garden on L as you walk from the stn towards town. Free parking on Queens Rd & Christchurch Rd.* **Sun 17 Sept (1.30-5.30). Adm £4, chd free.**

A wildlife and wildflower-friendly urban garden with many shrubs and climbers. At the front there are shrub borders and a gravel garden. The rear garden slopes upward from the house and is accessed by steps. There are seven ponds, old fruit trees, a vine-house, raised beds planted with perennials and dwarf shrubs, a wild area and a brickery for alpines. Animal sculpture trail for children. Two cafés 5 mins walk away.

The Manor, Moreton-in-Marsh

62 RADNORS

Wheatstone Lane, Lydbrook, GL17 9DP. Mrs Mary Wood, 01594 861690, mary.wood37@btinternet.com. *In the Wye Valley on the edge of the Forest of Dean. From Lydbrook, go down through village towards the River Wye. At the T junction turn L into Stowfield Rd. Wheatstone Ln (300m) is 1st turning L after the white cottages. Radnors is at end of lane.* **Visits by arrangement 6 Apr to 31 Oct for groups of up to 10. Adm £5, chd free. We offer a range of home-made cakes with tea, coffee or home-made elderflower cordial.**

5 acre hillside woodland garden in AONB on bank above the River Wye. Focus on wildlife with naturalistic planting and weeds, some left for specific insects/birds. It has many paths, a wooded area, wildflower area, flower beds and borders, lawns, stumpery, fernery, vegetable beds and white garden. Of particular interest is the path along a disused railway line, and the summer dahlias.

63 RAMBLERS

Lower Common, Aylburton, Lydney, GL15 6DS. Jane & Leslie Hale. *1½m W of Lydney. Off A48 Gloucester to Chepstow Rd. From Lydney through Aylburton, out of de-limit turn R signed Aylburton Common, ¾m along lane.* **Sun 7 May (1.30-5). Adm £5, chd free. Home-made teas.**

Peaceful medium sized country garden with informal cottage planting, herbaceous borders and small pond looking through hedge windows onto wildflower meadow and mature apple orchard. Some shade loving plants and topiary. Large productive vegetable garden. Past winner of The English Garden magazine's Britain's Best Gardener's Garden competition.

64 RICHMOND PAINSWICK RETIREMENT VILLAGE

Stroud Road, Painswick, Stroud, GL6 6UL. Richmond Villages/ bupa. *Approx 5m E of Stroud. Just outside Painswick village.* **Thur 11, Sat 13, Thur 25, Sat 27 May, Thur 8, Sat 10, Thur 22, Sat 24 June (9-3). Adm £5, chd free. Light refreshments.**

Situated on the southern slopes of Painswick this 4 acre retirement village boasts formal lawns and borders planted for year-round interest. A varied mix of herbaceous and shrubs, with many areas of interest inc a wildflower meadow with fruit trees that combine to attract an abundance of wildlife. The gardens are a blaze of colour throughout. Car parking, WC, cafe serving tea/coffee plus light bites, and a restaurant. There are gentle slopes in wildflower meadow and around some areas of the village.

65 ROCKCLIFFE

Upper Slaughter, Cheltenham, GL54 2JW. Mr & Mrs Simon Keswick, www.rockcliffegarden.co.uk. *2m from Stow-on-the Wold. 1½m from Lower Swell on B4068 towards Cheltenham. Leave Stow-on-the-Wold on B4068 through Lower Swell. Continue on B4068 for 1½m. Rockcliffe is well signed on R.* **Wed 31 May, Wed 28 June (10-5). Adm £7.50, chd free. Home-made teas in car park. Donation to Kate's Home Nursing.**

Large traditional English garden of 8 acres inc pink garden, white and blue garden, herbaceous borders, rose terrace, large walled kitchen garden and greenhouses. Pathway of topiary birds leading up through orchard to stone dovecot. Featured in many books and magazines. 2 wide stone steps through gate, otherwise good wheelchair access. Regret no dogs.

66 ◆ SEZINCOTE

Moreton-in-Marsh, GL56 9AW. Mrs D Peake, 01386 700444, enquiries@sezincote.com, www.sezincote.co.uk. *3m SW of Moreton-in-Marsh. From Moreton-in-Marsh turn W along A44 towards Evesham; in 1½m (just before Bourton-on-the-Hill) turn L, by stone lodge with white gate.* **For NGS: Sun 9 July (11-5). Adm £7.50, chd £2.50. Home-made teas. For other opening times and information, please phone, email or visit garden website. Donation to a local charity.**

Exotic, oriental water garden by Repton and Daniell, with lake, pools and meandering stream, banked with massed perennials. Large semi-circular orangery, formal Indian garden, fountain, temple and unusual trees of vast size in lawn and wooded park setting. House in Indian manner designed by Samuel Pepys Cockerell. Garden on slope with gravel paths, so not all areas wheelchair accessible.

67 SHEEPEHOUSE COTTAGE

Stepping Stone Lane, Painswick, Stroud, GL6 6RX. Mr Russ & Mrs Jackie Herbert, 01452 813229, sheepehouse@hotmail.com. *3m N of Stroud off the A46 in Painswick. From A46 turn into Stamages Lane (signed Painswick Car Park). Pass car park, keep straight on Stamages Lane. Stepping Stone Lane to bottom of hill follow until you find parking signs.* **Sun 26 Mar, Sun 3 Sept (11-5). Adm £5, chd free. Light refreshments. Visits also by arrangement 1 Apr to 1 Oct for groups of 6 to 15.**

Lovely country, hillside garden with outstanding views across the Cotswolds. Variety of mixed borders and garden 'rooms' inc perennials, herbaceous and evergreens. Cottage garden with formal pond and arbour. Wildflower orchard with mown paths and many spring bulbs. Greenhouse and raised vegetable beds. Wildlife pond. Much of garden can be seen from the paved and level pathways around house. Other areas wheelchair and pushchair accessible across grassy slopes.

68 SOUTH LODGE

Church Road, Clearwell, Coleford, GL16 8LG. Andrew & Jane MacBean, 01594 837769, southlodgegarden@btinternet.com, www.southlodgegarden.co.uk. *2m S of Coleford. Off B4228. Follow signs to Clearwell. Garden on L of castle driveway. Please park on rd in front of church or in village. No parking on castle drive.* **Sat 1 Apr, Sun 7 May, Sat 10 June (1-5). Adm £4.50, chd free. Home-made teas. Visits also by arrangement 1 Apr to 24 June for groups of 15+.**

Peaceful country garden in 2 acres with stunning views of surrounding countryside. High walls provide a backdrop for rambling roses, clematis, and honeysuckles. Organic garden with large variety of perennials, annuals, shrubs and specimen trees with year-round colour. Vegetable garden, wildlife and formal ponds. Rustic pergola planted with English climbing roses and willow arbour in gravel garden. Gravel paths and steep grassy slopes. Assistance dogs only.

GROUP OPENING

69 STANTON VILLAGE GARDENS

Stanton, nr Broadway, WR12 7NE. *3m S of Broadway. Off B4632, between Broadway (3m) & Winchcombe (6m).* **Sun 11 June (1-6). Combined adm £8, chd free. Home-made teas in several gardens around the village. Donation to Village charities.**

A selection of gardens open in this picturesque, unspoilt Cotswold village. Many houses border the street with long gardens hidden behind. Gardens vary, from those with colourful herbaceous borders, established trees, shrubs and vegetable gardens to tiny cottage gardens. Some also have attractive, natural water features fed by the stream which runs through the village. Plants for sale. Free parking. Church also open. The Mount Inn is open for lunch. An NGS visit not to be missed in this gem of a Cotswold village. Regret gardens not suitable for wheelchair users due to gravel drives.

70 ◆ STANWAY FOUNTAIN & WATER GARDEN

Stanway, Cheltenham, GL54 5PQ. The Earl of Wemyss & March, 01386 584528, office@stanwayhouse.co.uk, www.stanwayfountain.co.uk. *9m NE of Cheltenham. 1m E of B4632 Cheltenham to Broadway rd on B4077 Toddington to Stow-on-the-Wold rd.* **For NGS: Sun 21 May, Sun 13 Aug (2-5). Adm £7, chd £3. Home-made teas in Stanway Tea Room. For other opening times and information, please phone, email or visit garden website.**

20 acres of planted landscape in early C18 formal setting. The restored canal, upper pond and fountain have recreated one of the most interesting Baroque water gardens in Britain. Striking C16 manor with gatehouse, tithe barn and church. The garden features Britain's highest fountain at 300ft, and it is the world's highest gravity fountain. It runs at 2.45pm and 4.00pm for 30 mins each time. Partial wheelchair access in garden, some flat areas, able to view fountain and some of garden. House is not suitable for wheelchairs.

71 ◆ SUDELEY CASTLE & GARDENS

Winchcombe, GL54 5JD. Lady Ashcombe, 01242 604244, enquiries@sudeley.org.uk, www.sudeleycastle.co.uk. *8m NE Cheltenham, 10 m from M5 J9. SatNavs use GL54 5LP. Free parking.* **For NGS: Fri 22 Sept (10.30-3). Adm £10, chd £6. Light refreshments. For other opening times and information, please phone, email or visit garden website.**

Sudeley Castle features 10 magnificent gardens, each with its own unique style and design. Surrounded by striking views of the Cotswold Hills, each garden reflects the fascinating 1000 yr history of the Castle. This series of elegant gardens is set among the Castle and atmospheric ruins and inc a knot garden, Queen's garden and Tudor physic garden. Admission is for garden only. Sudeley Castle remains the only private castle in England to have a queen buried within the grounds - Queen Katherine Parr, the last and surviving wife of King Henry VIII who lived and died in the castle. A circular route around the gardens is wheelchair accessible although some visitors may require assistance from their companion.

72 TRENCH HILL

Sheepscombe, GL6 6TZ. Celia & Dave Hargrave, 01452 814306, celia.hargrave@btconnect.com. *1½m E of Painswick. From Cheltenham A46 take 1st turn signed Sheepscombe, follow for approx 1½m. Or from the Butcher's Arms in Sheepscombe (with it on R) leave village and take lane signed for Cranham.* **Sun 12, Sun 19 Feb (11-4); Sun 19 Mar, Sun 9, Mon 10 Apr, Sun 7 May (11-6). Every Wed 7 June to 28 June (2-6). Sun 16 July, Sun 10 Sept (11-6). Adm £5, chd free. Home-made teas. Gluten & dairy free usually available. 2024: Sun 11, Sun 18 Feb. Visits also by arrangement 13 Feb to 31 Aug for groups of up to 35. If coming by coach this must be approved by the owners because of size.**

Approx 3 acres set in a small woodland with panoramic views. Variety of herbaceous and mixed borders, rose garden, extensive vegetable plots, wildflower areas, plantings of spring bulbs with thousands of snowdrops and hellebores, woodland walk, two small ponds, waterfall and larger conservation pond. Interesting wooden sculptures, many within the garden. Cultivated using organic principles. Children's play area. Mostly wheelchair accessible but some steps and slopes.

GROUP OPENING

73 TUFFLEY GARDENS

Tuffley Lane GL4 0DT, Gloucester, GL4 0DT. Martyn & Jenny Parker. *3m S Gloucester. Follow arrows off St. Barnabas R'about. On street parking.* **Sun 28 May (11-4). Combined adm £5, chd £1. Home-made teas in St Barnabas Church Hall.**

A number of mature suburban gardens of different styles and sizes. Some corner plots, lawned or gravelled, lots of colourful flowers, shrubs, trees, baskets and tubs. Circular route one mile in total. Close to Robinswood Hill Country park, 250 acres of open countryside with viewpoint, pleasant walks and waymarked nature trails.

74 NEW 11 UPPER WASHWELL

Painswick, Stroud, GL6 6QY. Stella Barber. *From A46 Cheltenham/ Stroud road take B4073 Gloucester St. At the junction turn R into Pullens Rd then L into Upper Washwell. Parking in Stamages Lane village car park.* **Sat 10 June (9-3). Adm £5, chd free.**

Delightful and secluded. This is a cottage style garden, inc a pond, gravelled area, and box hedging. With many interesting nooks and crannies.

75 ◆ UPTON WOLD

Moreton-in-Marsh, GL56 9TR. Mr & Mrs I R S Bond, www.uptonwold.co.uk. *4½m W of Moreton-in-Marsh on A44. From Moreton/Stow ½m past A424 turn R opp deer sign to road into fields then L at mini Xrds. From Evesham 1m past B4081 C/Campden Xrds turn L at end of stone wall to road into fields then as above.* **Sun 16 Apr (10-5). Adm £15, chd free. Home-made teas.**

One of the secret gardens of the Cotswolds, Upton Wold has commanding views, yew hedges, herbaceous walk, vegetable, pond and woodland gardens, and a labyrinth. An abundance of unusual and interesting plants, shrubs and trees. National Collections of juglans and pterocarya. A garden of interest to any garden and plant lover. Snowdrop walks from end of February to end of March. Details on website.

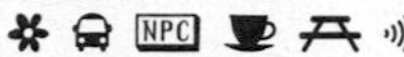

76 WEIR REACH

The Rudge, Maisemore, Gloucester, GL2 8HY. Sheila & Mark Wardle, weirreach@gmail.com. *3m NW of Gloucester. Turn into The Rudge by White Hart Pub. Parking 100yds from garden.* **Sun 11, Wed 14 June (11-5). Adm £5, chd free. Home-made teas. Visits also by arrangement in June for groups of 10 to 30.**

Country garden by River Severn. Approx 2 acres, half cultivated with herbaceous beds and mixed borders plus fruit and vegetable cages. Clematis and acers, stone ornaments, small sculptures, bonsai collection. Planted rockery with waterfall and stream connect 2 ponds. Large specimen koi pond borders patio. Meadow with specimen and fruit trees leading to river and country views.

77 WESTAWAY

Stockwell Lane, Cleeve Hill, Cheltenham, GL52 3PU. Liz & Ian Ramsay, 01242 672676, lizmramsay@gmail.com. *5m NW of Cheltenham. Off the B4632 Cheltenham to Winchcombe road at Cleeve Hill. Parking available adjacent to garden.* **Wed 26 Apr (2-5). Adm £5, chd free. Home-made teas. Visits also by arrangement in June for groups of 5 to 30.**

Hillside 1½ acre garden situated on the Cotswold escarpment with spectacular views across the Severn Vale. Interesting solutions to the challenges of gardening on a gradient, reflecting the local topography. Mixed shrub and herbaceous borders, bog garden, orchard, small arboretum and several wildflower areas. Landscaping inc extensive terracing with grass banks.

78 ◆ WESTONBIRT SCHOOL GARDENS

Tetbury, GL8 8QG. Holfords of Westonbirt Trust, 01666 881373, jbaker@holfordtrust.com, www.holfordtrust.com. *3m SW of Tetbury. Please enter through main school gates on A433 - some SatNavs will send you via a side entrance where there will be no access - main school gates only please.* **For NGS: Sun 16 July (11-4.30). Adm £7.50, chd free. For other opening times and information, please phone, email or visit garden website.**

20 acres. Former private garden of Robert Holford, founder of Westonbirt Arboretum. Formal Victorian gardens inc walled Italian garden now restored with early herbaceous borders and exotic border. Rustic walks, lake, statuary and grotto. Rare, exotic trees and shrubs. Beautiful views of Westonbirt House open with guided tours to see fascinating Victorian interior on designated days of the year. There are gravelled paths in some areas, grass in others and wheelchair users are limited to downstairs part of the house due to evacuation protocols.

79 NEW WINSTONE GLEBE

Croft Lane, Winstone, Cirencester, GL7 7LN. Amelia Baalack. *6.5 m N Cirencester. Leave A417 at Elkstone & Winstone turning L into Winstone. At the Xrds go straight on branch L & Winstone Glebe is on the L.* **Fri 9 June (10.30-2.30). Adm £6, chd free. Tea.**

The garden and grounds wrap around the house. Formally laid out topiary, a herbaceous border influenced by Gertrude Jekyll's planting style. The front lawn area blends flawlessly with the surrounding landscape thanks to the ha-ha. Restoration of the garden is in progress by the current owner, inc a newly established border of beautiful fragrant roses. The garden has 2 access points for wheelchair users & is on a gentle grass slope.

80 NEW WOODCHESTER PARK HOUSE

Nympsfield, Stonehouse, GL10 3UN. Robin & Veronica Bidwell. *Nr Nailsworth. Off the road joining Nailsworth & Nympsfield (Tinkley Lane). Around 3m from Nailsworth, turn R down next turning after the Thistledown campsite.* **Sun 9 July (12-5). Adm £6, chd free. Tea.**

This partially walled garden of approx 3 acres incorporates extensive herbaceous borders, a yew walk, a rose covered belvedere overlooking a large pond, a woodland garden, rose walk, vegetable garden and terrace. Wide variety of plants and settings. Wheelchair access to most parts of the garden but there are steep slopes to be negotiated.

81 WORTLEY HOUSE

Wortley, Wotton Under Edge, GL12 7QP. Simon & Jessica Dickinson, 01453 843174, jessica@wortleyhouse.co.uk. *On Wortley Rd 1m S of Wotton-under-Edge. Grand entrance on L as you enter Wortley coming from Wotton.* **Tue 25 Apr, Tue 20 June (2-5). Adm £15, chd free. Pre-booking essential, please visit www.ngs.org.uk for information & booking. Home-made teas included in the admission price. Visits also by arrangement 25 Apr to 1 Sept for groups of up to 30.**

A diverse garden of over 20 acres created during the last 30 yrs by the current owners. Inc a walled garden, pleached lime avenues, nut walk, potager, ponds, Italian garden, arbour, shrubberies and wildflower meadows. Strategically placed follies, urns and statues enhance extraordinary vistas throughout. The stunning surrounding countryside is incorporated into the garden with views up the steep valley that are such a feature in this part of Gloucestershire. Picnics are welcome, and a map and plant list are available. Wheelchair access to most areas, golf buggy also available.

9,500 patients and their families are being supported across four Horatio's gardens thanks to our annual donations

HAMPSHIRE

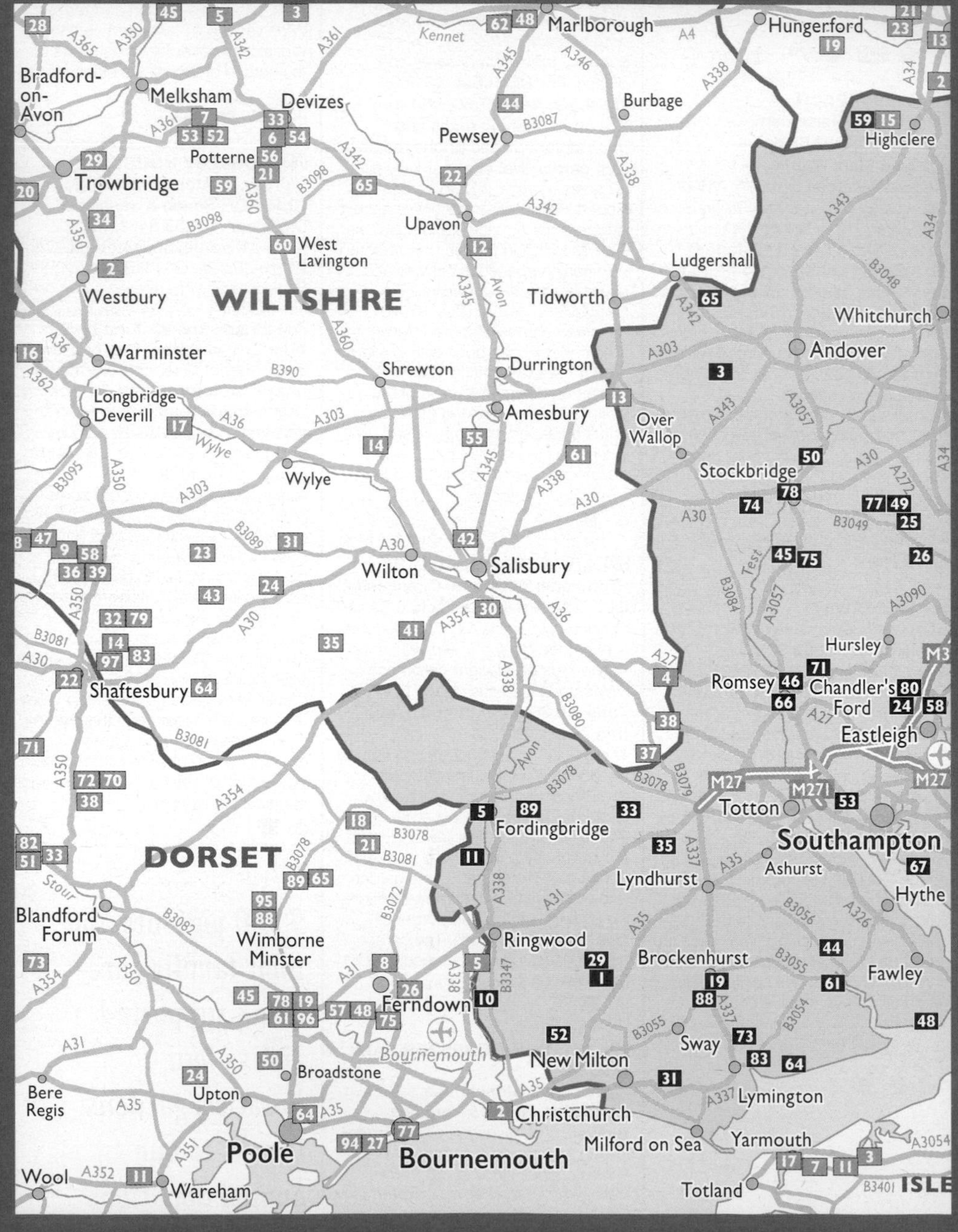

WILTSHIRE
DORSET
Marlborough
Hungerford
Bradford-on-Avon
Melksham
Devizes
Burbage
Highclere
Pewsey
Potterne
Trowbridge
Upavon
West Lavington
Ludgershall
Westbury
Tidworth
Whitchurch
Warminster
Shrewton
Durrington
Andover
Longbridge Deverill
Amesbury
Over Wallop
Wylye
Stockbridge
Wilton
Salisbury
Hursley
Shaftesbury
Romsey
Chandler's Ford
Eastleigh
Totton
Fordingbridge
Southampton
Ashurst
Lyndhurst
Blandford Forum
Hythe
Wimborne Minster
Ringwood
Brockenhurst
Fawley
Ferndown
Sway
New Milton
Broadstone
Bere Regis
Upton
Lymington
Christchurch
Milford on Sea
Yarmouth
Poole
Bournemouth
Wool
Wareham
Totland
ISLE

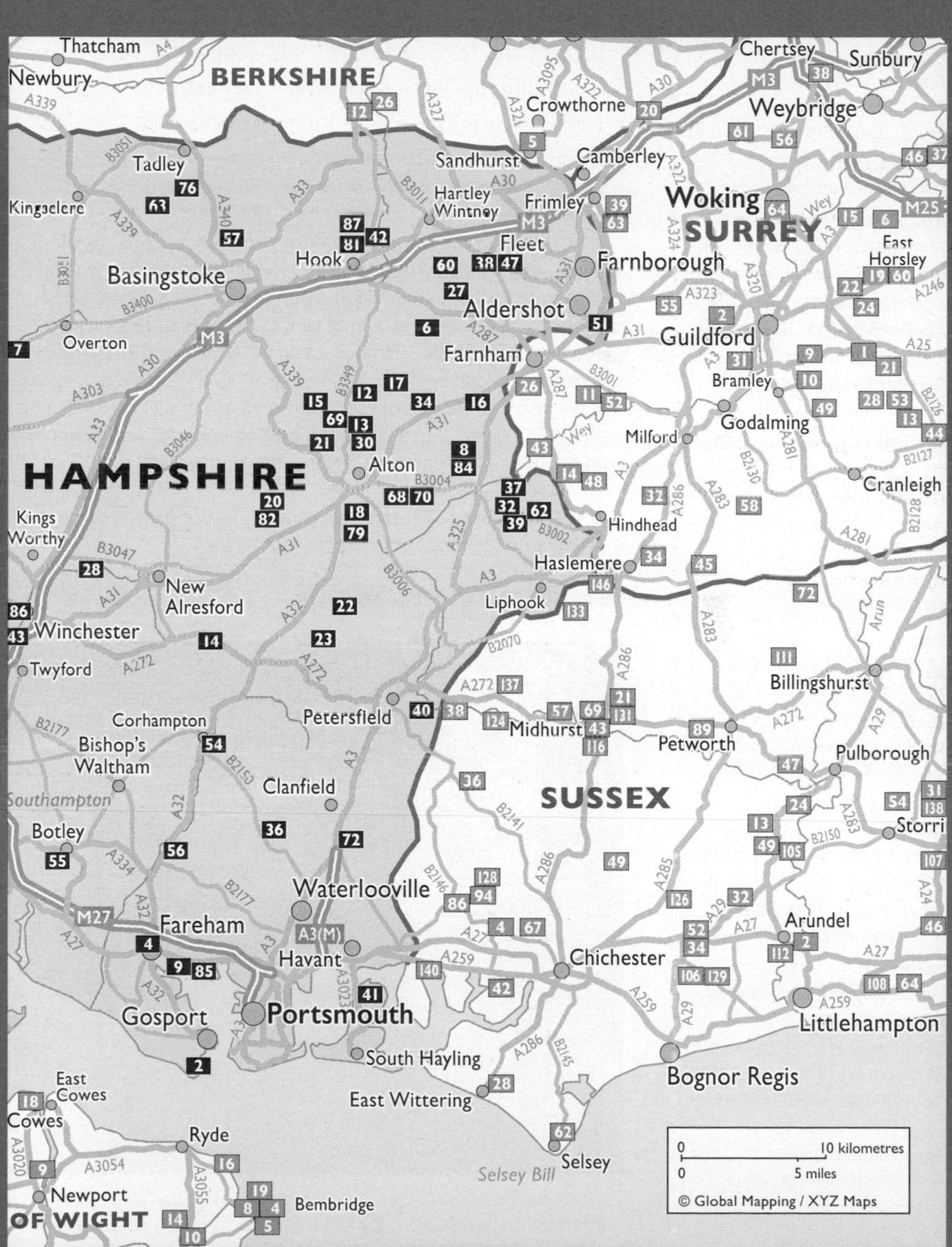
BERKSHIRE
SURREY
HAMPSHIRE
SUSSEX
OF WIGHT
Thatcham
Newbury
Tadley
Kingsclere
Basingstoke
Overton
Hook
Hartley Wintney
Sandhurst
Crowthorne
Camberley
Frimley
Fleet
Farnborough
Aldershot
Farnham
Woking
Chertsey
Sunbury
Weybridge
East Horsley
Guildford
Bramley
Godalming
Milford
Cranleigh
Hindhead
Haslemere
Liphook
Alton
Kings Worthy
New Alresford
Winchester
Twyford
Corhampton
Bishop's Waltham
Southampton
Botley
Petersfield
Clanfield
Midhurst
Petworth
Billingshurst
Pulborough
Storri
Waterlooville
Fareham
Havant
Gosport
Portsmouth
South Hayling
East Wittering
Chichester
Arundel
Littlehampton
Bognor Regis
Selsey
Selsey Bill
East Cowes
Cowes
Ryde
Newport
Bembridge
0 10 kilometres
0 5 miles
© Global Mapping / XYZ Maps

Volunteers

County Organiser
Mark Porter 07814 958810
markstephenporter@gmail.com

County Treasurer
Fred Fratter 01962 776243
fred@fratter.co.uk

Publicity
Pat Beagley 01256 764772
pat.beagley@ngs.org.uk

Social Media - Facebook
Mary Hayter 07512 639772
mary.hayter@ngs.org.uk

Social Media - Twitter
Louise Moreton 07943 837993
louise.moreton@ngs.org.uk

Booklet Co-ordinator
Mark Porter (as above)

Assistant County Organisers

Central
Sue Cox 01962 732043
suealex13@gmail.com

Central West
Kate Cann 01794 389105
kategcann@gmail.com

East
Linda Smith 01329 833253
linda.ngs@btinternet.com

North
Cynthia Oldale 01420 520438
c.k.oldale@btinternet.com

North East
Lizzie Powell 01420 23185
lizziepowellbroadhatch@gmail.com

North West
Adam Vetere 01635 268267
adam.vetere@ngs.org.uk

South
Barbara Sykes 02380 254521
barandhugh@aol.com

South West
Elizabeth Walker 01590 677415
elizabethwalker13@gmail.com

West
Jane Wingate-Saul 01725 519414
jane.wingatesaul@ngs.org.uk

Hampshire is a large, diverse county. The landscape ranges from clay/gravel heath and woodland in the New Forest National Park in the south west, across famous trout rivers – the Test and Itchen – to chalk downland in the east, where you will find the South Downs National Park.

Our open gardens are spread right across the county and offer a very diverse range of interest for both the keen gardener and the casual visitor.

We have many new gardens for you to visit - for example, we are delighted to welcome the village group of gardens at Headley (12 of them) in the east of the county as well as the wonderful spring garden at Lepe House on the south coast. Coles, also in the east, is a historic garden being extensively renovated by its new owners and we also welcome our second group of allotments in Hook in the north, which along with a sustainable community garden, is part of a plan to create a parkland centre for health and wellbeing.

You will be assured of a warm welcome by all our garden owners and we hope you enjoy your visits.

Below: Bay Tree House, Crawley Gardens

@HampshireNGS @HantsNGS 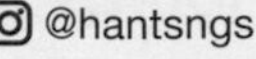

OPENING DATES

All entries subject to change. For latest information check **www.ngs.org.uk**

Extended openings are shown at the beginning of the month.

Map locator numbers are shown to the right of each garden name.

February

Snowdrop Festival

Sunday 5th
◆ Chawton House 18

Sunday 12th
Little Court 49

Monday 13th
Little Court 49

Sunday 19th
Bramdean House 14
Little Court 49

Monday 20th
Little Court 49

March

Monday 20th
Little Court 49

Monday 27th
Bere Mill 7

Wednesday 29th
Beechenwood Farm 6

April

Every Wednesday
Beechenwood Farm 6

Sunday 2nd
Durmast House 29
NEW Lepe House Gardens 48

Friday 7th
Crawley Gardens 25

Sunday 9th
Pylewell Park 64
Twin Oaks 80

Monday 10th
Crawley Gardens 25
Twin Oaks 80

Wednesday 12th
Abbotsfield 1

Thursday 13th
Appleyards 5

Friday 14th
Appleyards 5

Saturday 15th
Appleyards 5
21 Chestnut Road 19

Sunday 16th
Appleyards 5
21 Chestnut Road 19
Old Thatch & The Millennium Barn 60
Tylney Hall Hotel 81

Sunday 23rd
◆ Spinners Garden 73
Terstan 78

Friday 28th
Bluebell Wood 13

Saturday 29th
Bluebell Wood 13

Sunday 30th
Brick Kiln Cottage 15
The Cottage 24
Pylewell Park 64

May

Every Wednesday
Beechenwood Farm 6

Monday 1st
Beechenwood Farm 6
Bere Mill 7
The Cottage 24

Sunday 7th
The Cottage 24
Walhampton 83

Monday 8th
The Cottage 24

Friday 12th
NEW ◆ Furzey Gardens 35

Saturday 13th
◆ Alverstoke Crescent Garden 2
NEW ◆ Furzey Gardens 35
NEW Hart's Green Garden 38
Manor Lodge 55

Sunday 14th
NEW Coles 23
The House in the Wood 44
Manor Lodge 55
Tylney Hall Hotel 81

Tuesday 16th
NEW Hart's Green Garden 38

Thursday 18th
Appleyards 5

Friday 19th
Appleyards 5

Saturday 20th
Appleyards 5
Bisterne Manor & Stable Family Home Trust 10
21 Chestnut Road 19
NEW Maggie's Southampton 53

Sunday 21st
Appleyards 5
21 Chestnut Road 19
The Dower House 27
How Park Barn 45
Little Court 49
NEW Maggie's Southampton 53

Monday 22nd
Little Court 49

Saturday 27th
Abbotsfield 1
Spitfire House 74

Sunday 28th
Amport & Monxton Gardens 3
Manor House 54
Romsey Gardens 66
The Thatched Cottage 79
Twin Oaks 80

Monday 29th
Amport & Monxton Gardens 3
Bere Mill 7
Romsey Gardens 66
Spitfire House 74
The Thatched Cottage 79
Twin Oaks 80

June

Saturday 3rd
Ferns Lodge 31
Froyle Gardens 34
NEW Ladybower 47
Winchester College 86

Sunday 4th
Ferns Lodge 31
Froyle Gardens 34
◆ The Hospital of St Cross 43
NEW Ladybower 47
Winchester College 86

Wednesday 7th
Beechenwood Farm 6

Thursday 8th
Stockbridge Gardens 75

Friday 9th
Walden 82

Saturday 10th
21 Chestnut Road 19
NEW Headley Village Gardens 39
5 Oakfields 58
1 Povey's Cottage 63

Sunday 11th
21 Chestnut Road 19
NEW Headley Village Gardens 39
108 Heath Road 40
Manor House 54
5 Oakfields 58
Old Thatch & The Millennium Barn 60
1 Povey's Cottage 63
Shalden Park House 69
South View House 72
Stockbridge Gardens 75
Tylney Hall Hotel 81
Walden 82

Wednesday 14th
Redenham Park House 65

Thursday 15th
Appleyards 5
Tanglefoot 77

Friday 16th
Appleyards 5

Saturday 17th
Appleyards 5
Yew Hurst 89

Sunday 18th
Appleyards 5
Fritham Lodge 33
Longstock Park Water Garden 50
Tanglefoot 77
Terstan 78
Yew Hurst 89

Saturday 24th
Twin Oaks 80

Sunday 25th
Bramdean House 14
Broadhatch House 16
Durmast House 29
Tadley Place 76
The Thatched Cottage 79
Twin Oaks 80
Wicor Primary School Community Garden 85

Monday 26th
Broadhatch House 16

Tuesday 27th
Broadhatch House 16

Thursday 29th
NEW Mill House 57

July

Saturday 1st
Angels Folly 4
NEW Frey Elna 32
Hook Cross Allotments 42
26 Lower Newport Road 51
NEW Perrymead 62
'Selborne' 68

Sunday 2nd
Angels Folly 4
NEW Frey Elna 32
Hook Cross Allotments 42
26 Lower Newport Road 51
NEW Mill House 57
NEW Perrymead 62
'Selborne' 68

Thursday 6th
Crawley Gardens 25

Saturday 8th
21 Chestnut Road 19
15 Rothschild Close 67

Sunday 9th
21 Chestnut Road 19
Crawley Gardens 25
15 Rothschild Close 67
1 Wogsbarne Cottages 87

Monday 10th
1 Wogsbarne Cottages 87

Saturday 15th
Angels Folly 4
8 Birdwood Grove 9

Sunday 16th
Angels Folly 4
8 Birdwood Grove 9

Thursday 20th
Tanglefoot 77

Saturday 22nd
The Island 46
1 Povey's Cottage 63

Sunday 23rd
Bleak Hill Nursery & Garden 11
The Island 46
1 Povey's Cottage 63
Tanglefoot 77
Terstan 78

Monday 24th
Bleak Hill Nursery & Garden 11

August

Saturday 5th
Church House 20

Sunday 6th
Church House 20
The Homestead 41
'Selborne' 68

Monday 7th
'Selborne' 68

Saturday 12th
Angels Folly 4

Sunday 13th
Bleak Hill Nursery & Garden 11

Monday 14th
Bleak Hill Nursery & Garden 11

Saturday 19th
Twin Oaks 80
Wheatley House 84

Sunday 20th
Twin Oaks 80
Wheatley House 84

Sunday 27th
NEW Coles 23
The Thatched Cottage 79

Monday 28th
Bere Mill 7
Bleak Hill Nursery & Garden 11
The Thatched Cottage 79

September

Saturday 2nd
Bumpers 17
15 Rothschild Close 67
Woodpeckers Care Home 88

Sunday 3rd
Blounce House 12
Bumpers 17
Meon Orchard 56
15 Rothschild Close 67
Woodpeckers Care Home 88

Sunday 10th
Bramdean House 14
Terstan 78

Friday 22nd
Redenham Park House 65

October

Sunday 1st
Wheatley House 84

Sunday 22nd
NEW Coles 23

Friday 27th
NEW ◆ Furzey Gardens 35

Saturday 28th
NEW ◆ Furzey Gardens 35

February 2024

Sunday 11th
Little Court 49

Monday 12th
Little Court 49

Sunday 18th
Little Court 49

Monday 19th
Little Court 49

By Arrangement

Arrange a personalised garden visit with your club, or group of friends, on a date to suit you. See individual garden entries for full details.

Abbotsfield 1
Angels Folly 4
Appleyards 5
Beechenwood Farm 6
Bere Mill 7
Binsted Place 8
8 Birdwood Grove 9
Blounce House 12
Bluebell Wood 13
Brick Kiln Cottage 15
Broadhatch House 16
Church House 20
Clover Farm 21
Colemore House Gardens 22
NEW Coles 23
The Cottage 24
The Deane House 26
The Down House 28
Durmast House 29
Fairbank 30
Hambledon House 36
Hanging Hosta Garden 37
108 Heath Road 40
The Homestead 41
Little Court 49
The Old Rectory 59
1 Povey's Cottage 63
15 Rothschild Close 67
Silver Birches 70
Tanglefoot 77
Terstan 78
The Thatched Cottage 79
Trout Cottage, Stockbridge Gardens 75
Twin Oaks 80
Wheatley House 84

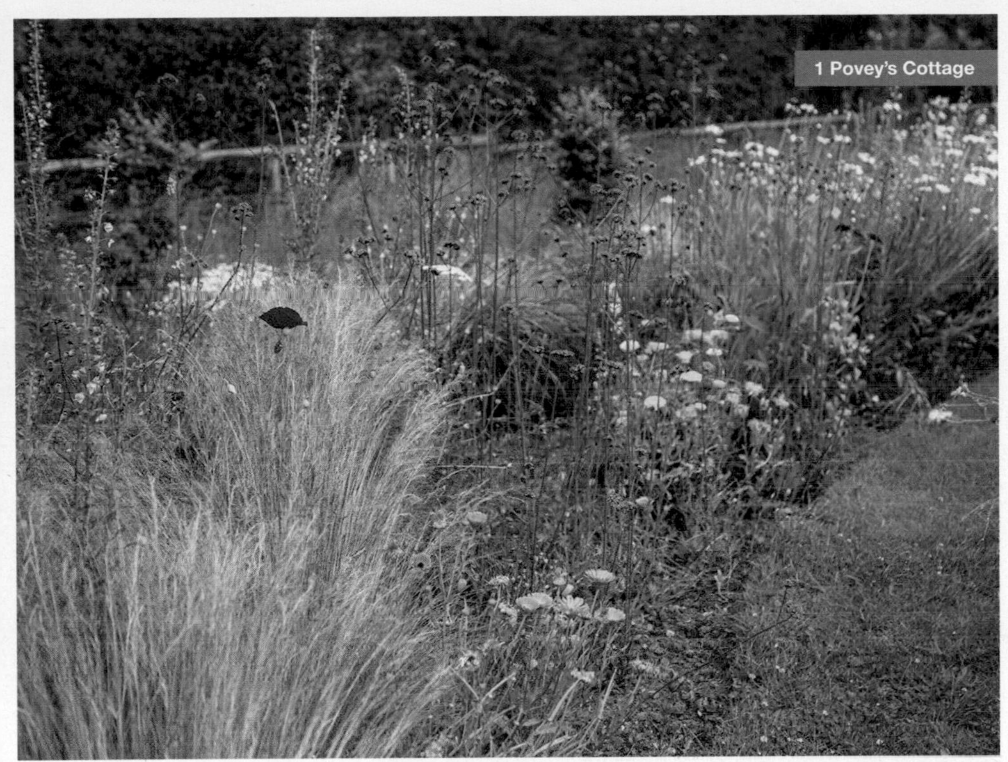

1 Povey's Cottage

THE GARDENS

1 ABBOTSFIELD

Bennetts Lane, Burley, Ringwood, BH24 4AT. Anne Blackman & Paul Moors, 01425 403305, a.blackman292@btinternet.com.

1m E of Burley village centre. Turn L in village centre. Parking at the White Buck, 300 metres walk from garden (mostly level access) & Mill Lawn Car Park, Mill Ln, 400 metres walk up steep hill. **Wed 12 Apr, Sat 27 May (10.30-4.30). Adm £3.50, chd free. Home-made teas. Visits also by arrangement 12 Apr to 27 May for groups of up to 12.**

A ½ acre, New Forest garden set in a borrowed landscape of mature native trees, full of birdsong in spring. Planted for year-round interest but best in spring for the varied hue and texture of the foliage of trees, shrubs and ferns forming a foil for bright spring bulbs, early herbaceous and flowering shrubs. A large fish pond, and kitchen garden with some raised bed for vegetables and dahlias. Summerhouse for teas if wet. Disabled parking for three cars at garden. Wheelchair access to most of the garden on accessible paths and lawns.

2 ◆ ALVERSTOKE CRESCENT GARDEN

Crescent Road, Gosport, PO12 2DH. Gosport Borough Council, www.alverstokecrescentgarden.co.uk.

1m S of Gosport. From A32 & Gosport follow signs for Stokes Bay. Continue alongside bay to small r'about, turn L into Anglesey Rd. Crescent Garden signed 50yds on R. **For NGS: Sat 13 May (10-4). Adm by donation. Home-made teas. For other opening times and information, please visit garden website.**

Restored Regency ornamental garden designed to enhance fine crescent (Thomas Ellis Owen 1828). Trees, walks and flowers lovingly maintained by community and council partnership. A garden of considerable local historic interest highlighted by impressive restoration and creative planting. Adjacent to St Mark's churchyard, worth seeing together. Heritage, history and horticulture, a fascinating package. Plant sale. Green Flag Award.

Our annual donations to Parkinson's UK meant 7,000 patients are currently benefiting from support of a Parkinson's Nurse

GROUP OPENING

3 AMPORT & MONXTON GARDENS

Amport and Monxton, SP11 8AY. *3m SW of Andover. Turn off the A303 signed East Cholderton from the E, or Thruxton village from the W. Follow signs to Amport. Car parking in field next to Amport village green.* **Sun 28, Mon 29 May (2-5.30). Combined adm £6, chd free. Home-made teas.**

BRIDGE COTTAGE
John & Jenny Van de Pette.

CORNER COTTAGE
15 Sarson. Ms Jill McAvoy.

WHITE GABLES
David & Coral Eaglesham.

Monxton and Amport are two pretty villages linked by Pill Hill Brook. Visitors have three gardens to enjoy. Bridge Cottage, a 2 acre haven for wildlife with the banks of the trout stream and lake planted informally with drifts of colour, a large vegetable garden, fruit cage, small mixed orchard and arboretum with specimen trees. Corner Cottage is a delightful cottage garden with a serpentine gravel path winding between gravel borders, clipped box hedging and old fashioned roses. A cottage style garden at White Gables with a collection of trees, along with old roses and herbaceous plants. Very popular large plant sale at Bridge Cottage. Preserves and chutneys for sale, along with an artist selling cards (10% to NGS). As always choc ices for sale at Bridge Cottage (100% to NGS). Amport has a lovely village green, so come early and bring a picnic, or refreshments available in a marquee. No wheelchair access to White Gables and partial access to Corner Cottage.

4 ANGELS FOLLY

15 Bruce Close, Fareham, PO16 7QJ. Teresa & John Greenwood, 07545 242654, tgreenwood@ntlworld.com, www.facebook.com/angelsfolly. *¾m NNE of Fareham Town Centre. M27 W leave J10 under M27 bridge in the RH-lane, do a U-turn. At r'about, take 3rd exit, across T-lights. Ist R Miller Drive. 2nd R Somervell Drive. 1st R Bruce Close.* **Sat 1, Sun 2, Sat 15, Sun 16 July (10.30-5). Adm £3.50, chd free. Home-made teas. Evening opening Sat 12 Aug (4-9). Adm £5, chd free. Wine. Visits also by arrangement 1 July to 1 Sept for groups of up to 20.**

The garden has a number of secluded areas each with their own character inc a Mediterranean garden, decking with raised beds and a seating area with a living wall. An arched folly, bench and fish pond leads to a raised planting bed and fireplace adjacent to a summerhouse. There is a wide range of colourful plants, hanging baskets and a lower secluded decked area with a planted gazebo and statue.

5 APPLEYARDS

Bowerwood Road, Fordingbridge, SP6 3BP. Bob & Jean Carr *½m from Fordingbridge on B3078. After church & houses, 400yds on L as road climbs after bridge. Parking for 8 cars only. No parking on narrow road.* **Thur 13, Fri 14, Sat 15, Sun 16 Apr, Thur 18, Fri 19, Sat 20, Sun 21 May, Thur 15, Fri 16, Sat 17, Sun 18 June (12-6). Adm £5, chd free. Pre-booking essential, please phone 01425 657631 or email bob.carr.rtd@gmail.com for information & booking. Light refreshments. Visits also by arrangement 18 Mar to 1 Sept for groups of 10 to 30.**

2 acre sloping south facing garden, newly restored, overlooking pasture. 100+ trees, sloping lawns and paths though wooded sections with massed daffodils and bluebells in spring. Newly planted rhododendrons in wooded area. Herbaceous beds, two rose beds, shrubberies, two wildlife ponds, orchard, sloping rockery beds, soft fruit cages and greenhouse.

6 BEECHENWOOD FARM

Hillside, Odiham, Hook, RG29 1JA. Mr & Mrs M Heber-Percy, 07944 162419, beechenwood@gmail.com. *5m SE of Hook. Turn S into King St from Odiham High St. Turn L after cricket ground for Hillside. Take 2nd R after 1½m, modern house ½m.* **Every Wed 29 Mar to 7 June (2-5). Mon 1 May (2-5). Adm £5, chd free. Home-made teas. Visits also by arrangement 29 Mar to 7 June.**

2 acre garden in many parts. Lawn meandering through woodland with drifts of spring bulbs. Rose pergola with steps, pots with spring bulbs and later aeoniums. Fritillary and cowslip meadow. Walled herb garden with pool and exuberant planting. Orchard inc white garden and hot border. Greenhouse and vegetable garden. Rock garden extending to grasses, ferns and bamboos. Shady walk to belvedere. 8 acre copse of native species with grassed rides. Assistance available with gravel drive and avoidable shallow steps.

7 BERE MILL

London Road, Whitchurch, RG28 7NH. Rupert & Elizabeth Nabarro, 07703 161074, rupertnab@gmail.com, www.beremillfarm.co.uk/garden. *9m E of Andover, 12m N of Winchester. In centre of Whitchurch, take London Rd at r'about. Uphill 1m, turn R 50yds beyond The Gables on R. Drop-off point for disabled at garden by side of butchery.* **Mon 27 Mar, Mon 1, Mon 29 May, Mon 28 Aug (1-5). Adm £7.50, chd free. Home-made teas. Visits also by arrangement 5 Feb to 15 Oct. Fixed fee of £400 for groups up to 50, call to discuss larger groups.**

Garden built around early C18 mill on idyllic isolated stretch of the River Test east of Whitchurch. Gardens have been built incrementally over 30 yrs with extensive bulbs planting; herbaceous borders with magnolia, irises, and tree peonies; summer and autumn borders; a traditional orchard and two small arboretums, one specialising in Japanese shrubs and trees. The garden aims to complement the natural beauty of the site and to incorporate elements of oriental garden design and practice. The working mill was where Portals first made paper for the Bank of England in 1716. Unfenced and unguarded rivers and streams. Wheelchair access unless very wet.

8 BINSTED PLACE

River Hill, Binsted Road, Binsted, Alton, GU34 4PQ. Max & Catherine Hadfield, 01420 23146, catherine.hadfield1@icloud.com. *At eastern edge of Binsted Village on Binsted Rd. 1m from Jolly Farmer Pub in Blacknest. 1½m from A325. Parking limited, but safe on-road parking outside the property.* **Visits by arrangement 5 June to 15 Sept for groups of up to 30. Home-made teas.**

Binsted Place, a C17 farmhouse with attractive local stone outbuildings, is surrounded by a series of garden rooms covering approx 1½ acres, enclosed by yew hedges and old walls. It is very traditional in style and inc many roses, pergolas, herbaceous borders, lily pond and a productive vegetable garden and orchards. Step-free disabled access to most of the garden.

9 8 BIRDWOOD GROVE

Downend, Fareham, PO16 8AF. Jayne & Eddie McBride, 01329 280838, jayne.mcbride@ntlworld.com. *M27 J11, L lane slip to Delme r'about, L on A27 to Portchester, over 2 T-lights, completely around small r'about, Birdwood Grove 1st L.* **Sat 15, Sun 16 July (12.30-5). Adm £3.50, chd free. Home-made teas. Visits also by arrangement July & Aug for groups of up to 20.**

The subtropics in Fareham! This small garden is influenced by the flora of Australia and New Zealand and inc many indigenous species and plants that are widely grown down under. The four climate zones; arid, temperate, lush and fertile and, a shady fernery, are all densely planted to make the most of dramatic foliage, from huge bananas to towering cordylines. Wheelchair access over short gravel path. Not suitable for mobility scooters.

10 BISTERNE MANOR & STABLE FAMILY HOME TRUST

Bisterne, Ringwood, BH24 3BN. Mr & Mrs Hallam Mills. *2½m S of Ringwood on B3347 Christchurch Rd, 500yds past church on L. Entrance signed Stable Family Home Trust on L (blue sign), just past lodge. Disabled parking signed.* **Sat 20 May (1-5). Adm £7.50, chd free. Tea.**

Two spectacular gardens in one visit, linked by history. The original commercial walled garden to the C16 Manor House (not open) now used by the Stable Family Home Trust, a charity, founded in 1980, supporting adults with learning difficulties, will be open alongside the Manor garden providing a wonderful 4 acre combination. There will be a band playing live music. Wheelchair access to a level garden with wide gravel paths.

11 BLEAK HILL NURSERY & GARDEN

Braemoor, Bleak Hill, Harbridge, Ringwood, BH24 3PX. Tracy & John Netherway, www.bleakhillplants.co.uk. *2½m S of Fordingbridge. Turn off A338 at Ibsley. Go through Harbridge village to T-junction at top of hill, turn R for ¼m.* **Sun 23 July (2-5); Mon 24 July (11-3); Sun 13 Aug (2-5); Mon 14 Aug (11-3); Mon 28 Aug (2-5). Adm £4, chd free. Home-made teas. No refreshments on 24 July & 14 Aug, welcome to bring a picnic.**

Enjoy this ¾ acre garden, pass through the moon gate to reveal the billowing borders contrasting against a seaside scene with painted beach huts and a boat on the gravel. Herbaceous borders complemented by a spectacular tropical border fill the garden with colour wrapping around a pond and small stream. Greenhouses with cacti and sarracenias. Vegetable patch and small wildflower meadow. Small adjacent nursery.

Bumpers

12 BLOUNCE HOUSE

Blounce, South Warnborough, Hook, RG29 1RX. Tom & Gay Bartlam, 01256 862234, tomb@thbartlam.co.uk. *In hamlet of Blounce, 1m S of South Warnborough on B3349 from Odiham to Alton.* **Sun 3 Sept (1-4). Adm £4, chd free. Home-made teas. Visits also by arrangement 4 June to 17 Sept for groups of 6+.**

A 2 acre garden surrounding a classic Queen Anne house (not open). Mixed planting to give interest from spring to late autumn. Herbaceous borders with a variety of colour themes. In later summer an emphasis on dahlias, salvias and grasses.

13 BLUEBELL WOOD

Stancombe Lane, Bavins, New Odiham Road, Alton, GU34 5SX. Mrs Jennifer Ospici, bavinsbnb@hotmail.com, www.bavins.co.uk. *On the corner of Stancombe Ln & the B3349 2½m N of Alton.* **Fri 28, Sat 29 Apr (11-4). Adm £10, chd free. Light refreshments. Visits also by arrangement 24 Apr to 5 May for groups of 5 to 20.**

Unique 100 acre ancient bluebell woodland. If you are a keen walker you will have much to explore on the long meandering paths and rides dotted with secluded seats. The wood is very challenging for those with mobility problems. Refreshments will be served in an original rustic building and inc soups using natural woodland ingredients.

14 BRAMDEAN HOUSE

Bramdean, Alresford, SO24 0JU. Mr & Mrs E Wakefield, garden@bramdeanhouse.com. *4m S of Alresford; 9m E of Winchester; 9m W of Petersfield. In centre of village on A272. Entrance opp sign to the church. Parking is usually signed across the road from entrance.* **Sun 19 Feb (1-3.30); Sun 25 June (11.30-4); Sun 10 Sept (1-3.30). Adm £6, chd free. Home-made teas. Donation to Bramdean Church.**

Beautiful 5 acre garden best known for its mirror image herbaceous borders and 1 acre walled garden. Also carpets of spring bulbs, especially snowdrops and a large and unusual collection of plants and shrubs giving year-round interest. The walled garden features prize-winning vegetables, fruit and flowers. Features inc fine snowdrops, a large collection of old fashioned sweet peas, an expansive collection of nerines, a boxwood castle and the nation's tallest sunflower 'Giraffe'. Visits also by arrangement for groups of 5+ (non-NGS). Assistance dogs only.

15 BRICK KILN COTTAGE

The Avenue, Herriard, nr Alton, RG25 2PR. Barbara Jeremiah, barbara@klca.co.uk. *4m NE of Alton. A339 Basingstoke to Alton, 7m out of Basingstoke turn L along The Avenue, past Lasham Gliding Club on R, then past Back Ln on L & take next track on L, one field later.* **Sun 30 Apr (12-5). Adm £4.50, chd free. Home-made teas. Visits also by arrangement 6 May to 30 June for groups of up to 30.**

Bluebell woodland garden in 2 acres with a perimeter woodland path inc treehouse, pebble garden, billabong, stumpery, ferny hollow, bug palace, waterpool, shepherd's hut and a traditional cottage garden filled with herbs. The garden is maintained using eco-friendly methods as a haven for wild animals, butterflies, birds, bees and English bluebells. New feature children's reading area. Wildlife friendly garden in a former brick works. A haven in the trees. Gallery of textiles.

16 BROADHATCH HOUSE

Bentley, Farnham, GU10 5JJ. Bruce & Lizzie Powell, 07799 031044, lizziepowellbroadhatch@gmail.com. *4m NE of Alton. Turn off A31 (Bentley bypass) through village, then L up School Ln. R to Perrylands, after 300yds drive on R.* **Sun 25, Mon 26, Tue 27 June (2-5.30). Adm £5, chd free. Home-made teas. Visits also by arrangement May & June.**

3½ acre garden set in lovely Hampshire countryside with views to Alice Holt. Divided into different areas by yew hedges and walled garden. Focussing on as long a season as possible on heavy clay. Two reflective pools help break up lawn areas; lots of flower borders and beds; mature trees. Working greenhouses and vegetable garden. Wheelchair access with gravel paths and steps in some areas.

17 BUMPERS

Sutton Common, Long Sutton, Hook, RG29 1SJ. Stella Wildsmith. *From village of Long Sutton, turn up Copse Ln, immed opp duck pond. Follow lane for 1½m to top of steep hill, house on L.* **Sat 2, Sun 3 Sept (2-5). Adm £5, chd free. Cream teas.**

Large country garden with beautiful views spread over 2 acres with mixed herbaceous and shrub borders and laid out in a series of individual areas. Some interesting sculptures and water features with informal paths through the grounds and a number of places to sit and enjoy the views. Visitors with wheelchairs, please park at front of house.

18 ◆ CHAWTON HOUSE

Chawton, Alton, GU34 1SJ. Chawton House, 01420 541010, info@chawtonhouse.org, www.chawtonhouse.org. *2m S of Alton. Take the Gosport Rd opp Jane Austen's House museum towards St Nicholas Church. Property is at the end of this road on the L. Parking is signed at the end of the road & in Chawton village.* **For NGS: Sun 5 Feb (10-3.30). Adm £5, chd free. Home-made teas. For other opening times and information, please phone, email or visit garden website.**

Snowdrops and spring flowering bulbs are scattered through this 15 acre listed English landscape garden. Sweeping lawns, wilderness, terraces, fernery and shrubbery walk surround the Elizabethan manor house. The walled garden designed by Edward Knight surrounds a rose garden, borders, orchard, vegetable garden and 'Elizabeth Blackwell' herb garden based on her book 'A Curious Herbal' of 1737-39. Hot and cold drinks, wine, light lunches, cream teas, home-made cakes and local ice creams available in our tea shed on the main drive and in The Old Kitchen Tearoom at the house.

19 21 CHESTNUT ROAD

Brockenhurst, SO42 7RF. Iain & Mary Hayter, www.21-chestnut-rdgardens.co.uk. *New Forest. Please use village car park. Limited parking for less mobile in road. Leave M27 J2, follow Heavy Lorry Route. Mainline station less than 10 min walk.* **Sat 15, Sun 16 Apr, Sat 20, Sun 21**

May, Sat 10, Sun 11 June, Sat 8, Sun 9 July (11.30-5). Adm £5, chd free. Home-made teas.
A ⅓ acre garden brimming with ideas and inspiration for different aspects and conditions, designed sympathetically with nature and wildlife. Ponds, mixed shrub and perennial borders, kitchen and fruit garden with wildflower area, all adorned with statues, arches and plenty of seating, perfect for enjoying home-made refreshments. Art exhibition of owners paintings with a donation to NGS from any sales made on open days. Plants, bug houses and bird boxes usually available to purchase.

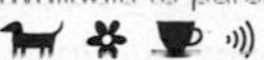

20 CHURCH HOUSE

Trinity Hill, Medstead, Alton, GU34 5LT. Mr Paul & Mrs Alice Beresford, 01420 562592, churchhouse4@gmail.com. *5m WSW of Alton. From A31 Four Marks follow signs to Medstead for 1½m to village centre, turn R into Church Ln/Trinity Hill. From N on A339, R at Bentworth Xrds & continue via Bentworth to Medstead.* **Sat 5 Aug (1-6); Sun 6 Aug (1-5). Adm £5, chd £1. Home-made teas. Visits also by arrangement 15 Apr to 10 Sept for groups of 10 to 32.**
A colourful 1 acre garden, set within a wide variety of mature trees and shrubs. Long, sweeping, colour themed mixed borders give lots of ideas for planting in sun and shade. Contrasting features and textures throughout the garden are enhanced by interesting sculptures. Espaliered fruit trees, a woodland area, small greenhouse and roses in different settings all contribute to this much loved garden. For further information see Facebook, search Church House Garden Medstead. Wheelchair access via gravel drive to flat lawned garden. No access to some paths and patio.

21 CLOVER FARM

Shalden Lane, Shalden, Alton, GU34 4DU. Tom & Sarah Floyd, 01420 86294. *Approx 3m N of Alton in the village of Shalden. Take A339 out of Alton. After approx 2m turn R up Shalden Ln. At top, turn sharp R next to church sign.* **Visits by arrangement June to Sept for groups of 20+. Adm £10, chd free. Tea & cake inc.**
3 acre garden with far-reaching views. Herbaceous borders and sloping lawns down to reflection pond, wildflower meadow, lime avenue, rose and kitchen garden and ornamental grass area.

22 COLEMORE HOUSE GARDENS

Colemore, Alton, GU34 3RX. Mr & Mrs Simon de Zoete, 01420 588202, simondezoete@gmail.com. *4m S of Alton, off A32. Approach from N on A32, turn L in to Shell Ln, ¼m S of East Tisted. Go under bridge, keep L until you see Colemore Church. Park on verge of church.* **Visits by arrangement May to July for groups of 10 to 40.**
4 acres in lovely unspoilt countryside, featuring rooms containing many unusual plants and different aspects with a spectacular arched rose walk, water rill, mirror pond, herbaceous and shrub borders. Newly designed by David Austin roses, an octagonal garden with 25 different varieties. Explore the interesting arboretum, grass gardens and thatched pavilion. Every yr the owners seek improvement and the introduction of new, interesting and rare plants. We propagate and sell plants, many of which can be found in the garden. Some are unusual and not readily available elsewhere. For private visits, we endeavour to give a conducted tour and try to explain our future plans, rationale and objectives.

✲

23 NEW COLES

Basing Dean, Privett, Alton, GU34 3PH. Mike & Shuna MacKillop-Hall, 07423 541105, mikemackillophall@gmail.com. *6m NW of Petersfield. From Petersfield, come up Bell Hill/Stoner Hill & turn L at top of hill via High Cross; from Winchester on A272 turn L at West Meon Xrds on to A32; after 2m turn R at The Angel.* **Sun 14 May (10.30-7.30); Sun 27 Aug (10.30-8); Sun 22 Oct (10.30-5.30). Adm £10, chd free. Light refreshments. Visits also by arrangement 3 Apr to 26 Nov for groups of 10+.**
Country Life called it 'a great post-war garden'; historic 25 acre creation being renovated; dazzling spring rhododendrons and azaleas with bluebells; Japanese acers among ancient oak woodland and arboretum for autumn colour; koi, wildlife and Japanese ponds; lawns and contemporary beds of grass and perennials around stunning modern house (not open). Illustrated talk at 11.30am and 2.30pm by owner who restored Old Alresford House gardens. Light refreshments inc home-made teas, soup and sourdough bread, vegetarian tarts and salads, wine. Easy access around the house; an all-terrain wheelchair is needed for the whole garden.

24 THE COTTAGE

16 Lakewood Road, Chandler's Ford, Eastleigh, SO53 1ES. Hugh & Barbara Sykes, 02380 254521, barandhugh@aol.com. *Leave M3 J12, follow signs to Chandler's Ford. At King Rufus on Winchester Rd, turn R into Mordon Ave, then 3rd road on L.* **Sun 30 Apr, Mon 1, Sun 7, Mon 8 May (1.30-5.30). Adm £4.50, chd free. Home-made teas. Visits also by arrangement Apr & May.**
The house was built in 1905, but the ¾ acre garden has been designed, planted and cared for since 1950 by two keen garden loving families. Azaleas, camellias, trilliums and erythroniums under old oaks and pines. Herbaceous cottage style borders with many unusual plants for year-round interest. Bog garden, ponds, kitchen garden. Bantams, bees and birdsong with over 35 bird species noted. Wildlife areas. NGS sundial for opening for 30 yrs. Childrens' quiz. 'A lovely tranquil garden', Anne Swithinbank. Hampshire Wildlife Trust Wildlife Garden Award. Honey from our garden hives for sale.

The National Garden Scheme searches the length and breadth of England, Wales, Northern Ireland and the Channel Islands for the very best private gardens

GROUP OPENING

25 CRAWLEY GARDENS

Crawley, Winchester, SO21 2PR. F J Fratter, 01962 776243, fred@fratter.co.uk. *5m NW of Winchester. Between B3049 (Winchester - Stockbridge) & A272 (Winchester - Andover). Parking throughout village & in field at Tanglefoot.* **Fri 7, Mon 10 Apr (2-5.30). Combined adm £8, chd free. Thur 6, Sun 9 July (2-5.30). Combined adm £10, chd free. Home-made teas in the village hall (April & 6 July) & Little Court (9 July).**

BAY TREE HOUSE
Julia & Charles Whiteaway.
Open on all dates

LITTLE COURT
Mrs A R Elkington.
Open on all dates
(See separate entry)

PAIGE COTTAGE
Mr & Mrs T W Parker.
Open on all dates

TANGLEFOOT
Mr & Mrs F J Fratter.
Open on Thur 6, Sun 9 July
(See separate entry)

Crawley is a pretty period village nestling in chalk downland with thatched houses, C14 church and village pond. The spring gardens are Bay Tree House, Little Court and Paige Cottage; all of the gardens are open in the summer; providing varied seasonal interest with traditional and contemporary approaches to landscape and planting. Most of the gardens have beautiful country views and other gardens can be seen from the road. Bay Tree House has bulbs, wild flowers, a Mediterranean garden, pleached limes, a rill and contemporary borders of perennials and grasses. Little Court is a 3 acre country garden with carpets of spring bulbs, herbaceous borders and a large meadow. Paige Cottage is a 1 acre traditional English garden surrounding a period thatched cottage (not open) with bulbs and wild flowers in spring and old climbing roses in summer. Tanglefoot has colour themed borders, herb wheel, kitchen garden, Victorian boundary wall supporting trained fruit and a large wildflower meadow. Plants from the garden for sale at Little Court and Tanglefoot.

26 THE DEANE HOUSE

Sparsholt, Winchester, SO21 2LR. Mr & Mrs Richard Morse, 07774 863004, chrissiemorse7@gmail.com. *3½m NW of Winchester. Turn off A3049 Stockbridge Rd, onto Woodman Ln, signed Sparsholt. Turn L at 1st cottage on L, cream with green gables, come to top of the drive. Plentiful parking.* **Visits by arrangement Mar to Sept for groups of 10+. Home-made teas, lunches, wine & canapés.**

A beautiful 4 acre rural garden, with vineyard nestled on a gentle south facing slope, has been landscaped to draw the eye from one gentle terraced lawn to another with borders merging with the surrounding countryside and vines. Featuring a good selection of specimen trees, a walled garden, prairie planting and herbaceous borders. Millennium avenue of tulip trees. Water features and sculptures. Sorry no dogs. Although the garden is on the side of a hill there is always a path to avoid steps.

27 THE DOWER HOUSE

Church Lane, Dogmersfield, Hook, RG27 8TA. Anne-Marie & Richard Revell. *3½m E of Hook. Turn N off A287. For SatNav please use RG27 8SZ.* **Sun 21 May (2-4.30). Adm £5, chd free. Home-made teas.**

6 acres inc bluebell wood with large and spectacular collection of rhododendrons, azaleas, magnolias and other flowering trees and shrubs; set in parkland with fine views over 20 acre lake.

28 THE DOWN HOUSE

Itchen Abbas, SO21 1AX. Jackie & Mark Porter, 07814 958810, markstephenporter@gmail.com. *5m E of Winchester on B3047. 5th house on R after the Itchen Abbas village sign, if coming on B3047 from Kings Worthy. 500 metres on L after The Plough pub if coming on B3047 from Alresford.* **Visits by arrangement 4 Feb to 12 Mar for groups of 6+. Adm £10, chd free. Home-made teas inc.**

A 2 acre garden laid out in rooms overlooking the Itchen Valley, adjoining the Pilgrim's Way. In winter come and see the snowdrops, aconites and crocus, plus borders of dogwoods, willow stems and white birches. A garden of structure with pleached hornbeams, rope-lined fountain garden, formal box-edged potager, a yew lined avenue and walks in adjoining meadows.

29 DURMAST HOUSE

Bennetts Lane, Burley, BH24 4AT. Mr & Mrs P E G Daubeney, 01425 402132, philip@daubeney.co.uk, www.durmasthouse.co.uk. *5m SE of Ringwood. Off Burley to Lyndhurst*

Clover Farm

Rd, nr White Buck Hotel, C10 road. **Sun 2 Apr, Sun 25 June (2-5). Adm £5, chd free. Home-made teas. Visits also by arrangement 1 Apr to 12 Aug.**
Designed by Gertrude Jekyll, Durmast has contrasting hot and cool colour borders, formal rose garden edged with lavender and a long herbaceous border. Many old trees, Victorian rockery and orchard with beautiful spring bulbs. Rare azaleas; Fama, Princeps and Gloria Mundi from Ghent. Features inc rose bowers with rare French roses; Eleanor Berkeley, Psyche and Reine Olga de Wurtemberg. Many old trees inc cedar and Douglas firs. Wheelchair access on stone and gravel paths.

30 FAIRBANK

Old Odiham Road, Alton, GU34 4BU. Jane & Robin Lees, 01420 86665, j.lees558@btinternet.com. *1½m N of Alton. From S, past Sixth Form College, then 1½m beyond road junction on R. From N, turn L at Golden Pot & then 50yds turn R. Garden 1m on L before road junction.* **Visits by arrangement May to Sept for groups of up to 30. Adm £4, chd free. Home-made teas.**
The planting in this large garden reflects our interest in trees, shrubs, fruit and vegetables. A wide variety of herbaceous plants provide colour and are placed in sweeping mixed borders that carry the eye down the long garden to the orchard and beyond. Near the house (not open), there are rose beds and herbaceous borders, as well as a small formal pond. There is a range of acers, ferns and unusual shrubs and 60 different varieties of fruit, together with a large vegetable garden. Wheelchair access with uneven ground in some areas.

31 FERNS LODGE

Cottagers Lane, Hordle, Lymington, SO41 0FE. Sue Grant, www.fernslodge.co.uk. *Approx 5½m W of Lymington. From Silver St turn into Woodcock Ln, 100 metres to Cottagers Ln, parking in field by opp garden. From A337 turn into Everton Rd & drive approx 1½m, Cottagers Ln on R.* **Sat 3, Sun 4 June (2-5). Adm £5, chd free. Home-made teas & cream teas.**
4 acres of wildlife garden inc a ½ acre cottage garden with a heady mix of colour and scent surrounding a Victorian lodge with winding brick paths. Azalea, fig, sweet pea, roses, agapanthus, salvia and foxgloves jostle for attention. While honeysuckle, clematis, passionflower and jasmine lead you to the big garden in restoration filled with rhododendron, tree ferns and mature trees of all description. Wheelchair access to many areas.

32 NEW FREY ELNA

Church Lane, Headley, Bordon, GU35 8PJ. Christine Leonard. *Coming from Alton on B3002. Pass Headley Church on L, take 1st turn L onto Curtis Ln, then 2nd turning L Church Ln. Frey Elna is 1st house on L. Please follow signs.* **Sat 1, Sun 2 July (10.30-5). Combined adm with Perrymead £5, chd free. Home-made teas. Opening with Headley Village Gardens on Sat 10, Sun 11 June.**
⅓ acre divided into four rooms. Enter by the front gate to the welcome garden in blue and yellow, through an archway to the main herbaceous and shrub garden; pass through the next archway to the vegetable garden and greenhouse; final stop is the entertaining room with conservatory, gazebo, patio and where tea and cakes are served. Wheelchair access to all areas.

33 FRITHAM LODGE

Fritham, SO43 7HH. Sir Chris & Lady Powell. *6m N of Lyndhurst. 3m NW of M27 J1 (Cadnam). Follow signs to Fritham.* **Sun 18 June (2-4). Adm £5, chd free. Home-made teas.**
A walled garden of 1 acre in the heart of the New Forest, set within 18 acres surrounding a house that was originally a Charles I hunting lodge (not open). Herbaceous and blue and white mixed borders, pergolas and ponds. A box hedge enclosed parterre of roses, fruit and vegetables. Visitors will enjoy the ponies, donkeys, sheep and old breed hens on their meadow walk to the woodland and stream.

GROUP OPENING

34 FROYLE GARDENS

Lower Froyle, Froyle, GU34 4LG. *5m NE of Alton. Access to Froyle from A31 between Alton & Farnham at Bentley, or at Hen & Chicken Inn. Park at Recreation Ground in Lower Froyle. Map provided. Additional signed parking in Upper Froyle.* **Sat 3, Sun 4 June (2-5.30). Combined adm £10, chd free. Home-made teas in the village hall & picnics welcome on the recreation ground, Lower Froyle.**

ALDERSEY HOUSE
Nigel & Julie Southern.

DAY COTTAGE
Nick & Corinna Whines.

NEW **2 HIGHWAY COTTAGE**
Faith Richards & Gordon Mitchell.

OLD BREWERY HOUSE
Vivienne & John Sexton.
D

OLD COURT
Sarah & Charlie Zorab.

WALBURY
Ernie & Brenda Milam.

WARREN COTTAGE
Gillian & Jonathan Pickering.

WELL LANE CORNER
Mark & Sue Lelliott.

You will certainly receive a warm welcome as Froyle Gardens open their gates again this yr, enabling visitors to enjoy a wide variety of gardens, which have undergone further development since last yr and will be looking splendid. Froyle is a beautiful village with many old and interesting buildings. Our gardens harmonise well with the surrounding landscape and most have spectacular views. The gardens themselves are diverse with rich planting. You will see greenhouses, water features, vegetables, roses, clematis and wildflower meadows. Lots of ideas to take away with you, along with plants to buy and delicious teas served in the village hall. Close by is a playground with a zip wire where children can let off steam. There is also an exhibition of richly embroidered historic vestments in the Church in Upper Froyle (separate donation). No wheelchair access to Day Cottage and on request at Warren Cottage. Long drive to Old Court. Gravel drive at 2 Highway Cottage.

35 NEW ◆ **FURZEY GARDENS**
School Lane, Minstead, Lyndhurst, SO43 7GL. Minstead Trust, furzey.manager@furzey-gardens.org, www.furzey-gardens.org. *From M27 J1 take A337 heading S for 1m, then R into Minstead Rd. Follow signs for Furzey Gardens. Please note: Limited parking.* **For NGS: Fri 12, Sat 13 May, Fri 27, Sat 28 Oct (9.30-4). Adm £9, chd free. Pre-booking essential, please visit www.ngs.org.uk for information & booking. Light refreshments. For other opening times and information, please email or visit garden website.**
Nestled in the New Forest, 10 acre stunning spring display of unusual rhododendrons, azaleas, camellias and magnolias, some planted over 100 yrs ago. More recent planting of specimen acers, liquidambars and sorbus create blazing autumn glory. Children delight in the winding paths and thatched structures leading to over 30 fairy doors! Regular garden maintenance by people with learning disabilities. Tearoom open daily from 10am to 4pm serving light lunches, cakes and a selection of drinks. Membership holders, coaches, picnic hampers and cream teas need to be pre-booked through garden website. Main paths accessible only. A wheelchair and an electric mobility scooter can be pre-booked with a week's notice. Assistance dogs only.

36 **HAMBLEDON HOUSE**
East Street, Hambledon, PO7 4RX. Capt & Mrs David Hart Dyke, 02392 632380, dianahartdyke@gmail.com. *8m SW of Petersfield, 5m NW of Waterlooville. In village centre, driveway leading to house in East St. Do not go up Speltham Hill even if advised by SatNav.* **Visits by arrangement Apr to Oct. Home-made teas.**
3 acre partly walled plantsman's garden for all seasons. Large borders filled with a wide variety of unusual shrubs and perennials with imaginative plant combinations culminating in a profusion of colour in late summer. Hidden, secluded areas reveal surprise views of garden and village rooftops. Planting a large central area, which started in 2011, has given the garden an exciting new dimension.

37 **HANGING HOSTA GARDEN**
Narra, Frensham Lane, Lindford, Bordon, GU35 0QJ. June Colley & John Baker, 01420 489186, hanginghostas@btinternet.com. *Approx 1m E of Bordon. From the A325 at Bordon take the B3002, then B3004 to Lindford. Turn L into Frensham Lane, 3rd house on L.* **Visits by arrangement 9 July to 24 July for groups of up to 20. Adm £3.50, chd free.**
This garden is packed with almost 2000 plants. The collection of over 1700 hosta cultivars is one of the largest in England. Hostas are displayed at eye level to give a wonderful tapestry of foliage and colour. Islamic garden, waterfall and stream garden, cottage garden. Talks given to garden clubs.

NPC

GROUP OPENING

38 NEW **HART'S GREEN GARDEN**
Edenbrook Country Park, Pale Lane, Fleet, RG27 8DH. Hart District Council, www.hart.gov.uk/Harts-green-garden. *Off A323 close to M3 South Bridge. Disabled parking only at Edenbrook Country Park, Pale Ln. Other parking at Hart's Sports Centre, GU51 5HS. Optional free minibus 5 min drive or 20 min stunning country walk (or cycle).* **Sat 13, Tue 16 May (2-5). Combined adm £6.50, chd free. Pre-booking essential, please visit www.ngs.org.uk for information & booking. Teas & cake.**

NEW **HART DISTRICT ALLOTMENTS**
Katy Sherman.

NEW **MINDING THE GARDEN**
Hart Voluntary Action

Hart's Green Garden forms a vibrant part of Edenbrook Country Park's active area. Minding The Garden and Hart District Allotments. They act as a beacon of sustainable community gardening; this project supports Hart District Council's aspiration to establish Edenbrook Country Park as an award-winning centre of excellence for health and wellbeing. Shared facilities and spaces inc a shelter, bike racks, compost WC and communal planting areas which will encourage connections between all site users. Wheelchair access to all areas.

GROUP OPENING

39 NEW **HEADLEY VILLAGE GARDENS**
All Saints Church Centre, High Street, Headley, GU35 8PW. *From the B3002 turn R into the High St. The car park is opp the church. Please buy combined ticket in the marquee.* **Sat 10, Sun 11 June (10.30-5). Combined adm £10, chd free. Lunches & teas at All Saints Church Centre. Cream teas in some gardens.**

NEW **ARFORD LODGE**
Jenny & Nick Record.

NEW **CHERRYCROFT**
Robin & Marjorie Hall.

NEW **CLOVER HOUSE**
Gail Cookson.

NEW **2 EASHING COTTAGES**
Liz & Mike Pennick.

NEW **FREY ELNA**
Christine Leonard.
(See separate entry)

NEW **THE HOLLIES**
Alison & Richard Kemp.

NEW **LITTLE BENIFOLD**
Pamela Williams.

NEW **THE OLD RECTORY**
Phyllida & Robin Smeeton.

NEW **PERRYMEAD**
Helen & Anne Kempster.
(See separate entry)

NEW **THE SQUARE HOUSE**
Mike Regan.

NEW **TY NEWYDD**
Jane & Hywel Bowen-Perkins.

NEW **VICTORIAN DREAMS**
Shabby Kay.

Headley offers a huge range of interesting and varied gardens inc meadowland, woodlands, mixed family gardens, sunny borders and shade-loving plants. There is everything one could wish for, courtyards, beautiful mature trees, borders, greenhouses, vegetable gardens, ponds, formal gardens, and a wild cottage garden. The group has a long experience with offering visitors a warm welcome and a great day out.

40 108 HEATH ROAD
Petersfield, GU31 4EL. Mrs Karen Llewelyn, 01730 269541, k.llewelyn@btinternet.com. *A3 N & S take A272 exit signed Midhurst. Take 1st exit from r'about, 1st R into Pullens Ln (B2199) & then 6th road on R into Heath Rd.* **Sun 11 June (2-5.30). Adm £4, chd free. Home-made teas. Visits also by arrangement 4 June to 17 Sept.**
2/3 acre garden close to town centre and Heath Pond. Greenhouse and succulent collection. Tropical plants, acers, small woodland walk. 30 metre long border with shade loving plants inc many hostas and ferns. Patio garden, seasonal pots and late summer herbaceous border. Mixed planted driveway borders. Lots to see. Wheelchair access after a 5 metre sloping gravel drive.

41 THE HOMESTEAD
Northney Road, Hayling Island, PO11 0NF. Stan & Mary Pike, 02392 464888, jhomestead@aol.com, www.homesteadhayling.co.uk. *3m S of Havant. From A27 Havant & Hayling Island r'about, travel S over Langstone Bridge & turn immed L into Northney Rd. Car park entrance on R after Langstone Hotel.* **Sun 6 Aug (2-5.30). Adm £4, chd free. Home-made teas. Visits also by arrangement June to Sept for groups of 12+.**
1 1/4 acre garden surrounded by working farmland with views to Butser Hill and boats in Chichester Harbour. Trees, shrubs, colourful herbaceous borders and small walled garden with herbs, vegetables and trained fruit trees. Large pond and woodland walk with shade-loving plants. A quiet and peaceful atmosphere with plenty of seats to enjoy the vistas within the garden and beyond. Extensive range of plants for sale. Wheelchair access with some gravel paths.

42 HOOK CROSS ALLOTMENTS
Reading Road, Hook, RG27 9DB. Hook Allotment Association. *Northern edge of Hook village on B3349 Reading Rd, 900 metres N of A30 r'about. Concealed entrance track is on RHS at foot of hill opp a farm entrance, straight after turns to B & M Fencing & Searle's Ln.* **Sat 1, Sun 2 July (1-5). Adm £4, chd free.**
5 1/4 acre community run allotments overlooking Hook village. More than 100 plots showcasing different vegetables, fruit and flower growing styles. Plot holder demonstrations of how to grow your own. Community orchard, wildflower meadow, beetle banks, wildlife friendly gardening information. Plants and refreshments for sale.

Froyle Gardens

© Jo Whitworth

43 ◆ THE HOSPITAL OF ST CROSS

St Cross Road, Winchester, SO23 9SD. The Hospital of St Cross & Almshouse of Noble Poverty, 01962 851375, porter@hospitalofstcross.co.uk, www.hospitalofstcross.co.uk. *½m S of Winchester. From city centre take B3335 (Southgate St & St Cross Rd) S. Turn L immed before The Bell Inn. If on foot follow riverside path S from Cathedral & College, approx 20 mins.* **For NGS: Sun 4 June (2-5). Adm £4, chd free. Light refreshments in the Hundred Men's Hall in the Outer Quadrangle. For other opening times and information, please phone, email or visit garden website.**

The Medieval Hospital of St Cross nestles in water meadows beside the River Itchen and is one of England's oldest almshouses. The tranquil, walled Master's Garden, created in the late C17 by Bishop Compton, now contains colourful herbaceous borders, old fashioned roses, interesting trees and a large fish pond. The Compton Garden has unusual plants of the type he imported when Bishop of London. Wheelchair access, but surfaces are uneven in places.

44 THE HOUSE IN THE WOOD

Beaulieu, SO42 7YN. Victoria Roberts. *New Forest. 8m NE of Lymington. Leaving the entrance to Beaulieu Motor Museum on R (B3056), take next R signed Ipley Cross. Take 2nd gravel drive on RH-bend, approx ½m.* **Sun 14 May (1.30-5). Adm £6, chd free. Cream teas.**

Peaceful 12 acre woodland garden with continuing progress and improvement. Very much a spring garden with tall, glorious mature azaleas and rhododendrons in every shade of pink, orange, red and white, interspersed with acers and other woodland wonders. A magical garden to get lost in with many twisting paths leading downhill to a pond and a more formal layout of lawns around the house (not open). Used in the war to train the Special Operations Executive.

45 HOW PARK BARN

Kings Somborne, Stockbridge, SO20 6QG. Kate & Chris Cann. *2m from Stockbridge, just outside Kings Somborne. A3057 from Stockbridge, turn R into Cow Drove Hill, then follow NGS signs.* **Sun 21 May (2-5). Adm £5, chd free. Home-made teas.**

A 2 acre country garden in elevated position with spectacular views over the Test Valley. Large borders of naturalistic planting and shrubs. Sweeping lawns with some slopes and a large natural wildlife pond form a tranquil setting within the landscape of C17 listed barn (not open). Adjacent to the Clarendon and Test Ways. Gravel drive and some slopes, but wheelchair access to most of the garden.

46 THE ISLAND

Greatbridge, Romsey, SO51 0HP. Mr & Mrs Christopher Saunders-Davies. *1m N of Romsey on A3057. Entrance at bridge. Follow drive 100yds. Car park on RHS.* **Sat 22, Sun 23 July (2-5). Adm £5, chd under 12 yrs free. Home-made teas.**

6 acres both sides of the River Test. The main garden has herbaceous and annual borders, fruit trees, rose pergola, lavender walk and extensive lawns. An arboretum planted in the 1930s by Sir Harold Hillier contains three ponds, shrubs and specimen trees providing interest throughout the yr. No dogs allowed. Disabled WC.

47 NEW LADYBOWER

47 Connaught Road, Fleet, GU51 3LR. Muriel Pratt. *Central Fleet. Exit J4A of M3 & follow directions to Fleet, pass train station & turn L at 2nd T-lights on to Kings Rd, Connaught Rd is 3rd on the R.* **Sat 3, Sun 4 June (2-5). Adm £5, chd free. Home-made teas.**

This garden of over ¼ of an acre was created 8 yrs ago. Over three terraces it has a large acer bed, several herbaceous borders and a large pond surrounded by rockery. Still developing it is a lovely oasis of peace and calm in the centre of a thriving market town. Floral displays from roses, rhododendrons, salvias, with an additional collection of hostas. Wheelchair access over flat patio and slopes to two of the three terraces. Parking on drive by prior arrangement, email muriel.pratt@ntlworld.com.

48 NEW LEPE HOUSE GARDENS

Lepe, Exbury, Southampton, SO45 1AD. Michael & Emma Page, www.lepe.org.uk. *New Forest. ½m from Lepe Country Park, 2m from Exbury Gardens. Entrance to drive through gates on S-side of Lepe Rd. What3words app - belong.nurses.highlight.* **Sun 2 Apr (1-4). Adm £5, chd free. Light refreshments.**

This 14 acre spring woodland garden was laid out in 1893. An embarkation point for D-Day, the lighthouse in the garden now marks the entrance to the Beaulieu River. Distinct areas inc walled garden with camellias, coastal walk overlooking the Solent, woodland with mature magnolias and rhododendrons, arboretum with drifts of spring bulbs, wildlife ponds plus formal areas with a wishing well!

49 LITTLE COURT

Crawley, Winchester, SO21 2PU. Mrs A R Elkington, 01962 776365, elkslc@btinternet.com. *5m NW of Winchester. Between B3049 (Winchester - Stockbridge) & A272 (Winchester - Andover), 400yds from either pond or church.* **Sun 12, Mon 13, Sun 19, Mon 20 Feb, Mon 20 Mar (2-5); Sun 21, Mon 22 May (2-5.30). Adm £5, chd free. Home-made teas in Crawley Village Hall. 2024: Sun 11, Mon 12, Sun 18, Mon 19 Feb. Opening with Crawley Gardens on Fri 7, Mon 10 Apr, Thur 6, Sun 9 July. Visits also by arrangement 10 Feb to 6 Aug.**

This walled, sheltered naturalistic garden is worth visiting in all seasons, and is specially exciting in spring. There are many established flower beds and climbers, and a traditional kitchen garden. Also free range bantams, a field of English wild flowers and many butterflies. There are informal seats to relax on both within the garden and to the surrounding countryside, each with a good view. Plants grown in the garden are for sale.

50 LONGSTOCK PARK WATER GARDEN

Leckford, Stockbridge, SO20 6EH. Leckford Estate Ltd, part of John Lewis Partnership, www.leckfordestate.co.uk. *4m S of Andover. From Leckford village on A3057 towards Andover, cross the river bridge & take 1st turning*

L signed Longstock. **Sun 18 June (1-4). Adm £10, chd £2.**
Famous water garden with extensive collection of aquatic and bog plants set in 7 acres of woodland with rhododendrons and azaleas. A walk through the park leads to National Collections of *Buddleja* and *Clematis viticella*; arboretum and herbaceous border at Longstock Park Nursery. Refreshments at Longstock Park Farm Shop and Nursery (last orders at 3.30pm). Assistance dogs only.

NPC

51 26 LOWER NEWPORT ROAD

Aldershot, GU12 4QD. Pete & Angie Myles. *Nr to the Aldershot junction of the A331. Parking is normally arranged with the factory opp 'Jondo' & the Salvation Army. Signage in place on the day if available. We are 100 metres away from the McDonalds drive through.* **Sat 1, Sun 2 July (11-4). Adm £3, chd free. Light refreshments.**
A T-shaped town garden full of ideas, split into four distinct sections; a semi-enclosed patio area with pots and water feature; a free-form lawn with a tree fern, perennials, bulbs and shrubs and over 200 varieties of hosta; secret garden with a 20ft x 6ft raised pond, exotic planting backdrop and African carvings; and a potager garden with vegetables, roses, cannas and plant storage, and sales.

52 ◆ MACPENNYS WOODLAND GARDEN & NURSERIES

Burley Road, Bransgore, Christchurch, BH23 8DB. Mr & Mrs T M Lowndes, 01425 672348, office@macpennys.co.uk, www.macpennys.co.uk. *6m SE of Ringwood, 5m NE of Christchurch. From Crown Pub Xrd in Bransgore take Burley Rd, following sign for Thorney Hill & Burley. Entrance ½m on R.* **For opening times and information, please phone, email or visit garden website.**
4 acre woodland garden originating from worked out gravel pits in the 1950s, offering interest year-round, but particularly in spring and autumn. Attached to a large nursery that offers for sale a wide selection of homegrown trees, shrubs, conifers, perennials, hedging plants, fruit trees and bushes. Tearoom offering home-made cakes, afternoon tea (pre-booking required) and light lunches, using locally sourced produce wherever possible. Nursery closed Christmas through to the New Year. Partial wheelchair access on grass and gravel paths. Can be bumpy with tree roots and, muddy in winter.

53 NEW MAGGIE'S SOUTHAMPTON

101 Tremona Road, Southampton, SO16 6HT. Maggie's Southampton. *Next to Southampton General Hospital Blue car park on Tremona Rd. There is also up to 2 hrs free parking on side roads around the centre.* **Sat 20, Sun 21 May (1.30-4.30). Adm by donation. Home-made teas.**
Designed by Sarah Price and inspired by the New Forest ecology. Paths take visitors through birch trees under-planted with mosses, ferns, and woodland flora. Long molinia grasses reach out between drifts of heathland flowers. Apple and Tibetan cherry trees bring colour and springtime enjoyment. The reflective surfaces of the centre mirror the small garden and give the space a more expansive feel.

54 MANOR HOUSE

Church Lane, Exton, SO32 3NU. Tina Blackmore. *Off A32, just N of Corhampton. Pass The Shoe Inn on your L, go to the end of the road to a T-junction, turn R & Manor House is immed on the L, just below the church.* **Sun 28 May, Sun 11 June (2-4.30). Adm £5, chd free. Home-made teas.**
An enchanting 1 acre mature walled garden set in the Meon Valley. Yew hedges and flint walls divide the garden into rooms. The white garden planted with hydrangeas and roses. Herbaceous borders with colourful cottage favourites; delphiniums, roses, geraniums and salvias. A secluded, highly productive walled vegetable garden, a parterre with fountain, woodland and meadow with wild flowers.

55 MANOR LODGE

Brook Lane, Botley, Southampton, SO30 2ER. Gary & Janine Stone. *6m E of Southampton. From A334 to the W of Botley village centre, turn into Brook Ln. Manor Lodge is ½m on the R. Limited disabled parking. Continue past Manor Lodge to parking (signed).* **Sat 13, Sun 14 May (2-5). Adm £4, chd free. Home-made teas.**
Close to Manor Farm Country Park, this mid-Victorian house (not open), set in 1½ acres is the garden of an enthusiastic plantswoman. A garden still in evolution with established areas and new projects, a mixture of informal and formal planting, woodland and wildflower meadow areas, large established and new specimen shrubs and trees. Perennials and planting combinations for year-round interest. Largely flat with hard paving, but some gravel and grass to access all areas.

56 MEON ORCHARD

Kingsmead, North of Wickham, PO17 5AU. Doug & Linda Smith, 01329 833253, meonorchard@btinternet.com. *5m N of Fareham. From Wickham take A32 N for 1½ m. Turn L at Roebuck Inn. Garden in ½m. Park on verge or in field N of property.* **Sun 3 Sept (2-5.30). Adm £5, chd free. Home-made teas.**
2 acre garden designed and constructed by current owners. An exceptional range of rare, unusual and architectural plants inc National Collection of Eucalyptus. Dramatic foliage plants from around the world, see plants you have never seen before! Bananas, tree ferns, cannas, gingers, palms and perennials dominate; streams and ponds plus an extensive range of planters complete the display. Visitors are welcome to explore the 20 acre meadow and ½ mile of the River Meon frontage attached to the garden. Extra big plant sale of the exotic and rare. Garden fully accessible by wheelchair, reserved parking.

NPC

In 2022, 20,000 people were supported by Perennial caseworkers and prevent teams thanks to our funding

Corner Cottage, Amport & Monxton Gardens

57 NEW MILL HOUSE

Vyne Road, Sherborne St John, Basingstoke, RG24 9HU. Harry & Devika Clarke. *2m N of Basingstoke. From Basingstoke take the A340 N. 400 metres beyond the hospital, turn R for Sherborne St John. Go through, past a red phone box & 400 metres up a hill. As it crests, on L, take track to Mill House.* **Thur 29 June, Sun 2 July (2-6). Adm £5, chd free. Home-made teas.**

Set in a private valley, the 3 acres of domestic fruit, vegetables, wild flowers and casual planting are arranged around a power generating watermill. Sustainability and low maintenance lie at its heart, to fit with modern life. Open views, set with mature trees and livestock, give a tranquil sense of space, with moving and static water framed in an undulating landscape. Partial wheelchair access. Park in car park and follow disabled access signs.

58 5 OAKFIELDS

Boyatt Wood, Eastleigh, SO50 4RP. Martin & Margaret Ward. *7m S of Winchester. M3 J12, follow signs to Eastleigh. 3rd exit at r'about into Woodside Ave, 2nd R into Bosville, 2nd R onto Boyatt Ln, 1st R to Porchester Rise & 1st L into Oakfields.* **Sat 10, Sun 11 June (2-5). Adm £4, chd free. Home-made teas.**

A ⅓ acre garden full of interesting and unusual plants in predominately woodland beds and, rambling roses cascading from birch trees. A pond with rockery, waterfall and flower beds formed from the intermittent winter streams, accommodate moisture loving plants. Colourful mixed herbaceous border and a terrace with architectural agapanthus and fuchsia. Wheelchair access over hard paths and some gravel.

59 THE OLD RECTORY

East Woodhay, Newbury, RG20 0AL. David & Victoria Wormsley, 07801 418976, victoria@wormsley.net. *6m SW of Newbury. Turn off A343 between Newbury & Highclere to Woolton Hill. Turn L to East End, continue ¾ m beyond East End. Turn R, garden opp St Martin's Church.* **Visits by arrangement for groups of 20+. Adm £11, chd free. Home-made teas.**

A classic English country garden of about 2 acres surrounding a Regency former rectory (not open). Formal lawns and terrace provide tranquil views over parkland. A large walled garden with topiary and roses full of interesting herbaceous plants and climbers for successional interest. Also a Mediterranean pool garden, orchard, wildflower meadow and fruit garden. Explore and enjoy.

GROUP OPENING

60 OLD THATCH & THE MILLENNIUM BARN

Sprats Hatch Lane, Winchfield, Hook, RG27 8DD. Jillian Ede, www.old-thatch.co.uk. *3m W of Fleet. 1½m E of Winchfield Stn, follow NGS signs. Turning to Sprats Hatch Ln is opp the The Barley Mow pub.* **Sun 16 Apr, Sun 11 June (2-6). Combined adm £5, chd free. Home-made teas. Pimms if hot & mulled wine if cool.**

THE MILLENNIUM BARN
Mr & Mrs G Carter.

OLD THATCH
Jill Ede.

Who could resist visiting Old Thatch, a chocolate box thatched cottage (not open), featured on film and TV, a smallholding with a 5 acre garden and woodland alongside the Basingstoke Canal (unfenced). A succession of spring bulbs, a profusion of wild flowers, perennials and homegrown annuals pollinated by our own bees and fertilised by the donkeys, who await your visit. Over 30 named clematis and rose cultivars. Children enjoy our garden quiz, adults enjoy tea and home-made cakes. Arrive by narrow boat! Trips on 'John Pinkerton' may stop at Old Thatch on NGS days www.basingstoke-canal.org.uk. Also Accessible Boating shuttle from Barley Mow wharf, approx every 45 mins. Entrance to the garden is an additional cost. Signed parking for Blue Badge holders: please use entrance by the red telephone box. Paved paths and grass slopes give access to the whole garden.

61 ◆ PATRICK'S PATCH

Fairweather's Garden Centre, High Street, Beaulieu, SO42 7YB. Patrick Fairweather, 01590 612307, info@fairweathers.co.uk, www.fairweathers.co.uk. *SE of New Forest at head of Beaulieu River. Leave M27 at J2 & follow signs for Beaulieu Motor Museum. Go up the one-way High St & park in Fairweather's car park on the L.* **For opening times and information, please phone, email or visit garden website.**

Model kitchen garden with a full range of vegetables, trained top and soft fruit and herbs. Salads in succession used as an educational project for all ages. Maintained by volunteers, primary school children and a head gardener. We run a series of fun educational gardening sessions for children as well as informal workshops for adults. Open daily by donation from dawn to dusk. Wheelchair access on gravelled site.

62 NEW PERRYMEAD

May Close, Headley, GU35 8LR. Helen & Anne Kempster. *From Crabtree Ln (B3002), turn into Liphook Rd. About 500yds along Liphook Rd is May Close which is a dirt track by the side of the allotments.* **Sat 1, Sun 2 July (10.30-5). Combined adm with Frey Elna £5, chd free. Home-made teas at Frey Elna. Opening with Headley Village Gardens on Sat 10, Sun 11 June.**

A garden built for wildlife. Mixed borders of perennials, grasses, trees and shrubs chosen for bees, birds and butterflies. Bug boxes, nest boxes and feeders. Walk over the bridge to the wildlife pond and explore our wildlife seating area. Wander round the garden and find nooks and crannies, sitting areas, patios and secret spaces where you can relax and enjoy the peace of this tranquil garden.

63 1 POVEY'S COTTAGE

Stoney Heath, Baughurst, Tadley, RG26 5SN. Jonathan & Sheila Richards, 01256 850633, smrichards3012@icloud.com, www.facebook.com/onepoveyscottage. *Between villages of Ramsdell & Baughurst, 10 min drive from Basingstoke. Take A339 out of Basingstoke, direction Newbury. Turn R off A339 towards Ramsdell, then 4m to Stoneyheath. Pass under overhead power cables, take next L into unmade road & Povey's is 1st on the R.* **Sat 10, Sun 11 June, Sat 22, Sun 23 July (1-5). Adm £5, chd free. Home-made teas. Visits also by arrangement for groups of 15+.**

Herbaceous borders, trees and shrubs, a small orchard, greenhouses, fruit cage, vegetable garden and a small wildflower meadow area. Beehives in one corner of the garden and chickens in another corner. A feature of the garden is an unusual natural swimming pond. Plant sales and sales of home-made 'Bee pottery'. Wheelchair access across flat grassed areas, but no hard pathways.

64 PYLEWELL PARK

South Baddesley, Lymington, SO41 5SJ. Lady Elizabeth Teynham. *Coast road 2m E of Lymington. From Lymington follow signs for Car Ferry to Isle of Wight, continue for 2m to South Baddesley.* **Sun 9, Sun 30 Apr (2-5). Adm £5, chd free.**

A large parkland garden laid out in 1890. Enjoy a walk along the extensive informal grass and moss paths, bordered by fine rhododendron, magnolia and azalea. Wild daffodils bloom in March and carpets of bluebells in late April. Large lakes feature giant gunnera and are home to magnificent swans. Distant views across the Solent to the Isle of Wight. Pylewell House and surrounding garden is private, and the glasshouses, swimming pool and other outbuildings are not open to visitors. Lovely day out for families and dogs. Bring your own tea or picnic and wear suitable footwear for muddy areas.

65 REDENHAM PARK HOUSE

Redenham Park, Andover, SP11 9AQ. Lady Clark. *Approx 1½m from Weyhill on the A342 Andover to Ludgershall road.* **Wed 14 June, Fri 22 Sept (2-5). Adm £6, chd £2.50. Home-made teas & cream teas in the thatched pool house.**

Redenham Park built in 1784. The garden sits behind the house (not open). The formal rose garden is planted with white flowered roses. Steps lead up to the main herbaceous borders, which peak in late summer. A calm green interlude, a gate opens into gardens with espaliered pears, apples, mass of scented roses, shrubs and perennial planting surrounds the swimming pool. A door opens onto a kitchen garden.

GROUP OPENING

66 ROMSEY GARDENS

Romsey, SO51 8LD. *Town Centre. Use Lortemore Place public car park SO51 8LD (free on Suns), which is directly next to King John's Garden. All gardens are within easy walking distance of each other.* **Sun 28, Mon 29 May (10.30-4.30). Combined adm £8, chd free. Home-made teas at King John's Garden.**

KING JOHN'S GARDEN
Friends of King John's Garden & Test Valley Borough Council, www.facebook.com/KingJohnsGarden.

LA SAGESSE CONVENT
Fiona Jenvey, 01794 830206, reception@wisdomhouseromsey.co.uk, www.wisdomhouseromsey.org.uk.

4 MILL LANE
Miss J Flindall.

THE NELSON COTTAGE
Margaret Prosser.

OLD THATCHED COTTAGE
Genevieve & Derek Langford.

King John's House is a fascinating listed C13 house (not open Sun) and the garden inc some pre-1700 period planting and a Victorian courtyard. The majestic C12 Romsey Abbey is the backdrop to 4 Mill Lane, a garden described by Joe Swift as 'the best solution for a long thin garden with a view'. The C15 listed Old Thatched Cottage on Mill Lane has a typical cottage garden with hollyhocks, wisteria and roses; it features a variety of shrubs, fruit cordons, rockery, water features, rain water harvesting and a small chicken coop. Nelson Cottage on Cherville St, formally a pub with a ½ acre garden, has a variety of perennial plants and shrubs with a wild grass meadow bringing the countryside into the town. La Sagesse Convent immediately to the south of Romsey Abbey has an unusual seven circuit meditation labyrinth, a rose garden and a walled garden. No wheelchair access at 4 Mill Lane.

67 15 ROTHSCHILD CLOSE

Southampton, SO19 9TE. Steve Campion, 07968 512773, Spcampion@talktalk.net. *3m from M27 J8. From M27 J8 follow A3025 towards Woolston. After cemetery r'about L to Weston, next r'about 2nd exit Rothschild Close.* **Sat 8, Sun 9 July, Sat 2, Sun 3 Sept (11-5). Adm £3, chd free. Pre-booking essential, please visit www.ngs.org.uk for information & booking. Home-made teas. Visits also by arrangement 1 July to 16 Sept for groups of 5 to 12.**

Small, 8 metre x 8 metre, modern city garden incorporating family living with lush tropical foliage. A large range of unusual tropical style plants inc tree ferns, bananas, cannas, gingers and dahlias. The garden has flourished over the last 6 yrs into a tropical oasis with many plants grown from seeds and cuttings. Tropical plant enthusiasts do come to have a cuppa in the jungle! Adjacent to the River Solent and Royal Victoria Country Park.

68 'SELBORNE'

Caker Lane, East Worldham, Alton, GU34 3AE. Mary Trigwell-Jones, www.worldham.org. *2m SE of Alton. On B3004 at Alton end of the village of East Worldham, nr The Three Horseshoes pub. Please note: 'Selborne' is the name of the house, it is not in the village of Selborne. Parking signed.* **Sat 1, Sun 2 July, Sun 6, Mon 7 Aug (2-5). Adm £4, chd free. Home-made teas.**

Described as a 'happy garden', this much-loved ½ acre mature cottage style garden provides visitors with surprises around every corner and views across farmland. Productive 60 yr old orchard of named varieties, densely planted borders, shrubs and climbers, especially clematis. Metal and stone sculptures enhance the borders. Containers. Bug mansion. Enjoy tea in the shade of the orchard. Summerhouses and conservatory provide shelter. Wheelchair access to parts of the garden with some gravel paths.

69 SHALDEN PARK HOUSE

The Avenue, Shalden, Alton, GU34 4DS. Mr & Mrs Michael Campbell. *4½m NW of Alton. B3349 from Alton or M3 J5 onto B3349. Turn W at Kapadokya Restaurant (formerly The Golden Pot pub) signed Herriard, Lasham, Shalden. Entrance ¼m on L. Disabled parking on entry.* **Sun 11 June (2-5). Adm £5, chd free. Light refreshments.**

Shalden Park House welcomes you to our 4 acre garden to enjoy a stroll through the arboretum, herbaceous borders, rose garden, kitchen garden and wildlife pond. Refreshments will be served from the pool terrace.

70 SILVER BIRCHES

Old House Gardens, East Worldham, nr Alton, GU34 3AN. Jenny & Roger Bateman, 07464 696245/01420 88307, Jennyabateman@gmail.com. *2m E of Alton off B3004. Turn L signed Wyck & Binstead, then 1st R into Old House Gardens.* **Visits by arrangement May to Aug for groups of 10 to 30. Adm £5, chd free. Home-made teas.**

½ acre garden to be featured in new series of ITV's 'Love Your Garden with Alan Titchmarsh' in spring 2023. Winding paths lead through colourful mixed borders to fish pond, stream, rockery and summerhouse. Rose garden with arbour. Planting designed for year-round colour and interest using foliage as well as flowers. Many sitting areas and some unusual plants. Partial wheelchair access.

71 ◆ SIR HAROLD HILLIER GARDENS

Jermyns Lane, Ampfield, Romsey, SO51 0QA. Hampshire County Council, 01794 369318, info.hilliers@hants.gov.uk, www.hants.gov.uk/hilliergardens. *2m NE of Romsey. Follow brown tourist signs off M3 J11, or off M27 J2, or A3057 Romsey to Andover.* **For opening times and information, please phone, email or visit garden website.**

Established by the plantsman Sir Harold Hillier, this 180 acre garden holds a unique collection of 12,000 different hardy plants from across the world. It inc the famous Winter Garden, Magnolia Avenue, Centenary Border, Himalayan Valley, Gurkha Memorial Garden, Magnolia Avenue, spring woodlands, Hydrangea Walk, fabulous autumn colour, 14 National Collections and over 600 champion trees. The Centenary Border is one of the longest double mixed border in the country, a feast from early summer to autumn. Celebrated

Winter Garden is one of the largest in Europe. Electric scooters and wheelchairs for hire (please pre-book). Accessible WC and parking. Registered assistance dogs only.

♿ ✿ 🚌 NPC ☕

72 SOUTH VIEW HOUSE

60 South Road, Horndean, Waterlooville, PO8 0EP. James & Victoria Greenshields. *Between Horndean & Clanfield. From N A3 towards Horndean, R at T-junction to r'about. From S A3 B2149 to Horndean, continue on A3 N to r'about, 1st exit into Downwood Way, 3rd L into South Rd. House is 3rd on R. Park in road.* **Sun 11 June (12-4). Adm £3.50, chd free. Light refreshments.**

A fusion of traditional and contemporary designs across a ½ acre site. The formal cottage garden to the front inc a large herbaceous border, small woodland garden and formal topiary. The cleverly designed garden to the rear features pond, alpine garden, mini fruit orchard, chickens and greenhouse and, for entertaining, a patio with bar (not open) and a large summerhouse. Well worth a visit. Garden is accessed along a gravel drive on a gentle slope.

♿ ✿ ☕))))

73 ◆ SPINNERS GARDEN

School Lane, Pilley, Lymington, SO41 5QE. Andrew & Vicky Roberts, 07545 432090, info@spinnersgarden.co.uk, www.spinnersgarden.co.uk. *New Forest. 1½ m N of Lymington off A337.* **For NGS: Sun 23 Apr (1.30-5). Adm £6, chd free. Cream teas. For other opening times and information, please phone, email or visit garden website.**

Peaceful woodland garden overlooking the Lymington valley with many rare and unusual plants. Drifts of trilliums, erythroniums and anemones light up the woodland floor in early spring. The garden continues to be developed with new plants added to the collections and the layout changed to enhance the views. The house was rebuilt in 2014 to reflect its garden setting. Andy will take groups of 15 on tours of the hillside with its woodland wonders and draw attention to the treats at their feet.

🐕 🚌 ☕))))

In 2022, our donations to Carers Trust meant that 456,089 unpaid carers were supported across the UK

The Old Rectory, Stockbridge Gardens

74 SPITFIRE HOUSE

Chattis Hill, Stockbridge, SO20 6JS. Tessa & Clive Redshaw. *2m from Stockbridge. Follow the A30 W from Stockbridge for 2m. Go past the Broughton/Chattis Hill Xrds, do not follow SatNav into Spitfire Ln, take next R to the Wallops & then immed R again to Spitfire House.* **Sat 27 May (2-5); Mon 29 May (11-4). Adm £5, chd free. Home-made teas on 27 May only. Visitors welcome to bring a picnic on 29 May.**

A country garden situated high on chalk downland. On the site of a WW11 Spitfire assembly factory with Spitfire tethering rings still visible. This garden has wildlife at its heart and inc fruit and vegetables, a small orchard, wildlife pond, woodland planting and large areas of wildflower meadow. Wander across the downs to be rewarded with extensive views. Wheelchair access with areas of gravel and a slope up to wildflower meadow.

GROUP OPENING

75 STOCKBRIDGE GARDENS

Stockbridge, SO20 6EX. *9m W of Winchester. On A30, at the junction of A3057 & B3049. All gardens & parking on High St.* **Thur 8, Sun 11 June (2-5). Combined adm £10, chd free. Home-made teas at St Peter's Church, High St.**

LITTLE WYKE
Mrs Mary Matthews.

THE OLD RECTORY
Robin Colenso & Chrissie Quayle.

SHEPHERDS HOUSE
Kim & Frances Candler.

TROUT COTTAGE
Mrs Sally Milligan, sally@sallymilligan.co.uk.
Visits also by arrangement 12 June to 16 July for groups of up to 10.

Four gardens will open this yr in Stockbridge, offering a variety of styles and character. Little Wyke, next to the Town Hall has a long mature town garden with mixed borders and fruit trees. Trout Cottage is a small walled garden, which will inspire those with small spaces and little time to achieve tranquillity and beauty. Full of approx 180 plants flowering for almost 10 months of the yr. The Old Rectory has a partially walled garden with formal pond, fountain and planting near the house (not open) with a stream-side walk under trees, many climbing and shrub roses and a woodland area. Shepherds House on Winton Hill with herbaceous borders, a kitchen garden and a belvedere overlooking the pond. Plant sale at the church with teas on the church lawn. Wheelchair access to all gardens. Gravel path at Shepherds House.

76 TADLEY PLACE

Church Lane, Baughurst, Tadley, RG26 5LA. Lyn & Ronald Duncan. *10 min drive from Basingstoke, nr Tadley.* **Sun 25 June (2-5). Adm £6, chd free. Light refreshments.**

Tadley Place is a Tudor manor house (not open) dating from the C15. The gardens surround the house and inc formal areas, a large kitchen garden and access to a bluebell wood with a pond.

77 TANGLEFOOT

Crawley, Winchester, SO21 2QB. Mr & Mrs F J Fratter, 01962 776243, fred@fratter.co.uk. *5m NW of Winchester. Between B3049 (Winchester - Stockbridge) & A272 (Winchester - Andover). Lane beside Crawley Court (Arqiva). For SatNav use SO21 2QA. Parking in adjacent mown field.* **Thur 15, Sun 18 June, Thur 20, Sun 23 July (2-5.30). Adm £5, chd free. Self-service drinks & biscuits. Open nearby Terstan on 18 June, 23 July only. Opening with Crawley Gardens on Thur 6, Sun 9 July. Visits also by arrangement 12 June to 28 July.**

Developed by owners since 1976, Tanglefoot's ½ acre garden is a blend of influences, from Monet-inspired rose arch and small wildlife pond to Victorian boundary wall with trained fruit trees. Highlights inc a raised lily pond, herbaceous bed (a riot of colour later in the summer), herb wheel, large productive kitchen garden and unusual flowering plants. In contrast to the garden, a 2 acre field with views over the Hampshire countryside has been converted into walk-through spring and summer wildflower meadows with mostly native trees and shrubs. The meadow contains plenty of ox-eye daisy, bedstraw, scabious, vetch, knapweed, bees, butterflies and much more; it has delighted many visitors. Visitors can picnic in the field, but there are no facilities. Plants from the garden for sale. Wheelchair access with narrow paths in vegetable area.

78 TERSTAN

Longstock, Stockbridge, SO20 6DW. Penny Burnfield, paburnfield@gmail.com, www.pennyburnfield.wordpress.com. *¾m N of Stockbridge. From Stockbridge (A30) turn N to Longstock at bridge. Garden ¾m on R.* **Sun 23 Apr, Sun 18 June, Sun 23 July, Sun 10 Sept (2-5). Adm £5, chd free. Home-made teas. Open nearby Tanglefoot on 18 June, 23 July only. Visits also by arrangement Apr to Sept for groups of 10+.**

A garden for all seasons, developed over 50 yrs into a profusely planted, contemporary cottage garden in peaceful surroundings. There is a constantly changing display in pots, starting with tulips and continuing with many unusual plants. Features inc gravel garden, water features, cutting garden, showman's caravan and live music. Wheelchair access with some gravel paths and steps.

79 THE THATCHED COTTAGE

Church Road, Upper Farringdon, Alton, GU34 3EG. Mr David & Mrs Cally Horton, 01420 587922, dwhorton@btinternet.com. *3m S of Alton off A32. From A32, take road to Upper Farringdon. At top of the hill turn L into Church Rd, follow round corner, past Masseys Folly (large red brick building) & we are the 1st house on the R.* **Sun 28, Mon 29 May, Sun 25 June, Sun 27, Mon 28 Aug (2-5). Adm £6, chd free. Home-made teas. Visits also by arrangement May to Sept for groups of 10 to 40. Donation to Cardiac Rehab.**

A 1½ acre garden hidden behind a C16 thatched cottage (not open). Borders burst with cottage garden plants and a pond provides the soothing sound of water. A pergola of roses, clematis and honeysuckle leads to many fine specimen trees. A productive garden with fruit trees, fruit cage and raised vegetable beds. Enjoy the wildflower area and gypsy caravan. Fully accessible by wheelchair after a short gravel drive.

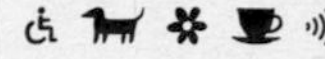

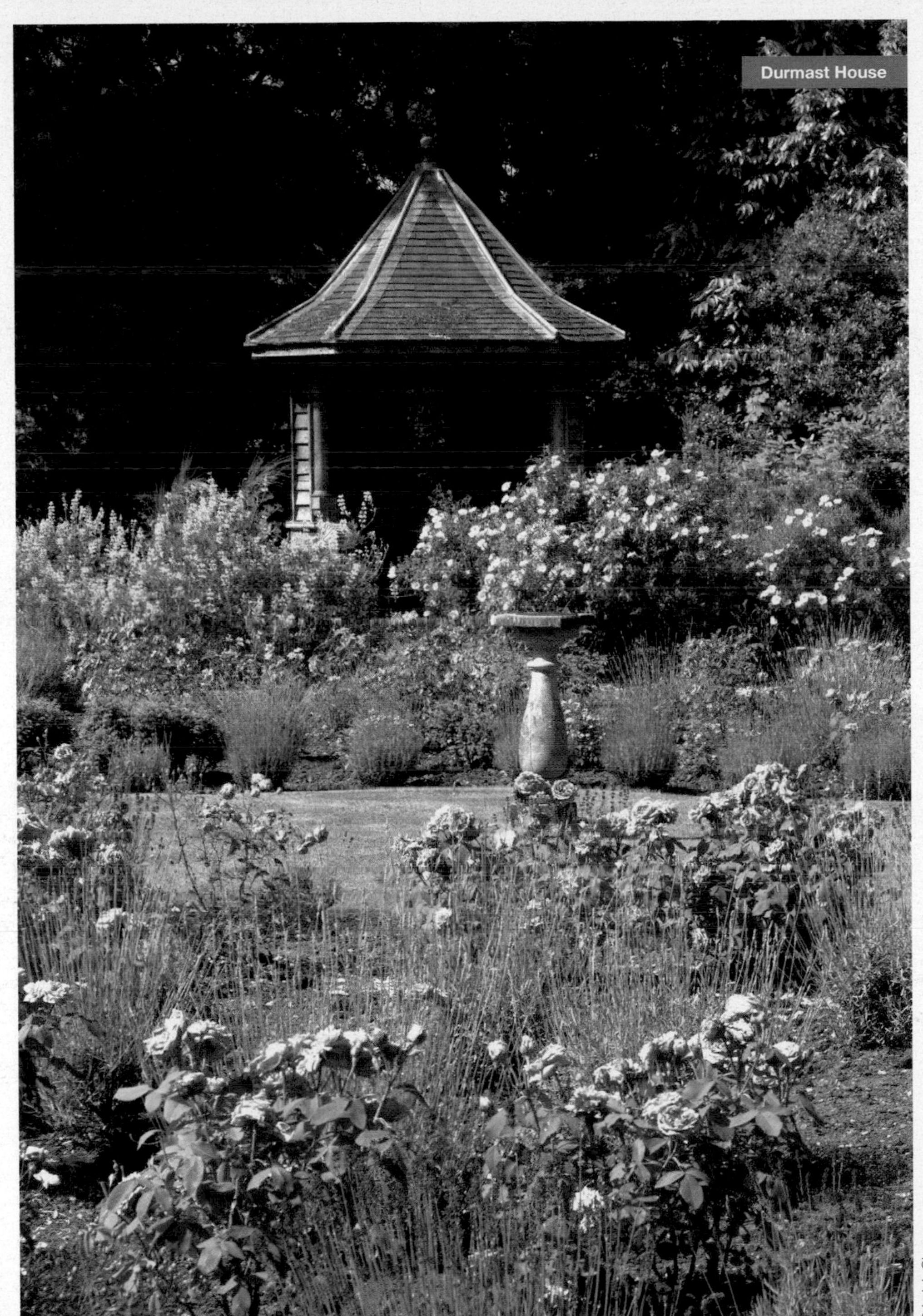
Durmast House

80 TWIN OAKS

13 Oakwood Road, Chandler's Ford, Eastleigh, SO53 1LW. Syd & Sue Hutchinson, 02380 907517, syd@sydh.co.uk, www.facebook.com/twinoaksngs. *Leave M3 J12. Follow signs to Chandlers Ford onto Winchester Rd. After ½m turn R into Hiltingbury Rd. After approx ½m turn L into Oakwood Rd.* **Sun 9, Mon 10 Apr, Sun 28, Mon 29 May, Sat 24, Sun 25 June, Sat 19, Sun 20 Aug (1-5). Adm £4, chd free. Home-made teas. Visits also by arrangement 11 Apr to 18 Aug for groups of 10 to 30.**

Evolving suburban water garden. Enjoy spring colour from azaleas, rhododendrons and bulbs then summer colour from perennials, water lilies and tropical plants. A lawn meanders between informal beds and ponds, and bridges lead to a tranquil pergola seating area overlooking a wildlife pond. A rockery is skirted by a stream and a waterfall tumbles into a large lily pond home to dragonflies. Aviary.

81 TYLNEY HALL HOTEL

Ridge Lane, Rotherwick, RG27 9AZ. Elite Hotels, 01256 764881, sales@tylneyhall.com, www.tylneyhall.co.uk. *3m NW of Hook. From M3 J5 via A287 & Newnham, M4 J11 via B3349 & Rotherwick.* **Sun 16 Apr, Sun 14 May, Sun 11 June (10-5). Adm £5, chd free. Cream teas in the Chestnut Suite from 12pm.**

Large garden of 66 acres with extensive woodlands and beautiful vista. Fine avenues of wellingtonias; rhododendrons and azaleas, Italian garden, lakes, large water and rock garden, dry stone walls originally designed with assistance of Gertrude Jekyll. Partial wheelchair access.

82 WALDEN

Common Hill, Medstead, Alton, GU34 5LZ. Terri & Neil Burman. *5m S of Alton. From N on A31 turn R into Boyneswood Rd signed Medstead. At small Xrds go R into Roedowns Rd. At T-junction at village green turn L. After church, Common Hill is 1st turning on L.* **Fri 9, Sun 11 June (1-5). Adm £5, chd free. Home-made teas.**

2 acre sloping garden on chalk, designed by the owners with panoramic views towards Winchester. Restored 60ft rockery with many alpine species and a pond. Early summer colour with hardy perennials, peonies, roses, mature shrubs and fruit trees. Interesting sculptures

Bramdean House

enhance the garden whilst recycled objects add interest and humour. Parking for disabled visitors in front of house on paved driveway. The garden is mainly grass with some slopes.

83 WALHAMPTON

Beaulieu Road, Walhampton, Lymington, SO41 5ZG. Walhampton School Trust Ltd. *1m E of Lymington. From Lymington follow signs to Beaulieu (B3054) for 1m & turn R into main entrance at 1st school sign, 200yds after top of hill.* **Sun 7 May (2-6). Adm £6, chd free. Light refreshments. Donation to St John's Church, Boldre.**

Glorious walks through large C18 landscape garden surrounding magnificent mansion (now a school). Visitors will discover three lakes, serpentine canal, climbable prospect mount, period former banana house and orangery, fascinating shell grotto, plantsman's glade and Italian terrace by Peto (c1907), drives and colonnade by Mawson (c1914) with magnificent views to the Isle of Wight. David Hill will give talks at the fountain every 30 mins to groups of 20 people, followed by a guided tour of the grounds. Wheelchair access with gravel paths and some slopes. Regret, no dogs allowed.

84 WHEATLEY HOUSE

Wheatley Lane, Kingsley, Bordon, GU35 9PA. Mrs Michael Adlington, 01420 23113, adlingtons36@gmail.com. *4m E of Alton, 5m SW of Farnham. Between Binsted & Kingsley. Take A31 to Bentley, follow sign to Bordon. After 2m, R at The Jolly Farmer pub towards Binsted, 1m L & follow signs to Wheatley.* **Sat 19, Sun 20 Aug (1.30-5.30). Home-made teas. Sun 1 Oct (2-5). Light refreshments. Adm £5, chd free. Visits also by arrangement Apr to Oct for groups of 10+.**

Situated on a rural hilltop with panoramic views over Alice Holt Forest and the South Downs. The owner admits to being much more of an artist than a plantswoman, but has had great fun creating this 1½ acre garden full of interesting and unusual planting combinations. The sweeping mixed borders and shrubs are spectacular with colour throughout the season, particularly in late summer. The black and white border, now with bright red accents, is very popular with visitors. Plants for sale. Selection of home-made, local crafts in the barn and a variety of artwork and sculptures in the garden. Wheelchair access with care on lawns, good views of garden and beyond from terrace.

85 WICOR PRIMARY SCHOOL COMMUNITY GARDEN

Portchester, Fareham, PO16 9DL. Louise Moreton, www.wicor.hants.sch.uk. *Halfway between Portsmouth & Fareham on A27. Turn S at Seagull Pub r'about into Cornaway Ln, 1st R into Hatherley Drive. Entrance to school is almost opp. Parking on site, pay at main gate.* **Sun 25 June (12-4). Adm £3.50, chd free. Home-made teas.**

As shown on BBC Gardeners' World in 2017. Beautiful school gardens tended by pupils, staff and community gardeners. Wander along Darwin's path to see the coastal garden, Jurassic garden, orchard, tropical bed, stumpery, wildlife areas, allotments and apiary, plus one of the few camera obscuras in the south of England. Wheelchair access to all areas over flat ground.

86 WINCHESTER COLLEGE

College Street, Winchester, SO23 9NA. The College, www.winchestercollege.org. *Entrance to the College is via the Porter's Lodge on College St, a short walk from Winchester City Centre. Note there is very limited parking nr the college.* **Sat 3 June, Sun 4 June (11-4). Adm £10, chd free. Home-made teas. Open nearby The Hospital of St Cross on 4 June only.**

See the historic gardens of Winchester College. These inc the main quad with herb garden, herbaceous borders and splendid climbing hydrangea; War Cloister designed by Herbert Baker, 'Bethesda' a soft cottage style garden; 'Meads' a stunning walled cricket ground surrounded by magnificent plane trees; and the Warden's Garden with Regency border, chalk stream and wilder woodland area.

87 1 WOGSBARNE COTTAGES

Rotherwick, RG27 9BL. Miss S & Mr R Whistler. *2½m N of Hook. M3 J5, M4 J11, A30 or A33 via B3349.* **Sun 9, Mon 10 July (2-5). Adm £4, chd free. Home-made teas.**

Small traditional cottage garden with a 'roses around the door' look, much photographed for calendars, jigsaws and magazines. Mixed flower beds and borders. Vegetables grown in abundance. Ornamental pond and alpine garden. Views over open countryside to be enjoyed whilst you take afternoon tea on the lawn. The garden has been open for the NGS for more than 30 yrs. Wheelchair access with some gravel paths.

88 WOODPECKERS CARE HOME

Sway Road, Brockenhurst, SO42 7RX. Mr Charles Hubberstey. *New Forest. Signed from A337. Sway Rd from village centre, past petrol station, then school & Woodpeckers on R.* **Sat 2, Sun 3 Sept (11-5). Adm £4, chd free. Home-made teas.**

A vibrant and colourful garden surrounds our care home. We have active involvement from our residents who enjoy the wide paths, whether in a wheelchair or strolling on foot. The courtyard area, small orchard and woodland, all look particularly beautiful in spring and late summer with views through neighbouring fields. An amazing bug house, chainsaw sculpture and vibrant plantings will wow you!

89 YEW HURST

Blissford, Fordingbridge, SP6 2JQ. Mr & Mrs Henry Richardson. *From Fordingbridge B3078 towards Brook. After approx 1m take R turn to Blissford. Downhill to water splash & Yew Hurst is 3rd house on the L after splash.* **Sat 17, Sun 18 June (2-6). Adm £5, chd free. Home-made teas.**

Set in a glorious position in the New Forest this is the home of a keen gardener and plant lover. A small but perfectly formed garden has lots to offer as it wraps around the house (not open) with every part having been designed to take advantage of the space. Each area has been given a sense of purpose whilst there is a flow from one part to another.

HEREFORDSHIRE

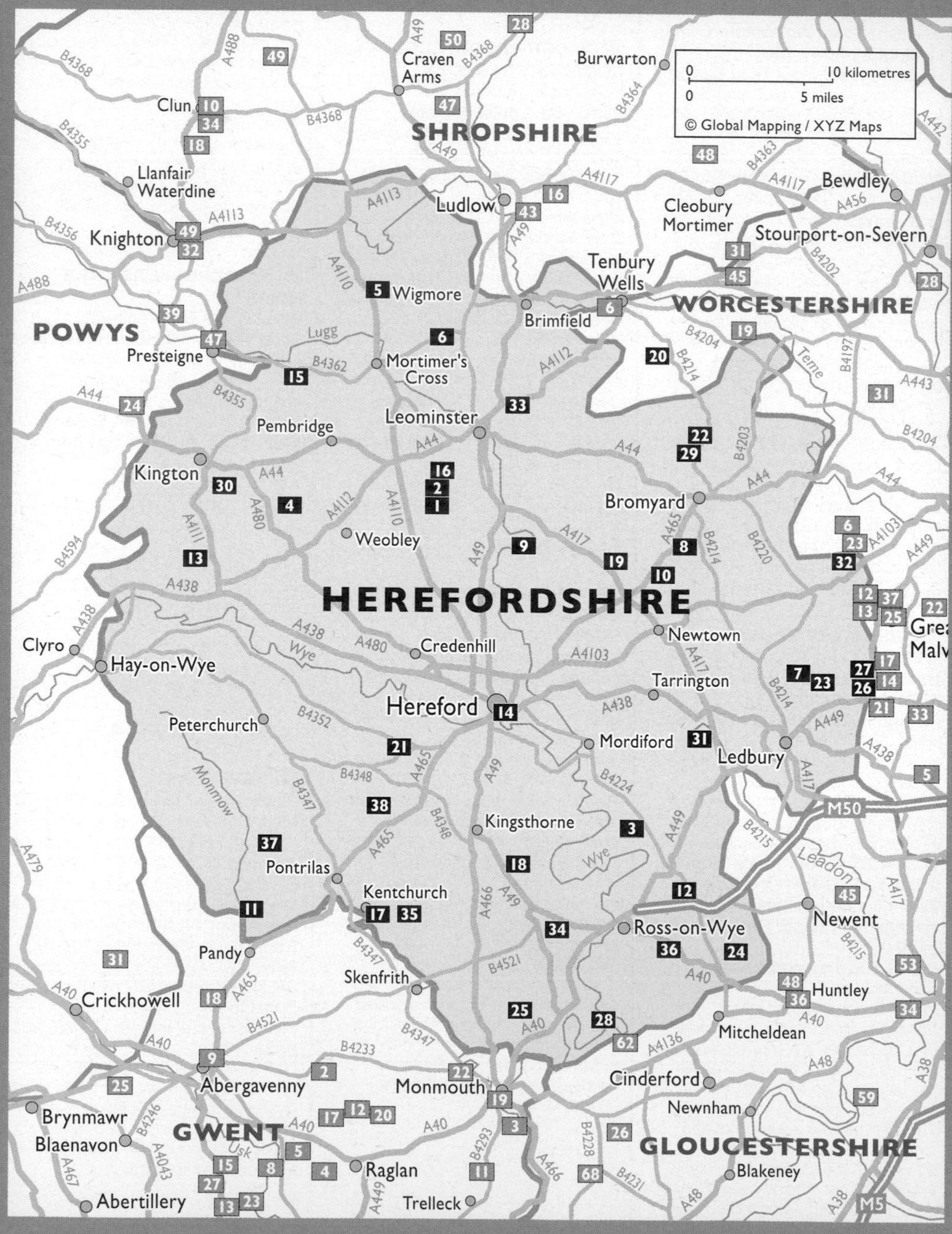

Herefordshire is essentially an agricultural county, characterised by small market towns, black and white villages, fruit and hop orchards, meandering rivers, wonderful wildlife and spectacular, and often remote, countryside (a must for keen walkers).

As a major region in the Welsh Marches, Herefordshire has a long and diverse history, as indicated by the numerous prehistoric hill forts, medieval castles and ancient battle sites. Exploring the quiet country lanes can lead to many delightful surprises.

For garden enthusiasts the National Garden Scheme offers a range of charming and interesting gardens ranging from informal cottage plots to those of grand houses with parterres, terraces and parkland. Widely contrasting in design and plantings they offer inspiration and innovative ideas.

National collections of Asters and Siberian Iris can be found at The Picton Garden and Aulden Farm, respectively; and, for Galanthophiles, Ivycroft will not disappoint. The numerous specialist nurseries offer tempting collections of rare and unusual plants.

You can always be sure of a warm welcome at a National Garden Scheme open garden.

Volunteers

County Organiser
Lavinia Sole
07880 550235
lavinia.sole@ngs.org.uk

County Treasurer
Angela Mainwaring
01981 251331
angela.mainwaring@ngs.org.uk

Booklet Coordinator
Chris Meakins
01544 370215
christine.meakins@btinternet.com

Booklet Distribution
Lavinia Sole (As above)

Social Media
Naomi Grove
naomi.grove@ngs.org.uk

Assistant County Organisers
Sue Londesborough
01981 510148
slondesborough138@btinternet.com

Penny Usher
01568 611688
pennyusher@btinternet.com

@NGSHerefordshire
@HerefordNGS

Left: Cuckhorn Estate

OPENING DATES

All entries subject to change. For latest information check www.ngs.org.uk

Extended openings are shown at the beginning of the month.

Map locator numbers are shown to the right of each garden name.

January

Thursday 26th
Ivy Croft 16

February

Snowdrop Festival

Every Thursday
Ivy Croft 16

Sunday 12th
◆ The Picton Garden 27

Saturday 18th
Wainfield 34

Sunday 26th
◆ The Picton Garden 27

March

Thursday 2nd
Ivy Croft 16

Saturday 11th
◆ Ralph Court Gardens 29

Sunday 12th
◆ Ralph Court Gardens 29

Friday 17th
◆ The Picton Garden 27

Saturday 25th
Wainfield 34

Sunday 26th
◆ Perrycroft 26
Whitfield 38

Wednesday 29th
Ivy Croft 16

Friday 31st
Coddington Vineyard 7

April

Every Wednesday
Ivy Croft 16

Saturday 1st
Coddington Vineyard 7
◆ Stockton Bury Gardens 33

Sunday 2nd
Bury Court Farmhouse 5

Monday 3rd
◆ Moors Meadow Gardens 22

Saturday 8th
◆ The Picton Garden 27

Friday 14th
Dinmore Gardens 9

Saturday 15th
Dinmore Gardens 9

Sunday 16th
Lower Hope 19

Sunday 23rd
Lower House Farm 20

Saturday 29th
Aulden Farm 2
Ivy Croft 16

Sunday 30th
Aulden Farm 2
Ivy Croft 16

May

Every Wednesday
Ivy Croft 16

Monday 8th
◆ Moors Meadow Gardens 22

Saturday 13th
◆ The Picton Garden 27

Friday 19th
Wainfield 34

Saturday 20th
Wainfield 34

Sunday 21st
Lower Hope 19
Lower House Farm 20

Saturday 27th
NEW Cuckhorn Estate 8

Sunday 28th
NEW Cuckhorn Estate 8
Sheepcote 31

June

Every Wednesday
Ivy Croft 16

Monday 5th
◆ Moors Meadow Gardens 22

Friday 9th
Kentchurch Court 17

Saturday 10th
Kentchurch Court 17
◆ The Picton Garden 27

Sunday 11th
Brockhampton Cottage 3
Broxwood Court 4
Grendon Court 12
Lower House Farm 20
◆ Perrycroft 26

Saturday 17th
Aulden Arts and Gardens 1
Byecroft 6
Ivy Croft 16

Sunday 18th
Aulden Arts and Gardens 1
Byecroft 6
Ivy Croft 16
Whitfield 38

Tuesday 20th
Hillcroft 15

Wednesday 21st
Hillcroft 15

Thursday 22nd
Hillcroft 15

Friday 23rd
◆ Hereford Cathedral Gardens 14

Friday 30th
Wainfield 34
NEW Well House 35

July

Every Wednesday
Ivy Croft 16

Saturday 1st
◆ Ralph Court Gardens 29
Wainfield 34
NEW Well House 35

Sunday 2nd
◆ Ralph Court Gardens 29

Monday 3rd
◆ Moors Meadow Gardens 22

Wednesday 5th
The Laskett Gardens 18

Friday 7th
Kentchurch Court 17

Saturday 8th
Kentchurch Court 17
◆ The Picton Garden 27

Thursday 13th
NEW White House Farm 37

Friday 14th
NEW White House Farm 37

Sunday 16th
◆ Rhodds Farm 30

Thursday 20th
Herbfarmacy 13

Friday 21st
Herbfarmacy 13

Saturday 22nd
Herbfarmacy 13

August

Every Wednesday
Ivy Croft 16

Saturday 5th
Millfield House 21

Sunday 6th
Millfield House 21

Monday 7th
◆ Moors Meadow Gardens 22

Friday 11th
Kentchurch Court 17

Saturday 12th
Kentchurch Court 17

Wednesday 16th
◆ The Picton Garden 27

Sunday 20th
Poole Cottage 28

Saturday 26th
◆ The Garden of the Wind at Middle Hunt House 11

Sunday 27th
◆ The Garden of the Wind at Middle Hunt House 11
Old Grove 25

September

Every Wednesday
Ivy Croft 16

Saturday 2nd
Aulden Farm 2
Dovecote Barn 10
Ivy Croft 16

Sunday 3rd
Aulden Farm 2
Dovecote Barn 10
Ivy Croft 16

Saturday 9th
Dovecote Barn 10

Sunday 10th
Brockhampton Cottage 3
Dovecote Barn 10
Grendon Court 12
Lower Hope 19

Wednesday 13th
◆ The Picton Garden 27

Saturday 16th
NEW Cuckhorn Estate 8

Sunday 17th
Bury Court Farmhouse 5
NEW Cuckhorn Estate 8

Sunday 24th
◆ Perrycroft 26

October

Every Wednesday to Wednesday 11th
Ivy Croft 16

Saturday 7th
Coddington Vineyard 7
◆ Ralph Court Gardens 29

Sunday 8th
◆ Ralph Court Gardens 29

Monday 16th
◆ The Picton Garden 27

November

Saturday 25th
◆ Ralph Court Gardens 29

Sunday 26th
◆ Ralph Court Gardens 29

February 2024

Sunday 25th
◆ The Picton Garden 27

By Arrangement

Arrange a personalised garden visit with your club, or group of friends, on a date to suit you. See individual garden entries for full details.

Aulden Farm 2
Bury Court Farmhouse 5
Byecroft 6
Coddington Vineyard 7
Hillcroft 15
Ivy Croft 16
Lower House Farm 20
Millfield House 21
Old Colwall House 23
The Old Corn Mill 24
Poole Cottage 28
Shuttifield Cottage 32
Wainfield 34
Weston Hall 36

Our 2022 donation supports 1,700 Queen's Nurses working in the community in England, Wales, Northern Ireland, the Channel Islands and the Isle of Man

The Old Corn Mill

© Carole Drake

THE GARDENS

GROUP OPENING

1 AULDEN ARTS AND GARDENS

Aulden, Leominster, HR6 0JT. www.auldenfarm.co.uk/auldenarts. *4m SW of Leominster. From Leominster, take Ivington/Upper Hill rd, ¾m after Ivington church turn R signed Aulden. From A4110 signed Ivington, take 2nd R signed Aulden.* **Sat 17, Sun 18 June (1-5). Combined adm £8, chd free. Home-made cake & ice cream.**

AULDEN FARM
Alun & Jill Whitehead.
(See separate entry)

HONEYLAKE COTTAGE
Jennie & Jack Hughes.

OAK HOUSE
Bob & Jane Langridge.

NEW **OLD BARN AT LOWER HYDE**
Bobbie Fennick.

Lost in the back lanes of Herefordshire, 4 neighbours looking to celebrate art and our gardens, which vary from traditional to more zany – as does our art! Honeylake Cottage; mature peaceful garden that encourages wildlife with a pond and numerous nest boxes. Cottage flowers predominate interspersed with fragrant roses, and a traditional vegetable garden - the garden inspires artwork using a variety of media. Oak House is ⅔ acre of garden, with an increased emphasis on more wild areas. Aiming to reduce mowing and encourage pollinators. The garden is divided into several areas inc borders, ponds, chickens and a vegetable plot. Photography and textiles displayed throughout the garden. The Old Barn at Lower Hyde is a barn conversion in progress, with workshops for wood and stone in the setting of an evolving wildlife habitat. With panoramic views of a developing wildflower meadow and surrounding hills, the workshop will be open for the first time to view stone and letter carving.

2 AULDEN FARM

Aulden, Leominster, HR6 0JT. Alun & Jill Whitehead, 01568 720129, web@auldenfarm.co.uk, www.auldenfarm.co.uk. *4m SW of Leominster. From Leominster take Ivington/Upper Hill rd, ¾m after Ivington church turn R signed Aulden. From A4110 signed Ivington, take 2nd R signed Aulden.* **Sat 29, Sun 30 Apr, Sat 2, Sun 3 Sept (11-4). Adm £5, chd free. Home-made teas. Opening with Aulden Arts and Gardens on Sat 17, Sun 18 June. Visits also by arrangement 17 Apr to 17 Sept. Minimum cost for groups wanting refreshments £90.**
Informal country garden, thankfully never at its Sunday best! 3 acres planted with wildlife in mind. Emphasis on structure and form, with a hint of quirkiness, a garden to explore with eclectic planting. Irises thrive around a natural pond, shady beds and open borders, seats abound, feels mature but ever evolving. Our own ice cream and home-burnt cakes, Lemon Chisel a speciality!

3 BROCKHAMPTON COTTAGE

Brockhampton, HR1 4TQ. Peter Clay. *8m SW of Hereford. 5m N of Ross-on-Wye on B4224. In Brockhampton take road signed to B Crt nursing home, pass N Home after ¾m, go down hill & turn L. Car park 500yds downhill on L in orchard.* **Sun 11 June, Sun 10 Sept (11-4). Adm £5, chd free. Open nearby Grendon Court. Picnic parties welcome by the lake.**
Created from scratch in 1999 by the owner and Tom Stuart-Smith, this beautiful hilltop garden looks south and west over miles of unspoilt countryside. On one side a woodland garden and 5 acre wildflower meadow, on the other side a Perry pear orchard and in valley below: lake, stream and arboretum. The extensive borders are planted with drifts of perennials in the modern romantic style. Allow 1½ hrs. Visit Grendon Court (11-4) after your visit to us.

4 BROXWOOD COURT

Broxwood, nr Pembridge, Leominster, HR6 9JJ. Richard Snead-Cox & Mike & Anne Allen. *From Leominster follow signs to Brecon A44/A4112. After approx 8m, just past Weobley turn off, go R to Broxwood/Pembridge. After 2m straight over Xrds to Lyonshall. 500yds on L over cattle grid.* **Sun 11 June (11-5.30). Adm £5, chd free. Home-made teas.**
Impressive 29 acre garden and arboretum, designed in 1859 by W. Nesfield. Magnificent yew hedges, long avenue of cedars and Scots pines. Spectacular view of Black Mountains, sweeping lawns, rhododendrons, gentle walks to summerhouse, chapel and lakes. Rose garden, mixed borders, rill, gazebo, sculpted benches and fountain. White and coloured peacocks, ornamental duck. Wildlife meadow in progress. Wheelchair access some gravel, but mostly lawn. Gentle slopes. Disabled WC.

♿ ✿ 🚌 ☕ ⛱

5 BURY COURT FARMHOUSE

Ford Street, Wigmore, Leominster, HR6 9UP. Margaret & Les Barclay, 01568 770618, l.barclay@zoho.com. *10m from Leominster, 10m from Knighton, 8m from Ludlow. On A4110 from Leominster, at Wigmore turn R just after shop & garage. Follow signs to parking & garden.* **Sun 2 Apr, Sun 17 Sept (2-5). Adm £5, chd free. Home-made teas. Visits also by arrangement 1 Feb to 30 Oct (excl Jun, Jul & Aug) for groups of up to 30.**
¾ acre garden surrounds the 1820's stone farmhouse (not open). The courtyard contains a pond, mixed borders, fruit trees and shrubs, with steps up to a terrace which leads to lawn and vegetable plot. The main garden (semi-walled) is on two levels with mixed borders, greenhouse, pond, mini-orchard, many spring flowers, and wildlife areas. Year-round colour. Mostly accessible for wheelchairs by arrangement.

6 BYECROFT

Welshman's Lane, Bircher, Leominster, HR6 0BP. Sue & Peter Russell, 01568 780559, peterandsuerussell@btinternet.com, www.byecroft.weebly.com. *6m N of Leominster. From Leominster take B4361. Turn L at T-junction with B4362. ¼m beyond Bircher village turn R at war memorial into Welshman's Ln, signed Bircher Common.* **Sat 17, Sun 18 June (2-5.30). Adm £4, chd free. Home-made teas. Visits also by arrangement May to July for**

groups of 10+.
A compact garden stuffed full of interesting and unusual plants chosen not just for their flowers but also for their foliage. Pergola smothered in old-fashioned roses, herbaceous borders full of things you may not have seen before, formal pond, lots of pots, vegetable garden, wildflower orchard, soft fruit area. Sue and Peter take particular pride in their plant sales table. Wheelchair access, most areas accessible with assistance. Some small steps.

7 CODDINGTON VINEYARD

Coddington, HR8 1JJ. Sharon & Peter Maiden, 01531 641817, sgmaiden@yahoo.co.uk, www.coddingtonvineyard.co.uk. *4m NE of Ledbury. From Ledbury to Malvern A449, follow brown signs to Coddington Vineyard.* **Fri 31 Mar, Sat 1 Apr (11-4); Sat 7 Oct (11-3). Adm £4, chd free. Wine, home-made ice cream & our own apple juice available. Visits also by arrangement 31 Mar to 30 Sept for groups of 10 to 50.**

5 acres inc 2 acre vineyard, listed farmhouse, threshing barn and cider mill. Garden with terraces, wildflower meadow, woodland with massed spring bulbs, large pond with wildlife, stream garden with masses of primula and hosta. Hellebores and snowdrops, hamamelis and parottia. Azaleas followed by roses and perennials. Lots to see all year. In spring, the gardens are a mass of bulbs.

8 NEW CUCKHORN ESTATE

Stoke Lacy, Bromyard, HR7 4HB. Jane & Roland Horton. *What3words app - risks.sweeten. producers. A465 to Stoke Lacy. House is on outskirts of village. Follow NGS signs.* **Sat 27, Sun 28 May, Sat 16, Sun 17 Sept (11-5). Adm £5, chd free. Light refreshments.**

An acre of gentle quintessential English flower beds and blossoms. A delightful mix of lupins, euphorbia, roses, peonies, trees and shrubs make this garden a riot of colour. The gardens are built around a natural pond, lawns add a vibrant touch of emerald green in contrast to the flower-filled beds. Regret no dogs.

GROUP OPENING

9 DINMORE GARDENS

Dinmore, Hereford, HR1 3JR. *8m N of Hereford, 8m S of Leominster. From Hereford on A49, turn R at bottom of Dinmore Hill towards Bodenham, gardens 1m on L. From Leominster on A49, L onto A417, 2m turn R & through Bodenham, following NGS signs to garden.* **Fri 14, Sat 15 Apr (11-5). Combined adm £7.50, chd free. Home-made teas at Southbourne garden.**

HILL HOUSE
Guy & Pippa Heath.

PINE LODGE
Frank Ryding.

SOUTHBOURNE
Graham & Lavinia Sole.

3 neighbouring gardens with panoramic views over Bodenham Lakes to the Malvern Hills and Black Mountains. Hill House: a 2 acre garden on the lower slopes of Dinmore Hill, bordering the Bodenham Lake Nature Reserve. Pine Lodge: 2½ acres of woodland featuring most of Britain's native trees. Goblins are thought to live here! Southbourne: 2 acre, terraced garden with herbaceous beds, shrubs, pond and ornamental woodland. Tickets sold at Hill House.

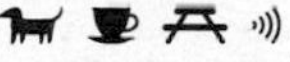

10 DOVECOTE BARN

Stoke Lacy, Bromyard, HR7 4HJ. Gill Pinkerton & Adrian Yeeles. *4m S of Bromyard on A465. Turn into lane running alongside Stoke Lacy Church. Parking in 50 metres.* **Sat 2, Sun 3, Sat 9, Sun 10 Sept (11-4). Adm £5, chd free.**

2 acre organic, wildlife friendly, garden in the unspoilt Lodon Valley. Year-round interest: ornamental vegetable and fruit gardens, peaceful and romantic pond area, copse with spring and autumn colour, winter walk, wildflower meadow, and stunning planting on late summer prairie banks. The C17 barn, framed by cottage garden beds, has views over the garden towards the Malvern Hills. Gravel paths.

♿ ✿

11 ◆ THE GARDEN OF THE WIND AT MIDDLE HUNT HOUSE

Walterstone, Hereford, HR2 0DY. Rupert & Antoinetta Otten, 01873 860600, gardenofthewind@gmail.com, www.gardenofthewind.co.uk. *4m W of Pandy, 17m S of Hereford, 10m N of Abergavenny. A465 to Pandy, W towards Longtown, turn R at Clodock Church, 1m on R. Disabled parking available. SatNav may take you via a different route but indicates arrival at adjacent farm.* **For NGS: Sat 26, Sun 27 Aug (2-6). Adm £6, chd free. For other opening times and information, please phone, email or visit garden website.**

A modern garden using swathes of herbaceous plants and grasses, surrounding stone built farmhouse and barns with stunning views of the Black Mountains. Special features: rose border, hornbeam alley, formal parterre and water rill and fountains, William Pye water feature, architecturally designed greenhouse and RIBA bridge, vegetable gardens. Carved lettering and sculpture throughout, garden covering about 4 acres. Garden seating throughout the site on stone, wood and metal benches inc some with carved lettering. Water rill and fountains. Spectacular views. Partial wheelchair access but WC facilities not easily accessible due to gravel.

12 GRENDON COURT

Upton Bishop, Ross-on-Wye, HR9 7QP. Mark & Kate Edwards. *3m NE of Ross-on-Wye. M50, J3. Hereford B4224 Moody Cow Pub, 1m open gate on R. From Ross. A40, B449, Xrds R Upton Bishop. 100yds on L by cream cottage.* **Sun 11 June, Sun 10 Sept (11-4). Adm £5, chd free. Home-made teas. Open nearby Brockhampton Cottage.**

A contemporary garden designed by Tom Stuart-Smith. Planted on two levels, a clever collection of mass-planted perennials and grasses of different heights, textures and colour give all year interest. The upper walled garden with a sea of flowering grasses makes a highlight. Picnics welcome in carpark field. Wheelchair access possible but some gravel.

13 HERBFARMACY

The Field, Eardisley, Hereford, HR3 6NB. Paul Richards, www.herbfarmacy.com. *Take A438 from Hereford to Willersley where Brecon rd turns L, cont straight on for 1m on A4111 to Eardisley & Kington. Turn off L by Tram Inn then turn R 3 times. Farm is near end of No Through Road on L, parking in field on the 1st R corner in the lane.* **Thur 20, Fri 21, Sat 22 July (10-4). Adm £5, chd free. Light refreshments.**

A 4 acre organic herb farm overlooking the Wye valley with views to the Black Mountains. Featured on BBC Countryfile, crops are grown for use in herbal skincare and medicinal products. Colourful plots of echinacea, marshmallow, mullein and calendula will be on show as well as displays on making products. Herbfarmacy products on sale in the shop. Wheelchair access possible with assistance when dry but some ground rough and some slopes.

14 ◆ HEREFORD CATHEDRAL GARDENS

Hereford, HR1 2NG. Dean of Hereford Cathedral, 01432 374251, events@herefordcathedral.org, www.herefordcathedral.org/events. *Centre of Hereford. The ticket desk is in the cathedral car park, please approach the desk to sign in with staff.* **For NGS: Fri 23 June (10-3.30). Adm £5, chd free. For other opening times and information, please phone, email or visit garden website.**

Explore the Chapter House garden, Cloister garden, the Canon's garden with plants with ecclesiastical connections and roses, the private Dean's Garden and the Bishop's Garden with fine trees and an outdoor chapel. These are open session where visitors can explore the gardens at their own pace, with Gardeners and Guides based within the gardens to share the history of the site. Partial wheelchair access. For more information please visit website or contact the team in advance.

15 HILLCROFT

Coombes Moor, Presteigne, LD8 2HY. Liz O'Rourke & Michael Clarke, 01544 262795, lorconsulting@hotmail.co.uk. *North Herefordshire, 10m from Leominster. Coombes Moor is under Wapley Hill on the B4362, between Shobdon & Presteigne. Garden on R just beyond Byton Cross when heading W.* **Tue 20, Wed 21, Thur 22 June (11-5). Adm £4, chd free. Home-made teas. Visits also by arrangement 19 June to 7 July.**

The garden is part of a 5 acre site on the lower slopes of Wapley Hill in the beautiful Lugg Valley. The highlight in mid summer is the romantic rose walk, combining over sixty roses with mixed herbaceous planting set in an old cider apple orchard. In addition to the garden area around the house, there is a secret garden, wildflower meadow and a vegetable and fruit area.

16 IVY CROFT

Ivington Green, Leominster, HR6 0JN. Roger Norman, 01568 720344, ivycroft@homecall.co.uk, www.ivycroftgarden.co.uk. *3m SW of Leominster. From Leominster take Ryelands Rd to Ivington. Turn R at church, garden ¾m on R. From A4110 signed Ivington, garden 1¾m on L.* **Every Thur 26 Jan to 2 Mar (10-4). Every Wed 29 Mar to 11 Oct (10-4). Sat 29, Sun 30 Apr (11-4). Home-made teas. Open nearby Aulden Farm. Sat 17, Sun 18 June (11-4). Home-made teas. Open nearby Aulden Arts and Gardens. Sat 2, Sun 3 Sept (11-4). Home-made teas. Open nearby Aulden Farm. Adm £5, chd free. Visits also by arrangement.**

Now over 25 yrs old, the garden shows signs of maturity, inc some surprising trees. A very wide range of plants is displayed, blending with countryside and providing habitat for wildlife. The cottage is surrounded by borders, raised beds, trained pear trees and containers giving all year interest. Paths lead to the wider garden inc mixed borders, vegetables framed with espalier apples. Snowdrops. Partial wheelchair access.

17 KENTCHURCH COURT

Kentchurch, HR2 0DB. Mr J Lucas-Scudamore, 01981 240228, enquiry@kentchurchcourt.co.uk, www.kentchurchcourt.co.uk. *12m SW of Hereford. From Hereford A465 towards Abergavanny, at Pontrilas turn L signed Kentchurch. After 2m fork L, after Bridge Inn. Garden opp church.* **Fri 9, Sat 10 June, Fri 7, Sat 8 July, Fri 11, Sat 12 Aug (11-4). Adm £7.50, chd free. Home-made teas.**

Kentchurch Court is sited close to the Welsh border. Formal rose garden, traditional vegetable garden redesigned with colour, scent and easy access. Walled garden and herbaceous borders, rhododendrons and wildflower walk. Extensive collection of mature trees and shrubs. Stream with habitat for spawning trout and a deer park at the heart of the estate. First opened for the NGS in 1927. Wheelchair access, some slopes, shallow gravel.

18 THE LASKETT GARDENS

Much Birch, HR2 8HZ. Perennial, www.perennial.org.uk. *Approx 7m from Hereford; 7m from Ross. On A49, midway between Ross-on-Wye & Hereford, turn into Laskett Ln towards Hoarwithy. The drive is approx 350yds on L.* **Wed 5 July (10.30-4.30). Adm £12.**

The Laskett Gardens are the largest private formal gardens to be created in England since 1945 consisting of 4 acres of stunning garden rooms inc rose and knot garden, fountains, statuary, topiary and a Belvedere from which to view the gardens on high. Partial wheelchair access.

19 LOWER HOPE

Lower Hope Estate, Ullingswick, Hereford, HR1 3JF. Mrs Clive Richards, www.lowerhopegardens.co.uk. *5m S of Bromyard. A465 N from Hereford, after 6m turn L at Burley Gate onto A417 towards Leominster. After approx 2m take 3rd R to Lower Hope. After ½m garden on L. Disabled parking available.* **Sun 16 Apr, Sun 21 May, Sun 10 Sept (2-5). Adm £7.50, chd £2. Light refreshments.**

Outstanding 5 acre garden with wonderful seasonal variations. Impeccable lawns with herbaceous borders, rose gardens, white garden, Mediterranean, Italian and Japanese gardens. Natural streams, man-made waterfalls, bog gardens. Woodland with azaleas and rhododendrons with lime avenue to lake with wild flowers and bulbs. Glasshouses with exotic plants and breeding butterflies. Tickets bought on the gate will be cash only. Wheelchair access to most areas.

20 LOWER HOUSE FARM
Vine Lane, Sutton, Tenbury Wells, WR15 8RL. Mrs Anne Durston Smith, 07891 928412, adskyre@outlook.com, www.kyre-equestrian.co.uk. *3m SE of Tenbury Wells; 8m NW of Bromyard. From Tenbury take A4214 to Bromyard. After approx 3m turn R into Vine Lane, then R fork to Lower House Farm.* **Sun 23 Apr, Sun 21 May, Sun 11 June (11-4). Adm £5, chd free. Home-made teas. Visits also by arrangement 1 Apr to 15 Oct.**
Award-winning country garden surrounding C16 farm-house (not open) on working farm. Herbaceous borders, roses, box-parterre, productive kitchen and cutting garden, spring garden, ha-ha allowing wonderful views. Wildlife pond. Walkers and dogs can enjoy numerous footpaths across the farm land. Autumn colour and dahlias. Home to Kyre Equestrian Centre with access to safe rides and riding events.

21 MILLFIELD HOUSE
Eaton Bishop, Hereford, HR2 9QS. Angela & Richard Mainwaring, 01981 251331, angela.mainwaring@ngs.org.uk. *4m SW of Hereford. A465 W to B4349 Hay on Wye. Approx 2m L in Clehonger B4349 to Kingstone. l00yds after 40mph sign R into lane 'unsuitable for long or wide vehicles'. Follow yellow signs.* **Sat 5, Sun 6 Aug (11-5). Adm £5, chd free. Home-made teas for visitors to help themselves. Visits also by arrangement 24 July to 25 Aug for groups of 15 to 30.**
1½ acre informal country garden shared with wildlife. The terrace and vegetable garden are enclosed by pergolas, espalier fruit trees and raised borders. Grass paths connect the borders, ponds, and wildflower meadow while formal and native hedges and trees enclose and divide the different areas. Work is progressing on new borders for 2023.

Brockhampton Cottage

22 ◆ MOORS MEADOW GARDENS

Collington, Bromyard, HR7 4LZ. Ros Bissell, 01885 410318/07942 636153, moorsmeadow@hotmail.co.uk, www.moorsmeadow.co.uk. *4m N of Bromyard, on B4214. ½m up lane follow yellow arrows.* **For NGS: Mon 3 Apr, Mon 8 May, Mon 5 June, Mon 3 July, Mon 7 Aug (10-5). Adm £7, chd £1.50. For other opening times and information, please phone, email or visit garden website.**

7 acre organic hillside garden with a vast amount of species, many rarely seen, an emphasis on working with nature to create a wildlife haven. With intriguing features and sculptures it is an inspiration to the garden novice as well as the serious plantsman. Wander through fernery, grass garden, extensive shrubberies, herbaceous beds, meadow, dingle, pools and kitchen garden. Resident Blacksmith. Huge range of unusual and rarely seen plants from around the world. Unique home-crafted sculptures.

23 OLD COLWALL HOUSE

Old Colwall, Malvern, WR13 6HF. Mr & Mrs Roland Trafford-Roberts, 07764 537181, garden1889@aol.com. *3m NE of Ledbury. From Ledbury, turn L off A449 to Malvern towards Coddington. Signed from 2½m along lane. Signed from Colwall & Bosbury. Visitors will need to walk from car park to garden.* **Visits by arrangement 31 Mar to 30 Sept for groups of 10+. Adm £5, chd free.**

Early C18 garden on a site owned by the Church till Henry VIII. Brick-walled lawns on various levels. The heart is the yew walk, a rare survival from the 1700s: 100 yds long, 30ft high, cloud clipped, and with a church aisle-like quality inside. Later centuries have brought a summerhouse, water garden, and rock gardens. Fine trees, inc enormous veteran yew; fine views. Steep in places. C18 yew walk and walled gardens.

24 THE OLD CORN MILL

Aston Crews, Ross-on-Wye, HR9 7LW. Mrs Jill Hunter, 01989 750059, mjhunterross@outlook.com, www.theoldcornmillgarden.com. *5m E of Ross-on-Wye. A40 Ross to Gloucester. Turn L at T-lights at Lea Xrds onto B4222 signed Newent, Garden ½m on L. Parking for disabled down drive. For SatNavs use HR9 7LA.* **Visits by arrangement 1 Feb to 20 Oct for groups of up to 50. Home-made teas inc in adm. Adm £5, chd free.**

C17 water mill nestling in a wooded valley. Winding paths lead through the meadows, woodland glades and orchards. Snowdrops in Feb, daffodils in Mar, tulips in Apr, orchids in May. Do come and enjoy the calm of this very natural valley garden. Pretty good cakes too.

25 OLD GROVE

Llangrove, Ross-On-Wye, HR9 6HA. Ken & Lynette Knowles. *Between Ross & Monmouth. 2m off A40 at Whitchurch. Disabled parking at house otherwise follow signs for parking in nearby field.* **Sun 27 Aug (2-6). Adm £5, chd free. Home-made teas.**

1½ acre garden plus two fields. South west facing with unspoilt views. Lots of mixed beds with plenty of late summer colour and many unusual plants. Large collection of dahlias and salvias; wildlife pond, masses of pots; interesting trees inc a new 15ft tree sculpture. Seating throughout and a large well fenced area where visiting canines can stretch their legs. Wheelchair access, gentle slopes, some steps but accessible.

Grendon Court

26 ◆ PERRYCROFT

Jubilee Drive, Upper Colwall, Malvern, WR13 6DN. Gillian & Mark Archer, 07858 393767, info@perrycroft.co.uk, www.perrycroft.co.uk. *Between Malvern & Ledbury. On B4232 between British Camp & Wyche cutting. Park in Gardiners Quarry pay & display car park on Jubilee Drive, short walk to the garden.* **For NGS: Sun 26 Mar (12-4); Sun 11 June (12-5); Sun 24 Sept (12-4). Adm £5, chd free. Home-made teas in the Coach House. For other opening times and information, please phone, email or visit garden website.**

Perrycroft is an Arts and Crafts house designed by CFA Voysey in 1895. Situated high on the Malvern Hills, with magnificent views to the south and west accross Herefordshire and into Wales, the garden has been developed over the past 23 yrs. Yew and box hedges, topiary, mixed and herbaceous borders, shrubs and roses, meadows, ponds and woodland combine to create a relaxed and varied garden with traditional and contemporary planting styles.

27 ◆ THE PICTON GARDEN

Old Court Nurseries, Walwyn Road, Colwall, WR13 6QE. Mr & Mrs Paul Picton, 01684 540416, oldcourtnurseries@btinternet.com, www.autumnasters.co.uk. *3m W of Malvern. On B4218 (Walwyn Rd) N of Colwall Stone. Turn off A449 from Lodbury or Malvern onto the B4218 for Colwall.* **For NGS: Sun 12, Sun 26 Feb, Fri 17 Mar (11-4); Sat 8 Apr, Sat 13 May, Sat 10 June, Sat 8 July, Wed 16 Aug, Wed 13 Sept, Mon 16 Oct (11-5). Adm £5, chd free. 2024: Sun 25 Feb. For other opening times and information, please phone, email or visit garden website. Donation to Plant Heritage.**

1½ acres west of Malvern Hills. Bulbs and a multitude of woodland plants in spring. Interesting perennials and shrubs in Aug. In late Sept and early Oct colourful borders display the National Plant Collection of Michaelmas daisies, backed by autumn colouring trees and shrubs. Many unusual plants to be seen, inc more than 100 different ferns and over 300 varieties of snowdrop. National Plant Collection of autumn-flowering asters and an extensive nursery that has been growing them since 1906.

28 POOLE COTTAGE

Coppett Hill, Goodrich, Ross-on-Wye, HR9 6JH. Jo Ward-Ellison & Roy Smith, 07718 229813, jo@ward-ellison.com. *5m from Ross on Wye, 7m from Monmouth. Above Goodrich Castle in Wye Valley AONB. Goodrich signed from A40 or take B4234 from Ross. No parking close to garden on 20th August. Park in village & follow signs. Shuttle service available to garden or walk 10-15 mins up from village.* **Sun 20 Aug (10.30-5.30). Adm £5, chd free. Home-made teas. Visits also by arrangement July & Aug for groups of up to 20. Parking for up to 5 cars only at the garden for by arrangement visitors.**

A modern country garden in keeping with the local natural landscape. Home to designer Jo Ward-Ellison the 2 acre garden is predominately naturalistic in style with a contemporary feel. A long season of interest with grasses and later flowering perennials. Some steep slopes, steps and uneven paths. Features inc a small orchard, a pond loved by wildlife and kitchen garden with fabulous views.

29 ◆ RALPH COURT GARDENS

Edwyn Ralph, Bromyard, Hereford, HR7 4LU. Mr & Mrs Morgan, 01885 483225, ralphcourtgardens@aol.com, www.ralphcourtgardens.co.uk. *1m from Bromyard. From Bromyard follow the Tenbury rd for approx 1m. On entering the village of Edwyn Ralph take 1st turning on R towards the church.* **For NGS: Sat 11, Sun 12 Mar, Sat 1, Sun 2 July, Sat 7, Sun 8 Oct, Sat 25, Sun 26 Nov (10-3.30). Adm £12, chd £8. Light refreshments. For other opening times and information, please phone, email or visit garden website.**

All tickets must be purchased through the garden website or at the gate. 13 amazing gardens set in the grounds of a gothic rectory. A family orientated garden with a twist, incorporating an Italian Piazza, an African Jungle, Dragon Pool, Alice in Wonderland and the elves in their conifer forest and our new section 'The Monet Garden'. These are just a few of the themes within this stunning garden. Overlooking the Malvern Hills 120 seater licensed restaurant. Offering a good selection of daily specials, delicious Sunday roasts, afternoon tea and our scrumptious homemade cakes. All areas ramped for wheelchair and pushchair access. Some grass areas, without help can be challenging during wet periods.

30 ◆ RHODDS FARM

Lyonshall, Kington, HR5 3LW. Richard & Cary Goode, 01544 340120, cary.goode@russianaeros.com, www.rhoddsfarm.co.uk. *1m E of Kington. From A44 take small turning S just E of Penrhos Farm, 1m E of Kington. Continue 1m, garden straight ahead (a little further than SatNav sends you).* **For NGS: Sun 16 July (11-5). Adm £7, chd free. Light refreshments. Self service. For other opening times and information, please phone, email or visit garden website.**

Created by the owner, a garden designer, over the past 16 yrs, the garden contains an extensive range of interesting plants. Formal garden with dovecote and 100 white doves, mixed borders, double herbaceous borders of hot colours, large gravel garden, three ponds, arboretum, perennial and wildflower meadows and 13 acres of woodland. A natural garden that fits the setting with magnificent views. Interesting and unusual trees, shrubs and perennials. A natural garden on a challenging site. A number of sculptures by different artists. No pesticides used.

31 SHEEPCOTE

Putley, Ledbury, HR8 2RD. Tim & Julie Beaumont. *5m W of Ledbury off the A438 Hereford to Ledbury Rd. Passenger drop off; parking 200 yds.* **Sun 28 May (2-5.30). Adm £5, chd free. Light refreshments.**

⅓ acre garden taken in hand from 2011 retaining many quality plants, shrubs and trees from earlier gardeners. Topiary holly, box, hawthorn, privet and yew formalise the varied plantings around the croquet lawn and gravel garden; beds with heathers, azaleas, lavender surrounded by herbaceous perennials and bulbs; small ponds; kitchen garden with raised beds.

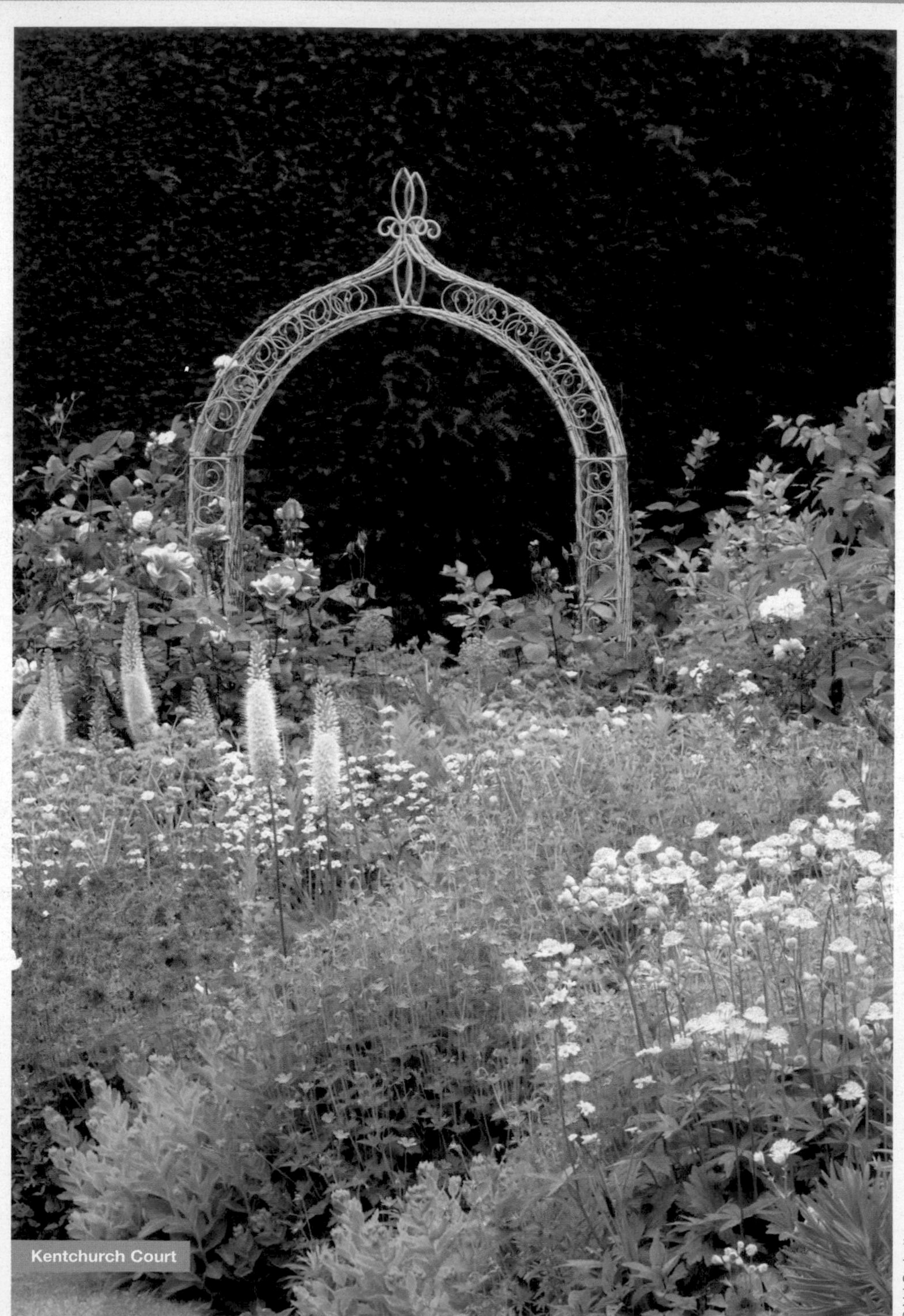
Kentchurch Court

© Val Corbett

32 SHUTTIFIELD COTTAGE

Birchwood, Storridge, WR13 5HA. Mr & Mrs David Judge, 01886 884243, judge.shutti@btinternet.com. *15m E of Hereford. Turn L off A4103 at Storridg opp the Church to Birchwood. After 1¼m L down steep tarmac drive. Please park on roadside at the top of the drive but drive down if walking is difficult (150 yards).* **Visits by arrangement Apr to Sept. Adm £5, chd free. Home-made teas.**

Superb position and views. Unexpected 3 acre plantsman's garden, extensive herbaceous borders, primula and stump bed, many unusual trees, shrubs, perennials, colour themed for all year interest. Anemones, bluebells, rhododendrons, azaleas and camelias are a particular spring feature. Large old rose garden with many spectacular climbers. Small deer park, vegetable garden and 20 acre wood. Wildlife ponds, wild flowers and walks in 20 acres of ancient woodland.

33 ◆ STOCKTON BURY GARDENS

Kimbolton, HR6 0HA. Raymond G Treasure, 07880 712649, twstocktonbury@outlook.com, www.stocktonbury.co.uk. *2m NE of Leominster. From Leominster to Ludlow on A49 turn R onto A4112. Gardens 300yds on R.* **For NGS: Sat 1 Apr (11-4.30). Adm £9, chd £5. Home-made teas in Tithe Barn Cafe. For other opening times and information, please phone, email or visit garden website.**

Superb, sheltered 4 acre garden with colour and interest from Apr until the end of Sept. Extensive collection of plants, many rare and unusual set amongst medieval buildings. Features pigeon house, tithe barn, grotto, cider press, auricula theatre, pools, secret garden, garden museum and rill, all surrounded by unspoilt countryside. Garden and Cafe are also open Wed-Sunday. See website. Garden museum and Roman hoard. Home made teas, light refreshments, wine in Tithe Barn Café. Partial wheelchair access. A fit able bodied companion is advisable for wheelchair users.

34 WAINFIELD

Peterstow, Ross-On-Wye, HR9 6LJ. Nick & Sue Helme, 01989 730663. *From the A49 between Ross-on-Wye & Hereford. Take the B4521 to Skenfrith/ Abergavenny. Wainfeild is 50yds on R.* **Sat 18 Feb, Sat 25 Mar, Fri 19, Sat 20 May, Fri 30 June, Sat 1 July (10.30-4). Adm £5, chd free. Home-made teas. Visits also by arrangement 1 Feb to 28 July for groups of 25 to 50.**

3 acre informal, wildlife garden inc rose garden, fruit trees, climbing roses and clematis. Delightful pond with waterfall. Climbing roses, many different clematis and honeysuckle. In spring, tulips, bluebells, crocuses and grasses followed by lush summer planting. Fruit walk with naturalised cowslips all set in an open area of interesting, unusual mature trees and sculptures. Wheelchair access on even ground

35 NEW WELL HOUSE

Garway Hill, Hereford, HR2 8RT. Mrs Betty Lovering. *12m SW Hereford A465 from Hereford turn L at Pontrilas onto B4347 turn 1st R & immed L signed Orcop & Garway Hill, 3m to Bagwyllydiart then follow signs.* **Fri 30 June, Sat 1 July (10-4.30). Adm £5, chd free. Pre-booking essential, please visit www.ngs.org.uk for information & booking. Light refreshments.**

Hillside garden, approx 1 acre views over Orcop on Garway Hill. This garden has been created from scratch since 1998 on a steep bank with steps and narrow paths and is ongoing with nature in mind. Herbaceous beds, mature trees, shrubs, wide collection of acers and unusual plants, veg garden, polytunnel, bog garden seating areas, on a working small holding with Dutch Spotted Sheep. Lovely views lots of seating. Secret areas.

36 WESTON HALL

Weston-under-Penyard, Ross-on-Wye, HR9 7NS. Mr P Aldrich-Blake, 01989 562597, aldrichblake@btinternet.com. *1m E of Ross-on-Wye. On A40 towards Gloucester.* **Visits by arrangement Apr to Sept for groups of 5 to 30. Light refreshments for modest extra charge. Adm £5, chd free.**

6 acres surrounding Elizabethan house (not open). Large walled garden with herbaceous borders, vegetables and fruit, overlooked by Millennium folly. Lawns and mature and recently planted trees and shrubs, with many unusual varieties. Orchard, ornamental ponds and lake. 4 generations in the family, but still evolving year on year. Wheelchair access to walled garden only.

38 WHITFIELD

Wormbridge, HR2 9BA. Mr & Mrs Edward Clive, www.whitfield-hereford.com. *8m SW of Hereford. The entrance gates are off the A465 Hereford to Abergavenny rd, ½m N of Wormbridge Postcode for SatNav HR2 9DG.* **Sun 26 Mar, Sun 18 June (2-5). Adm £5, chd free. Home-made teas.**

Parkland, wild flowers, ponds, walled garden, many flowering magnolias (species and hybrids), 1780 ginkgo tree, 1½m woodland walk with 1851 grove of coastal redwood trees. Dogs on leads welcome. Delicious teas. Partial access for wheelchair users, some gravel paths and steep slopes.

Our 2022 donation enabled 7,000 people affected by cancer to benefit from the garden at Maggie's in Oxford

HERTFORDSHIRE

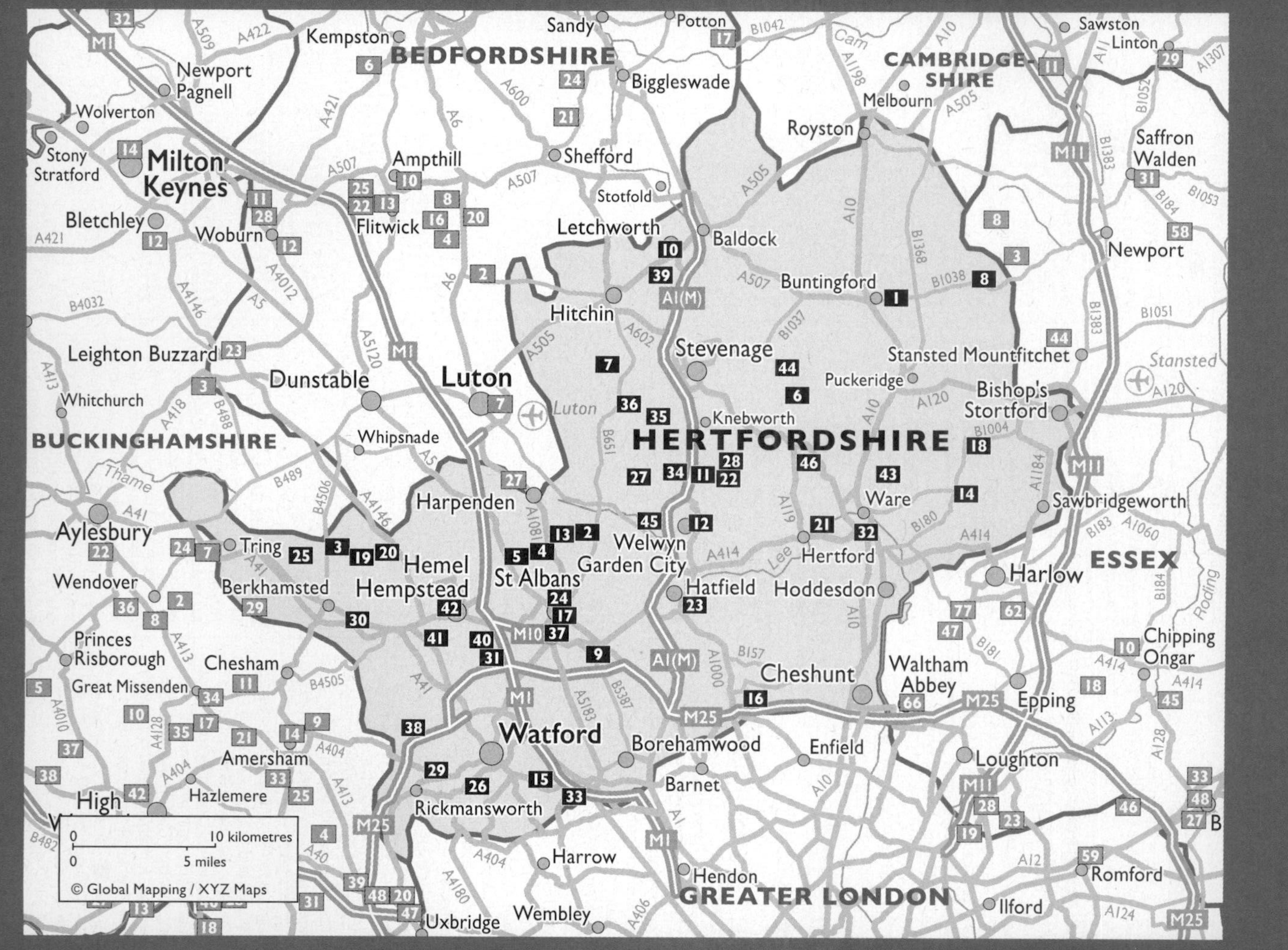

Hertfordshire is full of surprises. We have wonderful transport links to London and our neighbours around the M25. This has always been part of our appeal and why so many have come here, both historically and recently to live, work and enjoy the space and countryside close to the city.

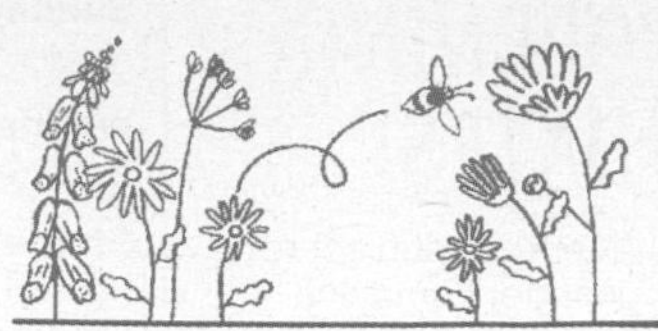

It is why the county is peppered with medieval deer park and woodland, great estates, smaller country estates with parkland, manor houses and some wonderful historic landscapes associated with celebrated designers. We are lucky that Hatfield House, St Pauls Walden Bury and Ashridge all open for us. We are also home to the first pioneer Garden Cities, Letchworth and then Welwyn Garden City, embedding the ideal of a garden and open space as central to urban development.

Don't be tempted to whiz through. If you dive into any of our sprawling urban areas or tiny rural hamlets you can find gardens that offer everything from contemporary garden design to rewilded spaces, from biodiverse wildflower meadows to productive potagers, from ancient woodlands to tropical oases. There is inspiration for everyone. We also have some exciting new gardens to add to our wonderful list this year. Amongst these there are two community gardens which are a first for us.

Many open 'by arrangement' so get in touch if you need help arranging a private group tour.

We raised a record amount last year thanks to you our visitors, volunteers and gardens owners. We look forward to another record year in 2023. Come and be surprised by Hertfordshire.

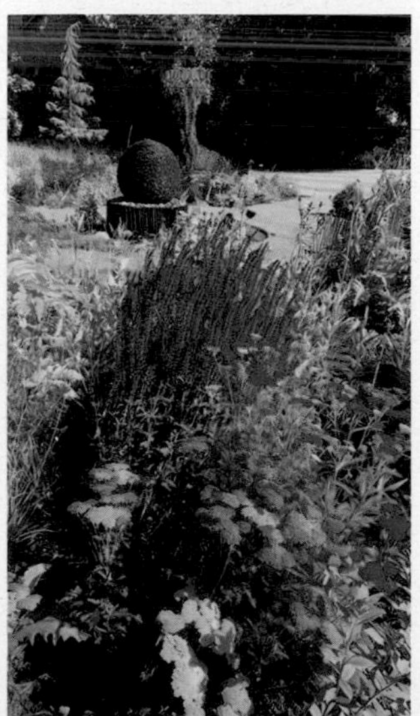

Left: The Lodge

Volunteers

County Organisers
Bella Stuart-Smith 07710 099132
bella.stuart-smith@ngs.org.uk

Kate Stuart-Smith 07551 923217
kate.stuart-smith@ngs.org.uk

County Treasurer
Peter Barrett 01442 393508
peter.barrett@ngs.org.uk

Publicity
Shubha Allard
shubha.allard@ngs.org.uk

Photography
Lucy Standen 07933 261347
lucy.standen@ngs.org.uk

Julie Meakins 07899 985324
meakinsjulie@gmail.com

Social Media - Facebook
Anastasia Rezanova
anastasia.rezanova@ngs.org.uk

Social Media - Twitter
Mark Lammin 07966 625559
markjango@msn.com

Social Media - Instagram
Rebecca Fincham
rebecca.fincham@ngs.org.uk

Radio (Position vacant)

Booklet Coordinator
Janie Nicholas 07973 802929
janie.nicholas@ngs.org.uk

New Gardens
Julie Wise 07759 462330
julie.wise@ngs.org.uk

Group Tours
Sarah Marsh 07813 083126
sarah.marsh@ngs.org.uk

Assistant County Organisers
Parul Bhatt
parul.bhatt@ngs.org.uk

Kerrie Lloyd-Dawson 07736 442883
kerrie.lloyddawson@ngs.org.uk

Barbara Goult 07712 131414
barbara.goult@ngs.org.uk

Tessa Birch 07721682481
tessa.birch@ngs.org.uk

@HertfordshireNGS 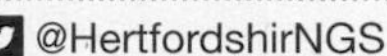 @HertfordshirNGS 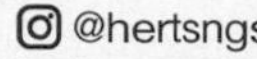 @hertsngs

OPENING DATES

All entries subject to change. For latest information check **www.ngs.org.uk**

Extended openings are shown at the beginning of the month.

Map locator numbers are shown to the right of each garden name.

February

Snowdrop Festival

Wednesday 1st
8 Gosselin Road 21

Friday 3rd
8 Gosselin Road 21

Saturday 4th
Walkern Hall 44

Sunday 5th
Walkern Hall 44

Tuesday 7th
◆ Benington Lordship 6

Wednesday 8th
8 Gosselin Road 21

March

Saturday 18th
◆ Hatfield House West Garden 23

Wednesday 22nd
8 Gosselin Road 21

Saturday 25th
Walkern Hall 44

Sunday 26th
Amwell Cottage 2
Walkern Hall 44

April

Sunday 2nd
◆ St Paul's Walden Bury 36

Monday 10th
10 Cross Street 10

Sunday 16th
Alswick Hall 1

Wednesday 19th
8 Gosselin Road 21

Sunday 23rd
Serendi 39

Sunday 30th
Beesonend House 5
Patchwork 30
Pie Corner 31

May

Saturday 13th
NEW Sunnyside Rural Trust Trail 41

Sunday 14th
39 Firs Wood Close 16
◆ St Paul's Walden Bury 36

Saturday 20th
The Manor House, Ayot St Lawrence 27
Morning Light 29

Sunday 21st
The Manor House, Ayot St Lawrence 27
Morning Light 29

Sunday 28th
◆ Benington Lordship 6
15 Gade Valley Cottages 20
The Pines 32

Monday 29th
◆ Ashridge House 3
43 Mardley Hill 28

June

Every day from Saturday 24th
42 Falconer Road 15

Friday 2nd
16 Langley Crescent 24

Saturday 3rd
16 Langley Crescent 24

Sunday 4th
NEW Brambley Hedge 7
NEW Gaddesden House 19
9 Tannsfield Drive 42

Thursday 8th
NEW Danesbury Fernery 11

Saturday 10th
NEW Danesbury Fernery 11

Sunday 11th
◆ St Paul's Walden Bury 36

Friday 16th
NEW Eastmoor Lodge 13
1 Elia Cottage 14
Rustling End Cottage 35

Saturday 17th
Rustling End Cottage 35

Sunday 18th
NEW Eastmoor Lodge 13
1 Elia Cottage 14
Serge Hill Gardens 40
Thundridge Hill House 43

Friday 23rd
28 Fishpool Street 17
NEW 54 Reynards Road 34

Saturday 24th
NEW The Lodge 25

Sunday 25th
28 Fishpool Street 17
NEW 54 Reynards Road 34
St Stephens Avenue Gardens 37

Friday 30th
2 Barlings Road 4

July

Every day to Sunday 2nd
42 Falconer Road 15

Saturday 1st
2 Barlings Road 4

Sunday 2nd
9 Tannsfield Drive 42
Woodhall Park 46

Saturday 8th
Gable House 18

Saturday 15th
NEW Sarratt Community Garden 38

Sunday 16th
Greenwood House 22
NEW Sarratt Community Garden 38

Sunday 23rd
15 Gade Valley Cottages 20

Sunday 30th
35 Digswell Road 12

August

Every day from Saturday 26th
◆ The Celebration Garden 9

Saturday 5th
12 Longmans Close 26

Sunday 6th
12 Longmans Close 26
Reveley Lodge 33

Sunday 20th
Patchwork 30

Wednesday 23rd
8 Gosselin Road 21

September

Every day to Sunday 10th
◆ The Celebration Garden 9

Sunday 3rd
St Stephens Avenue Gardens 37

Sunday 24th
Alswick Hall 1
The Pines 32

Wednesday 27th
8 Gosselin Road 21

October

Wednesday 4th
8 Gosselin Road 21

Sunday 29th
42 Falconer Road 15

By Arrangement

Arrange a personalised garden visit with your club, or group of friends, on a date to suit you. See individual garden entries for full details.

Our 2022 donation to ABF The Soldier's Charity meant 372 soldiers and veterans were helped through horticultural therapy

THE GARDENS

1 ALSWICK HALL

Hare Street Road, Buntingford, SG9 0AA. Mike & Annie Johnson, www.alswickhall.com/gardens. *1m from Buntingford on B1038. From the S take A10 to Buntingford, drive into town & take B1038 E towards Hare Street Village. Alswick Hall is 1m on R.* **Sun 16 Apr, Sun 24 Sept (12-4). Adm £8, chd free. Hog roast and home-made teas.**
Listed Tudor House with 5 acres of landscaped gardens set in unspoiled farmland. 2 well established natural ponds with rockeries. Herbaceous borders, shrubs, woodland walk and wildflower meadow with a fantastic selection of daffodils, tulips, camassias and crown imperials. Later in the year, enjoy the spectacular dahlia beds and late season planting, formal beds and orchard. Licensed bar, hog roast, teas and delicious home-made cakes. A plant stall and various trade stands as well as children's entertainment. Good access for disabled with lawns and wood chip paths. Slight undulations.

2 AMWELL COTTAGE

Amwell Lane, Wheathampstead, AL4 8EA. Colin & Kate Birss. *½m S of Wheathampstead. From St Helen's Church, Wheathampstead turn up Brewhouse Hill. At top L fork (Amwell Lane), 300yds down lane, park in field opp.* **Sun 26 Mar (2-5). Adm £5, chd free. Home-made teas.**
Informal garden of approx 2½ acres around C17 cottage. Large orchard of mature apples, plums and pear, planted with daffodils and laid out with paths. Extensive lawns with borders, framed by tall yew hedges and old brick walls. A large variety of roses, stone seats with views, woodland pond, greenhouse, vegetable garden with raised beds and fire-pit area. Wheelchair access, gravel drive.

♿ 🐕 ✿ ☕ 🔊

3 ◆ ASHRIDGE HOUSE

Berkhamsted, HP4 1NS. E F Corporate Education Ltd, 01442 843 491, tickets@ashridge.hult.edu, www.ashridgehouse.org.uk. *3m N of Berkhamsted. A4251, 1m S of Little Gaddesden.* **For NGS: Mon 29 May (9-6). Adm £5, chd £2.50. Ashridge House only accepts payments by card. Options to dine inside & out in the courtyard, for both light refreshments & substantial meals. For other opening times and information, please phone, email or visit garden website.**
The gardens cover 190 acres forming part of the Grade II Landscape of Ashridge Park. Based on designs by Humphry Repton, who presented the Red Book detailing his designs in 1813, and modified by Jeffry Wyatville. A collection of small, secluded gardens, and a large lawn area leading to avenues of trees. In late May, a highlight of the garden is Rhododendron Walk. Once a monastic site then home to Henry VIII and his children. One of Repton's finest gardens with influences from the Bridgewater dynasty, comprising colourful formal bedding, a rosary, shrubberies, topiary, stunning rhododendrons and tree lined avenues leading to an arboretum.

4 2 BARLINGS ROAD

Harpenden, AL5 2AN. Liz & Jim Machin. *1m S of Harpenden. Take A1081 S from Harpenden, after 1m turn R into Beesonend Ln, bear R into Burywick to T-junc. Turn R and follow signs.* **Evening opening Fri 30 June, Sat 1 July (5-8). Adm £5, chd free. Wine.**
This garden has been developed over many years to provide a haven of privacy and peace in an urban setting. Mature trees, shrubs, climbers and perennials give year-round structure and colour. Island beds invite inspection from several perspectives. The tranquil courtyard garden uses plants from a limited colour palette. A shady corner and formal pond add extra interest.

5 BEESONEND HOUSE

Beesonend Lane, Harpenden, AL5 2AB. John & Sarah Worth. *Turn off A1081(turn R from Harpenden & L from St Albans). Keep L up Beesonend Ln, past cottages & derelict stables on the L. Beesonend House is on L next to white stones.* **Sun 30 Apr (11-4.30). Adm £5, chd free. Light refreshments.**
4 garden areas. East garden lawn with herbaceous borders, pleached hedges, Japanese-themed border sitting under eucalyptus tree. South garden laid lawn with herbaceous borders with semi-mature Lebanese cedar. South-West facing garden inc pond and orchard. North garden comprises circular lawn, pots, roses and herbaceous border. Vegetable garden with raised borders and a greenhouse.

Our donation of £2.5 million over five years has helped fund the new Y Bwthyn NGS Macmillan Specialist Palliative Care Unit in Wales. 1,900 inpatients have been supported since its opening

6 ◆ BENINGTON LORDSHIP

Stevenage, SG2 7BS. Mr & Mrs R Bott, 01438 869668, garden@beningtonlordship.co.uk, www.beningtonlordship.co.uk. *4m E of Stevenage. In Benington Village, next to church. Signs off A602.* **For NGS: Tue 7 Feb (11-4); Sun 28 May (2-5). Adm £7, chd free. Light refreshments in February, served in tea room. In May, refreshments in village hall, adjacent to garden. For other opening times and information, please phone, email or visit garden website.**

7 acre site dating back to Saxon times which inc the ruin of a C12 Norman keep and C19 neo Norman folly. Garden comprises rose garden, walled kitchen garden with wildflower meadow, vegetables and bantams, orchard and lakes. Spectacular snowdrops, spring bulbs and herbaceous borders. Unspoilt panoramic views over surrounding parkland.

7 NEW BRAMBLEY HEDGE

1 Chequers Lane, Preston, Hitchin, SG4 7TX. Lynda & Steve Woodward. *3 m S of Hitchin, Herts. From A602, Three Moorhens r'about take exit to Gosmore, pass through Gosmore and continue for about 2 m into Preston. Chequers Le is 1st on R, by white fence.* **Sun 4 June (12-5.30). Adm £4, chd free. Home-made teas.**

Brambley Hedge covers an area of 0.25 acres approximately. The front garden is mainly set to shrubs and small trees. The rear garden is sections or 'rooms' planted in a cottage style, with around 40 varieties of roses, popular and some less common perennials. There is also a vegetable plot and numerous pots for annuals. The owner is a keen amateur wood turner and some of his work will be on show. The paths are mainly flat and level, some may be slightly narrow.

8 BRENT PELHAM HALL

Brent Pelham, Buntingford, SG9 0HF. Alex & Mike Carrell, catherine.cocks@enactorsupport.com. *From Buntingford take the B1038 E for 5m. From Clavering take the B1038 W for 3m.* **Visits by arrangement Apr to Sept for groups of 10+. Adm £8, chd free. Light refreshments in the Estate Office.**

Surrounding a beautiful Grade I listed property, the gardens consist of 12 acres of formal gardens, redesigned in 2007 by the renowned landscaper Kim Wilkie. With 2 walled gardens, a potager, walled kitchen garden, greenhouses, orchard and a new double herbaceous border, there is lots to discover. The further 14 acres of parkland boast lakes and wildflower meadows. All gardened organically. Access by wheelchair to most areas of the garden, inc paths of paving, gravel and grass.

9 ◆ THE CELEBRATION GARDEN

North Orbital Road, St Albans, AL2 1DH. Aylett Nurseries Ltd, 01727 822255, info@aylettnurseries.co.uk, www.aylettnurseries.co.uk. *The Celebration Garden is adjacent to Garden Centre. Aylett Nurseries is on the eastbound carriageway of A414, S of St Albans, between Park Street r'about & London Colney r'about. Turn L on entry & drive through green gates at end of car park.* **For NGS: Daily Sat 26 Aug to Sun 10 Sept (10-4). Adm by donation. Light refreshments in Tea Tent by the garden offering sandwiches, cakes, ice creams with hot/cold drinks, further refreshments in the Café in the Garden Centre. For other opening times and information, please phone, email or visit garden website.**

The Celebration Garden is sited next to our famous Dahlia Field. Dahlias are planted amongst other herbaceous plants and shrubs. We also have a wildflower border complete with insect hotel. The garden is open all year to visit, during the garden centre opening hours, but it is especially spectacular from July to early autumn when the dahlias are in flower. Annual Autumn Festival held in September. Refreshments in our Seasons Café in the Garden Centre available year-round. Wheelchair access to the Celebration Garden via grass paths.

10 10 CROSS STREET

Letchworth Garden City, SG6 4UD. Renata & Colin Hume, www.cyclamengardens.com. *Nr town centre. From A1(M) J9 signed Letchworth, across 2 r'abouts, R at 3rd, across next 3 r'abouts L into Nevells Rd, 1st R into Cross St.* **Mon 10 Apr (2-6). Adm £5, chd free. Home-made teas.**

A garden with mature fruit trees is planted for interest throughout the year. The structure of the garden evolved around 3 circles- 2 grass lawns and a wildlife pond. Mixed borders connect the different levels of the garden. A newt pond.

11 NEW DANESBURY FERNERY

North Ride, Welwyn, AL6 9RD. Welwyn Hatfield Borough Council, www.danesburyfernery.org.uk. *Exit Junc 6 off A1M signed to Welwyn Village and follow yellow signs from there. 1½ m from junc 6. Danesbury Fernery is located in Danesbury Nature Reserve. Street parking available nearby.* **Evening opening Thur 8, Sat 10 June (5.30-8.30). Adm £5, chd free. Wine.**

Danesbury Fernery was designed and built in 1859. The circular grassy site contains a Pulhamite (an artificial rock) grotto and cascade, Victorian water cistern and a rustic bridge over a plant river! Now planted with a myriad of ferns, perennials and a wild flower chalk bank. It was a derelict site that has been restored since 2015 by volunteers. We were awarded a Conservation Award in 2018.

12 35 DIGSWELL ROAD

Welwyn Garden City, AL8 7PB. Adrian & Clare de Baat, 01707 324074, adrian.debaat@ntlworld.com, www.adriansgarden.org. *½ m N of Welwyn Garden City centre. From the Campus r'about in city centre take N exit just past the Public Library into Digswell Rd. Over the White Bridge, 200yds on L.* **Sun 30 July (2-5.30). Adm £5, chd free. Home-made teas. Visits also by arrangement July to Oct for groups of 10 to 20. Adm incl refreshments.**

Town garden of around a ⅓ acre with naturalistic planting inspired by the Dutch garden designer, Piet Oudolf. The garden has perennial borders plus a small meadow packed with herbaceous plants and grasses. The contemporary planting gives way to the exotic, inc a succulent bed and under mature trees, a lush jungle garden inc bamboos, bananas, palms and tree ferns. Daisy Roots Nursery will be selling plants. Grass paths and gentle slopes to all areas of the garden.

13 NEW EASTMOOR LODGE

East Common, Harpenden, AL5 1DA. Ekaterina and Julian Gilbert, kate2304g@yahoo.com. *1m S of Harpenden, turn L onto Limbrick Rd. After 400 yards, turn R onto East Common Rd. After 100 yds turn L. The house is the 3rd on the L. For parking use the car park opp Bamville Cricket club.* **Fri 16, Sun 18 June (1-5). Adm £5. Home-made teas. Visits also by arrangement 10 May to 30 June for groups of 5 to 15.**

Beautiful modern garden of half an acre featuring an extensive collection of David Austin roses, various perennials, dwarf azaleas, rhododendrons and beautiful hydrangeas. Within a woodland setting of giant redwood and other conifer trees. Not suitable for wheelchairs due to steps.

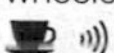

14 1 ELIA COTTAGE

Nether Street, Widford, Ware, SG12 8TH. Margaret & Hugh O'Reilly, 01279 843324, hughoreilly56@yahoo.co.uk. *B1004 from Ware towards Much Hadham down dip towards MH at Xrd take RH into Nether St. 8m W of Bishop's Stortford on B1004 through Much Hadham at Widford sign turn L. B180 from Stanstead Abbots.* **Fri 16 June (12-5.30); Sun 18 June (12.30-5.30). Adm £5, chd free. Light refreshments. Visits also by arrangement 18 Feb to 30 Sept for groups of 5 to 16.**

A ⅓ acre garden reflecting the seasons. Snowdrops, hellebores and crocus welcome visitors in spring Followed later in June by clematis and roses uneven paths with pond, cascade water features, stream with Monet-style bridge. Plenty of seats and two summerhouses. Steep nature of garden means we are sorry no wheelchair access. NEW 2022: ½ acre land opp, with wildflower meadow.

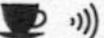

15 42 FALCONER ROAD

Bushey, Watford, WD23 3AD. Mrs Suzette Fuller, 077142 94170. *M1 J5 follow signs for Bushey. From London A40 via Stanmore towards Watford. From Watford via Bushey Arches, through to Bushey High St, turn L into Falconer Rd, opp St James church.* **Daily Sat 24 June to Sun 2 July (12-6). Evening opening Sun 29 Oct (7-9.30). Adm £4, chd free. Mulled Wine on 29th October. Visits also by arrangement May to Sept for groups of up to 10.**

Enchanting, magical and unusual Victorian style space. Bird cages and chimney pots feature, plus a walk through conservatory with many plants. Children so very welcome. Winter viewing for fairyland lighting, for all ages. Bring a torch.

16 39 FIRS WOOD CLOSE

Potters Bar, EN6 4BY. Val & Peter Mackie. *Potters Bar, High Street (A1000) fork R towards Cuffley, along the Causeway. Immediately before the T-lights/Chequers pub, R down Coopers Ln Rd. ½ m on L Firs Wood Close.* **Sun 14 May (11-5). Adm £5, chd free. Home-made teas.**

Situated within Northaw Park, a 3 acre garden made up of a woodland area, a field inc a wildflower meadow and a more formal ½ acre around the house. Raised vegetable beds and Victorian greenhouse. Choice of seating areas inc a breeze house, sunken fire pit area and colourful patio benches under the pergola covered with wisteria. Tranquil views up to Northaw village. Wheelchair access, flat access from road to back garden and then on grass.

17 28 FISHPOOL STREET

St Albans, AL3 4RT. Jenny & Antony Jay. *28 is at the Cathedral end of Fishpool St. Entrance to the garden is through the Lower Red Lion car park.* **Evening opening Fri 23 June (6-8). Wine. Sun 25 June (2-5). Home-made teas. Adm £6, chd free.**

Sculpted box and yew hedging and a C17 Tripe House feature strongly in this tranquil oasis set in the vicinity of St Albans Cathedral. Gravel paths lead to a lawn surrounded by late flowering sustainable herbaceous perennial borders and a relaxed woodland retreat. Imaginative planting in all areas offer unique perspectives. Plants are on sale. Not suitable for wheelchairs due to differing levels.

18 GABLE HOUSE

Church Lane, Much Hadham, SG10 6DH. Tessa & Keith Birch. *Situated in the picturesque & historic village of Much Hadham. Follow signs to the church. Continue around the R bend & past several white cottages. Gable House driveway is immed on R. Please park in the High St.* **Sat 8 July (2-6). Adm £5, chd free. Home-made teas.**

Enjoy this colourful and walled village garden of just under an acre. Surrounding the house on 3 sides, the design is both formal and naturalistic, with structural clipped evergreens and wildlife friendly grassy areas. The planting displays abundant year-round borders, with an emphasis on strong colour and varied texture. Features inc a woodland walk and a cuttings garden. Matched funding of all sales goes to Brain Tumour Research. Wheelchair access, level with easy access.

Amwell Cottage

© Mike Howes

19 NEW GADDESDEN HOUSE

Nettleden Road, Little Gaddesden, Berkhamsted, HP4 1PP. David & Jilly Scriven. *3½ m N of Berkhamsted. SW of Little Gaddesden village on Nettleden Rd. Look out for farm gate with yellow signs.* **Sun 4 June (11-6). Adm £5, chd free. Home-made teas.**

10 acre garden with lovely views onto the NT Golden Valley. Variety of established wildflower meadows, 30m mixed border planted with bulbs, grasses, 'prairie' plants, perennials, unusual annuals, shrubs and topiarised holly trees, kitchen garden with raised beds and bay topiary, an orchard, rhododendrons, interesting trees and woodland. Lots to see and places to wander through. All grass pathways, some uneven. Garden mainly flat with some slightly sloping paths. Woodland & back field meadows on steep hill.

20 15 GADE VALLEY COTTAGES

Dagnall Road, Great Gaddesden, Hemel Hempstead, HP1 3BW. Bryan Trueman. *3m N of Hemel Hempstead. Follow B440 N from Hemel Hempstead. Through Water End. Go past turning for Great Gaddesden on L. Park in village hall car park on R. Gade Valley Cottages on R (short walk).* **Sun 28 May, Sun 23 July (1.30-5). Adm £4.50, chd free. Home-made teas.**

Medium sized sloping rural garden. Patio, lawn, borders and pond. Paths lead through a woodland area emerging by wildlife pond and sunny border. A choice of seating offers views across the beautiful Gade Valley or quiet shady contemplation with sounds of rustling bamboos and bubbling water. Many acers, hostas and ferns in shady areas. Hemerocallis, dahlias, iris, crocosmia and phlox found in sun.

21 8 GOSSELIN ROAD

Bengeo, Hertford, SG14 3LG. Annie Godfrey & Steve Machin, www.daisyroots.com. *Take B158 from Hertford signed to Bengeo. Gosselin Rd 2nd R after White Lion Pub.* **Wed 1, Fri 3, Wed 8 Feb, Wed 22 Mar, Wed 19 Apr, Wed 23 Aug, Wed 27 Sept, Wed 4 Oct (1-4). Adm £5, chd free.**

Daisy Roots nursery garden acts as trial ground and show case for perennials and ornamental grasses grown there. Over 150 varieties of snowdrop in February, deep borders packed with perennials and grasses later in the year. Small front garden with lots of foliage interest. Regret no dogs.

22 GREENWOOD HOUSE

2a Lanercost Close, Welwyn, AL6 0RW. David & Cheryl Chalk. *1½ m E of Welwyn village, close to J6 on A1(M). Off B197 in Oaklands. Opp North Star pub turn into Lower Mardley Hill. Park here unless you are disabled as there is very limited parking at house (10 min walk). Take Oaklands Rise to top of the hill & take L fork.* **Sun 16 July (2-5). Adm £5, chd free. Home-made teas.**

Secluded garden of ⅓ acre surrounds our contemporary home and backs on to a beautiful ancient woodland. Garden has been transformed in last 12 years, with mature trees providing a natural backdrop to the many shrubs and perennials which provide interest and colour throughout the seasons. Access to rear garden across pebble paths and some steps.

23 ◆ HATFIELD HOUSE WEST GARDEN

Hatfield, AL9 5HX. The Marquess of Salisbury, 01707 287010, r.ravera@hatfield-house.co.uk, www.hatfield-house.co.uk. *Pedestrian Entrance to Hatfield Park is opp Hatfield Railway Stn, from here you can obtain directions to the gardens. Free parking is available, please use AL9 5HX with a SatNav.* **For NGS: Sat 18 Mar (11-4). Adm £9, chd £4.50. For other opening times and information, please phone, email or visit garden website. Donation to another charity.**

Visitors can enjoy the spring bulbs in the lime walk, sundial garden and view the famous Old Palace garden, childhood home of Queen Elizabeth I. The adjoining woodland garden is at its best in spring with masses of naturalised daffodils and bluebells. Beautifully designed gifts, jewellery, toys and much more can be found in the Stable Yard shops. Visitors can also enjoy relaxing at the River Cottage Restaurant which serves a variety of delicious foods throughout the day. There is a good route for wheelchairs around the West garden and a plan can be picked up at the garden kiosk.

24 16 LANGLEY CRESCENT

St Albans, AL3 5RS. Jonathan Redmayne. *½ m S of St Albans city centre. J21a M25, follow B4630. At mini-r'about by King Harry Pub turn L. At big r'about turn R along A4147, then R at mini-r'about. Langley Cres is 2nd L.* **Evening opening Fri 2 June (6-9). Wine. Sat 3 June (2.30-5.30). Home-made teas. Adm £5, chd free.**

A compact walled garden set on a slope with a wide range of unusual herbaceous perennials, shrubs and trees, both edible and ornamental, and a gentle ambience. The rear garden is divided into 2 parts; the lower section comprises extensive herbaceous beds inc a collection of Phlomis and leads to a secluded area for quiet contemplation beside a wildlife pond surrounded by rambling roses. Access by wheelchair to most areas of the garden, inc paved paths, grass and brick terrace beside pond.

25 NEW THE LODGE

New Ground Road, Aldbury, Tring, HP23 5SF. Tim & Tess Alps. *0.5m SW of Aldbury Village. Follow Newground Rd from the A4251 (Berkhamsted/Tring rd) towards Aldbury, across canal & rail bridges; the Lodge is the 1st house on the L. Parking in the field beyond the house.* **Sat 24 June (2-6). Adm £5, chd free. Home-made teas. Teas, coffees, soft drinks and cakes.**

A garden designed to attract pollinating insects. 1 acre of formal garden, inc pond, water features, sculptures, woodland walk) bordered by 3 acres of wildflower meadows, an orchard and vegetable garden. All-round views of the Chilterns and Ashridge Forest. The Lodge is part of a group of buildings which once formed a Victorian Isolation Hospital. There are many large trees still from that era in the garden which lined the original road down to the wards. Step-free route possible around the garden incl mown paths through meadows. Some bark, gravel and pebble paths and sloping lawn. No access to deck.

The Manor House, Ayot St Lawrence

© Anna Omiotek

26 12 LONGMANS CLOSE
Byewaters, Watford, WD18 8WP. Mark Lammin, www.instagram.com/hertstinytropicalgarden. *Leave M25 J18 (A404) & follow Rickmansworth/ Croxley Green then A412 to Watford. Follow signs for Watford & Croxley Business Parks & then NGS signs. Park on neighbouring roads if you can.* **Sat 5, Sun 6 Aug (2-6). Adm £4, chd free. Home-made teas.**
Hertfordshire's Tiny Tropical Garden. See how dazzling colour, scent, lush tropical foliage, trickling water and clever use of pots in a densely planted small garden can transport you to the tropics. Stately bananas and canna rub shoulders with delicate lily and roses amongst a large variety of begonia, hibiscus, ferns and houseplants in a tropical theme more often associated with warmer climes.

27 THE MANOR HOUSE, AYOT ST LAWRENCE
Welwyn, AL6 9BP. Rob & Sara Lucas. *4m W of Welwyn. 20 mins J4 A1M. Take B653 Wheathampstead. Turn into Codicote Rd follow signs to Shaws Corner. Parking in field, short walk to garden. A disabled drop-off point is available at the end of the drive.* **Sat 20, Sun 21 May (11-5). Adm £7.50, chd free. Home-made teas. Visitors paying at the gate on the day will be charged £8.**
A 6 acre garden set in mature landscape around Elizabethan Manor House (not open). 1 acre walled garden inc glasshouses, fruit and vegetables, double herbaceous borders, rose and herb beds. Herbaceous perennial island beds, topiary specimens. Parterre and temple pond garden surround the house. Gates and water features by Arc Angel. Garden designed by Julie Toll. Home-made cakes, tea and coffee. Produce for sale.

D

28 43 MARDLEY HILL
Welwyn, AL6 0TT. Kerrie & Pete, www.agardenlessordinary.blogspot.co.uk. *5m N of Welwyn Garden City. On B197 between Welwyn & Woolmer Green, on crest of Mardley Hill by bus stop for Arriva 300/301.* **Mon 29 May (1-5). Adm £5, chd free. Home-made teas.**
An unexpected garden created by plantaholics and packed with unusual plants. Focus on foliage and long season of interest. Various areas: alpine bed; sunny border; deep shade; white-stemmed birches and woodland planting; naturalistic stream, pond and bog; chicken house and potted vegetables; potted exotics. Seating areas on different levels. Homegrown plants for sale. Featured in Garden News and Garden Answers.

29 MORNING LIGHT

7 Armitage Close, Loudwater, Rickmansworth, WD3 4HL. Roger & Patt Trigg, 01923 774293, roger@triggmail.org.uk. *From M25 J18 take A404 towards Rickmansworth, after ¾m turn L into Loudwater Ln, follow bends, then turn R at T-junc & R again into Armitage Close.* **Plant sales Sat 20 & Sun 21 May (10-4). Visits also by arrangement Apr to Sept for groups of up to 20.**

South facing plantsman's garden, densely planted with mainly hardy and tender perennials and shrubs in a shady environment. Astilbes and phlox prominently featured. The garden inc island beds, pond, chipped paths and raised deck. Tall perennials can be viewed advantageously from the deck. Large conservatory (450 sq ft) stocked with sub-tropicals.

30 PATCHWORK

22 Hall Park Gate, Berkhamsted, HP4 2NJ. Jean & Peter Block, 01442 864731, patchwork2@btinternet.com. *3m W of Hemel Hempstead. Entering E side of Berkhamsted on A4251, turn L 200yds after 40mph sign.* **Sun 30 Apr, Sun 20 Aug (2-5). Adm £5, chd free. Light refreshments. Visits also by arrangement 1 Apr to 29 Oct for groups of 5 to 30.**

¼ acre garden with lots of year-round colour, interest and perfume, particularly on opening days. Sloping site containing rockeries, 2 small ponds, herbaceous borders, island beds with bulbs in spring and dahlias in summer, roses, fuchsias, hostas, begonias, patio pots and tubs galore - all set against a background of trees and shrubs of varying colours. Seating and cover from the elements.

31 PIE CORNER

Millhouse Lane, Bedmond, WD5 0SG. Bella & Jeremy Stuart-Smith, 07710099132, piebella1@gmail.com. *Between Watford & Hemel Hempstead. 1½m from J21 M25. 3m from J8 of M1. Go to the centre of Bedmond. Millhouse Ln is opp the shops. Entry is 50m down Millhouse Ln. Parking in field.* **Sun 30 Apr (2-5). Adm £6, chd free. Home-made teas. Visits also by arrangement May & June for groups of 15 to 30.**

A garden designed to complement the modern classical house. Formal borders near the house, pond and views across lawn towards valley. More informal shrub plantings with bulbs edge the woodland. A dry garden leads through new meadow planting to the vegetable garden. Enjoy blossom, bulbs, wild garlic, bluebells and young rhododendron planting in late spring. Wheelchair access to all areas on grass or gravel paths except the formal pond where there are steps.

32 THE PINES

58 Hoe Lane, Ware, SG12 9NZ. Peter Laing. *Approx ½m S of Ware centre. At S (top) end of Hoe Ln, close to Hertford Rugby Club (car parking) and opp Pinewood School. Look for prominent white gateposts with lions.* **Sun 28 May, Sun 24 Sept (2-5.30). Adm £5, chd free. Home-made teas.**

Plantsman's garden with many unusual plants, created by present owner over 30 yrs. 1 acre, E/W axis so much shade, sandy soil over chalk but can grow ericaceous plants. Front garden formal with fountain. Main garden mature trees, herbaceous borders and island beds. Features inc gravel garden, pergola and obelisk with moss rose "William Lobb". Among many specimen trees, the rare Kashmir cypress. Wheelchairs access ok unless recent heavy rain.

33 REVELEY LODGE

88 Elstree Road, Bushey Heath, WD23 4GL. Reveley Lodge Trust, www.reveleylodge.org. *3½m SE of Watford & 2m N of Stanmore.*

Serge Hill

From A41 take A411 signed Bushey & Harrow. At mini-r'about take 2nd exit into Elstree Rd. Garden ½ m on l. Disabled parking only onsite. **Sun 6 Aug (1-5). Adm £5, chd free. Home-made teas.**
2½ acre garden surrounding a Grade II listed Victorian house. Restored conservatory with collection of tender and exotic plants. Rose garden planted with many old roses. Variety of perennial and annual planting inc medicinal and tropical plants. Vegetable and cut flower garden and beehives. Teas and delicious home-made cakes. Variety of plant and art and craft stalls. Live music. Partial wheelchair access.

34 NEW 51 REYNARDS ROAD

Welwyn, AL6 9TP. Jackie and David Naylor. *Reynards Rd is off the Welwyn to Codicote Rd (B656). 1 m NW of Welwyn village.* **Evening opening Fri 23 June (5-8.30). Wine. Sun 25 June (2-5.30). Home-made teas. Adm £5, chd free.**
An enthusiast's garden of approx 2 acres with large perennial borders, pergolas, an orchard and extensive trees and hedging set at the end of Reynards Road bordering onto fields. A large naturalistic pond, a small woodland walk with ferns and a Mediterranean border add to the mix. There is also a variety of hard landscaping and many different seating areas within the garden. Most areas accessible by wheelchair.

35 RUSTLING END COTTAGE

Rustling End, Codicote, SG4 8TD. Julie & Tim Wise, www.rustlingend.com. *1m N of Codicote. From B656 turn L into '3 Houses Ln' then R to Rustling End. House 2nd on L.* **Evening opening Fri 16, Sat 17 June (5-8.30). Adm £5, chd free. Wine.**
Meander through our wildflower meadow to a cottage garden with contemporary planting. Behind lumpy hedges explore a garden managed for wildlife. Natural planting provides an environment for birds, small mammals and insects. Our terrace features drought tolerant low maintenance plants. A flowery mead surrounds the formal pond and an abundant floral vegetable garden provides produce for the summer. Hens in residence.

36 ◆ ST PAUL'S WALDEN BURY

Whitwell, Hitchin, SG4 8BP. The Bowes Lyon family, stpaulswalden@gmail.com, www.stpaulswaldenbury.co.uk. *5m S of Hitchin. On B651; ½m N of Whitwell village. From London leave A1(M) J6 for Welwyn (not Welwyn Garden City). Pick up signs to Codicote, then Whitwell.* **For NGS: Sun 2 Apr, Sun 14 May, Sun 11 June (2-7). Adm £7.50, chd free. Home-made teas. For other opening times and information, please email or visit garden website. Donation to St Paul's Walden Charity.**
Spectacular formal woodland garden, Grade I listed, laid out 1720, covering over 50 acres. Long rides lined with clipped beech hedges lead to temples, statues, lake and a terraced theatre. Seasonal displays of daffodils, cowslips, irises, magnolias, rhododendrons, lilies. Wildflowers are encouraged. This was the childhood home of the late Queen Mother. Children welcome. Dogs on leads. Pre-booking available online at ngs.org.uk. Good wheelchair access to part of the garden. Steep grass slopes in places.

GROUP OPENING

37 ST STEPHENS AVENUE GARDENS

St Albans, AL3 4AD. *1m S of St Albans City Centre. From A414 take A5183 Watling St. At mini-r'about by St Stephens Church/King Harry Pub take B4630 Watford Rd. St Stephens Ave is 1st R.* **Sun 25 June, Sun 3 Sept (2.30-5.30). Combined adm £6, chd free. Home-made teas.**

20 ST STEPHENS AVENUE
Heather & Peter Osborne.

30 ST STEPHENS AVENUE
Carol & Roger Harlow.

2 gardens of similar size and the same aspect, developed in totally different ways. No. 20 is designed for year-round colour, fragrance and structure. Dense planting makes it impossible to see from one end of the garden to the other, enticing visitors to explore. Paths meander through and behind the colour coordinated borders, giving access to all parts. Varied plant habitats inc cool shade, hot dry gravel and lush pondside displays. New borders are developed or areas replanted every year. Seating is in both sun and shade, a large conservatory provides shelter. No. 30 has a southwest facing graveled front garden that has a Mediterranean feel. Herbaceous plants, such as sea hollies and achilleas, thrive in the poor, dry soil. Clipped box, beech and hornbeam in the back garden provide a cool backdrop for the strong colours of the herbaceous planting. A gate beneath a beech arch frames the view to the park beyond. Plants for sale at June opening only.

38 NEW SARRATT COMMUNITY GARDEN

The Green, Sarratt, Rickmansworth, WD3 6AT. Sarratt Post Office Stores, 07759777475, floragarveygardening@icloud.com, m.facebook.com/sarrattcommunitygarden/. *Directly behind Sarratt Post Office Stores, in the centre of Sarratt Village.* **Sat 15, Sun 16 July (10.30-5). Adm £5, chd free. Home-made teas. Visits also by arrangement 1 Apr to 2 Sept for groups of 10 to 30.**
A thriving, productive community garden, growing fruit, veg and cut flowers for local people. This garden is run on organic principles, with a passion for the environment and education. There will be refreshments available, and there is space to sit for around 40 people. Parking is available in the village (extra in the KGV playing fields if needed). Wheeled access to refreshment area only. Paths in the garden unsuitable.

Our 2022 donation to Hospice UK meant that 911 NHS and Social Care staff accessed individual counselling support

39 SERENDI
22 Hitchin Road, Letchworth Garden City, SG6 3LT.
Valerie, 07548776809, valerie.aitken@ntlworld.com, www.instagram.com/serendigarden/?hl=en. *1m from city centre. A1 J9 signed Letchworth on A505. At 2nd r'about take 1st exit Hitchin A505. Straight over T-lights. Garden 1m on R.* **Sun 23 Apr (11-5). Adm £5, chd free. Home-made teas. Other refreshments available. Visits also by arrangement 25 Apr to 30 Oct for groups of up to 25.**
A mass of tulips and other spring bulbs. Many different areas within a well designed garden: silver birch grove, a 'dribble of stones', rill, contemporary knot garden, dry area with alliums. Later in the year an abundance of roses climbing 5 pillars, perennials, grasses and dahlias. A greenhouse for over wintering and a Griffin glasshouse with ginger lily, tibochina, aeoniums, pelargoniums and more. Regional Finalist, The English Garden's The Nation's Favourite Gardens. Gravel entrance, driveway and paths.

GROUP OPENING

40 SERGE HILL GARDENS
Serge Hill Lane, Bedmond, WD5 0RT. *½m E of Bedmond. Go to Bedmond & take Serge Hill Ln, where you will be directed past the lodge & down the drive.* **Sun 18 June (1-5). Combined adm £12, chd free. Pre-booking essential, please visit www.ngs.org.uk for information & booking. Home-made teas.**

THE BARN
Sue & Tom Stuart-Smith, www.tomstuartsmith.co.uk/our-work/toms-garden.
D

SERGE HILL
Kate Stuart-Smith.

2 large country gardens a short walk from each other. Tom and Sue Stuart-Smith's garden at The Barn has an enclosed courtyard with tanks of water, herbaceous perennials and shrubs tolerant of generally dry conditions. To the N there are views over the 5 acre wildflower meadow. The West Garden is a series of different gardens overflowing with bulbs, herbaceous perennials, and shrubs. There is also an exotic prairie planted from seed in 2011. The plant library, a collection of over 1000 herbaceous plants will also be open. Next door at Serge Hill, there is a lovely walled garden with a large foster and pearson greenhouse, orderly rows of vegetables, and disorderly self-seeded annuals and perennials. The walls are crowded with climbers and shrubs. From here you emerge to a meadow, a mixed border and a wonderful view over the ha-ha to the park and woods beyond.

GROUP OPENING

41 NEW SUNNYSIDE RURAL TRUST TRAIL
Two Waters Road, Hemel Hempstead, HP3 9BY.
Sunnyside Rural Trust Charity, www.sunnysideruraltrust.org.uk/. *The entrance & car park to Sunnyside Hemel Hempstead are found off the Two Waters Rd, behind the K2 restaurant accessed by the rd to its L.* **Sat 13 May (11-4). Combined adm £7, chd free. Home-made teas Cafe & Farm shop at Hemel Hempstead and Northchurch sites. The cafes offer hot & cold drinks, barista coffees, cakes & snacks from the farm shop. The Sunnyside Up cafe has vegetarian and vegan lunch options.**

NEW SUNNYSIDE RURAL TRUST- BERKHAMSTEAD
Sunnyside Rural Trust Charity,

NEW SUNNYSIDE RURAL TRUST- HEMEL HEMPSTEAD
Sunnyside Rural Trust Charity,

NEW SUNNYSIDE RURAL TRUST- NORTHCHURCH
Sunnyside Rural Trust Charity,

Sunnyside Rural Trust is a therapeutic horticultural charity and social enterprise, which supports young people and adults with learning disabilities with training and employment at 3 garden sites in Dacorum, SW Hertfordshire. The three 6 acre gardens are located along the Bulbourne valley, just off the A4251 and close to the Chilterns AONB. The Northchurch site is a market garden using 'no dig' methods, nature trail with wildlife pond and small animal farm whilst at Sunnyside Berkhamsted there is a vegetable growing area and extensive cane and espalier fruit garden, with a woodland and hazel coppice, each busy with wildlife. The 3rd site - Hemel Food Garden is a Green Flag award winning site that is also a plant nursery growing peat free perennial and annual bedding for domestic and commercial clients. They have an orchard, quiet garden space and market garden, growing a wide range of vegetables for their own use. For a day out why not drop in to all 3 sites under a combined ticket. A select range of perennial plants, veg seedlings and herb plants will be available for sale as well as home made chutneys and jams from our own produce. Assisted wheelchair access where there are woodland & grass paths.

42 9 TANNSFIELD DRIVE
Hemel Hempstead, HP2 5LG.
Peter & Gaynor Barrett, 01442 393508, peteslittlepatch@virginmedia.com, www.peteslittlepatch.co.uk. *Approx. 1m NE of Hemel Hempstead town centre & 2m W of J8 on M1. From M1 J8, cross r'about to A414 to Hemel Hempstead. Under footbridge, cross r'about then 1st R across dual c'way to Leverstock Green Rd. On to High St Green. L into Ellingham Rd then follow signs.* **Sun 4 June, Sun 2 July (1.30-4.30). Adm £4, chd free. Home-made teas. Visits also by arrangement 5 June to 20 Aug for groups of 6 to 12.**
A town garden to surprise. Dense planting together with the ever-present sound of water, create an intimate and welcoming oasis of calm. Narrow paths divide, leading visitors on a voyage of discovery of the garden's many features. The owners regularly experiment with the planting scheme which ensures the 'look' of the garden changes from year to year. Dense planting together with simple water features, stone statues, metal sculptures, wall art and mirrors can be seen throughout the garden. As a time and cost saving experiment all hanging baskets are planted with hardy perennials most of which are normally used for ground cover.

43 THUNDRIDGE HILL HOUSE

Cold Christmas Lane, Ware, SG12 0UE. Christopher & Susie Melluish, 01920 462500, c.melluish@btopenworld.com. *2m NE of Ware. ¾m from Maltons off the A10 down Cold Christmas Ln, Xing bypass.* **Sun 18 June (11-5.30). Adm £5, chd free. Cream teas. Visits also by arrangement May to Sept for groups of 15+.**
Well-established garden of approx 2½ acres; good variety of plants, shrubs and roses, attractive hedges. Visitors often ask for the unusual yellow-only bed. Several delightful places to sit. Wonderful views in and out of the garden especially down to the Rib Valley to Youngsbury, visited briefly by Lancelot 'Capability' Brown. 'A most popular garden to visit'. Fine views down to Thundridge Old Church.

44 WALKERN HALL

Walkern, Stevenage, SG2 7JA. Mrs Kate de Boinville. *4m E of Stevenage. Turn L at War Memorial as you leave Walkern, heading for Benington (immed after small bridge). Garden 1m up hill on R.* **Sat 4, Sun 5 Feb (12-4); Sat 25, Sun 26 Mar (12-5). Adm £5, chd free. Home-made teas. Warming home-made soup and cakes also available.**
Walkern Hall is essentially a winter woodland garden. Set in 8 acres, the carpet of snowdrops and aconites is a constant source of wonder in Jan/Feb. This medieval hunting park is known more for its established trees such as the tulip trees and a magnificent London plane tree which dominates the garden. Following on in March and April is a stunning display of daffodils. and other spring bulbs. There is wheelchair access but quite a lot of gravel. No disabled WC.

45 WATEREND HOUSE

Waterend Lane, Wheathampstead, St Albans, AL4 8EP. Mr & Mrs J Nall-Cain, 07736 880810, sj@nallcain.com. *2m E of Wheathampstead. Approx 10 mins from J4 of A1M. Take B653 to Wheathampstead, past Crooked Chimney Pub, after ½m turn R into Waterend Ln. Cross river, house is immed. on R.* **Visits by arrangement 6 Feb to 15 Nov for groups of 20 to 25. Adm £6. Home-made teas.**
A hidden garden of 4 acres sets off an elegant Jacobean Manor House (not open). Steep grass slopes and fine views of glorious countryside. Large quantities of spring bulbs, formal flint-walled garden. Roses, peonies and irises in the summer. Formal beds and lots of colour throughout the year. Mature specimen trees, ponds, formal vegetable garden and chickens. Meditation garden. Hilly garden. Wheelchair access to lower gardens only.

46 WOODHALL PARK

Woodhall Park, Watton-at-Stone, Hertford, SG14 3NQ. Woodhall Estate, 01920830286, reservations@woodhallestate.co.uk, www.woodhallestate.co.uk. *Woodhall Estate, near Watton at Stone, Hertfordshire. Use postcode SG14 3NQ and enter Woodhall Park via the A602. The entrance will be signposted.* **Sun 2 July (12-4). Adm £6, chd £3. Light refreshments.**
Stunning, recently-restored C18 , 6 acre Walled Gardens at Woodhall Estate The gardens have been meticulously restored (2020-21), designed by Lady Caroline Egremont. The formal Upper Garden with symmetrical shrub and herbaceous borders, and the Lower Garden laid out to wildflower meadows and fruit trees, inc a large collection of rare and specially grafted Hertfordshire apples. After seeing the gardens, visitors are welcome to walk and picnic in the beautiful C18 park. Long and short routes will be marked. Whilst wheelchair access to the gardens is possible, the area is rural and the ground may be uneven.

8 Gosselin Road

© Suzie Gibbons

ISLE OF WIGHT

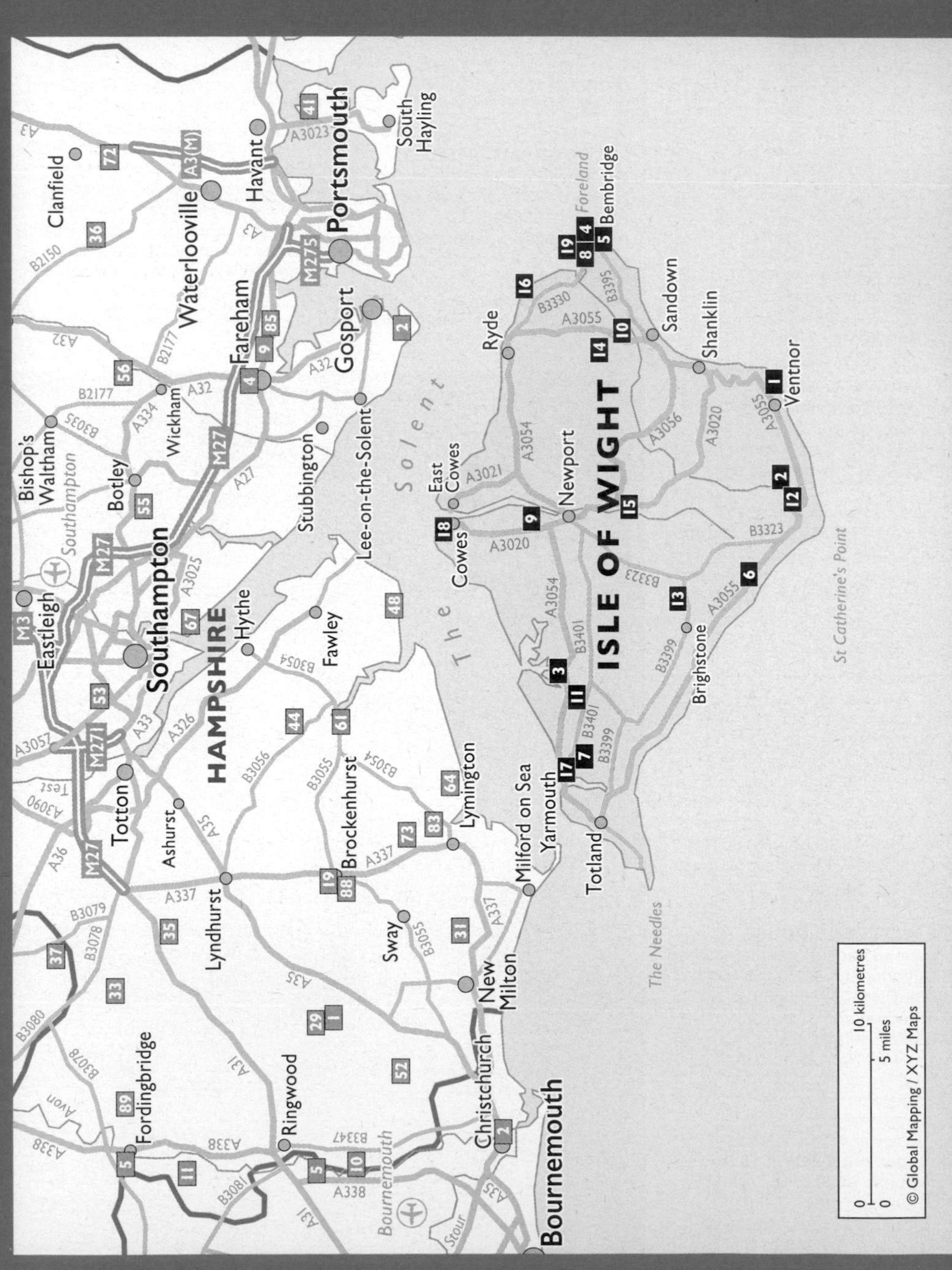

The island is a very special place to those who live and work here and to those who visit and keep returning. We have a range of natural features, from a dramatic coastline of cliffs and tiny coves to long sandy beaches.

Inland, the grasslands and rolling chalk downlands contrast with the shady forests and woodlands. Amongst all of this beauty nestle the picturesque villages and hamlets, many with gardens open for the National Garden Scheme. Most of our towns are on the coast and many of the gardens have wonderful sea views.

The one thing that makes our gardens so special is our climate. The moderating influence of the sea keeps hard frosts at bay, and the range of plants that can be grown is therefore greatly extended.

Conservatory plants are planted outdoors and flourish. Pictures taken of many island gardens fool people into thinking that they are holiday snaps of the Mediterranean and the Canaries.

Our gardens are very varied and our small enthusiastic group of garden owners are proud of their gardens, whether they are small town gardens or large manor gardens, and they love to share them for the National Garden Scheme.

Volunteers

County Organiser
Jane Bland, 01983 874592
jane.bland@ngs.org.uk

County Treasurer
Jennie Fradgley 01983 730805
Jennie.Fradgley@ngs.org.uk

Booklet Co-ordinator
Jane Bland (as above)

Assistant County Organisers
Sally Parker
01983 612495
sallyparkeriow@btinternet.com

Joanna Truman
01983 873822
joanna.truman@ngs.org.uk

Below: Crab Cottage

OPENING DATES

All entries subject to change. For latest information check **www.ngs.org.uk**

Map locator numbers are shown to the right of each garden name.

May

Sunday 7th
1 Union Road 18

Sunday 14th
Morton Manor 10

Saturday 20th
Niton Gardens 12

Sunday 21st
Goldings 7
Niton Gardens 12
Thorley Manor 17

June

Sunday 4th
East Dene 6
◆ Nunwell House 14

Saturday 10th
Ashcliff 1

Sunday 11th
Ashcliff 1
Seaview West End Gardens 16

Sunday 18th
Darts 4
Northcourt Manor Gardens 13
Yew Tree Lodge 19

Saturday 24th
Ashknowle House 2

Sunday 25th
Ashknowle House 2

July

Saturday 22nd
NEW Rookley Gardens 15

Sunday 23rd
Dove Cottage 5
NEW Rookley Gardens 15

August

Sunday 13th
Crab Cottage 3

Sunday 20th
Morton Manor 10

Sunday 27th
1 Union Road 18

By Arrangement

Arrange a personalised garden visit with your club, or group of friends, on a date to suit you. See individual garden entries for full details.

Crab Cottage 3
NEW 16 Harbour Strand 8
NEW 23 Horsebridge Hill 9
Morton Manor 10
Ningwood Manor 11
Northcourt Manor Gardens 13
1 Union Road 18

THE GARDENS

1 ASHCLIFF

The Pitts, Bonchurch, nr Ventnor, PO38 1NT. Judi & Sid Lines. *From A3055 follow signs for Bonchurch, then for Bonchurch pond & park in village or continue, following signs for parking in Bonchurch Shute.* **Sat 10, Sun 11 June (2-5). Adm £6, chd free. Home-made teas.**
The garden was started from a blank canvas 20 yrs ago and now contains many diverse areas of interest inc areas of sun and shade. Plantings are of interesting and unusual perennials, shrubs and trees over approx 1 acre blending into the natural landscape, part of which is a cliff.

2 ASHKNOWLE HOUSE

Ashknowle Lane, Whitwell, Ventnor, PO38 2PP. Mr & Mrs K Fradgley. *4m W of Ventnor. Take the Whitwell Rd from Ventnor or Godshill. Turn into unmade lane next to Old Rectory. Field parking. Disabled parking at house.* **Sat 24, Sun 25 June (12-4). Adm £5, chd free. Home-made teas.**
The mature garden of this Victorian house (not open) has great diversity inc woodland walks, water features, colourful beds and borders. The large, well maintained kitchen garden is highly productive and boasts a wide range of fruit and vegetables grown in cages, tunnels, glasshouses and raised beds. Diversely planted and highly productive orchard inc protected cropping of strawberries, peaches and apricots. Propagation areas for trees, shrubs and crops.

3 CRAB COTTAGE

Mill Road, Shalfleet, PO30 4NE. Mr & Mrs Peter Scott, 07768 065756, mencia@btinternet.com. *4½m E of Yarmouth. At New Inn, Shalfleet, turn into Mill Rd. Continue 400yds to end of metalled road, drive onto unmade road through open NT gates. After 100yds pass Crab Cottage on L. Park opp on grass.* **Sun 13 Aug (11-5). Adm £4, chd free. Home-made teas. Visits also by arrangement May to Sept.**
1¼ acres on gravelly soil. Part glorious views across croquet lawn over Newtown Creek and Solent, leading through wildflower meadow to hidden water lily pond, secluded lawn and woodland walk. Part walled garden protected from westerlies with mixed borders, leading to terraced sunken garden with ornamental pool and pavilion, planted with exotics, tender shrubs and herbaceous perennials. Croquet, plant sales and excellent tea and cake. Wheelchair access over gravel and uneven grass paths.

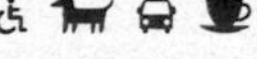

4 DARTS

Darts Lane, Bembridge, PO35 5YH. Joanna Truman. *Up the hill from Bembridge Harbour, turn L after the Co-op down Love Ln & take 1st L. The house is painted grey.* **Sun 18 June (12.30-5). Combined adm with Yew Tree Lodge £6, chd free. Home-made teas at Yew Tree Lodge.**
Classic walled garden with roses, climbers, shrubs, two large raised vegetable beds and running water feature. Work in progress continues! Access to garden via two raised steps.

5 DOVE COTTAGE

Swains Lane, Bembridge, PO35 5ST. Mr James & Mrs Alex Hearn. *Head E on Lane End Rd, passing Lane End Court Shops on L. Take 3rd turning on L onto Swains Ln. Dove Cottage is on R.* **Sun 23 July (12-4.30). Adm £6, chd free. Light refreshments.**
An enclosed garden with lawn and woodland area. A central path leads through the woodland setting which

has mature variagated shrubs and perennials, leading to the swimming pool area and tennis court.

6 EAST DENE

Atherfield Green, Ventnor, PO38 2LF. Marc & Lisa Morgan-Huws. *From A3055 Military Rd, turn into Southdown. At end of road turn L on to Atherfield Rd & the garden is ahead on the R.* **Sun 4 June (11-4). Adm £4, chd free. Light refreshments.**

Planted from 2 acres of farmland in the 1970s, the garden is structured around a variety of mature trees. Taken over by brambles and nettles prior to our arrival 8 yrs ago, the garden is continuing to be developed, whilst still being a haven for wildlife. Occupying a windswept coastal location with areas of woodland, fruit trees, mature pond and informal planting. Alpacas. Although the garden has level access, the ground is unmade.

7 GOLDINGS

Thorley Street, Thorley, Yarmouth, PO41 0SN. John & Dee Sichel. *E of Yarmouth. Follow directions for Thorley from Yarmouth/Newport road or from Wilmingham Ln. Then follow NGS signs. Parking shared with Thorley Manor.* **Sun 21 May (2-5). Combined adm with Thorley Manor £5, chd free.**

A country garden with many focuses of interest. A newly planted orchard already producing cider apples in large amounts. A small, but productive vegetable garden and a well maintained lawn with shrub borders and roses. A microclimate has been created by the adjustment of levels to create a series of terraced areas for planting.

8 NEW 16 HARBOUR STRAND

Bembridge, PO35 5NP. Charles & Suise Evans, 07887 508879, susiejoyevans@gmail.com. *Entrance in Embankment Rd, opp Bembridge Sailing Club, behind bus stop. Adjacent to Bembridge Harbour.* **Visits by arrangement 2 May to 5 May & 5 June to 9 June for groups of up to 6. Adm £5. Light refreshments.**

Small seaside garden tucked away with hot sunny main area leading down to shaded side garden and small courtyard. A good example of the seclusion that can be achieved between two busy roads.

9 NEW 23 HORSEBRIDGE HILL

Newport, PO30 5TJ. Paul Sheaf, 07840 887369, pasgardenservices@hotmail.co.uk. *1½ m N of Newport. On main Newport to Cowes road, past the hospital, first layby on L side of road as you travel up the hill.* **Visits by arrangement 30 Apr to 15 Oct for groups of up to 15. Adm £5, chd free. Light refreshments.**

A creative pallet of colour to enjoy in every season. Many unusual plants inc trees, shrubs, climbers, herbaceous and container planting. A pond and stream beautifully designed to look natural which can be viewed from the boardwalk. Pergola, trellises, greenhouse and work/potting shed. Areas to sit and enjoy the gardens tranquillity, and lighting throughout for evening visits.

Ashcliff

© Heather Edwards

Dove Cottage

© Lucy Hooper

10 MORTON MANOR

Morton Manor Road, Brading, Sandown, PO36 0EP. Mr & Mrs G Godliman, 07768 605900, patricia.godliman@yahoo.co.uk. *Off A3055 5m S of Ryde, just out of Brading. At Yarbridge T-lights turn into The Mall. Take next L into Morton Manor Rd.* **Sun 14 May, Sun 20 Aug (11-4). Adm £5, chd free. Light refreshments. Visits also by arrangement Apr to Oct.**

A colourful garden of great plant variety. Mature trees inc many acers with a wide variety of leaf colour. Early in the season a display of rhododendrons, azaleas and camellias and later hydrangeas and hibiscus. Ponds, sweeping lawns, roses set on a sunny terrace and much more to see in this extensive garden surrounding a picturesque C16 manor house (not open). Wheelchair access over gravel driveway.

11 NINGWOOD MANOR

Station Road, Ningwood, nr Newport, PO30 4NJ. Nicholas & Claire Oulton, 07738 737482, claireoulton@gmail.com. *Nr Shalfleet. From Newport, turn L opp the Horse & Groom Pub. Ningwood Manor is 300-400yds on the L. Please use 2nd set of gates.* **Visits by arrangement 1 May to 1 Aug for groups of up to 30. Adm £5. Light refreshments.**

A 3 acre, landscaped country garden divided into several rooms: a walled courtyard, croquet lawn, white garden and kitchen garden. They flow into each other, each with their own gentle colour schemes; the exception to this is the croquet lawn garden which is a riot of colour, mixing oranges, reds, yellows and pinks. Much new planting has taken place over the last few yrs. The owners have several new projects underway, so the garden is a work in progress. Features inc a vegetable garden with raised beds and a small summerhouse, part of which is alleged to be Georgian. Please inform us at time of booking if wheelchair access is required.

GROUP OPENING

12 NITON GARDENS

Niton, PO38 2AZ. *5m W of Ventnor. Parking at football ground (Blackgang Rd), in village & Allotment Rd car park. Tickets & maps from the library in the heart of the village on Niton Gardens open days only.* **Sat 20, Sun 21 May (11.30-4.30). Combined adm £6, chd free. Home-made teas at Tillington Villa & Winfrith.**

12 GREENLYDD CLOSE
Mrs G Rollfe.

NEW **HOLLY BANK**
Lynda Paice.

PUCKASTER CORNER
Mr & Mrs Ian McCallum.

SPRING COTTAGE
Mr & Mrs Neil White.

TALSA
Frances Pritchard.

TILLINGTON VILLA
Paul & Catherine Miller.

WINFRITH
Mrs Janet Tedman.

Niton is a delightful village with a busy community spirit, blessed with lovely churches, two pubs (one of which was renowned for smuggling), PO and shops, lovely walks and bridleways, school, football and cricket pitches and recreation ground. The gardens of this walk are very varied in both style and size and are full of colour, fragrance and interest; from cottage and country gardens to vegetable plot and havens for wildlife. The gardens are situated both in the heart of the village and the undercliff. We do hope you will enjoy them all.

13 NORTHCOURT MANOR GARDENS

Main Road, Shorwell, Newport, PO30 3JG. Mr & Mrs J Harrison, 01983 740415, john@northcourt.info, www.northcourt.info. *4m SW of Newport. On entering Shorwell from Newport, entrance at bottom of hill on R. If entering from other directions head through village in direction of Newport. Garden on the L, on bend after passing the church. Coaches, park opp church.* **Sun 18 June (11.30-5). Adm £7, chd free. Home-made teas. Visits also by**

arrangement 30 Mar to 15 Oct for groups of 5 to 50. Introductory talk & tour.
15 acre garden surrounding large C17 manor house (not open). Boardwalk along jungle garden. Stream and bog garden. A large variety of plants enjoying the different microclimates. Large collection of camellias and magnolias. Woodland walks. Tree collection. Salvias and subtropical plantings for autumn drama. Productive walled kitchen garden. Numerous shrub roses. Picturesque wooded valley around the house. Bathhouse and snail mount leading to terraces. 1 acre walled garden. The house celebrated its 407th yr anniversary. A plantsman' garden. Wheelchair access on main paths only, some paths are steep and uneven, and the terraces are hilly.

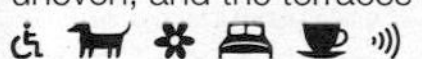

14 ◆ NUNWELL HOUSE

Coach Lane, Brading, PO36 0JQ. Mr & Mrs S Bonsey, www.nunwellhouse.co.uk. *3m S of Ryde. Signed off A3055 as you arrive at Brading from Ryde & turn into Coach Ln.* **For NGS: Sun 4 June (1-4.30). Adm £5, chd free. Home-made teas. For other opening times and information, please visit garden website.**
6 acres of tranquil and beautifully set formal and shrub gardens and old fashioned shrub roses prominent. Exceptional Solent views over historic parkland and Brading Haven from the terraces. Small arboretum and walled garden with herbaceous borders. House developed over 5 centuries and full of architectural interest.

GROUP OPENING

15 NEW ROOKLEY GARDENS

Rookley, Newport, PO30 3BJ. *4m S of Newport. From the main Newport to Sandown road, take turning for Rookley at Blackwater.* **Sat 22, Sun 23 July (12-4). Combined adm £6, chd free. Home-made teas at Corris.**

NEW **BLACKWATER MILL RESIDENTIAL HOME**
Debbie Webb.

CORRIS
Mr David & Mrs Helen Crook.

OAKDENE
Mr Tim Marshall.

NEW **OLD SCHOOL COTTAGE**
Mr Nigel Palmer.

NEW **THE OLD STABLES**
Mrs Susan Waldron.

A varied group of gardens diverse in size, design and interest. Set within the area of Rookley Village and extending towards Blackwater these gardens surround a well kept and lively village. Features inc lakes and grassy pathways with colourful borders and places to rest, to beautiful designs displaying a riot of colour, shape and form, plus fruit and vegetables, wild flowers and bees.

GROUP OPENING

16 SEAVIEW WEST END GARDENS

Salterns Road, Seaview, PO34 5AH. *Along E coast of Ryde. Follow signs for Seaview from Ryde or Brading along A3055. Parking in village or Duver Rd. Tickets & map available at each garden.* **Sun 11 June (12.30-4.30). Combined adm £6, chd free. Light refreshments at Northbank Hotel, Circular Road.**

ARMADALE
Mrs Mary Johnston-Taylor.

RED CROSS COTTAGE
Katie Barnfather & Stephen Jones.

SALTERNS COTTAGE
Susan & Noël Dobbs.

Enjoy a variety of garden styles set in the popular village resort of Seaview. Colourful borders of herbaceous, perennials and annual bedding, trees, shrubs and climbers, grasses and water features. Raised vegetable beds, glasshouses and potager, plus colourful sunny patios with a touch of the exotic.

17 THORLEY MANOR

Thorley, Yarmouth, PO41 0SJ. Mr & Mrs Anthony Blest. *1m E of Yarmouth. From Bouldnor take Wilmingham Ln, house ½m on L.* **Sun 21 May (2-5). Combined adm with Goldings £5, chd free. Home-made teas.**
Mature informal gardens of over 3 acres surrounding manor house (not open). Garden set out in a number of walled rooms, perennial and colourful self-seeding borders, shrub borders, lawns, large old trees and an unusual island lawn, all seamlessly blending into the surrounding farmland. The delightful cottage garden of Goldings is open a short walk away.

18 1 UNION ROAD

Cowes, PO31 7TW. Mr & Mrs B Hicks, georgeous1@gmail.com. *Take the Cowes Rd (A3020) as far as Northwood T-lights (by the car garage), bear L & follow signs for Northwood House. Parking at Northwood House, PO31 8AZ. 3 min walk from the car park.* **Sun 7 May, Sun 27 Aug (11-5). Adm £4, chd free. Home-made teas. Visits also by arrangement May to Sept for groups of up to 20.**
A small, south facing town garden with views of the Solent and plenty of sea air. Planting is tropical and takes full advantage of the many hours of sunshine here on the Isle of Wight. A combination of unusual plants from drier, arid environments, with lush planting in the zones towards the bottom of the garden from temperate climates. Specimen trees throughout the garden and beautiful walled garden area. There is a two level pond featuring wildlife and Mediterranean planting, as well as the working area of the garden featuring grapevines and olive, apple, fig, citrus and walnut trees. There is also a living agapanthus wall. Finally, there is a herb garden within the outdoor kitchen area. Garden designed by Helen Elks-Smith in 2019.

19 YEW TREE LODGE

Love Lane, Bembridge, PO35 5NH. Jane Bland. *On the Bembridge circular one way system, take 1st L after Co-op into Love Ln. Darts is on the L, Yew Tree Lodge is facing you when Love Ln turns sharply to the R.* **Sun 18 June (12.30-5). Combined adm with Darts £6, chd free. Home-made teas.**
Yew Tree Lodge garden is flanked by mature oak trees and well established gardens, its main axis being north south. It is divided into separate areas which take into account available natural light and soil type. I have been gardening here for 10 yrs, during this time the garden has altered a great deal. As well as flowering plants there is a vegetable garden, fruit cages and a cool greenhouse. Paved path connecting the front and back gardens and paved access to decking area.

KENT

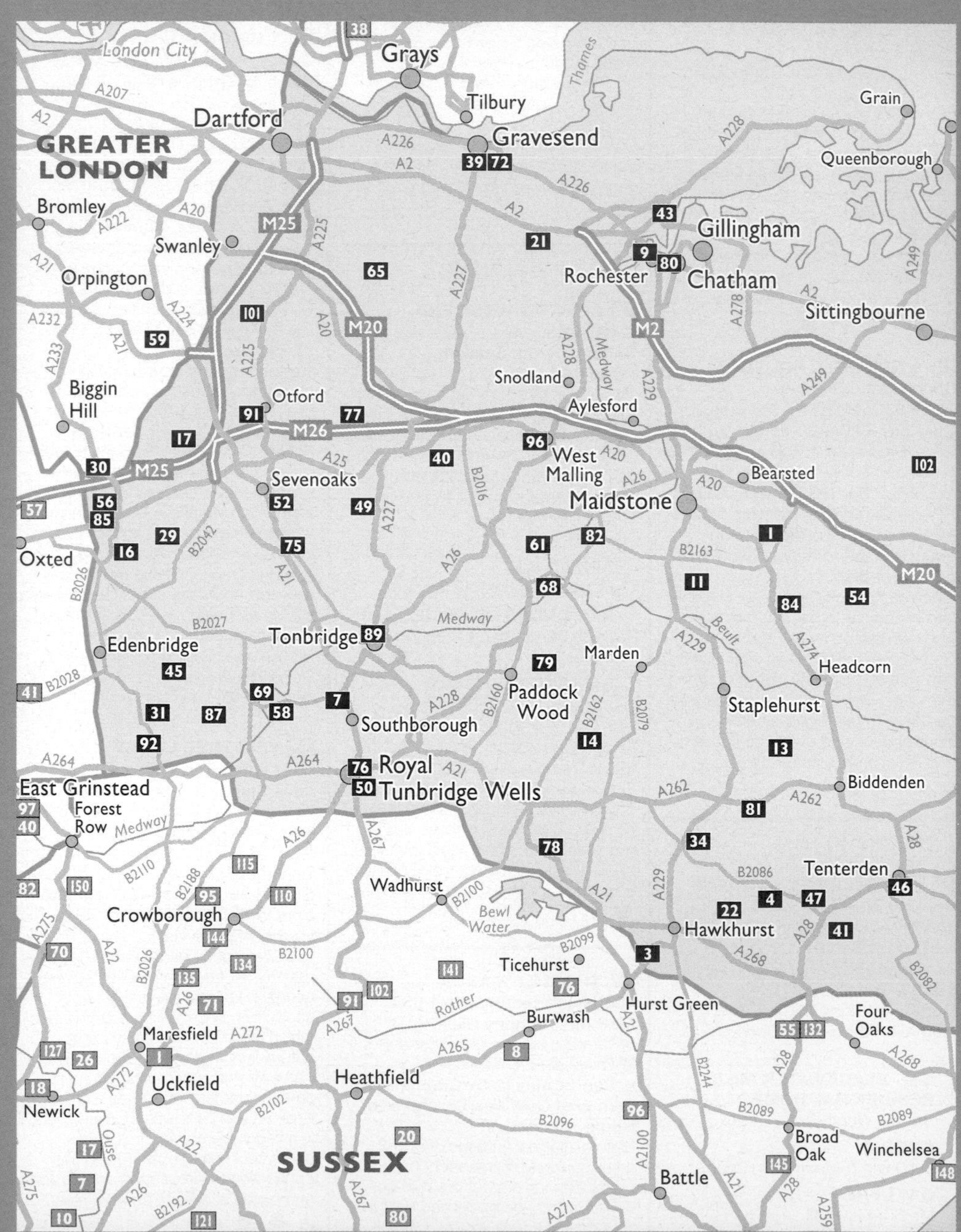
London City
Grays
Tilbury
Dartford
Gravesend
GREATER LONDON
Grain
Queenborough
Bromley
Swanley
Gillingham
Rochester
Chatham
Orpington
Sittingbourne
Snodland
Biggin Hill
Otford
Aylesford
West Malling
Sevenoaks
Bearsted
Maidstone
Oxted
Tonbridge
Marden
Edenbridge
Headcorn
Paddock Wood
Staplehurst
Southborough
Royal Tunbridge Wells
East Grinstead
Biddenden
Forest Row
Tenterden
Wadhurst
Bewl Water
Crowborough
Hawkhurst
Ticehurst
Hurst Green
Burwash
Four Oaks
Maresfield
Uckfield
Heathfield
Newick
Broad Oak
Winchelsea
SUSSEX
Battle
Thames
Medway
Beult
Rother
Ouse

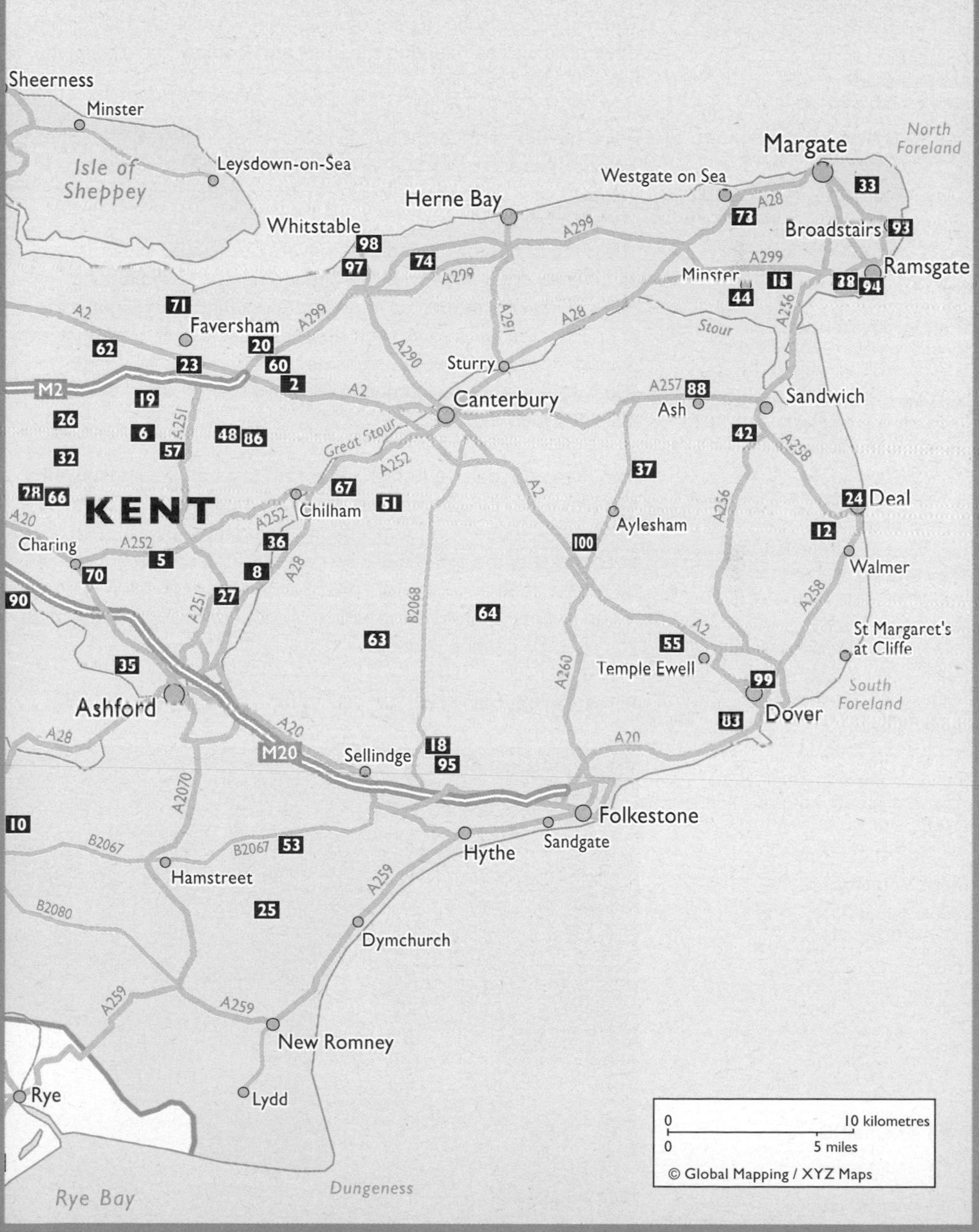
Sheerness
Minster
Isle of Sheppey
Leysdown-on-Sea
Margate
North Foreland
Westgate on Sea
Herne Bay
Whitstable
Broadstairs
Ramsgate
Minster
Stour
Faversham
Sturry
Canterbury
Sandwich
Ash
Great Stour
KENT
Chilham
Deal
Aylesham
Charing
Walmer
St Margaret's at Cliffe
Temple Ewell
South Foreland
Ashford
Dover
Sellindge
Folkestone
Sandgate
Hythe
Hamstreet
Dymchurch
New Romney
Rye
Lydd
Rye Bay
Dungeness
0 10 kilometres
0 5 miles
© Global Mapping / XYZ Maps

Volunteers

County Organiser
Jane Streatfeild 01342 850362
janestreatfeild@btinternet.com

County Treasurer
Andrew McClintock 01732 838605
andrew.mcclintock@ngs.org.uk

Publicity
Susie Challen
susie.challen@ngs.org.uk

Booklet Advertising
Nicola Denoon-Duncan
01233 758600
nicola.denoonduncan@ngs.org.uk

Booklet Co-ordinator
Ingrid Morgan Hitchcock
01892 528341
ingrid@morganhitchcock.co.uk

Booklet Distribution
Diana Morrish 01892 723905
diana.morrish@ngs.org.uk

Assistant County Organisers
Jacqueline Anthony 01892 518879
jacquelineanthony7@gmail.com

Clare Barham 01580 241386
clarebarham@holepark.com

Pam Bridges
pam.bridges@ngs.co.uk

Mary Bruce 01795 531124
mary.bruce@ngs.org.uk

Kate Dymant 07766 201806
katedymant@hotmail.co.uk

Neil Dymant 07836 781391
neildymant@btinternet.com

Andy Garland
andy.garland@bbc.co.uk

Virginia Latham 01303 862881
virginia.latham@ngs.org.uk

Sian Lewis
sian.lewis@ngs.org.uk

Diana Morrish (as above)

Nicola Talbot 01342 850526
nicola@falconhurst.co.uk

@KentNGS
@NGSKent
@nationalgardenschemekent

Famously known as 'The Garden of England', it is no wonder that over a quarter of Kent has been designated as Areas of Outstanding Natural Beauty.

The unique landscapes here offer ancient woodlands and rolling downs, as well as haunting marshes and iconic white cliffs along the 350-mile coastline.

Being the oldest county too means that Kent is filled with fascinating historic sites and buildings, several of which open their gardens for the NGS such as Ightham Mote and Sissinghurst Castle.

There are also many enchanting private gardens to explore, of all different sizes and styles. Inspiration can be found among expansive acres as well as in exquisite planting in the smallest of spaces. We have several group openings as well, offering a wonderful way to explore a seaside town or a country village. You will be warmly welcomed in all, often with a cup of tea and a delicious slice of cake!

If you are after garden design ideas and inspiration, or just want to escape into the countryside, there is a huge range of gardens in Kent to choose from. A selection of gardens are only open 'by arrangement'; the owners will be delighted to see you but please do phone or email to arrange a visit first. Always check ngs.org.uk for up to date opening information. We look forward to welcoming you over the coming season.

Below: Old Bladbean Stud

OPENING DATES

All entries subject to change. For latest information check **www.ngs.org.uk**

Map locator numbers are shown to the right of each garden name.

January

Thursday 19th
Spring Platt 84

Wednesday 25th
Spring Platt 84

Sunday 29th
Spring Platt 84

February

Snowdrop Festival

Thursday 2nd
Spring Platt 84

Saturday 4th
Knowle Hill Farm 54

Sunday 5th
Knowle Hill Farm 54
Spring Platt 84

Saturday 11th
Copton Ash 23

Sunday 19th
Copton Ash 23
◆ Doddington Place 26

March

Sunday 5th
Haven 44

Wednesday 15th
Haven 44

Sunday 19th
Copton Ash 23
◆ Godinton House & Gardens 35
Godmersham Park 36
Stonewall Park 87

Sunday 26th
Copton Ash 23
◆ Great Comp Garden 40
◆ Mount Ephraim Gardens 60

Thursday 30th
◆ Ightham Mote 49

April

Sunday 2nd
◆ Cobham Hall 21
◆ Doddington Place 26
Nettlestead Place 61

Sunday 9th
Copton Ash 23
Haven 44

Monday 10th
Haven 44

Thursday 13th
◆ Boldshaves 10

Friday 14th
Oak Cottage and Swallowfields Nursery 63

Saturday 15th
NEW ◆ Beech Court Gardens 5
The Knoll Farm 53
Oak Cottage and Swallowfields Nursery 63

Sunday 16th
Balmoral Cottage 4
NEW ◆ Beech Court Gardens 5
Frith Old Farmhouse 32
◆ Hole Park 47
The Knoll Farm 53

Thursday 20th
The Old Rectory, Otterden 66

Friday 21st
The Old Rectory, Otterden 66

Saturday 22nd
Bilting House 8

Sunday 23rd
Bilting House 8

Saturday 29th
Bishopscourt 9

Sunday 30th
Copton Ash 23
Frith Old Farmhouse 32
Stonewall Park 87

May

Monday 1st
Eagleswood 28
Haven 44

Sunday 7th
Balmoral Cottage 4
1 Brickwall Cottages 13
Meadow Wood 58
43 The Ridings 74
The Silk House 79

Tuesday 9th
◆ Riverhill Himalayan Gardens 75

Wednesday 10th
◆ Hole Park 47
Oak Cottage and Swallowfields Nursery 63
◆ Scotney Castle 78

Thursday 11th
Oak Cottage and Swallowfields Nursery 63

Friday 12th
NEW ◆ Squerryes Court 85

Saturday 13th
Avalon 2

Sunday 14th
Avalon 2
◆ Boughton Monchelsea Place 11
Frith Old Farmhouse 32
◆ Godinton House & Gardens 35
The Orangery 67
Whitstable Joy Lane Gardens 97

Saturday 20th
Bilting House 8
Haven 44
Little Gables 56

Sunday 21st
Balmoral Cottage 4
Bilting House 8
Copton Ash 23
Little Gables 56
12 The Meadows 59
St Clere 77

Wednesday 24th
Great Maytham Hall 41

Sunday 28th
Copton Ash 23
Downs Court 27
Eagleswood 28
Lacey Down 55
Old Bladbean Stud 64
Windy Ridge 99

Monday 29th
Falconhurst 31

June

Saturday 3rd
Churchfield 18
West Court Lodge 95

Sunday 4th
◆ Belmont 6
Churchfield 18
Nettlestead Place 61
Tram Hatch 90
West Court Lodge 95
West Malling Early Summer Gardens 96

Thursday 8th
◆ Mount Ephraim Gardens 60
NEW Pett Place 70

Friday 9th
NEW Pett Place 70

Saturday 10th
The Coach House 20
Ivy Chimneys 50
Little Gables 56

Sunday 11th
The Coach House 20
Downs Court 27
Godmersham Park 36
Little Gables 56
Lords 57
Old Bladbean Stud 64
Whitstable Town Gardens 98

Wednesday 14th
◆ Hole Park 47
◆ Riverhill Himalayan Gardens 75
Upper Pryors 92

Thursday 15th
◆ Goodnestone Park Gardens 37

Friday 16th
◆ Godinton House & Gardens 35

Saturday 17th
Avalon 2
Falconhurst 31

Sunday 18th
Arnold Yoke 1
Avalon 2
Chevening 17
Downs Court 27
Haven 44
◆ The World Garden at Lullingstone Castle 101

Monday 19th
Norton Court 62

Tuesday 20th
Chapel House Estate 15
Norton Court 62

Wednesday 21st
Great Maytham Hall 41

Saturday 24th
Bishopscourt 9
Pheasant Barn 71
Sir John Hawkins Hospital 80
Womenswold Gardens 100

Sunday 25th
◆ Boughton Monchelsea Place 11
Deal Town Gardens 24
Old Bladbean Stud 64
Pheasant Barn 71
Sir John Hawkins Hospital 80
Smiths Hall 82
Womenswold Gardens 100

Wednesday 28th
Pheasant Barn 71

Thursday 29th
Pheasant Barn 71

July

Saturday 1st
Eureka 30
The Garden Gate 33
NEW 95 High Street 46
Hoppickers East 48
NEW 3 Stanley Cottages 86

Sunday 2nd
Bidborough Gardens 7
Eureka 30
Goddards Green 34
Gravesend Garden for Wildlife 39
NEW 95 High Street 46
Pheasant Barn 71
Tram Hatch 90

Monday 3rd
Pheasant Barn 71

Saturday 8th
Hammond Place 43

Sunday 9th
Hammond Place 43
Haven 44
Old Bladbean Stud 64
Pheasant Barn 71
Sweetbriar 88
Tonbridge School 89

Monday 10th
Pheasant Barn 71

Saturday 15th
50 Pine Avenue 72

Sunday 16th
Lacey Down 55
◆ Quex Gardens 73
Windy Ridge 99

Tuesday 18th
◆ Knole 52

Saturday 22nd
Eureka 30

Sunday 23rd
Eureka 30
Old Bladbean Stud 64
South Dover Garden Trail 83

Saturday 29th
The Orangery 67
The Watch House 93

Sunday 30th
Haven 44
The Orangery 67
The Watch House 93

August

Wednesday 2nd
Hoppickers East 48

Sunday 6th
The Silk House 79

Saturday 12th
Eureka 30
NEW Grove Manor 42

Sunday 13th
Eureka 30
NEW Grove Manor 42

Saturday 19th
Avalon 2

Sunday 20th
Avalon 2
Sweetbriar 88
Tram Hatch 90

Monday 28th
Haven 44

September

Saturday 2nd
Eureka 30
Knowle Hill Farm 54

Sunday 3rd
Eureka 30
Knowle Hill Farm 54

Wednesday 6th
◆ Emmetts Garden 29

Friday 8th
NEW Churchmans Farm 19

Saturday 9th
NEW Churchmans Farm 19
NEW The Copper House 22

Sunday 10th
NEW The Copper House 22
104 Grange Road 38
12 West Cliff Road 94

Wednesday 13th
◆ Chartwell 16
◆ Penshurst Place & Gardens 69

Thursday 14th
◆ Goodnestone Park Gardens 37

Sunday 17th
◆ Sissinghurst Castle Garden 81
Sweetbriar 88

Friday 22nd
◆ Godinton House & Gardens 35

Sunday 24th
◆ Doddington Place 26
Haven 44

October

Sunday 1st
◆ Mount Ephraim Gardens 60

Wednesday 4th
◆ Hever Castle & Gardens 45

Sunday 15th
◆ Hole Park 47

Thursday 19th
◆ Ightham Mote 49

Sunday 29th
◆ Great Comp Garden 40
Haven 44

November

Sunday 12th
Haven 44

February 2024

Saturday 3rd
Knowle Hill Farm 54

Sunday 4th
Knowle Hill Farm 54

Saturday 10th
Copton Ash 23

Sunday 18th
Copton Ash 23

In 2022 the National Garden Scheme donated £3.11 million to our nursing and health beneficiaries

Chapel House Estate

By Arrangement

Arrange a personalised garden visit with your club, or group of friends, on a date to suit you. See individual garden entries for full details.

Arnold Yoke 1
Avalon 2
Badgers 3
Bilting House 8
Boundes End, Bidborough Gardens 7
Brewery Farmhouse 12
1 Brickwall Cottages 13
Cacketts Farmhouse 14
Churchfield 18
The Coach House 20
NEW The Copper House 22
Copton Ash 23
Dean House 25
Downs Court 27
Eagleswood 28
Falconhurst 31
Frith Old Farmhouse 32
Goddards Green 34
Gravesend Garden for Wildlife 39
Haven 44
Kenfield Hall 51
Knowle Hill Farm 54
Lords 57
The Old Rectory, Fawkham 65
The Old Rectory, Otterden 66
The Orangery 67
Parsonage Oasts 68
Pheasant Barn 71
43 The Ridings 74
18 Royal Chase 76
Spring Platt 84
Tram Hatch 90
Troutbeck 91
West Court Lodge 95
Yokes Court 102

THE GARDENS

1 ARNOLD YOKE

Back Street, Leeds, Maidstone, ME17 1TF. Richard & Patricia Stileman, 07968 787950, richstileman@btinternet.com. *5m E of Maidstone. Fm M20 J8 take A20 Lenham R to B2163 to Leeds. Thru Leeds R into Horseshoes Ln, 1st R into Back St. House ¾ m on L. Fm A274 follow B2163 to Langley L into Horseshoes Ln 1st R Back St.* **Sun 18 June (2-5). Adm £6, chd free. Home-made teas. Visits also by arrangement 4 June to 23 July for groups of 10 to 20.**

The garden is really a ½ acre 'room', bounded by yew to the South; an ivy wall to the West; a rock garden to the North; and the house to the East. There is a paradise garden and pond in the middle of the 'room', and profusely planted borders on the South and West sides. Wheelchair access from the car park and around most parts of the garden.

2 AVALON

57 Stoney Road, Dunkirk, ME13 9TN. Mrs Croll, avalongarden8@gmail.com. *4m E of Faversham, 5m W of Canterbury, 2½m E of J7 M2. M2 J7 or A2 E of Faversham take A299, first L, Staplestreet, then L, R past Mt Ephraim, turn L, R. From A2 Canterbury, turn off Dunkirk, bottom hill turn R, Staplestreet then R, R. Park in side roads.* **Sat 13, Sun 14 May, Sat 17, Sun 18 June, Sat 19, Sun 20 Aug (11-4). Adm £6, chd free. Visits also by arrangement 13 May to 31 Aug for groups of 5 to 15.**

½ acre woodland garden, for all seasons, on north west slope, something round every corner, views of surrounding countryside. Collections of roses, hostas and ferns plus rhododendrons, shrubs, trees, vegetables, fruit, unusual plants, flowers for local shows. Planted by feeling, making it a reflective space and plant lovers' garden. Plenty of seating for taking in the garden and resting from lots of steps.

3 BADGERS

Bokes Farm, Horns Hill, Hawkhurst, Cranbrook, TN18 4XG. Bronwyn Cowdery, 01580 754178, cowderyfamily@btinternet.com. *On the border of Kent & East Sussex. In centre of Hawkhurst, follow A229 in the direction of Hurst Green. Pass the The Wealden Advertiser. At sharp L bend turn R up Horns Hill. Drive slowly up Horns Hill.* **Visits by arrangement 21 Aug to 21 Sept for groups of 10 to 25. Adm £5, chd free. Home-made teas.**

The garden comes into its own from end of August, when the Tropical Garden is in full growth. There is also a walled Italian style garden, small Japanese area and woodland with ponds and a waterfall. The Tropical Garden has a wide variety of Palms, Bananas, Gingers, Eucomis and Dahlias and has a network of paths threading through for you to explore and immerse yourself in the Tropics.

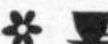

4 BALMORAL COTTAGE

The Green, Benenden, Cranbrook, TN17 4DL. Charlotte Molesworth, thepottingshedholidaylet@gmail.com. *Few 100 yds down unmade track to W of St George's Church, Benenden.* **Sun 16 Apr (12-5), open nearby Hole Park. Sun 7, Sun 21 May (12-5). Adm £6, chd £2.50. Refreshments available at Hole Park (separate additional price).**

An owner created and maintained garden now 40yrs mature. Varied, romantic and extensive topiary form the backbone for mixed borders. Vegetable garden, organically managed. Particular attention to the needs of nesting birds and small mammals lend this artistic plantswoman's garden a rare and unusual quality. No hot borders or dazzling dahlias here.

5 NEW ◆ BEECH COURT GARDENS

Canterbury Road, Challock, Ashford, TN25 4DJ. Chloë & Garth Oates, www.beechcourtgardens.co.uk. *5m N of Ashford, Faversham 6m, Canterbury 9m. W of Xrds A251/A252, off the Lees.* **For NGS: Sat 15, Sun 16 Apr (10.30-5). Adm £5, chd free. Home-made teas. For other opening times and information, please visit garden website.**

Beech Court Gardens are well established, planted in the 1930s. Exceptional collection of Acers and Rhododendron, along with numerous firs, oaks and pines. Many rare and unusual trees grow here and are home to many species of bird. The Gardens have been enhanced with island beds, climbing roses and hydrangeas. Please visit the gallery on our website.

6 ◆ BELMONT

Belmont Park, Throwley, Faversham, ME13 0HH. Harris (Belmont) Charity, 01795 890202, administrator@belmont-house.org, www.belmont-house.org. *4½m SW of Faversham. A251 Faversham-Ashford. At Badlesmere, brown tourist signs to Belmont.* **For NGS: Sun 4 June (12-5). Adm £5, chd free. Light refreshments. Home-made cakes. For other opening times and information, please phone, email or visit garden website.**

Belmont House is surrounded by large formal lawns that are landscaped with fine specimen trees, a pinetum and a walled garden containing long borders, wisteria and large rose border. Across the drive there is a second walled kitchen garden, restored in 2001 to a design by Arabella Lennox-Boyd. Inc vegetable and herbaceous borders, hop arbours and walls trained with a variety of fruit.

GROUP OPENING

7 BIDBOROUGH GARDENS

Bidborough, Tunbridge Wells, TN4 0XB. *3m N of Tunbridge Wells, between Tonbridge & Tunbridge Wells W off A26. Take B2176 Bidborough Ridge signed to Penshurst. Take 1st L into Darnley Drive, then 1st R into St Lawrence Ave, no. 2.* **Sun 2 July (1-5). Combined adm £6, chd free. Home-made teas at Boundes End inc gluten & dairy free options. Donation to Hospice in the Weald.**

BOUNDES END
Carole & Mike Marks, 01892 542233, carole.marks@btinternet.com, www.boundesendgarden.co.uk.
Visits also by arrangement 12 June to 31 Aug for groups of up to 20. Teas inc in adm.

Boundes End, Bidborough Gardens

NEW **4 THE CRESCENT**
Mrs Ann Tyler.

SHEERDROP
Mr John Perry.

The Bidborough gardens (collect garden list from Boundes End, 2 St Lawrence Avenue) are in a small village at the heart of which are The Kentish Hare pub (book in advance), the church, village store and primary school. Partial wheelchair access, some gardens have steps. Please phone if you have access queries

8 BILTING HOUSE

nr Ashford, TN25 4HA. Mr John Erle-Drax, 07764 580011, johnerledrax@gmail.com. *5m NE of Ashford. A28, 5m E from Ashford, 9m S from Canterbury. Wye 1½m.* **Sat 22, Sun 23 Apr, Sat 20, Sun 21 May (1.30-5.30). Adm £7, chd free. Cream teas. Visits also by arrangement 1 Apr to 30 July for groups of 10+.**
6 acre garden with ha-ha set in beautiful part of Stour Valley. Wide variety of rhododendrons, azaleas and ornamental shrubs. Woodland walk with spring bulbs. Mature arboretum with recent planting of specimen trees. Rose garden and herbaceous borders. Conservatory.

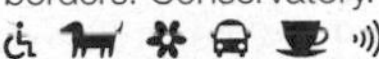

9 BISHOPSCOURT

24 St Margaret's Street, Rochester, ME1 1TS. Kelly Thomas (Head Gardener). *Central Rochester, nr castle & cathedral. On St Margaret's St at junction with Vines Lane.* **Sat 29 Apr, Sat 24 June (11-2). Adm £5, chd free. Home-made teas.**
Bishopscourt has been home to Bishops of Rochester since 1920 with the walled garden providing a secluded oasis in the town. The garden has been continuously developed for a decade with lawns, mature trees, hedging, rose garden, gravel garden, mixed herbaceous borders, meadow, glasshouse, vegetable garden and secret garden with a raised 'lookout' offering views of the castle and the river. WC inc disabled.

10 ◆ BOLDSHAVES

Woodchurch, Kent, nr Ashford, TN26 3RA. Mr & Mrs Peregrine Massey, 01233 860283, masseypd@hotmail.co.uk, www.boldshaves.co.uk. *Between Woodchurch & High Halden off Redbrook St. From centre of Woodchurch, with church on L & Bonny Cravat/Six Bells pubs on R, 2nd L down Susan's Hill, then 1st R after ½m before L after a few 100 yrds to Boldshaves. Park as indicated.* **For NGS: Thur 13 Apr (2-6). Adm £7.50, chd free. Home-made teas in the Cliff Tea House (weather permitting), otherwise in the Barn. For other opening times and information, please phone, email or visit garden website. Donation to Kent Minds.**
7 acre garden developed over past 30 yrs, partly terraced, south facing, with wide range of ornamental trees and shrubs, walled garden, Italian garden, Diamond Jubilee garden, Camellia Dell, herbaceous borders (inc flame bed, red borders and rainbow border), vegetable garden, bluebell walks in April, woodland and ponds; wildlife haven renowned for nightingales and butterflies. Home of the Wealden Literary Festival. Grass paths.

11 ◆ BOUGHTON MONCHELSEA PLACE

Church Hill, Boughton Monchelsea, Maidstone, ME17 4BU. Mr & Mrs Dominic Kendrick, 01622 743120, mk@boughtonplace.co.uk, www.boughtonplace.co.uk. *4m SE of Maidstone. For SatNav use ME17 4HP. What3words app - couch.blocks. picked. From Maidstone follow A229 S for 3½m to T-lights at Linton Xrds, turn L onto B2163, house 1m on R; or take J8 off M20 & follow Leeds Castle signs to B2163, house 5½m on L.* **For NGS: Sun 14 May, Sun 25 June (2-5.30). Adm £5, chd £1. Home-made teas. Picnics in parking field only. Cash payment only. For other opening times and information, please phone, email or visit garden website.**

150 acre estate mainly park and woodland, spectacular views over own deer park and the Weald. Grade I manor house (not open). Courtyard herb garden, intimate walled gardens, box hedges, herbaceous borders, orchard. Planting is romantic rather than manicured. Terrace with panoramic views, bluebell woods, wisteria tunnel, David Austin roses, traditional greenhouse and kitchen garden. Visit St. Peter's Church next door to see the huge stained glass Millennium Window designed by renowned local artist Graham Clark and the tranquil rose garden overlooking the deer park of Boughton Place.

12 BREWERY FARMHOUSE

182 Mongeham Road, Great Mongeham, Deal, CT14 9LR. Mr David & Mrs Maureen Royston-Lee, 07966 202243, david@davidroystonlee.com. *On the corner of Mongeham Rd & Northbourne Rd. Opp the village green, with a white picket fence at the front of the house, entrance to the gardens are through the Brewery yard at the back of the house - entrance on Northbourne Rd.* **Visits by arrangement 15 May to 28 July. Please contact us to discuss. Very welcome to bring your own refreshments. Adm £6, chd free.**

An established country walled garden divided into a series of 'rooms'. Roses, clematis, poppies abound in herbaceous borders and against walls and trellises.With water features, fruit tree pergola and a chinese pagoda.

13 1 BRICKWALL COTTAGES

Frittenden, Cranbrook, TN17 2DH. Mrs Sue Martin, 01580 852425, suemartin41@icloud.com, www.geumcollection.co.uk. *6m NW of Tenterden. E of A229 between Cranbrook & Staplehurst & W of A274 between Biddenden & Headcorn. Park in village & walk along footpath opp school.* **Sun 7 May (2-5). Adm £5, chd free. Home-made teas. Visits also by arrangement 10 Apr to 5 June for individuals & groups of up to 30.**

The garden is a secluded oasis in the centre of the village. It is filled with a wide range of plants, inc many trees, shrubs, perennials and over 100 geums which make up the National Collection. In an effort to attract more wildlife some areas of grass have been left unmown, and a new butterfly and moth 'meadow' was created during lockdown to replace the main nursery area. Some paths are narrow and wheelchairs may not be able to reach far end of garden.

NPC

14 CACKETTS FARMHOUSE

Haymans Hill, Horsmonden, TN12 8BX. Mr & Mrs Lance Morrish, 07831 432528, diana.morrish@hotmail.co.uk. *Take B2162 from Horsmonden towards Marden. 1st R into Haymans Hill, 200yds 1st L, drive immed to R of Little Cacketts.* **Visits by arrangement 1 June to 3 Aug for groups of 10+. £10 adm inc home-made teas, chd free.**

1½ acre garden surrounding C17 farmhouse (not open). Walled garden, bog garden and ponds, woodland garden with unusual plants, bug hotel. 4 acre hayfield with self planted wildflowers.

15 CHAPEL HOUSE ESTATE

Thorne Hill, Ramsgate, CT12 5DS. Chapel House Estate, www.chapelhouseestate.co.uk. *3m W of Ramsgate. From Canterbury take A253 towards Ramsgate, from Sandwich take A256 towards Ramsgate. At Sevenscore roundabout take slip road turn R, Cottington Rd, follow Chapel House Estate signs.* **Tue 20 June (10-4). Adm £6, chd £3. Light refreshments in the Walled Courtyard.**

Beautifully proportioned and designed garden show-casing established grasses, mixed perennials and unusual shrubs all set within an architectural framework of clipped yew. Coffee, tea and cakes served from The walled courtyard. 35 acre estate, with Arts and Craft style gardens, a wild apple orchard and C13 Chapel House. Helipad just off the Orchard.

16 ◆ CHARTWELL

Mapleton Road, Westerham, TN16 1PS. National Trust, 01732 868381, chartwell@nationaltrust.org.uk, www.nationaltrust.org.uk/chartwell. *4m N of Edenbridge, 2m S of Westerham. Fork L off B2026 after 1½m.* **For NGS: Wed 13 Sept (10-4.30). Adm £8, chd free. Light refreshments. For other opening times and information, please phone, email or visit garden website.**

Informal gardens on hillside with glorious views over Weald of Kent. Water features and lakes together with red brick wall built by Sir Winston Churchill, former owner of Chartwell. Lady Churchill's rose garden. Avenue of golden roses runs down the centre of a must see productive kitchen garden. Hard paths to Lady Churchill's rose garden and the terrace. Some steep slopes and steps.

17 CHEVENING

nr Sevenoaks, TN14 6HG. The Board of Trustees of the Chevening Estate, www.cheveninghouse.com. *4m NW of Sevenoaks. Turn N off A25 at Sundridge T-lights on to B2211; at Chevening Xrds 1½m turn L.* **Sun 18 June (2-5). Adm £8, chd £1. Home-made teas. Local ice cream, picnic area.**

The pleasure grounds of the Earls Stanhope at Chevening House are today characterised by lawns and wooded walks around an ornamental lake. First laid out between 1690 and 1720 in the French formal style, in the 1770s a more informal English design was introduced. In the early C19 lawns, parterres and a maze were established, a lake was created from the ornamental canal and basin, and many specimen trees were planted to shade woodland walks. Expert-guided group tours of the park and gardens can sometimes be arranged with the Estate Office when the house is unoccupied. Gentle slopes, gravel paths throughout.

18 CHURCHFIELD

Pilgrims Way, Postling, Hythe, CT21 4EY. Chris & Nikki Clark, 01303 863558, coulclark@hotmail.com. *2m NW of Hythe. From M20 J11 turn S onto A20. 1st L after ½m on bend take road signed Lyminge. 1st L into Postling.* **Sat 3, Sun 4 June (12-5). Combined adm with West Court Lodge £8, chd free. Home-made teas in Postling Village Hall. Visits also by arrangement 30 Apr to 30 Sept for groups of up to 35.**

At the base of the Downs, springs rising in this garden form the source of the East Stour. Two large ponds are home to wildfowl and fish and the banks have been planted with drifts of primula, large leaved herbaceous bamboo and ferns. The rest of the 5 acre garden is a Kent cobnut platt and vegetable garden, large grass areas and naturally planted borders and woodland. Postling Church open for visitors.

19 NEW CHURCHMANS FARM

Stalisfield Rd, Faversham, ME13 0HA. Mary & Andrew Bruce. *2½m S of Faversham. From A2 in Faversham turn S along Brogdale Rd for 2m, bear L at fork towards Stalisfield. Farm is ½m on R. Parking in farmyard.* **Fri 8, Sat 9 Sept (11-4). Adm £6, chd free. Tea. Picnics welcome by lake.**

Dry valley garden with paths round adjoining lake and fields with encouraging wildlife a priority. Chalk and flint soil and limited shade so dahlias and salvias in abundance. Large expanse of lawn with rose beds and herbaceous border. Wheelchair access to main garden only.

20 THE COACH HOUSE

Kemsdale Road, Hernhill, Faversham, ME13 9JP. Alison & Philip West, 07801 824867, alison.west@kemsdale.plus.com. *3m E of Faversham. At J7 of M2 take A299, signed Margate. After 600 metres take 1st exit signed Hernhill, take 1st L over dual carriageway to T-junction, turn R & follow yellow NGS signs.* **Sat 10 June (11-5). Sun 11 June (11-5), open nearby Lords. Adm £6, chd free. Cream teas. Visits also by arrangement June to Sept. Refreshments can be requested when booking.**

The ¾ acre garden has views over surrounding fruit-producing farmland. Sloping terraced site and island beds with year-round interest, a pond room, herbaceous borders containing bulbs, shrubs, perennials and a tropical bed. The different areas are connected by flowing curved paths. Unusual planting on light sandy soil where wildlife is encouraged. Some garden accessible to wheelchairs but some slopes. Seating available in all areas.

21 ◆ COBHAM HALL

Cobham, DA12 3BL. Mr D Standen (Bursar), 01474 823371, www.cobhamhall.com. *3m W of Rochester, 8m E of M25 J2. Ignore SatNav directions to Lodge Ln. Entrance drive is off Brewers Rd, 50 metres E from Cobham/ Shorne A2 junction.* **For NGS: Sun 2 Apr (2-5). Adm £4.50, chd free. Light refreshments in the Gilt Hall. For other opening times and information, please phone or visit garden website.**

1584 brick mansion (open for tours) and parkland of historical importance, now a boarding and day school for girls. Some herbaceous borders, formal parterres, drifts of daffodils, C17 garden walls, yew hedges and lime avenue. Humphry Repton designed 50 hectares of park. Most garden follies restored in 2009. Film location for BBC's Bleak House series and films by MGM and Universal. ITV serial The Great Fire. CBBC filmed serial 1 and 2 of Hetty Feather. Gravel and slab paths through gardens. Land uneven, many slopes. Stairs and steps in Main Hall. Please call in advance to ensure assistance.

22 NEW THE COPPER HOUSE

Hinksden Road, Benenden, Cranbrook, TN17 4LE. Eleanor Cochrane, eleanor.cochrane@btinternet.com. *Located close to Hinksden Dairy.* **Sat 9, Sun 10 Sept (11-3). Adm £5, chd free. Pre-booking essential, please visit www.ngs.org.uk for information & booking. Cream teas. Visits also by arrangement 4 July to 12 Sept for groups of 10 to 15. £7.50 inc tour. Regret no refreshments.**

The Copper House garden is a modern flower garden successionally planted to provide interest and colour throughout the season. Mixed planting of annuals, perennials, bulbs and shrubs. Small wildflower meadow and orchard. Three small ponds to encourage wildlife. Recently renovated woodland ponds. Borders, views, gravel garden, water features.

23 COPTON ASH

105 Ashford Road, Faversham, ME13 8XW. Drs Tim & Gillian Ingram, 01795 535919, coptonash@yahoo.co.uk, www.coptonash.plus.com. *½m S of A2, Faversham on A251. On A251 Faversham to Ashford rd. Opp E bound J6 with M2. Park in nearby laybys, single yellow lines are Mon to Fri restrictions.* **Sat 11 Feb (12-4). Sun 19 Feb (12-4), open nearby Doddington Place. Sun 19, Sun 26 Mar, Sun 9 Apr (12-5). Sun 30 Apr (12-5), open nearby Frith Old Farmhouse. Sun 21 May (12-5). Sun 28 May (12-5), open nearby Eagleswood. Adm £5, chd free. Home-made teas. 2024: Sat 10, Sun 18 Feb. Visits also by arrangement 13 Feb to 30 June.**

Garden grown out of a love and fascination with plants. Contains wide collection inc many rarities and newly introduced species raised from wild seed. Special interest in woodland flowers, snowdrops and hellebores with flowering trees and shrubs of spring. Refreshed Mediterranean plantings to adapt to a warming climate. Raised beds with choice alpines and bulbs. Alpine and dryland plant nursery. Gravel drive, shallow step by house and some narrow grass paths.

National Garden Scheme gardens are identified by their yellow road signs and posters. You can expect a garden of quality, character and interest, a warm welcome and plenty of home-made cakes!

Arnold Yoke

GROUP OPENING

24 DEAL TOWN GARDENS

Deal, CT14 6EB. *Signs from all town car parks, maps & tickets at all gardens.* **Sun 25 June (11-5). Combined adm £5, chd free. Home-made teas at 53 Sandown Road.**

NEW **61 COLLEGE ROAD**
Andrew Tucker.

4 GEORGE ALLEY
Lyn Freeman & Barry Popple.

16 ST ANDREW'S ROAD
Martin Parkes & Paul Green.

5 ST PATRICK'S CLOSE
Chris & Geoff Hobbs-East.

53 SANDOWN ROAD
Robin Green & Ralph Cade.

NEW **14 SUTHERLAND ROAD**
Joan Bull.

88 WEST STREET
Lyn & Peter Buller.

Start from any town car park (signs from here). 88 West Street: A non water cottage garden with perennials, shrubs, clematis and roses and shade garden. 4 George Alley: A pretty alley leads to a secret garden with courtyard, leading to a vibrant cottage garden with summerhouse. 16 St Andrew's Road: A tropical themed walled urban garden with large sun-drenched borders, lawn and family seating area. 53 Sandown Road: A south west facing garden with garden rooms for entertainment. Decking terraces surround the house with pots of summer flowering bulbs. 5 St Patrick's Close: South-facing, rented garden, featuring roses, herbs and scented flowers in a range of containers and pots. The summerhouse hosts a OO gauge railway which extends into the garden. 14 Sutherland Rd: Three small gardens in one with intense planting for all year interest and support for wildlife.61 College Rd: Small garden filled with colour and a profusion of plants.

25 DEAN HOUSE

Newchurch, Romney Marsh, TN29 0DL. Jaqui Bamford, 07480 150684, jaquibamford@gmail.com. *Between the village of Newchurch & New Romney. Leave B2067 at Bilsington Xrds SE towards New Romney. Continue for 2.9m to S bend. Garden on R after bend. From New Romney leave A259 NE onto St Marys Rd. Continue for 3.7m. Garden on L.* **Visits by arrangement 12 June to 17 Sept. Adm £5, chd free. Tea.**

A garden created over 20 yrs, from an old farmyard, featuring mature trees, shady areas, wildlife pond, paths, secluded sitting areas, sun-drenched gravel beds and herbaceous borders designed for pollinators. Extensive views across Romney Marsh. Visitors can explore the field to which the garden leads to see a rewilding project in its infancy. Large collection of cacti and succulents. The garden planting attracts a number of different bumblebee species inc one of the rarer species - the Ruderal bumblebee. Photos of the garden and surrounding area in different seasons are on display throughout the garden. Laminated A5 or A6 prints can be made up to order. Wheelchair must be operable on flat gravel and grass areas. Stepping stone paths and some seating will not be accessible.

26 ◆ DODDINGTON PLACE
Church Lane, Doddington, Sittingbourne, ME9 0BB. Mr & Mrs Richard Oldfield, 07596 090849, enquiries@doddingtonplacegardens.co.uk, www.doddingtonplacegardens.co.uk. *6m SE of Sittingbourne. From A20 turn N opp Lenham or from A2 turn S at Teynham or Ospringe (Faversham), all 4m.* **For NGS: Sun 19 Feb (11-4); Sun 2 Apr, Sun 24 Sept (11-5). Adm £9, chd £3. For other opening times and information, please phone, email or visit garden website.**
10 acre garden, wide views; trees and cloud clipped yew hedges; woodland garden with azaleas and rhododendrons; Edwardian rock garden; formal garden with mixed borders. A flint and brick late C20 gothic folly; newly installed at the end of the Wellingtonia walk, a disused pinnacle from the southeast tower of Rochester Cathedral. Snowdrops in February. Chelsea Fringe Events. Wheelchair access possible to majority of gardens except rock garden.

27 DOWNS COURT
Church Lane, Boughton Aluph, Ashford, TN25 4EU. Mr Bay Green, 07984 558945, bay@baygee.com. *4m NE of Ashford. From A28 Ashford or Canterbury, after Wye Xrds take next turn NW to Boughton Aluph Church signed Church Ln. Fork R at pillar box, garden only drive on R. Park in field. Disabled parking in drive.* **Sun 28 May, Sun 11, Sun 18 June (2-5). Adm £5, chd free. Visits also by arrangement 28 May to 16 July.**
3 acre downland garden on alkaline soil with fine trees, mature yew and box hedges, mixed borders. Shrub roses and rose arch pathway, small parterre. Sweeping lawns and lovely views over surrounding countryside.

28 EAGLESWOOD
Slade Road, Warren Street, Lenham, ME17 2EG. Mike & Edith Darvill, 01622 858702, mike.darvill@btinternet.com. *approx. 12m E of Maidstone. E on A20 nr Lenham, L into Hubbards Hill for approx 1m then 2nd L into Slade Rd. Garden 150yds on R. Coaches permitted by prior arrangement.* **Mon 1, Sun 28 May (11-5). Adm £5, chd free. Light refreshments. Visits also by arrangement Apr to Oct for groups of 5+. Guided tours can be arranged. Donation to Demelza House Hospice.**
2 acre plant enthusiasts garden. Wide range of trees and shrubs (many unusual), herbaceous material and woodland plants grown to give year-round interest.

29 ◆ EMMETTS GARDEN
Ide Hill, Sevenoaks, TN14 6BA. National Trust, 01732 751507, emmetts@nationaltrust.org.uk, www.nationaltrust.org.uk/emmetts-garden. *5m SW of Sevenoaks. 1½m S of A25 on Sundridge Ide Hill Rd. 1½m N of Ide Hill off B2042.* **For NGS: Wed 6 Sept (10-4.30). Adm £8, chd free. Light refreshments in the Old Stables. For other opening times and information, please phone, email or visit garden website.**
5 acre hillside garden, with the highest tree top in Kent, noted for its fine collection of rare trees and flowering shrubs. The garden is particularly fine in spring, while a rose garden, rock garden and extensive planting of acers for autumn colour extend the interest throughout the season. Hard paths to the Old Stables for light refreshments and WC. Some steep slopes. Volunteer driven buggy available for lifts up steepest hill.

30 EUREKA
Westerham Hill, TN16 2HR. Gordon & Suzanne Wright. *Park at Westerham Heights Garden Centre TN16 2HW off A233, 200m from Eureka. Parking at house for those with walking difficulties.* **Sat 1, Sun 2, Sat 22, Sun 23 July, Sat 12, Sun 13 Aug, Sat 2, Sun 3 Sept (11-4). Adm £5, chd free. Home-made teas.**
Approx 1 acre garden with a blaze of colourful displays in perennial borders and the eight cartwheel centre beds. Sculptures, garden art, chickens, lots of seating and stairs to a viewing platform. Many quirky surprises at every turn. 100s of annuals in tubs and baskets, a David Austin shrub rose border and a free Treasure Trail with prizes for all children. Garden art inc 12ft Blacksmith's made red dragon, a horse's head carved out of a 200yr old yew tree stump and a 10ft dragonfly on a reed. Wheelchair access to most of the garden.

31 FALCONHURST
Cowden Pound Road, Markbeech, Edenbridge, TN8 5NR. Mr & Mrs Charles Talbot, 01342 850526, nicola@falconhurst.co.uk, www.falconhurst.co.uk. *3m SE of Edenbridge. B2026 at Queens Arms pub turn E to Markbeech. 2nd drive on R before Markbeech village.* **Mon 29 May, Sat 17 June (11-5). Adm £6, chd free. Home-made teas. Visits also by arrangement 22 May to 30 Sept for groups of 15 to 40. Refreshments available by arrangement.**
4 acre garden with fabulous views devised and cared for by the same family for 170 yrs. Deep mixed borders with old roses, peonies, shrubs and a wide variety of herbaceous and annual plants; ruin garden; walled garden; interesting mature trees and shrubs; kitchen garden; wildflower meadows with woodland and pond walks. Market garden, pigs, orchard chickens; lambs in the paddocks.

32 FRITH OLD FARMHOUSE
Frith Road, Otterden, Faversham, ME13 0DD. Drs Gillian & Peter Regan, 01795 890556, peter.regan@cantab.net. *½m off Lenham to Faversham rd. From A20 E of Lenham follow signs Eastling;. after 4m L into Frith Rd. From A2 in Faversham turn S (Brogdale Rd); cont 7m (thro' Eastling), R into Frith Rd Limited parking.* **Sun 16 Apr (11-4.30). Sun 30 Apr (11-4.30), open nearby Copton Ash. Sun 14 May (11-4.30). Adm £6, chd free. Visits also by arrangement 19 Mar to 31 July for groups of up to 40. Please email to book.**
A riot of plants growing together as if in the wild, developed over nearly 50 years. No formal beds, but several hundred interesting (and some very unusual) plants. trees and shrubs chosen for year-round appeal. Special interest in bulbs and woodland plants. Visitor comments - 'one of the best we have seen, natural and full of treasures', 'a plethora of plants', 'inspirational', 'a hidden gem'. Altered habitat areas to increase the range of plants grown. Areas for wildlife. Unusual trees and shrubs.

33 THE GARDEN GATE
Northdown Park, Northdown Park Road, Margate, Kent, CT9 3TP. The Garden Gate Project Ltd, www.thegardengateproject.co.uk. *Located within Northdown Park, opp Friends Corner on Northdown Park Rd B2052 between Margate & Broadstairs, nr Northdown House.* **Sat 1 July (2-5). Adm £5, chd free. Light refreshments. inc wood fired pizzas with toppings from the garden will be on sale.**
The Garden Gate is a community garden based in Northdown Park, growing a mixture of plants, flowers and vegetables using organic methods. We also have a wildlife pond, two polytunnels, a shade house, some coppiced woodland and a green roof on one of our buildings. The garden is flat and on one level with grass or wood chip paths.

34 GODDARDS GREEN
Angley Road, Cranbrook, TN17 3LR. John & Linde Wotton, 01580 715507, jpwotton@gmail.com, www.goddardsgreengarden.com. *½m SW of Cranbrook. On W of Angley Rd. (A229) at junction with High St, opp War Memorial.* **Sun 2 July (12-4). Adm £5, chd free. Home-made teas. Visits also by arrangement Apr to Sept (excl Aug) for groups of 10 to 50.**
Gardens of about 5 acres, surrounding beautiful 500+yr old clothier's hall (not open), laid out in 1920s and redesigned since 1992 to combine traditional and modern planting schemes. fountain and rill, water garden, fern garden, mixed borders of bulbs, perennials, shrubs, trees and exotics; birch grove, grass border, pond, kitchen garden, meadows, arboretum and mature orchard. Some slopes and steps, but most areas (though not the toilets) are wheelchair accessible. Disabled parking is reserved near the house.

35 ◆ GODINTON HOUSE & GARDENS
Godinton Lane, Ashford, TN23 3BP. The Godinton House Preservation Trust, 01233 643854, info@godintonhouse.co.uk, www.godintonhouse.co.uk. *1½m W of Ashford. M20 J9 to Ashford. Take A20 towards Charing & Lenham, then follow brown tourist signs.* **For NGS: Sun 19 Mar, Sun 14 May, Fri 16 June, Fri 22 Sept (1-5). Adm £8, chd free. Home-made teas. Ticket office serves takeaway refreshments, check website for tearoom opening times. For other opening times and information, please phone, email or visit garden website.**
12 acres complement the magnificent Jacobean house. Terraced lawns lead through herbaceous borders, rose garden and formal lily pond to an intimate Italian garden and large walled garden with delphiniums, potager and cut flowers. March/April the 3 acre wild garden is a mass of daffodils, fritillaries, primroses and other spring flowers. Delphinium Festival (10 June - 25 June). Garden workshops and courses throughout the year. Partial wheelchair access to ground floor of house and most of gardens.

36 GODMERSHAM PARK
Godmersham, CT4 7DT. Mrs Fiona Sunley. *5m NE of Ashford. Off A28, midway between Canterbury & Ashford.* **Sun 19 Mar, Sun 11 June (1-5). Adm £7.50, chd free.**
24 acres of restored wilderness and formal gardens set around C18 mansion (not open). Topiary, rose garden, herbaceous borders, walled kitchen garden and recently restored Italian and swimming pool gardens. Superb daffodils in spring and roses in June. Historical association with Jane Austen. Also visit the Heritage Centre. Deep gravel paths.

37 ◆ GOODNESTONE PARK GARDENS
Wingham, Canterbury, CT3 1PL. Julian Fitzwalter, 01304 840107, office@goodnestone.com, www.goodnestoneparkgardens.co.uk. *6m SE of Canterbury. Village lies S of B2046 from A2 to Wingham. Brown tourist signs off B2046. Use CT3 1PJ for SatNav.* **For NGS: Thur 15 June, Thur 14 Sept (10-4). Adm £8, chd free. Light refreshments at The Old Dairy Cafe. Café tel: 01304 695098. For other opening times and information, please phone, email or visit garden website.**
One of Kent's outstanding gardens and the favourite of many visitors. 14 acres with views over parkland. Something special year-round from snowdrops and spring bulbs to the famous walled garden with old fashioned roses and kitchen garden. Outstanding trees and woodland garden with cornus collection and hydrangeas later. Two arboreta and a contemporary gravel garden.

38 104 GRANGE ROAD
Ramsgate, CT11 9PX. Anne-Marie Nixey. *Enter Ramsgate on A299, cont on A255. At r'about take 2nd exit London Rd. Continue for less than 1m to the r'about & turn L onto Grange Rd or straight ahead down West Cliff Rd.* **Sun 10 Sept (12-5). Combined adm with 12 West Cliff Road £5, chd free. Home-made teas.**
Fruit, chickens, vegetables and lots of flowers! A series of parts all of which intertwine nature with a traditional town garden. Reclaimed materials, food production, sustainability, water conservation and wildlife friendly.

39 GRAVESEND GARDEN FOR WILDLIFE
68 South Hill Road, Windmill Hill, Gravesend, DA12 1JZ. Judith Hathrill, 07810 550991, judith.hathrill@live.com. *On Windmill Hill 0.6m from Gravesend town centre, 1.8m from A2. From A2 take A227 towards Gravesend. At T-lights with Cross Ln turn R then L at next T-lights, following yellow NGS signs. Park in Sandy Bank Rd or Rouge Ln.* **Sun 2 July (12-5). Adm £5, chd free. Home-made teas. Visits also by arrangement 3 June to 27 Aug for groups of up to 20.**
Cottage garden dedicated to nurturing wildlife. Borders packed with flowers for pollinators. 3 ponds, visited by amphibians, dragonflies and birds. Small wildflower lawn. Vegetables all grown in containers to demonstrate gardening in a small space. Information and leaflets about gardening for wildlife always available.

40 ◆ GREAT COMP GARDEN
Comp Lane, Platt, nr Borough Green, Sevenoaks, TN15 8QS. Great Comp, 01732 885094, office@greatcompgarden.co.uk, www.greatcompgarden.co.uk. *7m E of Sevenoaks. 2m from Borough Green Station. Accessible from M20 & M26 motorways. A20 at Wrotham Heath, take Seven Mile Lane, B2016; at 1st Xrds turn R; garden on L ½m.* **For NGS: Sun 26 Mar, Sun 29 Oct (10-5). Adm £8.50, chd £3. For other opening times and information, please phone, email or**

The Copper House

visit garden website.
Skilfully designed 7 acre garden of exceptional beauty. Spacious setting of well maintained lawns and paths lead visitors through plantsman's collection of trees, shrubs, heathers and herbaceous plants. Early C17 house (not open). Magnolias, hellebores and snowflakes (leucojum), hamamellis and winter flowering heathers are a great feature in the spring. A great variety of perennials in summer inc salvias, dahlias and crocosmias. Tearoom open daily for morning coffee, home-made lunches and cream teas. Most of garden accessible to wheelchair users. Disabled WC.

41 GREAT MAYTHAM HALL
Maytham Road, Rolvenden, Tenterden, TN17 4NE. The Sunley Group. *3m from Tenterden. Maytham Rd off A28 at Rolvenden Church, ½m from village on R. Designated parking for visitors.* **Wed 24 May, Wed 21 June (1-4). Adm £7.50, chd free.**
Lutyens designed gardens famous for having inspired Frances Hodgson Burnett to write The Secret Garden (pre Lutyens). Parkland, woodland with bluebells. Walled garden with herbaceous beds and rose pergola. Pond garden with mixed shrubbery and herbaceous borders. Interesting specimen trees. Large lawned area, rose terrace with far reaching views.

42 NEW GROVE MANOR
The Street, Woodnesborough, Sandwich, CT13 0NH. Deborah Anderson. *1½m SW of Sandwich. From Sandwich drive to W'borough on Woodnesborough Road. From C'bury A257 through Ash & Marshborough. From Dover/Thanet A256 to Foxborough Hill.* **Sat 12, Sun 13 Aug (11-5). Adm £5, chd free. Home-made teas.**
Grove Manor, a Victorian farmhouse is surrounded by gardens reflecting its past. Close to the house are lawns, herbaceous beds and a shady fernery. Home to moorhens and ducks, the farm pond has recently been dredged and replanted. The moat is a haven for wildlife and borders the orchard and composting area (the engine room). Beyond are the newly established veg and cutting beds.

43 HAMMOND PLACE
High Street, Upnor, Rochester, ME2 4XG. Paul & Helle Dorrington. *3m NE of Strood or at A2. J1 take A289 twds Grain at r'about follow signs to Gillingham. After 2nd r'about take 1st L following signs to Upnor & Upnor Castle. Park in free car park & continue by foot to the High St.* **Sat 8, Sun 9 July (11-4). Adm £4, chd free. Home-made teas.**
A small garden in a historically interesting village growing an eclectic mix of flowers, fruit and vegetables around a Scandinavian style house. Greenhouse, Pond, Sauna Hut.

44 HAVEN

22 Station Road, Minster, Ramsgate, CT12 4BZ. Robin Roose-Beresford, 01843 822594, robin.roose@hotmail.co.uk. *Off A299 Ramsgate Rd, take Minster exit from Manston r'bout, straight rd, R fork at church is Station Rd.* **Sun 5, Wed 15 Mar, Sun 9, Mon 10 Apr, Mon 1, Sat 20 May, Sun 18 June, Sun 9, Sun 30 July, Mon 28 Aug, Sun 24 Sept, Sun 29 Oct, Sun 12 Nov (10-4). Adm £5, chd free. Visits also by arrangement 1 Feb to 1 Nov. Visits can be arranged for any day of the year with 24 hours notice.**

Award winning 300ft garden, designed in the Glade style, similar to Forest gardening but more open and with use of exotic and unusual trees, shrubs and perennials, with wildlife in mind, devised and maintained by the owner, densely planted in a natural style with stepping stone paths. Two ponds (one for wildlife, one for fish with water lilies), gravel garden, rock garden, fernery, Japanese garden, cactus garden, hostas and many exotic, rare and unusual trees, shrubs and plants inc tree ferns and bamboos and year-round colour. Greenhouse cactus garden. Collection of Bromeliads Treefern grove.Many Palms.

45 ◆ HEVER CASTLE & GARDENS

Edenbridge, TN8 7NG. Hever Castle Ltd, 01732 865224, info@hevercastle.co.uk, www.hevercastle.co.uk. *3m SE of Edenbridge. Between Sevenoaks & East Grinstead off B2026. Signed from J5 & J6 of M25, A21, A264.* **For NGS: Wed 4 Oct (10.30-4.30). For other opening times and information, please phone, email or visit garden website.**

Romantic double-moated castle, the childhood home of Anne Boleyn, set in 150 acres of formal and natural landscape. Topiary, Tudor herb garden, magnificent Italian garden with classical statuary, sculpture and fountains. 38 acre lake, yew and water mazes. Walled rose garden with over 5000 roses, 110 metre-long herbaceous border. Fine autumn colour displays in Oct. Partial wheelchair access.

46 95 HIGH STREET

Tenterden, TN30 6LB. Judy & Chris Older. *Nearest parking is Bridewell car park - Saturday visitors will have to pay but free on Sundays. Drive to far end. Venue on R.* **Sat 1, Sun 2 July (12-5). Adm £5, chd free. Light refreshments.**

Small town house garden divided into 3 rooms. Developed since June 2021 from a blank canvas, now well-stocked with perennials and summer flowers to capacity. Colourful and traditional. Plenty of seating. Teas available in Tenterden. Two small steps.

47 ◆ HOLE PARK

Benenden Road, Rolvenden, Cranbrook, TN17 4JB. Mr & Mrs Edward Barham, 01580 241344, info@holepark.com, www.holepark.com. *4m SW of Tenterden. Midway between Rolvenden & Benenden on B2086. Follow brown tourist signs from Rolvenden.* **For NGS: Sun 16 Apr, Wed 10 May, Wed 14 June, Sun 15 Oct (11-6). Adm £10, chd £2.50. Home-made teas. Picnics only in picnic site & car park please. For other opening times and information, please phone, email or visit garden website.**

Hole Park is proud to stand amongst the group of gardens which first opened in 1927 soon after it was laid out by my great grandfather. Our 15 acre garden is surrounded by parkland and contains fine yew hedges, large lawns with specimen trees, walled gardens, pools and mixed borders combined with bulbs, rhododendrons and azaleas. Massed bluebells in woodland walk, standard wisterias, orchids in flower meadow and glorious autumn colours make this a garden for all seasons. Wheelchairs are available for loan and may be reserved. Please email info@holepark.com.

48 HOPPICKERS EAST

Hogbens Hill, Selling, Faversham, ME13 9QZ. Katherine Pickering. *Signs from HogbensHill. From A251 signed Selling 1m, then NGS signs.* **Sat 1 July (11-5). Combined adm with 3 Stanley Cottages £6, chd free. Wed 2 Aug (11-5). Adm £5, chd free.**

A new garden, started in autumn 2020 on the edge of a field. No dig principles. Emphasis on plants for pollinators and other insects.New this year a mizmaze in meadow. Flat grass paths.

49 ◆ IGHTHAM MOTE

Mote Road, Ivy Hatch, Sevenoaks, TN15 0NT. National Trust, 01732 810378, ighthammote@nationaltrust.org.uk, www.nationaltrust.org.uk/ightham-mote. *Nr Ivy Hatch: 6m E of Sevenoaks; 6m N of Tonbridge; 4m SW of Borough Green. E from Sevenoaks on A25 follow brown sign R along coach road. W from Borough Green on A25 follow brown sign L along coach rd. N from Tonbridge on A227 follow brown sign L along High Cross rd.* **For NGS: Thur 30 Mar, Thur 19 Oct (10-4). Adm £17, chd £8.50. Cream teas at the Mote Café. For other opening times and information, please phone, email or visit garden website.**

Lovely 14 acre garden surrounding a picturesque medieval moated manor house c1320, open for NGS since 1927. Herbaceous borders, lawns, C18 cascade, fountain pools, courtyards and a cutting garden provide formal interest; while the informal lakes, stream, pleasure grounds, stumpery/fernery, dell and orchard complete the sense of charm and tranquillity. Estate walks open to dogs all year, regret no dogs in main garden. Please check NT Ightham Mote website for access details, prices and last admission. Access guide available at visitor reception.

50 IVY CHIMNEYS

Mount Sion, Tunbridge Wells, TN1 1TW. Laurence & Christine Smith. *At the end of Tunbridge Wells High St, with Pizza Express on the corner, turn L up Mount Sion. Ivy Chimneys is a red brick Queen Anne house at the top of the hill on the R.* **Sat 10 June (11-5). Adm £5, chd free.**

Town centre garden with herbaceous borders and masses of roses set on three layers of lawns enclosed in an old walled garden. This property also boasts a large vegetable garden with flowers for cutting and chickens for eggs! The property dates back to the late 1600's. Parking in public carparks near the Pantiles. Winner of best back garden in Tunbridge Wells in Bloom competition. Pantiles cafes nearby.

51 KENFIELD HALL

Kenfield, Petham, Canterbury, CT4 5RN. Barnaby & Camilla Swire, kenfieldhallgarden@gmail.com. *Petham nr Canterbury. Continue along Kenfield Rd, down the hill & up the other side. Pass the farm & look out for signs.* **Visits by arrangement May to July for groups of up to 30. Adm £10, chd free.**

The 8 acre gardens benefit from views within the AONB and its diverse wildlife, consisting of a sunken formal garden, lawns, mixed borders, herbaceous borders, spring bulbs, a Japanese water garden, orchards, an organic vegetable garden with glasshouses, a wildflower meadow, a rose garden and woodland garden.

52 ◆ KNOLE

Knole, Sevenoaks, TN15 0RP. Lord Sackville, 01732 462100, knole@nationaltrust.org.uk, www.nationaltrust.org.uk/knole. *1½m SE of Sevenoaks. Leave M25 at J5 (A21). Park entrance S of Sevenoaks town centre off A225 Tonbridge Rd (opp St Nicholas Church). For SatNav use TN13 1HX.* **For NGS: Tue 18 July (11-4). Adm £5, chd £2.50. For other opening times and information, please phone, email or visit garden website.**

Lord Sackville's private garden at Knole is a magical space, featuring sprawling lawns, a walled garden, an untamed wilderness area and a medieval orchard. Access is through the beautiful Orangery, off Green Court, where doors open to reveal the secluded lawns of the 26 acre garden and stunning views of the house. Last entry at 3.30pm and closes at 4pm. Please book in advance using Knole's website. Refreshments are available in the Brewhouse Café. Bookshop and shop in Green Court. Wheelchair access via the bookshop into the Orangery. Some paths may be difficult in poor weather. Assistance dogs are allowed in the garden.

53 THE KNOLL FARM

Giggers Green Road, Aldington, Ashford, TN25 7BY. Lord & Lady Aldington. *The Postcode leads to Goldenhurst, our drive entrance is opp & a little further down hill.* **Sat 15, Sun 16 Apr (12-4). Adm £8, chd free. Donation to Bonnington Church.**

10 acres of woodland garden with over 100 camellias, acer japonica, and a growing collection of specimen pines and oaks; bluebell wood; formal elements; Flock of Jacob Sheep around lake; long views across Romney Marsh. Picnics welcome;. Paths throughout but the whole garden is on a slope and clay can be slippery.

54 KNOWLE HILL FARM

Ulcombe, Maidstone, ME17 1ES. The Hon Andrew & Mrs Cairns, 01622 850240, elizabeth@knowlehillfarm.co.uk, www.knowlehillfarmgarden.co.uk. *7m SE of Maidstone. From M20 J8 follow A20 towards Lenham for 2m. Turn R to Ulcombe. After 1½m, L at Xrds, after ½m 2nd R into Windmill Hill. Past Pepper Box Pub, ½m 1st L to Knowle Hill.* **Sat 4, Sun 5 Feb (11-3). Light refreshments. Sat 2, Sun 3 Sept (2-5). Home-made teas. Adm £6, chd free. 2024: Sat 3, Sun 4 Feb. Visits also by arrangement Feb to Sept.**

2 acre garden created over nearly 40 years on south facing slope below the Greensand Ridge. Spectacular views. Snowdrops and hellebores, many tender plants, china roses, agapanthus, salvias and grasses flourish on light soil. Topiary continues to evolve with birds at last emerging. Lavender ribbons hum with bees. Pool enclosed in small walled white garden. A green garden was completed in 2018. Access only for 35 seater coaches. Some steep slopes.

55 LACEY DOWN

22 Canterbury Road, Lydden, Dover, CT15 7ER. Wendy & Nick Smith. *2m NW of Dover. From A2 N take Lydden turn. On main road through village just before end of 30 speed limit. From A2 S 2nd exit Whitfield r'about. Follow London Rd through Temple Ewell, under railway bridge.* **Sun 28 May, Sun 16 July (1-5). Combined adm with Windy Ridge, Dover Road £5, chd free. Home-made teas.**

A botanical artist's garden, long and very narrow, on a sloping site. Borders are loosely themed. There is a tiny pond and a keyhole vegetable garden. Paths are mainly gravel and narrow. Steps access the garden and continue within. The site was derelict and the budget small when the garden was begun in 2014. Tiny front garden. Not wheelchair accessible. Children welcome. Windy Ridge CT15 5EH.

56 LITTLE GABLES

Holcombe Close, Westerham, TN16 1HA. Mrs Elizabeth James. *Centre of Westerham. Off E side of London Rd A233, 200yds from The Green. Please park in public car park. No parking available at house.* **Sat 20, Sun 21 May, Sat 10, Sun 11 June (1.30-4.30). Adm £5, chd free. Home-made teas.**

¾ acre plant lover's garden extensively planted with a wide range of trees, shrubs, perennials etc, inc many rare varieties. Collection of climbing and bush roses. Large pond with fish, water lilies and bog garden. Fruit and vegetable garden. Large greenhouse.

57 LORDS

Sheldwich, Faversham, ME13 0NJ. John Sell CBE & Barbara Rutter, 01795 536900, john@sellwade.co.uk. *On A251 4m S of Faversham & 3½m N of Challock Xrds. From A2 or M2 take A251 towards Ashford. ½m S of Sheldwich church find entrance lane on R adjacent to wood.* **Sun 11 June (2-5). Adm £6, chd free. Home-made teas. Visits also by arrangement 1 Apr to 10 June for groups of 12 to 20. Charge TBC when arranging & will inc refreshments & guided visit.**

C18 walled garden. Mediterranean terrace and citrus standing. Flowery meadow under apples, pears, quince, crab apple and medlar. Grass tennis court. A cherry orchard grazed by Jacob sheep. Pleached hornbeams, clipped yew hedges and 120 ft. high tulip tree. Fruit, vegetables, herbs, salads. New exciting sculpture this year. Some gravel paths.

58 MEADOW WOOD

Penshurst, TN11 8AD. Mr & Mrs James Lee. *1¼m SE of Penshurst, 5m SW of Tonbridge. On B2176 in direction of Bidborough.* **Sun 7 May (12-5). Adm £10, chd free. Home-made teas. Entry price inc tea.**
Mature 1920s garden, on edge of woods, originally part of the Victorian Swaylands Estate. Long southerly views over the Weald, and interesting mature trees and shrubs, azaleas, rhododendrons, naturalised bulbs and bluebells in woodlands with mown walks. New lake.

59 12 THE MEADOWS

Chelsfield, Orpington, BR6 6HS. Mr Roger & Mrs Jean Pemberton. *3m from J4 on M25. 10 mins walk from Chelsfield station. Exit M25 at J4. At r'about 1st exit for A224, next r'about 3rd exit - A224, ½m, take 2nd L, Warren Rd. Bear L into Windsor Drive. 1st L The Meadway, follow signs to garden.* **Sun 21 May (11-5). Adm £5, chd free. Light refreshments.**
Front garden Mediterranean style gravel with sun loving plants. Rear ¾ acre garden in two parts. Semi-formal Japanese style area with tea house, two ponds, one Koi and one natural (lots of spring interest). Acers, grasses, huge bamboos etc and semi wooded area, children's path with 13ft high giraffe, Sumatran tigers and lots of points of interest. Children and well behaved dogs more than welcome. Designated children's area. Adults only admitted if accompanied by responsible child! Silver award for garden with the LGS. Winner of first prize for our back garden from Bromley in Bloom. Wheelchair access to all parts except small stepped area at very bottom of garden.

60 ◆ MOUNT EPHRAIM GARDENS

Hernhill, Faversham, ME13 9TX. Mr & Mrs E S Dawes & Mr W Dawes, 01227 751496, info@mountephraimgardens.co.uk, www.mountephraimgardens.co.uk. *3m E of Faversham. From end of M2, then A299 take slip rd 1st L to Hernhill, signed to gardens.* **For NGS: Sun 26 Mar (11-4), open nearby Copton Ash. Thur 8 June, Sun 1 Oct (11-4). Adm £8, chd £3. For other opening times and information, please phone, email or visit garden website.**

Mount Ephraim is a privately-owned family home set in 10 acres of terraced Edwardian gardens with stunning views over the Kent countryside. Highlights inc a Japanese rock and water garden, arboretum, unusual topiary and a spectacular grass maze plus many mature trees, shrubs and spring bulbs. Partial wheelchair access; top part manageable, but steep slope. Disabled WC. Full access to tea room.

61 NETTLESTEAD PLACE

Nettlestead, Maidstone, ME18 5HA. Mr & Mrs Roy Tucker, www.nettlesteadplace.co.uk. *6m W/SW of Maidstone. Turn S off A26 onto B2015 then Im on L, Nettlestead Court Farm after Nettlestead Church.* **Sun 2 Apr, Sun 4 June (1.30-4.30). Adm £7, chd free. Light refreshments.**

C13 manor house in 10 acre plantsman's garden. Large formal rose garden. Large herbaceous garden of island beds with rose and clematis walkway leading to a recently planted garden of succulents. Fine collection of trees and shrubs; sunken pond garden, a maze of Thuja, terraces, bamboos, glen garden, Acer lawn. Young pinetum adjacent to garden. Sculptures. Maze and beautiful countryside views. Collection of Shona and other sculptures. Gravel and grass paths. Most of garden accessible (but not sunken pond garden). Large steep bank and lower area accessible with some difficulty.

62 NORTON COURT

Teynham, Sittingbourne, ME9 9JU. Tim & Sophia Steel. *Off A2 between Teynham & Faversham. L off A2 at Esso garage into Norton Lane; next L into Provender Ln; L signed Church for car park.* **Mon 19, Tue 20 June (2-5). Adm £10, chd free. Home-made teas inc in adm.**

10 acre garden within parkland setting. Mature trees, topiary, wide lawns and clipped yew hedges. Orchard with mown paths through wildflowers. Walled garden with mixed borders and climbing roses. Pine tree walk. Formal box and lavender parterre. Treehouse in the Sequoia. Church open, adjacent to garden. Flat ground except for 2 steps where ramp is provided.

63 OAK COTTAGE AND SWALLOWFIELDS NURSERY

Elmsted, Ashford, TN25 5JT. Martin & Rachael Castle. *6m NW of Hythe. From Stone St (B2068) turn W opp the Stelling Minnis turning. Follow signs to Elmsted. Turn L at Elmsted village sign. Limited parking at house, further parking at Church (7mins walk).* **Fri 14, Sat 15 Apr, Wed 10, Thur 11 May (11-4). Adm £5, chd free. Cream teas.**

Get off the beaten track and discover this beautiful ½ acre cottage garden in the heart of the Kent countryside. This plantsman's garden is filled with unusual and interesting perennials, inc a wide range of salvias. There is a small specialist nursery packed with herbaceous perennials. Auriculas displayed in traditional theatres in April.

64 OLD BLADBEAN STUD

Bladbean, Canterbury, CT4 6NA. Carol Bruce, www.oldbladbeanstud.co.uk. *6m S of Canterbury. From B2068, follow signs into Stelling Minnis, turn R onto Bossingham Rd, then follow yellow NGS signs through single track lanes.* **Sun 28 May, Sun 11, Sun 25 June, Sun 9, Sun 23 July (2-6). Adm £6, chd free. Cream teas.**

Romantic walled rose garden with 90+ old fashioned rose varieties, tranquil yellow and white garden, square garden with a tapestry of self sowing perennials and Victorian style greenhouse, 300ft long colour schemed symmetrical double borders and an organic fruit garden. Maintained entirely by the owner, the gardens were designed to be managed as an ornamental ecosystem. Please see the garden website for more information.

65 THE OLD RECTORY, FAWKHAM

Valley Road, Fawkham, Longfield, DA3 8LX. Karin & Christopher Proudfoot, 01474 707513, keproudfoot@gmail.com. *1m S of Longfield. Midway between A2 & A20, on Valley Rd 1½m N of Fawkham Green, 0.3m S of Fawkham church, opp sign for Gay Dawn Farm/Corinthian Sports Club. Parking on drive only. Not suitable for coaches.* **Visits by arrangement in Feb for groups of up to 20. Adm £5, chd free. Home-made teas.**

1½ acres with impressive display of long-established naturalised snowdrops and winter aconites; collection of over 130 named snowdrops. Garden developed around the snowdrops over 35yrs, inc hellebores, pulmonarias and other early bulbs and flowers, with foliage perennials, shrubs and trees, also natural woodland. Gentle slope, gravel drive, some narrow paths.

66 THE OLD RECTORY, OTTERDEN

Bunce Court Road, Faversham, ME13 0BY. Mrs Gry Iverslien, 07734 538272, gry@iverslien.com. *North Downs. Postcode take you to a property called Bunce Ct. we are 600 yds further on just past Cold Harbour lane.* **Thur 20, Fri 21 Apr (11-4). Adm £5, chd free. Pre-booking essential, please visit www.ngs.org.uk for information & booking. Light refreshments. Visits also by arrangement 3 Apr to 30 Sept for groups of 5 to 30.**

4 acre woodland garden with numerous large Rhododendrons and Camellias, mass of spring bulbs with a formal rose garden, cutting garden and large hydrangea beds throughout the garden. The garden also has a large wildlife pond and a variety of trees of interest.

67 THE ORANGERY

Mystole, Chartham, Canterbury, CT4 7DB. Rex Stickland & Anne Prasse, 01227 738348, rex.mystole@btinternet.com. *5m SW of Canterbury. Turn off A28 through Shalmsford St. In 1½m at Xrds turn R downhill. Continue, ignoring roads on L & R. Ignore drive on L (Mystole House only). At sharp R bend in 600yds turn L into private drive.* **Sun 14 May, Sat 29, Sun 30 July (1-5). Adm £5, chd free. Home-made teas. Visits also by arrangement Mar to Sept.**

1½ acre gardens around C18 orangery, now a house (not open). Magnificent extensive herbaceous border and impressive ancient wisteria. Large walled garden with a wide variety of shrubs, mixed borders and unusual specimen trees. Water features and intriguing collection of modern sculptures in natural surroundings. Splendid views from the terrace over ha-ha to the lovely Chartham Downs. Ramps to garden.

68 PARSONAGE OASTS

Hampstead Lane, Yalding, ME18 6HG. Edward & Jennifer Raikes, 01622 814272, jmraikes@parsonageoasts.plus.com. *6m SW of Maidstone. On B2162 between Yalding village & stn, turn off at Boathouse Pub. over lifting bridge, cont 100 yds up lane. House & car park on L.* **Visits by arrangement 25 Mar to 1 Oct for groups of up to 30. Adm inc cream tea. Adm £7, chd free.**

Our garden has a lovely position on the bank of the River Medway. Typical Oast House (not open) often featured on calendars and picture books of Kent. 70yr old garden now looked after by grandchildren of its creator. ¾ acre garden with walls, daffodils, crown imperials, shrubs, clipped box and a spectacular magnolia. Small woodland on river bank. Best in spring, but always something to see. Local Boathouse pub (5mins) and several nice places for a picnic in the garden. Unfenced river bank. Gravel paths. Bee hives.

69 ◆ PENSHURST PLACE & GARDENS

Penshurst, TN11 8DG. Lord & Lady De L'Isle, 01892 870307, contactus@penshurstplace.com, www.penshurstplace.com. *6m NW of Tunbridge Wells. SW of Tonbridge on B2176, signed from A26 N of Tunbridge Wells.* **For NGS: Wed 13 Sept (10-5). Adm £12, chd £7.30. Light refreshments at The Porcupine Pantry. For other opening times and information, please phone, email or visit garden website.**

11 acres of garden dating back to C14. The garden is divided into a series of rooms by over a mile of yew hedge. Profusion of spring bulbs, formal rose garden and famous peony border. Woodland trail and arboretum. Year-round interest. Toy museum. Some paths not paved and uneven in places; own assistance will be required. 2 wheelchairs available for hire.

70 NEW PETT PLACE

Pett Lane, Charing, Ashford, TN27 0DS. Mr & Mrs P Hurst. *6m NW of Ashford. From A20 turn N into Charing High St. At end turn R into Pett Lane towards Westwell. Pett Place is about ½m on the L.* **Thur 8, Fri 9 June (11-4). Adm £7.50, chd free. Home-made teas.**

Walled gardens covering nearly 4 acres. A garden of pleasing vistas and secret places with an idyllic rose garden and wildly romantic planting throughout. Remains of a ruined medieval chapel is a charming feature beside the manor house (not open), which was re-fronted about 1700 and which Pevsner describes as 'presenting grandiloquently towards the road'. Rose garden, Victorian glassouses, long mixed border, specimen trees, ponds.

71 PHEASANT BARN

Church Road, Oare, ME13 0QB. Paul & Su Vaight, 07843 739301, suvaight46@gmail.com. *2m NW of Faversham. Entering Oare from Faversham, turn R at Three Mariners pub towards Harty Ferry. Garden 400yds on R, before church. Parking on roadside.* **Sat 24, Sun 25, Wed 28, Thur 29 June, Sun 2, Mon 3, Sun 9, Mon 10 July (11-4). Adm £7, chd free. Pre-booking essential, please visit www.ngs.org.uk for information & booking. Visits also by arrangement 20 May to 22 July for groups of up to 20.**

Series of smallish gardens around award-winning converted farm buildings in beautiful situation overlooking Oare Creek. Main area is nectar rich planting in formal design with a contemporary twist inspired by local landscape. Also vegetable garden, dry garden, water features, wildflower meadow and labyrinth. July optimum for wildflowers. Kent Wildlife Trust Oare Marshes Bird Reserve within 1m. Two village inns serving lunches/dinners, booking recommended.

72 50 PINE AVENUE

Gravesend, DA12 1QZ. Tony Threlfall. *10 mins from the A2. Leave the A2 at the Gravesend East exit and continue on Valley Drive to the T junction at the end of Valley Drive where you turn L onto Old Rd East. The next turn on the R is Pine Ave.* **Sat 15 July (11-4). Adm £5, chd free. Light refreshments.**

The property sits on a steep ¼ acre plot with a swimming pool to the rear surrounded by tropical planting. Steps lead down to the lawn with mixed borders, and on to the middle patio with more tropical planting. Steps then lead down to the woodland area and then out to the pond. Not suitable for small children due to deep water.

73 ◆ QUEX GARDENS

Quex Park, Birchington, CT7 0BH. Powell-Cotton Museum, 01843 842168, enquiries@powell-cottonmuseum.org, www.powell-cottonmuseum.org. *3m W of Margate. Follow signs for Quex Park on approach from A299 then A28 towards Margate, turn R into B2048 Park Lane. Quex Park is on L.* **For NGS: Sun 16 July (11-4). Adm £2.50, chd £2.50. Light refreshments in Felicity's Café & Quex Barn. For other opening times and information, please phone, email or visit garden website.**

10 acres of woodland and gardens with fine specimen trees unusual on Thanet, spring bulbs, wisteria, shrub borders, old figs and mulberries, herbaceous borders. Victorian walled garden with cucumber house, long glasshouses, cactus house, fruiting trees. Peacocks, dovecote, chickens, bees, woodland walk, wildlife pond, children's maze, croquet lawn, picnic grove, lawns and fountains. Head Gardener and team will be available on the day for a chat and to answer questions. Garden almost entirely flat with tarmac paths. Sunken garden has sloping lawns to the central pond.

74 43 THE RIDINGS

Chestfield, Whitstable, CT5 3QE. David & Sylvie Buat-Menard, 01227 500775, sylviebuat-menard@hotmail.com. *Nr Whitstable. From M2 heading E cont onto A299. In 3m take A2990. From r'about on A2990 at Chestfield, turn onto Chestfield Rd, 5th turning on L onto Polo Way which leads into The Ridings.* **Sun 7 May (11-4). Adm £5, chd free. Light refreshments. Visits also by arrangement for groups of 8 to 20.**

Delightful small garden brimming with interesting plants both in the front and behind the house. Many different areas. Dry gravel garden in front, raised beds with alpines and bulbs and borders with many unusual perennials and shrubs. The garden ornaments are a source of interest for visitors. The water feature will be of interest for those with a tiny garden as planted with carnivorous plants. Many alpine troughs and raised beds as well as dry shade and mixed borders all in a small space.

75 ◆ RIVERHILL HIMALAYAN GARDENS
Riverhill, Sevenoaks, TN15 0RR. The Rogers Family, 01732 459777, info@riverhillgardens.co.uk, www.riverhillgardens.co.uk. *2m S of Sevenoaks on A225. Leave A21 at A225 & follow signs for Riverhill Himalayan Gardens.* **For NGS: Tue 9 May, Wed 14 June (10-5). Adm £11, chd £7. Light refreshments. For other opening times and information, please phone, email or visit garden website.**
Beautiful hillside garden, privately owned by the Rogers family since 1840. Spectacular rhododendrons, azaleas and fine specimen trees. Edwardian Rock Garden with extensive fern collection, Rose Walk and Walled Garden with sculptural terracing. Bluebell walks. Extensive views across the Weald of Kent. Hedge maze, adventure playground and den building. Café serves speciality coffee, light lunches and cakes. Plant sales and quirky shed shop selling beautiful gifts and original garden ornaments. Disabled parking. Easy access to café, shop and tea terrace. Accessible WC.

76 18 ROYAL CHASE
Tunbridge Wells, TN4 8AY. Eithne Hudson, 07985 305941, eithne.hudson@gmail.com. *Situated at the top of Mount Ephraim & The Common. Leave A21 at Southborough, follow signs for Tunbridge Wells. After St John's Rd take R turn at junction on Common & sharp R again. From S, follow A26 through town up over Common. Last turn on L.* **Visits by arrangement 20 May to 1 July for groups of up to 15. Please email to arrange a group visit. Adm £5, chd free. Home-made teas.**
Town garden of a ¼ of acre on sandy soil which was originally Common and woodland. Large lawn with island beds planted with roses and herbaceous perennials and a small pond with lilies and goldfish.The owner is a florist and has planted shrubs and annuals specifically with flower arranging in mind. Roses and island beds in semi woodland setting in central Tunbridge Wells. Access through right hand side gate.

77 ST CLERE
Kemsing, Sevenoaks, TN15 6NL. Mr Simon & Mrs Eliza Ecclestone, www.stclere.com. *6m NE of Sevenoaks. 1m E of Seal on A25, turn L signed Heaverham. In Heaverham turn R signed Wrotham. In 75yds straight ahead marked Private Rd; 1st L to house.* **Sun 21 May (2-5). Adm £5, chd £1. Home-made teas in the Garden Room.**
4 acre garden, full of interest. Formal terraces surrounding C17 mansion (not open), with beautiful views of the Kent countryside. Herbaceous and shrub borders, productive kitchen and herb gardens, lawns and rare trees. Some gravel paths and small steps.

78 ◆ SCOTNEY CASTLE
Lamberhurst, TN3 8JN. National Trust, 01892 893820, scotneycastle@nationaltrust.org.uk, www.nationaltrust.org.uk/scotneycastle. *6m SE of Tunbridge Wells. On A21 London - Hastings, brown tourist signs. Bus: (Mon to Fri) 256 Tunbridge Wells to Wadhurst Autocar service via Lamberhurst alight Lamberhurst Green.* **For NGS: Wed 10 May (10-4.30). Adm £14, chd £7. Cream teas in the Courtyard tea-room. Lunches, snacks, & hot & cold drinks also available. For other opening times and information, please phone, email or visit garden website.**
The medieval moated Old Scotney Castle lies in a peaceful wooded valley on the Kent/Sussex border. In the 1830s its owner, Edward Hussey III, set about building a new house, partially demolishing the Old Castle to create a romantic ruin, the centrepiece of his visionary landscape. Manual wheelchairs and individual mobility scooters are available to borrow.

79 THE SILK HOUSE
Lucks Lane, Rhoden Green, nr Paddock Wood, Tunbridge Wells, TN12 6PA. Adrian & Silke Barnwell, www.silkhousegarden.com. *Rhoden Green. The Silk House is ½m from Queen St on sharp bend near Fish Lake. Also approached from Maidstone Rd. Parking at house or in lane when full.* **Sun 7 May, Sun 6 Aug (10-4). Adm £7.50, chd £3.50. Light refreshments in the garden - usually served from the tea house.**
Japanese stroll garden with Koi and wildlife ponds, canal pond, unusual acers, cherries, large bamboos, topiary and rare evergreen trees. Zen area. Secluded and calming. Small woodland walk, busy kitchen garden, dry garden and apiary. 2 acres overall. Designed, built and maintained entirely by owners. Bonsai classes and Japanese garden design talks available for groups. Best Japanese stroll garden in the South East! Spectacular spring cherry blossom and autumnal acers. Structural planting looks good year-round. Picnics welcome for an additional charitable donation of £10, please take all your litter home. Sorry no dogs. Disabled parking at the house. Wheelchairs have assisted side access by request avoiding steps. Some steps, banks, bridges, deep water ponds.

80 SIR JOHN HAWKINS HOSPITAL
High Street, Chatham, ME4 4EW. Sir John Hawkins Hospital, www.hawkinshospital.org.uk. *On the N side of Chatham High St, on the border between Rochester & Chatham. Leave A2 at J1 & follow signs to Rochester. Pass Rochester Station & turn L at main junction T-lights, travelling E towards Chatham.* **Sat 24, Sun 25 June (11-5). Adm £3, chd free. Cream teas.**
Built on the site of Kettle Hard - part of Bishop Gundulph's Hospital of St Bartholomew, the Almshouse is a square of Georgian houses dating from the 1790s. A delightful small secluded garden overlooks the River Medway, full of vibrant and colourful planting. A lawn with cottage style borders leads to the riverside and a miniature gnome village captivates small children. Disabled access via stairlift, wheelchair to be carried separately.

9,500 patients and their families are being supported across four Horatio's gardens thanks to our annual donations

81 ◆ SISSINGHURST CASTLE GARDEN
Biddenden Road, Sissinghurst, Cranbrook, TN17 2AB. National Trust, 01580 710700, sissinghurst@nationaltrust.org.uk, www.nationaltrust.org.uk/ sissinghurst-castle-garden. *2m NE of Cranbrook, 1m E of Sissinghurst on Biddenden Rd (A262), see our website for more information.* **For NGS: Sun 17 Sept (11-5.30). Adm £15, chd £7.50. Light refreshments in Coffee Shop or Restaurant. For other opening times and information, please phone, email or visit garden website.**
Historic, poetic, iconic; a refuge dedicated to beauty. Vita Sackville-West and Harold Nicolson fell in love with Sissinghurst Castle and created a world renowned garden. More than a garden, visitors can also find Elizabethan and Tudor buildings, find out about our history as a Prisoner of War Camp and see changing exhibitions. Picnic spot in the vegetable garden, picnics are not allowed in the formal garden. Free welcome talks and estate walks leaflets. Café, restaurant, gift, secondhand book and plant shops are open from 10am-5.30pm. Some areas unsuitable for wheelchair access due to narrow paths and steps.

82 SMITHS HALL
Lower Road, West Farleigh, ME15 0PE. Mr S Norman. *3m W of Maidstone. A26 towards Tonbridge, turn L into Teston Ln B2163. At T-junction turn R onto Lower Rd B2010. Opp Tickled Trout pub.* **Sun 25 June (11-5). Adm £5, chd free. Home-made teas. Donation to Heart of Kent Hospice.**
Delightful 3 acre gardens surrounding a beautiful 1719 Queen Anne House (not open). Lose yourself in numerous themed rooms: sunken garden, scented old fashioned rose walk, formal rose garden, intense wild flowers, peonies, deep herbaceous borders and specimen trees. Gravel paths.

GROUP OPENING

83 SOUTH DOVER GARDEN TRAIL
Dover, South, CT17 9PU. *Queens Avenue gardens are at the Dover end of B2011, signed from mini r'about at Elms Vale Road. 69, Capel Street, Capel is off B2011, signed from Battle of Britain memorial.* **Sun 23 July (11-4). Combined adm £5, chd free. Home-made teas at 69, Capel Street.**

69 CAPEL STREET
John & Jenny Carter.

1 QUEEN'S AVENUE, DOVER
Alison Whitnall.

2 QUEENS AVENUE
Deborah Gasking.

Three gardens each designed and carefully maintained by their owners. 2, Queens Road consists of numerous terraces each with a different identity and creative use of salvaged/recycled materials. 1, Queens Road is an oasis of calm with clever use of colour and texture. A curving lawn and well positioned key plants make it appear much wider than it is. 69, Capel Street is a contemporary cottage garden with central wisteria-clad arbour surrounded by distinctly different areas and imaginative features. Delicious teas served here!

84 SPRING PLATT
Boyton Court Road, Sutton Valence, Maidstone, ME17 3BY. Mr & Mrs John Millen *5m SE of Maidstone. From A274 nr Sutton Valence follow yellow NGS signs.* **Thur 19, Wed 25, Sun 29 Jan, Thur 2, Sun 5 Feb (10.30-3). Adm £5, chd free. Pre-booking essential, please phone 01622 843383, email j.millen@talktalk.net or visit www.kentsnowdrops.com for information & booking. Light refreshments inc home-made soup & bread. Visits also by arrangement 19 Jan to 5 Feb for groups of 8 to 50.**
1 acre garden under continual development with panoramic views of the Weald. Over 650 varieties of snowdrop grown in tiered display beds with spring flowers in borders. An extensive collection of alpine plants in a large greenhouse. Vegetable garden, natural spring fed water feature and a croquet lawn.

85 NEW ◆ SQUERRYES COURT
Squerryes, Westerham, TN16 1SJ. Henry Warde, www.squerryes.co.uk. *½m W of Westerham. Signed from A25.* **For NGS: Fri 12 May (11-5). Adm £8, chd free. Light refreshments. For other opening times and information, please visit garden website.**
15 acres of garden, lake and woodland surrounding beautiful C17 manor house (not open). Lovely throughout the seasons from the spring bulbs to later-flowering borders. Cenotaph commemorating General Wolfe. C18 dovecote. Lawns, yew hedges, ancient trees, parterres, azaleas, roses.

86 NEW 3 STANLEY COTTAGES
Perry Wood, Selling, Faversham, ME13 9SF. Marguerite Heath. *2.7m N of Chilham. With 'The Sondes' on your L go through Selling Continue straight over 2 junctions. Turn first L towards Perry Wood. Take the first L. Immed after Home Farm (R) turn into small lane on the L.* **Sat 1 July (11-5). Combined adm with Hoppickers East £6, chd free. Light refreshments.**
This small garden presents in two parts. The 'Studio' garden is bordered by bluebell woods and a rose hedge. It inc a wildlife pond, lawns, shrubs, perennials and vegetable beds and is enhanced by natural wood and fused glass features. The 'Home' garden is a sheltered haven where beech hedges enclose more formal planting, structures with climbing roses and water features. Wheelchair accessible.

87 STONEWALL PARK
Chiddingstone Hoath, nr Edenbridge, TN8 7DG. Mr & Mrs Fleming. *4m SE of Edenbridge. Via B2026. ½ way between Markbeech & Penshurst. Plenty of parking which will be signed.* **Sun 19 Mar, Sun 30 Apr (2-5). Adm £7, chd free. Tea. Donation to Sarah Matheson Trust & St Mary's Church, Chiddingstone.**
Even from the driveway you can see a vast amount of self-seeded daffodils, leading down to a romantic woodland garden in historic setting, featuring

species such as rhododendrons, magnolias, azaleas and bluebells. You can see those in full flower on our May opening day. You can also discover a range of interesting trees, sandstone outcrops, wandering paths and lakes. The woods can be very wet and muddy, please don't forget your boots!

88 SWEETBRIAR

69 Chequer Lane, Ash, nr Sandwich, CT3 2AX. Miss Louise Dowle & Mr Steven Edney. *8m from Canterbury, 3m from Sandwich. Turn off A257 into Chequer Lane, 100 metres from junction on the L.* **Sun 9 July, Sun 20 Aug, Sun 17 Sept (10.30-4.30). Adm £5, chd free. Home-made teas.**

Exotic jungle garden of Steve Edney and Louise Dowle, multi gold medal winners at RHS Chelsea and Hampton court flower shows. Gold award from Kent Wildlife Trust.

NPC

89 TONBRIDGE SCHOOL

High Street, Tonbridge, TN9 1JP. The Governors. *Maps & guides available for visitors. The school can be found at the North end of Tonbridge High St. Parking signed off London Rd (B245 Tonbridge-Hildenborough) in the Tonbridge Sports Centre car park.* **Sun 9 July (10.30-3.30). Adm £6, chd free. Cream teas.**

In front of and behind Tonbridge School you will find the four main gardens that you can visit: Front of School Gardens, The Garden of Remembrance, Smythe Library Garden as well as the newly created Barton Science Centre Garden. Tea/coffee and cake available. Head Gardener available to chat. Toilets on site. All gardens can be accessed via wheelchair and gardeners will be on hand to help and guide.

Our annual donations to Parkinson's UK meant 7,000 patients are currently benefiting from support of a Parkinson's Nurse

© Leigh Clapp

Upper Pryors

90 TRAM HATCH

Barnfield Road, Charing Heath, Ashford, TN27 0BN. Mrs P Scrivens, 07835 758388, Info@tramhatch.com, www.tramhatchgardens.co.uk. *10m NW of Ashford. A20 towards Pluckley, over motorway then 1st R to Barnfield. At end, turn L past Barnfield, Tram Hatch ahead.* **Sun 4 June, Sun 2 July, Sun 20 Aug (12.30-5). Adm £5, chd free. Home-made teas. Visits also by arrangement 2 Apr to 1 Oct.**

Meander your way off the beaten track to a mature 3 acre garden changing through the seasons. You will enjoy a garden laid out in rooms. Large selection of trees, vegetable, rose and gravel gardens, colourful containers. River Stour and the Angel of the South enhance your visit. Please come and enjoy, then relax in our lovely garden room for tea. Ample parking and WC facilities available. The garden is totally flat, apart from a very small area which can be viewed from the lane.

91 TROUTBECK

High Street, Otford, Sevenoaks, TN14 5PH. Mrs Jenny Alban Davies, albandavies@me.com. *3m N of Sevenoaks. At W end of village.* **Visits by arrangement in June for groups of 8 to 20. Adm £10, chd free. Light refreshments.**

3 acre garden surrounded by branches of River Darent which has been developed over 30 years. It combines the informal landscape of the river, a pond and an area of wild meadow with views to the North Downs. The garden has extensive structural planting using box and yew. Knots and topiary shapes are displayed throughout. All main areas of the garden can be accessed by wheelchair.

92 UPPER PRYORS

Butterwell Hill, Cowden, TN8 7HB. Mr & Mrs S G Smith. *4½m SE of Edenbridge. From B2026 Edenbridge-Hartfield, turn R at Cowden Xrds & take 1st drive on R.* **Wed 14 June (11-5.30). Adm £5, chd free. Home-made teas.**

8 acres of English country garden surrounding C16 house - a garden of many parts; colourful profusion, interesting planting arrangements, lawns, mature woodland, water and a terrace on which to appreciate the view, and tea.

93 THE WATCH HOUSE

7 Thanet Road, Broadstairs, CT10 1LF. Dan Cooper & John McKenna, www.frustratedgardener.com. *Town Centre Location. Off Broadstairs High St on narrow side rd. At Broadstairs station, cont along High St (A255) towards sea front. Turn L at Terence Painter Estate Agent then immed turn R.* **Sat 29, Sun 30 July (12-4). Adm £5, chd free.**

Adjoining a historic fishermen's cottage, two small courtyard gardens shelter an astonishing array of unusual plants. Thanks to a unique microclimate the east-facing garden is home to a growing collection of exotics, chosen principally for exuberant, colourful, jungly foliage. In the west-facing courtyard, a garden room leads onto a terrace where flowering plants jostle for space. Within a few mins walk of Viking Bay, The Dickens Museum and Bleak House. Teas available in town.

94 12 WEST CLIFF ROAD

Ramsgate, CT11 9JW. Brian Daubney. *Access down Ivy Ln off Addington St.* **Sun 10 Sept (12-5). Combined adm with 104 Grange Road £5, chd free. Home-made teas at 104 Grange Road.**

Relaxed gravel garden hidden behind a Regency townhouse, shingle laid on chalk. Supporting olive, figs and birch trees, combining Mediterranean influences with indigenous planting, Loudon meets Jarman. Front garden maintains flowers in bloom each month of the year.

95 WEST COURT LODGE

Postling Court, The Street, Postling, nr Hythe, CT21 4EX. Mr & Mrs John Pattrick, 01303 863285, pattrickmalliet@gmail.com. *2m NW of Hythe. From M20 J11 turn S onto A20. Immed 1st L. After ½m on bend take rd signed Lyminge. 1st L into Postling.* **Sat 3, Sun 4 June (12-5). Combined adm with Churchfield £8. Home-made teas in Postling village hall. Visits also by arrangement 15 Apr to 1 Sept for groups of up to 35.**

South facing 1 acre walled garden at the foot of the North Downs, designed in two parts: main lawn with large sunny borders and a romantic woodland glade planted with shadow loving plants and spring bulbs, small wildlife pond. Lovely C11 church will be open next to the gardens.

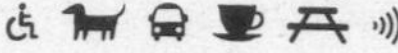

GROUP OPENING

96 WEST MALLING EARLY SUMMER GARDENS

West Malling, ME19 6LW. *On A20, nr J4 of M20. Park (Ryarsh Lane & Station) in West Malling. Please start your visit either at Brome House or Went House. Parking available at New Barns Cottages.* **Sun 4 June (12-5). Combined adm £8, chd free. Home-made teas at New Barns Cottages only. Donation to St Mary's Church, West Malling.**

ABBEY BREWERY COTTAGE
Dr David & Mrs Lynda Nunn.

BROME HOUSE
John Pfeil & Shirley Briggs.

LUCKNOW, 119 HIGH STREET
Ms Jocelyn Granville.

NEW BARNS COTTAGES
Mr & Mrs Anthony Drake.

TOWN HILL COTTAGE
Mr & Mrs P Cosier.

WENT HOUSE
Alan & Mary Gibbins.

West Malling is an attractive small market town with some fine buildings. Enjoy six lovely gardens that are entirely different from each other and cannot be seen from the road. Brome House and Went House have large gardens with specimen trees, old roses, mixed borders, attractive kitchen gardens and garden features inc a coach house, Roman temple, fountain and parterre. Town Hill Cottage is a walled town garden with mature and interesting planting. Abbey Brewery Cottage is a recent jewel-like example of garden restoration and development. Wheelchair access to Brome House and Went House only.

GROUP OPENING

97 WHITSTABLE JOY LANE GARDENS

Tickets and map from 19 Joy Lane, Whitstable, CT5 4LT. www.facebook.com/whitstableopengardens. *Off A299, or A290. Down Borstal Hill, L by garage into Joy Lane.* **Sun 14 May**

(10-5). Combined adm £6, chd free.
Enjoy a day visiting a dozen or so eclectic gardens along Joy Lane and neighbourhood by the sea. From Arts & Crafts villas to mid century bungalows, some productive, others wildlife friendly, large or compact, we garden on heavy clay, are prone to northerly winds and hope to inspire those new to gardening with our ingenuity and style. To find out more: visit Whitstable Gardens on Facebook. Plant stalls at 19 Joy Lane.

GROUP OPENING

98 WHITSTABLE TOWN GARDENS

Whitstable, CT5 1HJ. www.facebook.com/whitstableopengardens. *Tickets & map at: Stream Walk (CT5 1HJ), Umbrella Centre (CT5 1DD) & The Guinea 31 Island Wall (CT5 1EW).* **Sun 11 June (10-5). Combined adm £6, chd free. Light refreshments.**
From fishermen's yards to formal gardens, the residents of Whitstable are making the most of the mild climate. Enjoy contemporary gardens, seaside gardens, rose gardens, gravel gardens, designers' gardens and wildlife friendly plots, both large and small. Stream Walk is at the heart of our community, ideal for those without gardening space of their own. We toil on heavy clay soils and are prone to northerly winds. By opening, we hope to encourage those new to gardening with our ingenuity and style, rather than rolling acres. To find out more see Whitstable Gardens on Facebook. Refreshments at Stream Walk Community Garden NGS plant sale at The Guinea No 31 Island Wall.

99 WINDY RIDGE

Dover Road, Guston, Dover, CT15 5EH. Mr David & Mrs Marianne Slater. *1m N of Dover. From Dover Castle take Guston Rd next to Coach Park and follow for ¾m. Please use car park.* **Sun 28 May, Sun 16 July (1-5). Combined adm with Lacey Down £5, chd free. Home-made teas.**
Well established series of separate areas in ⅓ acre plot developed since 2008. Shrubs, herbaceous borders, lawns, pergola with wisteria, trees in pots, productive raised beds, soft and top fruit. Self sufficient in water (in most years) due to extensive rain harvesting, gravity fed to vegetable area. Paths smooth but narrow in places with some low steps to some seating areas. Lacey Down CT15 7ER.

GROUP OPENING

100 WOMENSWOLD GARDENS

Womenswold, Canterbury, CT4 6HE. Mrs Maggie McKenzie. *6m S of Canterbury, midway between Canterbury & Dover. Take B2046 for Wingham at Barham Xover. Turn 1st R, following signs.* **Sat 24, Sun 25 June (1-5). Combined adm £5, chd free. Home-made teas in the garden at Brambles, Womenswold.**
A diverse variety of cottage gardens in an idyllic situation in an unspoilt hamlet, mostly surrounding C13 Church. Cottage garden with variety of old climbing and shrub roses, clematis, vegetable bed and beehives; a garden in a setting of a traditional C17 thatched cottage; colourful garden with ponds, waterfalls, tropical area with many rare plants and a large collection of agapanthus; a 2 acre plantsman's garden partially created in old chalk quarry with vegetables, orchard, alpines, polytunnel with tender fruit. A very picturesque terraced cottage garden, with unusual plants, feature pond with lovely views of the church. Easy walking distance between gardens. North Downs Way runs through village. Many unusual plants for sale. Teas in lovely restful garden with home-made cakes. Produce stall; Church open. Parking in field a short walk from village. Disabled car park behind church. Most gardens have good wheelchair access although some areas may be inaccessible. Minibus available.

101 ◆ THE WORLD GARDEN AT LULLINGSTONE CASTLE

Eynsford, DA4 0JA. Mrs Guy Hart Dyke, 01322 862114, info@lullingstonecastle.co.uk, www.lullingstonecastle.co.uk. *1m from Eynsford. Over Ford Bridge in Eynsford Village. Follow signs to Roman Villa. Keep Roman Villa immed on R then follow Private Rd to Gatehouse.* **For NGS: Sun 18 June (11-5). Adm £10, chd £5. Light refreshments. For other opening times and information, please phone, email or visit garden website.**
The World Garden is located within the 2 acre, C18 Walled Garden in the stunning grounds of Lullingstone Castle, where heritage meets cutting-edge horticulture. The garden is laid out in the shape of a miniature map of the world. Thousands of species are represented, all planted out in their respective beds. The World Garden Nursery offers a host of horticultural and homegrown delights, to reflect the unusual and varied planting of the garden. Wheelchairs available upon request.

NPC

102 YOKES COURT

Coal Pit Lane, Frinsted, Sittingbourne, ME9 0ST. John & Kate Leigh Pemberton, 01795 830210, leighpems@btinternet.com. *2.4m from Doddington. Turn off Old Lenham Rd to Hollingbourne. 1st R at Torry Hill Chestnut Fencing, then 1st L to Torry Hill. 1st L into Coal Pit Ln.* **Visits by arrangement May to Aug. Adm £4.50, chd free.**
3 acre garden surrounded by countryside. Hedges and herbaceous borders set in open lawns. Rose beds. New prairie planting. Serpentine walkway through wildflowers. Walled vegetable garden.

The National Garden Scheme searches the length and breadth of England, Wales, Northern Ireland and the Channel Islands for the very best private gardens

LANCASHIRE

Merseyside, Greater Manchester

Welcome to Lancashire

Lancashire is a large county with diverse landscapes and includes the two metropolitan areas of Merseyside and Greater Manchester. In the west, Lancashire has a 137 mile long coastline stretching from Morecambe Bay in the north to the River Mersey in the south. The east of the county is bounded by the Pennine hills. We boast two Areas of Outstanding Natural Beauty including the Forest of Bowland and Beacon Fell together with the beautiful Ribble Valley.

We have a wide range of gardens for you to visit including some designed by Chelsea gold medal winners to wonderful private gardens designed and lovingly tended by their dedicated owners There are village gardens, city gardens, gardens surrounded by wonderful rural scenery, community gardens doing inspiring work, themed gardens including Japanese and Spanish inspired gardens. This year we have even more interest with 20 new gardens, 7 returning gardens and 15 group gardens which provide excellent value. Something for everyone.

Our gardeners look forward to welcoming you!

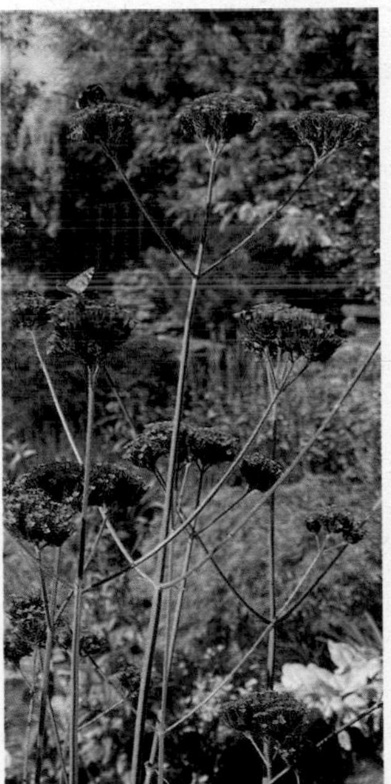

Left: 2 Pheasant Field, Hale Village Gardens

Visits by Arrangement
Roger Craine
07595 229700
roger.craine@ngs.org.uk

Talks
Maureen Sawyer & Sue Beacon
01695 574628

Assistant County Organisers
Sandra Curl 01704 893713
peter.curl@btinternet.com

Margaret & Geoff Fletcher
01704 567742
margaret.fletcher@ngs.org.uk

Deborah Jackson
deborah.jackson@ngs.org.uk

John & Jennifer Mawdsley
01704 564708

Carole Ann & Stephen Powers
(as above)

Claire Spendlove 01524 727770
claire@lavenderandlime.co.uk

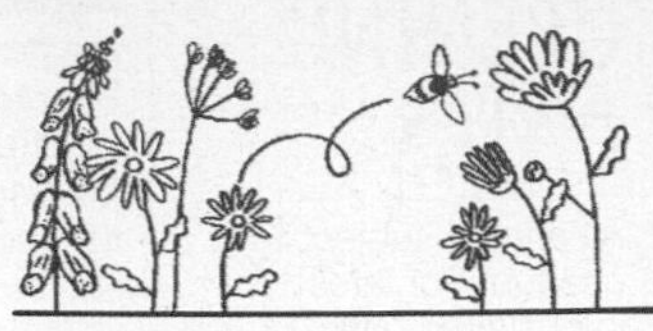

Volunteers

County Organiser
Marian & Brian Jones
01695 574628
marianandbrian.jones@ngs.org.uk

County Treasurer
Peter Curl
01704 893713
peter.curl@ngs.org.uk

Publicity
Christine Ruth
07740 438994
caruthchris@aol.com

Social Media
Carole Ann Powers
01254 824903
chows3@icloud.com

Instagram/Radio
Tania Craine
tania.craine@ngs.org.uk

Twitter
John Spendlove
07746 378194
john.spendlove@ngs.org.uk

Photographers
Norman Rigby
01704 840329
norman.rigby@ngs.org.uk

Laurie Lissett
07786 631280
laurielissett@hotmail.co.uk

Booklet Co-ordinator
Barbara & Richard Farbon
01772 600750
barbaraandrichard.farbon@ngs.org.uk

Booklet Distribution
Claire Spendlove
01524 727770 claire@lavenderandlime.co.uk

John & Jennifer Mawdsley
01704 564708

@Lancsngs  @lancsngs  @ngslancashire

OPENING DATES

All entries subject to change. For latest information check **www.ngs.org.uk**

Map locator numbers are shown to the right of each garden name.

February

Snowdrop Festival

Sunday 12th
Weeping Ash Garden 59

Sunday 19th
Weeping Ash Garden 59

Sunday 26th
Weeping Ash Garden 59

April

Saturday 15th
Dale House Gardens 13

Sunday 16th
Dale House Gardens 13

Sunday 23rd
Derian House Children's Hospice 15
Mill Green School 41

May

Monday 1st
◆ The Ridges 52

Sunday 7th
Kington Cottage 33

Saturday 20th
Halton Park House 29

Sunday 21st
Halton Park House 29
Holly House 32
Maggie's, Manchester 35
Parkers Lodges 49
NEW Rainbag Cottage 51

Saturday 27th
Dent Hall 14
Mill Barn 39
NEW Northcote 44

Sunday 28th
Bretherton Gardens 4
Dent Hall 14
Hightown Gardens 31
Mill Barn 39
Mill Croft 40
NEW Northcote 44

June

Saturday 3rd
Mill Barn 39

Sunday 4th
NEW 14 Elms Avenue 19
45 Grey Heights View 26
Mill Barn 39

Saturday 10th
136 Buckingham Road 7
Calder House Lane Gardens 8
NEW Ellesmere Park Gardens 18
Mill Barn 39
NEW The North Tithebarn 43
Old Hollows Farm 46

Sunday 11th
Ainsdale & Birkdale Gardens 1
136 Buckingham Road 7
Calder House Lane Gardens 8
31 Cousins Lane 10
79 Crabtree Lane 11
Didsbury Village Gardens 16
Green Farm Cottage 23
Kington Cottage 33
Mill Barn 39
NEW The North Tithebarn 43
Old Hollows Farm 46

Saturday 17th
Dale House Gardens 13
Giles Farm 21
The Old Vicarage 47
Turton Tower Kitchen Garden 57
Warton Gardens 58
Willow Wood Hospice Gardens 60

Sunday 18th
Bretherton Gardens 4
Canning Georgian Quarter 9
Dale House Gardens 13
Giles Farm 21
Warton Gardens 58

Saturday 24th
5 Crib Lane 12

Sunday 25th
Bridge Inn Community Farm 5
5 Crib Lane 12
Dutton Hall 17
NEW Greenfield Hall 24
Mill Croft 40
Ormskirk and Aughton Gardens 48
Woolton Village June Gardens 62

July

Saturday 1st
NEW Oaklands 45

Sunday 2nd
Freshfield Gardens 20
Glynwood House 22
NEW Oaklands 45

Saturday 8th
NEW 2 Brookside 6
Hale Village Gardens 28
Higher Bridge Clough House 30

Sunday 9th
79 Crabtree Lane 11
Hale Village Gardens 28
Higher Bridge Clough House 30
Kington Cottage 33
Parkers Lodges 49

Sunday 16th
Ainsdale & Birkdale Gardens 1
Bretherton Gardens 4
33 Greenhill Road 25
146 Mather Avenue 38
Moss Park Allotments 42
◆ The Ridges 52
Southlands 55
Woolton Village July Gardens 61

Saturday 22nd
Arevinti 2
NEW 47 The Crescent 56

Sunday 23rd
Arevinti 2
Holly House 32
NEW Lark Rise 34
Maggie's, Oldham 36
Maghull Station 37
Sefton Park July Gardens 53
NEW 47 The Crescent 56

Sunday 30th
The Growth Project 27

August

Saturday 5th
NEW 2 Brookside 6

Sunday 13th
Derian House Children's Hospice 15
Kington Cottage 33
Plant World 50
Weeping Ash Garden 59

Sunday 20th
45 Grey Heights View 26

September

Saturday 2nd
Ashton Walled Community Gardens 3

Sunday 3rd
33 Greenhill Road 25
146 Mather Avenue 38
NEW Rainbag Cottage 51

Saturday 9th
Halton Park House 29

Sunday 10th
Halton Park House 29
Sefton Park September Gardens 54
Weeping Ash Garden 59

October

Sunday 29th
Weeping Ash Garden 59

November

Saturday 4th
Arevinti 2

Sunday 5th
Arevinti 2

By Arrangement

Arrange a personalised garden visit with your club, or group of friends, on a date to suit you. See individual garden entries for full details.

In 2022 we donated £450,000 to Marie Curie which equates to 23,871 hours of community nursing care

THE GARDENS

GROUP OPENING

1 AINSDALE & BIRKDALE GARDENS

14 Saxon Road, Southport, PR8 2AX. Mrs Margaret Fletcher. *1-4m S of Southport. Gardens signed from A565 & A5267.* **Sun 11 June, Sun 16 July (11-5). Combined adm £6, chd free. Home-made teas at 14 Saxon Rd, 45 Stourton Rd & 115 Waterloo Rd.**

23 ASHTON ROAD
PR8 4QE. John & Jennifer Mawdsley.

14 SAXON ROAD
PR8 2AX. Margaret & Geoff Fletcher, 01704 567742, margaret.fletcher@ngs.org.uk. **Visits also by arrangement May to Aug for groups of 10 to 50.**

45 STOURTON ROAD
PR8 3PL. Pat & Bill Armstrong.

115 WATERLOO ROAD
PR8 4QN. Antony & Rebecca Eden.

A group formed to showcase some of the beautiful gardens to visit around the Victorian seaside town of Southport. They range in size and design from a walled garden featuring tender perennials, a flower arranger's garden of infinite detail and surprises and a garden of rooms leading to a spectacular fruit and vegetable plot. We welcome the return of 115 Waterloo Rd, a family garden reclaimed from neglect featuring unusual shrubs and perennials. Not all gardens are wheelchair friendly due to narrow paths or gravel.

2 AREVINTI

1 School Court, Ramsbottom, Bury, BL0 0SD. Lavinia Tod, 01706 822474, arevintgarden@gmail.com. *4m N of Bury. Exit at J1 turn R onto A56 signed Ramsbottom continue straight until yellow signs.* **Sat 22, Sun 23 July (11-4). Home-made teas in the garden room. Sat 4, Sun 5 Nov (1-5). Adm £4, chd free. Cakes & mulled wine available for Nov dates. Visits also by arrangement 15 May to 17 Sept for groups of 8 to 20. Inc tea,coffee & cakes.**

The main garden has a roofed pergola with clematis growing around it and a raised pond. The Chinese garden laid with gravel has two water features, stone dragons and buddhas. There is a working miniature railway in the garden shed. The church yard contains a wildlife pond, herbaceous borders and a fairy house for children. Local history information in folders and boards to read. Nov dates will offer visitors a unique Remembrance experience featuring over 2000 handmade poppies. Wear red if you can.

3 ASHTON WALLED COMMUNITY GARDENS

Pedders Lane, Ashton-On-Ribble, Preston, PR2 1HL. Annie Wynn, 07535 836364, letsgrowpreston@gmail.com, www.letsgrowpreston.org. *W of Preston. From M6 J30 head towards Preston turn R onto Blackpool Rd, continue for 3.4m, turn L onto Pedders Lane & next R onto the park. Entrance to walled garden is 50 metres on L.* **Sat 2 Sept (10-3). Adm £5, chd free. Light refreshments. Visits also by arrangement Mar to Sept for groups of 6 to 25. Refreshments for groups can be inc for an additional £1.50.**

Formal raised beds within a walled garden, a peace garden and an edible garden. The formal part of the garden uses plants predominantly from just 3 families, rose, geranium and aster. It is punctuated by grasses and has been designed to demonstrate how diverse and varied plants can be from just the one family. Practical demonstrations and talking tour about the work that Let's Grow Preston does within the PR postcode inc improving mental and physical wellbeing and working with the 41 food hubs of Preston.

GROUP OPENING

4 BRETHERTON GARDENS

Bretherton, Leyland, PR26 9AD. *8m SW of Preston. Between Southport & Preston, from A59, take B5247 towards Chorley for 1m. Gardens signed from South Rd (B5247) & North Rd (B5248). Maps & tickets at all gardens.* **Sun 28 May, Sun 18 June, Sun 16 July (12-5). Combined adm £6, chd free. Home-made teas at Bretherton Congregational Church from 12 noon on all dates, also light lunches on 16 July.**

HAZEL COTTAGE
PR26 9AN. John & Kris Jolley, 01772 600896, jolley@johnjolley.plus.com. **Visits also by arrangement 1 May to 15 Oct for groups of 10 to 40.**

◆ HAZELWOOD
PR26 9AY. Jacqueline Iddon & Thompson Dagnall, 01772 601433, jacquelineiddon@gmail.com, www.jacquelineiddon.co.uk.

OWL BARN
PR26 9AD. Richard & Barbara Farbon.

PALATINE, 6 BAMFORDS FOLD
PR26 9AL. Alison Ryan.

PEAR TREE COTTAGE
PR26 9AS. John & Gwenifer Jackson.

A group of contrasting gardens spaced across an attractive award winning village with a conservation area. Hazelwood is a 1½ acre plant lover's paradise with herbaceous and mixed borders, kitchen and cutting garden, alpine house, sculpture gallery, pond and plant nursery. At the sculpture gallery there are usually demonstrations at 2pm. Hazel Cottage's former Victorian orchard plot has evolved into a series of themed spaces while the adjoining land has a natural pond, meadow and developing native woodland. Owl Barn has herbaceous borders filled with cottage garden and hardy plants, a productive kitchen garden with fruit, vegetables and cut flowers. There are two ponds with water features and secluded seating areas. The garden was featured in Lancashire Life in 2018. Palatine is a garden of three contrasting spaces started in 2019 around a modern bungalow to attract wildlife and give year-round interest. Pear Tree Cottage has informal beds featuring ornamental grasses, a solar powered greenhouse watering system, a profusion of fruits, a pond, mature trees and a fine backdrop of the West Pennine Moors. On some openings there is live music in the garden. Palatine has full wheelchair access and the majority of Hazelwood is accessible, the other gardens have varying degrees of accessibility due to narrow paths or gravel.

5 BRIDGE INN COMMUNITY FARM

Moss Side, Formby, Liverpool, L37 0AF. www.bridgeinncommunityfarm.co.uk. *7m S of Southport. Formby bypass A565, L onto Moss Side.* **Sun 25 June (11-4). Adm £3.50, chd free. Light refreshments.**

Bridge Inn Community Farm was established in 2010 in response to a community need. Our farm sits on a beautiful 4 acre smallholding with views looking out over the countryside. We provide a quality service of training in a real life work environment and experience in horticulture, conservation and animal welfare.

Hazelwood, Bretherton Gardens

6 NEW 2 BROOKSIDE

Old Langho, Blackburn, BB6 8AP. Mr Peter Lumsden. *3m W of Whalley. Leave A59 at Northcote Manor, onto Northcote Lane. At the bottom, after 1m, turn R at T-junction, and 2 Brookside is on L.* **Sat 8 July, Sat 5 Aug (10-6). Adm £4, chd free. Home-made teas.**
A recently created rural cottage garden, designed to support birds and other wildlife, and in particular, providing year-round flowers for foraging bees and butterflies. Features inc 3 beehives, pond, raised beds for vegetables, cordon fruit trees, perennial flower beds, a small lawned area, water features, and an enclosed patio with variety of plants in pots. Seating areas located throughout.

7 136 BUCKINGHAM ROAD

Maghull, L31 7DR. Debbie & Mark Jackson. *7m N of Liverpool. End M57/M58, take A59 towards Ormskirk. Turn L after car superstore onto Liverpool Rd Sth, cont' on past Meadows pub, 3rd R into Sandringham Rd, L into Buckingham Rd.* **Sat 10, Sun 11 June (12-5). Combined adm with The North Tithebarn £5, chd free. Home-made teas. Donation to The Marina Dalglish Appeal.**
A suburban garden overflowing with cottage garden plants creating a haven for bees and butterflies. Wisteria covered pergola provides shade and roses complement the planting. A pond and water feature along with an array of planters complete the garden. Relax and enjoy our home-made cakes.

GROUP OPENING

8 CALDER HOUSE LANE GARDENS

Calder House Lane, Bowgreave, Preston, PR3 1ZE. Mrs Margaret Richardson, 07867 848218, marg254@btinternet.com. *7m N of M6 J32. 7m S of M6 J33. From J32, A6, turn R on to B6340 1m. From J33 follow A6 for 7m, turn L onto Cock Robin Lane, Catterall turn L at the end, B6340. ½m turn R into Calder House Lane.* **Sat 10 June (10.30-5); Sun 11 June (12-5). Combined adm £5, chd free. Home-made teas in the Friends Meeting House, adjacent to the gardens.** Visits also by arrangement 10 June to 12 Aug for groups of 6+.

1 CALDER HOUSE COTTAGE
Paul Hafren & Gaynor Gee.

2 CALDER HOUSE COTTAGE
Phil & Sarah Schofield.

3 CALDER HOUSE COTTAGE
Margaret & Mick Richardson.

The gardens range from a small satellite to a large (relatively new) garden featuring over 100 hosta varieties. They are mainly cottage gardens in style, each having its own features and interest. The small front gardens have all been made to 3 very different designs. No 1 has a satellite garden at the top of the track, a small but beautiful use of the space, making a tranquil haven. No 2's garden is divided into 3 rooms, a cottage garden, a secluded seating area and finally a greenhouse with vegetables and fruit. A newly built garden room adds to this lovely garden. No 3 has the largest garden, planted with perennials, trees, roses, hostas (over100 varieties mainly in pots) and shrubs with a separate fruit and cut flower garden and greenhouse. Plus a gravel area with seating and a pond.

GROUP OPENING

9 CANNING GEORGIAN QUARTER

27 Canning Street, Liverpool, L8 7NN. Ms Waltraud Boxall. *No 86 bus runs every 12 mins from city centre. Ask for bus stop Back Canning St. Gardens signed both sides of Parliament St.* **Sun 18 June (1-5). Combined adm £6, chd free. Light refreshments & home-made teas will be available at Canning St, Gambier Terrace & at the Community Gardens.**

27 CANNING STREET
L8 7NN. Mrs Waltraud Boxall.

NEW **EL JARDIN DE LA NUESTRA SENORA**
L8 7NL. R.C. Archdiocese of Liverpool.

10 GAMBIER TERRACE
L1 7BG. Mr Paul Murphy.

NEW **GRAPES COMMUNITY FOOD GARDEN**
L8 1XE. Squash Liverpool, admin@squashliverpool.co.uk.

NEW **PAKISTAN ASSOCIATION LIVERPOOL WELLBEING GARDEN**
L8 2TF. Pakistan Association Liverpool.

NEW **THE SQUASH CAFE GARDEN**
L8 8EQ. Squash Liverpool, www.squashliverpool.co.uk.

Two contrasting private gardens behind the Georgian homes in the Canning quarter open their gates this year. And they are joined by a vibrant community garden in the creative hub of nearby Windsor Street, the Squash project back garden, and the nearby Pakistan Association garden. Plus there's a unique opportunity for the public to experience the tranquillity of the Spanish garden, El Jardin de la Nuestra Senora, alongside the St Philip Neri Church. The Georgian Quarter of Liverpool has many important and historic buildings inc Liverpool Cathedral, the largest cathedral and religious building in Britain. The St Philip Neri church garden has a fascinating history, and was created on a WW2 bomb site.

10 31 COUSINS LANE

Ardilea 31 Cousins Lane, Rufford, Ormskirk, L40 1TN. Brenda & Roy Caslake. *From M6 J27, follow signs for Parbold then Rufford. Turn L onto the A59. Turn R at Hesketh Arms Pub. 4th turn.* **Sun 11 June (10-4). Adm £3.50, chd free. Cream teas at Rufford Cricket Club adjacent to garden.**
Rooms within rooms, discreet corners punctuated with an open vista to the cricket ground. This cottage garden welcomes the bees and butterflies. Frogs, toads and newts are resident and the blackbird and thrush lead a vocal chorus whilst wren and robin flit. A potpourri of planting with scents pleasing to the senses. A small, tranquil oasis. Take time to re-energise with a glass of prosecco and a cream tea whilst enjoying the best of English pastimes.

In 2022, our donations to Carers Trust meant that 456,089 unpaid carers were supported across the UK

I1 79 CRABTREE LANE
Burscough, L40 0RW. Sandra & Peter Curl, 01704 893713, peter.curl@btinternet.com, www.youtube.com/watch?v=CW7S8nB_iX8. *3m NE of Ormskirk. A59 Preston - Liverpool Rd. From N before bridge R into Redcat Lane signed for Martin Mere. From S over 2nd bridge L into Redcat Lane after ¾m L into Crabtree Lane.* **Sun 11 June, Sun 9 July (11-4). Adm £5, chd free. Home-made teas. Visits also by arrangement June & July.**
¾ acre year-round plantsperson's garden with many rare and unusual plants. Herbaceous borders and island beds, pond, rockery, rose garden, and autumn hot bed. Many stone features built with reclaimed materials. Shrubs and rhododendrons, Koi pond and waterfall, hosta and fern walk. Gravel garden with Mediterranean plants. Patio surrounded by shrubs and raised alpine bed. Trees give areas for shade loving plants. Many beds replanted recently. Many stone buildings and features. Flat grass and bark paths.

I2 5 CRIB LANE
Dobcross, Oldham, OL3 5AF. Helen Campbell. *5m E of Oldham. From Dobcross head towards Delph on Platt Lane, Crib Lane opp Dobcross Band Club. V limited parking for disabled visitors only100 metres up steep narrow lane, all other visitors to park in the village.* **Sat 24, Sun 25 June (1-4). Adm £3.50, chd free. Home-made teas.**
A well loved and well used family garden - challenging as on a high terraced hillside on tip site. Visited by deer, hares and the odd cow! Of additional interest are wildlife ponds, wildflower areas, a polytunnel, four beehives with 250,000 bees, an art gallery and garden sculptures. Areas re thought annually, dug up and changed. Local honey and cards for sale.

I3 DALE HOUSE GARDENS
off Church Lane, Goosnargh, Preston, PR3 2BE. Caroline & Tom Luke. *2½ m E of Broughton. M6 J32 signed Garstang Broughton, T-lights R at Whittingham Lane, 2½m to Whittingham. At PO turn L into Church Lane. Garden between nos 17 & 19.* **Sat 15, Sun 16 Apr, Sat 17, Sun 18 June (10-4). Adm £3.50, chd free. Home-made teas. Donation to St Francis School, Goosnargh.**
½ acre tastefully landscaped gardens comprising of limestone rockeries, well-stocked herbaceous borders, raised alpine beds, well-stocked koi pond, lawn areas, greenhouse and polytunnel, patio areas, specialising in alpines, rare shrubs and trees, large collection of unusual bulbs, Secret Garden. Year-round interest. Large indoor budgerigar aviary with 300+ budgies. Gravel path, lawn areas.

I4 DENT HALL
Colne Road, Trawden, Colne, BB8 8NX. Mr Chris Whitaker-Webb & Miss Joanne Smith. *10 min from end of M65. Turn L at end of M65. Follow A6068 for 2m; just after 3rd r'about turn R down B6250. After 1½m, in front of church, turn R, signed Carry Bridge. Keep R, follow road up hill, garden on R after 300yds.* **Sat 27, Sun 28 May (12-5). Adm £4, chd free. Home-made teas.**
Nestled in the oldest part of Trawden villlage and rolling Lancashire countryside, this mature and evolving country garden surrounds a 400 yr old grade II listed hall (not open); featuring a parterre, lawns, herbaceous borders, shrubbery, wildlife pond with bridge to seating area and a hidden summerhouse in a woodland area. Plentiful seating throughout. Some uneven paths and gradients.

I5 DERIAN HOUSE CHILDREN'S HOSPICE
Chancery Road, Chorley, PR7 1DH. Gareth Elliott, www.derianhouse.co.uk. *2m from Chorley town centre. From B5252 pass Chorley Hospital on L, at r'about 1st exit to Chancery Lane. Hospice on L after 0.4 m. Parking is available around the building & on the road. SatNav directions not always accurate.* **Sun 23 Apr, Sun 13 Aug (10-4.30). Adm £3, chd free. Light refreshments.**
The Chorley-based children's hospice provides respite and end-of-life care to more than 450 children and young people from across the North West and South Cumbria. The gardens at the hospice help to create an atmosphere of relaxation, tranquillity and joy. Distinct areas inc the seaside garden, the sensory garden, the memorial garden and the Smile Park adventure playground. Family activities. Home-made cakes and plant sale. Gardens and refreshments area are wheelchair friendly, accessible parking and WC facilities also available.

GROUP OPENING

I6 DIDSBURY VILLAGE GARDENS
Tickets from 68 Brooklawn Drive M20 3GZ or any garden, Didsbury, Manchester, M20 6RW. *5m S of Manchester. From M60 J5 follow signs to Northenden. Turn R at T-lights onto Barlow Moor Rd to Didsbury. From M56 follow A34 to Didsbury.* **Sun 11 June (12-5). Combined adm £6, chd free. Home made teas at 68 Brooklawn Drive.**

68 BROOKLAWN DRIVE
M20 3GZ. Anne & Jim Britt, www.annebrittdesign.com.

3 THE DRIVE
M20 6HZ. Peter Clare & Sarah Keedy.

1 OSBORNE STREET
M20 2QZ. Richard & Teresa Pearce-Regan.

5 PARKFIELD ROAD SOUTH, FLAT 1
M20 6DA. Kath & Rob Lowe.

40 PARRS WOOD AVENUE
M20 5ND. Tom Johnson.

38 WILLOUGHBY AVENUE
M20 6AS. Simon Hickey.

A wonderful collection of 6 very beautiful and individual gardens to visit in the attractive suburb of Didsbury. There is so much to see and inspire in these gardens, from our traditional cottage gardens with alliums, roses, clematis and topiary, to the crisp lines of two beautifully planted contemporary gardens with white rendered walls, pale porcelain paving and stylish outdoor living and dining ideas. Other gardens offer a diverse range of water features, inspirational shade planting and organic wildlife friendly planting ideas - all within our relatively small garden spaces which are a feast to the eye and full of detail to imitate at home. Wheelchair access to some gardens.

79 Crabtree Lane

17 DUTTON HALL

Gallows Lane, Ribchester, PR3 3XX. Mr & Mrs A H Penny, www.duttonhall.co.uk. *2m NE of Ribchester. Signed from B6243 & B6245 Directions on website also. What3words app - copying. botanists.mostly.* **Sun 25 June (1-5). Adm £7.50, chd free. Home-made teas. Donation to Plant Heritage.** An increasing range of unusual trees and shrubs have been added to the existing collection of old fashioned roses, inc rare and unusual varieties and Plant Heritage National Collection of Pemberton Hybrid Musk roses. Formal garden at front with backdrop of C17 house (not open). Analemmatic Sundial, pond, meadow areas all with extensive views over Ribble valley. Plant Heritage plant stall with unusual varieties for sale.

GROUP OPENING

18 NEW ELLESMERE PARK GARDENS

35 Westminster Road, Eccles, Manchester, M30 9FE. Enid Noronha. *4m S from central Manchester. From A56 Stretford take Edge Lane (A5154) ¾m then bear L onto Wilbraham Rd (A6010) for ¾m then L onto Westminster Rd.* **Sat 10 June (12.30-5). Combined adm £6, chd free. Cream teas at 35 Ellesmere Rd, hot dogs at 20 Stafford Rd and tea & biscuits at 11 Westminster Rd.**

35 ELLESMERE ROAD
M30 9FE. Enid Noronha.

NEW **20 STAFFORD ROAD**
M30 9HW. Mrs Paula Gibson.

11 WESTMINSTER ROAD
M30 9HF. George & Lynne Meakin.

The group consists of 3 diverse suburban gardens in adjoining streets in Eccles. 35 Ellesmere Road is a cottage style garden with walkways, palisade, greenhouse and raised beds with vegetables and herbs. Lawns with box hedges. Mature fruit and other trees. At 11 Westminster Road there is a pretty front garden with topiary chickens, well-stocked with perennials, mature trees, and box hedging. The garden is divided by a trellis and rose arch which

Glynwood House

separates the flower beds and lawn from the fruit growing area, and there are a large number of fuchsias grown in pots. Finally at 20 Stafford Road the garden features mature trees and shrubs alongside many recent additions. An ongoing project involves upcycling items the garden owners find interesting that others have discarded. The owners like to put bold colours and statements in wherever they can to enable colour year-round. Some areas with limited access.

19 NEW 14 ELMS AVENUE

Lytham St Annes, FY8 5PW. Mr Mark Fenton, www.markfentondesigns.co.uk. *Access from A584 running between Preston and Blackpool. 0.6m W of Lowther Gardens & Theatre.* **Sun 4 June (11-4.30). Adm £3, chd free. Light refreshments.**
Richly planted garden recently developed from a blank canvas to suit a Victorian property. Incorporating a fish pool, seating for relaxing and entertaining, a vegetable plot, a potting shed, and a greenhouse with cacti and succulents collection. The multitude of features and uses have been cleverly and effectively incorporated into a modest, difficult space by the Landscape Architect owner. Single step between main areas of the garden. Greenhouse and vegetable garden are not accessible by wheelchair.

GROUP OPENING

20 FRESHFIELD GARDENS

West Lane, Freshfield & Formby, L37 7BA. *6m S of Southport. Paradise Lane, Brewery Lane, Kenton Close & the West Lane gardens are close together in Freshfield. Buttermere Cl is off Ennerdale Rd approx 1m away from the other gardens. Cable St is off Church Rd.* **Sun 2 July (11-4). Combined adm £6, chd free. Home-made teas at West Lane, Brewery Lane, Kenton Close & Cable St. Light refreshments at Buttermere Close.**

33 BREWERY LANE
L37 7DY. Sue & Dave Hughes.

4 BUTTERMERE CLOSE
L37 2YB. Marilyn & Alan Tippett.

NEW **60 CABLE STREET**
L37 3LX. Mrs Kate Branigan.

18 KENTON CLOSE
L37 7EA. Alex & Peter Reed.

NEW **28 PARADISE LANE**
L37 7EJ. Mrs Lynette Marsden.

5 WEST LANE
L37 7AY. Ian & Kathy Taphouse.

6 WEST LANE
L37 7BA. Laurie & Sue Lissett.

7 coastal suburban gardens near to Formby sand dunes and NT Nature Reserve, home to the red squirrel. Soil is sandy and well drained.
All the gardens feature colourful herbaceous borders with a mixture of shrubs and perennials, and inc many roses and hydrangeas. The group gardens demonstrate a variety of styles and designs and some inc fruit and vegetables areas. Features inc raised beds, greenhouses, rockeries, ponds, arches, pergolas and basket/container displays.

21 GILES FARM

Four Acre Lane, Thornley, Preston, PR3 2TD. Kirsten & Phil Brown, 07925 603246, phil.brown32@aol.co.uk. *3m NE of Longridge. Pass through Longridge & follow signs for Chipping. After 2m turn R at the old school up Hope Lane. Turn L at the top and continue to the farm where there is ample parking.* **Sat 17, Sun 18 June (12-5). Adm £5, chd free. A wide selection of home-made cakes & sandwiches. Visits also by arrangement 5 June to 21 July for groups of 15+. Guided tour and hand drawn garden map provided for groups.**
Nestled high on the side of Longridge Fell, with beautiful long-reaching views across the Ribble Valley, the gardens surround the old farmhouse and buildings. Ever evolving, they inc an acre of perennial wildflower meadows, wildlife ponds, woodland areas and cottage gardens. There are plentiful areas to sit and take in the views. There are steps and uneven surfaces in the gardens.

22 GLYNWOOD HOUSE

Eyes Lane, Bretherton, Leyland, PR26 9AS. Terry & Sue Riding. *Once in Bretherton, turn onto Eyes Lane by the War Memorial. After 50 yds take the R fork, after 100 yds follow the road around to the L. Glynwood House is a further 200 yds on the L.* **Sun 2 July (11-5). Adm £4, chd free.**
Set in a peaceful rural location with spectacular views. Highlights of this ¾ acre garden inc colour themed mixed borders with rare and unusual plants, wildlife pond with cascade water feature and meandering paths through a woodland shaded area.
A patio garden, surrounded by a large pergola, features a new kinetic sycamore seed sculpture. Previous finalist in Daily Mail National Garden competition. Access available to most of the garden apart from sleeper paths in the woodland walk area.

23 GREEN FARM COTTAGE

42 Lower Green, Poulton-le-Fylde, FY6 7EJ. Eric & Sharon Rawcliffe. *500yds from Poulton-le-Fylde village. M55 J3 follow A585 Fleetwood. T- lights turn L. Next lights bear L A586. Poulton 2nd set of lights turn R Lower Green. Cottage on L.* **Sun 11 June (10-5). Adm £4, chd free. Home-made teas.**
½ acre well established formal cottage garden. Featuring a koi pond and paths leading to different areas. Lots of climbers and rose beds. Packed with plants of all kinds. Many shrubs and trees. Well laid out lawns. Collections of unusual plants. A surprise round every corner. Said by visitors to be a real hidden jewel.

24 NEW GREENFIELD HALL

Greenfield Road, Colne, BB8 9PE. Mr Mike & Mrs Lesley Pusey. *From r'about at end of M65 take A6068 signed Nelson & Colne. Take 1st exit A58 then 2nd L down Phillips Lane, follow to R into Greenfield Rd. Parking opp Industrial Units.* **Sun 25 June (12-5). Adm £4, chd free. Home-made teas.**
Set in a green oasis on the edge of Colne, C16 Greenfield Hall (not open) has a quirky informal garden featuring old and English roses and a wide variety of perennials, shrubs and young trees. A raised pond is home to ghost carp and goldfish. The garden has several different planting areas. It is divided by the beck, crossed by a medieval bridge, and showcases work by renowned sculptor Clare Bigger.

25 33 GREENHILL ROAD
Mossley Hill, Liverpool, L18 6JJ. Tony Rose. *From L lane at end of M62 follow Ring Road A5058 (S). Follow A5058 through 1st island. Keep to L lane through 2nd island then take B5180 for about ½m to signage.* **Sun 16 July, Sun 3 Sept (12-5). Combined adm with 146 Mather Avenue £4, chd free. Light refreshments at 146 Mather Avenue. Tea & cake.**
The central attraction in the garden is a raised water stream over cobbles running into a raised pond (reservoir). On two sides planted into the ground are mature bamboo and other plants around a golden Buddha statue. The remainder of the garden is hard paved in varying materials. There is no lawn. Throughout the garden the embellishment is with many types of plants in pots, exotic and unusual.

26 45 GREY HEIGHTS VIEW
Off Eaves Lane, Chorley, PR6 0TN. Barbara Ashworth. *From Wigan/ Coppull B5251. At town centre Xrds go straight across. At r'about across to Lyons Ln & follow NGS signs. From M61, J8 follow signs A6 Town Centre to Lyons Ln then signed from here.* **Sun 4 June, Sun 20 Aug (10-5). Adm £3, chd free. Cream teas.**
A small suburban garden with cottage garden style planting inc fruit trees. Heavily planted with a profusion of perennials, roses and clematis. No repeat planting. Inc a greenhouse, small vegetable and fruit area. An abundance of recycling and space saving ideas. Backdrop of Healey Nab, and a stone's throw from the Leeds - Liverpool Canal.

27 THE GROWTH PROJECT
Kellett Street Allotments, Rochdale, OL16 2JU. Karen Hayday, 07464 546962, k.hayday@hourglass.org.uk, www.facebook.com/Hourglass.org.uk/ *From A627 M. R A58 L Entwistle Rd then R Kellett St.* **Sun 30 July (11.30-3.30). Adm £3.50, chd free. Home-made lunches & afternoon tea served in the Victorian style 'Woodland Green' railway station & signal box. Visits also by arrangement 6 July to 27 Oct Only open on Thursdays each week. Donation to The Growth Project.**
The project covers over an acre and inc a huge variety of organic vegetables, a wildlife pond, insect hotels, formal flower and Japanese garden, potager and enchanted woodland garden. See the mock Elizabethan straw bale build and station, stroll down the pergola walk to the wildflower meadow and orchard. Jams and preserves, flowers, craft items and cakes to buy. The new attraction this year is the Japanese garden. The Growth Project is a partnership between Hourglass and Rochdale and District Mind. Guides are on hand to give horticultural advice and show you round. No disabled WC, ground can be uneven.

GROUP OPENING

28 HALE VILLAGE GARDENS
2 Pheasant Field, Hale Village, Liverpool, L24 5SD. Roger & Tania Craine. *6m S of M62 J6. Take A5300, A562 towards L'pool, then A561 and then L for Hale. From S L'pool head for the airport then L sign for Hale. The 82A bus from Widnes/Runcorn to L'pool has village stops.* **Sat 8, Sun 9 July (12-5). Combined adm £5, chd free. Cream teas at 2 Pheasant Field & cold drinks & cakes at 33 Hale Rd.**

33 ARKLOW DRIVE
L24 5RW. Rebecca & Ross Firman.

54 CHURCH ROAD
L24 4BA. Norma & Ray Roe.

4 CHURCH ROAD
L24 4BA. Ms Chris Chesters.

2 PHEASANT FIELD
L24 5SD. Roger & Tania Craine.

WHITECROFT, 33 HALE ROAD
L24 5RB. Mrs Donna & Mr Robert Richards.

The delightful village of Hale is set in rural South Merseyside between Widnes and Liverpool Airport. It is home to the cottage, sculpture and grave of the famous giant known as the Childe of Hale. Five gardens of various sizes have been developed by their present owners. There are 3 gardens (a large skilfully landscaped one and two beautiful contemporary ones) to the west of the village and 2 delightful gardens in Church Rd to the east of the village.

29 HALTON PARK HOUSE
Halton Park, Halton, Lancaster, LA2 6PD. Mr & Mrs Duncan Bowring. *7min drive from both J34 & 35 M6. On Park Lane, approx 1½m from Halton or Caton. Park Lane accessed either from Low Rd or High Rd out of Halton. From Low Rd turn into Park Lane through pillars over cattle grid.* **Sat 20, Sun 21 May, Sat 9, Sun 10 Sept (11-4). Adm £5, chd free. Home-made teas. BBQ over lunch time.**
Approx 6 acres of garden, with gravel paths leading through large mixed herbaceous borders, terraces, orchard and terraced vegetable beds. Wildlife pond and woodland walk in dell area, extensive lawns, greenhouse and herb garden. Plenty of places to sit and rest. Gravel paths (some sloping) access viewing points over the majority of the garden. Hard standing around the house.

30 HIGHER BRIDGE CLOUGH HOUSE
Coal Pit Lane, Rossendale, BB4 9SB. Karen Clough. *4m from Rawtenstall. Take A681 into Waterfoot. Turn L onto the B6238. Approx ½m turn R onto Shawclough Rd. Follow the road up until you reach the yellow signs.* **Sat 8, Sun 9 July (12-5). Adm £4, chd free. Cream teas, selection of cakes & drinks available.**
Nestled within Rossendale farmland in an exposed site the garden is split by a meandering stream. There are raised mixed borders with shrubs and perennials enclosed by a stone wall, with loose planting of shrubs and water loving plants around the stream. A local farmer once told me 'you won't grow 'owt up 'ere...' so the challenge is on. The Rossendale area provides some excellent walking and rambling sites. Rawtenstall has the famous East Lancashire railway, which is worth a visit.

Our 2022 donation enabled 7,000 people affected by cancer to benefit from the garden at Maggie's in Oxford

GROUP OPENING

31 HIGHTOWN GARDENS
Alt Road, Hightown, Liverpool, L38 3RH. Dr Jan Proctor. *11m N of Liverpool & 11m S of Southport. M 57/58 Join A5758 keep L, at r'about take 2nd exit A565. Turn 2nd L onto B5193/Orrell Hill Lane Turn R Moss Lane. Turn L onto Alt Rd, then R to Kerslake Way & follow NGS signs from here.* **Sun 28 May (1-5). Combined adm £5, chd free. Home-made teas at The Cross House.**

11 BLUNDELL AVENUE
L38 9ED. Karen Rimmer.

11 BLUNDELL ROAD
L38 9ED. Joyce Batey.

75 BLUNDELL ROAD
L38 9EF. Shirley & Phil Roberts.

NEW **THE CROSS HOUSE**
L38 3RH. Dr Jan Proctor.

A welcoming group of 4 gardens in the village of Hightown. Visit a plantswoman's quirky garden full of surprises, an informal organic wildlife garden with literary themed areas. We see the return of a garden on 2 levels surrounded by well established impressive trees and shrubs and the addition of a new garden recently developed as a labour of love.

32 HOLLY HOUSE
Moor Road, Croston, Leyland, PR26 9HP. Mrs Elaine Raper. *Situated on the A581, approx 200 metres after the Highfield pub if coming from Leyland area.* **Sun 21 May, Sun 23 July (11-5). Adm £5, chd free. Light refreshments.**
Large garden, approx an acre of perennial beds with mature trees. Pond with decking area leading to wooded area beyond with rill. Two wisteria clad pergolas, parterre with adjacent gazebo. Mostly accessible to the main parts of the garden. Wood behind pond not accessible.

33 KINGTON COTTAGE
Kirkham Road, Treales, Preston, PR4 3SD. Mrs Linda Kidd, 01772 683005. *M55 J3. Take A585 to Kirkham, exit Preston St, L into Carr Lane to Treales village. Cottage on L.* **Sun 7 May, Sun 11 June, Sun 9 July, Sun 13 Aug (10-5). Adm £4, chd free. Home-made teas. Visits also by arrangement 1 May to 1 Sept for groups of 8 to 30. Escorted tour by garden owner.**
Nestling in the beautiful village of Treales this generously sized Japanese garden has many authentic and unique Japanese features, alongside its 2 ponds linked by a stream. The stroll garden leads down to the tea house garden. The planting and the meandering pathways blend together to create a tranquil meditative garden in which to relax. New additions complete four Japanese garden styles. Wheelchair access to some areas, uneven paths.

34 NEW **LARK RISE**
515 Burnley Lane, Chadderton, Oldham, OL9 0BW. Wendy & Ian Connor. *Less than 3m from Oldham town centre. Leave A627(M) at J1-Elk Mill R'about. Take the exit for Middleton & the house is ½m on L before the T-lights. Turn R at lights for car parking at St Matthew's Church.* **Sun 23 July (12-5). Combined adm with Maggie's, Oldham £5, chd free. Home-made teas.**
Lark Rise is a ½ acre suburban garden. The garden consists of several long borders in both sun and shade planted mainly with perennials and some shrubs/small decorative trees. There is an apple archway leading to a box parterre with a centre piece decorative ironwork water feature. There are several seating areas and home made teas will be served. There is a short steep sloping driveway but access to the garden can be gained without steps through a side path.

35 MAGGIE'S, MANCHESTER
Kinnaird Road, Manchester, M20 4QL. Ruth Tobi, www.maggies.org/manchester. *At the end of Kinnaird Rd which is off Wilmslow Rd opp the Christie Hospital.* **Sun 21 May (12-4). Adm £4, chd free. Cream teas. Light refreshments.**
The architecture of Maggie's Manchester, designed by world-renowned architect Lord Foster, is complemented by gardens designed by Dan Pearson, Best in Show winner at Chelsea Flower Show. Combining a rich mix of spaces, inc the working glasshouse and vegetable garden, the garden provides a place for both activity and contemplation. The colours and sensory experience of nature becomes part of the Centre through micro gardens and internal courtyards, which relate to the different spaces within the building. Wheelchair access to most of the garden from the front entrance.

36 MAGGIE'S, OLDHAM
The Royal Oldham Hospital, Rochdale Road, Oldham, OL1 2JH. Maggie's Centres, www.maggies.org/oldham. *Maggie's is in the grounds of the Royal Oldham Hospital, next door to A&E. It is the wooden building on stilts & the garden lies underneath the building.* **Sun 23 July (12-5). Combined adm with Lark Rise £5, chd free. Home-made teas.**
The garden is framed by enclosing walls. The building 'floating' aloft is like a drop curtain to the scene, creating a picture window effect. The trees soar upwards filling the volume of space. A woodland understorey weaves between the structure of the numerous white birch and crispy bark of the pine trunks. The garden could be described as an ornamental woodland.

37 MAGHULL STATION
Station Road, Maghull, Liverpool, L31 3DE. Merseyrail. *7 m N of Liverpool. Follow A59 into Maghull turning R at 2nd set of T-lights (opp Maghull Town Hall) Follow rd over canal bridge & station is approx ½m.* **Sun 23 July (11-4.30). Combined adm with 47 The Crescent £4, chd free.**
Named RHS North West In Bloom Best Station in 2021. Filled with herbaceous plants, shrubs, rockery, hanging baskets, troughs and large planters tumbling with a wide variety of bedding plants - a wonderful sight for commuters arriving in Maghull. Disabled parking spaces available.

In 2022, 20,000 people were supported by Perennial caseworkers and prevent teams thanks to our funding

38 146 MATHER AVENUE

Allerton, Liverpool, L18 7HB. Barbara Peers. *3m S from end of M62. On the R of Mather Ave heading out of town: on the same side as Tesco and 800 metres beyond.* **Sun 16 July, Sun 3 Sept (12-5). Combined adm with 33 Greenhill Road £4, chd free. Light refreshments. Tea and cake.**

A small tree fringed town garden with two ponds, one for fish and one for wildlife. A large woodpile is left to encourage insects as are the nettles along the back wall. One bed allowed to self seed with wildflowers. No pesticides and only home-made compost used. A multitude of plants encroach on the ever decreasing patch of lawn. Paths around give access.

39 MILL BARN

Goosefoot Close, Samlesbury, Preston, PR5 0SS. Chris Mortimer, 07742 924124, millbarnchris@gmail.com. *6m E of Preston. From M6 J31 2½m on A59/A677 B/burn. Turn S. Nabs Head Lane, then Goosefoot Lane.* **Sat 27, Sun 28 May, Sat 3, Sun 4, Sat 10, Sun 11 June (11-5). Adm £5, chd free. Cream teas. Picnics welcome. Visits also by arrangement 1 May to 30 July for groups of up to 30. A donation of £20 is requested if the number of visitors is less than 4.**

The unique and quirky garden at Mill Barn is a delight, or rather a series of delights. Along the River Darwin, through the tiny secret grotto, past the suspension bridge and view of the fairytale tower, visitors can stroll past folly, sculptures, lily pond, and lawns, enjoy the naturally planted flowerbeds, then enter the secret garden and through it the pathways of the wooded hillside beyond. A garden developed on the site of old mills gives a fascinating layout which evolves at many levels. The garden jungle provides a smorgasbord of flowers to attract insects throughout the season. Children enjoy the garden very much. Partial wheelchair access.

40 MILL CROFT

Spout Lane, Wennington, Lancaster, LA2 8NX. Linda Ashworth & Carl Hunter. *From J34 of M6 take A683 towards Kirkby Lonsdale. Turn R on B6480 towards Bentham. Pass through Wray & the narrowing in the road at Wennington, then take the 1st L onto Spout Lane.* **Sun 28 May, Sun 25 June (10-4.30). Adm £4, chd free. Light refreshments. Home-made cakes, inc gluten-free & vegan.**

A rural garden extending to about 3¼ acres with a summer flowering meadow, stream, trees, shrubs, herbaceous planting, vegetable plot and stumpery. Various seating areas to rest and enjoy local wildlife and the surrounding views. An ongoing experiment to discover what will thrive in heavy clay with occasional flooding, exposure to wind and in a frost pocket. New projects always underway.

41 MILL GREEN SCHOOL

Lansbury Avenue, St Helens, WA9 1BU. George McClellan. *2 m from St Helens Town Centre (next to Parr Library). From St Helens Rugby Club follow Linkway East then Parr St for 1m, L onto Ashcroft St then Parr Stocks Rd for ½m. R into Fleet Lane for 200 yds, then L into Landsbury Ave.* **Sun 23 Apr (11-4). Adm £4, chd free. Light refreshments.**

Set in 4 acres the garden features 2 orchards with numerous types of fruit trees, an apple arch with espaliered trees, an allotment, pond, formal garden, a wildlife garden and wilderness area, a grass labyrinth and an annual wildflower meadow.

42 MOSS PARK ALLOTMENTS

Lesley Road, Stretford, Manchester, M32 9EE. Allison Sterlini 07778 931029. *3m SW of Manchester. From M60 J7 (Manchester), A56 (Manchester), A5181 Barton Rd, L onto B5213 Urmston Lane, ½m L onto Lesley Rd signed Stretford Cricket Club. Parking at 2nd gate.* **Sun 16 July (11-4). Adm £5, chd free. Gorgeous array of home-made cakes.**

Moss Park is a stunning, award winning allotment site in Stretford, Manchester. Wide grass paths flanked by pretty flower borders give way to a large variety of well-tended plots bursting with ideas to try at home, from insect hotels to unusual fruits and vegetables. Take tea and cake on the lawn outside the quirky society clubhouse that looks like a beamed country pub. WC facilities. Partial wheelchair access.

43 NEW THE NORTH TITHEBARN

Lunt Lane, Liverpool, L29 7WL. Mrs Janis & Mr Howard Sleeman. *From Switch Island follow A5758 turn R onto B5422 Brickwall Lane signposted Sefton. In Sefton turn L onto Lunt rd & cont into Lunt Village. Barn is signposted in village.* **Sat 10, Sun 11 June (12-5). Combined adm with 136 Buckingham Road £5, chd free. Home-made teas at 136 Buckingham Road.**

Nestled in the village of Lunt this 2 acre garden comprises of ½ an acre of woodland, and an impressive 25 metre pond surrounded by perennial planting. Several other ponds, orchard, wildflower area and vegetable beds along with mature trees, rhododendrons and 'hosta hill' provide plenty of interest in a tranquil setting.

44 NEW NORTHCOTE

Bambers Lane, Blackpool, FY4 5LH. Mrs Anne Lesniak. *3 min drive from the M55 exit. Access from Bambers Lane,3rd house on R. 2nd entrance. Access lane narrow with passing points, ample parking by house.* **Sat 27, Sun 28 May (11-4). Adm £5, chd free. Light refreshments and wine.**

Northcote garden is a 1 acre well established and constantly evolving garden that has mature trees and shrubs, large lawns, perennial borders, central beds, fairy garden, a rose bed and lily pond. Spring brings magnificent magnolias, cherry blossom and a crocus pathway running through the central lawn. Garden is mainly flat. Areas with steps can usually be accessed from other areas.

Our donation of £2.5 million over five years has helped fund the new Y Bwthyn NGS Macmillan Specialist Palliative Care Unit in Wales. 1,900 inpatients have been supported since its opening

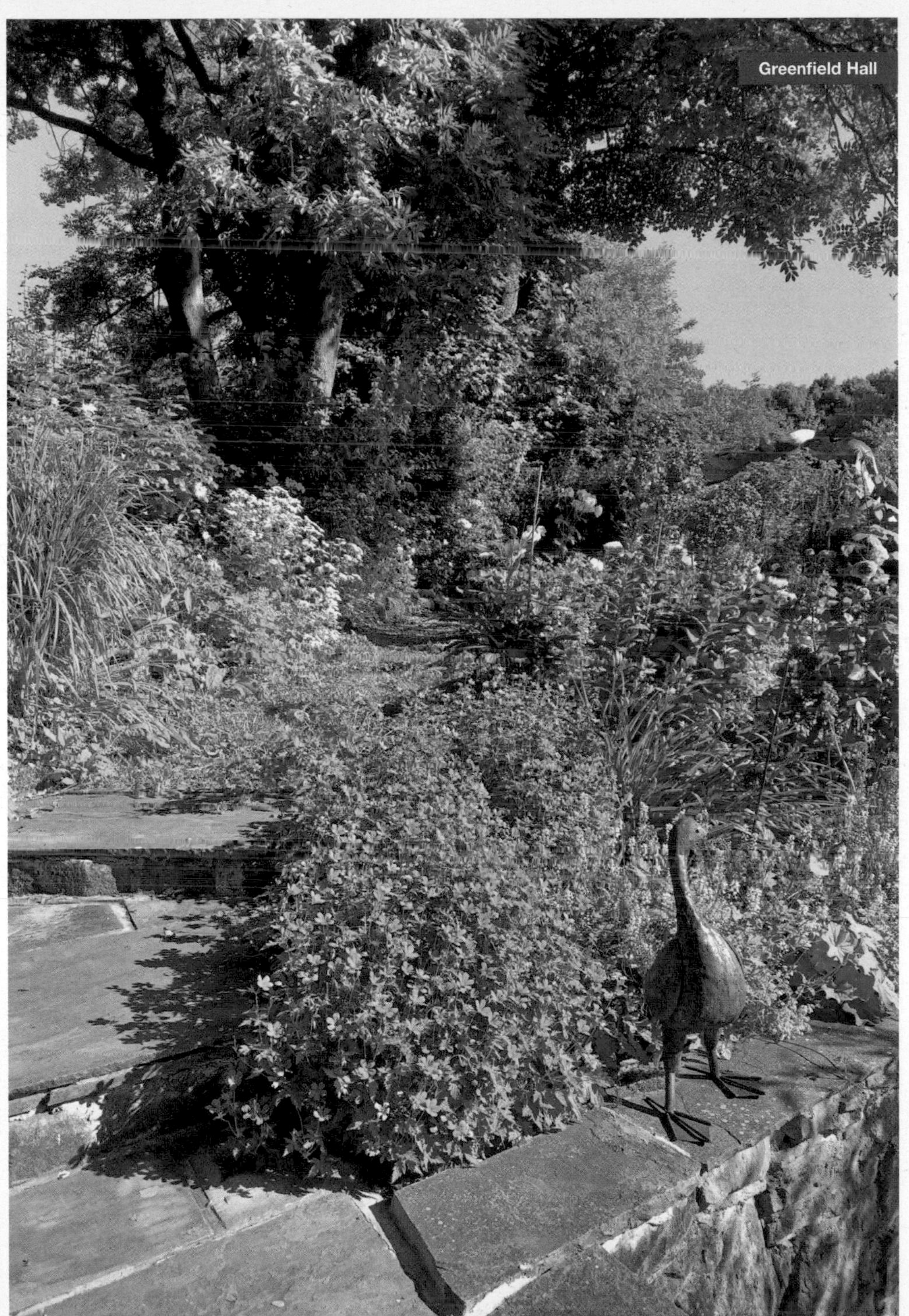

Greenfield Hall

45 NEW **OAKLANDS**

Flash Lane, Rufford, Ormskirk, L40 1SW. Mr John & Mrs Jane Hoban. *Pls park in the village hall car park which is situated on the L at the end of Flash Lane.* **Sat 1, Sun 2 July (10-4.30). Adm £4, chd free. Home-made teas.**

Oaklands is a woodland garden full of mature trees,acers,camellias and rhododendrons. We have over the years created beautiful pathways and seating areas flanked with ferns and hostas. The garden also has several water features and large areas of lawn. Most areas of our garden are wheelchair accessible but all the pathways are through the woodlands so it can be quite uneven.

46 **OLD HOLLOWS FARM**

Old Hollow Lane, Banks, Southport, PR9 8DU. Janet Baxter. *5m N of Southport. A565 r'about take turning for Banks. Straight through Banks. New Lane Pace Rd ends on bend, turn L on farm rd into Old Hollow Lane. Parking in yard at lane end.* **Sat 10, Sun 11 June (11-5). Adm £5, chd free. Home-made teas.**

The garden is in 2 locations, a large garden to the back of the property plus a small cottage garden by Auntie May's Cottage. Originally a waste ground and pit 18 yrs ago, now a peaceful tranquil garden on the edge of the Ribble Estuary. Patio area, garden room, large lawned areas, trees, large herbaceous border and arboretum. The wildlife pond has a large reed bed. Walks along the estuary embankment. Bee demonstration plus local crafts and artisans. No disabled WC facilities. Level in main area only.

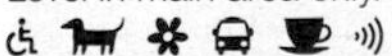

47 **THE OLD VICARAGE**

Church Road, Astley, Manchester, M29 7FS. Susan & Neil Kinsella, 07748 323222, susan.kinsella@hotmail.co.uk. *Halfway between Leigh & Worsley off A580 East Lancs Rd. Opp the site of the old St Stephens Astley CE Church & next to Dam House, 250 yds from the mini r'about junction at Church Rd (A5082) & Manchester Rd (A572). Close to the Bull's Head.* **Sat 17 June (10-4). Adm £4, chd free. Home-made teas. Visits also by arrangement 24 June to 2 Sept.**

Recently restored medium sized gardens of Grade II* vernacular Georgian vicarage comprising both formal and informal planting and statuary as the garden narrows to follow the woodland stream which runs through it. There are 6 separate areas/rooms which gradually merge into the surrounding trees ending several hundred yards from the frontage to the property. Most of the garden is flat with some gravel paths.

GROUP OPENING

48 **ORMSKIRK AND AUGHTON GARDENS**

Ormskirk, L39 1LF. Brian & Marian Jones. *13m N of Liverpool. From the end of M57 take A59 & follow signs for Ormskirk or from M58 J3 take A570 to Ormskirk.* **Sun 25 June (11-4). Combined adm £5, chd free. Home-made teas.**

NEW **65 COUNTY ROAD**
L39 1QG. Mrs Sandra Kuberski.

GORSE HILL NATURE RESERVE
Holly Lane, L39 7HB. Jonathan Atkins (Reserve Manager), www.nwecotrust.org.uk.

72 LUDLOW DRIVE
L39 1LF. Marian & Brian Jones.

Ormskirk is an historic market town in W Lancashire. The gardens are very diverse with something for everyone. 72 Ludlow Drive is a beautiful town garden overflowing with exuberant planting. Comprising a gravel garden, colourful herbaceous and shrub borders, raised shade and rose borders with many old and new roses and clematis. A large glasshouse with a collection of succulents, aeoniums, cacti and pelargoniums. There is also a raised pond and alpine troughs. 65 County Road is a cottage style garden with mature trees, perennials and shrubs, a central lawned area leads to a raised patio. Pebble beds and small sculptures are dotted around the garden. Gorse Hill Nature Reserve situated on a sandstone ridge offers spectacular views across the Lancashire Plain. There is a 5 acre wildflower meadow in summer brimming with a wide variety of wildflowers and grasses. The meadows pond is patrolled by dragonflies and damselflies.

49 **PARKERS LODGES**

28 Lodge Side, Bury, BL8 2SW. Keith Talbot. *2m W of M66 J2. From A58 then B6196 Ainsworth Rd turn R onto Elton Vale Rd, drive past the sports club onto a small estate, & follow the road round to car park.* **Sun 21 May, Sun 9 July (10-5). Adm £5, chd free. Home-made teas.**

Parkers Lodges is set on the site of a demolished Victorian Mill that was used for bleaching. Since the houses were completed in 2014 a small group of volunteers have been working to transform what was a jungle into a manicured wild space that still allows the local wildlife to flourish. Set in 12 acres with 2 lakes with 50 percent of them open for a leisurely stroll round.

50 **PLANT WORLD**

Myerscough College, St Michaels Road, Bilsborrow, Preston, PR3 0RY. 01995 642264, tmelia@myerscough.ac.uk, www.myerscough.ac.uk/commercial-services-equine-events/plant-world-gardens. *From M6 take J32 and head N for 5m up the A6. Turn L onto St Michaels Rd take 2nd entrance into the college signed Plant World.* **Sun 13 Aug (10-4.30). Adm £3.50, chd free. Tea.**

A gardener's paradise with an acre of RHS Gold Award winning gardens and glasshouses, inc tropical, temperate and desert zones. Themed and herbaceous borders, pinetum, fruit garden, bog garden, pond and woodland garden. Many plants labelled for identification purposes. Plant sales with many rare and unusual specimens. Myerscough Tearooms overlook the stunning gardens serving a wide range of delicious locally produced cakes and sandwiches. Indoor and outdoor seating. Guided tours throughout the day. Dogs welcome in the gardens, sales area and the café. Partial access for wheelchairs. Paths are grass and it can be wet in some areas.

51 NEW **RAINBAG COTTAGE**

Carr Moss Lane, Halsall, Ormskirk, L39 8RZ. Mr Mick & Mrs Sue Beacon. *8m NW of Ormskirk. Take A570 from Ormskirk to Southport for approx 4m, L into Gorsuch Lane A5147. After 1½m pass church R*

into Carr Moss Ln which becomes single track, house sign on L after 2m. **Sun 21 May, Sun 3 Sept (10.30-4). Adm £4, chd free. Home-made teas. Selection of home-made cakes.**
A magical ½ acre garden in a rural setting with open views to the surrounding countryside and adjoining Rainbag Wood. Oriental themed stroll garden with wildlife ponds and moon gate. Herbaceous borders and woodland walk. Small flower meadow. Fruit trees. Greenhouse. Wormery. Bug pad. Various seating areas. Garden ornaments. Projects under development inc vegetable and cut flower garden.

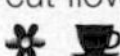

52 ◆ THE RIDGES
Weavers Brow (cont. of Cowling Rd), Limbrick, Chorley, PR6 9EB. Mr & Mrs J M Barlow, 01257 279981, barbara@barlowridges.co.uk, www.bedbreakfast-gardenvisits.com. *2m SE of Chorley town centre. From M6 J27 & M61 J8. Follow signs for Chorley A6 then signs for Cowling & Rivington. Passing Morrison's up Brook St, mini r'about 2nd exit, Cowling Brow. Pass Spinners Arms on L. Garden on R.* **For NGS: Mon 1 May, Sun 16 July (11-5). Adm £5, chd free. Light refreshments. For other opening times and information, please phone, email or visit garden website.**
3 acres, inc old walled orchard garden, cottage-style herbaceous borders, with perfumed rambling roses and clematis through fruit trees. Arch leads to formal lawn, surrounded by natural woodland, shrub borders and specimen trees with contrasting foliage. Woodland walks and dell. Natural looking stream, wildlife ponds. Walled water feature with Italian influence, and walled herb garden. Classical music played. Some gravel paths and woodland walks not accessible.

GROUP OPENING

53 SEFTON PARK JULY GARDENS
Liverpool, L17 1AS. Sefton Park Allotments Society. *From end of M62 take A5058 Queens Drive S through Allerton to Sefton Park. Circle the park and follow yellow signs. Maps available at all gardens.* **Sun 23 July (12-5). Combined adm £6, chd free. Home-made teas.**

6 CROXTETH GROVE
L8 0RX. Stuart Speeden.

FERN GROVE COMMUNITY GARDEN
L8 0RY. Liverpool City Council.

NEW **10 NORMANTON AVENUE**
L17 4JJ. Colette Howard.

SEFTON PARK ALLOTMENTS
Greenbank Drive, L17 1AS. Sefton Park Allotments Society.

SEFTON VILLA
L8 3SD. Patricia Williams.

A brand new garden under development joins the group this year. Two existing private gardens full of peak summer colour, plus nearly 100 productive allotments, and the urban oasis of a community garden are splendid examples of gardening in the city. Plants and produce for sale. Look out for the Liverpool City's 'Allotment of the Year 2022' at Sefton Park Allotments, along with a developing orchard and children's plot. Children's activities and bee demonstration at Fern Grove Community Garden. Classical music 2-4pm at Sefton Park Allotments.

GROUP OPENING

54 SEFTON PARK SEPTEMBER GARDENS
Sydenham Avenue, Sefton Park, Liverpool, L8 3SD. *From end of M62, take A5058 Queens Drive S through Allerton to Sefton Park & follow yellow signs. Parking roadside and in Sefton Park.* **Sun 10 Sept (12-5). Combined adm £6, chd free. Home-made teas.**

PARKMOUNT
L17 3BP. Jeremy Nicholls.

37 PRINCE ALFRED ROAD
L15 8HH. Jane Hammett.

NEW **8 SYDENHAM AVENUE**
L17 3AX. Dinah Dosser, 07956 740710, Dinah.dossor@talktalk.net.

17 SYDENHAM AVENUE
L17 3AU. Fatima Aabbar-Marshall.

A new garden planted and tended specifically for birds and wildlife joins the group of beautiful private gardens planted for late summer colour and interest. The secret garden in Prince Alfred Road is a hidden sandstone walled space with glorious planting, a lovely greenhouse, summerhouse and bothy. Gorgeous planting at Parkmount and in both Sydenham Avenue gardens.

55 SOUTHLANDS
12 Sandy Lane, Stretford, M32 9DA. Maureen Sawyer & Duncan Watmough, 0161 283 9425, moe@southlands12.com, www.southlands12.com. *3m S of Manchester. Sandy Lane (B5213) is situated off A5181 (A56) ¼m from M60 J7.* **Sun 16 July (12-5.30). Adm £5, chd free. Home-made teas. Cake-away service (take a slice of your favourite cake home). Visits also by arrangement 5 June to 27 Aug for groups of 5 to 20. Larger groups welcome but refreshments are not available for over 20.**
This beautiful multi-award winning garden inc a newly planted Mediterranean terrace, an ornamental garden with stunning herbaceous borders, organic kitchen potager with large greenhouse and tranquil woodland garden. Fabulous container plantings throughout add to its continued appeal as a garden that is 'Absolutely wonderful to walk through' and 'an unforgettable experience' for visitors. Artist's work on display.

Our 2022 donation supports 1,700 Queen's Nurses working in the community in England, Wales, Northern Ireland, the Channel Islands and the Isle of Man

56 NEW 47 THE CRESCENT

Maghull, Liverpool, L31 7BL. Mrs Nikki Fennah. *7m N of Liverpool. End M57/M58, take A59 Ormskirk. L at Aldi onto Liverpool Rd South, then 2nd into The Crescent. From Ormskirk, follow A59 to Maghull, turn R at town hall L at Meadows pub, 4th R.* **Sat 22 July (11-4.30). Adm £3, chd free. Sun 23 July (11-4.30). Combined adm with Maghull Station £4, chd free. Home-made teas.**

A garden with full borders, cleverly split into 'rooms', inc a gravel garden by the summerhouse and wildlife pebble pond, surrounded by heleniums and other perennials, a corner arbour overlooking the fish pond, a cut flower garden inc a secluded seat and small natural pond, greenhouse, courtyard garden and apple tree/passion flower walkway, edged with successional perennial planting.

57 TURTON TOWER KITCHEN GARDEN

Tower Drive, Turton, Bolton, BL7 0HG. Nancy Walsh. *1½m from Edgeworth. Turton Tower is signposted off the B6391 between Bolton & Edgworth.* **Sat 17 June (11-4). Adm £3, chd free. Light refreshments at The Woodland Café.**

The garden is set in the historic Turton Tower grounds and was originally the kitchen garden in Victorian times. A group of enthusiastic volunteers began to restore the heavily overgrown two thirds of an acre garden in 2008. Over the years the volunteers have created a tranquil and much visited garden. The garden today consists of raised vegetable beds and soft fruit areas but most of the garden is divided into smaller feature gardens. Turton Tower's history is reflected in its Tudor and Victorian beds. More contemporary gardens inc The White Garden and Japanese Garden. There are also two long herbaceous borders with a large variety of perennials. An area of the garden is shaded and we have an interesting variety of ferns and shade loving plants in our stumperies. Plants, cards and jam will be for sale. The paths are wheelchair friendly but in some places are moderately sloped.

Green Farm Cottage

GROUP OPENING

58 WARTON GARDENS

Warton, LA5 9PJ. 01524 727770, claire@lavenderandlime.co.uk. *3m J35 M6, 1m N of Carnforth. 5 gardens on Main Street span approx ¾m, 1 garden is just off Main St in Church Hill Ave. Allotments off Back Lane nr school down footpath.* **Sat 17, Sun 18 June (11-5). Combined adm £6, chd free. Home-made teas at 109/111 Main Street.**

BRIAR COTTAGE
LA5 9PT. Mr Bendall.

2 CHURCH HILL AVENUE
LA5 9NU. Mr & Mrs J Street.

9 MAIN STREET
LA5 9NR. Mrs Sarah Baldwin.

80 MAIN STREET
LA5 9PG. Mrs Yvonne Miller.

107 MAIN STREET
LA5 9PJ. Becky Hindley, www.erdabotanicals.co.uk.

111 MAIN STREET
LA5 9PJ. Mr & Mrs J Spendlove, 07939 343287, claire@onezeronine.co.uk. **Visits also by arrangement 19 June to 22 Sept for groups of 6 to 20. Pre-booking essential.**

WARTON ALLOTMENT HOLDERS
LA5 9QU. Mrs Jill Slaughter.

The 6 gardens and allotments are spread across the village and offer a wide variety of planting and design ideas. Our group offers a contemporary garden skilfully planted on limestone pavement, a flower picking garden, a mature cottage style garden with unusual herbaceous planting, trees and productive vegetable/soft fruit section, 21 allotments, a large garden comprised of a series of rooms each with a different atmosphere and 2 new gardens which have been created in the last 2 yrs and are 'work in progress'. We would like to encourage families and will be offering activities at some gardens throughout our openings. Warton is the birthplace of the medieval ancestors of George Washington, the family coat of arms can be seen in St Oswald's Church. The ruins of the Old Rectory (English Heritage) is the oldest surviving building in the village. Ascent of Warton Crag (AONB), provides panoramic views across Morecambe Bay to the Lakeland hills beyond.

59 WEEPING ASH GARDEN

Bents Garden & Home, Warrington Road, Glazebury, WA3 5NS. John Bent, www.bents.co.uk. *15m W of Manchester. Next to Bents Garden & Home, just off the A580 East Lancs Rd at Greyhound r'about nr Leigh. Follow brown 'Garden Centre' signs.* **Sun 12, Sun 19, Sun 26 Feb, Sun 13 Aug, Sun 10 Sept, Sun 29 Oct (11-4). Adm by donation.**

Created by retired nurseryman and photographer John Bent, Weeping Ash is a garden of year-round interest with a beautiful display of early snowdrops. Broad sweeps of colour lend elegance to this stunning garden which is much larger than it initially seems with hidden paths and wooded areas creating a sense of natural growth. Bents Garden & Home offers a choice of dining destinations inc The Fresh Approach Restaurant, Caffe nel Verde and a number of al fresco dining options.

60 WILLOW WOOD HOSPICE GARDENS

Willow Wood Close, Mellor Road, Ashton-Under-Lyne, OL6 6SL. www.willowwood.info. *3mins from J23 M60. Exit onto A6140 towards Ashton-u-Lyne turn R onto Manchester Rd/A635 then slight R onto Park Parade/A635 keep R, stay on A635 turn L onto Mellor Rd L onto Willow Wood Close.* **Sat 17 June (12-5). Adm £4, chd free. Cakes, light snacks, hot & cold beverages, cocktails.**

The sensory gardens, redesigned and maintained by two former National Trust gardeners and a wonderful team of volunteers, comprise a series of formal and informal gardens. These spaces contain lush, romantic planting design, filled with scent, texture, studies in colour combinations and sculptural form, water features, a kitchen garden, wildlife friendly planting and two woodland areas. The Sensory and Tranquil gardens are wheelchair accessible. Please check the Willow Wood website for further information.

GROUP OPENING

61 WOOLTON VILLAGE JUNE GARDENS

Hillside Drive, Woolton, Liverpool, L25 5NR. *6m SE of Liverpool city centre. Gardens are best accessed by car or bus (75, 78, 81, 89). Nearest train stn is Hunts Cross.* **Sun 25 June (12-5). Combined adm £5, chd free. Home-made teas.**

GREEN RIDGES, RUNNYMEDE CLOSE
L25 5JU. Sarah & Michael Beresford.

23 HILLSIDE DRIVE
L25 5NR. Bruce & Fiona Pennie.

MERRICK
L25 5NS. Kerry & Tony Marson.

Three suburban gardens located in the beautiful historic village of Woolton featuring C17 buildings that act as a backdrop to the public space plant displays maintained by Woolton In Bloom, for whom the gardens also open. The gardens feature herbaceous borders, water features, tropical plants, greenhouses, kitchen garden, and garden sculptures.

GROUP OPENING

62 WOOLTON VILLAGE JULY GARDENS

Woolton, Liverpool, L25 8QF. *7m S of Liverpool. Woolton Rd B5171 or Menlove Ave A562 follow signs for Woolton.* **Sun 16 July (12-5). Combined adm £5, chd free. Light refreshments at 16 Layton Close and 89 Oakwood Road.**

16 LAYTON CLOSE
L25 9NJ. Bob Edisbury.

71 MANOR ROAD
L25 8QF. John & Maureen Davies.

89 OAKWOOD ROAD
L26 1XD. Sean Gargan.

6 RONALDSWAY
L26 0XE. Mr Charles Hughes.

Woolton village has won multiple Britain in Bloom and North West in Bloom awards. The gardens are all different reflecting their owners' individuality and style. All within a short walk or drive from each other. Wheelchair access to some gardens.

LEICESTERSHIRE & RUTLAND

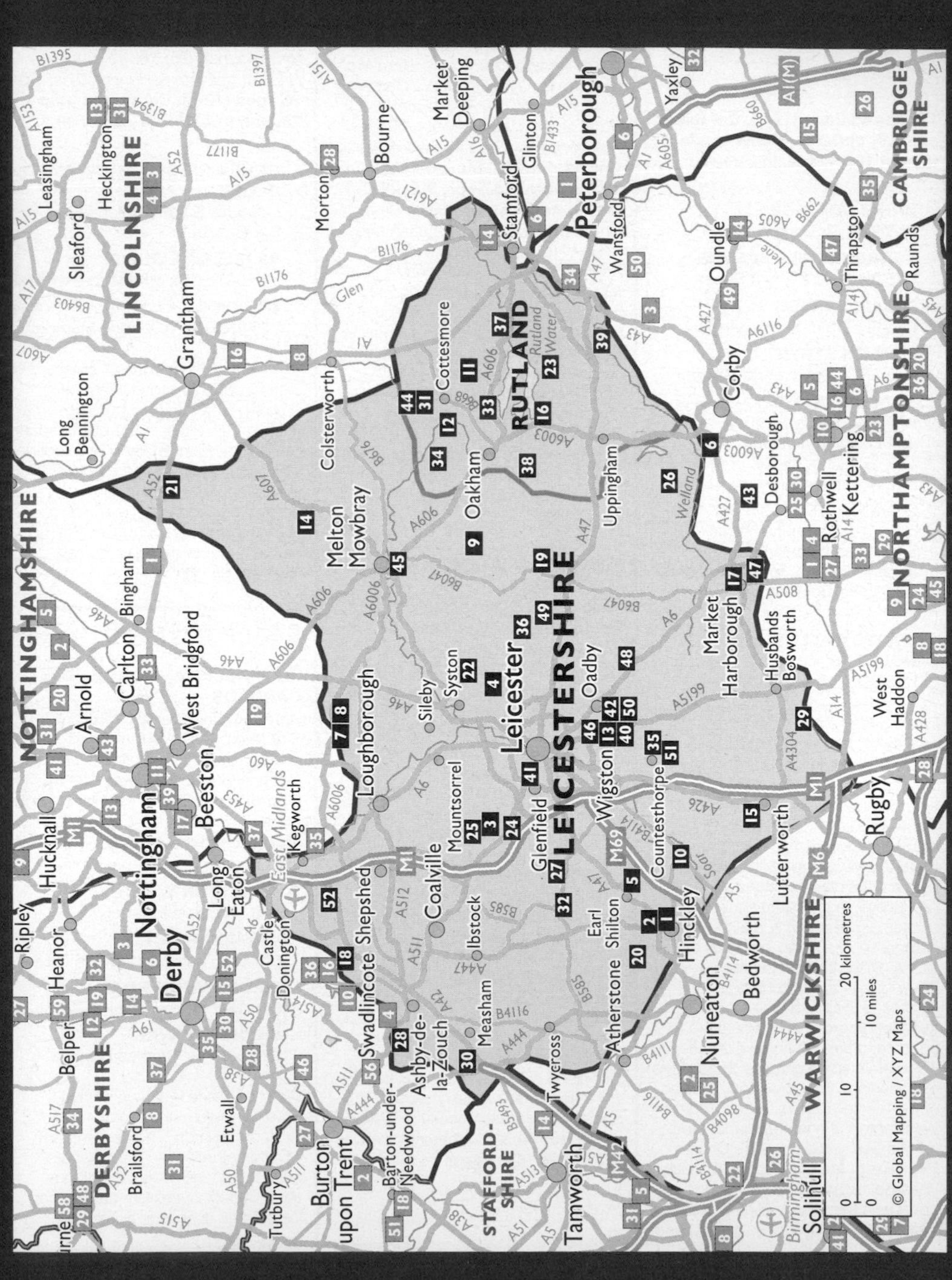

Leicestershire is a landlocked county in the Midlands with a diverse landscape and fascinating heritage providing a range of inspiring city, market town and village gardens.

Our gardens include Victorian terraces that make the most of small spaces and large country houses with historic vistas. We have an arboretum with four champion trees and a city allotment with over 100 plots. We offer something for everyone, from the serious plants person to the casual visitor.

You'll receive a warm welcome at every garden gate. Visit and get ideas for your own garden or to simply enjoy spending time in a beautiful garden. Most gardens sell plants and offer tea and cake, many are happy to take group bookings. We look forward to seeing you soon!

'Much in Little' is Rutland's motto. They say small is beautiful and never were truer words said.

Rutland is rural England at its best. Honey-coloured stone cottages make up pretty villages nestling amongst rolling hills; the passion for horticulture is everywhere you look, from stunning gardens to the hanging baskets and patio boxes showing off seasonal blooms in our two attractive market towns of Oakham and Uppingham.

There's so much to see in and around Rutland whatever the time of year, including many wonderful National Garden Scheme gardens.

Below: St Wolston's House

@NGSLeicestershire

@LeicsNGS

@leicestershire_ngs

@rutlandngs

@rutlandngs

Volunteers

Leicestershire

County Organiser
Pamela Shave 01858 575481
pamela.shave@ngs.org.uk

County Treasurer
Martin Shave 01455 556633
martin.shave@ngs.org.uk

Publicity
Carol Bartlett 01616 261053
carol.bartlett@ngs.org.uk

Booklet Co-ordinator (Leicestershire & Rutland)
Sharon Maher 01162 711680
sharon.maher@ngs.org.uk

Social Media
Zoe Lewin 07810 800 007
zoe.lewin@ngs.org.uk

Talks
Karen Gimson 07930 246974
k.gimson@btinternet.com

Group Visit Co-ordinator
Judith Boston 07740 945332
judith.boston@ngs.org.uk

Assistant County Organisers
Judith Boston (as above)

Janet Rowe 01162 597339
janetnandrew@btinternet.com

Gill Hadland 01162 592170
gillhadland1@gmail.com

Roger Whitmore 07837579672
whitmorerog@outlook.com

Rutland

County Organisers
Sally Killick 07799 064565
sally.killick@ngs.org.uk

Lucy Hurst 07958 534778
lucy.hurst@ngs.org.uk

County Treasurer
Sandra Blaza 01572 770588
sandra.blaza@ngs.org.uk

Publicity
Lucy Hurst (as above)

Social Media
Nicola Oakey 07516 663358
nicola.oakey@ngs.org.uk

OPENING DATES

All entries subject to change. For latest information check **www.ngs.org.uk**

Extended openings are shown at the beginning of the month.

Map locator numbers are shown to the right of each garden name.

February

Snowdrop Festival

Sunday 19th
The Acers 1

Saturday 25th
Hedgehog Hall 19
Westview 48

Sunday 26th
Hedgehog Hall 19
Westview 48

March

Sunday 5th
Tresillian House 45

Sunday 26th
Gunthorpe Hall 16

April

Saturday 1st
Oak Cottage 28

Sunday 2nd
Oak Cottage 28

Sunday 23rd
Quaintree Hall 38

Sunday 30th
Tresillian House 45
Westbrooke House 47

May

Monday 1st
NEW 2 Manor Farm Mews 22

Sunday 7th
Burrough Hall 9

Saturday 13th
Hedgehog Hall 19
8 Hinckley Road 20
Mountain Ash 25
Westview 48

Sunday 14th
Hedgehog Hall 19
8 Hinckley Road 20
Mountain Ash 25
The Old Vicarage, Whissendine 34
Westview 48

Saturday 20th
Goadby Marwood Hall 14

Sunday 21st
Fox Cottage 12

Sunday 28th
The Old Vicarage, Burley 33
Westbrooke House 47

June

Every Wednesday
Stoke Albany House 43

Saturday 3rd
28 Gladstone Street 13
NEW The Old Rectory 32

Sunday 4th
Dairy Cottage 10
28 Gladstone Street 13
Manton Gardens 23

Saturday 10th
88 Brook Street 7
109 Brook Street 8
The Paddocks 36
The Secret Garden at Wigston Framework Knitters Museum 40

Sunday 11th
88 Brook Street 7
109 Brook Street 8
Nevill Holt Hall 26
The Old Hall 31
4 Packman Green 35
The Paddocks 36
NEW St Wolstan's House 42
The Secret Garden at Wigston Framework Knitters Museum 40
15 The Woodcroft 52

Friday 23rd
Brickfield House 6

Saturday 24th
The Old Barn 30
NEW The Secret Garden, Glenfield Hospital 41

Sunday 25th
The Old Barn 30
59 Thistleton Road 44
Tresillian House 45
Westbrooke House 47

Wednesday 28th
The Old Vicarage, Burley 33

July

Every Friday, Saturday and Sunday from Friday 28th
221 Markfield Road 24

Every Wednesday
Stoke Albany House 43

Saturday 1st
8 Hinckley Road 20
Oak Tree House 29

Sunday 2nd
12 Hastings Close 18
8 Hinckley Road 20
Oak Tree House 29

Saturday 8th
Wigston Gardens 50

Sunday 9th
Barkby Hall 4
4 Packman Green 35
Prebendal House 37
Wigston Gardens 50
Willoughby Gardens 51

Saturday 15th
The Secret Garden at Wigston Framework Knitters Museum 40

Sunday 16th
Green Wicket Farm 15
The Secret Garden at Wigston Framework Knitters Museum 40

Wednesday 19th
Green Wicket Farm 15

Saturday 22nd
NEW The Secret Garden, Glenfield Hospital 41

August

Every Sunday
Honeytrees Tropical Garden 21

Every Friday, Saturday and Sunday to Sunday 20th
221 Markfield Road 24

Sunday 6th
182 Ashby Road 2

Saturday 19th
The Old Barn 30

Sunday 20th
The Old Barn 30

Sunday 27th
Tresillian House 45

September

Sunday 3rd
Washbrook Allotments 46

Sunday 10th
Redhill Lodge 39

Sunday 17th
Exton Hall 11
Westview 48

Saturday 23rd
The New Barn 27

Sunday 24th
The New Barn 27

October

Sunday 8th
Hammond Arboretum 17

Sunday 29th
Tresillian House 45

Our 2022 donation to ABF The Soldier's Charity meant 372 soldiers and veterans were helped through horticultural therapy

By Arrangement

Arrange a personalised garden visit with your club, or group of friends, on a date to suit you. See individual garden entries for full details.

THE GARDENS

1 THE ACERS

10 The Hills, Hinckley, LE10 1NA. Mr Dave Baggott. *Off B4668 out of Hinckley. Turn into Dean Rd then 1st on the R.* **Sun 19 Feb (10.30-4). Adm £5, chd free. Home-made teas.**

Medium sized garden, with a Japanese theme inc a zen garden, Japanese tea house, koi pond, more than 20 different varieties of acers, many choice alpines, trilliums, cyclamen, erythroniums, cornus, hamamelis and dwarf conifers. Over 150 different varieties of Snowdrops in spring. Large greenhouse.

2 182 ASHBY ROAD

Hinckley, LE10 1SW. Ms Lynda Blower. *Hinckley is SW Leicestershire near to the border with Warwickshire, with access off the M69 or A5. From Hinckley town centre on the B4667 on R 400yds before the A47/A447 jnc. Parking in the road nearby.* **Sun 6 Aug (11-5). Adm £3.50, chd free. Light refreshments.**

120ft cottage style garden packed with perennials and a small pond to attract frogs and other wildlife. A bog garden area complete with irrigation system. Kitchen garden with a selection of seasonal vegetables. Greenhouse and potting shed and plenty of seating to sit and view different aspects of the garden.

3 BANK COTTAGE

90 Main Street, Newtown Linford, Leicester, LE6 0AF. Jan Croft, 01530 244865/07429 159910, gardening91@icloud.com. *6m NW Leicester. 2½m from M1 J22. From Leicester on L after Markfield Ln. Just before a public footpath sign if coming via Markfield Ln or Anstey or on the R just after public footpath sign if coming via Warren Hill.* **Visits by arrangement Mar to Aug. Adm £3, chd free. Light refreshments by arrangement.**

Traditional cottage garden set on different levels leading down to the River Lin. Providing colour all yr-round but at its prettiest in the Spring. Aconites, snowdrops, blue and white bells, primroses, alliums, aquilegia, poppies, geraniums, roses, honeysuckle, philadelphia, acers, wisteria, autumn crocus, perennial sweet pea, perennial sunflowers, some fruit trees, small pond. Full of wildlife.

4 BARKBY HALL

Barkby, LE7 3QB. Mr & Mrs A J Pochin. *Entrance is by the lodge on the Queniborough Rd, postcode LE7 3QJ. From Barkby village follow long brick wall until a white cottage, turn in & proceed down drive.* **Sun 9 July (11-6). Adm £5, chd free. Light refreshments.**

The current garden is going through an extensive restoration and renovation, with formal gardens of circa 8 acres. Areas of interest inc the rose garden, fragrant garden, white garden, Japanese garden, Camellia houses, 1 acre walled kitchen garden under production. Mixture of herbaceous borders, azalea areas, woodland walk planted with rhododendrons, orchard, formal lawns, parkland. Large areas are accessible by wheelchair along gravel paths. Some areas have inclines, and other areas have raised viewing points up steps.

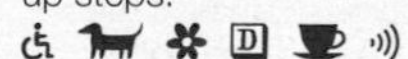

5 BARRACCA

Ivydene Close, Earl Shilton, LE9 7NR. Mr John & Mrs Sue Osborn, 01455 842609, susan.osborn1@btinternet.com, www.barraccagardens.co.uk. *10m W of Leicester. From A47 after entering Earl Shilton, Ivydene Close is 4th on L from Leicester side of A47.* **Visits by arrangement 20 Feb to 31 July for groups of up to 50. Home-made cakes & tea inc. Adm £8, chd free.**

1 acre garden with lots of different areas, silver birch walk, wildlife pond with seating, apple tree garden, Mediterranean planted area and lawns surrounded with herbaceous plants and shrubs. Patio area with climbing roses and wisteria. There is also a utility garden with greenhouses, vegetables in beds, herbs and perennial flower beds, lawn and fruit cage. Part of the old gardens owned by the Cotton family who used to open approx 9 acres to the public in the 1920's. Partial wheelchair access.

6 BRICKFIELD HOUSE

Rockingham Road, Cottingham, Market Harborough, LE16 8XS. Simon & Nicki Harker. *If using SatNav, postcode will take you into Cottingham Village but garden is ½m from Cottingham on Rockingham Rd/B670 towards Rockingham.* **Fri 23 June (12-4). Adm £5, chd free. Home-made teas.**

Lovely views overlooking the Welland Valley, this 2 acre garden has been developed from a sloping brickyard rubbish plot over 35 yrs. There is a kitchen garden, orchard, herbaceous borders filled with roses, perennials and shrubs, paths and many pots, filled with succulents, herbs and colourful annuals.

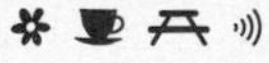

Our 2022 donation to Hospice UK meant that 911 NHS and Social Care staff accessed individual counselling support

7 88 BROOK STREET

Wymeswold, LE12 6TU. Adrian & Ita Cooke, 01509 880155, itacooke@btinternet.com. *4m NE of Loughborough. From A6006 Wymeswold turn S by church onto Stockwell, then E along Brook St. Roadside parking on Brook St.* **Sat 10, Sun 11 June (2-5). Combined adm with 109 Brook Street £6, chd free. Home-made teas at 109 Brook Street. Visits also by arrangement in June for groups of 10 to 40.**

The garden is set on a hillside, which provides lovely views across the village, and comprises three distinct areas: firstly, a cottage style garden; then a series of water features inc a stream and a 'champagne' pond; and finally at the top there is a vegetable plot, small orchard and wildflower meadow. The ponds attract great crested and common newts, frogs, toads and grass snakes.

8 109 BROOK STREET

Wymeswold, LE12 6TT. Maggie & Steve Johnson, 07973 692931, steve@brookend.org, www.brookend.org. *4m NE of Loughborough. From A6006 Wymeswold turn S onto Stockwell, then E along Brook St. Roadside parking along Brook St. Steep drive with limited disabled parking at house.* **Sat 10, Sun 11 June (2-5). Combined adm with 88 Brook Street £6, chd free. Home-made teas inc gluten free option. Visits also by arrangement in June for groups of 10+.**

South facing, 3/4 acre, gently sloping garden with views to open country. Modern garden with mature features. Patio with roses and clematis, wildlife and fish ponds, mixed borders, vegetable garden, orchard, hot garden and woodland garden. Something for everyone. Optional tour of rain water harvesting. Some gravel paths. Drive to top of drive and ask for assistance.

9 BURROUGH HALL

Somerby Road, Burrough on the Hill, Melton Mowbray, LE14 2QZ. Richard & Alice Cunningham. *Somerby Rd, Burrough on the Hill. Close to B6047. 10 mins from A606. 20 mins from Melton Mowbray.* **Sun 7 May (2-5). Adm £5, chd free. Home-made teas.**

Burrough Hall was built in 1867 as a classic Leicestershire hunting lodge. The garden, framed by mature trees and shrubs, was extensively redesigned by garden designer George Carter in 2007. The garden continues to develop. This family garden designed for all generations to enjoy is surrounded by magnificent views across High Leicestershire. In addition to the garden there will be a small collection of vintage and classic cars on display. Gravel paths and lawn.

10 DAIRY COTTAGE

15 Sharnford Road, Sapcote, LE9 4JN. Mrs Norah Robinson-Smith, 01455 272398, nrobinsons@yahoo.co.uk. *9m SW of Leicester. Sharnford Rd joins Leicester Rd in Sapcote to B4114 Coventry Rd. Follow NGS signs at both ends.* **Sun 4 June (11-4.30). Adm £4, chd free. Home-made teas. £2.00. Visits also by arrangement 7 May to 30 July for groups of 10 to 30.**

From a walled garden with colourful mixed borders to a potager approached along a woodland path, this mature cottage garden combines extensive perennial planting with many unusual shrubs and specimen trees. With colourful herbaceous borders and unusual small trees and shrubs this quintessential cottage garden is divided into several rooms. May need to access over a gravel drive.

11 EXTON HALL

Cottesmore Road, Exton, LE15 8AN. Viscount & Viscountess Campden, www.extonpark.co.uk. *Exton, Rutland. 5m E of Oakham. 8m from Stamford off A1 (A606 turning).* **Sun 17 Sept (2-5). Adm £5, chd free. Light refreshments.**

Extensive park, lawns, specimen trees and shrubs, lake, private chapel and C19 house (not open). Pinetum, woodland walks, lakes, ruins, dovecote and formal herbaceous garden. Coffee trailer. Whilst there is wheelchair access, areas of the garden are accessible along grass or gravel paths which, weather dependent, may make access difficult.

12 FOX COTTAGE

Woodside, Ashwell, Oakham, LE15 7LX. Mr & Mrs D Pettifer. *No access via Woodside. Follow signs for parking.* **Sun 21 May (11.30-4.30). Adm £5, chd free. Light refreshments in the garden with live music.**

A 2 acre country garden on the edge of the village with sweeping views across the surrounding open fields. Walks through the wonderful ancient oaks and ash underplanted with spring bulbs and lawns surrounded by rampant cow parsley. South facing terrace with seasonal planting.

13 28 GLADSTONE STREET

Wigston Magna, LE18 1AE. Chris & Janet Huscroft. *4m S of Leicester. Off Wigston by-pass (A5199) follow signs off McDonalds r'about.* **Sat 3, Sun 4 June (11-5). Adm £3.50, chd free. Home-made teas. Opening with Wigston Gardens on Sat 8, Sun 9 July.**

Our mature 70'x15' town garden is divided into rooms and bisected by a pond with a bridge. It is brimming with unusual hardy perennials, inc collections of ferns and hostas. David Austin roses chosen for their scent feature throughout, inc a 30' rose arch. A shade house with unusual hardy plants and a Hosta Theatre. Regular changes to planting. Wigston Framework Knitters Museum and garden nearby - open Sunday afternoons. Parts of the garden can be viewed, narrow paths and step limit full access.

14 GOADBY MARWOOD HALL

Goadby Marwood, Melton Mowbray, LE14 4LN. Mr & Mrs Westropp, 01664 464202, vwestropp@gmail.com. *4m NW of Melton Mowbray. Between Waltham-on-the-Wolds & Eastwell, 8m S of Grantham. Plenty of parking.* **Sat 20 May (10.30-5.30). Adm £5, chd £1. Visits also by arrangement Apr to Oct.**

Redesigned in 2000 by the owner based on C18 plans. A chain of five lakes (covering 10 acres) and several ironstone walled gardens all interconnected. Lakeside woodland walk. Planting for yr round interest. Landscaper trained under plantswoman Rosemary Verey at Barnsley House. Beautiful C13 church open. Water, walled gardens. Gravel paths and lawns.

15 GREEN WICKET FARM

Ullesthorpe Road, Bitteswell, Lutterworth, LE17 4LR. Mrs Anna Smith, 01455 552646, greenfarmbitt@hotmail.com. *2m NW of Lutterworth J20 M1. From Lutterworth follow signs through Bitteswell towards Ullesthorpe. Garden situated behind Bitteswell Cricket Club. Use this as a landmark rather than relying totally on SatNav.* **Sun 16, Wed 19 July (2-5). Adm £5, chd free. Light refreshments. Visits also by arrangement 26 June to 23 Sept for groups of 10 to 30.**

A fairly formal garden on an exposed site surrounded by open fields. Mature trees enclosing the many varied plants chosen to give all round year interest. Many unusual hardy plants grown along side good reliable old favourites present a range of colour themed borders. Grass and gravel paths allow access to the whole garden. Disabled parking areas.

16 GUNTHORPE HALL

Gunthorpe, Oakham, LE15 8BE. Tim Haywood, 01572 737514, ask for lettings. *A6003 between Oakham & Uppingham; 1½ m S of Oakham, up drive between gate lodges.* **Sun 26 Mar (2-5). Adm £5, chd free. Light refreshments.**

Large garden in a country setting with extensive views across the Rutland landscape with the carpets of daffodils being a key seasonal feature. The Stable Yard has been transformed into a series of parterres. The kitchen garden and borders have all been rejuvenated over recent years. Gravel path allows steps to be avoided.

17 HAMMOND ARBORETUM

Burnmill Road, Market Harborough, LE16 7JG. The Robert Smyth Academy, www.hammondarboretum.org.uk. *15m S of Leicester on A6. From High St, follow signs to The Robert Smyth Academy via Bowden Ln to Burnmill Rd. Park in 1st entrance on L.* **Sun 8 Oct (2-4.30). Adm £6, chd free. Cream teas. Provided by the Academy Parents Association.**

A site of just under 2½ acres containing an unusual collection of trees and shrubs, many from Francis Hammond's original planting dating from 1913 to 1936 whilst headmaster of the school. Species from America, China and Japan with malus and philadelphus walks and a moat. Proud owners of three champion trees identified by national specialist and 37 which are the best in Leicestershire. Walk plans available. Some steep slopes.

18 12 HASTINGS CLOSE

Breedon-On-The-Hill, Derby, DE73 8BN. Mr & Mrs P Winship. *5m N from Ashby de la Zouch. Follow NGS signs in the Village. Parking around the Village Green. Please do not park in the close due to limited parking.* **Sun 2 July (1-5). Adm £3.50, chd free. Light refreshments.**

A medium sized prairie garden, organically managed and planted in the Piet Oudolf style. Also many roses and a wide range of perennials. The back garden has a small 'white' garden with box hedging and more colourful style perennial borders. The garden has narrow paths and steps so is not suitable for wheelchairs.

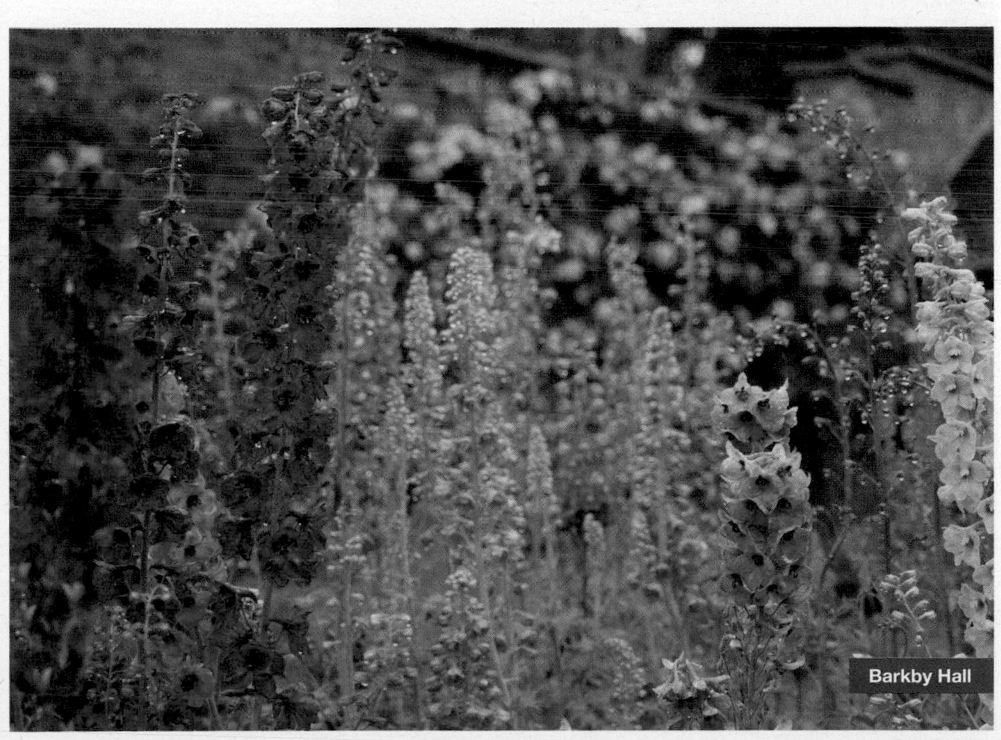

Barkby Hall

19 HEDGEHOG HALL

Loddington Road, Tilton on the Hill, LE7 9DE. Janet & Andrew Rowe. *8m W of Oakham. 2m N of A47 on B6047 between Melton & Market Harborough. Follow yellow NGS signs in Tilton towards Loddington. Disabled parking on road outside the White House 20 yds past our entrance.* **Sat 25, Sun 26 Feb (11-4). Sat 13, Sun 14 May (11-4). Light refreshments. Adm £5, chd free. Open nearby Westview.**

½ acre organically managed plant lover's garden. Steps leading to three stone walled terraced borders filled with shrubs, perennials, bulbs and a patio over looking the valley. Lavender walk, herb border, beautiful spring garden, colour themed herbaceous borders. Courtyard with collection of hostas and acers and terrace planted for yr-round interest with topiary and perennials. Snowdrop collection. Regret, no wheelchair access to terraced borders.

20 8 HINCKLEY ROAD

Stoke Golding, Nuneaton, CV13 6DU. John & Stephanie Fraser. *3m NW of Hinckley. Approach from any direction into village then follow NGS signs. Please park roadside with due consideration to other residents properties.* **Sat 13 May (12-4); Sun 14 May (12-3); Sat 1 July (12-4); Sun 2 July (12-3). Adm £3.50, chd free. Light refreshments.**

A small SSW garden with water features to add interest to the colourful and some unusual perennials inc climbers to supplement the trees and shrubs in May and July. Garden established and recently re-established by current owner to provide seating for a variety of views of the garden. Many perennials in the garden are represented in the plants for sale.

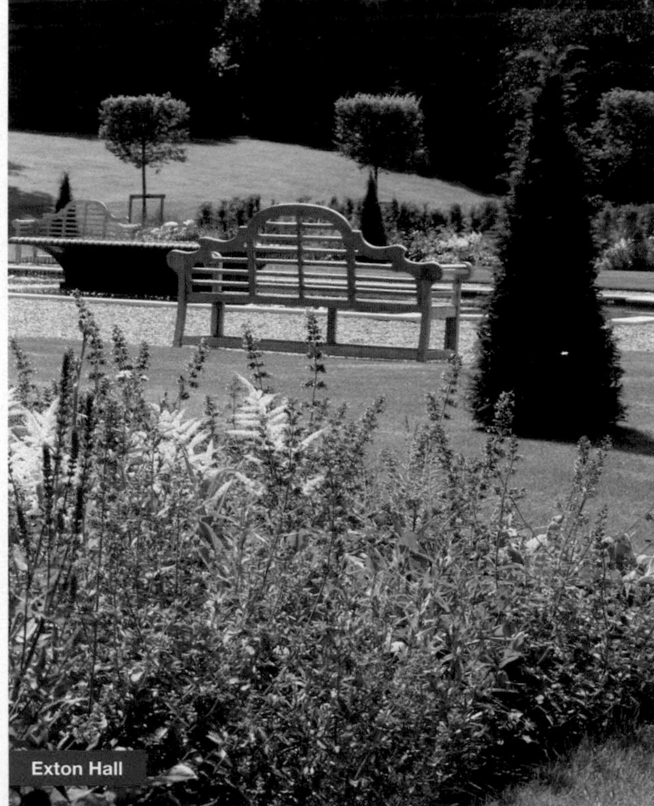

Exton Hall

21 HONEYTREES TROPICAL GARDEN

85 Grantham Road, Bottesford, NG13 0EG. Julia Madgwick & Mike Ford, 01949 842120, Julia_madgwick@hotmail.com, www.facebook.com/HoneytreesTropicalGarden. *Bottesford, Leicestershire. 7m E of Bingham on A52. Turn into village. Garden is on L on slip road behind hedge going out of village towards Grantham. Parking on grass opposite property.* **Every Sun 6 Aug to 27 Aug (11-4). Adm £4, chd free. Home-made teas. Vegan & gluten free cakes. Visits also by arrangement July & Aug for groups of 10+.**

Tropical and exotic with a hint of jungle! Raised borders with different themes from lush foliage to arid cacti. Exotic planting as you enter the garden gives way on a gentle incline to surprises, inc glasshouses dedicated to various climatic zones interspersed with more exotic planting, ponds and a stream. Representation of over 20 yrs plant hunting. Treehouse and viewing platform to view tree ferns from above whilst being among the canopy of the trees. There are some steps and ramps. Gravel and bark in certain areas but wheelchair access to most parts of the garden.

22 NEW 2 MANOR FARM MEWS

Main Street, Queniborough, Leicester, LE7 3EA. Mrs Jo Dolan. *7½m NE of Leicester. Off A46 follow signs for Queniborough. Entrance opp St Mary's Church on Main Street. No parking in Manor Farm Mews.* **Mon 1 May (11-4). Adm £5, chd free. Light refreshments.**

A modern, contemporary formal garden of approx ¼ acre that rises gently up a shallow slope. Large overflowing borders filled with a variety of bird, bee and butterfly friendly planting, surround formal lawns, topiary, hedging and shady retreats inc a small woodland and a summerhouse. There are collections of roses, salvias, agapanthus and a wide variety of year round bulbs.

GROUP OPENING

23 MANTON GARDENS

Oakham, LE15 8SR. *3m N of Uppingham & 3m S of Oakham. Manton is on S shore of Rutland Water ¼m off A6003. Please park carefully in village.* **Sun 4 June (1-5). Combined adm £7, chd free. Home-made teas in the Village Hall.**

22 LYNDON ROAD
Chris & Val Carroll.

MANTON GRANGE
Anne & Mark Taylor.

NEW **MANTON HOUSE**
Elisabeth & Robert Braddock.

MANTON LODGE
Emma Burnaby-Atkins, 01572 737258, info@mantonlodge.co.uk, www.mantonlodge.co.uk.

SHAPINSAY
Tony & Jane Bews.

NEW **WALNUT HOUSE**
Alison Crossley.

6 gardens in small village on S shore of Rutland Water. Manton Grange - 2½ acre garden with interesting trees, shrubs and herbaceous borders, inc a rose garden, water features, a lime tree walk and clematis pergola. 22 Lyndon Rd - a beautiful combination of cottage garden and unusual plants in overflowing borders, hanging baskets and decorative pots. Manton Lodge - steeply sloping garden with wonderful views and colourful beds of shrubs, roses and perennials, with ornamental pond and terrace. Shapinsay - ⅔ acre garden with mature trees and framed views, inc woodland walk, perennial borders, island shrub borders and stream linking numerous ponds. Manton House - walled garden with south facing borders, octagonal garden and kitchen garden. 20 St. Mary's Road - small, beautifully designed garden with roses and box hedging. Wheelchair access to Manton Grange only.

24 221 MARKFIELD ROAD

Groby, Leicester, LE6 0FT. Jackie Manship. *From M1 J22 take A50 towards Leicester. In approx 3m at the T-lights junction with Lena Drive turn L. Parking available along this road, no parking on A50.* **Every Fri, Sat and Sun 28 July to 20 Aug (10-4). Adm £5, chd free. Light refreshments.**
A south facing plot of land nestled between the village of Groby and Markfield approx 1 acre in size. The hidden treasures are deceptive from the front of the property which sits on one of the main trunk roads out of Leicester. Packed with interest and created over the last 20 yrs from a dishevelled overgrown plot you will be presented with a garden full of delight.

25 MOUNTAIN ASH

140 Ulverscroft Lane, Newtown Linford, LE6 0AJ. Mike & Liz Newcombe, 01530 242178, mjnew12@gmail.com. *7m SW of Loughborough, 7m NW of Leicester, 1m NW of Newtown Linford. Head ½m N along Main St towards Sharpley Hill, fork L into Ulverscroft Ln & Mountain Ash, is about ½m along on the L. Parking is along the opp verge.* **Sat 13, Sun 14 May (11-5). Adm £6, chd free. Home-made teas. Visits also by arrangement May to Sept for groups of 15 to 50.**
2 acre garden with stunning views across Charnwood countryside. Near the house are patios, lawns, water feature, flower and shrub beds, fruit trees, soft fruit cage, greenhouses and vegetable plots. Lawns slope down to a gravel garden, large wildlife pond and small areas of woodland with walks through many species of trees. Over 50 garden statues and ornaments. Many places to sit and relax. On the May open days we have stalls by Wildwater Ponds offering sales and advice, and others selling quality plants from a local nursery, gift stationery, vintage tools and books. Only the top part of the garden around the house is reasonably accessible by wheelchair.

26 NEVILL HOLT HALL

Drayton Road, Nevill Holt, Market Harborough, LE16 8EG. Mr David Ross. *5m NE of Market Haborough. Signed off B664 at Medbourne.* **Sun 11 June (12-4). Adm £6, chd free. Home-made teas in restaurant marquee, provided & served by Nevill Holt Opera. Donation to another charity.**
As well as being a private home, the C13 estate of Nevill Holt Hall is also home to Nevill Holt Opera. In addition to having an award winning theatre, the grounds also boast 3 walled gardens and an impressive modern sculpture collection. When we open in June this year the opera festival will be in full swing - the gardens climaxing - the perfect backdrop to Nevill Holts' country summerhouse opera. Nevill Holt Opera Event partner with the RHS.

27 THE NEW BARN

Newbold Road, Desford, Leicester, LE9 9GS. A Nichols & R Pullin. *8m from J21 M1 & 12m from J22 M1.* **Sat 23, Sun 24 Sept (11-4). Adm £5, chd free. Light refreshments. Donation to Plant Heritage.** A large, sloping garden of ¾ acre with wonderful views. Enter via a side gate and follow a narrow path opening up into what feels like a large secret garden, different levels and spaces add to this illusion enabling you to admire the planting and views. This is a well-established plantsman's garden with many unusual plants, inc a collection of asters, Benton irises and geranium phaeum.

NPC

28 OAK COTTAGE

Well Lane, Blackfordby, Swadlincote, DE11 8AG. Colin & Jenny Carr. *Blackfordby, just over 1m from (& between) Ashby-de-la-Zouch or Swadlincote. From Ashby-de-la-Zouch take Moira Rd, turn R on Blackfordby Ln. As you enter Blackfordby, turn L to Butt Ln & quickly R to Strawberry Ln. Park then it's a 2 min walk to Well Ln entrance.* **Sat 1, Sun 2 Apr (10-4). Adm £4.50, chd free. Light refreshments.**
½ acre garden set around Blackfordby's "hidden" listed thatched cottage, which itself is more than 300 years old. In total there are 3.4 acres of paddocks, front and rear gardens to explore with extensive displays of hellebores, snakes heads and mature magnolias in Spring. The lower paddock has been planted with 450 native trees as part of the National Forest Freewoods scheme, with a drainage pond created at its base. The central swathe is being developed with wildflowers. At the top of the rear garden there is a chicken run, old and new orchards and a peach house.

29 OAK TREE HOUSE

North Road, South Kilworth, LE17 6DU. Pam & Martin Shave. *15m S of Leicester. From M1 J20, take A4304 towards Market Harborough. At North Kilworth turn R, signed South Kilworth. Garden on L after approx 1m.* **Sat 1 July (11-4.30); Sun 2 July (11-2.30). Adm £5, chd free. Home-made teas.**

⅔ acre beautiful country garden full of colour, formal design, softened by cottage style planting. Modern sculptures. Large herbaceous borders, vegetable plots, pond, greenhouse, shady area, colour-themed borders. Extensive collections in pots, home to everything from alpines to trees. Trees with attractive bark. Many clematis and roses. Dramatic arched pergola. Constantly changing garden. Access to patio and greenhouse via steps.

30 THE OLD BARN

Rectory Lane, Stretton-En-Le-Field, Swadlincote, DE12 8AF. Gregg & Claire Mayles, 07870 160318, greggmayles@gmail.com. *On A444, 1½m from M42/A42 J11. Rectory Ln is a concealed turn off the A444, surrounded by trees. Look out for brown 'church' signs. Postcode in SatNav usually helps.* **Sat 24, Sun 25 June, Sat 19, Sun 20 Aug (11-5). Adm £5, chd free. Cream teas made by local WI. Visits also by arrangement May to Aug for groups of 10+.**

2 acres in leafy hamlet, lots of interest. Main garden has lawns, many colourful shrubs and tree-lined cobbled paths. Walled garden with fishpond, pergola with climbers and cottage garden planting. Rewilding meadow with paths and open views. Always new things to find! Lots of wildlife, inc our Peafowl. Plenty of drinks, cakes and seats to enjoy them! Redundant old church close, open to visitors. Craft stalls and a treasure hunt around the garden for children. Main garden fully wheelchair accessible. Walled garden partial access due to gravel. Meadow is accessible, but uneven.

31 THE OLD HALL

Main Street, Market Overton, LE15 7PL. Mr & Mrs Timothy Hart. *6m N of Oakham. 6m N of Oakham; 5m from A1via Thistleton. 10m E from Melton Mowbray.* **Sun 11 June (2-6). Adm £6, chd free. Home-made teas with Hambleton Bakery cakes.**

Set on a southerly ridge. Stone walls and yew hedges divide the garden into enclosed areas with herbaceous borders, shrubs, and young and mature trees. There are interesting plants flowering most of the time. In 2020 a Japanese Tea House was added at the bottom of the garden. Partial wheelchair access. Gravel and mown paths. Return to house is steep. It is, however, possible to just sit on the terrace.

32 NEW THE OLD RECTORY

Main Street, Newbold Verdon, Leicester, LE9 9NN. Mr & Mrs Gianni and Kate De Fraja. *10m W of Leicester. On the village Main Street, just before the turning for the church. Coordinates to main entrance: 52.6293, -1.34424. Bus stop 'Old White Swan' on lines 152 153 159.* **Sat 3 June (11-4). Adm £7, chd free. Home-made teas.**

The garden surrounds the Georgian Old Rectory. It contains many species of trees, ranging in age from centuries-old to very young saplings. In early June, the rose garden is about to flower, and so the garden's principal attractions are the rhododendrons, the wisteria, the hostas, and a sea of cow parsley in the woods around the gravel path. Recent additions are a greenhouse and a box parterre. The garden is accessed via a gravel drive. Wheelchair access is possible with some attention.

33 THE OLD VICARAGE, BURLEY

Church Road, Burley, Oakham, LE15 7SU. Jonathan & Sandra Blaza, 01572 770588, sandra.blaza@googlemail.com, www.theoldvicarageburley.com. *1m NE of Oakham. In Burley just off B668 between Oakham & Cottesmore. Church Rd is opp village green.* **Sun 28 May (11-5). Home-made teas. Evening opening Wed 28 June (5-8.30). Wine. Adm £6, chd free. Visits also by arrangement 25 May to 25 June for groups of 20+. Coaches welcome.**

The Old Vicarage is a relaxed country garden, planted for year-round interest and colour. There are lawns and borders, a lime walk, rose gardens and a sunken rill garden with an avenue of standard wisteria. The walled garden produces fruit, herbs, vegetables and cut flowers. There are two orchards, an acer garden and areas planted for wildlife inc woodland, a meadow and a pond. Some gravel and steps between terraces.

34 THE OLD VICARAGE, WHISSENDINE

2 Station Road, Whissendine, LE15 7HG. Prof Peter & Dr Sarah Furness, www.rutlandlordlieutenant.org/garden. *Up hill from St Andrew's church, first L in Station Rd.* **Sun 14 May (1.30-4.30). Adm £6, chd free. Home-made teas in St Andrew's Church, Whissendine (next door!).**

⅔ acre packed with variety. Terrace with topiary, a formal fountain courtyard and raised beds backed by gothic orangery. Herbaceous borders surround main lawn. Wisteria tunnel leads to raised vegetable beds and large ornate greenhouse, four beehives, Gothic hen house plus rare breed hens. Hidden white walk, unusual plants. New Victorian style garden room. The gravel drive is hard work for wheelchair users. Some areas are accessible only by steps.

35 4 PACKMAN GREEN

Countesthorpe, Leicester, LE8 5WS. Roger Whitmore & Shirley Jackson. *5m S of Leicester. Garden is close to village centre pass bank of shops off Scotland way.* **Sun 11 June, Sun 9 July (11-4). Adm £2, chd free. Home-made teas.**

A small town house garden packed with hardy perennials surrounded by climbing roses and clematis. A pond and rose arch complete the picture to make it an enclosed peaceful haven filled with colour and scent. Lots of inexpensive plants for sale besides a refreshing cuppa and slice of cake.

36 THE PADDOCKS

Main Street, Hungarton, LE7 9JY. John & Helen Galyer, 01162 595230, Michael.c.martin@talk21.com. *8m E of Leicester. Follow NGS signs in village.* **Sat 10, Sun 11 June (11-5). Adm £5, chd free. Home-made teas. Visits also by arrangement for groups of 20+.**

2 acre garden with mature and specimen trees, rhododendrons, azaleas, magnolia grandiflora, wisterias. Two lily ponds and stream.

Three lawn areas surrounded by herbaceous and shrub borders. Woodland walk. Pergola with clematis and roses, hosta collection and two rockeries, fern bed. Large well established semi permanent plant stall in aid of local charities. Huge magnolia grandiflora and hydrangea petiolaris. Rhododendrum spinney. Partial wheelchair access due to steep slopes at rear of garden. Flat terrace by main lawn provides good viewing area.

37 PREBENDAL HOUSE

Crocket Lane, Empingham, LE15 8PW. Matthew & Rebecca Eatough. *5m E of Oakham. Facing the church on Church St, we are through the large gates on R.* **Sun 9 July (1.30-5). Adm £6, chd free. Home-made teas.**

There has been a house on the site since the C11. The present house was built in 1688 owned by the diocese of Lincoln until the mid C19 and then absorbed into the Normanton Park Estate. The present owners have been in residence since 2016. The garden is a combination of open parklands, yew tree walks, herbaceous borders and a large formal C18 walled garden. The majority of the garden is accessible by wheelchair.

38 QUAINTREE HALL

Braunston, LE15 8QS.
Mrs Caroline Lomas. *Braunston, nr Oakham. 2m W of Oakham, in the centre of the village of Braunston in Rutland on High St.* **Sun 23 Apr (12-4.30). Adm £5, chd free. Home-made teas.**

An established garden surrounding the medieval hall house (not open) inc a formal box parterre to the front of the house, a woodland walk, formal walled garden with yew hedges, a small picking garden and terraced courtyard garden with conservatory. A wide selection of interesting plants can be enjoyed, each carefully selected for its specific site by the knowledgeable garden owner. The gravel drive can cause difficulty for wheelchairs.

39 REDHILL LODGE

Seaton Road, Barrowden, Oakham, LE15 8EN. Richard & Susan Moffitt, 07894 064789, www.m360design.co.uk. *Redhill Lodge is 1m from village of Barrowden along Seaton Rd.* **Sun 10 Sept (12-5.30). Adm £6, chd free. Home-made teas. Visits also by arrangement 2 Apr to 31 Oct.**

Bold contemporary design with formal lawns, grass amphitheatre and turf viewing mound, herbaceous borders and rose garden. Prairie style planting showing vibrant colour in late summer. Also natural swimming pond surrounded by Japanese style planting, bog garden and fernery. New for 2022 is a Japanese style stroll garden.

42 NEW ST WOLSTAN'S HOUSE

Church Nook, Wigston Magna, LE18 3RA. Mr Kevin De-Voy & Mr Stephen Walker. *On corner of Church Nook & Bull Head St opp St Wistan's church. No parking at garden. Public car parks, all within 5 min walk on Frederick St, Junction Rd & Paddock St.* **Sun 11 June (12-5). Adm £5, chd free. Home-made teas.**

Approximately ½ acre garden divided into a series of garden rooms, with formal and informal planting and featuring specimen Sequoia sempervirens and Cedrus libani. Inc formal white garden, upper garden, rose garden, sunken Italian garden, lawn, mixed border, paths with rose and wisteria pergola and laburnum arch, Edwardian conservatory, Tuscan garden and well garden with raised beds.

2 Manor Farm Mews

40 THE SECRET GARDEN AT WIGSTON FRAMEWORK KNITTERS MUSEUM

42-44 Bushloe End, Wigston, LE18 2BA. 07814 042889, chris.huscroft@tiscali.co.uk, www.wigstonframeworkknitters.org.uk. *4m S of Leicester. On A5199 Wigston bypass, follow yellow signs to Paddock St public car park. yellow signs onto Long St to All Saints Church turn R onto Bushloe End, Museum on R.* **Sat 10, Sun 11 June, Sat 15, Sun 16 July (11-5). Adm £3.50, chd free. Home-made teas. Visits also by arrangement 12 June to 18 Aug weekdays only for groups of up to 20.**

Victorian walled garden approx 70'x80' with traditional cottage garden planting, referred to as Higgledy Piggledy because of the random planting. Managed by a group of volunteers with much replanting during past 2 years. Garden located in the grounds of a historic museum, (an extra charge applies). A unique garden in the centre of Wigston which still retains an air of tranquillity. Cobbled area before garden, gravel paths through main parts of the garden.

41 NEW THE SECRET GARDEN, GLENFIELD HOSPITAL

Groby Road, Leicester, LE3 9QP. Karen James. *Upon arrival to the Glenfield Hospital, please follow the blue directional signage.* **Sat 24 June, Sat 22 July (10.30-3.30). Adm £3.50, chd free. Light refreshments at The Secret Cafe Garden.**

Set within the grounds of the Glenfield Hospital, the Secret Garden is 1 acre in size, hidden behind the walls of a Victorian Walled garden which has been lovingly designed and restored for the benefit of all those who visit it and in consideration of the rich history and heritage of the garden and the wider Leicester Frith site. All main pathways are wheelchair accessible.

43 STOKE ALBANY HOUSE

Desborough Road, Stoke Albany, Market Harborough, LE16 8PT. Mr & Mrs A M Vinton, www.stokealbanyhouse.co.uk. *4m E of Market Harborough. Via A427 to Corby, turn to Stoke Albany, R at the White Horse (B669) garden ½m on the L.* **Every Wed 7 June to 26 July (2-4.30). Adm £6, chd free. Donation to Marie Curie Cancer Care.**

4 acre country house garden; fine trees and shrubs with wide herbaceous borders and sweeping striped lawn. Good display of bulbs in spring, roses June and July. Walled grey garden; nepeta walk arched with roses, parterre with box and roses. Mediterranean garden. Heated greenhouse, potager with topiary, water feature garden and sculptures.

44 59 THISTLETON ROAD

Market Overton, Oakham, LE15 7PP. Wg Cdr Andrew Stewart JP, 07876 377397, stewartaj59@gmail.com. *Market overton. 6m NE of Oakham.* **Sun 25 June (12-5). Adm £5, chd free. Home-made teas at local Bowls Club, a short walk away. Visits also by arrangement June & July for groups of 5 to 20.**

Over the last 15 yrs, the owners have transformed 1.8 acres of bare meadow into a wildlife friendly garden full of colour and variety. Against a backdrop of mature trees, the garden inc a small kitchen garden, rose pergola, large pond, shrubbery with 'Onion Day Bed', orchard, wildflower meadow, a small arboretum, a woodland walk and large perennial borders a riot of colour. Small area of shingle to access main path, woodland walk inaccessible to wheelchairs.

45 TRESILLIAN HOUSE

67 Dalby Road, Melton Mowbray, LE13 0BQ. Mrs Alison Blythe, 01664 481997, alisonblythe@tresillianhouse.com, www.tresillianhouse.com. *½m S of Melton Mowbray centre. Situated on B6047 Dalby Rd, S of Melton town centre. (Melton to Gt Dalby/ Market Harborough rd). Parking on site.* **Sun 5 Mar, Sun 30 Apr, Sun 25 June, Sun 27 Aug, Sun 29 Oct (11-4). Adm £5, chd free. Light refreshments. Visits also by arrangement 5 Mar to 30 Sept for groups of up to 30. Parking available up to 20 seater coach or up to 14 cars.**

¾ acre garden re-established by current owner. Beautiful blue cedar trees, specimen tulip tree. Variety of trees, plants and bushes reinstated. Original bog garden and natural pond. Koi pond maturing well; glass garden room holds exhibitions and recitals. Vegetable plot. Cowslips and bulbs in Springtime. Unusual plants, trees and shrubs. Contemporary area added 2020. Quiet and tranquil oasis. Small Art Exhibition hosted by local artists. Wild Water Ponds will be at the garden. They are specialists in pond set ups, maintenance and planting and have replanted the natural pond at Tresillian House. Relax in June and August with cream tea listening to live traditional jazz. Keep warm in October with stew and dumplings or soup. Slate paths, steep in places but manageable.

46 WASHBROOK ALLOTMENTS

Welford Road (opposite Brinsmead Rd), Leicester, LE2 6FP. Sharon Maher. *Approx 2½m S of Leicester, 1½ m N of Wigston. No onsite parking. Welford Rd (opposite Brinsmead Rd) difficult to park on. Please use nearby side roads & Pendlebury Drive (LE2 6GY).* **Sun 3 Sept (11-3). Adm £3.50, chd free. Home-made teas.**

A hidden oasis. There are over 100 whole, half and quarter plots growing a wide variety of fruit, vegetables and flowers. Meadows, Anderson shelters and a composting toilet. You can also purchase fresh produce by donation. Circular route around the site is uneven in places but is suitable for wheelchairs and mobility scooters.

47 WESTBROOKE HOUSE

52 Scotland Road, Little Bowden, Market Harborough, LE16 8AX. Bryan & Joanne Drew. *½m S Market Harborough. From Northampton Rd follow NGS arrows & park in public car park or on nearby roads - not on the road directly opp the entrance. Regret, no disabled parking on property.* **Sun 30 Apr, Sun 28 May, Sun 25 June (10-4.30). Adm £5, chd free. Cream teas.**

Westbrooke House is a late Victorian property built in 1887. The gardens comprise 6 acres in total and are approached through a tree lined driveway of mature limes and giant redwoods. Key features are walled flower garden, walled kitchen garden, fernery, lower garden, wildlife pond, spring garden, lawns, woodland paths and a meadow with a wildflower area, ha-ha and hornbeam avenue.

D

48 WESTVIEW

1 St Thomas's Road, Great Glen, Leicester, LE8 9EH. Gill & John Hadland, 01162 592170, gillhadland1@gmail.com. *7m S of Leicester. Take either r'about from A6 into village centre then follow NGS signs. Please park in Oaks Rd.* **Sat 25, Sun 26 Feb, Sat 13, Sun 14 May (11-4), open nearby Hedgehog Hall. Sun 17 Sept (11-4). Adm £3.50, chd free. Home-made teas as well as home-made soup & rolls served at the Feb openings. Visits also by arrangement Feb to Sept for groups of 10 to 20.**

Organically managed small walled cottage garden with yr round interest. Rare and unusual plants, many grown from seed. Formal box parterre, courtyard garden, alpines, herbaceous borders, woodland areas with unusual ferns, small wildlife pond, greenhouse, vegetables, fruit and herbs. Collection of snowdrops. Recycled materials used to make quirky garden ornaments and water feature. Restored Victorian outhouse functions as a garden office and houses a collection of old garden tools and ephemera.

49 THE WHITE HOUSE FARM

Ingarsby, Houghton-on-the-Hill, LE7 9JD. Pam & Richard Smith, 0116 259 5448, Pamsmithtwhf@aol.com. *7m E of Leicester. 12m W of Uppingham. Take A47 from Leicester through Houghton-on-the-Hill towards Uppingham. 1m after Houghton, turn L (signed Tilton). After 1m turn L (signed Ingarsby), garden is 1m further on.* **Visits by arrangement June to Aug for groups of 20 to 40. Adm £6, chd free. Home-made teas.**

Former Georgian farm in 2 acres of country garden. Beautiful views. Box, yew and beech hedges divide a cottage garden of gaily coloured perennials and roses; a formal herb garden; a pergola draped with climbing plants; an old courtyard with roses, shrubs and trees. Herbaceous borders lead to pools with water lilies and informal cascade. Orchard, wild garden and lake. Home for lots of wildlife.

GROUP OPENING

50 WIGSTON GARDENS

Wigston, LE18 3LF. Zoe Lewin. *Just S of Leicester off A5199.* **Sat 8, Sun 9 July (11-5). Combined adm £6, chd free. Home-made teas at Little Dale Wildlife Garden, 9 Wicken Rise & 28 Gladstone Street.**

28 GLADSTONE STREET
Chris & Janet Huscroft.
(See separate entry)

2A HOMESTEAD DRIVE
Mrs Sheila Bolton.

LITTLE DALE WILDLIFE GARDEN
Zoe Lewin & Neil Garner, www.facebook.com/zoesopengarden.

'VALLENVINA' 6 ABINGTON CLOSE
Mr Steve Hunt.

9 WICKEN RISE
Sharon Maher & Michael Costall.

Wigston Gardens consists of five relatively small gardens all within a 2m radius of each other. There is something different to see at each garden from traditional and formal to a taste of the exotic via wildflowers, upcycling, interesting artefacts, prairie style planting, wildlife pond, fruit and vegetables. A couple of the gardens are within walking distance of one another but you will need to travel by car to visit all of the gardens in the group or it will make for quite a long walk and you'll need your comfy shoes.

GROUP OPENING

51 WILLOUGHBY GARDENS

Willoughby Waterleys, LE8 6UD. *9m S of Leicester. From A426 heading N turn R at Dunton Bassett lights. Follow signs to Willoughby. From Blaby follow signs to Countesthorpe. 2m S to Willoughby.* **Sun 9 July (11-5). Combined adm £6, chd free. Light refreshments in the Village Hall.**

2 CHURCH FARM LANE
Valerie & Peter Connelly.

FARMWAY
Eileen Spencer, 01162 478321, eileenfarmway9@msn.com.
Visits also by arrangement July & Aug for groups of up to 25.

HIGH MEADOW
Phil & Eva Day.

JOHN'S WOOD
John & Jill Harris.

KAPALUA
Richard & Linda Love.

3 ORCHARD ROAD
Diane Brearley.

3 YEW TREE CLOSE
Emma Clanfield.

Willoughby Waterleys lies in the South Leicestershire countryside. 7 gardens will be open. John's Wood is a 1½ acre nature reserve planted to encourage wildlife. 2 Church Farm Lane has been professionally designed with many interesting features. Farmway is a plant lovers garden with many unusual plants in colour themed borders. 3, Orchard Road is a small south facing garden packed with interesting features. High Meadow has been evolving over 13yrs. Inc mixed planting and ornamental vegetable garden. Kapalua has an interesting planting design incorporating views of open countryside. 3 Yew Tree Close is a wrap around garden that naturally creates a series of rooms with cottage garden style borders.

52 15 THE WOODCROFT

Diseworth, Derby, DE74 2QT. Nick & Sue Hollick. *The Woodcroft is off The Green, parking on The Woodcroft.* **Sun 11 June (11-5). Adm £4, chd free. Home-made teas.**

⅓ acre garden developed over 43 yrs with mature choice trees and shrubs, old and modern shrub roses, fern garden, wildlife garden inc barrel pond, alpine troughs, seasonal containers and mixed herbaceous borders, planted with a garden designer's eye and a plantsman's passion. Recently constructed gazebo overlooking new wildlife pond. Three steps from the upper terrace to the main garden.

LINCOLNSHIRE

Lincolnshire is a county shaped by a rich tapestry of fascinating heritage, passionate people and intriguing traditions; a mix of city, coast and countryside.

The city of Lincoln is dominated by the iconic towers of Lincoln Cathedral. The eastern seaboard contains windswept golden sands and lonely nature reserves. The Lincolnshire Wolds is a nationally important landscape of rolling chalk hills and areas of sandstone and clay, which underlie this attractive landscape.

To the south is the historic, religious and architectural heritage of The Vales, with river walks, the fine Georgian buildings of Stamford and historic Burghley House. In the east the unqiue Fens landscape thrives on an endless network of waterways inhabited by an abundance of wildlife.

Beautiful gardens of all types, sizes and designs are cared for and shared by their welcoming owners. Often located in delightful villages, a visit to them will entail driving through quiet roads often bordered by verges of wild flowers.

Lincolnshire is rural England at its very best. Local heritage, beautiful countryside walks, aviation history and it is the home of the Red Arrows.

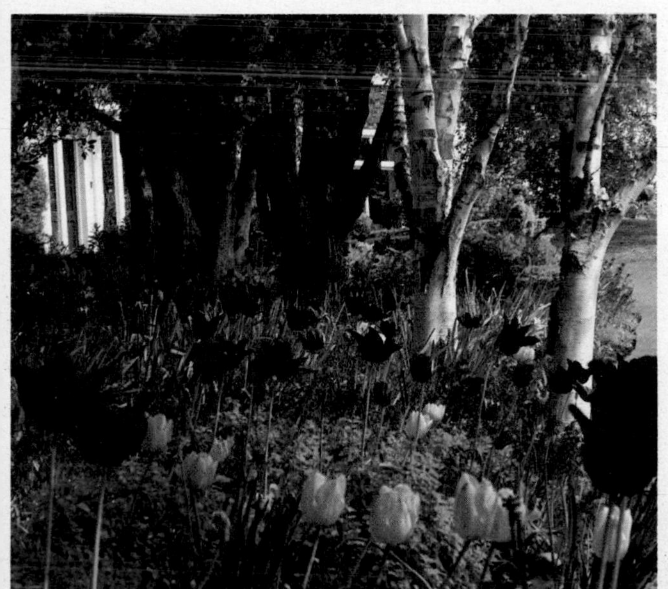

Volunteers

County Organisers
Lesley Wykes
01673 860356
lesley.wykes@ngs.org.uk

County Treasurer
Kate Richardson
07496 550516
kate.richardson@ngs.org.uk

Social Media
Diane Puncheon
01427 800008
dianepuncheon@yahoo.com

Publicity
Margaret Mann
01476 585905
marg_mann2000@yahoo.com

Tricia Elliott 01427 788517
t.elliott575@gmail.com

Assistant County Organisers
Karen Bourne 07860 504047
karen@karenwrightpr.com

Sally Grant 01205 750486
sallygrant50@btinternet.com

Sylvia Ravenhall 01507 526014
sylvan@btinternet.com

Robert Bailey- Scott 07860 833905
rbaileyscott@aol.com

Talks Co-ordinator
Neil Timm
01472 398092
neilfernnursery@gmail.com

@LincolnshireNGS
@LincsNGS

Left: Dunholme Lodge

OPENING DATES

All entries subject to change. For latest information check **www.ngs.org.uk**

Map locator numbers are shown to the right of each garden name.

Extended openings are shown at the beginning of each month.

April

Saturday 1st
◆ Burghley House Private South Gardens 6
The Manor House 19
Oasis Garden - Your Place 24

Sunday 2nd
◆ Burghley House Private South Gardens 6

Friday 7th
◆ Easton Walled Gardens 8

Sunday 9th
◆ Mill Farm 22

Monday 10th
◆ Mill Farm 22

Sunday 16th
Ashfield House 2
Woodlands 35

Sunday 30th
Dunholme Lodge 7
66 Spilsby Road 32

May

Sunday 7th
Woodlands 35

Saturday 13th
Oasis Garden - Your Place 24

Saturday 20th
23 Accommodation Road 1
Willoughby Road Allotments 34

Sunday 21st
23 Accommodation Road 1
Fydell House 10

Sunday 28th
Aswarby House 3
Aswarby Park 4
2 Mill Cottage 21

June

Saturday 10th
23 Accommodation Road 1

Sunday 11th
23 Accommodation Road 1
Aubourn Hall 5
The Fern Nursery 9
Manor Farm 18
The Old Vicarage 25
Old White House 26
Shangrila 31
Springfield 33

Sunday 18th
Little Ponton Hall 16
Millstone House 23
Woodlands 35

Saturday 24th
Home Farm 14

Sunday 25th
Hackthorn Hall 12
Home Farm 14
Ludney House Farm 17

July

Sunday 2nd
Dunholme Lodge 7

Thursday 20th
The Secret Garden of Louth 30

Sunday 23rd
Ludney House Farm 17
The Secret Garden of Louth 30
Yew Tree Farm 36

Thursday 27th
The Secret Garden of Louth 30

Sunday 30th
The Secret Garden of Louth 30

August

Thursday 3rd
The Secret Garden of Louth 30

Saturday 5th
◆ Gunby Hall and Gardens 11

Sunday 6th
The Fern Nursery 9
Fydell House 10
The Secret Garden of Louth 30

Thursday 10th
The Secret Garden of Louth 30

Sunday 13th
The Secret Garden of Louth 30

Thursday 17th
The Secret Garden of Louth 30

Sunday 20th
The Secret Garden of Louth 30
Willoughby Road Allotments 34
Woodlands 35

Thursday 24th
The Secret Garden of Louth 30

Sunday 27th
The Secret Garden of Louth 30

September

Saturday 16th
Inley Drove Farm 15

Sunday 17th
Inley Drove Farm 15
Woodlands 35

By Arrangement

Arrange a personalised garden visit with your club, or group of friends, on a date to suit you. See individual garden entries for full details.

23 Accommodation Road 1
Ashfield House 2
Fydell House 10
23 Handley Street 13
Home Farm 14
Inley Drove Farm 15
Ludney House Farm 17
Marigold Cottage 20
The Old Vicarage 25
Overbeck 27
The Plant Lover's Garden 28
Pottertons Nursery 29
The Secret Garden of Louth 30
Woodlands 35

Pottertons Nursery

THE GARDENS

1 23 ACCOMMODATION ROAD

Horncastle, LN9 5AS. Mr & Mrs D Chapman, davechapman1953@outlook.com. *Horncastle. From turning off Lincoln Rd A158 onto Accommodation Rd we are situated approx 600yards on R.* **Sat 20, Sun 21 May, Sat 10, Sun 11 June (11-4). Adm £3, chd free. Visits by arrangement 13 May to 17 June.**

We have a medium sized garden which mixes flowers with fruit, places to sit and admire the fish pond and flowers. We have a range of iris, perennials, auriculas and alpines. Plenty to see and enjoy. Limited wheelchair access to decking area.

2 ASHFIELD HOUSE

Lincoln Road, Branston, Lincoln, LN4 1NS. John & Judi Tinsley, 07977 505682, john@tinsleyfarms.co.uk. *3m S of Lincoln on B1188. N outskirts of Branston on the B1188 Lincoln Rd. Signed 'Tinsley Farms - Ashfield'. Nr bus stop, follow signs down drive.* **Sun 16 Apr (11-4). Adm £5, chd free. Light refreshments. Visits also by arrangement Apr to Oct for groups of 20 to 30.**

See 140 flowering cherries, 30 magnolias. Many thousands of spring bulbs, sweeping lawns and lake. Beautiful naturally landscaped garden with some superb mature trees as well as a fascinating arboretum. One of the best flowering cherry displays in the area.

3 ASWARBY HOUSE

Aswarby, Sleaford, NG34 8SE. Penny & James Herdman. *Past church on R of rd. 300 yds from the gates of Aswarby Park. Plenty of parking available on the roadside & gravel paths around the garden.* **Sun 28 May (2-5). Combined adm with Aswarby Park £8, chd free. Home-made teas in Aswarby Park.**

New garden of an acre planted 4 years ago in the grounds of a handsome C18 house and Coachhouse. It has a partial walled garden, wildflower meadow surrounded by ornamental grasses and a 30 metre long herbaceous border. With 2 box parterres, and woodland shrubs, it has stunning views over ancient ridge and furrow grassland. This garden would complement your visit to Aswarby Park. Gravel paths.

4 ASWARBY PARK

Aswarby, Sleaford, NG34 8SD. Mr & Mrs George Playne, www.aswarbyestate.co.uk. *5m S of Sleaford on A15. Take signs to Aswarby. Entrance is straight ahead by Church through black gates.* **Sun 28 May (2-5). Combined adm with Aswarby House £8, chd free. Home-made teas.**

Formal and woodland garden in a parkland setting of approx 20 acres. Yew trees form a backdrop to borders and lawns surrounding the house which is a converted stable block. Walled garden contains a greenhouse with a muscat vine, which is over 300 yrs old. Large display of daffodils, snowdrops and climbing roses in season. Partial wheelchair access on gravel paths and drives.

5 AUBOURN HALL

Harmston Road, Aubourn, Lincoln, LN5 9DZ. Mr & Mrs Christopher Nevile, www.aubourngardens.com. *7m SW of Lincoln. Signed off A607 at Harmston & off A46 at Thorpe on the Hill.* **Sun 11 June (2-5). Adm £6, chd free. Home-made refreshments available.**

Approx 9 acres. Lawns, mature trees, shrubs, roses, mixed borders, rose garden, large prairie and topiary garden, spring bulbs, woodland walk and ponds. C11 church adjoining. Access to garden is fairly flat and smooth. Depending on weather some areas may be inaccessible to wheelchairs. Parking in field.

6 ◆ BURGHLEY HOUSE PRIVATE SOUTH GARDENS

Stamford, PE9 3JY. Burghley House Preservation Trust, 01780 752451, burghley@burghley.co.uk, www.burghley.co.uk. *1m E of Stamford. From Stamford follow signs to Burghley via B1443.* **For NGS: Sat 1, Sun 2 Apr (10.30-3.30). Adm £6, chd free. Pre-booking essential, please visit www.ngs.org.uk for information & booking. For other opening times and information, please phone, email or visit garden website.**

The Private South Gardens at Burghley House will open for the NGS with spectacular spring bulbs in a park like setting with magnificent trees. Relish the opportunity to enjoy Capability Brown's famous lake and summerhouse. Entry via Garden Kiosks. Fine Food Market and TVR Car Rally. Admission charge is a special pre-book price only via NGS website. Visitors paying at the gate on the day will be charged a Gardens ticket price. Wheelchair access via gravel paths.

7 DUNHOLME LODGE

Dunholme, Lincoln, LN2 3QA. Hugh & Lesley Wykes. *4m NE of Lincoln. Turn off A46 towards Welton at the r'about. After ½m turn L up long private road. Garden at top.* **Sun 30 Apr, Sun 2 July (11-5). Adm £5, chd free. Light refreshments.**

5 acres. Spring bulb area, shrub borders, fern garden, natural pond, wildflower area, orchard and vegetable garden. Developing 2 acre arboretum. RAF Dunholme Lodge Museum and War Memorial within the grounds. Craft stalls.Ukulele Band. Vintage vehicles. Most areas wheelchair accessible but some loose stone and gravel.

8 ◆ EASTON WALLED GARDENS

Easton, NG33 5AP. Sir Fred & Lady Cholmeley, 01476 530063, info@eastonwalledgardens.co.uk, www.visiteaston.co.uk. *7m S of Grantham. 1m from A1, off B6403.* **For NGS: Fri 7 Apr (11-4). Adm £9, chd £4.50. For other opening times and information, please phone, email or visit garden website.**

A 400 yr old, restored, 12 acre garden set in the heart of Lincolnshire. Home to snowdrops, sweet peas, roses and meadows. The River Witham meanders through the gardens, teeming with wildlife. Other garden highlights inc a yew tunnel, turf maze and cut flower gardens. The Courtyard offers shopping, plant sales, a coffee room offering a selection of teas and coffees and beauty salon. Applestore Tearoom offers hot and cold drinks, savoury snacks, cakes and ice creams. Regret no wheelchair access to lower gardens but tearoom, shop, upper gardens and facilities are all accessible.

Little Ponton Hall

9 THE FERN NURSERY

Grimsby Road, Binbrook, Market Rasen, LN8 6DH. Neil Timm, www.fernnursery.co.uk. *On B1203 from Market Rasen. On the Grimsby rd from Binbrook Square 400m.* **Sun 11 June, Sun 6 Aug (11-4). Adm £3, chd free. Home-made teas in the bowling pavilion, and cakes.**

The garden has been designed as a wildlife garden with a number of features of interest to both visitors and wildlife, helped by having a natural stream running through the garden, which supplies water to a pond and water features. Visitors can also enjoy rock features, acid beds, and a sheltered winter garden with a sundial at its centre. From which a path leads to a small wood with the main fern collection. In addition there is a semi formal herb garden and bowling green, where they will often see a game being played, large shrubs, a bank of drought tolerant plants and herbaceous perennials, while steps, seats, a gazebo, bridge and many other features. Partial wheelchairs access, gravel paths.

10 FYDELL HOUSE

South Square, Boston, PE21 6HU. Boston Preservation Trust, 01205 351520, info@fydellhouse.org.uk, www.bostonpreservationtrust.com/fydell-garden.html. *Central Boston down South St. Through the Market Sq, past Boots the Chemist. One way street by Guildhall. There are 3 car parks within 200 yds of the house. Disabled parking in council car park opp the house.* **Sun 21 May, Sun 6 Aug (10-4). Adm £4, chd free. Light refreshments. Visits also by arrangement 21 May to 6 Aug.**

Within 3 original red brick walls a formal garden was been created in 1995. Yew buttresses, arbours and four parterres use dutch themes. The borders contain herbaceous plants and shrubs. The N facing border holds shade loving plants. There is a mulberry and walnut tree. The astrolabe was installed in 1997. A Victorian rockery is built from slag from ironworks in Boston. Walled garden Astrolabe Parterres formal borders topiary of box and yew. Wheelchair access is along the south alleyway from the front to the back garden.

11 ◆ GUNBY HALL AND GARDENS

Spilsby, PE23 5SS. National Trust, 01754 890102, gunbyhall@nationaltrust.org.uk, www.nationaltrust.org.uk/gunby-hall. *2½ m NW of Burgh-le-Marsh. 7m W of Skegness. On A158. Signed off Gunby r'about.* **For NGS: Sat 5 Aug (10-4). Adm £9, chd £4.50. For other opening times and information, please phone, email or visit garden website.**

Pre-booking advisable via the National Trust website to ensure we have a parking space for you. 8 acres of formal and walled gardens. Old roses, herbaceous borders, herb garden and kitchen garden with fruit trees and vegetables. Greenhouses, carp pond and sweeping lawns. Tennyson's Haunt of Ancient Peace. House built by Sir William Massingberd in 1700. Wheelchair access in gardens and with Gunby's dedicated wheelchair on ground floor of house. There are steps into the house.

12 HACKTHORN HALL
Hackthorn, Lincoln, LN2 3PQ. Mr & Mrs William Cracroft-Eley, www.hackthorn.com. *6m N of Lincoln. Follow signs to Hackthorn. Approx 1m off A15 N of Lincoln.* **Sun 25 June (1-5). Adm £5, chd free. Home-made teas at Hackthorn Village Hall.**
Formal and woodland garden, productive and ornamental walled gardens surrounding Hackthorn Hall and church extending to approx 15 acres. Parts of the formal gardens designed by Bunny Guinness. The walled garden boasts a magnificent Black Hamburg vine, believed to be 2nd in size to the vine at Hampton Court. Partial wheelchair access, gravel paths, grass drives.

13 23 HANDLEY STREET
Heckington, Nr Sleaford, NG34 9RZ. Stephen & Hazel Donnison, 01529 460097, donno5260@gmail.com. *Heckington. A17 from Sleaford, turn R into Heckington. Follow rd to the Green. L, follow rd past Church, R into Cameron St. At end of this rd L into Handley St.* **Visits by arrangement 15 May to 26 Aug for groups of 5 to 30. Adm £3, chd free. Home-made teas.**
Compact, quirky garden, large fish pond. Further 5 small wildlife ponds. Small wooded area and Jurassic style garden with tree ferns. Densely planted flower borders featuring penstemons. Large patio with seating.

14 HOME FARM
Little Casterton Road, Ryhall, Stamford, PE9 4HA. Steve & Karen Bourne, 07860 504047, karen@karenwrightpr.com. *1½m N of Stamford. Off A6121 at mini-r'about, towards Little Casterton & Tolethorpe Hall.* **Sat 24, Sun 25 June (1-5). Adm £5, chd free. Home-made teas. Visits also by arrangement June & July for groups of 10 to 25.**
A colourful and fragrant range of roses, herbaceous borders and an avenue of lavender and cherry trees established 12 years ago, alongside a woodland walk on the nine acre site. Plantings encourage wildlife. Wildflower meadow and Mediterranean borders introduced more recently to suit drier conditions. Orchard with local varieties of fruit trees, and a fruit cage of soft fruits.

15 INLEY DROVE FARM
Inley Drove, Sutton St James, Spalding, PE12 0LX. Francis & Maisie Pryor, 01406 540088, maisietaylor7@gmail.com, www.pryorfrancis.wordpress.com/. *5 m S of Holbeach. Just off rd from Sutton St James to Sutton St Edmund. 2m S of Sutton St James. Look for yellow NGS signs on double bend.* **Sat 16, Sun 17 Sept (11-5). Adm £6, chd free. Home-made teas. Visits also by arrangement 1 June to 30 July for groups of 10 to 30.**
Over 3 acres of Fenland garden and meadow plus 6½ acre wood developed over 25yrs. Garden planted for colour, scent and wildlife. Double mixed borders and less formal flower gardens all framed by hornbeam hedges. Unusual shrubs and trees, inc fine stand of Black Poplars, vegetable garden, woodland walks and orchard. Wheelchair access, some gravel and a few steps but mostly flat grass.

16 LITTLE PONTON HALL
Grantham, NG33 5BS. Bianca & George McCorquodale, www.littlepontonhallgardens.org.uk. *2m S of Grantham. ½m E of A1 at S end of Grantham bypass.* **Sun 18 June (11-4). Adm £8, chd free. Light refreshments.**
3 to 4 acre garden. River walk. Spacious lawns with cedar tree over 200 yrs old. Formal walled kitchen garden and listed dovecote, with herb garden. Victorian greenhouses with many plants from exotic locations. Wheelchair access on hard surfaces, unsuitable on grass. Disabled WC.

17 LUDNEY HOUSE FARM
Ludney, Louth, LN11 7JU. Jayne Bullas, 07733 018710, jayne@theoldgatehouse.com. *Between Grainthorpe & Conisholme.* **Sun 25 June, Sun 23 July (1.30-4). Adm £7, chd £3. Home-made teas. Cakes inc in adm. Visits also by arrangement 7 May to 24 Sept.**
A beautiful garden lovingly developed over the last 20 yrs with several areas of formal and informal planting inc a pond which attracts a wonderful variety of wildlife. There is an excellent mix of trees, shrubs, perennials, rose garden and wildflower area. There are plenty of seats positioned around to sit and enjoy a cuppa and piece of cake! Wheelchair access to most parts.

18 MANOR FARM
Horkstow Road, South Ferriby, Barton-upon-Humber, DN18 6HS. Geoff & Angela Wells. *3m from Barton-upon-Humber on A1077, turn L onto B1204, opp Village Hall.* **Sun 11 June (11-5). Combined adm with Springfield £6, chd free. Home-made teas.**
A garden which is much praised by visitors. Set within approx 1 acre with mature shrubberies, herbaceous borders, gravel garden and pergola walk. Rose bed, white garden and fernery. Many old trees with preservation orders. Wildlife pond set within a paddock.

19 THE MANOR HOUSE
Manor House Street, Horncastle, LN9 5HF. Mr Michael & Dr Marilyn Hieatt. *Manor House St runs off the Market Sq in middle of Horncastle, beside St Mary's Church. The Manor House is approx 100m from the Market Sq (on R).* **Sat 1 Apr (1-4.30). Adm £5, chd free.**
An informal spring garden and orchard bordered by the River Bain, hidden in the middle of Horncastle. The garden inc a short section of the C3/4 wall that formed part of a Roman fort (Scheduled Ancient Monument) with the remnants of an adjacent medieval well. Restricted wheelchair access (some parts not accessible).

In 2022 the National Garden Scheme donated £3.11 million to our nursing and health beneficiaries

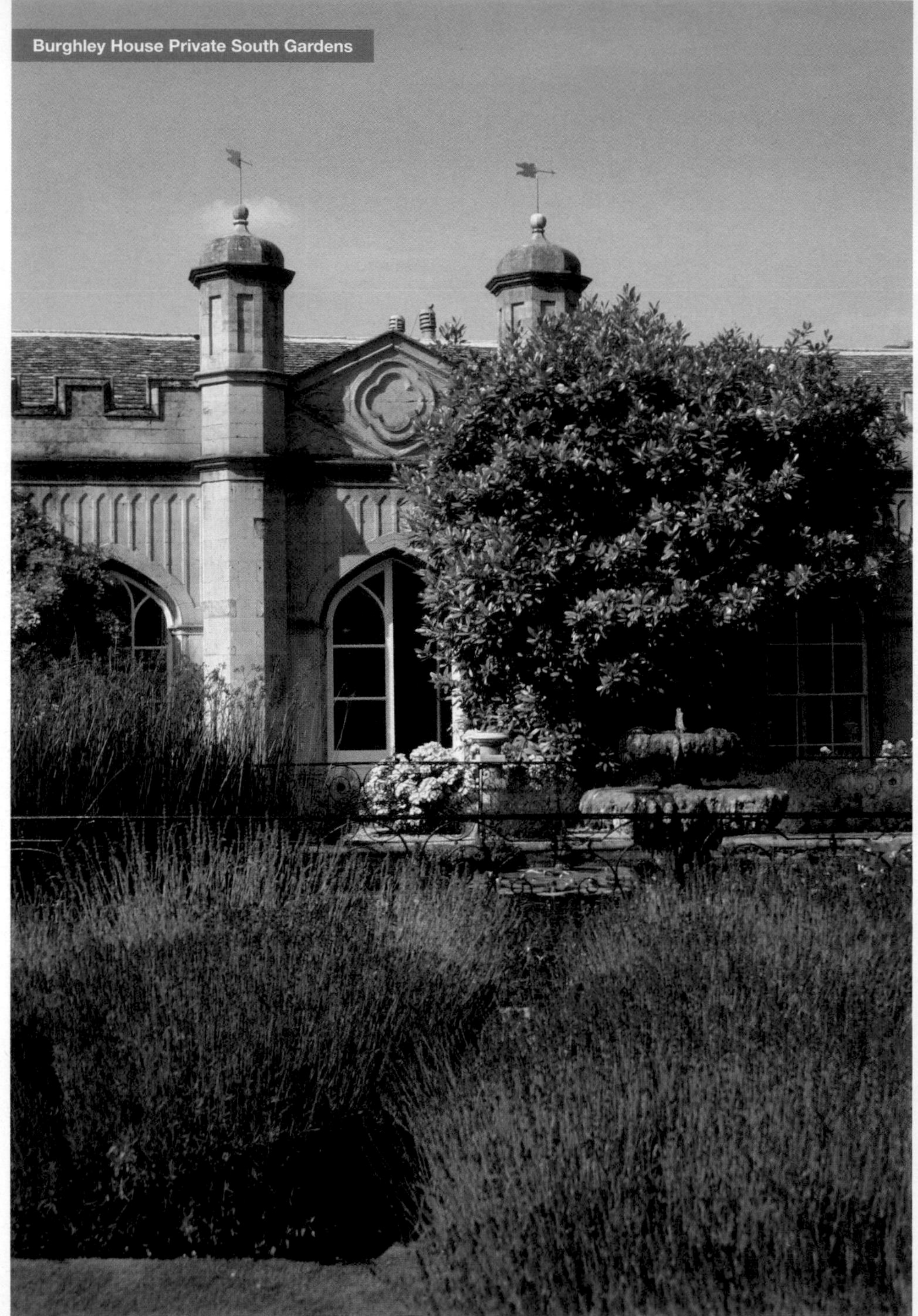

Burghley House Private South Gardens

20 MARIGOLD COTTAGE

Hotchin Road, Sutton-on-Sea, LN12 2NP. Stephanie Lee & John Raby, 01507 442151, marigoldlee@btinternet.com, www.rabylee.uk/marigold/. *16m N of Skegness on A52. 7m E of Alford on A1111. 3m S of Mablethorpe on A52. Turn off A52 on High St at Cornerhouse Cafe. Follow rd past playing field on R. Rd turns away from the dunes. House 2nd on L.* **Visits by arrangement 21 Apr to 15 Sept. Adm by donation.**

Slide open the Japanese gate to find secret paths, lanterns, a circular window in a curved wall, water lilies in pots and a gravel garden, vegetable garden and propagation area. Take the long drive to see the sea. Back in the garden, find a seat, enjoy the birds and bees. We face the challenges of heavy clay and salt ladened winds but look for unusual plants not the humdrum for these conditions. Most of garden accessible to wheelchairs along flat, paved paths.

21 2 MILL COTTAGE

Barkwith Road, South Willingham, Market Rasen, LN8 6NN. Mrs Jo Rouston & Mr Andrew Deaton. *5m E of Wragby. On A157 turn R at pub in East Barkwith then immed L to S Willingham. Cottage 1m on L.* **Sun 28 May (11-4). Adm £5, chd free. Home-made teas. Home-made cakes and savouries also available.**

A garden of several defined spaces, packed with interesting features, unusual plants and well placed seating areas, created by garden designer Jo. Original engine shed, working well, raised beds using local rock. Clipped box, alpines, roses, themed summerhouses. Greenhouses and newly installed raised beds. Living Sofa, woven metal and turf tree seat. 2 outdoor BBQ areas. Log and bottle wall. Partial wheelchair access. Gravel at far end of garden. Steps down to main greenhouse.

22 ◆ MILL FARM

Caistor Road, Grasby, Caistor, DN38 6AQ. Mike & Helen Boothman, 01652 628424, boothmanhelen@gmail.com, www.millfarmgarden.co.uk. *3m NW of Caistor on A1084. Between Brigg & Caistor. From Cross Keys pub towards Caistor for approx 200yds. Do not go into Grasby village.* **For NGS: Sun 9, Mon 10 Apr (11-4). Adm £5, chd free. Light refreshments. Hot soup and sausage rolls. For other opening times and information, please phone, email or visit garden website.**

Opening earlier than usual to enable visitors to enjoy the hellebores, spring bulbs and fresh spring growth. A much loved garden by visitors maintained to a high standard by the owners. Over 3 acres of garden with many diverse areas. Formal frontage with shrubs and trees. The rear is a plantsman haven with a diverse mix of perennials. Herbaceous beds with different grasses and hardy perennials. Small nursery on site with homegrown plants available. Open by arrangement for groups. Wheelchair access, mainly grass, but with some gravelled areas.

23 MILLSTONE HOUSE

7 Mill Lane, Gosberton, Spalding, PE11 4NN. Mrs J Chatterton. *From the main rd by the war memorial, turn into Mill Ln where you will see signs for Millstone House. Garden is 100 yds on the R.* **Sun 18 June (12-5). Adm £4, chd free.**

Hidden behind a privet hedge is a beautiful wrap-around garden with a wealth of herbaceous plants, trees and shrubs.There is a fabulous display of pots ferns and a small water feature.

24 OASIS GARDEN - YOUR PLACE

Rear of Your Place, 236 Wellington Street, Grimsby, DN32 7JP. Grimsby Neighbourhood Church, www.yourplacegrimsby.com. *Enter Grimsby (M180) over flyover, along Cleethorpes Rd. Turn R into Victor St, Turn L into Wellington St. Your Place is on R on junction of Wellington St & Weelsby St.* **Sat 1 Apr, Sat 13 May (11-3). Adm £3, chd free. Light refreshments.**

The multi award winning Oasis Garden, Your Place, has been described as one of the 'Most inspirational garden in the 6 counties of the East Midlands', is approximately 1½ acres and nestles in the heart of Great Grimsby's East Marsh Community. A working garden producing 15k plants per year, grown by local volunteers of all ages and abilities. Lawns, fruit, vegetable, perennial and annual beds.

25 THE OLD VICARAGE

Low Road, Holbeach Hurn, PE12 8JN. Mrs Liz Dixon-Spain, 01406 424148, lizdixonspain@gmail.com. *2m NE of Holbeach. Turn off A17 N to Holbeach Hurn, past post box in middle of village, 1st R at war memorial into Low Rd. Old Vicarage is on R approx 400yds Parking in grass paddock.* **Sun 11 June (1-5). Combined adm with Old White House £5, chd free. Visits also by arrangement Apr to Sept for groups of up to 30.**

2 acres of garden with 150yr old tulip, plane and beech trees: borders of shrubs, roses, herbaceous plants. Shrub roses and herb garden in old paddock area, surrounded by informal areas with pond and bog garden, wildflowers, grasses and bulbs. Small fruit and vegetable gardens. Kids love exploring winding paths through the wilder areas. Garden is managed environmentally. Fun for kids! Wheelchair access, gravel drive, some paths, mostly grass access.

26 OLD WHITE HOUSE

Baileys Lane, Holbeach Hurn, PE12 8JP. Mrs A Worth. *2m N of Holbeach. Turn off A17 N to Holbeach Hurn, follow signs to village, cont through, turn R after Rose & Crown pub at Baileys Ln.* **Sun 11 June (1-5). Combined adm with The Old Vicarage £5, chd free. Home-made teas.**

1½ acres of mature garden, featuring herbaceous borders, roses, patterned garden, herb garden and walled kitchen garden. Large catalpa, tulip tree that flowers, ginko and other specimen trees. Flat surfaces, some steps, wheelchair access to all areas without using steps.

The National Garden Scheme searches the length and breadth of England, Wales, Northern Ireland and the Channel Islands for the very best private gardens

27 OVERBECK

46 Main Street, Scothern, LN2 2UW. John & Joyce Good, 01673 862200, jandjgood@btinternet.com. *4m E of Lincoln. Scothern signed from A46 at Dunholme & A158 at Sudbrooke. Overbeck is E end of Main St.* **Visits by arrangement 15 May to 7 Aug for groups of 5 to 30. Adm £5, chd free. Light refreshments.**

Situated in an attractive village this approx 2⁄3 acre garden is a haven for wildlife. Long herbaceous borders and colour themed island beds with some unusual perennials. Hosta border, gravel bed with grasses, fernery, interesting range of trees, numerous shrubs, small stumpery, climbers inc roses and clematis, recently developed parterre and large prolific vegetable and fruit area.

28 THE PLANT LOVER'S GARDEN

Bourne, PE10 0XF. Danny & Sophie, 07850 239393, plantloversgarden@outlook.com. *Situated on the edge of beautiful S Lincolnshire close to the borders of Cambridgeshire & Rutland in the East of England.* **Visits by arrangement Mar to Aug for groups of up to 30. Adm £4, chd free. Light refreshments. Available by prior arrangement.**

The Plant Lovers Garden is a verdant space of colour, form and style, with brimming raised beds and packed herbaceous borders, peppered throughout with clipped topiary balls. With an ever changing interest and a variety of blooms throughout it's opening months of March to August. This inspirational garden is a plant lover's delight. Private visits to larger groups welcomed. Most areas accessible by wheelchair.

29 POTTERTONS NURSERY

Moortown Road, Nettleton, Caistor, LN7 6HX. Rob & Jackie Potterton, www.pottertons.co.uk. *1m W of Nettleton. From A46 at Nettleton turn onto B1205 (Moortown). Nursery 1¼m, turn by edge of wood.* **Visits by arrangement 1 Mar to 1 Oct for groups of 10 to 55. Adm £3, chd free. Home-made teas. Home-made light refreshments also available.**

5 acre garden of alpine rockeries, raised beds and troughs. Discover our pools and waterfall which are in a tranquil setting. We also boast a tufa bed, crevice garden and woodland beds which inc alpines, bulbs and woodland plants. which offer interest from March to September. We hope to arrange for a special open day and plant fair in May, see web site for details. Wheelchair access: mixed grass surfaces.

30 THE SECRET GARDEN OF LOUTH

68 Watts Lane, Louth, LN11 9DG. Jenny & Rodger Grasham *½m S of Louth town centre. For SatNav & to avoid opening/closing gate on Watts Ln, use postcode LN119DJ this is Mount Pleasant Ave, leads straight to our house front.* **Every Thur and Sun 20 July to 27 Aug (11-4). Adm £3, chd free. Pre-booking essential, please phone 07977 318145, email sallysing@hotmail.co.uk or visit www.facebook.com/thesecretgardenoflouth for information & booking. Home-made teas. Visits also by arrangement 20 July to 27 Aug.**

Blank canvas of 1⁄5 acre in early 90s. Developed into lush, colourful, exotic plant packed haven. A whole new world on entering from street. Exotic borders, raised exotic island, long hot border, ponds, stumpery. Intimate seating areas along garden's journey. Facebook page - The Secret Garden of Louth. Children can find where the frogs are hiding! Butterflies and bees but how many different types? Feed the fish, find Cedric the spider, Simon the snake, Colin the Crocodile and more.

31 SHANGRILA

Little Hale Road, Great Hale, Sleaford, NG34 9LH. Marilyn Cooke & John Knight. *On B1394 between Heckington & Helpringham.* **Sun 11 June (11-5). Adm £4.50, chd free. Home-made teas.**

Approx 3 acre garden with sweeping lawns long herbaceous borders, colour themed island beds, hosta collection, lavender bed with seating area, topiary, acer collection Japanese garden. Wheelchair access to all areas.

32 66 SPILSBY ROAD

Boston, PE21 9NS. Rosemary & Adrian Isaac. *From Boston town take A16 towards Spilsby. On L after Trinity Church. Parking on Spilsby Rd.* **Sun 30 Apr (11-4). Adm £5, chd free. Home-made teas.**

1⅓ acre with mature trees, moat, Venetian Folly, summerhouse and orangery, lawns and herbaceous borders. Children's Tudor garden house, gatehouse and courtyard. Wide paths most paths suitable for wheelchairs.

33 SPRINGFIELD

Main Street, Horkstow, Barton-Upon-Humber, DN18 6BL. Mr & Mrs G Allison. *4m from Barton on Humber. Take the A1077 towards Scunthorpe & in South Ferriby bear L onto the B1204, after 2m Springfield is on the hillside on L.* **Sun 11 June (11-5). Combined adm with Manor Farm £6, chd free.**

This beautiful hillside garden on the edge of the Wolds was renovated and redesigned in 2011 from an overgrown state. It features many shrubs and perennials with a rose pergola and stunning views over the Ancholme Valley.

34 WILLOUGHBY ROAD ALLOTMENTS

Willoughby Road, Boston, PE21 9HN. Willoughby Road Allotments Association. *Entrance is adjacent to 109 Willoughby Rd. Street Parking only.* **Sat 20 May, Sun 20 Aug (10.30-4). Adm £3.50, chd free. Light refreshments.**

Set in 5 acres the allotments comprise 60 plots growing fine vegetables, fruit, flowers and herbs. There is a small orchard and wildflower area and a community space adjacent. Grass paths run along the site. Several plots will be open to walk round. There will be a seed and plant stall. Artwork created by Bex Simon situated on site. Small orchard and wildflower beds Community area with kitchen and disabled WC. Large Polytunnel with raised beds inside and out. Accessible for all abilities.

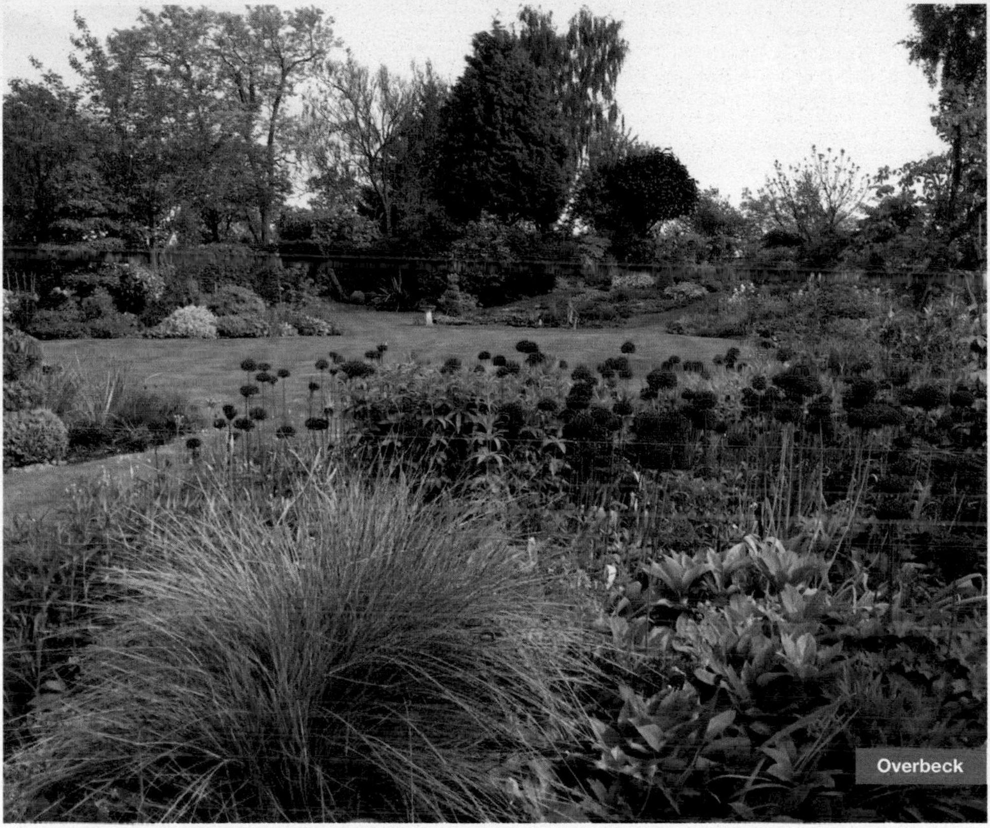

Overbeck

35 WOODLANDS
Peppin Lane, Fotherby, Louth, LN11 0UW. Ann & Bob Armstrong, 01507 603586, annbobarmstrong@btinternet.com, www.woodlandsplants.co.uk. *2m N of Louth off A16 signed Fotherby. Please park on R verge opp allotments & walk approx 350 yds to garden. If full, please park considerately in the village. No parking at garden. Please do not drive beyond designated area.* **Sun 16 Apr, Sun 7 May, Sun 18 June, Sun 20 Aug, Sun 17 Sept (10.30-4.30). Adm £4, chd free. Home-made teas. Visits also by arrangement.**
A lovely mature woodland garden where a multitude of unusual plants are the stars, many of which are available from the well-stocked RHS listed nursery. During 2022 several small areas have been redesigned and a great deal of fresh planting has taken place. Award winning professional artist's studio/gallery open to visitors. Specialist collection of Codonopsis for which Plant Heritage status has been granted. Wheelchair access possible to some areas with care. We leave space for parking at the house for those who cannot manage the walk from the car park.

36 YEW TREE FARM
Westhorpe Road, Gosberton, Spalding, PE11 4EP. Robert & Claire Bailey-Scott. *Nr Spalding. Enter the village of Gosberton. Turn into Westhorpe Rd, opp The Bell Inn, cont for approx. 1½ m. Property is 3rd on R after bridge.* **Sun 23 July (11-5). Adm £5, chd free. Home-made teas.**
A lovely country garden, 1½ acres. Large herbaceous and mixed borders surround well kept lawns. Wildlife pond with two bog gardens, woodland garden, shaded borders containing many unusual plants. Stunning reflective pool surrounded by ornamental grasses. Orchard, wildflower meadow and vegetable plot. See our video on YouTube 'Bailey-Scott - Gardens of the Year 2018'.

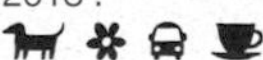

LONDON

HERTFORDSHIRE
BUCKINGHAMSHIRE
BERKSHIRE
SURREY
River Thames
London Heathrow
EN4
EN5
Barnet
N20
Southgate
N14
NW7
HA7
HA8
Edgware
N12
N11
Finchley
N3
N10
N22
N2
N8
N6
N19
HA6
UB9
HA5
HA
HA3
NW4
Hendon
NW9
NW11
Ruislip
Harrow
HA1
HA4
HA2
HA9
NW
NW2
Hampstead
NW3
NW5
N7
Islington
UB10
HA0
Wembley
Northolt
Uxbridge
UB5
UB6
NW10
NW6
NW8
NW1
UB8
Hillingdon
UB4
UB
W10
W9
Bayswater
W1
W13
Ealing
W7
W5
W3
W12
W11
W2
W8
UB1
West Drayton
Southall
UB3
UB7
UB2
Hammersmith
W14
Westminster
SW1
W6
W4
SW7
SW5
SW3
SW10
Brentford
TW8
TW5
TW7
SW8
Battersea
TW6
Hounslow
TW9
SW13
SW6
SW11
TW4
TW3
SW14
SW4
Richmond
Wandsworth
TW14
TW1
SW15
SW18
TW
SW
SW12
TW2
TW10
Feltham
Twickenham
TW13
Teddington
Wimbledon
SW17
TW12
TW11
SW19
SW16
KT2
Hampton
SW20
KT8
KT1
CR4
Mitcham
Kingston
KT
KT3
SM4
KT5
KT7
KT6
KT4
SM3
SM1
SM
Sutton
SM6
KT9
SM2
SM5
Purley
Coulsdon
CR5

ESSEX
KENT
Chingford
Edmonton
Woodford Green
Tottenham
Walthamstow
Ilford
Romford
Upminster
Stratford
Barking
Rainham
London City
Thamesmead
Greenwich
Peckham
Lewisham
Eltham
Bexleyheath
Bexley
Sidcup
Chislehurst
Bromley
Orpington
Addington
Biggin Hill
River Thames
EN1
EN3
N9
E4
N18
IG8
N17
E17
E18
IG5
IG6
IG7
RM4
RM5
RM1
RM3
RM6
RM2
RM7
RM11
RM12
RM14
RM8
RM10
RM9
RM13
IG4
IG2
IG3
IG1
IG11
N15
N16
E11
E10
E12
F5
F7
E8
E9
E15
E2
E3
E13
E6
E1
E14
E16
SE28
DA18
DA17
SE16
SE10
SE7
SE8
SE2
SE18
DA8
SE15
SE14
SE3
DA7
DA16
DA1
DA6
SE4
SE13
SE22
SE12
SE9
DA15
DA5
SE21
SE23
SE6
DA14
SE26
SE19
SE20
BR1
BR7
BR3
SE25
BR5
BR2
CR0
BR4
BR6
CR2
TN14
TN16
IG
RM
E
SE
DA
BR
CR
TN
DON
0
5
10 kilometres
0
5 miles
© Global Mapping / XYZ Maps

Volunteers

County Organiser
Penny Snell
01932 864532
pennysnellflowers@btinternet.com

County Treasurer
Joanna Gray
joanna.gray@ngs.org.uk

Publicity
Sonya Pinto
07779 609715
sonya.pinto@ngs.org.uk

Booklet Co-ordinator
Sue Phipps
07771 767196
sue@suephipps.com

Booklet Distributor
Joey Clover
020 8870 8740
joey.clover@ngs.org.uk

Social Media
Position Vacant
For details please contact
Sue Phipps (as above)

Assistant County Organisers

Central London
Eveline Carn
07831 136069
evelinecbcarn@icloud.com

Clapham & surrounding area
Sue Phipps (as above)

Croydon & outer S London
Christine Murray & **Nicola Dooley**
07889 888204
christineandnicola.gardens@gmail.com

Dulwich & surrounding area
Clive Pankhurst
07941 536934
alternative.ramblings@gmail.com

E London
Teresa Farnham
07761 476651
farnhamz@yahoo.co.uk

Finchley & Barnet
Debra Craighead 07415 166617
dcraighead@icloud.com

Hackney
Philip Lightowlers
020 8533 0052
plighto@gmail.com

Hampstead
Joan Arnold
020 8444 8752
joan.arnold40@gmail.com

Northwood, Pinner, Ruislip & Harrow
Brenda White
020 8863 5877
brenda.white@ngs.org.uk

NW London
Susan Bennett & Earl Hyde
020 8883 8540
suebearlh@yahoo.co.uk

Outer NW London
James Duncan Mattoon
07504 565612

Outer W London
Sarah Corvi
07803 111968
sarah.corvi@ngs.org.uk

SE London
Janine Wookey
07711 279636
j.wookey@btinternet.com

SW London
Joey Clover (as above)

W London, Barnes & Chiswick
Siobhan McCammon
07952 889866
siobhan.mccammon@gmail.com

@LondonNGS
@LondonNGS
@londonngs

From the tiniest to the largest, London gardens offer exceptional diversity. Hidden behind historic houses in Spitalfields are exquisite tiny gardens, while on Kingston Hill there are 9 acres of landscaped Japanese gardens.

The oldest private garden in London boasts 5 acres, while the many other historic gardens within these pages are smaller – some so tiny there is only room for a few visitors at a time – but nonetheless full of innovation, colour and horticultural excellence.

London allotments have attracted television cameras to film their productive acres, where exotic Cape gooseberries, figs, prizewinning roses and even bees all thrive thanks to the skill and enthusiasm of city gardeners.

The traditional sit comfortably with the contemporary in London – offering a feast of elegant borders, pleached hedges, topiary, gravel gardens and the cooling sound of water – while to excite the adventurous there are gardens on barges and green roofs to explore.

The season stretches from April to October, so there is nearly always a garden to visit somewhere in London. Our gardens opening this year are the beating heart of the capital just waiting to be visited and enjoyed.

LONDON GARDENS LISTED BY POSTCODE

Inner London postcodes

E and EC London

Spitalfields Gardens, E1
26 College Gardens, E4
Lower Clapton Gardens, E5
84 Lavender Grove, E8
London Fields Gardens, E8
17 Greenstone Mews, E11
37 Harold Road, E11
Aldersbrook Gardens, E12
28a Worcester Road, E17
83 Cowslip Road, E18
68 Derby Road, E18
15 Hillcrest Road, E18
Victoria Lodge, E18
The Inner and Middle Temple Gardens, EC4

N and NW London

26 Arlington Avenue, N1
Arlington Square Gardens, N1
Barnsbury Group, N1
4 Canonbury Place, N1
De Beauvoir Gardens, N1
57 Huntingdon Street, N1
King Henry's Walk Garden, N1
5 Northampton Park, N1
7 St Paul's Place, N1
6 Thornhill Road, N1
7 Deansway, N2
12 Lauradale Road, N2
24 Twyford Avenue, N2
32 Highbury Place, N5
10 Furlong Road, N7
1a Hungerford Road, N7
12 Fairfield Road, N8
11 Park Avenue North, N8
35 Weston Park, N8
Princes Avenue Gardens, N10
5 St Regis Close, N10
25 Springfield Avenue, N10
33 Wood Vale, N10
Golf Course Allotments, N11
11 Shortgate, N12
2 Conway Road, N14
70 Farleigh Road, N16
53 Manor Road, N16
15 Norcott Road, N16
21 Gospatrick Road, N17
36 Ashley Road, N19
21 Oakleigh Park South, N20
Ally Pally Allotments, N22
Railway Cottages, N22
Garden of Medicinal Plants, Royal College of Physicians, NW1
The Holme, NW1
36 Park Village East, NW1
28a St Augustine's Road, NW1
93 Tanfield Avenue, NW2
Marie Curie Hospice, Hampstead, NW3
27 Nassington Road, NW3
1A Primrose Gardens, NW3
The Mysteries of Light Rosary Garden, NW5
Torriano Community Garden, NW5
Highwood Ash, NW7
92 Hampstead Way, NW11
100 Hampstead Way, NW11
74 Willifield Way, NW11

SE and SW London

Garden Barge Square at Tower Bridge Moorings, SE1
The Garden Museum, SE1
Lambeth Palace, SE1
24 Grove Park, SE5
41 Southbrook Road, SE12
13 Waite Davies Road, SE12
15 Waite Davies Road, SE12
Choumert Square, SE15
Lyndhurst Square Garden Group, SE15
Court Lane Group, SE21
103 & 105 Dulwich Village, SE21
38 Lovelace Road, SE21
4 Cornflower Terrace, SE22
174 Peckham Rye, SE22
4 Piermont Green, SE22
58 Cranston Road, SE23
Forest Hill Garden Group, SE23
39 Wood Vale, SE23
5 Burbage Road, SE24
28 Ferndene Road, SE24
Stoney Hill House, SE26
Cadogan Place South Garden, SW1
Eccleston Square, SW1
Spencer House, SW1
Brixton Water Lane Gardens, SW2
51 The Chase, SW4
Royal Trinity Hospice, SW4
35 Turret Grove, SW4
152a Victoria Rise, SW4
1 Fife Road, SW14
40 Chartfield Avenue, SW15
39 Hazlewell Road, SW15
61 Arthur Road, SW19
97 Arthur Road, SW19
Paddock Allotments & Leisure Gardens, SW20

W and WC London

Rooftopvegplot, W1
Warren Mews, W1
Hyde Park Estate Gardens, W2
118b Avenue Road, W3
34 Buxton Gardens, W3
41 Mill Hill Road, W3
65 Mill Hill Road, W3
Zen Garden at Japanese Buddhist Centre, W3
Chiswick Mall Gardens, W4
Maggie's West London, W6
27 St Peters Square, W6
Temple Lodge Club, W6
1 York Close, W7
Edwardes Square, W8
57 St Quintin Avenue, W10
Arundel & Elgin Gardens, W11
Arundel & Ladbroke Gardens, W11
57 Tonbridge House, WC1H

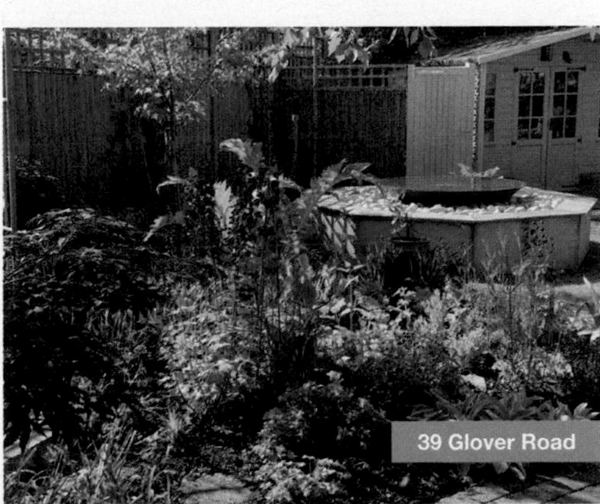
39 Glover Road

Outer London postcodes

37 Crescent Road, BR3
20 Hazelmere Road, BR5
Tudeley House, BR7
Oak Farm/Homestead, EN2
Theobald's Farmhouse, EN2
West Lodge Park, EN4
190 Barnet Road, EN5
45 Great North Road, EN5
9 Trafalgar Terrace, HA1
42 Risingholme Road, HA3
31 Arlington Drive, HA4
4 Manningtree Road, HA4
39 Glover Road, HA5
24 Moor Park Road, HA6
Horatio's Garden, HA7
53 Lady Aylesford Avenue, HA7
34 Barn Rise, HA9
26 Hillcroft Crescent, HA9
Perth Road Garden, IG2
Haven House Hospice, IG8
7 Woodbines Avenue, KT1
The Watergardens, KT2
52A Berrylands Road, KT5
15 Catherine Road, KT6
Hampton Court Palace, KT8
5 Pemberton Road, KT8
61 Wolsey Road, KT8
239A Hook Road, KT9
40 Ember Lane, KT10
9 Imber Park Road, KT10
Kew Green Gardens, TW9
28 Taylor Avenue, TW9
31 West Park Road, TW9
Dragon's Dream, UB8
Maggie's, SM2
Petersham House, TW10
16 Links View Road, TW12
106 Station Road, TW12
9 Warwick Close, TW12
70 Wensleydale Road, TW12
4 Hanworth Road, TW13

OPENING DATES

All entries subject to change. For latest information check www.ngs.org.uk

March

Sunday 26th
4 Canonbury Place, N1

April

Sunday 2nd
Royal Trinity Hospice, SW4

Saturday 15th
NEW Haven House Hospice, IG8

Sunday 16th
Edwardes Square, W8
Maggie's West London, W6
39 Wood Vale, SE23

Wednesday 19th
39 Wood Vale, SE23

Thursday 20th
◆ Hampton Court Palace, KT8

Sunday 23rd
Arundel & Ladbroke Gardens, W11
51 The Chase, SW4
84 Lavender Grove, E8
Petersham House, TW10

Tuesday 25th
51 The Chase, SW4

Sunday 30th
37 Crescent Road, BR3
5 St Regis Close, N10

May

Monday 1st
King Henry's Walk Garden, N1

Sunday 7th
7 Deansway, N2
27 St Peters Square, W6
NEW Theobald's Farmhouse, EN2
The Watergardens, KT2

Saturday 13th
21 Gospatrick Road, N17

Sunday 14th
Eccleston Square, SW1
Forest Hill Garden Group, SE23
Garden Barge Square at Tower Bridge Moorings, SE1
92 Hampstead Way, NW11
100 Hampstead Way, NW11
Highwood Ash, NW7
15 Norcott Road, N16
Princes Avenue Gardens, N10
42 Risingholme Road, HA3
West Lodge Park, EN4

Monday 15th
Lambeth Palace, SE1

Thursday 18th
57 Huntingdon Street, N1

Sunday 21st
61 Arthur Road, SW19
Arundel & Elgin Gardens, W11
10 Furlong Road, N7
Kew Green Gardens, TW9
Royal Trinity Hospice, SW4
Stoney Hill House, SE26

Saturday 27th
Cadogan Place South Garden, SW1
16 Links View Road, TW12
70 Wensleydale Road, TW12

Sunday 28th
36 Ashley Road, N19
De Beauvoir Gardens, N1
32 Highbury Place, N5
NEW 15 Hillcrest Road, E18
Kew Green Gardens, TW9
16 Links View Road, TW12
70 Wensleydale Road, TW12

Monday 29th
36 Ashley Road, N19

June

Saturday 3rd
The Mysteries of Light Rosary Garden, NW5
41 Southbrook Road, SE12
Zen Garden at Japanese Buddhist Centre, W3

Sunday 4th
31 Arlington Drive, HA4
Barnsbury Group, N1
40 Chartfield Avenue, SW15
Choumert Square, SE15
NEW 20 Hazelmere Road, BR5
NEW 39 Hazlewell Road, SW15
1a Hungerford Road, N7
London Fields Gardens, E8
Lower Clapton Gardens, E5
27 Nassington Road, NW3
11 Park Avenue North, N8
41 Southbrook Road, SE12
25 Springfield Avenue, N10
13 Waite Davies Road, SE12
15 Waite Davies Road, SE12
61 Wolsey Road, KT8
Zen Garden at Japanese Buddhist Centre, W3

Tuesday 6th
26 Arlington Avenue, N1

Friday 9th
Chiswick Mall Gardens, W4

Saturday 10th
Hyde Park Estate Gardens, W2
Maggie's, SM2
Spitalfields Gardens, E1

Sunday 11th
97 Arthur Road, SW19
51 The Chase, SW4
Chiswick Mall Gardens, W4
26 College Gardens, E4
4 Cornflower Terrace, SE22
Dragon's Dream, UB8
40 Ember Lane, KT10
39 Glover Road, HA5
37 Harold Road, E11
9 Imber Park Road, KT10
36 Park Village East, NW1
174 Peckham Rye, SE22
5 Pemberton Road, KT8
4 Piermont Green, SE22
6 Thornhill Road, N1

Wednesday 14th
The Inner and Middle Temple Gardens, EC4

Friday 16th
239A Hook Road, KT9

Saturday 17th
5 Northampton Park, N1
Rooftopvegplot, W1
7 St Paul's Place, N1
NEW Warren Mews, W1
Zen Garden at Japanese Buddhist Centre, W3

Sunday 18th
Arlington Square Gardens, N1
NEW 118b Avenue Road, W3
Brixton Water Lane Gardens, SW2
37 Crescent Road, BR3
103 & 105 Dulwich Village, SE21
Lyndhurst Square Garden Group, SE15
1A Primrose Gardens, NW3
Rooftopvegplot, W1
Zen Garden at Japanese Buddhist Centre, W3

Saturday 24th
Marie Curie Hospice, Hampstead, NW3
Paddock Allotments & Leisure Gardens, SW20
Victoria Lodge, E18

Sunday 25th
83 Cowslip Road, E18
70 Farleigh Road, N16
28 Ferndene Road, SE24
1 Fife Road, SW14
38 Lovelace Road, SE21
Royal Trinity Hospice, SW4
5 St Regis Close, N10
28 Taylor Avenue, TW9

Torriano Community Garden, NW5
Tudeley House, BR7
31 West Park Road, TW9

July

Saturday 1st
26 Hillcroft Crescent, HA9

Sunday 2nd
Aldersbrook Gardens, E12
NEW Ally Pally Allotments, N22
52A Berrylands Road, KT5
Court Lane Group, SE21
12 Fairfield Road, N8
NEW 4 Hanworth Road, TW13
26 Hillcroft Crescent, HA9
Railway Cottages, N22
57 St Quintin Avenue, W10
11 Shortgate, N12
57 Tonbridge House, WC1H
152a Victoria Rise, SW4

Sunday 9th
15 Catherine Road, KT6
84 Lavender Grove, E8
74 Willifield Way, NW11
33 Wood Vale, N10
7 Woodbines Avenue, KT1

Monday 10th
Garden of Medicinal Plants, Royal College of Physicians, NW1

Sunday 16th
24 Twyford Avenue, N2

Saturday 22nd
106 Station Road, TW12

Sunday 23rd
68 Derby Road, E18
NEW Horatio's Garden, HA7
53 Lady Aylesford Avenue, HA7
57 St Quintin Avenue, W10
106 Station Road, TW12
93 Tanfield Avenue, NW2
35 Turret Grove, SW4

Saturday 29th
34 Buxton Gardens, W3
The Holme, NW1
Perth Road Garden, IG2

Sunday 30th
34 Barn Rise, HA9
34 Buxton Gardens, W3
58 Cranston Road, SE23
45 Great North Road, EN5
The Holme, NW1
NEW 21 Moor Park Road, HA6
5 St Regis Close, N10

August

Saturday 12th
9 Warwick Close, TW12

Sunday 13th
45 Great North Road, EN5
4 Manningtree Road, HA4

Saturday 19th
1 York Close, W7

Sunday 20th
51 The Chase, SW4
35 Weston Park, N8
28a Worcester Road, E17
1 York Close, W7

Saturday 26th
The Holme, NW1

Sunday 27th
The Holme, NW1

September

Saturday 2nd
NEW 9 Trafalgar Terrace, HA1

Sunday 3rd
2 Conway Road, N14
Golf Course Allotments, N11
24 Grove Park, SE5
41 Mill Hill Road, W3
65 Mill Hill Road, W3
42 Risingholme Road, HA3

Saturday 9th
Temple Lodge Club, W6

Sunday 10th
12 Lauradale Road, N2
53 Manor Road, N16
Oak Farm/Homestead, EN2
28a St Augustine's Road, NW1

Tuesday 12th
◆ The Garden Museum, SE1

Sunday 17th
Royal Trinity Hospice, SW4

October

Sunday 22nd
The Watergardens, KT2
West Lodge Park, EN4

By Arrangement

Arrange a personalised garden visit with your club, or group of friends, on a date to suit you. See individual garden entries for full details.

36 Ashley Road, N19
NEW 118b Avenue Road, W3
190 Barnet Road, EN5
52A Berrylands Road, KT5
5 Burbage Road, SE24
51 The Chase, SW4
7 Deansway, N2
40 Ember Lane, KT10
45 Great North Road, EN5
17 Greenstone Mews, E11
NEW 20 Hazelmere Road, BR5
1a Hungerford Road, N7
57 Huntingdon Street, N1
9 Imber Park Road, KT10
Kew Green Gardens, TW9
84 Lavender Grove, E8
Maggie's West London, W6
41 Mill Hill Road, W3
65 Mill Hill Road, W3
The Mysteries of Light Rosary Garden, NW5
21 Oakleigh Park South, N20
1A Primrose Gardens, NW3
28a St Augustine's Road, NW1
27 St Peters Square, W6
57 St Quintin Avenue, W10
5 St Regis Close, N10
41 Southbrook Road, SE12
93 Tanfield Avenue, NW2
Tudeley House, BR7
24 Twyford Avenue, N2
152a Victoria Rise, SW4
West Lodge Park, EN4
74 Willifield Way, NW11
33 Wood Vale, N10

20 Hazelmere Road

GROUP OPENING

ALDERSBROOK GARDENS, E12
Wanstead, E12 5ES. *Empress Ave is a turning off Aldersbrook Rd. Bus 101 from Manor Park or Wanstead Stns. From Manor Park Stn take the 3rd R off Aldersbrook Rd. From Wanstead, drive past St Gabriel's Church, take 6th turning on L.* **Sun 2 July (12-5). Combined adm £10, chd free. Tea at 1 Clavering Road & 4 Empress Avenue.**

19 BELGRAVE ROAD
Gill Usher.

1 CLAVERING ROAD
Theresa Harrison.

4 EMPRESS AVENUE
Ruth Martin.

21 PARK ROAD
Theresa O'Driscoll & Barry Reeves.

13 ST MARGARET'S ROAD
Mrs Gerti Ashton.

Five different gardens situated on the Aldersbrook Estate between Wanstead Park and Wanstead Flats. 21 Park Road is a colour themed garden with evergreen shrubs for year-round structure and a vine covered pergola leading to a vegetable area. 13 St Margaret's Road is a town garden with many interesting, reclaimed features. 1 Clavering Road is an end of terrace garden where incremental space to the side has been adapted to create a kitchen garden and chicken coop, excess produce is eagerly received by five resident hens. Borders and beds contain a variety of planting. At 4 Empress Avenue a largish garden is divided in two with a cutting bed, soft fruit and wildlife areas inc pond and two mixed borders: one with hot colours and one with white planting. The diverse garden at 19 Belgrave Road is eclectic and busting at the seams with trees and plants. Gill is a ceramicist who uses the plants in her work in her garden studio.

NEW **ALLY PALLY ALLOTMENTS, N22**
Alexandra Palace Way, N22 7BB. Peter Campbell. *Entry gate in a tall wooden fence, nr the footpath to Springfield Ave. Bus W3 to Alexandra Palace Garden Centre stop. Car parking nearby in Dukes Ave N10. Blue badge holder spaces in The Grove car park. No cars on site.* **Sun 2 July (1-4.30). Adm £5, chd free. Home-made teas. Open nearby Railway Cottages.**
Situated on a south facing hillside alongside Alexandra Park, this 140 plot allotment site has the feel of an urban village, with spectacular views towards central London and, on a clear day, the North Downs. Its diverse community of gardeners grow fruit, vegetables and flowers, with an emphasis on promoting biodiversity. Clearly marked trails will enable you to explore the site and meet plotholders. Plants, produce and used tools on sale.

26 ARLINGTON AVENUE, N1
N1 7AX. Thomas Blaikie, www.instagram.com/thomas_blaikie. *South Islington. Off New North Rd via Arlington Ave or Linton St. Buses: 21, 76, 141, 271.* **Evening opening Tue 6 June (6-8). Adm £5, chd free. Wine. Opening with Arlington Square Gardens on Sun 18 June.**
A friend once said 'Your garden is like a tiny corner of some much grander horticultural vision'. I assume it was a compliment. Somehow, over 150 plants approx are crammed into the minute space. Visitors always like my slender willow tree, *Salix exigua*. In June 'a roses evening', the highlights should be a good selection of old roses, climbing and shrub, big dramatic alliums, lupins and tall spires of verbascum.

31 ARLINGTON DRIVE, HA4
Ruislip, HA4 7RJ. John & Yasuko O'Gorman. *Tube: Ruislip, then bus H13 to Arlington Drive, or 15 min walk up Bury St, opp Millar & Carter. Please note: No parking on Arlington Drive. Ample parking on all roads south of Arlington Drive.* **Sun 4 June (2-5). Adm £4, chd free. Home-made teas.**
Cottage garden at heart with a wonderful oriental influence. Traditional cottage garden favourites have been combined with Japanese plants, a reflection of Yasuko's passion for plants and trees of her native Japan. Acers, peonies, rhododendrons and flowering cherries underplanted with hostas, ferns and hellebores, create a lush exotic scheme. Emphasis on structure and texture.

GROUP OPENING

ARLINGTON SQUARE GARDENS, N1
N1 7DP.
www.arlingtonassociation.org.uk.
South Islington. Off New North Rd via Arlington Ave or Linton St. Buses: 21, 76, 141, 271. **Sun 18 June (2-5.30). Combined adm £10, chd free. Home-made teas at St James' Vicarage, 1A Arlington Square.**

15 ARLINGTON AVENUE
Armin Eiber & Richard Armit.

26 ARLINGTON AVENUE.
Thomas Blaikie.
(See separate entry)

21 ARLINGTON SQUARE
Alison Rice.

25 ARLINGTON SQUARE
Michael Foley.

27 ARLINGTON SQUARE
Geoffrey Wheat & Rev Justin Gau.

30 ARLINGTON SQUARE
James & Maria Hewson.

5 REES STREET
Gordon McArthur & Paul Thompson-McArthur.

ST JAMES' VICARAGE, 1A ARLINGTON SQUARE
John & Maria Burniston.

Behind the early Victorian facades of Arlington Square and Arlington Avenue are eight contrasting town gardens. One is the spacious garden of St James' Vicarage, with its impressive herbaceous border and mature London plane trees. The other seven are smaller and remarkably diverse. From city chic to organic wildlife haven, they reflect the diverse tastes and interests of their owners, who know each other through the community gardening of Arlington Square. It is hard to believe you are minutes from the bustle of the City of London.

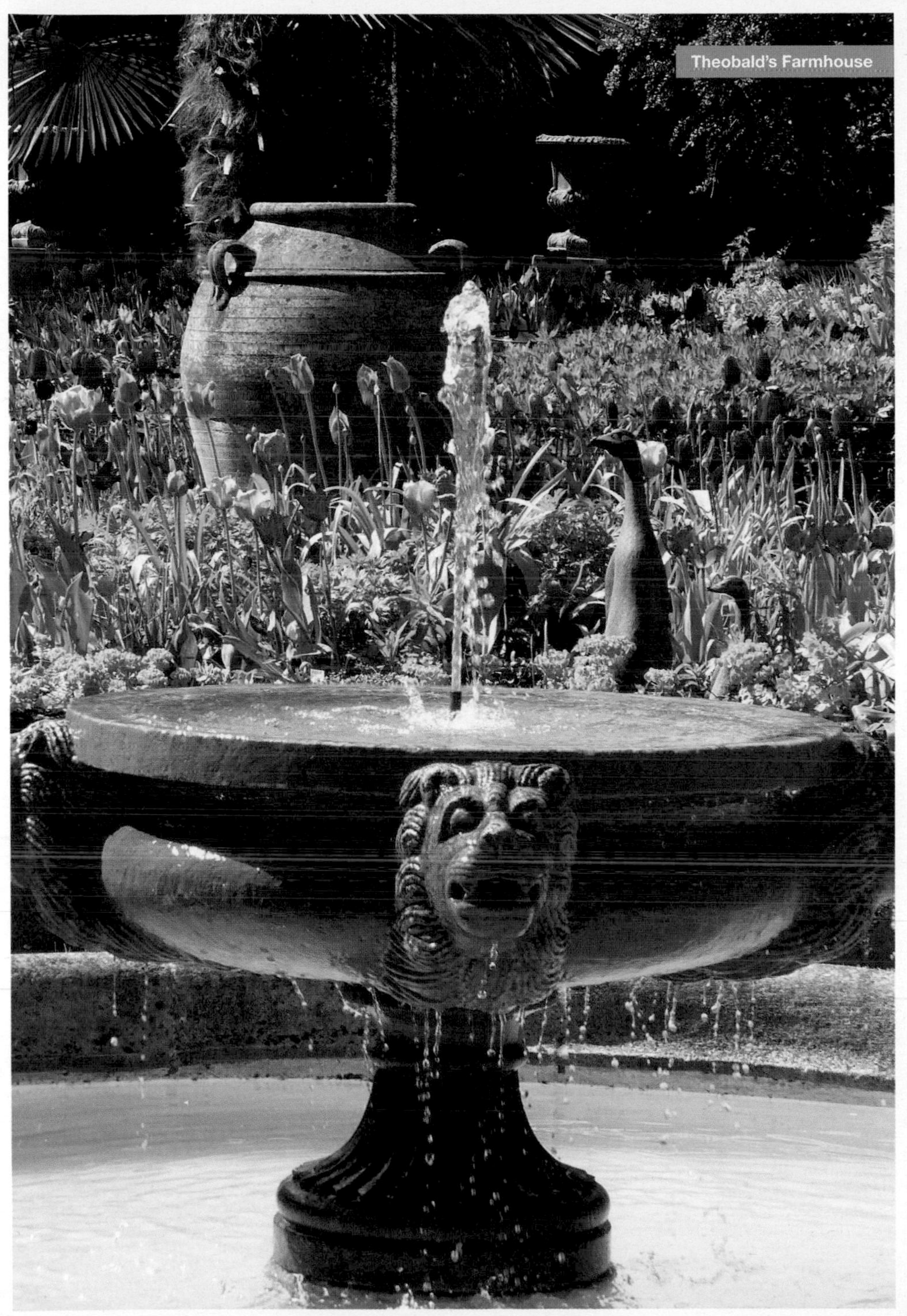

Theobald's Farmhouse

190 Barnet Road

61 ARTHUR ROAD, SW19

Wimbledon, SW19 7DN. Daniela McBride. *5 mins from St. Mary's Church. Tube: Wimbledon Park, then 8 min walk. Train: Wimbledon, 18 min walk.* **Sun 21 May (2-6). Adm £5, chd free. Home-made teas.** This steeply, sloping garden comprises woodland walks, filled with flowering shrubs and ferns. Azaleas, rhododendrons and acers bring early season colour. In early summer the focus is the many roses grown around the garden. Partial wheelchair access to top lawn and terrace only, steep slopes elsewhere.

97 ARTHUR ROAD, SW19

Wimbledon, SW19 7DP. Tony & Bella Covill. *Wimbledon Park Tube Stn, then 200yds uphill on R.* **Sun 11 June (2-6). Adm £5, chd free. Light refreshments.** ½ acre garden of an Edwardian house (not open). Garden established over 25 yrs with a large variety of mature plants, shrubs and sloping lawns become a wildflower meadow. Ponds and fountains encourage an abundance of wildlife and a bird haven, and our new gravel garden is planted to attract butterflies. A peaceful place with much colour, foliage and texture. Partial wheelchair access.

ARUNDEL & ELGIN GARDENS, W11

Kensington Park Road, Notting Hill, W11 2JD. Residents of Arundel Gardens & Elgin Crescent, www.instagram.com/arundelelgin. *Entrance opp 174 Kensington Park Rd. Nearest tube within walking distance: Ladbroke Grove (5 mins) or Notting Hill (10 mins). Buses: 52, 452, 23, 228 all stop opp garden entrance.* **Sun 21 May (2-6). Adm £4, chd free. Home-made teas.** A friendly and informal garden square. One of the best preserved gardens of the Ladbroke estate with mature and rare trees, plants and shrubs laid out according to the original Victorian design of 1862. The larger garden inc a topiary hedge, a rare mulberry tree, a pergola and benches from which vistas can be enjoyed. The central hedged garden is an oasis of tranquillity with extensive and colourful herbaceous borders. All looked after by Gardener, Paul Walsh. Play areas for young children.

ARUNDEL & LADBROKE GARDENS, W11

Kensington Park Road, Notting Hill, W11 2PT. Arundel & Ladbroke Gardens Committee, www. arundelladbrokegardens.co.uk. *Entrance on Kensington Park Rd, between Ladbroke & Arundel Gardens. Tube: Notting Hill Gate or Ladbroke Grove. Buses: 23, 52, 228, 452. Alight at stop for Portobello Market/Arundel Gardens.* **Sun 23 Apr (2-6). Adm £4, chd free.**

Home-made teas.
This private communal garden is one of the few that retains its attractive mid Victorian design of lawns and winding paths. A woodland garden at its peak in spring with rhododendrons, flowering dogwoods, early roses, bulbs, ferns and rare exotics. Live music on the lawn during tea. Playground for small children. Wheelchair access with a few steps and gravel paths to negotiate.

36 ASHLEY ROAD, N19

N19 3AF. Alan Swann & Ahmed Farooqui, swann.alan@googlemail.com, www.instagram.com/space36garden. *Between Stroud Green & Crouch End. Tube: Archway or Finsbury Park. Train: Crouch Hill. Buses: 210 or 41 from Archway to Hornsey Rise. W7 from Finsbury Park to Heathville Rd. Car: Free parking in Ashley Rd at weekends.* **Sun 28, Mon 29 May (2-6). Adm £5, chd free. Home-made teas. Visits also by arrangement 1 May to 17 Sept for groups of 5 to 35. Proceeds for group visits shared with Sunnyside Community Garden.**
A lush town garden rich in textures, colour and forms. At its best in late spring as Japanese maple cultivars display great variety of shape and colour whilst ferns unfurl fresh, vibrant fronds over a tumbling stream and wisteria, and clematis burst into flower on the balcony and pergola. A number of microhabitats inc fern walls, bog garden, stream and ponds, rockeries, alpines and shade plantings. Young ferns and plants propagated from the garden for sale. Pop-up café with a variety of cakes, biscuits, gluten free and vegan options, and traditional home-brewed ginger beer. Seating areas around the garden.

NEW 118B AVENUE ROAD, W3

Acton, W3 8QG. Gareth Sinclair, 07713 020002, londonwildlifegarden@yahoo.com. *Garden located behind the low-rise block of flats of 118. Go on to the grounds of 118, look to the R & go through some gates to arrive at 118b.* **Sun 18 June (12-6). Adm £3.50, chd free. Tea. Open nearby Zen Garden at Japanese Buddhist Centre. Visits also by arrangement 1 May to 1 Oct for groups of up to 15.**
A secret garden, tucked away off the main street. A surprisingly large back garden acquired 3 yrs ago, built on very poor soil and builders' rubble but already looking mature. The ethos of the garden is to encourage wildlife with its meadow, orchard, two ponds and 40 trees. You will find frogs, toads, newts, many invertebrates, dragonflies, stag beetles, nesting garden birds and invariably foxes.

34 BARN RISE, HA9

Wembley, HA9 9NJ. Gita Gami. *500 metres from Wembley Park Stn. Turn L out of station & head to end of road towards T-lights. Cross at T-lights to a precinct of shops & turn L towards Barn Rise.* **Sun 30 July (2-5). Adm £5, chd free. Home-made teas.**
A delightful lush garden, an oasis next to Fryent Country Park, delights the senses of a Mediterranean as well as an English garden, amidst the busy Wembley Park with the view of The Wembley Stadium Arc. Various palms, ferns and *Cornus controversa* 'Variegata'.

190 BARNET ROAD, EN5

The Upcycled Garden, Arkley, Barnet, EN5 3LF. Hilde Wainstein, 07949 764007, hildewainstein@hotmail.co.uk. *1m S of A1, 2m N of High Barnet Tube Stn. Garden located on corner of A411 Barnet Rd & Meadowbanks cul-de-sac. Nearest tube: High Barnet, then 107 bus, Glebe Ln stop. Ample unrestricted roadside parking. NB Do not park half on pavement.* **Visits by arrangement 1 May to 15 Sept for groups of up to 30. Adm £5, chd free. Home-made teas & gluten free options.**
The upcycled garden. Garden designer's walled garden, 90ft x 36ft. Modern design and year-round interest. Flowing herbaceous drifts around central pond. Trees, shrubs herbaceous, bulbs. Upcycled containers, recycled objects, home-made sculptures. Handmade copper pipe trellis divides space into contrasting areas. Garden continues evolving as planted areas are expanded. Gravel garden. Home-made jams and plants for sale, all propagated from the garden. Wheelchair access with single steps within garden.

♿ ✻ ☕ ·))

GROUP OPENING

BARNSBURY GROUP, N1

N1 1BX. *Tube: King's Cross, Caledonian Rd or Angel. Train: Caledonian Rd & Barnsbury. Buses: 17, 91, 259 to Caledonian Rd. Buses 4, 19, 30, 38, 43 to St Mary's Church, Upper St, for Lonsdale Square.* **Sun 4 June (2-6). Combined adm £10, chd free. Individual garden £3 each. Home-made teas at 57 Huntingdon Street.**

◆ BARNSBURY WOOD
London Borough of Islington, ecologycentre@islington.gov.uk.

57 HUNTINGDON STREET.
Julian Williams.
(See separate entry)

2 LONSDALE SQUARE
Jenny Kingsley.

36 THORNHILL SQUARE
Anna & Christopher McKane.

Discover four contrasting spaces within Barnsbury's Georgian squares and terraces. 57 Huntingdon Street is a secluded garden with silver birches, ferns, perennials, grasses and container ponds to encourage wildlife. 36 Thornhill Square, a 120ft garden with a country atmosphere, is filled with roses, specimen trees and perennials. A collection of bonsai and planters surrounds the patio. 2 Lonsdale Square is a charming small cobblestoned garden with herbaceous beds and apple trees, structured with box and yew hedges. Climbing roses and star jasmine add delicious scent; planters overflow with lavender and pansies. Barnsbury Wood is London's smallest nature reserve, a hidden gem and one of Islington's few sites of semi-mature woodland, a tranquil oasis of wild flowers and massive trees minutes from Caledonian Rd. Wildlife information available. The gardens reveal what can be achieved with the right plants in the right conditions, surmounting difficulties of dry walls and shade. Plants for sale at 36 Thornhill Square.

52A BERRYLANDS ROAD, KT5
Surbiton, KT5 8PD. Dr Tim & Mrs Julia Leunig, t.leunig@lse.ac.uk. *2m E of Kingston-upon-Thames. A3 to Tolworth; A240 (towards Kingston) for approx 1m, then R into Berrylands Rd (after Fire Stn). 52A on R after Xrds.* **Sun 2 July (2-5). Adm £4, chd free. Visits also by arrangement.**
Professionally designed. Bold verticality: eucalyptus, ironwood tree, nine bamboos around stream and pond. South shaped lawn, flowing water, tear shaped Indian sandstone patio, clipped hebes, choisya and bay, slender cypress. Exuberance: large tetrapanax leaves, bonkers *Lobelia tupa* flowers, majestic *Magnolia grandiflora*, many climbing roses, day lilies and star jasmine in abundance. Come see!
·)))

GROUP OPENING

BRIXTON WATER LANE GARDENS, SW2
Brixton, SW2 1QB. *Tube: Brixton. Train: Herne Hill, both 10 mins. Buses: 3, 37, 196 or 2, 415, 432 along Tulse Hill.* **Sun 18 June (2-5.30). Combined adm £5, chd free. Tea.**

60 BRIXTON WATER LANE
Caddy & Chris Sitwell.

62 BRIXTON WATER LANE
Daisy Garnett & Nicholas Pearson.

Two 90ft gardens backing onto Brockwell Park with original apple trees from the old orchard. No. 60 has a large garden with floral borders and a mature wisteria covering the house. Strong colour comes from laburnum and lilac under which teas will be served. No. 62 is a country garden with exuberant borders of soft colours, a productive greenhouse and a mass of pots on the terrace. Plenty of colour from old fashioned roses, peonies and other perennials.

5 BURBAGE ROAD, SE24
Herne Hill, SE24 9HJ. Crawford & Rosemary Lindsay, 020 7274 5610, rl@rosemarylindsay.com, www.rosemarylindsay.com. *Nr junction with Half Moon Ln. Herne Hill & N Dulwich Train Stns, 5 min walk. Buses: 3, 37, 40, 68, 196, 468.* **Visits by arrangement 13 Mar to 30 June.**
The garden of a member of The Society of Botanical Artists and regular writer for Hortus journal. 150ft x 40ft with large and varied range of plants, many unusual. Herb garden, packed herbaceous borders for sun and shade, climbing plants, pots, terraces, lawns. Immaculate topiary. Gravel areas to reduce watering. All the box has been removed because of attack by blight and moth and replaced with suitable alternatives to give a similar look. A garden that delights from spring through to summer. Limited plants for sale.

34 BUXTON GARDENS, W3
Black&Bold, Acton, W3 9LQ. Alex Buxton. *10-15 min walk from: Acton Town Stn (Piccadilly Line), Acton Main Train Stn, Acton Central Train Stn, West Acton Stn (Central Line).* **Sat 29, Sun 30 July (3-7). Adm £5, chd free. Home-made teas.**
Have you been to the Black & Bold yet? Traditional elements of a garden reimagined at Black & Bold. Darkest reds, purple and copper spill over the rockery, the cottage garden and on the banks of its ponds which are fed by a waterfall, even the smallest meadow is black. Black & Bold the 1000ft square garden which is different, extraordinary and wild!

CADOGAN PLACE SOUTH GARDEN, SW1
Sloane Street, Chelsea, SW1X 9PE. The Cadogan Estate, www.cadogan.co.uk. *Entrance to garden opp 97 Sloane St.* **Sat 27 May (10-4). Adm £5, chd free. Cream teas.**
Many surprises, unusual trees and shrubs are hidden behind the railings of this large London square. The first square to be developed by architect Henry Holland for Lord Cadogan at the end of C18, it was then called the London Botanic Garden. Mulberry trees planted for silk production at end of C17. Cherry trees, magnolias and bulbs are outstanding in spring. Beautiful 300 yr old black mulberry tree (originally planted to produce silk, but incorrect variety!). Area of medicinal plants in honour of Sir Hans Sloane. This was once the home of the Royal Botanic Garden. Now featuring a bug hotel and children's playground.

4 CANONBURY PLACE, N1
N1 2NQ. Mr & Mrs Jeffrey Tobias. *Canonbury. Highbury & Islington Tube & Train Stns. Bus 271 to Canonbury Square. Located in old part of Canonbury Place, off Alwyne Villas, in a cul-de-sac.* **Sun 26 Mar (2.30-5.30). Adm £4, chd free. Home-made teas.**
Come along and join us on the very first day of British Summer Time! Enjoy the romance of early spring flowering in our historic, secluded 100ft garden with its architectural stonework, sculpture, and ancient hidden fountain, all echoing the 1780 house. Mostly old pots, interesting shrubs and climbers, along with spectacular mature trees and plenty of seating. Artisan pastries and sourdough bread from the legendary Dusty Knuckle Bakery on sale (as supplied to Fortnum & Masons and Ottolenghi). Wonderful home-made cakes.

15 CATHERINE ROAD, KT6
Surbiton, KT6 4HA. Malcolm Simpson & Stefan Gross. *5-10 min walk from Surbiton Stn. Public transport is recommended. Please note limited free car parking in immed area at weekends.* **Sun 9 July (2-6). Combined adm with 7 Woodbines Avenue £7, chd free.**
A well loved town garden approx 40ft x 70ft with deep borders of shrubs and perennial planting, sculptures, a small folly and a walled area.

40 CHARTFIELD AVENUE, SW15
Putney, SW15 6HG. Sally Graham. *Train & Tube: 10 min walk from Putney Train Stn & 15 mins from East Putney Tube Stn. Buses: 14, 37, 93, 85, 39 stop at end of Chartfield Ave on Putney Hill. Some free parking at weekends.* **Sun 4 June (2.30-5). Combined adm with 39 Hazlewell Road £7, chd free. Tea & cakes.**
London garden built from scratch on a sloping site, inc planting more than 30 trees. Inspired by the borrowed landscape with swathes of perennials around a circular lawn leading to a more structural feel around the terrace. New kitchen garden, gravel garden and pergola for this yr. Several shaded seating areas under trees, planted 10 yrs ago.

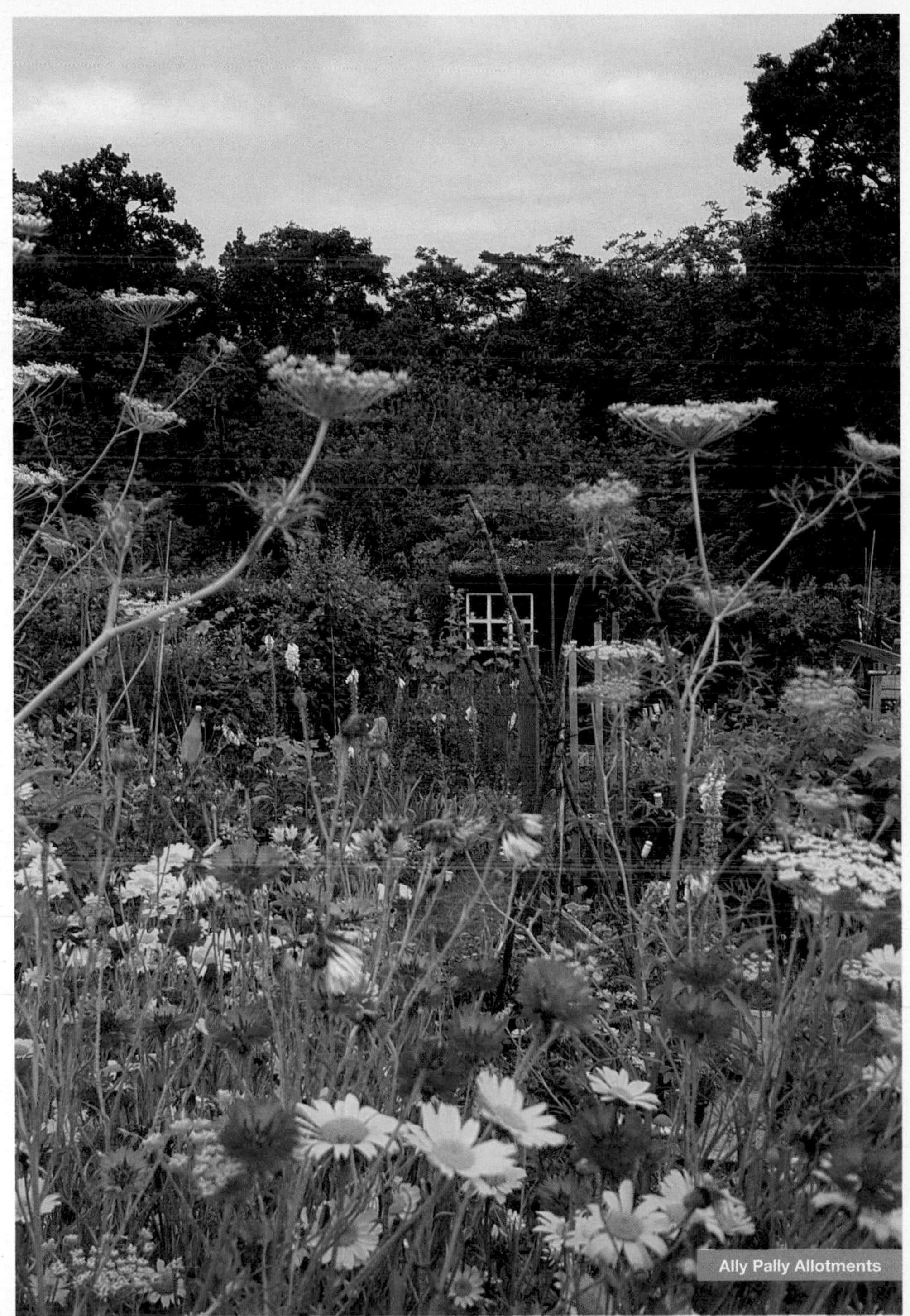

Ally Pally Allotments

51 THE CHASE, SW4
SW4 0NP. Mr Charles Rutherfoord & Mr Rupert Tyler, 07841 418399, charles@charlesrutherfoord.net, www.charlesrutherfoord.net. *Off Clapham Common Northside. Tube: Clapham Common. Buses: 137, 452, 77, 87, 345, 37.* **Sun 23 Apr (12-5). Evening opening Tue 25 Apr (5.30-8). Sun 11 June, Sun 20 Aug (12-5). Adm £4.50, chd free. Light refreshments. Visits also by arrangement 20 Apr to 12 Oct for groups of 10 to 30.**
Member of the Society of Garden Designers, Charles has created the garden over 40 yrs. In 2015 the main garden was remodelled, to much acclaim. Spectacular in spring, when 2000 tulips bloom among camellias, irises and tree peonies. Scented front garden. Rupert's geodetic dome shelters seedlings, succulents and subtropicals. Roses, brugmansia, hibiscus, and dahlias later in the season. Drought tolerant planting.

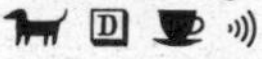

GROUP OPENING

CHISWICK MALL GARDENS, W4
Chiswick Mall, Chiswick, W4 2PS. *By car: Hogarth r'about turn down Church St or A4 (W), turn L to Eyot Gardens. Tube: Stamford Brook or Turnham Green, walk under A4 to river. Buses: 110, 190, 267, H91.* **Evening opening Fri 9 June (6-8). Combined adm £12.50, chd free. Wine at Field House & Swan House. Sun 11 June (1-5.30). Combined adm £15, chd free.**

CEDAR HOUSE
Stephanie & Philippe Camu.
Open on Sun 11 June

FIELD HOUSE
Rupert King, www.fieldhousegarden.co.uk.
Open on all dates

LONGMEADOW
Charlotte Fraser.
Open on all dates

MILLER'S COURT
Miller's Court Tenants Ltd.
Open on Sun 11 June

THE OLD VICARAGE
Eleanor Fein.
Open on Sun 11 June

ST PETERS WHARF
Barbara Brown.
Open on Sun 11 June

SWAN HOUSE
Mr & Mrs George Nissen.
Open on all dates

Gardens on or near the River Thames, three will open on 9 June and seven will open on 11 June. You will see a riverside garden in an artists' complex, several large walled gardens, and a communal garden on the riverbank with well planted borders.

CHOUMERT SQUARE, SE15
Choumert Grove, Peckham, SE15 4RE. The Residents. *Off Choumert Grove. Trains from London Victoria, London Bridge, London Blackfriars, Clapham Junction to Peckham Rye; buses 12, 36, 37, 63, 78, 171, 312, 345. Free car park in Choumert Grove.* **Sun 4 June (1-6). Adm £4, chd free. Savoury items, home-made teas & Pimms. Donation to St Christopher's Hospice.**
About 46 mini gardens with maxi planting in Shangri-la situation that the media has described as a floral canyon, which leads to small communal secret garden. The day is primarily about gardens and sharing with others our residents' love of this little corner of the inner city, but it is also renowned for its demonstrable community spirit and a variety of stalls and live music. Wheelchair access; one tiny step to raised paved area in the communal garden.

26 COLLEGE GARDENS, E4
Chingford, E4 7LG. Lynnette Parvez. *2m from Walthamstow. 15 min walk Chingford Train Stn. Bus 97 from Walthamstow Central Tube Stn, alight at College Gardens, then short walk downhill.* **Sun 11 June (2-5). Adm £4, chd free. Home-made teas.**
Large suburban garden, approx ⅔ acre. Sun terrace leads to established borders and a variety of climbing roses. Beyond this, wildlife pond and lawn, small woodland walk with spring plants and wildlife. A further area to the end with raised vegetable beds and a fruit cage.

2 CONWAY ROAD, N14
Southgate, N14 7BA. Eileen Hulse. *Buses: 121 & W6 from Palmers Green or Southgate to Broomfield Park stop. Walk up Aldermans Hill, R into Ulleswater Rd, 1st L into Conway Rd.* **Sun 3 Sept (2-6). Adm £4.50, chd free. Home-made teas. Open nearby Golf Course Allotments.**
A passion nurtured from childhood for growing unusual plants has culminated in two contrasting gardens. The original, calming with lawn, pond and greenhouse, complements the adjoining Mediterranean terraced rooms with pergolas clothed in exotic climbers, vegetable beds and cordon fruit. Tumbling achocha, figs, datura and *Rosa banksiae* mingle creating a horticultural adventure.

4 CORNFLOWER TERRACE, SE22
East Dulwich, SE22 0HH. Clare Dryhurst. *5 mins walk from 363 & 63 bus stop at bottom of Forest Hill Rd. Turn into Dunstans Rd, then 2nd on L. Stn: Peckham Rye or Honor Oak Park.* **Sun 11 June (2-5). Adm £3.50, chd free.**
This tiny terrace country cottage feel garden has had a rejuvenating rest last yr. While maintaining its roses, ferns and jasmine around the patio and delicate solar fountain, this yr will see simpler, more easily maintained planting, while retaining that welcoming feel as a calm place of relaxation to revitalise the senses.

GROUP OPENING

COURT LANE GROUP, SE21
Dulwich Village, SE21 7EA. *Buses: P4, 12, 40, 176, 185 to Dulwich Library 37. Train Stn: North Dulwich, then 12 min walk. Ample free parking in Court Ln.* **Sun 2 July (2-5.30). Combined adm £8, chd free. Tea & cakes at 122 Court Lane. Savouries at 164 Court Lane.**

122 COURT LANE
Jean & Charles Cary-Elwes.

148 COURT LANE
Mr Anthony & Mrs Sue Wadsworth.

164 COURT LANE
James & Katie Dawes.

No. 164 was recently redesigned to create a more personal and intimate space with several specific zones. A modern terrace and seating area leads onto a lawn with abundant borders and a beautiful mature oak. A rose arch leads to the vegetable beds and greenhouse. No. 122 has a countryside feel, backing onto Dulwich Park with colourful herbaceous borders. and unusual plants. No.148 a spacious garden, developed over 20 yrs backs on to Dulwich Park. From the wide sunny terrace surrounded by tall, golden bamboos and fan palms, step down into a garden designed with an artist's eye for colour, form, texture and flow, creating intimate spaces beneath mature trees and shrubs with colourful perennials. Live jazz on the terrace, a children's trail, plant sales and a wormery demonstration at No.122.

83 COWSLIP ROAD, E18
South Woodford, E18 1JN. Fiona Grant. *5 min walk from Central Line tube. Close to exit for A406.* **Sun 25 June (2-6). Adm £5, chd free. Home-made teas.**
80ft long wildlife friendly garden. On two levels at rear of Victorian semi. Patio has a selection of containers with a step down to the lawn past a pond full of wildlife. Flowerbeds stuffed with an eclectic mix of perennials. Ample seating on patio and under an ancient pear tree.

58 CRANSTON ROAD, SE23
Forest Hill, SE23 2HB. Mr Sam Jarvis & Mr Andres Sampedro. *12 min walk from nearest train stns Forest Hill or Honor Oak. Nearest bus stops: Stanstead Rd/Colfe Rd (185, 122), Kilmorie Rd (185, 171) or Brockley Rise/Cranston Rd (122, 171).* **Sun 30 July (12-5.30). Adm £4, chd free. Teas, cakes, tortilla & sangria.**
An exotic style plant lover's garden features a modern landscaped path and carefully curated subtropical planting. Vivid evergreens inc palms, cordyline, loquat and cycad provide structure and year-round interest. Tree ferns, bananas and tetrapanax add to the striking foliage, while cannas, dahlias and agapanthus provide vibrant pops of colour against the black painted boundaries.

122 Court Lane, Court Lane Group

37 CRESCENT ROAD, BR3
Beckenham, BR3 6NF. Myra & Tim Bright. *15 min walk from Beckenham Junction Stn. 227 bus route from Shortlands Stn. Street parking in adjacent roads.* **Sun 30 Apr, Sun 18 June (2-5.30). Adm £4, chd free. Home-made teas.**
A romantic garden. In spring tulips abound amongst euphorbias and forget-me-nots. In summer, an abundance of geraniums, foxgloves and allium fill the front borders. Go past the potting shed, under the vine arch you arrive in the garden filled with herbaceous plants, roses and clematis, a magnet for bees and butterflies. Gravel paths lead to secret corners with seating.

GROUP OPENING

DE BEAUVOIR GARDENS, N1
100 Downham Road, Hackney, N1 5BE. *Tube stations: Highbury & Islington, then 30 bus; Angel, then 38, 56 or 73 bus; Bank, then 21, 76 or 141 bus. 10 min walk from Dalston Junction Train Stn. Street parking.* **Sun 28 May (2-6). Combined adm £8, chd free. Home-made teas at 100 Downham Road.**

100 DOWNHAM ROAD
Ms Cecilia Darker.

64 LAWFORD ROAD

NEW **72 MORTIMER ROAD**
Mr Stephen King.

NEW **10 UFTON GROVE**
Ms Lynn Brooks.

Four gardens to explore in De Beauvoir, a leafy enclave of Victorian villas near to Islington and Dalston. The area boasts some of Hackney's keenest gardeners and a thriving garden club. 10 Ufton Grove is a modern garden with a sunken pond, sculpture, mirrors and dramatic but easy-to-maintain planting. 72 Mortimer Road is a secluded romantic garden with colourful and wildflower planting. 100 Downham Road features giant echiums, two green roofs and a pond with a miniature Giverny-style bridge. 64 Lawford Road is a small cottage style garden with old fashioned roses, espaliered apples and scented plants.

7 DEANSWAY, N2
East Finchley, N2 0NF. Joan Arnold & Tom Heinersdorff, 07850 764543, joan.arnold40@gmail.com. *Hampstead Garden Suburb. From East Finchley Tube Stn exit along the Causeway to East End Rd, then L down Deansway. From Bishops Ave, head N up Deansway towards East End Rd, close to the top.* **Sun 7 May (12-6). Adm £5, chd free. Home-made teas. Visits also by arrangement Apr to Aug for groups of 10 to 20.**
A garden of stories, statues, shapes and structures surrounded by trees and hedges. Bird friendly, cottage style with scented roses, clematis, mature shrubs, a weeping mulberry and abundant planting. Containers, spring bulbs and grapevine provide year-round colour. Developing secret, shady and wild woodland area with ferns and hostas. Cakes, gluten free cakes and plants for sale (cash preferred).

68 DERBY ROAD, E18
South Woodford, E18 2PS. Mrs Michelle Greene. *Nearest tube: South Woodford (Central Line). Buses: 20 & 179 to Chelmsford Rd. Close to Epping Forest & the Waterworks r'about.* **Sun 23 July (1-5). Adm £4, chd free. Tea.**
Old fashioned summer garden with two ponds, one for wildlife and one for goldfish. Tiny orchard, vegetable beds and rockery. Highlights inc American pokeweed and Himalayan honeysuckle. Many unusual plants with small plants for sale.

DRAGON'S DREAM, UB8
Grove Lane, Uxbridge, UB8 3RG. Chris & Meng Pocock. *Garden is in a small lane very close to Hillingdon Hospital. Parking available next door. Buses from Uxbridge Tube Stn: U1, U3, U4, U5, U7. Buses from West Drayton: U1, U3.* **Sun 11 June (2-5). Adm £5, chd free. Home-made teas.**
This is an unusual and secluded garden with two contrasting sections divided by a brick shed that has been completely covered by a rampant wisteria and a climbing hydrangea and rose. Other highlights inc rare dawn redwood tree and a huge *Gunnera manicata*. More features inc a Romneya poppy, ferns, rose and herb beds, tree peonies, acers and a pond. Also available Malaysian curry puffs and café gourmand (tea or coffee, plus bite-sized cakes).

GROUP OPENING

103 & 105 DULWICH VILLAGE, SE21
SE21 7BJ. *Train: North Dulwich or West Dulwich then 10-15 min walk. Tube: Brixton then P4 bus, alight Dulwich Picture Gallery stop. Street parking.* **Sun 18 June (2-5). Combined adm £10, chd free. Home-made teas at 103 Dulwich Village. Donation to Link Age Southwark.**

103 DULWICH VILLAGE
Mr & Mrs N Annesley.

105 DULWICH VILLAGE
Mr & Mrs A Rutherford.

Two Georgian houses with large gardens, 3 min walk from Dulwich Picture Gallery and Dulwich Park. 103 Dulwich Village is a country garden in London with a long herbaceous border, lawn, pond, roses and fruit and vegetable gardens. 105 Dulwich Village is a very pretty garden with many unusual plants, old fashioned roses, fish pond and water garden. Excellent cakes. Amazing collection of plants for sale from both gardens. Please bring your own bags for plants. Music provided by the Colomb Street Ensemble Wind Band.

ECCLESTON SQUARE, SW1
Pimlico, SW1V 1NP. The Residents of Eccleston Square, www.ecclestonsquaregardens.com. *Off Belgrave Rd nr Victoria Stn, parking allowed on Suns.* **Sun 14 May (2-5). Adm £5, chd free. Home-made teas.**
Planned by Cubitt in 1828, the 3 acre square is subdivided into mini gardens with camellias, iris, ferns and containers. Dramatic collection of tender climbing roses and 20 different forms of tree peonies. National Collection of Ceanothus inc more than 70 species and cultivars. Notable important additions of tender plants being grown and tested. World collection of ceanothus, tea roses and tree peonies.

EDWARDES SQUARE, W8
South Edwardes Square, Kensington, W8 6HL. Edwardes Square Garden Committee, www.edwardes-square-garden.co.uk. *Tube: Kensington High St & Earls Court. Buses: 9, 10, 27, 28, 31, 49 & 74 to Odeon Cinema. Entrance in South Edwardes Square.* **Sun 16 Apr (12-5). Adm £5, chd free. Home-made teas.**
One of London's prettiest secluded garden squares. 3½ acres laid out differently from other squares with serpentine paths by Agostino Agliothe, an Italian artist and decorator who lived at No.15 from 1814-1820. This quiet oasis is a wonderful mixture of rolling lawns, mature trees and imaginative planting. Children's play area. WC. Wheelchair access through main gate, South Edwardes Square.

40 EMBER LANE, KT10
Esher, KT10 8EP. Sarah & Franck Corvi, 07803 111968, sarah.corvi@ngs.org.uk. *½m from centre of Esher. From the A307, turn into Station Rd which becomes Ember Ln.* **Sun 11 June (1-5). Combined adm with 9 Imber Park Road £6, chd free. Home-made teas. Visits also by arrangement 19 June to 30 June for groups of up to 20.**
A contemporary family garden designed and maintained by the owners with distinct areas for outdoor living. A 70ft east facing plot where the lawn has been mostly removed to make space for the owner's love of plants and several ornamental trees.

12 FAIRFIELD ROAD, N8
N8 9HG. Christine Lane. *North London. Tube: Finsbury Park & then W3 (Weston Park stop) or W7 (Crouch End Broadway), alternatively Archway & then 41 bus (Crouch End Broadway), then a short walk.* **Sun 2 July (2-5.30). Adm £4, chd free.**
A tranquil garden created on two levels behind a large Victorian house. There are a variety of trees, shrubs and flowers, as well as succulents, palms and bamboos. Sculptures and statues add drama and plants in pots create interest and height. The patio, with its table and chairs overlooks the lawn, flower beds, cobbled garden and fern area, then up to the upper garden where there is more seating. Crouch End offers an abundance of cafés and restaurants.

70 FARLEIGH ROAD, N16
Stoke Newington, N16 7TQ. Mr Graham Hollick. *Short walk from junction of Stoke Newington High St & Amhurst Rd.* **Sun 25 June (10-6). Adm £4, chd free. Home-made teas.**
A diverse garden in a Victorian terrace with an eclectic mix of plants, many in vintage pots reflecting the owner's interests. A small courtyard leads onto a patio surrounded by pots followed by a lawn flanked by curving borders. At the rear is a paved area with raised beds containing vegetables.

28 FERNDENE ROAD, SE24
Herne Hill, SE24 0AB. Mr David & Mrs Lynn Whyte, www.instagram.com/dsw_garden. *Buses 68, 468, 42. A 5 min walk from Denmark Hill. Train stns: Herne Hill, Denmark Hill, Loughborough Junction. All 15 min walk. House overlooks Ruskin Park. Free parking.* **Sun 25 June (1-5.30). Adm £4, chd free. Home-made teas.**
It is all about structure and careful planting in this dramatically sloping south south east facing garden, 30 x 18 metres. A lively blend of perennials and shrubs shows definite Kiwi influences. The kitchen garden with raised beds and soft fruits is wonderfully secluded. Lower-level planting has a coastal feel. Upper-level has a hot colour border. Borrowed views of mature trees and big skies set it off. The garden office/summerhouse, constructed from sustainably sourced materials, has a rubble roof to attenuate water runoff. Recycling and re-use of materials.

1 FIFE ROAD, SW14
East Sheen, SW14 7EW. Mr & Mrs J Morgan. *Bus routes 33, 337 stops are a 15 min walk away on Upper Richmond Rd. Mortlake Train Stn SWT approx 20 mins away.* **Sun 25 June (1-6). Adm £5, chd free. Home-made teas.**
Dominated by two large cedar trees the garden has matured since it was redesigned in 2013. A maturing oak tree stands in the wild garden with fruit trees and 'dead hedge'. A productive vegetable garden and greenhouse is managed with a no dig regime. Lawns and borders with perennials and shrubs, sunny terrace close to the house.

GROUP OPENING

FOREST HILL GARDEN GROUP, SE23
Forest Hill, SE23 3DE. *Off S Circular Rd (A205) behind Horniman Museum & Gardens. Forest Hill Stn, 10 min walk. Buses 176,185,197, 356, P4.* **Sun 14 May (2-5.30). Combined adm £6, chd free. Home-made teas at 53 Ringmore Rise. Donation to St Christopher's Hospice & Marsha Phoenix Trust.**

THE COACH HOUSE, 3 THE HERMITAGE
Pat Rae.

27 HORNIMAN DRIVE
Rose Agnew.

NEW **35 NETHERBY ROAD**
Mr & Mrs K MacLennan.

53 RINGMORE RISE
Valerie Ward.

Four charming gardens (one new this yr) spread across the highest hill in South East London, nr Horniman Museum. Some with stunning views over the Downs and all in a short walk of each other. Enjoy the intricate combinations and zonal planting of a plantswoman's garden on three levels from a patio area offering roses and honeysuckle and fantastic views across London. The upper terrace is bordered on one side with woodland planting and the other with drought resistant plants. Wrapped round an C18 coach house, is an artist's walled courtyard with her own sculptures, fountain, birdbath and decorative plants in pots for year-round interest. Check the amazing dark wisteria. Relax amid a delightful embroidery of vibrant colours in a garden that's a haven of peace and harmony with truly breathtaking views over the Downs. A fruitful garden that takes you back to a time when this area was rich in orchards. Plum, cherry and apple trees spread their branches, and strawberries tumble out of pots. Plants for sale at 27 Horniman Drive.

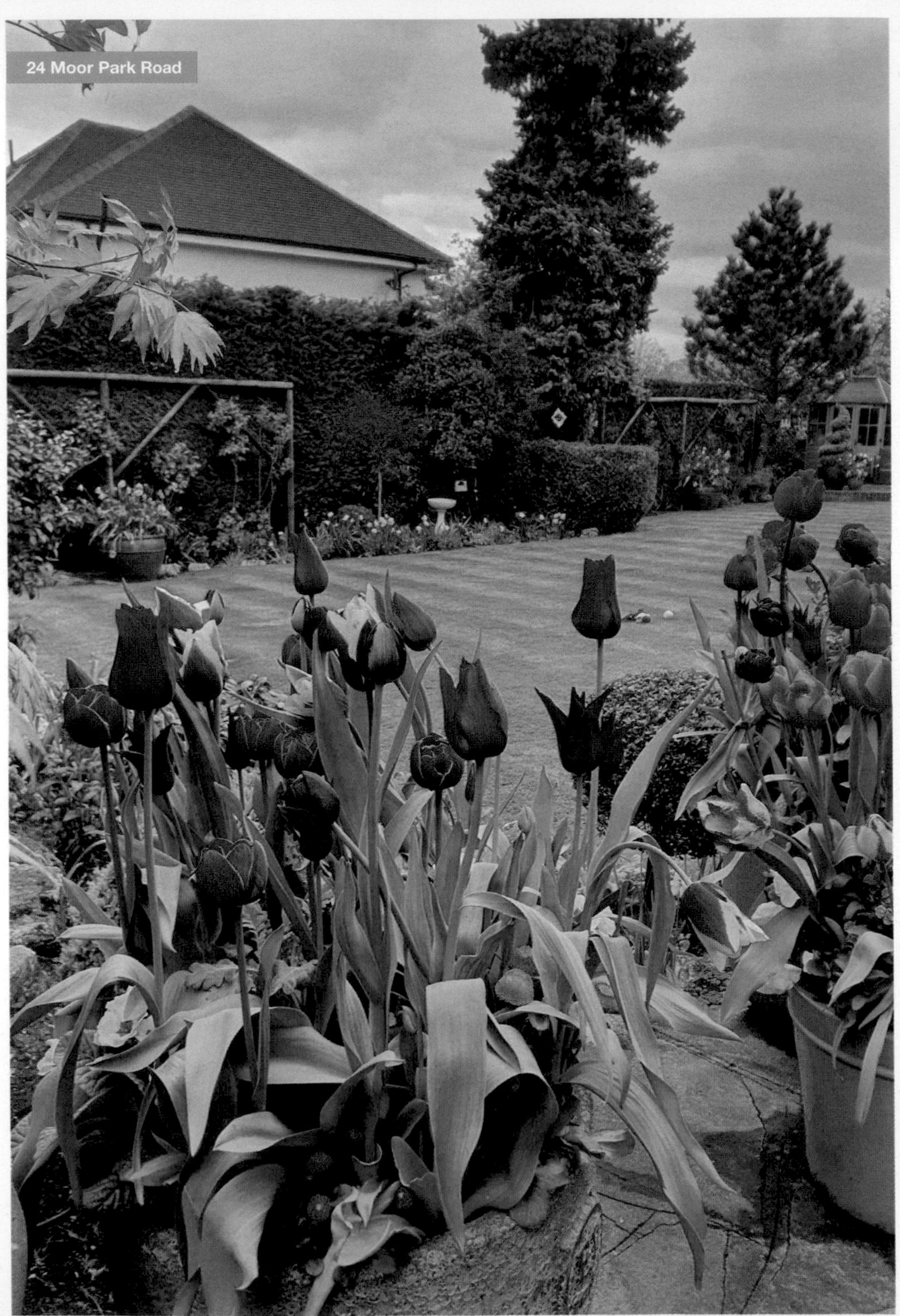
24 Moor Park Road

10 FURLONG ROAD, N7

N7 8LS. Gavin & Nicola Ralston. *Tube & Train: Highbury & Islington, 3 min walk along Holloway Rd, 2nd L. Furlong Rd runs between Holloway Rd & Liverpool Rd. Buses on Holloway Rd: 43, 271, 393; other buses 4, 19, 30.* **Sun 21 May (2-5.30). Adm £5, chd free. Home-made teas.**

A green oasis in the heart of a densely populated area, 10 Furlong Road is an open, sunny garden of considerable size for its urban location. Its new owners, previously at Canonbury House, extensively remodelled the garden in 2019, building on a foundation of trees, shrubs and roses, adding herbaceous planting, seating, winding brick path, pergola, raised vegetable bed and wildflower circle. Plenty of seating dotted around the garden.

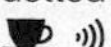

GARDEN BARGE SQUARE AT TOWER BRIDGE MOORINGS, SE1

31 Mill Street, SE1 2AX. Mr Nick Lacey, towerbridgemoorings.org. *5 min walk from Tower Bridge. Mill St off Jamaica Rd, between London Bridge & Bermondsey Stns, Tower Hill also nearby. Buses: 47, 188, 381, RV1.* **Sun 14 May (2-5). Adm £5, chd free. Home-made teas.**

Series of seven floating barge gardens connected by walkways and bridges. Gardens have an eclectic range of plants for year-round seasonal interest. Marine environment: suitable shoes and care needed. Small children must be closely supervised.

◆ THE GARDEN MUSEUM, SE1

5 Lambeth Palace Road, SE1 7LB. The Garden Museum, 020 7401 8865, info@gardenmuseum.org.uk, www.gardenmuseum.org.uk. *Lambeth side of Lambeth Bridge. Tube: Lambeth North, Vauxhall, Waterloo. Buses: 507 Red Arrow from Victoria or Waterloo Train & Tube Stns, also 3, 77, 344, C10.* **For NGS: Tue 12 Sept (10-5). Adm £7, chd free. Light refreshments. For other opening times and information, please phone, email or visit garden website.**

At the heart of the Garden Museum is our courtyard garden, designed by Dan Pearson as an 'Eden' of rare plants inspired by John Tradescant's journeys as a plant collector. Taking advantage of the sheltered, warm space, Dan has created a green retreat in response to the bronze and glass architecture, conjuring up a calm, reflective atmosphere. Visitors will also see a permanent display of paintings, tools, ephemera and historic artefacts; a glimpse into the uniquely British love affair with gardens. The Garden Café is an award-winning restaurant. The museum is accessible for wheelchair users via ramps and access lift.

GARDEN OF MEDICINAL PLANTS, ROYAL COLLEGE OF PHYSICIANS, NW1

11 St Andrews Place, Regents Park, NW1 4LE. Royal College of Physicians of London, garden.rcplondon.ac.uk. *Opp the SE corner of Regents Park. Tubes: Great Portland St & Regent's Park. Garden is one block N of station exits, on Outer Circle opp SE corner of Regent's Park. There is no access via Peto Place.* **Mon 10 July (10-4). Adm £5, chd free.**

We have over 1,000 different plants connected with the history of plants in medicine. These inc plants named after physicians, plants which make modern medicines as well as those used for millennia, plants which cause epidemics, plants used in different medical traditions from all the continents of the world and plants from the College's Pharmacopoeia of 1618. Guided tours will be offered throughout the day by physicians explaining the uses of the plants, their histories and other stories. Books about the plants in the medicinal garden will be on sale alongside free leaflets. All the plants are labelled with their botanical names. Entry to the garden is at far end of St Andrews Place. Accessible paths around the garden. Some slopes. Wheelchair lift for WC. No parking on site.

39 GLOVER ROAD, HA5

Pinner, HA5 1LQ. Prakash & Joanne Daswani. *2 tube stops N of Harrow on the Hill on Metropolitan line. 12 min walk/3 min drive from Pinner Tube Stn (Metropolitan line). Turn L into Marsh Rd 350yds, then R into Eastcote Rd 900yds, then L into Rosecroft Walk 250yds, then R at T-junction. Free parking.* **Sun 11 June (3-7). Adm £5, chd free. Cream teas & home-made teas.**

Newly landscaped compact space, sensitively filled with varied elements that complement one another, to generate a sense of union with nature and an instant welcome to sit and gaze. Range of plants of all types across numerous beds inc long established, towering trees alongside new specimens. Multiple seating areas, large deck, steel gazebo and octagonal water feature offer alternative vistas. Wheelchair access via side gate onto newly paved 1.2 metre wide path; temporary ramps to patio, then lawn and main planting beds.

GOLF COURSE ALLOTMENTS, N11

Winton Avenue, N11 2AR. GCAA Haringey, www.golfcourseallotments.co.uk. *Junction of Winton Ave & Blake Rd. Tube: Bounds Green. Buses: 102, 184, 299 to Sunshine Garden Centre, Durnsford Rd. Through park to Bidwell Gardens. Straight on up Winton Ave. New LTN on Blake Rd, beware cameras. No parking on site.* **Sun 3 Sept (1-4.30). Adm £4, chd free. Home-made teas & light lunches. Open nearby 2 Conway Road.**

Large, long established allotment with over 200 plots, some organic. Maintained by culturally diverse community growing a wide variety of fruit, vegetables and flowers enjoyed by bees. With picturesque corners, quirky sheds and tours of best plots and newly created plot 147, a jewel garden not to be missed. A visit feels like a holiday in the countryside. Healthy fresh allotment produce, chutneys, jams and honey for sale (cash only). Wheelchair access to main paths only. Gravel and uneven surfaces. WC inc disabled.

21 GOSPATRICK ROAD, N17
Tottenham, N17 7EH. Matthew Bradby. *London Zone 3. Tube: Turnpike Ln or Wood Green Stns. Train: Bruce Grove Stn. Bus routes: 144, 217, 231, 444 to Gospatrick Rd, or 123, 243 to Waltheof Ave, or 318 to Gt Cambridge Rd.* **Sat 13 May (1-5). Adm £4, chd free. Light refreshments.**
Diverse 40 metre plot with contrasting areas. Ornamental garden dominated by large weeping willow giving shade over fan palms, camellias, ferns and climbers. Fruit and herb garden with large Japanese banana, grapevine, olive, bay and mimosa trees. Greenhouse with succulent plants, and two ponds. Mainly organic, this is a very tranquil and welcoming garden.

45 GREAT NORTH ROAD, EN5
Barnet, EN5 1EJ. Ron & Miriam Raymond, 07880 500617, ron.raymond91@yahoo.co.uk. *1m S of Barnet High St, 1m N of Whetstone. Tube: Midway between High Barnet, Totteridge, & Whetstone Stns. Buses: 34, 234, 263, 326, alight junction Great North Rd & Lyonsdown Rd. 45 Great North Rd is on the corner of Cherry Hill.* **Sun 30 July, Sun 13 Aug (1.30-5.30). Adm £3.50, chd free. Home-made teas. Wine. Visits also by arrangement 1 Aug to 20 Aug for groups of 5 to 20.**
45 Great North Road is designed to give a riot of colour in late summer. The 90ft x 90ft cottage style front garden is packed with shrubs and perennials. Tiered stands line the side entrance with over 64 pots displaying a variety of flowering and foliage plants. The rear garden inc nearly 100 tubs and hanging baskets. Small pond surrounded by tiered beds. Magnificent named Tuberous begonia. Children's fun trail for 3-6yr olds and adult garden quiz with prizes. Partial wheelchair access.

17 GREENSTONE MEWS, E11
Wanstead, E11 2RS. Mrs T Farnham, 07761 476651, farnhamz@yahoo.co.uk. *Tube: Snaresbrook or Wanstead, 5 min walk. Buses: 101, 308, W12, W14 to Wanstead High St. Greenstone Mews is accessed via Voluntary Place which is off Spratt Hall Rd.* **Visits by arrangement 26 May to 3 Sept for groups of 6 to 8. Adm £5. Tea or coffee inc. Sandwiches & cake on request.**
Tiny slate paved garden, approx 15ft sq. Sunken reused bath now a fish pond surrounded by climbers clothing fences underplanted with hardy perennials, herbs, vegetables, shrubs and perennials grown from cuttings. Height provided by established palm and iron arch. Ideas aplenty for small space gardening. Regret garden unsuitable for children. Wheelchair access through garage but limited turning space.

24 GROVE PARK, SE5
Camberwell, SE5 8LH. Clive Pankhurst, www.alternative-planting.blogspot.com. *Chadwick Rd end of Grove Park. Peckham Rye or Denmark Hill Stns, both 10 min walk. Good street parking.* **Sun 3 Sept (11-4.30). Adm £5, chd free. Home-made teas.**
An inspiring exotic jungle of lush big, leafed plants and Southeast Asian influences transport you to the tropics. Huge hidden garden created from derelict land that had been the bottom halves of two neighbouring gardens gives the wow factor and unexpected size. Lawn and lots of hidden corners give spaces to sit and enjoy. Renowned for delicious home-made cake and plant sale.

92 HAMPSTEAD WAY, NW11
Hampstead Garden Suburb, NW11 7XY. Ann & Tom Lissauer. *In square set back on Hampstead Way, between Finchley Rd & Meadway. Buses 13,102 & 460 on Finchley Rd, getting off at Temple Fortune Ln & walk down Hampstead Way to the square. Golders Green Tube.* **Sun 14 May (2-6). Combined adm with 100 Hampstead Way £10, chd free. Home-made teas in the square between the two gardens.**
An informal garden, interesting year-round, combining wildlife friendly planting with a passion for plants. Different areas provide a variety of habitats. These inc a wildlife pond and, where there used to be lawns, there are now meadows with mown paths and wild flowers. Shaded and sunny beds offer opportunities to grow a wide range of interesting plants.

100 HAMPSTEAD WAY, NW11
Hampstead Garden Suburb, NW11 7XY. S & J Fogel. *In square set back on Hampstead Way, between Finchley Rd & Meadway. Buses 13,102 & 460 on Finchley Rd, getting off at Temple Fortune Ln & walk down Hampstead Way to the square. Golders Green tube.* **Sun 14 May (2-6). Combined adm with 92 Hampstead Way £10, chd free. Home-made teas on the square between the two gardens.**
Corner cottage garden with a variety of viewing perspectives and featuring sculpture and planting in recycled objects (pallets, sinks, dustbins, mattress on wheels, wine boxes, poles, chimneys). The garden comprises a number of rooms inc a formal parterre, wooded area, walkway, small meadow and formal lawn.

SPECIAL EVENT

◆ HAMPTON COURT PALACE, KT8
East Molesey, KT8 9AU. Historic Royal Palaces, www.hrp.org.uk. *Follow brown tourist signs on all major routes. Junction of A308 with A309 at foot of Hampton Court Bridge. Traffic is heavy around Hampton Court. Please leave plenty of time, the tour will start promptly at 6pm & will not be able to wait.* **For NGS: Evening opening Thur 20 Apr (6-8). Adm £15. Pre-booking essential, please visit www.ngs.org.uk for information & booking. Wine. For other opening times and information, please visit garden website. Donation to Historic Royal Palaces.**
Take the opportunity to join a special National Garden Scheme private tour after the wonderful historic gardens have closed to the public. Spring walk during the tulip festival in the remarkable gardens of Hampton Court Palace. Wheelchair access over some un-bound gravel paths.

NPC

NEW 4 HANWORTH ROAD, TW13
Feltham, TW13 5AB. Maria Vail. *5 mins walk from Feltham Train Stn. Opp Feltham Stn into Hanworth Rd. House on R after the school. Entrance through side gate on Cardinal Rd.* **Sun 2 July (1-4). Adm £4.50, chd free. Light refreshments.**
Appropriately designed garden

surrounding a Victorian villa. Pollarded lime trees in the front garden and mature palm trees with colourful under planting. Many tender and unusual Mediterranean plants. Lawn with flower filled urns and pots, rockery and colourful borders. An eclectic mix of plants.

37 HAROLD ROAD, E11

Leytonstone, E11 4QX. Dr Matthew Jones Chesters. *Tube: Leytonstone, exit L subway, 5 min walk. Train: Leytonstone High Road, 5 min walk. Buses: 257 & W14. Parking at station or limited on street.* **Sun 11 June (1-5). Adm £4, chd free. Home-made teas.**

50ft x 60ft pretty corner garden arranged around seven fruit trees. Fragrant climbers, woodland plants and shade tolerant fruit along north wall. Fastigiate trees protect raised vegetable beds and herb rockery. Long lawn bordered by roses and perennials on one side, prairie plants on the other. Patio with raised pond, palms and rhubarb. Planting designed to produce fruit, fragrance and lovely memories. Garden map with plant names and planting plans inc.

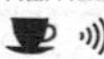

NEW HAVEN HOUSE HOSPICE, IG8

High Road, Woodford Green, IG8 9LB. *Woodford Green. Turn off Woodford Green High St following signs. Park at rugby club. Nearest tube station Woodford. From station bus 179 or W13. Alight at Chingford Ln.* **Sat 15 Apr (12-5). Adm £4, chd free. Light refreshments at café on site.**

Celebrating 20 yrs, this hidden historic garden started by accident when Viscount Montgomery planted walnut and chestnut trees. Two weeks later, Sir Winston Churchill added a red oak. There are over 128 trees many of them planted by Prime Ministers and Presidents. Woodland walk. Carpets of daffodils and spring flowering shrubs.

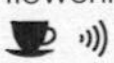

NEW 20 HAZELMERE ROAD, BR5

Petts Wood, Orpington, BR5 1PB. Victoria & Matthew Stephens, 07866 715125, matthewjrstephens@gmail.com. *A 10 min walk from Petts Wood Stn (22 mins direct from London Bridge / 35 mins from London Victoria). Free parking on street. Adjacent to Petts Wood NT woodland (free).* **Sun 4 June (2-5.30). Adm £4, chd free. Home-made teas. Visits also by arrangement 3 June to 24 Sept for groups of 5 to 15.**

A recently designed contemporary garden which was part built by the owners and features perennial planting with a mix of trees, blending into the borrowed landscape. Gravel pathways from the generous terrace lead you on a journey through the planting with a secret garden over the brook that runs through the garden. Clipped hedging and contemporary features punctuate the planting and pathways.

84 Lavender Grove, London Fields Gardens

NEW 39 HAZLEWELL ROAD, SW15

Putney, SW15 6LS. Kath Brown. *Train & Tube: 12 min walk from Putney Train Stn & 18 mins from East Putney Tube Stn. Buses: 14, 37, 93, 85 & 39 stop on Putney Hill nr St John's Ave. No parking restrictions on Sun.* **Sun 4 June (2.30-5). Combined adm with 40 Chartfield Avenue £7, chd free. Light refreshments.**

An 80x40ft town garden with a surprisingly secluded feel provided by the shared landscape of gardens and mature trees. A naturalistic planting scheme of trees, shrubs and perennials in curved beds surround the lawn. The sunny patio, bordered by lavender and climbing roses, looks toward a water feature set in a perennial bed. At the rear a small shady patio is framed by trees and woodland planting.

32 HIGHBURY PLACE, N5

N5 1QP. Michael & Caroline Kuhn. *Highbury Fields. Tube & Train: Highbury & Islington. Buses: 4, 19, 30, 43, 271, 393 to Highbury Corner. 3 min walk up Highbury Place which is opp station.* **Sun 28 May (2-5.30). Adm £5, chd free. Home-made teas.**

An 80ft garden behind a C18 terrace house (not open). An upper York stone terrace leads to a larger terrace surrounded by overfilled beds of cottage garden style planting. Further steps lead to a lawn by a rill and a lower terrace. A willow tree dominates the garden; amelanchiers, fruit trees and dwarf acers, winter flowering cherry, lemon trees and magnolia.

HIGHWOOD ASH, NW7

Highwood Hill, Mill Hill, NW7 4EX. Mrs P Gluckstein. *Totteridge & Whetstone on Northern line, then bus 251 stops outside Rising Sun/ Mill Hill stop. By car: A5109 from Apex Corner to Whetstone. Garden opp The Rising Sun pub at top of hill on the R.* **Sun 14 May (2-5.30). Adm £5, chd free. Home-made teas.**

Created over the last 56 yrs, this 3¼ acre garden features rolling lawns, two large interconnecting ponds with koi, herbaceous and shrub borders and a modern gravel garden. A country garden in London for all seasons with many interesting plants and sculptures. May should be perfect for the camellias, rhododendrons and azaleas. Partial access for wheelchairs, lowest parts too steep.

NEW 15 HILLCREST ROAD, E18

South Woodford, Woodford, E18 2JL. Roger Hammond. *Accessible by tube & bus.* **Sun 28 May (2-6). Adm £5, chd free.**

Lovely walled town garden divided into three areas. Near the house is a large York stone, brick and hardwood terrace, with plenty of spaces to sit and relax. Next section is a checkerboard of grass and brick squares with a beautiful robinia 'false acacia' at one end, giving shade in summer. Leading from this is a circular path, heavily planted all round, with a whitebeam in the centre.

26 HILLCROFT CRESCENT, HA9

Wembley Park, HA9 8EE. Gary & Suha Holmyard, 07773 691331. *½m from Wembley Park Stn. If held on Wembley event day, we can provide free parking permits. Turn R out of Wembley Park Stn walking down Wembley Park Drive, turn L into Manor Drive, Hillcroft Cres is 2nd on R.* **Sat 1, Sun 2 July (11-4). Adm £5, chd free. Home-made teas & soya option.**

Small cottage front garden with arched entrance with fuchsias, wisteria, yucca, canna, hydrangeas, roses and lilies. Rear garden approx 70ft x 80ft with summerhouse, arbours and water features. Planting inc fig, apple, pear, plum, olive and soft fruit. Ten different flower beds each holding its own particular interest inc lilies, unusual evergreen shrubs, exotics, clematis and roses. Plenty of seating around the garden. Wheelchair access through side gate via driveway.

THE HOLME, NW1

Inner Circle, Regents Park, NW1 4NT. Lessee of The Crown Commission. *In centre of Regents Park on The Inner Circle. Within 15 min walk from Great Portland St or Baker St Tube Stns, opp Regents Park Rose Garden Cafe.* **Sat 29, Sun 30 July, Sat 26, Sun 27 Aug (2.30-5.30). Adm £5, chd free.**

4 acre garden currently undergoing some restructuring. Set within Regent's Park, highlights inc sweeping lakeside lawns, an extensive rock garden with waterfall, stream and pool, and a formal garden with fountain and pool. Gardeners will help wheelchair users to negotiate gravel paths and steps.

103 Dulwich Village

239A HOOK ROAD, KT9
Chessington, KT9 1EQ. Mr & Mrs D St Romaine, www.gardenphotolibrary.com. *4m S of Kingston. A3 from London, turn L at Hook underpass onto A243. Garden 300yds on L. Parking opp in park or on road, no restrictions at night. Buses K4, 71, 465 from Kingston & Surbiton to North Star pub.* **Evening opening Fri 16 June (7.30-10). Adm £6, chd free. Wine.**
Patio with raised vegetable bed and garden room. A central path of standard hollies, urns and box balls leads to a water feature at the end of the garden. Rectangular beds with rose covered obelisks form a grid either side of central axis. Enclosed on three sides by sun and shade borders, shrubs and perennials. All cleverly lit to enhance the atmosphere in the evenings. Wheelchair access with one low step.

NEW **HORATIO'S GARDEN, HA7**
Royal National Orthopaedic Hospital, Brockley Hill, Stanmore, HA7 4LP. Horatio's Garden, www.horatiosgarden.org.uk. *Enter the Hosptial via Wood Ln, Aspire entrance if arriving by car.* **Sun 23 July (2-5). Adm £5, chd free. Home-made teas.**
Opened in Sept 2020, Horatio's Garden London & South East located at the Royal National Orthopaedic Hospital, Stanmore is designed by Tom Stuart-Smith. The garden is on one level with smooth paths throughout ensuring that it is easily accessible to patients in beds and wheelchairs. The essential design features inc a social space, private areas for patients to seek solitude or share with a family member or friend, the calming sound of flowing water, a garden room, a garden therapy area and a greenhouse. Throughout the garden the planting is designed to supply colour all year-round, whilst wildlife has been encouraged with bird and butterfly boxes. The Head Gardener Ashley Edwards will be on hand to answer any plant questions and give guided tours of this unique sanctuary. The whole site is designed for wheelchairs.

1A HUNGERFORD ROAD, N7
N7 9LA. David Matzdorf, davidmatzdorf@blueyonder.co.uk, growingontheedge.net/index.php. *Between Camden Town & Holloway. Tube: Caledonian Rd. Buses: 17, 29, 91, 253, 259, 274, 390 & 393. Parking free on Suns.* **Sun 4 June (1-6). Adm £4.50, chd free. Visits also by arrangement Apr to Oct for groups of up to 6.**
Unique eco house with walled, lush front garden in modern exotic and woodland style, densely planted with palms, acacia, bamboo, ginger lilies, bananas, ferns, yuccas, abutilons and unusual understory plants. Floriferous and ambitious green roof resembling Mediterranean or Mexican hillside, planted with yuccas, dasylirions, nolinas, agaves, cacti, aloes, cistus, euphorbias, grasses, sedums and herbs. Sole access to roof is via built in vertical ladder. Garden and roof each 50ft x 18ft (15 metres x 6 metres).

57 HUNTINGDON STREET, N1
Barnsbury, Islington, N1 1BX. Julian Williams, 07759 053001, julianandroman@me.com. *Train: Caledonian Rd & Barnsbury. Tube: Kings Cross or Highbury & Islington. Buses: Caledonian Rd 17, 91, 259, 274; Hemingford Rd 153 from Angel.* **Evening opening Thur 18 May (6-8). Adm £4, chd free. Wine. Opening with Barnsbury Group on Sun 4 June. Visits also by arrangement 10 Mar to 30 June for groups of up to 10.**
A secluded woodland garden room below an ash canopy and framed by timber palisade supporting roses, hydrangea and ivy. An understorey of silver birch and hazel provides the setting for shade loving ferns, perennials and grasses. Oak pathways offset the informal planting and lead to a tranquil central space with bench seating and two container ponds to encourage wildlife.

National Garden Scheme gardens are identified by their yellow road signs and posters. You can expect a garden of quality, character and interest, a warm welcome and plenty of home-made cakes!

GROUP OPENING

HYDE PARK ESTATE GARDENS, W2
Kendal Street, W2 2AN. Church Commissioners for England, www.hydeparkestate.com. *The Hyde Park Estate is bordered by Sussex Gardens, Bayswater Rd & Edgware Rd. Nearest tube stations inc Marble Arch, Paddington & Edgware Rd.* **Sat 10 June (10-4). Combined adm £6, chd free. Pre-booking essential, please visit www.ngs.org.uk for information & booking.**

CONISTON COURT
DEVONPORT
NEW **GLOUCESTER SQUARE GARDEN**
NEW **OXFORD SQUARE GARDEN**
THE QUADRANGLE
REFLECTIONS 2020
THE WATER GARDENS

Unique opportunity to visit seven Central London gardens usually only seen by residents. These gardens only open to the public for the NGS. Each garden planted sympathetically to reflect the surroundings and to support biodiversity. The gardens on the Hyde Park Estate are owned and managed by the Church Commissioners for England and play a key part in the environmental and ecological strategy on the Hyde Park Estate. The Estate covers 90 acres of which 12½ % is 'green', not only with the garden spaces but by installing planters on unused paved areas, green roofs on new developments and olive trees throughout Connaught Village. For 2023 we are delighted to inc Gloucester Square garden as part of the opening with the kind co-operation of the Garden Committee. Wheelchair access to most gardens. There are some steps to the upper levels of The Water Gardens.

9 IMBER PARK ROAD, KT10
Esher, KT10 8JB. Jane & John McNicholas, 07867 318655, jane_mcnicholas@hotmail.com. *½m from centre of Esher. From the A307, turn into Station Rd which becomes Ember Ln. Go past Esher Train Stn on R. Take 3rd road on R into Imber Park Rd.* **Sun 11 June (1-5). Combined adm with 40 Ember Lane £6, chd free. Home-made teas. Visits also by arrangement 19 June to 30 June for groups of up to 20.**
An established cottage style garden, always evolving, designed and maintained by the owners who are passionate about gardening and collecting plants. The garden is south facing with well-stocked, large, colourful herbaceous borders containing a wide variety of perennials, evergreen and deciduous shrubs, a winding lawn area and a small garden retreat.

SPECIAL EVENT

THE INNER AND MIDDLE TEMPLE GARDENS, EC4
Crown Office Row, Inner Temple, EC4Y 7HL. The Honourable Societies of the Inner & Middle Temples, www.innertemple.org.uk/www.middletemple.org.uk. *Entrance: Main Garden Gate on Crown Office Row, access via Tudor St Gate or Middle Temple Ln Gate.* **Wed 14 June (11.30-3). Adm £55, chd free. Pre-booking essential, please visit www.ngs.org.uk for information & booking. Light refreshments.**
Inner Temple Garden is a haven of tranquillity and beauty with sweeping lawns, unusual trees and charming woodland areas. The well known herbaceous border shows off inspiring plant combinations from early spring through to autumn. The award-winning gardens of Middle Temple are comprised of a series of courtyards. Adm inc conducted tour of the gardens by Head Gardeners. Light lunch in Middle Temple Hall, please advise of any dietary requirements. Please advise in advance if wheelchair access is required.

GROUP OPENING

KEW GREEN GARDENS, TW9
Kew, TW9 3AH. virginiagodfrey69@gmail.com. *NW side of Kew Green. Tube: Kew Gardens. Train Stn: Kew Bridge. Buses: 65, 110. Entrance via riverside.* **Sun 21 May (2-5). Combined adm £8, chd free. Evening opening Sun 28 May (6-8). Combined adm £10, chd free. Teas in St Anne's Church (21 May). Wine (28 May). Visits also by arrangement 29 May to 15 June.**

65 KEW GREEN
Giles & Angela Dixon.

67 KEW GREEN
Lynne & Patrick Lynch.

69 KEW GREEN
John & Virginia Godfrey.

71 KEW GREEN
Mr & Mrs Jan Pethick.

73 KEW GREEN
Sir Donald & Lady Elizabeth Insall.

The long gardens run for 100yds from the back of historic houses on Kew Green down to the Thames towpath. Together they cover nearly 1½ acres, and in addition to the style and structures of the individual gardens they can be seen as one large space, exceptional in London. The borders between the gardens are mostly relatively low and the trees and large shrubs in each contribute to viewing the whole, while roses and clematis climb between gardens giving colour to two adjacent gardens at the same time.
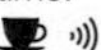

KING HENRY'S WALK GARDEN, N1
11c King Henry's Walk, N1 4NX. Friends of King Henry's Walk Garden, www.khwgarden.org.uk. *Buses: 21, 30, 38, 56, 141, 277. Behind adventure playground on KHW, off Balls Pond Rd.* **Mon 1 May (2-4.30). Adm £4, chd free. Home-made teas. Donation to Friends of KHW Garden.**
Vibrant ornamental planting welcomes the visitor to this hidden oasis leading into a verdant community garden with secluded woodland area, beehives, wildlife pond, wall trained fruit trees, and plots used by local residents to grow their own fruit and vegetables. Live music. Disabled access WC.

53 LADY AYLESFORD AVENUE, HA7
Stanmore, HA7 4FG. Jadon. *About 15 min walk from Stanmore Stn, off Uxbridge Rd, close to St John Church. H12 & 340 bus stops are 5 min walk from garden. Limited free parking nearby.* **Sun 23 July (11-5). Adm £5, chd free. Tea.**
A compact tropical fusion garden developed over the past 4 yrs, set in a 20 yr old development of a Battle of Britain, RAF base. This delightful gem of a corner garden with water features and a stunning display of plants and flowers, shows what can be achieved, even in a small space. Colours and textures blend effortlessly to create a harmonious space with exceptional attention to detail.

LAMBETH PALACE, SE1
Lambeth Palace Road, SE1 7JU. The Church Commissioners, www.archbishopofcanterbury.org. *Entrance via Main Gatehouse (Morton's Tower) facing Lambeth Bridge. Stn: Waterloo. Tube: Westminster, Vauxhall all 10 min walk. Buses: 3, C10, 77, 344, 507.* **Evening opening Mon 15 May (5-8). Adm £6, chd free. Wine.**
Lambeth Palace has one of the oldest and largest private gardens in London. It has been occupied by Archbishops of Canterbury since 1197. Formal courtyard boasts historic White Marseilles fig planted in 1556. Parkland style garden features mature trees, woodland and native planting. There is a formal rose terrace, summer gravel border, scented chapel garden and active beehives. Please note: Gates will open at 5pm, last entry is 7pm and garden closes at 8pm. Garden Tours will be available. Wheelchair access with ramped path to rose terrace. Disabled WC.

12 LAURADALE ROAD, N2
Fortis Green, N2 9LU. David Gilbert & Mary Medyckyj, www.davidgilbertart.com/garden. *300 metres from 102 & 234 bus stops. 500 metres from 43 & 134 bus stops. 10 min walk from East Finchley Tube Stn. Look for NGS signs.* **Sun 10 Sept (1-6). Adm £5, chd free. Home-made teas.**

Exotic, huge, featuring tropical and Mediterranean zone plants, now maturing well and extended with recent developments. Dramatic, architectural planting inc bananas, large tree ferns and rare palms weave along curving stone paths, culminating in a paradise garden. A modern take on the rockery embeds glacial boulders amid dry zone plants inc many succulents. Sculptures by artist owner.

84 LAVENDER GROVE, E8
Hackney, E8 3LS. Anne Pauleau, 07930 550414, a.pauleau@hotmail.co.uk. *Short walk from Haggerston or London Fields Train Stns.* **Sun 23 Apr, Sun 9 July (2-5). Adm £4, chd free. Home-made teas. Opening with London Fields Gardens on Sun 4 June. Visits also by arrangement 1 Apr to 18 Oct.**
Courtyard garden with tropical backdrop of bamboos and palms, foil to clipped shrubs leading to wilder area. The cottage garden with mingling roses, lilies, alliums, grasses, clematis, poppies, star jasmine and jasmine. A very highly scented garden with rampant ramblers and billowing vegetation enchanting all senses. Tulips and daffodils herald spring. Fiery crocosmias and dahlias trumpet late summer. Children's quiz offered with prize on completion.

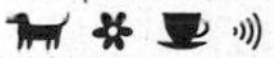

16 LINKS VIEW ROAD, TW12
Hampton Hill, TW12 1LA. Guy & Virginia Lewis. *5 min walk from Fulwell Stn. On 281, 267, 285 & R70 bus routes.* **Sat 27, Sun 28 May (2-4.30). Adm £5, chd free. Open nearby 70 Wensleydale Road. Home-made teas with gluten free options.**
A surprising garden featuring acers, hostas and fern collection, and other unusual shade loving plants. Many climbing roses, clematis and herbaceous border, and pots of exotic plants. Rockery and folly with shell grotto and waterfall to small pond and bog garden. Lawn and formal pond. Wild area with chickens and summerhouse. Greenhouse with succulent collection. A veranda with pelargonium collection. One very friendly dog. Wheelchair access with assistance.

GROUP OPENING

LONDON FIELDS GARDENS, E8
Hackney, E8 3JW. *London Fields. 7 min walk from 67, 149, 242, 243 bus stop Middleton Rd, 10 mins from 30, 38, 55 stops on Dalston Ln. 7 mins from Haggerston Train Stn or 10 min walk through London Fields from Mare St buses.* **Sun 4 June (2-5.30). Combined adm £10, chd free. Home-made teas at 84 Middleton Road & 65 Shrubland Road.**

84 LAVENDER GROVE.
Anne Pauleau.
(See separate entry)

53 MAPLEDENE ROAD
Tigger Cullinan.

55 MAPLEDENE ROAD
Amanda & Tony Mott.

84 MIDDLETON ROAD
Penny Fowler.

92 MIDDLETON ROAD
Mr Richard & Dr Louise Jarrett.

65 SHRUBLAND ROAD
Ms Jackie Cahoon.

A fascinating and diverse collection of gardens in London Fields within easy walking distance of each other. At 84 Lavender Grove there are twin south facing gardens, a courtyard with tropical backdrop, and a highly scented, romantic cottage garden. 92 Middleton Road is elegant and serene with a circular theme inc roses, acers and examples of stone lettering. At 84 Middleton lies an unusually large secret garden where you can wander down meandering woodland paths and forget you are in London. The other three are north facing with much the same space but totally different styles. 65 Shrubland Road is a designer's garden with exuberant planting and materials for discrete zoning. 53 Mapledene Road is an established plantaholic's garden in five sections with not a spare unplanted inch. No. 55 has a Moorish-inspired terrace leading to a wildlife garden with plants chosen to attract birds, bees and butterflies.

38 LOVELACE ROAD, SE21
Dulwich, SE21 8JX. José & Deepti Ramos Turnes. *Midway between West Dulwich & Tulse Hill Stns. Buses: 2, 3 & 68.* **Sun 25 June (12-4.30). Adm £4, chd free. Home-made teas.**
This gem of a garden has an all white front and a family friendly back. The garden slopes gently upwards with curving borders, packed with an informal mix of roses, perennials and annuals. The garden is designed to be an easy to maintain oasis of calm at the end of a busy day. There are several smile inducing features like a dragon, the Cheshire cat and a stream.

GROUP OPENING

LOWER CLAPTON GARDENS, E5
Lower Clapton, E5 0RL. *12 min walk from Hackney Central, Hackney Downs or Homerton Stns. Buses 38, 55, 106, 242, 253, 254 or 425, alight Lower Clapton Rd or Powerscroft Rd. Street parking.* **Sun 4 June (2-5). Combined adm £7, chd free.**

8 ALMACK ROAD
Philip Lightowlers.

NEW **10 ALMACK ROAD**
Mr Ben Myhill.

77 RUSHMORE ROAD
Penny Edwards.

Lower Clapton is an area of mid Victorian terraces sloping down to the River Lea. These gardens reflect their owner's tastes and interests. New this year is 10 Almack Road, a long garden with architectural plants like trachycarpus palms and cordylines, a large pond and much York stone and London brick. Next door is 8 Almack Road, a similar space but divided into two rooms, one cool and peaceful the other with hot colours, succulents and greenhouse. 77 Rushmore Road has a fruit and vegetable garden and wildlife pond.

65 MILL HILL ROAD, W3
Acton, W3 8JF. Anna Dargavel, 07802 241965, annadargavel@mac.com. *Tube: Acton Town, turn R, Mill Hill Rd 2nd R off Gunnersbury Ln.* **Sun 3 Sept (2-5). Combined adm with 41 Mill Hill Road £6, chd free. Visits also by arrangement May to Sept for groups of 5 to 15.**
Garden designer's garden. A secluded and tranquil space, paved with changes of level and borders. Sunny areas, topiary, a greenhouse and interesting planting combine to provide a wildlife haven. A pond and organic principles are used to promote a green environment and give a stylish walk to a studio at the end of the garden.

NEW **24 MOOR PARK ROAD, HA6**
Northwood, HA6 2DJ. Nikki & Tony Bello. *Nearest tube station Northwood. Garden is on the corner of Moor Park Rd & Grove Rd.* **Sun 30 July (2-5.30). Adm £5, chd free. Light refreshments.**
Immaculately designed and maintained garden. Many unusual features inc a collection of acers and olives in interesting pots and containers. Large collection of quirky topiary in both the front and back garden. Colourful border of summer perennials is mixed with seasonal bedding. Nooks and crannies, and a collection of wind chimes and other surprises.

THE MYSTERIES OF LIGHT ROSARY GARDEN, NW5
St Dominic's Priory (the Rosary Shrine), Southampton Road, Kentish Town, NW5 4LB. Raffaella Morini on behalf of the Church & Priory, 07778 526434, garden@raffaellamorini.com. *Entrance to the garden is from Alan Cheales Way on the RHS of the church, next to the school.* **Sat 3 June (1.30-5.30). Adm £4.50, chd free. Home-made teas. Visits also by arrangement 1 Apr to 1 Sept for groups of up to 20.**
A small walled garden behind the Priory Church of Our Lady of the Rosary and St Dominic, commissioned by the Dominican Friars as a spiritual and meditative space representing the 'Mysteries of Light' of the Holy Rosary. The sandstone path marks out a Rosary with black granite beads, surrounded by flowers traditionally associated with the Virgin Mary: roses, lilies, iris, periwinkle, columbine. The garden is fully accessible with a stone path and a wheelchair friendly gravel path.

27 NASSINGTON ROAD, NW3
Hampstead, NW3 2TX. Lucy Scott-Moncrieff. *From Hampstead Heath Train Stn & bus stops at South End Green, go up South Hill Pk, then Parliament Hill, R into Nassington Rd.* **Sun 4 June (2-6). Adm £5, chd free. Light refreshments.**
Double width town garden planted for colour and to support wildlife. Spectacular ancient wisteria, prolific roses; herbs and unusual fruit and salad crops in with the flowers. The main feature is a large eco pond with colourful planting in and out of the water and lots of minibeasts. Pots and planters, arches, bowers, view of allotments and very peaceful location give a rural feel in the city. Live music from the Secret Life Sax Quartet from 3.30pm. Cakes inc lemon drizzle made with lemons from the garden and gluten free cakes; teas inc rose hips from the garden, but also real tea.

15 NORCOTT ROAD, N16
Stoke Newington, N16 7BJ. Amanda & John Welch. *Buses: 67, 73, 76, 106, 149, 243, 393, 476, 488. Clapton & Rectory Rd Train Stns. Be aware of current traffic calming measures.* **Sun 14 May (2-6). Adm £3.50, chd free. Home-made teas.**
A large walled garden developed by the present owners over 40 yrs with pond, aged fruit trees and an abundance of herbaceous plants, many available in our plant sale, resumed after last yr's very limited Sept post drought sale. We have plenty of room for people to sit, relax and enjoy their tea.

5 NORTHAMPTON PARK, N1
N1 2PP. Andrew Bernhardt & Anne Brogan. *Backing on to St Paul's Shrubbery, Islington. 5 min walk from Canonbury Train Stn, 10 mins from Highbury & Islington Tube (Victoria Line) Buses: 73, 30, 56, 341, 476.* **Sat 17 June (1-6). Combined adm with 7 St Paul's Place £7, chd free. Wine, prosecco & strawberries.**
Early Victorian south facing walled garden (1840s), saved from neglect and developed over the last 28 yrs. Cool North European blues, whites and greys moving to splashes of red and orange Mediterranean influence. The contrast of the cool garden shielded by a small park creates a sense of seclusion from its inner London setting.

OAK FARM/HOMESTEAD, EN2
Cattlegate Road, Crews Hill, Enfield, EN2 9DS. Genine & Martin Newport. *5 mins from M25 J24 & J25. Follow yellow NGS signs. Few mins walk from Crews Hill Stn. If coming via train to Crews Hill Stn, on the main road turn L (opp Warmadams).* **Sun 10 Sept (12-4). Adm £6, chd free. Teas served in the barn.**
In the heart of Crews Hill is our 3 acre garden and meadow. It has been reclaimed over 30 yrs from pig farm to relaxed planting, a haven for wildlife. Walled garden leads to vegetable plot, greenhouse, and orchard. Woodland walk, lawns and stone ornaments. Martin built the house, the brick walls and metal work. Abundant cyclamen in woodland in Sept. Arboretum and lovely country views. Wheelchair access to gardens over grass.

21 OAKLEIGH PARK SOUTH, N20
N20 9JS. Carol & Robin Tullo, 07909 901731, robin.tullo@btinternet.com. *Totteridge & Whetstone Tube Stn (Northern Line), 15 min walk or 251 bus. Oakleigh Park Train Stn, 10 min walk. Also buses 34 & 125 from High Rd. Plenty of street parking.* **Visits by arrangement 1 Apr to 1 Oct. Adm £4, chd free. Tea & cake on request.**
A late spring opening. A mature 200ft garden framed by a magnificent 100 yr old ash tree. Path leads to a pond area fed by a natural spring within landscaped terraced paving. Beyond is a herb and vegetable area, orchard with bulbs and wild flowers and the working part of the garden. A mix of sunny borders, pond marginals and woodland shade areas with seating. Level wheelchair access to terrace and lawn. Path to pond area, but raised levels beyond.

PADDOCK ALLOTMENTS & LEISURE GARDENS, SW20
51 Heath Drive, Raynes Park, SW20 9BE. Paddock Horticultural Society. *Buses 57, 131, 200 to Raynes Park Stn, then 10 min walk or bus 163. Bus 152 to Bushey Rd, 7 min walk. Bus 413, 5 min walk from Cannon Hill Ln. Street parking.* **Sat 24 June (12-5). Adm £4, chd free. Light refreshments.**
An allotment site not to be missed, over 150 plots set in 5½ acres. Our tenants come from diverse communities growing a wide range of flowers, fruits and vegetables. Some plots are purely organic, others resemble English country gardens. Winner of London in Bloom Best Allotment on four occasions. Plants and produce for sale. Ploughman's lunch available. Wheelchair access over paved and grass paths, mainly level.

11 PARK AVENUE NORTH, N8
Crouch End, N8 7RU. Mr Steven Buckley & Ms Liz Roberts. *Buses: 144, W3, W7. Tube: Finsbury Park or Turnpike Ln. Train: Hornsey or Alexandra Palace.* **Sun 4 June (11.30-5.30). Adm £4, chd free. Home-made teas. Open nearby 25 Springfield Avenue.**
An award-winning, exotic 250ft garden, much developed in 2022. Dramatic foliage, spiky and lush dominates with the focus on dragon trees, palms, aloes, agaves, dasylirions, aeoniums, tree ferns, nolinas, cycads, bamboos, yuccas, bananas, cacti, puyas and hardy succulents. Trees inc orange, peach, *Cussonia spicata* and Szechuan pepper. Vegetables grow in oak raised beds and a glasshouse.

36 PARK VILLAGE EAST, NW1
Camden Town, NW1 7PZ. Christy Rogers. *Tube: Mornington Cres or Camden Town, 7 mins. Opp railway, just S of Mornington St bridge. Free parking on Sun.* **Sun 11 June (2-6). Adm £6, chd free. Home-made teas.**
A large peaceful garden behind a sympathetically modernised John Nash house. Relandscaped in 2014, retaining the original mature sycamores and adding hornbeam hedges dividing a woodland area and orchard from a central large lawn, mixed herbaceous border and rose bank. Children enjoy an artificial grass slide. Musical entertainment is provided by young local musicians. Wheelchair access via grass ramp down from driveway to main garden (steeper than wheelchair regulations).

174 PECKHAM RYE, SE22
East Dulwich, SE22 9QA. Mr & Mrs Ian Bland. *Station: Peckham Rye. Buses: 63, 363,12,197, 484. Overlooks Peckham Rye Common from Dulwich side.* **Sun 11 June (2.30-5.30). Adm £4, chd free. Home-made teas. Open nearby 4 Piermont Green. Donation to St Christopher's Hospice.**
This oasis, the size of a tennis court has been opening for over 25 yrs and is currently full of contrasting foliage, heights and colours. Among the usual suspects you will find rarer pseudowintera and panax, dodonaea, 'T. Rex', various arisaema's, dracunculus, a baby oven's wattle and a giant fleece flower. The plant sale is popular or just sit on the lawn with a cup of tea and the tamed cakes. Wheelchair access via side alley.

5 PEMBERTON ROAD, KT8
East Molesey, KT8 9LG. Armi Maddison. *Please enter the garden down the side path to R of house.* **Sun 11 June (2-5). Adm £4, chd free. Light refreshments.**
An artist's sheltered and secluded gravel garden, designed alongside our new build in 2015. Many grasses, pink, blue and white planting with occasional pops of bright colour, a galvanised drinking trough with bullrushes and water lilies, a large mature central acer tree, combine with several sitting areas to extend our living space into this fabulous outdoor room. Modern house and artist studio border the garden on three sides with large sliding doors making the garden our sheltered outside room.

PERTH ROAD GARDEN, IG2
110 Perth Road, Gant's Hill, Ilford, IG2 6AS. Mark Kenny. *5-10 mins from the M11 & the North Circular. Gant's Hill Tube & Ilford Train Stns are 15 min walk away.* **Sat 29 July (1-4). Adm £4, chd free.**
A suburban garden of a typical Ilford terrace house. The garden is long, and the pathways are small. Be prepared to disappear in its jungle like structure. Populated by a range of plants which form a rich tapestry combining the tropical, the Mediterranean, Japan and the UK. Cannas, bananas, buddleias and acers fight it out for space alongside a multiplicity of other plant types. A garden whose journey can be best described as Quaker Meeting Hall to Baroque Cathedral.

PETERSHAM HOUSE, TW10
Petersham Road, Petersham, Richmond, TW10 7AA. Francesco & Gael Boglione, www.petershamnurseries.com. *Station: Richmond, bus 65 to Dysart. Entry to garden off Petersham Rd, through Petersham Nurseries. Parking very limited on Church Ln.* **Sun 23 Apr (11-4). Adm £7, chd free. Light refreshments.**
Broad lawn with large topiary and generously planted double borders. Productive vegetable garden with chickens. Adjoins Petersham Nurseries with extensive plant sales, shop and café serving lunch, tea and cake.

4 PIERMONT GREEN, SE22
East Dulwich, SE22 0LP. Janine Wookey. *Triangle of green facing Peckham Rye at the Honor Oak Rd end. Stations: Peckham Rye & Honor Oak. Buses: 63 & 363 (pass the door) & 12. No parking on green, but free parking on side streets nearby.* **Sun 11 June (1.30-5). Adm £4.50, chd free. Home-made teas. Open nearby 174 Peckham Rye.**
This odd shaped enclosed garden is veering towards a more relaxed feel, aka 'easier to maintain', though it means sacrificing some lawn. A purple elder tree produces blossom for cordial. Global warming helps the banana grove flourish. A gravel garden is low maintenance with *Crambe maritima* and crimson dierama in June, and *Althaea cannabina* later. A pebble mosaic under the mulberry tree is new. Live music. Wheelchair access with a couple of front steps to negotiate.

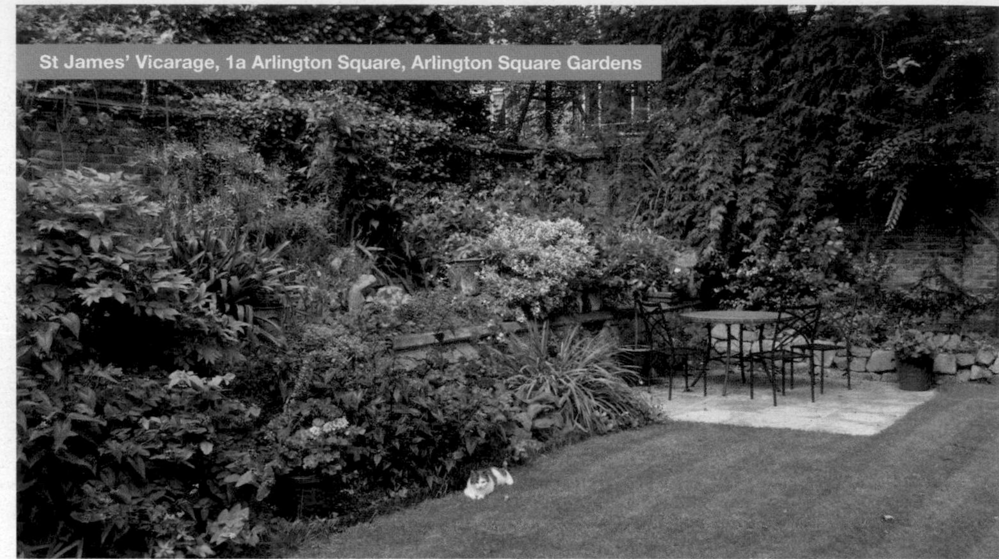
St James' Vicarage, 1a Arlington Square, Arlington Square Gardens

1A PRIMROSE GARDENS, NW3
Hampstead, NW3 4UJ. Debra Craighead, dcraighead@me.com. *Belsize Park. Convenient from Belsize & Chalk Farm Tube Stns (5 mins) or Swiss Cottage (12 mins). Also buses, 168, 268 & C11. Free parking on Suns.* **Sun 18 June (2-6). Adm £5, chd free. Tea or coffee inc. Visits also by arrangement 20 May to 7 Oct for groups of 10 to 35.**
Hidden oasis in the heart of Belsize Park, cool and relaxing. Planted with various microclimates owing to surrounding buildings, limited sun and desire for privacy. Mirrors help create sense of intrigue. An oasis of calm, it features grasses, perennial borders and a water feature in a sunny spot while shade loving species reside elsewhere. Bird friendly with a sedum rooftop attracting bees and butterflies. Cakes, desserts and refills for sale.

GROUP OPENING

PRINCES AVENUE GARDENS, N10
N10 3LS. *Muswell Hill. Buses: 43 & 134 from Highgate Tube Stn; also W7, 102, 144, 234, 299. Princes Ave opp M&S in Muswell Hill Broadway & The Village Green pub in Fortis Green Rd.* **Sun 14 May (12-6). Combined adm £5, chd free. Home-made teas.**

17 PRINCES AVENUE
Patsy Bailey & John Rance.

28 PRINCES AVENUE
Ian & Viv Roberts.

In a beautiful Edwardian avenue in the heart of Muswell Hill Conservation Area, two very different gardens reflect the diverse lifestyles of their owners. The peaceful garden at No. 17 is designed for relaxing and entertaining. Although south facing it is shaded by large surrounding trees, among which is a ginkgo. The garden features a superb hosta and fern display. No. 28 is a well established traditional garden reflecting the charm typical of the era. Mature trees, shrubs, mixed borders and woodland garden creating an oasis of calm just off the bustling broadway. Live music at 17 Princes Avenue by the Secret Life Sax Quartet, from 4.30 to 5.30pm.

GROUP OPENING

RAILWAY COTTAGES, N22
2 Dorset Road, N22 7SL. *Tube: Wood Green, 10 min walk. Train: Alexandra Palace, 3 mins. Buses: W3, 184, 3 mins. Free parking in local streets on Suns.* **Sun 2 July (2-5.30). Combined adm £5, chd free. Home-made teas at 2 Dorset Road. Open nearby Ally Pally Allotments.**

2 DORSET ROAD
Jane Stevens.

4 DORSET ROAD
Mark Longworth.

14 DORSET ROAD
Cathy Brogan.

22 DORSET ROAD
Mike & Noreen Ainger.

24A DORSET ROAD
Eddie & Jane Wessman.

A row of historical railway cottages, tucked away from the bustle of Wood Green nr Alexandra Palace, takes the visitor back in time. The tranquil country style garden at 2 Dorset Road flanks three sides of the house. Clipped hedges contrast with climbing roses, clematis, honeysuckle, abutilon, grasses and ferns. Trees inc mulberry, quince, fig, apple and a mature willow creating an interesting shady corner with a pond. There is an emphasis on scented flowers that attract bees and butterflies and the traditional medicinal plants found in cottage gardens. No. 4 is a pretty secluded garden (accessed through the rear of No. 2) and sets off the sculptor owners figurative and abstract work. There are three front gardens open for view. No. 14 is an informal, organic, bee friendly garden, planted with fragrant and useful herbs, flowers and shrubs. No. 22 is nurtured by the grandson of the original railway worker occupant. A lovely place to sit and relax and enjoy the varied planting. No. 24a reverts to the potager style cottage garden with raised beds overflowing with vegetables and flowers. Popular plant sale.

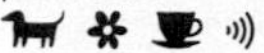

42 RISINGHOLME ROAD, HA3
Harrow, HA3 7ER. Brenda White. *Buses: 258, 340,182,140 Salvatorian College/St. Joseph's Catholic Church, Wealdstone. Tube/train: Harrow & Wealdstone Stn (10 min walk or bus). Road opp the Salvatorian College.* **Sun 14 May, Sun 3 Sept (12-4). Adm £5, chd free. Home-made teas.**
A stunning garden packed with plants, divided into different areas inc a tropical themed area, a Mediterranean area filled with roses and topiary, a raised bed vegetable garden, a large aviary and summerhouse. Feature plants at different times of the yr inc acers, azaleas, camellias, ferns, hydrangeas, rhododendrons, palms, cannas and grasses as well as box hedging and topiary.

ROOFTOPVEGPLOT, W1
122 Gt Titchfield Street, W1W 6ST. Miss Wendy Shillam, 020 7637 0057, coffeeinthesquare@me.com, www.rooftopvegplot.com. *Fitzrovia. Located on the 5th floor, flat roof of a private house. Ring the doorbell marked Shillam & Smith to be let into the building.* **Sat 17 June, Sun 18 June (11-5). Adm £5, chd free. Pre-booking essential, please visit www.ngs.org.uk for information & booking. Home-made teas. Open nearby Warren Mews on 17 June only.**
A nutritional garden, where fruit and vegetables grow amongst complementary flowers in 6 inches of soil, in raised beds on a flat roof. This is a tiny garden, so tours are restricted to six visitors. Home-made cakes, and growing and nutritional tips from Wendy Shillam, a keen environmentalist with an extensive knowledge of green nutrition.

ROYAL TRINITY HOSPICE, SW4
30 Clapham Common North Side, SW4 0RN. Royal Trinity Hospice, www.royaltrinityhospice.org.uk. *Tube: Clapham Common. Buses: 35, 37, 345,137 (37 & 137 stop outside).* **Sun 2 Apr, Sun 21 May, Sun 25 June, Sun 17 Sept (10.30-4.30). Adm £3, chd free. Light refreshments.**
Royal Trinity's beautiful, award-winning gardens play an important therapeutic role in the life and function of Royal Trinity Hospice. Over the yrs, many people have enjoyed our gardens and today they continue to be enjoyed by patients, families and visitors alike. Set over nearly 2 acres, they offer space for quiet contemplation, family fun and make a great backdrop for events. Wheelchair access via ramps and pathways.

28A ST AUGUSTINE'S ROAD, NW1
Camden Town, NW1 9RN. Ricky Patel, rickypatelfilm@gmail.com. *10-12 min walk from Camden Stn, 15 min walk from Kings Cross Stn, 15 min walk from Caledonian Rd Stn.* **Sun 10 Sept (2-6). Adm £5, chd free. Tea. Visits also by arrangement in Sept.**
Step from a wide, quiet street into a tropical paradise! This young garden away from busy Camden is only 4 yrs old but looking lush and well established thanks to the plants that moved here with the new owner inc bananas, ginger, cannas, tetrapanax and bamboos. A green fingered, self-taught plant lover's garden with an emphasis on impressive foliage and interesting texture.

7 ST PAUL'S PLACE, N1
Islington, N1 2QE. Mrs Fiona Atkins. *50 metres W & to the N of the junction of St Paul's Rd & Essex Rd in Islington.* **Sat 17 June (1-6). Combined adm with 5 Northampton Park £7, chd free. Cream teas.**
The main feature of the walled garden is a pond with natural planting to encourage pollinators and birds. It is a working garden; many of the plants have been grown from seed in the small greenhouse, compost is made from waste and leaves from surrounding trees and water is collected in water butts for the pond. A large terrace features a white wisteria and container planting.

27 ST PETERS SQUARE, W6
British Grove, W6 9NW. Oliver & Gabrielle Leigh Wood, 07810 677478, oliverleighwood@hotmail.com. *Tube to Stamford Brook, exit station & turn S down Goldhawk Rd. At T-lights continue ahead into British Grove. Entrance to garden at 50 British Grove, 100yds on L.* **Sun 7 May (2-6). Adm £5, chd free. Home-made teas. Visits also by arrangement 2 May to 4 Sept.**
This long, secret space, is a plantsman's eclectic semi-tamed wilderness. Created over the last 12 yrs it contains lots of camellias, magnolias and fruit trees. Much of the hard landscaping is from skips and the whole garden is full of other people's unconsidered trifles of fancy inc a folly and summerhouse.

57 ST QUINTIN AVENUE, W10
W10 6NZ. Mr H Groffman, 020 8969 8292. *Less than 1m from Ladbroke Grove or White City Tube Stn. Buses: 7, 70, 220 all to North Pole Rd.* **Sun 2, Sun 23 July (2-5.30). Adm £5, chd free. Home-made teas. Visits also by arrangement 14 July to 17 Sept.**
Award-winning 30ft x 40ft garden with a diverse selection of plants inc shrubs for foliage effects. Patio with colour themed bedding material. Focal points throughout. Clever use of mirrors and plant associations. New look front garden, new rear patio layout, new plantings for 2023 with a good selection of climbers and wall shrubs. This year's special display celebrates H.M. King Charles III Coronation.

In 2022 we donated £450,000 to Marie Curie which equates to 23,871 hours of community nursing care

5 ST REGIS CLOSE, N10

Alexandra Park Road, Muswell Hill, N10 2DE. Mrs S Bennett & Mr E Hyde, 020 8883 8540, suebearlh@yahoo.co.uk. *2nd L in Alexandra Park Rd coming from Colney Hatch Ln. 102 & 299 bus from Bounds Green tube to St Andrew's Church or102 from East Finchley. Buses 43 & 143 stop in Colney Hatch Ln. Short walk. Follow NGS signs. Parking on side roads inc coaches.* **Sun 30 Apr, Sun 25 June, Sun 30 July (2-6.30). Adm £5, chd free. Home-made teas, gluten free option & herbal teas. Visits also by arrangement Apr to Oct for groups of 10+. Short talk on history of garden.**

Cornucopia of sensual delights. Artist's garden famous for architectural features and delicious cakes. Baroque temple, pagodas, Raku tiled mirrored wall conceals plant nursery. American Gothic shed overlooks Liberace terrace and stairway to heaven. Maureen Lipman's favourite garden; combines colour, humour, trompe l'oeil with wildlife friendly ponds, waterfalls, weeping willow, lawns and abundant planting. A unique experience awaits! Unusual architectural features inc Oriental tea house overlooking carp pond. Mega plant sale and open studio with ceramics and cards (please bring cash). Gold winner of Haringey in Bloom 2022 'Best Back Garden'. Wheelchair access not suitable for everyone, please check with owners for details.

11 SHORTGATE, N12

North Finchley, N12 7JP. Jennifer O'Donovan. *Bus: 326, alight at the green on Southover, follow signs. Tube: Woodside Park, exit from northbound platform, follow signs. Parking: surrounding roads, not in Shortgate.* **Sun 2 July (1.30-5.30). Adm £5, chd free. Home-made teas.**

This corner plot is a spacious and elegant garden. It has sunny herbaceous beds, a neat vegetable patch, shade bearing trees with tranquil seating and a greenhouse busy with plants. Developed over 35 yrs, there wasn't even one tree when the owner arrived.

41 SOUTHBROOK ROAD, SE12

Lee, SE12 8LJ. Barbara Polanski, polanski101@yahoo.co.uk. *Southbrook Rd is situated off S Circular, off Burnt Ash Rd. Train: Lee & Hither Green, both 10 min walk. Bus: P273, 202. Please enter via side access.* **Sat 3, Sun 4 June (2-5.30). Adm £4, chd free. Home-made teas. Open nearby 15 Waite Davies Road on 4 June only. Visits also by arrangement May to July for groups of 4 to 30.**

Developed over 14 yrs, this large garden has a formal layout with wide mixed herbaceous borders full of colour, surrounded by mature trees, framing sunny lawns, a central box parterre and an Indian pergola. Ancient pear trees festooned in June with clouds of white Kiftsgate and Rambling Rector roses. Discover fish and damselflies in two lily ponds. Many sheltered places to sit and relax. Enjoy refreshments in a small classical garden building with interior wall paintings, almost hidden by roses climbing way up into the trees. Orangery, parterre, gazebo and wall fountain. Side access for standard wheelchairs. Gravel driveway and a few steps.

36 Thornhill Square, Barnsby Group

◆ SPENCER HOUSE, SW1
27 St James' Place, Westminster, SW1A 1NR. RIT Capital Partners, www.spencerhouse.co.uk. *From Green Park Tube Stn, exit on S side, walk down Queen's Walk, turn L through narrow alleyway. Turn R & Spencer House will be in front of you.* **For opening times and information, please visit garden website.**
Originally designed in the C18 by Henry Holland (son-in-law to Lancelot 'Capability' Brown), the garden was among the grandest in the West End. Restored since 1990 under the Chairmanship of Lord Rothschild, the garden with a delightful view of the adjacent Royal Park, now evokes its original layout with planting suggested by early C19 nursery lists.

GROUP OPENING

SPITALFIELDS GARDENS, E1
E1 6QE. *Nr Spitalfields Market. 10 min walk from Aldgate East Tube & 5 min walk from Liverpool St Stn. Train: Shoreditch High St, 3 min walk.* **Sat 10 June (10-4). Combined adm £16, chd free. Home-made teas at The Rectory (2 Fournier Street), Town House (5 Fournier Street), 29 Fournier Street & 31 Fournier Street.**

30 CALVIN STREET, FLAT 1
Susan Young.

20 FOURNIER STREET
Ms Charlie de Wet.

29 FOURNIER STREET
Juliette Larthe.

31 FOURNIER STREET
Tom Holmes.

21 PRINCELET STREET
Marianne & Nicholas Morse.

THE RECTORY, 2 FOURNIER STREET
Jack McCausland.

37 SPITAL SQUARE
Society for the Protection of Ancient Buildings.

21 WILKES STREET
Rupert Wheeler.

Discover a selection of courtyard gardens, some very small, behind fine C17 French Huguenots merchants' and weavers' houses in Spitalfields. Experience an architect designed garden in Wilkes St, two small courtyards in Fournier St, and a larger imaginatively created garden in Princelet St, among others. Each garden owner has adapted their particular urban space to complement a historic house: vegetables, herbs, vertical and horizontal beds, ornamental pots, statuary and architectural artefacts abound.

25 SPRINGFIELD AVENUE, N10
Muswell Hill, N10 3SU. Heather Hampson & Nigel Ragg. *From main r'about in Muswell Hill, down Muswell Hill towards Crouch End. Springfield Ave 1st on L.* **Sun 4 June (2-6). Adm £4, chd free. Home-made teas. Open nearby 11 Park Avenue North.**
No ordinary London garden! Be pleasantly surprised for a small city garden. Magical, packed with quirky ideas to stimulate conversation and imagination. Transformed from a steep grassy slope with love and hard work into three terraces, each with an individual atmosphere that leads to a peaceful summerhouse with a backdrop of mature trees in Alexandra Palace. Enjoy colour, fragrance and the sound of water. The summerhouse is somewhere to rest and view the owner's art works for sale in aid of the NGS. Delicious cakes too!

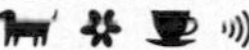

106 STATION ROAD, TW12
Hampton, TW12 2AS. Diane Kermack. *From Hampton Stn, turn L along Station Rd, past St Theodore's Church, opp the green. Access to garden via the garden gate on the adjacent side road Station Close.* **Sat 22, Sun 23 July (11-5). Adm £4, chd free. Light refreshments.**
An apiary garden with six busy hives which will be of special interest to beekeepers. Bushy front garden with vegetable plot and perimeter flower bed. Walk through past swimming pool, greenhouses, seating and many pots. Informal back garden featuring many beehives. Pond, chickens and informal flower beds.

STONEY HILL HOUSE, SE26
Rock Hill, SE26 6SW. Cinzia & Adam Greaves. *Off Sydenham Hill. Nearest train stations: Sydenham, Gipsy Hill or Sydenham Hill. Buses: To Crystal Palace, 202 or 363 along Sydenham Hill. House at end of cul-de-sac on L coming from Sydenham Hill.* **Sun 21 May (2-6). Adm £6.50, chd free. Home-made teas. Prosecco.**
Garden and woodland of approx 1 acre providing a secluded secret green oasis in the city. Paths meander through mature rhododendron, oak, yew and holly trees, offset by pieces of contemporary sculpture. The garden is on a slope and a number of viewpoints set at different heights provide varied perspectives. The planting in the top part of the garden is fluid and flows seamlessly into the woodland. Swings and woodland treehouse for entertainment of children and adults alike! We will have music playing throughout the afternoon. Dogs welcome if kept on a lead. Wheelchair access to the main lawn is via a series of shallow steps or grassy slope.

♿ 🐕 ✽ ☕ 🔊

93 TANFIELD AVENUE, NW2
Dudden Hill, NW2 7SB. Mr James Duncan Mattoon, 07504 565612. *Nr Dollis Hill, Willesden and Wembley. Nearest station: Neasden (Jubilee line), then 10 min walk; or various bus routes to Neasden Parade or Tanfield Ave.* **Sun 23 July (2-6). Adm £5, chd free. Home-made teas. Visits also by arrangement June to Sept for groups of up to 15.**
Intensely exotic Mediterranean and subtropical paradise garden! Sunny deck with implausible planting and panoramic views of Harrow and Wembley, plunges into incredibly exotic, densely planted oasis of delight, with two further seating areas engulfed by flowers, such as, acacia, colocasia, eryngium, hedychium, plumbago, salvias and hundreds more in vigorous competition! Birds and bees love it! Previous garden was Tropical Kensal Rise (Doyle Gardens), featured on BBC 2 Open Gardens and in Sunday Telegraph. This garden featured in Garden Week and Garden Answers magazine. Steep steps down to main garden.

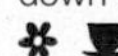

In 2022, 20,000 people were supported by Perennial caseworkers and prevent teams thanks to our funding

28 TAYLOR AVENUE, TW9
Kew, Richmond, TW9 4ED. Inma Lapena. *10 min walk from Kew Gardens Stn. Follow North Rd until Atwood Ave. Turn L & then follow to end where it becomes Taylor Ave. If driving turn R off the South Circular.* **Sun 25 June (2-6). Combined adm with 31 West Park Road £5, chd free. Tea.**
The garden is about 30 metres long and about 10 metres wide. It is an urban garden typical of any semi-detached house in London. There are borders on both sides with many varieties of plants.

TEMPLE LODGE CLUB, W6
51 Queen Caroline Street, Hammersmith, W6 9QL. The Rev. Peter van Breda, 020 8748 8388, Booking@templelodgeclub.com, templelodgeclub.com/about-us/#history. *Queen Caroline St is opp Broadway Shopping Centre which contains tube & bus stations. Piccadilly, District & Hammersmith & City lines. Buses: 9, 10, 27, 33, 419, 72, H91, 190, 211, 220, 267, 283, 295, 391, Fulham Palace Rd exit. Temple Lodge entrance via Gate Restaurant Courtyard.* **Sat 9 Sept (12-5). Adm £5, chd free. Light refreshments.**
Artist Sir Frank Brangwyn's former home, studio and walled garden. Discover an oasis in the heart of London. Perennials and grasses with long seasons of interest provide changing colour, movement and structure through seasons. Garden designed by Tom Ryder. Old carved stone unearthed and set aside during recent garden renovations are repurposed and are on display. Explore to the rhythmic sound of the flow form water feature. Temple Lodge Club is also a thriving and unique guesthouse and has a vegetarian restaurant (booking recommended). Wheelchair access through adjacent car park of St Vincent's Care Home. Garden does feature steps and sloped paths.
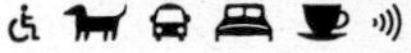

NEW **THEOBALD'S FARMHOUSE, EN2**
Burnt Farm Ride, Crews Hill, Enfield, EN2 9DY. Alison Green. *N Enfield, 1/4 inside M25 J24 or J25. Under 1/2 m from Crews Hill Stn. From Crews Hill Stn, down the hill (Cattlegate Rd) to sharp bend (Jollye's), then turn L into Burnt Farm Ride, from Enfield turn R on the bend. Garden 200yds along road on R.* **Sun 7 May (1.30-5). Adm £15, chd free. Pre-booking essential, please visit www.ngs.org.uk for information & booking. Tea, coffee & cakes on covered terrace.**
Award-winning, 2 acre organic Arts and Crafts garden created by designer owner Alison Green. The garden and its 1650s farmhouse now have 14 distinct gardens with colour themed garden rooms and borders, knot gardens, spiral land form, topiary, water gardens and wildflower meadow, woodlands and a vegetable garden. Colour themed design, exotics and unusual annuals and biennials for all year interest. Visits also by arrangement (non-NGS), please email alison.g.green@talk21.com.

6 THORNHILL ROAD, N1
Islington, N1 1HW. Janis Higgie. *Barnsbury. Tube: Angel or Highbury & Islington. Train: Caledonian & Barnsbury. Bus: to Liverpool Rd.* **Sun 11 June (12-5). Adm £4, chd free. Home-made teas.**
150ft Islington garden, designed and planted over the last 30 yrs. This fully accessible family garden draws inspiration from the owner's antipodean roots. A brick path guides visitors past lawns, raised beds, a water feature, fire bowl and a creative mix of plants from around the globe (inc kowhai and hoheria trees with many shade tolerant plants). This garden has a lot, even the kitchen sink! Completely wheelchair friendly.

57 TONBRIDGE HOUSE, WC1H
Flat 57, Tonbridge House, Tonbridge Street, WC1H 9PG. Sue Heiser. *S of Euston Rd. Tube: King's Cross & St Pancras Stn & Russell Sq. Behind Camden Town Hall off Judd St. Turn into Bidborough St which becomes Tonbridge St. Side entrance to garden, look for yellow posters.* **Sun 2 July (2-5.30). Adm £3.50, chd free. Home-made teas.**
Unexpected informal garden off the busy Euston Road, overlooked on all sides by tall buildings. Mixed planting, with seating, rockery, pergola and shady areas. Mature magnolia, sycamore, silver birches and holly and some long established shrubs and perennials inc ferns, hostas and heuchera. Small vegetable beds and herbs. Evolved over 42 yrs on a low budget with plenty of help from friends. Wheelchair access from street and throughout garden.

TORRIANO COMMUNITY GARDEN, NW5
Torriano Avenue, NW5 2ST. Mrs Elisa Puentes. *Kentish Town. Train station: Camden Road. 10 min walk along Camden Rd to Torriano Ave. Buses: 29, 253. Walk along Torriano Ave & garden is on R opp the primary school.* **Sun 25 June (2-6). Adm £3.50, chd free. Home-made teas.**
This beautiful 1 acre community garden, designed and built in 2003 by the residents helped by the Army, has amazing features with a variety of flowers. The design of the flower beds have the numbers 2003. This beautiful space is fully wheelchair accessible and open for all to enjoy even if you do not live on the estate. Our garden is maintained and cleaned by residents and is open all year-round.

NEW **9 TRAFALGAR TERRACE, HA1**
Harrow, HA1 3EU. George Reeve, www.instagram.com/thehillsjunglegarden. *Harrow-on-the-Hill Stn; 10 min walk via Churchfields & path to Trafalgar Terrace. Free roadside parking on West St; L onto Nelson Rd to end, L to 9 Trafalgar Terrace following NGS signs.* **Sat 2 Sept (1-5). Adm £5. Pre-booking essential, please visit www.ngs.org.uk for information & booking. Light refreshments.**
Small jungle and tropical themed garden with big leaf plants down to small succulents. The garden is a tranquil space with running water and views up to Harrow's famous St Mary's Church.

TUDELEY HOUSE, BR7
Royal Parade, Chislehurst, BR7 6NW. Mrs Bernadette & Mr Colin Katchoff, 07786 854943, katchoff@hotmail.com. *1m from the A20 at J3 of M25. 20 min walk from Chislehurst Stn. Buses 61, 160, 161,162, 269 & 273 to Chislehurst War Memorial. Limited parking in side roads.* **Sun 25 June (11-4). Adm £4, chd free. Home-made teas. Visits also by arrangement 30 Apr to 3 Sept for groups of up to 6.**
The layout of this Victorian town house garden has remained as shown in the original architect's plans of 1896. The current owners have recently restored the house; employing Jo Thompson, an RHS Chelsea Gold winner, to bring the garden up to date, whilst remaining sympathetic to a Victorian era town garden. Phase one and two, of three sections being restored, is completed. The garden is 95% flat. Some older paths (25% of garden) are narrow, so may be difficult for large mobility scooters.

35 TURRET GROVE
Clapham Old Town SW4 0ES. Wayne Amiel, www.turretgrove.com. *Off Rectory Grove. 10 min walk from Clapham Common Tube & Wandsworth Rd Mainline. Buses: 87, 137.* **Sun 23 July (10-5). Adm £5, chd free. Home-made teas.**
As featured on BBC Two Gardeners' World, 2018, this north facing garden shows what can be achieved in a small space (8 metres x 20 metres). The owner, who makes no secret of disregarding the rule book, describes this visual feast of intoxicating colours as Clapham meets Jamaica. This is gardening at its most exuberant, where bananas, bamboos, tree ferns and fire bright plants flourish beside the traditional. Children very welcome.

24 TWYFORD AVENUE, N2
N2 9NJ. Rachel Lindsay & Jeremy Pratt, 07930 632902, jeremypr@blueyonder.co.uk. *Twyford Ave runs parallel to Fortis Green, between East Finchley & Muswell Hill. Tube: Northern line to East Finchley. Buses: 102, 143, 234, 263 to East Finchley. Buses 43, 134, 144, 234 to Muswell Hill. Buses: 102 & 234 stop at end of road. Garden signed from Fortis Green.* **Sun 16 July (2-6). Adm £4, chd free. Home-made teas. Visits also by arrangement 24 July to 28 July.**

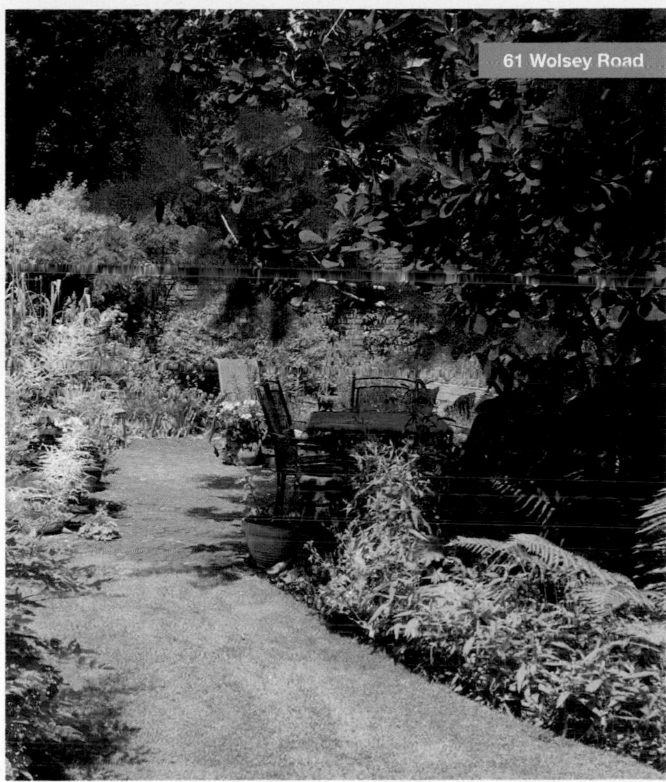
61 Wolsey Road

Sunny, 120ft south facing garden, planted for colour. Brick edged borders and overflowing containers packed with masses of traditional herbaceous and perennial cottage garden plants and shrubs. Shady area at rear evolving as much by happy accident as design. Some uneven ground. Water feature. Greenhouse bursting with cuttings. Many places to sit and think, chat or doze. Honey and bee products for sale.
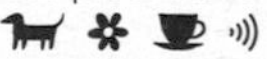

VICTORIA LODGE, E18
Hermitage Walk, Snaresbrook, E18 2BN. Lucy St Ville. *7 min walk from Snaresbrook Stn (Central line).* **Sat 24 June (2-5). Adm £3, chd £1. Tea.**
Enter this large London garden through a reclaimed iron pergola, past a colourful mixed bed to a relaxed fire pit. Next to a small rill created from cattle troughs, planted with irises and ferns, which leads to a walkway cut through mature rhododendrons. At the far end, sheltered by a high Victorian wall, is a wildlife garden inc a pond, bee friendly planting and secluded seating area. Most of the garden is accessible with gravel, brick and grass paths.

152A VICTORIA RISE, SW4
Clapham, SW4 0NW. Benn Storey, bennstorey@gmail.com, www.instagram.com/thenorthsouthgarden. *Entry via basement flat. Closest tube Clapham Common. Bus 77, 87, 137, 156, 345, 452.* **Sun 2 July (1-5). Adm £4.50, chd free. Visits also by arrangement 15 Mar to 10 Oct for groups of 5 to 20.**
In its 7th yr this terraced garden is 21 metres long by 8 metres wide. Planting ranges from the lush greens of the courtyard to the frothy, insect friendly plants of the main level to the espalier fruit trees and vegetables of the productive levels. A copper beech hedge hides a cozy arbour seat and fire pit at the top of the plot out of view from surrounding neighbours.

13 WAITE DAVIES ROAD, SE12

Lee, SE12 0NE. Janet Pugh. *Just off A205 South Circular. Lee Train Stn, 11 min walk. Hither Green & Grove Park Stns, 20 min walk. Bus route 261 stops close. Buses 202 & 160 stop nearby. Usually, free roadside parking.* **Sun 4 June (2-5.30). Combined adm with 15 Waite Davies Road £4.50, chd free. Home-made teas. Open nearby 41 Southbrook Road.**
A colourful and welcoming front garden leads you down steps to a bright and peaceful, private maturing garden. The gentle sound of a discreet water feature provides a calming atmosphere with attractive planting in a melange of pinks, purples and creams.

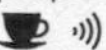

15 WAITE DAVIES ROAD, SE12

Lee, SE12 0NE. Will Jennings. *Just off the A205 South Circular. Lee Train Stn,11 min walk. Hither Green & Grove Park Stns, 20 min walk. Bus route 261 stops close. Buses 202 & 160 stop nearby. Usually, free roadside parking.* **Sun 4 June (2-5.30). Combined adm with 13 Waite Davies Road £4.50, chd free. Open nearby 41 Southbrook Road.**
Hidden behind a Victorian terrace is a tiny yet immaculately presented English garden. A huge Malvern Hills rambling rose gently billows over a willow coloured shed, foxgloves spring up beneath a young crab apple and tightly clipped box hedging frames the scene. Nearby, the soft pink, myrrh scented flowers of a climbing rose The Generous Gardener nods gracefully above a small, sunbaked stone patio.

NEW WARREN MEWS, W1

Fitzrovia, W1T 5NQ. Rebecca Hossack. *2 mins from Warren St Tube Stn. Entrance to garden on Warren St, W1T 5NQ.* **Sat 17 June (11-4). Adm £4, chd free. Home-made teas.**
Tucked away behind bustling Tottenham Court Road is the enchanting garden of Warren Mews. On first glance, it is impossible to tell that the plants in this verdant garden have no access to the earth. Warren Mews is a place where pots of paradise flowers, window boxes bursting with geraniums and containers of olive trees rule the street. Eclectic container planting with an Australian influence. The Rebecca Hossack Art Gallery is a 5 min walk away. Mews is fully accessible and entirely cobbled.

9 WARWICK CLOSE, TW12

Hampton, TW12 2TY. Chris Churchman. *2m W of Twickenham, 2m N of Hampton Court, overlooking Bushy Park. 100 metres from Hampton Open Air Swimming Pool.* **Sat 12 Aug (10.30-5). Adm £4, chd free. Home-made teas.**
A small suburban garden in South West London. Front garden with espaliered trees, featuring roses, lavender and stipa. Shade garden with rare ferns and herbaceous. Rear garden has new canal water feature, rectangular lawn with prairie style plantings crossed with subtropical species. Roof top allotment to garage (the garotment).

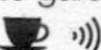

THE WATERGARDENS, KT2

Warren Road, Kingston-upon-Thames, KT2 7LF. The Residents' Association. *1m E of Kingston. From Kingston take A308 (Kingston Hill) towards London; after approx ½m turn R into Warren Rd. No. 57 bus along Coombe Lane West, alight at Warren Rd. Roadside parking only.* **Sun 7 May, Sun 22 Oct (1.30-4). Adm £5, chd free.**
Japanese landscaped garden originally part of Coombe Wood Nursery, planted by the Veitch family in the 1860s. Approx 9 acres with ponds, streams and waterfalls. Many rare trees, which in spring and autumn provide stunning colour. For the tree lover this is a must-see garden. Gardens attractive to wildlife. Major renovation and restoration have taken place over the past yr revealing a hitherto lost lake and waterfall. Restoration works ongoing.

70 WENSLEYDALE ROAD, TW12

Hampton, TW12 2LX. Mr Steve Pickering. *9 mins walk from Hampton Stn. From Hampton Stn, E on Station Rd towards Oldfield Rd, L onto Tudor Rd, R onto Wensleydale Rd, L to stay on Wensleydale Rd.* **Sat 27, Sun 28 May (1-5). Adm £4, chd free. Open nearby 16 Links View Road.**
Small traditional garden in classic layout with greenhouse, summerhouse, pergolas, patio and rockery with an emphasis on year-round colour. Many evergreen shrubs and perennials. Wisteria over pergola flowers in spring. David Austin Desdemona rose repeat flowering on the patio. Greenhouse containing succulents and many flowering plants. All created by the amateur gardener owners.

WEST LODGE PARK, EN4

Cockfosters Road, Hadley Wood, EN4 0PY. Beales Hotels, 020 8216 3904, janegray@bealeshotels.co.uk, www.bealeshotels.co.uk/westlodgepark. *1m S of Potters Bar. On A111. J24 from M25 signed Cockfosters.* **Sun 14 May (2-5); Sun 22 Oct (1-4). Adm £7, chd free. Visits also by arrangement 1 Apr to 29 Oct for groups of 10+.**
Open for the NGS for over 40 yrs, the 35 acre Beale Arboretum consists of over 800 varieties of trees and shrubs inc National Collections of Hornbeam cultivars (*Carpinus betulus*) Indian bean tree (*Catalpa bignonioides*) and Swamp Cypress (*Taxodium distichum*). Network of paths through good selection of conifers, oaks, maples and mountain ash, all specimens labelled. Stunning collection within the M25. Guided tours available. Breakfasts, morning coffee and biscuits, afternoon tea, restaurant lunches, light lunches, dinner all served in the hotel. Please see website for details.

NPC

31 WEST PARK ROAD, TW9

Kew, Richmond, TW9 4DA. Anna Anderson. *Close to the E side of Kew Gardens Stn. From Richmond bound exit from Kew Gardens Stn, West Park Rd is straight ahead & No.31 is the 2nd house on the LHS.* **Sun 25 June (2-6). Combined adm with 28 Taylor Avenue £5, chd free. Tea at 28 Taylor Avenue.**
Modern botanical garden with an oriental twist. Emphasis on foliage and an eclectic mix of unusual plants, a reflecting pool and willow screens. Dry bed, shady beds, mature trees and a private paved dining area with dappled light and shade.

35 WESTON PARK, N8

N8 9SY. Mrs Theresa & Mr Keith Rutter. *Tube to Finsbury Park then W3 bus (Weston Park stop) or W7 (Crouch End Broadway), or tube to Archway then 41 bus (Crouch End Broadway). Short walk from each one.* **Sun 20 Aug (2-5.30). Adm £4, chd free.**

Southeast facing large garden with a wide range of plants, shrubs and trees suited to varying conditions inc a bog garden. Summer colour in the densely planted beds and pots inc dahlias, cannas and salvias. Elements such as golden bamboo, phormiums, grasses and sculptures provide structure. A curving path leads up to an artist's studio.

74 WILLIFIELD WAY, NW11

NW11 6YJ. David Weinberg, 07956 579205, davidwayne@hotmail.co.uk. *Hampstead Garden Suburb. H2 bus from Golders Green will stop outside or take the 102,13 or 460 to Hampstead Way & walk up Asmuns Hill & turn R.* **Sun 9 July (1.30-5.30). Adm £5, chd free. Cream teas. Visits also by arrangement 28 May to 24 Sept.**

A very peaceful, traditional English country cottage garden. Borders full of herbaceous perennials and hydrangeas with a central rose bed surrounded by a box parterre with plenty of space to sit and take it all in and enjoy afternoon tea. Wheelchair access to patio area only.

61 WOLSEY ROAD, KT8

East Molesey, KT8 9EW. Jan & Ken Heath. *Less than 10 min walk from Hampton Court Palace & station, very easy to find.* **Sun 4 June (2-6). Adm £5, chd free. Home-made teas.**

Romantic, secluded and peaceful garden of two halves designed and maintained by the owners. Part is shaded by two large copper beech trees with woodland planting and fernery. The second reached through a beech arch with cottage garden planting, pond and wooden obelisks covered with roses. Beautiful octagonal gazebo overlooks pond, plus an oak framed summerhouse designed and built by the owners. Extensive seating throughout the garden to sit quietly and enjoy your tea and cake.

33 WOOD VALE, N10

Highgate, N10 3DJ. Mona Abboud, 020 8883 4955, monaabboud@hotmail.com, www.monasgarden.co.uk. *Tube: Highgate, 10 min walk. Buses: W3, W7 to top of Park Rd.* **Sun 9 July (1-5.30). Adm £4, chd free. Visits also by arrangement 13 May to 24 Sept for groups of 5+. Donation to Plant Heritage.**

This 100 metre long award-winning garden is home to the National Collection of Corokia along with a great number of other unusual Australasian, Mediterranean and exotic plants complemented by perennials and grasses which thrive thanks to 300 tons of topsoil, gravel and compost brought in by wheelbarrow. Emphasis on structure, texture, foliage and shapes brought alive by distinctive pruning.

39 WOOD VALE, SE23

Forest Hill, SE23 3DS. Nigel Crawley. *Entrance through Thistle Gates, 48 Melford Rd. Train stations: Forest Hill & Honor Oak Park. Victoria Stn to West Dulwich, then P4. Buses: 363 Elephant & Castle to Wood Vale/Melford Rd; 176, 185 & 197 to Lordship Ln/Wood Vale.* **Sun 16 Apr (1-5). Adm £4.50, chd free. Light refreshments. Evening opening Wed 19 Apr (5-7). Adm £6, chd free. Wine.**

Diverse garden dominated by a gigantic perry pear forming part of one of the East Dulwich orchards. The emphasis in the garden is on its inhabitants; white comfrey and pear blossom keeps the bees busy in the spring. Home to a variety of birds inc green woodpecker, dunnock and redwing. The clumps of narcissi around the old apple tree are a sight to see. Surprising green oasis in Forest Hill. Close to Sydenham Woods, Horniman Gardens and Camberwell Old Cemetery. Level wheelchair access, but rough terrain in the lane.

7 WOODBINES AVENUE, KT1

Kingston-upon-Thames, KT1 2AZ. Mr Tony Sharples & Mr Paul Cuthbert. *Take K2, K3, 71 or 281 bus. From Surbiton, bus stop outside Waitrose & exit bus Kingston University Stop. From Kingston, walk or bus from Eden St (opp Heals).* **Sun 9 July (12-5). Combined adm with 15 Catherine Road £7, chd free. Light refreshments.**

We have created a winding path through our 70ft garden with trees, evergreen structure, perennial flowers and grasses. Wide herbaceous borders, an ancient grapevine, a box hedge topiary garden, large silver birches and a hot summer terrace provide contrast.

28A WORCESTER ROAD, E17

Walthamstow, E17 5QR. Mark & Emma Luggie. *12 min walk from Blackhorse Road Tube Stn. Just off Blackhorse Ln. On street parking.* **Sun 20 Aug (12-5.30). Adm £4, chd free. Pre-booking essential, please visit www.ngs.org.uk for information & booking. Light refreshments.**

Typical Walthamstow terraced back garden, turned into a lush tropical oasis of foliage with a small stream and pond. Small and large leaves mix with floral accents and varying leaf textures from ferns, colocasia and bananas supported by a framework of larger established trees, jasmine, and tree ferns. A small gravel path leads you to a small seating area to reflect, or sit on the patio and enjoy.

1 YORK CLOSE, W7

Hanwell, W7 3JB. Tony Hulme & Eddy Fergusson. *By road only, entrance to York Close via Church Rd. Nearest station Hanwell Train Stn. Buses E3, 195, 207.* **Sat 19, Sun 20 Aug (2-6). Adm £7, chd free. Wine.**

Tiny, quirky, prize-winning garden extensively planted with eclectic mix inc hosta collection, many unusual and tropical plants. Plantaholics paradise. Many surprises in this unique and very personal garden.

ZEN GARDEN AT JAPANESE BUDDHIST CENTRE, W3

Three Wheels, 55 Carbery Avenue, Acton, W3 9AB. Reverend Prof K T Sato, www.threewheels.org.uk. *Tube: Acton Town, 5 min walk. 200yds off A406.* **Sat 3, Sun 4, Sat 17, Sun 18 June (2-5). Adm £3.50, chd free. Matcha tea £3. Open nearby 118b Avenue Road on 18 June only.**

Pure Japanese Zen garden (so no flowers) with 12 large and small rocks of various colours and textures set in islands of moss and surrounded by a sea of grey granite gravel raked in a stylised wave pattern. Garden surrounded by trees and bushes outside a cob wall. Oak framed wattle and daub shelter with Norfolk reed thatched roof. Talk on the Zen garden between 3-4pm. Buddha Room open to public.

NORFOLK

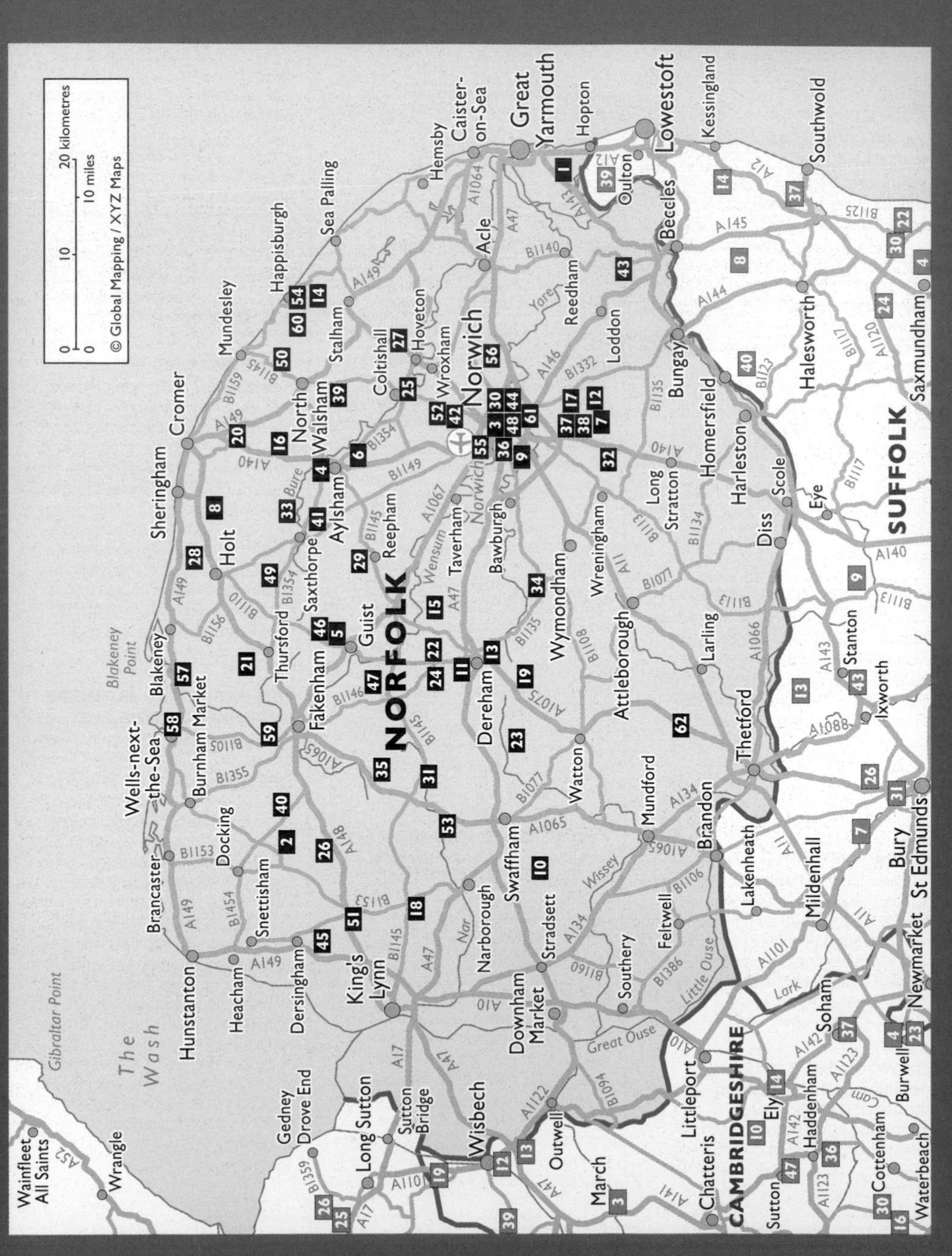
0 10 20 kilometres
0 10 miles
© Global Mapping / XYZ Maps
NORFOLK
SUFFOLK
CAMBRIDGESHIRE
The Wash
Gibraltar Point
Blakeney Point
Great Yarmouth
Caister-on-Sea
Hemsby
Hopton
Lowestoft
Kessingland
Southwold
Oulton
Beccles
Acle
Reedham
Loddon
Norwich
Wroxham
Hoveton
Coltishall
Stalham
Sea Palling
Happisburgh
Mundesley
North Walsham
Cromer
Sheringham
Holt
Aylsham
Saxthorpe
Reepham
Taverham
Bawburgh
Wymondham
Wreningham
Long Stratton
Bungay
Homersfield
Harleston
Halesworth
Saxmundham
Scole
Diss
Eye
Attleborough
Larling
Stanton
Ixworth
Thetford
Watton
Mundford
Brandon
Lakenheath
Mildenhall
Bury St Edmunds
Newmarket
Swaffham
Dereham
Guist
Thursford
Fakenham
Blakeney
Wells-next-the-Sea
Burnham Market
Brancaster
Docking
Snettisham
Hunstanton
Heacham
Dersingham
King's Lynn
Narborough
Stradsett
Downham Market
Southery
Feltwell
Littleport
Ely
Soham
Burwell
Haddenham
Cottenham
Waterbeach
Sutton
Chatteris
March
Outwell
Wisbech
Sutton Bridge
Long Sutton
Gedney Drove End
Wrangle
Wainfleet All Saints
Yare
Bure
Wensum
Nar
Wissey
Little Ouse
Great Ouse
Lark
Cam
A12
A47
A1064
A143
A145
A144
A146
A140
A149
A148
A1067
A11
A1066
A134
A1065
A1075
A10
A17
A1101
A1122
A1088
A1120
A1123
A142
A141
A52
A1065
B1140
B1332
B1135
B1149
B1354
B1145
B1159
B1156
B1110
B1105
B1355
B1153
B1454
B1146
B1077
B1108
B1113
B1134
B1117
B1123
B1125
B1106
B1386
B1160
B1094
B1359

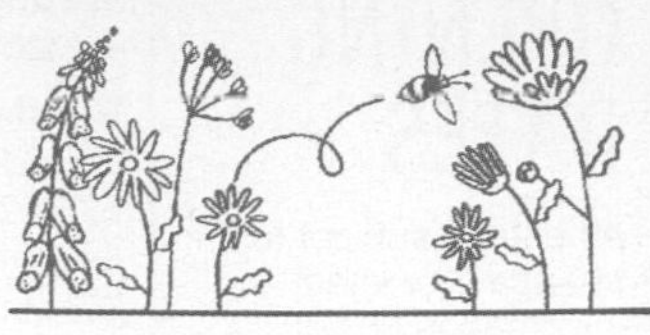

Norfolk is a lovely low-lying county, predominantly agricultural with an abundance of wildlife and a beautiful coastline.

Visitors come here because they are attracted not only to the peaceful and spacious countryside, but also to the medieval churches and historical houses. There is an extensive coastal footpath and a large network of rivers, together with the waterways of the Norfolk Broads. Norwich the capital is a fine city.

We are fortunate to have a loyal group of garden owners; Sandringham was one of the original gardens to open for the scheme in 1927 and has been supporting us continuously ever since.

Whilst many of our gardens have opened their gates for over half a century, others will be opening for the very first time. Located throughout the county, they range from those of the large estates and manor houses, to the smaller cottages, courtyards and town gardens, accommodating different styles of old and traditional, newly constructed and contemporary, designed and naturalistic.

So why not come and experience for yourself the county's rich tapestry of big skies, open countryside, attractive architecture and delightful gardens.

Below: Walcott House

Volunteers

County Organiser
Julia Stafford Allen
01760 755334
julia.staffordallen@ngs.org.uk

County Treasurer
Andrew Stephens OBE
07595 939769
andrew.stephens@ngs.org.uk

Publicity
Julia Stafford Allen
07778 169775
julia.staffordallen@ngs.org.uk

Social Media
Kenny Higgs 07791 429052
kenny.higgs@ngs.org.uk

Photographer
Simon Smith 01362 860530
simon.smith@ngs.org.uk

Booklet Co-ordinator
Juliet Collier 07986 607170
juliet.collier@ngs.org.uk

New Gardens Organiser
Fiona Black 01692 650247
fiona.black@ngs.org.uk

Group Talks & Visits
Graham Watts 01362 690065
graham.watts@ngs.org.uk

Assistant County Organisers
Jenny Clarke 01508 550261
jenny.clarke@ngs.org.uk

Nick Collier 07733 108443
nick.collier@ngs.org.uk

Sue Guest 01362 858317
guest63@btinternet.com

Sue Roe 01603 455917
sueroe8@icloud.com

Retty Wace 07876 648543
retty.wace@ngs.org.uk

@ngsnorfolk @Norfolkngs @norfolkngs

OPENING DATES

All entries subject to change. For latest information check **www.ngs.org.uk**

Map locator numbers are shown to the right of each garden name.

February

Snowdrop Festival

Sunday 5th
Lexham Hall 31

Sunday 12th
Lexham Hall 31

Saturday 18th
Horstead House 25

Sunday 19th
Bagthorpe Hall 2

Saturday 25th
◆ Hindringham Hall 21
◆ Raveningham Hall 43

Sunday 26th
Chestnut Farm 8

March

Saturday 18th
◆ East Ruston Old Vicarage 14

Sunday 26th
Gayton Hall 18
◆ Mannington Estate 33

April

Sunday 30th
Wretham Lodge 62

May

Monday 1st
Wretham Lodge 62

Sunday 7th
Kelling Hall 28

Sunday 14th
Chestnut Farm 8
Holme Hale Hall 23
Quaker Farm 42

Wednesday 17th
◆ Stody Lodge 49

Sunday 21st
Bolwick Hall 6
57 Ketts Hill 30
Lexham Hall 31
NEW 117 St Leonards Road 44
Warborough House 57

Sunday 28th
Blickling Lodge 4
Ferndale 17

June

Sunday 4th
The Old Rectory, Syderstone 40
Oulton Hall 41

Monday 5th
◆ Hoveton Hall Gardens 27

Wednesday 7th
NEW Erpingham House Farm 16

Thursday 8th
NEW Erpingham House Farm 16

Friday 9th
NEW Erpingham House Farm 16

Sunday 11th
Elsing Hall Gardens 15
High House Gardens 19
NEW The Norfolk Hospice, Tapping House 51

Sunday 18th
Bishop's House 3
Manor House Farm, Wellingham 35
Wells-Next-The-Sea Gardens 58
The White House 60

Friday 23rd
Silverstone Farm 47

Saturday 24th
47 Norwich Road 37
51 Norwich Road 38
NEW Old Manor Farmhouse 39

Sunday 25th
Ferndale 17
Highview House 20
NEW Old Manor Farmhouse 39

July

Sunday 2nd
Manor Farm, Coston 34
Walcott House 54
West Barsham Hall 59

Wednesday 5th
NEW The Walled Garden, Little Plumstead 56

Sunday 9th
Holme Hale Hall 23
Kerdiston Manor 29

Saturday 15th
Swafield Hall 50

Sunday 16th
4 Coach House Court 9
Dunbheagan 13
North Lodge 36
Swafield Hall 50

Wednesday 19th
Lexham Hall 31

Sunday 23rd
NEW Southgate House 48
Tudor Lodgings 53
NEW 1 Woodside Cottages 61

Wednesday 26th
Honeysuckle Walk 24

Friday 28th
Honeysuckle Walk 24

Sunday 30th
Dale Farm 11
Ferndale 17
The Long Barn 32
North Lodge 36
61 Trafford Way 52

Monday 31st
Honeysuckle Walk 24

August

Wednesday 2nd
Honeysuckle Walk 24

Saturday 5th
NEW Southgate House 48
NEW 1 Woodside Cottages 61

Sunday 6th
Brick Kiln House 7
NEW Southgate House 48
33 Waldemar Avenue 55
NEW 1 Woodside Cottages 61

Sunday 13th
NEW Bluebell Barn 5
Severals Grange 46

Sunday 20th
Chestnut Farm 8

Sunday 27th
Acre Meadow 1
NEW Cobweb Cottage 10

Monday 28th
Acre Meadow 1

September

Sunday 3rd
33 Waldemar Avenue 55

Sunday 10th
High House Gardens 19

October

Sunday 1st
Holme Hale Hall 23

Saturday 14th
◆ East Ruston Old Vicarage 14

By Arrangement

Arrange a personalised garden visit with your club, or group of friends, on a date to suit you. See individual garden entries for full details.

Acre Meadow 1
Blickling Lodge 4
Brick Kiln House 7
Chestnut Farm 8
Dale Farm 11
Dove Cottage 12
Dunbheagan 13
Highview House 20
Hoe Hall 22
Holme Hale Hall 23
Honeysuckle Walk 24
Horstead House 25
NEW Southgate House 48
Tudor Lodgings 53
West Barsham Hall 59

THE GARDENS

The White House

1 ACRE MEADOW

New Road, Bradwell, Great Yarmouth, NR31 9DU. Mr Keith Knights, 07476 197568, kk.acremeadow@gmail.com, www.acremeadow.co.uk. *Between Bradwell & Belton in arable surroundings. Take Belton & Burgh Castle turn (New Rd) from r'about at Bradwell on A143 Great Yarmouth to Beccles rd. 400 yds on R drive in very wide gateway. For SatNav use NR31 9JW.* **Sun 27, Mon 28 Aug (10-5). Adm £5, chd free. Pre-booking essential, please visit www.ngs.org.uk for information & booking. Light refreshments. Home-made cakes, tea & coffee. Visits also by arrangement 18 July to 30 Sept for groups of 8+.**

Dramatic, intensely planted mix of exotic and other late season plants, complementary and contrasting combinations of foliage and flowers, building into a hot colour and dark foliage crescendo. Alive with insects on sunny days. Planting inc brugmansia, dahlias, tall grasses, herbaceous perennials, aeoniums, and lots of cannas. Separate areas inc tea garden and wildlife pond.

2 BAGTHORPE HALL

Bagthorpe, Bircham, King's Lynn, PE31 6QY. Mr & Mrs D Morton. *3½ m N of East Rudham, off A148. Take turn opp The Crown in East Rudham. Look for white gates in trees, slightly set back from the rd. Other direction: triangle of grass, continue length of field, white gates on L.* **Sun 19 Feb (11-4). Adm £5, chd free. Light refreshments.**

A delightful circular walk which meanders through a stunning display of snowdrops naturally carpeting a woodland floor, and then returning through a walled garden.

3 BISHOP'S HOUSE

Bishopgate, Norwich, NR3 1SB. The Bishop of Norwich, www.dioceseofnorwich.org/gardens. *Located in the city centre near the Law Courts & The Adam & Eve Pub. Parking available at Town Centre car parks inc one by the Adam & Eve pub.* **Sun 18 June (1-4.30). Adm £5, chd free. Home-made teas.**

4 acre walled garden dating back to the C12. Extensive lawns with specimen trees. Borders with many rare and unusual shrubs. Spectacular herbaceous borders flanked by yew hedges. Rose beds underplanted with hosta, meadow labyrinth, organic kitchen garden, herb garden and bamboo walk. Popular plant sales. Wheelchair access, gravel paths and some slopes.

4 BLICKLING LODGE

Blickling, Norwich, NR11 6PS. Michael & Henrietta Lindsell, nicky@lindsell.co.uk. *½ m N of Aylsham. Leave Aylsham on Old Cromer rd towards Ingworth. Over hump back bridge & house is on R.* **Sun 28 May (12-5). Adm £6, chd free. Home-made teas. Cakes also available. Visits also by arrangement 2 Jan to 29 Dec for groups of 5+.**

Georgian house (not open) set in 17 acres of parkland inc cricket pitch, mixed border, walled kitchen garden, yew garden, woodland/water garden.

5 NEW BLUEBELL BARN

Lyng Hall Lane, Wood Norton, Dereham, NR20 5BJ. Rose and Phil Tweedie. *9 m SW of Holt. 9 m E of Fakenham. Take B1110 to Holt from Guist. Ignore the Xroads for Wood Norton and take next R onto Rectory Rd. Drive to the 30 mph signs, take next L onto gravel rd. Bluebell Barn is 1st L.* **Sun 13 Aug (11-5). Adm £4, chd free. Pre-booking essential, please visit www.ngs.org.uk for information & booking. Open nearby Severals Grange.**

Garden and meadow in total covering 2½ acres and newly planted by the current owners over the last 3 years. Colourful herbaceous borders with some specimen trees surround a circular lawn in the formal part of the garden. Informal meadow area with mown pathways through. Limited parking.

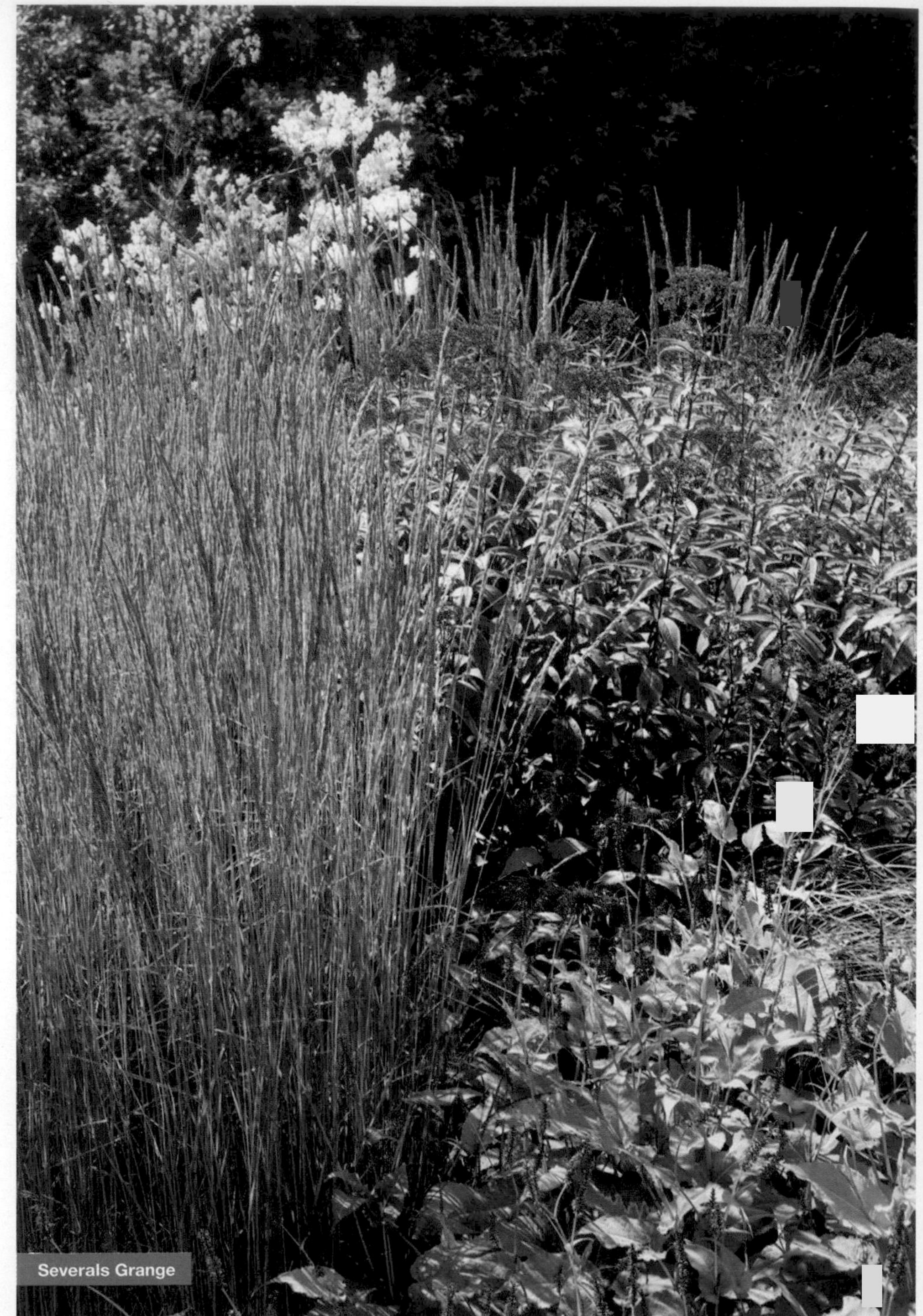

Severals Grange

6 BOLWICK HALL

Marsham, NR10 5PU. Mr & Mrs G C Fisher. *8m N of Norwich off A140. From Norwich, heading N on A140, just past Marsham take 1st R after Plough Pub, signed 'By Road' then next R onto private drive to front of Hall.* **Sun 21 May (1-5). Adm £6, chd free. Home-made teas. Refreshments in aid of Aldborough Primary School, Norfolk.**

Landscaped gardens and park surrounding a late Georgian hall. The original garden design is attributed to Humphrey Repton. The current owners have rejuvenated the borders, planted gravel and formal gardens and clad the walls of the house in old roses. Enjoy a woodland walk around the lake as well as as stroll through the working vegetable and fruit garden with its double herbaceous border. Please note that most of the paths are gravel.

7 BRICK KILN HOUSE

Priory Lane, Shotesham, Norwich, NR15 1UJ. Jim & Jenny Clarke, jennyclarke985@gmail.com. *6m S of Norwich. From Shotesham All Saints church Priory Ln is 200m on R on Saxlingham Rd.* **Sun 6 Aug (10-4). Adm £6, chd free. Home-made teas. Visits also by arrangement 1 June to 11 Sept for groups of up to 35.**

2 acre country garden with a large terrace, lawns and colourful herbaceous borders. There is a contemporary designed pergola garden, sculptures and a stream running through a diversely planted wood. Easy access to brick path.

8 CHESTNUT FARM

Church Road, West Beckham, Holt, NR25 6NX. Mr & Mrs John McNeil Wilson, 01263 822241, judywilson100@gmail.com. *2½m S of Sheringham. From A148 opp Sheringham Park entrance. Take the rd signed By Way To West Beckham, about ¾m to the garden by village sign. N.B. SatNav will take you to the pub!* **Sun 26 Feb (11-4); Sun 14 May, Sun 20 Aug (11-5). Adm £5, chd free. Light refreshments. Gluten Free options available. Visits also by arrangement 26 Feb to 10 Sept for groups of 5 to 40.**

Mature 3 acre garden developed over 60 years with collections of many rare and unusual plants and trees. 100+ varieties of snowdrops, drifts of crocus with seasonal flowering shrubs. Later, wood anemones, fritillary meadow, wildflower walk, pond, small arboretum, croquet lawn and colourful herbaceous borders. In May the handkerchief tree, cercis and camassias should be seen. Always something new. Good varied selection of spring bulbs. Wheelchair access tricky if wet.

9 4 COACH HOUSE COURT

Unthank Road, Norwich, NR4 7QR. Jackie Floyd. *Out of Norwich on the Unthank Rd for 1m, Coach House Court is situated on the L between Upton Rd and Judges Walk. To Norwich on A11 take Unthank Rd on the L. Coach House Court is 1m on the R.* **Sun 16 July (11-5). Adm £4, chd free. Open nearby North Lodge.**

Very small compact and secluded city courtyard garden with abundant trees, shrubs inc roses, climbers, perennials and container plants. The goal is to achieve consecutive flowering and all year interest. A new nature pond attracts a lot of birds and insects. A well trained and mature wisteria is a special feature. A small courtyard planted to the maximum. Paving to garden for wheelchair access.

10 NEW COBWEB COTTAGE

51 Shingham, Beachamwell, Swaffham, PE37 8AY. Sue Bunting. *5m from Swaffham. Shingham is next to Beachamwell. The cottage is 1st on the L after the Shingham village sign.* **Sun 27 Aug (10-5). Adm £4, chd free. Home-made teas.**

A small delightful cottage garden with mixed borders, a sunken greenhouse, pergola, and a wildlife pond. There is also a large productive ornamental kitchen garden with bees, chickens, treehouse, prairie planting and fruit cage. Plants for sale. Partial access with narrow gravel paths and lawn. All on one level.

11 DALE FARM

Sandy Lane, Dereham, NR19 2EA. Graham & Sally Watts, 01362 690065, grahamwatts@dsl.pipex.com. *16m W of Norwich. 12m E of Swaffham. From A47 take B1146 signed to Fakenham, turn R at T-junction, ¼m turn L into Sandy Ln (before pelican X-ing).* **Sun 30 July (11-5). Adm £6, chd free. Home-made teas. Visits also by arrangement 12 June to 23 July for groups of 15 to 45.**

2 acre plant lovers' garden with a large spring-fed pond. Over 1000 plant varieties in exuberantly planted borders with sculptures. Also, gravel, vegetable, nature and waterside gardens. Collection of 150 hydrangeas! Music during the day, some grass paths and gravel drive. Wide choice of plants for sale.

12 DOVE COTTAGE

Wolferd Green, Shotesham All Saints, Norwich, NR15 1YU. Sarah Cushion, 07706 566169, sarah.cushion815@btinternet.com. *From Norwich travel to far end of Poringland. Turn R by church into Shotesham rd, continue for approx 2½m, house pink cottage on L.* **Visits by arrangement 10 May to 31 Aug for groups of up to 30. Adm £5, chd free. Home-made teas.**

⅓ acre densely planted colourful cottage garden inc a small pond, summerhouse, and extensive views across the countryside. Profusion of roses. Some gravel areas.

13 DUNBHEAGAN

Dereham Road, Westfield, NR19 1QF. Jean & John Walton, 01362 696163, jandjwalton@btinternet.com. *2m S of Dereham. From Dereham turn L off A1075 into Westfield Rd by the Vauxhall garage/Premier food store. Straight ahead at Xrds into ln which becomes Dereham Rd. Garden on L.* **Sun 16 July (1-5). Adm £5, chd free. Home-made teas. Refreshments will be provided on the front lawn. Visits also by arrangement 10 June to 23 July for groups of 10+.**

Relax and enjoy the garden where the rare and unusual rub shoulders with the more recognisable plants in densely planted beds and borders in this ever-changing plantsman's garden. Lots of paths to explore. A riot of colour all summer. Lots of seating. Featured in Nick Bailey's book 365 Days of Colour. Several magazine articles. Wheelchair access via gravel driveway.

14 ◆ EAST RUSTON OLD VICARAGE

East Ruston, Norwich, NR12 9HN. Alan Gray & Graham Robeson, 01692 650432, erovoffice@btconnect.com, www.eastrustonoldvicarage.co.uk. *3m N of Stalham. Turn off A149 onto B1159 signed Bacton, Happisburgh. After 2m turn R 200yds N of East Ruston Church (ignore sign to East Ruston).* **For NGS: Sat 18 Mar, Sat 14 Oct (12-5.30). Adm £13, chd £2. Light refreshments. For other opening times and information, please phone, email or visit garden website.**

East Ruston Old Vicarage is a 32 acre garden with traditional borders and a modern landscape. We have many different types of gardens here including formal walled and rose gardens, which small delightful. We also have an exotic garden which should not be missed. Visitors must not miss our monumental fruit cage, Mediterranean garden and our containers are to die for in the warmer months. At East Ruston Old Vicarage we also have a meadow, parkland and a wonderful heritage orchard.

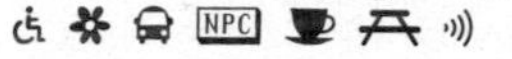

15 ELSING HALL GARDENS

Elsing Hall, Hall Road, Elsing, NR20 3DX. Patrick Lines & Han Yang Yap, www.elsinghall.com. *6km NW of Dereham. From A47 take the N Tuddenham exit. From A1067 take the turning to Elsing opp the Bawdeswell Garden Centre. Location details can be found on www.elsinghall.com.* **Sun 11 June (10-4). Adm £8, chd free. Home-made teas.**

C15 fortified manor house (not open) with working moat. 10 acre gardens and park surrounding the house. Significant collection of old roses, walled garden, formal garden, marginal planting, gingko avenue, viewing mound, moongate, interesting pinetum and terraced garden. Please note that there are very limited WC facilities available. Picnics allowed in the car park.

Our 2022 donation to Hospice UK meant that 911 NHS and Social Care staff accessed individual counselling support

GUIDED TOUR

16 NEW ERPINGHAM HOUSE FARM

Erpingham, Norwich, NR11 7QD. Buffy Willcox. *Turn off the A140 into Erpingham, signposted for caravans and camping. Continue ½m along the rd up a track with limetrees on either side.* **Evening opening Wed 7, Thur 8, Fri 9 June (6-8). Adm £15, chd free. Pre-booking essential, please visit www.ngs.org.uk for information & booking. Wine.**

The varieties of peonies grown have been carefully chosen to fill the whole season from early May to late June, with a generous selection of colours and beautiful scents. The peonies grown here benefit from being field grown in full sun and rich grade II loam soil and have been a successful part of the diversification on the farm.

17 FERNDALE

14, Poringland Road, Upper Stoke Holy Cross, Norwich, NR14 8NL. Dr Alan & Mrs Sheila Sissons. *4m S of Norwich. From Norwich or Poringland on B1332, take Stoke Rd at Railway Tavern r'about. Ferndale is 0.7m on L. Parking on road.* **Sun 28 May, Sun 25 June, Sun 30 July (11-5). Adm £4, chd free. Light refreshments.**

⅓ acre garden, paved area with seating surrounded by borders of shrubs and flowers. A pond with water feature. There is more seating in a second area, plus apple trees, soft fruit, vegetable plot, greenhouse and herb bed. Shingled area with water feature, rose arbour and low semi circular wall planted with flowers. There could be an accordionist playing. Wheelchair access 1m wide paved passageway to garden.

18 GAYTON HALL

Gayton, King's Lynn, PE32 1PL. Viscount & Viscountess Marsham. *6m E of King's Lynn. Off the B1145. At village sign take 2nd exit off Back St to entrance.* **Sun 26 Mar (11-4.30). Adm £6, chd free. Home-made teas.**

This rambling semi-wild garden, has over 2 miles of paths which meander through lawn, streams, bridges and woodland. Primulas, astilbes, hostas, lysichiton and gunnera grow along the water's edge. There is a good display of spring bulbs and a variety of unusual trees and shrubs, many labelled, have been planted over the years. Wheelchair access to most areas, gravel and grass paths.

19 HIGH HOUSE GARDENS

Blackmoor Row, Shipdham, Thetford, IP25 7PU. Sue & Fred Nickerson. *6m SW of Dereham. Take the airfield or Cranworth Rd off A1075 in Shipdham. Blackmoor Row is signed.* **Sun 11 June, Sun 10 Sept (12-5). Adm £5, chd free. Home-made teas.**

3 acre plantsman's garden developed and maintained by the current owners, over the last 40 years. Garden consists of colour themed herbaceous borders with an extensive range of perennials, box edged rose and shrub borders, woodland garden, pond and bog area, orchard and small arboretum. Plus large vegetable garden. Wheelchair access, gravel paths.

20 HIGHVIEW HOUSE

Norwich Road, Roughton, Norwich, NR11 8NA. Graham & Sarah Last, 07976 066896, grahamrc.last@gmail.com. *On A140 approx ½m N off Roughton village mini r'abouts. Parking on site is limited & therefore saved for wheelchair access. Main parking is signed, heading N of house on A140 on L side.* **Sun 25 June (12-4). Adm £6, chd free. Home-made teas. Visits also by arrangement 1 June to 23 July for groups of 20 to 80.**

2 acre garden featuring a large range of herbaceous borders. Long season of interest with a large range of perennial plants maintained by the owners. Over 120 salvia varieties feature through the planting of over 10,000 plants. Interesting garden structures inc a Japanese pagoda garden and large decked area. Pathways take you through each area of the garden. Suitable in most parts for wheelchairs. Football goals may be used by visitors at their own risk. For wheelchair access park on the main house driveway with access to pathway. Some slopes that will need to be accommodated.

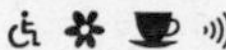

The Long Barn

21 ◆ HINDRINGHAM HALL
Blacksmiths Lane, Hindringham, NR21 0QA. Mr & Mrs Charles Tucker, 01328 878226, hindhall@btinternet.com, www.hindringhamhall.org. *7m from Holt/Fakenham/Wells. Turn off A148 at Crawfish Pub towards Hindringham. Drive 2m. Enter village of Hindringham. Turn L into Blacksmiths Ln after village hall.* **For NGS: Sat 25 Feb (10-4). Adm £9, chd £2.50. Light refreshments. Hot soup and sausage rolls as well as tea, coffee and cakes. For other opening times and information, please phone, email or visit garden website.**

A garden surrounding a Grade II* Tudor Manor House (not open) enveloped by a complete medieval moat plus 3 acres of fishponds. Working walled vegetable garden, Victorian nut walk, formal beds, bog and stream gardens. Something of interest throughout the year continuing well into autumn. Daffodils, spring wild garden, hellebore and primula, roses and clematis, autumn border. Some access for wheelchairs able to cope with gravel paths.

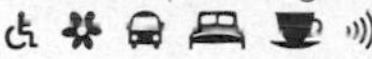

22 HOE HALL
Hall Road, Hoe, Dereham, NR20 4BD. Mr & Mrs James Keith, 01362 693169, vrkeith@hoehall.co.uk. *The garden is next to Hoe church in the village of Hoe.* **Visits by arrangement 15 May to 30 June for groups of up to 30. Adm £7, chd free. Tea and cake option available when booking.**

The main visual is a walled garden featuring a long white wisteria walk. This is set in the grounds of a Georgian rectory surrounded by parkland. The garden was redesigned in 1990 to incorporate climbers and herbaceous plants, with box parterres replacing the kitchen garden. There are espaliered fruit trees, and an old swimming pool with water lilies. Seating area and WC available in the walled garden.

23 HOLME HALE HALL
Holme Hale, Swaffham, Thetford, IP25 7ED. Mr & Mrs Simon Broke, 01760 440328, simon.broke@hotmail.co.uk. *2m S of Necton off A47. 1m E of Holme Hale village.* **Sun 14 May, Sun 9 July, Sun 1 Oct (12-4). Adm £7, chd free. Light refreshments. Cash payments only. Visits also by arrangement 3 Apr to 27 Oct.**

Walled kitchen garden designed by Arne Maynard and replanted in 2016/17. A soft palette of herbaceous plants inc some unusual varieties and provide a long season of interest. Greenhouse, vegetables, trained fruits, roses and topiary. 130 year old wisteria. Wildlife friendly with wildflower meadow and recently renovated Island pond. Historic buildings. Wheelchair access available to most areas.

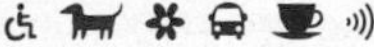

24 HONEYSUCKLE WALK

Litcham Road, Gressenhall, Dereham, NR20 4AR. Simon & Joan Smith, 01362 860530, simon.smith@ngs.org.uk. *17m W of Norwich. Follow B1146 & signs to Gressenhall Rural Life Museum then follow Litcham Rd into Gressenhall village. We are the 4th house on the L after passing the 30mph sign.* **Wed 26, Fri 28, Mon 31 July, Wed 2 Aug (2.30-5). Adm £15, chd free. Pre-booking essential, please visit www.ngs.org.uk for information & booking. Guided tour for 15 people starting at 2.30pm each day with afternoon tea inc in adm price. Visits also by arrangement 14 June to 4 Aug for groups of 10 to 20.**

2 acre woodland garden with 2 ponds, river and stream, 80 ornamental trees + collection of 400 different hydrangeas. Woodland access is by ¼ m grass track. Many rare hydrangea and hosta plants for sale.

25 HORSTEAD HOUSE

Mill Road, Horstead, Norwich, NR12 7AU. Mr & Mrs Matthew Fleming, 07771 655637, horsteadsnowdrops@gmail.com. *6m NE of Norwich on North Walsham Rd, B1150. Down Mill Rd opp the Recruiting Sergeant pub.* **Sat 18 Feb (11-4). Adm £5, chd free. Home-made teas. Visits also by arrangement 9 Feb to 28 Feb.**

Stunning display of beautiful snowdrops carpet the woodland setting with winter flowering shrubs. Another beautiful feature is the dogwoods growing on a small island in the River Bure, which flows through the garden. Small walled garden. Wheelchair access to main snowdrop area.

26 ◆ HOUGHTON HALL WALLED GARDEN

Bircham Road, Houghton, King's Lynn, PE31 6TY. The Cholmondeley Gardens Trust, 01485 528569, info@houghtonhall.com, www.houghtonhall.com. *11m W of Fakenham. 13m E of King's Lynn. Signed from A148.* **For opening times and information, please phone, email or visit garden website.**

The award-winning, 5 acre walled garden, beautifully divided into different areas designed by the Bannermans, inc a kitchen garden with espalier fruit trees. Glasshouses, a mediterranean garden, rose parterre, wisteria pergola and a spectacular double-sided herbaceous border. Many antique statues, fountains and contemporary sculptures. Gravel and grass paths. Electric buggies available in the walled garden.

27 ◆ HOVETON HALL GARDENS

Hoveton Hall Estate, Hoveton, Norwich, NR12 8RJ. Mr & Mrs Harry Buxton, 01603 784297, office@hovetonhallestate.co.uk, www.hovetonhallestate.co.uk. *8m N of Norwich. 1m N of Wroxham Bridge. Off A1151 Stalham Rd - follow brown tourist signs.* **For NGS: Mon 5 June (10.30-5). Adm £9, chd £4.50. Picnics are allowed for garden patrons only. For other opening times and information, please phone, email or visit garden website.**

15 acre gardens and woodlands taking you through the seasons. Mature walled herbaceous and kitchen gardens. Informal woodlands and lakeside walks. Nature Spy activity trail for our younger visitors. A varied events programme runs throughout the season. Please visit our website for more details. Light lunches and afternoon tea from our on site Garden Kitchen Cafe. The gardens are approx 75% accessible to wheelchair users. We offer a reduced entry price for wheelchair users and their carers.

28 KELLING HALL

Holt Road, Kelling, Holt, NR25 7EW. Mr & Mrs Widdowson, 01263 712201, hr@kelling-estate.co.uk, www.kelling-estate.co.uk. *2m NE of Holt, head through High St, follow Cromer Rd past BP petrol station, after 300 metres turn L onto Kelling Rd. Continue & follow signs.* **Sun 7 May (10.30-3.30). Adm £7.50, chd £3. Light refreshments catering and nearby Café at Holt Garden Centre (approx 0.3 mile) from Kelling Hall.**

The owners of The Kelling Estate and Holt Garden Centre invite you to walk the lower gardens and woods of Kelling Hall on their private estate. There is a wonderful bluebell walk, spring bulbs, Saxon/Roman ruins on an island, ponds, lakes and beautiful mature woods, as well as meadows with views to the sea. www.holtgardencentre.co.uk.

29 KERDISTON MANOR

Kerdiston, Norwich, NR10 4RY. Philip & Mararette Hollis. *1m from Reepham. Turn at Bawdswell garden centre towards Reepham after 3.8m turn L Smugglers Rd leaving Reepham surgery on L. L into Kerdiston Rd 1½m house on R.* **Sun 9 July (11-5). Adm £6, chd free. Light refreshments.**

2 acre tranquil garden surrounding Manor House (not open) that has been developed by the owners for over 30 years. Mature trees, colourful herbaceous borders, dell garden, potager style vegetable plot, pond, a 15 acre wild meadow walk and wonderful thatched C18 barn. Wheelchair access to teas and terrace over looking garden.

30 57 KETTS HILL

Norwich, NR1 4EX. Anne & John Farrow. *E of city centre. Follow the inner ring rd, A147, to the Barrack St r'about where it meets the B1140, Ketts Hill. The garden is situated 130 yds on L up the hill. Parking on Ketts Hill.* **Sun 21 May (11-5). Combined adm with 117 St Leonards Road £6, chd free. Light refreshments.**

A relaxed garden in an historic part of the city. There is a small lawned area, shrubs and perennial borders with a wide variety of plants, terraces and steep slopes leading to an elevated summerhouse with wonderful views of the cathedrals and city. The garden is approached via a steep gravelled drive and there are steep steps in the garden which means it is only suitable for the able bodied.

31 LEXHAM HALL

nr Litcham, PE32 2QJ. Mr & Mrs Neil Foster, www.lexhamestate.co.uk. *6m N of Swaffham off B1145. 2m W of Litcham.* **Sun 5, Sun 12 Feb (11-4). Adm £6, chd free. Sun 21 May, Wed 19 July (11-5). Adm £7.50, chd free. Light refreshments. Donation to Lexham Church (Feb only).**

Parkland with lake and river walks surround C17/18 Hall (not open). Formal garden with terraces, roses and mixed borders. Traditional working kitchen garden with crinkle-crankle wall. Year-round interest; woods and borders carpeted with snowdrops in February, rhododendrons, azaleas, camellias and magnolias in the 3 acre woodland garden in May, and July sees the walled garden borders at their peak.

32 THE LONG BARN
Flordon Road, Newton Flotman, Norwich, NR15 1QX. Mr & Mrs Mark Bedini. *6m S of Norwich along A140. Leave A140 in Newton Flotman towards Flordon. Exit Newton Flotman & approx 150 yds beyond 'passing place' on L, turn L into drive. Note that SatNav does not bring you to destination.* **Sun 30 July (10.30-4). Adm £6, chd free. Home-made teas.**
Beautiful informal garden with new haha creating infinity views across ancient parkland. Strong Mediterranean influence with plenty of seating in large paved courtyard under pollarded plane trees. Unusual wall-sheltered group of four venerably gnarled olive trees, figs and vines, then herbaceous borders, tiered lawns merging northwards into woodland garden. Walks through parkland to the River Tas - serene. Wheelchair drop off at front of house. Gradual lawn slope accesses upper tier of garden. WC requires steps.

33 ◆ MANNINGTON ESTATE
Mannington, Norwich, NR11 7BB. Lady Walpole, 01263 584175, admin@walpoleestate.co.uk, www.manningtonestate.co.uk. *18m NW of Norwich. 2m N of Saxthorpe via B1149 towards Holt. At Saxthorpe/Corpusty follow signs to Mannington.* **For NGS: Sun 26 Mar (11-4). Adm £6, chd free. Light refreshments in the Garden Tearooms. For other opening times and information, please phone, email or visit garden website.**
20 acres feature shrubs, lake and trees. Period gardens. Borders. Sensory garden. Extensive countryside walks and trails. Moated manor house and Saxon church with C19 follies. wildflowers and birds. The tearooms offer home-made locally sourced food with home-made teas. Wheelchair access, gravel paths, 1 steep slope.

34 MANOR FARM, COSTON
Coston Lane, Coston, Wymondham, NR9 4DT. Mr & Mrs J D Hambro. *10m W of Norwich. Off B1108 Norwich - Watton Rd. Take B1135 to Dereham at Kimberley. After approx 300yds sharp L bend, go straight over down Coston Ln, house & garden on L.* **Sun 2 July (12-5). Adm £6, chd free. Home-made teas.**
Wonderful 3 acre country garden set in larger estate. Several small garden rooms with both formal and informal planting. Climbing roses, walled kitchen garden, white, grass and late summer themes, classic herbaceous and shrub borders, box parterres and large areas of wildflowers. Large collection of sculptures dotted round the garden. Dogs welcome. Something for everyone. Wheelchair access, some gravel paths and steps.

35 MANOR HOUSE FARM, WELLINGHAM
Fakenham, King's Lynn, PE32 2TH. Robin & Elisabeth Ellis, 01328 838227, libbyelliswellingham@gmail.com, www.manor-house-farm.co.uk. *½m off A1065 7m W from Fakenham & 8m E. Swaffham. By the Church.* **Sun 18 June (11-5). Adm £6.50, chd free. Home-made teas.**
Charming 4 acre country garden surrounds an attractive farmhouse. Formal quadrants, 'hot spot' of grasses and gravel, small arboretum, pleached lime walk, vegetable parterre and rose tunnel. Unusual walled 'Taj' garden with old fashioned roses, tree peonies, lilies and formal pond. A variety of herbaceous plants. Small herd of Formosan Sika deer. Picnics allowed. Dogs on leads. Featured in Country Living, June 2022.

36 NORTH LODGE
51 Bowthorpe Road, Norwich, NR2 3TN. Bruce Bentley & Peter Wilson. *1½m W of Norwich City Centre. Turn into Bowthorpe Rd off Dereham Rd, garden150m on L. By bus: 21, 22, 23, & 24 from City centre, Old Catton, Heartsease, Thorpe & most of W Norwich. Parking available outside & room for bikes.* **Sun 16 July (11-5), open nearby 4 Coach House Court. Sun 30 July (11-5). Adm £5, chd free. Home-made teas. Cakes inc gluten-free and vegan.**
Delightful town garden surrounding Victorian Gothic Cemetery Lodge. Strong structure and vistas created by follies built by current owners, inc a classical temple, oriental water garden and formal ponds. Original 25m-deep well! Predominantly herbaceous planting. Carnivorous and succulent collection in hand-built conservatory. House extension won architectural award. Slide show of house and garden history. Wheelchair access possible but difficult. Sloping gravel drive followed by short, steep, narrow, brickweave ramp. WC not easily wheelchair accessible.

37 47 NORWICH ROAD
Stoke Holy Cross, Norwich, NR14 8AB. Anna & Alistair Lipp. *Heading S through Stoke Holy Cross on Norwich Rd, garden on R down private drive. No parking at property. Parking at sports ground car on L of Long Ln, 500 yds to garden.* **Sat 24 June (11-5). Combined adm with 51 Norwich Road £6, chd free. Home-made teas.**
Medium sized, west facing garden developed over 25 years with views over Tas Valley and water meadows. Vine and rose covered pergola, gravel beds, terrace with raised beds and shading magnolias, large greenhouse with exotic plants. Small meadow and informal pond. Gravel driveway to brick paths with slope to raised terrace allows majority of garden to be viewed by wheelchair users.

38 51 NORWICH ROAD
Stoke Holy Cross, Norwich, NR14 8AB. Mrs Vivian Carrington. *5m S of Norwich. From Norwich past Wildebeest restaurant on R, No 51 is on R. 50 yds from the junc with Long Ln. Parking at sports ground car park on L of Long Ln. 500 yds to garden.* **Sat 24 June (11-5). Combined adm with 47 Norwich Road £6, chd free.**
The garden has a variety of perennials, roses and annuals at the front with an asparagus bed and vegetables to the side and rear. There are cold frames and a small greenhouse. Behind the house there are further flower beds and a small lawn. Wheelchair access at the front and rear of garden.

39 NEW OLD MANOR FARMHOUSE

The Hill, Swanton Abbott, Norwich, NR10 5EA. Drs Paul and Sian Everden, 07768 376621, info@hostebarn.com. *Just outside Swanton Abbott on North Norfolk Coast. From Swanton Abbott take Long Common Ln until you see the car park sign on The Hill.* **Sat 24, Sun 25 June (9-4.30). Adm £5, chd free. Home-made teas.**

Originally a field surrounding a derelict C17 listed Farmhouse (not open). Winning the 1991 Graham Allen Conservation Award, the garden structure, sympathetic to the Dutch style of the house was laid down. Closed knot garden of box surrounded by pleached hornbeam, pollarded plane trees, beech and yew hedges divide areas and flank walks, herbaceous borders and lawns, clematis and rose walk to potager and paddock.

40 THE OLD RECTORY, SYDERSTONE

Creake Road, Syderstone, King's Lynn, PE31 8SF. Mr & Mrs Tom White. *Off B1454 , 8 miles W of Fakenham,. Access from Creake Rd, or side gate opp Village Hall on The St.* **Sun 4 June (11-5). Adm £5, chd free. Home-made teas.**

Charming Old Rectory garden designed in 1999 by Arne Maynard. There are lawns, box, hornbeam, yew and beach hedging, pleached crab apple, wisteria and climbing roses, a parterre of English shrub roses, herbaceous beds, a shrubbery and an orchard.

41 OULTON HALL

Oulton, Aylsham, NR11 6NU. Bolton Agnew. *4m NW of Aylsham. From Aylsham take B1354. After 4m turn L for Oulton Chapel, Hall ½m on R. From B1149 (Norwich/Holt rd) take B1354, next R, Hall ½m on R.* **Sun 4 June (1.30-5). Adm £6, chd free.**

C18 manor house (not open) and clocktower set in 6 acre garden with lake and woodland walks. Chelsea designer's own garden - herbaceous, Italian, bog, water, wild, verdant, sunken and parterre gardens all flowing from one tempting vista to another. Developed over 25 yrs with emphasis on structure, height and texture, with a lot of recent replanting in the contemporary manner.

42 QUAKER FARM

Quaker Lane, Spixworth, Norwich, NR10 3FL. Mr & Mrs Peter Cook. *3m N of Norwich. From Norwich take Buxton Rd towards Spixworth, through Old Catton. Turning signed Quaker Ln just over the NDR (A1270).* **Sun 14 May (11-4). Adm £5, chd free. Home-made teas.**

An acre of garden surrounding a traditional Norfolk farmhouse which has evolved over the last 40 plus yrs. Comprises garden rooms inc herbaceous borders, gravel garden and a variety of shrubs and trees. Also available is an extended circular walk on the farm through a bluebell wood.

43 ◆ RAVENINGHAM HALL

Raveningham, Norwich, NR14 6NS. Sir Nicholas & Lady Bacon, 01508 548480, sonya@raveningham.com, www.raveningham.com. *14m SE of Norwich. 4m from Beccles off B1136.* **For NGS: Sat 25 Feb (2-4). Adm £22, chd free. Pre-booking essential, please visit www.ngs.org.uk for information & booking. Guided Walk by Sir Nicholas Bacon around the gardens. Adm inc tea & cake. For other opening times and information, please phone, email or visit garden website.**

Traditional country house garden in glorious parkland setting. Restored Victorian conservatory, walled kitchen garden, herbaceous borders, newly planted stumpery. Arboretum est after 1987 gale, Millennium lake and sculpture by Susan Bacon. February sees large drifts of many different varieties of snowdrops.

44 NEW 117 ST LEONARDS ROAD

Norwich, NR1 4JF. John and Vanessa Trevelyan. *Down short access drive off St Leonards Road between nos 113C and 119.* **Sun 21 May (11-5). Combined adm with 57 Ketts Hill £6, chd free. 57 Ketts Hill will be serving home made teas.**

An established walled garden in an historic part of Norwich with packed flowerbeds, trees, shrubs, fruit and vegetables, planted and managed for year-round interest. Plant sales. This garden overlooks Ketts Heights which is an historic area of Norwich open to the public and with fine views across Norwich. Garden is level with no steps, but the access drive is gravel.

45 ◆ SANDRINGHAM GARDENS

Sandringham Estate, Sandringham, PE35 6EH. His Majesty The King, 01485 545408, visits@sandringhamestate.co.uk, www.sandringhamestate.co.uk. *Sandringham is 6 miles NE of King's Lynn and is signposted from the A148 Fakenham rd and the A149 Hunstanton rd. The Royal Park's postcode for SatNav is PE35 6AB.* **For opening times and information, please phone, email or visit garden website.**

Set in 25 hectares (60 acres) and enjoyed by the British Royal Family and their guests when in residence, the more formal gardens are open from April - October. The grounds have been developed in turn by each Monarch since 1863 when King Edward VII and Queen Alexandra purchased the Estate. Gravel paths are not deep, some long distances. Please contact us or visit the website for an Accessibility Guide.

46 SEVERALS GRANGE

Holt Road B1110, Wood Norton, NR20 5BL. Jane Lister, 01362 684206, hoecroft@hotmail.co.uk. *8m S of Holt, 6m E of Fakenham. 2m N of Guist on L of B1110. Guist is situated 5m SE of Fakenham on A1067 Norwich rd.* **For NGS: Sun 13 Aug (1-5). Adm £6, chd free. Pre-booking essential, please visit www.ngs.org.uk for information & booking. Home-made teas. Open nearby Bluebell Barn. For other opening times and information, please phone or email.**

The gardens surrounding Severals Grange are a perfect example of how colour, shape and form can be created by the use of foliage plants, from large shrubs to small alpines. Movement and lightness are achieved by interspersing these plants with a wide range of ornamental grasses, which are at their best in late summer. Splashes of additional colour are provided by a variety of herbaceous plants. Wheelchair access, some gravel paths but help can be provided.

47 SILVERSTONE FARM

North Elmham, Dereham, NR20 5EX. George Carter, 01362 668130, grcarter@easynet.co.uk, www.georgecartergardens.co.uk. *Nearer to Gateley than North Elmham. From North Elmham church head N to Guist. Take 1st L onto Great Heath Rd. L at T-junc. Take 1st R signed Gateley, Silverstone Farm is 1st drive on L by a wood.* **Evening opening Fri 23 June (6-8). Adm £6.50. Wine.**

Garden belonging to George Carter described by the Sunday Times as 'one of the 10 best garden designers in Britain'. 1830s farmyard and formal gardens in 2 acres. Inspired by C17 formal gardens, the site consists of a series of interconnecting rooms with framed views and vistas designed in a simple palette of evergreens and deciduous trees and shrubs such as available in that period. Books by the owner for sale. Level site, mostly wheelchair accessible.

48 NEW SOUTHGATE HOUSE

Southgate Lane, Norwich, NR1 2AQ. Mr Matthew Williams, matthew@createdesign.org. *Southgate Ln is located by the pedestrian X-ing on Bracondale. The house is found on the L at the bottom of the ln.* **Sun 23 July, Sat 5, Sun 6 Aug (10.30-4). Adm £5, chd free. Light refreshments. Open nearby 1 Woodside Cottages. Teas, coffees and cakes. Visits also by arrangement 5 June to 31 Aug for groups of 6 to 30.**

The former Harbour Master's house, the garden extends to about an acre, half of which is formed on a steep escarpment. Little of the original garden layout existed by 2018. However, we uncovered some historic maps and plans and have recreated the garden and Harbour Master's walks down the hillside. The garden now comprises a combination of garden rooms and styles and woodland hillside walks. Woodland garden, thematic gardens and a city vista.

49 ◆ STODY LODGE

Melton Constable, NR24 2ER. Mr & Mrs Charles MacNicol, 01263 860572, enquiries@stodyestate.co.uk, www.stodylodgegardens.co.uk. *16m NW of Norwich, 3m S of Holt. Off B1354. Signed from Melton Constable on Holt Rd. For SatNav NR24 2ER. Gardens signed as you approach.* **For NGS: Wed 17 May (1-5). Adm £9, chd free. Home-made teas. For other opening times and information, please phone, email or visit garden website.**

Spectacular gardens with 1 of the largest concentrations of rhododendrons and azaleas in East Anglia. Created in the 1920s, the gardens also feature magnolias, camellias, a variety of ornamental and specimen trees, late daffodils and bluebells. Expansive lawns and magnificent yew hedges. Woodland walks and 4 acre Water Garden filled with over 2,000 vividly-coloured azalea mollis. Home-made teas provided by selected local and national charities. Wheelchair access to most areas of the garden. Gravel paths with some uneven ground.

In 2022 the National Garden Scheme donated £3.11 million to our nursing and health beneficiaries

Elsing Hall Gardens

50 SWAFIELD HALL

Knapton Road, Swafield, North Walsham, NR28 0RP. Tim Payne & Boris Konoshenko, www.swafieldhall.co.uk. *Swafield Hall is approx ½m along Knapton Rd (also called Mundesley Rd) from its start in the village of Swafield. You need to add about 300 yds to the location given by most SatNavs.* **Sat 15, Sun 16 July (10-5). Adm £5, chd free. Light refreshments.**

C16 Manor House with Georgian additions (not open) set within 4 acres of gardens inc a parterre and various rooms inc a summer garden, orchard, cutting garden, pear tunnel, secret oriental garden (with 9 flower beds based on a Persian carpet), the Apollo Promenade of theatrical serpentine hedging, a duck pond and woodland walk. The whole garden is accessible by wheelchair.

51 NEW THE NORFOLK HOSPICE, TAPPING HOUSE

Wheatfields, Hillington, King's Lynn, PE31 6BH. The Norfolk Hospice, www.norfolkhospice.org.uk. *Take the A148 to Fakenham/Cromer. In Hillington turn R into Stn Rd, signed B1153 Grimston. Wheatfields is the 1st turning on the R after 160m. The Hospice is straight ahead at the end of the rd.* **Sun 11 June (9-5). Adm £4, chd free. Light refreshments.**

This hospice garden has been created and maintained by a team of volunteers. The site was purpose built and was opened in 2016. The gardens are still ongoing but many areas are well established and provide a peaceful and tranquil backdrop to residents, visitors and staff alike. Set in grounds of approx 2 acres, there is a variety of cottage garden plants, perennials and shrubs. A vegetable plot, the produce of which is used by the kitchen team, and a wildlife area with pond. A tree was planted to commemorate the Queen's Platinum Jubilee and to celebrate the coronation of King Charles III. Full wheelchair access means that our garden space is available for all to enjoy.

52 61 TRAFFORD WAY

Spixworth, Norwich, NR10 3QL. Mr & Mrs Colin Ryall. *Park in Spixworth Community car park in Crosswick Ln.* **Sun 30 July (10-5). Adm £4, chd free. Light refreshments.**

Small garden showing what can be achieved with careful planting and the use of pleached hornbeam trees. Gravel garden with flower beds and pot plants. Colourful herbaceous borders with roses.

53 TUDOR LODGINGS

Castle Acre, King's Lynn, PE32 2AN. Gus & Julia Stafford Allen, 01760 755334, jstaffordallen@btinternet.com. *Parking in field below the house.* **Sun 23 July (10.30-5.30). Adm £6, chd free. Light refreshments. Visits also by arrangement 1 July to 1 Sept.**

2 acre garden fronts a C15 flint house (not open), incorporates part of the Norman earthworks. A variety of planting in mixed borders, lawns, abstract topiary, and a C18 dovecote. Ornamental grasses, hot border and cutting garden with a productive fruit cage. Natural wild area inc a shepherd's hut and informal pond inhabited by ducks and poultry. Newly planted orchard with beehive. Partial wheelchair access, please ask for assistance. Disabled WC.

54 WALCOTT HOUSE

Walcott Green, Walcott, Norwich, NR12 0NU. Nick & Juliet Collier. *3m N of Stalham. Off the Stalham to Walcott Rd (B1159).* **Sun 2 July (10-4). Adm £6, chd free. Home-made teas.**

A 12 acre site with over 1 acre of formal gardens based on model C19 Norfolk farm buildings. Woodland and damp gardens, arboretum, vistas with tree lined avenues, woodland walks. Wheelchair access, small single steps to negotiate when moving between gardens in the yards.

55 33 WALDEMAR AVENUE

Hellesdon, Norwich, NR6 6TB. Sonja Gaffer & Alan Beal, www.facebook.com/Hellesdontropicalgarden/?modal=admin_todo_tour. *Waldemar Ave is situated approx 400 yds off Norwich ring road towards Cromer on A140.* **Sun 6 Aug, Sun 3 Sept (10-5). Adm £5, chd free. Home-made teas.**

A surprising and large suburban garden of many parts with an exciting mix of exotic and tropical plants combined with unusual perennials. A quirky palm-thatched Tiki hut is an eye catching feature. There is a wonderful treehouse draped in plants. You can sit by the pond which is brimming with wildlife and rare plants. A large collection of succulents will be on show and there will be plants to buy. Wheelchair access surfaces are mostly of lawn and concrete and are on one level.

Kelling Hall

56 NEW THE WALLED GARDEN, LITTLE PLUMSTEAD

Old Hall Road, Little Plumstead, Norwich, NR13 5FA. Little Plumstead Walled Garden Community Shop & Cafe, thewalledgardenshop.co.uk. *On arriving in Little Plumstead follow signs to The Walled Garden.* **Evening opening Wed 5 July (6-8). Adm £10. Pre-booking essential, please visit www.ngs.org.uk for information & booking.**

Richard Hobbs will be taking 2 guided walks for 20 people (prebooked in advance) around the recently restored Victorian walled garden with heritage apples and pears on beautifully restored brick walls. There are cutting beds, herbaceous and shrub areas together with a Victorian style glasshouse. There is a newly planted alpine area, a stumpery a range of herbs and some rare and unusual plants. Good wheelchair access throughout.

57 WARBOROUGH HOUSE

2 Wells Road, Stiffkey, NR23 1QH. Mr & Mrs J Morgan. *13m N of Fakenham, 4m E of Wells-Next-The-Sea on A149 in the centre of Stiffkey village. Parking is available & signed at garden entrance. Coasthopper bus stop outside garden. Please do not park on the main rd as this causes congestion.* **Sun 21 May (11-4). Adm £6, chd free. Home-made teas.**

7 acre garden on a steep chalk slope, surrounding C19 house (not open) with views across the Stiffkey valley and to the coast. Woodland walks, formal terraces, shrub borders, lawns and walled garden create a garden of contrasts. Garden slopes steeply in parts. Paths are gravel, bark chip or grass. Disabled parking allows access to garden nearest the house and teas.

GROUP OPENING

58 WELLS-NEXT-THE-SEA GARDENS

Wells-Next -The-Sea, NR23 1DP. *10m N of Fakenham. All gardens near Coasthopper 'Burnt St' or 'The Buttlands' bus stops. Car parking for all gardens in Market Ln area.* **Sun 18 June (11-5). Combined adm £6, chd free. Home-made teas in The Old Rectory Garden.**

NEW DAHLIA COTTAGE
Lynne & Jim Dowdy.

7 MARKET LANE
Wells-next-the-Sea. Hazel Ashley.

NORFOLK HOUSE
17 Burnt Street, Wells-next-the-Sea. Katrina & Alan Jackson.

NEW THE OLD RECTORY
Baroness Patricia Rawlings.

POACHER COTTAGE
15 Burnt Street, Wells-next-the-Sea. Roger & Barbara Oliver.

Wells-next-the-Sea is a small, friendly coastal town on the North Norfolk coast. Popular with families, walkers and bird watchers. The harbour has shops, cafes, fish and chips. Beach served by a narrow gauge railway. Fine parish church. The five town gardens, though small, demonstrate a variety of design and a wide selection of planting. Partial wheelchair access at all gardens.

59 WEST BARSHAM HALL

Fakenham, NR21 9NP. Mr & Mrs Jeremy Soames, 01328 863519, susannasoames@gmail.com. *3m N of Fakenham. From Fakenham take A148 to Cromer then L on B1105 to Wells. After ½m L again to Wells. After 1½m R The Barshams & West Barsham Hall.* **Sun 2 July (10.30-5.30). Adm £7.50, chd free. Light refreshments. Morning coffee and afternoon cream teas. Visits also by arrangement May to Sept.**

Large garden with lake, approx 10 acres. Mature yew hedging and sunken garden originally laid out by Gertrude Jeykll. Swimming pool garden, shrub borders, kitchen garden with herbaceous borders, fruit cage, cutting garden and bog garden. Separate old fashioned cottage garden also open. Some slopes and gravel paths.

60 THE WHITE HOUSE

The Street, Ridlington, North Walsham, NR28 9NR. Richard & Vanda Barker. *5m E of North Walsham, in village of Ridlington. Can be approached from North Walsham or from the B1159 Stalham to Bacton Rd. Off-road parking 5 minutes walk.* **Sun 18 June (11-4). Adm £5, chd free. Home-made teas. Home-made cakes and sandwiches also available.**

1 acre garden of former rectory, with colour-themed herbaceous and mixed borders, gravel borders, roses, wildlife pond, kitchen garden and fruit garden. Most of the garden is wheelchair-accessible with care. Access is through a gravel courtyard, and there are slopes.

61 NEW 1 WOODSIDE COTTAGES

Bracondale, Norwich, NR1 2AY. Wayne Waith. *From County Hall r'about, heading towards the City Centre on Bracondale, turn L at Conesford Drive and bear R towards white painted cottages. Parking at County Hall and then a short walk.* **Sun 23 July, Sat 5, Sun 6 Aug (11-5). Adm £4, chd free. Home-made teas. Open nearby Southgate House. Light refreshments are also served at nearby Southgate House.**

A very pretty, small cottage garden surrounding aptly named 1 Woodside Cottages near Norfolk's County Hall. It is packed with colourful plants both ornamental and edible and even the resident chickens are decorative. Fruit trees and vegetable grow side by side with perennial and annual flowers and a small pond supports a multitude of frogs each spring. A warm welcome awaits you. There are 2 shallow steps to negotiate, otherwise level.

62 WRETHAM LODGE

East Wretham, IP24 1RL. Mr Gordon Alexander & Mr Ian Salter. *6m NE of Thetford. A11 E from Thetford, L up A1075, L by village sign, R at Xrds then bear L.* **Sun 30 Apr, Mon 1 May (11-5). Adm £5, chd free. Tea at local church.**

10 acre garden surrounding former Georgian rectory (not open). In spring masses of species tulips, hellebores, fritillaries, daffodils and narcissi; bluebell walk and small woodland walk. Topiary pyramids and yew hedging lead to double herbaceous borders. Shrub borders and rose beds (home of the Wretham Rose). Traditionally maintained walled garden with fruit, vegetables and perennials.

NORTH EAST

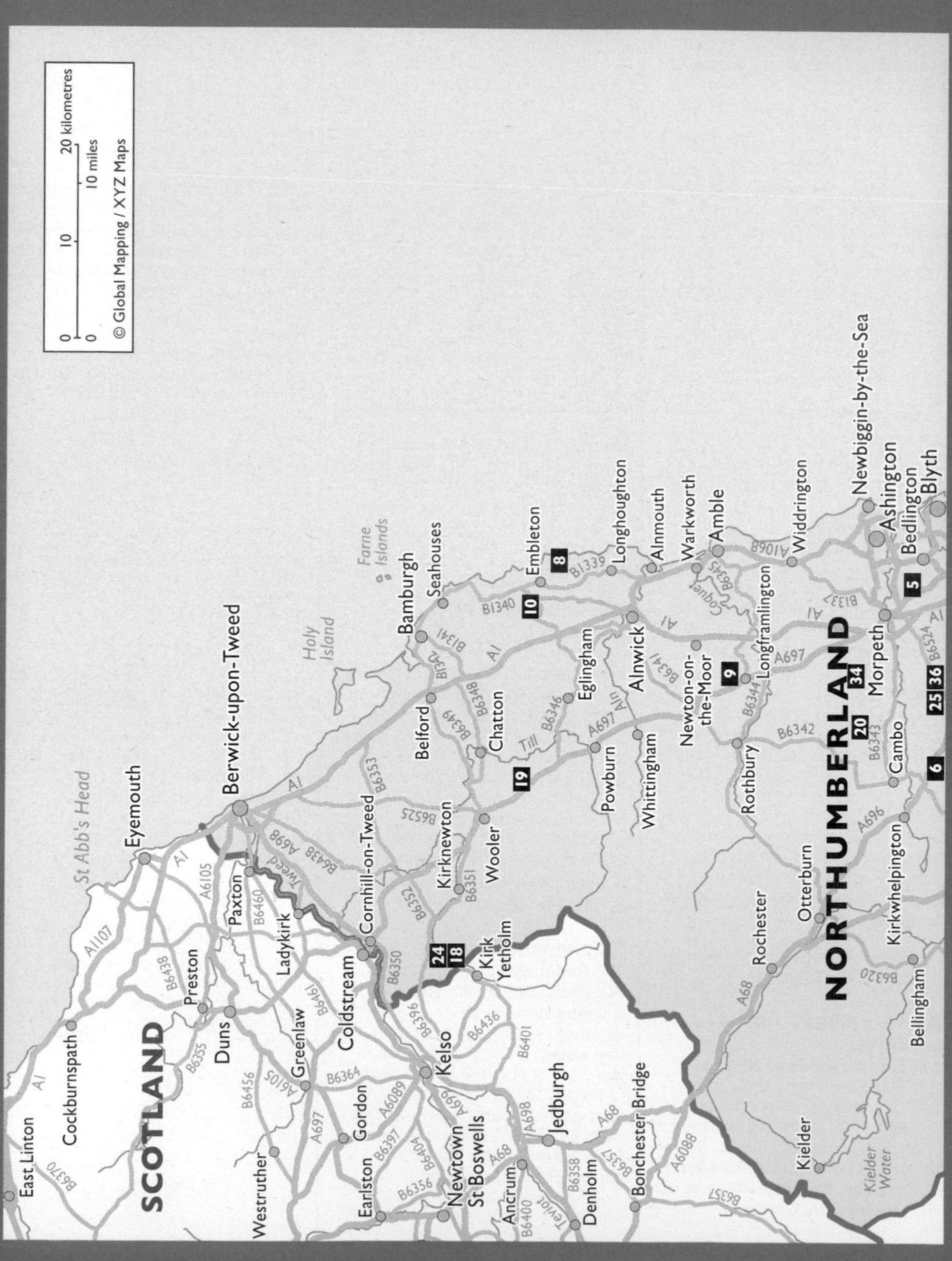

0 10 20 kilometres
0 10 miles
© Global Mapping / XYZ Maps
SCOTLAND
NORTHUMBERLAND
St Abb's Head
Eyemouth
Berwick-upon-Tweed
Holy Island
Farne Islands
Bamburgh
Seahouses
Embleton
Longhoughton
Alnmouth
Warkworth
Amble
Widdrington
Newbiggin-by-the-Sea
Ashington
Bedlington
Blyth
Morpeth
Longframlington
Newton-on-the-Moor
Alnwick
Eglingham
Chatton
Belford
Wooler
Kirknewton
Cornhill-on-Tweed
Coldstream
Kirk Yetholm
Powburn
Whittingham
Rothbury
Cambo
Kirkwhelpington
Otterburn
Rochester
Bellingham
Kielder
Kielder Water
Paxton
Ladykirk
Preston
Duns
Greenlaw
Cockburnspath
East Linton
Westruther
Gordon
Earlston
Kelso
Newtown St Boswells
Ancrum
Jedburgh
Denholm
Bonchester Bridge
Tweed
Till
Aln
Coquet
Teviot
A1
A1107
A6105
A698
A697
A68
A6088
A696
A699
A6089
A1068
B6438
B6355
B6456
B6364
B6397
B6356
B6404
B6400
B6358
B6357
B6436
B6401
B6396
B6350
B6352
B6351
B6525
B6353
B6349
B6348
B6346
B6341
B1341
B1342
B1340
B1339
B6345
B6344
B6342
B6343
B1337
B6524
B6320
B6461
B6460
B6370
8
10
9
19
24
18
20
34
25
36
5
6

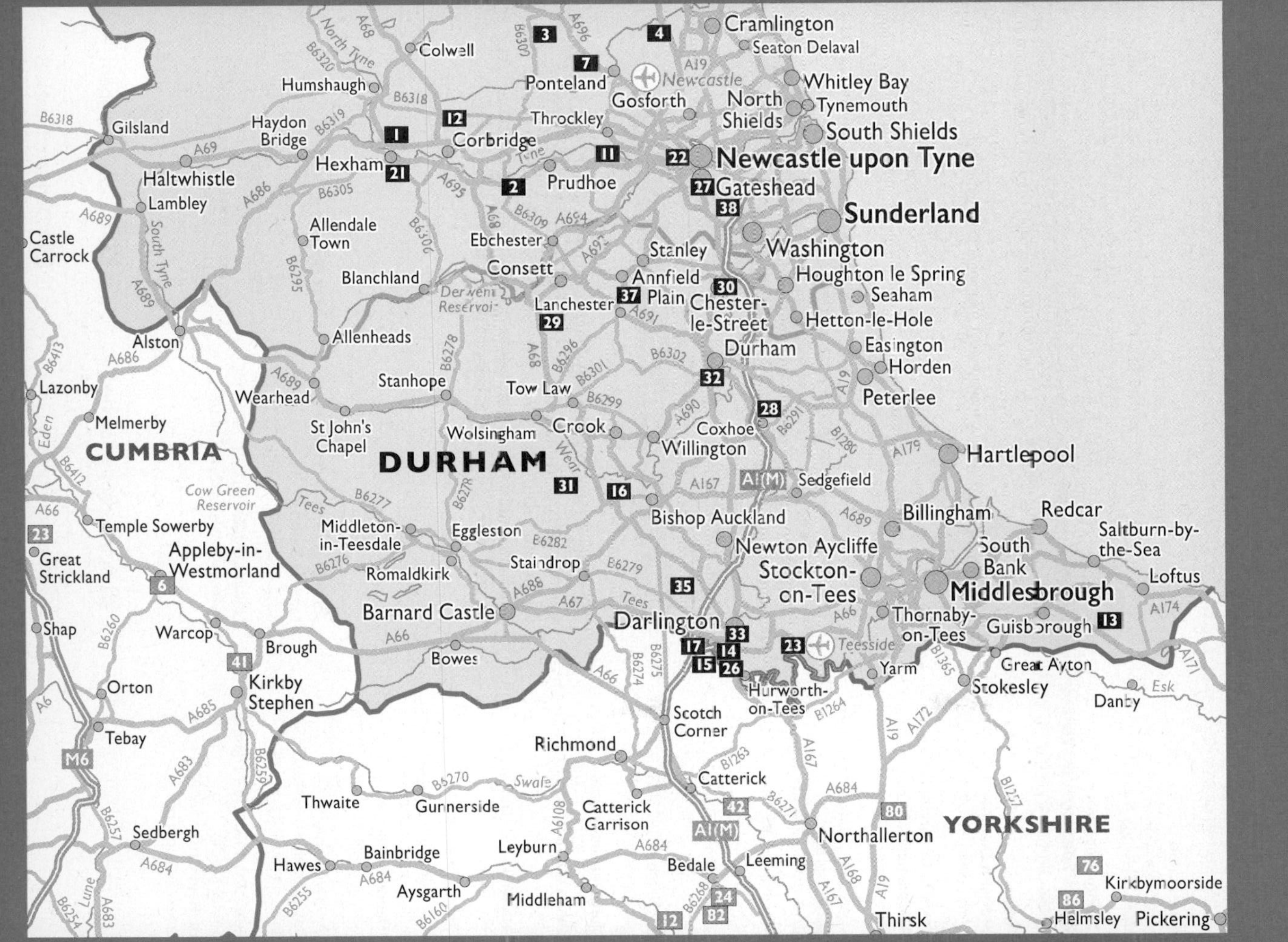
Cramlington
Seaton Delaval
Colwell
Humshaugh
Ponteland
Newcastle
Gosforth
Whitley Bay
North Shields
Tynemouth
South Shields
Gilsland
Haydon Bridge
Corbridge
Throckley
Hexham
Newcastle upon Tyne
Haltwhistle
Prudhoe
Gateshead
Lambley
Sunderland
Allendale Town
Ebchester
Castle Carrock
Stanley
Washington
Consett
Annfield Plain
Houghton le Spring
Blanchland
Derwent Reservoir
Seaham
Lanchester
Chester-le-Street
Hetton-le-Hole
Alston
Allenheads
Easington
Durham
Horden
Stanhope
Lazonby
Wearhead
Tow Law
Peterlee
Melmerby
St John's Chapel
Crook
Coxhoe
Wolsingham
Willington
Hartlepool
CUMBRIA
DURHAM
Sedgefield
Cow Green Reservoir
Billingham
Redcar
Bishop Auckland
Temple Sowerby
Middleton-in-Teesdale
Eggleston
Saltburn-by-the-Sea
Newton Aycliffe
South Bank
Great Strickland
Appleby-in-Westmorland
Staindrop
Stockton-on-Tees
Romaldkirk
Loftus
Middlesbrough
Barnard Castle
Darlington
Thornaby-on-Tees
Guisborough
Shap
Warcop
Brough
Teesside
Bowes
Yarm
Great Ayton
Orton
Kirkby Stephen
Hurworth-on-Tees
Stokesley
Danby
Scotch Corner
Tebay
Richmond
Catterick
Thwaite
Gunnerside
Catterick Garrison
Sedbergh
YORKSHIRE
Northallerton
Bainbridge
Leyburn
Hawes
Bedale
Leeming
Aysgarth
Middleham
Kirkbymoorside
Thirsk
Helmsley
Pickering
Tyne
North Tyne
South Tyne
Wear
Tees
Swale
Eden
Esk
Lune

Volunteers

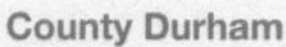

County Durham

Joint County Organisers
Iain Anderson
01325 778446
iain.anderson@ngs.org.uk

Aileen Little
01325 356691
aileen.little@ngs.org.uk

County Treasurer
Monica Spencer 01325 286215
monica.spencer@ngs.org.uk

Publicity
Margaret Stamper 01325 488911
margaretstamper@tiscali.co.uk

Booklet Co-ordinator
Sue Walker 01325 481881
walker.sdl@gmail.com

Assistant County Organisers
Sarah Garbutt
sarah.garbutt@ngs.org.uk

Helen Jackson
helen.jackson@ngs.org.uk

Gill Knights 01325 483210
gillianknights55@gmail.com

Gill Naisby 01325 381324
gillnaisby@gmail.com

Margaret Stamper (see above)

Sheila Walke 07837 764057
sheila.walke@ngs.org.uk

Sue Walker (see above)

Northumberland & Tyne and Wear County Organiser & Booklet Coordinator
Maureen Kesteven 01914 135937
maureen.kesteven@ngs.org.uk

County Treasurer
David Oakley 07941 077594
david.oakley@ngs.org.uk

Publicity, Talks Co-ordinator & Social Media
Liz Reid 01914 165981
liz.reid@ngs.org.uk

Assistant County Organisers
Maxine Eaton 077154 60038
maxine.eaton@ngs.org.uk

Natasha McEwen 07917 754155
natashamcewengd@aol.co.uk

Liz Reid (as left)

Susie White 07941 077594
susie@susie-white.co.uk

David Young 01434 600699
david.young@ngs.org.uk

County Durham: an unsung county.

County Durham lies between the River Wear and River Tees and is varied and beautiful.

Our National Garden Scheme open gardens can be found in the city, high up in the Dales and in the attractive villages of South Durham. Something different every week.

We are delighted to have favourite gardens opening again this year and we also welcome some new gardens, one of which is Willowburn Hospice at Lanchester and a new group in Langholm Crescent, Darlington. It is wonderful to be able to welcome new gardens and we look forward to them receiving support from visitors.

Northumberland is a county of ancient castles and wild coastline, of expansive views and big skies.

Northumberland is a rich and varied landscape of wild uplands where moorland birds nest, picturesque valleys and a coastline of castles and fishing villages. It's the perfect backdrop for historic as well as exciting contemporary gardens.

Some have been created over generations. A manor house garden associated with Gertrude Jekyll, a 19th century mansion set in 10 acres of walled gardens, a 14th century pele tower, and Jacobean house with fabulous views over the Tyne Valley. Others have been made over a number of years by their owners and are packed full of ideas. A town garden on a dramatically steep site, a country cottage filled with roses and lavender, an artist's garden by the sea.

Then there's newly designed gardens, one of which won a major award from the Society of Garden Designers in 2022, another redesigned with a gravel garden and stone rill. And for the specialist there's an outstanding collection of cacti that has been given a new home in this beautiful county.

@gardensopenforcharity   @NGSNorthumberl1

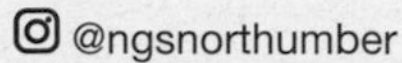
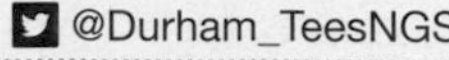

@ngsnorthumber   @Durham_TeesNGS

OPENING DATES

All entries subject to change. For latest information check **www.ngs.org.uk**

Map locator numbers are shown to the right of each garden name.

April

Sunday 23rd
Blagdon 4

May

Sunday 14th
Ravensford Farm 31

Sunday 21st
Lilburn Tower 19

Thursday 25th
Nags Head Farm 26

Sunday 28th
Ferndene House 11

June

Sunday 4th
The Beacon 2
2 Hillside Cottage 16

Sunday 11th
Coldcotes Moor Garden 7
Oliver Ford Garden 29
NEW Willow Burn Hospice 37

Saturday 17th
Bichfield Tower 3

Sunday 18th
Marie Curie Hospice 22
◆ Mindrum Garden 24
The Moore House 25

Saturday 24th
Fallodon Hall 10
Woodlands 38

Sunday 25th
Halton Castle 12
High Bank Farm 15
Holmelands 17
Old Quarrington Gardens 28
◆ Whalton Manor Gardens 36

Friday 30th
Hidden Gardens of Croft Road 14

July

Saturday 1st
Capheaton Hall 6
Craster's Hidden Gardens 8
Kirky Cottage 18

Sunday 2nd
The Beacon 2
Capheaton Hall 6
Hidden Gardens of Croft Road 14

Sunday 9th
2 Briarlea 5
Longwitton Hall 20
NEW Secret Gardens of Langholm Crescent 33

Sunday 16th
NEW Acomb High House 1

Friday 21st
Ferndene House 11

Sunday 30th
St Cuthbert's Hospice 32

August

Sunday 6th
Heather Holm 13

Sunday 13th
Walworth Gardens 35
Droomveld 9

Saturday 19th
Middleton Hall Retirement Village 23

By Arrangement

Arrange a personalised garden visit with your club, or group of friends, on a date to suit you. See individual garden entries for full details.

The Beacon 2
2 Briarlea 5
Coldcotes Moor Garden 7
Droomveld 9
Fallodon Hall 10
Ferndene House 11
2 Hillside Cottage 16
Kirky Cottage 18
Lilburn Tower 19
Loughbrow House 21
Old Quarrington Gardens 28
25 Park Road South 30
Ravensford Farm 31
Stanton Fence 34
Woodlands 38

Briarlea

THE GARDENS

1 NEW ACOMB HIGH HOUSE

The Green, Acomb, Hexham, NE46 4PH. Naomi Liller. *2 m N of Hexham. Around Hexham leave A69 signposted Acomb, in Acomb turn R up hill, past Miners Arms on R. Signed to parking area.* **Sun 16 July (12.30-4.30). Adm £6, chd free. Home-made teas.**

Within the remains of an old walled garden, featuring extensive perennial planting around a central rill and reflective pool. Also potager kitchen garden, wildflower meadow and herbaceous border. Designed by Matthew Wilson (attending on the day) and awarded Best Large Landscape and Garden 2022 Society of Garden Designers Awards. No chemicals and wildlife friendly. Art exhibition by Chris Partridge in garden.

2 THE BEACON

10 Crabtree Road, Stocksfield, NE43 7NX. Derek & Patricia Hodgson OBE, 01661 842518, patandderek@btinternet.com. *12m W of Newcastle. From A69 follow signs into village. Stn & cricket ground on L. Turn R into Cadehill Rd then 1st R into Crabtree Rd (cul-de-sac) Park on Cadehill.* **Sun 4 June, Sun 2 July (2-5.30). Adm £6, chd free. Home-made teas. Visits also by arrangement 29 May to 22 Aug for groups of 10+.**

This garden illustrates how to make a cottage garden on a steep site with loads of interest at different levels. Planted with acers, roses and a variety of cottage garden and formal plants. Water runs gently through it and there are tranquil places to sit and talk or just reflect. Stunning colour and plant combinations. Wildlife friendly - numerous birds, frogs, newts, hedgehogs. Haven for butterflies and bees. Owner available for entertaining group talks. Comedy/piano recital available at By Arrangement visits (£4 per person in aid of NSPCC).

3 BICHFIELD TOWER

Belsay, Newcastle Upon Tyne, NE20 0JP. Lesley & Stewart Manners, 07511 439606, lesleymanners@gmail.com, www.bitchfieldtower.co.uk. *Private rd off B6309, 4m N of Stamfordham & SW of Belsay village.* **Sat 17 June (1-4). Adm £8, chd free. Home-made teas in the carriage house garden & building.**

A 6 acre mature garden. Set around a Medieval Pele Tower, there is an impressive stone water feature, large trout lake, mature woodland, pear orchard, and 2 walled gardens. Extensive herbaceous borders, prairie borders and contemporary grass borders. Tennis court, croquet lawn and set. Delicious home-made teas provided by the 6th Morpeth Scout Group.

4 BLAGDON

Seaton Burn, NE13 6DE. Viscount Ridley. *5m S of Morpeth on A1. 8m N of Newcastle on A1, N on B1318, L at r'about (Holiday Inn) & follow signs to Blagdon. Entrance to parking area signed.* **Sun 23 Apr (1-4). Adm £6, chd free. Home-made teas.**

Unique 27 acre garden encompassing formal garden with Lutyens designed 'canal', Lutyens structures and walled kitchen garden. Valley with stream and various follies, quarry garden and woodland walks. Large numbers of ornamental trees and shrubs planted over many generations. National Collections of Acer, Alnus and Sorbus. Partial wheelchair access.

5 2 BRIARLEA

Hepscott, Morpeth, NE61 6PA. Richard & Carolyn Torr, 07470 391936, richardtorr6@gmail.com. *Entering Hepscott from A192, Briarlea is 2nd road on R. Please park considerately within the village. Please do not park on Briarlea.* **Sun 9 July (11-5). Adm £5, chd free. Light refreshments at 7 Thornlea. Refreshments approx 100m away, with further cactus collection and plant stall. Visits also by arrangement 5 June to 28 July for groups of 10 to 25. By arrangement adm inc tea/ coffee & home-made cake/scones.**

A plant lover's garden of just under ½ acre, rejuvenated and redeveloped over the last 6 yrs, featuring gravel areas, stream side planting and many herbaceous borders. Planting features hostas, grasses, roses, penstemons, ferns, a wide range of shrubs and other perennials. A vegetable area with raised beds inc greenhouses with an extensive collection of cacti and other succulents. Good access to most of garden, however, some steps and gravel areas.

6 CAPHEATON HALL

Capheaton, Newcastle Upon Tyne, NE19 2AB. William & Eliza Browne-Swinburne, 01913 758152, estateoffice@capheatonhall.co.uk, www.capheatonhall.co.uk. *Off A696 24m N of Newcastle. From S turn L off A696 onto Silver Hill rd signed Capheaton. From N, past Wallington/Kirkharle junction, turn R.* **Sat 1, Sun 2 July (10-5). Adm £10, chd free. Home-made teas.**

Set in parkland, Capheaton Hall has magnificent views over the Northumberland countryside. Formal ponds sit south of the house which has C19 conservatory and a walk to a Georgian folly of a chapel. The outstanding feature is the very productive walled kitchen garden, mixing colourful vegetables, espaliered fruit with annual and perennial flowering borders. Victorian glasshouse and conservatory.

7 COLDCOTES MOOR GARDEN

Ponteland, Northumberland, Newcastle Upon Tyne, NE20 0DF. Ron & Louise Bowey, 07798 532291, info@theboweys.co.uk. *Off A696 N of Ponteland. From S, leave Ponteland on A696 towards Jedburgh, after 1m take L turn marked 'Milbourne 2m'. After 400yds turn L into drive.* **Sun 11 June (11-5). Adm £8, chd free. Pre-booking essential, please visit www.ngs.org.uk for information & booking. Light refreshments. Visits also by arrangement 12 June to 20 Aug for groups of 20 to 50.**

The garden, landscaped grounds and woods cover around 15 acres. The wooded approach opens out to lawn areas surrounded by ornamental and woodland shrubs and trees. A courtyard garden leads to an ornamental walled garden, beyond which is an orchard, fruit and flower garden and rose arbour. To the south the garden looks out over a lake and field walks, with woodland walks to the west. Most areas are accessible though sometimes by circuitous routes or an occasional step. WC access involves three steps.

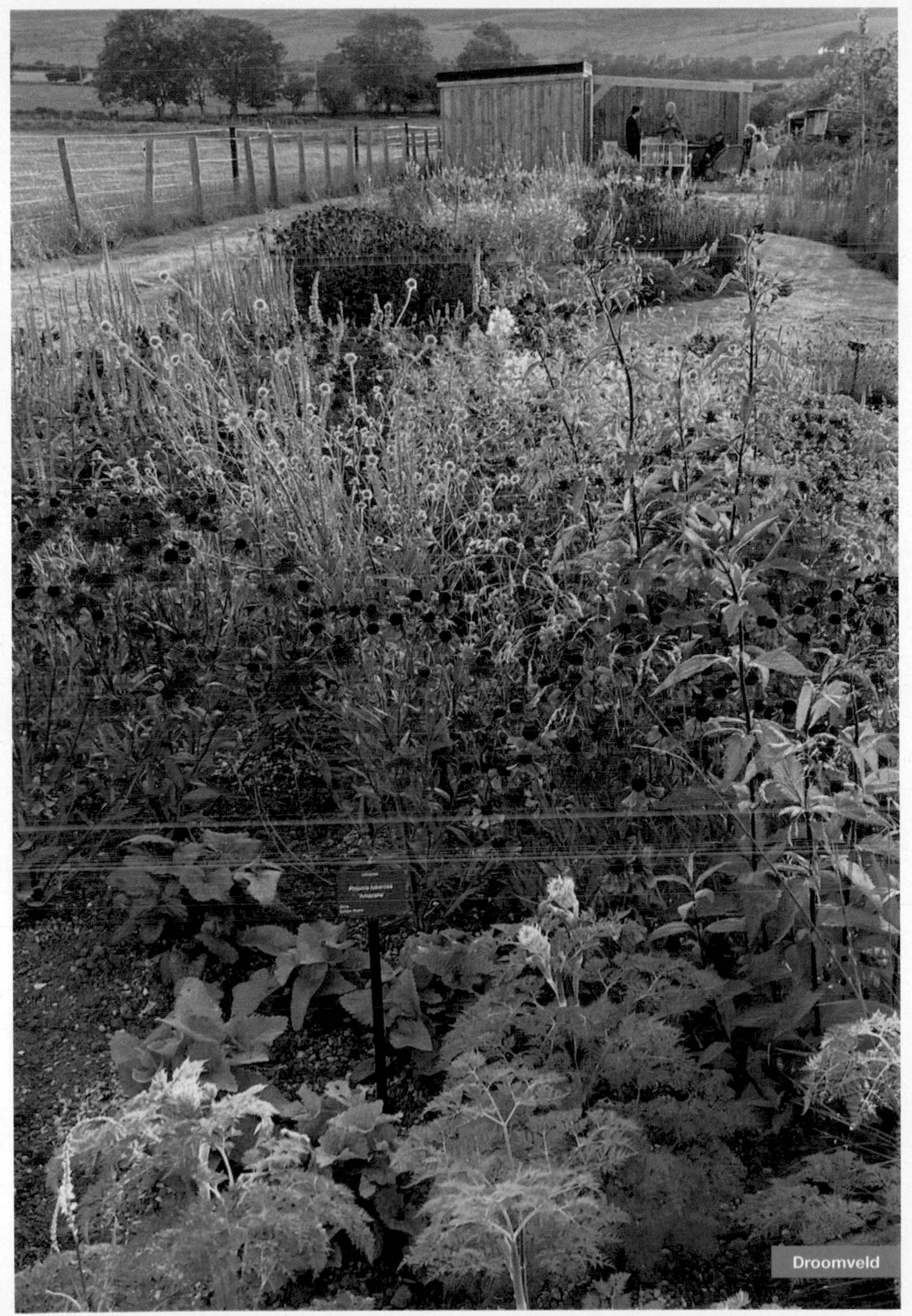

Droomveld

GROUP OPENING

8 CRASTER'S HIDDEN GARDENS

1, 2 and 5 Chapel Row, Craster, Alnwick, NE66 3TU. Sue Chapman, Gill Starkey, June Drage. *Park in the pay & display parking at entrance to village or nearby overspill field £4 pd. Walk to the harbour & turn L along Dunstanburgh Rd. Look for NGS sign on L before you reach the castle field. No parking within the village.* **Sat 1 July (11-4). Combined adm £5, chd free.**

In sight of Dunstanburgh Castle, in the picturesque village of Craster, sit 3 very different cottage gardens showing what can be achieved in a windy seaside setting. Formerly allotments attached to fishermen's cottages - No.1 is newly developed, has two ponds, colour-themed borders and prairie style planting. No. 2 has sea views, many perennial bee friendly plants, roses galore and a lovely herbaceous border. No. 5 is a ¼ acre, divided into several separate areas: woodland garden, pond, fruit and vegetables plus borders featuring many unusual perennials, shrubs, grasses and ferns. The garden is managed to encourage wildlife. There is a pub and 2 cafes in the village and picnic benches and ice cream vendor in the harbour area.

9 DROOMVELD

Christmas Farm, Longframlington, Morpeth, NE65 8DA. Maxine Eaton, 07715 460038, alittleplantcompany@gmail.com, www.alittleplantcompany.co.uk. *At the northern edge of the village on the A697, take the road E with signs for log cabins. After 1m turn L to Christmas Farm (signposted).* **Sun 13 August (11-4). Adm £5, chd free. Homemade teas using fresh produce from Christmas Farm. Visits by arrangement 7 July to 8 Sept for groups of up to 20. Adm £5, chd free.**

Nestled in the Northumberland hills, this small nursery garden has several borders to showcase different styles - cottage, naturalistic and woodland, with roses and a wildlife pool.

 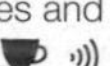

10 FALLODON HALL

Alnwick, NE66 3HF. Mr & Mrs Mark Bridgeman, 07765 296197, luciabridgeman@gmail.com, www.bruntoncottages.co.uk. *5m N of Alnwick, 2m off A1. From the A1 turn R onto the B6347 signed Christon Bank & Seahouses. Turn into the Fallodon gates after exactly 2m, at Xrds. Follow drive for 1m.* **Sat 24 June (2-5). Adm £6, chd free. Home-made teas. Visits also by arrangement for groups of 5+.**

Extensive, well established garden, with a hot greenhouse beside the bog garden. The late C17 walls of the kitchen garden surround cutting and vegetable borders and the fruit greenhouse. Natasha McEwen replanted the sunken garden from 1898 and the redesigned 30 metres border was planted in 2019. Woodlands, pond and arboretum with over 10 acres to explore. Grave of Sir Edward Grey, Foreign Secretary during WW1, famous ornithologist and fly fisherman, is in the woods near the pond and arboretum. The walls of the kitchen garden contain a fireplace built to heat the fruit trees of the Salkeld family, renowned for their gardening expertise in the C17. Partial wheelchair access.

Acomb High House

11 FERNDENE HOUSE

2 Holburn Lane Court, Holburn Lane, Ryton, NE40 3PN. Maureen Kesteven, 0191 413 5937, maureen.kesteven@ngs.org.uk, www.facebook.com/northeastgardenopenforcharity. *In Ryton Old Village, 8m W of Gateshead. Off B6317, on Holburn Lane. Park on street or in Co-op car park on High St, cross rd through Ferndene Park following yellow signs.* **Sun 28 May (12.30-4.30). Adm £6, chd free. Home-made teas. Pasta & prosecco. Fri 21 Jul (6-8). Adm £10, chd free. Wine & canapes inc in adm. Visits also by arrangement 12 Apr to 30 Aug for groups of 10+.**

¾ acre garden surrounded by trees. Informal areas of herbaceous perennials, formal box bordered area, meadow, wildlife pond, gravel and bog gardens (with boardwalk). Willow work. Early interest - hellebores, snowdrops, daffodils, bluebells and tulips. Summer interest from wide range of flowering perennials and meadow. 1½ acre mixed broadleaf wood with beck running through. Driveway, from which main borders can be seen, is wheelchair accessible but phone for assistance.

12 HALTON CASTLE

Corbridge, NE45 5PH. Hugh & Anna Blackett. *2m N of Corbridge turn E off the A68 onto the B6318 (Military Rd) towards Newcastle. Turn R onto drive after ¼m.* **Sun 25 June (11.30-4.30). Adm £6, chd free. Cream teas. Light lunches from 12pm.**

The terraced garden has stunning views over the Tyne Valley. Massive beech hedges give protection for herbaceous borders, lawns and shrubs. A box parterre is filled with fruit, vegetables and picking flowers. Paths lead through a recently created wildflower meadow garden. The Castle (not open) is a C14 Pele tower with Jacobean manor house attached beside a charming chapel with Norman origins. Partial wheelchair access.

13 HEATHER HOLM

Stanghow Road, Stanghow, Saltburn-By-The-Sea, TS12 3JU. Arthur & June Murray. *Stanghow is 5 km E of Guisborough on A171. Turn L at Lockwood Beck (signed Stanghow) Heather Holm is on the R past the Xrds.* **Sun 6 Aug (11-4). Adm £4, chd free. Home-made teas.**

Divided into several rooms, this is a garden of different aspects, formal, floral and architectural. The owners have a philosophy of colour throughout the year having many different species of lilies flowering from March to October. Sheltered by mature hedges, hostas and hydrangeas thrive. There are fruit trees, soft fruit bushes, a greenhouse, summerhouse and a large pond attracting wildlife. Regret no dogs.

GROUP OPENING

14 HIDDEN GARDENS OF CROFT ROAD

Darlington, DL2 2SD. Carol Pratt. *2m S of Darlington on A167. ¾m S from A167/A66 r'about between Darlington & Croft.* **Fri 30 June, Sun 2 July (1-5). Combined adm £6, chd free. Home-made teas at Oxney Flatts Farm (30 June) & Orchard Gardens (2 July).**

NAGS HEAD FARM
Jo & Ian Fearnley.
(See separate entry)

NEW COTTAGE
Jane & John Brown.

ORCHARD GARDENS
Gill & Neil Segger.

OXNEY COTTAGE
Gypsy Nichol.

OXNEY FLATTS FARM
Carol & Chris Pratt.

Five very different and interesting gardens, well named as 'Hidden Gardens' as they are not visible from the road. Oxney Cottage is a very pretty cottage garden with lawns, herbaceous borders and roses, colourful and varied unusual plants. Nags Head Farm has a wonderful rill running alongside a sloping garden with a variety of plants leading to a quiet, peaceful courtyard. There is a large vegetable garden in which stands a magnificent glasshouse with prolific vines. A woodland walk leads to a polytunnel, view point and wildflower meadow. Orchard Gardens is a large interesting garden of mixed planting, stump sculptures, colourful themed beds, fruit trees and different imaginative ornaments. Oxney Flatts has well-stocked herbaceous borders and a wildlife pond. New Cottage garden has mixed herbaceous flower beds, a fruit and vegetable garden and a newly created wildlife pond. Aycliffe Bee Keepers Association will have a stall in the garden of Nags Head Farm. There will be information about bees and bee keeping plus a small display hive where you can safely view bees in action. Honey will be on sale, subject to availability, and 10% of proceeds will be donated to NGS. Partial wheelchair access, grassy and mostly flat. Some areas of gravel. Dogs on a lead welcome in some gardens but not all.

15 HIGH BANK FARM

Cleasby Road, Stapleton, Darlington, DL2 2QE. Lesley Thompson. *Located on rd leading to Cleasby & Manfield from the Stapleton rd. Drive 300 yds, look out for a sign saying 'Nursery on a Farm', take turning & park in the car park.* **Sun 25 June (1-4). Combined adm with Holmelands £5, chd free. Home-made teas.**

The front garden has an abundance of roses, leading onto the orchard. The back garden has a lovely cottage garden, with a secluded summerhouse to relax in, together with paved area with beautiful pots where you can sit and listen to the trickling water from our lovely fishpond. Regret no dogs.

Our donation of £2.5 million over five years has helped fund the new Y Bwthyn NGS Macmillan Specialist Palliative Care Unit in Wales. 1,900 inpatients have been supported since its opening

16 2 HILLSIDE COTTAGE

Low Etherley, Bishop Auckland, DL14 0EZ. Mrs M Smith, 07789 366702, mary@maryruth.plus.com. *3m W of Bishop Auckland on B6282 and ½m E of Toft Hill. Follow the B6282 from Bishop Auckland. After the Etherley sign continue along the road for approx ¼m, the Cottage is down the track on the L near the 2nd footpath sign.* **Sun 4 June (1-5). Adm £5, chd free. Light refreshments. Visits also by arrangement 4 Feb to 31 Oct.**

In this half acre mature cottage garden grass paths lead you past a wide variety of beds with shrubs, grasses, perennials and trees (native and exotic inc a Wollemi Pine and a Ginko) to a wild area with a small pond and greenhouse with a collection of cacti and succulents. At the top is a larger pond with frogs, toad and newts, patios, vegetable beds, fruit trees and roses. Planting for wildlife. A variety of sculptures and lots of places to sit and admire the garden. Views over Weardale and the surrounding countryside. It is possible to negotiate most of the garden on grass paths.

17 HOLMELANDS

Cleasby, Darlington, DL2 2QY. Nicky & Clare Vigors, nickyvigors@btinternet.com. *Cleasby village. 1st open yard on R approaching village from Stapleton.* **Sun 25 June (1-4). Combined adm with High Bank Farm £5, chd free. Home-made teas at High Bank Farm.**

Semi formal garden with orchard and large vegetable patch. Herbaceous beds and climbing roses as well as standard roses. Laid out in a semi formal design to create a harmonious effect. Attractive mixed borders, formal topiary and lavender at the front of the house.

18 KIRKY COTTAGE

12 Mindrum Farm Cottages, Mindrum, TD12 4QN. Mrs Ginny Fairfax, 01890 850246, ginny@mindrumgarden.co.uk, mindrumestate.co.uk/kirky-garden. *6m SW of Coldstream. 9m NW of Wooler on B6352. 4m N of Yetholm village.* **Sat 1 July (11-5). Adm £5, chd free. Home-made teas. Visits also by arrangement.**

Ginny Fairfax has created Kirky Cottage Garden in the beautiful Bowmont valley surrounded and protected by the Border Hills. A gravel garden in cottage garden style, old roses, violas and others jostle with favourites from Mindrum. A lovely, abundant garden and, with Ginny's new and creative ideas, ever evolving. All plants propagated from garden. Regional Finalist, The English Garden's The Nation's Favourite Gardens 2021.

19 LILBURN TOWER

Alnwick, NE66 4PQ. Mr & Mrs D Davidson, 01668 217291, lilburntower@outlook.com. *3m S of Wooler. On A697.* **Sun 21 May (1.30-5.30). Adm £6, chd free. Home-made teas. Visits also by arrangement 21 May to 4 June for groups of 10+.**

10 acres of magnificent walled and formal gardens set above river; rose parterre, topiary, scented garden, Victorian conservatory, wildflower meadow. Extensive fruit and vegetable garden, large glasshouse with vines. 30 acres of woodland with walks. Giant lilies, meconopsis around pond garden. Rhododendrons and azaleas. Also ruins of Pele Tower, and C12 church. Partial wheelchair access.

20 LONGWITTON HALL

Longwitton, Morpeth, NE61 4JJ. Michael & Louise Spriggs. *2m N of Hartburn off B6343. Entrance at east end of Longwitton village.* **Sun 9 July (12-4). Adm £7, chd free. Home-made teas.**

6 acre historic site with glorious views to the south. Sheltered, mature garden, with specimen trees and acers, redeveloped with new borders. Circular rose garden, crescent shaped pool surrounded by foliage plants, 'standing stone' feature and a rhododendron and azalea glade leading to newly planted tunnel and cherry circle. Tree peonies and yew walk. Good wheelchair access.

21 LOUGHBROW HOUSE

Hexham, NE46 1RS. Mrs K A Clark, 01434 603351, patriciaclark351@hotmail.com. *1m S of Hexham on B6306. Dipton Mill Rd. Rd signed Blanchland, ¼m take R fork; then ¼m at fork, lodge gates & driveway at intersection.* **Visits by arrangement 1 Jan to 14 Dec. Adm £5, chd free.**

A real country house garden with sweeping, colour themed herbaceous borders set around large lawns. Unique Lutyens inspired rill with grass topped bridges and climbing rose arches. Part walled kitchen garden and paved courtyard. Bog garden with pond. New rose bed. Wildflower meadow with specimen trees. Woodland quarry garden with rhododendrons, azaleas, hostas and rare trees. Home-made jams and chutneys for sale.

22 MARIE CURIE HOSPICE

Marie Curie Drive, Newcastle Upon Tyne, NE4 6SS. www.mariecurie.org.uk/help/hospice-care/hospices/newcastle/about. *In West Newcastle just off Elswick Rd. At the bottom of a housing estate. Turning is between MA Brothers & Dallas Carpets.* **Sun 18 June (2-4.30). Adm by donation. Light refreshments in our Garden Café.**

The landscaped garden of the purpose-built Marie Curie Hospice overlooks the Tyne and Gateshead and offers a beautiful, tranquil place for patients and visitors to sit and chat. Rooms open onto a patio garden with gazebo and fountain. There are climbing roses, evergreens and herbaceous perennials. The garden is well maintained by volunteers. Come and see the work NGS funding helps make possible. Plant sale. The Hospice and gardens are wheelchair accessible.

23 MIDDLETON HALL RETIREMENT VILLAGE

Middleton St George, Darlington, DL2 1HA. www.middletonhallretirementvillage.co.uk. *From A67 D'ton/Yarm, turn at 2nd r'about signed to Middleton St George. Turn L at the mini r'about & immed R after the railway bridge, signed Low Middleton. Main entrance is ¼m on L.* **Sat 19 Aug (11-4). Adm £5, chd free. Home-made teas.**

Like those in our retirement community, the extensive grounds are gloriously mature, endlessly interesting and with many hidden depths. 45 acres of features beckon; natural woodland and parkland, Japanese, Mediterranean and Butterfly Gardens, putting green,

Longwitton Hall

© Susie White

allotments, ponds, wetland and bird hide. For more information please visit our website. All wheelchair accessible and linked by a series of woodland walks.

24 ◆ MINDRUM GARDEN

Mindrum, TD12 4QN. Mr & Mrs T Fairfax, 01890 850634, tpfairfax@gmail.com, www.mindrumestate.com. *6m SW of Coldstream, 9m NW of Wooler. Off B6352. Disabled parking close to house.* **For NGS: Sun 18 June (2-5). Adm £6, chd free. Home-made teas. For other opening times and information, please phone, email or visit garden website.**

A magical combination of old-fashioned roses and hardy perennials. Lawns surround the house with climbing roses, mature shrubs and trees throughout. Mindrum has evolved into a series of gardens inc a rose garden, limestone rock garden (steep slope), fishponds with terraced area, a woodland walk with mature pines, and wildflower garden by the river.

25 THE MOORE HOUSE

Whalton, Morpeth, NE61 3UX. Phillip & Filiz Rodger. *5m W of Morpeth & 4m N of Belsay on B6524. House in middle of village. Park in village & follow NGS signs.* **Sun 18 June (2-5). Adm £6, chd free. Home-made teas.**

2½ acre mature garden extensively but sympathetically redesigned by Sean Murray, BBC/RHS Great Chelsea Garden Challenge winner. Garden divided into sections. Wide use and mix of flowering perennials and grasses, Thyme walk, orchard, rose garden, Zen inspired courtyard. Decorative stonework around beds in patio area. Emphasis on scent, colour, texture and form with year-round interest.

26 NAGS HEAD FARM

Croft Road, Darlington, DL2 2SD. Jo & Ian Fearnley. *2m S of Darlington on A167. ¾m S from A167/A66 r'about between Darlington & Croft.* **Evening opening Thur 25 May (5-8). Adm £5, chd free. Light refreshments. Opening with Hidden Gardens of Croft Road on Fri 30 June, Sun 2 July.**

3 adjacent rural gardens invite you to spend an evening strolling around their late spring flower beds and wildlife ponds. New Cottage also has vegetable beds and spring borders, whilst Oxney Flatts has interesting planting and wide borders; both gardens have stunning wisteria displays. Nags Head Farm offers a vegetable plot with greenhouse and woodland walks leading to a viewpoint and a polytunnel. Partial wheelchair access over grass, mainly level but with some gravel.

27 ◆ NGS BUZZING GARDEN
East Park Road, Gateshead, NE9 5AX. Gateshead Council, maureen.kesteven@ngs.org.uk, www.gateshead.gov.uk/article/3958/Saltwell-Park. *Between Pets' Corner & Saltwell Towers. Pedestrian entrance from East Park Rd or car park in Joicey Rd. Open all year as part of the 55 acre Saltwell Park, known as The People's Park.* **For opening times and information, please email or visit garden website.**
A unique collaboration between the National Garden Scheme North East, Trädgårdsresan, Region Västra Götaland and Gateshead Council. It was funded by sponsorship and is a tribute to the importance of international friendship. The Swedish design reflects the landscape of West Sweden, with coast, meadow and woodland areas. Many of the plant species grow wild in Sweden, providing a welcoming vision for visitors and a feast for pollinators. Planted in 2019 the garden is maturing. Visitors can make a donation to the NGS. Refreshments available at the Saltwell Towers Café. Wide tarmac path around the garden and mown grass paths through the meadow, but much of the garden is loose gravel.

GROUP OPENING

28 OLD QUARRINGTON GARDENS
The Stables, Old Quarrington, Durham, DH6 5NN. John Little, 07967 267864, johndlittle10@gmail.com, facebook.com/thestablesOQ. *1m from J61 of A1(M). All vehicle access from Crow Trees Lane, Bowburn. SatNav may be misleading.* **Sun 25 June (11-4). Combined adm £6, chd free. Home-made teas at The Stables - also picnics and WC. Visits also by arrangement Apr to Oct for groups of 10+.**

13 HEUGH HALL ROW
Jacqueline Robson, facebook.com/OldQuarringtonGardens.

ROSE COTTAGE
Mr Richard & Mrs Ann Cowen.

THE STABLES
John & Claire Little, facebook.com/thestablesOQ.

Three very distinct gardens in the hamlet of Old Quarrington. The Stables is a large family garden full of hidden surprises and extensive views. The main garden is about an acre inc gravel garden, vegetable patch, orchard, play area, lawn and woodland gardens. There is a further 4 acres to explore which inc wildlife ponds, woodlands, meadows, hens, ducks and alpacas. Number 13 has been developed over 20 yrs with many small garden rooms inc a rose garden, a white garden, a small link garden which contains ferns, viburnum, birch trees and a *Cercidiphyllum japoniocum*. There is a pond and herbaceous beds which contain perennials and grasses. Many roses can be seen throughout the garden and a greenhouse devoted to a large Brown Turkey fig . The main garden at Rose Cottage is planted with wildlife friendly flowers, has slate paths and a pond that attracts 2 species of newts. To the rear there is a Mediterranean area and a woodland garden with stream beyond. Plant sales at 13 Heugh Hall Road.

Oxney Cottage

29 OLIVER FORD GARDEN

Longedge Lane, Rowley, Consett, DH8 9HG. Bob & Bev Tridgett, www.gardensanctuaries.co.uk. *5m NW of Lanchester. Signed from A68 in Rowley. From Lanchester take rd towards Sately. Garden will be signed as you pass Woodlea Manor.* **Sun 11 June (1-5). Adm £4, chd free. Light refreshments.**

A peaceful, contemplative 3 acre garden developed and planted by the owner and BBC Gardener of the Year as a space for quiet reflection. Arboretum specialising in bark, stream, wildlife pond and bog garden. Semi-shaded Japanese maple and dwarf rhododendron garden. Rock garden and scree bed. Insect nectar area, orchard and 1½ acre meadow. Annual wildflower area. Terrace and ornamental herb garden. The garden is managed to maximise wildlife and there are a number of sculptures around the garden.

30 25 PARK ROAD SOUTH

Chester le Street, DH3 3LS. Mrs A Middleton, 0191 388 3225, midnot2@gmail.com. *4m N of Durham. Located at S end of A167 Chester-le-St bypass rd. Precise directions provided when booking visit.* **Visits by arrangement 2 May to 1 Sept. Adm £3.50, chd free. Light refreshments.**

A stunning town garden with year-round interest. Herbaceous borders with unusual perennials, grasses, shrubs surrounding lawn and paved area. Courtyard planted with foliage and small front gravel garden. The garden owner is a very knowledgeable plantswoman who enjoys showing visitors around her inspiring garden. No minimum size of group. Plants for sale.

31 RAVENSFORD FARM

Hamsterley, Bishop Auckland, DL13 3NH. Jonathan & Caroline Peacock, 01388 488305, caroline@ravensfordfarm.co.uk. *7m W of Bishop Auckland. From A68 near Witton-le-Wear turn W to Hamsterley. On Open Day follow yellow NGS signs. Otherwise, turn left at W end of village & go to bottom of hill.* **Sun 14 May (2-5). Adm £5, chd free. Home-made teas. Visits also by arrangement Mar to Oct.**

In 1984 Ravensford Farm was a ruin in a field full of weeds, with just one tree. Today it is surrounded by a richly varied garden, with a wood, two ponds, an orchard, a Rhododendron dell and extensive flowerbeds. While children hunt for hidden animal figures, expert gardeners will have fun discovering unusual shrubs and trees, and all visitors will enjoy tea, home-made cakes and local music. Colour, variety, ponds, unusual trees and shrubs. Refreshments under cover in case of inclement weather. Some gravel, so assistance will be needed for wheelchairs. Assistance dogs only, please.

32 ST CUTHBERT'S HOSPICE

Park House Road, Durham, DH1 3QF. Paul Marriott, CEO, www.stcuthbertshospice.com. *1m SW of Durham City on A167. Turn into Park House Rd, the Hospice is on the L after bowling green car park. Parking available.* **Sun 30 July (10-5). Adm £5, chd free. Light refreshments.**

5 acres of mature gardens surround this CQC outstanding-rated Hospice. In development since 1988, the gardens are cared for by volunteers inc a Victorian-style greenhouse and large vegetable, fruit and cut flower area. Lawns surround smaller scale specialist planting, and areas for patients and visitors to relax. Woodland area with walks, sensory garden, and an 'In Memory' garden with stream. Plants and produce for sale. We are active participants in Northumbria in Bloom and Britain in Bloom, with several awards in recent years, inc overall winner in 2015, 2018 and 2019 for the Care/Residential/Convalescent Homes/Day Centre/Hospices category. Almost all areas are accessible for wheelchairs.

GROUP OPENING

33 NEW SECRET GARDENS OF LANGHOLM CRESCENT

Langholm Crescent, Darlington, DL3 7ST. Barbara-Anne Johnson. *All on street parking in the surrounding streets.* **Sun 9 July (1-5). Combined adm £5, chd free. Home-made teas at The Cottage, 20 Langholm Crescent.**

The Cottage Garden at 20 Langholm Crescent is a mature garden but new to the present owner. It is evolving, hidden from public view, has large lawns, borders, mature trees, small pond, rose garden and more. The 7 gardens behind the Langholm Terrace houses are individually delightful, with each making creative use of the limited space available.

34 STANTON FENCE

Stanton, Morpeth, NE65 8PP. Sir David Kelly, stantonfence@hotmail.com. *5m NW of Morpeth. Nr Stanton on the C144 between Pigdon & Netherwitton. OS map ref NZ 135890.* **Visits by arrangement Mar to Sept for groups of 15+. Visits to show seasonal changes possible for groups. Adm £10, chd free.**

Contemporary 4.7 acre country garden designed by Chelsea Gold Medal winner, Arabella Lennox-Boyd, in keeping with its rural setting. A strong underlying design unites the different areas from formal parterre and courtyard garden to orchard, wildflower meadows and woodland. Romantically planted rose covered arbours and long clematis draped pergola walk. Nuttery, greenhouse and greenhouse garden. Delightful views. Wheelchair access for chairs that can use mown paths as well as hard paving.

Our 2022 donation supports 1,700 Queen's Nurses working in the community in England, Wales, Northern Ireland, the Channel Islands and the Isle of Man

GROUP OPENING

35 WALWORTH GARDENS

Walworth, Darlington, DL2 2LY. Iain & Margaret Anderson, 01325 778446. *Approx 5m W of Darlington on A68 or ½m E of Piercebridge on A67. Follow brown signs to Walworth Castle Hotel. Just up the hill from the Castle entrance, follow NGS yellow signs down private track. Tickets & teas at Quarry End.* **Sun 13 Aug (1-4.30). Combined adm £6, chd free. Home-made teas at Quarry End.**

THE ARCHES
Stephen & Becky Street-Howard.

CASTLE BARN
Joe & Sheila Storey.

THE DOVECOTE
Tony & Ruth Lamb.

QUARRY END
Iain & Margaret Anderson.

Quarry End is a 1½ acre woodland garden developed over 20 yrs on the site of an ancient quarry. The garden offers a wide variety of mature trees, shrubs, perennials, a fernery and potager. It inc a reclaimed quarry woodland, an C18 ice house and offers lovely views over South Durham. The Dovecote has three themes; an English cottage garden featuring espalier fruit trees, and heavily scented tea roses, a Japanese garden with a Moon Gate leading to a collection of specimen Acer trees and a small Zen garden overlooking a Koi pond. The compact vegetable plot is wrapped round with a Grade II listed vented wall and has great space saving ideas. Castle Barn gardens are now well established with fruit trees, raised beds and colourful borders. This year a new herb garden and new grape vines have been developed. The Arches' expansive garden comprises an orchard, a large vegetable garden, a wild swimming pond, an ornamental garden and an arboretum. There will be plants for sale at Quarry End. Some areas of the Quarry End garden are not accessible for wheelchairs but most is grassy and flat.

36 ◆ WHALTON MANOR GARDENS

Whalton, Morpeth, NE61 3UT. Mr T R P S Norton, gardens@whaltonmanor.co.uk, www.whaltonmanor.co.uk. *5m W of Morpeth. On the B6524, the house is at E end of the village & will be signed.* **For NGS: Sun 25 June (11-4). Adm £8, chd free. Light refreshments in The Game Larder. For other opening times and information, please email or visit garden website.**

The historic Whalton Manor, altered by Sir Edwin Lutyens in 1908, is surrounded by 3 acres of magnificent walled gardens, designed by Lutyens with the help of Gertrude Jekyll. The gardens, developed by the Norton family since the 1920s inc extensive herbaceous borders, spring bulbs, 30yd peony border, rose garden, listed summerhouses, pergolas and walls festooned with rambling roses and clematis. Partial wheelchair access to main area but otherwise stone steps and gravel paths.

37 NEW WILLOW BURN HOSPICE

Howden Bank, Lanchester, DH7 0BF. Rachel Todd, www.willowburnhospice.org.uk. *Located in the Maiden View estate. If walking from Lanchester, there is a pedestrian entrance with signage for the Willows Café. Blue badge parking at the hospice, additional parking at St Bede's Catholic School (DH7 0RD), a 10-15 min uphill walk.* **Sun 11 June (10-3). Adm £3, chd free. Light refreshments in the Willows Café. A selection of refreshments, inc breakfast, lunch & baked goods are available.**

With a mix of both natural and cultivated areas, this beautiful garden is the setting for outstanding views from the hospice across the Derwent Valley. The gardens are maintained and cared for by volunteers and inc wildflowers, shrubs, perennials, and a large wooded area, as well as memorials and dedications for loved ones. The Willows Café is open throughout the duration of the open day. Lawn games, The Potting Shed Plant & Gift Shop. Pathways give access to view all the areas of the garden.

38 WOODLANDS

Peareth Hall Road, Springwell Village, Gateshead, NE9 7NT. Liz Reid, 07719 875750, liz.reid@ngs.org.uk, www.facebook.com/visitgarden. *3½m N Washington Galleries. 4m S Gateshead town centre. On B1288 turn opp Guide Post pub (NE9 7RR) onto Peareth Hall Rd. Continue for ½m passing 2 bus stops on L. 3rd drive on L past Highbury Ave.* **Sat 24 June (1.30-4.30). Adm £4, chd free. Home-made teas. Beer & wine also available. Visits also by arrangement 5 June to 31 July for groups of 10 to 30. Refreshments can be arranged on request.**

Mature garden on a site of approx one seventh acre - quirky, with tropical themed planting and Caribbean inspired bar. Also an area of cottage garden planting. A fun garden with colour year-round, interesting plants, informal beds and borders and pond area. On the 24-26 June 2022 Springwell Village plans a 'Forties Weekend' with many attractions. Eg: 2nd WW battle re-enactments and a military camp are planned on the nearby Bowes Railway (SAM) site. There are also plans for craft and other stalls and live music throughout the village.

Our 2022 donation to ABF The Soldier's Charity meant 372 soldiers and veterans were helped through horticultural therapy

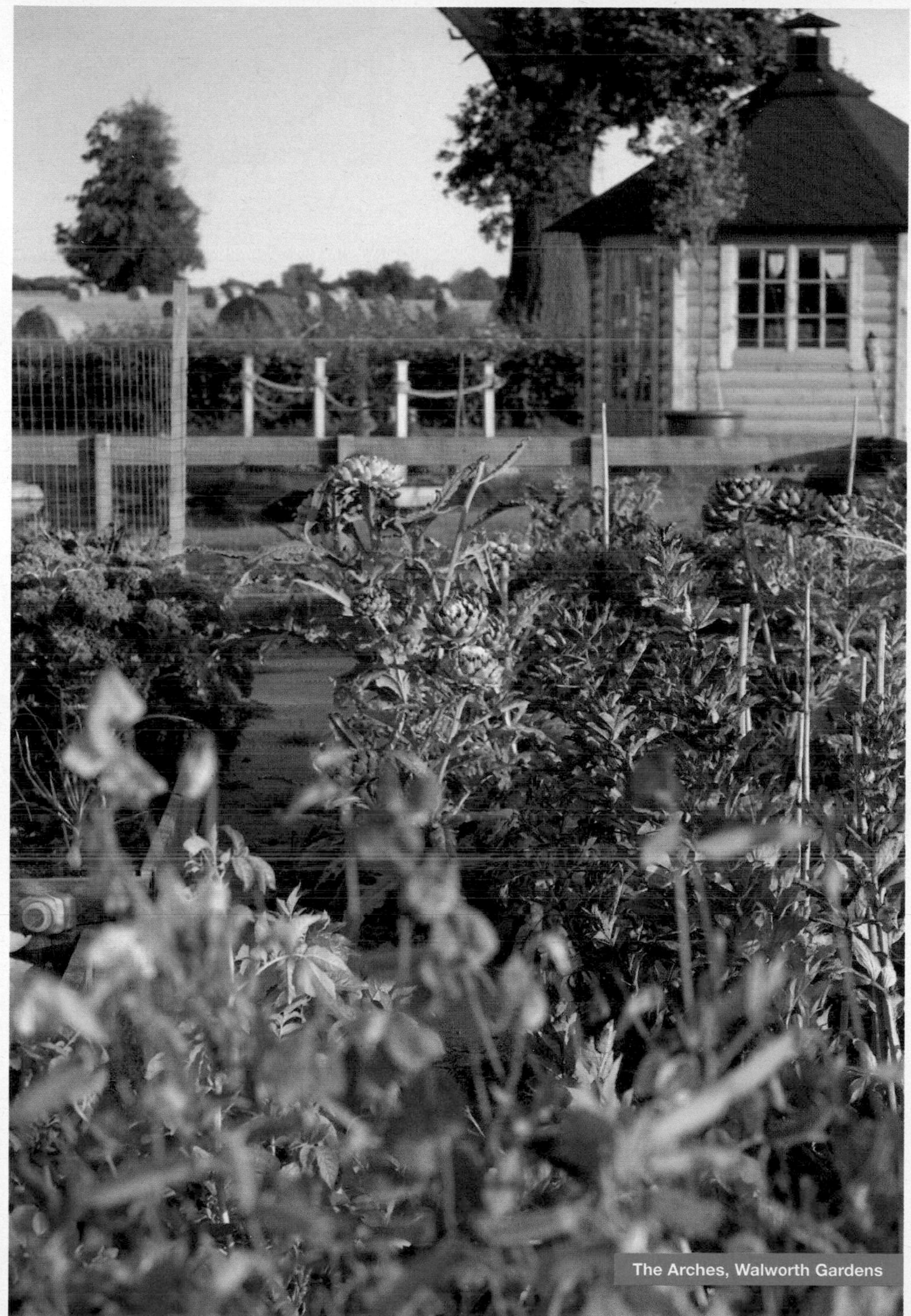

The Arches, Walworth Gardens

NORTHAMPTONSHIRE

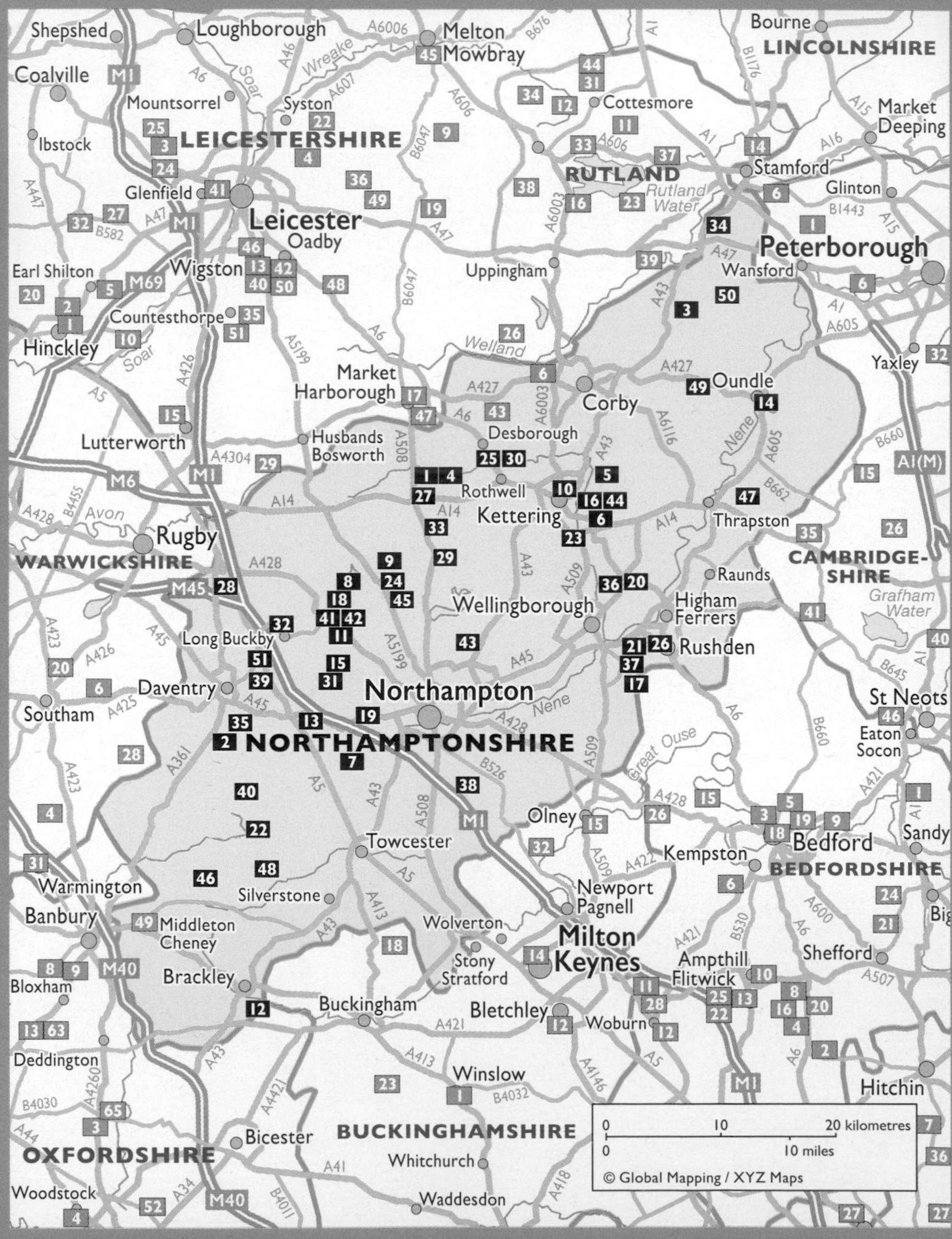

The county of Northamptonshire is famously known as the 'Rose of the Shires', but is also referred to as the 'Shire of Spires and Squires', and lies in the East Midlands area of the country bordered by eight other counties.

Take a gentle stroll around charming villages with thatch and stone cottages and welcoming inns. Wander around stately homes, discovering art treasures and glorious gardens open for the National Garden Scheme at Kelmarsh Hall, Holdenby House, Cottesbrooke Hall and Boughton House. In contrast visit some village groups, which include small imaginatively designed gardens. Explore historic market towns such as Oundle and Brackley in search of fine footwear, antiques and curiosities.

The serenity of our waterways with Woodcote Villa at Long Buckby Wharf will delight, and our winding country lanes and footpaths will guide you around a rural oasis, far from the pressures of modern living, where you can walk through swathes of snowdrops at Boughton House, view hellebores and spring flowers at 67/69 High Street, Finedon or The Old Vicarage at Norton through the seasons, to the late autumn colours of Briarwood.

Our first garden opens in February and the final opening occurs in October, giving a glimpse of gardens throughout the seasons.

Volunteers

County Organisers

David Abbott
01933 680363
david.abbott@ngs.org.uk

Gay Webster
01604 740203
gay.webster@ngs.org.uk

County Treasurer

David Abbott (as above)

Publicity

David Abbott (as above)

Photographer

Snowy Ellson
07508 218320
snowyellson@googlemail.com

Booklet Coordinator

William Portch
01536 522169
william.portch@ngs.org.uk

Talks

Elaine & William Portch
01536 522169
elaine.portch@yahoo.com

Assistant County Organisers

Amanda Bell
01327 860651
amanda.bell@ngs.org.uk

Lindsey Cartwright
01327 860056
lindsey.cartwright@ngs.org.uk

Jo Glissmann-Hill
07725 258755
joanna.glissmann-hill@ngs.org.uk

Philippa Heumann
01327 860142
pmheumann@gmail.com

Elaine & William Portch
(as above)

@Northants Ngs
@NorthantsNGS
@northantsngs

Left: Briarwood

OPENING DATES

All entries subject to change. For latest information check **www.ngs.org.uk**

Map locator numbers are shown to the right of each garden name.

February

Snowdrop Festival

Thursday 16th
136 High Street 21

Friday 17th
136 High Street 21

Saturday 18th
136 High Street 21

Sunday 19th
◆ Boughton House 5
136 High Street 21
The Old Vicarage 39

Sunday 26th
67-69 High Street 20

March

Sunday 26th
Woodcote Villa 51

April

Monday 10th
Titchmarsh House 47

Friday 14th
Ravensthorpe Nursery 42

Saturday 15th
Ravensthorpe Nursery 42

Sunday 16th
Briarwood 6
Flore Gardens 13
◆ Holdenby House & Gardens 24
Ravensthorpe Nursery 42
Rosi's Taverna 44

Sunday 23rd
◆ Cottesbrooke Hall Gardens 9
104 Irchester Road 26
The Old Vicarage 39

Sunday 30th
Great Brington Gardens 15

May

Monday 1st
Titchmarsh House 47

Sunday 7th
Guilsborough Gardens 18
NEW 83 Main Road 34

Friday 12th
Ravensthorpe Nursery 42

Saturday 13th
Ravensthorpe Nursery 42

Sunday 14th
Ravensthorpe Nursery 42

Sunday 21st
Badby Gardens 2
NEW Dene Lodge 10
Greywalls 17
136 High Street 21

Sunday 28th
Newnham Gardens 35
NEW Oaklea 37

Monday 29th
East Haddon Gardens 11

June

Saturday 3rd
Willow Cottage 49

Sunday 4th
67-69 High Street 20
1 Hinwick Close 23
Maidwell Gardens 33
Willow Cottage 49

Saturday 10th
◆ Lamport Hall 29
Titchmarsh House 47

Sunday 11th
Foxtail Lilly 14
Harpole Gardens 19
Hostellarie 25
16 Leys Avenue 30
Old Rectory, Quinton 38
Spratton Gardens 45

Saturday 17th
Flore Gardens 13

Sunday 18th
Flore Gardens 13
Kilsby Gardens 28
Rosearie-de-la-Nymph 43

Sunday 25th
Arthingworth Open Gardens 1
67-69 High Street 20
Little Brington Gardens 31
Rosearie-de-la-Nymph 43
Sulgrave Gardens 46
Weedon Lois & Weston Gardens 48

Friday 30th
Ravensthorpe Nursery 42

July

Saturday 1st
Ravensthorpe Nursery 42

Sunday 2nd
NEW Bugbrooke Gardens 7
104 Irchester Road 26
Ravensthorpe Gardens 41

Saturday 15th
◆ Evenley Wood Garden 12

Sunday 16th
NEW Oaklea 37

Sunday 23rd
Blatherwycke Estate 3
104 Irchester Road 26
Long Buckby Gardens 32

Sunday 30th
The Green Patch 16
Nonsuch 36

August

Sunday 6th
136 High Street 21

Sunday 20th
Highfields 22
Maidwell Gardens 33

Sunday 27th
104 Irchester Road 26

September

Sunday 3rd
Woodcote Villa 51

Sunday 10th
Briarwood 6
Rosi's Taverna 44

Saturday 16th
◆ Coton Manor Garden 8

Sunday 17th
◆ Kelmarsh Hall & Gardens 27

Friday 29th
Ravensthorpe Nursery 42

Saturday 30th
Ravensthorpe Nursery 42

October

Sunday 1st
◆ Boughton House 5
Ravensthorpe Nursery 42

February 2024

Sunday 25th
67-69 High Street 20

By Arrangement

Arrange a personalised garden visit with your club, or group of friends, on a date to suit you. See individual garden entries for full details.

NEW 16 Ace Lane, Bugbrooke Gardens 7
Bosworth House 4
Briarwood 6
23 Cotton End, Long Buckby Gardens 32
NEW Dene Lodge 10
Foxtail Lilly 14
The Green Patch 16
Greywalls 17
67-69 High Street 20
136 High Street 21
Hostellarie 25
104 Irchester Road 26
16 Leys Avenue 30
NEW 83 Main Road 34
Old West Farm 40
Ravensthorpe Nursery 42
Titchmarsh House 47
Willowbrook House 50
Wisteria House, Long Buckby Gardens 32
Woodcote Villa 51

THE GARDENS

GROUP OPENING

1 ARTHINGWORTH OPEN GARDENS

Arthingworth, nr Market Harborough, LE16 8LA. *6m S of Market Harborough. From Market Harborough via A508, after 4m take L to Arthingworth. From Northampton, A508 turn R just after Kelmarsh. Park cars in Arthingworth village & tickets for sale in the village hall.* **Sun 25 June (1.30-6). Combined adm £7, chd free. Home-made teas by residents of the village at Bosworth House & village hall.**

Come and celebrate the 10th Arthingworth NGS Open Gardens Anniversary. You will see around ten gardens, from artisan to large gardens, from natural to manicured ones, from annuals to perennials. We promise you an afternoon of discovery. The owners of the gardens will even give you some of the best tips. So come and enjoy, visiting our gardens and sitting down for a cup of tea at one of our two pop-up tearooms. St Andrew's Church, Grade II* listed will be open and the village is next to the national cycle path. Wheelchair access to some gardens.

GROUP OPENING

2 BADBY GARDENS

Badby, Daventry, NN11 3AR. *3m S of Daventry on E-side of A361.* **Sun 21 May (1-5). Combined adm £6, chd free. Home-made teas in St Mary's Church.**

CORNER HOUSE
Philip Turvil.

NEW **THE GLEBE**
Linda Clow.

NEW **RONKSWOOD COTTAGE**
Siobhan & David Bullock.

SHAKESPEARES COTTAGE
Jocelyn Hartland-Swann & Pen Keyte.

SOUTHVIEW COTTAGE
Alan & Karen Brown.

NEW **SPRINGFIELD HOUSE**
Chris & Linda Lofts.

Delightful hilly village with attractive old houses of golden coloured Hornton stone, set around a C14 church and two village greens (no through traffic). There are six gardens of differing sizes and styles inc three opening for the first time; a wisteria-clad thatched cottage with a sloping garden and modern sculptures; an elevated garden with views over the village and beyond; a 1/4 acre garden with mature planting and extensive new landscaping and planting in an 'informal formality', a new garden with a comparatively modern house that will develop over the next few yrs as the owner continues working with it; a garden which has been remodelled over the last 2 yrs with lawns, herbaceous beds, a pond, and a small vegetable and soft fruit area; a pretty cottage garden, also elevated and with lovely views. We look forward to welcoming you to our lovely village!

3 BLATHERWYCKE ESTATE

Blatherwycke, Peterborough, PE8 6YW. Mr George, Owner & S Bonney, Head Gardener. *Blatherwycke is signed off the A43 between Stamford & Corby. Follow road through village & the gardens entrance is next to the large river bridge.* **Sun 23 July (11-4). Adm £5, chd free. Home-made teas.**

Blatherwycke Hall was demolished in the 1940s and its gardens lost. In 2011 the renovation of the derelict 4 acre walled gardens started. So far, a large kitchen garden, wall trained fruit trees, extensive herbaceous borders, parterre, wildflower meadows, tropical bed, shrub borders and large arboretum have been planted. Restoration of the crinkle crankle wall and new steps is now completed. Wheelchair access over grass and gravel paths, some slopes and steps.

4 BOSWORTH HOUSE

Oxendon Road, Arthingworth, nr Market Harborough, LE16 8LA. Mr & Mrs C E Irving-Swift, 01858 525202, cirvingswift@gmail.com. *From the phone box, when in Oxendon Rd, take the little lane with no name, 2nd to the R.* **Visits by arrangement Apr to Oct for groups of 10 to 25. One hour tour by the owner. Home-made teas. Adm £12, chd free.**

Approx 3 acres, almost completely organic garden and paddock with fabulous panoramic views and magnificent wellingtonia. The garden also inc herbaceous borders, orchard, cottage garden with greenhouse, vegetable garden, herbs and strawberries and spinney. Bring a bag for any cuttings. Partial wheelchair access.

5 ◆ BOUGHTON HOUSE

Geddington, Kettering, NN14 1BJ. Duke of Buccleuch & Queensberry, KT. *3m NE of Kettering. From A14, 2m along A43 Kettering to Stamford, turn R into Geddington, house entrance 1½m on R. What3words app - crispier.sensible. maps.* **For NGS: Sun 19 Feb, Sun 1 Oct (1-5). Adm £6, chd £3. Pre-booking essential, please phone 01536 515731, email info@boughtonhouse.co.uk or visit www.boughtonhouse.org.uk for information & booking. Tea. For other opening times and information, please phone, email or visit garden website.**

The Northamptonshire home of the Duke and Duchess of Buccleuch. The garden opening inc opportunities to see the historic walled garden and herbaceous border, and the sensory and wildlife gardens. The wilderness woodland will open for visitors to view the spring flowers or the autumn colours. As a special treat the garden originally created by Sir David Scott (cousin of the Duke of Buccleuch) will also be open. Designated disabled parking. Gravel around house, please see our accessibility document for further information.

In 2022, our donations to Carers Trust meant that 456,089 unpaid carers were supported across the UK

16 Ace Lane, Bugbrooke Gardens

6 BRIARWOOD
4 Poplars Farm Road, Barton Seagrave, Kettering, NN15 5AF. William & Elaine Portch, 01536 522169, elaine.portch@yahoo.com, www.briarwoodgarden.com. *1½m SE of Kettering Town Centre. J10 off A14 turn onto Barton Rd (A6) towards Wicksteed Park. R into Warkton Ln, after 200 metres R into Poplars Farm Rd.* **Sun 16 Apr, Sun 10 Sept (10-4). Combined adm with Rosi's Taverna £6.50, chd free. Light lunches & refreshments. Visits also by arrangement Apr to Oct for groups of 10 to 30. Adm inc refreshments.**
A garden for all seasons with quirky original sculptures and many faces. Firstly, a south aspect lawn and borders containing bulbs, shrubs, roses and rare trees with year-round interest; hedging, palms, climbers, a wildlife, fish and lily pond, terrace with potted bulbs and unusual plants in odd containers. Secondly, a secret garden with garden room, small orchard, raised bed potager and greenhouse. Good use of recycled and repurposed materials throughout the garden, inc a unique self-build garden cabin, sculpture and planters.

GROUP OPENING

7 NEW BUGBROOKE GARDENS
Ace Lane, Bugbrooke, Northampton, NN7 3PQ. *Located 1m E of the A5, 3m S of M1 J16. Four gardens on Ace Ln & Orchard House is on Church Ln. SatNav for parking NN7 3RG.* **Sun 2 July (1-5.30). Combined adm £6, chd free. Home-made teas in Church Community Café.**

NEW **2 ACE LANE**
Nina Swain.

NEW **16 ACE LANE**
Steve & Kate, kate@qdl.co.uk. **Visits also by arrangement for groups of up to 6.**

NEW **DE FERRERS**
John Hammons.

NEW **ORCHARD HOUSE**
Sarah Spiers.

Bugbrooke is a large, picturesque village with its historic centre close to St. Michael's Church and the Millennium Green where the interesting and contrasting gardens are located. The circular route to enjoy the gardens takes in the church, crosses the brook, up to the High St with its splendid ironstone houses and then back towards the community café on Church Ln with its own garden where refreshments will be served. Each garden has a distinct character and they have been created, planted and maintained by their friendly and enthusiastic owners. They inc a garden of themed rooms structured by hedges with water features and sculptures, with a wide variety of plants; a calm and meditative garden hidden behind a lovely stone wall with sitting areas; a dramatic artistic garden with a striking and beautifully planted feature border; and a garden surrounding an Arts and Crafts house with lush planting and creative upcycled home-made features. We look forward to welcoming you to our gardens.

8 ◆ COTON MANOR GARDEN
Coton, Northampton, NN6 8RQ. Mr & Mrs Ian Pasley-Tyler, 01604 740219, pasleytyler@cotonmanor.co.uk, www.cotonmanor.co.uk. *10m N of Northampton, 11m SE of Rugby. From A428 & A5199 follow tourist signs.* **For NGS: Sat 16 Sept (11.30-5). Adm £10, chd free. Light refreshments & home-made teas at Stableyard Café. For other opening times and information, please phone, email or visit garden website.**
10 acre garden set in peaceful countryside with old yew and holly hedges and extensive herbaceous borders, containing many unusual plants. One of Britain's finest throughout the season, the garden is at its most magnificent in Sept and is an inspiration as to what can be achieved in late summer. Adjacent specialist nursery with over 1000 plant varieties propagated from the garden. Partial wheelchair access as some paths are narrow and the site is on a slope.

9 ◆ COTTESBROOKE HALL GARDENS
Cottesbrooke, Northampton, NN6 8PF. Mr & Mrs A R Macdonald-Buchanan, 01604 505808, welcome@cottesbrooke.co.uk, www.cottesbrooke.co.uk. *Cottesbrooke Hall is 3m off A14, J1. Follow brown signs S towards Northampton & in Creaton turn L onto Violet Ln. What3words app - expand.allow.curly.* **For NGS: Sun 23 Apr (2-5.30). Adm £12, chd £4. Home-made teas. Please visit www.cottesbrooke.co.uk to pre-book your tickets. For other opening times and information, please visit garden website. Donation to All Saints Church, Cottesbrooke.**
Award-winning gardens by Geoffrey Jellicoe, Dame Sylvia Crowe, James Alexander-Sinclair, and more recently Arne Maynard and Angel Collins. Formal gardens and terraces surround Queen Anne house with extensive vistas onto the lake and C18 parkland containing many mature trees. Wild and woodland gardens, a short distance from the formal areas, are exceptional in spring. Partial wheelchair access, please call ahead to discuss, as paths are grass, stone and gravel. Access map identifies best route.

10 NEW DENE LODGE
257 Rockingham Road, Kettering, NN16 9JE. Mr & Mrs R Pooley, 01536 481012, denelodge@btinternet.com. *N of Kettering Town Centre on A6003. From J7 of A14 take A43 towards Stamford, at next r'about take A6003 towards town centre, then over 1 r'about. Entrance opp Cotswold Ave. Parking by Beeswing Pub & Xrd.* **Sun 21 May (11-5). Adm £5, chd free. Home-made teas. Visits also by arrangement 7 May to 1 Oct for groups of 10 to 30.**
An established garden sloping downward from east to west with many trees, shrubs, roses, perennials and bulbs. Small pond, terrace with wisteria covered pergola leads past the rock garden to the lower patio, a larger pond with pergola, decking, climbing roses, jasmine and clematis. A further lawn leads through an arch to the vegetable garden with fruit trees and soft fruit, bushes and climbers. Concrete and paved patios and paths to bottom of the garden.

GROUP OPENING

11 EAST HADDON GARDENS

East Haddon, Northampton, NN6 8BT. *A few hundred yds off the A428 (signed) between M1 J18 (8m) & Northampton (8m). On-road parking. Strictly no parking in Priestwell Court, St Andrews Rd or on the properties.* **Mon 29 May (1-5). Combined adm £6, chd free. Home-made teas at St Mary's Church.**

BRAEBURN HOUSE
Judy Darby.

LIMETREES
Barry & Sally Hennessey.

LINDEN HOUSE
Hilary Van den Boogaard.

NEW **9 PRIESTWELL COURT**
Emma Forbes.

SADDLER'S COTTAGE
Val Longley.

THARFIELD
Julia Farnsworth.

TOWER COTTAGE
John Benson.

The pretty village of East Haddon dates back to the Norman invasion. The oldest surviving building is St Mary's, a C12 church. The village has many thatched cottages built in the local honey-coloured ironstone. Other features inc a thatched village pump and fire station which used to house a hand drawn pump and is now used as the bus shelter. The gardens opening this yr are a mixture of small, mature and family gardens with some having beautiful views across rolling hills. They are bursting with rare and unusual plants, various shrubs, climbers, roses and perennials. Borders galore, vegetable beds, espaliered fruit trees, and woodland areas. Plants propagated from the garden and raised from seed will be on sale inc some rarities at Limetrees. The village is less than 10 mins from Coton Manor Gardens. Partial wheelchair access at Tower Cottage.

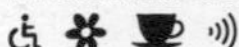

12 ◆ EVENLEY WOOD GARDEN

Evenley, Brackley, NN13 5SH. Whiteley Family, 07788 207428, info@evenleywoodgarden.co.uk, www.evenleywoodgarden.co.uk. *¾m S of Brackley. Turn off at Evenley r'about on A43 & follow signs within the village to the garden which is situated off the Evenley & Mixbury road.* **For NGS: Sat 15 July (10-4). Adm £7.50, chd £1. Light refreshments. For other opening times and information, please phone, email or visit garden website.**

Please come and celebrate summer in the woods when the roses and lillies will have taken over from the azaleas and rhododendrons. A wonderful opportunity to see all that has been developed in the woods since Timothy Whiteley acquired them in 1980 with the continuation of his legacy. Morning tea or coffee, lunch with a glass of wine and home-made cakes available in the café. Please take care as all paths are grass.

GROUP OPENING

13 FLORE GARDENS

Flore, Northampton, NN7 4LQ. *Situated between the towns of Daventry & Northampton & 2m from J16 of the M1.* **Sun 16 Apr (2-6). Combined adm £6, chd free. Sat 17, Sun 18 June (11-6). Combined adm £8, chd free. Home-made teas in Chapel School Room (Apr). Morning coffee & teas in Church & light lunches & teas in Chapel School Room (June). Donation to All Saints Church & United Reform Church, Flore (June).**

24 BLISS LANE
John Miller.
Open on Sun 16 Apr

THE CROFT
John & Dorothy Boast.
Open on all dates

THE GARDEN HOUSE
Gary & Julie Moinet.
Open on Sat 17, Sun 18 June

THE OLD BAKERY
John Amos & Karl Jones.
Open on all dates

PRIVATE GARDEN OF BLISS LANE NURSERY
Christine & Geoffrey Littlewood.
Open on all dates

ROCK SPRINGS
Tom Higginson & David Foster.
Open on all dates

RUSSELL HOUSE
Peter Pickering & Stephen George, 01327 341734, peterandstephen@btinternet.com, www.RussellHouseFlore.com.
Open on all dates

1 YEW TREE GARDENS
Mr & Mrs Martin Millard.
Open on all dates

Flore gardens have been open since 1963 as part of the Flore Flower Festival. The partnership with the NGS started in 1992 with openings every yr since. Flore is an attractive village with views over the Upper Nene Valley. We have a varied mix of gardens, large and small. They have all been developed by friendly, enthusiastic and welcoming owners who are in their gardens when open. Our gardens range from the traditional to the eccentric providing year-round interest. Some have been established over many yrs and others have been developed more recently. There are greenhouses, gazebos, summerhouses and seating opportunities to rest while enjoying the gardens. The village Flower Festival is our main event on the same two days as the June opening and garden tickets will be valid for both days. Partial wheelchair access to most gardens, some assistance may be required.

14 FOXTAIL LILLY

41 South Road, Oundle, PE8 4BP. Tracey Mathieson, 01832 274593, foxtaillilly41@gmail.com, www.foxtail-lilly.co.uk. *1m from Oundle town centre. From A605 at Barnwell Xrds take Barnwell Rd, 1st R to South Rd.* **Sun 11 June (11-5). Adm £4.50, chd free. Tea & cakes. Visits also by arrangement May to July for groups of 10 to 50.**

A cottage garden where perennials and grasses are grouped creatively together amongst gravel paths, complementing one another to create a natural look. Some unusual plants and quirky oddities create a different and colourful informal garden. Lots of flowers for cutting and a shop in the barn. New meadow pasture turned into new cutting garden. Plants, flowers and gift shop.

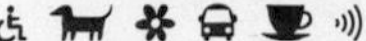

GROUP OPENING

15 GREAT BRINGTON GARDENS

Northampton, NN7 4JJ. *7m NW of Northampton. Off A428 Rugby Rd. From Northampton, turn 1st L past main gates of Althorp. Details & maps available at free car park, NN7 4HY.* **Sun 30 Apr (11-4). Combined adm £6, chd free. Home-made teas.**

FOLLY HOUSE
Sarah & Joe Sacarello.

ROSE COTTAGE
David Green & Elaine MacKenzie.

THE STABLES
Mrs J George.

SUNDERLAND HOUSE
Mrs Margaret Rubython.

THE WICK
Ray & Sandy Crossan.

YEW TREE HOUSE
Mrs Joan Heaps.

Great Brington is proud of over 25 yrs association with the NGS and arguably one of the most successful one day scheme events in the county. The gardens are situated in a circular and virtually flat walk around the village. Our gardens provide inspiration and variety; continuing to evolve each yr and designed, planted and maintained by their owners. The village on the Althorp Estate is particularly picturesque and well worth a visit, predominantly local stone with thatched houses, a church of historic interest, and walkers can extend their walk to view Althorp House. Visitors will enjoy a warm welcome with plants for sale, tea, coffee and cake available from 11pm-4pm. There are also secluded spots for artists to draw, sketch and paint. Small coaches for groups welcome by prior arrangement only, please email friends_of_stmarys@btinternet.com. Wheelchair access to some gardens.

16 THE GREEN PATCH

Valley Walk, Kettering, NN16 0LU. Grey Lindley, 07967 638710, grey.lindley@groundwork.org.uk, www.greenpatch.org.uk. *NE of Kettering Town Centre. Signed from A4300 Stamford Rd, on the junction of Valley Walk & Margaret Rd.* **Sun 30 July (10-4). Adm £5, chd free. Light refreshments. Visits also by arrangement 17 June to 15 Oct.**

The Green Patch is a 2½ acre, Green Flag award-winning community garden, nestled in the heart of England. We have hens, ducks, beehives, ponds, children's play area, orchard and so much more. We rely on our wonderful volunteers to make our friendly and magical garden the warm and welcoming place it is. Run by the environmental charity Groundwork Northamptonshire. Vegetable seedlings, herbs, and apple trees for sale. Tea, coffee and biscuits available. Bring a blanket for a picnic. Wheelchair access and disabled WC facilities.

17 GREYWALLS

Farndish, nr Wellingborough, NN29 7HJ. Mrs P M Anderson, 01933 353495, greywalls@dbshoes.co.uk. *2½m SE of Wellingborough. A609 from Wellingborough, B570 to Irchester, turn to Farndish by cenotaph. House adjacent to church.* **Sun 21 May (12-4). Combined adm with 136 High Street £6, chd free. Home-made teas. Visits also by arrangement for groups of 10 to 30.**

Greywalls is an old vicarage set in 2 acres of relaxed country gardens planted for year-round interest, featuring mature specimen trees, an impressive Banksia rose, three large ponds, many stone features, wildflower meadow and a Highgrove inspired stumpery. The borders feature unusual plants and there is an outside aviary. Strictly no dogs.

GROUP OPENING

18 GUILSBOROUGH GARDENS

Guilsborough, NN6 8RA. *10m NW of Northampton. 10m E of Rugby. Between A5199 & A428. J1 off A14. Car parking in field on Cold Ashby Rd NN6 8QN. Please park in the car park to avoid congestion in the village centre.* **Sun 7 May (1-5). Combined adm £7.50, chd free. Home-made teas in the village hall.**

NEW **THE CUTTING PATCH**
Suzanne McGlasson & Tina Hazle, www.instagram.com/sue_mcglasson_.

DRIPWELL HOUSE
Mr J W Langfield & Dr C Moss.

FOUR ACRES
Mark & Gay Webster, www.instagram.com/gardenatfouracres.

THE GATE HOUSE
Mike & Sarah Edwards.

GOWER HOUSE
Ann Moss.

THE OLD HOUSE
Richard & Libby Seaton Evans.

THE OLD VICARAGE
John & Christine Benbow.

Enjoy a warm welcome in this village with its very attractive rural setting of rolling hills and reservoirs. We have a group of seven contrasting gardens for you to visit this yr. Several of us are passionate about growing fruit and vegetables, and walled kitchen gardens and a potager are an important part of our gardening. Plants both rare and unusual from our plantsmen's gardens are for sale, a true highlight here. A garden new this year is dedicated to growing cut flowers. No wheelchair access at Dripwell House and The Gate House.

National Garden Scheme gardens are identified by their yellow road signs and posters. You can expect a garden of quality, character and interest, a warm welcome and plenty of home-made cakes!

GROUP OPENING

19 HARPOLE GARDENS

Harpole, NN7 4BX. *On A45 4m W of Northampton towards Weedon. Turn R at The Turnpike Hotel into Harpole. Village maps given to all visitors.* **Sun 11 June (1-6). Combined adm £5, chd free. Home-made teas at The Close.**

BRYTTEN COLLIER HOUSE
Heather Jennings.

CEDAR COTTAGE
Spencer & Joanne Hannam.

THE CLOSE
Michael Orton-Jones.

19 MANOR CLOSE
Caroline & Eamonn Kemshed.

THE OLD DAIRY
David & Di Ballard.

Harpole is an attractive village nestling at the foot of Harpole Hills with many houses built of local sandstone. Visit us and delight in a wide variety of gardens of all shapes, sizes and content. You will see luxuriant lawns, mixed borders with plants for sun and shade, mature trees, shrubs, herbs, alpines, water features and tropical planting. We have interesting and quirky artifacts dotted around, garden structures and plenty of seating for the weary.

20 67-69 HIGH STREET

Finedon, NN9 5JN. Mary & Stuart Hendry, 01933 680414, sh_archt@hotmail.com. *6m SE Kettering. Garden signed from A6 & A510 junction.* **Sun 26 Feb (11-3); Sun 4, Sun 25 June (2-5.30). Adm £3.50, chd free. Soup & roll in Feb (inc in adm). Home-made teas in June. 2024: Sun 25 Feb. Visits also by arrangement Feb to Sept.**

⅓ acre rear garden of C17 cottage (not open). Early spring garden with snowdrops and hellebores, summer and autumn mixed borders, many obelisks and containers, kitchen garden, herb bed, rambling roses and at least 60 different hostas. All giving varied interest from Feb through to Oct. Large selection of home-raised plants for sale (all proceeds to NGS).

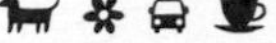

21 136 HIGH STREET

Irchester, NN29 7AB. Ade & Jane Parker, jane692@btinternet.com. *200yds past the church on the bend as you leave the village going towards the A45. Please park on High St. Disabled parking only in driveway.* **Thur 16, Fri 17, Sat 18, Sun 19 Feb (12-3). Adm £3.50, chd free. Sun 21 May (12-4). Combined adm with Greywalls £6, chd free. Sun 6 Aug (11-4). Adm £3.50, chd free. Light refreshments (Feb & Aug). Home-made teas at Greywalls (May only). Visits also by arrangement.**

Large garden developed by the current owners over the past 20 yrs. Various different planting habitats inc areas designed for shade, sun and pollinator friendly sites. Wildlife pond attracting large range of birds, insects and other creatures into the garden. Alpine houses, planted stone sinks and raised beds. Seasonally planted tubs adding bold summer colour. Wheelchair access mainly over grass with some gravel pathways.

22 HIGHFIELDS

Adstone, Towcester, NN12 8DS. Rachel Halvorsen. *The village of Adstone is midway between Banbury & Northampton, about 15m from each.* **Sun 20 Aug (2-5.30). Adm £5, chd free. Cream teas.**

Two sheltered mixed courtyard gardens with vibrant colours; new rill winds between lawn borders; formal walled garden with central aquaglobe. Plantsman's garden. Collection of succulents. Cream teas on patio overlooking panoramic views with ha-ha to 15 miles of unbroken countryside, lake and 100 acre Plumpton wood. Walk down to lake across a couple of fields to see wildlife and water lilies.

23 1 HINWICK CLOSE

Kettering, NN15 6GB. Mrs Pat Cole-Ashton. *J9 A14 A509 Kettering. At Park House r'about take 4th exit to Holdenby. Hinwick Close 3rd exit on R. From Kettering A509, at Park House r'about take the 1st exit to Holdenby, Hinwick Close 3rd exit on R.* **Sun 4 June (12-4.30). Adm £3.50, chd free. Home-made teas & savouries.**

In the past 14 yrs Pat and Snowy have transformed this space into a wildlife haven. The garden has numerous influences; seaside, woodland and English country garden. Ponds and waterfalls add to the delights. Vintage signs, numerous figures and seating areas at different vantage points are dotted throughout the garden.

24 ◆ HOLDENBY HOUSE & GARDENS

Holdenby House, Holdenby, Northampton, NN6 8DJ. Mr & Mrs James Lowther, 01604 770074, office@holdenby.com, www.holdenby.com. *7m NW of Northampton. Off A5199 or A428 between East Haddon & Spratton.* **For NGS: Sun 16 Apr (11-4). Adm £6, chd £4. Adm subject to change. Cream teas & light refreshments. For other opening times and information, please phone, email or visit garden website.**

Holdenby has a historic Grade I listed garden. The inner garden inc Rosemary Verey's renowned Elizabethan Garden and Rupert Golby's Pond Garden and long borders. There is also a delightful walled kitchen garden. Away from the formal gardens, the terraces of the original Elizabethan Garden are still visible, one of the best preserved examples of their kind. Accessible, but contact garden for further details. Assistant dogs only.

25 HOSTELLARIE

78 Breakleys Road, Desborough, NN14 2PT. Stella Freeman, 01536 760124, stelstan78@outlook.com. *6m N of Kettering. 5m S of Market Harborough. From church & war memorial turn R into Dunkirk Ave, then 3rd R. From cemetery L into Dunkirk Ave, then 4th L.* **Sun 11 June (2-5). Combined adm with 16 Leys Avenue £4, chd free. Home-made teas & a gluten-free option. Visits also by arrangement 1 June to 30 July for groups of 10 to 25.**

Hostellarie is a long town garden divided into rooms. Lawns and grass paths lead you through varied colour borders, ponds and water features. Courtyard garden shaded by an old clematis has over 40 different hostas, there are even more mature specimen hostas in the north facing bed and other shady spots. Roses, clematis, cottage and gravel borders and a welcoming relaxing atmosphere.

26 104 IRCHESTER ROAD

Rushden, NN10 9XQ. Mr Jason Richards, 07957 811173, jrrushden@gmail.com. *From Irchester the house is at the top of the hill just after the green, by the Welcome Inn on the L. From Rushden Town Centre the house is on the R after Knuston Drive.* **Sun 23 Apr, Sun 2, Sun 23 July, Sun 27 Aug (2-5.30). Adm £5, chd free. Light refreshments. Visits also by arrangement Apr to Sept.**

An Edwardian home set in ½ acre with different levels of patio areas and surrounded by trees. A rose garden with English rose varieties everywhere inc climbers on walls, a rose frame and a very long pergola winding its way down to a pear tree with seating and borders. In each area of the garden there are seating places to eat or just enjoy different perspectives. The owner has developed an environmentally friendly rose spray which will be on sale. The garden is on a hill with wheelchair access to all main areas with non-slip ramps. Please contact us for disabled parking priority.

27 ◆ KELMARSH HALL & GARDENS

Main Road, Kelmarsh, Northampton, NN6 9LY. The Kelmarsh Trust, 01604 686543, marketing@kelmarsh.com, www.kelmarsh.com. *Kelmarsh is 5m S of Market Harborough & 11m N of Northampton. From A14, exit J2 & head N towards Market Harborough on the A508.* **For NGS: Sun 17 Sept (10-4). Adm £6, chd £3.50. Light lunches, cream teas & cakes in Sweet Pea's Tearoom. For other opening times and information, please phone, email or visit garden website.**

Kelmarsh Hall is an elegant Palladian house set in glorious Northamptonshire countryside with highly regarded gardens, which are the work of Nancy Lancaster, Norah Lindsay and Geoffrey Jellicoe. Hidden gems inc an orangery, sunken garden, long border, rose gardens and, at the heart of it all, a historic walled garden. Highlights throughout the seasons inc fritillaries, tulips, roses and dahlias. Beautiful interiors brought together by Nancy Lancaster in the 1930s, in a Palladian style hall designed by James Gibbs. The recently restored laundry and servants' quarters in the Hall are open to the public, providing visitors the incredible opportunity to experience life 'below stairs'. Blue badge disabled parking close to the Visitor Centre entrance. Paths are loose gravel, wheelchair users advised to bring a companion.

GROUP OPENING

28 KILSBY GARDENS

Middle Street, Kilsby, Rugby, CV23 8XT. *5m SE of Rugby. 6m N of Daventry on A361.* **Sun 18 June (1-5.30). Combined adm £8, chd free. Home-made teas at Kilsby Village Hall (1-5).**

Kilsby's name has long been associated with Stephenson's famous railway tunnel and an early skirmish in the Civil War. The houses and gardens of the village offer a mixture of sizes and styles, which reflect its development through time. We welcome you to test the friendliness for which we are renowned and visit the gardens opening in the village this year!

29 ◆ LAMPORT HALL

Northampton, NN6 9HD. Lamport Hall Preservation Trust, 01604 686272, engagement@lamporthall.co.uk, www.lamporthall.co.uk. *For SatNav please use postcode NN6 9EZ. Exit J2 of the A14. Entry through the gate flanked by swans on the A508.* **For NGS: Sat 10 June (10-4). Adm £7, chd £4. Light refreshments in The Stables Café. For other opening times and information, please phone, email or visit garden website.**

Home of the Isham family for over 400 yrs, the extensive herbaceous borders complement the Elizabethan bowling lawns, together with topiary from the 1700s. The 2 acre walled garden is full of colour, with 250 rows of perennials. Another highlight is the famous Lamport Rockery, among the earliest in England and home of the world's oldest garden gnome. Wheelchair access on gravel paths within the gardens.

30 16 LEYS AVENUE

Desborough, NN14 2PY. Keith & Beryl Norman, 01536 760950, bcn@stainer16.plus.com. *6m N of Kettering, 5m S of Market Harborough. From church & War Memorial turn R into Dunkirk Ave & 5th R into Leys Ave.* **Sun 11 June (2-5). Combined adm with Hostellarie £4, chd free. Tea. Visits also by arrangement May to July for groups of 10 to 30.**

A town garden with two water features, plus a stream and a pond flanked by a 12ft clinker-built boat. There are six raised beds and an area of sweet peas. A patio lined with acers has two steps down to a gravel garden with paved paths. Mature trees and acers give the garden year-round structure and interest. Wheelchair access down two steps from patio to main garden.

GROUP OPENING

31 LITTLE BRINGTON GARDENS

Main Street, Little Brington, Northampton, NN7 4HS. *Off road parking can be found at the end of Folly Ln, Little Brington, NN7 4JR.* **Sun 25 June (11.30-4). Combined adm £7.50, chd free. Home-made teas.**

1 FERMOY COURT
Hilary & Chris Moore.

1 FOLLY LANE
Nick & Kerstin Banham.

MANOR FARM HOUSE
Rob Shardlow.

14 PINE COURT
Chris & Judy Peck.

ROCHE COTTAGE
Malcolm & Susan Uttley.

STONECROFT
Peter & Jenny Holman.

Six wonderful gardens in the historic village of Little Brington, nestled in the Northamptonshire farmlands will open for the NGS. They offer a variety of shapes, sizes and styles, from more formal planting to relaxed cottage gardens, along with a variety of vegetable gardens on show too. Lots of colours on display with a mix of both familiar and unusual plants and flowers. Refreshments will be served and the village pub Saracens Head will also be open, book early as the pub is very popular. Come and enjoy the wonderful gardens and the pretty village setting. Full or partial wheelchair access to most gardens. Some have gravel driveways.

GROUP OPENING

32 LONG BUCKBY GARDENS

Northampton, NN6 7RE. *8m NW of Northampton, midway between A428 & A5. Long Buckby is signed from A428 & A5. 10 mins from J18 M1. Long Buckby Train Stn is ½m from centre of the village. Free car park at Cotton End on B5385. Shuttle bus between farthest gardens.* **Sun 23 July (1-6). Combined adm £7, chd free. Home-made teas, ice creams & cold drinks.**

3 COTTON END
Roland & Georgina Wells.

4 COTTON END
Sue & Giles Baker.

23 COTTON END
Lynnette & Malcolm Cannell, 01327 844069, lynnettecannell68@gmail.com. **Visits also by arrangement 1 May to 1 Oct for groups of 10 to 15.**

THE GROTTO
Andy & Chrissy Gamble.

3A KNUTSFORD LANE
Tim & Jan Hunt.

10 LIME AVENUE
June Ford.

NEW **15 SYERS GREEN CLOSE**
Shona Mcnamee.

WISTERIA HOUSE
David & Clare Croston, 07771 911892, dad.croston@gmail.com. **Visits also by arrangement 1 May to 1 Oct for groups of up to 15. Combined visit with Woodcote Villa and/or 23 Cotton End.**

WOODCOTE VILLA
Sue & Geoff Woodward.
(See separate entry)

Nine gardens in the historic villages of Long Buckby and Long Buckby Wharf, inc a new garden and a reopening one. The gardeners have been busy creating new borders, features and extensively replanting, offering our visitors something new to enjoy. We welcome visitors old and new to see our wonderful variety of gardens which offer something for everyone. The gardens in the group vary in size and style, from courtyard and canal side to cottage garden, some are established and others evolving. They inc water features, pergolas, garden structures, but the stars are definitely the plants. Bursting with colour, visitors will find old favourites and the unusual, used in a variety of ways inc, trees, shrubs, perennials, climbers, annuals, fruit and vegetables. Of course, there will be teas and plants for sale to complete the visit. Come and see us for a friendly welcome and a good afternoon out. Full or partial wheelchair access to all gardens.

GROUP OPENING

33 MAIDWELL GARDENS

Draughton Road, Maidwell, Northampton, NN6 9JF. *From Market Harborough turn L off the A508 (or from Northampton turn R) & look for the yellow signs.* **Sun 4 June, Sun 20 Aug (10-5). Combined adm £6, chd free. Home-made teas at Wyatts.**

NIGHTINGALE COTTAGE
Ken & Angela Palmer.

ROSENHILL
Mr Ivan & Mrs Diana Barrett.

NEW **WYATTS**
Mr Colin & Mrs Amanda Goddard.

A friendly village with horses exercising daily and often sheep being herded through. The gardens opening are varied from a completely organic garden with many flowering plants and vegetables to a small terraced garden with hidden paths, sculptures and obelisks made from prunings. Wheelchair access to two gardens, partial access to Nightingale Cottage.

34 NEW 83 MAIN ROAD

Collyweston, Stamford, PE9 3PQ. Rosemary & Robert Fromm, 07597 684816, rrfromm@msn.com. *On A43 3½m SW of Stamford. 1m from A43/A47 r'about. Three doors from The Collyweston Slater pub.* **Sun 7 May (10-4). Adm £4, chd free. Visits also by arrangement Mar to May for groups of up to 15.** Wildlife friendly garden with many small trees, bushes and perennials. Spring bulbs and primroses proliferate underneath the trees, also a laburnum arch. Gravel paths and stone steps give access to the sloping site which is just under a ¼ acre. The garden is in the process of being made even more wildlife friendly, with many plants being added to flower to feed bees and butterflies. Refreshments at The Collyweston Slater, three doors away; very good food, recommended.

GROUP OPENING

35 NEWNHAM GARDENS

Newnham, Daventry, NN11 3HF. *2m S of Daventry on B4037 between the A361 & A45. Continue to the centre of the village & follow signs for the car park, just off the main village green.* **Sun 28 May (11-5). Combined adm £6, chd free. Light lunches & cakes in village hall.**

THE BANKS
Sue & Geoff Chester.

THE COTTAGE
Jacqueline Minor, www.instagram.com/newnhamngs.

HILLTOP
David & Mercy Messenger.

STONE HOUSE
Pat & David Bannerman.

WREN COTTAGE
Mr & Mrs Judith Dorkins.

You are so welcome again to Newnham. Five plant-lover gardens set in our beautiful old village nestled in the gentle rolling hills of west Northamptonshire. The gardens, set around charming traditional village houses, are packed with spring colour, horticultural delights and lovely views. Spend the day with us enjoying the gardens, buying at our ever-popular plant sale, strolling around the village lanes and visiting our C14 church. Why not treat yourself to a tasty light lunch and scrumptious cakes and refreshments in the village hall (several times!). We look forward to seeing you. The old village is hilly in parts. While most gardens are accessible for wheelchairs, some have steps or narrow paths in part.

Our 2022 donation enabled 7,000 people affected by cancer to benefit from the garden at Maggie's in Oxford

Old Rectory, Quinton

36 NONSUCH

11 Mackworth Drive, Finedon, Wellingborough, NN9 5NL. David & Carrie Whitworth. *Off Wellingborough Rd (A510) onto Bell Hill, then to Church Hill, 2nd L after church, entrance on the L as you enter Mackworth Drive.* **Sun 30 July (1-5). Adm £4, chd free. Drinks, cakes & biscuits.**

⅓ acre country garden within a conservation boundary stone wall. Mature trees, enhanced by many rare and unusual shrubs and perennial plants. A garden for all seasons with several seating areas. Wheelchair access on a level site with paved and gravel paths.

37 NEW OAKLEA

84 Wollaston Road, Irchester, NN29 7DF. Nicola Wilson-Brown. *3m SE of Wellingborough. Leave Wellingborough via A509 heading towards Wollaston, turn L at r'about onto B570 signed Irchester Country Park. Continue to end of road, turn L at r'about into Wollaston Rd.* **Sun 28 May, Sun 16 July (12-5). Adm £4, chd free. Tea.**

An interesting and deceptive garden featuring courtyard area with seasonal pots, troughs and baskets. Borders well-stocked with shrubs and herbaceous perennials together with climbing plants and roses giving colour throughout the growing season. Two lawned areas, a recently added pond planted with pond plants and stocked with fish. Several seating areas and a vegetable garden with potting shed. Garden is all on one level.

38 OLD RECTORY, QUINTON

Preston Deanery Road, Quinton, Northampton, NN7 2ED. Alan Kennedy & Emma Wise, www.garden4good.co.uk. *M1 J15, 1m from Wootton towards Salcey Forest. House is next to the church. On-road parking in village. Please note parking on village green is prohibited.* **Sun 11 June (10-4). Adm £10, chd free. Pre-booking essential, please visit www.ngs.org.uk for information & booking. Hot drinks & cake (all day). Lunches to pre-order via garden website (11am-2pm).**

A beautiful contemporary 3 acre rectory garden designed by multi-award-winning designer, Anoushka Feiler. Taking the Old Rectory's C18 history and its religious setting as a key starting point, the main garden at the back of the house has been divided into six parts; a kitchen garden, glasshouse and flower garden, a woodland menagerie, a pleasure garden, a park and an orchard. Elements of C18 design such as formal structures, parterres, topiary, long walks, occasional seating areas and traditional craft work have been introduced, however with a distinctly C21 twist through the inc of living walls, modern materials and features, new planting methods and abstract installations. Pop-up shop selling plants, garden produce, local

Mill Hollow Barn, Sulgrave Gardens

honey, home-made bread and bath products. Wheelchair access with gravel paths.

39 THE OLD VICARAGE

Daventry Road, Norton, Daventry, NN11 2ND. Barry & Andrea Coleman. *Norton is approx 2m E of Daventry, 11m W of Northampton. From Daventry follow signs to Norton for 1m. On A5 N from Weedon follow road for 3m, take L turn signed Norton. On A5 S take R at Xrds signed Norton, 6m from Kilsby. Garden is R of All Saints Church.* **Sun 19 Feb (11-3); Sun 23 Apr (1-5). Adm £5, chd free. Soup & bread in Feb (inc in adm). Home-made teas in orangery (Apr).**

The vicarage days bequeathed dramatic and stately trees to the modern garden. The last 40 yrs of evolution and the happy accidents of soil-type, and a striking location with lovely vistas have shaped the garden around all the things that makes April so thrilling, inc prodigious sweeps of primulas. In Feb walk through adjacent fields and the churchyard to see snowdrops in profusion. The interesting and beautiful C14 church of All Saints will be open to visitors.

40 OLD WEST FARM

Little Preston, Daventry, NN11 3TF. Mr & Mrs G Hoare, caghoare@gmail.com. *7m SW Daventry, 8m W Towcester, 13m NE Banbury. ¼ m E of Preston Capes on road to Maidford. Last house on R in Little Preston with white flagpole. Beware, the postcode applies to all houses in Little Preston.* **Visits by arrangement May & June for groups of up to 35. Adm £5, chd free. Home-made teas.**

Large rural garden developed over the past 40 yrs on a very exposed site, planted with hedges and shelter. Roses, shrubs and borders aiming for year-round interest. Partial wheelchair access over grass.

GROUP OPENING

41 RAVENSTHORPE GARDENS

Ravensthorpe, NN6 8ES. *7m NW of Northampton. Signed from A428. Please start & purchase tickets at Ravensthorpe Nursery.* **Sun 2 July (1.30-5.30). Combined adm £7, chd free. Home-made teas in village hall.**

CORNERSTONE
Lorna Jones.

NEW **MANOR VIEW**
Viv & David Rees.

QUIETWAYS
Russ Barringer.

RAVENSTHORPE NURSERY
Mr & Mrs Richard Wiseman.
(See separate entry)

TREETOPS
Ros Smith.

Attractive village in Northamptonshire uplands near to Ravensthorpe reservoir and Top Ardles Wood Woodland Trust which have bird watching and picnic opportunities. Established and developing gardens set in beautiful countryside displaying a wide range of plants, many available from the nursery that now only opens on NGS open days. Offering inspirational planting, quiet contemplation, beautiful views, water features and gardens encouraging wildlife. Disabled WC in village hall.

42 RAVENSTHORPE NURSERY

6 East Haddon Road, Ravensthorpe, NN6 8ES. Mr & Mrs Richard Wiseman, 01604 770548, ravensthorpenursery@hotmail.com. *7m NW of Northampton. First property on L approaching from A428.* **Fri 14, Sat 15, Sun 16 Apr, Fri 12, Sat 13, Sun 14 May, Fri 30 June, Sat 1 July, Fri 29, Sat 30 Sept, Sun 1 Oct (11-5). Adm £5, chd free. Tea & home-made cake. Opening with Ravensthorpe Gardens on Sun 2 July. Visits also by arrangement 14 Apr to 1 Oct for groups of 15+.**

Approx 1 acre garden wrapped around the nursery with beautiful views. Planted with many unusual shrubs and herbaceous perennials over the last 30+ yrs to reflect the wide range of plants produced. Plants for sale. Wheelchair access with gradual slope to garden and nursery.

43 ROSEARIE-DE-LA-NYMPH

55 The Grove, Moulton, Northampton, NN3 7UE. Peter Hughes, Mary Morris, Irene Kay, Steven Hughes & Jeremy Stanton. *N of Northampton town. Turn off A43 at small r'about to Overstone Rd. Follow NGS signs in village. The garden is on the Holcot Rd out of Moulton.* **Sun 18, Sun 25 June (11-5). Adm £5, chd free. Home-made teas.**

We have been developing this romantic garden for over 15 yrs and now have over 1800 roses, inc English, French and Italian varieties. Many unusual water features and specimen trees. Roses, scramblers and ramblers climb into trees, over arbours and arches. Collection of 140 Japanese maples. Mostly flat wheelchair access via a standard width doorway.

44 ROSI'S TAVERNA

20 St Francis Close, Barton Seagrave, Kettering, NN15 5DT. Rosi & David Labrum. *Approx 2m from J10 of the A14. Head towards Barton Seagrave, going through 3 sets of T-lights. At the 4th set turn R onto Warkton Ln. At the r'about turn L. Take the next 2 R turns into St Francis Close.* **Sun 16 Apr, Sun 10 Sept (10-4). Combined adm with Briarwood £6.50, chd free.**

An established medium south facing town garden. Kitchen garden intermingles with flowers, shrubs, plenty of fruit trees, soft fruits and vegetables in raised beds. Cacti and succulents hold high importance in the greenhouse. A covered koi carp pond and a pyramid water feature are key elements in the garden. A unique taverna with mosaic flooring offers perfect shelter from sun, wind and drizzle.

Our 2022 donation to Hospice UK meant that 911 NHS and Social Care staff accessed individual counselling support

GROUP OPENING

45 SPRATTON GARDENS

Smith Street, Spratton, NN6 8HP. *6½ m NNW of Northampton. To find car park, please follow yellow NGS signs from outskirts of village from A5199 or Brixworth Rd. In Spratton Hall School car park.* **Sun 11 June (11-5). Combined adm £7, chd free. Home-made teas.**

 **COTFIELD**
Christina Warner-Keogh.

THE COTTAGE
Mrs Judith Elliott.

DALE HOUSE
Fiona & Chris Cox.

28 GORSE ROAD
Lee Miller.

11 HIGH STREET
Philip & Frances Roseblade.

MULBERRY COTTAGE
Kerry Herd.

NORTHBANK HOUSE
Helen Millichamp.

OLD HOUSE FARM
Susie Marchant.

STONE HOUSE
John Forbear.

VALE VIEW
John Hunt.

WALTHAM COTTAGE
Norma & Allan Simons.

NEW **25 YEW TREE LANE**
Mrs Lise Beynon.

As well as attractive cottage gardens alongside old Northampton stone houses, Spratton also has unusual gardens, inc those showing good use of a small area; those dedicated to encouraging wildlife with views of the surrounding countryside; newly renovated gardens and those with new planting; courtyard garden; gravel garden with sculpture; mature gardens with fruit trees and herbaceous borders. Refreshments available in Norman St Andrew's Church. The King's Head Pub will be open, lunch reservations recommended.

GROUP OPENING

46 SULGRAVE GARDENS

Banbury, OX17 2RP. *8m NE of Banbury. Just off B4525 Banbury to Northampton road, 7m from J11 off M40. Car parking at church hall.* **Sun 25 June (2-6). Combined adm £6, chd free. Home-made teas at Rectory Farm.**

NEW **THE CHESTNUTS**
Mrs Mel Kirkpatrick.

NEW **HANGLANDS**
Clive & Donna Nicholls.

MILL HOLLOW BARN
David & Judith Thompson.

NEW **OLD FORGE**
Paul Shanley & Brian Jopling.

RECTORY FARM
Charles & Joanna Smyth-Osbourne, 01295 760261, sosbournejm@gmail.com.

VINECROFT
Claire & Jon Sadler.

THE WATERMILL
Mr & Mrs T Frost.

WOOTTON HOUSE
Zoe & Richard McCrow.

Sulgrave is a small historic village having recently celebrated its strong American connections as part of the 150 yrs of the signing of the Treaty of Ghent. Eight gardens opening; Rectory Farm has lovely views, a rill, well and planted arbours. Mill Hollow Barn, a large garden with lakes, streams, ponds and many rare and interesting trees, shrubs and perennials. The Watermill, a contemporary garden designed by James Alexander-Sinclair, set around a C16 watermill and mill pond. A quirky garden at Wootton House with a good selection of rare and interesting plants. The Chesnuts, a small garden, but with huge interest, where every plant tells a story. Hanglands is a modern garden, designed for entertaining. Vinecroft shows what can be done on a typical site, and Old Forge is a medium sized 'country meets modern' garden with lawn and curved borders with a variety of plants. An award-winning community owned and run village shop will be open.

47 TITCHMARSH HOUSE

Chapel Street, Titchmarsh, NN14 3DA. Sir Ewan & Lady Harper, 01832 732439, ewan@ewanh.co.uk, www.titchmarsh-house.co.uk. *2m N of Thrapston. 6m S of Oundle. Exit A14 at junction signed A605, Titchmarsh signed as turning E towards Oundle & Peterborough.* **Mon 10 Apr (12.30-5); Mon 1 May (2.30-5.30); Sat 10 June (12.30-5). Adm £6, chd free. Visits also by arrangement Apr to June for groups of 5+.**

The gardens are spread over 4½ acres with formal and wild areas, a hedged quiet garden, and inc collections of naturalised bulbs, irises, peonies, magnolias, shrub roses, cherries, malus and a number of rare trees, expanded and planted by the existing owners over the past 55 yrs. Refreshments inc: BBQ 12–2pm and teas 2.30–4.45pm (Apr). To celebrate our 58th Wedding Anniversary on the 1st May we are offering our first 100 entrants a glass of bubbly. Teas also available (May). BBQ and teas at the village fete (June). Wheelchair access to most of the garden without using steps. No dogs.

GROUP OPENING

48 WEEDON LOIS & WESTON GARDENS

Weedon Lois, Towcester, NN12 8PJ. *7m W of Towcester. 7m N of Brackley. Turn off A43 at Towcester towards Abthorpe & Wappenham & turn R in Wappenham for Weedon Lois. Or turn off A43 at Brackley, follow signs to Helmdon & Weston.* **Sun 25 June (1-6). Combined adm £6, chd free. Home-made teas at The Chapel, Weston.**

THE GARDENER'S COTTAGE
Mrs Sitwell.

4 HELMDON ROAD
Mrs S Wilde.

8A HIGH STREET
Mr & Mrs J Archard-Jones.

LOIS WEEDON HOUSE
Lady Greenaway.

MIDDLETON HOUSE
Mark & Donna Cooper.

OLD BARN
John & Iris Gregory.

POST COTTAGE
Jane Kellar.

RIDGEWAY COTTAGE
Jonathan & Elizabeth Carpenter.

4 VICARAGE RISE
Ashley & Lindsey Cartwright.

Our two small attractive villages combine to offer an interesting range of gardens. Everything from large and formal to small and intimate. We have wonderful herbaceous borders, far reaching views, fruit and vegetable gardens, cutting edge modern design and traditional cottage garden planting. So, we hope you will join us for our open day, enjoy looking round our gardens, visit the plant stalls and tuck into our famous home-made teas.

49 WILLOW COTTAGE

55 Upper Benefield, Upper Benefield, Peterborough, PE8 5AL. Mrs Nathalie Tarbuck. *From Oundle on A427 parking is located at Upper Benefield Cricket Club, 1st turning on the L as you enter the village. The garden is located opp Hill Farm. It is a short walk to the garden.* **Sat 3 June (1.30-4.30); Sun 4 June (10-2). Adm £5, chd free. Home-made teas at Upper Benefield Cricket Club.**
Willow Cottage is wrapped in a beautiful cottage style garden. Wander this ¼ acre site and discover its hidden gems. Planted to be a haven for pollinators and wildlife. Herbaceous borders burst with a wide variety of plants. A large pond, several water features, established and newly planted trees. Meander through rose arches and find a spot to perch to soak up the stunning views. Partial wheelchair access via a slope to enter the garden.

50 WILLOWBROOK HOUSE

76 Park Street, Kings Cliffe, Peterborough, PE8 6XN. Dr Robert E Stebbings, 03330 116577, bob@stebbing-cons.ndonet.com.
E of village on road to Wansford. Turn off A43 at Bulwick signed Kings Cliffe. Also signed from A47 just W of A1. **Visits by arrangement Feb to Sept for groups of up to 8.**
Original walled ⅕ acre garden with Victorian fruit trees, is over 250 yrs old. New garden created by ecologist in 2018 to inc native and exotic plants to encourage wildlife. Bulbs and flowering scented shrubs from the New Year and continue through the yr. Garden inc gravel areas with acid and alkaline rockeries and meadow. About 20 magnolias, unusual trees and shrubs, and fernery.

51 WOODCOTE VILLA

Old Watling Street, Long Buckby Wharf, Long Buckby, Northampton, NN6 7EW. Sue & Geoff Woodward, geoff.and.sue@btinternet.com.
2m NE of Daventry, just off A5. From M1 J16, take Flore by-pass, turn R at A5 r'about for approx 3m. From Daventry follow Long Buckby signs but turn L at A5 Xrds. From M1 J18 signed Kilsby, follow A5 S for approx 6m. **Sun 26 Mar, Sun 3 Sept (11-5). Adm £4, chd free. Home-made teas. Gluten free option. Opening with Long Buckby Gardens on Sun 23 July. Visits also by arrangement 4 Mar to 25 Aug for groups of 12 to 36.**
In a much admired location, this stunning canal side garden has a large variety of plants, styles, structures and unusual bygones. Bulbs and hellebores abound in Mar, colourful planting in July/Sept, many pots, all set against a backdrop of trees and shrubs in themed areas. Places to sit and watch the narrowboats and wildlife. Plants for sale (cash only). Sorry no WC. Wheelchair access via ramp at entrance to garden.

Holdenby House & Gardens

NOTTINGHAMSHIRE

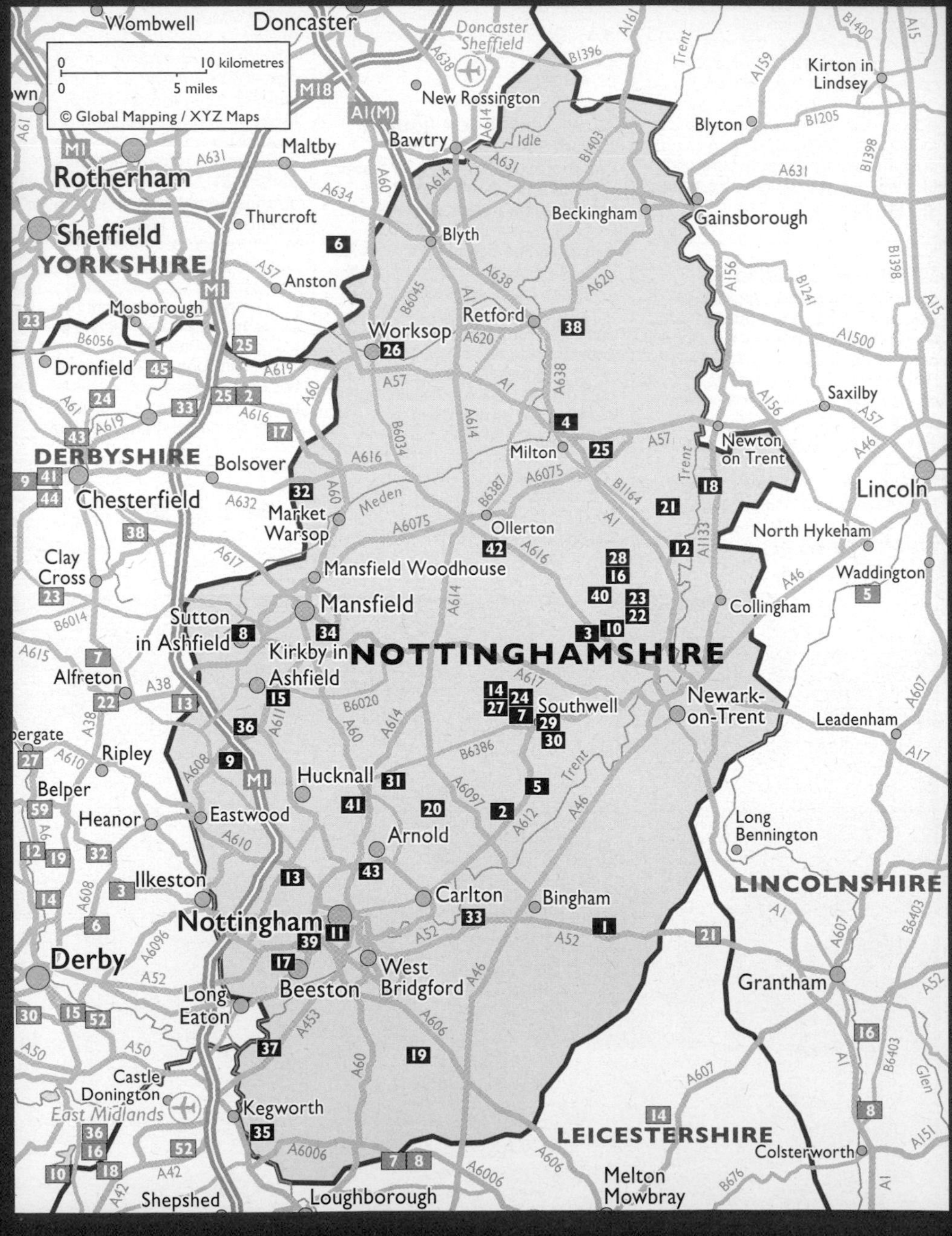

Nottinghamshire is best known as Robin Hood country. His legend persists and his haunt of Sherwood Forest, now a nature reserve, contains some of the oldest oaks in Europe. The Major Oak, thought to be 800 years old, still produces acorns.

Civil War battles raged throughout Nottinghamshire, and Newark's historic castle bears the scars. King Charles I surrendered to the Scots in nearby Southwell after a night at The Saracen's Head, which is still an inn today. Southwell is also home to the famous Bramley apple, whose descendants may be found in many Nottinghamshire gardens.

We have groups of cottage gardens, and we have historic houses with links to famous gardeners from the past. We have gardens which showcase environmental concerns, we have gardens full of rare exotic imports, and we have artists' gardens of inspiration. All will offer a warm welcome, and most will offer a brilliant tea.

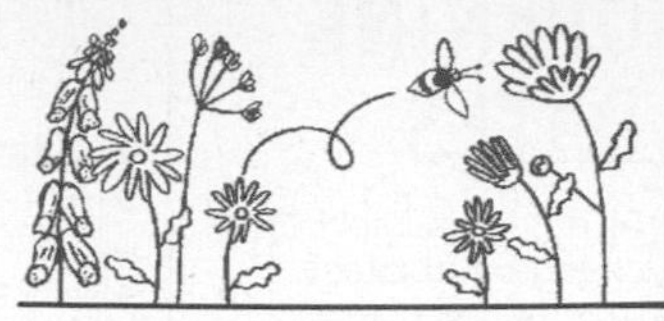

Volunteers

County Organiser
Georgina Denison
01636 821385
georgina.denison@ngs.org.uk

County Treasurer
Nicola Cressey
01159 655132
nicola.cressey@gmail.com

Publicity
Julie Davison
01302 719668
julie.davison@ngs.org.uk

Social Media
Malcolm Turner
01159 222831
malcolm.turner14@btinternet.com

Booklet Co-ordinators
Malcolm & Wendy Fisher
0115 966 4322
wendy.fisher111@btinternet.com.

Assistant County Organisers
Judy Geldart
01636 823832
judygeldart@gmail.com

Beverley Perks
01636 812181
perks.family@talk21.com

Mary Thomas
01509 672056
nursery@piecemealplants.co.uk

Andrew Young
01623 863327
andrew.young@ngs.org.uk

@National Garden Scheme Nottinghamshire

@nottsngs

Left: The Palace Garden of Southwell Minster

OPENING DATES

All entries subject to change. For latest information check **www.ngs.org.uk**

Map locator numbers are shown to the right of each garden name.

February

Snowdrop Festival

Saturday 11th
1 Highfield Road 13

Sunday 12th
Church Farm 4

Sunday 26th
NEW Norwood Park 24

March

Sunday 12th
Upper Grove Farm 40

April

Sunday 16th
◆ Felley Priory 9

Saturday 22nd
Capability Barn 2
Oasis Community Gardens 26

Sunday 23rd
Capability Barn 2
1 Highfield Road 13

Sunday 30th
NEW 160 Southwell Road West 34

May

Saturday 13th
Sutton Bonington Gardens 35

Sunday 14th
Sutton Bonington Gardens 35

Thursday 18th
Rhubarb Farm 32

Saturday 20th
Church House 5
The Old Vicarage 27

Sunday 21st
6 Hope Street 17
Ivy Bank Cottage 18
◆ Norwell Nurseries 23
Wellow Village Gardens 42

Sunday 28th
East Meets West 8
NEW Normanton Gardens 21
NEW The Palace Garden of Southwell Minster 29

Monday 29th
East Meets West 8

June

Sunday 4th
Capability Barn 2
Keyworth Gardens 19
Park Farm 30

Saturday 10th
Hill's Farm 14
The Old Vicarage 27
NEW Thyme House 38

Sunday 11th
Hollinside 15
Patchings Art Centre 31

Saturday 17th
NEW Clyde House 7
NEW Thyme House 38

Sunday 18th
10 Harlaxton Drive 11
NEW Sycamores House 36
Thrumpton Hall 37
6 Weston Close 43

Thursday 22nd
Rhubarb Farm 32

Saturday 24th
NEW Thyme House 38

Sunday 25th
Norwell Gardens 22

Wednesday 28th
Norwell Gardens 22

July

Saturday 1st
The Old Vicarage 27

Sunday 2nd
5 Burton Lane 1
Hopbine Farmhouse, Ossington 16
Ossington House 28

Saturday 8th
NEW Caunton Manor 3
Oasis Community Gardens 26

Sunday 9th
NEW Caunton Manor 3

Sunday 16th
10 Harlaxton Drive 11

Thursday 20th
Rhubarb Farm 32

Saturday 22nd
NEW 8 Church Lane 6
NEW Clyde House 7

Sunday 23rd
NEW 8 Church Lane 6

Saturday 29th
Floral Media 10

Sunday 30th
5a High Street 12

August

Saturday 5th
NEW 8 Church Lane 6

Sunday 6th
NEW 8 Church Lane 6
The Old Vicarage 27

Sunday 13th
University Park Gardens 39

Thursday 17th
Rhubarb Farm 32

Sunday 27th
East Meets West 8

Monday 28th
5 Burton Lane 1
East Meets West 8

September

Sunday 10th
Oak Barn Exotic Garden 25

Sunday 24th
◆ Norwell Nurseries 23

By Arrangement

Arrange a personalised garden visit with your club, or group of friends, on a date to suit you. See individual garden entries for full details.

5 Burton Lane 1
Capability Barn 2
NEW Church Farm, Normanton Gardens 21
NEW 8 Church Lane 6
10 Harlaxton Drive 11
5a High Street 12
Home Farm House, 17 Main Street, Keyworth Gardens 19
6 Hope Street 17
38 Main Street 20
Oasis Community Gardens 26
The Old Vicarage 27
NEW The Palace Garden of Southwell Minster 29
Park Farm 30
Rhubarb Farm 32
Riseholme, 125 Shelford Road 33
Rose Cottage, Keyworth Gardens 19
NEW Waxwings & Goldcrest 41
6 Weston Close 43

In 2022 we donated £450,000 to Marie Curie which equates to 23,871 hours of community nursing care

THE GARDENS

1 5 BURTON LANE

Whatton in the Vale, NG13 9EQ. Ms Faulconbridge, 01949 850942, jpfaulconbridge@hotmail.co.uk, www.ayearinthegardenblog. wordpress.com. *3m E of Bingham. Follow signs to Whatton from A52 between Bingham & Elton. Garden nr Church in old part of village. Follow yellow NGS signs.* **Sun 2 July, Mon 28 Aug (11-4.30). Adm £4, chd free. Home-made teas. Cakes for special dietary requirements available. Visits also by arrangement 1 June to 10 Sept for groups of up to 20.**

Modern organic cottage garden which is productive, highly decorative and wildlife friendly. It is full of colour and scent from spring to autumn. Several distinct areas, inc fruit and vegetables. Large beds filled with over 500 varieties of plants with paths through so you can wander and get close. Also features seating, gravel garden, pond, shade planting, pergola with grapevine, wildflower lawn. Attractive village with walks.

2 CAPABILITY BARN

Gonalston Lane, Hoveringham, NG14 7JH. Malcolm & Wendy Fisher, 01159 664322, wendy.fisher111@btinternet.com, www.capabilitybarn.com. *8m NE of Nottingham. A612 from Nottingham through Lowdham. Take 1st R into Gonalston Lane. Garden is 1m on L.* **Sat 22, Sun 23 Apr, Sun 4 June (11-4.30). Adm £4.50, chd free. Home-made teas. Visits also by arrangement 2 May to 9 June for groups of 20+. By arrangement adm inc home-made teas.**

Imaginatively planted large country garden with something new each year. April brings displays of daffodils, hyacinths and tulips along with erythroniums, brunneras and primulas. Wisteria, magnolia, rhodos and apple blossom greet May/June. A backdrop of established trees, shrubs and shady paths give a charming country setting. Large vegetable/ fruit gardens with orchard/meadow completes the picture.

3 NEW CAUNTON MANOR

Manor Road, Caunton, Newark, NG23 6AD. Sir John & Lady Peace. *6m from Newark & Southwell. Approach off the A616 signposted Caunton. Entry is through 2 large wrought iron gates opp the playing field next to Dean Hole sch.* **Sat 8, Sun 9 July (1-4). Adm £6, chd free. Teas.**

The gardens date back to the early C18. There is a kitchen garden, orchard, glasshouses and a well garden which all have paths around as well as an arboretum and walled garden. The gardens evolve each year so you may find a project in progress.

4 CHURCH FARM

Church Lane, West Drayton, Retford, DN22 8EB. Robert & Isabel Adam. *5m S of Retford. A1 exit Markham Moor. A638 Retford 500 yds signed West Drayton. 3/4 m, turn R, into Church Lane, 1st R past church. Ample parking in farm yard.* **Sun 12 Feb (10.30-4). Adm £4, chd free. Light refreshments served from 11.30am.**

The garden is essentially a spring garden with a small woodland area which is carpeted with many snowdrops, aconites and cyclamen which have seeded into the adjoining churchyard, with approx. 180 named snowdrops growing in island beds, along with hellebores and daffodils. Limited amount of snowdrops are for sale.

5 CHURCH HOUSE

Hoveringham, NG14 7JH. Alex & Sue Allan. *6m E of Nottingham. Off A612 from Southwell or Nottingham - To the R of the Church & Church Hall in the centre of Hoveringham village.* **Sat 20 May (12-4). Combined adm with The Old Vicarage £7, chd free.**

Stunning small gem of a garden, beautifully planted with immaculate hostas which greet you along the gravelled path leading past a mini pseudo-roof garden at eye level. Relax in an oasis of cottage garden style planting where the chickens, small pond, vegetable beds, charming auricula theatre and espalier fruit trees combine to create a delightfully peaceful atmosphere.

6 NEW 8 CHURCH LANE

Letwell, Worksop, S81 8DE. Vicky Bennett, 07795 058490, ywv.bennett64@gmail.com. *Turn off the B6463 onto Ramper Rd (signposted Letwell). At the end follow the 90 degree R bend. Then Church Lane forks off to the R. No 8 is the last property on the L opp church.* **Sat 22, Sun 23 July, Sat 5, Sun 6 Aug (12-4). Adm £6, chd free. Pre-booking essential, please visit www.ngs.org. uk for information & booking. Home-made teas. Visits also by arrangement 23 July to 5 Aug for groups of 5 to 20. Between midday & 4pm, limited parking.**

An award-winning, professionally designed garden in 1/3 acre. The free-flowing lawn is surrounded by colourful, naturalistic borders with shrubs, grasses and perennials. An informal path runs through a colourful gravel garden, planted with aromatics and grasses leading to the wildlife pond and sunken garden with oak pergola. Short woodland walk. Ample seating. Disabled parking on drive.

7 NEW CLYDE HOUSE

23 Westgate, Southwell, NG25 0JN. John & Lucy Murkett. *Pass Southwell Minister on your L on Westgate. Garden entrance signed with NGS yellow arrow on L into a footpath between 21 & 23 Westgate. First gate on the R.* **Sat 17 June, Sat 22 July (12-4). Adm £5, chd free. Home-made teas.**

Secret, intriguing, organic walled garden in the centre of Southwell, first opened for NGS in 1983 now reopening! Part formal to mirror the wisteria clad Georgian house with C13/14 listed stone arch. Sweeping lawn bottom of formal steps with long and circular herbaceous borders. Through to beautiful wildflower meadow, wildlife pond, fruit, veg, cut flower beds with greenhouse and compost bins. Wheelchair access to lower garden limited.

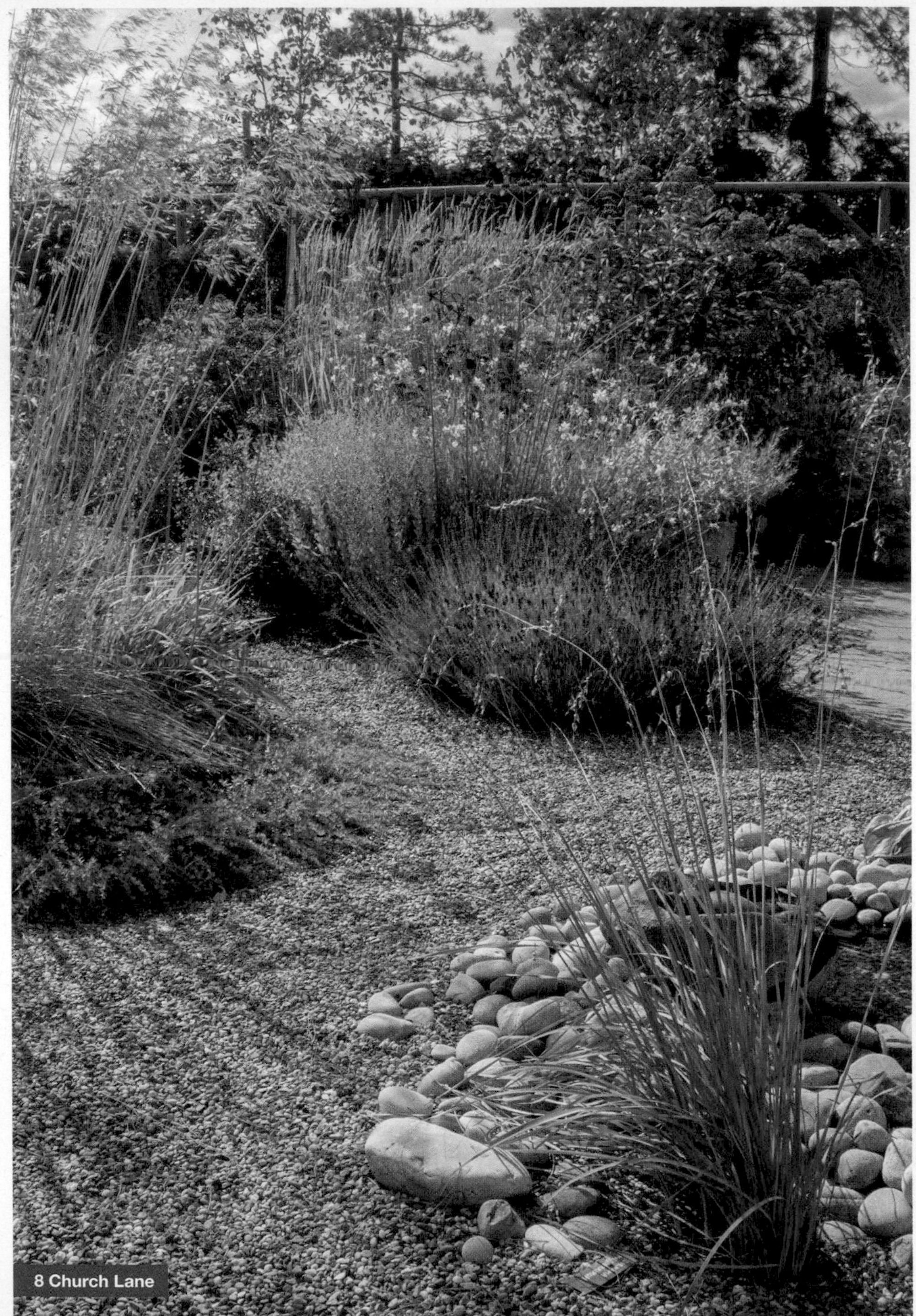

8 Church Lane

8 EAST MEETS WEST
85 Cowpes Close, Sutton-In-Ashfield, NG17 2BU. Kate & Mel Calladine. *Close to Quarrydale School entrance to Carsic Housing Estate. The Cl has limited parking res. for those with blue badges. More parking on Stoneyford Rd, NG17 2DU. Cross on zebra & walk down jitty into Cowpes Cl following NGS signposting.* **Sun 28, Mon 29 May, Sun 27, Mon 28 Aug (12-4). Adm £3, chd free. Pre-booking essential, please visit www.ngs.org.uk for information & booking.**
Our very small but lovely garden combines the tranquillity of the Orient with the colour of a traditional English cottage garden. East: We have a number of sizable acers, bamboos and Japanese lanterns. A stream flows past a cloud tree into a pond with goldfish and water lilies. West: A trompe d'oeil arch creates a magical garden illusion. A 'rainbow' flower bed. No steps from pavement to garden. Cobbled path allows wheelchair access to full length of garden.

9 ◆ FELLEY PRIORY
Underwood, NG16 5FJ. The Brudenell Family, 01773 810230, michelle@felleypriory.co.uk, www.felleypriory.co.uk. *8m SW of Mansfield. Off A608 ½m W M1 J27.* **For NGS: Sun 16 Apr (10-4). Adm £6.50, chd free. Light refreshments. For other opening times and information, please phone, email or visit garden website.**
Garden for all seasons with yew hedges and topiary, snowdrops, hellebores, herbaceous borders and rose garden. There are pergolas, a white garden, small arboretum and borders filled with unusual trees, shrubs, plants and bulbs. The grass edged pond is planted with primulas, bamboo, iris, roses and eucomis. Bluebell woodland walk. Orchard with extremely rare daffodils.

10 FLORAL MEDIA
Norwell Road, Caunton, Newark, NG23 6AQ. Mr & Mrs Steve Routledge, 01636 636283, info@floralmedia.co.uk, www.floralmedia.co.uk. *Take Norwell Rd from Caunton. Approx ½m from Caunton on L.* **Sat 29 July (10-4). Adm £4, chd free. Home-made teas.**
A beautifully well maintained country garden. Beds overflowing with a variety of roses, shrubs and flowers. A gravel/oriental garden, cutting gardens, vegetable beds, Flower Farm supplying British grown stems to florists/farm shops. Long sweeping borders surrounding the main lawn leading to the wildflower meadows where you will find an interesting garden retreat. A horticulturists haven. Excellent facilities, refreshments, plenty of parking. Often live music in the garden from a local folk group of musicians. A good range of plants available for sale. Full wheelchair access inc disabled WC.

11 10 HARLAXTON DRIVE
Lenton, Nottingham, NG7 1JA. Jan Brazier, 07968 420046, jan-28b@hotmail.com. *W of Nottingham city centre. From Nottingham centre, follow Derby Rd signs (A52), past St Barnabas Cathedral & just after Canning Circus, take 3rd L From M1 J25, take A52 Nottingham & after 7m, 5th R after Savoy cinema.* **Sun 18 June, Sun 16 July (11-4.30). Adm £5, chd free. Home-made teas. Visits also by arrangement June to Aug for groups of 10 to 25.**
City centre oasis, a short walk from the centre of Nottingham. Garden presented on three levels, separated by steep steps with handrails. The top terrace overlooks a large koi pond surrounded by bog plants, marginals and herbaceous perennials. Seating areas on second terrace under mature beech trees. On third level, a summerhouse as well as a small pond and densely-planted borders. Free on-street parking.

12 5A HIGH STREET
Sutton-on-Trent, NG23 6QA. Kathryn & Ian Saunders, 07827 920236, kathrynsaunders.optom@gmail.com. *6m N of Newark. Leave A1 at Sutton on Trent , follow Sutton signs. L at Xrds. 1st R turn (approx 1m) onto Main St. 2nd L onto High St. Garden 50 yds on R. Park on rd.* **Sun 30 July (1-4). Adm £5, chd free. Home-made teas. Visits also by arrangement 24 June to 31 July for groups of 15+.**
Manicured lawns are the foil for this plantsman's garden. Vistas lead past succulents to tropical areas and vibrant herbaceous beds. Ponds run through the plot, leading to woodland walks and a dedicated fernery (180+ varieties, with magnificent tree ferns). Topiary links all the different planting areas to great effect. Around 1000 named varieties of plants with interesting plant combinations.

13 1 HIGHFIELD ROAD
Nuthall, Nottingham, NG16 1BQ. Richard & Sue Bold. *4mins from J26 of the M1- 4m NW of Nottingham City. From J26 of the M1 take the A610 towards Notts, R lane at r'about, take turn-off to Horsendale. Follow rd to Woodland Dr, then 2nd R is Highfield Rd.* **Sat 11 Feb, Sun 23 Apr (10-4). Adm £4, chd free. Light refreshments.**
Visit us in Feb to see our collection of 600+ snowdrop varieties, with 300 varieties in the garden and many more in show benches. Our spring garden has lots of colour with many rare and unusual plants - miniature narcissus, acers, aconites and hellebores A good selection of unusual pots and garden ornaments. Many snowdrop varieties and other plants available for sale.

14 HILL'S FARM
Edingley, NG22 8BU. John & Margaret Hill. *Hill's Farm wildflower meadow is a short drive of ½m towards Edingley & turn R at brow of the hill as signed into tarmac farmyard for parking. Adm tickets at Old Vicarage.* **Sat 10 June (12-4). Combined adm with The Old Vicarage £7, chd free.**
Created with passion to produce a delightful walk - 6 acres of Nottinghamshire hay meadow carpeted with colourful wildflowers - ragged robin, yellow rattle, pyramidal orchids. 40+ acres of this mixed organic farm were taken out of arable production 2006 to create traditional hay meadows - hay taken in July is fed to the native beef shorthorn cattle. See some of the cows in adjoining fields.

In 2022, our donations to Carers Trust meant that 456,089 unpaid carers were supported across the UK

15 HOLLINSIDE

252 Diamond Avenue, Kirkby-In-Ashfield, Nottingham, NG17 7NA. Sue & Bob Chalkley. *1m E of Kirkby in Ashfield at the Xrds of the A611 & B6020. Please park away from the busy junction.* **Sun 11 June (1-5). Adm £3, chd free. Light refreshments.**

A formal front garden with terraced lawns and borders lead to a shaded area with ferns, camellias, roses, hydrangeas and other flowering shrubs. The rear garden has a wildlife pond, a wildflower meadow, a summerhouse and a victorian style greenhouse. There are topiary box and yews and a box parterre planted with roses. Majority of garden is suitable for wheelchairs. Disabled parking near house only by prior arrangement.

16 HOPBINE FARMHOUSE, OSSINGTON

Main Street, Ossington, NG23 6LJ. Mr & Mrs Geldart. *From A1 N take exit marked Carlton, Sutton-on-Trent, Weston etc. At T-Junction turn L to Kneesall. Drive 2m to Ossington. In village park on R in crewyard.* **Sun 2 July (2-5). Combined adm with Ossington House £6, chd free. Home-made teas in The Hut, Ossington.**

A cottage garden, half of which is walled, with roses, rambling clematis, and a varied border with unusual planting. The south garden has a long herbaceous border with many interesting varieties. A full central bed has favourite salvias, euphorbias and veronicastrum. Some narrow paths.

17 6 HOPE STREET

Beeston, Nottingham, NG9 1DR. Elaine Liquorish, 0115 9253239, eliquorish@outlook.com. *From M1 J25, A52 for Nottm. After 2 r'abouts, turn R for Beeston at The Nurseryman (B6006). Beyond hill, turn R into Bramcote Dr. 3rd turn on L into Bramcote Rd, then immed R into Hope St.* **Sun 21 May (1.30-5.30). Adm £4, chd free. Home-made teas. Visits also by arrangement 29 Apr to 12 Aug for groups of 5 to 20.**

A small garden packed with a wide variety of plants providing flower and foliage colour year-round. Collections of alpines, bulbs, mini, small and medium size hostas (60+), ferns, grasses, carnivorous plants, succulents, perennials, shrubs and trees. A pond and a greenhouse with subtropical plants. Troughs and pots. Home-made crafts. Shallow step into garden, into greenhouse and at rear. No wheelchair access to plant sales area or the back garden.

18 IVY BANK COTTAGE

The Green, South Clifton, Newark, NG23 7AG. David & Ruth Hollands. *12m N of Newark. From S, exit A46 N of Newark onto A1133 towards Gainsborough. From N, exit A57 at Newton-on-Trent onto A1133 towards Newark.* **Sun 21 May (1-5). Adm £3.50, chd free. Home-made teas.**

A traditional cottage garden; herbaceous borders, fruit trees inc a Nottinghamshire Medlar, vegetable plots and many surprises inc a stumpery, a troughery, dinosaur footprints and even fairies! Many original features: pigsties, double privy and a wash house. Children can search for animal models and explore inside the shepherd's van. Seats around and a covered refreshment area.Live music.

GROUP OPENING

19 KEYWORTH GARDENS

Keyworth, Nottingham, NG12 5AA. Graham & Pippa Tinsley. *Keyworth is a large village between the A60 & A606 7m S of Nottingham. Home Farm is near the church, Rose Cottage about ½m down Nottingham Rd & Ash Grove is off Beech Ave. Approx 15 min walk between each garden. Maps at each garden.* **Sun 4 June (12-5). Combined adm £6, chd free. Home-made teas at Home Farm and Rose Cottage.**

NEW **4 ASH GROVE**
Linda & Paul Truman.

HOME FARM HOUSE, 17 MAIN STREET
Graham & Pippa Tinsley, 0115 9377122, Graham_Tinsley@yahoo.co.uk, www.homefarmgarden.wordpress.com.
Visits also by arrangement May to Sept for groups of up to 30.

ROSE COTTAGE
Richard & Julie Fowkes, 0115 9376489, richardfowkes@yahoo.co.uk.
Visits also by arrangement May to Aug for groups of 10 to 25.

Home Farm ceased to work as a farm in 1949 and since then, a garden has evolved. The old paddock is now a wild garden of unmown grass with perennials and wildlife ponds surrounded by maturing trees. The cart shed is rebuilt as a pergola with roses and the old farm garden and orchard are divided by high hedges to create hidden places to explore. Perhaps the turf mound is an ancient monument? Rose Cottage: Densely-planted with colourful wildlife-friendly plants. Many features and different zones add unique interest. A wildlife stream meanders down between several ponds and bog areas. Some narrow paths. Ash Grove's inspiration comes from the land of the rising sun. This unique gem of a tiny garden skilfully combines many interesting features whilst retaining a sense of magic and tranquillity. Rose Cottage: Art studio open with paintings and art cards designed by Julie on sale. At Home Farm, interesting and unusual perennials for sale by Piecemeal Plants (nursery@piecemealplants.co.uk).

20 38 MAIN STREET

Woodborough, Nottingham, NG14 6EA. Martin Taylor & Deborah Bliss, 07999 786097. *About 8m NE of Nottingham centre. Turn off Mapperley Plains Rd at sign for Woodborough. Alternatively, follow signs to Woodborough off A6097 (Epperstone bypass). Property is between Park Av & Bank Hill.* **Visits by arrangement 8 May to 25 June for groups of up to 15. Light refreshments. to be discussed at booking.**

Started as a lawn at the front and the back and 125 junipers and one rosebush. Now a varied ⅓ acre. Bamboo fenced Asian species area with traditional outdoor wood fired Ofuro bath, herbaceous border, raised species rhododendron bed, vegetables, greenhouse, pond area and art studio and terrace. The art studio will be open

GROUP OPENING

21 NEW NORMANTON GARDENS

South Street, Normanton-On-Trent, Newark, NG23 6RQ. *Leave A1 at Sutton Carlton/Normanton-on-Trent junction. Turn L onto B1164 in Carlton. In Sutton-on-Trent turn R at Normanton sign. Go through Grassthorpe, turn L at Normanton sign.* **Sun 28 May (1-4). Combined adm £6, chd free. Home-made teas in Village Hall.**

NEW **CHURCH FARM**
Fiona Nelson, 01636 821340, geoff-fi.nelson@hotmail.co.uk.
Visits also by arrangement 27 May to 30 June.

LAVENDER COTTAGE
Margaret & Hedley Harper.

NEW **ORMONDE COTTAGE**
Lara Haynes.

NEW **WOODYARD HOUSE**
Jane & Craig Sutherland.

Lavender Cottage is new cottage style garden with herbaceous borders, mixed fruit and ornamental trees, and raised vegetable beds. Started in 2021 and planted with mixed perennial beds, bulbs and patio pots. Church Farm has cottage garden style rear and front gardens, with some lawn, perennials, shrubs, ornamental trees, ferns and grasses, shady borders under mature Bramley apple trees with hellebores, hostas, spring bulbs. Small pond with aquatic plants and goldfish, frogs and newts. Woodyard is a 1 acre garden with a variety of flowerbeds inc long borders, mixed borders, French bearded iris and lavender borders, a beautiful willow garden with Japanese wisteria on the front walls. There is a contained courtyard garden with roses and lavender. Yew topiary hedging and various fruit trees. There is also a large vegetable potager with greenhouse. Ormonde is a newly planted cottage and wildlife garden with emphasis on cutting beds for flowers set on quarter of an acre.

GROUP OPENING

22 NORWELL GARDENS

Newark, NG23 6JX. *6m N of Newark. Halfway between Newark & Southwell. Off A1 at Cromwell turning, take Norwell Rd at bus shelter. Or off A616 take Caunton turn.* **Sun 25 June (1-5). Evening opening Wed 28 June (6.30-9.30). Combined adm £6, chd free. Home-made teas in Village Hall (25 June) and Norwell Nurseries (28 June).**

CHERRY TREE HOUSE
Simon & Caroline Wyatt.

FAUNA FOLLIES
Lorraine & Roy Pilgrim.

JUXTA MILL
Janet McFerran.

◆ NORWELL NURSERIES
Andrew & Helen Ward.
(See separate entry)

NEW **PINFOLD COTTAGE**
Mrs Pat. Foulds.

ROSE COTTAGE
Mr Iain & Mrs Ann Gibson.

This is the 27th yr that Norwell has opened a range of different, very appealing gardens all making superb use of the beautiful backdrop of a quintessentially English countryside village. Inc a garden and nursery of national renown. To top it all there are a plethora of breathtaking village gardens showing the diversity that is achieved under the umbrella of a cottage garden description. The beautiful medieval church and its peaceful churchyard with grass labyrinth will be open for quiet contemplation.

23 ◆ NORWELL NURSERIES

Woodhouse Road, Norwell, NG23 6JX. Andrew & Helen Ward, 01636 636337, wardha@aol.com, www.norwellnurseries.co.uk. *6m N of Newark halfway between Newark & Southwell. Off A1 at Cromwell turning, take rd to Norwell at bus stop. Or from A616 take Caunton turn.* **For NGS: Sun 21 May, Sun 24 Sept (2-5). Adm £4, chd free. Home-made teas in Norwell Village Hall, Carlton Lane, on Sunday 25th June, other dates at the nurseries. Opening with Norwell Gardens on Sun 25, Wed 28 June. For other opening times and information, please phone, email or visit garden website.**

Jewel box of over 3,000 different, beautiful and unusual plants sumptuously set out in a one acre plantsman's garden inc shady garden with orchids, woodland gems, cottage garden borders, alpine and scree areas. Pond with opulently planted margins. Extensive herbaceous borders and effervescent colour themed beds. Sand beds showcase Mediterranean, North American and alpine plants. Nationally renowned nursery open with over 1,500 different rare plants for sale. Autumn opening features UK's largest collection of hardy chrysanthemums for sale and the National Collection of Hardy Chrysanthemums. New borders inc the National Collection Of Astrantias. Innovative sand beds. Grass paths, no wheelchair access to woodland paths.

NPC

24 NEW NORWOOD PARK

Halam Road, Southwell, NG25 0PF. Sir John Starkey, 01636 302099, events@norwoodpark.co.uk, www.norwoodpark.co.uk. *NW edge of Southwell. From Southwell follow brown signs to Norwood Park.* **Sun 26 Feb (11-4). Adm £7.50, chd free. Light refreshments at Norwood Park Golf Club. Hot apple punch served at the Temple.**

The grounds of Norwood Park date back to medieval times when they were part of a series of deer parks. A new garden on the south front of the C18 house was created in 2021 to showcase plants for all seasons. To the west a lime avenue lined with snowdrops and daffodils leads on to Mrs. Delaney's Path to the ornamental temple.

The National Garden Scheme searches the length and breadth of England, Wales, Northern Ireland and the Channel Islands for the very best private gardens

25 OAK BARN EXOTIC GARDEN
Church Street, East Markham, Newark, NG22 0SA. Simon Bennett & Laura Holmes, www.facebook.com/OakBarn1. *From A1 Markham Moor junction take A57 to Lincoln. Turn R at Xrds into E Markham. L onto High St & R onto Plantation Rd. Enter farm gates at T-Junction, garden located on L.* **Sun 10 Sept (1-5). Adm £3.50, chd free. Home-made teas.**
On entering the oak lych-style gate you will be met with the unexpected dense canopy of greenery and tropical foliage. The gravel paths wind under towering palms and bananas which are underplanted with cannas and gingers. On the lowest levels houseplants are bedded out from the large greenhouse to join the summer displays. They surround the Jungle Hut raised walkway.

26 OASIS COMMUNITY GARDENS
2a Longfellow Drive, Kilton Estate, Worksop, S81 0DE. Steve Williams, 07795 194957, Stevemark126@hotmail.com, www.oasiscommunitycentre.org. *Nottinghamshire. From Kilton Hill (leading to the Worksop hospital), take 1st exit to R (up hill) onto Kilton Cres, then 1st exit on R Longfellow Dr. Car Park off Dickens Rd (1st R).* **Sat 22 Apr, Sat 8 July (10-3). Adm £4, chd free. Visits also by arrangement Mar to Sept for groups of 5 to 25.**
Oasis Gardens is a community project transformed from abandoned field to an award winning garden. Managed by volunteers the gardens boast over 30 project areas, several garden enterprises and hosts many community events. Take a look in the Cactus Kingdom, the Children's pre-school play village, Wildlife Wonderland or check out a wonderful variety of trees, plants, seasonal flowers and shrubs. The Oasis Gardens hosts the first Liquorice Garden in Worksop for 100 years. The site hosts the 'Flowers for Life' project which is a therapeutic gardening project growing and selling cut flowers and floristry. There is disabled access from Longfellow Drive. From the town end there is a driveway after the first fence on the right next to house number 2.

27 THE OLD VICARAGE
Halam Hill, Halam, NG22 8AX. Mrs Beverley Perks. *1m W of Southwell. Please park diagonally into beech hedge on verge with speed interactive sign or in village - a busy road so no parking on roadside.* **Sat 20 May (12-4). Combined adm with Church House £7, chd free. Sat 10 June (12-4). Combined adm with Hill's Farm £7, chd free. Sat 1 July, Sun 6 Aug (12-4). Adm £5, chd free. Home-made teas at The Old Vicarage for all openings. Visits also by arrangement June to Aug for groups of 15+.**
An artful eye for design/texture/colour/ love of unusual plants/trees makes this a welcoming gem to visit. One time playground for children, this 2 acre hillside garden has matured over 26yrs into a much admired, landscape garden. New garden planting/design at the bottom with pond which attracts diverse wildlife. Beautiful C12 Church open only a short walk into the village or across field through attractively planted churchyard - rare C14 stained glass window. Guitar Duo Paul & Paul will be playing live at one of our openings. Short gravel drive at entrance - undulating levels as on a hillside - plenty of cheerful help available.

28 OSSINGTON HOUSE
Moorhouse Road, Ossington, Newark, NG23 6LD. Georgina Denison. *10m N of Newark, 2m off A1. From A1 N take exit marked Carlton, Sutton-on-Trent, Weston etc. At T-junction turn L to Kneesall. Drive 2m to Ossington. In village park in Hopbine crewyard on R.* **Sun 2 July (2-5). Combined adm with Hopbine Farmhouse, Ossington £6, chd free. Home-made teas in The Hut, Ossington.**
Vicarage garden redesigned in 1960 and again in 2014. Chestnuts, lawns, formal beds, woodland walk, poolside planting, orchard. Terraces, yews, grasses. Ferns, herbaceous perennials, roses and a new kitchen garden. Disabled parking available in drive to Ossington House.

29 NEW THE PALACE GARDEN OF SOUTHWELL MINSTER
Church Street, Southwell, NG25 0HD. Southwell Minster, 07890 387626, office@southwellminster.org.uk, www.southwellminster.org/theme/palace-gardens. *Centre of Southwell, Notts. Once at The Minster make your way to the South Door, opp the Archbishop's Palace, & follow the NGS yellow arrow & sign post down the footpath L for approx 40m to the garden entrance.* **Sun 28 May (11-4). Adm £5, chd free. Light refreshments in the Palace Gardens. Visits also by arrangement 3 Jan to 29 Dec. Inc tour of the gardens by Head Gardener.**
Set amongst magnificent ruins of The Archbishop's Palace the garden is packed with shrubs, flowering plants, mature trees, lawns and medieval herb garden. Sculptures/ literature provide links to the 'Leaves of Southwell', C13 stone leaf carvings - Minster's Chapter House. Separate grasslands (short walk) to Potwell Dyke are not to be missed - a spectacular array of wild flowers and many orchids. The Minster, churchyard & garden are accessible to wheelchairs. The grasslands are not accessible.

30 PARK FARM
Crink Lane, Southwell, NG25 0TJ. Ian & Vanessa Johnston, 01636 812195, v.johnston100@gmail.com. *1m SE of Southwell. From Southwell town centre go down Church St, turn R on to Fiskerton Rd & 200yds up hill turn R into Crink Lane. Park Farm is on 2nd bend.* **Sun 4 June (1-4.30). Adm £5, chd free. Home-made teas. Visits also by arrangement Apr to July for groups of up to 30. Guided visits (minimum 10 people) 50p extra per person.**
3 acre garden noted for its extensive variety of trees, shrubs and perennials, many rare or unusual. Long colourful herbaceous borders, rose arches, alpine/scree garden, large wildlife pond and area of woodland and acid loving plants. Spectacular views of the Minster across a wildflower meadow and ha-ha.

31 PATCHINGS ART CENTRE
Oxton Road, Calverton, Nottingham, NG14 6NU. Chas & Pat Wood, www.patchingsartcentre.co.uk. *N of Nottingham city take A614 towards Ollerton. Turn R on to B6386 towards Calverton & Oxton. Patchings is on L before turning to Calverton. Brown tourist directional signs.*

Sun 11 June (11-3). Adm £3, chd free. Light refreshments at Patchings Café.
Promoting the enjoyment of art. Established in 1988, Patchings is set in 50 acres with a visitor centre and galleries. The grounds are a haven for artists and a tranquil setting for visitors. The Patchings Artists' Trail - a walk through art history, famous paintings presented in glass take visitors through the centuries, meeting well known artists from the past along the way. Major exhibitions in two galleries. New Farmhouse gift shop, café and art materials and art book sale. Grass and compacted gravel paths with some undulations and uphill sections accessible to wheelchairs with help. Please enquire for assistance.

32 RHUBARB FARM

Hardwick Street, Langwith, nr Mansfield, NG20 9DR. Rhubarb Farm, 01623 741210, enquiries@rhubarbfarm.co.uk, www.rhubarbfarm.co.uk. *On NW border of Nottinghamshire in village of Nether Langwith. From A632 in Langwith, by bridge (single file traffic) turn up steep Devonshire Drive. N.B. Then turn off SatNav. Take 2nd L into Hardwick St. Rhubarb Farm at end. Parking to R of gates.* **Thur 18 May, Thur 22 June, Thur 20 July, Thur 17 Aug (10-3). Adm £2.50, chd free. Home-made teas in our on-site café, made by Rhubarb Farm volunteers. Visits also by arrangement May to Oct for groups of 10 to 15. At least a month's notice for group booking is appreciated.**

This 2 acre horticultural social enterprise provides training and volunteering opportunities to 60 ex-offenders, drug and alcohol misusers, older people, school students, people with mental and physical ill health and learning disabilities. Eight polytunnels, 100 hens, pigs, donkey and a Shetland pony. Forest school barn, willow dome and arch, large Keder polytunnel (bubblewrap walls), flower borders, small shop, pond, raised beds, comfrey bed and comfrey fertiliser factory, junk sculpture, Heath Robinson Feature. Chance to meet and chat with volunteers and staff. Main path suitable for wheelchairs but bumpy. Not all site accessible. Cafe & composting toilet wheelchair-accessible. Electric mobility scooter.

Felley Priory

33 RISEHOLME, 125 SHELFORD ROAD
Radcliffe on Trent, NG12 1AZ. John & Elaine Walker, 01159 119867, elaine.walker10@hotmail.co.uk. *4m E of Nottingham. From A52 follow signs to Radcliffe. In village centre take turning for Shelford (by Co-op). Approx ¾m on L.* **Visits by arrangement June & July for groups of 12 to 50. Adm £5, chd free. Home-made teas.**
Imaginative and inspirational is how the garden has been described by visitors. A huge variety of perennials, grasses, shrubs and trees combined with an eye for colour and design. Jungle area with exotic lush planting contrasts with tender perennials particularly salvias thriving in raised beds and in gravel garden with stream. Unique and interesting objects complement planting. Artwork, fairies and dragons feature in the garden!

34 NEW 160 SOUTHWELL ROAD WEST
Mansfield, NG18 4HB. Barrie & Lynne Jackson. *Approx half way between Mansfield & Rainworth on A1691. From A60 towards Mansfield, turn R at T-lights up Berry Hill Lane. At the T junction turn L. We are on the service road on the L. No parking on the service road.* **Sun 30 Apr (1-4.30). Adm £4, chd free. Home-made teas.**
The ⅓ acre garden was started from scratch in spring 2018. We have a range of growing environments inc a woodland garden, a border designed to cope with sun, large island beds, a vegetable garden and a wide range of climbers growing up a variety of structures. Our particular interests are Agapanthus, woodland plants inc Trillium, plus fruit and vegetables. A sloping site which is accessible with assistance. Some grass paths.

GROUP OPENING

35 SUTTON BONINGTON GARDENS
Main Street, Sutton Bonington, Loughborough, LE12 5PE. *2m SE of Kegworth (M1 J24). 6m NW of Loughborough. From A6 Kegworth, past University of Nottingham Sutton Bonington campus. From Loughborough, A6 then A6006. From Nottingham, A60 then A6006.* **Sat 13, Sun 14 May (2-6). Combined adm £6, chd free. Home-made teas at 118 Main Street.**

FORGE COTTAGE
Judith & David Franklin.

118 MAIN STREET
Alistair Cameron & Shelley Nicholls.

PIECEMEAL
Mary Thomas.

All 3 gardens are located at the north end of the village, near St Michael's Church and just a few minutes' walk from each other. Piecemeal has a tiny walled garden featuring a wide range of unusual shrubs, most in terracotta pots bordering narrow paths. Also climbers, perennials and even a few trees! Focus is on distinctive form, foliage shape and colour combination. Half-hardy and tender plants fill the conservatory. Forge Cottage garden, reclaimed from a blacksmith's yard and a little larger, has vibrant, curved herbaceous borders. 118 Main Street is large with well-established trees and varied planting as well as an orchard/wildflower meadow. Varied features inc Japanese-inspired seating area, redesigned patio with water feature, as well as more traditional established borders surrounding lawn and pond. Attractive greenhouse.

36 NEW SYCAMORES HOUSE
Salmon Lane, Annesley Woodhouse, Nottingham, NG17 9HB. David & Julie Scarle. *From M1 J27 follow Mansfield signs to Badger Box T-lights. Turn L. Straight 1km to hill crest. Gate on L just by 'No footway for 600 yds' sign.* **Sun 18 June (12.30-5). Adm £5, chd free. Light refreshments.**
Visitors will discover a 1⅓ acre garden that gently slopes from formal flower beds to well-stocked borders to informal woodland walks, with a number of secret pathways. There is also a large vegetable garden, polytunnel and fruit cage. David and Julie took on the garden as a retirement project but have not yet retired, so the garden is a work in progress. Julie is registered blind and particularly enjoys the sensory elements of the garden. Children's trail for under 5s. Gravel drive and paths, shallow steps nr house. Short grass slope to access informal garden. Wood chip on woodland walk, allotment and secret paths.

37 THRUMPTON HALL
Thrumpton, NG11 0AX. Miranda Seymour, www.thrumptonhall.com. *7m S of Nottingham. M1 J24 take A453 towards Nottingham. Turn L to Thrumpton village & cont to Thrumpton Hall.* **Sun 18 June (1.30-4.30). Adm £6, chd free. Home-made teas.**
2 acres inc lawns, rare trees, lakeside walks, flower borders, rose garden, new pagoda, and box-bordered sunken herb garden, all enclosed by C18 ha-ha and encircling a Jacobean house. Garden is surrounded by C18 landscaped park and is bordered by a river. Rare opportunity to visit Thrumpton Hall (separate ticket). Jacobean mansion, unique carved staircase, Great Saloon, State Bedroom, Priest's Hole.

38 NEW THYME HOUSE
Leverton Road, Retford, DN22 0DR. Colin Bowler. *Go up Spital Hill Retford after the Canalside workshops go over the canal humpback bridge, turn R. House has a brick wall at the gate.* **Sat 10, Sat 17, Sat 24 June (10.30-4). Adm £4, chd free. Pre-booking essential, please phone 01777 948620 for information & booking. Light refreshments.**
An opportunity to see an imaginative 2 acre garden in early development, started in 2018 from an open lawn. The garden consists of islands of unusual shrubs and perennials, with many original features such as a large sunken firepit, a moongate entranced pond area, and sculptures. Plenty of seating in shaded gazebos or sun. All grassed area, but no steps or steep hills.

39 UNIVERSITY PARK GARDENS
Nottingham, NG7 2RD. University of Nottingham, www.nottingham.ac.uk/estates/grounds/. *Approx 4m SW of Nottingham city centre & opp Queens Medical Centre. NGS visitors: Please purchase adm tickets online or in the Millennium Garden (in centre of campus), signed from N & W entrances to University Park & within internal road network.*

Sun 13 Aug (11.30-4). Adm £5, chd free. Light refreshments in Lakeside Pavilion Kiosk & Café. Also adjacent to the Millennium Garden.
University Park has many beautiful gardens inc the award-winning Millennium Garden with its dazzling flower garden, timed fountains and turf maze. Also the huge Lenton Firs rock garden, and the Jekyll garden. During summer, the Walled Garden is alive with exotic plantings. In total, 300 acres of landscape and gardens. Picnic area, café, walking tours, information desk, workshop, accessible minibus only to feature gardens within campus. Plants for sale in Millennium garden. Some gravel paths and steep slopes. Very large site.

40 UPPER GROVE FARM

Norwell Woodhouse, Newark, NG23 6NG. Kathryn Wiltshire. *A616 either from Newark (10m) or Ollerton (7m) at Xroads turn to Laxton R to Norwell 1st white farmhouse on L, look out for the big chimney pots!* **Sun 12 Mar (12-4). Adm £5, chd free. Light refreshments.**
We are a country garden around a farmhouse. We have a small orchard and informal beds of spring/summer flowers. Lots of daffodils. Most of garden accessible to wheelchairs, some gravel paths.

41 NEW WAXWINGS & GOLDCREST

Lamins Lane, Bestwood Village, Nottingham, NG6 8WS. Robert Carlyle, 07495 934449, robcarlyle@me.com. *A60 N of Redhill r'about. Take Lamins Lane turn. Single track road, limited passing places. Our properties are 1m along the lane on the L.* **Visits by arrangement 14 July to 14 Aug for groups of 20+. Adm £5, chd free. Light refreshments.**
Our six acre gardens for wildlife are open by invitation for groups. Orchard, meadows, prairie beds, woodland, stumpery, vegetable and fragrant gardens.

GROUP OPENING

42 WELLOW VILLAGE GARDENS

Potter Lane, Wellow, Newark, NG22 0EB. *12m NW of Newark. Wellow is on A616 between Ollerton & Newark approx 1m from Ollerton. Parking available around village green & as indicated by car park signs.* **Sun 21 May (1-5). Combined adm £5, chd free. Home-made teas. Donation to Wellow Memorial Hall.**

4 POTTER LANE
Anne & Fred Allsop.

TITHE BARN
Andrew & Carrie Young.

Wellow, formerly Wellah, from the number of wells, has a green and a famous maypole. Not so well known is the Wellow Dyke, which surrounds the village and can still be seen in some places. Now come and see some Wellow gardens: 4 Potter Lane: Created over more than 20 yrs, this series of gardens has a section of the Wellow Dyke at the top and has a well tended vegetable garden. Shade garden at side and colourful pleasure garden in centre. Tithe Barn: The approach to this garden between yew hedges sets the tone to the wide sweeps of lawn and generous terrace. Herbaceous beds, roses and rambling roses. Mature planting of shrubs and weeping trees. Woodland bank on Wellow Dyke open for the first time this year. Tithe Barn: Wheelchair access to the majority of the garden. No wheelchair access to patio or Wellow Dyke.

43 6 WESTON CLOSE

Woodthorpe, Nottingham, NG5 4FS. Diane & Steve Harrington, 0115 9857506, mrsdiharrington@gmail.com. *3m N of Nottingham. A60 Mansfield Rd. Turn R at T-lights into Woodthorpe Dr. 2nd L Grange Rd. R into The Crescent. R into Weston Close. Please park on The Crescent.* **Sun 18 June (1-5). Adm £3, chd free. Home-made teas. inc gluten free options. Visits also by arrangement June & July for groups of 10 to 35. Adm inc tea or coffee & home-made cake.**
Set on a substantial slope with three separate areas surrounding the house. Dense planting creates a full, varied yet relaxed display inc many scented roses, clematis and a collection of over 90 named mature hostas in the impressive colourful rear garden. Walls covered by many climbers. Home propagated plants for sale.

Norwood Park

OXFORDSHIRE

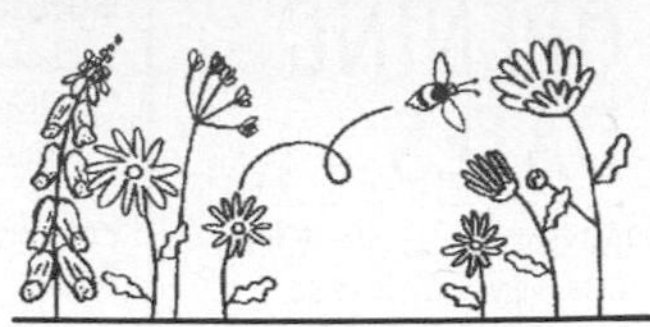

In Oxfordshire we tend to think of ourselves as one of the most landlocked counties, right in the centre of England and furthest from the sea.

We are surrounded by Warwickshire, Northamptonshire, Buckinghamshire, Berkshire, Wiltshire and Gloucestershire, and, like these counties, we benefit from that perfect British climate which helps us create some of the most beautiful and famous gardens in the world.

Many gardens open in Oxfordshire for the National Garden Scheme between spring and late-autumn. Amongst these are the perfectly groomed college gardens of Oxford University, and the grounds of stately homes and palaces designed by a variety of the famous garden designers such as William Kent, Capability Brown, Rosemary Verey, Tom Stuart-Smith and the Bannermans of more recent fame.

But we are also a popular tourist destination for our honey-coloured mellow Cotswold stone villages, and for the Thames which has its spring near Lechlade. More villages open as 'groups' for the National Garden Scheme in Oxfordshire than in any other county, and offer tea, hospitality, advice and delight with their infinite variety of gardens.

All this enjoyment benefits the excellent causes that the National Garden Scheme supports.

Volunteers

County Organiser
Marina Hamilton Baillie
01367 710486
marina.hamilton-baillie@ngs.org.uk

Treasurer
Tom Hamilton-Baillie
01367 710486
tom.hamilton-baillie@ngs.org.uk

Talks Co-Ordinator
Priscilla Frost 01608 811818
info@oxconf.co.uk

Dr David Edwards
07973 129473
david.edwards@ngs.org.uk

Social Media- Instagram
Dr Jill Edwards 07971 201352
jill.edwards@ngs.org.uk

Social Media & Marketing
John Fleming 01865 739327
john@octon.scot

Assistant County Organisers
Lynn Baldwin 01608 642754
elynnbaldwin@gmail.com

Dr David Edwards (as above)

Dr Jill Edwards (as above)

Penny Guy 01865 862000
penny.theavon@virginmedia.com

Michael & Pat Hougham
01865 890020
mike@gmec.co.uk
pat@gmec.co.uk

Lyn Sanders
01865 739486
sandersc4@hotmail.com

Paul Youngson 07946 273902
paulyoungson48@gmail.com

@NGSOxfordshire
@ngs_oxfordshire
@ngs_oxfordshire

Left: Home Close

OPENING DATES

All entries subject to change. For latest information check **www.ngs.org.uk**

Map locator numbers are shown to the right of each garden name.

February

Snowdrop Festival

Sunday 12th
23 Hid's Copse Road 34
6 High Street 35

Sunday 26th
Lime Close 44

March

Sunday 12th
◆ Waterperry Gardens 71

April

Sunday 2nd
Ashbrook House 1
Buckland Lakes 11

Friday 7th
Midsummer House 50
Sarsden Glebe 61

Monday 10th
Kencot Gardens 39

Sunday 16th
38 Leckford Road 43
Magdalen College 45

Saturday 22nd
Central North Oxford Gardens 14
Claridges Barn 20
Sandys House 60

Sunday 23rd
Broughton Poggs & Filkins Gardens 10
Central North Oxford Gardens 14
Claridges Barn 20
Sandys House 60

Saturday 29th
Central North Oxford Gardens 14

Sunday 30th
◆ Broughton Grange 9
Central North Oxford Gardens 14

May

Friday 5th
Midsummer House 50

Sunday 7th
9 Rawlinson Road 57
11 Rawlinson Road 58

Sunday 14th
◆ Blenheim Palace 4
Old Boars Hill Gardens 53

Sunday 21st
Church Farm Field 19
Headington Gardens 33
Lime Close 44
NEW Upper Bolney House 69
Westwell Manor 74

Wednesday 24th
Kingham Lodge 41

Friday 26th
Kidmore House Garden & Vineyard 40

Sunday 28th
Barton Abbey 3
Bolters Farm 5
Kings Cottage 42

Monday 29th
Bolters Farm 5
Kings Cottage 42

June

Friday 2nd
Midsummer House 50
Woolstone Mill House 77

Sunday 4th
Chestlion House 16
Friars Court 28
Steeple Aston Gardens 65
Wheatley Gardens 75
Whitehill Farm 76

Wednesday 7th
Claridges Barn 20

Thursday 8th
Wootton Gardens 78

Saturday 10th
Claridges Barn 20
Sandys House 60

Sunday 11th
Brize Norton Gardens 6
NEW Burford Gardens 12
Claridges Barn 20
Cumnor Village Gardens 21
Failford 26
Friars Court 28
Iffley Gardens 38
116 Oxford Road 54
Sandys House 60
West Oxford Gardens 73

Sunday 18th
◆ Broughton Grange 9
NEW Dean Manor 22
Middleton Cheney Gardens 49
The Priory, Charlbury 55

Monday 19th
NEW Dean Manor 22

Wednesday 21st
Church Farm Field 19

Thursday 22nd
◆ Stonor Park 66

Sunday 25th
◆ Broughton Castle 8
Chalkhouse Green Farm 15
Sibford Gardens 62

July

Saturday 1st
Stow Cottage Arboretum & Garden 67

Sunday 2nd
Dorchester Gardens 25
Green and Gorgeous 30

Friday 7th
Midsummer House 50

Sunday 9th
NEW Pusey House 56

Sunday 16th
113 Brize Norton Road 7

Sunday 23rd
Merton College Oxford Fellows' Garden 48

Sunday 30th
◆ Broughton Grange 9

August

Friday 4th
Midsummer House 50

Saturday 19th
Aston Pottery 2

Sunday 20th
Aston Pottery 2

September

Friday 1st
Midsummer House 50

Saturday 2nd
Christ Church Masters', Pocock & Cathedral Gardens 18

Sunday 3rd
Ashbrook House 1
Southbank 64

Sunday 10th
◆ Broughton Grange 9
Trinity College 68

Sunday 17th
Bolters Farm 5
Kings Cottage 42
◆ Waterperry Gardens 71

October

Friday 6th
Midsummer House 50

By Arrangement

Arrange a personalised garden visit with your club, or group of friends, on a date to suit you. See individual garden entries for full details.

Ashbrook House 1
Aston Pottery 2
Bolters Farm 5
Bush House 13
Carter's Yard, Sibford Gardens 62
Chalkhouse Green Farm 15
Chivel Farm 17
NEW Dean Manor 22
103 Dene Road 23
Denton House 24
Dorchester Manor House 46
Failford 26
Foxington 27
The Grange 29
Greenfield Farm 31
Greyhound House 32

Dean Manor

THE GARDENS

1 ASHBROOK HOUSE

Westbrook St, Blewbury, OX11 9QA. Mr & Mrs S A Barrett, 01235 850810, janembarrett@me.com. *4m SE of Didcot. Turn off A417 in Blewbury into Westbrook St. 1st house on R. Follow yellow signs for parking in Boham's Rd.* **Sun 2 Apr, Sun 3 Sept (2-5.30). Adm £5, chd free. Home-made teas. Visits also by arrangement for groups of 10+.**

The garden where Kenneth Grahame read Wind in the Willows to local children and where he took inspiration for his description of the oak doors to Badger's House. Come and see, you may catch a glimpse of Toad and friends in this 3½ acre chalk and water garden, in a beautiful spring line village. In spring the banks are a mass of daffodils and in late summer the borders are full of unusual plants.

2 ASTON POTTERY

Aston, Bampton, OX18 2BT. Mr Stephen Baughan, 01993 852031, info@astonpottery.co.uk, www.astonpottery.co.uk. *4m S of Witney. On the B4449 between Bampton & Standlake.* **Sat 19, Sun 20 Aug (9.30-5). Adm £5, chd free. Visits also by arrangement 19 Aug to 20 Aug.**

6 stunning borders set around Aston Pottery. 72 metre double hornbeam border full of riotous perennials. 80 metre long hot bank of alstroemeria, salvias, echinacea and knifophia. Quadruple dahlia border with over 600 dahlias, grasses and asters. Tropical garden with bananas, cannas and ricinus. Finally, 80 metres of 120 different annuals planted in four giant successive waves of over 6000 plants.

3 BARTON ABBEY

Steeple Barton, OX25 4QS. Mr & Mrs P Fleming. *8m E of Chipping Norton. On B4030, ½m from junc of A4260 & B4030.* **Sun 28 May (2-5). Adm £5, chd free. Home-made teas.**

15 acre garden with views from house (not open) across sweeping lawns and picturesque lake. Walled garden with colourful herbaceous borders, separated by established yew hedges and espalier fruit, contrasts with more informal woodland garden paths with vistas of specimen trees and meadows. Working glasshouses and fine display of fruit and vegetables.

4 ◆ BLENHEIM PALACE

Woodstock, OX20 1PX. His Grace the Duke of Marlborough, 01993 810530, customerservice@blenheimpalace.com, www.blenheimpalace.com. *8m N of Oxford. The S3 bus runs every 30 mins from Oxford Train Stn & Oxford's Gloucester Green Bus Stn to Blenheim. Oxford Bus Company No. 500 departs from Oxford Parkway & stops at Blenheim.* **For NGS: Sun 14 May (10-4). Adm £5, chd free. For other opening times and information, please phone, email or visit garden website.**

Blenheim Gardens, originally laid out by Henry Wise, inc the formal Water Terraces and Italian Garden by Achille Duchêne, Rose Garden, Arboretum, and Cascade. The Secret Garden offers a stunning garden paradise in all seasons. Blenheim Lake, created by Capability Brown and spanned by Vanburgh's Grand Bridge, is the focal point of over 2,000 acres of landscaped parkland. The Walled Gardens complex inc the Herb and Lavender Garden and Butterfly House. Other activities inc the Marlborough Maze, adventure play area, giant chess and draughts. Wheelchair access with some gravel paths, uneven terrain and steep slopes. Dogs allowed in parkland and East Courtyard only.

Kingham Lodge

5 BOLTERS FARM
Pudlicote Lane, Chilson, Chipping Norton, OX7 3HU. Robert & Amanda Cooper, 07778 476517, art@amandacooper.co.uk. *Centre of Chilson village. On arrival in the hamlet of Chilson, heading N, we are the last in an old row of cottages on R. Please drive past & park considerately on the L in the lane. Limited parking, car sharing recommended.* **Sun 28, Mon 29 May, Sun 17 Sept (2-5). Combined adm with Kings Cottage £8, chd free. Light refreshments. Gluten free options. Visits also by arrangement in June for groups of 5 to 20. Donation to Hands Up Foundation.**
A cherished old cottage garden restored over the last 16 yrs. Tumbly moss covered walls and sloping lawns down to a stream with natural planting and quirky character. Running water, weeping willows and a sense of peace.

GROUP OPENING

6 BRIZE NORTON GARDENS
Brize Norton, OX18 3LY. www.bncommunity.org/ngs. *3m SW of Witney. Brize Norton Village, S of A40, between Witney & Burford. Parking at Elderbank Hall. Coaches welcome with plenty of parking nearby. Tickets & maps available at Elderbank Hall & at each garden.* **Sun 11 June (1-6). Combined adm £7.50, chd free. Home-made teas in Elderbank Village Hall & at Grange Farm.**

BARNSTABLE HOUSE
Mr & Mrs P Butcher.

17 CHICHESTER PLACE
Mr & Mrs D Howard.

CHURCH FARM HOUSE
Philip & Mary Holmes,

CLUMBER
Mr & Mrs S Hawkins.

GRANGE FARM
Mark & Lucy Artus.

MILLSTONE
Bev & Phil Tyrell.

PAINSWICK HOUSE
Mr & Mrs T Gush.

ROSE COTTAGE
Brenda & Brian Trott.

95 STATION ROAD
Mr & Mrs P A Timms.

STONE COTTAGE
Mr & Mrs K Humphris.

Doomsday village on the edge of the Cotswold's offering a number of gardens open for your enjoyment. You can see a wide variety of planting inc ornamental trees and grasses, herbaceous borders, traditional fruit and vegetable gardens. Features inc a Mediterranean style patio, courtyard garden, terraced roof garden, water features; plus gardens where you can just sit, relax and enjoy the day. Plants for sale at individual gardens. A Flower Festival will take place in the Brize Norton St Britius Church. Partial wheelchair access to some gardens.

7 113 BRIZE NORTON ROAD
Minster Lovell, Witney, OX29 0SQ. David & Lynn Rogers. *Approx 2½m W of Witney. On main village rd (B4447), ½m N of A40 Witney Bypass intersection on the R. Parking at nearby Scout hut.* **Sun 16 July (2-5). Adm £5, chd free. Home-made teas.**
Mixed variety garden of over 1 acre inc quirky features for added interest. Lawns, meadow grass area, small woodland, native mixed hedging, wildlife pond, fish pond, tree ferns, acers, huge mix of plants and trees with emphasis on year-round colour. The whole garden has been developed with wildlife in mind. This garden will feature on a new series 'Love Your Garden' with Alan Titchmarsh. Gravel and grass surfaces.

8 ◆ BROUGHTON CASTLE
Banbury, OX15 5EB. Lord Saye and Sele, 01295 276070, info@broughtoncastle.com, www.broughtoncastle.com. *2½m SW of Banbury. On Shipston-on-Stour road (B4035).* **For NGS: Sun 25 June (2-4.30). Adm £10, chd free. Home-made teas. For other opening times and information, please phone, email or visit garden website.**
1 acre; shrubs, herbaceous borders, walled garden, roses, climbers seen against background of C14-C16 castle surrounded by moat in open parkland. House also open (additional charge).

9 ◆ BROUGHTON GRANGE
Wykham Lane, Broughton, Banbury, OX15 5DS. S Hester, 07791 747371, enquiries@broughtongrrange.com, www.broughtongrange.com. *¼m out of village. From Banbury take B4035 to Broughton. Turn L at Saye & Sele Arms Pub up Wykham Ln (one way). Follow road out of village for ¼m. Entrance on R.* **For NGS: Sun 30 Apr, Sun 18 June, Sun 30 July, Sun 10 Sept (10-5). Adm £10, chd free. Light refreshments. For other opening times and information, please phone, email or visit garden website.**
An impressive 25 acres of gardens and light woodland in an attractive Oxfordshire setting. The centrepiece is a large terraced walled garden created by Tom Stuart-Smith in 2001. Vision has been used to blend the gardens into the countryside. Good early displays of bulbs followed by outstanding herbaceous planting in summer. Formal and informal areas combine to make this a special site inc newly laid arboretum with many ongoing projects.

National Garden Scheme gardens are identified by their yellow road signs and posters. You can expect a garden of quality, character and interest, a warm welcome and plenty of home-made cakes!

GROUP OPENING

10 BROUGHTON POGGS & FILKINS GARDENS

Lechlade, GL7 3JH. www.filkins.org.uk. *3m N of Lechlade. 5m S of Burford. Just off A361 between Burford & Lechlade on the B4477. Map of the gardens available.* **Sun 23 Apr (2-6). Combined adm £8, chd free. Home-made teas at Filkins Village Hall. Ice creams at village shop.**

BROUGHTON POGGS MILL
Charlie & Avril Payne.

3 THE COACH HOUSE
Peter & Bea Berners-Price.

THE CORN BARN
Ms Alexis Thompson.

THE FIELD HOUSE
Peter & Sheila Gray.

FILKINS ALLOTMENTS
Filkins Allotments.

FILKINS HALL
Filkins Hall Residents.

LITTLE PEACOCKS
Colvin & Moggridge.

MUFFITIES
Arthur Parkinson.

PEACOCK FARMHOUSE
Pauline & Peter Care.

PIGEON COTTAGE
Lynne Savege.

PIP COTTAGE
G B Woodin.

THE TALLOT
Ms M Swann & Mr D Stowell.

TAYLOR COTTAGE
Mrs Ronnie Bailey.

12 gardens and flourishing allotments in these beautiful and vibrant Cotswold stone twin villages. Scale and character vary from the grand landscape setting of Filkins Hall, to the small but action packed Pigeon Cottage, Taylor Cottage, Muffities and The Tallot. Broughton Poggs Mill has a rushing mill stream with an exciting bridge; Pip Cottage combines topiary, box hedges and a fine rural view. In these and the other equally exciting and varied gardens horticultural interest abounds. Features inc Swinford Museum of Cotswolds tools and artefacts, and Cotswold Woollen Weavers. Many gardens have gravel driveways, but most are suitable for wheelchair access. Most gardens welcome dogs on leads.

11 BUCKLAND LAKES

nr Faringdon, SN7 8QW. The Wellesley Family. *3m NE of Faringdon. Buckland is midway between Oxford (14m) & Swindon (15m), just off the A420. Faringdon 3m, Witney 8m. Follow yellow signs which will lead you to driveway & car park by St Mary's Church.* **Sun 2 Apr (2-5). Adm £6, chd free. Home-made teas at Memorial Hall. Donation to Buckland Memorial Hall.**

Descend down wooded path to 2 large secluded lakes with views over undulating historic parkland, designed by Georgian landscape architect Richard Woods. Picturesque mid C18 rustic icehouse, cascade with iron footbridge, thatched boathouse and round house, and exedra. Many fine mature trees, drifts of spring bulbs and daffodils amongst shrubs. Norman church adjoins. Cotswold village. Children must be supervised due to large expanse of unfenced open water.

GROUP OPENING

12 NEW BURFORD GARDENS

Burford, OX18 4SJ. *Swan Lane, Burford. Parking in the public car park signed near the Church. Walk up Guildenford & cross Witney St. Up Pytts Ln & access to The Lodge for ticket sales is on the junc of Pytts Ln and Swan Ln.* **Sun 11 June (2-6). Combined adm £7.50, chd free. Home-made teas.**

NEW **CYGNET LODGE**
Alice Temple-Bruce.

NEW **33 FRETHERN CLOSE**
Miss Michelle Thornton.

NEW **THE LODGE**
Hugo and Sue Ashton.
D

NEW **4 SWAN LANE CLOSE**
David & Jan Cohen.

NEW **WALNUT TREE COTTAGE**
Lorna ans Sarah Millard.
D

This group comprise of 5 very different gardens located close together in the attractive Cotswold town of Burford. Their variety of size and character provides a wide range of planting and design styles in a beautiful setting. The Lodge, from where tickets and teas are available, has a well established family garden with a distinctive glasshouse at its centre, divided into several defined areas set on a sloping site. The compact, cottage-style garden at 4, Swan Close accommodates herbaceous perennials, grasses, vegetables and herbs, a mini orchard and water features. The owners of Cygnet Lodge have built a highly individualistic garden, comprising individual 'rooms', filled with unexpected detail. The newly designed and built garden on Walnut Tree Cottage's steep site is generously planted with roses, herbaceous perennials, woodland plants and herbs, with a backdrop of mature trees. This garden was designed by Susan Ashton (MSGD), who also owns The Lodge. Opening for the 1st time this year, 33 Frethern Close is tucked away in an unlikely part of Burford. It is an interesting small garden with many trees, shrubs and exotic plants.

13 BUSH HOUSE

Wigginton Road, South Newington, Banbury, OX15 4JR. Mr John Ainley, 07503 361050, rojoainley@btinternet.com. *In S Newington on A361 from Banbury to Chipping Norton, take 1st R to Wigginton, Bush House 1st house on the L in Wigginton Rd.* **Visits by arrangement 30 Jan to 29 Oct. Adm £5, chd free.**

Set in 8 acres, over 10 yrs a 2 acre garden has emerged. Herbaceous borders partner dual level ponds and stream. The terrace leads to a walled parterre framed by roses and wisteria. The orchard is screened by rose and vine covered wrought iron trellis. Kitchen gardens, greenhouses and fruit cage provide organically grown produce. Wildflowers sown in the orchard in Aug 2021. Stream and interconnecting ponds. Walled parterre and knot garden. 1000 native broadleaved trees planted 2006, 2011 and 2014. Gravel drive, a few small steps, and two gentle grass slopes on either side of the garden.

GROUP OPENING

14 CENTRAL NORTH OXFORD GARDENS

Leckford Road, Oxford, OX2 6HY. *City gardens situated in parallel streets; in Leckford Rd & Plantation Rd. Access from Kingston Rd or Woodstock Rd. Parking in Leckford Rd, Warnborough Rd & Farndon Rd.* **Sat 22, Sun 23, Sat 29, Sun 30 Apr (2-5). Combined adm £5, chd free. Tea at 50 Plantation Road.**

41 LECKFORD ROAD
Liz & Mark Jennings.

50 PLANTATION ROAD
Philippa Scoones.

Two very different town gardens. 50 Plantation Rd is a surprisingly wide and deep garden in two parts: lovely trees and planting and lots of pots; and an unusual water feature with stepping stones. Many variety of tulips. In Leckford Rd is an exquisitely designed garden for entertaining and relaxing. Hedged with pleached hornbeams is a foreground for lovely spring planting. Every area has vegetation suited to its microclimate. Plants for sale at 50 Plantation Road.

15 CHALKHOUSE GREEN FARM

Chalkhouse Green, Kidmore End, Reading, RG4 9AL. Mr J Hall, 01189 723631, chgfarm@gmail.com, www.chgfarm.com. *2m N of Reading, 5m SW of Henley-on-Thames. Situated between A4074 & B481. From Kidmore End take Chalkhouse Green Rd. Follow yellow signs.* **Sun 25 June (2-6). Adm £4, chd free. Home-made teas. Visits also by arrangement May to Sept for groups of 5 to 15.**

1 acre garden and open traditional farmstead. Herbaceous borders, herb garden, shrubs, old fashioned roses, trees inc medlar, quince and mulberries, walled ornamental kitchen garden and cherry orchard. Rare breed farm animals inc British White cattle, Suffolk Punch horses, donkeys, geese, chickens, ducks and turkeys. Plant and jam stall, donkey rides, swimming in covered pool, grass tennis court, trailer rides, farm trail, WWII bomb shelter, heavy horse and bee display. Partial wheelchair access.

Our annual donations to Parkinson's UK meant 7,000 patients are currently benefiting from support of a Parkinson's Nurse

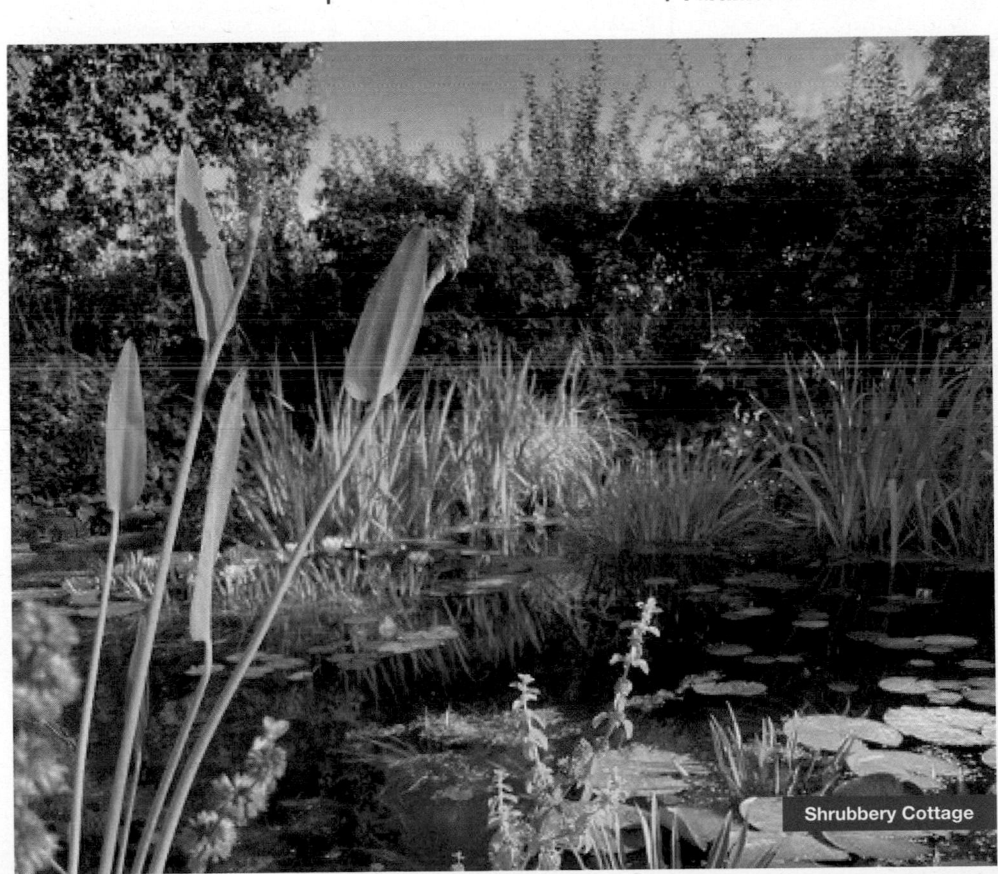

Shrubbery Cottage

16 CHESTLION HOUSE

Chestlion Lane, Bourton Road, Clanfield, Bampton, OX18 2PA. Louise Johnson. *3m S of Carterton. Chestlion Ln is off A4020 Bourton Rd in the village of Clanfield.* **Sun 4 June (2-6). Adm £6, chd free. Home-made teas. Open nearby Friars Court.**

The garden surrounding the house (not open), has been planted since the new owners took on the house 10yrs ago. The garden to the rear of the house, has approx ¾ acre of mixed planting, with many blowsy roses adorning the walls and soft romantic planting in borders. There is a circular meditative garden and a round pond along with a productive kitchen, cutting garden, greenhouse and herb garden. There is a boat house and pond outside the walled area and a newly planted wildflower meadow with a fruit tree avenue and a hornbeam avenue leading to the back entrance.

17 CHIVEL FARM

Heythrop, OX7 5TR. John & Rosalind Sword, 01608 683227, rosalind.sword@btinternet.com. *4m E of Chipping Norton. Off A361 or A44. Parking at Chivel Farm.* **Visits by arrangement 18 Feb to 30 Sept for groups of 5+.**

Beautifully designed country garden with extensive views, designed for continuous interest that is always evolving. Colour schemed borders with many unusual trees, shrubs and herbaceous plants. Small formal white garden and a conservatory.

18 CHRIST CHURCH MASTERS', POCOCK & CATHEDRAL GARDENS

St Aldate's, Oxford, OX1 1DP. Christ Church, www.chch.ox.ac.uk/gardens-and-meadows. *5 mins walk from Oxford city centre. Entry from St Aldate's through the Memorial Gardens, into Christ Church Meadow, then turn L into Masters' Garden gate after main visitor entrance. No parking available.* **Sat 2 Sept (10-4). Adm £5, chd free.**

3 walled gardens, not normally open to visitors, with herbaceous, shrub, Mediterranean and tropical borders. Inc the magnificent 'Jabberwocky' Tree, an Oriental Plane planted in the mid 1600s, as well as other links to Alice in Wonderland, St Frideswide and Harry Potter. Wheelchair access over gravel paths.

19 CHURCH FARM FIELD

Church Lane, Epwell, Banbury, OX15 6LD. Mrs D Castle. *7½m W of Banbury on N side of Epwell village.* **Sun 21 May, Wed 21 June (2-6). Adm £2, chd free. Light refreshments. Tea & cake.**

Woodland and an arboretum with wild flowers (planting started 1992). There are over 90 different trees and shrubs within the 4½ acres. Paths cut through trees for access to various parts of the garden. There is also a lawn tennis court and croquet lawn.

20 CLARIDGES BARN

Charlbury Road, Chipping Norton, OX7 5XG. Drs David & Jill Edwards. *3m SE of Chipping Norton. Take B4026 from Chipping Norton to Charlbury after 3m turn R to Dean, we are 200 metres on the R. Please park on the verge.* **Sat 22, Sun 23 Apr (11-5). Combined adm with Sandys House £5, chd free. Wed 7 June (11-5). Adm £5, chd free. Sat 10, Sun 11 June (11-5). Combined adm with Sandys House £5, chd free. Home-made teas. Also serving coffee & light lunches.**

3½ acres of family garden, wood and meadow hewn from a barley field on limestone brash. Situated on top of the Cotswolds, it is open to all weathers, but rewarding views and dog walking opportunities on hand. Large vegetable, fruit and cutting garden, wildlife pond and 5 cedar greenhouses, all loved by rabbits, deer and squirrel. Herbaceous borders and woodland gardens with gravel and flagged paths, divided by stone walls. Claridges Barn dates back to the 1600s, the cottage 1860s, converted about 30 yrs ago. Plants for sale.

GROUP OPENING

21 CUMNOR VILLAGE GARDENS

Leys Road, Cumnor, Oxford, OX2 9QF. *4m W of central Oxford. From A420, exit for Cumnor & follow B4017 into the village. Parking on rd & side rds, behind PO, or behind village hall in Leys Rd.* **Sun 11 June (2-5.30). Combined adm £5, chd free. Home-made teas in United Reformed Church Hall, Leys Road.**

6 HIGH STREET
Dr Dianne & Prof Keith Gull.
(See separate entry)

19 HIGH STREET
Janet Cross.

10 LEYS ROAD
Penny & Nick Bingham.

NEW **18 LEYS ROAD**
Philip James.

41 LEYS ROAD
Philip & Jennie Powell.

43 LEYS ROAD
Anna Stevens.

These 6 gardens show the wonderful variety of village green space with many different plants, shrubs, trees, vegetables, fruit and wild flowers. Two of the gardens are cottage style with unusual perennials and many shrubs, trees and vegetables. Another has a Japanese influence exhibiting many plants with black or bronze foliage and gravel areas. There are 2 courtyard gardens, one with extremes of light and shade that reflects the conditions the other cleverly packs in roses, perennials and vegetables. The largest has many rooms to explore and delight the imagination and harp music may be played. Wheelchair access to 6, 19 High Street, 18, 41 and 43 Leys Road. WC facilities in United Reformed Church Hall.

22 NEW DEAN MANOR

Dean, Chipping Norton, OX7 3LD. Mr & Mrs Johnny Hornby. *In Dean, 3 m SE of Chipping Norton. Follow A361 for 2 m. Turn R onto the lane and the garden is on the L.* **Sun 18, Mon 19 June (11-6). Adm £7, chd free. Light refreshments. Visits also by arrangement.**

The gardens at Dean Manor cover approx 6 acres. Stone walls are home to an abundant and varied selection of climbing/rambling roses, clematis and hydrangeas. The formal gardens inc complex yew hedging and herbaceous borders, kitchen and cutting garden, areas of wildflower meadow, an orchard and water gardens make up areas around the house. A spectacular new kitchen and cutting garden designed by Frances Rasch complete with new large glasshouse is currently in development, offering the opportunity to witness the beginnings of a major new chapter in the garden's rich history.

23 103 DENE ROAD
Headington, Oxford, OX3 7EQ. Steve & Mary Woolliams, 07778 617616, stevewoolliams@gmail.com. *S Headington, nr Nuffield. Dene Rd accessed from The Slade from the N, or from Hollow Way from the S. Both access roads are B4495. Garden on sharp bend.* **Visits by arrangement 15 Apr to 2 Sept for groups of 5 to 15. Adm £3.50, chd free. Home-made teas.**
A surprising eco-friendly garden with borrowed view over the Lye Valley Nature Reserve. Lawns, a wildflower meadow, pond and large kitchen garden are inc in a suburban 60ft x 120ft sloping garden. Fruit trees, soft fruit and mixed borders of shrubs, hardy perennials, grasses and bulbs, designed for seasonal colour. This garden has been noted for its wealth of wildlife inc a variety of birds, butterflies and other insects such as the rare Brown Hairstreak butterfly, the rare Currant Clearwing moth and the Grizzled Skipper.

24 DENTON HOUSE
Denton, Oxford, OX44 9JF. Mr & Mrs Luke, 01865 874440, waveney@jandwluke.com. *Nr Oxford. In a valley between Garsington & Cuddesdon.* **Visits by arrangement Apr to Sept. Adm £5, chd free.**
Large walled garden surrounds a Georgian mansion (not open) with shaded areas, walks, topiary and many interesting mature trees, large lawns, herbaceous borders and rose beds. The windows in the wall were taken in 1864 from Brasenose College Chapel and Library. Wild garden and a further walled fruit garden.

GROUP OPENING

25 DORCHESTER GARDENS
Dorchester-On-Thames, Wallingford, OX10 7HZ. *Off A4074 or A415 signed to Dorchester. Parking at Old Bridge Meadow, at SE end of Dorchester Bridge. Disabled parking at 26 Manor Farm Rd (OX10 7HZ). Maps available at the car park and at the Co-op in the High St.* **Sun 2 July (2-5). Combined adm £5, chd free. Tea in downstairs tearoom at Abbey Guesthouse. Last orders at 4.30pm.**

26 MANOR FARM ROAD
David & Judy Parker.

6 MONKS CLOSE
Leif & Petronella Rasmussen.

7 ROTTEN ROW
Michael & Veronica Evans.

3 contrasting gardens in a historic village surrounding the medieval abbey. 26 Manor Farm Rd (OX10 7HZ) was part of an old, larger, partially walled garden, which now has a formal lawn and planting, a vegetable garden, greenhouse and several interesting trees. From the yew hedge down towards the River Thames, is an apple orchard underplanted with spring bulbs. 6 Monks Close (OX10 7JA) is idyllic and surprising. A small spring-fed stream and sloping lawn surrounded by naturalistic planting runs down to a monastic fish pond. Bridges over this deep pond lead to the River Thames with steep banks. 7 Rotten Row (OX10 7LJ) is Dorchester's lawnless garden, a terrace with borders leads to a lovely geometric garden supervised by a statue of Hebe. Access is from the allotments. Due to deep water at 26 Manor Farm Road and 6 Monks Close children should be accompanied at all times. Wheelchair access to 26 Manor Farm Rd, partial access to other gardens. Assistance dogs only.

46 DORCHESTER MANOR HOUSE
Manor Farm Road, Dorchester-on-Thames, OX10 7HZ. Simon & Margaret Broadbent, 01865 340101, manor@dotoxon.uk. *8m SSE of Oxford. Off A4074 in village centre, next to Abbey.* **Visits by arrangement 5 June to 30 July. Home-made teas.**
2 acre garden in beautiful setting around Georgian house (not open) and medieval abbey. Spacious lawn leading to riverside copse of towering poplars with fine views of Dorchester Abbey. Terrace with rose and vine covered pergola around lily pond. Colourful herbaceous borders, small orchard and vegetable garden. Wheelchair access on gravel paths.

26 FAILFORD
118 Oxford Road, Abingdon, OX14 2AG. Miss R Aylward, 01235 523925, aylwardsdooz@hotmail.co.uk. *No. 118 is on the L of Oxford Rd when coming from Abingdon Town, or on the R when approaching from the N. Entrance to this garden is via 110 Oxford Rd.* **Sun 11 June (11-4). Combined adm with 116 Oxford Road £6, chd free. Home-made teas. Visits also by arrangement May to Sept for groups of 8 to 30.**
This town garden is an extension of the home divided into rooms both formal and informal. It changes every yr. Features inc walkways through shaded areas, arches, a beach, grasses, fernery, roses, topiaries, acers, hostas and heucheras.
Be inspired by the wide variety of planting, many unusual and quirky features, all within an area 570 sq ft.

27 FOXINGTON
Britwell Salome, Watlington, OX49 5LG. Ms Mary Roadnight, 01491 612418, mary@foxington.co.uk. *1 mile from Watlington. At Red Lion Pub take turning to Britwell Hill. After 350yds turn into drive on L.* **Visits by arrangement 10 Apr to 30 Sept for groups of 10 to 30. Adm £11, chd free. Home-made teas. Admission inc teas. Special dietary options by prior request.**
Stunning views to the Chiltern Hills provide a wonderful setting for this impressive garden, remodelled in 2009. Patio, heather and gravel gardens enjoy this view, whilst the back and vegetable gardens are more enclosed. A very large apple tree blew down in the spring of 2022 and has been replaced by a summerhouse. This has resulted in a major revision of the planting. There is an orchard and a flock of white doves. Well behaved dogs are welcome, but they must be kept on a lead at all times as there are many wild animals in the garden, wildflower meadow, wood and neighbouring fields. Wheelchair access throughout the garden on level paths with no steps.

28 FRIARS COURT

Clanfield, OX18 2SU. Charles Willmer, www.friarscourt.com. *4m N of Faringdon. On A4095 Faringdon to Witney rd. ½m S of Clanfield.* **Sun 4 June (2-5), open nearby Chestlion House. Sun 11 June (2-5). Adm £4, chd free. Cream teas.**

Over 3 acres of formal and informal gardens with flower beds, borders and specimen trees, lie within the remaining arms of a C16 moat which partially surrounds the large C17 Cotswold stone house. Bridges span the moat with water-lily-filled ponds to the front whilst beyond the gardens is a woodland walk. A museum about Friars Court is located in the old Coach House. A level path goes around part of the gardens. The museum is accessed over gravel.

29 THE GRANGE

1 Berrick Road, Chalgrove, OX44 7RQ. Mrs Vicky Farren, 01865 400883, vickyfarren@mac.com, thegrangegardener.com. *12m E of Oxford & 4m from Watlington, off B480. The entrance to The Grange is at the grass triangle between Berrick Rd & Monument Rd, by the pedestrian x-ing. GPS is not reliable in the final 200yds.* **Visits by arrangement 1 June to 27 Oct for groups of up to 35. Adm £10, chd free. Home-made teas.**

11 acres of gardens inc herbaceous borders and a large expanse of prairie with many grasses and a wildflower meadow. There is a lake with bridges and a planted island, a dry river bed and a labyrinth. A brook runs through the garden with a further pond, arboretum, old orchard and partly walled vegetable garden. Partial wheelchair access on many grass paths.

30 GREEN AND GORGEOUS

Little Stoke, Wallingford, OX10 6AX. Rachel Siegfried, www.greenandgorgeousflowers.co.uk. *3m S of Wallingford. Off B4009 between N & S Stoke, follow single track road down to farm.* **Sun 2 July (1-5). Adm £5, chd free. Home-made teas.**

6 acre working flower farm next to River Thames. Cut flowers (many unusual varieties) in large plots and polytunnels, planted with combination of annuals, bulbs, perennials, roses and shrubs, plus some herbs, vegetables and fruit to feed the workers! Flowers selected for scent, novelty, nostalgia, weather tolerance and naturalistic style. Floristry demonstrations, cut flowers, plant and craft stalls. Wheelchair access on short grass paths, with large concrete areas.

31 GREENFIELD FARM

Christmas Common, nr Watlington, OX49 5HG. Andrew & Jane Ingram, 01491 612434, andrew@andrewbingram.com. *4m from J5 M40, 7m from Henley. J5 M40, A40 towards Oxford for ½m, turn L signed Christmas Common. ¾m past Fox & Hounds Pub, turn L at Tree Barn sign.* **Visits by arrangement 15 May to 29 Sept for groups of 5 to 30. Adm £5, chd free.**

10 acre wildflower meadow surrounded by woodland, established 25 yrs ago under the Countryside Stewardship Scheme. Traditional

Barton Abbey

Chiltern chalkland meadow in beautiful peaceful setting with 100 species of perennial wildflowers, grasses and nine species of orchids. ½ m walk from parking area to meadow. Opportunity to return via typical Chiltern beechwood.

32 GREYHOUND HOUSE

The Street, Ewelme, Wallingford, OX10 6HU. Mrs Wendy Robertson, flowerfrond@gmail.com. *Greyhound House is on the main rd at the school end of Ewelme.* **Visits by arrangement 5 June to 29 Sept for groups of 10 to 50. Adm £6, chd free.**

2 acre hillside garden set in historic Chiltern village of Ewelme. Mixed herbaceous borders (inc salvias and unusual plants), ornamental grass walk, walled courtyard with dahlias and late summer flowering perennials, houseplant theatre, fernery, cutting garden, orchard and woodland.

GROUP OPENING

33 HEADINGTON GARDENS

Old Headington, Oxford, OX3 9BT. *2m E from centre of Oxford. At T-lights in the centre of Headington, travelling towards Oxford, turn R into Old High St. Parking on R- gardens further down in Old Headington.* **Sun 21 May (2-6). Combined adm £6, chd free. Home-made teas.**

THE COACH HOUSE
David & Bryony Rowe.

MONCKTON COTTAGE
Julie Harrod & Peter McCarter, 01865 751471, petermccarter@msn.com.

NEW **THE OLD POUND HOUSE**
Steve & Jane Cowls.

9 STOKE PLACE
Clive Hurst.

NEW **10 STOKE PLACE**
Tamasin and James Went.

Situated above Oxford, Headington is centred round an old village that is remarkable for its mature trees, high stone walls, narrow lanes and Norman church. The 5 gardens in Old Headington offer a contrast in styles and formality and there are traces of history: a partially cobbled courtyard for horses, a pound for stray animals, surviving rural surroundings and plaques honouring some of those who have lived here. One garden, 10 Stoke Place, is new this year. Partial wheelchair access to some gardens.

34 23 HID'S COPSE ROAD

Oxford, OX2 9JJ. Kathy Eldridge. *W of central Oxford, halfway up Cumnor Hill. Turn into Hids Copse Rd from Cumnor Hill. We are the last house on the R before T-junction.* **Sun 12 Feb (2-5). Combined adm with 6 High Street £3, chd free. Opening with West Oxford Gardens on Sun 11 June.**

The garden surrounds a house built in the early 30s (not open) on a ½ acre plot. Designed to be wildlife friendly, it has a kitchen garden, two ponds with frogs and newts, a wildflower meadow and woodland areas under many trees. Small winding paths lead to areas to sit and contemplate. The planting is dependent on the area within the garden, but a special treat are the snowdrops in early spring.

35 6 HIGH STREET

Cumnor, Oxford, OX2 9PE. Dr Dianne & Prof Keith Gull. *4m from central Oxford. Exit to Cumnor from the A420. In centre of village opp Post Office. Parking at back of Post Office.* **Sun 12 Feb (2-5). Combined adm with 23 Hid's Copse Road £3, chd free. Opening with Cumnor Village Gardens on Sun 11 June.**

Front, side and rear garden of a thatched cottage (not open). Front is partly gravelled and side courtyard has many pots. Rear garden overlooks meadows with old apple trees underplanted with ferns, wildlife pond, unusual plants, many with black or bronze foliage, planted in drifts and repeated throughout the garden. Planting has mild Japanese influence; rounded, clipped shapes interspersed with verticals. Snowdrops feature in Feb. There are two pubs in the village serving food; The Bear & Ragged Staff and The Vine. Former has accommodation. Wheelchair access to garden via gravel drive.

36 HOLLYHOCKS

North Street, Islip, Kidlington, OX5 2SQ. Avril Hughes, 01865 377104, ahollyhocks@btinternet.com. *3m NE of Kidlington. From A34, exit Bletchingdon & Islip. B4027 direction Islip, turn L into North St.* **Visits by arrangement 5 Feb to 30 Sept for groups of up to 25.**

Plantswoman's small Edwardian garden brimming with year-round interest, especially planted to provide winter colour, scent and snowdrops. Divided into areas with bulbs, May tulips, herbaceous borders, roses, clematis, shade and woodland planting especially trillium, podophyllum and arisaema, late summer salvias and annuals give colour. Large pots and troughs add seasonal interest and colour.

37 HOME CLOSE

29 Southend, Garsington, OX44 9DH. Mrs M Waud & Dr P Giangrande, 01865 361394, m.waud@btinternet.com. *3m SE of Oxford. N of B480, opp Garsington Manor.* **Visits by arrangement Apr to Sept. Adm £5, chd free. Please discuss refreshments when booking.**

2 acre garden with listed house (not open), listed granary and 1 acre mixed tree plantation with fine views. Unusual trees and shrubs planted for year round effect. Terraces, stone walls and hedges divide the garden and the planting reflects a Mediterranean interest. Vegetable garden and orchard.

Our donation of £2.5 million over five years has helped fund the new Y Bwthyn NGS Macmillan Specialist Palliative Care Unit in Wales. 1,900 inpatients have been supported since its opening

GROUP OPENING

38 IFFLEY GARDENS

Iffley, Oxford, OX4 4EF. *2m S of Oxford. Within Oxford's ring road, off A4158 Iffley road, from Magdalen Bridge to Littlemore r'about, to Iffley village. Map provided at each garden.* **Sun 11 June (2-6). Combined adm £5, chd free. Home-made teas in Church Hall and 25 Abberbury Road.**

17 ABBERBURY ROAD
Mrs Julie Steele.

25 ABBERBURY ROAD
Rob & Bridget Farrands.

NEW **50 CHURCH WAY**
William and Sarah Beaver.

86 CHURCH WAY
Helen Beinart & Alex Coren.

6 FITZHERBERT CLOSE
Eunice Martin.

Secluded old village with renowned Norman church. Visit several gardens ranging in variety and style. Mixed family gardens with shady borders and vegetables. Varied planting throughout the gardens inc herbaceous borders, shade loving plants, roses, fine specimen trees and plants in terracing. Features inc a small Japanese garden. Plant sales at several gardens. Wheelchair access to some gardens.

GROUP OPENING

39 KENCOT GARDENS

Kencot, Lechlade, GL7 3QT. *5m NE of Lechlade. E of A361 between Burford & Lechlade.* **Mon 10 Apr (2-6). Combined adm £6, chd free. Home-made teas at village hall.**

THE ALLOTMENTS
Amelia Carter Charity.

BELHAM HAYES
Mr Joseph Jones.

THE GARDENS
Mark & Jayne Hodds.

GREYSTONES
Mr Tom Valentine.

HILLVIEW HOUSE
Andrea Moss.

IVY NOOK
Gill & Wally Cox.

MANOR FARM
Henry & Kate Fyson.

WELL HOUSE
Janet & Richard Wheeler.

The 2023 Kencot Gardens group will consist of 7 gardens and the allotments. The Allotments tended by 9 people, growing a variety of vegetables, flowers and fruit. Belham Hayes, mature cottage garden, mixed herbaceous borders, two old fruit trees, vegetables. Emphasis on scent and colour coordination. Manor Farm, 2 acre walled garden with bulbs, wood anemones, fritillaries, mature orchards, pleached limewalk, ancient yew balls, revolving summerhouse. Chickens and pony. Well House, ⅓ acre garden, mature trees, hedges, miniature woodland glade, brook with waterfall. Spring bulbs, rockeries. Hill View, 2 acres lime tree drive, established trees, pond, wild flowers, shrubs, perennial borders, daffodils, aconites, vegetables A family garden. Greystones, stone animals and olive trees, climbers and pots adding colour. The Gardens 2 old apple trees, herbaceous beds, perennial borders, bulbs pots of seasonal flowers. Ivy Nook, medium sized garden, mixed borders, vegetable garden, greenhouse, small pond and magnolia tree. Wheelchair access to The Allotments, difficult due to step entrance and narrow paths. Other gardens maybe difficult due to gravel and uneven paths.

40 KIDMORE HOUSE GARDEN & VINEYARD

Chalkhouse Green Road, Kidmore End, Reading, RG4 9AR. Mr Stephen & Mrs Niamh Kendall, 07740 290990, niamhk@btinternet.com. *6m NNW of Reading, Emma Green/Reading end of the village. Parking in the field through gate signed Kidmore Vineyard.* **Fri 26 May (1.30-5). Adm £6, chd free. Home-made teas in the barn next to the Rose Garden. Visits also by arrangement 25 May to 21 June for groups of 8 to 20. Wednesday afternoons or early evenings.**
14 acres set in South Oxfordshire landscaped gardens with features inc ha-ha, walled garden with delphiniums, calla lilies and white wisteria, a rose garden and one hectar of vines. Garden sympathetic with a red-brick Queen Anne house (not open). Anecdotally, the large sweet chestnut in the lawn was planted with the house in 1680. Picnics welcome in the field. Partial wheelchair access with gravel pathways and some steps.

41 KINGHAM LODGE

West End, Kingham, Chipping Norton, OX7 6YL. Christopher Stockwell, 01608 658226, info@sculptureatkinghamlodge.com, www.sculptureatkinghamlodge.com. *From Kingham Village, West St turns into West End at tree in middle of road, bear R & you will see black gates for Kingham Lodge immed on L. Follow signs for parking. Do not park on street.* **Wed 24 May (10-5). Adm by donation. Pre-booking essential, please phone, email, or visit our website for information & booking. Light refreshments.**
Many ericaceous plants not normally seen in the Cotswolds grow on 5 acres of garden, planted over 2 decades. Big display of rhododendron, laburnum arch and azaleas. Formal 150 metre border, backed with trellis, shaded walks with multi-layered planting, an informal quarry pond, formal mirror pond, pergola, parterre and unique Islamic garden. Sculpture show from 20 to 29 May 2023. Islamic pavilion with fountains and rills. Disabled parking on gravel at entrance and level access to all areas of the garden.

42 KINGS COTTAGE

Pudlicote Lane, Chilson, Chipping Norton, OX7 3HU. Mr Michael Anderson. *S end of Chilson village. Entering Chilson from the S, off the B4437 Charlbury to Burford rd, we are the 1st house on the L. Please drive past & park considerately in centre of village.* **Sun 28, Mon 29 May, Sun 17 Sept (2-5). Combined adm with Bolters Farm £8, chd free. Light refreshments.**
An old row of cottages with mature trees and yew hedging. Over the last 7 yrs a new design and planting scheme has been started to reduce the areas of lawn, bring in planting to complement the house (not open) and setting, introduce new borders and encourage wildlife. A work in progress. Partial wheelchair access via gravel drive. Moderate slopes, grass and some sections of garden only accessible via steps.

43 38 LECKFORD ROAD

Oxford, OX2 6HY. Dinah Adams, 01865 511996, dinah_zwanenberg@fastmail.com. *Central Oxford. N on Woodstock Rd take 3rd L. Coming into Oxford on Woodstock Rd, 1st R after Farndon Rd. Some 2 hr parking nearby.* **Sun 16 Apr (2-5). Adm £5, chd free. Home-made teas. Visits also by arrangement 17 Apr to 31 Oct.**
Behind the rather severe facade of a Victorian town house (not open) is a very protected long walled garden with mature trees. The planting reflects the varied levels of shade and inc rare and unusual plants. The trees in the front garden deserve attention. The back garden is divided into 3 distinct parts, each with a very different character. Amongst other things there is a hornbeam roof.

44 LIME CLOSE

35 Henleys Lane, Drayton, Abingdon, OX14 4HU. M C de Laubarede, mail@mclgardendesign.com, www.mclgardendesign.com. *2m S of Abingdon. Henleys Ln is off main road through Drayton. Please Note: When visiting Lime Close, please respect local residents & park considerately.* **Sun 26 Feb (2-5); Sun 21 May (2-5.30). Adm £5, chd free. Cream teas. Visits also by arrangement 15 Feb to 31 Oct for groups of 10+. Garden closed from mid July to mid September. Donation to International Dendrology Society.**
4 acre mature plantsman's garden with rare trees, shrubs, roses and bulbs. Mixed borders, raised beds, pergola, topiary and shade borders. Herb garden by Rosemary Verey. Listed C16 house (not open). Cottage garden by MCL Garden Design, planted for colour, an iris garden with 100 varieties of tall bearded irises. Winter bulbs. New arboretum with rare exotic trees and shrubs from Asia and America. The garden is flat and mostly grass with gravel drive and paths.

Woolstone Mill House

45 MAGDALEN COLLEGE

Oxford, OX1 4AU. Magdalen College, www.magd.ox.ac.uk. *Oxford. Entrance in High St.* **Sun 16 Apr (10-7). Adm £8, chd free. Light refreshments in the Old Kitchen Bar.**
60 acres inc deer park, college lawns, numerous trees 150-200 yrs old; notable herbaceous and shrub plantings. Magdalen meadow where purple and white snake's head fritillaries can be found is surrounded by Addison's Walk, a tree lined circuit by the River Cherwell developed since the late C18. Ancient herd of 60 deer. Press bell at the lodge for porter to provide wheelchair access.

47 MEADOW COTTAGE

Christmas Common, Watlington, OX49 5HR. Mrs Zelda Kent-Lemon, 01491 613779, zelda_kl@hotmail.com. *1m from Watlington. Coming from Oxford M40 to J6. Turn R & go to Watlington. Turn L up Hill Rd to top. Turn R after 50yds, turn L into field.* **Visits by arrangement for groups of 10+. Home-made teas.**
1¾ acre garden adjoining ancient bluebell woods, created by the owner from 1995 onwards. Many areas to explore inc a professionally designed vegetable garden, many composting areas, wildflower garden and pond, old and new fruit trees, many shrubs, much varied hedging and a tall treehouse which children can climb under supervision. Indigenous trees and C17 barn (not open). Come and see the wonderful snowdrops in Feb then stunning aconites and during the month of May visit the bluebell woodland. Partial wheelchair access over gravel driveway and lawns.

48 MERTON COLLEGE OXFORD FELLOWS' GARDEN

Merton Street, Oxford, OX1 4JD. Merton College, 01865 276310. *Merton St runs parallel to High St about halfway down.* **Sun 23 July (10-5). Adm £6.50, chd free.**
Ancient mulberry, said to have associations with James I. Specimen trees, long mixed border, recently established herbaceous bed. View of Christ Church meadow.

GROUP OPENING

49 MIDDLETON CHENEY GARDENS

Middleton Cheney, Banbury, OX17 2ST. *3m E of Banbury. From M40 J11 follow A422 signed Middleton Cheney. Parking at nursery school car park, next to 19 Glovers Ln, OX17 2NU. Map available at all gardens.* **Sun 18 June (1-6). Combined adm £8, chd free. Home-made teas at Peartree House.**

CROFT HOUSE
Richard & Sandy Walmsley.

NEW **6 GLOVERS LANE**
Mike Scarlett.

NEW **LEXTON HOUSE**
Matthew & Diane Tims.

5 LONGBURGES
Mr D Vale.

38 MIDWAY
Margaret & David Finch.

PEARTREE HOUSE
Roger & Barbara Charlesworth.

14 QUEEN STREET
Brian & Kathy Goodey.

2A RECTORY LANE
Debbie Evans.

Large village with C13 church (open) with renowned pre-Raphaelite stained glass. 8 open gardens with a variety of sizes, styles and maturity. Of the smaller gardens, one contemporary garden contrasts formal features with colour-filled borders and exotic plants. A mature small front and back garden is planted profusely with a feel of an intimate haven. A garden that has evolved through family use, features rooms and dense planting. A steeply-pitched garden on three levels has patio, ponds, herbaceous borders and acer collection. One garden is designed to support nature, providing an environment to attract bird and insects. A new garden inc a formal terrace surrounded by hedges and fig trees with steps down to a secluded lower space and small vegetable garden. A larger garden new to its owners, is being replanted with cottage-style plants and vegetable garden. Another has an air of mystery with hidden corners and an extensive water feature weaving its way throughout the garden.

50 MIDSUMMER HOUSE

Woolstone, Faringdon, SN7 7QL. Penny Spink. *7m W & 7m S of Faringdon. Woolstone is a small village off B4507, below Uffington White Horse Hill. Take rd towards Uffington from the White Horse Pub.* **Fri 7 Apr, Fri 5 May (2-5). Fri 2 June (2-5), open nearby Woolstone Mill House. Fri 7 July, Fri 4 Aug, Fri 1 Sept, Fri 6 Oct (2-5). Adm £5, chd free. Home-made teas.**

Midsummer Garden is designed by Justin Spink for his parents, Anthony and Penny Spink, in 2015. In 2018 Mike and Ann Collins continued designing and planting the garden. Over the road there is a newly planted arboretum and nature reserve established 2019. Rare and unusual plants. Picnics welcome in the field opp. Wheelchair access over short gravel drive at entrance.

51 MILL BARN

25 Mill Lane, Chalgrove, OX44 7SL. Pat Hougham, 01865 890020, Gmec@outlook.com. *12m E of Oxford. Chalgrove is 4m from Watlington off B480. Mill Barn is in Mill Ln, W of Chalgrove, 300yds S of Lamb Pub.* **Visits by arrangement 1 May to 1 Oct. Adm £6, chd free. Cream teas.**

Mill Barn has an informal cottage garden with a variety of flowers throughout the seasons. Rose arches and a pergola lead to a vegetable plot surrounded by a cordon of fruit trees, all set in a mill stream landscape. The Manor garden has a lake and wildlife areas, mixed shrubs and herbaceous beds that surround the C15 Grade I listed Manor House (not open).

52 MONKS HEAD

Weston Road, Bletchingdon, OX5 3DH. Sue Bedwell, 01869 350155, bedwell615@btinternet.com. *Approx 4m N of Kidlington. From A34 take B4027 to Bletchingdon, turn R at Xrds into Weston Rd.* **Visits by arrangement 7 Jan to 5 Nov. Home-made teas.**

Plantaholics' garden for year-round interest. Bulb frame and alpine area, greenhouse. Changes evolving all the time, becoming more of a wildlife garden.

GROUP OPENING

53 OLD BOARS HILL GARDENS

Jarn Way, Boars Hill, Oxford, OX1 5JF. *3m S of Oxford. From S ring road towards A34 at r'about follow signs to Wootton & Boars Hill. Up Hinksey Hill take R fork. 1m R into Berkley Rd to Old Boars Hill.* **Sun 14 May (2-5.30). Combined adm £5, chd free. Home-made teas.**

TALL TREES
David Clark.

YEW COTTAGE
Michael Edwards, 018765 735180.
Visits also by arrangement 1 May to 20 May for groups of up to 25.

2 gardens located in a wooded conservation area favoured by walkers. Tall Trees is surrounded by pots, more acid soil for rhododendrons and gravel gardens. The garden at Yew Tree Cottage is a delightful mixture of plants to give year-round colour and the owners will display their vintage cars, one has been on TV.

54 116 OXFORD ROAD

Abingdon, OX14 2AG. Mr & Mrs P Aylward, 01235 523925, aylwardsdooz@hotmail.co.uk. *North Abingdon. 116 Oxford Rd is on the R if coming from A34 N exit, or on the L after Picklers Hill turn if approaching from Abingdon town centre.* **Sun 11 June (11-4). Combined adm with Failford £6, chd free. Home-made teas. Visits also by arrangement May to Sept for groups of up to 30.**

This young town garden brings alive the imagination of its creators. It will inspire both the enthusiast and the beginner. The use of recycled materials, lots of colour and architectural plants. A folly/greenhouse is just one of many quirky features. Raised beds, rockeries, specimen plants, beds of roses and hostas. Pushes the boundaries of conventional gardening. There is something for everyone!

55 THE PRIORY, CHARLBURY

Church Lane, Charlbury, OX7 3PX. Dr D El Kabir & Colleagues. *6m SE of Chipping Norton. Large Cotswold village on B4022 Witney-Enstone Rd,*

near St Mary's Church. **Sun 18 June (2-5). Adm £6, chd free.**
1½ acre of formal terraced topiary gardens with Italianate features. Foliage colour schemes, shrubs, parterres with fragrant plants, old roses, water features, sculpture and inscriptions aim to produce a poetic, wistful atmosphere. Formal vegetable and herb garden. Arboretum of over 3 acres borders the River Evenlode and inc wildlife garden and pond. Partial wheelchair access.

56 NEW PUSEY HOUSE
Faringdon, SN7 8QB. Head Gardener, Lloyd Newman. *1 m off the A420 between Oxford & Swindon.* **Sun 9 July (11-4). Adm £8, chd free. Home-made teas.**
Walled and kitchen gardens surrounded by C18 landscape park with lake. Much has been recently renovated, some areas still in progress. Great view of the Berkshire Downs. Mostly gravel paths, some bridges and grass to cross. Dogs on leads welcome.

57 9 RAWLINSON ROAD
Oxford, OX2 6UE. Ramnique Lall. *¾ m N of Oxford City Centre. Rawlinson Rd runs between Banbury & Woodstock rds midway between Oxford City Centre & Summertown shops.* **Sun 7 May (1-5). Adm £5, chd free. Open nearby 11 Rawlinson Road.**
Townhouse garden with structured disarray of roses. Terrace of stone inlaid with brick and enclosed by Chinese fretwork balustrade, chunky brick and oak pergola covered with roses, wisteria and clematis; potted topiary. Until autumn the garden is delightfully replete with aconites, lobelias, phloxes, daisies and meandering clematis. Adm £7 for visiting both 9 & 11 Rawlinson Road.

58 11 RAWLINSON ROAD
Oxford, OX2 6UE. Emma Chamberlain. *Central N Oxford. Halfway between Central Oxford & Summertown, off Banbury Rd.* **Sun 7 May (1-5). Adm £5, chd free. Tea. Open nearby 9 Rawlinson Road.**
A S facing 120ft walled Victorian town garden created from scratch since 2010. Struggle with heavy clay and poor soil was instructive. Interesting mix of herbaceous and shrubs with particular interest in late tulips, grasses and unusual plants. Great emphasis on colour and succession planting with paths running through the beds. New bed done in 2022 and being extended in 2023. Sculpture and colour paths. Adm £7 for visiting both 9 and 11 Rawlinson Road. Wheelchairs possible and disabled visitors welcome.

59 NEW RECTORY FARMHOUSE
Church Enstone, Chipping Norton, OX7 4NN. Andrew Hornung & Sally Coles, 01608 677374, giussanese@yahoo.co.uk. *Entrance on unmade rd behind The Crown Inn.* **Visits by arrangement 3 June to 9 July for groups of up to 30. Adm £5, chd free. Home-made teas.**
A large mixed garden on three main levels, inc fruit trees, soft fruit, vegetable garden. About 100 varieties of rose, particularly climbers, many unusual, some rare plants. The garden inc many different settings, ranging from gravel areas to pondside plantings. Newer, less developed areas alongside well established plantings.

60 SANDYS HOUSE
Bull Hill, Chadlington, Chipping Norton, OX7 3ND. Jane Bell. *Centre of Chadlington village accessible via A361 (Burford-Chipping Norton) or A44 (Oxford). At Xrds turn down Bull Hill & enter garden on L through wooden gates. Parking on surrounding streets but not at property.* **Sat 22, Sun 23 Apr, Sat 10, Sun 11 June (11-5). Combined adm with Claridges Barn £5, chd free. Teas, coffees and light lunches available.**
Bulbs, perennials and shrubs create a colourful mix in this informal cottage-style garden to the rear of the former Sandys Arms pub. There is a strong emphasis on wildlife-friendly planting, and a series of small ponds encourage bio-diversity. The garden inc a mature summer-flowering Magnolia Grandiflora against the house and tender annuals are displayed in the main greenhouse.

Aston Pottery

61 SARSDEN GLEBE
Churchill, Chipping Norton, OX7 6PH. Mr & Mrs Rupert Ponsonby. *Situated between Sarsden and Churchill. Sarsden Glebe is ¼ m S of Churchill on the road to Sarsden. Turn in to the drive by a small lodge.* **Fri 7 Apr (12.30-4.30). Adm £7.50, chd free. Home-made teas.**
A rectory garden and park laid out by Humphrey Repton and his son George. Terraced formal garden; wild garden with spring bulbs and mature oaks; and walled kitchen garden. Wild garden with a sea of blue and white anemone blanda interspersed with fritillaries and daffodils in spring. Most of the garden can be accessed in a wheelchair.

GROUP OPENING

62 SIBFORD GARDENS
Sibford Ferris, OX15 5RE. *7m W of Banbury. Nr the Warwickshire border, S of B4035, in centre of Sibford Ferris village at T-junc & additional gardens nr the Xrds & Wykham Arms Pub in Sibford Gower. Parking in both villages.* **Sun 25 June (2-6). Combined adm £7, chd free. Home-made teas at Sibford Gower Village Hall (opp the church).**

BUTTSLADE HOUSE
James & Sarah Garstin.

CARTER'S YARD
Sue & Malcolm Bannister, 01295 780365, sebannister@gmail.com.
Visits also by arrangement May to Sept for groups of 7 to 25.

COPPERS
Mr Andrew & Mrs Chris Tindsley.

HOME CLOSE
Graham & Carolyn White.

THE LONG HOUSE
Jan & Diana Thompson.

NEW **SHRUBBERY COTTAGE**
Nic Durrant.

2 charming small villages of Sibford Gower and Sibford Ferris, off the beaten track with thatched stone cottages. 6 contrasting gardens comprising a truly traditional plants woman's cottage garden,a woodland setting with shade garden, the other a private garden of a renowned landscape architect. New for 2023 an Artists garden with planting designed to inspire her works, a revised garden of small rooms and the gardens of an early C20 Arts and Crafts house.A fantastic collection bursting with bloom, structural intrigue, interesting planting and some rather unusual plants.

63 SOUTH NEWINGTON HOUSE
South Newington, OX15 4JW. Mr & Mrs David Swan, 07711 720135, claire_ainley@hotmail.com. *6m SW of Banbury. South Newington is between Banbury & Chipping Norton. Take Barford Rd off A361, 1st L after 100yds in between oak bollards. For SatNav use OX15 4JL.* **Visits by arrangement 30 Jan to 29 Oct. Adm £5, chd free. Home-made teas.**
Meandering tree lined drive leads to 2 acre garden. Herbaceous borders designed for year-round colour. Organic garden with established beds and rotation planting scheme. Orchard full of fruit trees with pond encouraging wildlife. Walled parterre planted for seasonal colour. A family garden with a small menagerie, all beautifully designed to blend seamlessly into the environment; a haven for all. Some gravel paths, otherwise full wheelchair access.

64 SOUTHBANK
Back Lane, Epwell, Banbury, OX15 6LF. Alan Cooper, 01295 780577, alan.cooper2019@gmail.com. *1st bungalow on L after Church Ln on L. Epwell is midway between Banbury & Shipston on Stour. Turn at yellow national Garden Scheme sign for Epwell. Then 500yds turn R to village. 1m turn into square. Exit via L corner by church, approx 200yds Southbank.* **Sun 3 Sept (1-6). Adm £4, chd free. Home-made teas in the village hall. Visits also by arrangement Apr to Oct for groups of up to 12.**
Front and back garden evolved over 20 yrs with several mature trees. Very densely planted with many rare and unusual plants. The front inc a rockery, mixed herbaceous border, peony and geranium beds. The back has various borders inc a sub-tropical bed, greenhouse, small vegetable plot and wildflower hedgerow.

GROUP OPENING

65 STEEPLE ASTON GARDENS
Steeple Aston, OX25 4SF. Richard and Daphne Preston. *14m N of Oxford, 9m S of Banbury. ½m E of A4260. Parking at village hall, OX25 4SF.* **Sun 4 June (2-6). Combined adm £7, chd free. Home-made teas in village hall.**

ACACIA COTTAGE
Jane & David Stewart.

THE CHURCH ALLOTMENTS
Oxford Diocese.

COMBE PYNE
Chris & Sally Cooper.

NEW **33 GRANGE PARK**
P McKenna.

THE LONGBYRE
Mr & Mrs V Billings.

THE OLD POST OFFICE
Mr & Mrs John Adriaanse.

THE POUND HOUSE
Mr & Mrs R Clarke.

PRIMROSE GARDENS
Richard & Daphne Preston, 01869 340512, richard.preston5@btopenworld.com.
Visits also by arrangement 15 May to 21 July for groups of 10 to 35.

Steeple Aston, often considered the most easterly of the Cotswold villages, is a beautiful stone-built village with gardens that provide a wide range of interest. A stream meanders down the hill as the landscape changes from sand to clay. The eight open gardens inc small floriferous cottage gardens, large, landscaped gardens, natural woodland areas, ponds and bog gardens, themed borders and allotments. The gardens provide something for everyone from a courtyard garden to a former walled kitchen garden and so much in between. Allotments are wheelchair accessible with care, Primrose Gardens and Longbyre have gravel and many of the gardens are situated on a slope.

66 ◆ STONOR PARK
Stonor, Henley-On-Thames, RG9 6HF. Lady Ailsa Stonor, 01491 638587, administrator@stonor.com, www.stonor.com. *4 m from Henley on Thames. Stonor is located between the M4 (J8/J9) & the M40 (J6) on the B480 Henley-on-Thames to Watlington rd. If you are approaching Stonor on the M40 from the E, please exit at J6 only.* **For NGS: Thur 22 June (10-5). Adm £6.50, chd free. For other opening times and information, please phone, email or visit garden website.**
Surrounded by dramatic swooping valleys and nestled within an ancient deer park, you will find the gardens at Stonor, which date back to Medieval times. Visitors love the serenity of the our C17 walled, Italianate Pleasure Garden and herbaceous perennial borders beyond.

67 STOW COTTAGE ARBORETUM & GARDEN
Junction Road, Churchill, Chipping Norton, OX7 6NP. Tom Heywood-Lonsdale. *2½ m SW of Chipping Norton, off the B4450. Parking: use postcode OX7 6NP & parking will be signed from William Smith Close.* **Sat 1 July (1-5). Adm £5, chd free.**
The arboretum and garden cover approx 15 acres with extensive views towards Stow-on-the-Wold and beyond. Stow Cottage (not open) is surrounded by the garden with trees and large deep borders with many rare shrubs as well as roses and daphnes. The arboretum began in 2009 and has been extensively developed over the years. There is an array of 500 trees inc different oaks, sorbus and limes as well as many magnolias, dogwoods, walnuts, birches and liquidambars.

68 TRINITY COLLEGE
Broad Street, Oxford, OX1 3BH. Kate Burtonwood, Head Gardener, www.trinity.ox.ac.uk. *Central Oxford. Entrance in Broad St.* **Sun 10 Sept (10-4). Adm £8, chd free. Home-made teas in dining hall.**
Historic, listed Oxford College gardens, on a site dating back to C12th. The park-like front quadrangle has mature specimen trees inc Catalpa and Cedrus. Formal courtyards house seasonal pot displays. New planting areas have been created in the Woodland and Library Quad in 2021 following extensive redevelopment of college buildings. The private President's Gardens will be open for viewing. Further restoration is underway through 2023. A focus on sustainable plantings. Most of site is accessible for wheelchair users. Some uneven paths and gravel at points. Accessible facilities are provided.

69 NEW UPPER BOLNEY HOUSE
Upper Bolney Road, Harpsden, Henley-On-Thames, RG9 4AQ. Anna and Richard Wilson. *2m S of Henley on Thames, between Shiplake and Harpsden. Only put postcode into SatNav, or you may be directed to Lower Shiplake. ½m up private rd, turn off Woodlands Rd.* **Sun 21 May (11-5.30). Adm £6, chd free. Light refreshments.**
The 4 acre garden has been extensively developed over the last 15 years. There are lawns with herbaceous borders. The terraces around the house inc planted areas and yew topiary. Various interesting areas lie beyond; a rose garden, wildflower meadow, rhododendron bed, tropical area, Victorian stumpery, kitchen garden and greenhouse. Some garden statuary and several sitting areas and benches.

70 UPPER GREEN
Brill Road, Horton cum Studley, Oxford, OX33 1BU. Susan & Peter Burge, 01865 351310, sue.burge@ndm.ox.ac.uk, www.uppergreengarden.co.uk. *6½m NE of Oxford. Enter village, turn R up Horton Hill. At T-junc turn L into Brill Rd. Upper Green 250yds on R, 2 gates before pillar box. Roadside parking.* **Visits by arrangement Feb to Oct for groups of up to 30. Adm £5. Tea.**
Mature ½ acre wildlife friendly garden packed with interest and colour throughout the yr. Over 1500 plants in garden database. Wildlife friendly. Herbaceous borders, gravel bed, rock bed, ferns, hot border, potager, bog and pond. Snowdrop collection. Alpines. Apple trees support rambling roses. Spectacular views. Sorry, no dogs.

71 ◆ WATERPERRY GARDENS
Waterperry, Wheatley, OX33 1JZ. The School of Philosophy and Economic Science, 01844 339254, office@waterperrygardens.co.uk, www.waterperrygardens.co.uk. *7½m from Oxford city centre. From E M40 J8, from N M40 J8a. Follow brown tourist signs. For SatNav please use OX33 1LA.* **For NGS: Sun 12 Mar (10-5); Sun 17 Sept (10-5.30). Adm £9.95, chd free. Light refreshments in the teashop (10-5). For other opening times and information, please phone, email or visit garden website.**
8 acres of beautifully landscaped ornamental gardens featuring a spectacular 200ft herbaceous border. New developments underway inc the replanting of the Mary Spiller rose garden, planting the new winter border and creating a new bluebell copse. The formal garden is particularly neatly designed and colourful, with a small knot garden, herb border and wisteria tunnel. Newly redesigned walled garden, river walk, statues and pear orchard. Riverside walk may be inaccessible to wheelchair users if very wet.

72 WAYSIDE
82 Banbury Road, Kidlington, OX5 2BX. Margaret & Alistair Urquhart, 01865 460180, alistairurquhart@ntlworld.com. *5m N of Oxford. On R of A4260 travelling N through Kidlington.* **Visits by arrangement June to Sept for groups of up to 25. Adm £4, chd free. Tea.**
¼ acre garden shaded by mature trees. Mixed border with some rare and unusual plants and shrubs. A climber clothed pergola leads past a dry gravel garden to the woodland garden with an extensive collection of hardy ferns. Conservatory and large fern house with a collection of unusual species of tree ferns and tender exotics. Partial wheelchair access.

9,500 patients and their families are being supported across four Horatio's gardens thanks to our annual donations

GROUP OPENING

73 WEST OXFORD GARDENS

Cumnor Hill, Oxford, OX2 9HH. *Take Botley interchange off A34 from N or S. Follow signs for Oxford & then turn R at Botley T-lights, opp MacDonalds & follow National Garden Scheme yellow signs. Street parking.* **Sun 11 June (2-5). Combined adm £5, chd free. Home-made teas at 10 Eynsham Road.**

10 EYNSHAM ROAD
Jon Harker.

NEW **3 HALLIDAY LANE**
John and Viccy Fleming, 07710 578174, garden@octon.scot.
Visits also by arrangement Mar to July for groups of up to 20.

23 HID'S COPSE ROAD
Kathy Eldridge.
(See separate entry)

86 HURST RISE ROAD
Ms P Guy & Mr L Harris.

6 SCHOLAR PLACE
Helen Ward.

86 Hurst Rise Road, a small garden abounding in perennials, roses, clematis and shrubs displayed in layers and at different levels around a circular lawn and path with sculptural features. Delightful colour. 10 Eynsham Road in Botley, is a New Zealander's take on an English-style garden inc many rose varieties, white garden, herbaceous borders and pond. 23 Hid's Copse Road, a ½ acre, wildlife friendly plot under many trees with two ponds, a wildflower meadow, kitchen garden and contemplative areas. 6 Scholar Place, an urban garden on a gently upward slope with a selection of small trees, shrubs and perennials. There is a small pond. Steps up from the patio lead to the main garden. There are also planted areas to the front and side of the house. NEW for 2023 - 3 Halliday Lane, North Hinksey is a beautiful town garden with a stunning alpine crevice garden at the front complemented by a well laid out and interesting back garden. Gravel steps at 6 Scholar Place, pebble path at 86 Hurst Rise Road and narrow bark paths to the back of 3 Halliday Lane.

74 WESTWELL MANOR

Westwell, nr Burford, OX18 4JT. Mr Thomas Gibson. *2m SW of Burford. From A40 Burford-Cheltenham, turn L ½m after Burford r'about signed Westwell. After 1½m at T-junc, turn R & Manor is 2nd house on L.* **Sun 21 May (2-6). Adm £5, chd free. Light refreshments. Tea, biscuits, cake and other stands open in the village. Donation to Aspire.**

7 acres surrounding old Cotswold manor house (not open) with knot garden, potager, shrub roses, herbaceous borders, topiary, earth works, moonlight garden, auricula ladder, rills and water garden.

GROUP OPENING

75 WHEATLEY GARDENS

High Street, Wheatley, OX33 1XX. 07813 339480, echess@hotmail.co.uk. *5m E of Oxford. Leave A40 at Wheatley, turn into High St. Gardens at W end of High St, S side.* **Sun 4 June (2-6). Combined adm £5, chd free. Cream teas. Visits also by arrangement 12 Mar to 30 Sept.**

BREACH HOUSE GARDEN
Liz Parry.

THE MANOR HOUSE
Mrs Elizabeth Hess.

THE STUDIO
Ann Buckingham.

3 adjoining gardens in historic Wheatley are: Breach House Garden with many shrubs, perennials, a contemporary reflective space and a wild meadow with ponds; The Studio with walled garden and climbing roses, clematis, herbaceous borders, vegetables and fruit trees; The Elizabethan Manor House (not open) is a romantic oasis with formal box walk, herb garden, rose arches and old rose shrubbery. Various musical events. Wheelchair access with assistance due to gravel paths, 2 shallow steps and grass.

76 WHITEHILL FARM

Widford, Burford, OX18 4DT. Mr & Mrs Paul Youngson, 01993 822894, paulyoungson48@gmail.com. *1m E of Burford. From A40 take rd signed Widford. Turn R at bottom of hill, 1st house on R with ample car parking.* **Sun 4 June (2-6). Adm £5, chd free. Home-made teas. Visits also by arrangement June to Sept.**

2 acres of hillside gardens and woodland with spectacular views overlooking Burford and Windrush valley. Informal plantsman's garden built up by the owners over 25 yrs. Herbaceous and shrub borders, ponds and bog area, old fashioned roses, ground cover, ornamental grasses, bamboos and hardy geraniums. Large cascade water feature, pretty tea patio and wonderful Cotswold views.

77 WOOLSTONE MILL HOUSE

Woolstone, Faringdon, SN7 7QL. Mr & Mrs Justin Spink. *7m W of Wantage. 7m S of Faringdon. Woolstone is a small village off B4507, below Uffington White Horse Hill.* **Fri 2 June (2-5). Adm £5, chd free. Tea. Open nearby Midsummer House.**

Redesigned by new owner, garden designer Justin Spink in 2020, this 1½ acre garden has large mixed perennial beds, and small gravel, cutting, kitchen and bog gardens. Topiary, medlars and old fashioned roses. Treehouse with spectacular views to Uffington White Horse and White Horse Hill. C18 millhouse and barn (not open). Partial wheelchair access.

GROUP OPENING

78 WOOTTON GARDENS

Wootton, OX13 6DP. *Wootton is 3m SW of Oxford. From Oxford ring road S, take turning to Wootton. Parking for all gardens at the Bystander Pub & on the road.* **Thur 8 June (1-5.30). Combined adm £5, chd free. Home-made teas.**

13 AMEY CRESCENT
Sylv & Liz Gleed.

57 BESSELSLEIGH ROAD
John & Lin Allen.

3 HOME CLOSE
Katherine Schomberg.

14 HOME CLOSE
Kev & Sue Empson.

35 SANDLEIGH ROAD
Hilal Baylav Inkersole.

5 inspirational small gardens, all with very different ways of providing

personal joy. 13 Amey Crescent is a gravel garden with grasses and colourful prairie plants. The garden has a wildlife pond, alpine house and troughs. 3 Home Close is laid to lawn with perennial borders, homegrown annuals and a small space for vegetables. Mature shrubs, trees and garden arches give height to the garden. 14 Home Close is a garden containing mainly shrubs with an unusually shaped lawn, a pond, raised bed and a vegetable garden. 57 Besselsleigh Road is a large family garden with many unusual plants alongside shrubs, fruit trees and vegetable plot, with a polytunnel and greenhouse. 35 Sandleigh Road is a mature garden, laid to lawn on two levels. The plants are mainly homegrown from cuttings. The garden is brimming with vibrant flowers, pondside planting and glass art. Partial wheelchair access, some gravel paths.

♿ ✿ ☕))

Pusey House

SHROPSHIRE

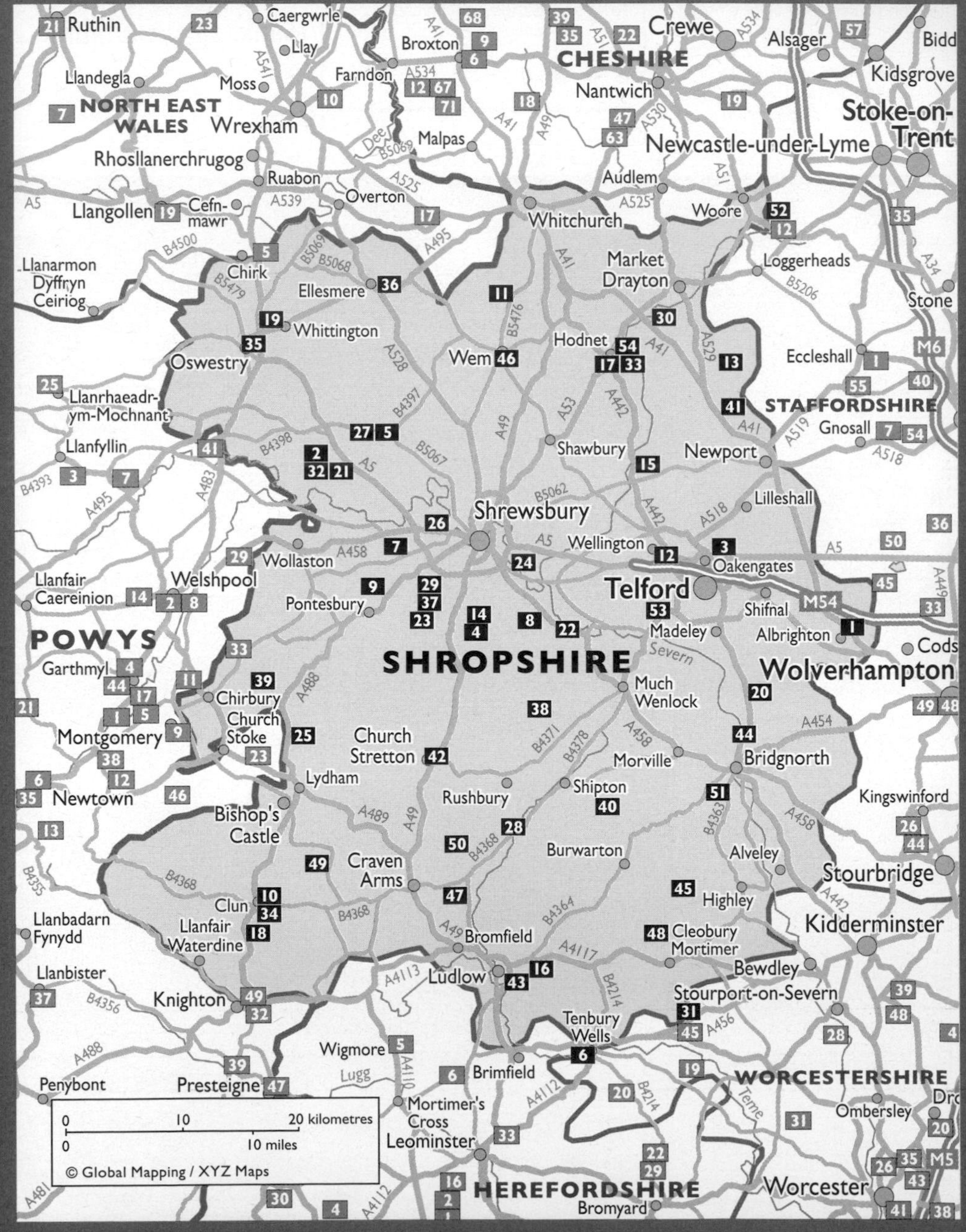

Ruthin
Caergwrle
Llay
Broxton
Crewe
Alsager
CHESHIRE
Kidsgrove
Llandegla
Moss
Farndon
Nantwich
Stoke-on-Trent
NORTH EAST WALES
Wrexham
Malpas
Newcastle-under-Lyme
Rhosllanerchrugog
Ruabon
Audlem
Overton
Llangollen
Cefn-mawr
Whitchurch
Woore
Chirk
Loggerheads
Llanarmon Dyffryn Ceiriog
Ellesmere
Market Drayton
Stone
Whittington
Hodnet
Oswestry
Wem
Eccleshall
Llanrhaeadr-ym-Mochnant
STAFFORDSHIRE
Llanfyllin
Gnosall
Shawbury
Newport
Lilleshall
Shrewsbury
Wellington
Wollaston
Oakengates
Llanfair Caereinion
Welshpool
Telford
Pontesbury
Shifnal
POWYS
Madeley
Albrighton
Cods
Garthmyl
SHROPSHIRE
Wolverhampton
Severn
Chirbury
Much Wenlock
Church Stoke
Montgomery
Church Stretton
Morville
Bridgnorth
Newtown
Lydham
Shipton
Rushbury
Kingswinford
Bishop's Castle
Burwarton
Alveley
Craven Arms
Stourbridge
Clun
Highley
Llanfair Waterdine
Kidderminster
Llanbadarn Fynydd
Bromfield
Cleobury Mortimer
Llanbister
Ludlow
Bewdley
Knighton
Stourport-on-Severn
Tenbury Wells
Wigmore
Lugg
Brimfield
Penybont
Presteigne
WORCESTERSHIRE
Teme
Mortimer's Cross
Ombersley
Leominster
Worcester
HEREFORDSHIRE
Bromyard
0 10 20 kilometres
0 10 miles
© Global Mapping / XYZ Maps

Welcome to our wonderful Shropshire gardens, set in a predominantly rural county with stunning countryside, fascinating geology, and a wealth of historic towns, castles and hillforts.

In 2023 we have 54 gardens opening, including 13 new or returning gardens, ranging from small cottage or town plots, to large country estates. Gardens are seen as being increasingly important to our wellbeing, and the wellbeing of the planet through encouraging wildlife, so what better than a garden visit to find inspiration for your own plot, whatever its size? As entry to most gardens is £6 or less, with children free, all the family can enjoy a great day out whilst raising essential funds for our nursing and health charities. To help plan your visits, photos and extended descriptions can be found on our website, www.ngs.org.uk.

In addition to published garden openings, we expect to have a number of 'pop-up' events advertised on the website and through social media. To make sure you don't miss out, why not join our supporters' email list by emailing Andy to receive regular updates throughout the season. Or why not become more involved through hosting your own Great British Garden Party? Look for information on the website or contact Angela for further details.

If you love gardens and have a little time to spare, why not consider joining our friendly team of volunteers? We couldn't open our gardens without their support with publicity, distributing leaflets or attending garden open days. For ideas on how you could help, please contact Sheila. Finally, we are always on the lookout for new gardens to open for us, so if you would like to chat about this, Andy would love to hear from you.

Volunteers

County Organiser
Andy Chatting
07546 560615
andy.chatting@ngs.org.uk

Treasurer
Elaine Jones 01588 650323
elaine.jones@ngs.org.uk

Volunteers Co-ordinator
Sheila Jones 01743 244108
smaryjones@icloud.com

Publicity
Douglas Wood 07768 058730
douglas.woods@ngs.org.uk

Facebook/Twitter
Vicky Kirk 01743 821429
vicky.kirk@ngs.org.uk

Instagram
John Butcher 07817 443837
butchinoz@hotmail.com

Booklet Co-ordinator
Fiona Chancellor 01952 507675
fiona.chancellor@ngs.org.uk

Assistant County Organisers
Angela Woolrich
angelawoolrich@hotmail.co.uk

Lynne Beavan
lynne.beavan@googlemail.com

Ruth Dinsdale
ruth.dinsdale@ngs.org.uk

Sue Griffiths
sue.griffiths@btinternet.com

Above: The Mill House

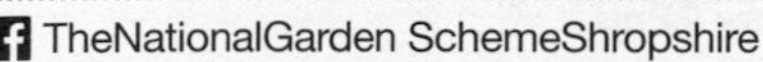

TheNationalGarden SchemeShropshire

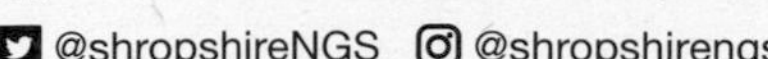

@shropshireNGS @shropshirengs

OPENING DATES

All entries subject to change. For latest information check **www.ngs.org.uk**

Map locator numbers are shown to the right of each garden name.

February

Snowdrop Festival

Sunday 19th
Millichope Park 28

March

Wednesday 29th
NEW Hundred House Hotel 20

April

Sunday 16th
Edge Villa 9

Sunday 23rd
Westwood House 51

Wednesday 26th
NEW Hundred House Hotel 20

Sunday 30th
Millichope Park 28

May

Monday 1st
Ruthall Manor 40

Wednesday 3rd
Neen View 31

Sunday 7th
Henley Hall 16
Longner Hall 24
Oteley 36

Wednesday 10th
◆ Goldstone Hall Gardens 13
Neen View 31

Saturday 13th
The Paddock 37
Rorrington Lodge 39

Sunday 14th
The Ferns 10
NEW The Leasowes, Cound 22
Longden Manor 23

Saturday 20th
Kinton Grove 21
Ruthall Manor 40

Sunday 21st
Brownhill House 5
Kinton Grove 21
Ruthall Manor 40
Stanley Hall Gardens 44

Tuesday 23rd
Brownhill House 5

Saturday 27th
Beaufort 3
NEW ◆ Burford House Gardens 6

Sunday 28th
NEW ◆ Burford House Gardens 6
◆ Walcot Hall 49

Monday 29th
◆ Walcot Hall 49

June

Saturday 3rd
NEW Tower House 47
Windy Ridge 53

Sunday 4th
Windy Ridge 53

Saturday 10th
Ruthall Manor 40
Westhope College 50

Sunday 11th
The Gardeners Lodge 12
Ruthall Manor 40
Sunningdale 46

Wednesday 14th
◆ Goldstone Hall Gardens 13

Thursday 15th
Four Winds 11

Friday 16th
NEW The Mill House 27

Saturday 17th
Cruckfield House 7
NEW Horatio's Garden 19

Sunday 18th
Preen Manor 38

Wednesday 21st
NEW Hundred House Hotel 20

Saturday 24th
Grooms Cottage 15
The Secret Gardens at Steventon Terrace 43

Sunday 25th
The Bramleys 4
NEW 1 Grange Court 14
◆ Hodnet Hall Gardens 17

Tuesday 27th
Brownhill House 5

July

Saturday 1st
The Albrighton Trust Moat & Gardens 1
NEW 1 Scotsmansfield 42

Sunday 2nd
Upper Marshes 48

Wednesday 5th
◆ Goldstone Hall Gardens 13

Saturday 8th
Lower Brookshill 25

Sunday 9th
Lower Brookshill 25
NEW 17 Mortimer Road 30

Saturday 15th
Ruthall Manor 40
153 Willoughbridge 52

Sunday 16th
NEW Eaton Mascott Hall 8
Ruthall Manor 40
Stottesdon Village Open Gardens 45
153 Willoughbridge 52

Wednesday 19th
NEW Hundred House Hotel 20

Thursday 20th
Appledore 2
Offcot 32

Friday 21st
Appledore 2
Offcot 32

Saturday 22nd
Appledore 2
Offcot 32

Sunday 23rd
Oswestry Gatacre Allotments & Gardens Association 35

August

Sunday 13th
The Ferns 10

Wednesday 16th
◆ Goldstone Hall Gardens 13

Sunday 20th
Edge Villa 9

Sunday 27th
Sambrook Manor 41

September

Saturday 2nd
NEW ◆ Burford House Gardens 6

Sunday 3rd
NEW ◆ Burford House Gardens 6

Tuesday 5th
◆ Wollerton Old Hall 54

Wednesday 13th
◆ Goldstone Hall Gardens 13

Saturday 16th
Ruthall Manor 40

Sunday 17th
Ruthall Manor 40

October

Friday 6th
Appledore 2
Offcot 32

Saturday 7th
Appledore 2
Offcot 32

Sunday 15th
Millichope Park 28

By Arrangement

Arrange a personalised garden visit with your club, or group of friends, on a date to suit you. See individual garden entries for full details.

The Albrighton Trust Moat & Gardens 1
Brownhill House 5
Cruckfield House 7
Edge Villa 9
The Hollies 18
Kinton Grove 21
NEW The Leasowes, Cound 22
Merton 26
Moat Hall 29
NEW The Old Rectory, Hodnet 33
The Old Vicarage, Clun 34
Oswestry Gatacre Allotments & Gardens Association 35
Ruthall Manor 40
Sambrook Manor 41
The Secret Gardens at Steventon Terrace 43
Sunningdale 46
NEW Tower House 47
153 Willoughbridge 52
Windy Ridge 53

Our 2022 donation enabled 7,000 people affected by cancer to benefit from the garden at Maggie's in Oxford

THE GARDENS

1 THE ALBRIGHTON TRUST MOAT & GARDENS

Blue House Lane, Albrighton, Wolverhampton, WV7 3FL. Stephen Jimson, 01902 372441, moat@albrightontrust.org.uk, www.albrightontrust.org.uk. *Off A41 & adjacent to motorway network. We are very nr Albrighton train stn & RAF Cosford base.* **Sat 1 July (11-3). Adm £5, chd free. Home-made teas. Visits also by arrangement 1 Apr to 1 July.**

The Albrighton Trust gardens are designed around the remains of a C13 fortified manor house and an ancient moat, offering excellent recreational and educational opportunities for anyone wanting to visit and enjoy this award winning outdoor space. The gardens are wheelchair accessible, there are disabled toilets and a personal care room with hoist and couch.

2 APPLEDORE

Kynaston, Kinnerley, Oswestry, SY10 8EF. Lionel parker. *Just off A5 on Wolfshead r'about (at Oswestry end of Nesscliffe bypass) towards Knockin. Then1st L towards Kinnerley. Follow NGS signs from this road.* **Thur 20, Fri 21, Sat 22 July, Fri 6, Sat 7 Oct (10-4). Combined adm with Offcot £6, chd free. Light refreshments.**

Appledore adjoins neighbouring garden Offcot. It has gone through extensive changes and is managed by Tom Pountney from Offcot. Both gardens are opening together this year so that visitors can see a developing 'edible garden' with 6 large beds for a wide range of fruit and vegetables, all grown organically. To encourage pollinators, the garden also has a wildlife pond and a small wildflower meadow.

3 BEAUFORT

Coppice Drive, Moss Road Wrockwardine Wood, Telford, TF2 7BP. Mike King, www.carnivorousplants.uk.com. *Approx 2m N from Telford town centre. From Asda Donnington, turn L at lights on Moss rd, ⅓m, turn L into Coppice Drive. 4th Bungalow on L with solar panels.* **Sat 27 May (10-5). Adm £5, chd free. Light refreshments.**

If carnivorous plants are your thing, then come and visit our National Collection of Sarracenia (pitcher plants); also over 100 different Venus flytrap clones (Dionaea muscipula), Sundews (Drosera) and Butterworts (Pinguicula) - over 6000 plants in total. Large greenhouses at Telford's first carbon negative house; a great place to visit - kids will love it! Visitors are asked to wear a face covering in the greenhouses, please. Regret, greenhouses are not wheelchair accessible.

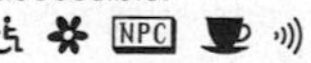

4 THE BRAMLEYS

Condover, Shrewsbury, SY5 7BH. Toby & Julie Shaw. *3m S of Shrewsbury. Through Condover towards Dorrington. Pass village hall follow road round, cross the bridge, in approx 100 metres there is a drive on L. Parking limited, please park at school & walk down to garden.* **Sun 25 June (11-5). Combined adm with 1 Grange Court £6, chd free. Home-made teas.**

A large country garden extending to 2 acres with a variety of trees and shrubs, herbaceous borders and a woodland with the Cound Brook flowing through. A courtyard oasis welcomes you as you enter the garden with far reaching views over open countryside. Wheelchair access around most of the garden, although not for the woodland area.

5 BROWNHILL HOUSE

Ruyton XI Towns, SY4 1LR. Roger & Yoland Brown, 01939 261121, brownhill@eleventowns.co.uk, www.eleventowns.co.uk. *9m NW of Shrewsbury on B4397. On the B4397 in the village of Ruyton XI Towns.* **Sun 21, Tue 23 May, Tue 27 June (10-6.30). Adm £5, chd free. Pre-booking essential, please visit www.ngs.org.uk for information & booking. Home-made teas. Visits also by arrangement 15 Apr to 15 July.**

A unique 2 acre hillside garden with many steps and levels bordering River Perry. Visitors can enjoy a wide variety of plants and styles from formal terraces to woodland paths. The Good Garden Guide said 'It has to be seen to be believed'. The lower areas are for the sure-footed while the upper levels with a large kitchen garden and glasshouses have many places to sit and enjoy the views. Kit cars on show.

Our 2022 donation to ABF The Soldier's Charity meant 372 soldiers and veterans were helped through horticultural therapy

6 NEW ◆ BURFORD HOUSE GARDENS

Burford House, Burford, Tenbury Wells, WR15 8HQ. British Garden Centres, 01584 810777, pbenson@britishgardencentres.com, www.britishgardencentres.com/burford-house-garden-centre. *1m W of Tenbury Wells on A456. Follow signs for Burford House Garden Centre off the A456.* **For NGS: Sat 27, Sun 28 May, Sat 2, Sun 3 Sept (10-4.30). Adm £5, chd free. Cream teas inside the wonderful Burford House on NGS open days. For other opening times and information, please phone, email or visit garden website.**

The garden is currently being restored by the Garden Angels (volunteers) under the leadership of the newly appointed Head Gardener. 4 acres of sweeping lawns bordered by the River Teme, and serpentine borders, set in the beautiful Teme Valley, around an elegant Georgian House (open for teas). Designed by the late, great plantsman, John Treasure and featuring over 100 varieties of clematis in addition to a myriad of plants in wonderful combinations and colours. Garden centre offers a comprehensive range of plants, gifts and outdoor leisure. Well behaved dogs on lead welcomed.

7 CRUCKFIELD HOUSE

Shoothill, Ford, SY5 9NR. Geoffrey Cobley, 01743 850222. *5m W of Shrewsbury. A458 from Shrewsbury, turn L towards Shoothill.* **Sat 17 June (2-5). Adm £8, chd free. Home-made teas. Visits also by arrangement 17 June to 15 July.**

An artist's romantic 3 acre garden, formally designed, informally and intensively planted with a great variety of unusual herbaceous plants. Nick's garden, with many species' trees, shrubs and wildflower meadow, surrounds a lake with bog and moisture-loving plants. Ornamental kitchen garden. Rose and peony walk. Courtyard fountain garden, large shrubbery and extensive clematis collection. Extensive topiary, and lily pond.

8 NEW EATON MASCOTT HALL

Eaton Mascott, Cross Houses, Shrewsbury, SY5 6HG. *About 7m S of Shrewsbury on A458 2nd R past Cross Houses, thereafter follow signs.* **Sun 16 July (11-4.30). Adm £5, chd free. Home-made teas.**

A 6 acre garden consisting of 1½ acre walled garden with extensive rose collection. Emphasis on symmetry with pergolas, water features and cross views. Bamboo walk and acer collection, specimen trees. 4½ acre woodland garden with walks and glimpses of parkland and surrounding countryside, ferns, azaleas and rhododendrons. Wheelchair access possible if good wheels and accompanied by helper.

♿ ☕

9 EDGE VILLA

Edge, nr Yockleton, Shrewsbury, SY5 9PY. Mr & Mrs W F Neil, 01743 821651, billfneil@me.com. *6m SW of Shrewsbury. From A5 take either A488 signed to Bishops Castle or B4386 to Montgomery for approx 6m then follow NGS signs.* **Sun 16 Apr, Sun 20 Aug (2-5). Adm £5, chd free. Home-made teas. Visits also by arrangement 1 Apr to 30 Aug for groups of 10+.**

Two acres nestling in South Shropshire hills. Self-sufficient vegetable plot. Chickens in orchard, foxes permitting. Large herbaceous borders. Dewpond surrounded by purple elder, irises, candelabra primulas and dieramas. Large selection of fragrant roses. Wendy house for children. Extensive plant sale in aid of NGS charities. Book sale in aid of NGS charities. Visitors can stay in the garden for as long as they like. Some German and French spoken. Some gravel paths.

10 THE FERNS

Newport Street, Clun, SY7 8JZ. Andrew Dobbin. *Enter Clun from Craven Arms. Take 2nd R signed into Ford St. At T-junction turn R for parking in the Memorial Hall car park (100 yds). Retrace steps to T junction. The Ferns is on L.* **Sun 14 May, Sun 13 Aug (12-6). Adm £5, chd free.**

A formal village garden of ¾ acre, approached via a drive lined with crab apple and pear trees. On the right is the autumn garden, giving fine views of the surrounding hills. From the front courtyard garden a path leads through double herbaceous borders full of late summer colour, to further rooms, of yew, beech and box. There is also a rear courtyard with tender exotics. Lily pond and statuary.

11 FOUR WINDS

Gilberts Lane, Whixall, Whitchurch, SY13 2PR. Lynne Beavan. *3m N of Wem. B5476 through Quina Brook turn L onto Coton Park/ Gilbert's Lane follow rd bearing L at junction Whitchurch B5476 turn R past Bull & Dog take 1st L past social centre then L Gilbert's Lane.* **Thur 15 June (10.30-4.30). Adm £5, chd free. Home-made teas.**

The traditional garden around the bungalow is backed by 150yr old oak trees. In the adjoining paddock, a garden is being planted out, now 3-4 yrs old. Fruit trees, ornamental trees silver birches, shrub beds, hedges, pergolas with roses and wisteria, perennial beds, alliums making a delightful garden for wildlife and family. An ongoing evolving project for the owner.

12 THE GARDENERS LODGE

2 Roseway, Wellington, TF1 1JA. Amanda Goode, www.lovegrowshereweb.wordpress.com/. *1½m (4 mins) from J7 (M54). B5061 Holyhead Rd 2nd L after NT 'Sunnycroft'. (New Church Rd) We are the cream house on corner of NCR & Roseway.* **Sun 11 June (11-4.30). Adm £5, chd free. Home-made teas.**

There was just one tree in the garden when the current owner purchased The Gardener's Lodge, the ground having been cleared, ready to sell as a building plot. However the owner had other plans for it: the garden is now eclectically divided, arranged and planted into areas: Mediterranean, Cottage, Indian etc but, seamlessly, each section merges and leads into the next developing idea. A small urban garden with seating, water features and shade.

13 ◆ GOLDSTONE HALL GARDENS

Goldstone, Market Drayton, TF9 2NA. John Cushing, 01630 661202, enquiries@goldstonehall.com, www.goldstonehall.com. *5m N of Newport on A41. Follow brown signs from Hinstock. From Shrewsbury A53, R for A41 Hinstock & follow brown signs & NGS signs.* **For NGS: Wed 10 May, Wed 14 June, Wed 5 July, Wed 16 Aug, Wed 13 Sept (11-4.30). Adm £7.50, chd free. Cream teas in award winning oak framed pavilion with**

Hundred House Hotel

cakes created by our talented chef. For other opening times and information, please phone, email or visit garden website.
5 acres with highly productive beautiful kitchen garden. Unusual vegetables and fruits - alpine strawberries; heritage tomatoes, salad, chillies, celeriac. Roses in walled garden from May; Double herbaceous in front of old English garden wall at its best July and August; sedums and roses stunning in September. Winner of the prestigious Good Hotel Guide's Editor's Choice Award for Gardens. Lawn aficionados will enjoy the stripes, Extensive kitchen garden, Box hedge growing sign. Majority of garden can be accessed on gravel and lawns.

14 NEW 1 GRANGE COURT

Condover, Shrewsbury, SY5 7BU. Vicky Barnes. *5m S of Shrewsbury. If travelling N on A49 turn R in Dorrington and follow Station Road to Condover. Avoid turning R off A49 further N. If travelling S on A49 just follow signs for Condover then for Golf Course.* **Sun 25 June (11-5). Combined adm with The Bramleys £6, chd free.**
A love of nature at the heart of the garden, this small space celebrates the beauty of plants and the wonder of soil. Roses. Herbaceous. Tiny meadows and hedgerows. Wildlife pond. No dig beds. Regenerative. Water conserving, composting, organic. RHS Gold Award winner designed 2006. From 2020 a productive potager showcases the front of the property. A nearby roadside verge is managed as a meadow.

15 GROOMS COTTAGE

Waters Upton, Telford, TF6 6NP. Joanne & Andi Butler, www.instagram.com/groomscottagegarden. *No parking at the property due to restricted access; parking available on the verge opp the church or in the village hall car park. The garden will be signed nearby.* **Sat 24 June (10-5). Adm £5, chd free. Pre-booking essential, please visit www.ngs.org.uk for information & booking. Light refreshments.**
A cottage garden around ½ an acre in size comprising mixed herbaceous borders and an abundance of English roses. Highlights inc water features, pergolas, oriental garden, hosta garden, alpine and herb beds and a productive vegetable garden and greenhouse. Many established trees and shrubs and shady plant area. Wheelchair access, some gravel areas not accessible. Parking on the road, with wheeled access down the tarmac drive.

16 HENLEY HALL

Henley, Ludlow, SY8 3HD. Helen & Sebastian Phillips, www.henleyhallludlow.com. *2m E of Ludlow. 1½m from A49. Take A4117 signed to Clee Hill & Cleobury Mortimer. On reaching Henley, the gates to Henley Hall are on R when travelling towards Clee Hill.* **Sun 7 May (10-5). Adm £6, chd free. Home-made teas.**

Henley Hall offers a mixture of formal and informal gardens. The historic elements inc a formal lawn, stone staircase with balustrades, a ha-ha and an ornate stone arched bridge with decorative weirs over the river Ledwyche, created in 1874. Other highlights inc walkways along the banks of the river and woodland paths. The listed Orangery will serve as a 'pop-up' tea room at the garden opening. Some parts of the garden are accessible by wheelchair.

17 ◆ HODNET HALL GARDENS

Hodnet, Market Drayton, TF9 3NN. Sir Algernon & The Hon Lady Heber-Percy, 01630 685786, secretary@hodnethall.com, www.hodnethallgardens.org. *5½m SW of Market Drayton. 12m NE Shrewsbury. At junction of A53 & A442. Tickets available on the gate, on the day.* **For NGS: Sun 25 June (10-5). Adm £9, chd free. Light refreshments in the Garden Restaurant. For other opening times and information, please phone, email or visit garden website.**

The 60+ acres of Hodnet Hall Gardens are amongst the finest in the country. There has been a park and gardens at Hodnet for many hundreds of years. Magnificent forest trees, ornamental shrubs and flowers planted to give interest and colour from early Spring to late Autumn. Woodland walks alongside pools and lakes, home to abundant wildlife. Productive walled kitchen garden and historic dovecot. For details please see website and Facebook page. Maps are available to show access for our less mobile visitors.

18 THE HOLLIES

Rockhill, Clun, SY7 8LR. Pat & Terry Badham, 01588 640805, patbadham@btinternet.com. *10m W of Craven Arms. 8m S of Bishops Castle. From A49 Craven Arms take B4368 to Clun. Turn L onto A488 continue for 1½m. Bear R signed Treverward. After 50 yds turn R at Xrds, property is 1st on L.* **Visits by arrangement June to Aug. Adm £5, chd free.**

A garden of 2 acres at 1200ft. Features inc kitchen garden, large island beds, borders with perennials, shrubs, grasses, specimen bamboos and trees. Birch grove and wildlife dingle with stream, rain permitting! Refreshments can be arranged at a local cafe in Clun

1 Scotsmanfield

(daytime hours). We leave signs out on the road so anyone passing can visit on spec! If the gates are shut, then we are out for the day. Wheelchair access is available to the majority of the garden over gravel and grass.

19 NEW **HORATIO'S GARDEN**

The Robert Jones & Agnes Hunt Orthopaedic Hospital, Gobowen, Oswestry, SY10 7AG. Horatio's Garden, www.horatiosgarden.org.uk. *From A5 follow signs to Orthopaedic Hospital; park in pay & display car park by hospital. Once parked, facing main hospital entrance turn R at mini r'about & walk to end of internal road to grey gate.* **Sat 17 June (2-5). Adm £5, chd free. Home-made teas.**

Beautifully designed by Bunny Guinness and delightfully planted, Horatio's Garden Midlands opened to great acclaim in September 2019. Partly funded by the National Garden Scheme, the garden offers a therapeutic place of peace for patients, their loved ones and NHS staff spending time in the Midland Centre for Spinal Injuries. Featuring raised beds, beautiful specimen trees, multi-purpose garden room, gorgeous glasshouse, and a pretty rill running the length of the garden, this is a horticultural sanctuary like no other. Very good access throughout.

20 NEW **HUNDRED HOUSE HOTEL**

Bridgnorth Road, Norton, Telford, TF11 9EE. Henry Phillips, reservations@hundredhouse.co.uk, www.hundredhouse.co.uk. *Situated on the A442 in the village of Norton. Midway between Bridgnorth & Telford.* **Wed 29 Mar, Wed 26 Apr, Wed 21 June, Wed 19 July (10-6). Adm by donation. Light refreshments.**

An acre of gardens crafted over 35 years, complete with sculptures, stonework and working herb and kitchen garden. From Mar onwards, you will find over 5000 bulb flowers inc tulips, hyacinths, daffodils, alliums, blossoming trees and dancing pond flowers. Summer brings a variety of flowers, inc David Austin Roses, giant agapanthus, dahlias, clary, foxgloves, fuchsias and arum lilies. Please see website for restaurant booking and information.

21 **KINTON GROVE**

Kinton, SY4 1AZ. Tim & Judy Creyke, 01743 741263, judycreyke@icloud.com. *Off A5, between Shrewsbury & Oswestry, approx 1m from Nesscliffe. Coming from Shrewsbury to Oswestry on the A5, take the 3rd exit off the Wolfshead r'about (at the end of the Nesscliffe bypass) & follow directions to Kinton, approx 1m.* **Sat 20, Sun 21 May (11-5). Adm £5, chd free. Home-made teas. Visits also by arrangement 1 Apr to 1 Oct.**

A garden of ¾ acre surrounding a Georgian house. The garden features well filled herbaceous borders, roses, a gravel area, raised vegetable beds, and a wide range of interesting trees and shrubs. Hedges, inc some as old as the house, divide up the garden. Lovely views across the Breidden hills and plenty of pleasant places to sit, whilst enjoying cream tea and cakes. Wheelchair access is possible, but on uneven, narrow paths.

22 NEW **THE LEASOWES, COUND**

Cound, Shrewsbury, SY5 6AF. Robert & Tricia Bland, 07802 667636, rjbbland1@outlook.com. *7m SE of Shrewsbury. On the A458 SE of Shrewsbury, 500m to the E of the Riverside Inn adjoining Oakleys, the lawnmower business.* **Sun 14 May (10.30-5). Adm £6, chd free. Light refreshments. Visits also by arrangement Apr to Nov for groups of 15 to 25.**

Within the super structure of mature larches and interconnected ponds, a wide range of rhododendrons and azaleas have been planted as well as shade appreciating trees and shrubs over 10 acres. A large offering of roses of all descriptions provide colour throughout the formal gardens.

23 **LONGDEN MANOR**

Plealey, SY5 0XL. Karen Lovegrove. *From Shrewsbury take C150 Longden Road. On entering Longden Village, opp The Tankerville Arms & Village Shop on L, turn R into Manor Ln.* **Sun 14 May (10-4). Adm £5, chd free. Home-made teas.**

Large estate garden with lots of character and interest: woodland walks and grass paths; humorous topiary; wide variety of specimen trees; rhododendrons and azaleas; wildflowers; restored water garden; panoramic vistas of surrounding countryside; holly garden. Also, giant Jenca, skittles and croquet - a great place for all the family to visit. Topiary trail for children inc giraffe, pelican, otter, kingfisher, shark, gorilla and a goose.

24 **LONGNER HALL**

Atcham, Shrewsbury, SY4 4TG. Mr & Mrs R L Burton, www.longner.co.uk. *4m SE of Shrewsbury. From M54 follow A5 to Shrewsbury, then B4380 to Atcham. From Atcham take Uffington Rd, entrance ¼m on L.* **Sun 7 May (2-5). Adm £6, chd free. Home-made teas.**

A long drive approach through parkland designed by Humphry Repton. Walks lined with golden yew through extensive lawns, with views over Severn Valley. Borders containing roses and herbaceous shrubs, also ancient yew wood. Enclosed 1 acre walled garden open to NGS visitors. Mixed planting, garden buildings, tower and game larder. Short woodland walk around old moat pond which is not suitable for wheelchairs.

25 **LOWER BROOKSHILL**

Nind, nr Lydham, SY5 0JW. Patricia & Robin Oldfield, 01588 650137, robin.oldfield@live.com. *3m N of Lydham on A488. Take signed turn to Nind & after ½m sharp L & follow narrow rd for another ½m. Drive very slowly & use temporary passing places.* **Sat 8, Sun 9 July (1-5). Adm £6, chd free. Cream teas.**

10 acres of hillside garden and woods at 950ft within the AONB. Begun in 2010 from a derelict and overgrown site, cultivated areas now rub shoulders with the natural landscape using fine borrowed views over and down a valley. Inc brookside walks, a 'pocket' park, four ponds (inc a Monet lily pond), mixed borders and lawns, cottage garden (with cottage) and wildflowers. Lots of lovely picnic spots.

26 MERTON

Shepherds Lane, Bicton, Shrewsbury, SY3 8BT. David & Jessica Pannett, 01743 850773, jessicapannett@hotmail.co.uk. *3m W of Shrewsbury. Follow B4380 from Shrewsbury past Shelton for 1m Shepherd's Ln turn L garden signed on R or from A5 by pass at Churncote r'about turn towards Shrewsbury 2nd turn L Shepherds Ln.* **Visits by arrangement June to Sept for groups of up to 25. Adm £5, chd free.**
Mature ½ acre botanical garden with a rich collection of trees and shrubs inc unusual conifers from around the world. Hardy perennial borders with seasonal flowers and grasses plus an award winning collection of hosta varieties in a woodland setting. Outstanding gunneras in a waterside setting with moisture loving plants. Tea and biscuits £1 or visitors may bring own refreshments. Wheelchair access, level paths and lawns.

27 NEW THE MILL HOUSE

Mill Lane, Ruyton XI Towns, Shrewsbury, SY4 1LR. Debbie Sargent & Jeffrey Ewin. *SE of Ruyton X1Towns. Turn S off B4397 at eastern end of Ruyton X1 Towns by Platt Bridge. Follow narrow lane ½m. Mill House is where tarmac finishes.* **Fri 16 June (11-4). Adm £5, chd free. Home-made teas.**
Approx 4 acres of garden situated alongside the River Perry. Herbaceous borders inc roses, peonies and climbers. A terrace behind the house, sloping lawns, and unusual trees in a naturalised Arboretum. Mill pond. Vegetable garden. Wildlife friendly. Uneven ground and some sloping pathways. Adjacent to Arboretum with national collection of Betula.

28 MILLICHOPE PARK

Munslow, SY7 9HA. Mr & Mrs Frank Bury, www.millichopepark.com. *8m NE of Craven Arms. off B4368 Craven Arms to Bridgnorth Rd. Nr Munslow then follow yellow signs.* **Sun 19 Feb (2-5). Sun 30 Apr (2-6). Tea at the Hall (Apr only). Sun 15 Oct (2-5). Adm £6, chd free.**
Historic landscape gardens covering 14 acres with lakes, and cascades dating from C18, woodland walks and wildflowers. Snowdrops feature in February, bluebells and violas in May, and autumn colours in October.

29 MOAT HALL

Annscroft, Shrewsbury, SY5 8AZ. Martin & Helen Davies, 01743 860216, helenatthefarm@hotmail.co.uk. *Take the Longden road from Shrewsbury to Hook a Gate. Our lane is 2nd on R after Hook a Gate & before Annscroft. Single track lane for ½m.* **Visits by arrangement Mar to Sept for groups of 10+. Adm £6, chd free. Teas £4.**
1 acre garden around an old farmhouse within a dry moat which can be walked around. Well organised and extensive kitchen garden, fruit garden and orchard for self-sufficiency. Colourful herbaceous borders; stumpery; many interesting stone items inc troughs, cheese weights, staddle stones some uncovered in the garden. Plenty of seating areas on the lawns for enjoying the garden and a homemade tea. Wheelchair access: mostly lawn with one grass ramp and one concrete ramp; kitchen garden has 2' wide paved paths.

30 NEW 17 MORTIMER ROAD

Buntingsdale, Market Drayton, TF9 2EP. Martin & Stella Clifford-Jones. *What3Words - undivided/orbited/apprehend Access only via Tern Hill A41. North on A41, as you enter Tern Hill take first R into Hedley Way, or going S turn L into Hedley Way after mobile home sales site.* **Sun 9 July (2-5). Adm £5, chd free. Home-made teas.**
A wildlife-friendly garden with plantaholic owners. There are three ponds, a meadow area, and a huge range of plants from Enchanter's Nightshade to carnivorous pitcher plants. Most importantly, there are several delightful places to sit whilst enjoying your tea and cake! Wheelchair access to some of the garden.

31 NEEN VIEW

Neen Sollars, Cleobury Mortimer, DY14 9AB. Ian & Chris Ferguson. *3m from the Market Town of Cleobury Mortimer. From A456 past Mamble, turn R at Neen Sollars, follow lane to Live & Let Live for parking. From Cleobury take the Tenbury Rd, turn L & L again then up the hill to Neen View gardens.* **Wed 3, Wed 10 May (12-5). Adm £5, chd free.**
A woodland garden with stunning panoramic views of the Clee Hills and Teme Valley. Mature trees inc: azaleas, camellias and rhododendrons. Wildlife ponds and wildflower meadow. There are plenty of seats to enjoy the views. Walking around you will find a ruined Bothy and Japanese stream garden.

32 OFFCOT

Kynaston, Kinnerley, SY10 8EF. Tom Pountney. *Just off A5 on Wolfshead r'about (at Oswestry end of Nesscliffe bypass) towards Knockin. Then take the 1st L towards Kinnerley. Follow NGS signs from this road.* **Thur 20, Fri 21, Sat 22 July, Fri 6, Sat 7 Oct (10-4). Combined adm with Appledore £6, chd free. Home-made teas.**
A cottage garden with lots of winding pathways leading to different focal points. The garden is packed with a wide range of evergreen and deciduous trees and shrubs and underplanted with herbaceous perennials. There is a natural looking pond with a running stream feeding into it. A haven for wildlife. So many different areas to see and enjoy inc the garden bar.

33 NEW THE OLD RECTORY, HODNET

Hearne Lane, Hodnet, Market Drayton, TF9 3NG. Sarah Riley, 01630 685671, rileys55@icloud.com. *¼m NW of the village of Hodnet. SatNav takes you to the end of Hearne Ln. Bear L, on the single track road, up the hill towards the trees.* **Visits by arrangement Mar to Sept for groups of up to 20. Adm £5, chd free. Tea.**
3 ½ acres of gardens, created over 45 yrs by John and Elizabeth Ravenscroft. Formal gardens, a walled garden, large pool with stepping stones, lawns and wooded areas. Collections of azaleas, magnolias, peonies. After being somewhat neglected by the current owners for a couple of years as work on the house took priority, the garden is now being brought back in phases.

34 THE OLD VICARAGE, CLUN

Vicarage Road, Clun, SY7 8JG. Peter & Jay Upton, 01588 640775, jay@salopia.plus.com. *16m NW of Ludlow. Over bridge at Clun heading towards Knighton (parking in public car park by bridge); walk up*

17 Mortimer Road

to church, turn L into Vicarage Rd; house on R next to church. **Visits by arrangement Apr to Aug for groups of up to 30. Adm £5, chd free.**

A revived, old vicarage garden: a slow retrieval and recovery revealing a wealth of features and plants chosen by plantsmen vicars: buddleia globosa fascinates bees and butterflies; glorious oriental poppies fascinate visitors. 'The Tree', an enormous Leyland cypress (5th biggest girth in the world); the most dramatic feature is a formal wisteria allee with alliums - a symphony of mauve and purple. Refreshments in Clun centre.

35 OSWESTRY GATACRE ALLOTMENTS & GARDENS ASSOCIATION

Lloyd Street, Oswestry, SY11 1NL. Graham Mitchell, 01691 654961, gsfmitchell@gmail.com, gatacre.wordpress.com. *2 parts to the allotments either side of Liverpool Road. From Whittington r'about on A5 turn towards Oswestry B4580. In ½ m go across staggered junction (L then R). Continue 100y then bear sharp L. After 200y straight over r'about, then 3rd on L.* **Sun 23 July (11-4.30). Adm £5, chd free. Tea, coffee & cakes. Visits also by arrangement June to Aug for groups of 5 to 10.**

Two adjacent, large allotment communities in the heart of Oswestry; well supported and well-loved by local residents. Vast array of fruit, flowers, vegetables in every shape and form grown on the allotments. Allotment holders will be on-hand to talk about their produce and give advice. Entry ticket covers both sides of the allotments. The main pathways are wheelchair friendly, but the small paths tend to be rather steep or narrow. There are disabled WC on one side of the site.

36 OTELEY

Ellesmere, SY12 0PB. Mr RK Mainwaring, www.oteley.com. *1m SE of Ellesmere. Entrance out of Ellesmere past Mere, opp Convent nr to A528/495 junction.* **Sun 7 May (11-4). Adm £6, chd free. Light refreshments.**

10 acres running down to The Mere. Walled kitchen garden, architectural features, many old interesting trees. Rhododendrons, azaleas, wild woodland walk and views across Mere to Ellesmere. First opened in 1927 when the National Garden Scheme started. Beautiful setting on the Mere.

37 THE PADDOCK

Annscroft, Shrewsbury, SY5 8AN. Ian Ross. *On road from Annscroft to Hanwood/Plealey 200 yds from the Xrds. Parking is at The Farriers SY5 8AN - on the road from Annscroft to Exfords Green.* **Sat 13 May (11-5). Adm £6, chd free. Home-made teas.**

A mini arboretum in which 70 trees are tagged - some of which are rare. The collection inc 12 different Acers and a cluster of 6 Gingko Bilobas. A waterfall cascades through three levels. As the new owner since 2001 I see myself as the custodian of an impressive small garden. Wheelchair access is available, by paths and grass, to most of the garden.

38 PREEN MANOR

Church Preen, SY6 7LQ. Mr & Mrs J Tanner, 07971 955609, katytanner@msn.com. *6m W of Much Wenlock; For SatNav, please use postcode SY6 7LF. From A458 Shrewsbury to Bridgnorth Rd turn off at Harley and follow signs to Kenley & Church Preen. From B4371 Much Wenlock to Church Stretton Rd go via Hughley to Church Preen.* **Sun 18 June (2-6). Adm £6, chd free. Home-made teas.**

Fine historical and architectural 6 acre garden on site of Cluniac priory and former Norman Shaw mansion. Compartmentalised with large variety of garden rooms, inc kitchen parterre and fernery. Formal terraces with fine yew and hornbeam hedges have panoramic views over parkland to Wenlock Edge. Dell, woodland walks and specimen trees. Gardens also open by appointment (not NGS).

39 RORRINGTON LODGE

Rorrington, Chirbury, SY15 6BX. Adrian & Samantha Boyes. *Between Chirbury & Marton. Signed from Marton & Wotherton.* **Sat 13 May (2-5). Adm £5, chd free. Home-made teas.**

Hillside garden with spectacular 360 views of the surrounding countryside and Welsh Hills. Interesting collection of trees and shrubs in their spring beauty lovely places to wander with some very steep banks but accessible around most of the garden. Attractive parterre-style planting around the house and pool. Plenty of places to sit, have a cup of tea and admire the views. Teas in aid of local charities. Wheelchair access around most of the garden except steep banks.

40 RUTHALL MANOR

Ditton Priors, WV16 6TN. Mr & Mrs G T Clarke, 01746 712608, clrk608@btinternet.com. *7m SW of Bridgnorth. At Ditton Priors Church take road signed Bridgnorth. then 2nd L. Garden 1m.* **Mon 1, Sat 20, Sun 21 May, Sat 10, Sun 11 June, Sat 15, Sun 16 July, Sat 16, Sun 17 Sept (12.30-5.30). Adm £5, chd free. Home-made teas until 4:30 pm. Visits also by arrangement Apr to Sept.**

Offset by a mature collection of specimen trees, the garden is divided into intimate sections, carefully linked by winding paths. The front lawn flanked by striking borders, extends to a gravel, art garden and ha-ha. Clematis and roses scramble through an eclectic collection of wrought-iron

Windy Ridge

work, unique pottery and secluded seating. A stunning horse pond with primulas, iris and bog plants. Lots of lovely shrubs to see. Jigsaw puzzle sale. Donation to NGS from refreshments. Wheelchair access to most parts.

41 SAMBROOK MANOR

Sambrook, TF10 8AL. Mrs E Mitchell, 01952 550256, eileengran@hotmail.com . *Between Newport & Ternhill, 1m off A41. In the village of Sambrook.* **Sun 27 Aug (12-5). Adm £5, chd free. Home-made teas. Visits also by arrangement for groups of 10+.**

Deep, colourful, well planted borders offset by sweeping lawns surrounding an early C18 manor house (not open). Wide ranging herbaceous planting with plenty of roses to enjoy; the arboretum below the garden, with views across the river, has been further extended with new trees. The waterfall and Japanese garden are now linked by a pretty rill. Lovely garden to visit for all the family. Woodland area difficult for wheelchairs.

42 NEW 1 SCOTSMANSFIELD

Burway Road, Church Stretton, SY6 6DP. Peter Vickers & Hilary Taylor. *Scotsmansfield is on L, coming up Burway Rd. Accessible, by foot, up steep hill from Church Stretton. No parking at Scotsmansfield & v. limited on Burway Rd. Parking in Church Stretton (main car parks off Easthope Rd, SY6 6BL).* **Sat 1 July (11-5). Adm £5, chd free. Light refreshments.**

Terraced garden of 3/4 acre, renewed over a decade after long neglect. Attached to E wing of 'Scotsmansfield' (not open), built 1908. Celebration of favourite plants, with colour, shape, texture, scent, light, shade and plentiful wildlife. Trees, fernery, lily pond, mixed borders, yew hedges, lavish roses. Areas of tranquillity and intimacy, occasions of drama and long views of surrounding woods and hills. Limited access via steep gravelled paths to some areas. Wheelchair users will need support from a competent and fit assistant!

GROUP OPENING

43 THE SECRET GARDENS AT STEVENTON TERRACE

Steventon Terrace, Steventon New Road, Ludlow, SY8 1JZ. Kevin & Carolyn Wood, 01584 876037, carolynwood2152@yahoo.co.uk. *Gardens are located behind row of terraced cottages. Easily accessible from A49; on-street parking; Park & Ride stops outside the garden.* **Sat 24 June (12.30-4.30). Combined adm £5, chd free. Home-made teas & ice cream. Visits also by arrangement June to Sept.**

Very secret gardens behind a row of Victorian terraced cottages in Ludlow: 1/2 acre plot south facing garden developed over 30 years with the love of gardening divided up into different sections rose garden, herbaceous borders, koi fish, chickens, polytunnel, Mediterranean garden. Heart of England in Bloom chairman's award. Tropical garden, chickens, large polytunnel greenhouse large fishpond.

44 STANLEY HALL GARDENS

Bridgnorth, WV16 4SP. Mr & Mrs M J Thompson. *1/2 m N of Bridgnorth. Leave Bridgnorth by N gate B4373; turn R at Stanley Lane. Pass Golf Course Club House on L & turn L at Lodge.* **Sun 21 May (2-6). Adm £5, chd free. Home-made teas.**

Georgian landscaped drive with rhododendrons, fine trees in parkland setting, woodland walks and fish ponds. Restored ice house. Dower House (Mr and Mrs C Wells): 4 acres of specimen trees, contemporary sculpture, walled vegetable garden and potager; South Lodge (Mr Tim Warren) Hillside cottage garden. Wheelchair access to the main gardens.

GROUP OPENING

45 STOTTESDON VILLAGE OPEN GARDENS

Stottesdon, DY14 8TZ. *In glorious S Shropshire near Cleobury Mortimer (A4117/B4363). 30m from Birmingham (M5/42), 15m E of Ludlow (A49/4117) & 10m S of Bridgnorth (A458/442) Stottesdon is between Clee Hill & the Severn Valley. NGS Signed from B4363.* **Sun 16 July (12-4). Combined adm £5, chd free. Home-made teas in the Parish Church.**

Located in unspoilt countryside near the Clee Hills, up to 8 gardens and the heritage church in Stottesdon village are open to visitors. Several places have stunning views. Some gardens feature spaces for outdoor living. Many are traditional or more modern 'cottage gardens', containing fruit, vegetables and livestock. There are contrasting vegetable gardens inc one devoted to permaculture principles and one to growing championship winners. Take teas and refreshments In the Norman church and book on the day to join a unique guided tour of the historic Tower, Bells and Turret Clock. A garden-related competition to be held and be judged by garden visitors. Dogs on leads please. Heritage Church open - tower tours. Lunches may be available at The Fighting Cocks pub - call on 01746 718270 to pre-book (essential). Most gardens have some wheelchair access. Those gardens not suitable for wheelchair access will be listed on our gardens guide.

46 SUNNINGDALE

9 Mill Street, Wem, SY4 5ED. Mrs Susan Griffiths, 01939 236733, sue.griffiths@btinternet.com. *Town centre. Wem is on B5476. Parking in public car park Barnard St. The property is opp the purple house below the church. Some on street free parking on the High St. Coaches can drop off at the main gate.* **Sun 11 June (10.30-4). Adm £4, chd free. Home-made teas. Visits also by arrangement 28 Jan to 3 Dec for groups of 5 to 60.**

1/2 acre town garden. Wildlife haven for a variety of birds inc nesting gold crests. A profusion of excellent nectar rich plants means that butterflies and other pollinators are in abundance. Interesting plantings with carefully collected rare plants and unusual annuals means there is always something new to see. Large perennial borders, with exotic climbers, designed as an all yr-round garden. Koi pond and natural stone waterfall rockery. Antique and modern sculpture. Sound break yew walkway. Wheelchair access the garden is on the level but with a number of steps mostly around the pond area; paths are mainly gravel or flags; flat lawn.

1 Grange Court

47 NEW **TOWER HOUSE**
Bache, Craven Arms, SY7 9LN. Lady Spicer, 01584 861692, nicspicer@yahoo.com. *7m NW of Ludlow. B4365 Ludlow to Much Wenlock. Turn L after 3m (Bache, Burley), garden at top of hill after 1½m. B4368 Craven Arms to Bridgnorth. Turn R at Xrds after 1½m immed fork L, garden after ¾m.* **Sat 3 June (11-5). Adm £5, chd free. Home-made teas. Visits also by arrangement 1 Apr to 16 July for groups of up to 40.**
Tower House, a folly, built on the orders of Benjamin Flounders in (we think) 1838 is a twin to his more well known folly on Callow Hill. The garden of just under 2 acres is designed with the panoramic views of the Corvedale in mind. Created gradually over 50yrs by Priscilla, when her painting allows, has formal and informal flowers, vegetables and a small wooded area in the quarry. Rough paths.

48 **UPPER MARSHES**
Catherton Common, Hopton Wafers, nr Cleobury Mortimer, DY14 0JJ. Jo & Chris Bargman. *3m NW of Cleobury Mortimer. From A4117 follow signs to Catherton. Property is on common land 100yds at end of track.* **Sun 2 July (12-5). Adm £5, chd free. Home-made teas.**
Commoner's stone cottage and 3 acre small holding. 800' high. Garden has been developed to complement its unique location on edge of Catherton common with herbaceous borders, vegetable plot, herb garden. Short walk down to a spring fed wildlife pond. Plenty of seats to stop and take in the tranquillity. Optional circular walk across Wildlife Trust common to SSI field. Various animals and poultry.

49 ◆ **WALCOT HALL**
Lydbury North, SY7 8AZ. Mr & Mrs C R W Parish, 01588 680570, enquiries@walcothall.com, www.walcothall.com. *4m SE of Bishop's Castle. B4385 Craven Arms to Bishop's Castle, turn L by The Powis Arms in Lydbury North.* **Sun 28, Mon 29 May (1.30-5). Adm £5, chd free. Home-made teas.**
Arboretum planted by Lord Clive of India's son, Edward in1800. Cascades of rhododendrons and azaleas amongst specimen trees and pools. Fine views of Sir William

Chambers' Clock Towers, with lake and hills beyond. Walled kitchen garden, dovecote, meat safe, ice house and mile-long lakes. Russian wooden church, grotto and fountain; tin chapel. Relaxed borders and rare shrubs. Lakeside replanted and water garden at western end re-established. Outstanding Ballroom where excellent teas are served.

50 WESTHOPE COLLEGE

Westhope, Craven Arms, SY7 9JL. Anne Dyer, www.westhope.org.uk. *At Westhope, off B4368 Craven Arms/Bridgnorth Road. From A49 at Craven Arms, turn L (from Shrewsbury) or R (from Ludlow) onto B4368. L to Westhope after 3.4m. On B4368 from Bridgnorth, through Diddlebury, R after 2m See yellow NGS sign.* **Sat 10 June (10-4). Adm £5, chd free. Hot and cold drinks available.**

Anne Dyer is opening her garden at Westhope College where more than 3,500 common orchids and other ancient hay meadow plants live in the meadow; a delight to see! There are wildflowers, wildlife and a lovely woodland walk with a stream and ponds. The now restored walled garden is full of produce and flowers. The bright front garden welcomes visitors to Westhope College. Part of the gardens first opened for the National Garden Scheme in the 1980s. The large main garden has been rewilded for over 30 years. Good wheelchair access throughout some of the garden (please enquire).

51 WESTWOOD HOUSE

Oldbury, Bridgnorth, WV16 5LP. Hugh & Carolyn Trevor-Jones. *Take the Ludlow Road B4364 out of Bridgnorth. Past the Punch Bowl Inn, turn 1st L, Westwood House signed on R.* **Sun 23 Apr (2-5). Adm £5, chd free. Home-made teas.**

A country garden, well designed and planted around the house, particularly known for its tulips and use of colour. Sweeping lawns offset by deeply planted mixed borders; pool garden and lawn tennis court; kitchen and cutting garden, with everything designed to attract wildlife for organic growth. Far reaching views of this delightful corner of the county and woodland walks to enjoy. Reasonable wheelchair access around the house, but gravel paths and some steps.

52 153 WILLOUGHBRIDGE

Market Drayton, TF9 4JQ. John Butcher & Sarah Berry, 07817 443837, butchinoz@hotmail.com, www.instagram.com/_thebeegardener_. *Approx ½m from the Dorothy Clive Garden, turn onto Minn Bank from A51, 3rd drive on L. Parking will be signed.* **Sat 15, Sun 16 July (1-5). Adm £5, chd free. Pre-booking essential, please visit www.ngs.org.uk for information & booking. Cream teas. Visits also by arrangement May to Sept for individuals & groups of up to 40.**

Small and beautiful garden packed full of scented insect friendly plants. The garden inc a long cottage-prairie garden full of climbers, English roses, grasses and herbaceous perennials, a woodland-jungle garden consisting of huge scented tree lilies, tree ferns, acers, bamboo and bananas, two green roofs, three bug hotels and a large Victorian greenhouse full of traditional and exotic plants. A wide selection of home-made cakes will be available, as well as honey to buy collected from John's bees.

53 WINDY RIDGE

Church Lane, Little Wenlock, TF6 5BB. George & Fiona Chancellor, 01952 507675, fiona.chancellor@ngs.org.uk. *2m S of Wellington. Follow signs for Little Wenlock from N (J7, M54) or E (off A5223 at Horsehay). Parking signed. Do not rely on SatNav.* **Sat 3, Sun 4 June (12-5). Adm £6, chd free. Home-made teas. Visits also by arrangement June to Sept for groups of 10+.**

Universally admired for its structure, inspirational planting and balance of texture, form and all-season colour, the garden more than lives up to its award-winning record. Developed over 30 years, 'open plan' garden rooms display over 1000 species (mostly labelled) in a range of colour-themed planting styles, beautifully set off by well-tended lawns, plenty of water and fascinating sculpture. Wheelchair access, some gravel paths but help available.

54 ◆ WOLLERTON OLD HALL

Wollerton, Market Drayton, TF9 3NA. Lesley & John Jenkins, 01630 685760, info@wollertonoldhallgarden.com, www.wollertonoldhallgarden.com. *4m SW of Market Drayton. On A53 between Hodnet & A53-A41 junction. Follow brown signs.* **For NGS: Tue 5 Sept (11-5). Adm £9.50, chd free. Light refreshments in our excellent cafe. For other opening times and information, please phone, email or visit garden website.**

4 acre garden created around C16 house (not open). Formal structure creates variety of gardens each with own colour theme and character. Planting is mainly of perennials many in their late summer/early autumn hues, particularly the asters. Winner of many awards and nationally acclaimed. Ongoing lectures by Gardening Celebrities inc Chris Beardshaw, and other garden designers and personalities. We will suspend the booking system for the NGS day. Partial wheelchair access.

Our 2022 donation supports 1,700 Queen's Nurses working in the community in England, Wales, Northern Ireland, the Channel Islands and the Isle of Man

SOMERSET, BRISTOL AREA & SOUTH GLOUCESTERSHIRE incl BATH

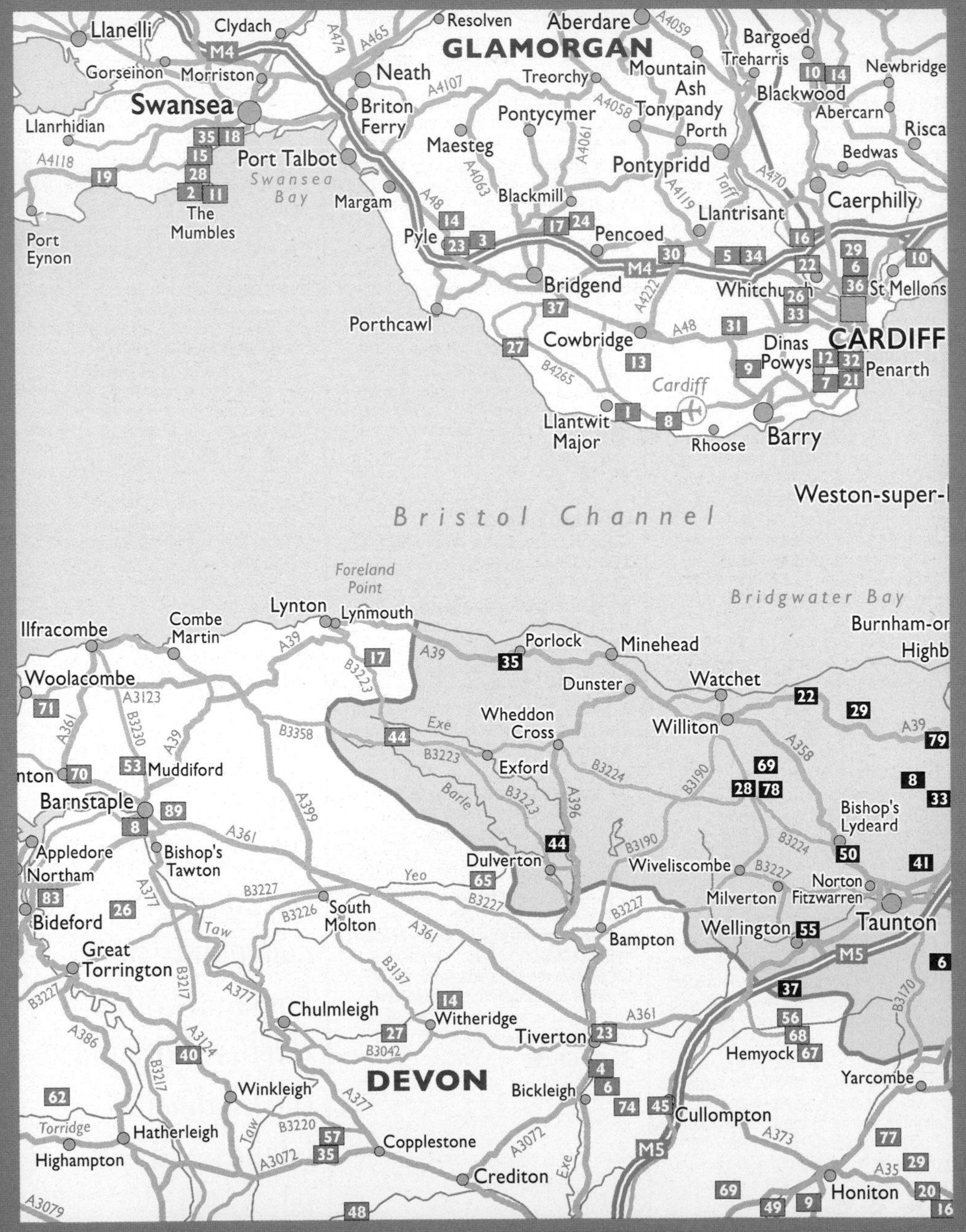

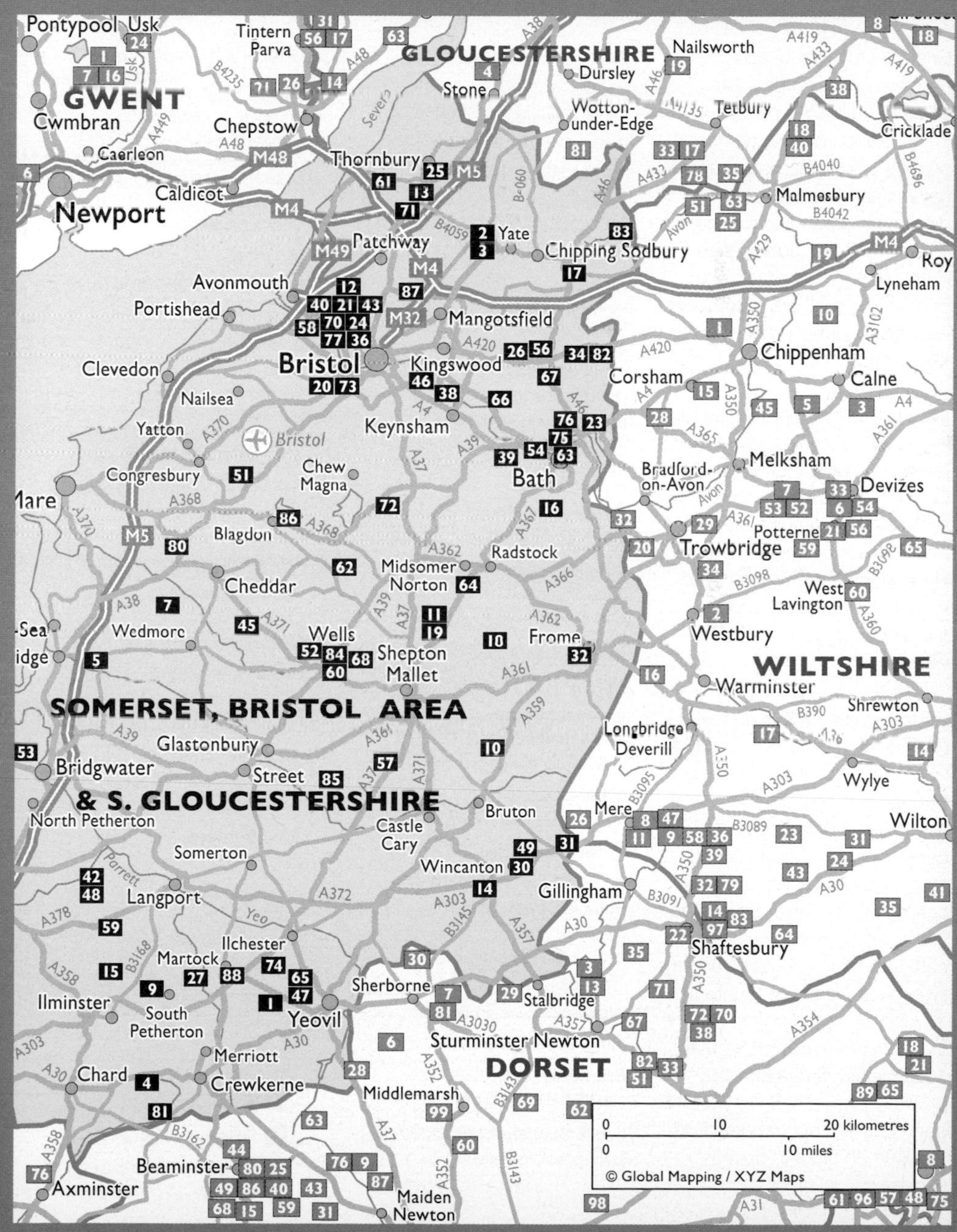
GLOUCESTERSHIRE
GWENT
SOMERSET, BRISTOL AREA
& S. GLOUCESTERSHIRE
WILTSHIRE
DORSET
Newport
Bristol
Bath
Chepstow
Thornbury
Yate
Chipping Sodbury
Kingswood
Keynsham
Wells
Shepton Mallet
Frome
Trowbridge
Chippenham
Glastonbury
Street
Bridgwater
Yeovil
Sherborne
Shaftesbury
Gillingham
Wincanton
Chard
Crewkerne
Ilminster
Beaminster
Axminster
0 10 20 kilometres
0 10 miles
© Global Mapping / XYZ Maps

Somerset, Bristol, Bath and South Gloucestershire make up a National Garden Scheme 'county' of captivating contrasts, with castles and countryside and wildlife and wetlands, from amazing cities to bustling market towns, coastal resorts and picturesque villages.

Bristol's stunning location and famous landmarks offer a wonderful backdrop to our creative and inspiring garden owners who have made tranquil havens and tropical back gardens in urban surroundings. The surrounding countryside is home to gardens featuring contrasting mixtures of formality, woodland, water, orchard and kitchen gardens.

Bath is a world heritage site for its Georgian architecture and renowned for its Roman Baths. Our garden visitors can enjoy the secluded picturesque gardens behind the iconic Royal Crescent Hotel in the heart of the city, or venture further afield and explore the hidden gems in nearby villages.

Somerset is a rural county of rolling hills such as the Mendips, the Quantocks and Exmoor National Park contrasted with the low-lying Somerset Levels. Famous for cheddar cheese, strawberries and cider; agriculture is a major occupation. It is home to Wells, the smallest cathedral city in England, and the lively county town of Taunton.

Visitors can explore more than 100 diverse gardens, mostly privately owned and not normally open to the public ranging from small urban plots to country estates.

Somerset Volunteers

County Organiser
Laura Howard 01460 282911
laura.howard@ngs.org.uk

County Treasurer
Jill Wardle 07702 274492
jill.wardle@ngs.org.uk

Publicity
Roger Peacock
roger.peacock@ngs.org.uk

Social Media
Janet Jones 01749 850509
janet.jones@ngs.org.uk

Lisa Prior 07773 440147
lisa.prior@ngs.org.uk

Presentations
Dave & Prue Moon 01373 473381
davidmoon202@btinternet.com

Booklet Co-ordinator
John Simmons 07855 944049
john.simmons@ngs.org.uk

Booklet Distributor
Laura Howard (see above)

Assistant County Organisers
Jo Beaumont 07534 777278
jo.beaumont@ngs.org.uk

Marsha Casely 07854 882616
marsha.casely@ngs.org.uk

Kirstie Dalrymple 07772 170537
kirstie.dalrymple@ngs.org.uk

Patricia Davies-Gilbert
01823 412187
pdaviesgilbert@gmail.com

John & Sue Denmark 01643 863513
johnandsuedenmark@ngs.org.uk

Janet Jones (as above)

Sue Lewis 07885 369280
sue.lewis@ngs.org.uk

Judith Stanford 01761 233045
judith.stanford@ngs.org.uk

Bristol Area Volunteers

County Organiser
Su Mills 01454 615438
su.mills@ngs.org.uk

County Treasurer
Harsha Parmar 07889 201185
harsha.parmar@ngs.org.uk

Publicity
Myra Ginns 07766 021616
myra.ginns@ngs.org.uk

Booklet Co-ordinator
John Simmons 07855 944049
john.simmons@ngs.org.uk

Booklet Distributor
John Simmons (as above)

Assistant County Organisers
Tracey Halladay
07956 784838
thallada@icloud.com

Frances Stewart
07463 776244
frances.stewart@ngs.org.uk

Helen Hughesdon
07793 085267
helen.hughesdon@ngs.org.uk

Jeanette Parker
01454 299699
jeanette_parker@hotmail.co.uk

John Burgess
07795 466513

Irene Randow
01275 857208
irene.randow@sky.com

Karl Suchy 07873 588540
karl.suchy@icloud.com

@visitsomersetngs
@SomersetNGS
@ngs_bristol_s_glos_somerset

OPENING DATES

All entries subject to change. For latest information check **www.ngs.org.uk**

Extended openings are shown at the beginning of the month.

Map locator numbers are shown to the right of each garden name.

January

Sunday 29th
Rock House 61

February

Snowdrop Festival

By arrangement
1 Birch Drive 13

Thursday 2nd
Elworthy Cottage 28

Sunday 5th
Rock House 61

Thursday 9th
◆ East Lambrook Manor Gardens 27

Sunday 12th
Greystones 36

Tuesday 14th
Elworthy Cottage 28

Thursday 23rd
Elworthy Cottage 28

Sunday 26th
Algars Manor 2
Algars Mill 3

Saturday 11th
Blackmore House 14

March

Tuesday 7th
◆ Hestercombe Gardens 41

Sunday 19th
Elworthy Cottage 28
Rock House 61

Saturday 25th
Forest Lodge 31
Lower Shalford Farm 49

Sunday 26th
Nynehead Court 55
Rock House 61

April

Every Thursday
Model Farm 53

Saturday 1st
NEW 6 Green Lane 34
Weir Cottage 82

Sunday 2nd
NEW 6 Green Lane 34
Weir Cottage 82

Monday 10th
Elworthy Cottage 28
Stoneleigh Down 71

Saturday 15th
NEW Skool Beanz Children's Allotment 65
◆ The Walled Gardens of Cannington 79

Sunday 16th
Fairfield 29
◆ The Walled Gardens of Cannington 79
Watcombe 80
◆ The Yeo Valley Organic Garden at Holt Farm 86

Tuesday 18th
Elworthy Cottage 28

Wednesday 19th
◆ Greencombe Gardens 35

Sunday 23rd
Algars Manor 2
Algars Mill 3
Greystones 36

Tuesday 25th
Elworthy Cottage 28

May

Every Thursday
Model Farm 53

Sunday 7th
◆ Milton Lodge 52
NEW Stoke Bishop Gardens 70

Friday 12th
◆ Stoberry Garden 68

Saturday 13th
◆ East Lambrook Manor Gardens 27
Hillcrest 42
Japanese Garden Bristol 46
◆ Stoberry Garden 68

Sunday 14th
Hillcrest 42
The Red Post House 59
◆ Stoberry Garden 68
Watcombe 80
Wayford Manor 81
The Yews 87

Tuesday 16th
Elworthy Cottage 28

Thursday 18th
Barford House 8

Saturday 20th
Forest Lodge 31
Lower Shalford Farm 49

Sunday 21st
Penny Brohn UK 58

Tuesday 23rd
Elworthy Cottage 28

Saturday 27th
The Hayes 39

Sunday 28th
81 Coombe Lane 21
Court House 22
Elworthy Cottage 28
The Hayes 39
4 Haytor Park 40

Tuesday 30th
Elworthy Cottage 28

June

Every Thursday
Model Farm 53

Friday 2nd
Stoneleigh Down 71

Saturday 3rd
Babbs Farm 5
Badgworth Court Barn 7
Coleford House 18

Sunday 4th
Babbs Farm 5
Badgworth Court Barn 7
Lydeard House 50
◆ Milton Lodge 52
Rock House 61
Stoneleigh Down 71

Tuesday 6th
◆ Hestercombe Gardens 41

Wednesday 7th
Mathlin Cottage 51
Pennard House 57

Thursday 8th
Watcombe 80

Friday 9th
NEW Caisson House 16
NEW Tregunter 76

Saturday 10th
NEW 10 Flingers Lane 30
Hillside 43
The Old Rectory, Doynton 56

Sunday 11th
NEW 10 Flingers Lane 30
Japanese Garden Bristol 46
Mathlin Cottage 51
Nynehead Court 55
NEW Tregunter 76

Thursday 15th
◆ Special Plants 67

Saturday 17th
Batcombe House 10
Westbrook House 85

Sunday 18th
Frome Gardens 32
42 Silver Street 64
Stogumber Gardens 69

Saturday 24th
Hanham Court 38

Sunday 25th
Coombe Cottage 20
Crete Hill House 24
Hanham Court 38
Sunnymead 73
Vexford Court 78

Tuesday 27th
Elworthy Cottage 28

July

Every Thursday
Model Farm 53

Saturday 1st
165 Newbridge Hill 54
NEW Tintinhull Gardens 74

Sunday 2nd
Honeyhurst Farm 45
◆ Milton Lodge 52
165 Newbridge Hill 54
NEW Tintinhull Gardens 74
◆ University of Bristol Botanic Garden 77
Yews Farm 88

Monday 3rd
Honeyhurst Farm 45
The Royal Crescent Hotel & Spa 63

Tuesday 4th
The Royal Crescent Hotel & Spa 63

Saturday 8th
Babbs Farm 5
Doynton House 26
◆ East Lambrook Manor Gardens 27

Sunday 9th
Babbs Farm 5
Doynton House 26

Monday 10th
◆ Berwick Lodge 12

Tuesday 11th
Elworthy Cottage 28

Saturday 15th
Goathurst Gardens 33

Sunday 16th
Daggs Allotments 25
Goathurst Gardens 33
Hangeridge Farmhouse 37
Nynehead Court 55

Wednesday 19th
◆ Greencombe Gardens 35

Thursday 20th
◆ Special Plants 67

Saturday 22nd
The Rib 60

Sunday 23rd
Benter Gardens 11
Court House 22
Stowey Gardens 72

Tuesday 25th
Elworthy Cottage 28

Berwick Lodge

August

Every Thursday
Model Farm 53

Saturday 12th
NEW Skool Beanz Children's Allotment 65

Tuesday 15th
Elworthy Cottage 28

Thursday 17th
◆ Special Plants 67

Monday 28th
Elworthy Cottage 28

September

Every Thursday
Model Farm 53

Saturday 2nd
NEW 10 Flingers Lane 30

Sunday 3rd
NEW 10 Flingers Lane 30

Saturday 9th
Batcombe House 10

Sunday 10th
College Barn 19
Yews Farm 88

Saturday 16th
Babbs Farm 5
Blackmore House 14
NEW The Kitchen Garden at Martock Workspace 47
◆ The Walled Gardens of Cannington 79

Sunday 17th
Babbs Farm 5
◆ The Walled Gardens of Cannington 79

Thursday 21st
◆ Special Plants 67

October

Every Thursday
Model Farm 53

Saturday 7th
◆ The Yeo Valley Organic Garden at Holt Farm 86

Thursday 19th
◆ Special Plants 67

November

Every Thursday
Model Farm 53

By Arrangement

Arrange a personalised garden visit with your club, or group of friends, on a date to suit you. See individual garden entries for full details.

Abbey Farm 1
Avalon 4
Badgers Holt 6
NEW The Barn 9
Batcombe House 10
1 Birch Drive 13
Blackmore House 14
Bradon Farm 15
Camers 17
81 Coombe Lane 21
Court View 23
Elworthy Cottage 28
NEW 10 Flingers Lane 30
Hangeridge Farmhouse 37
The Hayes 39
4 Haytor Park 40
Hillcrest 42
NEW Hollam House 44
Honeyhurst Farm 45
Knoll Cottage, Stogumber Gardens 69
Lane End House 48
Lucombe House, Stoke Bishop Gardens 70
165 Newbridge Hill 54
Nynehead Court 55
Penny Brohn UK 58
The Red Post House 59
Rock House 61
Rose Cottage 62
42 Silver Street 64
South Kelding 66
Stoneleigh Down 71
NEW Trafalgar House 75
Vexford Court 78
Watcombe 80
Well Cottage 83
Wellfield Barn 84

THE GARDENS

1 ABBEY FARM

Montacute, TA15 6UA. Elizabeth McFarlane, 01935 823556, abbey.farm64@gmail.com. *4m from Yeovil. Follow A3088, take slip rd to Montacute, turn L at T-junction into village. Turn R between church & King's Arms (no through rd).* **Visits by arrangement 10 May to 27 Sept for groups of 10 to 30. Adm £6, chd free. Home-made teas. Details on request.**

2½ acres of mainly walled gardens on sloping site provide the setting for Cluniac Medieval Priory gatehouse. Interesting plants inc roses, shrubs, grasses, clematis. Herbaceous borders, white garden, gravel garden. Small arboretum. Pond for wildlife - frogs, newts, dragonflies. Fine mulberry, walnut and monkey puzzle trees. Seats for resting.

2 ALGARS MANOR

Station Rd, Iron Acton, BS37 9TB. Mrs B Naish. *9m N of Bristol, 3m W of Yate/Chipping Sodbury. Turn S off Iron Acton bypass B4059, past village green & White Hart pub, 200yds, then over level Xing. No access from Frampton Cotterell via lane; ignore SatNav. Parking at Algars Manor.* **Sun 26 Feb (1-4). Light refreshments. Sun 23 Apr (1-5). Home-made teas. Combined adm with Algars Mill £7, chd free.**

2 acres of woodland garden beside River Frome, mill stream, native plants mixed with collections of 60 magnolias and 70 camellias, rhododendrons, azaleas, eucalyptus and other unusual trees and shrubs. Daffodils, snowdrops and other early spring flowers.

3 ALGARS MILL

Frampton End Rd, Iron Acton, Bristol, BS37 9TD. Mr & Mrs John Wright. *9m N of Bristol, 3m W of Yate/Chipping Sodbury. (For directions see Algars Manor).* **Sun 26 Feb (1-4). Light refreshments. Sun 23 Apr (1-5). Home-made teas. Combined adm with Algars Manor £7, chd free.**

2 acre woodland garden bisected by River Frome; spring bulbs, shrubs; very early spring feature (Feb-Mar) of wild Newent daffodils. 300-400yr-old mill house (not open) through which millrace still runs.

4 AVALON

Moor Lane, Higher Chillington, Ilminster, TA19 0PT. Dee & Tony Brook, 07506 688191, dee1jones@hotmail.com. *Just off the A30 between Crewkerne and Chard. From A30 take turning signed to Chillington opp Swandown Lodges. Take 2nd L down Coley Lane and 1st L Moor Lane. Avalon is the large pink house. Parking limited, so please car share if possible.* **Visits by arrangement 30 May to 20 Aug for groups of up to 40. Adm £5, chd £2. Home-made teas. Gluten & dairy free cakes available if requested in advance.**

Secluded hillside garden with wonderful views as far as Wales. The lower garden has large herbaceous borders, a sizeable wildlife pond and 2 greenhouses filled with RSA succulents. The middle garden has mixed borders, wild spotted orchids on the lawn, allotment area and a small orchard. The upper garden has a spring fed water course with ponds, and many terraces with different planting schemes. Partial wheelchair access across lower lawns & side paths. Steep slope & gravel paths. Wheelchairs will require to be pushed & attended at all times.

5 BABBS FARM

Westhill Lane, Bason Bridge, Highbridge, TA9 4RF. Sue & Richard O'Brien, www.babbsfarm.co.uk. *1½m E of Highbridge, 1½m SSE of M5 exit 22. Turn into Westhill Lane off B3141 (Church Rd), 100yds S of where it joins B3139 (Wells-Highbridge rd).* **Sat 3, Sun 4 June, Sat 8, Sun 9 July, Sat 16, Sun 17 Sept (2-5). Adm £7, chd free. Home-made teas.**

1½ acre plantsman's garden on Somerset Levels, gradually created out of fields surrounding old farmhouse over last 30 yrs and still being developed. Trees, shrubs and herbaceous perennials planted with an eye for form and shape in big flowing borders. Plants for sale if circumstances allow.

6 BADGERS HOLT

Frost Street, Thurlbear, Taunton, TA3 5BA. Mr Neil Jones & Ms Sharon Bradford, 07734 318425 or 07814 842811, Neil.jones881@gmail.com. *SatNav will not bring you directly to the house. Exact details will be given on booking.* **Visits by arrangement 1 June to 2 July for groups of up to 12. Adm £4, chd free. Home-made teas. Refreshments by prior arrangement.**

Quintessential English Cottage garden, boasting beautiful views over the Blackdown Hills. The garden owners have designed this wildlife friendly garden of just under ½ an acre from a grassy site over the last 4 yrs. Full of interesting perennials, shrubs, and a large selection of roses. The addition of a wildlife pond and good sized vegetable patch completes the garden. Regret no dogs.

7 BADGWORTH COURT BARN

Notting Hill Way, Stone Allerton, Axbridge, BS26 2NQ. Trish & Jeremy Gibson, www.instagram.com/theoldbarngardeners. *4m SW of Axbridge. Turn off A38 in Lower Weare, signed Wedmore, Weare. Continue 1.1m, past rd on R (Badgworth Arena). Continue 0.1m, to 'No Footway for 600 yds' sign. Garden & car park will be signed.* **Sat 3, Sun 4 June (2-5.30). Adm £5, chd free. Home-made teas.**

This 1 acre plot surrounds old stone barn buildings. A small orchard leads to a colourful part-walled garden with areas of perennial meadow and multi-stemmed trees. Gently curving beds are flanked by a more formal oak pergola walk. The planting is a relaxed contemporary mix. In the atmospheric courtyard, planting is more established and leads on to a new sand garden in front of the barns. Main garden is level and accessible. Courtyard and drive areas are loose stone chippings and can be difficult for some wheelchairs. No accessible WC.

8 BARFORD HOUSE

Spaxton, Bridgwater, TA5 1AG. Donald & Bee Rice. *4½m W of Bridgwater. Midway between Enmore & Spaxton.* **Thur 18 May (2-5). Adm £6.50, chd free. Home-made teas.**

Secluded walled garden contains wide borders, kitchen garden beds, shrubs and fruits trees; lawns lead to a 6 acre woodland garden of camellias, rhododendrons, azaleas and magnolias. Stream-side gardens feature candelabra primulas, ferns, foxgloves and lily-of-the-valley among veteran pines and oaks, and some rarer trees. Partial wheelchair access. Some areas of woodland garden inaccessible.

9 NEW THE BARN

Water Street, Barrington, Ilminster, TA19 0JR. Debbie & Mal Eustice, 01460 52614, debbie.robins@btinternet.com. *Enter Barrington from B3168, following main rd through village, past Barrington Boar pub. At old school building turn L into Water St. House on R after 1st pair of cottages.* **Visits by arrangement 27 May to 30 June for groups of up to 20.Pls contact owner for specific dates available. Adm £4, chd free. Light refreshments.**

0.3 acre village cottage garden, close to NT Barrington Court, created in last 3 yrs from challenging, steeply sloping site. Massed informal planting for wildlife and pollinators. Mediterranean courtyard with water feature. Lower terrace with wildlife pond, mid level lawn and cottage borders. Upper level with rose clad pergola, small cutting garden and greenhouse. Multiple seating areas.

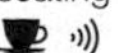

10 BATCOMBE HOUSE

Gold Hill, Batcombe, Shepton Mallet, BA4 6HF. Libby Russell, libby@mazzullorussellandscapedesign.com, www.mazzullorussellandscapedesign.com. *In centre of Batcombe, 3m from Bruton. Parking between Batcombe House & church at centre of village will be clearly marked.* **Sat 17 June, Sat 9 Sept (2-6). Adm £7.50, chd free. Cream teas. Visits also by arrangement 15 May to 17 Sept for groups of 15+.**

Plantswoman's and designer's garden of two parts – one a riot of colour through kitchen terraces, potager leading to wildflower orchard; the other a calm contemporary amphitheatre with large herbaceous borders and interesting trees and shrubs. Always changing. Dogs are allowed but on a lead.

GROUP OPENING

11 BENTER GARDENS

Benter, Oakhill, Radstock, BA3 5BJ. instagram.com/al.gardening. *Between the villages of Chilcompton, Strat-on-Fos & Oakhill, narrow lanes. A37 from Bristol & Shepton Mallet, turn to gardens by village shop in Gurney Slade. Follow lane for approx 1m, turn R at grass triangle before incline. From Bath A367, after Strat-on-Fos 2nd R to Benter.* **Sun 23 July (10-5). Combined adm £6, chd free. Delicious home-made teas and cakes.**

COLLEGE BARN
Alex Crossman & Jen Weaver.
(See separate entry)

FIRE ENGINE HOUSE
Patrick & Nicola Crossman.

Two contrasting gardens in a beautiful, peaceful and tranquil secluded valley setting, surrounded by woodland. A profusion of herbaceous borders add colour with interest in each garden. The garden at Fire Engine House is mature and established, with lawns, generous borders and narrow, enticing paths; through a garden door and tumble-down bothy is a small orchard with specimen trees. At College Barn a wildlife pond has been created in recent years at the woodland edge with native planting. Perennial borders, prairie planting, grasses, cottage garden planting, specimen trees, wildlife pond, woodland walk. Please wear stout shoes for the woodland walk.

12 ◆ BERWICK LODGE

Berwick Drive, Bristol, BS10 7TD. Sarah Arikan, 0117 958 1590, info@berwicklodge.co.uk, www.berwicklodge.co.uk. *Leave M5 at J17. A4018 towards Bristol West. 2nd exit on r'about, straight on at mini r'about. At next r'about by Old Crow pub do 360° turn back down A4018, follow sign saying BL off to L.* **For NGS: Mon 10 July (11-3). Adm £5, chd free. Cream teas. For other opening times and information, please phone, email or visit garden website.**

Berwick Lodge, named Bristol's hidden gem by its customers, is an independent hotel with beautiful gardens on the outskirts of Bristol. Built in 1890, this Victorian Arts & Crafts property is set within 18 acres, of which 4 acres are accessible and offer a peaceful garden for use by its visitors. The gardens enjoy pretty views across to Wales, and continue to evolve. Created by Head Gardener Robert Dunster, an ex Royal gardener who worked for Prince Charles at Highgrove, it features an elegant water fountain, Victorian summerhouse, orchard, wildflower meadow, beehives, pond, wisteria clad pergola plus a thriving house martin colony.

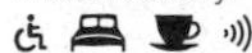

13 1 BIRCH DRIVE

Alveston, Bristol, BS35 3RQ. Myra Ginns, 01454 415396, myra.ginns@ngs.org.uk. *14m N of Bristol. Alveston on A38 Bristol to Gloucester. Just before traffic lights turn left into David's Lane. At end, L then R onto Wolfridge Ride. Birch Drive 2nd on R.* **Visits by arrangement in Feb for groups of up to 10. Adm £5, chd free.**

A newly planted garden with particular interest in the spring. A wide range of bulbs will be in flower, along with unusual named varieties of anemone, hellebore, hepatica and crocus. The garden's main feature are the snowdrops, collected since 2008. Many named varieties, flowering between November and March, will interest snowdrop collectors. The garden will also be available to book by arrangement in February of 2024. Ramp to decked area gives a view of the garden.

♿ ✿

14 BLACKMORE HOUSE

Holton Street, Holton, Wincanton, BA9 8AN. Mrs Lisa Prior, 07773 440147, lisa.prior@ngs.org.uk, www.instagram.com/lisaprior. *Just off the Wincanton junction of the A303. In the centre of Holton village, 2 doors down from Holton Village Hall, round the corner from The Old Inn. Limited street parking.* **Sat 11 Feb (11-3); Sat 16 Sept (1-5). Adm**

Caisson House

£4.50, chd free. Home-made teas in the garden or Holton Village Hall if the weather is inclement. Visits also by arrangement 1 Feb to 16 Sept for groups of 6+. Tea & cake can be booked in advance.
A hidden terraced cottage garden with dry stone walls and gravel paths, designed for year-round interest complementing the Georgian listed house. Wide borders packed full of perennials and self-seeded surprises. 2 intensely scented 10ft daphnes, sarcocca, and a plethora of hellebores in winter invite you outside while most gardens sleep. Roses, rudbeckia and helianthus bring bright colours to autumn.

15 BRADON FARM

Isle Abbotts, Taunton, TA3 6RX. Mr & Mrs Thomas Jones, deborahjstanley@hotmail.com. *Take turning to Ilton off A358. Bradon Farm is 1½m out of Ilton on Bradon Lane.* **Visits by arrangement for groups of 10+. Adm £7, chd free. Home-made teas.**
Classic formal garden demonstrating the effective use of structure with parterre, knot garden, pleached lime walk, formal pond, herbaceous borders, orchard and wildflower planting.

16 NEW CAISSON HOUSE

Combe Hay, Bath, BA2 7EF. Amanda Honey. *3m S of Bath. Take A367 from centre of Bath towards Radstock, at 2nd r'about take 1st exit, follow signs to Combe Hay. 1st L after Wheatsheaf pub, marked No Through Rd. Entrance 1st on R.* **Fri 9 June (11-4). Adm £7, chd free. Pre-booking essential, please visit www.ngs.org.uk for information & booking. Home-made teas.**
This is a wonderfully eclectic garden set in the most beautiful English countryside around a gorgeous Georgian house built in 1815. It is a mix of herbaceous borders, topiaries, ponds and rills, a walled garden with fruit trees, greenhouses, flower and vegetable beds.There are wildflower meadows surrounding the garden and the disused Somerset Coal Canal running through the property. Great variety of species and biodiversity, inc native orchids.

17 CAMERS

Badminton Road, Old Sodbury, Bristol, BS37 6RG. Mr & Mrs Michael Denman, 01454 327929, jodenman@btinternet.com, www.camers.org. *2m E of Chipping Sodbury. Entrance in Chapel Lane off A432 at Dog Inn. Enter through the field gate and drive to the top of the fields to park next to the garden.* **Visits by arrangement May & June for groups of 20+. Adm £6, chd free. Refreshments by arrangement.**
Elizabethan farmhouse (not open) set in 4 acres of constantly developing garden and woodland with spectacular views over Severn Vale. Garden full of surprises, formal and informal areas planted with wide range of species to provide year-round interest. Parterre, topiary, Japanese garden, bog and prairie areas, white and hot gardens, woodland walks. Some steep slopes.

18 COLEFORD HOUSE

Underhill, Coleford, Radstock, BA3 5LU. Mr James Alexandroff. *5m from Frome. Coleford House is opp Kings Head Pub in Lower Coleford with black wrought iron gates just before bridge over river. Parking in field 100 metres away.* **Sat 3 June (10.30-4). Adm £7.50, chd free. Home-made teas.**

The River Mells flows through this picturesque garden with large lawns, wildflower planting, ornamental pond, woodland, substantial herbaceous borders, walled garden, arboretum/orchard, kitchen garden, vegetable garden, bat house and orangery. A collection of classic 1960's Aston Martin cars will be on the lawn. Artwork will also be for sale. Most of garden is wheelchair friendly.

19 COLLEGE BARN

Benter, Oakhill, Radstock, BA3 5BJ. Alex Crossman & Jen Weaver, instagram.com/al.gardening. *Between the villages of Chilcompton, Stratton on the Fosse and Oakhill, narrow lanes. A37, Bristol to Shepton Mallet, take lane E in Gurney Slade by village shop. Follow lane for approx 1m, turn at grass triangle before incline. From Bath A367, 1m after Strat-on-Fos, R to Benter.* **Sun 10 Sept (10-5). Adm £5, chd free. Delicious teas with home made cakes. Opening with Benter Gardens on Sun 23 July.**

Created over the last 7 years, the garden at College Barn draws upon its surroundings of meadows and woodland, with hazel and hornbeam hedges and naturalistic plantings of perennials in large blocks or in matrices with ornamental grasses. Intimate walled garden filled with vegetables, herbs, flowers and fruit. Wildlife pond created in recent years at the woodland edge with native planting. A peaceful and tranquil secluded valley setting, surrounded by woodland. Please wear stout shoes for the woodland walk.

20 COOMBE COTTAGE

161 Long Ashton Road, Long Ashton, Bristol, BS41 9JQ. Peter & Sheena Clark. *3m SW of Bristol city centre. Exit off A370 (Long Ashton by-pass) onto B3128. L at junction with Long Ashton Rd (sign post Long Ashton & opp gateway to Ashton Court). Garden 100 yds on R opp field.* **Sun 25 June (1-4). Combined adm with Sunnymead £5, chd free.**

Georgian cottage overlooking field; fronted with flagstone paved pathway behind hedge and containing ericaceous beds; linking with small walled side garden laid to lawn with herbaceous borders; arch leads to rear paved patio with old water cistern; beyond is hedged lawn bordered by rockery, small pond and beds stocked with flowers and shrubs.

The Barn

21 81 COOMBE LANE

Stoke Bishop, Bristol, BS9 2AT. Karl Suchy, karl.suchy@icloud.com. *4m from Bristol city centre. J17 of M5, then follow A4018, direction Bristol for approx 2m.Turn R onto Canford Rd A4162. After 0.8m turn L onto Coombe Lane. Destination on R.* **Sun 28 May (12-5). Adm £5, chd free. Home-made teas. Visits also by arrangement 1 May to 7 July for groups of up to 40. Admission of £7.50, inc tea/beverage & cake.**

Hidden Victorian walled garden. You'll encounter substantial mixed borders containing traditional and contemporary planting. Large lawns, numerous seating areas, summerhouse and French inspired patio with coppiced lime trees. Parterre and large raised Koi pond surrounded by bananas and tree ferns. Access to parterre and Koi pond might be difficult for wheelchair due to narrow gravel path, main part of garden is accessible.

22 COURT HOUSE

East Quantoxhead, TA5 1EJ. Mr & Mrs Hugh Luttrell. *12m W of Bridgwater. Off A39, house at end of village past duck pond. Enter by Frog St (Bridgwater/Kilve side from A39). Car park £1 in aid of church.* **Sun 28 May, Sun 23 July (2-5). Adm £6, chd free. Cream teas.**

Lovely 5 acre garden, trees, shrubs (many rare and tender), herbaceous and 3 acre woodland garden with spring interest and late summer borders. Traditional kitchen garden (chemical free). Views to sea and Quantocks. Gravel, stone and some mown grass paths.

23 COURT VIEW

Solsbury Lane, Batheaston, Bath, BA1 7HB. Maria & Jeremy Heffer, 07786 668278, maria.heffer@btinternet.com, www.thebathgreenhouse.com. *3m E of Bath. In Batheaston High St take turning on L signed Northend, St Catherine. At top of rise L into Solsbury Lane. Court View is 2nd driveway on L. Parking for 8 cars. Public car park in the village.* **Visits by arrangement July & Aug for groups of 10 to 30. Adm £5, chd free. Refreshments by arrangement.**

2 acre south facing gardens with ⅓ acre devoted to cut flowers and foliage. Colourful mix of annuals, biennials and perennials. Spectacular views from terraced lawns, box parterre, small orchard and meadow area. A floral experience for garden lovers, artisan florists, flower arrangers and anyone interested in the revival of beautiful, diverse and locally grown British cut flowers.

24 CRETE HILL HOUSE

Cote House Lane, Durdham Down, Bristol, BS9 3UW. John Burgess. *2m N of Bristol city centre, 3m S J16 M5. A4018 Westbury Rd from city centre, L at White Tree r'about, R into Cote Rd, continue into Cote House Lane across the Downs. 2nd house on L. Parking on street.* **Sun 25 June (12-5). Adm £5, chd free. Home-made teas.**

C18 house in hidden corner of Bristol. Mainly SW facing garden, 80'x40', with shaped lawn, heavily planted traditional mixed borders - shrub, rose, clematis and herbaceous. Pergola with climbers, terrace with pond, several seating areas. Shady walled garden. Roof terrace (44 steps) with extensive views. A couple of seating areas and roof terrace not accessible by wheelchair. Whole garden can be viewed.

25 DAGGS ALLOTMENTS

High Street, Thornbury, BS35 2AW. Thornbury Town Trust. *15m N of Bristol. Park in free car park off Chapel St.* **Sun 16 July (2-5). Adm £5, chd free.**

Thornbury is a historic market town and has been a regular winner of awards in the RHS Britain in Bloom competition. 120 plots, all in cultivation, many organic, inc vegetables, soft fruit, herbs and flowers for cutting. Narrow, steep, grass paths between plots. A plot holder will be available to answer any questions you may have about the plots, cultivation techniques and varieties grown. Short talk on the history of Daggs since 1546. Coffee shops, cafes and pubs are available nearby in Thornbury High St.

26 DOYNTON HOUSE

Bury Lane, Doynton, Bristol, BS30 5SR. Frances & Matthew Lindsey-Clark. *5m S of M4 J18, 6m N of Bath, 8m E of Bristol. Doynton is NE of Wick (turn off A420 opp Bath Rd) and SW of Dyrham (signed from A46). Doynton House is at S end of Doynton village, opp Culleysgate/Horsepool Lane. Park in signed field.* **Sat 8, Sun 9 July (2-5). Adm £7.50, chd free. Home-made teas.**

A variety of garden areas separated by old walls and hedges. Mixed borders, lawns, wall planting, parterre, rill garden, walled vegetable garden, cottage beds, pool garden, dry gardens, peach house and greenhouse. Bees, chickens and meadow area. We can recommend our local pub, the Cross House, just a short stroll across our parking field. Paths are of hoggin, stone and gravel. The grade of the gravel makes it a hard push in places but all areas are just about wheelchair accessible.

27 ◆ EAST LAMBROOK MANOR GARDENS

Silver Street, East Lambrook, TA13 5HH. Mike Werkmeister, 01460 240328, enquiries@eastlambrook.com, www.eastlambrook.com. *2m N of South Petherton. Follow brown tourist signs from A303 South Petherton r'about or B3165 Xrds with lights N of Martock.* **For NGS: Thur 9 Feb, Sat 13 May, Sat 8 July (10-5). Adm £6.50, chd free. Tea and cake. For other opening times and information, please phone, email or visit garden website.**

The quintessential English cottage garden created by C20 gardening legend Margery Fish. Plantsman's paradise with contemporary and old-fashioned plants grown in a relaxed and informal manner to create a remarkable garden of great beauty and charm. With noted collections of snowdrops, hellebores and geraniums and the excellent specialist Margery Fish Plant Nursery. Partial wheelchair access.

Our 2022 donation to Hospice UK meant that 911 NHS and Social Care staff accessed individual counselling support

28 ELWORTHY COTTAGE
Elworthy, Taunton, TA4 3PX. Mike & Jenny Spiller, 01984 656427, mike@elworthy-cottage.co.uk, www.elworthy-cottage.co.uk. *12m NW of Taunton. On B3188 between Wiveliscombe & Watchet.* **Thur 2, Tue 14, Thur 23 Feb, Sun 19 Mar, Mon 10, Tue 18, Tue 25 Apr, Tue 16, Tue 23, Sun 28, Tue 30 May, Tue 27 June, Tue 11, Tue 25 July, Tue 15, Mon 28 Aug (11-4.30). Adm £4.50, chd free. Home-made teas. Visits also by arrangement Feb to Aug.**
1 acre plantsman's garden in tranquil setting. Island beds, scented plants, clematis, unusual perennials and ornamental trees and shrubs to provide year-round interest. In spring, pulmonarias, hellebores and more than 350 varieties of snowdrops. Planted to encourage birds, bees and butterflies. Lots of birdsong, wildflower areas and developing wildflower meadow, decorative vegetable garden, living willow screen. Seats for visitors to enjoy views of the surrounding countryside. Garden attached to plantsman's nursery, open at the same time.

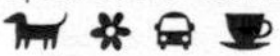

29 FAIRFIELD
Stogursey, Bridgwater, TA5 1PU. Lady Acland Hood Gass. *7m E of Williton. 11m W of Bridgwater. From A39 Bridgwater to Minehead road turn N. Garden 1½m W of Stogursey on Stringston road. No coaches.* **Sun 16 Apr (2-5). Adm £5, chd free. Home-made teas.**
Woodland garden with many interesting bulbs inc naturalised anemones, fritillaria with roses, shrubs and fine trees. Paved maze. Views of Quantocks. The ground is flat and should be accessible, around half the paths are grass so may be more difficult when wet.

30 NEW 10 FLINGERS LANE
Wincanton, BA9 9LE. Yseult Ogilvie, 07736 609789, yseultogilvie@hotmail.com. *Situated on the L side of Flingers Lane to the N of the B3081. Additional parking on the High St or around the Memorial Hall.* **Sat 10, Sun 11 June (10.30-5); Sat 2, Sun 3 Sept (10.30-4). Adm £5, chd free. Home-made teas. Visits also by arrangement 1 Apr to 3 Sept for groups of 5 to 10. Pls contact by email & allow for 2 days notice.**
A secluded third of an acre secret walled garden hidden behind the High Street in Wincanton on a no through road. Box hedges, topiary, a formal kitchen garden with fruit trees, a fernery, glasshouses, and a lawn flanked by herbaceous borders and shrubs. Designed by an architect, the striking layout leads the visitor from one notional 'room' to another.

31 FOREST LODGE
Pen Selwood, BA9 8LL. James & Lucy Nelson, forestlodgegardens.co.uk. *1½m N of A303, 3m E of Wincanton. Leave A303 at B3081 (Wincanton to Gillingham rd), up hill to Pen Selwood, L towards church. ½m, garden on L - low curved wall and sign saying Forest Lodge Stud.* **Sat 25 Mar, Sat 20 May (12-3.30). Combined adm with Lower Shalford Farm £10, chd free. Home-made teas. Donation to Heads Up Wells, Balsam Centre Wincanton.**
3 acre mature garden with many camellias and rhododendrons in May. Lovely views towards Blackmore Vale. Part formal with pleached hornbeam allée and rill, part water garden with lake. Wonderful roses in June. Unusual spring flowering trees such as *Davidia involucrata* and many beautiful cornus. Interesting garden sculpture. Good plant interest year-round due to acidic greensand soil (hamamelis, Daphne, magnolia, lovely rhododendrons, camellias and flowering trees in the spring) and south west facing slope. Beautiful in all seasons and good structure which underpins the planting. Wheelchair access to front garden only, however much of garden viewable from there.

GROUP OPENING

32 FROME GARDENS
Frome, BA11 4HR. *61 Nunney Rd, A362 Badcox up Nunney Rd approx 275m on L before Dommetts Ln opp Portland Rd. Enter side gate R of house. 84 Weymouth Rd, Badcox turn opp The Artisan Restaurant. Pls park considerately.* **Sun 18 June (1-5). Combined adm £6, chd free.**

61 NUNNEY ROAD
Mrs Caroline Toll.

84 WEYMOUTH ROAD
Amanda Relph.

A warm welcome awaits at 2 differing town gardens. 61 Nunney Road's well cared for garden welcomed a new owner in 2019 who has redesigned the garden with a terrace, lawn, shrubs, and flowerbeds around old established fruit trees; 2 small deep ponds, a wild corner encouraging wildlife into the garden, vegetable patch and greenhouse. The owner considers herself an untidy gardener, she falls for plants then finds the right place for them! In contrast 84 Weymouth Road, seasoned NGS openers, open their garden designed by and celebrating the life of Simon Relph. Architectural designed metal arches and geometrical raised beds surround a central raised deep pond. Sculpture by their son Alex Relph, and other artists, feature throughout. Mature shrubs surround the walled garden; fern area, roses and camellias, enhance the central courtyard. Raised vegetable beds, greenhouse and soft fruits hide at the far end of the garden. Very different and interesting styles of design and planting complement each other in these two hidden town gardens. Very small step from terrace to lawn at 61 Nunney Road; 3 small steps at 84 Weymouth Road.

♿

GROUP OPENING

33 GOATHURST GARDENS
Goathurst, Bridgwater, TA5 2DF. *4m SW of Bridgwater, 2½m W of N Petherton. Parking available at Halswell House. Very limited parking at the Temple of Pan.* **Sat 15, Sun 16 July (2-6). Combined adm £7.50, chd free. Home-made teas at Halswell House.**

NEW HALSWELL HOUSE
Mrs Oksana Kadatskaya.

THE LODGE
Sharon & Richard Piron.

OLD ORCHARD
Mr Peter Evered.

NEW THE TEMPLE OF PAN
Peter Strivens & Tessa Shaw.

Four very different gardens, two new this year, in the picturesque village of Goathurst on the edge of the Quantock Hills. The Lodge and Old Orchard are beautiful examples of quintessentially English cottage gardens approx 30

Walters Farm

metres apart in a rural village setting. The gardens at Halswell Park are a history of garden design in England, with a new knot garden created to reflect an original C16 design. The garden of the Temple of Pan is centred on an C18 baroque folly built as part of the pleasure gardens for Halswell House. The Lodge, Old Orchard and Halswell knot gardens have wheelchair access to most areas. The gardens at The Temple of Pan are steep in some places.

34 NEW 6 GREEN LANE

Marshfield, Chippenham, SN14 8JW. Richard Whish. *West end of Marshfield. Look for yellow sign off A420. Park in layby on R. 2nd yellow sign to Green Lane. Weir Cottage approx ½ a mile away.* **Sat 1, Sun 2 Apr (11-3). Combined adm with Weir Cottage £5, chd free. Home-made teas at Weir Cottage.**

Traditional borders that merge with open countryside; extensive fields planted with native trees, mostly deciduous. Designed to attract wildlife. An arboretum featuring native trees.

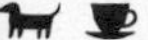

35 ◆ GREENCOMBE GARDENS

Porlock, Minehead, TA24 8NU. Greencombe Garden Trust, 01643 862363, info@greencombe.org, www.greencombe.org. *W of Porlock below the wooded slopes of Exmoor. Take A39 to west end of Porlock and turn onto B3225 to Porlock Weir. Drive ½ m, turn L at Greencombe Gardens sign. Go up drive, parking signed.* **For NGS: Wed 19 Apr, Wed 19 July (2-6). Adm £7, chd £1. Cream teas. For other opening times and information, please phone, email or visit garden website. Donation to Plant Heritage.**

Organic woodland garden of international renown, Greencombe stretches along a sheltered hillside and offers outstanding views over Porlock Bay. Moss-covered paths meander through a collection of ornamental plants that flourish beneath a canopy of oaks, hollies, conifers and chestnuts. Camellias, rhododendrons, azaleas, lilies, roses, clematis, and hydrangeas blossom among 4 National Collections. Champion English Holly tree (*Ilex aquifolium*), one of the largest and oldest in the UK. Giant rhododendrons species and exceptionally large camellias. A millennium chapel hides in the mossy banks of the wood. A moon arch leads into a walled garden. Spectacular views onto Porlock Bay.

36 GREYSTONES

Hollybush Lane, Bristol, BS9 1JB. Mrs P Townsend. *2m N of Bristol city centre, close to Durdham Down in Bristol, backing onto the Botanic Garden. A4018 Westbury Rd, L at White Tree r'about, L into Saville Rd, Hollybush Lane 2nd on R. Narrow lane, parking limited, recommended to park in Saville Rd.* **Sun 12 Feb, Sun 23 Apr (11-4). Adm £4, chd free. Light refreshments.**

Peaceful garden with places to sit and enjoy a quiet corner of Bristol. Interesting courtyard, raised beds, large variety of conifers and shrubs leads to secluded garden of contrasts - sunny beds with olive tree and brightly coloured flowers to shady spots, with acers, hostas and ferns. Snowdrops, hellebores, spring bulbs, naturalised daffodils. Rambling roses, small orchard, espaliered pears. Paved footpath provides level access to all areas.

The Kitchen Garden at Martock Workspace

37 HANGERIDGE FARMHOUSE

Wrangway, Wellington, TA21 9QG. Mrs J M Chave, 07812 648876, hangeridge@hotmail.co.uk. *2m S of Wellington. Off A38 Wellington bypass signed Wrangway. 1st L towards Wellington monument, over motorway bridge 1st R.* **Sun 16 July (2-5). Adm £4, chd free. Home-made teas. Visits also by arrangement 7 May to 6 Aug for groups of 10 to 30.**

Rural fields and mature trees surround this 1 acre informal garden offering views of the Blackdown and Quantock Hills. Magnificent hostas and heathers, colourful flower beds, cascading wisteria and roses and a trickling stream. Relax with home-made refreshments on sunny or shaded seating admiring the views and birdsong.

38 HANHAM COURT

Ferry Road, Hanham Abbots, BS15 3NT. Hanham Court Gardens, www.hanhamcourtgardens.co.uk. *5m E of Bristol centre, 9m W of Bath centre. E-A431 from Bath, through Willsbridge, L at mini r'about, Court Farm Rd-1m Entrance sharp L bend. W-A420 from Bristol. A431-1m. R at mini r'about Memorial Rd 1m. Entrance on Court Farm Rd sharp R bend.* **Sat 24, Sun 25 June (11-4.30). Adm £10, chd free. Cream teas.**

Hanham Court Gardens is a deeply romantic and enchanting idyll, hidden in a rural bowl, containing a rich mix of bold formal topiary and blowsy planting, water, woodland, orchard, meadow and kitchen garden with emphasis on scent, structure and romance, set amid a remarkable cluster of manorial buildings between Bath and Bristol.

39 THE HAYES

Newton St. Loe, Bath, BA2 9BU. Jane Giddins, 01225 873592, jeturner@btinternet.com. *From the Globe Pub on A4 take Pennyquick which is signed to Odd Down. Take 1st turning R which is just after the bend. At top of hill turn R then L.* **Sat 27, Sun 28 May (2-5). Adm £6, chd free. Home-made teas. Visits also by arrangement 1 May to 15 Sept for groups of 20+.**

Stunning in all seasons. 1 acre garden on edge of Duchy of Cornwall village. Herbaceous borders and formal lawns and terraces; informal garden of trees and long grass, bulbs and meadow flowers; formal potager and greenhouse; small orchard with espalier apple trees. Tulips, wisteria, alliums, foxgloves, gladioli, dahlias, asters. Wonderful views. All of garden can be accessed in a wheelchair but some grassy inclines.

40 4 HAYTOR PARK

Bristol, BS9 2LR. Mr & Mrs C J Prior, 07779 203626, p.l.prior@gmail.com. *3m NW of Bristol city centre. From A4162 Inner Ring Rd take turning into Coombe Bridge Ave, Haytor Park is 1st on L. Please no parking in Haytor Park.* **Sun 28 May (1.30-5). Adm £4, chd free. Open nearby 81 Coombe Lane. Opening with Stoke Bishop Gardens on Sun 7 May. Visits also by arrangement May to Aug for groups of 10 to 30.**

A journey of discovery, under many arches, around a lovingly tended plot, full of year-round interest at every turn. A wildlife pond, spaces to sit and quirky features, a bicycle wheel screen, rusty bed trellis, so many plants to find. There are dragons for children to discover and maybe a prize to win.

41 ◆ HESTERCOMBE GARDENS
Cheddon Fitzpaine, Taunton, TA2 8LG. Hestercombe Gardens Trust, 01823 413923, info@hestercombe.com, www.hestercombe.com. *3m N of Taunton, less than 6m from J25 of M5. Follow brown daisy signs. SatNav postcode TA2 8LQ.* **For NGS: Tue 7 Mar (10-4.30). Adm £14.85, chd £7.45. Light refreshments at the Stables with indoor & covered courtyard seating. Discount/prepaid vouchers not valid on NGS charity days. Evening opening Tue 6 June (6-8). Adm £30. Pre-booking essential, please visit www.ngs.org.uk for information & booking. For other opening times and information, please phone, email or visit garden website.**
Magnificent Georgian landscape garden designed by artist Coplestone Warre Bampfylde, a contemporary of Gainsborough and Henry Hoare of Stourhead. Victorian terrace and shrubbery and an exquisite example of a Lutyens/Jeykll designed formal garden; C17 water garden opening 2022. Enjoy 50 acres of woodland walks, temples, terraces, pergolas, lakes and cascades. Restaurant and café, restored watermill and barn, contemporary art gallery. Four centuries of garden design. A limited number of tickets have been made available for a private tour of the gardens on the evening of the 6th June. After an introductory talk by Head Gardener, Claire Greenslade, there will be the opportunity to explore the gardens, accompanied by Claire. This will be followed by refreshments. Gravel paths, steep slopes, steps. An all-access route is shown on the guide map, & visitors can phone to pre-book an all-terrain 'tramper' vehicle.

42 HILLCREST
Curload, Stoke St Gregory, Taunton, TA3 6JA. Charles & Charlotte Sundquist, 01823 490852, chazfix@gmail.com. *At top of Curload. From A358 turn L along A378, then branch L to North Curry & Stoke St Gregory. L ½m after Willows & Wetlands centre. Hillcrest is 1st on R with parking directions.* **Sat 13, Sun 14 May (1-5). Adm £5, chd free. Home-made teas. Visits also by arrangement 1 Apr to 30 July for groups of up to 30. Guided tour inc but groups welcome to explore on their own.**
Boasting stunning views of the Somerset Levels, Burrow Mump and Glastonbury Tor this 6 acre garden offers plenty of interest, inc a standing stone. Enjoy woodland walks, varied borders, flowering meadow and several ponds. There are greenhouses, orchards, a new produce garden and newly built large gravel garden. Garden is mostly level with gentle sloping paths down through meadow.

43 HILLSIDE
35 Westbury Hill, Westbury-on-Trym, Bristol, BS9 3AG. Stephanie Pritchett, hillsideoldstables@gmail.com, www.hillsideoldstables.com/ourgardens. *3 m from J17 M5 or Bristol City Centre. We are next door (downhill) from the Post Office Tavern pub on Westbury Hill. Access will be via the large green gates next to the pub.* **Sat 10 June (10-4). Adm £5, chd free. Home-made teas.**
A hidden Grade II listed period gem dating to 1715 with large and established formal walled gardens and family gardens in ½ acre. There are many different areas to explore – magnificent trees c 500 years old, pond and kitchen herb garden, rockery, rose and cutting gardens, orchard and seasonal interest borders, wildflower meadow and established kitchen garden and fruit bushes.

44 NEW HOLLAM HOUSE
Dulverton, TA22 9JH. Annie Prebensen, 01398 323445, annie@hollam.co.uk. *From Dulverton Bridge, go straight and bear R at the chemist. At the garage, Hollam Lane is the 2nd L turning immed after the garage, it is between a cottage & music shop.* **Visits by arrangement 17 Apr to 29 June for groups of 8 to 25. Monday through Thursday. No large coaches. Adm £7, chd free. Light refreshments by arrangement.**
Extending over 5 acres, this sloping Exmoor garden inc ponds, a water garden, woodland planting, borders and meadow areas. There are magnificent mature trees and old rhododendrons. Spring highlights are the thousands of tulips and other bulbs as well as flowering shrubs and trees; magnolia, cornus and viburnum among others. Not suitable for those of limited mobility or small children. No dogs.

45 HONEYHURST FARM
Honeyhurst Lane, Rodney Stoke, Cheddar, BS27 3UJ. Don & Kathy Longhurst, 01749 870322, donlonghurst@btinternet.com, www.ciderbarrelcottage.co.uk. *4m E of Cheddar. From Wells (A371) turn into Rodney Stoke signed Wedmore. Pass church on L and continue for almost 1m. Car park signed. From Cheddar (A371) turn R signed Wedmore, through Draycott to car park.* **Sun 2, Mon 3 July (2-5). Adm £4, chd free. Home-made teas. Visits also by arrangement May to Aug for groups of 10 to 50.**
⅔ acre part walled rural garden with babbling brook and 4 acre traditional cider orchard, with views. Specimen hollies, copper beech, paulownia, yew and poplar. Pergolas, arbour and numerous seats. Mixed informal shrub and perennial beds with many unusual plants. Many pots planted with shrubs, hardy and half-hardy perennials. Level, grass and some shingle.

46 JAPANESE GARDEN BRISTOL
13 Glenarm Walk, Brislington, Bristol, BS4 4LS. Martin Fitton, www.facebook.com/japanesegardenbristol. *A4 Bristol to Bath. A4 Brislington, at Texaco Garage at bottom of Bristol Hill turn into School Rd & immed R into Church Parade. Car park 1st turn on R or proceed to Glenarm Walk.* **Sat 13 May, Sun 11 June (10-4). Adm £5, chd free. Pre-booking essential, please visit www.ngs.org.uk for information & booking. Home-made teas.**
As you walk through the gate you will be welcomed by Japanese Koi. Then take a step to another level to the relaxing Japanese garden room and tea house surrounded by acers and cloud trees. Walk past Buddha corner into the Bonsai and Zen water feature area. Continue to a Japanese courtyard through a gate to a peaceful Zen rock garden. There you will find seating to enjoy the serene atmosphere. The owner will provide a brief talk about his garden to visitors before they enter, on the hour. Please note steps to different levels means the garden is unsuitable for disabled access.

47 NEW THE KITCHEN GARDEN AT MARTOCK WORKSPACE

Thorne, Yeovil, BA21 3PZ. Mrs Lara Honnor. *From A303 take turning for Stoke Sub Hamdon/ Martock & follow Stoke rd. Garden is on L. If coming from Martock, with church on R, turn L at r'about & garden is on 2nd R after the rec/tennis courts.* **Sat 16 Sept (11-5). Adm £5, chd free. Light refreshments at The Hub Cafe, Martock Workspace.**

This no dig garden has been created by Lara Honnor, who previously worked for acclaimed Somerset No Dig Market Gardener Charles Dowding. Inspired by the traditional Victorian Kitchen Garden designs Lara has transformed a 500sq metre area of dirt into an abundant and thriving vegetable and cut flower garden in just 6 months, providing produce for the Hub Cafe opp and for locals to buy. The Hub Cafe sells refreshments with seating outside under a canopy. Florist and jewellery maker open. The paths inside are made of woodchip, but the main path that crosses through the middle is wide enough for wheelchairs. Cafe & car park accessible.

48 LANE END HOUSE

Curload, Stoke St Gregory, Taunton, TA3 6JA. Eric & Veronica Martin, 07817 518655, va.martin@gmail.com. *Taunton A358 turn L on A378, fork L to North Curry & Stoke St Gregory. L ½m after Willows & Wetlands Centre into Curload. 1st house on L.* **Visits by arrangement 9 Apr to 30 June for groups of up to 20. Limited parking available for max 6 cars. Adm £5, chd free.**

Our garden has grown over the past 6 yrs with the addition of a large number of specimen trees and spring bulbs planted into what was once a rough grazing field. Separate areas inc a vegetable and fruit patch, orchard and wildlife pond sitting alongside grasses and nectar rich flowers, apiary, flower borders, seating areas and a number of unique sculptures. Level gardens with gravel courtyard.

49 LOWER SHALFORD FARM

Shalford Lane, Charlton Musgrove, Wincanton, BA9 8HE. Mr & Mrs David Posnett. *Lower Shalford is 2m NE of Wincanton. Leave A303 at Wincanton go N on B3081 towards Bruton. Just beyond Otter Garden Centre turn R Shalford Lane, garden is ½m on L. Parking opp house.* **Sat 25 Mar (10-3); Sat 20 May (10-4). Combined adm with Forest Lodge £10, chd free. Light refreshments.**

Fairly large open garden with extensive lawns and wooded surroundings with drifts of daffodils in spring. Small winterbourne stream running through with several stone bridges. Walled rose/parterre garden, hedged herbaceous garden and several ornamental ponds.

50 LYDEARD HOUSE

West Street, Bishops Lydeard, Taunton, TA4 3AU. Mrs Vaun Wilkins. *5m NW of Taunton. A358 Taunton to Minehead, do not take 1st 3 R turnings to Bishops Lydeard but 4th signed to Cedar Falls. Follow signs to Lydeard House.* **Sun 4 June (2-5). Adm £5, chd free. Light refreshments.**

4 acre garden with C18 origins and many later additions. Sweeping lawns, lake overhung with willows, canal running parallel to Victorian rose-covered pergola, along with box parterre, chinoiserie-style garden, recent temple folly and walled vegetable garden plus wonderful mature trees. Plants for sale. Children must be supervised because of very deep water. Deep gravel paths and steps may cause difficulty for wheelchairs but most features accessible by lawn.

51 MATHLIN COTTAGE

School Road, Wrington, Bristol, BS40 5NB. Tony & Sally Harden. *10m SW of Bristol, midway between A38 at Redhill or Lower Langford & A370 at Congresbury.* **Wed 7, Sun 11 June (2-5.30). Adm £5, chd free. Home-made teas.**

A cottage garden accessed by numerous shallow steps with wonderful views of the Mendip hills. Enjoy the front garden border with pergola and walkway before entering the back, more cottage style space full of bee/butterfly friendly plants. There is a small pond, greenhouse with tomatoes, cucumbers and chilies alongside a salad/soft fruit plot. Plenty of seating and places to relax as well as a large specimen of Paul's Himalayan Musk over a new walkway to enjoy.

52 ◆ MILTON LODGE

Old Bristol Road, Wells, BA5 3AQ. Simon Tudway Quilter, 01749 679341, www.miltonlodgegardens.co.uk. *½m N of Wells. From A39 Bristol-Wells, turn N up Old Bristol Rd; car park 1st gate on L signed.* **For NGS: Sun 7 May, Sun 4 June, Sun 2 July (2-5). Adm £5. Tea. Children under 14 free entry to garden. Discount/prepaid vouchers are not valid on the National Garden Scheme charity days. For other opening times and information, please phone or visit garden website.**

A must for garden lovers. Sloping land transformed into architectural terraces capitalising on views of Wells Cathedral & Vale of Avalon. The Grade II terraced garden was restored to its former glory by the owner's parents who moved here in 1960, replacing the orchard with a collection of ornamental trees. Herbaceous borders, blooming roses, inc Gertrude Jekyll, flourish next to well tended lawns. A serene, relaxing atmosphere within the garden succeeds the ravages of two World Wars. Cross over Old Bristol Rd to our 7 acre woodland garden, The Combe, a natural peaceful contrast to the formal garden of Milton Lodge, well worth a visit.

53 MODEL FARM

Perry Green, Wembdon, Bridgwater, TA5 2BA. Dave & Roz Young, 01278 429953, daveandrozontour@hotmail.com, www.modelfarm.com. *4m from J23 of M5. Follow Brown signs from r'about on A39 2m W of Bridgwater.* **Every Thur 6 Apr to 30 Nov (10-4). Adm £5, chd free. Tea, coffee, squash.**

4 acres of flat gardens to S of Victorian country house. Created from a field in last 13 yrs and still being developed. A dozen large mixed flower beds planted in cottage garden style with wildlife in mind. Wooded areas, mixed orchard, lawns, wildflower meadows and wildlife pond. Plenty of seating throughout the gardens with various garden sculptures by Somerset artists.

54 165 NEWBRIDGE HILL

Bath, BA1 3PX. Helen Hughesdon, 07793 085267, thefragrantlife@hotmail.com. *On the western fringes of Bath (A431), 100m on L after Apsley Rd. Several bus routes go to Newbridge Hill.* **Sat 1, Sun 2 July (10-5). Adm £5, chd free. Light refreshments. Visits also by arrangement June and Sept for groups of 8 to 24. Admission price for groups include cream tea.**

Incorporating 'Sculpture to Enhance a Garden' at its July opening this south facing garden on the edge of the city has unusual and exotic plants, vegetable garden, wildlife pond, greenhouse, treehouse and swing, late summer colour, fabulous views and a sunny terrace overlooking the garden where light lunches, cream teas and home-made cakes are served. Pieces of sculpture are for sale.

55 NYNEHEAD COURT

Nynehead, Wellington, TA21 0BN. Nynehead Care Ltd, 01823 662481, nyneheadcare@aol.com, www.nyneheadcourt.co.uk. *1½m N of Wellington. M5 J26 B3187 towards Wellington. R on r'about marked Nynehead & Poole, follow lane for 1m, take Milverton turning at fork, turning into Chipley Rd.* **Sun 26 Mar, Sun 11 June, Sun 16 July (2-5). Adm £5.50, chd free. Cream teas in Orangery. Visits also by arrangement 9 Jan to 18 Dec for groups of 10 to 35.**

Nynehead Court was the home of the Sandford family from 1590-1902. The 14 acres of gardens are noted for specimen trees, and there will be a garden tour with the Head Gardener at 2.15pm. Nynehead is now a private residential care home; for the latest Covid precautions pls contact us or view our Facebook page for details. Partial wheelchair access: cobbled yards, gentle slopes, chipped paths, liable to puddle during or after rain. Please wear suitable footwear.

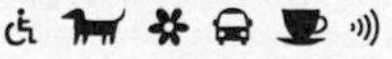

56 THE OLD RECTORY, DOYNTON

18 Toghill Lane, Doynton, Bristol, BS30 5SY. Edwina & Clive Humby, www.tumblr.com/doyntongardens. *At heart of village of Doynton, between Bath & Bristol. Parking in Bury Lane, just after junction with Horsepool Lane. Car parking is signed & charges may apply due to another local event being held that day.* **Sat 10 June (11-4). Adm £5, chd free. Home-made teas by WI.**

Doynton's Grade II-listed Georgian Rectory's walled garden and extended 15 acre estate. Renovated over 12 yrs, it sits within AONB. Garden has diversity of modern and traditional elements, fused to create an atmospheric series of garden rooms. Large landscaped kitchen garden featuring canal, vegetable plots, fruit cages and treehouse.

57 PENNARD HOUSE

East Pennard, Shepton Mallet, BA4 6TP. Martin Dearden. *5m S of Shepton Mallet on A37. 1m W of A37 Turn R at the top of the hill, or L if coming from the south.* **Wed 7 June (12-4). Adm £6, chd free. Home-made teas.**

Two delightful separate gardens. Both with extensive lawns, mature trees, rose beds, rustic topiary, a Victorian spring fed swimming pool and ponds. Garden layout dates from 1835 when the Napier family enlarged the main house and acquired the house next door. Next-door, Pennard Plants, will also open for plant purchasing. On offer is a selection of edible plants, fruit trees, herbs and seeds. The gardens are on a slope, but accessible with assistance.

58 PENNY BROHN UK

Chapel Pill Lane, Pill, BS20 0HH. Penny Brohn UK, 07951 033495, fundraising@pennybrohn.org.uk, www.pennybrohn.org.uk. *4m W of Bristol. Off A369 Clifton Suspension Bridge to M5 (J19 Gordano Services). Follow signs to Penny Brohn UK and to Pill/Ham Green (5 mins).* **Sun 21 May (10-4). Adm £5, chd free. Hot & cold drinks and home-made cakes are available. Visits also by arrangement 31 May to 31 July for groups of 5 to 15.**

3½ acre tranquil garden surrounds Georgian mansion with many mature trees, wildflower meadow, flower garden and cedar summerhouse. Fine views from historic gazebo overlooking the River Avon. Courtyard gardens with water features. Garden is maintained by volunteers and plays an active role in the Charity's 'Living Well with Cancer' approach. Plants, teas, music and plenty of space to enjoy a picnic. Gift shop. Tours of centre to find out more about the work of Penny Brohn UK. Some gravel and grass paths.

59 THE RED POST HOUSE

Fivehead, Taunton, TA3 6PX. The Rev Mervyn & Mrs Margaret Wilson, 01460 281 558, margaretwilson426@gmail.com. *10m E of Taunton. On the corner of A378 and Butcher's Hill, opp garage. From M5 J25, take A358 towards Langport, turn L at T-lights at top of hill onto A378. Garden is at Langport end of Fivehead.* **Sun 14 May (2-5). Adm £5, chd free. Home-made teas. Visits also by arrangement 1 Jan to 1 Dec for groups of up to 20.**

⅓ acre walled garden with shrubs, borders, trees, circular potager, topiary. We combine beauty and utility. Further 1½ acres, lawn, orchard and vineyard. Plums, 40 apple and 20 pear, walnut, quince, medlar, mulberry, fig. Mown paths, longer grass. Views aligned on Ham Hill. Summerhouse with sedum roof, belvedere. Garden in its present form has been developed over last 20 yrs. Paths are gravel and grass, belvedere is not wheelchair accessible. Dogs on leads.

60 THE RIB

St. Andrews Street, Wells, BA5 2UR. Paul Dickinson & David Morgan-Hewitt. *Wells City Centre, adjacent to east end of Wells Cathedral & opp Vicars Close. There is absolutely no parking at or very near this city garden. Visitors should use one of the 5 public car parks and enjoy the 10-15 minutes stroll through the city to The Rib.* **Sat 22 July (12-5). Adm £5, chd free.**

The Rib is one of the few houses in England that can boast a cathedral and a sacred well in its garden. Whilst the garden is compact, it delivers a unique architectural and historical punch. Long established trees, interesting shrubs and more recently planted mixed borders frame the view in the main garden. Ancient walled orchard and traditionally planted cottage garden. Lunch, tea and WC facilities available in Wells marketplace or Wells Cathedral. Slightly bumpy but short gravel drive and uneven path to main rear garden. 2-3 steps up to orchard and cottage gardens. Grass areas uneven in places.

61 ROCK HOUSE

Elberton, BS35 4AQ. Mr & Mrs John Gunnery, 01454 413225. *10m N of Bristol. From Old Severn Bridge on M48 take B4461 to Alveston. In Elberton, take 1st turning L and then immed R. SatNav starts at top of village, come down hill to Littleton turning on R, R again.* **Sun 29 Jan, Sun 5 Feb, Sun 19, Sun 26 Mar, Sun 4 June (11-4). Adm £5, chd free. Visits also by arrangement.**

2 acre garden. Woodland vistas with swathes of snowdrops and carpets of daffodils, some unusual. Spring flowers, cottage garden plants and climbing roses in season. Old yew tree, maturing cedar tree, pond. The garden will also be open to visitors on Sunday 28th January and Sunday 5th February 2024 between 11am and 4pm.

62 ROSE COTTAGE

Smithams Hill, East Harptree, Bristol, BS40 6BY. Bev & Jenny Cruse, 01761 221627, bandjcruse@gmail.com. *5m N of Wells, 15m S of Bristol. From B3114 turn into High St in EH. L at Clock Tower & immed R into Middle St, up hill for 1m. From B3134 take EH rd opp Castle of Comfort, continue 1½m. Car parking in field opp cottage.* **Visits by arrangement Apr to June for groups of 10+. Pls confirm visitor numbers 2 weeks prior to visit. Adm £5.50, chd free. Home-made teas.**

Organically gardened and planted to encourage wildlife. Bordered by a stream and mixed hedges, our acre of hillside cottage garden is carpeted with seasonal bulbs, primroses and hellebores in spring, roses and hardy geraniums in summer. Plenty of seating areas to enjoy the panoramic views over Chew Valley. Opp our cottage, in the corner of the field, our wildlife area with pond develops with interest. Full of spring and summer colour.

63 THE ROYAL CRESCENT HOTEL & SPA

16 Royal Crescent, Bath, BA1 2LS. Topland Ltd (The Royal Crescent Hotel and Spa), 01225 823 333, info@royalcrescent.co.uk, www.royalcrescent.co.uk. *Walk through front door of hotel, carry straight on, exit via back door to gardens. Nearest car park is in Charlotte St. Lansdown P & R bus stop nearby.* **Mon 3, Tue 4 July (9.30-4). Adm £5, chd free. Light refreshments. Booking essential for lunch or afternoon tea.**

1 acre of secluded gardens sits waiting for you behind this iconic hotel, lovingly curated by our gardeners. Gently winding lavender paths take you across the beautiful lawns, with various stunning floral displays along the route. There are many beautiful and varied aspects of the garden to enjoy, inc rescue hedgehogs, rose bushes, and interesting statues. Garden map and virtual tour available on website.

64 42 SILVER STREET

Midsomer Norton, Radstock, BA3 2EY. Andrew King & Kevin Joint, kingandrew@talk21.com. *For open day only parking is at Norton Hill School, Charlton Rd, BA3 4AD. From town centre, follow the B3355 Silver St for 0.4m, turn L into Charlton Rd, School car park is100m on R.* **Sun 18 June (11-4). Adm £5, chd free. Home-made teas. For group visits please enquire re refreshments. Visits also by arrangement May to July for groups of 10 to 30. Limited street parking for By Arrangement visits only, pls car share.**

Half acre plantsman's garden divided into 7 distinct zones, developed from scratch by the owners over the last 10 yrs. Features inc a gravel garden with formal rill and pond, mixed/herbaceous borders in cool colours, Iris garden (bearded irises), elliptical lawn with hot colour planting, informal area with woodland character and experimental bank with intermingled mixed herbaceous planting. The herbaceous planting on the bank was inspired by the Merton Borders at Oxford Botanic Gardens. The garden inc various structures made from green oak, inc an arbour and pergola incorporating timbers from an old bus shelter at Bath. Level wheelchair access to most of the garden, enter via separate side entrance - pls ask at main entrance on arrival.

♿ ✽ ☕

65 NEW SKOOL BEANZ CHILDREN'S ALLOTMENT

Little Sammons Allotments, Chilthorne Domer, BA22 8RB. South Somerset County Council. *Situated between the Carpenters Arms & the village school adjacent to Tintinhull rd. Parking either at pub or at The Rec just off Main St next to the school.* **Sat 15 Apr, Sat 12 Aug (11-5). Adm £3, chd free. Home-made teas at The Rec at Chilthorne Domer.**

Skool Beanz is a gardening club for children, created by Lara Honnor, to encourage children to enjoy gardening. The allotment has a huge cut flower dahila bed, vegetable area, fruit trees and bushes, a rainwater collecting station, 'Muddy Buddy' compost heap, quiet wildlife garden, secret den, polytunnel and plenty of seating and tables for upcycling garden arts & crafts. Lara contributed to Charles Dowding's No Dig Children's Gardening Book sharing tips she has learnt from Skool Beanz on teaching the joys of gardening to children.

66 SOUTH KELDING

Brewery Hill, Upton Cheyney, Bristol, BS30 6LY. Barry & Wendy Smale, 0117 9325145, wendy.smale@yahoo.com. *Halfway between Bristol and Bath. Upton Cheyney lies ½m up Brewery Hill off A431 just outside Bitton. Detailed directions & parking arrangements given when appt made. Restricted access means pre-booking essential.* **Visits by arrangement 1 Mar to 30 Oct for groups of up to 30. Adm £10, chd free. Home-made teas. Adm price includes refreshments & tour by owner.**

7 acre hillside garden offering panoramic views from its upper levels, with herbaceous and shrub beds, prairie-style scree beds, orchard, native copses and small, labelled arboretum grouped by continents. Large wildlife pond, boundary stream and wooded area featuring shade and moisture-loving plants. Due to slopes and uneven terrain this garden is unsuitable for disabled access.

67 ◆ SPECIAL PLANTS

Greenway Lane, Cold Ashton, SN14 8LA. Derry Watkins, 01225 891686, derry@specialplants.net, www.specialplants.net. *6m N of Bath. From Bath on A46, turn L into Greenways Lane just before r'about with A420.* **For NGS: Thur 15 June, Thur 20 July, Thur 17 Aug, Thur 21 Sept, Thur 19 Oct (10.30-5). Adm £5, chd free. Home-made teas. For other opening times and information, please phone, email or visit garden website.**

Architect-designed ¾ acre hillside garden with stunning views. Started autumn 1996. Exotic plants. Gravel gardens for borderline hardy plants. Black and white (purple and silver) garden. Vegetable garden and orchard. Hot border. Lemon and lime bank. Annual, biennial and tender plants for late summer colour. Spring fed ponds. Bog garden. Woodland walk. Allium alley. Free list of plants in garden.

68 ◆ STOBERRY GARDEN

Stoberry Park, Wells, BA5 3LD. Frances & Tim Young, 01749 672906, stay@stoberry-park.co.uk, www.stoberryhouse.co.uk. *½m N of Wells. From Bristol - Wells on A39, L into College Rd & immed L through Stoberry Park, signed.* **For NGS: Fri 12, Sat 13, Sun 14 May (1.30-5). Adm £5, chd free. Home-made teas. Discount/prepaid vouchers not valid on NGS charity day. For other opening times and information, please phone, email or visit garden website.**

With breathtaking views over Wells Cathedral, this 6 acre family garden is planted sympathetically within its landscape providing stunning combinations of vistas accented with wildlife ponds and water features. In comparison 1½ acre walled garden, full of interesting planting. Colour and interest for every season; spring bulbs, irises, salvias, a newly planted wildflower area, a new modern rockery. Interesting sculpture artistically integrated in the garden.

GROUP OPENING

69 STOGUMBER GARDENS

Station Road, Stogumber, TA4 3TQ. *11m NW of Taunton. 3m W of A358. Signed to Stogumber, W of Crowcombe. Village maps given to all visitors.* **Sun 18 June (2-6). Combined adm £6, chd free. Home-made teas in the Village Hall.**

HIGHER KINGSWOOD
Fran & Tom Vesey.

KNOLL COTTAGE
Elaine & John Leech,
01984 656689,
john@Leech45.com,
www.knoll-cottage.co.uk.
Visits also by arrangement June to Sept for groups of 5 to 30.

POUND HOUSE
Barry & Jenny Hibbert.

WICK BARTON
Sara & Russ Coward.

4 delightful and very varied gardens in a picturesque village on the edge of the Quantocks. 2 surprisingly large gardens near village centre, plus 2 very large gardens on outskirts of village, with many rare and unusual plants. Conditions range from waterlogged clay to well-drained sand. Features inc courtyard, ponds, bog gardens, rockery, extensive mixed beds, vegetable and fruit gardens, and a collection of over 80 different roses. Fine views of surrounding countryside from some gardens. Wheelchair access to main features of all gardens.

GROUP OPENING

70 NEW STOKE BISHOP GARDENS

Stoke Bishop, Bristol, BS9 1DD. Malcolm Ravenscroft, Pat Prior, Magda Goss. *All gardens are within the BS9 postal code area of Bristol.* **Sun 7 May (1-5). Combined adm £7.50, chd free. Home-made teas in Lucombe House, which is the only venue offering teas and cake.**

4 HAYTOR PARK
Mr & Mrs C J Prior.
(See separate entry)

LUCOMBE HOUSE
Malcolm Ravenscroft,
01179 682494,
famrave@gmail.com.
Visits also by arrangement 1 May to 1 Oct Pls contact owner at least two weeks prior to visit.

1 SUNNYSIDE
Mrs Magda Goss.

Three very different gardens, all within ½m of each other in the BS9 area of Bristol. 4 Haytor Park: A journey of discovery, under many arches, around a lovingly tended plot, full of year-round interest at every turn. A wildlife pond, spaces to sit and quirky features, a bicycle wheel screen, rusty bed trellis, so many plants to find. Lucombe House: For tree lovers of all ages! In addition to the 260 yr old Lucombe Oak, registered as one of the most significant trees in the UK, there are over 30 mature English trees planted together with ferns and bluebells to create an urban woodland - plus a newly designed Arts & Crafts front garden. A new path through the woodland has been completed providing a behind the scenes look at the garden. Sunnyside: C17 cottage in heart of Stoke Bishop with part walled, cottage style front garden. Garden sculptures dominated by large magnolia, perennials, roses and spring bulbs. Courtyard garden with open studio at rear.

71 STONELEIGH DOWN
Upper Tockington Road, Tockington, Bristol, BS32 4LQ. Su & John Mills, 01454 615438, susanlmills@gmail.com. *12m N of Bristol. On LH side of Upper Tockington Rd when travelling from Tockington towards Olveston. Set back from road up gravel drive. Parking in village.* **Mon 10 Apr, Fri 2, Sun 4 June (1-5). Adm £6, chd free. Home-made teas. Visits also by arrangement 8 Apr to 11 June for groups of 16 to 30.**
Approaching ⅔ acre, the south facing garden has curved gravel pathways around an S-shaped lawn that connects themed areas: exotic border; summer walk; acers; oriental pond; winter garden; spring garden. On a level site, it has been densely planted with trees, shrubs, perennials and bulbs for year-round interest. Plenty of places to sit. Steps into courtyard.

GROUP OPENING

72 STOWEY GARDENS
Stowey, Bishop Sutton, Bristol, BS39 5TL. *10m W of Bath. Stowey Village on A368 between Bishop Sutton & Chelwood. From Chelwood r'about take A368 to Weston-s-Mare. At Stowey Xrds turn R to car park, 150yds down lane, ample off road parking opp Dormers. Limited disabled parking at each garden which will be signed.* **Sun 23 July (2-6). Combined adm £6, chd free. Home-made teas at Stowey Mead.**

DORMERS
Mr & Mrs G Nicol.

◆ MANOR FARM
Richard Baines & Alison Fawcett, 01275 332297.

STOWEY MEAD
Mr Victor Pritchard.

A broad spectrum of interest and styles developing year on year. But there is far more than this in these gardens; the visitor's senses will be aroused by the sights, scents and diversity of these gardens in the tiny, ancient village of Stowey. There are flower-packed beds, borders and pots, roses, topiary, hydrangeas, exotic garden, many unusual trees and shrubs, orchards, vegetables, ponds, specialist sweet peas, lawns, Stowey Henge, and a ha-ha. Abundant collections of mature trees and shrubs and ample seating areas, with glorious views from each garden. There is something of interest for everyone, all within a few minutes walk of car park at Dormers. Plant sales at Dormers. Well behaved dogs on short leads welcome. Wheelchair access restricted in places, many grassed areas in each garden.

73 SUNNYMEAD
153 Long Ashton Road, Long Ashton, Bristol, BS41 9JQ. Anna-Liisa & Ian Blanks-Walden (Colonel Rtd, MBE). *Located opp The Angel Inn, with a LALHS historic green plaque & large wooden gates; sat on the main Long Ashton Rd opp Church Lane junction. Parking on main road on single white line.* **Sun 25 June (1-4). Combined adm with Coombe Cottage £5, chd free. Home-made teas.**
Sunnymead is a medieval village property dating back to 1403, with a wrap around family garden (0.4 acre) inc a number of different 'garden rooms' and themes: 1700s pond and rockery, 200yr old box hedge (designed in the 1700s and planted in the 1800s), various period features, drought tolerant 'Christopher Lloyd' inspired exotic border, Finnish themed woodland area, ericaceous planting and more. Cream teas, cakes and other treats for sale.

GROUP OPENING

74 NEW TINTINHULL GARDENS
68 Queen Street, Tintinhull, Yeovil, BA22 8PQ. Felicity Down. *Turn off A303 towards Tintinhull. Follow garden signs & park at village hall BA22 8PY where tickets can be purchased.* **Sat 1, Sun 2 July (12-5). Combined adm £8, chd free. Home-made teas at village hall. £10 adm for visitors wishing to attend both days.**

NEW THE BIRCHES
Martin & Flora Wragg.

NEW 3 CHURCH STREET
Derek & Anita Mills.

NEW 28 CHURCH STREET
Geoff & Jo Fisher.

NEW FRANCIS COTTAGE
Joanna Hunt.

NEW FRANCIS HOUSE
Steve & Sue Creaney.

NEW 11 HEAD STREET
Dave & Sue Shorey.

NEW LAMB FARM
John & Marilyn Smith.

NEW LEACHES FARM
Cathie Brown.

NEW THE OLD DAIRY HOUSE
Geoff & Sarah Stone.

NEW 68 QUEEN STREET
Alan & Felicity Down.

NEW SOUTH VIEW
Patrick & Ruth Sullivan.

NEW WALTERS FARM
Ed & Anette Lorch.

These 12 beautiful gardens set in the conservation village of Tintinhull offer a full range of plants, many to encourage bees and butterflies. From enclosed courtyards to large country house landscapes, there is something for everyone. Roses climbing up hamstone walls and billowing over borders will encourage visitors to explore these well cared for gardens. As you wander through the village, some gardens are visible from the road. Scrumptious teas can be enjoyed in the Village Hall.

75 NEW TRAFALGAR HOUSE
29 Sion Hill, Bath, BA1 2UW. Mrs Judith Lywood, 07779 165075, judith.lywood@btinternet.com. *Top of Sion Hill opp allotments. Free street parking. Please car share where possible.* **Visits by arrangement May to Sept for groups of 15 to 50. Entry via back gate only. Adm £5, chd free. Refreshments available on request only.**
The garden has continual colour and features a fernery, gravel bed, climbing roses, clematis, also a woodland bed, cloud pruned yew and pots for every season. Spring bulbs start the season and dahlias and salvias end it. Across the path from the main garden the owner has created her dream of what a family allotment should be. As well as raised beds for easy harvesting and maintenance it affords a restful place to sit and enjoy the most glorious views over Bath. Allotment by front of the house, with espalier fruit trees,

raised beds, arches with clematis and sweet peas. A glorious view from the top of Sion hill, over the city of Bath. Partial access to main part of garden.

76 NEW TREGUNTER

Charlcombe Lane, Lansdown, Bath, BA1 5TT. Jane & Peter Mawle. *1m N of Bath city centre. From Lansdown Rd turn into Charlcombe Lane. At the junction with Richmond Rd turn L; house is 50m on R. Parking available higher up Charlcombe Lane or along Richmond Rd.* **Fri 9, Sun 11 June (12-5). Adm £5, chd free. Home-made teas.**

A Cotswold stone terraced garden, with spectacular views of Little Solsbury Hill, designed, organically planted and maintained by the owners. Flower borders are planted for year-round interest seamlessly moving between seasons. Unusual and rare plants, contemporary and classical planting, formal and natural ponds, large kitchen and herb garden, varied and deep herbaceous borders, yew and box topiary.

77 ◆ UNIVERSITY OF BRISTOL BOTANIC GARDEN

Stoke Park Road, Stoke Bishop, Bristol, BS9 1JG. 01174 282041, botanic-gardens@bristol.ac.uk, botanic-garden.bristol.ac.uk. *Located in Stoke Bishop ¼m W of Durdham Downs & 1m from city centre. After crossing the Downs to Stoke Hill, Stoke Park Rd is 1st on R.* **For NGS: Sun 2 July (10-5). Adm £9, chd free. Light refreshments, provided by local deli. For other opening times and information, please phone, email or visit garden website. Donation to University of Bristol Botanic Garden.**

Exciting and contemporary award winning Botanic Garden with dramatic displays illustrating collections of Mediterranean flora, rare native, useful plants (inc European and Chinese herbs) and those that illustrate plant evolution. Large floral displays illustrating pollination in flowering plant and evolution. Glasshouses are home to giant Amazon water lily, tropical fruit, medicinal plants, orchids, cacti and unique sacred lotus collection. Wheelchair available to borrow from Welcome Lodge on request. Wheelchair friendly primary route through garden inc glasshouses, accessible WCs.

78 VEXFORD COURT

Higher Vexford, Lydeard St Lawrence, Taunton, TA4 3QF. Dr David Yates, 01984 656735, yatesdavid135@gmail.com. *The best approach is from B3224 into Sheepstealing Lane. Take L after ¼m & continue to the bottom of the hill (0.9m). Turn L up Higher Vexford Farm drive & Vexford Court entrance on the R.* **Sun 25 June (10-5). Adm £5, chd free. Tea and soft drinks available. Visits also by arrangement 10 June to 18 Sept for groups of 10 to 25.**

This large garden surrounds 3 converted barns. 1½ acres of grounds slopes gently downhill to a small stream. Roses around the courtyard entrance lead to herbaceous beds and beyond to trees and shrubs. Amongst the interesting trees a large *Liriodendron tulipifera* tree blooms in early summer and there are azaleas, camellias, catalpa, ginko, Wollemi Pine and Persian Ironwood trees.

East Lambrook Gardens

© Ellen Rooney

79 ◆ THE WALLED GARDENS OF CANNINGTON

Church Street, Cannington, TA5 2HA. Bridgwater College, 01278 655042, walledgardens@btc.ac.uk, www.canningtonwalledgardens.co.uk.

Part of Bridgwater & Taunton College Cannington Campus, 3m NW of Bridgwater. On A39 Bridgwater-Minehead rd - at 1st r'about in Cannington 2nd exit, through village. War memorial, 1st L into Church St then 1st L. **For NGS: Sat 15, Sun 16 Apr, Sat 16, Sun 17 Sept (10-4). Adm £7.50, chd free. Light refreshments. For other opening times and information, please phone, email or visit garden website.**

Within the grounds of a medieval priory, the Walled Gardens of Cannington are a gem waiting to be discovered! Classic and contemporary features inc hot herbaceous border, blue garden, sub-tropical walk and Victorian style fernery, amongst others. Botanical glasshouse where arid, sub-tropical and tropical plants can be seen. Tea room, plant nursery, gift shop, also events throughout the year. Gravel paths.

80 WATCOMBE

92 Church Road, Winscombe, BS25 1BP. Peter & Ann Owen, 01934 842666, peterowen449@btinternet.com.

12m SW of Bristol, 3m N of Axbridge. 100 yds after yellow signs on A38 turn L, (from S), or R. (from N) into Winscombe Hill. After 1m reach The Square. Watcombe on L after 150yds. **Sun 16 Apr, Sun 14 May, Thur 8 June (2-5). Adm £5, chd free. Home-made teas. Gluten-free available. Visits also by arrangement 16 Apr to 9 July.**

¾ acre mature Edwardian garden with colour-themed, informally planted herbaceous borders. Strong framework separating several different areas; pergola with varied wisteria, unusual topiary, box hedging, lime walk, pleached hornbeams, cordon fruit trees, 2 small formal ponds and growing collection of clematis. Many unusual trees and shrubs. Small vegetable plot. Some steps but most areas accessible by wheelchair with minimal assistance.

81 WAYFORD MANOR

Wayford, Crewkerne, TA18 8QG.

3m SW of Crewkerne. Turn N off B3165 at Clapton, signed Wayford or S off A30 Chard to Crewkerne rd, signed Wayford. **Sun 14 May (2-5). Adm £6, chd £3. Home-made teas.**

The mainly Elizabethan manor (not open) mentioned in C17 for its 'fair and pleasant' garden was redesigned by Harold Peto in 1902. Formal terraces with yew hedges and topiary have fine views over W Dorset. Steps down between spring-fed ponds past mature and new plantings of magnolia, rhododendron, maples, cornus and, in season, spring bulbs, cyclamen and giant echium. Primula candelabra, arum lily and gunnera around lower ponds.

Stoneleigh Down

82 WEIR COTTAGE

Weir Lane, Marshfield, SN14 8NB. Ian & Margaret Jones. *Opp Weir Farm. 7m NE of Bath. Look for yellow sign on A420 between Cold Ashton & Chippenham. Weir Lane is at the E end of the High St & approx ½ m away from Green Lane.* **Sat 1, Sun 2 Apr (11-3). Combined adm with 6 Green Lane £5, chd free. Home-made teas.**

South facing garden of ¼ acre. High stone walls to the north protect against Marshfield weather. Limestone soil. We think the garden is best in early spring with daffodils and primroses. Sat 1st Apr guided walk around the outskirts of the village led by Cotswold Wardens. Boots advised. Sun 2nd Apr guided history walk in the village. Both walks meet 10am in Market Place.

83 WELL COTTAGE

Little Badminton, Tetbury, GL9 1AB. Miranda Beaufort, 01454 218729. *Signs from both ends of Little Badminton. Ignore no entry signs. Parking in Badminton Park.* **Visits by arrangement Apr to Sept for groups of 10 to 20. Adm £7, chd free. Light refreshments on request.**

The garden is on 4 levels. The bottom garden was planted 6 yrs ago, helped by the owner's husband, the late Duke of Beaufort. It is mostly planted with David Austin roses, which feature in their catalogue. Also many clematis, which come alive after the first flush of roses have finished. Over the road are wild flowers, picking borders and greenhouses. Wheelchair access on bottom 2 levels and over the road.

84 WELLFIELD BARN

Walcombe Lane, Wells, BA5 3AG. Virginia Nasmyth, 01749 675129. *½ m N of Wells. From A39 Bristol to Wells rd turn R at 30 mph sign into the narrow Walcombe Lane. Entrance at 1st cottage on R, parking signed.* **Visits by arrangement June & July for groups of 10 to 30. Pls confirm visitor numbers 2 weeks prior to visit. Adm £5.50, chd free. You are very welcome to bring your own picnic.**

Our garden is a haven for wildlife. 26yrs ago, a once bustling concrete farmyard grew into a tranquil ½ acre garden. Planned and created and still evolving today to provide enjoyment of colour and form for year-round enjoyment. Structured design integrates house, lawn and garden with the landscape. Wonderful views, ha-ha, specimen trees, mixed borders, hydrangea bed, hardy geraniums and roses. Formal sunken garden, grass walks with interesting young and semi-mature trees. Now tranquillity, sheep as neighbours, perfect peace. Our collection of hardy geraniums was featured on BBC Gardeners' World with Carol Klein. Moderate slopes in places, some gravel paths.

85 WESTBROOK HOUSE

West Bradley, BA6 8LS. Keith Anderson & David Mendel, www.instagram.com/keithbfanderson. *4m E of Glastonbury; 8m W of Castle Cary. From A361 at W Pennard follow signs to W Bradley (2m). From A37 at Wraxall Hill follow signs to W Bradley (2m).* **Sat 17 June (11-5). Adm £5, chd free. Donation to West Bradley Church.**

Garden developed since 2003 by a garden designer and a painter. 4 acres comprising 3 distinct gardens around house with mixed herbaceous and shrub borders leading to meadow and orchard with spring bulbs, species roses and lilacs.

86 ◆ THE YEO VALLEY ORGANIC GARDEN AT HOLT FARM

Bath Road, Blagdon, BS40 7SQ. Mr & Mrs Tim Mead, 01761 258155, visit@yeovalleyfarms.co.uk, www.yeovalley.co.uk. *12m S of Bristol. Off A368. Entrance approx ½ m outside Blagdon towards Bath, on L, then follow garden signs past dairy.* **For NGS: Sun 16 Apr, Sat 7 Oct (10-5). Adm £7, chd £2. Pre-booking essential, please visit www.ngs.org.uk for information & booking. Light refreshments. For other opening times and information, please phone, email or visit garden website.**

One of only a handful of ornamental gardens that is Soil Association accredited, 6½ acres of contemporary planting, quirky sculptures, bulbs in their thousands, purple palace, glorious meadow and posh vegetable patch. Great views, green ideas. Events, workshops and exhibitions held throughout the year - see website for further details. Level access to café. Around garden there are some grass paths and some uneven bark and gravel paths. Accessibility map available at ticket office.

87 THE YEWS

Harry Stoke Road, Stoke Gifford, Bristol, BS34 8QH. Dr Barbara Laue & Dr Chris Payne. *From A38/ Filton, take A4174 ring road to M32. L turn at Sainsbury R'about. Straight across next R'about. After 200 yds, sharp R turn into Harrystoke Rd. Parking in paddock.* **Sun 14 May (2-5). Adm £5, chd free. Home-made teas.**

Approx 1 acre, developed by present owners since 1987. Part of the old hamlet of Harrystoke, now surrounded by housing development. Formal area with pond, gazebo, herbaceous borders, clipped box and yew. 300 yr old yews, wedding cake tree, magnolias, eucalyptus, Indian bean tree, gingkoes and more. Vegetable garden, greenhouse, orchard and meadow. Spring bulbs and blossom.

88 YEWS FARM

East Street, Martock, TA12 6NF. Louise & Fergus Dowding, www.instagram.com/dowdinglouise. *Turn off main rd, Church St, at Market House/visitor's centre, onto East St, go past White Hart & PO on R. Yews Farm 150 yds on R, opp Foldhill Lane. Turn around if you get to Nag's Head.* **Sun 2 July, Sun 10 Sept (1.30-5). Adm £8, chd free. Home-made teas.**

Theatrical planting in large south facing walled garden. Sculptural planting for height, form, leaf and texture. Prolific healthy Box topiary. High maintenance pots. Self seeding hugely encouraged, especially in farmyard populated by hens and pigs. Working organic kitchen garden. Greenhouses bursting with summer vegetables. Organic orchard with heritage varieties and active cider barn – taste the difference with our homemade cider.

In 2022 the National Garden Scheme donated £3.11 million to our nursing and health beneficiaries

STAFFORDSHIRE

Birmingham & West Midlands

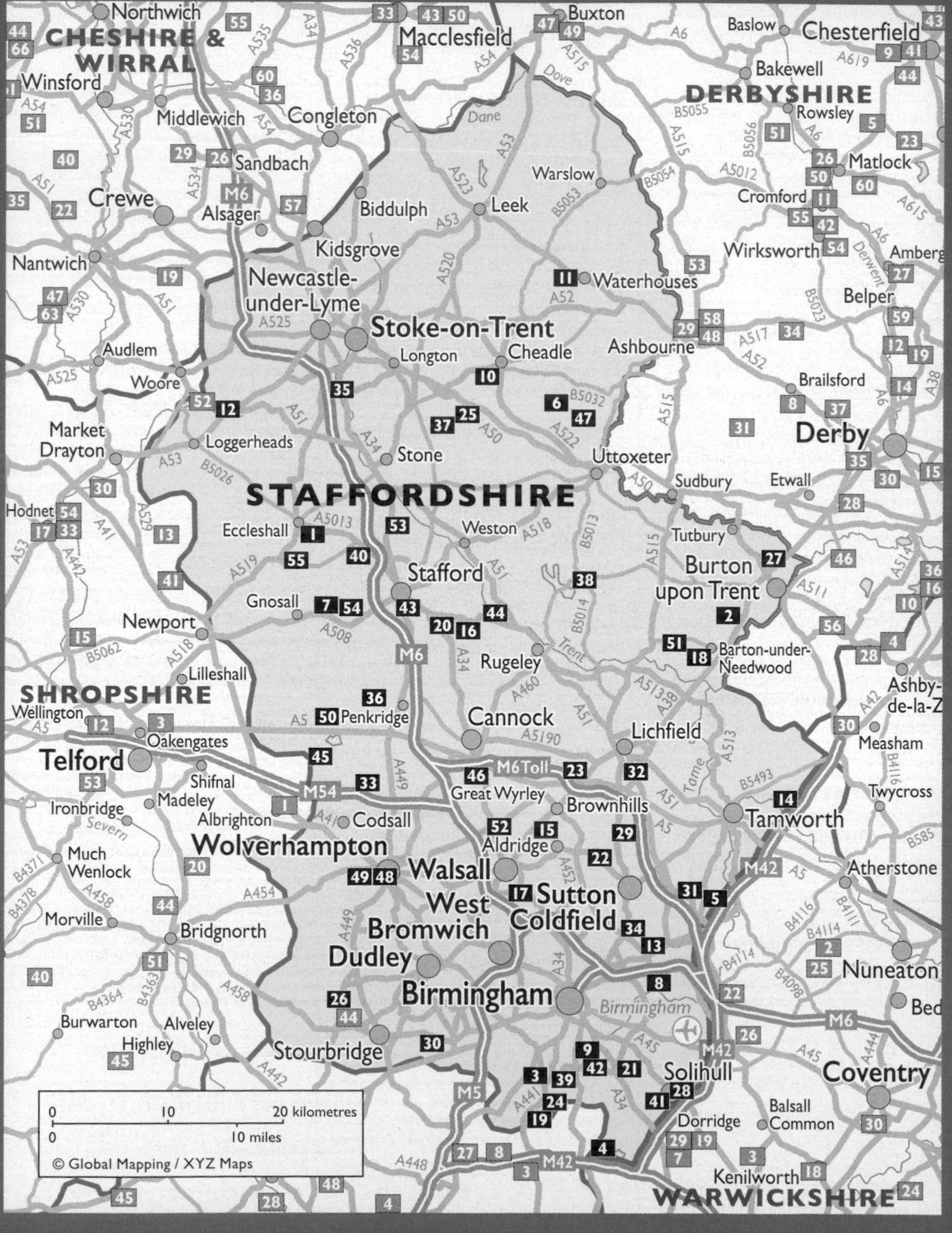

Staffordshire, Birmingham and part of the West Midlands is a landlocked 'county', one of the furthest from the sea in England and Wales.

It is a National Garden Scheme 'county' of surprising contrasts, from the 'Moorlands' in the North East, the 'Woodland Quarter' in the North West, the 'Staffordshire Potteries' and England's 'Second City' in the South East, with much of the rest of the land devoted to agriculture, both dairy and arable.

The garden owners enthusiastically embraced the National Garden Scheme from the very beginning, with seven gardens opening in the inaugural year of 1927, and a further thirteen the following year.

The county is the home of the National Memorial Arboretum, the Cannock Chase Area of Outstanding Natural Beauty and part of the new National Forest.

There are many large country houses and gardens throughout the county with a long history of garden-making and with the input of many of the well known landscape architects.

Today, the majority of National Garden Scheme gardens are privately owned and of modest size. However, a few of the large country house gardens still open their gates for National Garden Scheme visitors.

Volunteers

County Organiser
Anita & David Wright
01785 661182
davidandanita@ngs.org.uk

County Treasurer
Brian Bailey
01902 424867
brian.bailey@ngs.org.uk

Publicity
Ruth & Clive Plant
07591 886925
ruthandcliveplant@ngs.org.uk

Booklet Co-ordinator
Fiona Horwath
07908 918181
fiona.horwath@ngs.org.uk

Assistant County Organisers
Jane Cerone
01827 873205
janecerone@btinternet.com

Ken & Joy Sutton
07791 041189
kenandjoysutton@ngs.org.uk

Alison & Peter Jordan
01785 660819
alisonandpeterjordan@ngs.org.uk

@StaffsNGS

Left: 52 Elm Road

OPENING DATES

All entries subject to change. For latest information check **www.ngs.org.uk**

Map locator numbers are shown to the right of each garden name.

February

Snowdrop Festival

Sunday 12th
5 East View Cottages 14

March

Sunday 19th
NEW The Old Mission 36

Thursday 23rd
23 St Johns Road 43

Sunday 26th
Millennium Garden 32

April

Sunday 9th
'John's Garden' at Ashwood Nurseries 26

Saturday 22nd
Springfield Cottage 45

Thursday 27th
23 St Johns Road 43

In 2022 the National Garden Scheme donated £3.11 million to our nursing and health beneficiaries

May

Sunday 7th
Cats Whiskers 9

Sunday 14th
Brackencote 4
Millennium Garden 32

Thursday 18th
23 St Johns Road 43

Friday 19th
◆ Middleton Hall & Gardens 31
The Secret Garden 44

Saturday 20th
The Home Bee Garden 24

Sunday 21st
The Home Bee Garden 24
10 Paget Rise 38

Sunday 28th
Butt Lane Farm 7
Hamilton House 22
12 Meres Road 30
The Old Dairy House 35

Monday 29th
Bridge House 5
The Old Dairy House 35

June

Sunday 4th
The Bungalow, Wood Farm 6
The Garth 16
The Pintles 40
12 Waterdale 48
19 Waterdale 49

Thursday 8th
23 St Johns Road 43

Friday 9th
22 Greenfield Road 20

Saturday 10th
Hall Green Gardens 21
14 Longbow Close, Stretton 27

Sunday 11th
Ashcroft and Claremont 1
Brackencote 4
The Bungalow, Wood Farm 6
◆ Castle Bromwich Hall Gardens 8
Hall Green Gardens 21
3 Marlows Cottages 29
NEW The Old Mission 36
8 Rectory Road 41

Wednesday 14th
Bankcroft Farm 2

Friday 16th
NEW Hammerwich House Farm 23
Yarlet House 53

Saturday 17th
NEW Hammerwich House Farm 23

Sunday 18th
33 Gorway Road 17
Monarchs Way 33

Wednesday 21st
Bankcroft Farm 2

Saturday 24th
Springfield Cottage 45
Yew Trees 55

Sunday 25th
Cheadle Allotments 10
5 East View Cottages 14
The Garth 16
Marie Curie Hospice Garden 28
12 Meres Road 30
15 New Church Road 34
The Old Vicarage 37
Yew Trees 55

Wednesday 28th
Bankcroft Farm 2
5 East View Cottages 14
The Secret Garden 44

July

Sunday 2nd
Bournville Village 3
The Bungalow, Wood Farm 6
Grafton Cottage 18
12 Waterdale 48
19 Waterdale 49
25 Wolverhampton Road 52

Thursday 6th
23 St Johns Road 43
Yew Tree Cottage 54

Friday 7th
Grafton Cottage 18

Sunday 9th
The Bungalow, Wood Farm 6
NEW Izaak Walton Farm 25
76 Station Street 46
The Wickets 50

Thursday 13th
Yew Tree Cottage 54

Saturday 15th
Fifty Shades of Green 15

Sunday 16th
198 Eachelhurst Road 13
Fifty Shades of Green 15
Grafton Cottage 18

Thursday 20th
The Secret Garden 44
Yew Tree Cottage 54

Sunday 23rd
128 Green Acres Road 19
Yew Tree Cottage 54

Sunday 30th
Grafton Cottage 18

August

Saturday 5th
Springfield Cottage 45

Sunday 6th
Grafton Cottage 18

Friday 11th
Colour Mill 11

Saturday 12th
Colour Mill 11

Saturday 26th
Fifty Shades of Green 15

Sunday 27th
Fifty Shades of Green 15
The Wickets 50

September

Saturday 9th
Springfield Cottage 45

Sunday 10th
8 Rectory Road 41

October

Saturday 21st
◆ Dorothy Clive Garden 12

Sunday 22nd
◆ Dorothy Clive Garden 12

Sunday 29th
Wild Thyme Cottage 51

February 2024

Sunday 18th

5 East View Cottages 14

By Arrangement

Arrange a personalised garden visit with your club, or group of friends, on a date to suit you. See individual garden entries for full details.

Brackencote 4
Bridge House 5
Butt Lane Farm 7
5 East View Cottages 14
36 Ferndale Road, Hall Green Gardens 21
Fifty Shades of Green 15
The Garth 16
33 Gorway Road 17
Grafton Cottage 18
22 Greenfield Road 20
The Old Dairy House 35
Paul's Oasis of Calm 39
The Pintles 40
8 Rectory Road 41
120 Russell Road, Hall Green Gardens 21
56 St Agnes Road 42
23 St Johns Road 43
Truckle House, Croxden 47
19 Waterdale 49
The Wickets 50
Yew Tree Cottage 54
Yew Trees 55

In 2022 the National Garden Scheme donated £3.11 million to our nursing and health beneficiaries

THE GARDENS

GROUP OPENING

1 ASHCROFT AND CLAREMONT

Stafford Road, Eccleshall, ST21 6JP. *7m W of Stafford. J14 M6. At Eccleshall end of A5013 the garden is 100 metres before junction with A519. On street parking nearby. Note: Some SatNavs give wrong directions.* **Sun 11 June (2-5). Combined adm £4.50, chd free. Home-made teas at Ashcroft.**

ASHCROFT
Gillian Bertram.

26 CLAREMONT ROAD
Maria Edwards.

Weeping limes hide Ashcroft, a 1 acre garden of green tranquillity, rooms flow seamlessly around the Edwardian house. Covered courtyard with lizard water feature, sunken herb bed, kitchen garden, greenhouse, wildlife boundaries home to five hedgehogs increasing yearly. Deep shade border, woodland area and ruin with stone carvings and stained glass sculpture. Claremont is a master class in clipped perfection. An artist with an artist's eye has blurred the boundaries of this small Italianate influenced garden. Overlooking the aviary, a stone lion surveys the large pots and borders of vibrant planting, completing the Feng-Shui design of this beautiful all seasons garden. Tickets, teas and plants available at Ashcroft. Partial wheelchair access at Claremont.

♿ ✽ ☕

2 BANKCROFT FARM

Tatenhill, Burton-on-Trent, DE13 9SA. Mrs Penelope Adkins. *2m NW of Burton-on-Trent. 1m NW of Burton upon Trent take Tatenhill Rd off A38 on Burton/Branson flyover, 1m 1st house on L after village sign. Parking on farm.* **Wed 14, Wed 21, Wed 28 June (1.30-4.30). Adm £4, chd free.**

Lose yourself for an afternoon in our 1½ acre organic country garden. Arbour, gazebo and many other seating areas to view ponds and herbaceous borders, backed with shrubs and trees with emphasis on structure, foliage and colour. Productive fruit and vegetable gardens, wildlife areas and adjoining 12 acre native woodland walk. Picnics welcome. Gravel paths.

♿ 🚌

GROUP OPENING

3 BOURNVILLE VILLAGE

Birmingham, B30 1QY. Bournville Village Trust, www.bvt.org.uk. *Bournville. Gardens spread across 1,000 acre estate. Walks of up to 30 mins between some. Map supplied on day. Additional parking: Bournville Garden Centre, (B30 2AE) & Rowheath Pavilion, (B30 1HH).* **Sun 2 July (11-5). Combined adm £7, chd free. Light refreshments at various locations.**

5 BLACKTHORN ROAD
Mr & Mrs Andrew Christie.

103 BOURNVILLE LANE
Mrs Jennifer Duffy.

52 ELM ROAD
Mr Glenn & Mrs Jackie Twigg.

21 HIGH HEATH CLOSE
Dr Faint & Ms Dorward.

11 KESTREL GROVE
Mr Julian Stanton.

32 KNIGHTON ROAD
Mrs Anne Ellis & Mr Lawrence Newman.

MASEFIELD COMMUNITY GARDEN
Mrs Sally Gopsill, masefieldcommunitygarden.wordpress.com.

54 RAMSDEN CLOSE
Ms Lesley Pattenson.

Bournville Village is showcasing eight gardens. Bournville is famous for its large gardens, outstanding open spaces and of course its chocolate factory in a garden! We have a garden railway amongst bonsai, a considered recycling garden, a long garden with surprises at each glance, a small garden for easy maintenance, a produce grower's paradise, a little changed original style garden, a community garden and a recovered Japanese rockery garden. Free information sheet/map available on the day. Gardens spread across the 1,000 acre estate, with walks of up to 30 minutes between sites. For those with a disability, full details of access are available on the NGS website. Visitors with particular concerns with regards to access are welcome to call Bournville Village Trust on 0300 333 6540 or email: CommunityAdmin@bvt.org.uk. Music, singing and food available across a number of sites. Please check on the day.

♿ ✽ ☕ 🔊

4 BRACKENCOTE

Forshaw Heath Road, Earlswood, Solihull, B94 5JU. Mr & Mrs Sandy Andrews, 01564 702395 mob: 07710 107000, mjandrews53@hotmail.com. *From J3 M42 take exit signed to Forshaw Heath. At the T junction, turn L onto Forshaw Heath Rd, signed to Earlswood. Garden approx ½m on the L, before you get to Earlswood Nurseries.* **Sun 14 May, Sun 11 June (11-4). Adm £5, chd free. Pre-booking essential, please visit www.ngs.org.uk for information & booking. Light refreshments. Visits also by arrangement 1 May to 30 July for groups of up to 15.**

A beautiful country garden with stunning wildflower meadow, full of orchids in spring and early summer. The 1¼ acre plot is surrounded by mature trees, with herbaceous borders enclosing a large circular lawn. Beyond this, the garden inc a brick and turf labyrinth, a rockery, a pond and bog garden as well as a large raised vegetable plot and garden buildings.

5 BRIDGE HOUSE

Dog Lane, Bodymoor Heath, B76 9JD. Mr & Mrs J Cerone, 01827 873205, janecerone@btinternet.com. *5m S of Tamworth. From A446 at Belfry Island take A4091to Tamworth, after 1m turn R onto Bodymoor Heath Lane & continue 1m into village, parking in field opp garden.* **Mon 29 May (2-5). Adm £5, chd free. Home-made teas. Visits also by arrangement 6 May to 30 Sept for groups of 6 to 25.**

1 acre garden surrounding converted public house. Divided into smaller areas with a mix of shrub borders, azalea and fuchsia, herbaceous and bedding, orchard, kitchen garden with large greenhouse. Pergola walk, arch to formal garden with big fish pool, pond, bog garden and lawns. Some unusual carefully chosen trees. Kingsbury Water Park and RSPB Middleton Lakes Reserve located within a mile.

6 THE BUNGALOW, WOOD FARM

Great Gate, Nr Tean, Stoke-On-Trent, ST10 4HF. Mrs Dorothy Hurst. *Staffordshire. Arrive in Great Gate & follow yellow signs.* **Sun 4, Sun 11 June, Sun 2, Sun 9 July (11.30-5). Adm £5, chd free. Light refreshments.**

Unique 1 acre country cottage garden with stunning views of the Weaver Hills surrounded by farm land, the garden inc a Thai theme underground temple with water feature, a relaxing Japanese area also an area with a New Zealand and Mediterranean vibe all with varied planting, as you wader around you will find plenty of quiet and tranquil seating. It an experience that will lift your spirits.

7 BUTT LANE FARM

Butt Lane, Ranton, Stafford, ST18 9JZ. Pete Gough, 07975 928968, claire-pickering@hotmail.co.uk. *6m W of Stafford. Take A518 to Haughton, turn R Station Rd (signed Ranton) 2m turn L at Butt*

19 Waterdale

Ln. Or from Great Bridgeford head W B5405 for 3m. Turn L Moorend Ln, 2nd L Butt Ln. **Sun 28 May (10-4). Adm £5, chd free. Light refreshments inc home-made pizza & BBQ. Visits also by arrangement May to July for groups of 15 to 50.**
Entering from unspoilt farmland through small wooded area into multi award winning garden. The lawns are surrounded by varying floral beds, fruit and vegetable gardens. Different areas of interest can be found, greenhouse courtyard, outside kitchen and glass topped well. Many arts and craft stalls, vintage tractor display and live music. Several craft stalls to be in attendance.

8 ◆ CASTLE BROMWICH HALL GARDENS

Chester Road, Castle Bromwich, Birmingham, B36 9BT. Castle Bromwich Hall & Gardens Trust, 0121 749 4100, cbhallgardens@gmail.com, www.castlebromwichhallgardens.org.uk. *4m E of Birmingham centre. 1m J5 M6 (exit N only).* **For NGS: Sun 11 June (10.30-4.30). Adm £5, chd £2. For other opening times and information, please phone, email or visit garden website.**
10 acres of restored C17/18 walled gardens attached to a Jacobean manor (now a hotel), with a further 30 acres of historic parkland and nature reserve. Formal yew parterres, wilderness walks, summerhouses, holly maze, espaliered fruit and wild areas. Just minutes from J5 of M6. Cream tea in a box, cafe and picnics all season. Paths are either lawn or rough hoggin - sometimes on a slope. Most areas generally accessible, rough areas outside the walls difficult when wet.

9 CATS WHISKERS

42 Amesbury Rd, Moseley, Birmingham, B13 8LE. Dr Alfred & Mrs Michele White. *Opp back of Moseley Hall Hospital. Past Edgbaston Cricket ground straight on at r'about & up Salisbury Rd. Amesbury Rd, 1st on R.* **Sun 7 May (12.30-5). Adm £6, chd free. Light refreshments inc cake, wine & soft drinks.**
A plantsman's garden developed over the last 40 years but which has kept its 1923 landscape. The front garden whilst not particularly large is full of interesting trees and shrubs; the rear garden is on three levels with steps leading to a small terrace and further steps to the main space. At the end of the garden is a pergola leading to the vegetable garden and greenhouse.

10 CHEADLE ALLOTMENTS

Delphouse Road, Cheadle, Stoke-on-Trent, ST10 2NN. Cheadle Allotment Association. *On the A521 1m to the W of Cheadle town centre.* **Sun 25 June (1-5). Adm £5, chd free. Home-made teas inc in adm**
The allotments, which were opened in 2015, are located on the western edge of Cheadle (Staffs). There are 29 plots growing a variety of vegetables, fruits and flowers. A new addition in 2019 was a community area, with an adjacent wildlife area. A small orchard is currently being developed. Wheelchair access on all main paths.

11 COLOUR MILL

Winkhill, Leek, ST13 7PR. Jackie Pakes. *7m E of Leek. Follow A523 from either Leek or Ashbourne, look for NGS signs on the side of the main rd which will direct you down to Colour Mill.* **Fri 11, Sat 12 Aug (1.30-5). Adm £5, chd free. Home-made teas.**
1½ acre south facing garden, created in the shadow of a former iron foundry, set beside the delightful River Hamps. Informal planting in a variety of rooms surrounded by beautiful 7ft beech hedges. Large organic vegetable patch complete with greenhouse and polytunnel. Maturing trees provide shade for the interesting seating areas. River walk through woodland and willows. Herbaceous borders.

12 ◆ DOROTHY CLIVE GARDEN

Willoughbridge, Market Drayton, TF9 4EU. Willoughbridge Garden Trust, 01630 647237, info@dorothyclivegarden.co.uk, www.dorothyclivegarden.co.uk. *3m SE of Bridgemere Garden World. From M6 J15 take A53 W bound, then A51 N bound midway between Nantwich & Stone, near Woore.* **For NGS: Sat 21, Sun 22 Oct (10-4). Adm £5, chd £2. Home-made teas in the Tearooms. Vintage Afternoon Tea can be pre-booked. For other opening times and information, please phone, email or visit garden website.**
12 informal acres, inc superb woodland garden, alpine scree, gravel garden, fine collection of trees and spectacular flower borders. Renowned in May when woodland quarry is brilliant with rhododendrons. Waterfall and woodland planting. Laburnum arch in June. Creative planting has produced stunning summer borders. Large Glasshouse. Spectacular autumn colour. Much to see, whatever the season. The Dorothy Clive Tea Rooms will be open throughout the weekend during winter for refreshments, lunch and afternoon tea. Open all week in summer. Plant sales, gift room, picnic area and children's play area for a wide age range. Wheelchairs (inc electric) are available to book through the tea rooms. Disabled parking is available. Toilets on both upper and lower car parks.

13 198 EACHELHURST ROAD

Walmley, Sutton Coldfield, B76 1EW. Jacqui & Jamie Whitmore. *5mins N of Birmingham. M6 J6, A38 Tyburn Rd to Lichfield, continue to T-lights at Lidl & continue on Tyburn Rd, at island take 2nd exit to destination rd.* **Sun 16 July (12.30-4.30). Adm £3, chd free. Home-made teas.**
A long garden approx 210ft x 30ft divided by arches and pathways. Plenty to explore inc wildlife pond, corner arbour, cottage garden and hanging baskets leading to formal garden with box-lined pathways, well, stocked borders, gazebo then through to raised seating area, overlooking Pype Hayes golf course, with summerhouse bar and courtyard garden.

The National Garden Scheme searches the length and breadth of England, Wales, Northern Ireland and the Channel Islands for the very best private gardens

14 5 EAST VIEW COTTAGES

School Lane, Shuttington, nr Tamworth, B79 0DX. Cathy Lyon-Green, 01827 892244, cathyatcorrabhan@hotmail.com, www.ramblinginthegarden.wordpress.com. *2m NE of Tamworth. From Tamworth, Amington Rd or Ashby Rd to Shuttington. From M42 J11, B5493 for Seckington & Tamworth, 3m L turn to Shuttington. Pink house nr top of School Ln. Parking signed, disabled at house.* **Sun 12 Feb (12-4); Sun 25 June (1-5); Wed 28 June (12-4). Adm £4, chd free. Home-made teas. 2024: Sun 18 Feb. Visits also by arrangement February & 17 June to 16 July for groups of up to 30.**

Deceptive, quirky plantlover's garden, full of surprises and always something new. Informally planted themed borders, cutting beds, woodland and woodland edge, stream, water features, sitooterie, folly, greenhouses and many artefacts. Roses, clematis, perennials, potted hostas. Snowdrops and witch hazels in Feb. Seating areas for contemplation and enjoying home-made cake. 'Wonderful hour's wander'.

15 FIFTY SHADES OF GREEN

20 Bevan Close, Shelfield, Walsall, WS4 1AB. Annmarie & Andrew Swift, 07963 041402, annmarie.1963@hotmail.co.uk. *Walsall. M6 J10 take A454 to Walsall for 1.6m, L at Lichfield St for ⅓m, L onto A461 Lichfield Rd for 2m, at Co-op T-lights turn L onto Mill rd then follow yellow signs. Parking limited, some in Broad Lane.* **Sat 15, Sun 16 July (10-5); Sat 26 Aug (1.30-8.30); Sun 27 Aug (11-8.30). Adm £3.50, chd free. Home-made teas. Visits also by arrangement 13 May to 15 Oct for groups of up to 20.**

Our garden is 15m x11m taking several years to create and landscape one area at a time. We have many distinctive areas inc two ponds linked by a stream with a stone waterfall, many water features, places to sit, watch and relax. We encourage and welcome wildlife. Our planting style is varied and inc architectural plants for foliage, texture, over 40 trees. A calm garden of surprises, intrigue and discovery. Find us on facebook at 'Fifty Shades of Green'.

16 THE GARTH

2 Broc Hill Way, Milford, Stafford, ST17 0UB. Anita & David Wright, 01785 661182, anitawright1@yahoo.co.uk, www.anitawright.co.uk. *4½m SE of Stafford. A513 Stafford to Rugeley rd; at Barley Mow turn R (S) to Brocton; L after ½m.* **Sun 4,**

15 New Church Road

Sun 25 June (1-5). Adm £5, chd free. Cream teas. Visits also by arrangement June to Sept for groups of 20+.
½ acre garden of many levels on Cannock Chase AONB. Acid soil loving plants. Series of small gardens, water features, raised beds. Rare trees, island beds of unusual shrubs and perennials, many varieties of hosta and ferns. Varied and colourful foliage, Summerhouses, arbours and quiet seating to enjoy the garden. Ancient sandstone caves.

17 33 GORWAY ROAD

Walsall, WS1 3BE. Gillian Brooks, 07972 615501, Brooks.gillian@gmail.com. *M6 J9 turn N onto Bescot Rd, at r'about take Wallows L (A4148), in 2m at r'about, 1st exit Birmingham Rd, L to Jesson Rd, L to Gorway Rd.* **Sun 18 June (10.30-3.30). Adm £3.50, chd free. Home-made teas. Visits also by arrangement 2 May to 31 Aug.**
Cottage style Edwardian house garden. Late spring bulbs and roses. Willow tunnel, pond and rockery. Garden viewing is over flat grass and paths. Wheelchair access through garage.

18 GRAFTON COTTAGE

Bar Lane, Barton-under-Needwood, DE13 8AL. Margaret & Peter Hargreaves, 01283 713639, marpeter1@btinternet.com. *6m N of Lichfield. Leave A38 for Catholme S of Barton, follow sign to Barton Green, L at Royal Oak, ¼m.* **Sun 2, Fri 7, Sun 16, Sun 30 July, Sun 6 Aug (1-5). Adm £5, chd free. Home-made teas. Visits also by arrangement 2 June to 20 Aug. Minimum group adm £100 if less than 20 people. Donation to Alzheimer's Research UK.**
Designed by the gardeners to create a picturesque cottage garden. Old fashioned roses climb over trellis and arches laden with clematis, Hollyhocks adorn the front of the cottage a wide range of salvias dahlias and perennials. Colour themed borders with scent being insect friendly amphitheatre stream parterre and much more.The garden has attracted visitors for over 30 years.

19 128 GREEN ACRES ROAD

Kings Norton, Birmingham, B38 8NL. Mr & Mrs Mark Whitehouse. *From the M42 J2 take the A441 towards Birmingham take the 2nd exit at r'about continue along the A441 for 3m Green Acres Rd is on the R next to Spar.* **Sun 23 July (10.30-4.30). Adm £3, chd free. Light refreshments.**
A small garden with lots of character full of interest and abundance of colour, inc various seating areas to enjoy different aspects and views. The garden has many different types of perennials, Hostas, Roses and Dahlias. A fish pond, water feature plus many more points of interest. To access the garden, please follow signs to enter the garden from the rear.

20 22 GREENFIELD ROAD

Stafford, ST17 0PU. Alison & Peter Jordan, 01785 660819, alison.jordan2@btinternet.com. *3m S of Stafford. Follow the A34 out of Stafford towards Cannock. 2nd L onto Overhill Rd.1st R into Greenfield Rd.* **Evening opening Fri 9 June (7-9). Adm £6. Wine & nibbles. Visits also by arrangement 15 Apr to 30 July for groups of up to 20.**
Suburban garden,working towards all round interest. In spring bulbs and stunning azaleas and rhododendrons. June onwards perennials and grasses. A garden that shows being diagnosed with Parkinson's needn't stop you creating a peaceful place to sit and enjoy. You might be able to catch a glimpse of Pete's model railway running. Come in the evening and enjoy a glass of wine and live music. Flat garden but with some gravelled areas.

GROUP OPENING

21 HALL GREEN GARDENS

Hall Green, Birmingham, B28 8SQ. *Off A34, 4m from city centre, 6m from M42, J4. From City Centre start at 120 Russell Rd B28 8SQ. Alternatively, from M42 start at 638 Shirley Rd B28 9LB.* **Sat 10, Sun 11 June (1.30-5.30). Combined adm £6, chd free. Home-made teas at 111 Southam Road.**

42 BODEN ROAD
Mrs Helen Lycett.

36 FERNDALE ROAD
Mrs E A Nicholson, 0121 777 4921.
Visits also by arrangement 27 Feb to 1 Sept for groups of up to 30. Tea inc in adm price for booked visits.

120 RUSSELL ROAD
Mr David Worthington, 07552 993911, hildave@hotmail.com.
Visits also by arrangement May to July for groups of up to 30.

638 SHIRLEY ROAD
Dr & Mrs M Leigh.

111 SOUTHAM ROAD
Ms Val Townend & Mr Ian Bate.

Five very different large suburban gardens in leafy Hall Green with its beautiful mature trees and friendly residents. Our visitors love the unique atmosphere in each garden. It is such a perfect way to share a relaxing early summer saunter and pause for refreshments served by our brilliant catering team. We look forward to seeing you! 42 Boden Rd: Large restful garden, mature trees, cottage borders, seating areas and small vegetable area. 36 Ferndale Rd: Florist's large suburban garden, ponds and waterfalls, and fruit garden. 120 Russell Rd: Plantsman's garden, formal raised pond and hosta collection and unusual perennials, container planting. 111 Southam Rd: Mature garden with well defined areas inc ponds, white garden, rescue hens and a majestic cedar. 638 Shirley Rd: Large garden with herbaceous borders, vegetables and greenhouses. Unfortunately, there is only partial wheelchair access at some gardens.

9,500 patients and their families are being supported across four Horatio's gardens thanks to our annual donations

22 HAMILTON HOUSE
Roman Grange, Roman Road, Little Aston Park, Sutton Coldfield, B74 3GA. Philip & Diana Berry, www.hamiltonhousegarden.co.uk. *3m N of Sutton Coldfield. Follow A454 (Walsall Rd) & enter Roman Rd, Little Aston Park. Roman Grange is 1st L after church but enter rd via pedestrian gate.* **Sun 28 May (2-5). Adm £6, chd free. Home-made teas. Pimms.**
½ acre north facing English woodland garden in a tranquil setting, making the most of challenging shade, providing a haven for birds and other wildlife. Large pond with a stone bridge, pergolas, water features, box garden with a variety of roses. Interesting collection of rhododendrons, clematis, hostas, ferns, old English roses and stunning artworks in the garden. Join us for tea and cakes, listening to live music and admire the art of our garden.

23 NEW HAMMERWICH HOUSE FARM
Hall Lane, Hammerwich, Burntwood, WS7 0JP. Donna Harvey-Bailye. *3m from Lichfield Staffordshire. Farm gate to the L of Church Lane.* **Fri 16, Sat 17 June (11-6). Adm £5, chd free. Cream teas.**
Gardens in the grounds of an 1807 Staffordshire Farmhouse inc a large wildlife pond and an extensive collection of Roses. The garden has a greenhouse and vegetable garden to explore along with open fields overlooking Hammerwich Church to picnic in.

24 THE HOME BEE GARDEN
143 Redditch Road, Kings Norton, Birmingham, B38 8RH. Mr & Mrs Zoe Brady, www.thehomebeegarden.co.uk. *We are on the A441, 3m from the M42 Hopwood services or 5.4m from the city centre Birmingham.* **Sat 20, Sun 21 May (11-4.30). Adm £3, chd free. Home-made cakes, ploughman's lunch, coffee, soft drinks, teas, home-made fresh herbal garden tea.**
The Home Bee Garden is a large family garden, set within historic Kings Norton, Birmingham. Extensive kitchen garden growing a variety of seasonal vegetables, herbs, fruit, with seasonal herbaceous borders, bedding, war effort fruit trees, secret paths, interesting features and resident chickens. Each season offers a unique aspect, spring meadow, summer flowers and autumn colour.

25 NEW IZAAK WALTON FARM
Cresswell Lane, Cresswell, Stoke-On-Trent, ST11 9RE. Mr Mark & Mrs Veronica Swinnerton. *Entrance, Off Cresswell Old Lane, 100 metres from railway Xing.* **Sun 9 July (11-4). Adm £4, chd free. Light refreshments inc home made cakes & deli boards.**
Izaak Walton Farm garden has lots of small interesting areas. Flower borders, gravel gardens, shaded areas, vegetables and fruit trees, beautiful bug hotel made from a wine rack and oak pagoda. Water fountain and small greenhouse with cacti's. Interesting children's den, lawn and gravel paths and small grass meadow. Plenty of seating to enjoy light lunches or tea and cake.

26 'JOHN'S GARDEN' AT ASHWOOD NURSERIES
Ashwood Lower Lane, Ashwood, nr Kingswinford, DY6 0AE. John Massey, www.ashwoodnurseries.com. *9m S of Wolverhampton. 1m past Wall Heath on A449 turn R to Ashwood along Doctor's Lane. At T-junction turn L. Garden entrance off Main Car Park at Ashwood Nurseries.* **Sun 9 Apr (10-4). Adm £6.50, chd free.**
A stunning private garden adjacent to Ashwood Nurseries, it has a huge plant collection and many innovative design features in a beautiful canal-side setting. There are informal beds, woodland dells, a stunning rock garden, a unique ruin garden, an Anemone pavonina meadow and wildlife meadow. Fine displays of bulbs and Spring-flowering plants and a notable collection of Malus and Amelanchier. Tea Room, Garden Centre and Gift Shop at adjacent Ashwood Nurseries. Coaches are welcome by appointment only. Disabled access difficult if very wet. Sorry no disabled access to the wildlife garden.
NPC

27 14 LONGBOW CLOSE, STRETTON
Burton upon Trent, DE13 0XY. Debbie & Gavin Richards. *3m NW of Burton upon Trent in the village of Stretton. From A38 turn off at A5121 (Burton north) Follow signs to Stretton turning into Claymills Road, then L into Church Road, R into Bridge Street & R into Athelstan Way. Please park on Athelstan Way.* **Sat 10 June (11-4). Adm £3, chd free. Home-made teas.**
A plantwoman's garden with exotic evergreens, gorgeous smelling roses, a flower-filled sunny patio where refreshments can be enjoyed and a welcoming front garden. Its design features planting of herbaceous perennials, climbers and evergreen trees with a mauve, pink and purple colour theme.

28 MARIE CURIE HOSPICE GARDEN
Marsh Lane, Solihull, B91 2PQ. Mrs Do Connolly, www.mariecurie.org.uk/westmidlands. *Close to J5 M42 to E of Solihull Town Centre. M42 J5, travel towards Solihull on A41. Take slip toward Solihull to join B4025 & after island take 1st R onto Marsh Lane. Hospice is on R. Limited onsite parking - available for blue badge holders.* **Sun 25 June (11-4). Adm £4, chd free. Light refreshments at the hospice bistro.**
The gardens contain two large, formally laid out patients' gardens, indoor courtyards, a long border adjoining the car park and a wildlife and pond area. The volunteer gardening team hope that the gardens provide a peaceful and comforting place for patients, their visitors and staff.

29 3 MARLOWS COTTAGES
Little Hay Lane, Little Hay, Lichfield, WS14 0QD. Phyllis Davies. *4m S of Lichfield. Take A5127, Birmingham Rd. Turn L at Park Ln (opp Tesco Express) then R at T junction into Little Hay Ln, ½m on L.* **Sun 11 June (11-4). Adm £3, chd free. Home-made teas.**
Long, narrow, gently sloping cottage style garden with borders and beds containing abundant herbaceous perennials and shrubs leading to vegetable patch.

The Old Dairy House

30 12 MERES ROAD
Halesowen, B63 2EH. Nigel & Samantha Hopes, www.hopesgardenplants.co.uk. *4m from M5 J3. Follow the A458 out of Halesowen heading W for 2m, L turn onto Two Gates Ln just after Round of Beef pub, this becomes Meres Rd. House is on R.* **Sun 28 May, Sun 25 June (11-4). Adm £3.50, chd free. Light refreshments. Donation to Plant Heritage.**
A little oasis in the heart of the Black Country, with stunning views across to Shropshire. The borders are filled with interest and colour throughout the year. A small family garden, designed for young children and plantaholics alike, we hold a National Collection of Border Auriculas, with a passion for snowdrops, hellebores, epimedium, bearded iris, roscoea, cyclamen and kniphofia.

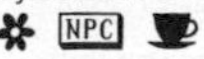

31 ◆ MIDDLETON HALL & GARDENS
Middleton, Tamworth, B78 2AE. Middleton Hall Trust, 01827 283095, enquiries@middleton-hall.co.uk, www.middleton-hall.co.uk. *4m S of Tamworth, 2m N of J9 M42. On A4091 between The Belfry & Drayton Manor.* **For NGS: Fri 19 May (11-4). Adm £8, chd £3. For other opening times and information, please phone, email or visit garden website.**
Our formal gardens form part of the 42 acre estate of the Grade II* Middleton Hall, the C17 home of naturalists Francis Willoughby and John Ray. The formal gardens are made up of a Walled Garden, lawns and an orchard. The Walled Garden contains a variety of herbaceous and seasonal planting with specimen plants that have a botanical and/or historical significance to our site. Bake 180 Coffee Shop will also be open serving lunches and other light refreshments. Walled Garden paths are gravel, Glade and Orchard paths are grass. Regret no wheelchair access to1st floor of the Hall and Nature Trail.

32 MILLENNIUM GARDEN
London Road, Lichfield, WS14 9RB. Carol Cooper. *1m S of Lichfield. Off A38 along A5206 towards Lichfield ¼m past A38 island towards Lichfield. Park in field on L. Yellow signs on field post.* **Sun 26 Mar, Sun 14 May (1-5). Adm £3.50, chd free. Home-made teas.**
2 acre garden with mixed spring bulbs in the woodland garden. In May the laburnum walk and wisteria arch are in full bloom in this English country garden. Designed with a naturalistic edge and with the environment in mind. A relaxed approach creates a garden of quiet sanctuary with the millennium bridge sitting comfortably, with its surroundings of lush planting and mature trees. Well-stocked borders give shots of colour to lift the spirit and the air fills with the scent of wisterias and climbing roses. A stress free environment awaits you at the Millennium Garden. Park in field then footpath round garden. Some uneven surfaces.

120 Russell Road

33 MONARCHS WAY

Park Lane, Coven, Wolverhampton, WV9 5BQ. Eileen & Bill Johnson, www.monarchsway-garden.co.uk. *From Port Ln (main road between Codsall & Brewood) turn E onto Park Ln. Monarchs Way is on the sharp bend on Park Ln. Limited parking, (further parking is available at other end of Park ln).* **Sun 18 June (11-4). Adm £6, chd free. Pre-booking essential, please visit www.ngs.org.uk for information & booking. Light refreshments.**

We bought a $1\frac{3}{4}$ acre bare, treeless blank canvas in 2010 with grass around three feet high! Since then we have designed a Tudor folly and jungle hut, built a pergola, excavated a lily pond with bog garden created a cottage garden, orchard, and vegetable garden. We have planted hundreds of conifer, evergreen, fruit and flowering trees, roses, hydrangeas, perennials and designed numerous flower beds. Wheelchair access is easy for most of the garden, on arrival request side gate for access. Minibuses can be accepted.

34 15 NEW CHURCH ROAD

Sutton Coldfield, B73 5RT. Owen & Lloyd Watkins. *$4\frac{1}{2}$ m NW of Birmingham city centre. Take A38M then A5127 to 6 Ways r'about, take 2nd exit then bear L Summer Rd then Gravelly Ln over Chester Rd to Boldmere Rd, 5th R New Church Rd.* **Sun 25 June (1-5). Adm £4, chd free. Home-made teas.**

A garden changing and evolving over 14 yrs, never standing still. An extended pond and bridge take centre stage with a recently added stream. Different areas inc a small stumpery, a dry shade area, large herbaceous borders, tree ferns a rockery and woodland area. Raised vegetable beds, chickens and a composting system. Wheelchair access is available, although some areas may not be accessible. Several steps, portable ramps available.

35 THE OLD DAIRY HOUSE

Trentham Park, Stoke-on-Trent, ST4 8AE. Philip & Michelle Moore, 07779 158394, olddairyhouse@hotmail.com. *S edge of Stoke-on-Trent. Next to Trentham Gardens. Off Whitmore Rd. Please follow NGS signs or signs for Trentham Park Golf Club. Parking in church car park.* **Sun 28 May (1.30-5.30); Mon 29 May (1-5). Adm £4.50, chd free. Home-made teas. Visits also by arrangement for groups of 15+.**

Grade II listed house which originally

formed part of the Trentham Estate forms backdrop to this 2 acre garden in parkland setting. Shaded area for rhododendrons, azaleas plus expanding hosta and fern collection. Mature trees, 'cottage garden', long borders and stumpery. Narrow brick paths in vegetable plot. Large courtyard area for teas. Wheelchair access - some gravel paths but lawns are an option.

36 NEW THE OLD MISSION

Bickford road, Whiston, Penkridge, ST19 5QH. Mr Jason & Mrs Laura Beet. *Postcode accurate for SatNav.* **Sun 19 Mar, Sun 11 June (10-4). Adm £6, chd free. Light refreshments.**

The old Mission is situated amongst farmland in the small hamlet of Whiston. The one acre garden boasts an ancient giant oak, many shrub and flower borders, statues, carvings and other artistic ornaments. An amazing swimming pool with seating and an Outdoor kitchen completes the scene. There are many shrub and climbing roses plus an Orchard, kitchen garden and soft fruit cages.

37 THE OLD VICARAGE

Fulford, nr Stone, ST11 9QS. Mike & Cherry Dodson. *4m N of Stone. From Stone A520 (Leek). 1m R turn to Spot Acre & Fulford, turn L down Post Office Terrace, past village green/Pub towards church. Parking signed on L.* **Sun 25 June (11-4). Adm £5, chd free. Home-made teas.**

1½ acres of formal sloping garden around Victorian house. Sit on the terrace or in the summerhouse to enjoy home-made cakes and tea amongst mature trees, relaxed herbaceous borders, roses and a small pond. Move to the organic vegetable garden with raised beds, fruit cage and very big compost heaps! In complete contrast, easy walk around the natural setting of a 2 acre reclaimed lake planted with native species designed to attract wildlife. Waterfall, jetty, fishing hut, acer and fern glade plus arboretum provide more interest. Children will enjoy meeting the chickens and horses. A garden of contrasts, with easy formality around the Victorian house and the accent on very natural planting around the lake, managed for wildlife and sustainability. Wheelchair access to most areas.

38 10 PAGET RISE

Paget Rise, Abbots Bromley, Rugeley, WS15 3EF. Mr Arthur Tindle. *4m W of Rugeley 6m S of Uttoxeter & 12m N of Lichfield. From Rugeley: B5013 E. At T junction turn R on B5014. From Uttoxeter take the B5013 S then B5014. From Lichfield take A515 N then turn L on B5234. In Abbots Bromley follow NGS yellow signs.* **Sun 21 May (11-4). Adm £3, chd free. Light refreshments.**

This medium sized split level garden has a strong Japanese influence. Rhododendrons and a wide range of flowering shrubs. Many bonsai-style Acer trees in shallow bowls occupy a central gravel area with stepping stones. The rear of the garden has a woodland feel with a fairy dell under the pine tree. A gem of a garden! Arthur hopes visitors will be inspired with ideas to use in their own garden. Arthur is a watercolour artist and will be displaying a selection of his paintings for sale.

39 PAUL'S OASIS OF CALM

18 Kings Close, Kings Heath, Birmingham, B14 6TP. Mr Paul Doogan, 0121 444 6943, gardengreen18@hotmail.co.uk. *4m from city centre. 5m from M42 J4. Take A345 to Kings Heath High St then B4122 Vicarage Rd. Turn L onto Kings Rd then R to Kings Close.* **Visits by arrangement 28 May to 31 Aug for groups of 5 to 15. Adm £3, chd free.**

Garden cultivated from nothing into a little oasis. Measuring 18ft x 70ft. It's small but packed with interesting and unusual plants, water features and seven seating areas. It's my piece of heaven. I have cultivated the council land in front of my house which has doubled the pleasure.

Our annual donations to Parkinson's UK meant 7,000 patients are currently benefiting from support of a Parkinson's Nurse

40 THE PINTLES

18 Newport Road, Great Bridgeford, Stafford, ST18 9PR. Peter & Leslie Longstaff, 01785 282582, peter.longstaff@ngs.org.uk. *From J14 M6 take A5013 towards Eccleshall, in Great Bridgeford turn L onto B5405 after 600 metres turn L onto Great Bridgeford Village Hall car park. The Pintles is opp the hall main doors.* **Sun 4 June (1-5). Adm £3.50, chd free. Home-made teas inc gluten free & reduced sugar options. Visits also by arrangement 10 June to 9 July for groups of 10 to 30.**

Located in the village of Great Bridgeford this traditional semi-detached house has a medium sized wildlife friendly garden designed to appeal to many interests. There are two greenhouses, 100s of cacti and succulents, vegetable and fruit plot, wildlife pond, weather station and hidden woodland shady garden. Plenty of outside seating to enjoy the home-made cakes and refreshments.

41 8 RECTORY ROAD

Solihull, B91 3RP. Nigel & Daphne Carter, 07527 475759, npcarter@blueyonder.co.uk. *Town centre. Located in the town centre: off Church Hill Rd, turn into Rectory Rd, bear L down the rd, house on the R 100 metres down.* **Sun 11 June (11-6). Adm £4, chd free. Home-made teas. Evening opening Sun 10 Sept (7-10). Adm £4. Wine. Visits also by arrangement 10 June to 10 Sept for groups of 5 to 20.**

Stunning town garden divided into areas with different features. A garden with unusual trees and intensively planted borders. Walk to the end of the garden and step into a Japanese themed garden complete with pond, fish and traditional style bridge. Sit on the shaded decking area. A garden to attract bees and butterflies with plenty of places to relax. Japanese themed features and stunning Moon gate.

42 56 ST AGNES ROAD

Moseley, Birmingham, B13 9PN. Michael & Alison Cullen, 07748 312541, alison.cullen1@mac.com. *3m from city centre. From Moseley T-lights take St Mary's Row which becomes Wake Green Rd. After ½m turn R into St Agnes Rd and L at the church. Number 56 is approx 200 metres on L.* **Visits by arrangement 22 July to 20 Aug for groups of up to 8. Adm £3.50, chd free. Home-made teas.**

Immaculately maintained, medium-sized, urban garden with curving borders surrounding a formal lawn punctuated with delicate acers and contemporary sculpture. Seating by a Victorian-style fish pond with a fountain and waterfall offers a peaceful setting to enjoy the tranquillity of this elegant garden.

43 23 ST JOHNS ROAD

Rowley Park, Stafford, ST17 9AS. Colin & Fiona Horwath, 07908 918181, fiona_horwath@yahoo.co.uk. *½m S of Stafford Town Centre. Just a few mins from J13 M6, towards Stafford. After approx 2m on the A449, turn L into St. John's Rd after bus-stop.* **Thur 23 Mar, Thur 27 Apr, Thur 18 May, Thur 8 June, Thur 6 July (2-5). Adm £4, chd free. Home-made teas. Visits also by arrangement 23 Mar to 31 July for groups of 20 to 60.**

Pass through the black and white gate of this Victorian house into a part-walled gardener's haven. Bulbs, alpines and shady woodlanders in Spring and masses of unusual herbaceous plants and climbers. Numerous Japanese maples. Many spots to sit in sun or shade. Steps and gravel in places. Gardener is keen Hardy Plant Society member and sows far too many seeds, so always something good for sale! Our outdoor kitchen and new revolving summerhouse are great for refreshments! Plenty of places to sit and enjoy the myriad plants.

44 THE SECRET GARDEN

3 Banktop Cottages, Little Haywood, ST18 0UL. Derek Higgott & David Aston. *5m SE of Stafford. A51 from Rugeley or Weston signed Little Haywood A513 Stafford Coley Ln, Back Ln R into Coley Gr. Entrance 50 metres on L.* **Fri 19 May, Wed 28 June, Thur 20 July (11-4). Adm £4, chd free. Home-made teas.**

Wander past the other cottage gardens and through the evergreen arch and there before you is a fantasy for the eyes and soul. Stunning garden approx ½ acre, created over the last 30+yrs. Strong colour theme of trees and shrubs, underplanted with perennials, 1000 bulbs and laced with clematis; other features inc water, laburnum and rose tunnel and unique buildings. Is this the jewel in the crown? Water feature and a warm Bothy for inclement days.

45 SPRINGFIELD COTTAGE

Kiddemore Green Road, Bishops Wood, Stafford, ST19 9AA. Mrs Rachel Glover. *A5 to Telford from Gailey Island. Turn L at sign for Boscobel House. Follow Ivetsey Bank Rd to Bishop's Wood. Turn 1st L Old Coach Road past church second white cottage on L.* **Sat 22 Apr, Sat 24 June, Sat 5 Aug, Sat 9 Sept (11-4). Adm £4, chd free. Home-made teas inc vegan options.**

Newly designed and landscaped plant enthusiasts cottage garden with unusual planting in areas, a tropical area inc Gunera's, Cannas, Hedichiums and palms, a rose garden, a herb garden and an established vegetable garden with large greenhouse. Fantastic views of the south Staffordshire countryside from all sides of the garden. Disabled parking for two cars, flat garden area and WC facilities.

46 76 STATION STREET

Cheslyn Hay, Walsall, WS6 7EE. Mr Paul Husselbee. *Located on B4156. Limited on road parking.* **Sun 9 July (12-5). Adm £3.50, chd free. Home-made teas.**

40 yrs of gardening on this site has produced a hosta filled courtyard which leads to a Mediterranean patio, steps down to small area with folly, formal gardens area with stream, gated courtyard then finally find the secret garden. A quirky garden with a surprise round every corner.

47 TRUCKLE HOUSE, CROXDEN

Uttoxeter, ST14 5JD. Mrs Hilary Hawksworth, 01889 507124, hilaryhawk@outlook.com. *7m NW of Uttoxeter. From Uttoxeter take B5030 towards Rocester. At the r'about by JCB HQ turn L onto Hollington Rd & follow brown signs to Croxden Abbey (+yellow NGS signs) Parking at Croxden Church 375yds past house.* **Visits by arrangement May to July for groups of 6 to 20. Adm £5, chd free. Home-made teas.**

A truly unique and tranquil setting for this rural ¾ acre garden, set against the impressive remains of Croxden Abbey. Lovingly created over the past 15 years to inc mixed herbaceous borders, dozens of roses, trees and shrubs, wildflower meadow, large natural pond with bridge over, vegetable patch. The garden offers a riot of colour and scent in summer but has year-round interest.

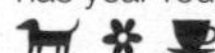

48 12 WATERDALE

Compton, Wolverhampton, WV3 9DY. Colin & Clair Bennett. *1½m W of Wolverhampton city centre. From Wolverhampton Ring Rd take A454 towards Bridgnorth for 1m. Waterdale is on the L off A454 Compton Rd West.* **Sun 4 June, Sun 2 July (11.30-5). Combined adm with 19 Waterdale £5, chd free.**

A riot of colour welcomes visitors to this quintessentially English garden. The wide central circular bed and side borders overflow with classic summer flowers, inc the tall spires of delphiniums, lupins, irises, campanula, poppies and roses. Clematis tumble over the edge of the decked terrace, where visitors can sit among pots of begonias and geraniums to admire the view over the garden.

49 19 WATERDALE

Compton, Wolverhampton, WV3 9DY. Anne & Brian Bailey, 01902 424867, m.bailey1234@btinternet.com. *1½m W of Wolverhampton city centre. From Wolverhampton Ring Rd take A454 towards Bridgnorth for 1m. Waterdale is on L off A454 Compton Rd West.* **Sun 4 June, Sun 2 July (11.30-5). Combined adm with 12 Waterdale £5, chd**

free. Home-made teas. Visits also by arrangement 6 June to 25 Aug for groups of 10 to 35.
A romantic garden of surprises, which gradually reveals itself on a journey through deep, lush planting, full of unusual plants. From the sunny, flower filled terrace, a ruined folly emerges from a luxuriant fernery and leads into an oriental garden, complete with tea house. Towering bamboos hide the way to the gothic summerhouse and mysterious shell grotto. Find us on facebook at 'Garden of Surprises'.

50 THE WICKETS

47 Long Street, Wheaton Aston, ST19 9NF. Tony & Kate Bennett, 01785 840233, ajtonyb@talktalk.net. *8m W of Cannock, 10m N of Wolverhampton, 10m E of Telford. M6 J12 W towards Telford on A5; 3m R signed Stretton; 150yds L signed Wheaton Aston; 2m L; over canal, garden on R or at Bradford Arms on A5 follow signs.* **Sun 9 July, Sun 27 Aug (1.30-4.30). Adm £4, chd free. Home-made teas. Visits also by arrangement July & Aug.**
There's a delight around every corner and lots of quirky features in this most innovative garden. Its themed areas inc a fernery, grasses bed, hidden gothic garden, succulent theatre, cottage garden beds and even a cricket match! It will certainly give you ideas for your own garden as you sit and have tea and cake. Afterwards, walk by the canal and enjoy the beautiful surrounding countryside. Wheelchair access - two single steps in garden and two gravel paths.

51 WILD THYME COTTAGE

Woodhouses, Barton Under Needwood, Burton-On-Trent, DE13 8BS. Ray & Michele Blundell. *B5016 Midway between villages of Barton & Yoxall. From A38 through Barton Village B5016 towards Yoxall. Or from A513 at Yoxall centre take Town Hill sign for Barton. Garden is at Woodhouses, approx 2m from either village.* **Sun 29 Oct (11-4.30). Adm £4, chd free. Tea.**
The garden extends to ⅓ acre surrounding a self built Oak framed house crafted 20 years ago. Now with Great Dixter porch and Lich Gate. The garden has over 120 Japanese Maples. but also features white bark birch, ornamental grasses, herbaceous perennials and established shrubs. Plants sales of Zantedeschia (Arum Lily's) and Hosta. Emphasis is on autumn colour and successive year long interest. Great Dixter inspired porch and garden entrance Oak Lychgate. Ornamental grasses. Rare trees and shrubs. Partial wheelchair access on gravel and grass.

52 25 WOLVERHAMPTON ROAD

Bloxwich, Walsall, WS3 2HB. Mr Mick Carter. *2m S of J11 of M6. Leave J11 of M6 take A462 to Willenhall, Warstone Rd. At first set of T-lights turn L onto B4210 Broad Ln at next set of T-lights turn L A424 then Ist R then 1st L is Wolv Rd.* **Sun 2 July (11-4). Adm £3, chd free. Cream teas.**
Garden packed with plants, flowering perennials, acers, shrubs, trees, wildlife pond with waterfall. A small Japanese garden with bamboo deer scarer water feature and Torre Gate. Eye catching Gothic Folly adorned with plants and flowers. Hand made raised vegetable beds.

53 YARLET HOUSE

Yarlet, Stafford, ST18 9SD. Mr & Mrs Nikolas Tarling. *2m S of Stone. Take A34 from Stone towards Stafford, turn L into Yarlet School & L again into car park.* **Fri 16 June (10-1). Adm £4, chd free. Home-made teas. Donation to Staffordshire Wildlife Trust.**
4 acre garden with extensive lawns, walks, lengthy herbaceous borders and traditional Victorian box hedge. Water gardens with fountain and rare lilies. Sweeping views across Trent Valley to Sandon. New peaceful Japanese garden for 2021. Victorian School Chapel and War Memorial. Garden chess board. Boules pitch. Yarlet School art display. Gravel paths.

54 YEW TREE COTTAGE

Podmores Corner, Long Lane, White Cross, Haughton, ST18 9JR. Clive & Ruth Plant, 07591 886925, pottyplantz@aol.com. *4m W of Stafford. Take A518 W Haughton, turn R Station Rd (signed Ranton) 1m, then turn R at Xrds ¼m on R.* **Thur 6 July (2-5); Thur 13, Thur 20 July (11-4); Sun 23 July (2-5). Adm £4, chd free. Home-made teas. Visits also by arrangement 3 July to 21 July for groups of 10+. Donation to Plant Heritage.**
Hardy Plant Society member's garden brimming with unusual plants. Year-round interest inc meconopsis, trillium and other shade lovers. ½ acre inc gravel, borders, veg and plant sales. National Collection Dierama featured on BBC Gardeners' World flowering first half July. Covered vinery for tea if weather is unkind, seats in the garden for sunny days. A series of different areas and interests. Cottage garden, shade garden, and National Collection of Dierama (Angels Fishing Rods) in a rural location with a working vegetable garden. A garden for plantaholics. Find us on facebook at 'Dierama Species in Staffordshire'. Partial wheelchair access, grass and paved paths, some narrow and some gravel.

NPC

55 YEW TREES

Whitley Eaves, Eccleshall, Stafford, ST21 6HR. Mrs Teresa Hancock, 07973 432077, hancockteresa@gmail.com. *7m from J14 M6. Situated on A519 between Eccleshall (2.2m) & Woodseaves, (1m). Traffic cones & signs will highlight the entrance.* **Sat 24, Sun 25 June (10-4). Adm £4, chd free. Pre-booking essential, please visit www.ngs.org.uk for information & booking. Light refreshments inc locally made cakes. Visits also by arrangement June to Aug for groups of 6+.**
1 acre garden divided into rooms by mature hedging, shrubs and trees enjoying views over the surrounding countryside. Large patio area with containers and seating. Other features inc pond, topiary, vegetable plot, hen run and wildlife area. During June and July there are 8 acres of natural wildflower meadow to enjoy before it is cut for hay.

In 2022 we donated £450,000 to Marie Curie which equates to 23,871 hours of community nursing care

SUFFOLK

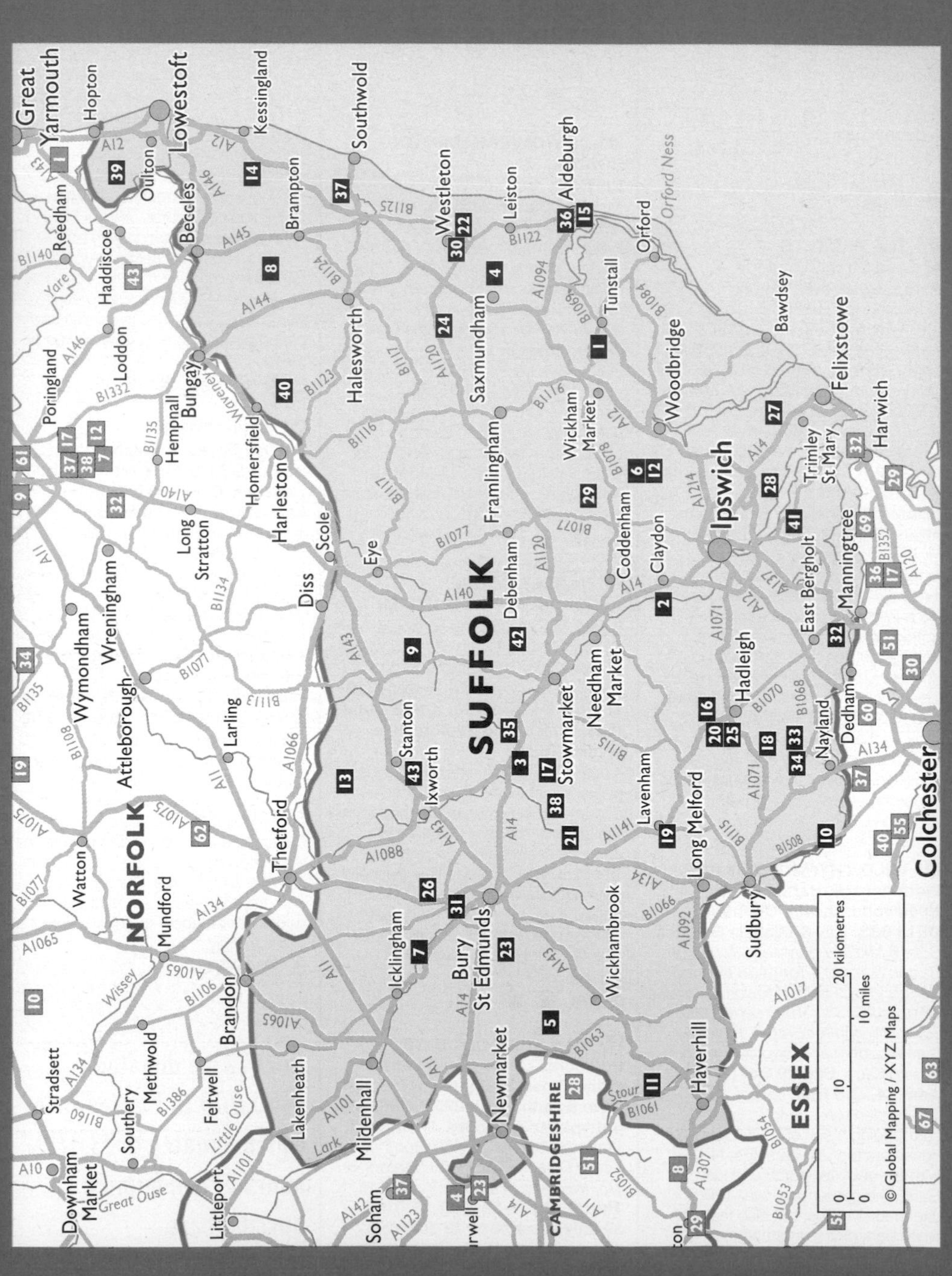
SUFFOLK
NORFOLK
ESSEX
CAMBRIDGESHIRE
Great Yarmouth
Hopton
Lowestoft
Kessingland
Southwold
Oulton
Beccles
Reedham
Haddiscoe
Brampton
Westleton
Leiston
Aldeburgh
Orford Ness
Orford
Tunstall
Bawdsey
Halesworth
Saxmundham
Loddon
Poringland
Bungay
Hempnall
Homersfield
Harleston
Wickham Market
Woodbridge
Felixstowe
Harwich
Trimley St Mary
Framlingham
Ipswich
Long Stratton
Scole
Eye
Debenham
Coddenham
Claydon
Diss
Wreningham
Wymondham
Attleborough
Larling
Stanton
Ixworth
Stowmarket
Needham Market
Hadleigh
East Bergholt
Manningtree
Dedham
Nayland
Lavenham
Long Melford
Sudbury
Colchester
Watton
Thetford
Mundford
Bury St Edmunds
Icklingham
Wickhambrook
Brandon
Methwold
Feltwell
Lakenheath
Mildenhall
Newmarket
Haverhill
Stradsett
Southery
Downham Market
Littleport
Soham
Great Ouse
Little Ouse
Lark
Wissey
Waveney
Yare
Stour
0 10 20 kilometres
0 10 miles
© Global Mapping / XYZ Maps

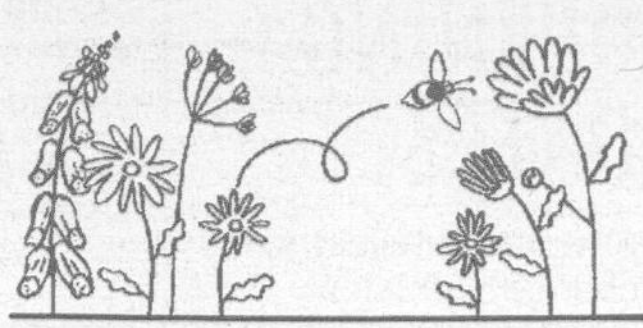

Suffolk has so much to offer – from charming coastal villages, ancient woodlands and picturesque valleys – there is a landscape to suit all tastes.

Keen walkers and cyclists will enjoy Suffolk's low-lying, gentle countryside, where fields of farm animals and crops reflect the county's agricultural roots.

Stretching north from Felixstowe, the county has miles of Heritage Coast set in an Area of Outstanding Natural Beauty. The Suffolk coast was the inspiration for composer Benjamin Britten's celebrated work, and it is easy to see why.

To the west and north of the county are The Brecks, a striking canvas of pine forest and open heathland, famous for its chalky and sandy soils – and one of the most important wildlife areas in Britain.

A variety of gardens to please everyone open in Suffolk, so come along on an open day and enjoy the double benefit of a beautiful setting and supporting wonderful charities.

© Jacqui Hirst

Volunteers

County Organiser
Jenny Reeve
01638 715289
jenny.reeve@ngs.org.uk

County Treasurer
Julian Cusack
01728 649060
julian.cusack@ngs.org.uk

Publicity
Jenny Reeve
(as above)

Social Media
Barbara Segall
01787 312046
barbara.segall@ngs.org.uk

Booklet Co-ordinator
Michael Cole
07899 994307
michael.cole@ngs.org.uk

Assistant County Organisers
Michael Cole
(as above)

Gillian Garnham 01394 448122
gill.garnham@btinternet.com

Wendy Parkes 01473 785504
wendy.parkes@ngs.org.uk

Barbara Segall
(as above)

Peter Simpson 01787 249845
peter.simpson@ngs.org.uk

@SuffolkNGS
@SuffolkNGS

Left: Gable House

OPENING DATES

All entries subject to change. For latest information check **www.ngs.org.uk**

Map locator numbers are shown to the right of each garden name.

February

Snowdrop Festival

Sunday 12th
◆ Blakenham Woodland Garden 2
Gable House 8
Great Thurlow Hall 11

Sunday 26th
Manor House 22

April

Sunday 2nd
Great Thurlow Hall 11

Sunday 16th
◆ The Place for Plants, East Bergholt Place Garden 32

Sunday 23rd
◆ Blakenham Woodland Garden 2

Wednesday 26th
◆ Somerleyton Hall Gardens 39

Sunday 30th
Moat House 23

May

Monday 1st
◆ The Red House 36

Sunday 7th
Finndale House 6
◆ Fullers Mill Garden 7
Grundisburgh House 12
The Old Rectory, Nacton 28
◆ The Place for Plants, East Bergholt Place Garden 32

Saturday 20th
The Lodge 21

Sunday 21st
The Priory 34

Saturday 27th
Pulham Cottage 35
◆ Wyken Hall 43

Sunday 28th
Bridges 3
Lavenham Hall 19
NEW Otley Hall 29
◆ Wyken Hall 43

June

Saturday 3rd
The Old Rectory, Ingham 26

Sunday 4th
Great Bevills 10

Sunday 11th
Ashe Park 1
Great Thurlow Hall 11
Holm House 17
NEW Oakley Cottage 24

Saturday 17th
The Old Rectory, Kirton 27

Sunday 18th
Hillside 16
5 Parklands Green 31

Saturday 24th
Lillesley Barn 20
Old Gardens 25

Sunday 25th
Heron House 15

July

Sunday 2nd
NEW Reydon Grange 37

Sunday 16th
Squires Barn 40

Sunday 23rd
The Old Rectory, Ingham 26

August

Saturday 5th
Gislingham Gardens 9

Sunday 6th
Gislingham Gardens 9

Sunday 20th
Henstead Exotic Garden 14

September

Sunday 24th
Bridges 3

October

Sunday 1st
◆ Fullers Mill Garden 7

Sunday 8th
◆ The Place for Plants, East Bergholt Place Garden 32

By Arrangement

Arrange a personalised garden visit with your club, or group of friends, on a date to suit you. See individual garden entries for full details.

By the Crossways 4
12 Chapel Farm Close, Gislingham Gardens 9
Dip-on-the-Hill 5
Finndale House 6
Gislingham Gardens 9
Helyg 13
Heron House 15
NEW 2 Holmwood Cottages 18
Ivy Chimneys, Gislingham Gardens 9
Lillesley Barn 20
The Lodge 21
Manor House 22
Moat House 23
Old Gardens 25
The Old Rectory, Kirton 27
The Old Rectory, Nacton 28
Paget House 30
5 Parklands Green 31
Polstead Mill 33
Pulham Cottage 35
Smallwood Farmhouse 38
Stone Cottage 41
Wood Farm, Gipping 42

In 2022, 20,000 people were supported by Perennial caseworkers and prevent teams thanks to our funding

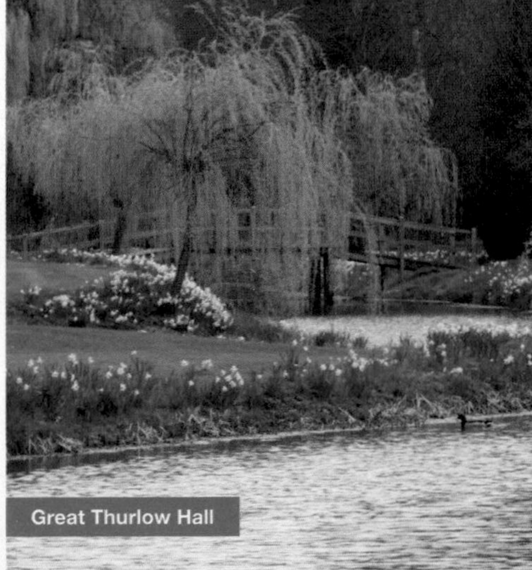
Great Thurlow Hall

THE GARDENS

1 ASHE PARK

Ivy Lodge Road, Campsea Ashe, Woodbridge, IP13 0QB. Mr Richard Keeling. *Using the postcode in SatNav will bring you to the entrance to Ashe Park on Ivy Lodge Rd. Drive through entrance signed Ashe Park, past the gate cottage on L & follow signs to car park.* **Sun 11 June (10.30-5.30). Adm £6.30, chd free.**

An old 12 acre garden comprised of different areas; a large yew hedge, ancient cedars of Lebanon, canal and other water features. Walled garden and wild areas. Garden planted in ruins of old house. Partial wheelchair access, some gravel paths and steps.

2 ◆ BLAKENHAM WOODLAND GARDEN

Little Blakenham, Ipswich, IP8 4LZ. M Blakenham, 07917612355, info@blakenhamfarms.com, www.blakenhamwoodlandgarden.org.uk. *4m NW of Ipswich. Follow signs at Little Blakenham, 1m off B1113 or go to Blakenham Woodland Garden web-site.* **For NGS: Sun 12 Feb, Sun 23 Apr (10-4). Adm £5, chd £3. Home-made teas. For other opening times and information, please phone, email or visit garden website.**

Beautiful 6 acre woodland garden with variety of rare trees and shrubs, Chinese rocks and a landscape spiral form. Lovely in spring with snowdrops, daffodils and camellias followed by magnolias and bluebells. Woodland Garden open from 1 March to 28 June. Suffolk Punches. Tosier Chocolate Stall.

3 BRIDGES

The Street, Woolpit, Bury St Edmunds, IP30 9SA. Mr Stanley Bates & Mr Michael Elles. *Through green coach gates marked Deliveries. From A14 take slip rd to Woolpit, follow signs to centre of village, rd curves to R. Bridges is on L & covered in Wisteria & opp Co-op.* **Sun 28 May, Sun 24 Sept (11-5). Adm £5, chd free. Home-made teas.**

C15 Grade II terraced house in the centre of a C12 Suffolk village with walled garden to the rear of the property. Additional land was acquired 20 years ago, and this garden was developed into formal and informal planting. The main formal feature is the Shakespeare Garden featuring the bust of Shakespeare, and the 'Umbrello' a recently constructed pavillion in an Italianate design. Statue of Shakespeare, The Umbrello, various surprises for children of all ages. Usually a Wind Quintet playing in the main garden.

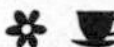

4 BY THE CROSSWAYS

Kelsale, Saxmundham, IP17 2PL. Mr & Mrs William Kendall, miranda@bythecrossways.co.uk. *2m NE of Saxmundham, just off Clayhills Rd. ½ m N of town centre, turn R to Theberton on Clayhills Rd. After 1½ m, 1st L to Kelsale, turn L immed after white cottage.* **Visits by arrangement in Sept for groups of up to 10. Adm £5, chd free.**

3 acre wildlife garden designed as a garden within a working organic farm where wilderness areas lie next to productive beds. Large semi-walled vegetable and cutting garden and a spectacular crinkle-crankle wall. Extensive perennial planting, grasses and wild and uneven areas. The garden is mostly flat, with paved or gravel pathways around the main house, a few low steps and extensive grass paths and lawns.

5 DIP-ON-THE-HILL

Ousden, Newmarket, CB8 8TW. Geoffrey & Christine Ingham, 07947 309900, gki1000@cam.ac.uk. *5m E of Newmarket; 7m W of Bury St Edmunds. From Newmarket: 1m from junc of B1063 & B1085. From Bury St Edmunds follow signs for Hargrave. Parking at village hall. Follow NGS sign at the end of the lane.* **Visits by arrangement for groups of up to 12. Adm £5, chd free.**

Approx 1 acre in a dip on a south facing hill based on a wide range of architectural/sculptural evergreen trees, shrubs and groundcover: pines; grove of Phillyrea latifolia; 'cloud pruned' hedges; palms; large bamboo; ferns; range of kniphofia and Croscosmia. Winner: 'Britain's Best Garden' 2018, Telegraph/ Yorkshire Tea; Gardeners' World, 11 March 2022. Visitors may wish to make an appointment when visiting gardens nearby.

6 FINNDALE HOUSE

Woodbridge Road, Grundisburgh, Woodbridge, IP13 6UD. Bryan & Catherine Laxton, 01473 735228, catherine@laxton-interiors.co.uk. *On B1079, 2m NW of Woodbridge. Parking in field (postcode IP13 6PU), blue badge parking by the house. Short walk to Grundisburgh House.* **Sun 7 May (11-5). Combined adm with Grundisburgh House £7, chd free. Home-made teas. Visits also by arrangement 20 Mar to 30 Sept for groups of 10 to 25.**

A Georgian house surrounded by 10 acres of garden and meadows which are bisected by a 'Monet' style bridge over the River Lark. The garden was designed 30 years ago and has been refreshed by recent additions. Many mature trees, colourful herbaceous borders, thousands of daffodils followed by tulips and alliums, roses in the summer and dahlias to round the year off. Productive kitchen garden. Wheelchair access: gravel drive.

7 ◆ FULLERS MILL GARDEN

West Stow, IP28 6HD. Perennial, 01284 728888, fullersmillgarden@perennial.org.uk, www.fullersmillgarden.org.uk. *6m NW of Bury St Edmunds. Turn off A1101 Bury to Mildenhall Rd, signed West Stow Country Park, go past Country Park continue for ¼ m, garden entrance on R. Sign at entrance. Follow yellow signs on all major routes.* **For NGS: Sun 7 May, Sun 1 Oct (11-5). Adm £7, chd free. Tea, coffee, cake and ice creams. For other opening times and information, please phone, email or visit garden website.**

An enchanting 7 acre garden on the banks of the River Lark. A beautiful site with light dappled woodland and a plantsman's paradise of rare and unusual shrubs, perennials and marginals planted with great natural charm. Euphorbias and lilies are a particular feature with the late flowering colchicums, inc many rare varieties, being of great interest in autumn. Plants are for sale in the garden. Gift shop and Bothy serving tea, coffee and soft drinks. Home-made cakes. Partial wheelchair access around garden.

8 GABLE HOUSE
Halesworth Road, Redisham, Beccles, NR34 8NE. Brenda Foster. *5m S of Beccles. Signed from A144 Bungay/Halesworth Rd.* **Sun 12 Feb (11-4). Adm £4.50, chd free. Home-made teas. Soup lunches.**
We have a large collection of snowdrops, cyclamen, hellebores and other flowering plants for the Snowdrop Day in February. Many bulbs and plants will be for sale. Greenhouses contain rare bulbs and tender plants. A 1 acre garden with lawns and scree with water feature. We have a wide range of unusual trees, shrubs, perennials and bulbs collected over the last fifty years.

GROUP OPENING

9 GISLINGHAM GARDENS
Mill Street, Gislingham, IP23 8JT. 01379 788737, alanstanley22@gmail.com. *4m W of Eye. Gislingham 2½m W of A140. 9m N of Stowmarket, 8m S of Diss. Disabled parking at Ivy Chimneys. Parking limited to disabled parking at Chapel Farm Close.* **Sat 5, Sun 6 Aug (11-4.30). Combined adm £5, chd free. Light refreshments at Ivy Chimneys. Wide range of tea, coffee & cordials; home-made cakes, a gluten free offering & some savouries. Visits also by arrangement Aug & Sept for groups of 10 to 25.**

12 CHAPEL FARM CLOSE
Ross Lee, 01379 788737, alanstanley22@gmail.com.
Visits also by arrangement Aug & Sept for groups of 10 to 25.

IVY CHIMNEYS
Iris & Alan Stanley, 01379 788737, alanstanley22@gmail.com.
Visits also by arrangement Aug & Sept for groups of 10 to 25.

2 varied gardens in a picturesque village with a number of Suffolk timbered houses. Ivy Chimneys is planted for year-round interest with ornamental trees, some topiary, exotic borders and fishpond set in an area of Japanese style. Wisteria draped pergola supports a productive vine. Also a separate ornamental vegetable garden. Small orchard on front lawn. New: 12 Chapel Farm Close is a tiny garden, exquisitely planted and an absolute riot of colour despite the garden's size, the owner has planted a Catalpa, a Cornus Florida Rubra and many unusual plants. It is a fine example of what can be achieved in a small space. Wheelchair access to Ivy Chimneys. Access for smaller wheelchairs only at 12 Chapel Farm Close.

10 GREAT BEVILLS
Sudbury Road, Bures, CO8 5JW. Mr & Mrs G T C Probert. *4m S of Sudbury. Just N of Bures on the Sudbury rd B1508.* **Sun 4 June (2-5.30). Adm £5, chd free. Home-made teas.**
Overlooking the Stour Valley the gardens surrounding an Elizabethan manor house are formal and Italianate in style with Irish yews and mature specimen trees. Terraces, borders, ponds and woodland walks. A short drive away from Great Bevills visitors may wish to also see the C13 St Stephen's Chapel with wonderful views of the Old Bures Dragon recently re-created by the owner. Woodland walks give lovely views over the Stour Valley. Gravel paths.

11 GREAT THURLOW HALL
Great Thurlow, Haverhill, CB9 7LF. Mr George Vestey. *12m S of Bury St Edmunds, 4m N of Haverhill. Great Thurlow village on B1061 from Newmarket; 3½m N of junc with A143 Haverhill/Bury St Edmunds rd.* **Sun 12 Feb (1-4). Light refreshments in the Church. Sun 2 Apr, Sun 11 June (2-5). Home-made teas in the Church. Adm £5, chd free.**
13 acres of beautiful gardens set around the River Stour, Masses of snowdrops in late winter are followed by daffodils and blossom around the riverside walk in spring. Herbaceous borders, rose garden and extensive shrub borders come alive with colour from late spring onwards, there is also a large walled kitchen garden and arboretum. Also open is the Curwen Print Study Centre, Art Studios and Gallery, located adjacent to Great Thurlow Hall will also be open to all garden visitors. Artists will be demonstrating fine art printmaking skills and the Gallery features prints by leading C20 Artists inc Paula Rego, Edward Bawden, Mark Hearld. www.curwenprintstudy.co.uk.

12 GRUNDISBURGH HOUSE
Woodbridge Road, Grundisburgh, Woodbridge, IP13 6UD. Mrs Linden Hibbert. *Not,as the postcode claims, at the junct. Just off B1079. Car parking in field IP13 6PU.* **Sun 7 May (11-5). Combined adm with Finndale House £7, chd free. The Grundisburgh Dog (pub) and cafes in the village. Also 3m from Woodbridge with numerous restaurants and cafes.**
3 acre garden wrapping around classic Georgian house (not open), highlights inc natural swimming pond, formal garden, spring bulbs and fruit blossom, roses, irises and hydrangeas. New projects are on-going but inc planting more hedging, ornamental trees and pleaching. There is a pop-up art gallery in the old coach house which is open intermittently throughout the year. Wheelchair access to paved terrace and path in the formal garden. Decking with 2 steps on swimming pond terrace. Remainder is gravel path.

13 HELYG
Thetford Road, Coney Weston, Bury St Edmunds, IP31 1DN. Jackie & Briant Smith, 01359 220106, briant.broadsspirituality@gmail.com. *From Barningham Xrds/shop turn off the B1111 towards Coney Weston & Knettishall Country Park. After approx 1m Helyg will be found on L behind some large willow trees.* **Visits by arrangement 17 Apr to 30 Sept for groups of 10 to 25. Adm £5, chd free. Home-made teas.**
A garden that combines planting and novel ideas to stimulate interest, with many seats for relaxation. There are different areas with their own character, naturalised spring bulbs,many hostas, roses, rhododendrons, camellias, irises, buddleias, fuchsias, azaleas, herbs, vegetables and exotics. Raised beds, 3 ponds and 4 water features supplement the plants and a recently added 'dry garden'. Many varied, colourful and inspirational areas as above in descriptions. Most of the garden is wheelchair accessible.

Wood Farm, Gipping

14 HENSTEAD EXOTIC GARDEN

Church Road, Henstead, Beccles, NR34 7LD. Andrew Brogan, www.hensteadexoticgarden.co.uk. *Equal distance between Beccles, Southwold & Lowestoft approx 5m. 1m from A12 turning after Wrentham (signed Henstead) very close to B1127.* **Sun 20 Aug (11-4). Adm £6, chd free. Home-made teas. Also available are coffees, local delicatessen made cakes, home made cheese scones & sausage rolls.**

2 acre exotic garden featuring 100 large palms, 20+ bananas and giant bamboo, some of biggest in the UK. Streams, 20ft tiered walkway leading to Thai style wooden covered pavilion. Mediterranean and jungle plants around 3 large ponds with fish. Winner Britain's Best Garden 2015 on ITV as voted by Alan Titchmarsh. Unique garden buildings, streams, waterfalls, rock walkways, different levels, Victorian grotto, giant compost toilet etc. Wheelchair access, can gain access to the garden but not the whole garden.

15 HERON HOUSE

Priors Hill Road, Aldeburgh, IP15 5EP. Mr & Mrs Jonathan Hale, 01728 452200, jonathanrhhale@aol.com. *At the SE junction of Priors Hill Rd & Park Rd. Last house on Priors Hill Rd on S side, at the junction where it rejoins Park Rd.* **Sun 25 June (2-5). Adm £5, chd free. Tea in conservatory attached to house. Visits also by arrangement 1 Mar to 28 Oct.**

2 acres with superb views over the North Sea, River Alde and marshes. Unusual trees, herbaceous beds, shrubs and ponds with a waterfall in large rock garden, and a stream and bog garden. Some half hardy plants in the coastal microclimate. Partial wheelchair access.

16 HILLSIDE

Union Hill, Semer, Ipswich, IP7 6HN. Mr & Mrs Neil Mordey. *Car park through field gate off A1141.* **Sun 18 June (11-4). Adm £5, chd free. Light refreshments.**

This garden in its historic setting of 10½ acres has sweeping lawns running down to a spring fed carp pond. The formal garden has island beds of mixed planting for a long season of interest. The wild area of meadow has been landscaped with extensive tree planting to complement the existing woodland. There is also a small walled kitchen garden and raised beds in the stable yard. Wheelchair access, most areas are accessible although the fruit and vegetable garden is accessed over a deep gravel drive.

17 HOLM HOUSE

Garden House Lane, Drinkstone, Bury St Edmunds, IP30 9FJ. Mrs Rebecca Shelley. *7m SE of Bury St Edmunds. Coming from the E exit A14 at J47, from W J46. Follow signs to Drinkstone, then Drinkstone Green. Turn into Rattlesden Rd & look for Garden House Ln on L. 1st house on L.* **Sun 11 June (10.30-5). Adm £7.50, chd free. Home-made teas.**

Approx 10 acres inc orchard and lawns with mature trees and clipped holm oaks; formal garden with topiary, parterre and mixed borders; rose garden; woodland walk with hellebores, camellias, rhododendrons and bulbs; lake set in wildflower meadow; cut flower garden with greenhouse; large kitchen garden with impressive greenhouse; Mediterranean courtyard with mature olive tree.

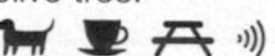

18 NEW 2 HOLMWOOD COTTAGES

Bower House Tye, Polstead, Colchester, CO6 5BZ. Neil Bradfield, 07887 516054, scuddingclouds2@gmail.com. *3m W of Hadleigh. Just off A1071, Boxford to Hadleigh Rd. Look for The Brewers Arms & just beyond, turn off A1071 onto Bower House Tye. Cottage 100yards along this rd.* **Visits by arrangement Apr to Oct for groups of 5 to 30. Not opening July & August. Adm £5, chd free. Donation to Plant Heritage.**

Plantsman's garden with many unusual plants. Hot, sunny borders and areas of dappled shade planted to explore flower colour, foliage, texture, form, autumn interest and design principles. Rich variety of perennials, grasses, species and shrub roses, ferns and woodland plants, matrix and gravel planting. Plus, summerhouse, several seats and a part of the National Collection, Engleheart Narcissus. Wheelchair access along some narrow paths & slightly uneven ground. Please note: there are 2 steps.

NPC

19 LAVENHAM HALL

Hall Road, Lavenham, Sudbury, CO10 9QX. Mr & Mrs Anthony Faulkner, www.katedenton.com. *Next to Lavenham's iconic church & close to High St. From church turn off the main rd down the side of churchyard (Potland Rd). Go down hill. Car Park on R after 100m.* **Sun 28 May (10.30-4.30). Adm £6, chd free.**

5 acre garden built around the ruins of the original ecclesiastical buildings on the site and the village's fishpond. Deep borders of herbaceous planting with sweeping vistas provide the perfect setting for the sculptures which Kate makes in her studio at the Hall and exhibits both nationally and internationally under her maiden name of Kate Denton. 40 garden sculptures by Kate Denton on display. There is a gallery in the grounds which displays a similar number of indoor sculptures and her working drawings. Mostly wheelchair accessible. Note gravel paths / slopes may limit access to certain areas. Follow signs for separate disabled parking.

20 LILLESLEY BARN

The Street, Kersey, Ipswich, IP7 6ED. Mr Karl & Mrs Bridget Allen, 07939 866873, bridgetinkerseybarn@gmail.com, www.instagram.com/bridgets_suffolk_garden/. *In village of Kersey, 2m NW Hadleigh. Driveway is 200m above 'The Bell' pub. Lillesley Barn is situated behind 'The Ancient Houses'. Parking is on st.* **Sat 24 June (11-5). Combined adm with Old Gardens £6, chd free. Home-made teas. Tea, coffee and home-made cakes available. Visits also by arrangement May to Sept £4 per garden; £6 entry if group visits both Lillesley Barn & Old Gardens.**

Dry gravel garden (inspired by the Beth Chatto Garden) inc variety of mediterranean plants, ornamental grasses and herbs. Large herbaceous borders, pleached hornbeam hedge, rose arbours and small orchard with poultry. The garden contains various species of trees inc birch, amelanchier and willow in less than an acre of garden bordered on 2 sides by fields. Meals available at The Bell Inn.

Oakley Cottage, Peasenhall

21 THE LODGE

Bury Road, Bradfield St Clare, Bury St Edmunds, IP30 0ED. Christian & Alice Ward-Thomas, 07768 347595, alice.baring@btinternet.com, www.instagram.com/alicewardthomas. *4m S of Bury St Eds. From N turn L up Water Ln off A134, at Xrds, turn R & go exactly 1m on R. From S turn R up Ixer Ln, R at T-junction, ½m on R. Before post box on R & next door to Lodge Farm.* **Sat 20 May (1-5). Adm £5, chd free. Visits also by arrangement Apr to Sept for groups of up to 30. We are also happy to open for individuals.**

A self-made country garden, surrounded by beautiful parkland established from blank canvas over 20yrs but recently remodelled. Gravel garden inc alpines and rockery, herbaceous borders, rose garden, large veg garden, tulip and wildlife meadows, extensive grassland with mown paths. Life is busy and help is minimal so please don't expect immaculacy! Ground uneven but wheelchair access is possible.

22 MANOR HOUSE

Leiston Road, Middleton, Saxmundham, IP17 3NS. Mandy Beaumont & Steve Thorpe, mandshome@btinternet.com. *Please use IP17 3NZ which takes you direct to parking. Once at parking (village recreation ground car park), the entrance to Manor House gardens is an 8-min walk away - down the side of a field.* **Sun 26 Feb (11-4). Adm £5, chd free. Light refreshments. Hot soup & roll. Visits also by arrangement 27 Feb to 30 Apr for groups of 6 to 20. Incl an informative tour by the owner/s.**

Just over an acre of garden on a triangular plot which was once the northern tip of a medieval green. Started from a wilderness in 2015 - now an ambulatory garden with seating to enjoy many new tree plantings, borders, meadows and a vegetable garden all created from scratch. We have focussed on interest for all seasons and there are many winter/spring flowering trees, shrubs and bulbs to enjoy. An area of the garden has been designated a 'Silent Space' with https://silentspace.org.uk/garden/manor-house/.

23 MOAT HOUSE

Little Saxham, Bury St Edmunds, IP29 5LE. Mr & Mrs Richard Mason, 01284 810941, suzanne@countryflowerssuffolk.com. *2m SW of Bury St Edmunds. Leave A14 at J42- leave r'about towards Westley. Through Westley village at Xroads. R towards Barrow/Saxham. After 1.3m turn L down track and follow signs.* **Sun 30 Apr (1-5). Adm £5, chd free. Home-made teas. Visits also by arrangement 30 Apr to 2 July for groups of 20 to 50.**

Set in a 2 acre historic and partially moated site. This tranquil mature garden has been developed over 20yrs. Bordered by mature trees the garden is in various sections inc a sunken garden, rose and clematis arbours, herbaceous borders with hydrangeas and alliums, small arboretum. A Hartley Botanic greenhouse erected and partere have been created. Each year the owners enjoy new garden projects. Secluded and peaceful setting, each year new additions and wonderful fencing.

24 NEW **OAKLEY COTTAGE**
Chapel Street, Peasenhall, Saxmundham, IP17 2JD. Mr Robin Brooks. *Centre of Peasenhall Village. On the A1120 in the centre of the village of Peasenhall, next door to the Weavers Tea-rooms.* **Sun 11 June (12.30-5). Adm £5, chd free.**
This little cottage garden is divided into 5 small areas - a formal-ish courtyard, 2 lawns, a vegetable plot, and what should be a rose garden once the roses get their act together. Persicaria, hardy geraniums, honesty, gaura, foxgloves; lots grown from cuttings and seeds, not counting David Austin's shy little beauties. Pergolas galore. Mostly flat, few steps, but grass and gravel to be negotiated.

25 OLD GARDENS
The Street, Kersey, Ipswich, IP7 6ED. Mr & Mrs David Anderson, 01473 828044, davidmander15@gmail.com. *10m W of Ipswich. Up hill, approx 100m from The Bell Inn Pub. On the same side of the rd.* **Sat 24 June (11-5). Combined adm with Lillesley Barn £6, chd free. Visits also by arrangement May to Sept. £4 per garden; £6 entry if group visits both Lillesley Barn & Old Gardens.**
Entered from The Street, a natural garden with wildflowers under a copper beech tree. To the rear, a formal garden designed by Cherry Sandford with a sculpture by David Harbour. Refreshments at Lillesley Barn. Meals available at The Bell Inn in the village.

26 THE OLD RECTORY, INGHAM
Ingham, Bury St Edmunds, IP31 1NQ. Mr J & Mrs E Hargreaves. *5m N of Bury St Edmunds & A14. Park in Churchyard, on R as you come N on A134 into Ingham.* **Sat 3 June, Sun 23 July (11-5). Adm £4, chd free. Cream teas. Range of home-made cakes and scones. Variety of teas, coffee and soft drinks.**
Come and enjoy teas, scones and cakes in a lovely setting of the large gardens of a Georgian Rectory. Formal parterre, lawns, and a wide variety of mature trees inc weeping willows and a Wellingtonia. We are building a vibrant display of colours, inc dahlias, geraniums and more, in insect-friendly borders; alliums and roses in June, and a lavender walk in July. Children friendly. Wheelchair access through 1m wide gate.

27 THE OLD RECTORY, KIRTON
Church Lane, Kirton, Ipswich, IP10 0PT. Mr & Mrs N Garnham, gill.garnham@btinternet.com. *Adjacent to Kirton parish church on corner of Church Ln & Burnt House Ln. Car parking available in Church Hall car park next door and on village green.* **Sat 17 June (10.30-4). Adm £5, chd free. Home-made teas in Kirton Methodist Chapel garden nearby with musical accompaniment. Visits also by arrangement in June for groups of up to 20.**
A 3 acre garden surrounding a Georgian rectory with numerous herbaceous beds in sun and shade, large child friendly lawn and a variety of mature trees. Tracy Barritt-Brown, who created the 'Willow Family' on Felixstowe seafront, will be exhibiting her willow sculptures throughout the garden for purchase or commission.

28 THE OLD RECTORY, NACTON
Nacton, IP10 0HY. Mrs Elizabeth & Mr James Wellesley Wesley, 01473 659673, tizyww@gmail.com. *The garden is down the rd to Nacton from the 1st A14 turn off after Orwell bridge going S/E. Parking on Church Rd opp Old Rectory driveway. This will be signed.* **Sun 7 May (10.30-5). Adm £5, chd free. Home-made teas. Visits also by arrangement 17 Apr to 7 Oct for groups of up to 30.**
Just under 2 acres of garden divided into areas for different seasons: mature trees and herbaceous borders, ample spring bulbs and blossom. Light soil so many self sown flowers. Damp area with emphasis on foliage (rheum, darmera, rodgersia, hydrangea). Still a work in progress after 33 years. We're looking at how to bring in more butterflies and deal with very dry conditions in most of the garden. Most areas are accessible for disabled visitors, however several grassy slopes and various different levels, so some energy needed to get everywhere.

29 NEW **OTLEY HALL**
Hall Lane, Otley, Ipswich, IP6 9PA. Mr Stephen & Mrs Louisa Flavell, www.otleyhall.co.uk. *Take Chapel Rd past the Otley village Post Office after 250 yds take 1st L onto Hall Ln.* **Sun 28 May (10-4). Adm £5, chd free. Light refreshments at Marthas Barn Cafe which has varied recipes using fresh seasonal produce. A Barista, with a fully stocked bar, offers fine home-made cakes baked on the premises. Advisory to book in advance.**
The 10 acres of both formal and informal gardens at Otley Hall are managed with an eye to nature. The gardens feature 3 medieval/ Elizabethan Garden recreations by Sylvia Landsberg, stew ponds and woodland. Wheelchair access to Martha's Barn Cafe.

30 PAGET HOUSE
Back Road, Middleton, Saxmundham, IP17 3NY. Julian & Fiona Cusack, 01728 649060, julian.cusack@btinternet.com. *From A12 at Yoxford take B1122 towards Leiston. Turn L after 1.2m at Middleton Moor. After 1m enter Middleton & drive straight ahead into Back Rd. Turn 1st R on Fletchers Ln for car park.* **Visits by arrangement 17 Apr to 10 Sept for groups of up to 50. Light refreshments. Gluten free & vegan options.**
The garden is designed to be wildlife friendly with wild areas meeting formal planting.There is an orchard and vegetable plot. There are areas of woodland, laid hedges, a pond supporting amphibians and dragonflies, wildflower meadow and an abstract garden sculpture by local artist Paul Richardson. In 2023 we will be making changes designed to increase our resilience to drought. We record over 40 bird species each year and a good showing of butterflies, dragonflies and wildflowers inc orchids. Wheelchair access: gravel drive and mown paths. Parking on drive by prior arrangement.

31 5 PARKLANDS GREEN
Fornham St Genevieve, Bury St Edmunds, IP28 6UH. Mrs Jane Newton, newton.jane@talktalk.net. *2m NW of Bury St Edmunds off B1106. Plenty of parking on the green.* **Sun 18 June (11-4). Adm**

£5, chd free. Home-made teas. Visits also by arrangement Apr to Sept for groups of up to 60.
1½ acres of gardens developed since the 1980s for all year interest. There are mature and unusual trees and shrubs and riotous herbaceous borders. Explore the maze of paths to find 4 informal ponds, a treehouse, the sunken garden, greenhouses and woodland walks.

32 ◆ THE PLACE FOR PLANTS, EAST BERGHOLT PLACE GARDEN

East Bergholt, CO7 6UP. Mr & Mrs Rupert Eley, 01206 299224, sales@placeforplants.co.uk, www.placeforplants.co.uk. *2m E of A12, 7m S of Ipswich. On B1070 towards Manningtree, 2m E of A12. Situated on the edge of East Bergholt.* **For NGS: Sun 16 Apr, Sun 7 May, Sun 8 Oct (1-5). Adm £8, chd free. Home-made teas. For other opening times and information, please phone, email or visit garden website.**
20 acre woodland garden originally laid out at the turn of the last century by the present owner's great grandfather. Full of many fine trees and shrubs, many seldom seen in East Anglia. A fine collection of camellias, magnolias and rhododendrons, topiary, and the National Collection of deciduous Euonymus. Partial wheelchair access in dry conditions - it is advisable to telephone before visiting.

NPC

33 POLSTEAD MILL

Mill Lane, Polstead, Colchester, CO6 5AB. Mrs Lucinda Bartlett, 07711 720418, lucyofleisure@hotmail.com. *Between Stoke by Nayland & Polstead on the River Box. From Stoke by Nayland take rd to Polstead - Mill Ln is 1st on L & Polstead Mill is 1st house on R.* **Visits by arrangement 19 May to 9 Sept for groups of 10 to 50. Adm £6, chd free. Light refreshments.**
The garden has been developed since 2002, it has formal and informal areas, a wildflower meadow and a large productive kitchen garden. The River Box runs through the garden and there is a mill pond, which gives opportunity for damp gardening, while much of the rest of the garden is arid and is planted to minimise the need for watering. Range of refreshments available inc full cream teas and light lunches. Featured in Secret Gardens of East Anglia. Partial wheelchair access.

34 THE PRIORY

Stoke by Nayland, Colchester, CO6 4RL. Mrs H F A Engleheart. *5m SW of Hadleigh. Entrance on B1068 to Sudbury (NW of Stoke by Nayland).* **Sun 21 May (2-5). Adm £5, chd free. Home-made teas.**
Interesting 9 acre garden with fine views over Constable countryside; lawns sloping down to small lakes and water garden; fine trees, rhododendrons and azaleas; walled garden; mixed borders and ornamental greenhouse. Wide variety of plants. Plant stall. Wheelchair access over most of garden, some steps.

35 PULHAM COTTAGE

Wetherden, Stowmarket, IP14 3LQ. Susanne & David Barker, 07882 849379, susanne@evolution-planning.co.uk. *Take the track signed 'Mutton Hall' & we are the red house on L.* **Sat 27 May (10.30-5). Adm £5, chd free. Pre-booking essential, please visit www.ngs.org.uk for information & booking. Home-made teas. Visits also by arrangement 6 May to 30 Sept for groups of 10 to 30.**
A relaxed cottage garden set in 1½ acres, created over the last 5 yrs. Divided into rooms using mixed hedging, hazel fencing and vintage iron gates and fencing. A clipped box parterre, a kitchen garden with flowers for cutting, orchard and wildflower areas to explore. Areas left for re-wildng to encourage moths, butterflies and bees. Lots of seating to relax and enjoy the vistas and views. Victorian style greenhouse, orchard and free range chickens. The Kings Arms is a lovely pub in Haughley with a garden.

36 ◆ THE RED HOUSE

Golf Lane, Aldeburgh, IP15 5PZ. Britten Pears Arts, 01728 451700, info@brittenpearsarts.org, brittenpearsarts.org/visit-us/the-red-house. *Top of Aldeburgh, approx. 1m from the sea. From A12, take A1094 to Aldeburgh. Follow the brown sign directing you towards the r'about, take 1st exit: B1122 Leiston Rd. Golf Ln is 2nd L, follow sign to 'The Red House'.* **For NGS: Mon 1 May (11-3). Adm £5, chd free. Light refreshments. For other opening times and information, please phone, email or visit garden website.**
The former home of the renowned British composer Benjamin Britten and partner, the tenor Peter Pears. The 5 acre garden provides an atmospheric setting for the house they shared and contain many plants loved by the couple. Mixed herbaceous borders, kitchen garden, contemporary planting and a summer tropical border. This is a garden-only visit with a cafe, shop and WC facilities available. The Red House inc the collections left by the 2 men and the Archive holds an extraordinary wealth of material documenting their lives. The garden offers a peaceful setting to this beautiful corner of Suffolk. For museum opening times, see website. Wheelchair access: brick, concrete, gravel paths and grass. Some areas are uneven, may require effort to navigate. Wheelchair available upon request.

37 NEW REYDON GRANGE

Mardle Road, Wangford, Beccles, NR34 8AU. Aileen and Bill Irving. *2m W of Southwold.* **Sun 2 July (11-4). Adm £5, chd free. Home-made teas. Coffee, cakes & biscuits also available.**
The approximately 3 acre gardens consists mainly of a series of mixed herbaceous and shrub borders, orchards, a rose garden, dwarf box cloud hedge, kitchen garden, lawns and a pond. We have been developing a romantic and colour co-ordinated style to the garden, with a number of internal vistas, and also grouping plants according to their environmental needs. There is a 6 acre wild meadow/field.

In 2022, our donations to Carers Trust meant that 456,089 unpaid carers were supported across the UK

Otley Hall

38 SMALLWOOD FARMHOUSE

Smallwood Green, Bradfield St George, Bury St Edmunds, IP30 0AJ. Mr & Mrs P Doe, 01449 736127, philipdoe142@btinternet.com. *S of A14, E of A134 on Bradfield St George to Felsham Rd. Disregard sign to Smallwood Green & follow the yellow signage.* **Visits by arrangement 15 May to 15 July for groups of 10 to 30. Adm £5, chd free. Light refreshments.**

The garden is a combination of traditional cottage planting and contemporary styles. At its heart, a C16 farmhouse provides the backdrop to a number of old English roses, a profusion of clematis and honeysuckle, and a variety of perennials. There are two natural ponds, a Mediterranean garden and productive kitchen garden, whilst paths wind through an ancient meadow and orchard. Partially wheelchair accessible, although not suitable in damp or wet weather.

39 ◆ SOMERLEYTON HALL GARDENS

Somerleyton, NR32 5QQ. Lord Somerleyton, 01502 734901, visitors@somerleyton.co.uk, www.somerleyton.co.uk. *5m NW of Lowestoft. From Norwich take the B1074, 7m SE of Great Yarmouth (A143). Coaches should follow signs to the rear W gate entrance.* **For NGS: Wed 26 Apr (11-4). Adm £8.45, chd £5.45. Light refreshments. For other opening times and information, please phone, email or visit garden website.**

12 acres of beautiful gardens contain a wide variety of magnificent specimen trees, shrubs, borders and plants providing colour and interest throughout the year. Sweeping lawns and formal gardens combine with majestic statuary and original Victorian ornamentation. Highlights inc the Paxton glasshouses, pergola, walled garden and yew hedge maze. House and gardens remodelled in 1840s by Sir Morton Peto. House created in Anglo-Italian style with lavish architectural features and fine state rooms. Most areas of the gardens are accessible, path surfaces are gravel/stone and can be difficult with small wheeled chairs.

40 SQUIRES BARN

St. Cross South Elmham, Harleston, IP20 0PA. Stephen & Ann Mulligan. *6m W of Halesworth, 6m E of Harleston & 7m S of Bungay. On New Rd between St Cross & St James. Parking in field opp. Yellow signs from A143 and other local rds.* **Sun 16 July (10.30-4). Adm £5, chd free. Home-made teas. Refreshments inc gluten free, vegan options and ice creams.**

A young and evolving garden of some 3 acres, inc an orchard, greenhouse and fruit cage, raised vegetable beds, large ornamental pond with water lilies, fish and waterfall, wildflower mound, island beds of mixed planting and a range of mature and younger trees. A recent addition is a swimming pool area. Views over surrounding countryside. Gravel drive with pool. Plant stall and art exhibition. The garden is largely grass with some slight slopes. Seating is available across the garden. One single flight of steps can be bypassed.

41 STONE COTTAGE

34 Main Road, Woolverstone, Ipswich, IP9 1BA. Mrs Jen Young, 01473 780156, j_s_young@icloud.com. *Take B1456 towards Shotley. When leaving Woolverstone Village the yellow signs will direct you to Stone Cottage.* **Visits by arrangement 24 Apr to 11 June for groups of up to 40. Adm £5. Home-made teas.**

An idyllic Suffolk country cottage, surrounded by a garden created and maintained by the owner from a derelict space into a beautiful, calming garden. Over 100 roses, unusual delphiniums, irises, spring bulbs and many more are planted together in interesting colour combinations that provide year-round interest and perfume. A lovely space to sit and relax. Areas of gravel paths.

42 WOOD FARM, GIPPING

Back Lane, Gipping, Stowmarket, IP14 4RN. Mr & Mrs R Shelley, 07809 503019, els@maritimecargo.com. *From A14 take A1120 to Stowupland. Turn L opp petrol stn & then turn R at T- junc. Follow for approx 1m turn L at Walnut Tree Farm, then imm R & follow for 1m along country ln. Wood Farm is on L.* **Visits by arrangement 15 May to 18 June for groups of 10 to 40. Adm £5, chd free. Home-made teas. Large Party Barn with facilities available.**

Wood Farm is an old farm with ponds, orchards and a magnificent 8 acre wildflower meadow (with mown paths) bordered with traditional hedging, trees and woodland. The large cottage garden was created in 2011 with a number of beds planted with flowers, vegetables and topiary. Wildlife is very much encouraged in all parts of the garden (particularly bees and butterflies). In June 2018 used by an international fashion brand as their A/W18 Photo Shoot. Partial wheelchair access.

43 ◆ WYKEN HALL

Stanton, IP31 2DW. Sir Kenneth & Lady Carlisle, 01359 250262, kenneth.carlisle@wykenvineyards.co.uk, www.wykenvineyards.co.uk. *9m NE of Bury St Edmunds. Along A143. Follow signs to Wyken Vineyards on A143 between Ixworth & Stanton.* **For NGS: Sat 27, Sun 28 May (10-5). Adm £5, chd free. For other opening times and information, please phone, email or visit garden website.**

4 acres around the old manor. The gardens inc knot and herb gardens, old-fashioned rose garden, kitchen and wild garden, nuttery, pond, gazebo and maze; herbaceous borders and old orchard. Woodland walk and vineyard nearby. Restaurant (booking 01359 250287), shop and vineyard. Farmers' Market Sat 9am-1pm.

National Garden Scheme gardens are identified by their yellow road signs and posters. You can expect a garden of quality, character and interest, a warm welcome and plenty of home-made cakes!

SURREY

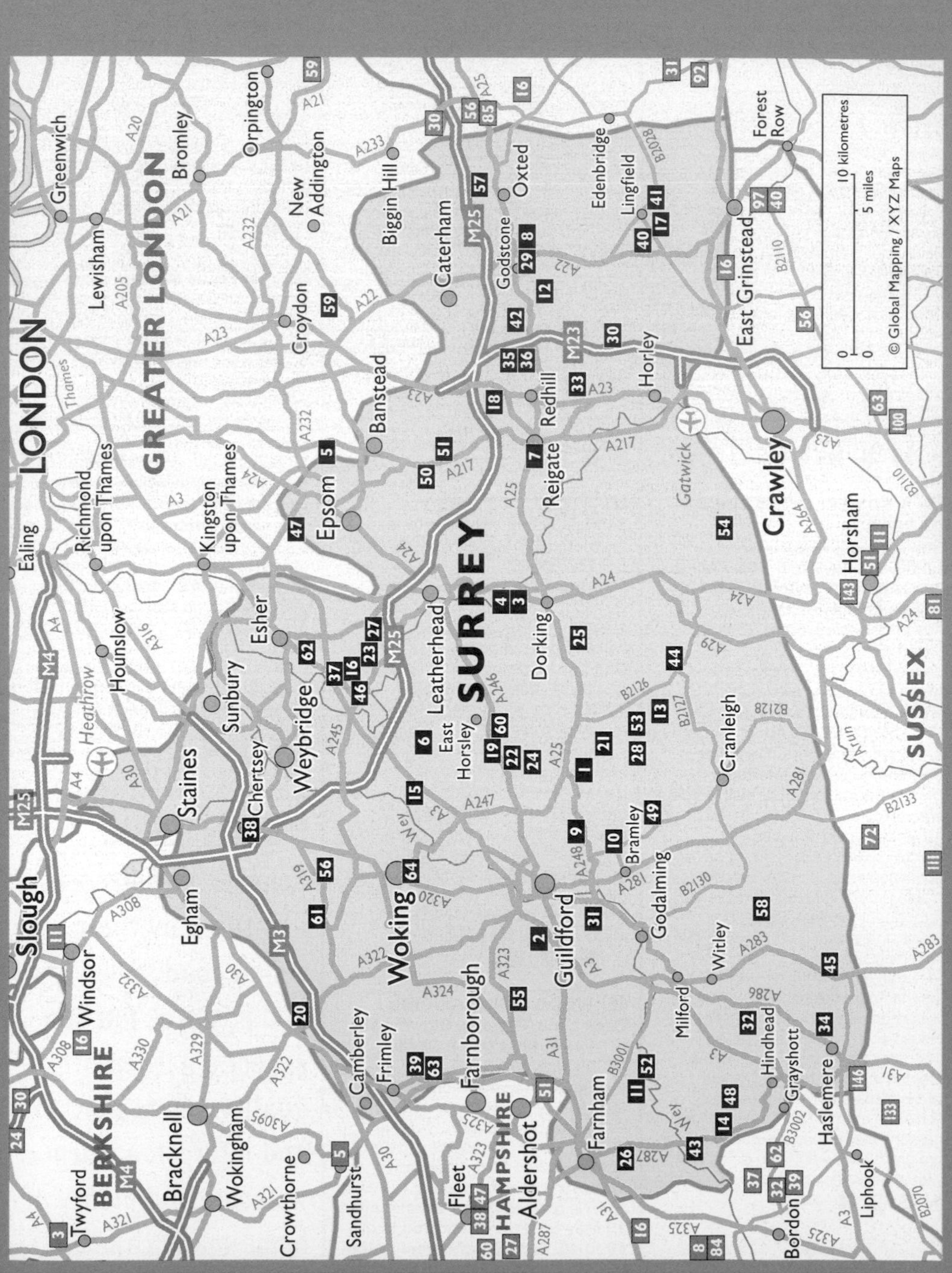

LONDON
GREATER LONDON
SURREY
BERKSHIRE
HAMPSHIRE
SUSSEX
Slough
Windsor
Twyford
Bracknell
Wokingham
Crowthorne
Sandhurst
Camberley
Frimley
Farnborough
Fleet
Aldershot
Farnham
Bordon
Liphook
Haslemere
Grayshott
Hindhead
Milford
Witley
Godalming
Guildford
Bramley
Cranleigh
Woking
Egham
Staines
Chertsey
Weybridge
Sunbury
Esher
Hounslow
Heathrow
Ealing
Richmond upon Thames
Kingston upon Thames
Epsom
Leatherhead
East Horsley
Dorking
Reigate
Redhill
Banstead
Croydon
Caterham
Godstone
Oxted
Biggin Hill
New Addington
Orpington
Bromley
Lewisham
Greenwich
Thames
Edenbridge
Lingfield
East Grinstead
Forest Row
Horley
Gatwick
Crawley
Horsham
Wey
Arun
0 5 miles
0 10 kilometres
© Global Mapping / XYZ Maps

As a designated Area of Outstanding Natural Beauty, it's no surprise that Surrey has a wealth of gardens on offer.

With its historic market towns, lush meadows and scenic rivers, Surrey provides the ideal escape from the bustle of nearby London.

Set against the rolling chalk uplands of the unspoilt North Downs, the county prides itself on extensive country estates with historic houses and ancient manors.

Surrey is the heartland of the National Garden Scheme at Hatchlands Park and the RHS at Wisley, both promoting a precious interest in horticulture. Surrey celebrates a landscape coaxed into wonderful vistas by great gardeners such as John Evelyn, Capability Brown and Gertrude Jekyll.

With many eclectic gardens to visit, there's certainly plenty to treasure in Surrey.

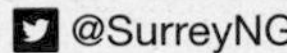 @surreyngs 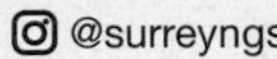 @SurreyNGS @surreyngs

Volunteers

County Organiser
Clare Bevan
07956 307546
clare.bevan@ngs.org.uk

County Treasurer
Nigel Brandon
020 8643 8686
nbrandon@ngs.org.uk

Booklet Co-ordinator
Annabel Alford-Warren
01483 203330
annabel.alford-warren@ngs.org.uk

Publicity
Sarah Wilson
07932 445868
sarah.wilson@ngs.org.uk

Social Media
Annette Warren
07790 045354
annette.warren@ngs.org.uk

Assistant County Organisers

Jan Brandon 020 8643 8686
janmbrandon@outlook.com

Penny Drew 01252 792909
penelopedrew@yahoo.co.uk

Angela Gilchrist 01306 884613
angela.gilchrist@ngs.org.uk

Joy Greasley 01342 837369
joy.greasley@ngs.org.uk

Di Grose 01883 742983
di.grose@ngs.org.uk

Annie Keighley 01252 838660
annie.keighley12@btinternet.com

Louise McAllister 07758 482881
louise.mcallister@ngs.org.uk

Jean Thompson 01483 425633
norney.wood@btinternet.com

Left: West Horsley Place

OPENING DATES

All entries subject to change. For latest information check **www.ngs.org.uk**

Extended openings are shown at the beginning of the month.

Map locator numbers are shown to the right of each garden name.

January

Wednesday 4th
Timber Hill 56

Wednesday 11th
Timber Hill 56

Monday 30th
Timber Hill 56

February

Snowdrop Festival

Every Monday to Monday 20th
Timber Hill 56

Tuesday 7th
◆ The Sculpture Park 48

Sunday 12th
◆ Gatton Park 18

Sunday 19th
Shieling 51

March

Daily from Monday 27th
◆ Vann 58

Monday 6th
Timber Hill 56

Sunday 26th
Albury Park 1

April

Daily to Sunday 2nd
◆ Vann 58

Sunday 9th
Coverwood Lakes 13

Monday 10th
Moleshill House 37
Shieling 51

Friday 14th
◆ Dunsborough Park 15

Saturday 15th
11 West Hill 59

Sunday 16th
Coverwood Lakes 13
11 West Hill 59

Monday 17th
Timber Hill 56

Friday 21st
Little Orchards 30

Sunday 23rd
◆ Hatchlands Park 22
41 Shelvers Way 50

Sunday 30th
The Garth Pleasure Grounds 17

May

Daily from Monday 8th to Sunday 14th
Chauffeur's Flat 8

Daily from Monday 1st to Sunday 7th
◆ Vann 58

Monday 1st
Coverwood Lakes 13
Shieling 51

Sunday 7th
◆ Crosswater Farm 14
The Garth Pleasure Grounds 17
NEW West Horsley Place 60

Monday 8th
◆ Crosswater Farm 14

Friday 12th
◆ Ramster 45

Saturday 13th
Hall Grove School 20
Lower House 32

Sunday 14th
Coverwood Lakes 13
NEW Harry Edwards Healing Sanctuary 21
Lower House 32
The Manor House 34
Westways Farm 61

Friday 19th
2 Knott Park House 27

Saturday 20th
NEW Cobbetts Corner 11
2 Knott Park House 27
Leigh Place 29
The White House 62

Sunday 21st
Chilworth Manor 9
Knowle Grange 28
Leigh Place 29
The Therapy Garden 55
The White House 62

Friday 26th
Little Orchards 30

Saturday 27th
Monks Lantern 38

Sunday 28th
15 The Avenue 5
Little Orchards 30
Timber Hill 56
◆ Titsey Place Gardens 57

June

Tuesday 6th
◆ The Sculpture Park 48

Saturday 10th
Bridge End Cottage 6
NEW The Old Vicarage 43

Sunday 11th
◆ Loseley Park 31
The Oast House 41
The Old Rectory 42
NEW The Old Vicarage 43

Saturday 17th
Lower House 32
The White House 62

Sunday 18th
Lower House 32
Timber Hill 56
The White House 62

Sunday 25th
NEW Fairmile Common Gardens 16
Macmillan Cancer Support Centre 33
NEW 1 Marsh Cottages 35
NEW 3 Marsh Cottages 36
The Nutrition Garden 39
◆ Titsey Place Gardens 57
Wildwood 63

July

Sunday 2nd
NEW Grace & Flavour CIC 19

Sunday 9th
High Clandon Estate Vineyard 24

Sunday 16th
◆ Titsey Place Gardens 57

Saturday 22nd
The White House 62

Sunday 23rd
The White House 62
NEW Woodpeckers 64

Sunday 30th
41 Shelvers Way 50

August

Sunday 13th
◆ Titsey Place Gardens 57

Sunday 20th
Pratsham Grange 44
NEW Woodpeckers 64

Saturday 26th
26 Rowden Road 47

Sunday 27th
26 Rowden Road 47

September

Sunday 3rd
Moleshill House 37
The Nutrition Garden 39
Randalls Allotments 46

Wednesday 6th
Little Orchards 30

Sunday 10th
NEW 11 Hillary Road 26
Little Orchards 30

Sunday 17th
Hill Farm 25
The Therapy Garden 55

Sunday 24th
NEW West Horsley Place 60

Saturday 30th
Hall Grove School 20

October

Sunday 1st
Albury Park 1

Wednesday 11th
Timber Hill 56

Sunday 15th
Coverwood Lakes 13

Wednesday 25th
Timber Hill 56

November

Wednesday 15th
Timber Hill 56

January 2024

Wednesday 3rd
Timber Hill 56

Wednesday 10th
Timber Hill 56

Wednesday 17th
Timber Hill 56

Wednesday 31st
Timber Hill 56

February 2024

Wednesday 7th
Timber Hill 56

Wednesday 14th
Timber Hill 56

Wednesday 21st
Timber Hill 56

By Arrangement

Arrange a personalised garden visit with your club, or group of friends, on a date to suit you. See individual garden entries for full details.

NEW 29 Applegarth Avenue 2
Ashcombe 3
Ashleigh Grange 4
15 The Avenue 5
Bridge End Cottage 6
Caxton House 7
2 Chinthurst Lodge 10
Coldharbour House 12
Coverwood Lakes 13
The Garth Pleasure Grounds 17
Heathside 23
Little Orchards 30
NEW 1 Marsh Cottages 35
NEW 3 Marsh Cottages 36
Moleshill House 37
Monks Lantern 38
The Nutrition Garden 39
Oaklands 40
26 Rowden Road 47
Shamley Wood Estate 49
41 Shelvers Way 50
Shieling 51
Springwood 52
Spurfold 53
Tanhouse Farm 54
Westways Farm 61
The White House 62
NEW Woodpeckers 64

THE GARDENS

1 ALBURY PARK

Albury, GU5 9BH. Trustees of Albury Estate. *5m SE of Guildford. From A25 take A248 towards Albury for ¼ m, then up New Rd, entrance to Albury Park immed on L.* **Sun 26 Mar, Sun 1 Oct (2-5). Adm £5, chd free. Home-made teas.**
14 acre pleasure grounds laid out in 1670s by John Evelyn for Henry Howard, later 6th Duke of Norfolk. ¼ m of terraces, fine collection of trees, lake and river. Wheelchair access over gravel path and slight slope.

2 NEW 29 APPLEGARTH AVENUE

Guildford, GU2 8LX. Mr Robert Avenell, robavenell@outlook.com. *Just off the A3 at Guildford. Follow directions for The Royal Surrey Hospital from the A3 along Egerton Rd. Turn L at the r'about by Kings College & Applegarth Ave is opp the shops on Park Barn Estate.* **Visits by arrangement in June for groups of up to 5. Adm £5, chd free.**
Not a large garden but somewhere to escape to and relax. The owner's love of vibrant colours is reflected in the garden which is full of cherished plants, glorious blooms and attractive scents. Walking through a series of outdoor rooms, you'll discover a mixture of trees, shrubs, herbaceous perennials and roses. Best seen in early summer with a cream tea.

3 ASHCOMBE

Chapel Lane, Westhumble, Dorking, RH5 6AY. Vivienne & David Murch, 01306 743062, murch@ashcombe.org. *2m N of Dorking. From A24 at Boxhill/Burford Bridge follow signs to Westhumble. Through village & take L drive by ruined chapel (1m from A24). Parking for up to 10 cars.* **Visits by arrangement June to Sept for groups of 10+. Home-made teas. Wine for eve visits.**
Plantaholics 1½ acre wildlife friendly sloping garden on chalk and flint. Enclosed ⅓ acre cottage garden with large borders of roses, delphiniums and clematis. Amphibian pond. Secluded decking and patio area with colourful acers and views over garden and Boxhill. Gravel bed of salvia and day lilies. House (not open) surrounded by banked flower beds and lawn leading to bee and butterfly garden.

4 ASHLEIGH GRANGE

off Chapel Lane, Westhumble, RH5 6AY. Clive & Angela Gilchrist, 01306 884613, ar.gilchrist@btinternet.com. *2m N of Dorking. From A24 at Boxhill/Burford Bridge follow signs to Westhumble. Through village & L up drive by ruined chapel (1m from A24). Sorry no access for coaches.* **Visits by arrangement May to July. Groups warmly welcomed. Home-made teas.**
Plant lover's chalk garden on 3½ acre sloping site in charming rural setting with delightful views. Many areas of interest inc rockery and water feature, raised ericaceous bed, prairie style bank, foliage plants, woodland walk, fernery and folly. Large mixed herbaceous and shrub borders planted for dry alkaline soil and widespread interest.

Our 2022 donation enabled 7,000 people affected by cancer to benefit from the garden at Maggie's in Oxford

5 15 THE AVENUE
Cheam, Sutton, SM2 7QA. Jan & Nigel Brandon, 020 8643 8686, janmbrandon@outlook.com. *1m SW of Sutton. By car: exit A217 onto Northey Av, 2nd R into The Ave. By train: 10 min walk from Cheam Stn. By bus: use 470.* **Evening opening Sun 28 May (4-8). Adm £8, chd £3. Wine. Visits also by arrangement 13 May to 2 July for groups of 10 to 50.**
A contemporary garden designed by RHS Chelsea Gold Medal Winner, Marcus Barnett. Four levels divided into rooms by beech hedging and columns: formal entertaining area, contemporary outdoor room, lawn and wildflower meadow. Over 100 hostas hug the house. Silver birch, cloud pruned box, ferns, grasses, tall bearded irises, contemporary sculptures. Regional finalist in the 2022 Garden of The Year competition as shown on Channel 4. Partial wheelchair access, terraced with steps; sloping path provides view of whole garden.

6 BRIDGE END COTTAGE
Ockham Lane, Ockham, GU23 6NR. Clare & Peter Bevan, 07956 307546, clare.bevan@ngs.org.uk, bridgeendcottage.co.uk. *Nr RHS Garden Wisley. At Wisley r'about turn L onto B2039 to Ockham/Horsley. After ½m turn L into Ockham Ln. House ½m on R. From Cobham go to Blackswan Xrds.* **Sat 10 June (1.30-6). Adm £6, chd free. Home-made teas in the garden room. Visits also by arrangement May to July for groups of up to 25.**
A 2 acre country garden with different areas of interest inc perennial borders, mature trees, pond and streams, small herb parterre, fruit trees and a vegetable patch. An adjacent 2 acre field was sown with perennial wildflower seed in May 2013 and has flowered each summer and will be of interest to anyone establishing a wildflower garden. Partial wheelchair access.

7 CAXTON HOUSE
67 West Street, Reigate, RH2 9DA. Bob Bushby, 01737 243158/07836 201740, bushbyybob@gmail.com. *On A25 towards Dorking, approx ¼m W of Reigate. Parking on road or past Black Horse on Flanchford Rd.* **Visits by arrangement 1 Mar to 4 Aug for groups of 10 to 75. Adm £10, chd free. Light refreshments inc.**
Lovely large spring garden with arboretum, two well-stocked ponds, large collection of hellebores and spring flowers. Pots planted with colourful displays, plants. Small Gothic Folly built by owner. Herbaceous borders with grasses, perennials and spring bulbs, parterre, bed with wild daffodils and prairie style planting in summer, and new wildflower garden in arboretum. Wheelchair access to most parts of the garden. Dogs on leads please.

8 CHAUFFEUR'S FLAT
Tandridge Lane, Tandridge, RH8 9NJ. Mr & Mrs Richins. *2m E of Godstone. 2m W of Oxted. Turn off A25 at r'about for Tandridge. Take 2nd drive on L past church. Follow arrows to circular courtyard. Do not use Jackass Ln even if your SatNav tells you to do so.* **Daily Mon 8 May to Sun 14 May (10-5). Adm £5, chd free. Home-made teas (Sats & Suns only).**
Enter a 1½ acre tapestry of magical secret gardens with magnificent views. Touching the senses, all sure footed visitors may explore the many surprises on this constantly evolving exuberant escape from reality. Imaginative use of recycled materials creates an inspired variety of ideas, while wild and specimen plants reveal an ecological haven.

9 CHILWORTH MANOR
Halfpenny Lane, Chilworth, Guildford, GU4 8NN. Mia & Graham Wrigley, www.chilworthmanorsurrey.com. *3½m SE of Guildford. From centre of Chilworth village turn into Blacksmith Ln. 1st drive on R on Halfpenny Ln.* **Sun 21 May (11-5). Adm £7.50, chd free. Pre-booking essential, please visit www.ngs.org.uk for information & booking. Home-made teas.**
The grounds of the C17 Chilworth Manor create a wonderful tapestry, a jewel of an C18 terraced walled garden, topiary, herbaceous borders, sculptures, mature trees and stew ponds that date back a 1000 yrs. A fabulous, peaceful garden for all the family to wander and explore or just to relax and enjoy! Perhaps our many visitors describe it best, 'Magical', 'a sheer delight', 'elegant and tranquil', 'a little piece of heaven', 'spiffing!'. There will be a garden or tree talk at 12.15pm, 1.15pm, 3.15pm, 4.15pm. Sorry, dogs are not allowed.

10 2 CHINTHURST LODGE
Wonersh Common, Wonersh, Guildford, GU5 0PR. Mr & Mrs M R Goodridge, 01483 535108, michaelgoodridge@ymail.com. *4m S of Guildford. From A281 at Shalford turn E onto B2128 towards Wonersh. Just after Waverley sign, before village, garden on R, via stable entrance opp Little Tangley.* **Visits by arrangement May to July for groups of 10 to 50. Adm £6, chd free. Home-made teas.**
1 acre enthusiast's atmospheric and tranquil garden divided into rooms with year-round interest. Herbaceous borders, dramatic white garden, specimen trees and shrubs, gravel garden with new water feature, small kitchen garden, fruit cage, two wells, ornamental ponds, herb parterre and millennium parterre garden. Wheelchair access with some avoidable gravel paths.

11 NEW COBBETTS CORNER
Tilford, Farnham, GU10 2AJ. Mr Nicholas Faulkner. *3m SE of Farnham. From Farnham take the B3001 towards Elstead, or go through Tilford. Cobbetts corner is on the corner of Tilford St & Charles Hill (B3001). 2m from Elstead & A3.* **Sat 20 May (2-5). Adm £4, chd free. Light refreshments.**
Cobbetts Corner has a garden for the owner to relax in, away from the busy world of work. It features colourful borders with minimal maintenance, a productive vegetable garden and greenhouse. It is also home for the owner's chickens together with a pond providing an oasis for wildlife. With a wildflower meadow, the garden is continually evolving to encourage eco diversity.

Our 2022 donation to ABF The Soldier's Charity meant 372 soldiers and veterans were helped through horticultural therapy

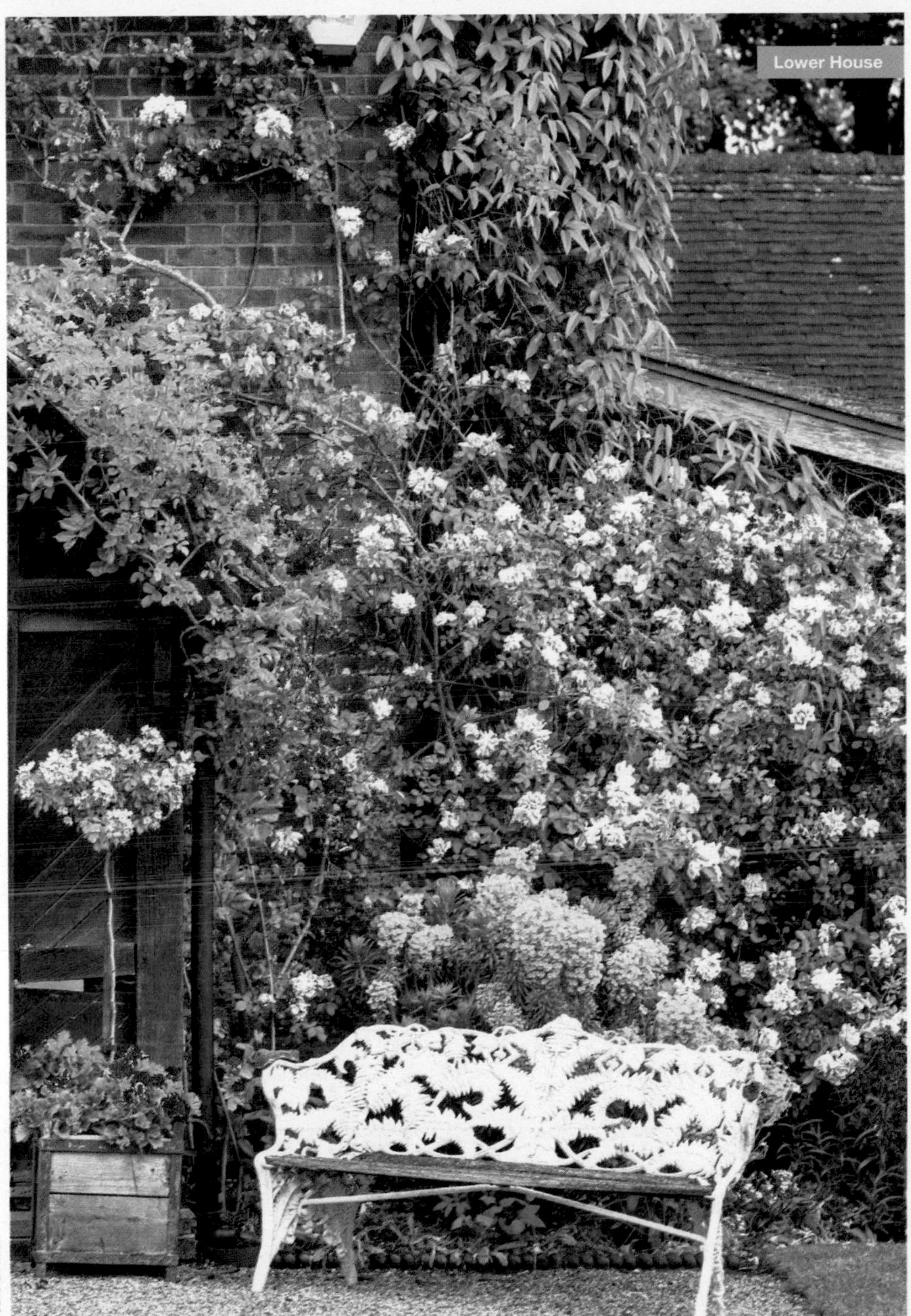

Lower House

© Julie Skelton

12 COLDHARBOUR HOUSE

Coldharbour Lane, Bletchingley, Redhill, RH1 4NA. Mr Tony Elias, 01883 742685, eliastony@hotmail.com. *Coldharbour Ln off Rabies Heath Rd, ½m from A25 at Bletchingley & ⅛m from Tilburstow Hill Rd. Park in field & walk down to house.* **Visits by arrangement Apr to Oct for groups of 10+. Adm £8, chd free. Coffee or tea & a slice of something sweet inc.**

This 1½ acre garden offers breathtaking views to the South Downs. Originally planted in the 1920s, it has since been adapted and enhanced. Several mature trees and shrubs inc a copper beech, a Canadian maple, magnolias, azaleas, rhododendrons, camellias, wisterias, *Berberis* 'Georgei', *Vitex agnus-castus*, fuchsias, hibiscus, potentillas, mahonias, a fig tree and a walnut tree.

13 COVERWOOD LAKES

Peaslake Road, Ewhurst, GU6 7NT. The Metson Family, 01306 731101, farm@coverwoodlakes.co.uk, www.coverwoodlakes.co.uk. *7m SW of Dorking. From A25 follow signs for Peaslake; garden ½m beyond Peaslake on Ewhurst Rd.* **Sun 9, Sun 16 Apr, Mon 1, Sun 14 May, Sun 15 Oct (11-5). Adm £7, chd free. Light refreshments. Visits also by arrangement Apr to Sept for groups of 20+.**

14 acre landscaped garden in stunning position high in the Surrey Hills with four lakes and bog garden. Extensive rhododendrons, azaleas and fine trees. 3½ acre lakeside arboretum. Marked trail through the 180 acre working farm with Hereford cows and calves, sheep and horses, extensive views of the surrounding hills. Light refreshments inc home produced beef and lamb burgers, gourmet coffee and home-made cakes from the Fillet & Bean Cafe (outdoor mobile kitchen). Sorry, dogs are not allowed.

14 ◆ CROSSWATER FARM

Crosswater Lane, Churt, Farnham, GU10 2JN. David & Susanna Millais, 01252 792698, sales@rhododendrons.co.uk, www.rhododendrons.co.uk. *6m S of Farnham, 6m NW of Haslemere. From A287 turn E into Jumps Rd ½m N of Churt village centre. After ¼m turn acute L into Crosswater Ln & follow signs for Millais Nurseries.* **For NGS: Sun 7, Mon 8 May (10-4.30). Adm £6, chd free. Home-made teas. For other opening times and information, please phone, email or visit garden website.**

Idyllic 5 acre woodland garden. Plantsman's collection of

2 Chinthurst Cottage

© Leigh Clapp

rhododendrons and azaleas inc rare species collected in the Himalayas and hybrids raised by the family. Everything from alpine dwarfs to architectural large leaved trees. Ponds, stream and companion plantings inc sorbus, magnolias and Japanese acers. Recent new plantings and wildflower meadows. Woodland garden and specialist rhododendron, azalea and magnolia plant centre. Grass paths may be difficult for wheelchairs after rain.

15 ◆ DUNSBOROUGH PARK

Ripley, GU23 6AL. Baron & Baroness Sweerts de Landas Wyborgh, events@dunsboroughpark.com, www.dunsboroughpark.com. *6m NE of Guildford. Entrance via Newark Ln, Ripley through Tudor-style gatehouses & courtyard up drive. Car park signposted. For SatNav use GU23 6BZ.* **For NGS: Fri 14 Apr (9.30-12 & 1-3.30). Adm £8, chd free. Pre-booking essential, please visit www.dunsboroughpark.com for information & booking. Charity teas for sale. For other opening times and information, please email or visit garden website.**

6 acre garden redesigned by Penelope Hobhouse and Rupert Golby. Garden rooms, lush herbaceous borders, standard wisteria, 70ft ginkgo hedge, potager and 300 yr old mulberry tree. Rose Walk, Italian Garden and Water Garden with folly bridge. April Tulip Festival: Formal borders and colourful informal display in meadow. June: Mid-summer herbaceous and roses. Sept: Autumn colour and dahlia displays. Wheelchair access over gravel paths and grass, cobbled over folly bridge.

GROUP OPENING

16 NEW FAIRMILE COMMON GARDENS

Portsmouth Road, Cobham, KT11 1BG. *2m NE of Cobham. On A307 Esher to Cobham Rd next to free car park by A3 bridge, at entrance to Waterford Close.* **Sun 25 June (2-5). Combined adm £10, chd free. Tea at Fairmile Lea.**

NEW 23 FAIRACRES
Miranda Filkins.

FAIRMILE LEA
Steven Kay.

MOLESHILL HOUSE
Penny Snell.
(See separate entry)

Three contrasting gardens inc many mature specimen trees. One with new water feature surrounding oak room (not open), vegetables and potager. Another large garden with formal pond, sanctuary garden and new features planned. The third, a romantic, naturalistic garden on the wild side, much photographed and published. Original features to inspire. Plants for sale at Moleshill House.

17 THE GARTH PLEASURE GROUNDS

The Garth, Newchapel Road, Lingfield, RH7 6BJ. Mr Sherlock & Mrs Stanley, ab_post@yahoo.com, www.oldworkhouse.webs.com. *In Lingfield, opp Barge Tiles on Newchapel Road. From A22 take B2028 by the Mormon Temple to Lingfield. The Garth is on the L after 1½m, opp Barge Tiles. Parking: Gun Pit Rd in Lingfield & limited space for disabled at Barge Tiles.* **Sun 30 Apr, Sun 7 May (1-5). Adm £8, chd free. Home-made teas. Visits also by arrangement 8 May to 30 July for groups of 5 to 30.**

Mature 9 acre Pleasure Grounds created by Walter Godfrey in 1919, present an idyllic setting surrounding the former parish workhouse refurbished in Edwardian style. The formal gardens, enchanting nuttery, a spinney with many mature trees and a pond attract wildlife. Wonderful bluebells in spring. The woodland gardens and beautiful borders are full of colour and fragrance for year-round pleasure. Many areas of interest, large specimen plants inc 500 yr old oak, many architectural features designed by Walter H Godfrey. Picnics are welcome. Partial wheelchair access in woodland, iris and secret gardens.

18 ◆ GATTON PARK

Reigate, RH2 0TW. Royal Alexandra & Albert School. *3m NE of Reigate. 5 mins from M25 J8 (A217) or from top of Reigate Hill, over M25 then follow sign to Merstham. Entrance off Rocky Ln accessible from Gatton Bottom or A23 Merstham.* **For NGS: Sun 12 Feb (12-5). Adm £7, chd free. Pre-booking essential, please phone 01737 649068, email events@gatton-park.org.uk or visit www.gattonpark.co.uk for information & booking. Light refreshments. For other opening times and information, please phone, email or visit garden website.**

Historic 260 acre estate in the Surrey Hills AONB. Capability Brown parkland with ancient oaks. Discover the Japanese garden, Victorian parterre and breathtaking views over the lake. Seasonal highlights inc displays of snowdrops and aconites in Feb. Ongoing restoration projects by the Gatton Trust. Bird hide open to see herons nesting. Free guided tours. A selection of hot and cold drinks, cakes and snacks. Plants for winter interest for sale. Adm: Adults £6 pre-book tickets only (£7 on the day).

19 NEW GRACE & FLAVOUR CIC

Ripley Lane, West Horsley, Leatherhead, KT24 6JW. Dr Gerard Robbins, www.graceandflavour.org. *At r'about on A246 in West Horsley, turn L (signed West Horsley & Ripley) into The Street. Follow road for ⅓m. Turn L into Ripley Ln. Garden is ⅓m on L, immed after BUPA Nursing Home.* **Sun 2 July (10.30-5). Adm £5, chd free. Home-made teas.**

Grace & Flavour is a community kitchen garden and 28 allotments. In 2009 it took over a derelict 3 acre walled garden and orchard, part of NT's Hatchlands Estate. It features vegetable beds, soft fruit cages, polytunnels, fruit tree avenue with lavender borders, cut flower beds, wildlife area and pond. It is free of chemicals and pesticides, and has a no dig policy. Wheelchair access to main paths, but not over some uneven side paths.

20 HALL GROVE SCHOOL

London Road (A30), Bagshot, GU19 5HZ. Mr & Mrs Graham, www.hallgrove.co.uk. *6m SW of Egham. M3 J3, follow A322 1m until sign for Sunningdale A30, 1m E of Bagshot, opp Longacres Garden Centre, entrance at footbridge. Ample car parking.* **Sat 13 May, Sat 30 Sept (2-5). Adm £5, chd free. Home-made teas.**

Formerly a small Georgian country estate, now a co-educational preparatory school. Grade II listed house (not open). Mature parkland with specimen trees. Historical features inc ice house and a recently restored walled kitchen garden with flower borders, fruit and children's vegetable plots. There is also a lake, woodland walks, rhododendrons, azaleas and acers. Live music at 3pm.

21 NEW HARRY EDWARDS HEALING SANCTUARY

Hook Lane, Burrows Lea, Shere, Guildford, GU5 9QQ. *Close to Shere & Gomshall. From the A25 take the turning to Shere Village & drive through the village. Crossover the railway bridge & turn L into Hook Ln. The Sanctuary is on the R & is clearly signed.* **Sun 14 May (11-5). Adm £5, chd free. Light refreshments.**

A place of serenity and peace nestled within the Surrey Hills, set in 30 acres of beautiful grounds that inc paddocks, bluebell woods, a rose garden with pond and waterfall, a meditation glade and labyrinth. Spend a while sitting on Cherry Tree Walk and enjoy the stunning Surrey Hills views.

22 ◆ HATCHLANDS PARK

East Clandon, Guildford, GU4 7RT. National Trust, 01483 222482, hatchlands@nationaltrust.org.uk, www.nationaltrust.org.uk/hatchlands-park. *4m E of Guildford. Follow brown signs to Hatchlands Park (NT).* **For NGS: Sun 23 Apr (10-5). Adm £11, chd £5.50. Adm subject to change. For other opening times and information, please phone, email or visit garden website.**

Garden and park designed by Repton in 1800. Follow one of the park walks to the stunning bluebell wood in spring (2½ km round walk over rough and sometimes muddy ground). In autumn enjoy the changing colours on the long walk. Partial wheelchair access to parkland with rough and undulating terrain, tracks and cobbled courtyard. Mobility scooter booking essential.

23 HEATHSIDE

10 Links Green Way, Cobham, KT11 2QH. Miss Margaret Arnott & Mr Terry Bartholomew, 01372 842459, m.a.arnott@btinternet.com. *1½m E of Cobham. Through Cobham A245, 4th L after Esso garage into Fairmile Ln. Straight on into Water Ln. Links Green Way 3rd turning on L.* **Visits by arrangement. Morning coffee, afternoon tea, or wine & canapés.**

Terraced, plants persons garden, designed for year-round interest. Gorgeous planting all set off by harmonious landscaping. Many urns and pots give seasonal displays. Several water features add tranquil sound. Stunning colour combinations excite. Dahlias and begonias a favourite. Beautiful Griffin Glasshouse housing the exotic. Many inspirational ideas. Situated 5 miles from RHS Garden Wisley.

24 HIGH CLANDON ESTATE VINEYARD

High Clandon, East Clandon, GU4 7RP. Sibylla & Bruce Tindale, www.highclandon.co.uk. *A3/Wisley junction, L for Ockham/Horsley for 2m to A246. R for Guildford for 2m, then 100yds past landmark Hatchlands NT, turn L into Blakes Ln, straight uphill through gates High Clandon to vineyard entrance. Extensive parking in our woodland area.* **Sun 9 July (11-4). Adm £7.50, chd free. Pre-booking preferred, please visit www.ngs.org.uk. Cream teas & home-made teas. Gold awarded English sparkling wine by the glass & bottle.**

Vistas, views, gardens, 1 acre wildflower meadow with rare butterflies, multi-gold award-winning English sparkling wine vineyard, and set in 12 acres of beautiful Surrey Hills, AONB. Panoramic views to London, water features, Japanese garden, truffière, apiary and sculptures. Twice winner Cellar Door of Year. Sculptures exhibition with over 180 works of art on show in gardens and vineyard. Pond usually has wild Mallard ducks and ducklings at this time of the yr. Access is good; all garden paths are lawns on firm ground on hard chalky substrate.

25 HILL FARM

Logmore Lane, Westcott, Dorking, RH4 3JY. Helen Thomas. *1m W of Dorking. Parking on Westcott Heath just past church. Entry to garden opp.* **Sun 17 Sept (11.30-4.30). Adm £5, chd free. Home-made teas.**

1¾ acre set in the magnificent Surrey Hills landscape. The garden has a wealth of different natural habitats to encourage wildlife, and planting areas which come alive through the different seasons. Features inc a wildlife pond, woodland walk, a tapestry of heathers, glorious late summer grasses and perennials, wildflower meadow, and a vegetable and cut flower garden. Greenhouse, and lime kiln in wildings area. A garden to be enjoyed by all. Everyone welcome. Sloping garden; most areas are accessible for wheelchair users. Paths are grass so care needed if very wet.

26 NEW 11 HILLARY ROAD

Farnham, GU9 8QY. Mrs Lenka Cooke. *1½ m from Farnham town centre. Please do not park in Hillary Rd, it is a narrow cul-de-sac. Please park on Brambleton Ave. The garden is within a short walking distance.* **Sun 10 Sept (11-3.30). Adm £5, chd free. Home-made teas.**

A ½ acre English style garden with large colourful herbaceous borders. Small pond with stream by a patio with container grown canna, brugmansia and citrus fruit. Garden divided by a rose arch leading to vegetable garden, fruit cage and greenhouse. Narrow access gate suitable for compact wheelchairs only. Uneven lawn and ramp to upper part of garden.

27 2 KNOTT PARK HOUSE
Wrens Hill, Oxshott, Leatherhead, KT22 0HW. Joanna Nixon. *10 mins from Oxshott village centre. From A244, which is off the A3, or S from Leatherhead take Wrens Hill, next to Bear Pub, continue c200 metres, take R fork, part of 1st big house on L.*
Fri 19, Sat 20 May (10-2). Adm £5, chd free.
A south facing terrace with far-reaching views is planted with herbs and lavender. Steps lead down to a lower area extending to ¼ acre. Wildlife friendly and maintained without pesticides. On four levels, it features pollinator loving plants, trees and shrubs. These inc irises, alliums, sedum, geums, acers, flowering currant and succulents. Scented flowering shrubs and climbers attract bees. Beehive.

Our 2022 donation to Hospice UK meant that 911 NHS and Social Care staff accessed individual counselling support

Moleshill House, Fairmile Common Gardens

© Leigh Clapp

28 KNOWLE GRANGE
Hound House Road, Shere, Guildford, GU5 9JH. Mr P R & Mrs M E Wood. *8m S of Guildford. From Shere (off A25), through village for ¾ m. After railway bridge, continue 1½ m past Hound House on R (stone dogs on gateposts). After 100yds turn R at Knowle Grange sign, go to end of lane.* **Sun 21 May (11-4.30). Adm £6, chd free. Light refreshments on front lawn.**
80 acres undulating landscape. 7 acre gardens. New features recently added. Small knot garden. Double herbaceous border. About seven various garden rooms with French, Japanese and English inspiration. A new clock tower garden. The one mile bluebell valley unicursal path which snakes through two valleys and a hill and inc the labyrinth upon a labyrinth.

29 LEIGH PLACE
Leigh Place Lane, Godstone, RH9 8BN. Mike & Liz McGhee. *Take B2236 Eastbourne Rd from Godstone village. Turn 2nd L onto Church Ln. Follow parking directions.* **Sat 20, Sun 21 May (10-4). Adm £5, chd free. Home-made teas.**
Leigh Place garden has 25 acres on greensand. Part of the Godstone Pond's with SSSI and lakeside paths, walled garden inc cutting garden, orchard and vegetable quadrant with greenhouses. Orchard, beehives and large rock garden. WC available. The walled garden has Breedon gravel paths suitable for wheelchairs and pushchairs.

30 LITTLE ORCHARDS
Prince Of Wales Road, Outwood, Redhill, RH1 5QU. Nic Howard, 01883 744020, hello@weloveplants.co. *A few hundred metres N of the Dog & Duck Pub.* **Fri 21 Apr, Fri 26, Sun 28 May, Wed 6, Sun 10 Sept (12-4). Adm £5, chd free. Home-made teas. Visits also by arrangement Apr to Sept for groups of 20+. Guided tour with designer & on hand garden advice.**
A magical, artistic, plantsman's paradise planted for year-round interest using a tapestry of foliage as well as flower interest. The garden is arranged as a series of connected areas that flow between two properties, the old gardener's cottage, and The Old Stables. Often compared to a mini Petersham Nursery by our visitors; we have a gift shop selling garden paraphernalia and a lovely café area.

31 ◆ LOSELEY PARK
Guildford, GU3 1HS. Mr & Mrs A G More-Molyneux, 01483 304440/405112, pa@loseleypark.co.uk, www.loseleypark.co.uk. *4m SW of Guildford. For SatNav please use GU3 1HS, Stakescorner Ln.* **For NGS: Sun 11 June (11.30-4.30). Adm by donation. Light refreshments. For other opening times and information, please phone, email or visit garden website.**
Delightful 2½ acre walled garden. Award-winning rose garden (over 1000 bushes, mainly old fashioned varieties), extensive herb garden, fruit and flower garden, white garden with fountain and spectacular organic vegetable garden. Magnificent vine walk, herbaceous borders, moat walk, ancient wisteria and mulberry trees.

32 LOWER HOUSE
Bowlhead Green, Godalming, GU8 6NW. Georgina Harvey. *1m from A3 Thursley/Bowlhead Green junction. Using A286 leave at Brook (6m from Haslemere & 3m from Milford). Follow Bowlhead Green & NGS signs for approx 2m. Good parking.* **Sat 13, Sun 14 May, Sat 17, Sun 18 June (1-5). Adm £6, chd free. Home-made teas.**
Established trees and shrubs shape the layout of the garden creating stunning vistas and quiet areas. Ericaceous azaleas, rhododendrons, camellias and magnolias are followed by roses to continue the colour until the herbaceous plants bloom. The white topiary garden with fish pond brings calm and the kitchen garden, greenhouses, orchard and fruit cage bring seasonal changes. Alternative routes to avoid steps for wheelchair users and some narrow paths.

33 MACMILLAN CANCER SUPPORT CENTRE
East Surrey Hospital, Canada Avenue, Redhill, RH1 5RH. East Surrey Macmillan Cancer Support Centre. *Follow signs for East Surrey Hospital. Drive past main hospital entrance, the Centre is opp E Entrance. Follow parking signs.* **Sun 25 June (11-4). Adm £3, chd free. Home-made teas. Open nearby 1 & 3 Marsh Cottages (15 mins away).**
The new Macmillan Cancer Support Centre offers holistic support to people affected by cancer. The two small gardens surrounding the state of art Centre provide a peaceful haven to visitors. The thoughtful planting ensures there is interest in the garden throughout the yr and offers a therapeutic retreat away from the hospital environment.

34 THE MANOR HOUSE
Three Gates Lane, Haslemere, GU27 2ES. Mr & Mrs Gerard Ralfe. *NE of Haslemere. From Haslemere centre take A286 towards Milford. Turn R after Museum into Three Gates Ln. At T-Junction turn R into Holdfast Ln. Car park on R.* **Sun 14 May (10-5). Adm £5, chd free. Home-made teas.**
Described by Country Life as 'The Hanging Gardens of Haslemere', the well established Manor House gardens are tucked away in a valley of the Surrey Hills. Set in 6 acres, it was one of Surrey's inaugural NGS gardens with fine views, an impressive show of azaleas, wisteria, beautiful trees underplanted with bulbs, enchanting water gardens and a magnificent rose garden.

35 NEW 1 MARSH COTTAGES
Chilmead Lane, Nutfield Marsh, Redhill, RH1 4ES. Debbie Koenraads, 07946 577151, debbie.koenraads@btconnect.com. *2½ m E of Redhill. Access via Chilmead Ln, off Cormongers Ln, Nutfield.* **Sun 25 June (11-5). Combined adm with 3 Marsh Cottages £5, chd free. Home-made teas. Open nearby Macmillan Cancer Support Centre. Visits also by arrangement June to Sept for groups of 5 to 10. Weekdays only.**
Established cottage garden, colour themed borders with year-round interest. Sunken patio, raised vegetable beds, greenhouse, potting shed and garden room attached to the house for tender plants. Views to the North Downs. Partial wheelchair access due to narrow paths.

36 NEW 3 MARSH COTTAGES

Chilmead Lane, Nutfield Marsh, Redhill, RH1 4ES. Mandy Winder, 07549 663291, mandyonthemarsh@gmail.com. *2½m E of Redhill. Access via Chilmead Ln, off Cormongers Ln, Nutfield.* **Sun 25 June (11-5). Combined adm with 1 Marsh Cottages £5, chd free. Home-made teas. Open nearby Macmillan Cancer Support Centre. Visits also by arrangement June to Sept for groups of 5 to 10. Weekdays only.**

Cottage style garden with mainly herbaceous perennials and brick paths leading to a circular lawn. The garden is arranged in rooms leading to a small cutting and vegetable garden and a shade garden to the rear of the property. There is an elevated deck with views over the Chaldon Hills and surrounding fields.

37 MOLESHILL HOUSE

The Fairmile, Cobham, KT11 1BG. Penny Snell, pennysnellflowers@btinternet.com, www.pennysnellflowers.co.uk. *2m NE of Cobham. On A307 Esher to Cobham Rd next to free car park by A3 bridge, at entrance to Waterford Close.* **Mon 10 Apr (1-4.30); Sun 3 Sept (2-5). Adm £5, chd free. Light refreshments. Open nearby Randalls Allotments on 3 Sept only. Opening with Fairmile Common Gardens on Sun 25 June. Visits also by arrangement 24 Apr to 31 Aug for groups of 10 to 50.**

Romantic, naturalistic garden on the wild side. Short woodland path leads from dovecote to beehives. Informal planting contrasts with formal topiary, box and garlanded cisterns. Colourful courtyard, conservatory, pond with fountain, pleached avenue, circular gravel garden replacing most of the lawn. Gipsy caravan garden, green wall and stumpery. Espaliered crab apples. Chickens and bees. Garden 5 mins from Claremont Landscape Garden, Painshill Park and Wisley, also adjacent to excellent dog walking woods.

38 MONKS LANTERN

Ruxbury Road, Chertsey, KT16 9NH. Mr & Mrs J Granell, 01932 569578, janicegranell@hotmail.com. *1m NW from Chertsey. M25 J11, signed A320/Woking. At r'about take 2nd exit A320/Staines, straight over next r'about. L onto Holloway Hill, R Hardwick Ln. ½m, R over motorway bridge, on Almners then Ruxbury Rd.* **Sat 27 May (2-5.30). Adm £6, chd free. Light refreshments. Visits also by arrangement June to Sept for groups of up to 20.**

A delightful garden with borders arranged with colour in mind; silvers and white, olive trees blend together with nicotiana and senecio. A weeping silver birch leads to the oranges and yellows of a tropical bed with large bottle brush, hardy palms and *Fatsia japonica*. Large rockery and an informal pond. There is a display of hostas, *Cytisus battandieri* and a selection of grasses in an island bed. Aviary with small finches. Workshop with handmade guitars, and paintings. Pond with ornamental ducks and fish. Music and wine. Wheelchairs welcome; two reserved parking spaces at entrance to garden, as gravel drive is not easy.

39 THE NUTRITION GARDEN

156A Frimley Green Road, Frimley Green, Camberley, GU16 6NA. Dr Trevor George RNutr, 07914 911410, t-george@hotmail.co.uk. *2m (5 mins) from J4 of the M3. From M3, follow signs for A331 towards Farnborough, then follow signs to Frimley Green (B3411). The house is down a long drive with telegraph poles at each end, almost opp the recreation ground.* **Sun 25 June, Sun 3 Sept (1-5). Adm £5, chd free. Pre-booking essential, please visit www.ngs.org.uk for information & booking. Light refreshments. Visits also by arrangement July & Aug for groups of up to 20. Guided tour of the garden & explanation of the plants can be inc.**

A garden designed by a registered nutritionist to produce and display a wide variety of edible plants inc fruits, vegetables, herbs and plants for infusions. There are trees, shrubs, tubers, perennials and annual plants. Over 100 types of edible plants and over 200 varieties are grown throughout the year. These inc unusual food plants, heritage varieties and unusual coloured varieties. Tea, coffee and light refreshments, infusions from plants growing in the garden also available. Wheelchair access on paved paths around fruit and vegetable beds. Other areas are step free around uneven grass lawn.

40 OAKLANDS

Eastbourne Road, Blindley Heath, Lingfield, RH7 6LG. Joy & Justin Greasley, 01342 837369, joy@greasley.me.uk. *S of Blindley Heath Village. From M25 take A22 to East Grinstead. Go straight on at Blindley Heath T-lights & after 800yds take sign on L to Nestledown Boarding Kennels. Oaklands is down the lane on the L. Limited parking.* **Visits by arrangement June to Aug for groups of 8 to 30. Adm £5, chd free. Home-made teas.**

A middle sized garden surrounded by natural woodland and featured in Surrey Life. Specimen trees, area of large planted pots, pond, folly, colourful herbaceous borders with many areas of interest and plenty of seating areas. Long curved pergola complemented by climbing plants and hanging baskets. Level access on paths to main features.

41 THE OAST HOUSE

Station Road, Lingfield, RH7 6EF. Mrs Andrea Watson. *In Station Rd look for sign to New Place Farm on the L. Park in field opp the drive entrance & walk 100 metres down the lane to The Oast House.* **Sun 11 June (11-5). Adm £6, chd free. Home-made teas.**

Traditional country garden with large croquet lawn, borders, old brick walls, courtyard garden, beds, rambling roses, vines, topiary and mature trees inc 35 yr old ginkgo. Approx 1½ acres. Views of Grade II* manor house and Grade I church. The Oast House itself is a building of local interest. Level wheelchair access with gravel paths.

In 2022 the National Garden Scheme donated £3.11 million to our nursing and health beneficiaries

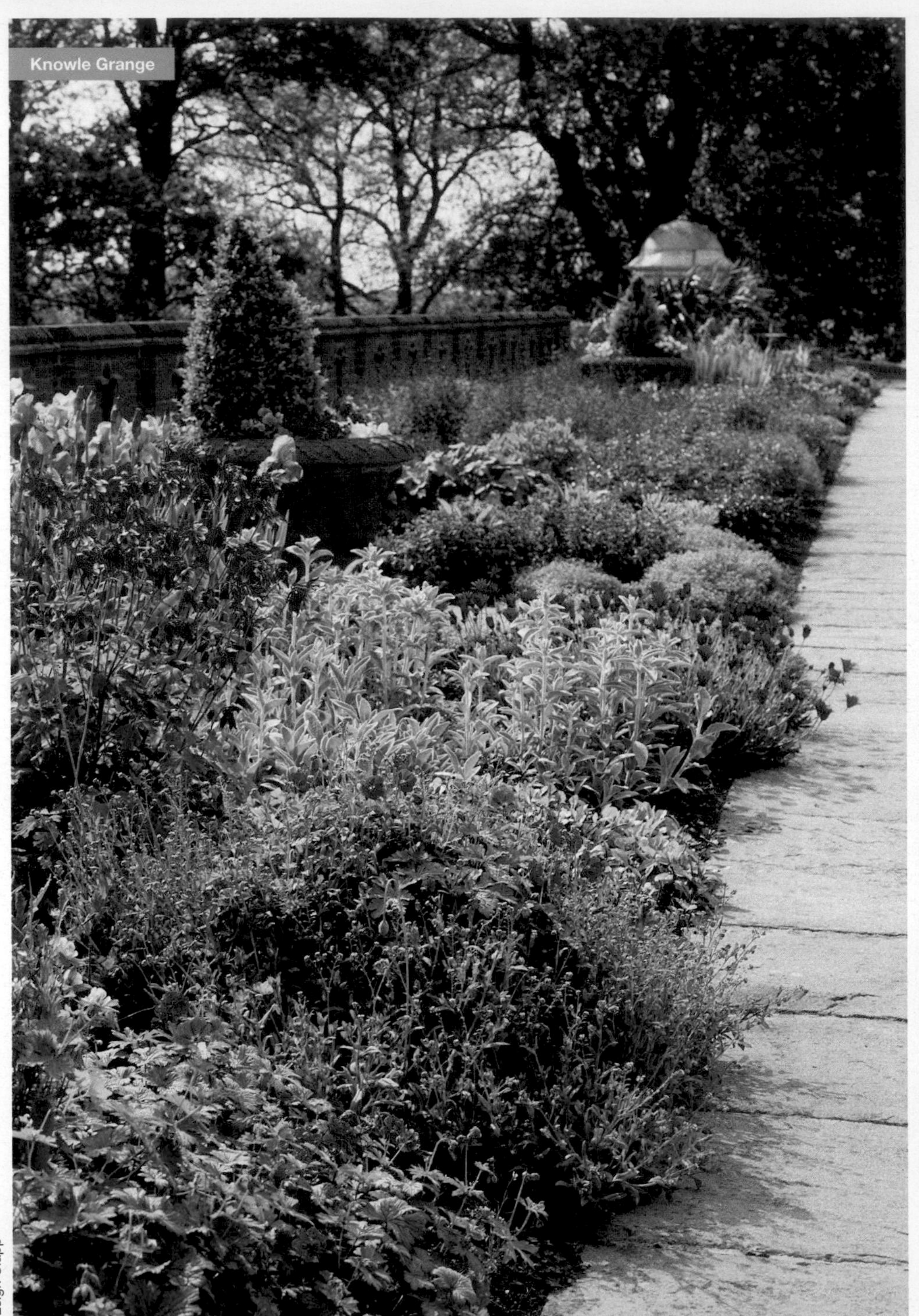
Knowle Grange

42 THE OLD RECTORY

Sandy Lane, Brewer Street, Bletchingley, RH1 4QW. Mr & Mrs A Procter. *Top of village nr The Red Lion pub, turn R into Little Common Ln, then R at Cross Rd into Sandy Ln. Parking nr house, disabled parking in courtyard.* **Sun 11 June (11-4). Adm £5, chd free. Home-made teas.**

Georgian Manor House (not open). Quintessential Italianate topiary garden, statuary, box parterres, courtyard with columns, water features, antique terracotta pots. New sunken water garden. Much of the 4 acre garden is the subject of ongoing reclamation. This inc the ancient moat, woodland with fine specimen trees, one of the largest tulip trees in the country. New water garden and tropical garden. Wheelchair access with gravel paths.

43 NEW THE OLD VICARAGE

The Street, Frensham, Farnham, GU10 3DU. Kate Smith, www.instagram.com/katesmithgardendesign. *3m S of Farnham, 4m N of Hindhead. Turn off A287 Farnham to Hindhead road at small village green at St Mary's School. Travel along The Street for ½m, the Old Vicarage is on the RHS, next to St Mary's Church.* **Sat 10 June (10.30-2.30); Sun 11 June (12.30-6). Adm £5, chd free. Light refreshments.**

12 acres of garden surrounding the Old Vicarage in Frensham. The garden consists of 2 acres of herbaceous borders and lawns adjacent to the house. Steep slopes lead to the less formal grounds with large pond and the River Wey. Mown paths cut through the water meadows.

44 PRATSHAM GRANGE

Tanhurst Lane, Holmbury St Mary, RH5 6LZ. Alan & Felicity Comber. *12m SE of Guildford, 8m SW of Dorking. From A25 take B2126, after 4m turn L into Tanhurst Ln. From A29 take B2126, before Forest Green turn R on B2126 then 1st R to Tanhurst Ln.* **Sun 20 Aug (12.30-5). Adm £6, chd free. Home-made teas.**

5 acre garden overlooked by Holmbury Hill and Leith Hill. Features inc two ponds joined by cascading stream, extensive scented rose and blue hydrangea beds. Also, herbaceous borders, cutting flower garden and two white beds. Partial wheelchair access with some steps, steep slopes (slippery when wet) and gravel paths. Deep ponds and a drop from terrace.

45 ◆ RAMSTER

Chiddingfold, GU8 4SN. Mrs R Glaister, 01428 654167, office@ramsterhall.com, www.ramsterevents.com. *12m S of Guildford. Ramster is on A283 1½m S of Chiddingfold, large iron gates on R, the entrance is signed from the main road.* **For NGS: Fri 12 May (10-5). Adm £9, chd £4. Light refreshments. For other opening times and information, please phone, email or visit garden website.**

A stunning, mature woodland garden set in over 20 acres, famous for its rhododendron and azalea collection and its carpets of bluebells in spring. Enjoy a peaceful wander down the grass paths and woodland walk, explore the bog garden with its stepping stones, or relax in the tranquil enclosed tennis court garden. The Tea House, by the entrance to the garden is open every day while the garden is open, serving sandwiches, home-made soup, cream teas, cakes and drinks. Sculpture Exhibition runs Friday 28th April until Monday 29th May. Wheelchair access to the Tea House and some paths in the garden.

Our 2022 donation supports 1,700 Queen's Nurses working in the community in England, Wales, Northern Ireland, the Channel Islands and the Isle of Man

46 RANDALLS ALLOTMENTS

Old Nursery Lane, Cobham, KT11 1JN. Cobham Garden Club, www.cobhamgardenclub.com. *3m W of Esher. Alongside Premier Garage (KT11 1JN) on A307 is Old Nurseries Ln next to common. Halfway down on RHS is a gate. Additional parking close to Moleshill House (KT11 1BG).* **Sun 3 Sept (11-3). Adm £4, chd free. Home-made teas. Open nearby Moleshill House.**

Well presented allotments in Cobham, growing unusual vegetables, fruit and salad items from different countries. Guided tours are available by the knowledgeable owners of the plots. If you have ever thought about 'Growing Your Own' this will inspire you. The Greenfingers Shop will be open offering plants, garden sundries, and composts for sale. Local honey will also be for sale. Wheelchair access: some uneven areas will need careful attention.

47 26 ROWDEN ROAD

Epsom, KT19 9PN. Mr Robert Stacewicz, robstacewicz@hotmail.com. *At Ruxley Ln take the turning at the Co-op on to Cox Ln. At the end of road, take a R on to Rowden Rd. The garden is on the R.* **Sat 26 Aug (12-5); Sun 27 Aug (12-4). Adm £8. Pre-booking essential, please visit www.ngs.org.uk for information & booking. Home-made teas. Visits also by arrangement 17 June to 8 Oct for groups of up to 15.**

A small 19 x 8 metre south facing garden, transformed from a plain lawn in spring 2020 to a plant lovers paradise. The owner is a garden designer, and the space is carefully planted with hardy exotics. Inspiration has come from world travels, in particular Southeast Asia and Indonesia. A large circular pond dominates the main garden, while raised beds home a succulent collection. Cactus and succulent border, wildlife pond, tropical water lilies, hardy banana plants, palm collection, planted succulent and alpine house, hardy bromeliads, exotic aquatic plants, small garden inspiration, exotic garden, unusual ferns. 15 min drive from Wisley. Sorry, no children due to deep water.

48 ◆ THE SCULPTURE PARK

Tilford Road, Churt, Farnham, GU10 2LB. Eddie Powell, 01428 605453, sian@thesculpturepark.com, www.thesculpturepark.com. *Corner of Jumps & Tilford Rd, Churt. Directly opp Bell & The Dragon pub. Use our car park or park in the pub, where refreshments are available.* **For NGS: Tue 7 Feb, Tue 6 June (10-5). Adm £10, chd £5. Pre-booking essential, please visit www.ngs.org.uk for information & booking. Senior rate for 65+ £5, please book chd ticket. For other opening times and information, please phone, email or visit garden website.**

This garden sculpture exhibition is set within an enchanting arboretum and wildlife inhabited water garden. You should set aside between 2-4 hrs for your visit as there are 2 miles of trail within 10 acres. Our displays evolve and diversify as the seasons pass with vivid and lush colours of the rhododendrons in May and June, to the enchanting frost in the depths of winter. 10 acres of arboretum and water gardens, hundreds of outdoor sculptures. Immersive experience - complete darkness! Approx one third of The Sculpture Park is accessible to wheelchairs. Disabled WC.

49 SHAMLEY WOOD ESTATE

Woodhill Lane, Shamley Green, Guildford, GU5 0SP. Mrs Claire Merriman, 07595 693132, claire@merriman.co.uk. *5m (15 mins) S of Guildford in village of Shamley Green. Entrance is approx ¼m up Woodhill Ln from centre of Shamley Green.* **Visits by arrangement 1 Mar to 24 Nov. Adm £8, chd free. Teas with gluten free options.**

A relative newcomer, this garden is worth visiting just for the setting! Sitting high on the North Downs, the garden enjoys beautiful views of the South Downs and is approached through a 10 acre deer park. Set within approx 3 acres, there is a large pond and established rose garden. More recent additions inc fire pits, vegetable patch, stream, tropical pergola and terraced wildflower lawn. Wheelchair access to most of garden. Step to access ground level WC.

50 41 SHELVERS WAY

Tadworth, KT20 5QJ. Keith & Elizabeth Lewis, 01737 210707, kandelewis@ntlworld.com. *6m S of Sutton off A217. 1st turning on R after Burgh Heath T-lights heading S on A217. 400yds down Shelvers Way on L.* **Sun 23 Apr, Sun 30 July (2-5.30). Adm £5, chd free. Home-made teas. Visits also by arrangement 1 Apr to 1 Sept for groups of 10+.**

In spring a myriad of small bulbs, specialist daffodils and an assortment of many pots of tulips. Choice perennials follow together with annuals to ensure colour well into Sept. A garden for all seasons.

51 SHIELING

The Warren, Kingswood, Tadworth, KT20 6PQ. Drs Sarah & Robin Wilson, 01737 833370, sarahwilson@doctors.org.uk. *Kingswood Warren Estate. Off A217, gated entrance just before church on S-bound side of dual carriageway after Tadworth r'about. ¾m walk from station. Parking on The Warren or by church on A217.* **Sun 19 Feb, Mon 10 Apr, Mon 1 May (11-3). Adm £5, chd free. Home-made teas. Visits also by arrangement May to July for groups of 10 to 25.**

1 acre garden restored to its original 1920s design. Formal front garden with island beds and shrub borders. Unusual large rock garden and mixed borders with collection of beautiful slug free hostas and uncommon woodland perennials. The rest is an interesting woodland garden with acid loving plants, a new shrub border and a stumpery. Plant list provided for visitors. Lots for children to do. Some narrow paths in back garden. Otherwise, resin drive, grass and paths easy for wheelchairs.

52 SPRINGWOOD

Farnham Road, Tilford, Farnham, GU10 2AU. Ali O'Connell, 07833 191460, allyoconnell@hotmail.com. *3m SE of Farnham. From Farnham take B3001 towards Elstead. After 3m look out for a small red post box on the R. Turn R down unmade drive. Parking in field adjacent to Springwood.* **Visits by arrangement in June for groups of 5 to 10. Light refreshments.**

A constantly evolving former Victorian market garden, now a family home. Beautiful views over the surrounding countryside and Hankley Common from the 1 acre walled garden, planted in a cottage garden style, with mixed borders of roses, lavenders, alliums and perennial flowers. Cutting flower patch, fruit and nut trees, established apple trees along with a new raised bed and vegetable plot. Features inc an original south facing, 30 metre Victorian lean-to greenhouse housing a 60 yr old grapevine and a heavily fruiting apricot tree along with another north facing 30 metre lean-to recently refurbished with beautiful original features.

53 SPURFOLD

Radnor Road, Peaslake, Guildford, GU5 9SZ. Mr & Mrs A Barnes, 01306 730196, spurfold@btinternet.com. *8m SE of Guildford. A25 to Shere then through to Peaslake. Pass village stores & L up Radnor Rd.* **Visits by arrangement May & June for groups of 15 to 35. Adm £6, chd free. Home-made teas or evening wine & nibbles.**

2½ acres, large herbaceous and shrub borders, formal pond with Cambodian Buddha head, sunken gravel garden with topiary box and water feature, terraces, beautiful lawns, mature rhododendrons, azaleas, woodland paths, and gazebos. Garden contains a collection of Indian elephants and other objets d'art. Topiary garden and formal lawn area.

54 TANHOUSE FARM

Rusper Road, Newdigate, RH5 5BX. Mrs N Fries, 01306 631334. *8m S of Dorking. On A24 turn L at r'about at Beare Green. R at T-junction in Newdigate, 1st farm on R approx ⅔m. Signed Tanhouse Farm Shop.* **Visits by arrangement 17 June to 26 Aug for groups of 8 to 50. Adm £5, chd free.**

Country garden created by owners since 1987. 1 acre of charming rambling gardens surrounding a C16 house (not open). Herbaceous borders, small lake and stream with ducks and geese. Orchard with wild garden with plentiful seats and benches to stop for contemplation.

55 THE THERAPY GARDEN
Manor Fruit Farm, Glaziers Lane, Normandy, Guildford, GU3 2DT. The Centre Manager, www.thetherapygarden.org. *SW of Guildford. Take A323 travelling from Guildford towards Aldershot, turn L into Glaziers Ln in centre of Normandy village, opp War Memorial. The Therapy Garden is 200yds on L.* **Sun 21 May, Sun 17 Sept (10-4). Adm £5, chd free. Light refreshments.**
The Therapy Garden is a horticulture and education charity that uses gardening to have a positive and significant impact on the lives of people facing challenges in life. In our beautiful and tranquil 2 acre garden we work to change lives for the better and we do this by creating a safe place to enjoy the power of gardening and to connect with nature. We are a working garden full of innovation with an on site shop selling plants and produce. BBQ, salads, sandwiches, teas, coffees and cakes are available with a selection of fun family garden related activities to take part in. Wheelchair access over paved pathways throughout most of garden, many with substantial handrails.

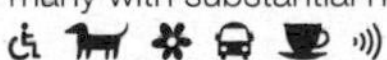

56 TIMBER HILL
Chertsey Road, Chobham, GU24 8JF. Nick & Lavinia Sealy, www.timberhillgarden.com. *4m N of Woking between Ottershaw & Chobham. 2½m E of Chobham & ⅓m E of Fairoaks airport on A319 (N side). 1¼m W of Ottershaw, J11 M25. (If approaching from Ottershaw the A319 is the Chobham Rd). See garden website for more detail.* **Wed 4, Wed 11 Jan (11.30-2.30). Every Mon 30 Jan to 20 Feb (11.30-2.30). Mon 6 Mar (11.30-2.30); Mon 17 Apr, Sun 28 May, Sun 18 June (2-5); Wed 11, Wed 25 Oct, Wed 15 Nov (11.30-2.30). Adm £7, chd £1. Pre-booking essential, please visit www.ngs.org.uk for information & booking. Self-service refreshments (Jan, Feb, Mar & Nov). Cream tea & home-made teas (Apr, May & June). 2024: Wed 3, Wed 10, Wed 17, Wed 31 Jan, Wed 7, Wed 14, Wed 21 Feb.**
16 acres of garden, park and woodland with views to North Downs. Enjoy winter walks, then a sea of snowdrops, crocuses and spring bulbs, followed by spectacular camellias. A new nature and wildlife trail in the making. Ten trees planted for the 'Queen's Green Canopy'. For further information, pop-ups, events, and opportunities see Timber Hill website or please telephone 01932 873875.

57 ◆ TITSEY PLACE GARDENS
Titsey, Oxted, RH8 0SA. The Trustees of the Titsey Foundation, 07889 052461, office@titsey.org, www.titsey.org. *3m N of Oxted. A25 between Oxted & Westerham. Follow brown heritage signs to Titsey Estate from A25 at Limpsfield or see website for directions. Please enter via Water Ln & the Pitchfont car park.* **For NGS: Sun 28 May, Sun 25 June, Sun 16 July, Sun 13 Aug (1-5). Adm £7.50, chd £2. Light refreshments. For other opening times and information, please phone, email or visit garden website.**
One of the largest surviving historic estates in Surrey. Magnificent ancestral home and gardens of the Gresham family since 1534. Walled kitchen garden restored early 1990s. Golden Jubilee rose garden. Etruscan summerhouse adjoining picturesque lakes and fountains. 15 acres of formal and informal gardens in an idyllic setting. Highly Commended in the 2019 Horticulture Week Custodian Awards. Tearoom serving delicious home-made cakes and selling local produce from 12.30-5pm on open days. Walks through the estate woodland are open year-round. Pedigree herd of Sussex Cattle roam the park. Last admissions to gardens at 4pm. Dogs on leads allowed in picnic area, car park and woodland walks only. Disabled car park alongside tea room.

9,500 patients and their families are being supported across four Horatio's gardens thanks to our annual donations

58 ◆ VANN
Hambledon, Godalming, GU8 4EF. Caroe Family, 01428 683413, vann@caroe.com, www.vanngarden.co.uk. *6m S of Godalming. A283 to Lane End, Hambledon. On NGS days only, follow yellow Vann signs for 2m. Please park in the field as signed, not in road. At other times follow the website instructions.* **For NGS: Daily Mon 27 Mar to Sun 2 Apr, Mon 1 May to Sun 7 May (10-4). Adm £8, chd free. For other opening times and information, please phone, email or visit garden website.**
5 acre, 2* English Heritage registered garden surrounding house dating back to 1542 with Arts and Crafts additions by W D Caröe inc a Bargate stone pergola. At the front, brick paved original cottage garden; to the rear a lake, yew walk with rill and Gertrude Jekyll water garden. Snowdrops and hellebores, spring bulbs, spectacular fritillaria in Feb/March. Island beds, crinkle crankle wall, orchard with wild flowers. Vegetable garden. Also open for by arrangement visits for individuals or groups. Cash only on the day. Some paths not suitable for wheelchairs due to uneven stone walkways. Please ring prior to visit to request disabled parking.

59 11 WEST HILL
Sanderstead, CR2 0SB. Rachel & Edward Parsons. *M25 J6, A22, 2.9m r'about 4th exit to Succombs Hill, R to Westhall Rd, at r'about 2nd exit to Limpsfield Rd, r'about 2nd exit on Sanderstead Hill 0.9m, sharp R to West Hill. Please park on West Hill.* **Sat 15, Sun 16 Apr (2-5). Adm £5, chd free. Home-made teas. Donation to British Hen Welfare Trust & Croydon Animal Samaritans.**
A hidden gem tucked away. A beautiful country cottage style garden set in ½ acre, designed by Sam Aldridge of Eden Restored. The garden flows through pathways, lawn, vegetable and play areas. Flower beds showcase outstanding tulips, informal seating areas throughout the garden allows you to absorb the wonderful garden, whilst observing our rescued chickens and rabbits!

60 NEW WEST HORSLEY PLACE

Epsom Road, West Horsley, Leatherhead, KT24 6AN. West Horsley Place Trust, www.westhorsleyplace.org. *5m E of Guildford. West Horsley Place is off the A246 between Guildford & Leatherhead. A 10 min drive from the A3/M25 intersection. Leave the A3 at J10.* **Sun 7 May, Sun 24 Sept (10-5). Adm £10, chd free. Light refreshments in the Place Farm Barn courtyard.**

West Horsley Place is set within a 380 acre estate. The garden adjacent to the Manor House dates back to the C15 and is approx 5 acres, completely surrounded by a wall over 300 yrs old. It has an ancient orchard, rose garden, interestingly striped formal lawns, an historic box hedge, many herbaceous borders, a magnificent white wisteria over 60ft tall and many wildflower areas. Accessible via a step-free route. Unpaved and the predominant surface is turf. Accessible ground floor WC suitable for wheelchair users.

61 WESTWAYS FARM

Gracious Pond Road, Chobham, GU24 8HH. Paul & Nicky Biddle, 01276 856163, nicolabiddle@rocketmail.com. *4m N of Woking. From Chobham Church proceed over r'about towards Sunningdale, 1st Xrds R into Red Lion Rd to junction with Mincing Ln.* **Sun 14 May (10.30-5). Adm £6, chd free. Home-made teas. Visits also by arrangement 24 Apr to 9 June for groups of 10 to 50. Eve visits on request.**

6 acre garden surrounded by woodlands planted in 1930s with mature and some rare rhododendrons, azaleas, camellias and magnolias, underplanted with bluebells, lilies and dogwood. Extensive lawns and sunken pond garden. Working stables and sand school. Lovely Queen Anne House (not open) covered with listed *Magnolia grandiflora*. Victorian design glasshouse. New planting round garden room.

62 THE WHITE HOUSE

21 West End Lane, West End, Esher, KT10 8LB. Lady Peteranne Hunt & Mr David John, 07850 367600, peajaya@btinternet.com. *From Esher to Cobham on the Portsmouth Rd take the 1st turning on the R into Hawskhill Way. At T-junction turn R & we are 5th house on R.* **Sat 20, Sun 21 May, Sat 17, Sun 18 June, Sat 22, Sun 23 July (11-5). Adm £4.50, chd free. Teas & cakes. Visits also by arrangement for groups of 5 to 15. Day or eve visits Mon 22-Wed 31 May, Mon 19-Fri 30 June, Mon 24-Mon 31 July.**

Although our garden is compact and on several levels it is dominated by a beautiful very old oak tree. The garden is full of colour with interesting plants and grasses. Modern water feature and a few sculptures add interest and further enhance the garden. There is a perfect peaceful spot to sit in the evening for a drink in the late sun. Patio access only for wheelchairs.

63 WILDWOOD

34 The Hatches, Frimley Green, Camberley, GU16 6HE. Annie Keighley. *3m S of Camberley. M3 J4 follow A325 to Frimley Centre, towards Frimley Green for 1m. Turn R by the green, R into The Hatches for on street parking.* **Sun 25 June (1-5). Adm £4.50, chd free. Pre-booking essential, please visit www.ngs.org.uk for information & booking. Home-made teas. Open nearby The Nutrition Garden.**

Groups love the hidden surprises in this romantic cottage garden with tumbling roses, topiary and scented *Magnolia grandiflora*. Enjoy discovering a sensory haven of sun and shade with wildlife pond, hidden dell, fernery and loggia. Secret cutting garden with raised beds, vegetables, fruit trees and potting shed patio.

64 NEW WOODPECKERS

Poplar Grove, Woking, GU22 7SD. Mr Janis Raubiška, 07478 025188, janis@grandiflorus.co.uk. *Next to Woking Leisure Centre. Please use Woking Leisure Centre car park (GU22 9BA) & follow signs for garden entrance. Coach parking available.* **Sun 23 July, Sun 20 Aug (12-5). Adm £5, chd free. Cream teas. Visits also by arrangement 1 July to 17 Sept for groups of 10 to 30.**

A journey through a horticultural designer's own garden where plants take centre stage in succession, providing year-round colour and interest. Three rooms, each with a different planting style and purpose; vibrant colours in the jewel's amphitheatre, leafy textures and pastels in the social room, glasshouse and grow your own productive area. Created in 2022, the garden has sustainability at its heart.

Our donation of £2.5 million over five years has helped fund the new Y Bwthyn NGS Macmillan Specialist Palliative Care Unit in Wales. 1,900 inpatients have been supported since its opening

Woodpeckers

SUSSEX

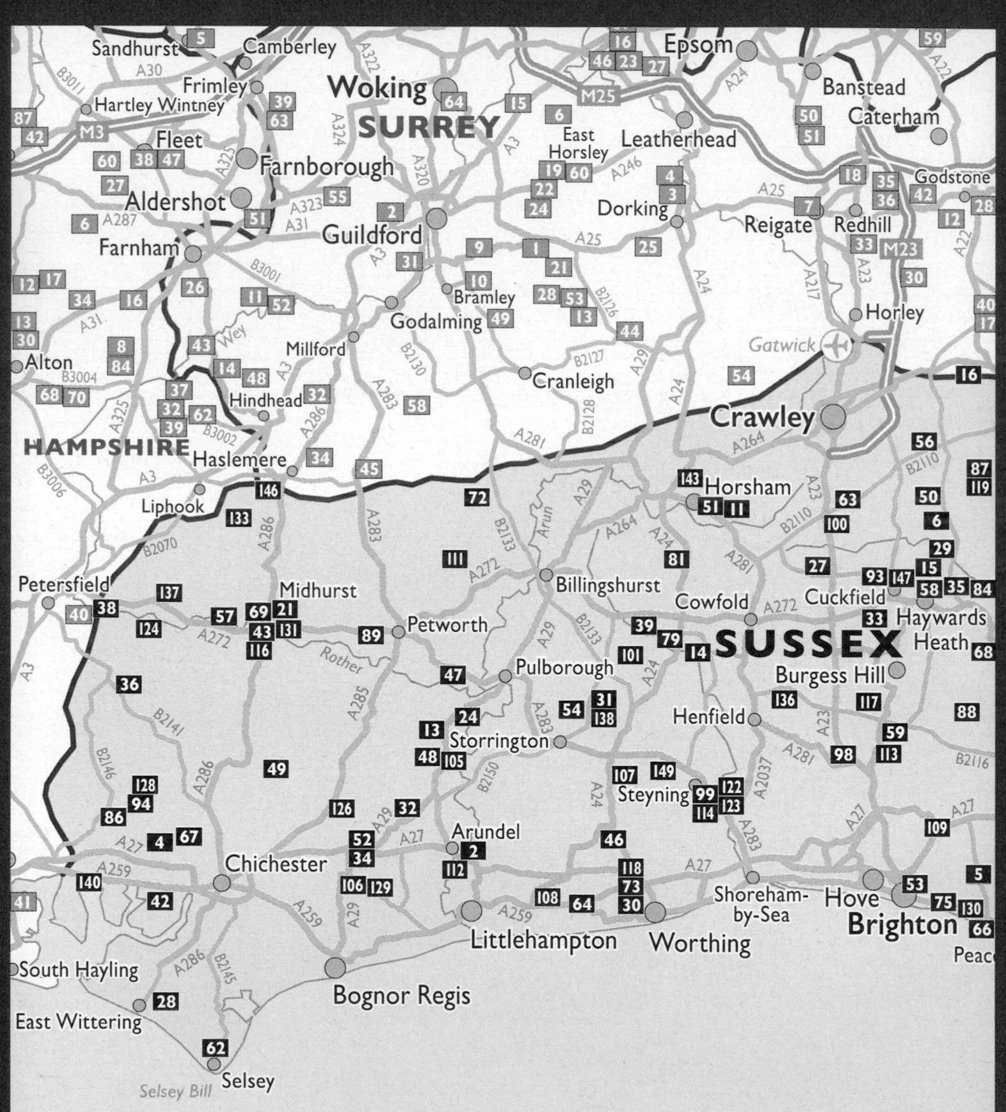

SURREY
HAMPSHIRE
SUSSEX
Sandhurst
Camberley
Frimley
Hartley Wintney
Woking
Epsom
Banstead
Caterham
Fleet
Farnborough
East Horsley
Leatherhead
Godstone
Aldershot
Guildford
Dorking
Reigate
Redhill
Farnham
Bramley
Godalming
Horley
Millford
Gatwick
Alton
Cranleigh
Hindhead
Crawley
Haslemere
Liphook
Horsham
Petersfield
Midhurst
Billingshurst
Cowfold
Cuckfield
Haywards Heath
Petworth
Pulborough
Burgess Hill
Henfield
Storrington
Steyning
Arundel
Chichester
Shoreham-by-Sea
Hove
Brighton
Littlehampton
Worthing
South Hayling
Bognor Regis
East Wittering
Selsey
Selsey Bill
Wey
Arun
Rother

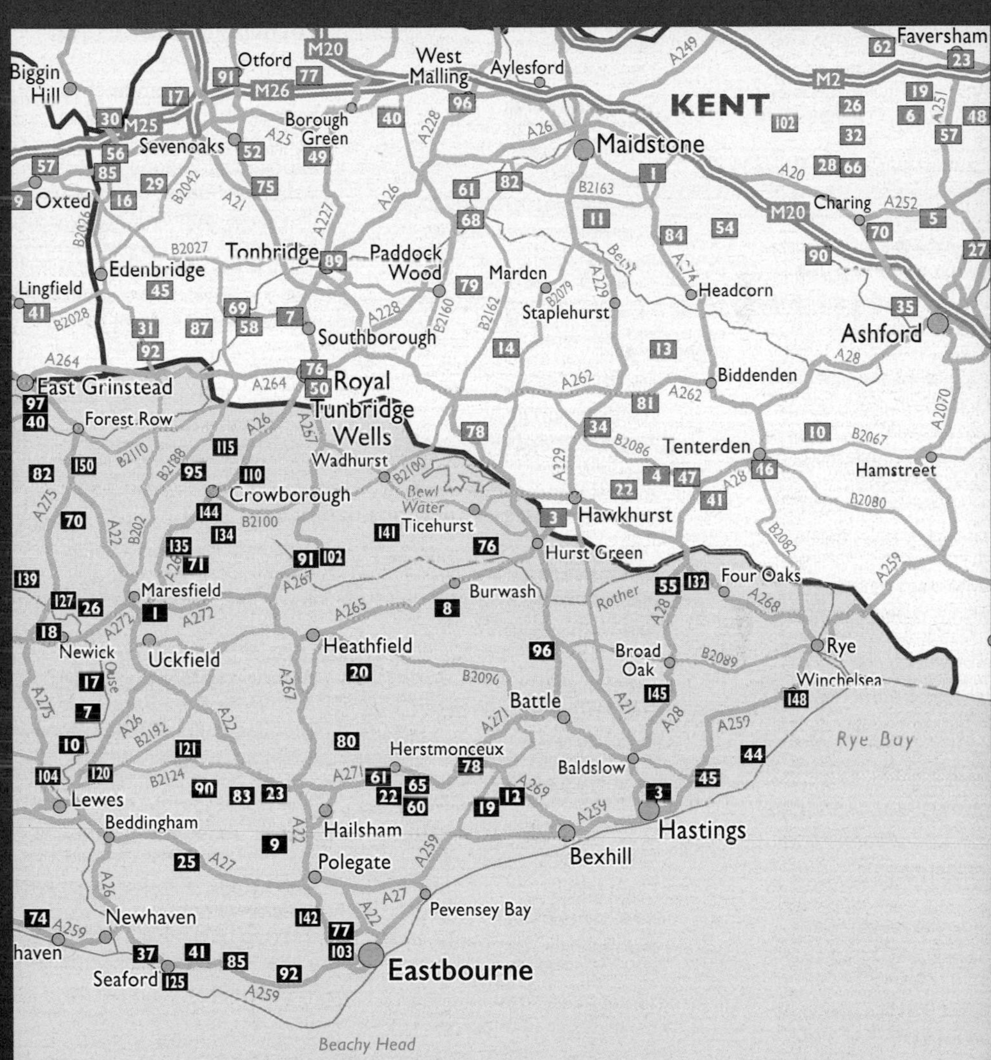
KENT
Faversham
Biggin Hill
Otford
West Malling
Aylesford
Sevenoaks
Borough Green
Maidstone
Oxted
Charing
Tonbridge
Paddock Wood
Edenbridge
Marden
Lingfield
Headcorn
Staplehurst
Southborough
Ashford
East Grinstead
Royal Tunbridge Wells
Biddenden
Forest Row
Wadhurst
Tenterden
Hamstreet
Crowborough
Bewl Water
Ticehurst
Hawkhurst
Hurst Green
Four Oaks
Maresfield
Burwash
Rother
Newick
Uckfield
Heathfield
Broad Oak
Rye
Winchelsea
Battle
Rye Bay
Herstmonceux
Baldslow
Lewes
Hastings
Beddingham
Hailsham
Bexhill
Polegate
Pevensey Bay
Newhaven
Seaford
Eastbourne
Beachy Head
0 10 kilometres
0 5 miles
© Global Mapping / XYZ Maps

East & Mid Sussex Volunteers

County Organiser, Booklet & Advertising Co-ordinator
Irene Eltringham-Willson
01323 833770
irene.willson@btinternet.com

County Treasurer
Andrew Ratcliffe 01435 873310
andrew.ratcliffe@ngs.org.uk

Publicity
Geoff Stonebanks 01323 899296
sussexeastpublicity@ngs.org.uk

Social Media
Nicki Conlon 07720 640761
nicki.conlon@ngs.org.uk

Photographer & Social Media
Lilly Carreras 07763 919602
lilly.carreras@ngs.org.uk

Booklet Distributor
Dr Denis Jones 01323 899452
sweetpeasd49@gmail.com

Assistant County Organisers
Jane Baker 01273 842805
jane.baker@ngs.org.uk

Michael & Linda Belton
01797 252984
belton.northiam@gmail.com

Shirley Carman-Martin
01444 473520
shirleycarmanmartin@gmail.com

Isabella & Steve Cass
07908 123524
oaktreebarn@hotmail.co.uk

Linda Field

Diane Gould 01825 750300
lavenderdgould@gmail.com

Aideen Jones 01323 899452
sweetpeasa52@gmail.com

Dr Denis Jones (as above)

Susan Laing 01892 770168
splaing@btinternet.com

Sarah Ratcliffe 01435 873310
sarah.ratcliffe@ngs.org.uk

Dianna Tennant 01892 752029
tennantdd@gmail.com

David Wright 01435 883149
david.wright@ngs.org.uk

@SussexNGSEast
@SussexNGS
@ngseastsussex

West Sussex Volunteers

County Organisers, Booklet & Advertising Co-ordinators
Maggi Hooper 07793 159304
maggi.hooper@ngs.org.uk

Meryl Walters 07766 761926
meryl.walters@ngs.org.uk

County Treasurer
Philip Duly 07789 050964
philipduly@tiscali.co.uk

Publicity
Kate Harrison 01798 817489
kate.harrison@ngs.org.uk

Social Media
Claudia Pearce 07985 648216
claudiapearce17@gmail.com

Photographer
Judi Lion 07810 317057
judi.lion@ngs.org.uk

Booklet Distributor
Lesley Chamberlain 07950 105966
chamberlain_lesley@hotmail.com

Assistant County Organisers
Teresa Barttelot 01798 865690
tbarttelot@gmail.com

Sanda Belcher 01428 723259
sandambelcher@gmail.com

Lesley Chamberlain
(as above)

Patty Christie 01730 813323
pattychristie49@gmail.com

Elizabeth Gregory 01903 892433
elizabethgregory1@btinternet.com

Carrie McArdle 01403 820272
carrie.mcardle@btinternet.com

Claudia Pearce (as above)

Fiona Phillips 07884 398704
fiona.h.phillips@btinternet.com

Susan Pinder 07814 916949
nasus.rednip@gmail.com

Teresa Roccia 07867 383753
teresa.m.roccia@gmail.com

Diane Rose 07789 565094
dirose8@me.com

@Sussexwestngs
@SussexWestNGS
@sussexwestngs

Sussex is a vast county with two county teams, one covering East and Mid Sussex and the other covering West Sussex.

Over 80 miles from west to east, Sussex spans the southern side of the Weald from the exposed sandstone heights of Ashdown Forest, past the broad clay vales with their heavy yet fertile soils and the imposing chalk ridge of the South Downs National Park, to the equable if windy coastal strip.

Away from the chalk, Sussex is a county with a largely wooded landscape with imposing oaks, narrow hedged lanes and picturesque villages. The county offers much variety and our gardens reflect this. There is something for absolutely everyone and we feel sure that you will enjoy your garden visiting experience - from rolling acres of parkland, country and town gardens, to small courtyards and village trails. See the results of the owner's attempts to cope with the various conditions, discover new plants and talk with the owners about their successes.

Many of our gardens are open by arrangement, so do not be afraid to book a visit or organise a visit with your local gardening or U3A group.

Should you need advice, please e-mail irene.willson@btinternet.com for anything relating to East and Mid Sussex or maggi.hooper@ngs.org.uk for anything in West Sussex.

OPENING DATES

All entries subject to change. For latest information check **www.ngs.org.uk**

Extended openings are shown at the beginning of the month.

Map locator numbers are shown to the right of each garden name.

January

Wednesday 25th
5 Whitemans Close 147

Friday 27th
5 Whitemans Close 147

Monday 30th
5 Whitemans Close 147

February

Snowdrop Festival

Every Thursday from Thursday 9th
The Old Vicarage 107

Every Thursday and Friday from Friday 10th to Friday 17th
Pembury House 113

Wednesday 1st
5 Whitemans Close 147

Friday 3rd
5 Whitemans Close 147

Saturday 4th
5 Whitemans Close 147

Monday 6th
5 Whitemans Close 147

Tuesday 7th
5 Whitemans Close 147

Wednesday 8th
5 Whitemans Close 147

Sunday 12th
Sandhill Farm House 124

Tuesday 14th
5 Whitemans Close 147

Wednesday 15th
◆ Highdown Gardens 64

Thursday 16th
5 Whitemans Close 147

Sunday 19th
◆ Bates Green 9
Manor of Dean 89

Thursday 23rd
NEW Buddington Farm 21

Friday 24th
Pembury House 113

Saturday 25th
NEW Buddington Farm 21

Sunday 26th
◆ Denmans Garden 34

March

Every Thursday
The Old Vicarage 107

Every Thursday and Friday to Friday 17th
Pembury House 113

Sunday 5th
Manor of Dean 89

Sunday 12th
The Old Vicarage 107

Saturday 18th
◆ Nymans 100

Sunday 19th
◆ Bates Green 9
47 Denmans Lane 35

Tuesday 21st
◆ Borde Hill Garden 15
Manor of Dean 89

Saturday 25th
Down Place 36
◆ King John's Lodge 76

Sunday 26th
47 Denmans Lane 35
Down Place 36

Friday 31st
Judy's Cottage Garden 73

April

Daily from Monday 24th | to Friday 28th
◆ Bateman's 8

Every Wednesday from Wednesday 19th
Fittleworth House 47

Every Thursday
The Old Vicarage 107

Saturday 1st
Butlers Farmhouse 22

Sunday 2nd
Butlers Farmhouse 22

Monday 10th
47 Denmans Lane 35
The Old Vicarage 107

Tuesday 11th
Peelers Retreat 112

Saturday 15th
Peelers Retreat 112

Sunday 16th
Newtimber Place 98
Penns in the Rocks 115

Tuesday 18th
Bignor Park 13

Friday 21st
The Garden House 53

Saturday 22nd
The Garden House 53
The Oast 102
Warnham Park 143

Sunday 23rd
Manor of Dean 89
The Oast 102

Tuesday 25th
Peelers Retreat 112

Wednesday 26th
Fairlight End 44
◆ Sheffield Park and Garden 127

Saturday 29th
Banks Farm 7
◆ King John's Lodge 76
Peelers Retreat 112

Sunday 30th
Banks Farm 7
◆ Denmans Garden 34
NEW The Hidden Garden 62
◆ King John's Lodge 76
Stanley Farm 133

May

Every Wednesday to Wednesday 10th
Fittleworth House 47

Every Thursday
The Old Vicarage 107

Monday 1st
47 Denmans Lane 35

Sunday 7th
Forest Ridge 50
Hollymount 71
Mountfield Court 96

Monday 8th
Stone Cross House 134

Tuesday 9th
Bignor Park 13
Peelers Retreat 112

Wednesday 10th
Cookscroft 28

Saturday 13th
96 Ashford Road 3
Cookscroft 28
Limekiln Farm 83
Peelers Retreat 112
The Warren, Crowborough Gardens 144

Sunday 14th
Champs Hill 24
Hammerwood House 57
28 Larkspur Way 81
Limekiln Farm 83
Penns in the Rocks 115
The Warren, Crowborough Gardens 144

Tuesday 16th
Manor of Dean 89
NEW Stroods 135

Wednesday 17th
Balcombe Gardens 6
Wych Warren House 150

Saturday 20th
96 Ashford Road 3
54 Elmleigh 43
Legsheath Farm 82

Sunday 21st
The Beeches 10
47 Denmans Lane 35
54 Elmleigh 43
Seaford Gardens 125

Tuesday 23rd
Peelers Retreat 112

Friday 26th
Holford Manor & Chailey Iris Garden 68
1 Pest Cottage 116

Saturday 27th
96 Ashford Road 3
Holford Manor & Chailey Iris Garden 68
◆ Knepp Castle 79
Peelers Retreat 112
◆ The Priest House 119
Skyscape 130

Sunday 28th
Copyhold Hollow 29
Foxglove Cottage 51
Holford Manor & Chailey Iris Garden 68
1 Pest Cottage 116
Skyscape 130
NEW Terwick House 137

Monday 29th
Copyhold Hollow 29
47 Denmans Lane 35

Wednesday 31st
◆ Highdown Gardens 64

June

Every Thursday
The Old Vicarage 107

Friday 2nd
NEW Hillside 66
Judy's Cottage Garden 73
Orchard Cottage 110

Saturday 3rd
Bradness Gallery 17
Harlands Gardens 58
Oaklands Farm 101
The Old Rectory 106
Orchard Cottage 110
The Shrubbery 129

Sunday 4th
Bradness Gallery 17
NEW Brickyard Farm Cottage 19
NEW Eastbourne Garden Trail 41
Harlands Gardens 58
◆ High Beeches Woodland and Water Garden 63
NEW Hillside 66
28 Larkspur Way 81
Offham House 104
The Old Rectory 106
Orchard Cottage 110
The Shrubbery 129

Tuesday 6th
Kitchenham Farm 78

Wednesday 7th
Fittleworth House 47

Friday 9th
Five Oaks Cottage 48

Saturday 10th
Five Oaks Cottage 48
Foxwood Barn 52
Jacaranda 72
Kitchenham Farm 78
Lordington House 86
Mayfield Gardens 91
Peelers Retreat 112

Sunday 11th
Deaks Lane Gardens 33
Down Place 36
Five Oaks Cottage 48
4 Hillside Cottages 67
Lordington House 86
Mayfield Gardens 91
Ringmer Park 120
Seaford Gardens 125
Town Place 139

Monday 12th
Down Place 36

Wednesday 14th
Fittleworth House 47
Kemp Town Enclosures: South Garden 75
NEW ◆ One Garden Brighton 109
Town Place 139

Thursday 15th
Butlers Farmhouse 22

Friday 16th
Bramley 18
Butlers Farmhouse 22
Holford Manor & Chailey Iris Garden 68

Saturday 17th
Alpines 1
Balcombe Gardens 6
Bramley 18
Durford Abbey Barn 38
54 Elmleigh 43
Grove Farm House 56
Holford Manor & Chailey Iris Garden 68
◆ King John's Lodge 76
Sandhill Farm House 124
Winchelsea's Secret Gardens 148

Sunday 18th
Balcombe Gardens 6
Bexhill-on-Sea Trail 12
Durford Abbey Barn 38
54 Elmleigh 43
Holford Manor & Chailey Iris Garden 68
Hollymount 71
◆ King John's Lodge 76
Sandhill Farm House 124
Selhurst Park 126

Tuesday 20th
Bignor Park 13
Peelers Retreat 112

Wednesday 21st
Fittleworth House 47
NEW Nightingale House 99
NEW 15 Penlands Rise 114
Town Place 139

Thursday 22nd
◆ Clinton Lodge 26

Friday 23rd
NEW 40 Michel Dene Road 92
Parsonage Farm 111
◆ St Mary's House Gardens 123
Wadhurst Park 141

Saturday 24th
Cookscroft 28
Luctons 87
Peelers Retreat 112
◆ The Priest House 119
◆ St Mary's House Gardens 123
Wadhurst Park 141

Sunday 25th
The Folly 49
Gorselands 54
Hassocks Garden Trail 59
Herstmonceux Parish Trail 61
Luctons 87
16 Musgrave Avenue 97
Town Place 139

Tuesday 27th
Luctons 87

Wednesday 28th
16 Musgrave Avenue 97
NEW Nightingale House 99
NEW 15 Penlands Rise 114

Friday 30th
16 Musgrave Avenue 97
Old Well Cottage 108

July

Every Thursday
The Old Vicarage 107

Saturday 1st
NEW Hill House 65
Hollist House 69
Waterworks & Friends 145
Wiston House 149

Sunday 2nd
Findon Place 46
Gorselands 54
NEW Hill House 65
NEW 255 Kings Drive 77
Town Place 139
NEW 50 Wannock Lane 142
Whitehanger 146

Tuesday 4th
16 Musgrave Avenue 97

Wednesday 5th
Foxglove Cottage 51

Friday 7th
16 Musgrave Avenue 97

Sunday 9th
The Beeches 10
Foxglove Cottage 51
Town Place 139

Monday 10th
16 Musgrave Avenue 97

Tuesday 11th
Peelers Retreat 112

Wednesday 12th
Fittleworth House 47

Thursday 13th
NEW 64 Cuckfield Crescent 30
16 Musgrave Avenue 97

Friday 14th
NEW 64 Cuckfield Crescent 30

Saturday 15th
Bagotts Rath 5
54 Elmleigh 43
The Jungle Garden 74
Oaklands Farm 101
Peelers Retreat 112

Sunday 16th
Bagotts Rath 5
NEW 64 Cuckfield Crescent 30
54 Elmleigh 43
4 Hillside Cottages 67
The Jungle Garden 74

Wednesday 19th
Fittleworth House 47

Thursday 20th
Cumberland House 31
Thakeham Place Farm 138

Saturday 22nd
South Grange 132

Sunday 23rd
Cumberland House 31
The Folly 49
Hollymount 71
South Grange 132
Thakeham Place Farm 138

Tuesday 25th
NEW Marchants Hardy Plants 90
Peelers Retreat 112

Wednesday 26th
◆ Herstmonceux Castle Estate 60
NEW Marchants Hardy Plants 90

Friday 28th
NEW Pitfield Barn Cut Flower Farm & Studio 117

Saturday 29th
NEW Pitfield Barn Cut Flower Farm & Studio 117

Sunday 30th
East Grinstead Gardens 40
NEW The Hidden Garden 62

August

Every Thursday
The Old Vicarage 107

Thursday 3rd
Rose Cottage 121

Saturday 5th
Rose Cottage 121

Sunday 6th
Penns in the Rocks 115
Whitehanger 146

Tuesday 8th
Peelers Retreat 112

Wednesday 9th
Fittleworth House 47
Kitchenham Farm 78

Saturday 12th
Kitchenham Farm 78
Peelers Retreat 112

Sunday 13th
Champs Hill 24
Colwood House 27

Thursday 17th
NEW Bourne Botanicals 16

Saturday 19th
Camberlot Hall 23
54 Elmleigh 43
Fairlight Hall 45
Holly House 70

Sunday 20th
Camberlot Hall 23
54 Elmleigh 43
Fairlight Hall 45
The Folly 49
Holly House 70

Tuesday 22nd
Peelers Retreat 112

Saturday 26th
Butlers Farmhouse 22

Sunday 27th
Butlers Farmhouse 22
Whitehanger 146

Monday 28th
47 Denmans Lane 35
Durrance Manor 39
Lindfield Jungle 84
The Old Vicarage 107

September

Every Thursday
The Old Vicarage 107

Saturday 2nd
NEW Bourne Botanicals 16

Sunday 3rd
Parsonage Farm 111

Wednesday 6th
Knightsbridge House 80

Friday 8th
Judy's Cottage Garden 73

Saturday 9th
54 Elmleigh 43
◆ King John's Lodge 76
Knightsbridge House 80

Sunday 10th
54 Elmleigh 43
NEW 255 Kings Drive 77
◆ Sussex Prairies 136

Wednesday 13th
NEW Ashling Park Estate 4

Saturday 16th
Sandhill Farm House 124

Sunday 17th
Sandhill Farm House 124

Wednesday 20th
NEW Ashling Park Estate 4

Saturday 23rd
Limekiln Farm 83
Peelers Retreat 112

Sunday 24th
◆ Bates Green 9
Limekiln Farm 83

Saturday 30th
Peelers Retreat 112

October

Every Thursday to Thursday 12th
The Old Vicarage 107

Sunday 1st
◆ High Beeches Woodland and Water Garden 63

Tuesday 10th
Peelers Retreat 112

Saturday 14th
Peelers Retreat 112

Sunday 22nd
◆ Bates Green 9

Sunday 29th
◆ Denmans Garden 34

In 2022 we donated £450,000 to Marie Curie which equates to 23,871 hours of community nursing care

By Arrangement

Arrange a personalised garden visit with your club, or group of friends, on a date to suit you. See individual garden entries for full details.

The Beeches 10
4 Ben's Acre 11
Black Barn 14
NEW Bourne Botanicals 16
NEW Brickyard Farm Cottage 19
Brightling Down Farm 20
Butlers Farmhouse 22
Camberlot Hall 23
Champs Hill 24
Colwood House 27
Cookscroft 28
Copyhold Hollow 29
Cosy Cottage, Seaford Gardens 125
Dale Park House 32
47 Denmans Lane 35
Dittons End, Eastbourne Garden Trail 41
Down Place 36
Driftwood 37
Durrance Manor 39
Eastfield Cottage 42
54 Elmleigh 43
Fairlight End 44
Findon Place 46
Fittleworth House 47
The Folly 49
Foxglove Cottage 51
Foxwood Barn 52
The Garden House 53
Hardwycke, Eastbourne Garden Trail 41
4 Hillside Cottages 67
Holford Manor & Chailey Iris Garden 68
Holly House 70
Hollymount 71
Jacaranda 72
NEW 36 Jellicoe Close, Eastbourne Garden Trail 41
The Jungle Garden 74
Legsheath Farm 82
Lindfield Jungle 84
The Long House 85
Lordington House 86
Luctons 87
Malthouse Farm 88
Manor of Dean 89
Mill Hall Farm 93

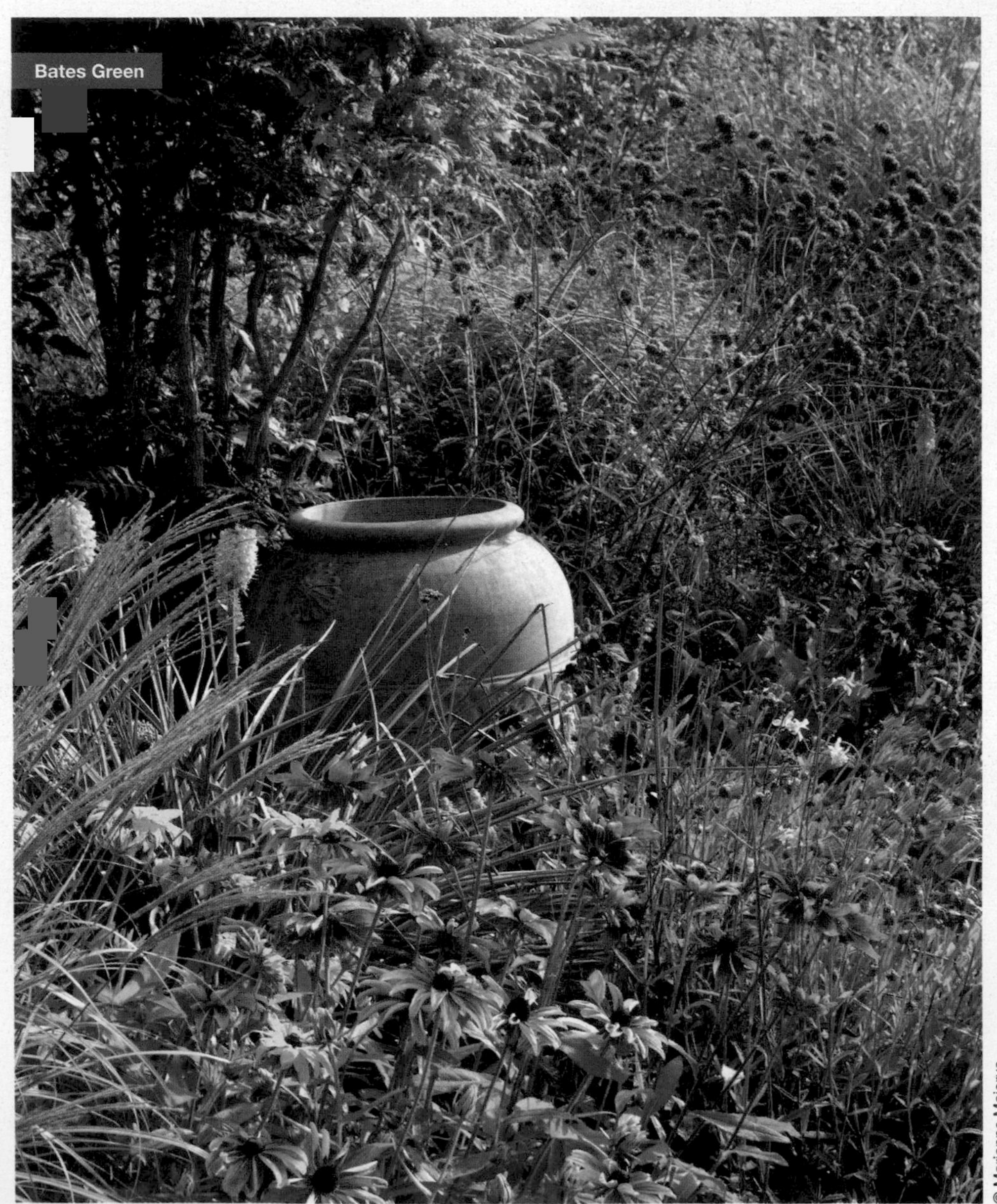

Bates Green

THE GARDENS

1 ALPINES

High Street, Maresfield, Uckfield, TN22 2EG. Ian & Cathy Shaw. *1½m N of Uckfield. Garden approx 150 metres N of Budletts r'about towards Maresfield. Blue Badge parking at garden, other parking in village.* **Sat 17 June (11-5). Adm £5, chd free. Home-made teas.**

A 1 acre garden with large and rampant mixed borders, a riot of colour and scent. Numerous new trees, orchard with beehives and meadow grasses. Wildlife pond with bog garden, developing stumpery and fernery. Vegetable patch with raised beds, fruit cage, pretty greenhouse full of edibles and bee garden. Wheelchair access largely on one level with a few steps round greenhouse and gravel paths.

2 ◆ ARUNDEL CASTLE & GARDENS

Arundel, BN18 9AB. Arundel Castle Trustees Ltd, 01903 882173, visits@arundelcastle.org, www.arundelcastle.org. *In the centre of Arundel, N of A27.* **For opening times and information, please phone, email or visit garden website.**

Ancient castle, family home of the Duke of Norfolk. 40 acres of grounds and gardens which inc hot subtropical borders, English herbaceous borders, stumpery, wildflower garden, two glasshouses with exotic fruit and vegetables, walled flower and organic kitchen gardens. C14 Fitzalan Chapel white garden.

3 96 ASHFORD ROAD

Hastings, TN34 2HZ. Lynda & Andrew Hayler. *From A21 (Sedlescombe Rd N) towards Hastings, take 1st exit on r'about A2101, then 3rd on L (approx 1m).* **Sat 13, Sat 20, Sat 27 May (1-4.30). Adm £3, chd free.**

Small (100ft x 52ft) Japanese inspired front and back garden. Full of interesting planting with many acers, azaleas and bamboos. Over 100 different hostas, many miniature. Lower garden with greenhouse and raised beds. Also, an attractive Japanese Tea House.

4 NEW ASHLING PARK ESTATE

West Ashling, Chichester, PO18 9DJ. Gail Gardner, 01243 967700, contact@ashlingpark.co.uk, ashlingpark.co.uk. *Located on the Funtington Rd, 5m W of Chichester. Through the large metal gates onto the private driveway to the vineyard.* **Wed 13, Wed 20 Sept (10-11.30am). Adm £8, chd free. Pre-booking essential, please visit www.ngs.org.uk for information & booking. Wine. Tea.**

A truly fascinating vineyard tour with wine tasting that starts promptly at 10am. First created in the C19, Ashling Park was cited on the Tithe map and has had a new lease of life when a vineyard was planted in 2017. Pinot Noir, Pinot Meunièr, Chardonnay and Bacchus grape varieties, chef's kitchen garden and 50 beehives.

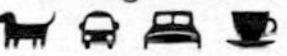

5 BAGOTTS RATH

5 Crescent Drive North, Woodingdean, BN2 6SP. Stephen & Den McDonnell-Daly. *A bus service from Brighton Stn & Brighton seafront stops outside the house, otherwise follow NGS signs & park in adjoining streets.* **Sat 15, Sun 16 July (12-5). Combined adm with The Jungle Garden £6, chd free. Home-made teas.**

A small garden on the Downs, near Brighton. Created on a challenging site with shallow chalk soil. The garden looks more mature than its 7 yrs of planning. The garden inc many of the elements that were needed to win the owner eight Chelsea gold medals and twice Best in Show for floral design. Water features, topiary, carnivorous plants, succulents and many pieces of sculptures. Enjoy refreshments at various seating areas throughout the garden.

GROUP OPENING

6 BALCOMBE GARDENS

3m N of Cuckfield on B2036, 3m S of J10A on M23. Individual gardens signed within Balcombe. **Wed 17 May, Sat 17, Sun 18 June (12-5). Combined adm £7.50, chd free. Home-made teas at Stumlet (17 May & 17 June) & at the tea venue for 46 Westup Farm Cottages (18 June).**

STUMLET
Oldlands Avenue, RH17 6LW. Max & Nicola Preston Bell.

46 WESTUP FARM COTTAGES
London Road, RH17 6JJ. Chris & Clare Cornwell, 01444 811891, chris.westup@btinternet.com. **Visits also by arrangement 17 Apr to 30 Sept for groups of up to 50.**

WINTERFIELD
Oldlands Avenue, RH17 6LP. Sue & Sarah Howe, 01444 811380, sarahjhowe_uk@yahoo.co.uk. **Visits also by arrangement 1 Mar to 16 July.**

Within the Balcombe AONB there are three quite different gardens that are full of variety and interest, which will appeal to plant lovers. Set amidst the countryside of the High Weald, 46 Westup Farm Cottages is a classic cottage garden with unique and traditional features linked by intimate paths through lush and subtle planting with pollinators and wildlife in abundance. In the village, Winterfield is a country garden packed with uncommon shrubs and trees, herbaceous borders, a summerhouse, pond and wildlife area. At nearby Stumlet the garden is restful, there are places to sit and enjoy a little peace, scent and colour. Redesigned to inc interesting plants for lasting enjoyment by different generations. A garden to watch develop in the future. Wheelchair access at Winterfield and Stumlet, also at the tea venue for 46 Westup Farm Cottages on 18 June only.

Our annual donations to Parkinson's UK meant 7,000 patients are currently benefiting from support of a Parkinson's Nurse

7 BANKS FARM
Boast Lane, Barcombe, Lewes, BN8 5DY. Nick & Lucy Addyman. *6m N of Lewes. From Barcombe Cross follow signs to Spithurst & Newick. 1st road on R into Boast Ln towards the Anchor Pub. At sharp bend carry on into Banks Farm.* **Sat 29, Sun 30 Apr (11-4). Adm £5, chd free. Home-made teas.**
9 acre garden set in rural countryside. Extensive lawns and shrub beds merge with the more naturalistic woodland garden set around the lake. An orchard, vegetable garden, ponds and a wide variety of plant species add to an interesting and very tranquil garden. Refreshments served outside, so may be limited during bad weather. Wheelchair access to the upper part of garden. Sloping grass paths in the lower part.

8 ◆ BATEMAN'S
Bateman's Lane, Burwash, TN19 7DS. National Trust, 01435 882302, batemans@nationaltrust.org.uk, www.nationaltrust.org.uk/batemans. *6m E of Heathfield. ½m S of A265 on road leading S at W end of Burwash, or N from Woods Corner (B2096). Pick up & drop off point available.* **For NGS: Daily Mon 24 Apr to Fri 28 Apr (10-5). Adm £15, chd £7. Light refreshments. For other opening times and information, please phone, email or visit garden website.**
Bateman's is an idyllic spot; a family home loved by Rudyard Kipling. Nestled in a shallow valley, the house and garden were a joy and an inspiration to him, from the formal borders and clipped yew hedges to romantic meadows with the meandering river flowing through. Highlights inc wild garden carpeted with spring bulbs below flowering trees and shrubs, orchard, vegetable garden, spring flowering borders, working watermill and an annual tulip and wallflower extravaganza. On site tearoom offering beverages, cream teas, cakes, hot and cold food. Most of the garden is accessible. There are some slopes and uneven paths.

9 ◆ BATES GREEN
Tye Hill Road, Arlington, BN26 6SH. John McCutchan, 01323 485151, john@bluebellwalk.co.uk, www.batesgreen.co.uk. *3½m SW of Hailsham & A22. Midway between the A22 & A27, 2m S of Michelham Priory. Bates Green is in Tye Hill Rd (N of Arlington village), 350yds S of Old Oak Inn. Ample parking on hard-standing verges.* **For NGS: Sun 19 Feb, Sun 19 Mar, Sun 24 Sept, Sun 22 Oct (10.30-4). Adm £6, chd £3. Pre-booking essential, please visit www.ngs.org.uk for information & booking. Home-made soup, cakes & scones, plus light lunches in a large insulated barn. For other opening times and information, please phone, email or visit garden website.**
This plantswoman's 2 acre tranquil garden provides interest through the seasons. Woodland garden created around a majestic oak tree. Middle garden peaks in late summer. Courtyard gardens with seasonal container displays. Front garden a winter and spring joy with narcissi, primroses, violets, early tulips and coloured stems of cornus and salix. Wildlife pond and conservation meadow. Gardened for nature. NGS spring visitors will be able to walk through the daffodil glade leading to a 24 acre ancient oak woodland. This wood has been owned by the McCutchan family for 100 yrs and is managed for conservation and diversity. Our autumn guests can enjoy spotting the many fungi abundant in the woods. Wheelchair access to most areas. Mobility scooters available to borrow free of charge. Accessible WC.

10 THE BEECHES
Church Road, Barcombe, Lewes, BN8 5TS. Sandy Coppen, 07771 655862, sand@thebeechesbarcombe.com, www.thebeechesbarcombe.com. *From Lewes, A26 towards Uckfield for 3m, turn L signed Barcombe. Follow road for 1½m, turn L signed Hamsey & Church. Follow road for approx ½m & parking will be in the church car park on LHS.* **Sun 21 May, Sun 9 July (1-5). Adm £6, chd free. Home-made teas. Visits also by arrangement June to Aug for groups of 10+.**
C18 walled garden with cut flowers, vegetables, salads and fruit. Separate orchard and rose garden. Herbaceous borders, a hot border and extensive lawns. A hazel walk is being developed and a short woodland walk. An old ditch has been made into a flowing stream with gunnera, ferns, tree ferns, hostas and a few flowers going into a pond. Wheelchair access without steps, but some ground is a little bumpy.

11 4 BEN'S ACRE
Horsham, RH13 6LW. Pauline Clark, 01403 266912, brian.clark8850@yahoo.co.uk. *E of Horsham. A281 via Cowfold, after Hilliers Garden Centre, turn R by Tesco on to St Leonards Rd, straight over r'about to Comptons Ln, next R Heron Way, 2nd L Grebe Cres, 1st L Ben's Acre.* **Visits by arrangement June to Aug for groups of 10 to 30. Adm £9, chd free. Home-made teas inc. Call to discuss wine & canapé options.**
Described as inspirational excellence in design and artistry. A visual delight on different levels, featuring ponds, rockery, summerhouse and arbour. Seating for your teas surrounded by borders full of harmonising perennials, roses, shrubs and more. See how a diverse small space created on different levels can be exciting and inspiring. Seating throughout the garden. 5 mins from Hilliers Garden Centre, 15 mins from Leonardslee and NT Nymans. Visit us on YouTube Pauline & Brian's Sussex Garden.

GROUP OPENING

12 BEXHILL-ON-SEA TRAIL
Bexhill & Little Common. Follow individual NGS signs to gardens from main roads. Tickets & maps at all gardens. **Sun 18 June (12-5). Combined adm £6, chd free. Home-made teas & light lunch at De Wilp.**

THE CLINCHES
Collington Lane East, TN39 3RJ. Val Kemm.

64 COLLINGTON AVENUE
TN39 3RA. Dr Roger & Ruth Elias.

DE WILP
Collington Lane East, TN39 3RJ. Stuart & Hazel Wood.

SMALL HOUSE
Sandhurst Lane, TN39 4RG. Veronika & Terry Rogers.

The Small House, returning to the trail, is a developing garden surrounded by mature trees and shrubs, built over three levels on a sloping site. Panoramic views of the whole garden from a beautiful terrace, descending into a garden of delightful island beds filled with interesting plants and flowering shrubs. Opening out into a wildflower and wildlife friendly garden. Beautiful vegetable beds and pond finish off this tranquil oasis. 64 Collington Avenue is a bit of a surprise as it incorporates the garden next door too, yrs in the making, it has a beautiful mix of herbaceous planting with structural shrubs, vegetables, and planting to encourage wildlife and stunning roses. De Wilp, a beautiful garden laid to lawn with beds containing a large variety of perennial and annual plants, vegetables, mature trees and shrubs. Also, the main tea venue for the trail. Adjacent to De Wilp, The Clinches is a mature cottage garden surrounded by trees, pond with fish, toads and newts. Partial wheelchair access to some gardens.

13 BIGNOR PARK

Pulborough, RH20 1HG. The Mersey Family, www.bignorpark.co.uk. *5m S of Petworth & Pulborough. Well signed from B2138. Nearest villages Sutton, Bignor & West Burton. Approach from the E, directions & map available on website.* **Tue 18 Apr, Tue 9 May, Tue 20 June (2-5). Adm £5, chd free. Home-made teas.**

11 acres of peaceful garden to explore with magnificent views of the South Downs. Interesting trees, shrubs, wildflower areas with swathes of daffodils in spring. The walled flower garden has been replanted with herbaceous borders. Temple, Greek loggia, Zen pond and unusual sculptures. Former home of romantic poet Charlotte Smith, whose sonnets were inspired by Bignor Park. Spectacular cedar of Lebanon and rare Lucombe oak. Wheelchair access to shrubbery and croquet lawn. Gravel paths in rest of garden and steps in stables quadrangle.

14 BLACK BARN

Steyning Road, West Grinstead, Horsham, RH13 8LR. Jane Gates, 07774 980819, jane.er.gates@gmail.com. *Between Partridge Green & A24 on Steyning Rd at West Grinstead. From A24 follow Steyning Rd signed West Grinstead for 1m, passing Park Ln & Catholic Church on the L, continue round 2 bends, Black Barn is on L. Parking limited so car share would be appreciated.* **Visits by arrangement 1 Apr to 29 Sept for groups of up to 20. Adm £5, chd free. Home-made teas.**

Black Barn Garden was designed and planted in Jan 2018. The site was cleared and only the mature trees remained. A hedgerow separating the garden from the field was removed and a large pond has been created in the southwest corner. A large gravel garden was created to the south of the barn and a terrace laid. The existing terrace at the back was increased and a large bed was created. Wheelchair access over grass only.

15 ◆ BORDE HILL GARDEN

Borde Hill Lane, Haywards Heath, RH16 1XP. Borde Hill Garden, 01444 450326, info@bordehill.co.uk, www.bordehill.co.uk. *1½m N of Haywards Heath. 20 mins N of Brighton, or S of Gatwick on A23 taking exit 10a via Balcombe.* **For NGS: Tue 21 Mar (10-5). Adm £10.50, chd £7. For other opening times and information, please phone, email or visit garden website.**

Rare plants and stunning landscapes make Borde Hill Garden the perfect day out for horticultural enthusiasts, families and those who love beautiful countryside. Enjoy tranquil outdoor rooms, woodland and parkland walks, playground, picnic areas, café and events throughout the season. Wheelchair access to 17 acres of formal garden.

16 NEW BOURNE BOTANICALS

The Bourne, Chesterfield Close, Furnace Wood Felbridge, East Grinstead, RH19 2PY. Jackie & Andy Doherty *A264 between Copthorne & Felbridge. 1⅙m from Felbridge. Parking in layby RH19 2QF on W bound A264 signed to Furnace Wood. Metro Bus 400. Enter Furnace Wood via footpath R of barrier, 8 min walk to Bourne Botanicals.* **Thur 17 Aug, Sat 2 Sept (11-5). Adm £8. Pre-booking essential, please phone 07785 562558 or email bournebotanic@outlook.com for information & booking. Home-made teas. Visits also by arrangement 1 Aug to 17 Sept for groups of 5 to 20.**

A lush setting of huge bananas jostling alongside gunnera and arid beds of agave, yucca and cacti. A diverse tropical look garden with a wildlife pond and stream inc bog beds of carnivorous plants. Palms, tree ferns, tetrapanax and many unusual plants and quirky touches all set in an acre of woodland. National Collection holder. Parking at property for disabled badge holders only.

17 BRADNESS GALLERY

Spithurst Road, Spithurst, Barcombe, BN8 5EB. Michael Cruickshank & Emma Burnett, www.bradnessgallery.com. *5m N of Lewes. Bradness Gallery lies midway between the villages of Barcombe & Newick on Spithurst Rd.* **Sat 3, Sun 4 June (11-5). Adm £5, chd free. Home-made teas.**

Welcome to this delightful and tranquil, mature, wild, wildlife garden with trees, scented shrubs, old roses, herbaceous borders and wild garden planting. A wooded stream flows along the bottom and two large ponds are home to wild ducks, dragonflies and frogs. Gravelled courtyard with raised planters. Bradness Gallery will be open showing a special exhibition of new garden paintings by Michael Cruickshank and Emma Burnett. Search The Bradness Garden on YouTube to watch a video of the garden.

18 BRAMLEY

Lane End Common, North Chailey, Lewes, BN8 4JH. Marcel & Lee Duyvesteyn. *From A272 North Chailey mini r'abouts take NE turn to A275 signed Bluebell Railway, after 1m turn R signed Fletching. Bramley is 300yds on R. Free car park opp & roadside parking.* **Fri 16, Sat 17 June (2-7). Adm £5, chd free.**

Rural 1 acre garden. Planting began in 2006 and inc an orchard avenue with wild flowers and long grassed areas and beehives. Vegetable garden, numerous formal and informal ornamental flowerbeds and borders. Planted for strong seasonal effect and structure. Interesting planting and design. SSSI nearby. Cream teas and Hindleap English Sparkling wine from the Bluebell Vineyard Estate.

19 NEW BRICKYARD FARM COTTAGE

Top Road, Hooe, Battle, TN33 9EJ. David & Grace Constable, 07740 407998, dc@constablespublishing.com. *From A269 turn R onto B2095, 2m on R. From A259 turn L onto B2095, 2½m on L.* **Sun 4 June (11-4.30). Adm £5, chd free. Home-made teas. Visits also by arrangement 1 May to 11 Sept.**

A 5 acre garden that we started from scratch 23 yrs ago. The large rockery around the pond is planted with miniature pines, behind is a small topiary garden leading into a pinetum and the follies which we added in the past few yrs. The long border leads into a rhododendron walk and out onto a planted parterre. Fruit and vegetables can also be seen in the fruit cages and orchard. Unique ruins in garden; also metal sculptures.

20 BRIGHTLING DOWN FARM

Observatory Road, Dallington, TN21 9LN. Val & Pete Stephens, 07770 807060, valstephens@icloud.com. *1m from Woods Corner. At Swan Pub, Woods Corner, take road opp to Brightling. Take 1st L to Burwash & almost immed, turn into 1st driveway on L.* **Visits by arrangement 9 June to 29 Sept for groups of 10 to 35. Thursdays & Fridays only. Adm £12.50, chd free. Home-made teas inc.**

The garden has several different areas inc a Zen garden, water garden, walled vegetable garden with two large greenhouses, herb garden, herbaceous borders and a woodland walk. The garden makes clever use of grasses and is set amongst woodland with stunning countryside views. Winner of the Society of Garden Designers award. Most areas of garden can be accessed with the use of temporary ramps.

21 NEW BUDDINGTON FARM

Buddington Lane, Easebourne, Midhurst, GU29 0QP. James & Catherine Renwick. *Coming out of Midhurst take the 1st L at the r'about towards Haslemere. Then follow the yellow NGS signs into Hollist Ln, Buddington Farm is on the R.* **Thur 23, Sat 25 Feb (2-5). Adm £5, chd free. Light refreshments.**

We are a working farm with wonderful daffodils and snowdrops by the farmhouse planted by my late mother-in-law who was a very keen gardener.

22 BUTLERS FARMHOUSE

Butlers Lane, Herstmonceux, BN27 1QH. Irene Eltringham-Willson, 01323 833770, irene.willson@btinternet.com, www.butlersfarmhouse.co.uk. *3m E of Hailsham. Take A271 from Hailsham, go through village of Herstmonceux, turn R signed Church Rd, then approx 1m turn R. Do not use SatNav!* **Sat 1, Sun 2 Apr, Thur 15, Fri 16 June (2-5). Adm £6, chd free. Sat 26, Sun 27 Aug (2-5). Adm £7, chd free. Home-made teas. Visits also by arrangement 20 Mar to 27 Oct.**

Lovely rural setting for 1 acre garden surrounding C16 farmhouse with views of South Downs. Pretty in spring with daffodils, hellebores and primroses. Come and see our meadow in June and perhaps spot an orchid or two. Quite a quirky garden with surprises round every corner inc a rainbow border, small pond, Cornish inspired beach corners, a poison garden and secret jungle garden. Plants for sale. Picnics welcome in June and Aug. Live jazz in Aug. Most of garden accessible by wheelchair.

23 CAMBERLOT HALL

Camberlot Road, Lower Dicker, Hailsham, BN27 3RH. Nicky & Paul Kinghorn, 07710 566453, nickykinghorn@hotmail.com. *500yds S of A22 at Lower Dicker, 4½m N of A27 Drusillas r'about. From A27 Drusilla's r'about through Berwick Stn to Upper Dicker & L into Camberlot Rd after The Plough pub, we are 1m on L. From A22 we are 500yds down Camberlot Rd on R.* **Sat 19, Sun 20 Aug (2-5). Adm £6, chd free. Home-made teas. Visits also by arrangement 30 June to 31 Aug for groups of 8 to 20.**

A 3 acre country garden with a lovely view across fields and hills to the South Downs. Created from scratch over the last 9 yrs with all design, planting and maintenance by the owner. Lavender lined carriage driveway, naturalistic border, vegetable garden, shady garden, 30 metre white border and dahlia garden. New part-walled garden and summerhouse with new planting. Wheelchair access over gravel drive and some uneven ground.

24 CHAMPS HILL

Waltham Park Road, Coldwaltham, Pulborough, RH20 1LY. Mrs Mary Bowerman, 01798 831205, info@thebct.org.uk, www.thebct.org.uk. *3m S of Pulborough. On A29 turn R to Fittleworth into Waltham Park Rd, garden 400 metres on R.* **Sun 14 May, Sun 13 Aug (2-5). Adm £5, chd free. Tea. Visits also by arrangement Apr to Sept for groups of 10+.**

A natural landscape, the garden has been developed around three disused sand quarries with far-reaching views across the Amberley Wildbrooks to the South Downs. A woodland walk in spring leads you past beautiful sculptures, against a backdrop of colourful rhododendrons and azaleas. In summer the garden is a colourful tapestry of heathers, which are renowned for their abundance and variety.

25 ◆ CHARLESTON

Firle, Lewes, BN8 6LL. The Charleston Trust, 01323 811626, www.charleston.org.uk. *Charleston is on the A27 signed halfway between Brighton & Eastbourne. From A27 follow narrow lane with deep ditches either side. Parking 200 metres from House entrance. Blue badge parking only 50 metres from house.* **For opening times and information, please phone or visit garden website.**

The Bloomsbury artists Vanessa Bell and Duncan Grant moved to Charleston in 1916. They transformed the walled vegetable plot into a quintessential painters' garden mixing Mediterranean influences with cottage garden planting. The garden is full of surprises inc a variety of sculpture, from classical forms to works by Quentin Bell, mosaics and tiled pools, an orchard and tranquil pond. Garden and grounds partially accessible. Gravel pathways, some uneven and narrow.

26 ◆ CLINTON LODGE

Fletching, TN22 3ST. Lady Collum, 01825 722952, garden@clintonlodge.com, www.clintonlodgegardens.co.uk. *4m NW of Uckfield. Clinton Lodge is situated in Fletching High St, N of The Griffin Inn. Off road parking provided, weather permitting. It is important visitors do not park in street. Parking available from 11am.*

For NGS: Thur 22 June (11-5.30). Adm £7, chd free. Home-made teas from 12pm. No lunches. For other opening times and information, please phone, email or visit garden website. Donation to local charities.
6 acre formal and romantic garden overlooking parkland with old roses, William Pye water feature, double white and blue herbaceous borders, yew hedges, pleached lime walks, medieval style potager, vine and rose allée, wildflower garden, small knot garden and orchard. Caroline and Georgian house (not open).

27 COLWOOD HOUSE

Cuckfield Lane, Warninglid, RH17 5SP. Mrs Rosy Brenan, 01444 461352. *6m W of Haywards Heath, 6m SE of Horsham. Entrance on B2115 Cuckfield Ln. From E, N & S, turn W off A23 towards Warninglid for ¾m. From W come through Warninglid village.* **Sun 13 Aug (1-5). Adm £7, chd free. Visits also by arrangement 1 May to 1 Sept for groups of 10+. Donation to Seaforth Hall, Warninglid.**
12 acres of garden with mature and specimen trees from the late 1800s, lawns and woodland edge. Formal parterre, rose and herb gardens. 100ft terrace and herbaceous border overlooking flower rimmed croquet lawn. Cut turf labyrinth and forsythia tunnel. Water features, statues and gazebos. Pets' cemetery. Giant chessboard. Lake with island and temple. Picnics welcome. Wheelchair access with gravel paths and some slopes.

28 COOKSCROFT

Bookers Lane, Earnley, Chichester, PO20 7JG. Mr & Mrs J Williams, 01243 513671, john@cookscroft.co.uk, www.cookscroft.co.uk. *6m S of Chichester. At end of Birdham Straight A286 from Chichester, take L fork to East Wittering B2198. 1m before sharp bend, turn L into Bookers Ln, 2nd house on L. Parking available.* **Wed 10, Sat 13 May (11-4). Evening opening Sat 24 June (5-9). Adm £5, chd free. Light refreshments (May). Wine & cheese (June). Visits also by arrangement Apr to Dec for groups of up to 30.**
A garden for all seasons which delights the visitor. Started in 1988, it features cottage, woodland and Japanese style gardens, water features and borders of perennials with a particular emphasis on southern hemisphere plants. Unusual plants for the plantsman to enjoy, many grown from seed. The differing styles of the garden flow together making it easy to wander anywhere. Wheelchair access over grass, bark paths and unfenced ponds.

29 COPYHOLD HOLLOW

Copyhold Lane, Borde Hill, Haywards Heath, RH16 1XU. Frances Druce, 01444 413265, frances.druce@yahoo.com. *2m N of Haywards Heath. Follow signs for Borde Hill Gardens. With Borde Hill Gardens on L over brow of hill, take 1st R signed Ardingly. Garden ½m. Please park in the lane.* **Sun 28, Mon 29 May (12.30-4). Adm £6, chd free. Home-made teas. Visits also by arrangement May & June.**
A different NGS experience in two north facing acres. The cottage garden surrounding C16 house (not open) gives way to slopes and steps up to woodland garden. Species primulas a particular interest of the owner. Stumpery. Not a manicured plot, but with a relaxed attitude to gardening, an inspiration to visitors.

Knepp Castle

30 NEW **64 CUCKFIELD CRESCENT**

Worthing, BN13 2EB. The Allen Family. *Approx 2m from seafront & Worthing town. Bus routes stop S of Stone Ln, within a 5 min walk. Plenty of parking in Rogate Rd. Please be mindful of residents.* **Thur 13 July (10.30-4); Fri 14 July (1-5); Sun 16 July (10.30-4). Adm £5, chd free. Home-made teas.**

Relax and unwind in this tranquil wildlife friendly town garden. Take a seat under the pergola, surrounded by a vast range of ferns and hostas. Walk along the short winding path enjoying the colour from the borders and, listen to the gentle sound of the water from the koi pond, as you wander through the rose covered arch into a grassed area with a wildlife pond and herbaceous border. Home-made teas inc vegan, gluten free, almond and soya milk options.

31 CUMBERLAND HOUSE

Cray's Lane, Thakeham, Pulborough, RH20 3ER. George & Jane Blunden. *At junction of Cray's Ln & The Street, nr the church.* **Thur 20, Sun 23 July (2-5.30). Combined adm with Thakeham Place Farm £8, chd free. Home-made teas at Thakeham Place Farm.**

A Georgian village house (not open), next to the C12 church with a beautiful, mature, ¾ acre English country garden comprising a walled garden laid out as a series of rooms with well-stocked flower beds, two rare ginkgo trees and yew topiary, leading to an informal garden with vegetable, herb and fruit areas, pleached limes and a lawn shaded by a copper beech tree. Wheelchair access through gate at right-hand side of house.

32 DALE PARK HOUSE

Madehurst, Arundel, BN18 0NP. Robert & Jane Green, 01243 814260, robertgreenfarming@gmail.com. *4m W of Arundel. Take A27 E from Chichester or W from Arundel, then A29 (London) for 2m, turn L to Madehurst & follow red arrows.* **Visits by arrangement 22 May to 10 July for groups of 10+. Adm £5, chd free. Home-made teas.**

Set in parkland, enjoying magnificent views to the sea. Come and relax in the large walled garden which features an impressive 200ft herbaceous border. There is also a sunken gravel garden, mixed borders, a small rose garden, dreamy rose and clematis arches, an interesting collection of hostas, foliage plants and shrubs, and an orchard and kitchen garden.

GROUP OPENING

33 DEAKS LANE GARDENS

Deaks Lane, Ansty, Haywards Heath, RH17 5AS. *3m W of Haywards Heath on A272. 1m E of A23. At r'about at junction of A272 & B2036 take exit A272 Bolney, then take immed R onto Deaks Ln. Park at Ansty Village Centre on R, where tickets for gardens are purchased.* **Sun 11 June (12-4). Combined adm £6, chd free. Light refreshments at Ansty Village Centre.**

BUTTERFLY HOUSE
Cindy & Roger Edmonston.

NEW **8 CROUCH FIELDS**
Danielle & Jason Chapman.

3 LAVENDER COTTAGES
Derry Baillieux.

Bradness Gallery

NUTBOURNE
David Miller.

THICKETS
Becky & Matt Morgan.

All five gardens are close together in the centre of Ansty, each bringing different designs and feel but all united in attracting bees and wildlife. Established and new build gardens with design and planting ideas to take away. Plenty of flowers and interest to see and enjoy.

34 ◆ DENMANS GARDEN

Denmans Lane, Fontwell, BN18 0SU. Gwendolyn van Paasschen. *5m from Chichester & Arundel. Off A27, ½m W of Fontwell r'about.* **For NGS: Sun 26 Feb, Sun 30 Apr, Sun 29 Oct (11-4). Adm £9, chd free. Pre-booking essential, please phone 01243 278950, email office@denmans.org or visit www.denmans.org for information & booking. Light refreshments. For other opening times and information, please phone, email or visit garden website.**

Created by Joyce Robinson, a brilliant pioneer in gravel gardening and former home of influential landscape designer, John Brookes MBE. Denmans is a Grade II registered post-war garden renowned for its curvilinear layout and complex plantings. Enjoy year-round colour, unusual plants, structure and fragrance in the gravel gardens, faux riverbeds, intimate walled garden, ponds and conservatory. On site there is a plant centre with unusual plants for sale, a gift shop and recently opened Midpines Café offering breakfast, lunch and a selection of sweet treats.

35 47 DENMANS LANE

Lindfield, Haywards Heath, RH16 2JN. Sue & Jim Stockwell, 01444 459363, jamesastockwell@aol.com, www.lindfield-gardens.co.uk/47denmans-lane. *Approx 1½m NE of Haywards Heath town centre. From Haywards Heath Train Stn follow B2028 signed Lindfield & Ardingly for 1m. At T-lights turn L into Hickmans Ln, then after 100 metres take 1st R into Denmans Ln.* **Sun 19, Sun 26 Mar, Mon 10 Apr, Mon 1, Sun 21, Mon 29 May (1-5). Adm £6, chd free. Mon 28 Aug (1-5). Combined adm with Lindfield Jungle £7, chd free. Home-made teas. Pre-booking preferred. Visits also by arrangement Mar to Sept for groups of 5+. Visits can be combined with Lindfield Jungle garden from mid-July.**

This beautiful and tranquil 1 acre garden was described by Sussex Life as a 'Garden Where Plants Star'. Created by the owners over the past 20 yrs, it is planted for interest throughout the yr. Spring bulbs are followed by azaleas, rhododendrons, roses and herbaceous perennials. The garden also has ponds, vegetable and fruit gardens. Plants and home-made jams for sale. Most of the garden accessible by wheelchair with some steep slopes.

36 DOWN PLACE

South Harting, Petersfield, GU31 5PN. Mrs David Thistleton-Smith, 01730 825374, selina@downplace.co.uk. *1m SE of South Harting. B2141 to Chichester, turn L down unmarked lane below top of hill.* **Sat 25, Sun 26 Mar, Sun 11, Mon 12 June (1.30-5.30). Adm £5, chd free. Home-made teas & cream teas. Visits also by arrangement 14 Mar to 1 July for groups of 15+.**

Set on the South Downs with panoramic views out to the undulating wooded countryside. A garden which merges seamlessly into its surrounding landscape with rose and herbaceous borders that have been moulded into the sloping ground. There is a well-stocked vegetable garden and walks shaded by beech trees, which surround the natural wildflower meadow where various native orchids flourish. The wildflower meadow is covered in natural wild daffodils in April, cowslips in May, and six varieties of wild orchids, plus various other wild flowers from the end of May through to early July. Substantial top terrace and borders accessible to wheelchairs.

37 DRIFTWOOD

4 Marine Drive, Bishopstone, Seaford, BN25 2RS. Geoff Stonebanks & Mark Glassman, 01323 899296, visitdriftwood@gmail.com, www.driftwoodbysea.co.uk. *A259 between Seaford & Newhaven. From Seaford, turn R into Bishopstone Rd, then immed L into Marine Drive, 2nd on R. Only park same side as house please, not on bend beyond the drive.* **Visits by arrangement 1 June to 11 Aug for groups of up to 20. Adm £6, chd free. Discuss catering options upon booking.**

Monty Don introduced Geoff's garden on BBC Gardeners' World saying, 'a small garden by the sea, full of character with inspired planting and design'. Francine Raymond wrote in her Sunday Telegraph feature, 'Geoff's enthusiasm is catching, he and his amazing garden deserve every visitor that makes their way up his enchanting garden path'. Check out our 170 5-star reviews on TripAdvisor, awarded three successive Certificates of Excellence and two successive Travellers Choice Awards. Selection of Geoff's home-made cakes, all served on vintage china, on trays, in the garden.

38 DURFORD ABBEY BARN

Petersfield, GU31 5AU. Mr & Mrs Lund. *3m from Petersfield. Situated on the S side of the A272 between Petersfield & Rogate, 1m from the junction with B2072.* **Sat 17, Sun 18 June (1-5.30). Adm £5, chd free. Home-made teas.**

A 1 acre plot with areas styled with cottage garden, prairie and shady borders set in the South Downs National Park with views of the Downs. Plants for sale. Partial wheelchair access due to quite steep grass slopes.

39 DURRANCE MANOR

Smithers Hill Lane, Shipley, RH13 8PE. Gordon Lindsay, 01403 741577, gallndsay@gmail.com. *7m SW of Horsham. A24 to A272 (S from Horsham, N from Worthing), turn W towards Billingshurst. Approx 1¾m, 2nd L Smithers Hill Ln signed to Countryman Pub. Garden 2nd on L.* **Mon 28 Aug (12-6). Adm £7, chd free. Home-made teas. Visits also by arrangement 1 May to 24 Sept.**

This 2 acre garden surrounding a medieval hall house (not open) with Horsham stone roof, enjoys uninterrupted views over a ha-ha of the South Downs and Chanctonbury Ring. There are many different gardens here, Japanese inspired gardens, a large pond, wildflower meadow and orchard, colourful long borders and vegetable garden. There is also a Monet style bridge over a pond with water lilies.

GROUP OPENING

40 EAST GRINSTEAD GARDENS

7m E of Crawley on A264 & 14m N of Uckfield on A22. For Allotments park just N of entrance, disabled parking on site. For 5 Nightingale Close & 27 Mill Way park at school in Mill Way, RH19 4DD then 5 min walk. Roadside parking at 35 Blount Ave. **Sun 30 July (1-5). Combined adm £6, chd free. Home-made teas.**

35 BLOUNT AVENUE
RH19 1JJ. Nicki Conlon.

IMBERHORNE ALLOTMENTS
RH19 1TX. Imberhorne Allotment Association, www.imberhorneallotments.org.

27 MILL WAY
RH19 4DD. Jeff Dyson, 01342 311504, jeffreydyson@btinternet.com.
Visits also by arrangement 1 Aug to 17 Sept for groups of up to 15.

5 NIGHTINGALE CLOSE
RH19 4DG. Carole & Terry Heather.

35 Blount Avenue, a contemporary design meets colourful and exuberant planting in this south facing town garden. Colour themed large borders, tropical planting and areas of colour are on show. Dahlias and a variety of interesting perennials feature in a front garden packed with colour. 5 Nightingale Close occupies a corner plot. Landscaped on two levels with mature planting around a naturalistic koi pond, productive fruit, and vegetable area, rose and herbaceous beds, and topiary trees. 27 Mill Way an established developing town garden with tranquil tropical planted rear garden. The front garden has a tropical theme interplanted with dahlias, canna, and salvias. Begonia planted hanging baskets not to be missed. Imberhorne Allotments consist of 80 plots with a diverse range of planting inc grapevines, assorted vegetables, fruit, flowers, and a community orchard, well worth a visit with produce on sale. Teas at 5 Nightingale Close and plant sale at 27 Mill Way. The Town Council hanging baskets and planting in the High St are not to be missed and have achieved a Gold Medal from South and South East in Bloom. For steam train fans, the Bluebell Railway starts nearby.

GROUP OPENING

41 NEW EASTBOURNE GARDEN TRAIL

Two gardens are nr Eastbourne Stn, one close to the Congress Theatre & two nr the Marina. Not a walking trail. Tickets & trail maps available at each garden. **Sun 4 June (1-5). Combined adm £7, chd free. Home-made teas.**

51 CARLISLE ROAD
BN21 4JR. E & N Fraser-Gausden.

DITTONS END
Southfields Road, BN21 1BZ. Mrs Frances Hodkinson, 01323 647163.
Visits also by arrangement Apr to Sept.

Highdown Gardens

HARDWYCKE
Southfields Road, BN21 1BZ. Lois Machin, 01323 729391, loisandpeter@yahoo.co.uk. **Visits also by arrangement.**

NEW **16 HARDY DRIVE**
BN23 6ED. Deb Cornford.

NEW **36 JELLICOE CLOSE**
BN23 6DD. Amanda Haines, 07956 322966, polegate@me.com. **Visits also by arrangement May to Sept.**

Two new gardens in Eastbourne join three other gardens that used to open independently to create a new garden trail. 51 Carlisle Road is a small walled south facing garden with a small pool and mixed beds full of profuse and diverse planting. Dittons End and Hardwycke are just two houses apart and are lovely, well maintained, small town front and back gardens. Dittons End has a small lawn area surrounded by packed borders and pots with lots of colour. Hardwcyke has many usual and unusual plants and many shrubs, inc 50 types of clematis. The gardens at 36 Jellicoe Close and 16 Hardy Drive are at the far end of Eastbourne, both developed on land reclaimed from the sea. 36 Jellicoe Close is a new front and rear garden created over the last 4 yrs on shingle/beach reclaimed land. Mediterranean, cottage-style garden with wildlife, ecology and sustainability at its heart. 16 Hardy Drive is a small urban back garden packed with plants dominated by a large palm tree and full borders.

42 EASTFIELD COTTAGE
Taylors Lane, Bosham, Chichester, PO18 8QQ. Mrs Jenifer Fox, 01243 572479, jenifox@waitrose.com. *3m W of Chichester. At Bosham r'about, take Delling Ln to T-junction, then take L turn, then 1st R into Taylors Ln. Eastfield Cottage is located on the R.* **Visits by arrangement 16 Apr to 25 June for groups of up to 10. Adm £4.50, chd free. Home-made teas.**
Year-round garden featuring a tulip display, wisteria and spring colour. Twenty rose varieties flower in June. Many other interesting shrubs and plants. The design inc a number of zones such as rose pergola, gravel garden, and well planted herbaceous borders. Restful and tranquil. Views to South Downs and within walking distance to Bosham waterfront. Garden designed for owner's wheelchair access.

43 54 ELMLEIGH
Midhurst, GU29 9HA. Wendy Liddle, 07796 562275, wendyliddle@btconnect.com. *¼m W of Midhurst off A272. Parking off road on marked grass area. Reserved disabled parking at top of drive, please phone on arrival for assistance.* **Sat 20, Sun 21 May, Sat 17, Sun 18 June, Sat 15, Sun 16 July, Sat 19, Sun 20 Aug, Sat 9, Sun 10 Sept (11-5). Adm £4, chd free. Home-made teas. Visits also by arrangement 20 May to 10 Sept for groups of 10 to 20. Donation to Chestnut Tree House, Canine Partners.**
⅕ acre property with terraced front garden, leading to a heavily planted rear garden with majestic 120 yr old black pines. Shrubs, perennials, packed with interest around every corner, providing all season colours. Many raised beds, numerous sculptures, vegetables in boxes, a greenhouse, pond and hedgehogs in residence. A large collection of tree lilies, growing 8-10ft. Child friendly. Come and enjoy the peace and tranquillity in this award-winning garden, enjoy our little bit of heaven. Not suitable for large electric buggies.

44 FAIRLIGHT END
Pett Road, Pett, Hastings, TN35 4HB. Chris & Robin Hutt, 07774 863750, chrishutt@fairlightend.co.uk, www.fairlightend.co.uk. *4m E of Hastings. From Hastings take A259 to Rye. At White Hart Beefeater turn R into Friars Hill. Descend into Pett village. Park in village hall car park, opp house.* **Wed 26 Apr (11-5). Adm £6, chd free. Home-made teas. Visits also by arrangement May to Sept for groups of 10 to 35. Donation to Pett Village Hall.**
Gardens Illustrated said 'The 18th century house is at the highest point in the garden with views down the slope over abundant borders and velvety lawns that are punctuated by clusters of specimen trees and shrubs. Beyond and below are the wildflower meadows and the ponds with a backdrop of the gloriously unspoilt Wealden landscape'. Wheelchair access with steep paths, gravelled areas and unfenced ponds.

45 FAIRLIGHT HALL
Martineau Lane, Hastings, TN35 5DR. Mr & Mrs David Kowitz, www.fairlighthall.co.uk. *2m E of Hastings. A259 from Hastings towards Dover & Rye, 2m turn R into Martineau Ln.* **Sat 19, Sun 20 Aug (10-4). Adm £8, chd free. Pre-booking preferred. Home-made teas.**
A welcome return to the NGS, a recently restored stunning garden in East Sussex. The formal gardens extend over 9 acres and surround the Victorian Gothic mansion (not open). Features semitropical woodland avenues, a huge contemporary walled garden with amphitheatre and two 110 metre perennial borders above and below ha-ha with far reaching views across Rye Bay. Home-made preserves for sale. Most of the garden can be viewed by wheelchair, please inform us when you arrive, so we can direct you to a disabled parking place.

46 FINDON PLACE
Findon, Worthing, BN14 0RF. Miss Caroline Hill, 01903 877085, hello@findonplace.com, www.findonplace.com. *Directly off A24 N of Worthing. Follow signs to Findon Parish Church & park through the 1st driveway on LHS.* **Sun 2 July (3-7). Adm £7, chd free. Pre-booking essential, please visit www.ngs.org.uk for information & booking. Cream teas. Visits also by arrangement 1 May to 8 Sept for groups of 15 to 40.**
Stunning grounds and gardens surrounding a Grade II listed Georgian country house (not open), nestled at the foot of the South Downs. The most glorious setting for a tapestry of perennial borders set off by Sussex flint walls. The many charms inc a yew allee, cloud pruned trees, espaliered fruit trees, a productive ornamental kitchen garden, rose arbours and arches, and a cutting garden.

In 2022, 20,000 people were supported by Perennial caseworkers and prevent teams thanks to our funding

47 FITTLEWORTH HOUSE

Bedham Lane, Fittleworth, Pulborough, RH20 1JH. Edward & Isabel Braham, 01798 865074, marksaunders66.com@gmail.com, www.racingandgreen.com. *2m E, SE of Petworth. Midway between Petworth & Pulborough on the A283 in Fittleworth, turn into lane by sharp bend signed Bedham. Garden is 50yds along on the L. Plenty of car parking space.* **Every Wed 19 Apr to 10 May (2-5). Wed 7, Wed 14, Wed 21 June, Wed 12, Wed 19 July, Wed 9 Aug (2-5). Adm £5, chd free. Home-made teas. Visits also by arrangement 19 Apr to 23 Aug for groups of 8 to 40. Weekdays only.**

3 acre tranquil, romantic, country garden, walled kitchen garden growing a wide range of fruit, vegetables and flowers inc a large collection of dahlias. Large glasshouse and old potting shed, mixed flower borders, roses, rhododendrons and lawns. Magnificent 115ft tall cedar overlooks wisteria covered Grade II listed Georgian house (not open). Wild garden, long grass areas and stream. The garden sits on a gentle slope, but is accessible for wheelchairs and buggies.

48 FIVE OAKS COTTAGE

Petworth, RH20 1HD. Jean & Steve Jackman. *5m S of Pulborough. SatNav does not work! To ensure best route, we provide printed directions, please email jeanjackman@hotmail.com or call 07939 272443.* **Fri 9, Sat 10, Sun 11 June (10-5). Adm £5. Pre-booking essential, please visit www.ngs.org.uk for information & booking. Home-made teas (cash only).**

An acre of delicate jungle surrounding an Arts and Crafts style cottage (not open) with stunning views of the South Downs. Our unconventional garden is designed to encourage maximum wildlife with a knapweed and hogweed meadow on clay attracting clouds of butterflies, plus two small ponds and lots of seating. An award-winning, organic garden with a magical atmosphere.

49 THE FOLLY

Charlton, Chichester, PO18 0HU. Joan Burnett & David Ward, 07711 080851, joankeirburnett@gmail.com, www.thefollycharlton.com. *7m N of Chichester & S of Midhurst off A286 at Singleton, follow signs to Charlton. Follow NGS parking signs. No parking in lane, drop off only. Parking nr the Fox Goes Free pub.* **Sun 25 June, Sun 23 July, Sun 20 Aug (2-5). Adm £5, chd free. Home-made teas. Visits also by arrangement 25 June to 20 Aug.**

Colourful cottage garden surrounding a C16 period house (not open), set in pretty downland village of Charlton, close to Levin Down Nature Reserve. Herbaceous borders well-stocked with a wide range of plants. Variety of perennials, grasses, annuals and shrubs to provide long season of colour and interest. Old well. Busy bees. Art Studio open to visitors. Partial wheelchair access with steps from patio to lawn. Visitors with mobility issues can be dropped off at the gate. No dogs.

50 FOREST RIDGE

Paddockhurst Lane, Balcombe, RH17 6QZ. Philip & Rosie Wiltshire, 07900 621838, rosiem.wiltshire@btinternet.com. *3m from M23, J10a. M23 J10a take B2036 to Balcombe. After 2/3m take 1st L onto B2110. After Worth School turn R into Back Ln. After 2m it becomes Paddockhurst Ln. Forest Ridge on R after 2¼m. Ignore SatNav to track!* **Sun 7 May (2-5.30). Adm £5, chd free. Home-made teas.**

A charming 4½ acre Victorian garden with far-reaching views, boasting the oldest Atlantic cedar in Sussex. The owners themselves are currently undertaking a major restoration: felling, planting and redesigning areas. Within the garden there is formal and informal planting, woodland dell and mini arboretum. Azaleas, rhododendrons and camellias abound, rare and unusual species. The major restoration project has opened up new vistas. Although the garden received the full force of Storm Eunice, it has now been cleared and along with a large amount of chopped timber, further vistas have opened up. A garden to watch over the coming yrs as the new plantings develop and the newly designed areas grow and flourish. A garden to explore! Please Note: There are several deep-water ponds.

51 FOXGLOVE COTTAGE

29 Orchard Road, Horsham, RH13 5NF. Peter & Terri Lefevre, 01403 256002, teresalefevre@outlook.com. *At Horsham Stn, over bridge, at r'about 3rd exit (Crawley), 1st R Stirling Way, at end turn L, 1st R Orchard Rd. From A281, take Clarence Rd, at end turn R, at end turn L Orchard Rd. Street parking.* **Sun 28 May, Wed 5, Sun 9 July (1-5). Adm £5, chd free. Home-made teas. Vegan, gluten & dairy free cake. Visits also by arrangement June & July for groups of 10+.**

A ¼ acre plantaholic's garden, full of containers, vintage finds and quirky elements. Paths intersect both sun and shady borders bursting with colourful planting and salvias in abundance! A beach hut summerhouse and deck are flanked by a water feature in a pebble circle. Two additional small ponds encourage wildlife. A surprise at every turn. The end of the garden is dedicated to a plant nursery. A recently added gravel garden for drought tolerant plants with a large, rusty metal arbour, called the Dome! Members of the Hardy Plant Society. A large selection of unusual plants for sale. Please check our Facebook page 'Foxglove-Cottage-Horsham' for further information.

52 FOXWOOD BARN

Little Heath Road, Fontwell, Arundel, BN18 0SR. Stephen & Teresa Roccia, 07867 383753, Teresa.m.roccia@gmail.com. *At the junction of Little Heath Rd & Dukes Rd, follow NGS signs. Separate drop off area for wheelchair access.* **Sat 10 June (11-4). Adm £5, chd free. Home-made teas. Visits also by arrangement 10 June to 31 July.**

From a green field site 4 yrs ago, this 1 acre garden has been developed into different areas inc a wildlife pond, rose and gravel garden, patio areas and a dry riverbed. There are herbaceous borders, wildflower areas and a small orchard and allotment. Picnic area adjacent to a paddock with donkeys and ponies.

53 THE GARDEN HOUSE

5 Warleigh Road, Brighton, BN1 4NT. Bridgette Saunders & Graham Lee, 07729 037182, contact@gardenhousebrighton.co.uk, www.gardenhousebrighton.co.uk. *1½m N of Brighton pier, Garden House is 1st L after Xrds, past the open market. Paid street parking. London Road Stn nearby. Buses 26 & 46 stopping at Bromley Rd.*

The Beeches

Fri 21, Sat 22 Apr (1.30-4.30). Adm £6, chd free. Home-made teas. Visits also by arrangement 31 Mar to 1 Oct for groups of 15 to 35. Tour & talk about the garden on request.

One of Brighton's secret gardens. We aim to provide year-round interest with trees, shrubs, herbaceous borders and annuals, fruit and vegetables, two glasshouses, a pond and rockery. A friendly garden, always changing with a touch of magic to delight visitors, above all it is a slice of the country in the midst of a bustling city. Plants for sale.

54 GORSELANDS

Common Hill, West Chiltington, Pulborough, RH20 2NL. Philip Maillou. *12m N of Worthing, approx 1m N of Storrington. Take either School Hill or Old Mill Drive from Storrington. 2nd L on to Fryern Rd to West Chiltington & continue on to Common Hill. Gorselands is on the R. Roadside parking.* **Sun 25 June, Sun 2 July (2-5). Adm £5, chd free. Home-made teas.**

Approx 3/4 acre garden featuring mature mixed borders, dahlia garden, fruit area, woodland area with giant redwood, camellias, azaleas, rhododendrons and two ponds. While wandering round the garden there will be gentle music by members of the Downland Ensemble. Also on display will be sculptures for sale by Uckfield sculptor Allan Mackenzie. Wheelchair access on sloping garden with short grass.

55 ◆ GREAT DIXTER HOUSE, GARDENS & NURSERIES

Northiam, TN31 6PH. Great Dixter Charitable Trust, 01797 253107, ticketoffice@greatdixter.co.uk, www.greatdixter.co.uk. *8m N of Rye. Off A28 in Northiam, follow brown signs.* **For opening times and information, please phone, email or visit garden website.**

Designed by Edwin Lutyens and Nathaniel Lloyd. Christopher Lloyd made the garden one of the most experimental and constantly changing gardens of our time, a tradition now being carried on by Fergus Garrett. Clipped topiary, wildflower meadows, the famous long border, pot displays, exotic garden and more. Spring bulb displays are of particular note. Please see garden website for accessibility information.

56 GROVE FARM HOUSE

Paddockhurst Road, Turners Hill, RH10 4SF. Mr & Mrs Piers Gibson. *1/4 m W of Turners Hill on B2110. The garden is on the edge of the village of Turners Hill; entrance is 1st on the L on Paddockhurst Rd (B2110), 1/4 m W of village Xrds.* **Sat 17 June (2-6). Adm £7, chd free.**

A welcome return to the NGS for this gorgeous, 4 acre classic terraced garden with views of South Downs. The garden inc a circular yew maze, ha-ha, lime walk, herb and vegetable gardens, shrub and herbaceous borders, and a pond in a woodland setting.

57 HAMMERWOOD HOUSE
Iping, Midhurst, GU29 0PF. Mr & Mrs M Lakin. *3m W of Midhurst. Take A272 from Midhurst, approx 2m outside Midhurst turn R for Iping. Over bridge, uphill to grassy junction, turn R. From A3 leave for Liphook, follow B2070, turn L for Milland & Iping.* **Sun 14 May (1-5). Adm £6, chd free. Home-made teas.**
Large south facing garden with lots of mature shrubs inc camellias, rhododendrons and azaleas. An arboretum with a variety of flowering and fruit trees. The old yew and beech hedges give a certain amount of formality to this traditional English garden. Tea on the terrace is a must with the most beautiful view of the South Downs. For the more energetic there is a woodland walk. Partial wheelchair access as garden is set on a slope.

GROUP OPENING

58 HARLANDS GARDENS
Penland Road, Haywards Heath, RH16 1PH. *Follow yellow signs from Balcombe Rd or Milton Rd & Bannister Way (Sainsbury's). Bus stop on Bannister Way (Route 30, 31, 39 & 80). Gardens within 5 min easy walk from train station & bus stop. Roadside parking is readily available. Park once to visit all gardens.* **Sat 3, Sun 4 June (12-4.30). Combined adm £5, chd free. Home-made teas.**

52 PENLAND ROAD
Karen & Marcel van den Dolder.

55 PENLAND ROAD
Steve & Lisa Williams.

72 PENLAND ROAD
Sarah Gray & Graham Delve.

5 SUGWORTH CLOSE
Lucy & Brian McCully.

NEW **57 TURNERS MILL ROAD**
Linda & Steve Mitchell.

Come and meet old favourites and make new friends as we welcome a new garden to our eclectic mix of inspiring town gardens. We comprise differing shapes and gradients, with gardens designed with both relaxation and socialisation in mind. We offer rich cottage style planting schemes, varied ponds, pretty courtyards, practical kitchen gardens and habitats for wildlife. Look out for interesting pots and carnivorous plants. Please check our Facebook page 'Harlands Gardens' for updated information. Our home-made cakes gifted by neighbours are always a delight and reflect our sense of community. The quaint English villages of Cuckfield and Lindfield are worthy of a visit. Nymans (NT) and Wakehurst (Kew) are also nearby. Partial wheelchair access to most gardens, no access to No. 72.

GROUP OPENING

59 HASSOCKS GARDEN TRAIL
Hassocks, BN6 8EY. *Close to A23, 6m N of Brighton, off A273. Gardens are between Stonepound Xrds & Keymer, N & S of B2116. Park on road or in free car parks in Hassocks. No parking in Parklands Rd. Trail walkable from Hassocks Stn.* **Sun 25 June (1-5). Combined adm £6, chd free. Home-made teas at St Edwards's Church, Lodge Lane, BN6 8NT.**

NEW **5 EWART CLOSE**
Jeannie & Alex Sheppard.

NEW **LODGE HOUSE**
Jo & Glen Welfare.

THE OLD THATCH
Melville Moss & Julie Latham.

PARKLANDS ROAD ALLOTMENTS
Tony Copeland & Jeannie Brooker.

Three contrasting gardens to suit many interests plus the allotments, along with a small strip of ancient woodland, make up this walkable trail. There is a young, modern garden around new build home (not open), a well established cottage garden at the 500 yr old, thatched cottage (not open) and a large wildlife garden close by, with the allotments a short walk away. The wildlife/cottage garden is wonderful but as to be expected, it is not manicured. There are some trip hazards, uneven steps, low branches and slippery areas, but as you explore this rich habitat the progress made by the owners in just 5 yrs is evident. The 55 allotments are not to be missed, a range of classic vegetables inc some exotic ones are grown. The allotmenteers will show you their plots and answer questions. There are spectacular views across to the Downs to Jack and Jill Windmills. Public WC available at Adastra Park, BN6 8QH. Dogs permitted at allotments only.

60 ◆ HERSTMONCEUX CASTLE ESTATE
Wartling Road, Hailsham, BN27 1RN. Bader College, Queen's University (Canada), 01323 833816, caroline.harber@queensu.ca, www.herstmonceux-castle.com. *Located between Herstmonceux & Pevensey on Wartling Rd. From Herstmonceux take A271 to Bexhill, 2nd R signed Castle. If using SatNav, enter Wartling Rd instead of postcode.* **For NGS: Wed 26 July (10-7). Adm £8, chd £3.50. Light refreshments in Chestnuts Tea Room. For other opening times and information, please phone, email or visit garden website.**
The Herstmonceux Castle Estate has formal gardens, woodland trails, meadows and lakes set around a majestic C15 moated castle. Features inc an avenue of ancient sweet chestnut trees and lakeside folly in the style of a Georgian house with a walled cottage garden. The gardens and grounds first opened for the NGS in 1927. Last admission 5pm. Partial wheelchair access to formal gardens.

GROUP OPENING

61 HERSTMONCEUX PARISH TRAIL
4m NE of Hailsham. Follow yellow NGS signs, maps given out at each venue. Ticket covers all gardens. **Sun 25 June (12-5). Combined adm £7, chd free. Light refreshments at Hill House. Teas (12-5) & BBQ lunch (12-2) at The Windmill.**

THE ALLOTMENTS, STUNTS GREEN
BN27 4PP. Nicola Beart.

COWBEECH HOUSE
BN27 4JF. Mr Anthony Hepburn.

1 ELM COTTAGES
BN27 4RT. Audrey Jarrett.

NEW **HILL HOUSE**
BN27 4RT. Maureen Madden & Terry Harland.
(See separate entry)

NEW **KERPSES**
BN27 4JG. Mrs Lynn & Mr Peter Maguire.

MERRIE HARRIERS BARN
BN27 4JQ. Lee Henderson.

THE WINDMILL
BN27 4RT. Windmill Hill Windmill Trust, **windmillhillwindmill.org.**

Seven gardens inc a historic windmill and allotments. In Cowbeech there are three gardens. Cowbeech House, a place to linger, has an exciting range of water features and sculpture in the garden dating back to 1731. Vintage car collection on site to view. Merrie Harriers Barn is a garden with sweeping lawn and open countryside beyond. Colourful herbaceous planting and a large pond with places to sit and enjoy the view. A developing garden that 7 yrs ago was agricultural land. Kerpes, new in 2023, is a delightful mixture of mature trees, shrubs, herbaceous borders, vegetable garden, ponds and a meadow. The Allotments at Stunts Green comprise 54 allotments growing a huge variety of traditional and unusual crops. In Herstmonceux there are three gardens to visit. 1 Elm Cottages is a lovely cottage garden packed full of edible and flowering plants you cannot afford to miss. Hill House, a developing garden with trees, shrubs, many perennials and roses. Wildlife pond with water lilies. Plenty of seating areas to sit and reflect and have tea. The historic Windmill is well worth a visit and where you can enjoy a BBQ lunch or tea.

Sandhill Farm House

© Judi Lion

62 NEW THE HIDDEN GARDEN

School Lane (behind Selsey Library), Selsey, Chichester, PO20 9EH. Paul Sadler. *Park behind Selsey Library or in front of the Academy School on School Ln. You will see The Bridge Support Centre, enter through the gate & The Hidden Garden is behind the centre.* **Sun 30 Apr, Sun 30 July (10.30-4.30). Adm £5, chd free. Tea.**

The Hidden Garden is a community gardening project encouraging local people to become involved with growing fruit, vegetables, herbs and flowers as well as providing spaces for wildlife to thrive. The garden is open to people of all ages and abilities organised by the Selsey Community Forum but looked after by a dedicated group of local volunteers.

63 ◆ HIGH BEECHES WOODLAND AND WATER GARDEN

High Beeches Lane, Handcross, Haywards Heath, RH17 6HQ. High Beeches Gardens Conservation Trust, 01444 400589, gardens@highbeeches.com, www.highbeeches.com. *5m NW of Cuckfield. On B2110, 1m E of A23 at Handcross.* **For NGS: Sun 4 June, Sun 1 Oct (1-5). Adm £10, chd free. Light refreshments. For other opening times and information, please phone, email or visit garden website.**

25 acres of enchanting, landscaped woodland and water gardens with spring daffodils, bluebells and azalea walks, many rare and beautiful plants, an ancient wildflower meadow and glorious autumn colours. Picnic area. National Collection of Stewartias.

NPC

64 ◆ HIGHDOWN GARDENS

33 Highdown Rise, Littlehampton Road, Goring-by-Sea, Worthing, BN12 6FB. Worthing Borough Council, 01273 263060, highdown.gardens@adur-worthing.gov.uk, www.highdowngardens.co.uk. *3m W of Worthing. Off the A259 approx 1m from Goring-by-Sea Train Stn.* **For NGS: Wed 15 Feb (10-4.30); Wed 31 May (10-8). Adm by donation. For other opening times and information, please phone, email or visit garden website.**

Highdown Gardens were created by Sir Frederick Stern. They are home to rare plants and trees, many grown from seed collected by Wilson, Farrer and Kingdon-Ward. A fully equipped glasshouse enables the propagation of this National Plant Collection. A visitor centre shares stories of the plants and people behind the gardens. A new accessible path leads to a sensory garden with a secret sea view. Accessible top pathway and lift to visitor centre, see garden website accessibility page for full details.

The Oast

© Bennet Smith

65 NEW HILL HOUSE

Windmill Hill, Hailsham, BN27 4RT. Maureen Madden & Terry Harland. *4m NE of Hailsham. Opp Windmill Hill PO. Parking for about 6 cars otherwise parking in adjacent streets.* **Sat 1, Sun 2 July (2-5). Adm £4, chd free. Cream teas. Opening with Herstmonceux Parish Trail on Sun 25 June.**

5 yrs ago the garden was just lawn and a pond which had been filled in. The pond now is a wildlife pond with water lilies, and the garden has three sections divided by a pergola and two arches with climbing roses and honeysuckle. Trees, shrubs and as many perennials as can fit in the flower beds have been planted, and roses abound in the rose bed. A garden very easy to walk around with plenty of seating areas to sit and reflect. Wheelchair access through side gate. No steps in main garden, but steps (not steep) at end of garden to greenhouse and vegetable patch.

66 NEW HILLSIDE

The Green, Rottingdean, Brighton, BN2 7HA. Nicky & Dave Boys. *5m E of Brighton on A259. Turn onto High St, which becomes The Green. Last house on L before RH-bend, opp Kipling Gardens. Seafront short & long stay car parks, 5 mins. Public transport from Brighton, Eastbourne & Woodingdean.* **Fri 2, Sun 4 June (12-5). Adm £6, chd free. Home-made teas.**

Historic Grade II* listed, flint walled garden with C18 barn and gazebo. Large chalk downland garden with naturalistic planting and mown pathways. Relaxed not manicured. Roses, shady trees, birds and bees give a secret garden feel. Herbaceous, woodland and poolside planting, willow walk and cottage garden. Behind the barn, ancient trees and beehives fringe a wild meadow with views to windmill.

67 4 HILLSIDE COTTAGES

Downs Road, West Stoke, Chichester, PO18 9BL. Heather & Chris Lock, 01243 574802, chlock@btinternet.com. *3m NW of Chichester. From A286 at Lavant, head W for 1½m, nr Kingley Vale.* **Sun 11 June, Sun 16 July (11-4). Adm £5, chd free. Home-made teas. Visits also by arrangement 1 June to 20 Aug.**

In a rural setting this stunning garden is densely planted with mixed borders and shrubs. Large collection of roses, clematis, fuchsias and dahlias, a profusion of colour and scent in a well maintained garden. Please visit www.ngs.org.uk for pop-up openings in June, July and August.

68 HOLFORD MANOR & CHAILEY IRIS GARDEN

Holford Manor Lane, North Chailey, Lewes, BN8 4DU. Martyn Price, 01444 471714, martyn@holfordmanor.com, www.chailey-iris.co.uk. *4½m SE of Haywards Heath. SE From Scaynes Hill on A272, after 1¼m turn R onto Holford Manor Ln.* **Fri 26, Sat 27, Sun 28 May, Fri 16, Sat 17, Sun 18 June (11-4). Adm £8, chd free. Home-made teas. Visits also by arrangement 22 May to 30 June for groups of 10+.**

5 acre garden and the largest iris nursery in the UK, Chailey Iris Gardens. The manor house gardens feature herbaceous borders, iris beds and formal parterre rose garden. Secret Chinese garden, tropical beds, cutting garden and a wildflower meadow. Ornamental pond, landscaped 'horse pond' with water lilies and a 2 acre lake and swamp garden. In the Chailey Iris Gardens walk amongst 1000s of iris in full bloom showcasing 700 varieties of bearded iris, as well as Japanese ensata iris, siberica iris and many rare species iris. Many iris available to buy on the day, and experts are on hand for help and advice. Brick or gravel paths give wheelchair access to most of the garden.

69 HOLLIST HOUSE

Hollist Lane, Easebourne, Midhurst, GU29 9RS. Mrs Caro & Mr Gavin Darlington. *Coming from Midhurst towards Petworth take 1st L on r'about at bottom of town signed Haslemere. Then follow large yellow NGS arrow signs.* **Sat 1 July (2-5.30). Adm £5, chd free. Home-made teas.**

Traditional English garden with natural spring pond and the River Rother running through acres of lawns, woods, herbaceous borders and kitchen garden.

70 HOLLY HOUSE

Beaconsfield Road, Chelwood Gate, Haywards Heath, RH17 7LF. Mrs Deirdre Birchell, 01825 740484, db@hollyhousebnb.co.uk, www.hollyhousebnb.co.uk. *7m E of Haywards Heath. From Nutley village on A22 turn off at Hathi Restaurant signed Chelwood Gate 2m. Chelwood Gate Village Hall on R, Holly House is opp.* **Sat 19, Sun 20 Aug (2-5). Adm £6, chd free. Home-made teas. Visits also by arrangement 8 May to 31 Aug.**

An acre of English garden providing views and cameos of plants and trees round every corner with many different areas giving constant interest. A fish pond and a wildlife pond beside a grassy area with many shrubs and flower beds. Among the trees and winding paths there is a cottage garden which is a profusion of colour and peace. Exhibition of paintings and cards by owner. Garden accessible by wheelchair in good weather, but it is not easy.

71 HOLLYMOUNT

Burnt Oak Road, High Hurstwood, Uckfield, TN22 4AE. Jonathan Hughes-Morgan, 07968 848418, jonnyhughesmorgan@gmail.com. *Exactly halfway between Uckfield & Crowborough, just off the A26. From A26 S of Crowborough or A272 between Uckfield & Buxted, follow sign to High Hurstwood. From N approx 2m down Chillies Ln take 1st L, from S 1½m up Hurstwood Rd take 2nd R, ½m up on L.* **Sun 7 May, Sun 18 June, Sun 23 July (2-6). Adm £7, chd free. Home-made teas. Visits also by arrangement for groups of 10+.**

A new 7 acre garden on sloping hillside. Four large ponds, a stream, waterfalls, terraced beds, interesting planting, some tropical, large rhododendrons, wildlife; alpacas, chickens, ducks, fish. The garden has a great variety of trees, two greenhouses, shepherds hut, summerhouse, gravel paths, kitchen garden, many decks and wildflower terracing to the rear. Very peaceful setting with great views.

D

72 JACARANDA

Chalk Road, Ifold, RH14 0UE. Brian & Barbara McNulty, 01403 751532, bmcn0409@icloud.com. *1m S of Loxwood. From A272/A281 take B2133 (Loxwood). ½m S of Loxwood take Plaistow Rd, then 3rd R into Chalk Rd. Follow signs for parking & garden. Wheelchair users can park in driveway.* **Sat 10 June (1-5). Adm £5, chd free. Home-made teas. Visits also by arrangement May to Sept for groups of up to 25.**

Created from scratch by the owners over the past 21 yrs, this is a plant lover's garden with curved borders featuring unusual shrubs, trees with interesting bark, roses, climbers and perennials. Hostas in pots are displayed in many places. The vegetable and cutting area feature a large raised bed, a greenhouse and a hanging potting bench. A garden for all seasons!

73 JUDY'S COTTAGE GARDEN

33 The Plantation, Worthing, BN13 2AE. Mrs Judy Gordon. *Salvington. A24 meets A27 at Offington r'about, turn into Offington Ln, 1st R into The Plantation.* **Fri 31 Mar, Fri 2 June, Fri 8 Sept (11-4). Adm £5, chd free. Home-made teas.**

A beautiful medium sized cottage garden with something of interest all year-round. The garden has several mature trees creating a feeling of seclusion. The informal beds contain a mixture of shrubs, perennials, cottage garden plants and spring bulbs. There are little hidden areas to enjoy, a small fish pond and other water features. There is also a pretty log cabin overlooking the garden.

74 THE JUNGLE GARDEN

Saltdean, Brighton, BN2 8PJ. Sue & Ray Warner, 01273 305137/07990 992985, suwarner.uk@gmail.com. *A259 between Rottingdean & Telscombe, turn into Saltdean (Longridge Ave) at T-lights. Rodmell Ave last turning on R, house ½m on R. Parking on the road outside the house or adjoining streets.* **Sat 15, Sun 16 July (12-5). Combined adm with Bagotts Rath £6, chd free. Home-made teas. Visits also by arrangement 1 July to 20 Aug for groups of 5 to 12. Refreshments on prior request.**

A fun jungle garden, created in 2012, but completely revamped in 2021 and returning to the NGS after a 3 yr break. Measuring 65ft x 36ft but appearing larger with winding paths that lead you through lush jungle and insect friendly planting, listen out for jungle sounds. Tea and home-made cakes served overlooking the garden. Exquisite planting of exotic plants.

75 KEMP TOWN ENCLOSURES: SOUTH GARDEN

Lewes Crescent, Brighton, BN2 1FH. Kemp Town Enclosures Ltd, kte.org.uk. *On S coast, 1m E of Brighton Palace Pier, ½m W of Brighton Marina. 2m from Brighton Stn. Bus No. 7, stop St Mary's Hall. M23/A23 to Brighton Palace Pier; L on A259. 5m S A27 on B2123, R on A259.* **Wed 14 June (12-5). Adm £7.50, chd free. Pre-booking essential, please visit www.ngs.org.uk for information & booking. Tours every 30 mins from 12pm to 3.30pm (2 groups of 15 per tour) inc. No refreshments, good coffee shops nearby.**

Unique historic Grade II listed Regency private town garden in a spectacular seaside setting. Enduring strong salty winds and thin chalk soil, the garden balances naturalistic planting and a more ordered look, relaxed not manicured. With several distinctive areas, this wonderful 5 acre garden combines open lawns, winding paths, trees, herbaceous and shrub borders and a shaded woodland garden. Celebrating its Bicentennial in 2023. Designed by Henry Phillips, the garden was a central feature of the Regency development of the Kemp Town Estate. Queen Victoria and Edward VII walked in these gardens. Ramped gate entry. Gravel winding paths give access to most of the garden. Relatively steep slopes on main lawn and in the garden tunnel.

Our 2022 donation supports 1,700 Queen's Nurses working in the community in England, Wales, Northern Ireland, the Channel Islands and the Isle of Man

76 ◆ KING JOHN'S LODGE

Sheepstreet Lane, Etchingham, TN19 7AZ. Jill Cunningham, 01580 819220, harry@kingjohnsnursery.co.uk, www.kingjohnsnursery.co.uk. *2m W of Hurst Green. Off A265 nr Etchingham. From Burwash turn L before Etchingham Church, from Hurst Green turn R after church, into Church Ln, which leads into Sheepstreet Ln after ½m, then L after 1m.* **For NGS: Sat 25 Mar, Sat 29, Sun 30 Apr, Sat 17, Sun 18 June, Sat 9 Sept (11-5). Adm £5, chd free. Home-made teas. For other opening times and information, please phone, email or visit garden website.**

4 acre romantic garden for all seasons. An ongoing family project since 1987 with new areas completed in 2020. From the eclectic shop, nursery and tearoom, stroll past wildlife pond through orchard with bulbs, meadow, rose walk and fruit according to the season. Historic house (not open) has broad lawn, fountain, herbaceous border, pond and ha-ha. Explore secret woodland with renovated pond and admire majestic trees and 4 acre meadows. Garden is mainly flat. Areas with steps can usually be accessed from other areas. Disabled WC.

77 NEW 255 KINGS DRIVE

Willingdon, Eastbourne, BN21 2UR. Barbara & Stephen Stroud. *On the A2021 between Eastbourne & Polegate, between Hampden Park & Decoy Drive on the LHS of Kings Drive travelling N out of Eastbourne.* **Sun 2 July, Sun 10 Sept (1-5). Adm £5, chd free. Light refreshments. Open nearby 50 Wannock Lane on 2 July only.**

½ acre tropical paradise. Lush jungle planting and unusual vibrant plants. Walk through dense planting with lots of secret nooks and crannies. Formal koi pond surrounded by giant cannas and gunnera. Two colourful borders lead you into the garden.

78 KITCHENHAM FARM

Kitchenham Road, Ashburnham, Battle, TN33 9NP. Amanda & Monty Worssam. *5m S of Battle. S of Ashburnham Place from A271 Herstmonceux to Bexhill road, take L turn 500 metres after Boreham St. Kitchenham Farm is 500 metres on L.* **Tue 6, Sat 10 June, Wed 9, Sat 12 Aug (2-5). Adm £5, chd free. Home-made teas.**

1 acre country house garden set amongst traditional farm buildings with stunning views over the Sussex countryside. Series of borders around the house and Oast House (not open). Lawns and mixed herbaceous borders inc roses and delphiniums. A ha-ha separates the garden from the fields and sheep. The garden adjoins a working farm. Wheelchair access to the garden. One step to WC.

79 ◆ KNEPP CASTLE

West Grinstead, Horsham, RH13 8LJ. Sir Charles & Lady Burrell, www.kneppsafaris.co.uk. *8m S of Horsham. Turning to Shipley off the A272. ½m to entrance on L. Follow driveway through the parkland. Follow signs to parking.* **For NGS: Sat 27 May (10-5.30). Adm £15, chd free. Pre-booking essential, please visit www.ngs.org.uk for information & booking. Light refreshments. For other opening times and information, please visit garden website.**

The Walled Garden at Knepp has been transformed into a garden for biodiversity. With designers Tom Stuart-Smith and James Hitchmough, we have applied some of the principles we've learned from rewilding the wider landscape to this small, confined space to create a mosaic of dynamic habitats for wildlife. The croquet lawn is now a riot of humps and hollows, hosting almost 900 species of plants.

D

80 KNIGHTSBRIDGE HOUSE

Grove Hill, Hellingly, Hailsham, BN27 4HH. Andrew & Karty Watson. *3m N of Hailsham, 2m S of Horam. From A22 at Boship r'about take A271, at 1st set of T-lights turn L into Park Rd, new road layout here, so you need to take a L turn approx 150 metres from the lights & drive for 2⅔m, garden on R.* **Wed 6, Sat 9 Sept (2-5). Adm £7, chd free. Home-made teas.**

Mature landscaped garden set in 5 acres of tranquil countryside surrounding Georgian house (not open). Several garden rooms, spectacular herbaceous borders planted in contemporary style. Lots of late season colour with grasses and some magnificent specimen trees; also partly walled garden. Note no card machine at this garden (poor phone signal). Wheelchair access to most of garden with gravel paths. Ask to park by house if slope from car park is too steep.

81 28 LARKSPUR WAY

Southwater, Horsham, RH13 9GR. Peter & Marjorie Cannadine. *From A24 take Worthing Rd signed Southwater, at 2nd r'about take 1st exit Blakes Farm Rd. At T-junction turn R, follow road round into Larkspur Way. At T-junction turn L, then R, No. 28 is on the L.* **Sun 14 May, Sun 4 June (2-5). Adm £4, chd £2. Home-made teas.**

Created for a new build this small garden is packed with big ideas for those looking for inspiration. A small front garden with pond, Mediterranean area and well-stocked border. The back garden being approx 15 x 10 metres, backing onto woodland is filled with cottage garden plants, containers, ornaments and more to inspire.

82 LEGSHEATH FARM

Legsheath Lane, nr Forest Row, RH19 4JN. Mr & Mrs M Neal, legsheath@btinternet.com. *4m S of East Grinstead. 2m W of Forest Row, 1m S of Weirwood Reservoir.* **Sat 20 May (2-4.30). Adm £5, chd free. Home-made teas. Visits also by arrangement 22 Apr to 9 Sept. Donation to Holy Trinity Church, Forest Row.**

Legsheath was first mentioned in Duchy of Lancaster records in 1545. It was associated with the role of Master of the Ashdown Forest. Set high in the Weald with far-reaching views of East Grinstead and Weirwood Reservoir. The garden covers 11 acres with a spring fed stream feeding ponds. There is a magnificent davidia, rare shrubs, embothrium and many different varieties of meconopsis and abutilons.

83 LIMEKILN FARM

Chalvington Road, Chalvington, Hailsham, BN27 3TA. Dr J Hester & Mr M Royle. *10m N of Eastbourne. Nr Hailsham. Turn S off A22 at Golden Cross & follow the Chalvington Rd for 1m. The entrance has white gates on LHS. Disabled parking space close to house, other*

parking 100 metres further along road. **Sat 13, Sun 14 May, Sat 23, Sun 24 Sept (2-5). Adm £7, chd free. Home-made teas in the Oast House.**
The garden was designed in the 1930s when the house was owned by Charles Stewart Taylor, MP for Eastbourne. It has not changed in basic layout since then. The planting aims to reflect the age of the C17 property (not open) and original garden design. The house and garden are mentioned in Virginia Woolf's diaries of 1929, depicting a particular charm and peace that still exists today. Flint walls enclose the main lawn, herbaceous borders and rose garden. Nepeta lined courtyard, Physic garden with talk at 3pm about medicinal plants, informal pond and specimen trees inc a very ancient oak. Many spring flowers and tree blossom. Prairie-style garden under development. Mostly flat access with two steps up to main lawn and herbaceous borders.

84 LINDFIELD JUNGLE

16 Newton Road, Lindfield, Haywards Heath, RH16 2ND. Tim Richardson & Clare Wilson, 01444 484132, info@lindfieldjungle.co.uk, www.lindfieldjungle.co.uk. *Approx 1½ m NE of Haywards Heath town centre. Take B2028 into Lindfield. Turn onto Lewes Rd (B2111), then Chaloner Rd & turn R into Chaloner Close (no parking in close or at garden). Garden is located at far end, use postcode RH16 2NH.* **Mon 28 Aug (1-5). Combined adm with 47 Denmans Lane £7, chd free. Visits also by arrangement 15 July to 30 Sept for groups of 5 to 12.**
A surprising, intimate garden, 17 x 8 metres. Transformed since 1999 into an atmospheric jungle oasis planted for tropical effect. Lush and exuberant with emphasis on foliage and hot colours. From the planter's terrace enjoy the winding path through lillies, cannas, ginger and bamboo, to the tranquil sundowner's deck over hidden pools. Tropical jungle effect planting.

85 THE LONG HOUSE

The Lane, Westdean, nr Seaford, BN25 4AL. Robin & Rosie Lloyd, 01323 870432, rosiemlloyd@gmail.com, www.thelonghousegarden.co.uk. *3m E of Seaford, 6m W of Eastbourne. From A27 follow signs to Alfriston then Litlington, Westdean 1m on L. From A259 at Exceat, L on Litlington Rd, ¼ m on R. Free parking in village. Coach parties go to Friston Forest car park.* **Visits by arrangement May to July for groups of 10+. Adm £8, chd free. Home-made teas.**
The Long House's 1 acre garden has become a favourite for private group visits, being compared for romance, atmosphere and cottage garden planting to Great Dixter and Sissinghurst. Lavenders, hollyhocks, roses, a wildflower meadow, a long perennial border, water folly and pond are just some of the features, and everyone says Rosie's home-made cakes are second to none. Situated on the South Downs Way in the South Downs National Park within the medieval village of Westdean and a stones throw from the Seven Sisters Country Park. Wheelchair access over gravel forecourt at entrance, some slopes and steps.

86 LORDINGTON HOUSE

Lordington, Chichester, PO18 9DX. Mr & Mrs John Hamilton, 01243 375862, hamiltonjanda@btinternet.com. *7m W of Chichester. On W side of B2146, ½ m S of Walderton, 6m S of South Harting. Enter through white railings on bend.* **Sat 10, Sun 11 June (2-5). Adm £5, chd free. Home-made teas. Visits also by arrangement 12 June to 27 Oct.**
Early C17 house (not open) and walled gardens in South Downs National Park. Clipped yew and box, lawns, borders and fine views. Vegetables, fruit and poultry in the kitchen garden. 100+ roses planted since 2008. Trees both mature and young. Lime avenue planted in 1973 to replace elms. Wildflowers in field outside walls, accessible from garden. Gardens overlook Ems Valley, farmland and wooded slopes of South Downs, all in AONB. Wheelchair access is possible, but challenging with gravel paths, uneven paving and slopes. No disabled WC.

87 LUCTONS

North Lane, West Hoathly, East Grinstead, RH19 4PP. Drs Hans & Ingrid Sethi, 01342 810085, ingrid@sethis.co.uk. *4m SW of East Grinstead, 6m E of Crawley. In centre of West Hoathly village, near church, Cat Inn & Priest House. Car parks in village & at school at weekend.* **Sat 24 June (1-5). Combined adm with The Priest House £7, chd free. Sun 25, Tue 27 June (1-5). Adm £5, chd free. Home-made teas. Visits also by arrangement 15 May to 29 Sept for groups of 10 to 30.**
'A garden with everything', 'lots of unusual plants', 'stunning herbaceous borders', are comments of NGS and overseas garden tours visitors. 2 acres of Gertrude Jekyll style garden with herbaceous borders, wildflower orchard, swathes of spotted orchids, pond, roses, chickens, soft fruit, vegetables, herb garden, vine house, greenhouses, croquet lawn and shrubberies.

88 MALTHOUSE FARM

Streat Lane, Streat, Hassocks, BN6 8SA. Richard & Helen Keys, 01273 890356, helen.k.keys@btinternet.com. *2m SE of Burgess Hill. From r'about between B2113 & B2112 take Folders Ln & Middleton Common Ln E (away from Burgess Hill); after 1m R into Streat Ln, garden is ½ m on R. Please park carefully as signed.* **Visits by arrangement Apr to Sept for groups of 10 to 40. Adm £13, chd free. Home-made teas inc.**
Rural 5 acre garden with stunning views to South Downs. Garden divided into separate rooms; box parterre and borders with glass sculpture, herbaceous and shrub borders, mixed border for seasonal colour and kitchen garden. Orchard with newly planted wild flowers leading to partitioned areas with grass walks, snail mound, birch maze and willow tunnel. Wildlife farm pond with planted surround. Wheelchair access mainly on grass with some steps (caution if wet).

Our 2022 donation to ABF The Soldier's Charity meant 372 soldiers and veterans were helped through horticultural therapy

89 MANOR OF DEAN
Tillington, Petworth, GU28 9AP. Mr & Mrs James Mitford, 07887 992349, emma@mitford.uk.com. *3m W of Petworth just off A272. A272 from Petworth towards Midhurst, pass Tillington & turn R onto New Rd follow NGS signs. From Midhurst on A272 towards Petworth, past Halfway Bridge, 3rd L onto New Rd follow NGS signs. Sorry no coaches due to parking.* **Sun 19 Feb (2-4); Sun 5 Mar (2-5); Tue 21 Mar (10.30-12.30); Sun 23 Apr (2-5); Tue 16 May (10.30-12.30). Adm £5, chd free. Home-made teas. Visits also by arrangement 1 Feb to 29 Sept for groups of 20+. Dates exc school holidays.**
Approx 3 acres of traditional English garden with extensive views of the South Downs. Herbaceous borders, early spring bulbs, bluebell woodland walk, walled kitchen garden with fruit, vegetables and cutting flowers. NB under long term programme of restoration, some parts of the garden may be affected.

90 NEW MARCHANTS HARDY PLANTS
2 Marchants Cottages, Mill Lane, Laughton, Lewes, BN8 6AJ. Hannah Fox & Henry Macaulay, www.marchantshardyplants.co.uk. *6m E of Lewes. From Laughton Xrds on B2124 (at Roebuck Inn), head E for ½m, at Xrds turn S for Ripe down Mill Ln; on R.* **Tue 25, Wed 26 July (10-5). Adm £6, chd free. Pre-booking essential, please visit www.ngs.org.uk for information & booking. Home-made teas.**
Atmospheric 2 acre garden and plant nursery with striking backdrop of the South Downs. Featured on Carol Klein's Great British Gardens TV series, the garden is imaginatively designed and sensitively planted with a rich tapestry of unusual herbaceous plants, and graceful grasses add to the interest. One of the country's leading independent nurseries, the team propagates all their plants. Agapanthus are a specialty, inc some developed and cultivated on site. Wheelchair access over grass paths with some slopes.

GROUP OPENING

91 MAYFIELD GARDENS
Mayfield, TN20 6AB. *10m S of Tunbridge Wells. Turn off A267 into Mayfield. Parking in the village, TN20 6BE & field parking at Hoopers Farm, TN20 6BD. A detailed map available at each garden.* **Sat 10, Sun 11 June (11-5). Combined adm £8, chd free. Home-made teas at Hoopers Farm & The Oast.**

HOOPERS FARM
Andrew & Sarah Ratcliffe.

MEADOW COTTAGE
Adrian & Mo Hope.

OAKCROFT
Nick & Jennifer Smith.

THE OAST
Mike & Tessa Crowe.
(See separate entry)

SOUTH STREET PLOTS
Val Buddle.

NEW **2 SOUTHVIEW VILLAS**
Andrew & Clementine Westwood.

SUNNYBANK COTTAGE
Eve & Paul Amans.

Mayfield is a beautiful Wealden village with tearooms, an old pub and many interesting historical connections. The gardens to visit are all within walking distance of the village centre. They vary in size and style inc colour themed, courtyard and cottage garden planting, wildlife meadows and fruit and vegetable plots. There are far-reaching, panoramic views over the beautiful High Weald.

92 NEW 40 MICHEL DENE ROAD
East Dean, Eastbourne, BN20 0JU. Stella Lockyer. *In East Dean village on the A259 between Seaford & Eastbourne. From Seaford on A259, turn L in the centre of East Dene into Michel Dene Rd & continue for about ½m. From Eastbourne turn R into Michel Dene Rd.* **Fri 23 June (11-4). Adm £5, chd free. Home-made teas.**
A medium sized sloping garden on chalk and flint in the South Downs National Park, with stunning views over the Downs to the sea. Informal flower beds display a love of colour in mainly herbaceous borders. A summerhouse overlooks a wild garden with pond, and a small vegetable plot. Several seating areas nestle in a riot of salvias, grasses, roses, penstemon and other perennial favourites.

93 MILL HALL FARM
Whitemans Green, Cuckfield, Haywards Heath, RH17 5HX. Kate & Jonathan Berry, 07974 115658, katehod@gmail.com. *2½m E of A23, junction with B2115. From S & E take Staplefield Rd, after 300 metres turn R into Burrell Cottages, then L. From W after 30mph & Webster Cottages turn L to Burrell Cottages, then L. Please ignore SatNav.* **Visits by arrangement 11 Apr to 31 Oct for groups of up to 30. Adm £5, chd free. Home-made teas. Donation to Plant Heritage.**
A 2½ acre, north facing, 10 yr old garden with long maturing view sloping down to pond and beyond with lilies, irises and sanguisorba, dragonflies and other wildlife. Long border with ornamental trees, herbaceous plants inc cornus, acer, peonies, daylily, phlox, hollyhock, brunnera, geum and potentillas, climbing and shrub roses and part of the National Collection of Noel Burr daffodils. Victorian underground water cistern guarded by a nymph and used for our watering system. Yr long changing colour enhanced by morning light and stunning sunsets. Deep pond. Sloping bumpy lawn and no hard paths. No wheelchair access to WC.

NPC

94 MITCHMERE FARM
Stoughton, Chichester, PO18 9JW. Neil & Sue Edden, 02392 631456, sue@mitchmere.co.uk. *5½m NW of Chichester. Turn off the B2146 at Walderton towards Stoughton. Farm is ¾m on L, ¼m beyond the turning to Upmarden. Please do not park on verge, follow signs for parking.* **Visits by arrangement Feb to Nov. Adm £5, chd free. Min charge £60. Tea, coffee & biscuits.**
1½ acre garden in lovely downland position. Unusual trees and shrubs growing in dry gravel, briefly wet most yrs when the Winterbourne rises and flows through the garden. Coloured stems, catkins, drifts of snowdrops and crocuses. Small collection of special snowdrops. Optional 10 min walk down the long field beside the river, across the sleeper bridge, take the path through the copse with snowdrops up the steps into the new wood, then into the meadow and

Old Cross Street Farm
© Judi Lion

back to the garden. Wellies advisable. Dogs on short leads. Wheelchair access over gravel and grass.

95 MOORLANDS

Friars Gate, Crowborough, TN6 1XF. Calum Love, 01892 611546, calumlove@hotmail.co.uk. *2m N of Crowborough. St Johns Rd to Friar's Gate, or turn L off B2188 at Friar's Gate signed Horder Hospital.* **Visits by arrangement Apr to Oct for groups of up to 20. Adm £5, chd free. Light refreshments.**

4 acres set in lush valley deep in Ashdown Forest. The many special trees planted 50 yrs ago make this garden an arboretum. Also, a water garden with ponds, streams and river; primulas, rhododendrons and azaleas. River walk with grasses and bamboos.

96 MOUNTFIELD COURT

Robertsbridge, TN32 5JP. Mr & Mrs Simon Fraser. *3m N of Battle. On A21 London-Hastings; ½m NW from Johns Cross.* **Sun 7 May (2-5). Adm £5, chd free. Home-made teas.**

3 acre wild woodland garden; bluebell lined walkways through exceptional rhododendrons, azaleas, camellias, and other flowering shrubs; fine trees and outstanding views. Stunning paved herb garden. Recently restored unique C18 walled garden.

97 16 MUSGRAVE AVENUE

East Grinstead, RH19 4BS. Carole & Bob Farmer. *7m E of Crawley on A264 & 14m N of Uckfield on A22. From N on A22 to town centre at r'about take London Rd/High St, R then L at mini r'about, Musgrave Ave 2nd on R. From S on A22 L at r'about, L at mini r'about, Musgrave Ave is 2nd R. Street parking.* **Sun 25, Wed 28, Fri 30 June, Tue 4, Fri 7, Mon 10, Thur 13 July (11-5). Adm £8. Pre-booking essential, please phone 01342 321015 or email bobandcarole@outlook.com for information & booking. Home-made teas inc.**

A colourful country cottage style garden on a ⅓ acre plot. A partly terraced garden with herbaceous borders, small terraces, roses, dahlias and collection of salvias, also a vegetable garden. Wheelchair access to front garden and patio at rear of house only. No motorised wheelchairs or scooters.

98 NEWTIMBER PLACE

Newtimber, BN6 9BU. Mr & Mrs Andrew Clay, 01273 833104, andy@newtimberholidaycottages.co.uk, www.newtimberplace.co.uk. *7m N of Brighton. From A23 take A281 towards Henfield. Turn R at small Xrds signed Newtimber in approx ½m. Go down Church Ln, garden is on L at end of lane.* **Sun 16 Apr (2-5.30). Adm £6, chd free. Home-made teas.**

Beautiful C17 moated house (not open). Gardens and woods full of bulbs and wild flowers in spring. Herbaceous border and lawns. Moat flanked by water plants. Mature trees, wild garden, ducks, and fish. Wheelchair access across lawn to parts of garden, tearoom and WC.

In 2022, our donations to Carers Trust meant that 456,089 unpaid carers were supported across the UK

Legsheath Farm

© Leigh Clapp

99 NEW NIGHTINGALE HOUSE
Twittenside, Penfold Way, Steyning, BN44 3TW. Lynne Broome. *6m NE of Worthing. Exit r'about on A283 at S end of Steyning bypass into Clays Hill Rd. Turn L at Welcome to Steyning sign then 2nd R into Penfold Way. Park in road & follow NGS signs.* **Wed 21, Wed 28 June (10.30-5). Combined adm with 15 Penlands Rise £7, chd free. Home-made teas.**
Recently redesigned and replanted cottage garden with many perennials and summer annuals. Super new greenhouse and a bespoke metal screen with climbing roses and clematis. The garden is still a work in progress, come and see how it is developing! Wheelchair access to a level garden.

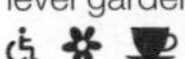

100 ◆ NYMANS
Staplefield Road, Handcross, RH17 6EB. National Trust, 01444 405250, nymans@nationaltrust.org.uk, www.nationaltrust.org.uk/nymans. *4m S of Crawley. On B2114 at Handcross signed off M23/A23 London-Brighton road. Metrobus 271 & 273 stop nearby.* **For NGS: Sat 18 Mar (10-5). Adm £15, chd £7.50. Adm subject to change. Light refreshments. For other opening times and information, please phone, email or visit garden website. Donation to Plant Heritage.**
One of NT's premier gardens with rare and unusual plant collections of national significance. In spring see blossom, bulbs and a stunning collection of subtly fragranced magnolias. The Rose Garden, inspired by Maud Messel's 1920s design, is scented by hints of old fashioned roses. The comfortable yet elegant house, a partial ruin, reflects the personalities of the creative Messel family. Some level pathways. See full access statement on Nymans website.

101 OAKLANDS FARM
Hooklands Lane, Shipley, Horsham, RH13 8PX. Zsa & Stephen Roggendorff, 01403 741270, zedrog@roggendorff.co.uk. *S of Shipley village. Off the A272 towards Shipley, R at Countryman Pub, follow yellows signs. Or N of A24 Ashington, off Billingshurst Rd, 1st R signed Shipley, garden 2m up the lane.* **Sat 3 June, Sat 15 July (11-5). Adm £6, chd free. Home-made teas. Visits also by arrangement 15 Apr to 27 Sept for groups of up to 25.**
Country garden designed by Nigel Philips in 2010. Oak lined drive leading to the house and farm opens out to an enclosed courtyard with pleached hornbeam and yew. The herbaceous borders are colourful throughout the yr. Vegetable garden with raised beds and greenhouse with white peach and vine. Wild meadow leading to orchard and views across the fields, full of sheep and poultry. Mature trees. Lovely Louise will be here with her special perennials for sale. Picnics welcome. Wheelchair access over gravel and brick paths, large lawn area and grassy paths.

102 THE OAST
Fletching Street, Mayfield, TN20 6TN. Mike & Tessa Crowe. *10m S of Tunbridge Wells. Turn off A267 into Mayfield. Parking along East St & in the village public car parks. Please do not park outside*

The Oast in Fletching St as it is very narrow. **Sat 22, Sun 23 Apr (11-5). Adm £5, chd free. Home-made teas. Opening with Mayfield Gardens on Sat 10, Sun 11 June.**
A south facing, gently sloping 1 acre garden with a lovely view, herbaceous borders and a ½ acre wildflower meadow. There is an interesting and varied selection of hardy and half-hardy plants, trees and shrubs, vegetables and fruit. Lots of colour with imaginative planting, a fine display of tulips and spring flowers. Homegrown plants for sale. Local nursery Rapkyns will also be selling plants.

103 OCKLYNGE MANOR

Mill Road, Eastbourne, BN21 2PG. Wendy & David Dugdill, 01323 734121, ocklyngemanor@hotmail.com, www.ocklyngemanor.co.uk. *Close to Eastbourne District General Hospital. Take A22 (Willingdon Rd) towards Old Town, turn L into Mill Rd by Hurst Arms Pub.* **Visits by arrangement Apr & May.**
A hidden oasis behind an ancient flint wall. Informal and tranquil, ½ acre chalk garden with sunny and shaded places to sit. Use of architectural and unusual trees. Rhododendrons, azaleas and acers in raised beds. Garden evolved over 20 yrs, maintained by owners. Georgian house (not open), former home of Mabel Lucie Attwell. Wheelchair access via short gravel path before entering garden. Brick path around perimeter.

104 OFFHAM HOUSE

The Street, Offham, Lewes, BN7 3QE. Mr & Mrs P Carminger & Mr S Goodman. *2m N of Lewes on A275. Offham House is on the main road (A275) through Offham between the filling station & Blacksmiths Arms.* **Sun 4 June (1-5). Adm £6, chd free. Home-made teas.**
Romantic garden with fountains, flowering trees, arboretum, double herbaceous border and long peony bed. 1676 Queen Anne house (not open) with well knapped flint facade. Herb garden and walled kitchen garden with glasshouses, cold frames, chickens, guinea fowl, sheep and ducks. A selection of pelargoniums and other plants for sale.

105 OLD CROSS STREET FARM

West Burton, Pulborough, RH20 1HD. Belinda & David Wilkinson, 01798 839373, belinda@westburton.com. *S of Petworth, nr the Roman Villa at Bignor. If travelling S on A29, continue & turn off at the signs for Bignor, Roman Villa. The house is in the centre of West Burton, not as indicated by the postcode! Parking in bottom of field.* **Visits by arrangement May to Oct for groups of 18 to 40. Adm £6, chd free. Home-made teas.**
A modern garden with a nod to traditional planting nestled in the ancient landscape of the South Downs. Despite its ancient buildings the garden was only designed and planted 15 yrs ago and enjoys many of the contemporary twists not usually found in such a landscape. An abundance of mass planting demonstrates the advantages of a limited planting palate. An enormous circular lawn, cloud hedging, an orchard, cutting garden, transformed farmyard, modern mass planting, year-round interest, a cottage garden, use of hedging, a raised formal pond, uses of different hard landscaping materials and planting of over 40 trees. Sorry, no picnics or WC.

106 THE OLD RECTORY

97 Barnham Road, Barnham, PO22 0EQ. Peter & Alexandra Vining *8m E of Chichester. Between Arundel & Chichester at A27 Fontwell junction, take A29 road to Bognor Regis. Turn L at next r'about onto Barnham Rd. 30 metres after speed camera arrive at garden.* **Sat 3, Sun 4 June (10-4). Combined adm with The Shrubbery £5, chd free. Pre-booking essential, please email theoldrectory97@gmail.com for information & booking. Home-made teas at The Shrubbery. Visits also by arrangement 5 June to 16 June for groups of 6 to 20. Combined adm with The Shrubbery.**
Scratch built in June 2019 after the 300m² garden was removed down to 40cm, then new topsoil and returfed. By summer 2021 the garden was well established. Some formal areas and a range of plants with acers, salvias, lilies, roses, cypresses and boxes. New flower bed, new trees and plants added for 2023. Garden now fully mature. No steps. Wheelchair access through 90cm wide entrance, please advise when booking.

107 THE OLD VICARAGE

The Street, Washington, RH20 4AS. Sir Peter & Lady Walters, 07766 761926, meryl.walters@ngs.org.uk, www.instagram.com/the_old_vicarage_washington. *2½m E of Storrington, 4m W of Steyning. From Washington r'about on A24 take A283 to Steyning. 500yds R to Washington. Pass Frankland Arms, R to St Mary's Church. Car park available, but no large coaches.* **Every Thur 9 Feb to 12 Oct (10.30-4.30). Pre-booking essential, please visit www.ngs.org.uk for information & booking. Self service refreshments on Thurs & picnics welcome. Sun 12 Mar, Mon 10 Apr, Mon 28 Aug (10.30-4.30). Purchase ticket in advance or at the gate on the day. Adm £7, chd free. Home-made teas. Visits also by arrangement 20 Feb to 29 Sept for groups of 10 to 30.**
Gardens of 3½ acres set around 1832 Regency house (not open). The front is formally laid out with topiary, wide lawn, mixed border and contemporary water sculpture. The rear features new and mature trees from C19, herbaceous borders, water garden and stunning uninterrupted views of the North Downs. The Japanese garden with waterfall and pond leads to a large copse, stream, treehouse and stumpery. Each yr 2000 tulips are planted for spring as well as another 2000 snowdrops and mixed bulbs throughout the garden. WC available. Wheelchair access to front garden, but rear garden is on a slope.

108 OLD WELL COTTAGE

High Street, Angmering, Littlehampton, BN16 4AG. Mr N Waters, www.instagram.com/nickwatersgardendesign. *Situated nr Angmering Manor Hotel & almost adjacent to the top of Weavers Hill in the High St. Look for the 'mushroom' shape tree! On road parking only, please be mindful of residents.* **Fri 30 June (10-2). Adm £5, chd free. Home-made teas.**
⅓ acre plot featuring topiary, formal areas and perennial borders. Framed within flint walls and surrounding the C16 to C18 cottage (not open) in the Angmering Conservation Area. Splendid holm oak and bay topiary trees, large espalier apple trees and a small kitchen garden. Lots of purples, whites and pinks.

109 NEW ◆ ONE GARDEN BRIGHTON
The Walled Garden, Stanmer Park, Brighton, BN1 9SE. hello@onegardenbrighton.com, www.onegardenbrighton.com. *Follow signs from A27 to Stanmer Park. Pay & display parking, follow signs for Patchway Car Park. By train to Falmer or by bus routes 5B, 23, 25 to University of Sussex, 30 min walk through park.* **Evening opening Wed 14 June (5-8). Adm £5, chd free. Light refreshments. For other opening times and information, please email or visit garden website.**
This will be an exclusive event for the NGS at this stunning 2½ acre garden which has recently been rediscovered and restored with a large National Lottery award. Visitors will have the chance to talk to the senior gardener, take guided tours with volunteers, and visit areas behind the scenes that are not normally open to the public, inc student areas and the propagation area. One Market offering a range of locally sourced goods, tea, cakes and plant sales. Gentle slopes, most paths suitable for wheelchair access.

110 ORCHARD COTTAGE
Boars Head Road, Boarshead, Crowborough, TN6 3GR. Jane Collins, 01892 653444, collinsjane1@hotmail.co.uk. *6m S of Tunbridge Wells, off A26. At T-junction, do not follow SatNav. Instead, turn L down dead end. Orchard Cottage is at the bottom of the hill on LHS.* **Fri 2, Sat 3, Sun 4 June (10-4). Adm £6, chd free. Home-made teas. Visits also by arrangement Mar to June for groups of 5 to 30.**
Mature 1½ acre plantaholic's garden with a large variety of trees and shrubs, perennials and bulbs, many unusual. Gardened organically. Mainly colour themed beds, planted informally. Small woodland, meadow and deep pond to encourage wildlife. Kitchen garden with raised beds. Hardy Plant Society member. Access via gravel drive with wide gently sloping grass paths suitable for wheelchairs and mobility scooters. Drop-off in drive by prior arrangement.

111 PARSONAGE FARM
Kirdford, nr Billingshurst, RH14 0NH. David & Victoria Thomas. *5m NE of Petworth. Located between Petworth & Wisborough Green in the village of Kirdford. Opp side of the road from the turn to Plaistow.* **Fri 23 June, Sun 3 Sept (2-6). Adm £7, chd free. Home-made teas.**
Major garden in beautiful setting developed over 30 yrs with fruit theme and many unusual plants. Formally laid out on grand scale with long vistas. C18 walled garden with borders in apricot, orange, scarlet and crimson. Topiary walk, pleached lime allée, tulip tree avenue, rose borders and vegetable garden with trained fruit. Turf amphitheatre, informal autumn shrubbery, yew cloisters and jungle walk.

112 PEELERS RETREAT
70 Ford Road, Arundel, BN18 9EX. Tony & Lizzie Gilks, 01903 884981, timespan70@tiscali.co.uk, www.timespanhistoricalpresentations.co.uk. *1m S of Arundel. At Chichester r'about take exit to Ford & Bognor Regis onto Ford Rd. We are situated close to Maxwell Rd.* **Tue 11, Sat 15, Tue 25, Sat 29 Apr, Tue 9, Sat 13, Tue 23, Sat 27 May (2-5); Sat 10 June (12-5); Tue 20, Sat 24 June, Tue 11, Sat 15, Tue 25 July, Tue 8, Sat 12, Tue 22 Aug, Sat 23, Sat 30 Sept, Tue 10, Sat 14 Oct (2-5). Adm £5, chd free. Home-made teas. Visits also by arrangement 8 July to 14 Oct for groups of 5 to 24.**
This inspirational space is a delight, with plenty of areas to sit and relax, enjoying delicious teas. Interlocking beds packed with year-round colour and scent, shaded by specimen trees, inventive water feature and rill, raised fish pond and a working Victorian fireplace.

113 PEMBURY HOUSE
Ditchling Road, Clayton, BN6 9PH. Nick & Jane Baker, www.pemburyhouse.co.uk. *6m N of Brighton, off A23. No parking at house. Use village green by church. Then follow signs across playing field to footpath by railway to back gate. Good public transport.* **Every Thur and Fri 10 Feb to 17 Feb, Thur and Fri 24 Feb to 17 Mar (10.30-12 & 2-3.30). Adm £10, chd free. Pre-booking essential, please visit www.ngs.org.uk for information & booking. Home-made teas inc.**
Depending on the vagaries of the season, hellebores and snowdrops are at their best in Feb and March. It is a country garden, tidy but not manicured. Work always in progress on new areas. Winding paths give a choice of walks through 3 acres of garden, which is in and enjoys views of the South Downs National Park. Wellies, macs and winter woollies advised. A German visitor observed 'this is the perfect woodland garden'. Year-round interest. Plants for sale.

114 NEW 15 PENLANDS RISE
Steyning, BN44 3PJ. Patsy Walton. *6m NE of Worthing. Exit r' about on A283 at S end of Steyning bypass into Clays Hill. Turn L by the Welcome to Steyning sign into Penlands Way & L by mailbox into Penlands Rise. No. 15 is on L. Park in road.* **Wed 21, Wed 28 June (10.30-5). Combined adm with Nightingale House £7, chd free. Home-made teas at Nightingale House.**
The cottage style small garden has evolved into a series of island beds with narrow paths in between. There are roses, clematis, many salvias and perennials. A collection of over 100 pots, many planted with hydrangeas, but the majority are a colourful mix of annuals and tender plants. There are steps into the garden, so not suitable for wheelchair access.

115 PENNS IN THE ROCKS
Groombridge, Tunbridge Wells, TN3 9PA. Mr & Mrs Hugh Gibson, www.pennsintherocks.co.uk. *7m SW of Tunbridge Wells. On B2188 Groombridge to Crowborough road, just S of Xrd to Withyham. For SatNav use TN6 1UX which takes you to the white drive gates, through which you should enter the property.* **Sun 16 Apr, Sun 14 May, Sun 6 Aug (2-6). Adm £7, chd free. Home-made teas.**
Large garden with spectacular outcrop of rocks, 140 million yrs old. Lake, C18 temple and woods. Daffodils, bluebells, azaleas, magnolia and tulips. Old walled garden with herbaceous borders, roses and shrubs. Stone sculptures by Richard Strachey. Part C18 house (not open) once owned by William Penn of Pennsylvania. Cash only on the day please. Restricted wheelchair access. No disabled WC.

Shepherds Cottage

116 1 PEST COTTAGE

Carron Lane, Midhurst, GU29 9LF. Jennifer Lewin. *W edge of Midhurst behind Carron Lane Cemetery. Free parking at recreation ground at top of Carron Ln. Short walk on woodland track to garden, please follow signs.* **Fri 26 May (2-6); Sun 28 May (1-5.30). Adm £4, chd free.**

This edge of woodland, architect's studio garden of approx ¾ acre sits on a sloping sandy site. Designed to support wildlife and biodiversity, a series of outdoor living spaces connected with informal paths through lightly managed areas, creates a charming secret world tucked into the surrounding common land. The garden spaces have made a very small house (not open) into a hospitable family home. Exhibition of architect's projects.

117 NEW PITFIELD BARN CUT FLOWER FARM & STUDIO

Chalkers Lane, Hurstpierpoint, Hassocks, BN6 9LR. Mrs Emma Martin, www.pitfieldbarn.co.uk. *Approach from B2117, Cuckfield Rd, turn into Chalkers Ln & we are the gate on the L just past the 30mph sign, close to Hurstpierpoint College.* **Fri 28, Sat 29 July (12-4). Adm £5, chd free. Home-made teas.**

Working flower farm, tidy but not manicured. We sow, grow and harvest seasonal, British cut flowers in approx 2 acres, without use of chemicals or pesticides, and all seedlings grown in peat free compost. We hand cut our flowers to order and are passionate about promoting British flowers. Tours explaining the British flower market, how the garden works and the philosophy inc. A converted barn and studio where various flower, art and craft workshops take place and café. Freshly cut bunches of seasonal flowers, local produce and gifts for sale. Wheelchair access over path from barn to second field and cut grass to other areas.

118 6 PLANTATION RISE

Worthing, BN13 2AH. Nigel & Trixie Hall, 01903 262206, trixiehall008@gmail.com. *2m from seafront on outskirts of Worthing. A24 meets A27 at Offington r'about. Turn into Offington Ln, 1st R into The Plantation, 1st R again into Plantation Rise. Parking on The Plantation only, short walk up to 6 Plantation Rise.* **Visits by arrangement 6 Mar to 12 Sept for groups of 6 to 20. Adm inc home-made teas. Please advise of specific dietary requirements upon booking.**

Clever use is made of evergreen shrubs, azaleas, rhododendrons and acers enclosing our 70' x 80' garden, enhancing the flower decked pergolas, folly and summerhouse, which overlooks the pond. Planting inc nine silver birches 'Tristis', which are semi pendula, plus a lovely combination of primroses, anemones and daffodils in spring and a profusion of roses, clematis and perennials in summer. Wheelchair access to patios only with a good view of the garden.

119 ◆ THE PRIEST HOUSE

North Lane, West Hoathly, RH19 4PP. Sussex Archaeological Society, 01342 810479, priest@sussexpast.co.uk, www.sussexpast.co.uk. *4m SW of East Grinstead, 6m E of Crawley. In centre of West Hoathly village, nr church, the Cat Inn & Luctons. Car parks in village.* **For NGS: Sat 27 May (10.30-5.30). Adm £2, chd free. Sat 24 June (10.30-5.30). Combined adm with Luctons £7, chd free. Home-made teas. For other opening times and information, please phone, email or visit garden website.**

C15 timber framed farmhouse with cottage garden on acid clay. Large collection of culinary and medicinal herbs in a small formal herb garden and mixed with perennials and shrubs in exuberant borders. Long established yew topiary and espalier apple trees provide structural elements. Traditional fernery and stumpery, recently enlarged with a small, secluded shrubbery and gravel garden. Be sure to visit the fascinating Priest House Museum, adm £1 for NGS visitors.

120 RINGMER PARK

Uckfield Road, Ringmer, Lewes, BN8 5RW. Deborah & Michael Bedford, www.ringmerpark.com. *On A26 Lewes to Uckfield road. 1½m NE of Lewes, 6m S of Uckfield.* **Sun 11 June (2-5). Adm £6, chd free. Home-made teas.**

The garden at Ringmer Park has been developed over the last 36 yrs as the owner's interpretation of a classic English country house garden. It extends over approaching 8 acres and comprises 15 carefully differentiated individual gardens and borders, which are presented to optimise the setting of the house (not open), close to the South Downs. See garden website for more detailed information.

121 ROSE COTTAGE

Laughton, Lewes, BN8 6BX. Paul Seaborne & Glenn Livingstone, www.pelhamplants.co.uk. *5m E of Lewes; 6m W of Hailsham. Signed from junction of Common Ln & Shortage Ln, ½m N of Roebuck Pub, Laughton.* **Thur 3, Sat 5 Aug (1-5). Adm £6.50, chd £3.50. Pre-booking essential, please visit www.ngs.org.uk for information & booking. Home-made teas.**

Rare opportunity to access nurseryman's private garden packed with unusual examples of herbaceous perennials and grasses. Informally planted 1 acre garden subdivided by strong structural shaped hedging and surrounding an old cottage (not open). Multiple densely planted borders with new plantings each yr. Pelham Plants nursery forms part of the 2 acre woodland edge site.

122 SAFFRONS

Holland Road, Steyning, BN44 3GJ. Tim Melton & Bernardean Carey, 07850 343516, tim.melton@btinternet.com, www.thetransplantedgardener.uk. *6m NE of Worthing. Exit r'about on A283 at S end of Steyning bypass into Clays Hill Rd. 1st R into Goring Rd, 4th L into Holland Rd. Park in Goring Rd & Holland Rd.* **Visits by arrangement 26 June to 15 Aug for groups of 10 to 30. Individuals can ask to join an existing group. Home-made teas.**

Planted with an artist's eye for contrasts and complementary colours. Vibrant late summer flower beds of salvias, eryngiums, agapanthus, grasses and lilies attract bees and butterflies. A broad lawn is surrounded by borders with maples, rhododendrons, hydrangeas and mature trees interspersed with ferns and grasses. The large fruit cage and vegetable beds comprise the productive area of the garden. One level garden with good access, except in very wet conditions.

123 ◆ ST MARY'S HOUSE GARDENS

Bramber, BN44 3WE. Roger Linton & Peter Thorogood, 01903 816205, info@stmarysbramber.co.uk, www.stmarysbramber.co.uk. *1m E of Steyning. 10m NW of Brighton in Bramber village, off A283.* **For NGS: Fri 23, Sat 24 June (2-5.30). Adm £8, chd free. Light refreshments. For other opening times and information, please phone, email or visit garden website.**

5 acres inc formal topiary, large prehistoric *Ginkgo biloba* and magnificent *Magnolia grandiflora* around enchanting timber-framed medieval house (not open for NGS). Victorian Secret Gardens inc splendid 140ft fruit wall with pineapple pits, Rural Museum, Terracotta Garden, Jubilee Rose Garden, King's Garden and circular Poetry Garden. Woodland walk and Landscape Water Garden. In the heart of the South Downs National Park. WC facilities. Wheelchair access with level paths throughout.

124 SANDHILL FARM HOUSE

Nyewood Road, Rogate, Petersfield, GU31 5HU. Rosemary Alexander. *4m SE of Petersfield. From A272 Xrds in Rogate take road S signed Nyewood & Harting. Follow road for approx 1m over small bridge. Sandhill Farm House on R, over cattle grid.* **Sun 12 Feb (12-4); Sat 17, Sun 18 June, Sat 16, Sun 17 Sept (2-5). Adm £5, chd free. Home-made teas.**

Front and rear gardens broken up into garden rooms inc small kitchen garden. Front garden with small woodland area, planted with early spring flowering shrubs, ferns and bulbs. White and green garden, large leaf border and terraced area. Rear garden has rose borders, small decorative vegetable garden, red border and grasses border. Snowdrop day on Sun 12 Feb. Home of author and principal of The English Gardening School.

D

GROUP OPENING

125 SEAFORD GARDENS

As there are several gardens on the trail, they will be signed from the A259 as the gardens are N & S of this road. NB Not a walking trail. Maps available at each garden. **Sun 21 May (12-5). Combined adm £7, chd free. Sun 11 June (12-5). Combined adm £8, chd free. Home-made teas at 34 Chyngton Road & Cosy Cottage.**

BURFORD
Cuckmere Road, BN25 4DE. Chris Kilsby.
Open on all dates

34 CHYNGTON ROAD
BN25 4HP. Dr Maggie Wearmouth & Richard Morland.
Open on all dates

COSY COTTAGE
69 Firle Road, BN25 2JA. Ernie & Carol Arnold, 07763 196343, ernie.whitecrane@gmail.com, www.facebook.com/CosyCottage69Garden.
Open on all dates
Visits also by arrangement 1 May to 17 July for groups of 5 to 15.

NEW **8 DOWNS ROAD**
BN25 4QL. Mr Phil & Mrs Julie Avery.
Open on Sun 11 June

LAVENDER COTTAGE
69 Steyne Road, BN25 1QH. Christina & Steve Machan.
Open on all dates

MADEHURST
67 Firle Road, BN25 2JA. Martin & Palo.
Open on all dates

SEAFORD ALLOTMENTS
Sutton Drove, BN25 3NQ. Peter Sudell.
Open on Sun 11 June

SEAFORD COMMUNITY GARDEN
East Street, BN25 1AD. Seaford Community Garden, www.seaford-sussex.co.uk/scg.
Open on Sun 21 May

On 21 May, six gardens open. Cosy Cottage, a cottage garden over three levels with ponds, flowers, vegetables and shrubs. Lavender Cottage a flint walled garden with a coastal and kitchen garden. Madehurst is a garden on different levels with interesting planting and seasonal interest. Burford, a mature garden with a mix of flowers, vegetables and an auricula theatre. The Community Garden provides an interesting space with flower and vegetable beds for members of the community to come together to share their gardening experience. Chyngton is a mature garden of shrubs, trees, herbaceous and vegetable beds. On 11 June, seven gardens will open inc five of the above, plus a new garden at 8 Downs Road, a delightful mix of shrubs, block planted herbaceous beds and a vegetable garden. Seaford Allotments offer a unique chance to see a variety of planting and colour. A site of 189 well maintained plots with a wildlife area, a dye bed and a compost WC. Some Allotment holders will be around on the day to answer any questions.

126 SELHURST PARK
Halnaker, Chichester, PO18 0LZ. Richard & Sarah Green, 01243 839310, mail@selhurstparkhouse.co.uk. *8m S of Petworth. 4m N of Chichester on A285.* **Sun 18 June (2-5). Adm £6, chd free. Home-made teas. Visits also by arrangement 5 June to 31 July for groups of 10 to 30.**
Come and explore the varied gardens surrounding a beautiful Georgian flint house (not open), approached by a chestnut avenue. The flint walled garden has a mature 160ft herbaceous border with unusual planting along with rose, hellebore and hydrangea beds. Pool garden with exotic palms and grasses divided from a formal knot and herb garden by espalier apples. Kitchen and walled fruit garden. Wheelchair access to walled garden, partial access to other areas.

127 ◆ SHEFFIELD PARK AND GARDEN
Uckfield, TN22 3QX. National Trust, 01825 790231, sheffieldpark@nationaltrust.org.uk, www.nationaltrust.org.uk/sheffieldpark. *10m S of East Grinstead. 5m NW of Uckfield; E of A275.* **For NGS: Wed 26 Apr (10-4.30). Adm £14.25, chd free. For other opening times and information, please phone, email or visit garden website.**
Magnificent, landscaped garden laid out in C18 by Capability Brown and Humphry Repton covering 120 acres (40 hectares). Further development in the early yrs of this century by its owner Arthur G Soames. Centrepiece is original lakes with many rare trees and shrubs. Beautiful at all times of the yr, but noted for its spring and autumn colours. National Collection of Ghent azaleas. Natural play trail for families on South Park. Large number of champion trees, 87 in total. Light refreshments available in the Coach House Café and The Shant. Last entry 4.30pm. Garden largely accessible for wheelchairs, please call for information.

128 SHEPHERDS COTTAGE
Milberry Lane, Stoughton, Chichester, PO18 9JJ. Jackie & Alan Sherling, 07795 388047, milberrylane@gmail.com, www.instagram.com/drjackieblackman. *9⅙m NW Chichester. Off B2146, next village after Walderton. Cottage is nr telephone box & beside St Mary's Church. No parking in lane beside house.* **Visits by arrangement Apr to Aug for groups of 10 to 35. Adm £6, chd free. Home-made teas.**
A compact terraced garden using the borrowed landscape of Kingley Vale in the South Downs. The south facing flint stone cottage (not open) is surrounded by a Purbeck stone terrace and numerous individually planted and styled seating areas. A small orchard under planted with meadow, lawns, yew hedges, amelanchier, cercis and drifts of wind grass provide structure and year-round interest. Many novel design ideas for a small garden. Ample seating throughout the garden to enjoy the views.

129 THE SHRUBBERY
140 Barnham Road, Barnham, PO22 0EH. John & Ros Woodhead *Between Arundel & Chichester. At A27 Fontwell junction take A29 road to Bognor Regis. Turn L at next r'about onto Barnham Rd. After ½m the garden is 30 metres after the speed camera.* **Sat 3, Sun 4 June (10-4). Combined adm with The Old Rectory £5, chd free. Pre-booking essential, please email johnrosw@sky.com for information & booking. Home-made teas. Visits also by arrangement 5 June to 16 June for groups of 6 to 20.**
¼ acre plot with mature trees, shrubs and colourful borders of mixed perennials. Features inc hosta, sculpture and areas of soft fruit.

130 SKYSCAPE
46 Ainsworth Avenue, Ovingdean, Brighton, BN2 7BG. Lorna & John Davies. *From Brighton take A259 coast road E, passing Roedean School on L. Take 1st L at r'about to Greenways & 2nd R into Ainsworth Ave. Skyscape at the top on R. No. 52 bus on Sat & No. 57 on Sun.* **Sat 27, Sun 28 May (1-5). Adm £5, chd free. Tea & cake served on the patio.**
250ft south facing rear garden on a sloping site with fantastic views of the South Downs and the sea. Garden created by owners over past 10 yrs. Orchard, flower beds, wildlife ponds and planting with bees in mind. Small, protected apiary in orchard. Full access to site via purpose built sloping path.

131 NEW **62 SNOWHILL**

Easebourne, Midhurst, GU29 9BL. Annie Bevan Lean, 01730 817493, infoanniel@gmail.com. *½m N from Midhurst/Easebourne Bridge. Front of house & lower garden front on to Wheelbarrow Castle at junction with Petworth Rd.* **Visits by arrangement 26 May to 11 Aug for groups of 5 to 10. Adm £5. Please discuss refreshments when booking.**

Partly shaded lower garden inc inherited shrubs and climbing roses. Many more plants constantly being planted. A side gate leads through a tiny courtyard and up stone steps to a garden professionally designed and completed in Oct 2021. Mixed planting of ornamental trees, flowering and ornamental shrubs and flowers was completed in March 2022. A gate leads to a wildflower meadow.

132 **SOUTH GRANGE**

Quickbourne Lane, Northiam, Rye, TN31 6QY. Linda & Michael Belton, 01797 252984, belton.northiam@gmail.com. *Between A268 & A28, approx ½m E of Northiam. From Northiam centre follow Beales Ln into Quickbourne Ln, or Quickbourne Ln leaves A286 approx ½m S of A28 & A286 junction. Disabled parking at front of house.* **Sat 22, Sun 23 July (11-5). Adm £6, chd free. Teas & light lunches made to order. Visits also by arrangement Apr to Oct for groups of 5+.**

Hardy Plant Society members' garden with wide variety of trees, shrubs, perennials, grasses and pots arranged into a complex garden display for year-round colour and interest. Raised vegetable beds, wildlife pond, water features, orchard, rose arbour, soft fruit cage and living gazebo. House roof runoff diverted to storage and pond. Small area of wild wood. An emphasis on planting for insects. We try to maintain nectar and pollen supplies and varied habitats for most of the creatures that we share the garden with, hoping that this variety will keep the garden in good heart. Home propagated plants for sale. Wheelchair access over hard paths through much of the garden, but steps up to patio and WC.

133 **STANLEY FARM**

Highfield Lane, Liphook, GU30 7LW. Bill & Emma Mills. *For SatNav please use GU30 7LN, which takes you to Highfield Ln & then follow NGS signs. Track to Stanley Farm is 1m.* **Sun 30 Apr (12-5). Adm £5, chd free. Home-made teas.**

1 acre garden created over the last 15 yrs around an old West Sussex farmhouse (not open), sitting in the midst of its own fields and woods. The formal garden inc a kitchen garden with heated glasshouse, orchard, espaliered wall trained fruit, lawn with ha-ha and cutting garden. A motley assortment of animals inc sheep, donkeys, chickens, ducks and geese. Bluebells flourish in the woods, so feel free to bring dogs and a picnic, and take a walk after visiting the gardens. Wheelchair access via a ramp to view main part of the garden. Difficult access to woods due to muddy, uneven ground.

134 **STONE CROSS HOUSE**

Alice Bright Lane, nr Crowborough, TN6 3SH. Mr & Mrs D A Tate. *1½m S of Crowborough Cross. At Crowborough T-lights (A26) turn S into High St & shortly R onto Croft Rd. Over 3 mini r'abouts to Alice Bright Ln. Garden on L at next Xrds.* **Mon 8 May (2-5). Adm £5, chd free. Home-made teas.**

Beautiful 9 acre country property (not open) with gardens containing a delightful array of azaleas, acers, rhododendrons and camellias, interplanted with an abundance of spring bulbs. The very pretty cottage garden has interesting examples of topiary and unusual plants. Jacob sheep graze the surrounding pastures. Gravel drive, but mainly flat and no steps. No WC.

135 NEW **STROODS**

Herons Ghyll, Uckfield, TN22 4DB. Geraldine Ogilvy. *Located on A26 opp Clay Studio.* **Tue 16 May (10-4). Adm £6, chd free. Pre-booking essential, please visit www.ngs.org.uk for information & booking. Home-made teas.**

Nestled in the beautiful hills of Sussex, the garden is a peaceful space where ancient trees, an orchard and olive grove team carefully with curated herbaceous borders and meandering wildflower meadows, all attracting bees, butterflies and birds. The walled kitchen garden delivers delicious, traditional vegetables, embroidered with flowers and shrubs, plus the occasional wandering chicken.

136 ◆ **SUSSEX PRAIRIES**

Dutch Barn Morlands Farm, Wheatsheaf Road (B2116), Henfield, BN5 9AT. Paul & Pauline McBride, 01273 495902, morlandsfarm@btinternet.com, www.sussexprairies.co.uk. *2m NE of Henfield on B2116 Wheatsheaf Rd (also known as Albourne Rd). Follow Brown Tourist signs indicating Sussex Prairie Garden.* **For NGS: Sun 10 Sept (1-5). Adm £10. Home-made teas. For other opening times and information, please phone, email or visit garden website.**

Exciting prairie garden of approx 8 acres planted in the naturalistic style using 60,000 plants and over 1,600 different varieties. A colourful garden featuring a huge variety of unusual ornamental grasses. Expect layers of colour, texture and architectural splendour. Surrounded by mature oak trees with views of Chanctonbury Ring and Devil's Dyke on the South Downs. Permanent sculpture collection and exhibited sculpture throughout the season. Rare breed sheep and pigs. Tropical entrance garden. Woodchip pathway at entrance. Soft woodchip paths in borders not accessible, but flat garden for wheelchairs and mobility scooters. Disabled WC.

137 NEW **TERWICK HOUSE**

Rogate, Petersfield, GU31 5BY. Mrs Fiona Dix. *2m N of Rogate village on road to Chithurst. Approx 1m E of Tullecombe Xrds on RHS. Roadside parking.* **Sun 28 May (2-5). Adm £6, chd free. Home-made teas.**

Wild woodland garden planted for spring interest some 40 yrs ago by plant collectors of rhododendron, azalea, camellia and acer, along winding steep paths to the south of the house. A collection of *R. yakushimanum* among many others; small pinetum and many specialist trees. Herbaceous planting is taking shape around the house (not open). Small potager garden off front driveway. Notable garden for the extensive range of rhododendron cultivars, rare and unusual trees and shrubs.

138 THAKEHAM PLACE FARM
The Street, Thakeham, Pulborough, RH20 3EP. Mr & Mrs T Binnington. *In the village of Thakeham, 3m N of Storrington. The farm is at the E end of The Street, where it turns into Crays Ln. Follow signs down farm drive to Thakeham Place.* **Thur 20, Sun 23 July (2-5.30). Combined adm with Cumberland House £8, chd free. Home-made teas.**
Set in the middle of a working dairy farm, the garden has evolved over the last 30 yrs. Taking advantage of its sunny position on free draining greensand, the borders are full of sun loving plants and grasses with a more formal area surrounding the farmhouse (not open). Lovely views across the farm to Warminghurst from the orchard.

139 TOWN PLACE
Ketches Lane, Freshfield, Sheffield Park, RH17 7NR. Anthony & Maggie McGrath, 01825 790221, mcgrathsussex@hotmail.com, www.townplacegarden.org.uk. *5m E of Haywards Heath. From A275 turn W at Sheffield Green into Ketches Ln for Lindfield. 1¾m on L.* **Sun 11, Wed 14, Wed 21, Sun 25 June, Sun 2, Sun 9 July (2-5). Adm £8, chd free. Visits also by arrangement 1 June to 7 July.**
A stunning 3 acre garden with a growing international reputation for the quality of its design, planting and gardening. Set round a C17 Sussex farmhouse (not open), the garden has over 400 roses, herbaceous borders, herb garden, white garden, topiary inspired by the sculptures of Henry Moore, an 800 yr old oak, potager, and a unique ruined Priory Church and Cloisters in hornbeam. Sorry, no dogs allowed, and no refreshments available. Picnics welcome. There are steps, but all areas can be viewed from a wheelchair.

140 TUPPENNY BARN
Main Road, Southbourne, PO10 8EZ. Maggie Haynes, 01243 377780, contact@tuppennybarn.co.uk, www.tuppennybarn.co.uk. *6m W of Chichester, 1m E of Emsworth. On Main Rd A259, corner of Tuppenny Ln. Disabled parking.* **Visits by arrangement 16 Jan to 7 Oct for groups of 12 to 30. Adm £5, chd free. Food intolerances & allergies catered for.**
An iconic, organic smallholding used as an outdoor classroom to teach children about the environment, sustainability and healthy food. 2½ acres packed with a wildlife pond, orchard with heritage top fruit varieties, two solar polytunnels, fruit cages, raised vegetables, herbs and cut flower garden. Willow provides natural arches and wind breaks. Bug hotel and beehives support vital pollinators. Most of the grounds are accessible for wheelchairs, but undulated areas are more difficult.

141 WADHURST PARK
Riseden Road, Wadhurst, TN5 6NT. Nicky Browne, wadhurstpark.co.uk. *6m SE of Tunbridge Wells. Turn R along Mayfield Ln off B2099 at NW end of Wadhurst. Then turn L by The Best Beech Inn & L at Riseden Rd. Please note, parking is limited.* **Fri 23 June (9.30-3.30); Sat 24 June (10-4). Adm £6, chd free. Home-made teas inc vegan & gluten-free cakes in the Common Room.**
The naturalistic gardens, designed by Tom Stuart-Smith, created on a C19 site, situated within a 2000 acre estate managed organically to enhance its wildlife, cultural heritage and beauty. The gardens invite the wider landscape in, while meadows and hedgerows, woodland trees and ground cover soften and frame views to hills and lake. We strive to garden with a greater respect for the natural world. Features inc restored Victorian orangery, naturalistic gardens planted with native species, potager, log hives and woodland walks. Due to uneven paths, please wear sensible shoes. Sorry no dogs. Wheelchair access to main features of garden, some surfaces inc grass, cobbles and steps.

The Long House

142 NEW **50 WANNOCK LANE**
Willingdon, Eastbourne, BN20 9SD. Chris & Nick Ireland. *Between Eastbourne & Polegate on the A2270. From Eastbourne (A2270) turn L into Gorringe Valley Rd, at give way turn R. From Polegate turn R into Gorringe Valley Rd.* **Sun 2 July (1-5). Adm £5, chd free. Open nearby 255 Kings Drive.**

A fully packed front garden with central large rogue Christmas tree. West facing 200ft x 55ft rear garden, with a wide selection of shrubs, herbaceous and perennials plus trees. There are two small ponds with fish, vegetable plot and greenhouses. Little nooks and seating areas provide interest and sunny patio. Light refreshments at 255 Kings Drive nearby.

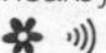

143 **WARNHAM PARK**
Robin Hood Lane, Warnham, Horsham, RH12 3RP. Mrs Caroline Lucas. *Turn in off A24 end of Robin Hood Ln, turn immed R by the large poster & follow signs.* **Sat 22 Apr (11-5). Adm £6, chd free. Soup & home-made teas.**

The garden is situated in the middle of a 200 acre Deer Park, which has a very special herd of Red Deer husbanded by the Lucas Family for over 150 yrs. Borders with traditional planting and a kitchen garden that is prolific most of the yr. The rest of the garden comprises different spaces, inc a small white garden, a Moroccan courtyard and a walled garden. There is also a woodland walk. Disabled WC.

GROUP OPENING

144 **THE WARREN, CROWBOROUGH GARDENS**
All gardens S of Crowborough town centre. On street parking at Redriff & Weavers. Hoath Cottage is 270yds from public car park, adjacent to main entrance of Waitrose. **Sat 13, Sun 14 May (2-5). Combined adm £6, chd free. Home-made teas at Weavers.**

HOATH COTTAGE
Croft Road, TN6 1DS. Frances Arrowsmith.

REDRIFF
Rannoch Road, TN6 1RA. John Mitchell.

WEAVERS
Glenmore Road East, TN6 1RE. Mr Graham James.

Three gardens south of Crowborough town centre. Redriff is a Japanese influenced, richly planted spring garden; bridge and seating with views overlooking fish pond. Weavers has rhododendrons, azaleas, stumpery and hidden wooded areas with natural and ornamental stone features. Hoath Cottage is an informal medium sized garden dedicated to supporting wildlife.

GROUP OPENING

145 **WATERWORKS & FRIENDS**
Broad Oak & Brede. *4 Waterworks Cottages, Brede, off A28 by church & opp Red Lion, ¾m at end of lane. Sculdown, B2089 Chitcombe Rd, W off A28 at Broad Oak Xrds. Start at either garden, a map will be provided.* **Sat 1 July (10.30-4). Combined adm £6, chd free. Light refreshments at Sculdown.**

SCULDOWN
TN31 6EX. Mrs Christine Buckland.

4 WATERWORKS COTTAGES
TN31 6HG. Mrs Kristina Clode, 07950 748097, kristinaclode@gmail.com, www.kristinaclodegardendesign.co.uk.
Visits also by arrangement 2 July to 30 Sept for groups of 10 to 20.
D

Foxglove Cottage

© Judi Lion

An opportunity to visit two unique gardens and discover the Brede Steam Giants 35ft Edwardian water pumping engines, and Grade II listed pump house located behind 4 Waterworks Cottages. Garden designer Kristina Clode has created her wildlife friendly garden at 4 Waterworks Cottages over the last 13 yrs. Delightful perennial wildflower meadow, pond, wisteria covered pergola and mixed borders packed full of unusual specimens with year-round interest and colour. Sculdown's garden is dominated by a very large wildlife pond formed as a result of iron-ore mining over 100 yrs ago. The stunning traditional cottage (not open) provides a superb backdrop for several colourful herbaceous borders and poplar trees. Plants for sale at 4 Waterworks Cottages. At Brede Steam Giants, WC available, but regret no disabled facilities and assistance dogs only (free entry, donations encouraged). Wheelchair access at Sculdown (park in flat area at top of field) and in the front garden of 4 Waterworks Cottages only.

146 WHITEHANGER

Marley Lane, Haslemere, GU27 3PY. Lynn & David Paynter, 07774 010901, lynn@whitehanger.co.uk. *3m S of Haslemere. Take A286 Midhurst Rd from Haslemere & after approx 2m turn R into Marley Ln (opp Hatch Ln). After 1m turn into drive shared with St Magnus Nursing & keep bearing L through the nursing home.* **Sun 2 July, Sun 6, Sun 27 Aug (10-3). Adm £6.50, chd £6.50. Pre-booking essential, please visit www.ngs.org.uk for information & booking. Visits also by arrangement 6 June to 10 Sept for groups of 8 to 30.**

Set in 6 acres on the edge of the South Downs National Park surrounded by NT woodland, this rural garden was started in 2012 when a new Huf house was built on a derelict site. Now there are lawned areas with beds of perennials, a serenity pool with koi carp, a wildflower meadow, a Japanese garden, a sculpture garden, a woodland walk, a large rockery and an exotic walled garden.

147 5 WHITEMANS CLOSE

Cuckfield, Haywards Heath, RH17 5DE. Shirley Carman-Martin. *1m N of Cuckfield. On B2036 signed Balcombe, Whitemans Close is 250yds from r'about on LHS. No parking in Whitemans Close. Buses stop at Whitemans Green, where there is also a large free car park.* **Wed 25, Fri 27, Mon 30 Jan, Wed 1, Fri 3, Sat 4, Mon 6, Tue 7, Wed 8, Tue 14, Thur 16 Feb (10.30-4.30). Adm £8, chd free. Pre-booking essential, please phone 01444 473520 or email shirleycarmanmartin@gmail.com for information & booking. Adm inc home-made teas.**

A garden visit for snowdrop and plant lovers. A relatively small cottage garden packed full of exciting and unusual winter plants, a large snowdrop collection, and also many hellebores and ferns. All snowdrops can be viewed from paths, so you can leave the wellies at home, but warm clothes are essential.

GROUP OPENING

148 WINCHELSEA'S SECRET GARDENS

Winchelsea, TN36 4EJ. *2m W of Rye, 8m E of Hastings.* **Sat 17 June (11-5). Combined adm £7, chd free. Home-made teas at Winchelsea New Hall.**

NEW **BURRIN HOUSE**
North Street. Charlotte Beecroft.

CLEVELAND PLACE
Sally & Graham Rhodda.

GILES POINT
Ant Parker & Tom Ashmore.

KING'S LEAP
Philip Kent.

MAGAZINE HOUSE
Susan Stradling.

THE ORCHARDS
Brenda & Ralph Courtenay.

PERITEAU HOUSE
Dr & Mrs Lawrence Youlten.

NEW **THE RECTORY**
Rev Jonathan & Mrs Shirley Meyer.

SOUTH MARITEAU
Robert Holland.

THE WELL HOUSE
Alice Kenyon.

Many styles, large and small, secret walled gardens, roses, herbaceous borders and more, in the beautiful setting of the Cinque Port of Winchelsea. Explore the town with its magnificent church. Check winchelsea.com for the latest information on tours of our famous medieval cellars. If you are bringing a coach please contact ryeview@gmail.com, 01797 226524. WC facilities available. Wheelchair access to four of the ten gardens.

149 WISTON HOUSE

Steyning Road, Wiston, BN44 3DD. Mr & Mrs R J Goring & Wilton Park, www.wistonestate.com. *1m NW Steyning. A24 Washington r'about, take A283 to Steyning. Driveway 2m on RHS.* **Sat 1 July (10.30-5.30). Adm £10, chd free. Home-made teas. Wiston Sparkling Wine.**

Nestled at the foot of the South Downs within a landscaped park, Wiston House has a Victorian garden under restoration. Features inc a conservatory, terraced lawns with herbaceous borders, a cascade, woodland garden, Italian parterre, wildflower garden, walled vegetable garden and Victorian greenhouses. Wiston House will be open for guests to walk through the ground floor. St Mary's Church will be open. Wheelchair access over gravel paths.

150 WYCH WARREN HOUSE

Wych Warren, Forest Row, RH18 5LF. Colin King & Mary Franck, 07852 272898, mlfranck@hotmail.com. *1m S of Forest Row. Proceed S on A22, track turning on L, 100 metres past 45mph warning triangle sign. Or 1m N of Wych Cross T lights, track turning on R. Go 400 metres across golf course till the end.* **Wed 17 May (2-5). Adm £5, chd free. Home-made teas. Visits also by arrangement 18 May to 20 Sept.**

6 acre garden in Ashdown Forest, AONB, much of it mixed woodland. Perimeter walk around property (not open). Delightful and tranquil setting with various aspects of interest providing sensory and relaxing visit. Lovely stonework, specimen trees, tulip display, bluebell woods, three ponds, herbaceous borders, greenhouse and always something new on the go. Plenty of space to roam and explore and enjoy the fresh air. Plants for sale and a great range of chutney and jams. Dogs on leads and children welcome. Partial wheelchair access by tarmac track to the kitchen side gate.

WARWICKSHIRE

For Birmingham & West Midlands see Staffordshire

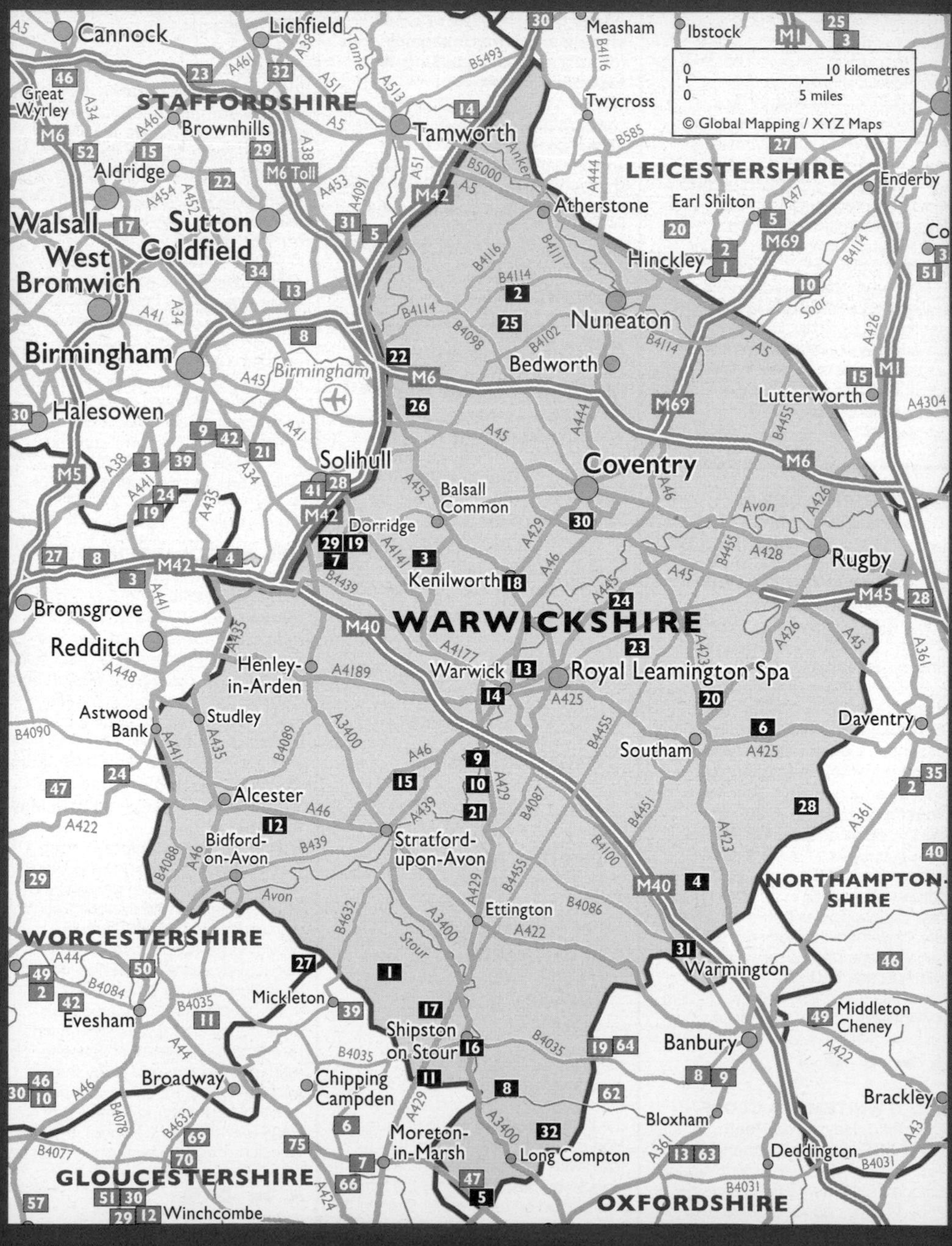

Set in the heart of England, Warwickshire is essentially a rural county, with many delightful villages and historic towns, amidst enchanting stretches of unspoilt countryside.

In the rolling hills of the south, Cotswold villages of honey-coloured stone nestle in the valleys and Stratford-upon-Avon is a mecca for tourists who flock to see Shakespeare's birthplace and enjoy his plays.

The ancient county town of Warwick with its great castle and elegant Leamington Spa lie at Warwickshire's centre, with Coventry and the more industrial landscapes to the north.

The Coventry Canal runs along the county's northern border, and the towns give way once again to open countryside. Warwickshire's gardens offer the visitor a tapestry of styles and settings, with locations ranging from picturesque villages and towns to nurseries and allotments. Grand country house gardens and charming tiny village ones are generously opened by their owners to support the National Garden Scheme Beneficiaries in their vital work.

So do please visit us. You will be assured of a warm welcome by our Garden Owners who take tremendous pleasure in sharing their gardens, knowledge and enthusiasm. And, of course, there is always the promise of a glorious afternoon tea to complete your day visiting our delightful gardens.

Volunteers

County Organiser
Liz Watson 01926 512307
liz.watson@ngs.org.uk

County Treasurer
Ian Roberts 01926 864181
ian.roberts@ngs.org.uk

Publicity
Lily Farrah 07545 560298
lily.farrah@ngs.org.uk

Social Media
Fiona Murphy
07754 943277
fiona.murphy@ngs.org.uk

Booklet Co-ordinator
Hazel Blenkinsop 07787 005290
hazel.blenkinsop@ngs.org.uk

Assistant County Organisers
James Coleman 07746 561570
james.coleman@talk21.com

Jenny Edwards 07884 177889
jenny.edwards@ngs.org.uk

Jane Redshaw 07803 234627
jane.redshaw@ngs.org.uk

Isobel Somers 07767 306673
ifas1010@aol.com

Left: Avening

@WarwickshireNGS   @WarksNGS

OPENING DATES

All entries subject to change. For latest information check **www.ngs.org.uk**

Extended openings are shown at the beginning of the month

Map locator numbers are shown to the right of each garden name.

February

Snowdrop Festival

Saturday 18th
◆ Hill Close Gardens 14

April

Every Saturday and Sunday from Friday 7th
◆ Bridge Nursery 6

Monday 10th
◆ Bridge Nursery 6

Sunday 16th
Broadacre 7

May

Every Saturday and Sunday
◆ Bridge Nursery 6

Monday 1st
◆ Bridge Nursery 6

Saturday 13th
6 Canon Price Road 9

Sunday 14th
6 Canon Price Road 9

Saturday 27th
Ilmington Gardens 17

Sunday 28th
Ilmington Gardens 17
Pebworth Gardens 27

Monday 29th
◆ Bridge Nursery 6
Pebworth Gardens 27

June

Every Saturday and Sunday
◆ Bridge Nursery 6

Sunday 4th
NEW 27 Knowle Wood Road 19
6 Rodborough Road 29

Saturday 10th
6 Canon Price Road 9

Sunday 11th
6 Canon Price Road 9
Honington Gardens 16
Maxstoke Castle 22
Styvechale Gardens 30

Friday 16th
Priors Marston Manor 28

Saturday 17th
Priors Marston Manor 28

Sunday 18th
Kenilworth Gardens 18
Packington Hall 26
Warmington Gardens 31
Whichford & Ascott Gardens 32

Sunday 25th
Blacksmiths Cottage 5

July

Every Saturday and Sunday
◆ Bridge Nursery 6

Sunday 2nd
Ansley Gardens 2

Saturday 8th
6 Canon Price Road 9
The Malt House 21

Sunday 9th
Avon Dassett Gardens 4
6 Canon Price Road 9
Old Arley Gardens 25

Saturday 22nd
Guy's Cliffe Walled Garden 13

August

Every Saturday and Sunday
◆ Bridge Nursery 6

Sunday 27th
Cedar House 10

Monday 28th
◆ Bridge Nursery 6

September

Every Saturday and Sunday to Sunday 24th
◆ Bridge Nursery 6

Sunday 10th
Burmington Grange 8

October

Saturday 28th
◆ Hill Close Gardens 14

By Arrangement

Arrange a personalised garden visit with your club, or group of friends, on a date to suit you. See individual garden entries for full details.

Admington Hall 1
NEW Avening 3
10 Avon Carrow, Avon Dassett Gardens 4
Broadacre 7
6 Canon Price Road 9
NEW Cedar House 10
Court House 11
The Croft House 12
Fieldgate, Kenilworth Gardens 18
The Hill Cottage 15
19 Leigh Crescent 20
The Motte 23
Oak House 24
Springfield House, Warmington Gardens 31

Gourdon

THE GARDENS

1 ADMINGTON HALL

Admington, Shipston-on-Stour, CV36 4JN. Mark & Antonia Davies, 01789 450279, adhall@admingtonhall.com. *6m NW of Shipston-on-Stour. From Ilmington, follow signs to Admington. Approx 2m, turn R to Admington by Polo Ground. Continue for 1m.* **Visits by arrangement May to Sept for groups of 20+. Adm £8, chd free. Home-made teas.**

A continually evolving 10 acre garden with an established structure of innovative planning and planting. A wide ranging collection of fine and mature specimen trees provide the essential core structure to this traditional country garden. Features inc a lush broad lawn, orchard, water garden, large walled garden, wildflower meadows and extensive modern topiary. This is a garden in motion. Wheelchair access to most parts of garden.

GROUP OPENING

2 ANSLEY GARDENS

Ansley, CV10 0QR. *Ansley is situated W of Nuneaton, adjacent to Arley. Ansley is directly off the B4114.* **Sun 2 July (1.30-6.30). Combined adm £6, chd free. Light refreshments. Cream teas & home-made cakes in specified gardens.**

1A BIRMINGHAM ROAD
Adrian & Heather Norgrove.

25 BIRMINGHAM ROAD
Pat & David Arrowsmith.

59 BIRMINGHAM ROAD
Joan & Peter McParland.

NEW **79 BIRMINGHAM ROAD**
Stephen Shaw.

NEW **HOOD LANE FARM**
David Kearns, www.hoodlanefarm.co.uk.

NEW **5 MARY CARROLL ROAD**
Derek Greedy.

35 NUTHURST CRESCENT
Roger & Heather Greaves.

1 PARK COTTAGES
Janet & Andy Down.

Ansley is a small ex-mining village situated in North Warwickshire. The 8 gardens open are a selection of different styles and offerings and this year there are three new gardens opening, from a very small traditional cottage garden crammed with flowers and pots to larger gardens and those maximising the amazing views of the countryside. There is one with a lush tropical feel, a country garden with mature plants and ancient roses and one boasting a glamping pod complete with large grounds to get lost in! 6 of the gardens are in the village with the opportunity to walk across the fields on public footpaths to the other two if you are able. (There is parking available at both if not!). The local Norman church will be open and the Morris Dancing Group will be entertaining visitors during the day.

3 NEW AVENING

Oldwich Lane East, Kenilworth, CV8 1NR. Helen Jones, 07894 321919, hello@mygardenoasis.co.uk, www.mygardenoasis.co.uk. *8 m SE of Solihull & 5 m W of Kenilworth. Avening is at the end of College Ln, a short single-track lane off Oldwich Ln E which runs between Chadwick End and Fen End.* **Visits by arrangement Apr to Sept for groups of up to 30.**

Designed in 2009, this is a ¾ acre garden oasis in the West Midlands. It has been designed to attract wildlife and has a typical cottage garden feel. Features inc 2 ponds joined by a waterfall, wildflower meadow, wide mixed shrub/herbaceous border with a stepping stone path running through it and small vegetable plot. There are no steps in the garden, although some of the paths are unsuitable for wheelchairs.

GROUP OPENING

4 AVON DASSETT GARDENS

Avon Dassett, CV47 2AE. *7m N of Banbury. From M40 J12 turn L & L again onto the B4100, following signs to Herb Centre & Gaydon. Take 2nd L into village (signed). Please park in cemetery car park at top of hill or where signed.* **Sun 9 July (2-6). Combined adm £5, chd free. Home-made teas. The Yew Tree pub will be open for lunch.**

10 AVON CARROW
Anna Prosser, 01295 690926, annaatthecarrow@btopenworld.com.
Visits also by arrangement Mar to Sept for groups of 5 to 20.

NEW **DASSETT HOUSE**
Mrs Sarah Rutherford, dassettdwelling@gmail.com.

THE EAST WING, AVON CARROW
Christine Fisher & Terry Gladwin.

HILL TOP FARM
Mr D Hicks.

THE OLD RECTORY
Lily Hope-Frost.

THE SNUG
Mrs Deb Watts.

Pretty Hornton stone village sheltering in the lee of the Burton Dassett hills, well wooded with parkland setting and The Old Rectory mentioned in Domesday Book. Wide variety of gardens inc kitchen gardens, cottage, gravel and tropical gardens. Range of plants inc alpines, herbaceous, perennials, roses, climbers and shrubs. The gardens are on/off the main road through the village. We would be grateful if visitors could park in designated areas and not along the main road in the village. For 2023, we hope to run a shuttle service from top to bottom of the village. Plant sales, historic church open and lunch available at the Yew Tree village pub, recently purchased by the Community, visit www.theyewtreepub.co.uk for details. Wheelchair access to most gardens.

5 BLACKSMITHS COTTAGE

Little Compton, Moreton-in-Marsh, GL56 0SE. Mrs Andrew Lukas. *Garden entrance is opp the village hall car park. Please park at village hall or in Reed College car park nearby.* **Sun 25 June (2-5.30). Adm £5, chd free. Home-made teas.**

A walled garden created in the last 8 years for all seasons with many fine and unusual trees, shrubs, plants and bulbs, set around a large lawn with a beautiful Aqualens fountain at its centre. Unusual varieties of clematis climb through shrubs and up walls. There is a productive vegetable garden, greenhouse and summerhouse. All designed and made by the owners to create a sense of magic. World Class Teas in the Village Hall opposite!

6 ◆ BRIDGE NURSERY

Tomlow Road, Napton, Southam, CV47 8HX. Christine Dakin & Philip Martino, 01926 812737, chris.dakin25@yahoo.com, www.bridge-nursery.co.uk. *3m E of Southam. Brown tourist sign at Napton Xrds on A425 Southam to Daventry Rd.* **For NGS: Every Sat and Sun 7 Apr to 24 Sept (10-4). Mon 10 Apr, Mon 1, Mon 29 May, Mon 28 Aug (10-4). Adm £3.50, chd free. Light refreshments. For other opening times and information, please phone, email or visit garden website.**

Clay soil? Don't despair. Here is an acre of garden with an exciting range of plants which thrive in hostile conditions. Grass paths lead you round borders filled with many unusual plants, a pond and bamboo grove complete with panda! A peaceful haven for wildlife and visitors. A visitor commented 'it is garden that is comfortable with itself.' Tea or coffee and biscuits gladly provided on request. Group visits welcome.

7 BROADACRE

Grange Road, Dorridge, Solihull, B93 8QA. John Woolman, 07818 082885, jw234567@gmail.com, www.broadacregarden.org. *Approx 3m SE of Solihull. On B4101 opp The Railway Inn. Plenty of parking.* **Sun 16 Apr (2-6). Adm £5, chd free. Home-made teas. Visits also by arrangement.**

Broadacre is a semi-wild garden, managed organically. Attractively landscaped with pools, lawns and trees, beehives, vegetable area and adjoining stream and wildflower meadows. Dorridge cricket club is on the estate. Lovely venue for a picnic. Dogs and children are welcome. Excellent country pub, The Railway Inn, at the bottom of the drive. Wheelchair access, improvement suggestions welcome.

8 BURMINGTON GRANGE

Cherington, Shipston-on-Stour, CV36 5HZ. Mr & Mrs Patrick Ramsay. *2m E of Shipston-on-Stour. Take Oxford Rd (A3400) from Shipston-on-Stour, after 2m turn L to Burmington, go through village & continue for 1m, turn L to Willington & Barcheston, on sharp L bend turn R over cattle grid.* **Sun 10 Sept (2-6). Adm £6, chd free. Tea.**

Interesting plantsman's garden extending to about 1½ acres, set in the rolling hills of the North Cotswolds with wonderful views over unspoilt countryside. The garden is well developed considering it was planted 20 years ago. Small vegetable and picking garden, beautiful sunken rose garden with herbaceous and shrub borders. Orchard and tree walk with unusual trees.

9 6 CANON PRICE ROAD

Nursery Meadow, Barford, CV35 8EQ. Mrs Marie-Jane Roberts, 07775 584336. *From A429 turn into Barford. Park on Wellesbourne Rd & walk into Nursery Meadow by red phone box. No 6 is R in 1st close. Disabled parking by house.* **Sat 13, Sun 14 May (2-4.30). Adm £3, chd free. Sat 10, Sun 11 June, Sat 8, Sun 9 July (2-4.30). Adm £5, chd free. Light refreshments in the garden room. Visits also by arrangement May to Sept.**

Unexpectedly large and mature garden with colour themed shrub and perennial plants immaculately grown. Separate areas for cut flowers, herbs, vegetables and 19 types of fruit, some fan trained. Spectacularly colourful patio pots, a small delightful rockery, pond and wildlife garden. With many seating areas New shrub and perennial garden for 2023. Hardy perennials for sale - please bring cash. Access via a single slab path that joins a wide path through the garden. No WC.

10 NEW CEDAR HOUSE

Wasperton, Warwick, CV35 8EB. Mr D Burbidge, 07836 532914, david.burbidge@burbidge.org.uk. *Turn off A429 into Wasperton. There is only 1 rd in the village. Continue for approx. a half mile.* **Sun 27 Aug (1-5.30). Adm £7.50, chd free. Home-made teas at Wasperton Village Hall. Visits also by arrangement for groups of 15 to 25.**

Large former vicarage garden, mature and exciting. Colourful herbaceous borders, collection of mature acer palmatums, walled garden, woodland glade and walk with mix of young and mature specimen trees, tranquil grass and bamboo area leading to the swimming pool garden. Several seating areas to enjoy the beautiful and peaceful setting. Interesting walks across the fields to the River Avon, Charlecote and Hampton Lucy. Some gravel and woodchip paths.

11 COURT HOUSE

Stretton-on-Fosse, GL56 9SD. Christopher White, 01608 663811, pennycourthouse@gmail.com. *Off A429 between Moreton-in-Marsh & Shipston-on-Stour. Located in centre of village, next to the church.* **Visits by arrangement 1 Mar to 1 Oct for groups of 10 to 40. Coaches to park at village hall (5 min walk). Weekdays only. Adm £8, chd free. Refreshments available on request.**

Most of the 4 acre garden lies to the rear of the house and on both sides of it. It consists of several different garden rooms, of unequal size and each with its own style. The range of planting is wide and is designed to achieve year-round interest. In some areas the effect is fully mature but in others we see more recently planted woody shrubs and trees as well as established herbaceous areas. Newly planted winter garden. Herbaceous borders, fernery, restored walled kitchen garden. Rose garden, pond area, paddocks established with wild flowers. Wheelchair access is not impossible, but difficult with a gravel drive.

12 THE CROFT HOUSE

Haselor, Alcester, B49 6LU. Isobel & Patrick Somers, 07767 306673, ifas1010@aol.com. *6m W of Stratford-upon-Avon, 2m E of Alcester, off A46. From A46 take Haselor turn. From Alcester take old Stratford Rd, turn L signed Haselor, then R at Xrds. Garden in centre of village. Please park considerately.* **Visits by arrangement 27 May to 11 June for groups of up to 30. Adm £4, chd free. Home-made teas.**

Wander through an acre of trees, shrubs and herbaceous borders densely planted with a designer's passion for colour and texture. Hidden areas invite you to linger. Gorgeous scented wisteria on 2 sides of the house. Organically managed, providing a haven for birds and other wildlife. Frog pond, treehouse, small vegetable plot and a few venerable old fruit trees from its days as a market garden.

Priors Marston Manor

13 GUY'S CLIFFE WALLED GARDEN

Coventry Road, Guy's Cliffe, Warwick, CV34 5FJ. Sarah Ridgeway, www.guyscliffewalledgarden.org.uk. *Behind Hintons Nursery in Guy's Cliffe. Guy's Cliffe is on the A429, between N Warwick & Leek Wootton.* **Sat 22 July (10-3.30). Adm £3.50, chd free.**

A Grade II listed garden of special historic interest, having been the kitchen garden for Guy's Cliffe House. The garden dates back to the mid-1700s. Restoration work started 9 yrs ago using plans from the early C19. The garden layout has already been reinstated and the beds, once more, planted with fruit, flowers and vegetables inc many heritage varieties. Glasshouses awaiting restoration. Original C18 walls. Exhibition of artefacts discovered during restoration. Wheelchair access through entrance of Hintons Nursery and paths through garden. WC with disabled access.

14 ◆ HILL CLOSE GARDENS

Bread and Meat Close, Warwick, CV34 6HF. Hill Close Gardens Trust, 01926 493339, centremanager@hcgt.org.uk, www.hillclosegardens.com. *Town centre. Follow signs to Warwick racecourse. Entry from Friars St onto Bread & Meat Close. Car park by entrance next to racecourse. 2 hrs free parking. Disabled parking outside the gates.* **For NGS: Sat 18 Feb (11-4); Sat 28 Oct (11-5). Adm £5, chd free. Light refreshments. For other opening times and information, please phone, email or visit garden website. Donation to Plant Heritage.**

Restored Grade II* Victorian leisure gardens comprising 16 individual hedged gardens, 8 brick summerhouses. Herbaceous borders, heritage apple and pear trees, C19 daffodils, over 100 varieties of snowdrops, many varieties of asters and chrysanthemums. Heritage vegetables. Plant Heritage border, auricula theatre, and Victorian style glasshouse. Children's garden. Wheelchair available. Please phone to book in advance.

NPC

15 THE HILL COTTAGE

Kings Lane, Snitterfield, Stratford-upon-Avon, CV37 0QA. Gillie & Paul Waldron, 07895 369387, info@thehillcottage.co.uk, www.thehillcottage.co.uk. *5 mins from M40, J15. Take A46 to Stratford. 1m take 2nd exit at r'about. 1m take L into Kings Ln, through S bends, house on R. Or from village, up White Horse Hill, R at T-junc, over A46, R into Kings Ln, 2nd on L.* **Visits by arrangement Apr to Sept for groups of up to 20. Adm £6, chd free. Home-made teas. £3.50pp.**

High on a ridge overlooking orchards and golf course with fabulous views to distant hills, this 2¼ acre garden, full of surprises, offers varied planting; sunny gravel with exotic specimens; cool, shady woodland with relaxed perennial groups; romantic green oak pond garden and stone summerhouse; pool with newts and dragonflies. Traditional glasshouse in walled kitchen garden with raised beds.

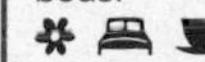

GROUP OPENING

16 HONINGTON GARDENS
Honington, Shipston-on-Stour, CV36 5AA. *1½m N of Shipston-on-Stour. Take A3400 towards Stratford-upon-Avon, then turn R signed Honington.* **Sun 11 June (2-5.30). Combined adm £6, chd free. Home-made teas.**

NEW **THE GARDEN HOUSE**
Mr & Mrs Andrew Sadleir.

HONINGTON HALL
B H E Wiggin.

MALT HOUSE RISE
Mr P Weston.

THE OLD HOUSE
Mrs D Beaumont.

ORCHARD HOUSE
Mr & Mrs Monnington.

NEW **THE ORCHARD, HOME FARM**
Mr Guy Winter.

NEW **ROSE COTTAGE**
Mr & Mrs Andrew Sadleir.

SHOEMAKERS COTTAGE
Christopher & Anne Jordan.

C17 village, recorded in Domesday, entered by old toll gate. Ornamental stone bridge over the River Stour and interesting church with C13 tower and late C17 nave after Wren. 8 contrasting gardens. One with extensive lawns and fine mature trees with river and garden monuments. A small, colourful and well-stocked garden, a developing garden with a hint of Japan, a typical cottage garden fit for a chocolate box cover, a 1 acre garden with informal mixed flower and vegetable beds. A secluded walled country garden with mixed perennial and annual planting, a structured cottage garden formally laid out with box hedging. A traditional orchard with mown pathways and active beehives.

GROUP OPENING

17 ILMINGTON GARDENS
Ilmington, CV36 4LA. *8m S of Stratford-upon-Avon. 8m N of Moreton-in-Marsh. 4m NW of Shipston-on-Stour off A3400. 3m NE of Chipping Campden.* **Sat 27, Sun 28 May (12.30-6). Combined adm £10, chd free. Cream teas in Ilmington Community Shop, Upper Green (Sat) & at the Village Hall (Sun). Donation to Shipston Home Nursing.**

THE BEVINGTONS
Mr & Mrs N Tustain.

THE DOWER HOUSE
Mr & Mrs M Tremellen.

FOXCOTE HILL
Mr & Mrs Michael Dingley.

FROG ORCHARD
Mr & Mrs Jeremy Snowden.

GRUMP COTTAGE
Mr & Mrs Martin Underwood.

ILMINGTON MANOR
Mr Martin Taylor.

OLD FOX HOUSE
Rob & Sarah Beebee.

RAVENSCROFT
Mr & Mrs Clasper.

STUDIO COTTAGE
Sarah Hobson.

Ilmington is an ancient hillside Cotswold village 2m from the Fosse Way with 2 good pubs. Start at Ilmington Manor (next to the Red Lion Pub); wander the 3 acre gardens with fish pond and then walk to the upper green to see Foxcote Hill's large gardens and Old Fox House. Next go up Grump St to Ravenscroft's large, sculpture filled, sloping vistas commanding the hilltop. Walk to nearby Frog Lane to view the cottage garden of Frog Orchard. Then go on to the Bevingtons many-chambered cottage garden at the bottom of Valanders Lane, by the church and manor ponds, and finally visit the Dower House in Back Street.

The Motte

GROUP OPENING

18 KENILWORTH GARDENS

Kenilworth, CV8 1BT. *Fieldgate Lane, off A452. Parking available at Abbey Fields or in town. Street parking on Fieldgate Ln (limited), Siddley Ave & Beehive Hill. Tickets & maps at all gardens. Transport is necessary to visit all the gardens.* **Sun 18 June (12-5.30). Combined adm £7, chd free. Home-made teas at St Nicholas Parochial Hall.**

BEEHIVE HILL ALLOTMENTS
Kenilworth Allotment Association.

34 CLARENDON ROAD
Barrie & Maggie Rogers.

FIELDGATE
Liz & Bob Watson, 01926 512307, liz.watson@ngs.org.uk.
Visits also by arrangement May to Sept for groups of 5 to 30.

14C FIELDGATE LANE
Sandra Aulton.

NEW **9 LAWRENCE GARDENS**
Leo Lewis & Judith Masson.

NEW **23 LINDSEY CRESCENT**
Mr John & Mrs Ruth Titley.

NEW **38 LINDSEY CRESCENT**
Mrs Christine Poole.

2 ST NICHOLAS AVENUE
Mr Ian Roberts.

NEW **2 SIDDLEY AVENUE**
Mr Simon & Mrs Celia Christie.

TALISMAN SQUARE COMMUNITY GARDEN
Discovery Properties and Cobalt Estates, www.facebook.com/FriendsofTalismanSquare/.

TREE TOPS
Joanna & George Illingworth.

14 WHITEHEAD DRIVE
James Coleman.

Kenilworth was historically a very important town in Warwickshire. It has one of England's best castle ruins, Abbey Fields and plenty of pubs and good cafes. The gardens open this year are spread out around the town. There are small, medium and large gardens with great variety - formal, relaxed and cottage styles with trees, shrubs, herbaceous borders, ponds and many wildlife friendly features - plus plenty of vegetables at the allotments. Several of the gardens have won Gold in the Kenilworth in Bloom garden competition.

19 NEW 27 KNOWLE WOOD ROAD

Dorridge, Solihull, B93 8JS. Mr Geoff Onyett. *2.5 m SE from M42 junc 5. From K&D Cricket Club on Station Rd follow Grove Rd. Take 3rd exit at the island. Park on the rd. Limited accessible parking on the front driveway (4 cars max).* **Sun 4 June (1-5.30). Combined adm with 6 Rodborough Road £4, chd free.**
This medium-sized urban garden with a countryside view has been redesigned by the owner along with redeveloping the house. In 2018 the garden was traditionally laid along rectangular lines. Over the last 4 years the beds have been reshaped and extended, the paths redesigned and areas of the garden defined with soft and hard landscaping.

20 19 LEIGH CRESCENT

Long Itchington, Southam, CV47 9QS. Tony Shorthouse, 01926 817192, tonyshorthouse19@gmail.com. *Off the Stockton Rd in Long Itchington village.* **Visits by arrangement 25 June to 30 Sept for groups of 5 to 10. Adm £5, chd free. Tea.**
Discover a tropical garden where dappled sunlight filters through the green canopy above and sounds of a waterfall fills the air. A tropical garden, hidden in a Warwickshire village, densely packed with rare and unusual species, many raised from seed.

21 THE MALT HOUSE

Charlecote, Warwick, CV35 9EW. Katriona & Rupert Collins. *10 mins from J15 M40, take the A429. About 2m along the A429 take R turn to Charlecote. Enter the village passing telephone box on L, The Malt House is approx 100m past this on the L.* **Sat 8 July (1.30-5). Adm £5, chd free. Home-made teas. There will also be cake for sale.**
Lovely Grade II listed house (not open) set in charming hamlet of Charlecote, on the edge of Charlecote House and deer park. Beautiful borders offering interest and colour throughout the yr and featuring a small pond which attracts an abundance of wildlife. Enjoy mixed borders brimming with agapanthus and beautiful pots. Walk around the vegetable plot and cutting patch and admire a large variety of dahlias. Hand tied bunches of flowers will be available to purchase.

22 MAXSTOKE CASTLE

Castle Lane, Coleshill, B46 2RD. Mr G M Fetherston-Dilke, www.maxstoke.com. *2½ m E of Coleshill. Located in between B'ham and Coventry. From B'ham on the B4114, take R turn down Castle Ln. The castle drive is 1¼ m on the R (the next turning after the golf course).* **Sun 11 June (11-5). Adm £9, chd free. Teas, home made cakes and savoury options will be available on site. With regret, we do not allow picnics. Donation to National Garden Scheme & other charities.**
Approx 5 acres of garden and grounds with roses, herbaceous plants, shrubs and trees in the courtyard and immediate surroundings of this C14 moated castle. Plant and gift stalls. Wheelchair access to the garden but not into the house.

23 THE MOTTE

School Lane, Hunningham, Leamington Spa, CV33 9DS. Margaret & Peter Green, 01926 632903, margaretegreen100@gmail.com. *5m E of Leamington Spa. 1m NW off B4455 (Fosse Way) at Hunningham Hill Xrds.* **Visits by arrangement 24 Apr to 31 Aug for groups of up to 30. Adm £5, chd free. Home-made teas by prior request.**
Plant lover's garden of about ⅓ acre, well-stocked with unusual trees, shrubs and herbaceous plants. Set in the quiet village of Hunningham with views over the River Leam and surrounding countryside. The garden has been developed over the last 20yrs and inc woodland, exotic and herbaceous borders, raised alpine bed, troughs, pots, wildlife pond, fruit and vegetable plot. Plant-filled conservatory. Wheelchair access to front and upper rear garden.

Our 2022 donation enabled 7,000 people affected by cancer to benefit from the garden at Maggie's in Oxford

24 OAK HOUSE

Waverley Edge, Bubbenhall, Coventry, CV8 3LW. Helena Grant, 07731 419685, helena.grant@btinternet.com. *15 mins from Leamington Spa via the Oxford Rd/A423 & the Leamington Rd/A445 & the A46. Spaces for 5 cars only.* **Visits by arrangement for groups of up to 12. Adm £5, chd free. Tea & cakes can be provided for a charge of £3 per person.**

Tucked away next to Waverley Wood Oak House, enjoy a walled garden that has been landscaped and extended over 30 years. The garden is split on 2 levels with 7 seating areas allowing for relaxed appreciation of every aspect of the garden, with peaceful places to sit, ponder and enjoy. A focal point is the large terracotta urn, over 60 years old, which delivers vertical interest and the summerhouse and arbour which face each other diagonally across the garden. The curved borders have a wide range of planting creating distinct areas which surround the lawn. There are over 80 different plant varieties giving year-round interest which is a haven for birds and wildlife. Wheelchair access only on level path which runs round the house.

GROUP OPENING

25 OLD ARLEY GARDENS

Ansley Lane, Arley, Coventry, CV7 8FT. Mrs Carolyn McKay. *2m off Tamworth Rd or ½ m from B'ham Rd. Enter Ansley Ln via B'ham Rd or Rectory Rd.* **Sun 9 July (2-6). Combined adm £5, chd free. Light refreshments in local church.**

7 ST WILFREDS COTTAGES
Mrs Sarah Shepherd.

21 ST WILFREDS COTTAGES
Miss Jo Carter.

24 ST WILFREDS COTTAGES
Mrs Carolyn McKay.

29 ST WILFREDS COTTAGES
Ms Maggie Eggar.

32 ST WILFREDS COTTAGES
Mr Ronald Whiting.

NEW **35 ST WILFREDS COTTAGES**
Mrs Justine Clee.

THE SCHOOL HOUSE
Mrs Pauline McAleese.

Old Arley is an ancient village which appears in the Domesday book. More recently it was a small mining community and St Wilfreds Cottages were built in 1907 to accommodate the mine's supervisors and their families. The gardens vary in style and size and can be easily reached one after the other on the lane. The planting is varied, ranging from traditional cottage gardens to more contemporary.

26 PACKINGTON HALL

Meriden, nr Coventry, CV7 7HF. Lord & Lady Guernsey, www.packingtonestate.co.uk. *Midway between Coventry & B'ham on A45. Entrance 400yds from Stonebridge r'about towards Coventry. For SatNav please use CV7 7HE or What3words app - dreamers.mandolin.portfolio.* **Sun 18 June (2-5). Adm £7.50, chd free. Home-made teas.**

Packington is the setting for an elegant Capability Brown landscape. Designed from 1751, the gardens inc a serpentine lake, impressive Cedars of Lebanon, Wellingtonias and a 1762 Japanese bridge. There is also a millennium rose garden, wildflower meadow, mixed terrace borders, woodland walks and a newly restored walled garden. Home-made teas on the terrace or in The Pompeiin Room if wet. Wheelchair access is possible but please note there are no paths in the garden. The gravelled terrace, where teas are served, is easily accessible.

National Garden Scheme gardens are identified by their yellow road signs and posters. You can expect a garden of quality, character and interest, a warm welcome and plenty of home-made cakes!

GROUP OPENING

27 PEBWORTH GARDENS

Stratford-upon-Avon, CV37 8XZ. www.pebworth.org/ngs-open-gardens. *7m SW of Stratford-upon-Avon. For parking & SatNav please use CV37 8XN.* **Sun 28, Mon 29 May (1-5.30). Combined adm £7, chd free. Home-made teas at Pebworth Village Hall. Coffee and soft drinks also available. Home-made cakes served by The Pebworth and District W.I.**

NEW **BROADLEAF FARM,**
Alix and Ben Hay.

FELLY LODGE
Maz & Barrie Clatworthy.

JASMINE COTTAGE
Ted & Veronica Watson.

THE KNOLL
Mr & Mrs K Wood.

MAPLE BARN
Richard & Wendi Weller.

MEON COTTAGE
David & Sally Donnison.

NEW **ORCHARD WILLOW**
Mr Roy Bowron.

PEBWORTH ALLOTMENTS
Les Madden.

Pebworth is a delightful village with thatched cottages and properties young and old. There are a variety of garden styles from cottage gardens to modern, walled and terraced gardens. Pebworth is topped by St Peter's Church which has a large ring of 10 bells, unusual for a small rural church. This year we have 7 gardens and the Pebworth Allotments opening, with scrumptious tea and cakes provided by the Pebworth WI in the village hall. The Pebworth Allotments have only been in existence a few years and residents have lovingly tended to them. Some of the Allotmenteers are real characters so do visit and have a chat, they have a wealth of knowledge. Pebworth has been featured on TV a number of times. Partial wheelchair access in some gardens. Ramp available for village hall.

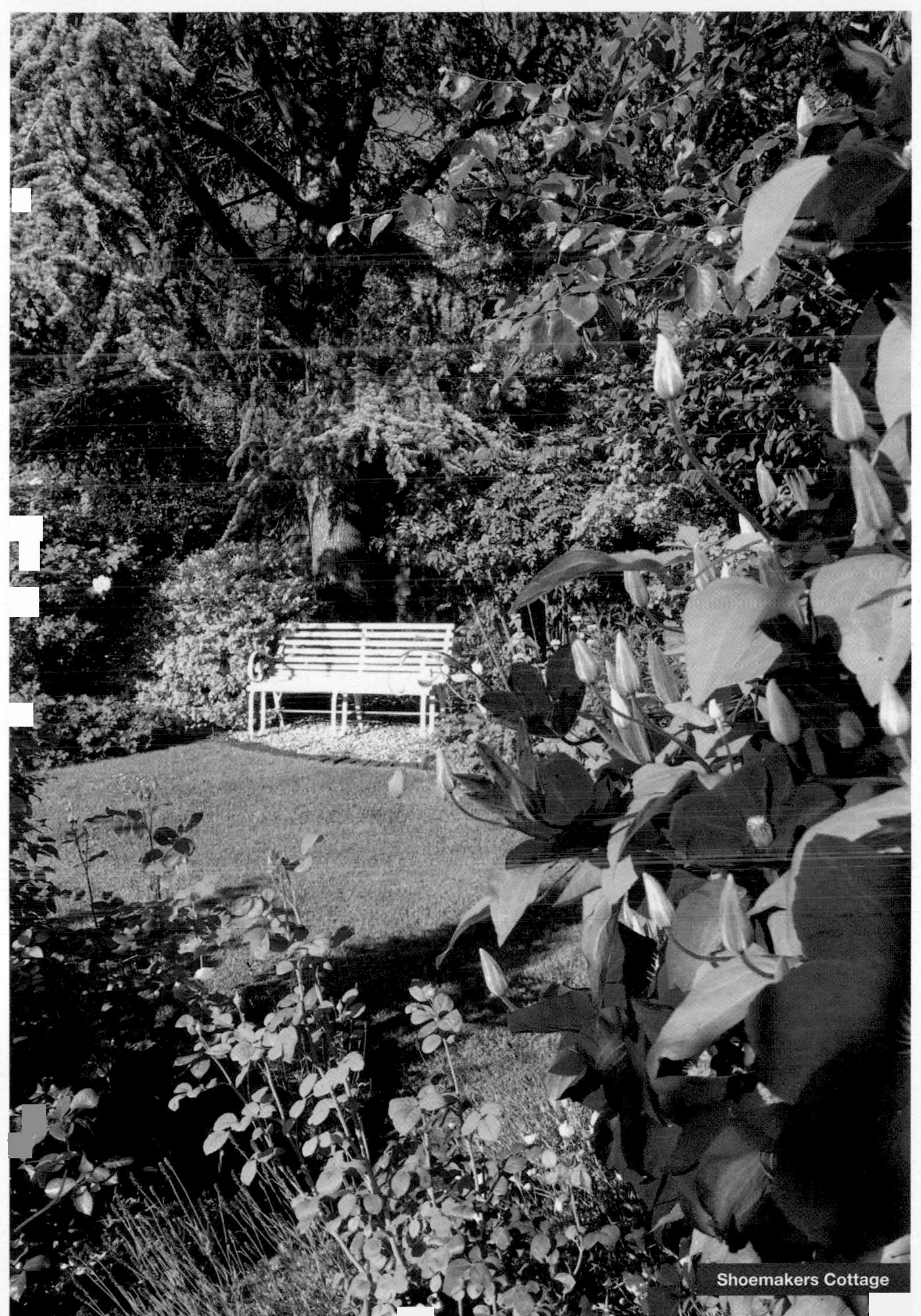

Shoemakers Cottage

St Wilfreds Cottages

28 PRIORS MARSTON MANOR
The Green, Priors Marston, CV47 7RH. Dr & Mrs Mark Cecil. *8m SW of Daventry. Off the A361 between Daventry & Banbury at Charwelton. Follow signs to Priors Marston, approx 2m. Arrive at T-junc with a war memorial on R. The manor will be on your L.* **Fri 16, Sat 17 June (10-6). Adm £7, chd free. Tea, coffee & home-made cakes. Sculpture on display. Friday 16th opening inc a Bonhams Valuation Day (11-2). Suggested donation £2 per item.**
Arrive in Priors Marston village and explore the manor gardens. Greatly enhanced by the present owners to relate back to a Georgian manor garden and pleasure grounds. Wonderful walled kitchen garden provides seasonal produce and cut flowers for the house. Herbaceous flower beds and a sunken terrace with water feature by William Pye. Lawns lead down to the lake and estate around which you can walk amongst the trees and wildlife with stunning views up to the house and garden aviary. Partial wheelchair access.

29 6 RODBOROUGH ROAD
Dorridge, Solihull, B93 8ED. Beryl & Peter Harrison. *Manor Rd end of Rodborough Rd. From Dorridge village centre go under the railway bridge and turn R into Manor Rd. At the small r'about turn L into Rodborough Rd. No. 6 is the 3rd house on the R.* **Sun 4 June (1-5.30). Combined adm with 27 Knowle Wood Road £4, chd free.**
A ¼ acre garden with wide borders and a large, central island bed. There is an extensive range of plants inc flowering evergreen shrubs, acers, cornus, roses and herbaceous perennials in a relaxed mingled style, designed to give year round interest and colour.

GROUP OPENING

30 STYVECHALE GARDENS

Knoll Drive, Coventry, CV3 5DE. styvechale-gardens.wixsite.com/ngs2020. *The gardens are located on the S side of Coventry close to A45. Tickets & map available on the day from The Margaret Roper Room St Thomas More Church Knoll Drive Coventry CV3 5DE.* **Sun 11 June (11-5). Combined adm £5, chd free.**

11 BAGINTON ROAD
Ken & Pauline Bond.

164 BAGINTON ROAD
Fran & Jeff Gaught.

59 THE CHESILS
John Marron & Richard Bantock.

66 THE CHESILS
Ami Samra & Rodger Hope.

96 THE CHESILS
Rebecca & Barry Preston.

NEW **102 THE CHESILS**
Mary & Peter Kinsella.

16 DELAWARE ROAD
Val & Roy Howells.

2 THE HIRON
Sue & Graham Pountney.

177 LEAMINGTON ROAD
Barry & Ann Suddens.

40 RANULF CROFT
Spencer & Sue Swain.

An eclectic mix of lovely, mature suburban gardens. Enjoy their variety from kitchen garden to a touch of the exotic, restful gardens to a riot of colourful borders. Roses, water features, ponds, shady areas and more. Something for everyone and plenty of ideas for you to take home. 2 of our gardens have been featured in garden magazines and one garden was awarded garden of the year by a national newspaper. Relax in the gardens where a warm friendly welcome will await you. Plant sales and refreshments in some gardens. Additional gardens will be open on the day.

GROUP OPENING

31 WARMINGTON GARDENS

Banbury, OX17 1BU. *5m NW of Banbury. Take B4100 N from Banbury, after 5m turn R across short dual carriageway into Warmington. From N take J12 off M40 onto B4100.* **Sun 18 June (1-5). Combined adm £7, chd free. Home-made teas at village hall. Sunday lunches available at the Plough Inn.**

GOURDON
Jenny Deeming.

GREENWAYS
Tim Stevens

HILL COTTAGE, SCHOOL LANE
Mr Mike Jones.

LANTERN HOUSE
Peter & Tessa Harborne.

THE MANOR HOUSE
Mr & Mrs G Lewis.

THE ORCHARD
Mike Cable.

SPRINGFIELD HOUSE
Jenny & Roger Handscombe, 01295 690286, jehandscombe@btinternet.com. **Visits also by arrangement 1 Apr to 22 July for groups of up to 12.**

1 THE WHEELWRIGHTS
Ms E Bunn.

Warmington is a charming historic village, mentioned in the Doomsday Book, situated at the NE edge of the Cotswolds in a designated AONB. There is a large village green with a pond overlooked by an Elizabethan Manor House (not open). There are other historic buildings inc St Michael's Church, The Plough Inn and Springfield House all dating from the C16 or before. There is a mixed and varied selection of gardens to enjoy during your visit to Warmington. These inc the formal knot gardens and topiary of The Manor House, cottage and courtyard gardens, terraced gardens on the slopes of Warmington Hill and orchards containing local varieties of apple trees. Some gardens will be selling homegrown plants. WC at village hall, along with delicious home-made cakes, and hot and cold drinks.

GROUP OPENING

32 WHICHFORD & ASCOTT GARDENS

Whichford & Ascott, Shipston-on-Stour, CV36 5PG. *6m SE of Shipston-on-Stour. For parking please use CV36 5PG. We have a large car park opp the church. Parking available at Wood House.* **Sun 18 June (1.30-5). Combined adm £7, chd free. Home-made teas.**

ASCOTT LODGE
Charlotte Copley.

ASCOTT RISE
Carol & Jerry Moore.

BELMONT HOUSE
Robert & Yoko Ward.

NEW **LENTICULARS**
Mrs Diana Atkins.

THE OLD RECTORY
Peter & Caroline O'Kane.

PLUM TREE COTTAGE
Janet Knight.

WHICHFORD HILL HOUSE
Mr & Mrs John Melvin.

THE WHICHFORD POTTERY
Jim & Dominique Keeling, www.whichfordpottery.com.

NEW **WOOD HOUSE**
Mrs Mandy James.

The gardens in this group reflect many different styles. The 2 villages are in an AONB, nestled within a dramatic landscape of hills, pasture and woodland, which is used to picturesque effect by the garden owners. Fine lawns, mature shrub planting and much interest plantsmen will all enrich your visit to our beautiful gardens. Many incorporate the inventive use of natural springs, forming ponds, pools and other water features. Classic cottage gardens contrast with larger and more classical gardens which adopt variations on the traditional English garden of herbaceous borders, climbing roses, yew hedges and walled enclosures. Partial wheelchair access as some gardens are on sloping sites.

WILTSHIRE

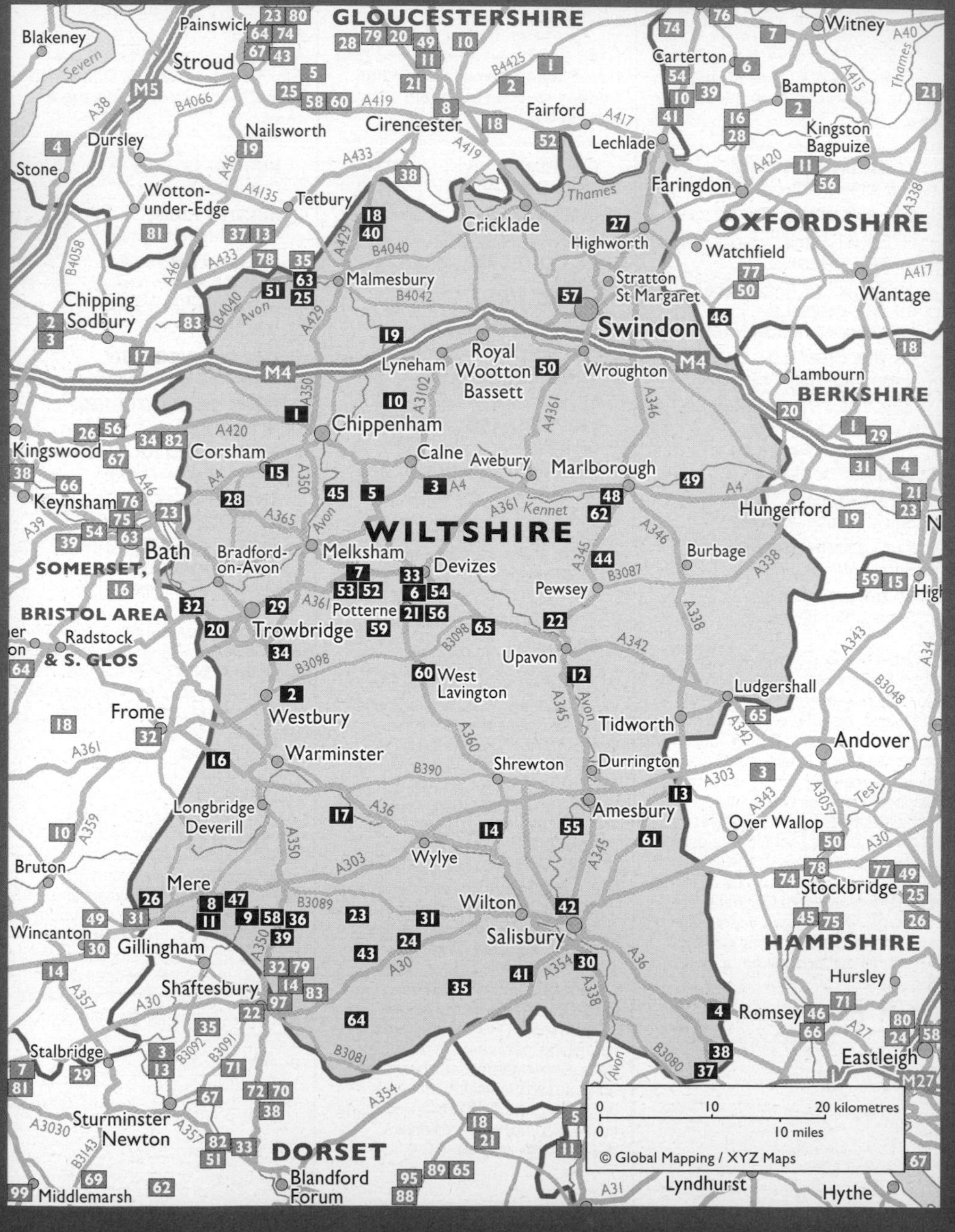
GLOUCESTERSHIRE
OXFORDSHIRE
BERKSHIRE
WILTSHIRE
SOMERSET, BRISTOL AREA & S. GLOS
HAMPSHIRE
DORSET
Swindon
Chippenham
Salisbury
Trowbridge
Devizes
Marlborough
Malmesbury
Melksham
Calne
Corsham
Bath
Bradford-on-Avon
Westbury
Warminster
Amesbury
Tidworth
Wilton
Mere
Stroud
Cirencester
Tetbury
Cricklade
Highworth
Royal Wootton Bassett
Lyneham
Wroughton
Avebury
Pewsey
Burbage
Upavon
West Lavington
Potterne
Shrewton
Durrington
Wylye
Longbridge Deverill
Gillingham
Shaftesbury
Frome
Andover
Hungerford
Romsey
Stockbridge
0 10 20 kilometres
0 10 miles
© Global Mapping / XYZ Maps

Wiltshire, a predominantly rural county, covers 1,346 square miles and has a rich diversity of landscapes, including downland, wooded river valleys and Salisbury Plain.

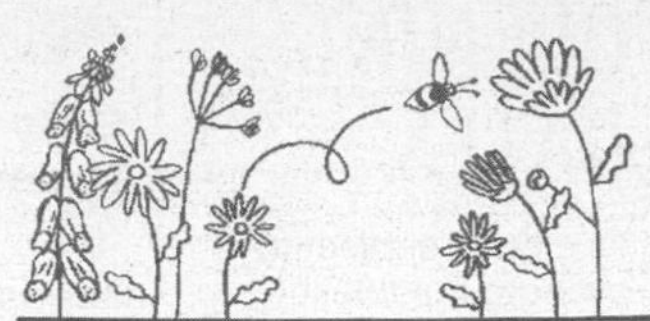

Chalk lies under two-thirds of the county with limestone to the north which includes part of the Cotswold Areas of Outstanding Natural Beauty. The county's gardens reflect its rich history and wide variety of environments.

Gardens opening for the National Garden Scheme include the Grade II listed Edwardian-style gardens of Hazelbury Manor and large privately owned gems such as Eastwell Manor, Chisenbury Priory and Cadenham Manor. We also admire less extravagant properties that are lovingly maintained by the owners such as Trymnells in Potterne and 1 Southview in Devizes.

The season opens with snowdrops at Westcroft, daffodils at Fonthill House, fabulous tulips at Blackland House and magnolias in spring at, amongst others, Broadleas House and Oare House. We also have some returning gardens including Allington Grange and Wellaway.

A wide selection of gardens, large and small, are at their peak in the summer. There are also three village openings which number a total of twenty-three gardens. These include a new one at both Dauntsey (The Old Rectory) and Landford Village (Foxhollow). If you are from a gardening club there are, in addition, a wide range of gardens open by arrangement totalling seventeen in all. There is something to delight the senses from January to September and one will always have a warm welcome wherever one goes.

Below: Falkners Cottage

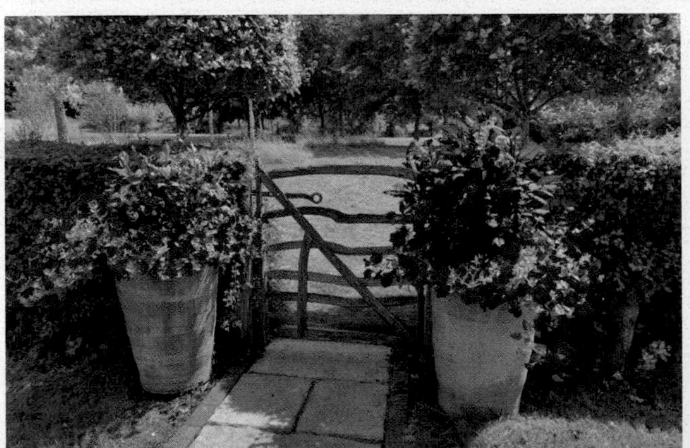

Volunteers

County Organiser
Amelia Tester 01672 520218
amelia.tester@ngs.org.uk

County Treasurer
Tony Roper 01249 447436
tony.roper@ngs.org.uk

Publicity & Booklet Co-ordinator
Tricia Duncan 01672 810443
tricia.duncan@ngs.org.uk

Social Media
Maud Peters 07595 266299
maudahpeters@gmail.com

Assistant County Organisers
Sue Allen 07785 294153
sue.allen@ngs.org.uk

Sarah Coate 01722 782365
sarah.coate@ngs.org.uk

Annabel Dallas 01672 520266
annabel.dallas@btinternet.com

Andy Devey 07810 641595
andy.devey@ngs.org.uk

Ros Ford 01380 722778
ros.ford@ngs.org.uk

Alex Graham 07906 146337
alex.graham@ngs.org.uk

Donna Hambly 07760 889277
donna.hambly@ngs.org.uk

Jo Hankey 01722 742472
jo.hankey@ngs.org.uk

Alison Parker 07786 985741
alison.parker@ngs.org.uk

@WiltshireNGS
@WiltshireNgs
@Wiltshirengs

OPENING DATES

All entries subject to change. For latest information check **www.ngs.org.uk**

Map locator numbers are shown to the right of each garden name.

Extended openings are shown at the beginning of the month.

January

Every Thursday from Thursday 12th
Westcroft 61

Saturday 28th
Westcroft 61

Sunday 29th
Westcroft 61

February

Snowdrop Festival

Every Thursday
Westcroft 61

Sunday 19th
Westcroft 61

Saturday 25th
Westcroft 61

Sunday 26th
Westcroft 61

March

Every Thursday to Thursday 9th
Westcroft 61

Sunday 19th
◆ Corsham Court 15
Fonthill House 23

April

Sunday 2nd
◆ Corsham Court 15

Sunday 16th
Broadleas House Gardens 6
NEW Brow Cottage 7
Seend House 52

Friday 21st
Blackland House 3

Saturday 22nd
◆ Iford Manor Gardens 32

Sunday 23rd
Allington Grange 1
Foxley Manor 25

Sunday 30th
Oare House 44
Waterdale House 58

May

Saturday 6th
Wellaway 59

Sunday 7th
Bush Farm 9
Knoyle Place 36
Lavender Gardens 38
Lower Lye 39
Wellaway 59

Saturday 13th
Julia's House Children's Hospice 33

Sunday 14th
Allington Grange 1
1 Southview 54
NEW Trymnells 56

Saturday 20th
NEW Winkelbury House 64

Sunday 21st
◆ Twigs Community Garden 57

Friday 26th
NEW King's Old Rectory 35

Saturday 27th
NEW Eastwell Manor 21

Sunday 28th
North Cottage 43
The Parish House 47

June

Saturday 3rd
Fovant House 24
The Old Vicarage 46

Sunday 4th
NEW The Chantry 11
Hyde's House 31

Wednesday 7th
Whatley Manor 63

Thursday 8th
Cadenham Manor 10

Friday 9th
Salthrop House 50

Saturday 10th
Landford Village Gardens 37
Salthrop House 50

Sunday 11th
Burton Grange 8
Chisenbury Priory 12
NEW Cholderton Estate 13
NEW Crudwell House 18
Hannington Village Gardens 27
Hazelbury Manor Gardens 28
Landford Village Gardens 37
Manor Farm 40

Saturday 17th
Hilperton House 29
Seend House 52
Seend Manor 53

Sunday 18th
Cortington Manor 17
Dauntsey Gardens 19
Hilperton House 29
Mawarden Court 42
North Cottage 43

Wednesday 21st
NEW Riverside House 49

Friday 23rd
The Old Mill 45

Saturday 24th
The Old Mill 45

Sunday 25th
Duck Pond Barn 20
Oare House 44

July

Sunday 2nd
Horatio's Garden 30
Teasel 55

Friday 14th
Salthrop House 50

Saturday 15th
Salthrop House 50

Sunday 16th
◆ Twigs Community Garden 57

Sunday 30th
Peacock Cottage 48
NEW Westwind 62

Horatio's Garden

By Arrangement

Arrange a personalised garden visit with your club, or group of friends, on a date to suit you. See individual garden entries for full details.

THE GARDENS

1 ALLINGTON GRANGE

Allington, Chippenham, SN14 6LW. Mrs Rhyddian Roper, www.allingtongrange.com. *2m W of Chippenham. Take A420 W from Chippenham. 1st R signed Allington Village, entrance I m up lane on L.* **Sun 23 Apr, Sun 14 May (2-5). Adm £7.50, chd free. Home-made teas.**

Informal country garden around C17 farmhouse (not open) with year-round interest and a diverse range of plants. Many early spring bulbs. Mixed and herbaceous borders. Pergola underplanted with shade loving plants and newly developed fountain garden. Walled potager. Wildlife pond with natural planting. Many alpine troughs, and increasing fern collection. Mainly level with ramp into potager. Dogs on leads.

2 BEGGARS KNOLL CHINESE GARDEN

Newtown, Westbury, BA13 3ED. Colin Little & Penny Stirling, 01373 823383, silkendalliance@talktalk.net. *1m SE of Westbury. Turn off B3098 at White Horse Pottery, up hill towards the White Horse for ¾m. Parking at end of drive for 10-12 cars.* **Visits by arrangement 1 June to 30 July for groups of up to 20. Adm £6, chd free. Tea & cake on request.**

A series of garden rooms are Chinese-style, separated by elaborate gateways inc moongate, with mosaic paths winding past pavilions, ponds and many rare Chinese trees, shrubs and flowers. Relatively new, a tranquil Islamic-style tiled garden influenced by NW China. Potager full of flowers and vegetables. Spectacular views to the Mendips. Map and garden tours by owners inc in ticket.

3 BLACKLAND HOUSE

Quemerford, Calne, SN11 8UQ. Polly & Edward Nicholson, www.bayntunflowers.co.uk. *Situated just off A4. Use Google Maps. Enter the grounds through the side entrance signed St. Peter's Church & Blackland Park Deliveries. Opp The Willows on Quemerford.* **Fri 21 Apr (2-5). Adm £10, chd free. Pre-booking essential, please visit www.ngs.org.uk for information & booking. Home-made teas. Vegan & gluten free cakes. Please bring cash for teas & plants. Donation to Dorothy House Hospice.**

A wonderfully varied 5 acre garden adjacent to River Marden (house not open). Formal walled productive and cutting garden, traditional glasshouses, rose garden and wide herbaceous borders. Interesting topiary, trained fruit trees, historic tulips and other unusual spring bulbs. Hand-tied bunches of flowers for sale. Partial wheelchair access, steps, grass and cobbles, wooden bridges.

♿ ✻ NPC ☕ 🔊

4 BLUEBELLS

Cowesfield, Whiteparish, Salisbury, SP5 2RB. Hilary Mathison, 01794 885738, hilary.mathison@icloud.com. *SW of Salisbury. On main A27 road from Salisbury to Romsey. 1½m SE of Whiteparish, on A27, 100-200 m inside Wiltshire county boundary.* **Visits by arrangement 19 Mar to 1 Oct for groups of 5 to 35. Parking for 10-12 cars max (no street parking). Adm £5, chd free. Cream teas. Wine & savoury bites for evening visits.**

Newly established 4 yr old garden on 1½ acre plot, with original deciduous woodland inc bluebells in season. Adjoining new-build contemporary house, so the rear courtyard reflects this. Other areas inc shady borders, winter garden, vegetable and fruit plot, formal and wildlife ponds. Large lawn surrounded by differently styled borders and large feature bed planted with white birch and cornus.

Our donation of £2.5 million over five years has helped fund the new Y Bwthyn NGS Macmillan Specialist Palliative Care Unit in Wales. 1,900 inpatients have been supported since its opening

5 ◆ BOWOOD WOODLAND GARDENS

Calne, SN11 9PG. The Marquis of Lansdowne, 01249 812102, houseandgardens@bowood.org, www.bowood.org. *3½m SE of Chippenham. Located off J17 M4 nr Bath & Chippenham. Entrance off A342 between Sandy Lane & Derry Hill Villages. Follow brown tourist signs.* **For opening times and information, please phone, email or visit garden website.**

This 30 acre woodland garden of azaleas, magnolias, rhododendrons and bluebells is one of the most exciting of its type in the country. From the individual flowers to the breathtaking sweep of colour, this is a garden not to be missed. With two miles of meandering paths, you will find hidden treasures at every corner. The Woodland Gardens are 2m from Bowood House and Garden.

6 BROADLEAS HOUSE GARDENS

Devizes, SN10 5JQ. Mr & Mrs Cardiff, instagram.com/broadleasgarden. *1m S of Devizes. Turn L from Hartmoor Rd onto Broadleas Park. Follow the road until you reach grassed area on R with red brick wall, stone pillars, grey gates & cattle grid, which is the entrance.* **Sun 16 Apr, Sun 3 Sept (2-5). Adm £10, chd free. Home-made teas.**

6 acre garden of hedges, perennial borders, walled rose garden, secret garden, bee garden and orchard stuffed with good plants. Well-stocked kitchen and herb garden. Mature collection of specimen trees inc oaks, magnolia, handkerchief, redwood and dogwood. Overlooked by the house and arranged above the valley the garden is crowded with magnolias, camellias, rhododendrons, azaleas, cornus, hydrangeas etc. Broadleas' professional beekeeper, Graham Davison, sells honey at open days. Wheelchair access to upper garden only, some gravel and narrow grass paths.

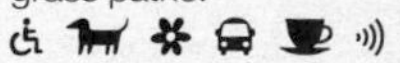

7 NEW BROW COTTAGE

Seend Hill, Seend, Melksham, SN12 6RU. James & Alexandra Gray, www.alexandragray.com. *2m W of Devizes on A361. Garden on L when coming from E, at top of Inmarsh Lane, but parking on green in village centre. Footpath along road or via fields as marked by yellow signs.* **Sun 16 Apr (1-5). Combined adm with Seend House £7, chd free. Home-made teas.**

Half acre contemporary cottage garden owned by a garden designer and created over the past 23 yrs. A garden of many harmonious parts inc lawns, well-stocked borders, topiary, potager, sunken pool garden, wildlife pond, short woodland walk, species bulb lawn and canopied dining area. Open with Seend House and connected by a field footpath offering fine views to Salisbury Plain.

8 BURTON GRANGE

Burton, Mere, BA12 6BR. Sue Phipps & Paddy Sumner, www.suephipps.com. *Take lane, signposted to Burton, on A303 just E of Mere bypass. After 400 yds follow rd past pond and round to L. Go past wall on R. Burton Grange entrance is in laurel hedge on R.* **Sun 11 June (11-5). Adm £5, chd free. Home-made teas.**

1½ acre garden, created from scratch since 2014. Lawns, borders, large ornamental pond, some gravel planting, vegetable garden and pergola rose garden, together with a number of wonderful mature trees. Artist's studio open.

9 BUSH FARM

West Knoyle, Mere, BA12 6AE. Lord & Lady Seaford, www.bisonfarm.co.uk. *From A303 turn off at Esso petrol station (Willoughby Hedge) follow L hand lane to West Knoyle (signed Bush Farm) through to end of village. Entrance on bend through woods to Bush Farm.* **Sun 7 May (2-5). Adm £6, chd free. Light refreshments. Bison burgers.**

Mature oak woodland glades jungliefied with climbing roses, clematis, honeysuckle and specimen trees. Bog garden with ferns, skunk cabbage and iris. Nearby lakeside walk with farm trail to see the bison and elk. Wildflowers everywhere. Bluebells in May. Flat woodland walks - not paved.

10 CADENHAM MANOR

Foxham, Chippenham, SN15 4NH. Victoria & Martin Nye, www.cadenham.com. *B4069 from Chippenham or M4 J17, turn R in Christian Malford & L in Foxham. On A3102 turn L from Calne or R from Lyneham at Xrds between Hilmarton & Goatacre. See map https://www.cadenham.com/contact.* **Thur 8 June (2-5). Adm £10, chd free. Home-made teas.**

This glorious 4 acre garden surrounds a listed C17 manor house and C16 dovecote. Divided by yew hedges and moats, its many rooms are furnished with specimen trees and fountains and statues to focus the eye. Known for its stunning displays of old roses, it also has bold swathes of spring bulbs, wisteria, bearded iris, a water garden in the old canal, plus extensive vegetable and herb gardens.

11 NEW THE CHANTRY

Church Street, Mere, Warminster, BA12 6DS. Mr & Mrs Richard Wilson, instagram.com/thechantrygarden. *In the centre of Mere, turn down Angel Lane & follow it to the Cemetery Car Park (RHS) & through it into the field at the Chantry. Pedestrian access is via Church Street.* **Sun 4 June (11-5). Adm £7.50, chd free. Home-made teas.**

Medieval chantry house with large landscaped gardens (inc cottage and walled gardens) ponds, orchard, fields and woodland. Historical connection with celebrated Dorset poet William Barnes who wrote his most famous work 'My Orchard in Linden Lea' about the garden.

12 CHISENBURY PRIORY

East Chisenbury, SN9 6AQ. Mr & Mrs John Manser, 07810 483984, peterjohnmanser@yahoo.com. *3m SW of Pewsey. Turn E from A345 at Enford then N to E Chisenbury, main gates 1m on R.* **Sun 11 June (2-5.30). Adm £5, chd free. Home-made teas. Visits also by arrangement May & June.**

Medieval Priory with Queen Anne face and early C17 rear (not open) in middle of 5 acre garden on chalk. Mature garden with fine trees within clump and flint walls, herbaceous borders, shrubs, roses. Moisture loving plants along mill leat, carp pond, orchard and wild garden, many unusual plants.

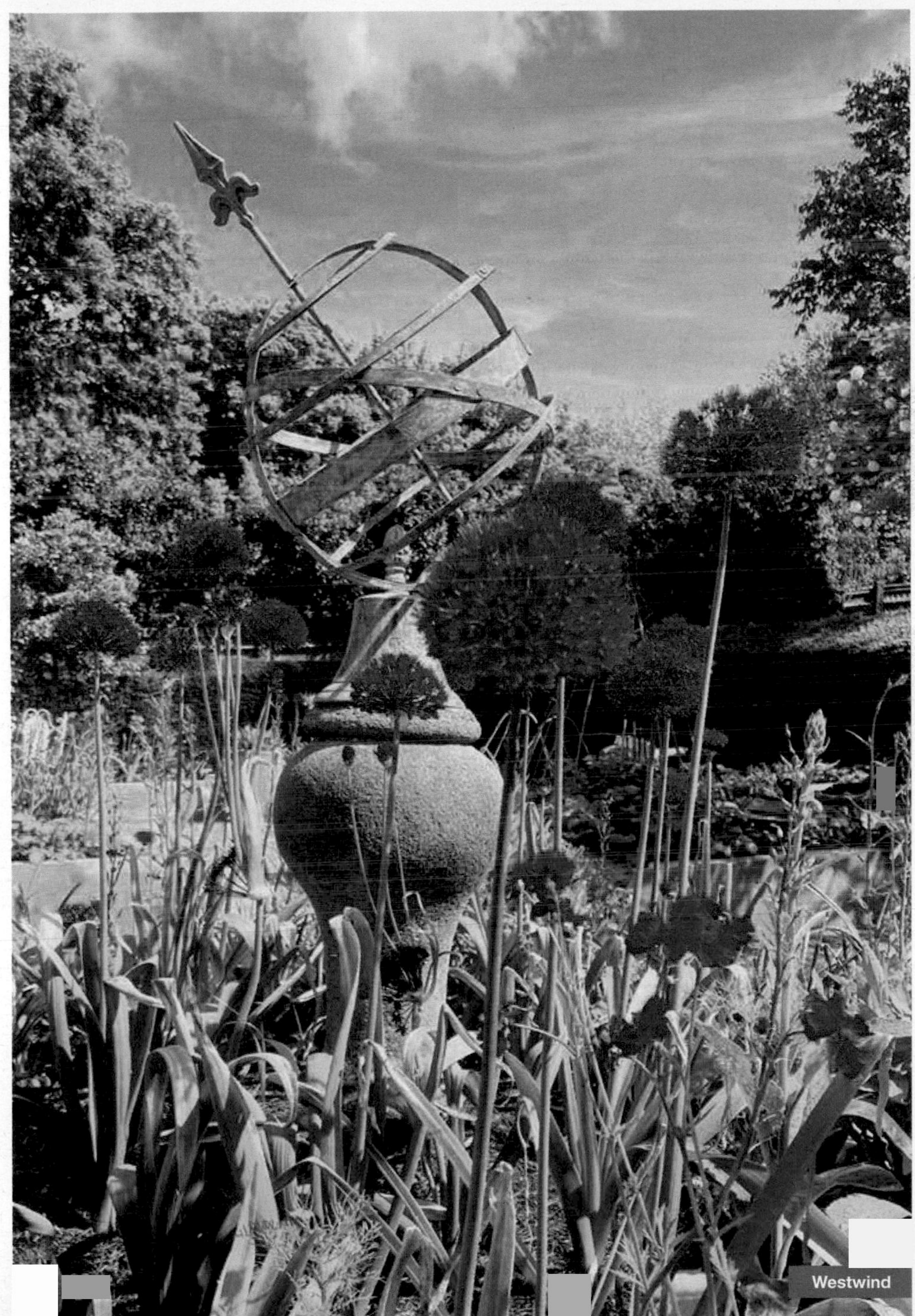

Westwind

13 NEW **CHOLDERTON ESTATE**
Kingsettle Stud, Cholderton Park, Cholderton, SP4 0DX. Henry & Felicity Edmunds, 07794 365419, office@kingsettlestud.co.uk, www.kingsettlestud.co.uk. *Follow signs from Cholderton Farm Shop on A338 (Tidworth to Salisbury rd) nr A303 junction.* **Sun 11 June (2-4.30). Adm £5, chd free. Light refreshments in the courtyard at Kingsettle Stud, the Estate's Grade II listed stable block. Teas, coffees and cake sale.**
A striking Victorian walled garden featuring a peony walk dating over 100 yrs old and a working kitchen garden. In the arboretum there is a wildlife garden with a maze and numerous tree species. There are no stairs, but ground is uneven.

14 COCKSPUR THORNS
Berwick St James, Salisbury, SP3 4TS. Stephen & Ailsa Bush, 01722 790445, stephenjdbush@gmail.com. *8m NW of Salisbury. 1m S of A303, on B3083 at S end of village of Berwick St James.* **Visits by arrangement 22 May to 24 July for groups of 5 to 20. Adm £5, chd free. Cream teas.**
2¼ acre garden, completely redesigned 22 yrs ago and developments since, featuring roses (particularly colourful in June), herbaceous border, shrubbery, small walled kitchen garden, secret pond garden, mature and unusual small trees, fruit trees and areas of wildflowers. Beech, yew and thuja hedgings planted to divide the garden. Small number of vines planted during early 2016.

15 ◆ **CORSHAM COURT**
Corsham, SN13 0BZ. Lord Methuen, 01249 701610, staterooms@corsham-court.co.uk, www.corsham-court.co.uk. *4m W of Chippenham. Signed off A4 at Corsham.* **For NGS: Sun 19 Mar (2-4.30); Sun 2 Apr (2-5.30). Adm £5, chd £2.50. For other opening times and information, please phone, email or visit garden website.**
Park and gardens laid out by Capability Brown and Repton. Large lawns with fine specimens of ornamental trees surround the Elizabethan mansion. C18 bath house hidden in the grounds. Spring bulbs, beautiful lily pond with Indian bean trees, young arboretum and stunning collection of magnolias. Wheelchair (not motorised) access to house, gravel paths in garden.

16 CORSLEY HOUSE
Corsley, Warminster, BA12 7QH. Glen Senk & Keith Johnson. *From Longleat on the A362 turning 1st R on Deep Lane, Corsley House is ¼m on R. The parkland entrance, where there is parking, is on R beyond the house.* **Sun 10 Sept (10-5). Adm £10, chd free. Cream teas.**
A garden full of surprises to reflect the eclectic nature of a Georgian home with a secret Jacobean facade. A unique sculpted wave lawn and a truly exceptional walled garden. Many well preserved ancient outbuildings such as a potting and apple storage shed and a granary built on staddle stones. All gloriously overlooking the NT's Cley Hill.

17 CORTINGTON MANOR
Corton, Warminster, BA12 0SY. Mr & Mrs Simon Berry. *5m S of Warminster. From Warminster, take rd thro Sutton Veny to Corton. Do not bear L into Corton but continue for ½m. From A303, take A36, L to Boyton, cross railway and R at T jnct. Continue for ¾m.* **Sun 18 June (2-5.30). Adm £8, chd free. Home-made teas.**
4 acres of wild and formal gardens surround rose clad C18 manor house. Herbaceous border, yew bays with Portuguese laurel line main lawn. Bank of white roses and foxgloves edges swimming pool. Sweet pea arch opens to cutting garden, vegetable garden and orchard divided by yew hedges. Lime avenue leads to river and pond from walled herb garden. Stable yard features pleached hornbeams and beech hedges.

18 NEW **CRUDWELL HOUSE**
Crudwell, Malmesbury, SN16 9EW. Mr & Mrs T Chichester-Kaner. *Off A429 leaving Crudwell to N next to Potting Shed. Crudwell is on A429 between Cirencester and Malmesbury. Leave M4 at J17. Parking at Manor Farm, Crudwell SN16 9ER (follow signs).* **Sun 11 June (11-5). Combined adm with Manor Farm £7, chd free. Home-made teas at Manor Farm, Crudwell.**
Entering past mature copper beech, chestnut and ash, the main garden to the side of the house faces west. Borders are a riot of colourful perennials flowering through spring and summer. A fragrant rose garden and part walled kitchen garden adjoin the property to the rear with a selection of fruit trees and asparagus beds.

GROUP OPENING

19 DAUNTSEY GARDENS
Church Lane, Dauntsey, Malmesbury, SN15 4HT. *5m SE of Malmesbury. Approach via Dauntsey Rd from Gt Somerford, 1¼m from Volunteer Inn Great Somerford.* **Sun 18 June (1-5). Combined adm £10, chd free. Home-made teas at Idover House.**

THE COACH HOUSE
Col & Mrs J Seddon-Brown.

DAUNTSEY PARK
Mr & Mrs Giovanni Amati, 01249 721777, enquiries@dauntseyparkhouse.co.uk.

THE GARDEN COTTAGE
Miss Ann Sturgis.

IDOVER HOUSE
Mr & Mrs Christopher Jerram.

THE OLD COACH HOUSE
Tony & Janette Yates.

THE OLD POND HOUSE
Mr & Mrs Stephen Love.

NEW **THE OLD RECTORY**
Mr Christopher & Mrs Claire Mellor-Hill, www.theoldrectorydauntsey.com.

This group of 7 gardens, centred around the historic Dauntsey Park Estate, ranges from the Classical C18 country house setting of Dauntsey Park, with spacious lawns, old trees and views over the River Avon, to mature country house gardens and traditional walled gardens. Enjoy the formal rose garden in pink and white, old fashioned borders and duck ponds at Idover House, and the quiet seclusion of The Coach House with its thyme terrace and gazebos, climbing roses and clematis. Here, mop-headed pruned *Crataegus prunifolia* line the drive. The Garden Cottage has a traditional walled kitchen garden with organic vegetables, orchard, woodland walk and yew topiary. Meanwhile the 2 acres at The Old Pond House are both clipped and unclipped! Large pond with lilies and fat carp, and look out for the giraffe and turtle. The Old Coach House is a small garden with perennial plants, shrubs and climbers. The Old Rectory is a 2 acre garden adjoining the River Avon with mature trees, roses and shrubs.

20 DUCK POND BARN

Church Lane, Wingfield, Trowbridge, BA14 9LW. Janet & Marc Berlin. *9m SW of Bath. On B3109 from Frome to Bradford on Avon, turn opp Poplars pub into Church Ln. Duck Pond Barn is at end of lane. Big field for parking.* **Sun 25 June, Sun 6 Aug (10-5). Adm £5, chd free. Home-made teas in front of the church outside the Barn.**

Garden of 1.6 acres with large duck pond, lawns, ericaceous beds, orchard, vegetable garden, big greenhouse, wood and wild areas of grass and trees with many wildflowers. Large dry stone wall topped with flowerbeds with rose arbour. 3 ponds linked by a rill in flower garden and large pergola in orchard. Set in farmland and mainly flat. In August there is an opportunity to see many interesting and rare succulents, Eucomis and agapanthus in flower.

21 NEW EASTWELL MANOR

Eastwell Road, Potterne, Devizes, SN10 5QG. Mr & Mrs Robert Hunt-Grubbe. *2 ½m S of Devizes. Stone pillars & black gates on W side of A360, 400 metres S of Potterne Xrds & George & Dragon. Entrance will be signposted.* **Sat 27 May (1.30-5). Adm £8, chd free. Home-made teas.**

Eastwell Manor was built by the family in 1570 and is representative of a small manorial estate with its own brew house, granary and stables. The garden, based on terraces, features a banqueting house against a backdrop of woodland, with views to Salisbury Plain over water. Eastwell boasts various specialist trees and mature palms. There are uneven steps and limited wheelchair access.

22 NEW FALKNERS COTTAGE

North Newnton, Pewsey, SN9 6LA. Anne & John Thompson-Ashby. *3m W Pewsey. SW from Pewsey on A345 direction Salisbury, exit 3 at Woodbridge r'about to Hilcott. Pass farm, next house on sharp L hand bend. Care turning R.* **Sat 2 Sept (1.30-5.30). Adm £6, chd free. Cream teas.**

This 5 acre site inc a formal garden with colourful borders, topiary and box framed parterre and courtyard 'rooms' with contemporary sculpture. Behind the thatched barn lies a kitchen garden with hens. There is a meadow with ornamental trees, a small lake with a summerhouse on an island, a woodland and a new wildlife pond. A short walk from the meadow is the C13 St James' Church. Dogs allowed on short leads in meadow only.

23 FONTHILL HOUSE

Tisbury, SP3 5SA. The Lord Margadale of Islay, www.fonthill.co.uk/gardens. *13m W of Salisbury. Via B3089 in Fonthill Bishop. 3m N of Tisbury.* **Sun 19 Mar (12-5). Adm £8, chd free. Sandwiches, quiches & cakes, soft drinks, tea & coffee. Bottled beer & wine.**

Wonderful woodland walks with daffodils, rhododendrons, azaleas, shrubs, bulbs. Magnificent views, formal gardens. The gardens have been extensively redeveloped under the direction of Tania Compton and Marie-Louise Agius. The formal gardens are being continuously improved with new designs, exciting trees, shrubs and plants. Gorgeous William Pye fountain and other sculptures. Partial wheelchair access.

24 FOVANT HOUSE

Church Lane, Fovant, Salisbury, SP3 5LA. Amanda & Noel Flint. *Approx 10m W of Salisbury. 6½m W of Wilton & 9.3m E of Shaftesbury. Take A30 to Fovant then head N through village and follow signs to St Georges Church.* **Sat 3 June (2-5.30). Adm £7, chd free. Home-made teas.**

Fovant House is a former Rectory set in about 3 acres of formal garden (house not open). In 2016 the garden was redesigned by Arabella Lennox Boyd. Garden inc 60m herbaceous border, terraces and parterre. A range of mature trees inc cedars, copper beech and lime. Majority of garden has easy wheelchair access subject to ground conditions being dry.

D

25 FOXLEY MANOR

Foxley, Malmesbury, SN16 0JJ. Richard & Louisa Turnor. *2m W of Malmesbury. 10 mins from J17 on M4. Turn towards Malmesbury/Cirencester, then towards Norton & follow signs for the Vine Tree, yellow signs from here.* **Sun 23 Apr (12-4). Adm £7, chd free. Home-made teas.**

Yew hedges divide lawns, borders, rose garden, lily pond and a newer wild area with a natural swimming pond shaded by a liriodendron. Views through large Turkey oaks to farmland beyond. Small courtyard gravel garden. Sculptures are sited throughout the gardens.

26 GASPER COTTAGE

Gasper Street, Gasper Stourton, Warminster, BA12 6PY. Bella Hoare & Johnnie Gallop, 07812 555 883, bella.hoare@icloud.com, www.youtube.com/channel/UCgar2Akygp4j3dWNGMI6i0A. *Nr Stourhead Gardens, 4m from Mere, off A303. Turn off A303 at B3092 Mere. Follow Stourhead signs. Go through Stourton. After 1m, turn R after phone box, signed Gasper. House 2nd on R going up hill. Parking past house on L, in field.* **Sun 20 Aug (11-5). Adm £5, chd free. Visits also by arrangement 1 May to 17 Sept for groups of 10 to 25. No more than 6 cars on the day - parking is limited. Pls car share.**

1½ acre garden, with views to glorious countryside. Luxurious planting of dahlias, grasses, asters, cardoons and more, inc new perennial planting combinations. Orchard with wildlife pond. Artist's studio surrounded by colour balanced planting and formal pond. Pergola with herb terrace. Several seating areas. Garden model railway.

In 2022 the National Garden Scheme donated £3.11 million to our nursing and health beneficiaries

GROUP OPENING

27 HANNINGTON VILLAGE GARDENS

Hannington, Swindon, SN6 7RP. *Off B4019 Blunsdon to Highworth Rd by the Freke Arms. Park as directed by signs, in the street where possible or opp Lushill House.* **Sun 11 June (11-5). Combined adm £8, chd free. Home-made teas in Hannington Village Hall.**

CHESTNUT HOUSE
Mary & Garry Marshall.

CHESTNUT VILLA
David Cornish.

GLEBE HOUSE
Charlie & Tory Barne.

LOWER FARM
Mr Piers & Mrs Jenny Martin.

LUSHILL HOUSE
John & Sasha Kennedy.

QUARRY BANK
Paul Minter & Michael Weldon.

22 QUEENS ROAD
Jan & Pete Willis.

ROSE COTTAGE
Mrs Ruth Scholes.

STEP COTTAGE
Mr & Mrs J Clarke.

SUNRISE VILLA
Roy George.

YORKE HOUSE GARDEN
Mr Miles & Mrs Cath Bozeat.

Hannington has a dramatic hilltop position on a Cotswold ridge overlooking the Thames Valley. A great variety of gardens, from large manor houses to small cottage gardens, many of which follow the brow of the hill and afford stunning views of the surrounding farmland. You will need lots of time to see all that is on offer in this beautiful historic village.

28 HAZELBURY MANOR GARDENS

Wadswick, Box, Corsham, SN13 8HX. Mr L Lacroix, 07505 543415, phylip@mac.com. *5m SW of Chippenham, 5m NE of Bath. From A4 at Box, A365 to Melksham, at Five Ways junction L onto B3109 towards Corsham; 1st L in ¼m, drive immed on R.* **Sun 11 June (11-5); Wed 13 Sept (10-4); Sun 17 Sept (11-5). Adm £7.50, chd free. Home-made teas. Visits also by arrangement May to Sept for groups of 5+.**

The C15 Manor house comes into view as you descend along the drive and into the Grade II landscaped Edwardian gardens. The extensive plantings that surround the house are undergoing considerable redevelopment by the owners and their head gardener. A wide range of organic horticulture is practiced in 8 acres of relaxed gardens.

29 HILPERTON HOUSE

The Knap, Hilperton, Trowbridge, BA14 7RJ. Chris & Ros Brown. *1½m NE of Trowbridge. Follow A361 towards Trowbridge and turn R at r'about signed Hilperton. House is next door to St Michael's Church in The Knap off Church St.* **Sat 17, Sun 18 June (2-6). Adm £5, chd free. Home-made teas.**

2½ acres well-stocked borders, small stream leading to large pond with fish, ducks, water lilies, waterfall and fountain. Fine mature trees inc unusual specimens. Walled fruit and vegetable garden, small woodland area. Rose walk and rose garden with roses and clematis. Interesting wood carvings, mainly teak. 170yr old vine in conservatory of Grade II listed house, c1705 (not open). Activity sheets for children. 'Fairy' tree is recent addition. Some gravel and lawns. Conservatory not wheelchair accessible. Path from front gate has uneven paving but can be bypassed on lawn.

30 HORATIO'S GARDEN

Duke of Cornwall Spinal Treatment Centre, Salisbury Hospital NHS Foundation Trust, Odstock Road, Salisbury, SP2 8BJ. Horatio's Garden Charity, www.horatiosgarden.org.uk. *1m from centre of Salisbury. Follow signs for Salisbury District Hospital. Please park in car park 10, which will be free to NGS visitors on the day.* **Sun 2 July (2-5). Adm £5, chd free. Tea & delicious cakes, made by Horatio's Garden volunteers, will be served in the Garden Room.**

Award winning hospital garden, opened in Sept 2012 and designed by Cleve West for patients with spinal cord injury at the Duke of Cornwall Spinal Treatment Centre. Built from donations given in memory of Horatio Chapple who was a volunteer at the centre in his school holidays. Low limestone walls, which represent the form of the spine, divide densely

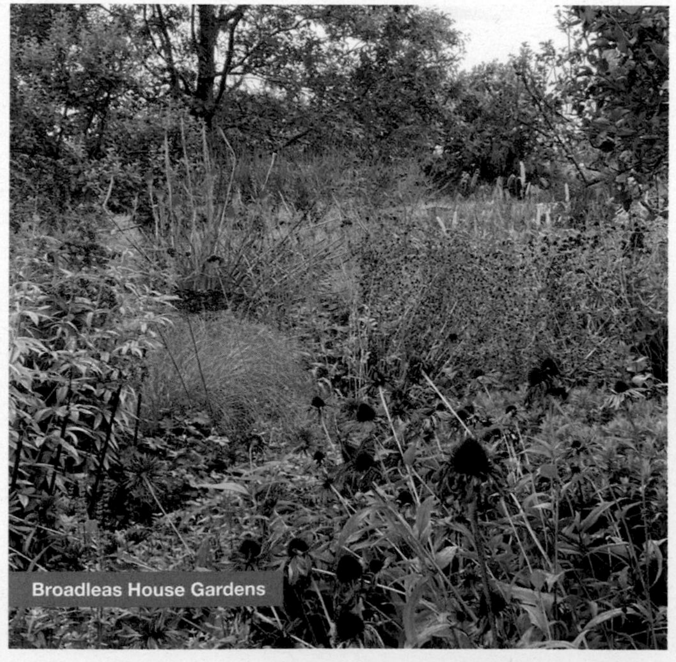

Broadleas House Gardens

planted herbaceous beds. Everything in the garden designed to benefit patients during their long stays in hospital. Garden is run by a Head Gardener and team of volunteers. SW Regional Winner, The English Garden's The Nation's Favourite Gardens 2019. Designer Cleve West has 8 RHS gold medals. At 3pm there will be a short talk from the Head Gardener about therapeutic gardens and the work of Horatio's Garden.

31 HYDE'S HOUSE

Dinton, SP3 5HH. Mr George Cruddas. *9m W of Salisbury. Off B3089 nr Dinton Church on St Mary's Rd.* **Sun 4 June (2-5). Adm £7, chd free. Home-made teas at Thatched Old School Room with outside tea tables.**

3 acres of wild and formal garden in beautiful situation with series of hedged garden rooms. Numerous shrubs, flowers and borders, all allowing tolerated wildflowers and preferred weeds, while others creep in. Large walled kitchen garden, herb garden and C13 dovecote (open). Charming C16/18 Grade I listed house (not open), with lovely courtyard. Every year varies. Free walks around park and lake. Steps, slopes, gravel paths and driveway.

32 ◆ IFORD MANOR GARDENS

Bradford-on-Avon, BA15 2BA. Mr Cartwright-Hignett, 01225 863146, info@ifordmanor.co.uk, www.ifordmanor.co.uk. *7m S of Bath. Off A36, brown tourist sign to Iford 1m. From Bradford-on-Avon or Trowbridge via Lower Westwood Village (brown signs). Please note all approaches via narrow, single track lanes with passing places.* **For NGS: Sat 22 Apr (11-4). Adm £8, chd £6. Pre-booking essential, please phone 01225 863146, email info@ifordmanor.co.uk or visit www.ifordmanor.co.uk for information & booking. Light refreshments in our cafe (no booking needed). Children under 10 will not be admitted to the garden. For other opening times and information, please phone, email or visit garden website.**

Harold Peto's former home, this 2½ acre terraced garden provides timeless inspiration. Influenced by his travels, particularly to Italy and Japan, Peto embellished the garden with a collection of classical statuary and architectural fragments. Steep steps link the terraces with pools, fountains, loggias, colonnades, urns and figures, with magnificent rural views across the Iford Valley. Afternoon teas, freshly baked cakes and a selection of savoury items, are served in the on site cafe, inc hot and cold drinks. Booking required for our restaurant serving lunch.

33 JULIA'S HOUSE CHILDREN'S HOSPICE

Bath Road, Devizes, SN10 2AT. Nicky Clack. *Situated just off A361 Bath Road in Devizes. Please note there is no parking available at hospice. We kindly ask visitors to use nearby Station Rd car park. For SatNav use SN10 1BZ.* **Sat 13 May (10-4). Adm £4, chd free. Home-made teas.**

Julia's House Children's Hospice provides respite care for children with life threatening or limiting conditions. Our garden has been designed so the children can experience and enjoy different sensory elements and is cared for by our volunteer gardeners. Explore how our children use the garden. Hospice tours will be available.

34 KETTLE FARM COTTAGE

Kettle Lane, West Ashton, Trowbridge, BA14 6AW. Tim & Jenny Woodall, 01225 753474, trwwoodall@outlook.com. *Kettle Lane is halfway between West Ashton T-lights & Yarnbrook r'about on S side of A350. Garden ½m down end of lane. Limited car parking.* **Visits by arrangement 19 June to 17 Sept for groups of up to 20. Adm inc home-made teas. Adm £10, chd free.**

Previously of Priory House, Bradford on Avon, which featured on Gardeners' World, September 2017, we have now created a new cottage garden, full of colour and style, flowering from June to Oct. Bring a loved one/friend to see the garden. One or two steps.

35 NEW KING'S OLD RECTORY

South Street, Broad Chalke, Salisbury, SP5 5DH. Sir Christopher & Lady Butcher. *10m SW of Salisbury. Take minor rd off A354 at Coombe Bissett. House on R with line of pleached limes in front. Enter through the Tudor arch opp Bulls Lane. Parking in church car park & also at Sports Hall.* **Fri 26 May (11-5). Adm £10, chd free.**

4 acres of formal and wild gardens transected by a meandering chalk stream (the Ebble). Fine mature trees and shrubs, inc unusual and very large specimens. Series of walled or hedged garden rooms. Wildflower orchard. Beautiful Grade II* listed house (not open), c1470 with C16 and C17 additions, gifted by King Henry VI for the endowment of King's College Cambridge. Wheelchair accessible.

36 KNOYLE PLACE

Holloway, East Knoyle, Salisbury, SP3 6AF. Lizzie de la Moriniere. *Turn off A350 into E Knoyle & follow signs to parking at Lower Lye. Walk 5 mins to Knoyle Place through village following signs.* **Sun 7 May (2-5). Adm £10, chd free. Home-made teas. Open nearby Lower Lye.**

Very beautiful and elegant garden created over 60 years by previous and current owners. Above the house there are several acres of mature rhododendron and magnolia woodland planting. Among the many different areas in this 9 acre garden is a box parterre, rose garden, vegetable garden and, around the house, a recently planted formal garden designed by Dan Combes. Sloping lawns and woodland paths, stone terrace.

The National Garden Scheme searches the length and breadth of England, Wales, Northern Ireland and the Channel Islands for the very best private gardens

GROUP OPENING

37 LANDFORD VILLAGE GARDENS

Landford, Salisbury, SP5 2AX. *From Salisbury take A36 then L through village. From S leave M27 at J2 onto A36 and R towards Nomansland. At Xrds take Forest Rd. Parking on L at The Gatehouse.* **Sat 10, Sun 11 June (2-6). Combined adm £7, chd free. Home-made teas at Bentley, Whitehorn Drive. Delicious home-made cakes and refreshments for sale.** Donation to local charities.

BENTLEY
Jacky Lumby.

FOREST COTTAGE
Norah Dunn.

NEW **FOXHOLLOW**
Kate Leevers & Martin Callaghan, 0780 1091080, martin@burnbridge.org.uk.
Visits also by arrangement July to Sept normally combined with 5 Whitehorn Drive.

THE GATEHOUSE
Mrs Jackie Beatham.

5 WHITEHORN DRIVE
Jackie & Barry Candler, 07999 548955, jackie.candler37@gmail.com.
Visits also by arrangement 1 June to 10 Sept for groups of 5 to 15. Talk, tour & refreshments inc in adm.

Landford is a small village set in the northern New Forest, famous for the beauty of its beech and oak trees and free roaming livestock. The 5 gardens range in size, planting conditions and setting. 4 are in the village, adjacent to Nomansland with its green grazed by New Forest ponies. The Gatehouse has formal herbaceous borders with exuberant planting and colour schemes in mixed herbaceous borders, looking out over paddocks to the forest. Bentley is a small garden, a green oasis containing a wide range of plants and raised vegetable borders. 5 Whitehorn Drive is a damp garden with lush colour and surprises in planting. Foxhollows is a brand new garden for 2023 featuring naturalistic planting and a short distance away in the forest itself is Forest Cottage, a garden with hidden delights around each corner, inc a field of wild orchids.

38 LAVENDER GARDENS

Giles Lane, Landford, Salisbury, SP5 2BG. Mr Michael Hayward, 07775 511884, newforestlavender@btinternet.com, www.newforestlavender.com. *11m from Southampton, 11m from Salisbury on A36. Village landmark on A36 Royal Jaipur Indian restaurant approx 1m turn L. From Southampton landmark The Shoe Inn approx 1m turn R.* **Sun 7 May (11-4). Adm £5, chd free. Cream teas. Light lunch available. Visits also by arrangement 15 Apr to 17 June for groups of 10 to 25. Refreshments available on request.**

Take time to wander through colourful borders. Be intrigued by plants you may not have seen before. Step back and admire a stunning colourful display of annual and perennial plants. Find areas en masse and specimens in mixed borders. Stroll around themed gardens, dahlia borders, a white garden along with colourful grasses. Spring bulbs and lily pond. A selection of different trees and shrubs. Wheelchair access from car park to garden and plant sales area. Gravel pathway runs through the garden. Disabled WC.

39 LOWER LYE

Holloway, East Knoyle, Salisbury, SP3 6AQ. Belinda & Andrew Scott. *From A350 exit at East Knoyle. Lower Lye is at upper southern end of village, follow signs to house. Parking on L entrance field.* **Sun 7 May (1.30-4). Adm £10, chd free. Pre-booking essential, please visit www.ngs.org.uk for information & booking. Open nearby Knoyle Place.**

The gardens at Lower Lye are exceptional and were completely reconfigured by the landscape architect Michael Balston 24 yrs ago. An extremely large garden, approx 9 acres placed on the brow of the hill with wonderful views across the valley and the magnificent King Alfred's Tower in the distance. More recently the gardens have been worked on by garden designers Tania Compton and Jane Hurst, who have given improved structure through replanting of the various borders. The gardens have been developed in a simplistic and modern style, which gives a sense of elegance whilst respecting the natural habitat There are numerous modern sculptures and water features placed throughout the garden which are spectacular and provide a number of individual focal points. The gardens inc herbaceous borders, mature trees, a meadow area and a vegetable and cutting garden. Coffee/tea and light refreshments will be served at the beautiful gardens just along the road at Knoyle Place, which will also be open throughout the afternoon. Some steps and slopes.

40 MANOR FARM

Crudwell, Malmesbury, SN16 9ER. Mr & Mrs J Blanch. *4m N of Malmesbury on A429. Heading N on A429 in Crudwell, turn R signed to Eastcourt. Farm entrance is on L 200m after end of speed limit sign.* **Sun 11 June (11-5). Combined adm with Crudwell House £7, chd free. Home-made teas. Delicious cakes and gluten free options.**

Set within Cotswold stone walls and a backdrop of Crudwell Church with a further 5 acres of parkland featuring Japanese maples and a Roman style summerhouse. The garden is divided by box and yew hedges to create different areas both formal and informal. Herbaceous borders with old fashioned roses are grown among fountains and extensive lawns.

41 MANOR HOUSE, STRATFORD TONY

Stratford Tony, Salisbury, SP5 4AT. Mr & Mrs Hugh Cookson, 01722 718496, lc@stratfordtony.co.uk, www.stratfordtony.co.uk. *4m SW of Salisbury. Take minor road W off A354 at Coombe Bissett. Garden on S after 1m. Or take minor road off A3094 from Wilton signed Stratford Tony & racecourse.* **Visits by arrangement 1 Feb to 15 Sept. Adm £10, chd free. Picnics and refreshments please enquire.**

Varied 4 acre garden with year-round interest. Formal and informal areas. Herbaceous borders, vegetable garden, parterre garden, orchard, shrubberies, roses, specimen trees, lakeside planting, winter colour and structure, many original contemporary features and places to sit and enjoy the downland views. Some gravel.

42 MAWARDEN COURT

Stratford Road, Stratford Sub Castle, SP1 3LL. Alastair & Natasha McBain. *2m WNW Salisbury. A345 from Salisbury, L at T-lights, opp St Lawrence Church.* **Sun 18 June (2-5). Adm £6, chd free. Home-made teas in pool pavilion.**

Mixed shrub and herbaceous borders set around C17 house (not open). Pergola walk, rose garden and white beam avenue down to River Avon, with pond pontoon and walk through newly planted woodland.

43 NORTH COTTAGE

Tisbury Row, Tisbury, SP3 6RZ. Jacqueline & Robert Baker, 01747 870019, baker_jaci@yahoo.co.uk. *12m W of Salisbury. From A30 turn N through Ansty, L at T-junction, towards Tisbury. From Tisbury take Ansty road. Car park entrance nr junction signed Tisbury Row.* **Sun 28 May, Sun 18 June (11-5). Adm £5, chd free. Light home-made lunches & teas.**

From car park, walk past vegetables and through the paddock to reach house and gardens. Although modest, there is much variety to find. Garden is divided with lots to explore as each part differs in style and feel, augmented with pottery and woodwork handmade by the owners. From the intimacy of the garden go out to see the orchard and ponds, walk the coppice wood and see the rest of the smallholding. As well as the owners' handiwork for sale (https://northcottagecreations.com/), there are exciting metalwork sculptures by Metal Menagerie. Please see NGS website for extra pop-up open days.

44 OARE HOUSE

Rudge Lane, Oare, nr Pewsey, SN8 4JQ. Sir Henry Keswick. *2m N of Pewsey. On Marlborough Rd (A345).* **Sun 30 Apr, Sun 25 June (2-6). Adm £8, chd free. Home-made teas in potting shed. Donation to The Order of St John.**

1740s mansion house later extended by Clough Williams Ellis in 1920s (not open). The original formal gardens around the house have been developed over the years to create a wonderful garden full of many unusual plants. Current owner is very passionate and has developed a fine collection of rarities. Garden is undergoing a renaissance but still maintains split compartments each with its own individual charm; traditional walled garden with fine herbaceous borders, vegetable areas, trained fruit, roses and grand mixed borders surrounding formal lawns. The magnolia garden is wonderful in spring with some trees dating from 1920s, together with strong bulb plantings. Large arboretum and woodland with many unusual and champion trees. In spring and summer there is always something of interest, with the glorious Pewsey Vale as a backdrop. Partial wheelchair access.

Winkelbury House

Brow Cottage

45 THE OLD MILL

Reybridge, Lacock, Chippenham, SN15 2PF. Trudy Clayton, 07908 152332, trudy@calibrehomes.co.uk, www.airbnb.co.uk/rooms/57819294039441 5288. *Limited parking in nearby field. Alternatively park in the Lacock car park & walk into Reybridge which will take about 20mins. Once there, follow the NGS signs.* **Fri 23, Sat 24 June (10-6). Adm £6, chd free. Home made savouries, light lunches, home baked cakes & cream teas. Ice creams. Hot & cold drinks.**

The Old Mill dates back to Doomsday and is surrounded by the River Avon. It lies in a very tranquil, picturesque setting, in the heart of beautiful Wiltshire countryside in the hamlet of Reybridge, which is within the parish of the NT village of Lacock. There are lots of areas to wander or to sit and relax, taking in and enjoying the wildlife all around you.

46 THE OLD VICARAGE

Church Lane, Ashbury, Swindon, SN6 8LZ. Mr David Astor. *In the village of Ashbury, next to the church.* **Sat 3 June (11-3). Adm £5, chd free. Light refreshments.**

Within The Vale of the White Horse, Ashbury is a picturesque village with thatched cottages and a friendly atmosphere. Nestled between the village pub and St Mary's church is the garden of The Old Vicarage. The garden surrounds the house and is made up of a floriferous walled garden, a wildflower meadow, an extensive kitchen garden, filled with flowers, as well as vegetables, plus more.

47 THE PARISH HOUSE

West Knoyle, Mere, BA12 6AJ. Philip & Alex Davies. *From E on A303 take signpost to West Knoyle at petrol station. Continue down hill and past church, take first R signposted Charnage, continue 1m to end of lane. Parking in field on L.* **Sun 28 May (2-5). Adm £5, chd free. Home-made teas.**

The garden at The Parish House is small and south facing surrounded by hedges protecting it from the exposure of a high wide open landscape beyond. The garden is laid out with gravel to encourage a self-seeding natural look but there are also roses and herbaceous perennials as well. The look is held together with topiary. Beyond this garden there is a larger area of lawn, trees and developing bulb cover (Camassia). Large conservatory greenhouse containing mature Muscat grapevine and pelargoniums. Small vegetable garden. Level but gravel surfaces, some lawn.

48 PEACOCK COTTAGE

Preshute Lane, Manton, Marlborough, SN8 4HQ. Ian & Clare Maurice. *1m to W of Marlborough along A4, turn L at Manton Village sign. Continue over the bridge and take 1st L following Car Park signs. If arriving through Manton village turn R by pub.* **Sun 30 July (2-5). Combined adm with Westwind £10, chd free.**

Set in rural surroundings, this expertly tended, undulating garden, sweeps up from the rear of the house and comprises herbaceous and mixed borders, parterre and exotic bed. Created in 1966 when 2 cottages and their gardens combined, and has been developed considerably over the last 15yrs. Disabled parking places identified in car park. Wheelchair access to garden is via lawns and grass paths.

49 NEW RIVERSIDE HOUSE

Axford, Marlborough, SN8 2HA. Hamish & Susie Watson. *4 m E of Marlborough. Parking in grass field next to Red Lion pub (adjoining gate to wildflower meadow & garden.).* **Wed 21 June (11-5). Adm £5, chd free. Light refreshments in the village hall opp house.**

Beautiful wisteria-clad village house (originally a farmhouse) dating back to C16, with William and Mary additions in late C17. Cottage style garden of over 1 acre, with many wild areas, mature trees and herbaceous borders. A charming rose garden and lawns sweeping down to a stream and small pond, with fine views to the south over water meadows to hills beyond.

50 SALTHROP HOUSE

Basset Down, Wroughton, Swindon, SN4 9QP. Sophie Conran. *Driving W on M4, exit at J16, take 1st exit signposted for Butterfly World, drive 2m, passing the golf course on your R, keep going up the hill, Salthrop House is the drive on the L.* **Fri 9, Sat 10 June, Fri 14, Sat 15 July, Fri 15, Sat 16 Sept (10-4). Adm £6.50, chd free. Light refreshments.**

Salthrop House is a manor house garden set on the edge of the Marlborough Downs. A tremendous variety of perennials, shrubs, borders and pots unfurl themselves across this romantic garden and surround the sweeping lawn. You can wind your way through the woodland paths, enjoy a moment of peace by the pond or visit the new greenhouse and kitchen garden.

51 SCOTS FARM

Pinkney, Malmesbury, SN16 0NZ. Mr & Mrs Martin Barrow, 07947 792919, ljikob@gmail.com, www.lawrencebarrow.com. *5m W of Malmesbury on B4040 towards Sherston. Take L at Pinkney Xrds, follow road over a bridge and up a hill.* **Visits by arrangement 1 June to 30 July for groups of up to 20. Adm £7, chd free. We offer tea & cake for group visits.**

Across 7 acres a colourful perennial border, a white garden, a Japanese Zen garden, a Mediterranean garden and follow a path through a wildflower meadow to a pottery studio and koi pond in this haven of tranquillity. Garden ware, pottery and sculptures on display.

52 SEEND HOUSE

High Street, Seend, Melksham, SN12 6NR. Maud Peters, www.instagram.com/maud_seendgarden. *In Seend village. Nr church & opp PO. Parking on village green.* **Sun 16 Apr (1-5). Combined adm with Brow Cottage £7, chd free. Sat 17 June (1-5). Combined adm with Seend Manor £10, chd free.**

Seend House is a Georgian house with 6 acres of gardens and paddocks. Framed with yew and box. Highlights inc cloud and rose garden, stream lavender, view of knot garden from above, fountain with grass border, walled garden as well as formal borders. Amazing view across the valley to Salisbury Plain.

53 SEEND MANOR

High Street, Seend, Melksham, SN12 6NX. Stephen & Amanda Clark, SeendManorGardens.com. *In centre of village, opp village green with car parking for garden visitors.* **Sat 17 June (1-5). Combined adm with Seend House £10, chd free. Home-made teas.**

Created over 20 yrs, a stunning walled garden with 4 quadrants evoking important parts of the owners' lives - England, China, Africa and Italy, with extensive trelliage, hornbeam hedges on stilts, cottage orne, temple, Chinese ting, grotto, fern walk, fountains, parterres, stone loggia and more. Kitchen garden. Folly ruin in woods. Courtyard garden. Extensive walled gardens, water features, garden structures, topiary and hedging and one of the best views in Wiltshire. Many gravel paths, so only wheelchairs with thick tyres.

54 1 SOUTHVIEW

Wick Lane, Devizes, SN10 5DR. Teresa Garraud, 01380 722936, tl.garraud@hotmail.co.uk. *From Devizes Mkt Pl go S (Long St). At r'about go straight over, at mini r'about turn L into Wick Ln. Continue to end of Wick Ln. Park in road or roads nearby.* **Sun 14 May (12-5), open nearby Trymnells. Sun 10 Sept (2-5). Adm £5, chd free. Visits also by arrangement 15 May to 30 Sept for groups of up to 25. Coffee/tea & biscuits available for small groups.**

An atmospheric and very long town garden, full of wonderful planting surprises at every turn. Densely planted with both pots near the house and large borders further up, it houses a collection of beautiful and often unusual plants, shrubs and trees many with striking foliage. Colour from seasonal flowers is interwoven with this textural tapestry. 'Truly inspirational' is often heard from visitors.

55 TEASEL

Wilsford, Amesbury, Salisbury, SP4 7BL. Ray Palmer. *2m SW of Amesbury in the Woodford Valley on western banks of R Avon which runs through the gardens.* **Sun 2 July (11-5). Adm £7, chd free. Home-made teas.**

The gardens at Teasel were originally laid out in the 1970s by James Mitchell the publisher who made his fortune from the worldwide best seller ' The Joy of Sex'. The gardens have been the subject of extensive restoration and new landscaping during 2020. This inc tropical gardens around the swimming pool terrace, long herbaceous borders, a lake and a ½ m riverside walk.

56 NEW TRYMNELLS

1a Coxhill Lane, Potterne, Devizes, SN10 5PH. Linda Smith. *Coxhill Lane is opp George & Dragon pub. Trymnells is 2nd property on L & is at the top of a gravel drive. Parking at Potterne Village Hall, some space on drive for those with limited mobility* **Sun 14 May (12-5). Adm £5, chd free. Home-made teas. Open nearby 1 Southview.**

Trymnells is a relatively new garden which was started approx 5 yrs ago. It is now maturing and has many trees, shrubs, hedges and perennials. The garden is steep and slopes up from the property and reaches a steep bank at the very top. Sleepers, fences, walls and plants have imposed a sense of structure with bulbs and perennials providing wonderful colour.

57 ◆ TWIGS COMMUNITY GARDEN

Manor Garden Centre, Cheney Manor, Swindon, SN2 2QJ. TWIGS, 01793 523294, twigs.reception@gmail.com, www.twigscommunitygardens.org.uk. *From Gt Western Way, under Bruce St Bridges onto Rodbourne Rd. 1st L at r'about, Cheney Manor Industrial Est. Through estate, 2nd exit at r'about. Opp Pitch & Putt. Signs on R to Manor Garden Centre.* **For NGS: Sun 21 May, Sun 16 July (12-4). Adm £4, chd free. Home-made teas. For other opening times and information, please phone, email or visit garden website.**

Delightful 2 acre community garden, created and maintained by volunteers. Features inc 7 individual display gardens, ornamental pond, plant nursery, Iron Age round house, artwork, fitness trail, separate kitchen garden site, Swindon beekeepers and the haven, overflowing with wildflowers. Excellent hot and cold lunches available at Olive Tree café within Manor Garden Centre adj to Twigs (pre booking required) 01793 533152. Most areas wheelchair accessible. Disabled WC at garden centre.

58 WATERDALE HOUSE

East Knoyle, SP3 6BL. Mr & Mrs Julian Seymour, 01747 830262. *8m S of Warminster. N of East Knoyle, garden signed from A350. Do not use SatNav.* **Sun 30 Apr (2-5). Adm £5, chd free. Home-made teas.**

In the event of inclement weather please check www.ngs.org.uk before visiting. 4 acre mature woodland garden with rhododendrons, azaleas, camellias, maples, magnolias, ornamental water, bog garden, herbaceous borders. Bluebell walk. Shrub border created by storm damage, mixed with agapanthus and half-hardy salvias. Difficult surfaces, sensible footwear essential as parts of garden can be very wet. Please keep to raked and marked paths in woodland.

59 WELLAWAY

Close Lane, Marston, SN10 5SN. Mrs P Lewis. *5m SW of Devizes. From A360, Devizes to Salisbury, R in Potterne just before George & Dragon pub. Through Worton. L at end of village signed to Marston, Close Lane ½m on L.* **Sat 6, Sun 7 May, Sat 16, Sun 17 Sept (1-5). Adm £6, chd free. Home-made teas.**

2 acre flower arranger's garden comprising herbaceous borders, orchard, vegetable garden, ornamental and wildlife ponds, lawns and naturalised areas. Planted since 1979 for year-round interest. Shrubberies and rose garden, other areas underplanted with bulbs or ground cover. Springtime particularly colourful with daffodils, tulips and hellebores. Extensive autumn colour.

60 WEST LAVINGTON MANOR

1 Church Street, West Lavington, SN10 4LA. Andrew Doman & Jordina Evins, andrewdoman01@gmail.com. *6m S of Devizes, on A360. House opp White St, where parking is available. See more details on Instagram westlavingtonmanor.* **Visits by arrangement 1 Jan to 24 Dec for groups of 10+. Adm £15, chd free. Donation to West Lavington Youth Club and Nestling Trust.**

A C15 manor house with spectacular 5 acre walled garden, established by Sir John Danvers, who brought the Italianate garden to the UK. Delightful aspects including Laburnum walk, replanted herbaceous border, an authentic Japanese garden, new Mulberry rotunda, orchard with 25 different apple and pear species, an arboretum with outstanding specimen trees and lake with duck house. Partial wheelchair access.

61 WESTCROFT

Boscombe Village, nr Salisbury, SP4 0AB. Lyn Miles, 01980 610877 (evenings), lynmiles@icloud.com, www.westcroftgarden.co.uk. *7m N/E Salisbury. On A338 from Salisbury, just past Boscombe & District Social Club. Park there or in field opp house, or where signed on day. Disabled parking only on drive.* **Every Thur 12 Jan to 9 Mar (11-4). Sat 28, Sun 29 Jan, Sun 19, Sat 25, Sun 26 Feb (11-4). Adm £4, chd free. Home-made soups, scrummy teas, inside or outside. Visits also by arrangement 3 Jan to 31 Mar for groups of 20+. Refreshments can be pre-booked for groups.**

Whilst overflowing with roses in June, in Jan and Feb the bones of this ⅔ acre galanthophile's garden on chalk are on show. Brick and flint walls, terraces, rustic arches, gates and pond add character. Drifts of snowdrops carpet the floor whilst throughout is a growing collection of well over 400 named varieties. Many hellebores, pulmonarias, grasses and seedheads add interest. Snowdrops (weather dependent) and snowdrop sundries for sale inc greetings cards, mugs, bags, serviettes, also chutneys and free range eggs.

62 NEW WESTWIND

Manton Drove, Manton, Marlborough, SN8 4HL. M G Campbell-Sharp, 07738 180759, westwindmanton@gmail.com. *1m W of Marlborough off A4. In Manton, turn L at pub & follow signs to car park at Peacock Cottage. Only 4 spaces available at Westwind for visitors with limited mobility.* **Sun 30 July (2-5). Combined adm with Peacock Cottage £10, chd free. Home-made teas. Visits also by arrangement 9 Apr to 8 Oct for groups of up to 15.**

An informal country garden set in 4 acres inc a 2½ acre meadow, home to a rich assortment of wildlife. Westwind has an abundance of mature trees, a cottage garden and beds full of herbaceous plants. 2022 welcomed the newly designed Armillary Garden which is home to 10 raised beds stuffed with colour from April to October. There will be a minibus to take visitors between the gardens or the walk is 0.4 mile part uphill.

63 WHATLEY MANOR

Easton Grey, Malmesbury, SN16 0RB. Christian & Alix Landolt, 01666 822888, reservations@whatleymanor.com, www.whatleymanor.com. *4m W of Malmesbury. From A429 at Malmesbury take B4040 signed Sherston. Manor 2m on L.* **Wed 7 June (2-6). Adm £8.50, chd free. Light refreshments in The Loggia Garden.**

12 acres of English country gardens with 26 distinct areas each with a strong theme based on colour,

scent or style. Original 1920s Arts & Crafts plan inspired the design and combines classic style with more contemporary touches, inc specially commissioned sculpture. Dogs must be on a lead at all times. Hotel also open for lunch and full afternoon tea.

64 NEW WINKELBURY HOUSE

Berwick St. John, Shaftesbury, SP7 0EY. Ian & Carrie Stewart. *5m E of Shaftesbury. From A30 follow sign to Berwick St John, after 1½m turn L into Woodlands Lane.* **Sat 20 May (2-5). Adm £7.50, chd free. Home-made teas.**

1½ acres with glorious views of surrounding countryside. The garden has evolved over last 6 yrs to inc kitchen garden with bothy and greenhouse, mown paths through informal areas, meadow planted with spring bulbs, 30m iris border, ha-ha, wildlife pond and bee friendly planting.

65 WUDSTON HOUSE

High Street, Wedhampton, Devizes, SN10 3QE. David Morrison, 07881 943213, djm@piml.co.uk. *Wedhampton lies on N side of A342 approx 4m E of Devizes. House is set back on E side of village street at end of drive with beech hedge on either side.* **Visits by arrangement 1 June to 16 Sept for groups of 6+. £10 per visitor for garden visit or £15 with light refreshments.**

The garden was started in 2010, following completion of the house. It consists of formal gardens round the house, perennial meadow, pinetum and arboretum. Nick Macer and James Hitchmough, who have pioneered the concept of perennial meadows, have been extensively involved in aspects of the garden, which is still developing.

Our 2022 donation to Hospice UK meant that 911 NHS and Social Care staff accessed individual counselling support

Riverside House

WORCESTERSHIRE

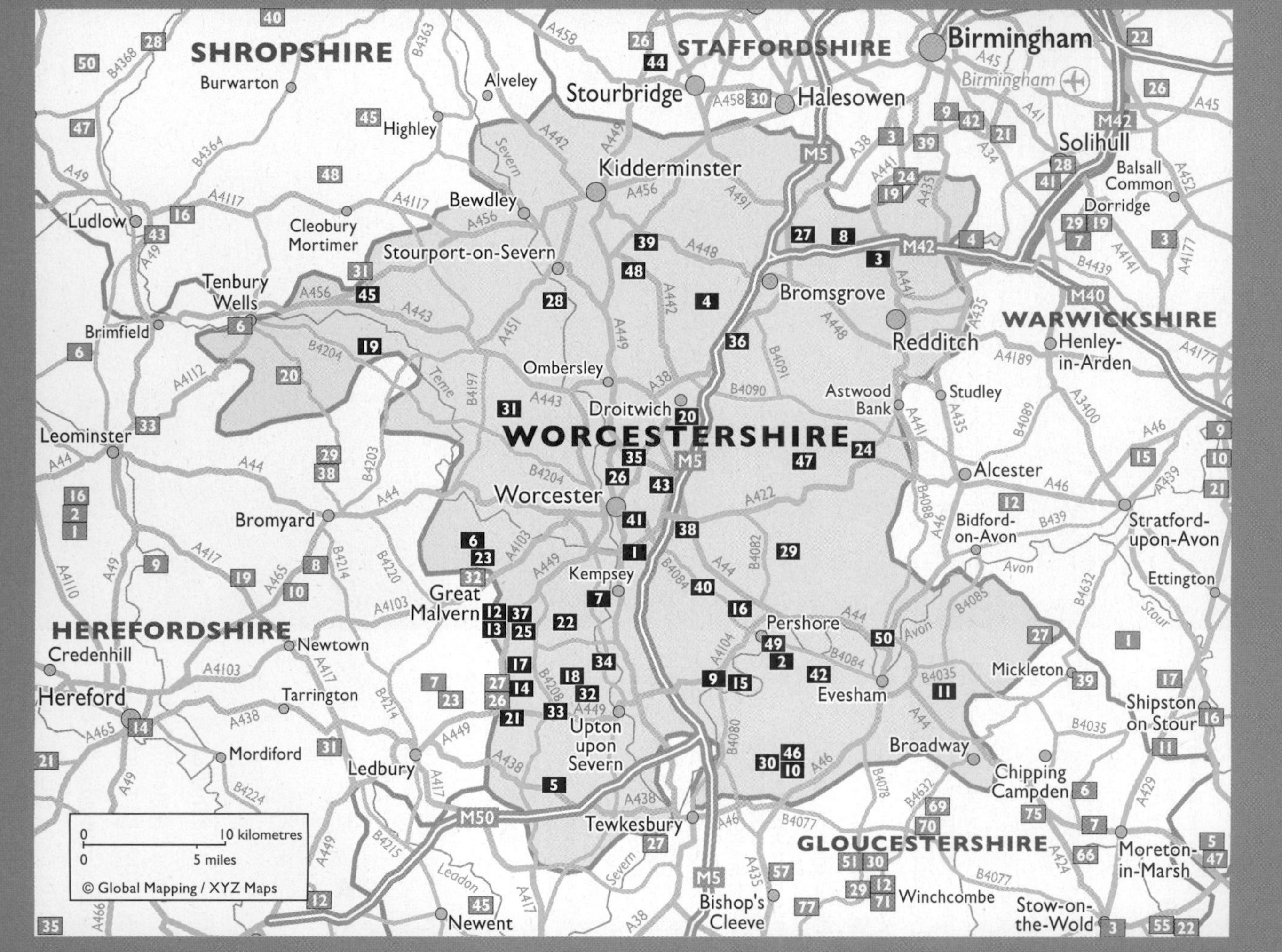

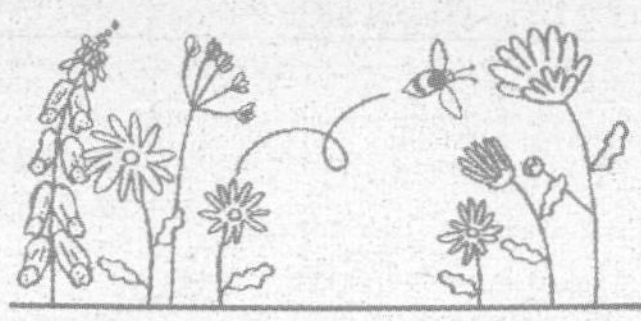

Worcestershire has something to suit every taste, and the same applies to its gardens.

From the magnificent Malvern Hills, the inspiration for Edward Elgar, to the fruit orchards of Evesham which produce wonderful blossom trails in the spring, and from the historic city of Worcester, with its 11th century cathedral and links to the Civil War, to the numerous villages and hamlets that are scattered throughout, there is so much to enjoy in this historic county.

Worcestershire is blessed with gardens created by both amateur and professional gardeners, and the county can boast properties with grounds of many acres to small back gardens of less than half an acre, even prize winning gardens, but all have something special to offer.

There are gardens with significant historical interest and some with magnificent views and a few that are not what you might consider to be a "typical" National Garden Scheme garden! We also have a number of budding artists within the Scheme, and a few display their works of art on garden open days.

Worcestershire's garden owners guarantee visitors beautiful gardens, some real surprises and a warm welcome.

© Jane Carr

Volunteers

County Organiser
David Morgan
01214 453595
meandi@btinternet.com

County Treasurer
Doug Bright
01886 832200
doug.bright@ngs.org.uk

Publicity
Pamela Thompson
01886 888295
peartree.pam@gmail.com

Social Media
Brian Skeys
01684 311297
brian.skeys@ngs.org.uk

Booklet Co-ordinator
Steven Wilkinson & Linda Pritchard
01684 310150
steven.wilkinson48412@gmail.com

Assistant County Organisers
Andrea Bright
01886 832200
andrea.bright@ngs.org.uk

Brian Bradford
07816 867137
bradf0rds@icloud.com

Lynn Glaze
01386 751924
lynnglaze@cmail.co.uk

Philippa Lowe
01684 891340
philippa.lowe@ngs.org.uk

Stephanie & Chris Miall
0121 445 2038
stephaniemiall@hotmail.com

Alan Nokes
alan.nokes@ngs.org.uk

@WorcestershireNGS
@WorcsNGS

Left: Conderton Manor

OPENING DATES

All entries subject to change. For latest information check **www.ngs.org.uk**

Map locator numbers are shown to the right of each garden name.

February

Snowdrop Festival

Sunday 19th
Brockamin 7
Warndon Court 43

March

Friday 17th
◆ Little Malvern Court 21

Saturday 18th
NEW Winchester House 50

Sunday 19th
Brockamin 7
NEW Winchester House 50

Friday 24th
◆ Little Malvern Court 21

Friday 31st
◆ Little Malvern Court 21

April

Saturday 1st
The Alpine Garden Society 2
Overbury Court 30
Whitcombe House 46
◆ Whitlenge Gardens 48

Sunday 2nd
◆ Whitlenge Gardens 48

Sunday 9th
The Dell House 14
◆ Spetchley Park Gardens 38

Monday 10th
NEW ◆ Cotswold Garden Flowers 11
The Dell House 14

Saturday 15th
NEW The Walled Garden & No. 53 41

Sunday 16th
◆ White Cottage & Nursery 47

Wednesday 19th
NEW The Walled Garden & No. 53 41

Saturday 22nd
The River School 35
NEW Winchester House 50

Sunday 23rd
NEW Winchester House 50

Saturday 29th
The Alpine Garden Society 2

Sunday 30th
Brockamin 7
The Folly 17
Nimrod, 35 Alexandra Road 25
◆ White Cottage & Nursery 47

May

Monday 1st
Nimrod, 35 Alexandra Road 25

Sunday 7th
The Folly 17
Hiraeth 20

Saturday 13th
Cowleigh Park Farm 13
Oak Tree House 27

Sunday 14th
Oak Tree House 27
Pear Tree Cottage 31

Saturday 20th
Madresfield Court 22
Ravelin 32
NEW Winchester House 50

Sunday 21st
2 Brookwood Drive 8
Ravelin 32
Rothbury 37
Warndon Court 43
NEW Winchester House 50

Saturday 27th
1 Church Cottage 9

Sunday 28th
1 Church Cottage 9
Millbrook Lodge 23
◆ White Cottage & Nursery 47

Monday 29th
NEW ◆ Cotswold Garden Flowers 11
Rothbury 37
◆ White Cottage & Nursery 47

Wednesday 31st
The Alpine Garden Society 2

June

Saturday 3rd
NEW Acorns Children's Hospice 1
Eckington Gardens 15
NEW The Walled Garden & No. 53 41

Sunday 4th
Brockamin 7
Eckington Gardens 15
3 Oakhampton Road 28

Wednesday 7th
NEW The Walled Garden & No. 53 41

Saturday 10th
Hanley Swan NGS Gardens 18

Sunday 11th
Birtsmorton Court 5
The Dell House 14
Hanley Swan NGS Gardens 18
The Old House 29
◆ White Cottage & Nursery 47
Wick Village 49

Saturday 17th
Alvechurch Gardens 3

Sunday 18th
Alvechurch Gardens 3
Rothbury 37
Warndon Court 43

Saturday 24th
Rest Harrow 33
Walnut Cottage 42
Westacres 44
◆ Whitlenge Gardens 48

Sunday 25th
Cowleigh Lodge 12
The Firs 16
Hiraeth 20
3 Oakhampton Road 28
The Old House 29
Rest Harrow 33
Walnut Cottage 42
Westacres 44
Whitcombe House 46
◆ Whitlenge Gardens 48

July

Saturday 1st
North Worcester Gardens 26
Wharf House 45

Sunday 2nd
The Firs 16
North Worcester Gardens 26
◆ Spetchley Park Gardens 38
Wharf House 45

Sunday 9th
2 Brookwood Drive 8
Millbrook Lodge 23

Saturday 15th
Cowleigh Park Farm 13

Sunday 23rd
Rothbury 37

Thursday 27th
The River School 35

Sunday 30th
NEW ◆ Cotswold Garden Flowers 11
Hiraeth 20
3 Oakhampton Road 28

August

Sunday 13th
Hiraeth 20

Saturday 26th
◆ Morton Hall Gardens 24
Westacres 44

Sunday 27th
Bridges Stone Mill 6
The Dell House 14
3 Oakhampton Road 28
Pear Tree Cottage 31
Westacres 44
Whitcombe House 46

Monday 28th
NEW ◆ Cotswold Garden Flowers 11
3 Oakhampton Road 28
Westacres 44

September

Saturday 2nd
Rest Harrow 33
◆ Whitlenge Gardens 48

Sunday 3rd
Rest Harrow 33
◆ Whitlenge Gardens 48

Saturday 16th
Ravelin 32
Rhydd Gardens 34

Sunday 17th
Brockamin 7
The Old House 29
Ravelin 32
Rhydd Gardens 34

October

Sunday 1st
NEW ◆ Cotswold Garden Flowers 11

Saturday 7th
Ravelin 32

Sunday 8th
The Dell House 14
Ravelin 32

In 2022 we donated £450,000 to Marie Curie which equates to 23,871 hours of community nursing care

By Arrangement

Arrange a personalised garden visit with your club, or group of friends, on a date to suit you. See individual garden entries for full details.

The Alpine Garden Society 2
Badge Court 4
5 Beckett Drive, North Worcester Gardens 26
Birtsmorton Court 5
Brockamin 7
2 Brookwood Drive 8
1 Church Cottage 9
Conderton Manor 10
Cowleigh Lodge 12
Cowleigh Park Farm 13
The Dell House 14
The Firs 16
The Folly 17
Hanley Swan NGS Gardens 18
Highways 19
Hiraeth 20
Mantoft, Eckington Gardens 15
Millbrook Lodge 23
Nimrod, 35 Alexandra Road 25
Oak Tree House 27
3 Oakhampton Road 28
Overbury Court 30
Pear Tree Cottage 31
Ravelin 32
Rest Harrow 33
Rhydd Gardens 34
The Tynings 40
Warndon Court 43
Westacres 44
Wharf House 45
Whitcombe House 46
19 Winnington Gardens, Hanley Swan NGS Gardens 18

THE GARDENS

1 NEW ACORNS CHILDREN'S HOSPICE

350 Bath Road, Worcester, WR5 3EZ. Acorns Children's Hospice. *On A38 approx 1¼ m S of Worcester. From M5 J7 towards Worcester then L at 1st island onto A4440. R at 2nd island onto A38 towards city centre for approx ½ m.* **Sat 3 June (11-4). Adm £4, chd free. Light refreshments.**

1½ acres designed for the benefit of children and young people with life limiting illnesses and their families. Inc play areas, raised beds for young people to plant vegetables, tranquil memorial garden with gentle stream and memorial pebbles for children who have passed away and an enclosed bedroom garden set aside for recently bereaved families. Acers, silver birch and other trees and shrubs feature. This is not a traditional National Garden Scheme garden but one that shows how gardens can have a positive effect on peoples' wellbeing. All areas wheelchair accessible.

♿ ☕

2 THE ALPINE GARDEN SOCIETY

Avon Bank, Wick, Pershore, WR10 3JP. The Alpine Garden Society, 01386 554790, ags@alpinegardensociety.net, www.alpinegardensociety.net. *Wick, ½ m from Pershore town. Take the Evesham Road from Pershore over the River Avon, cont for ½ m & turn R for Pershore College, the garden is the 1st entrance on the L.* **Sat 1, Sat 29 Apr (11-4); Wed 31 May (2-6). Adm £4, chd free. Light refreshments. Visits also by arrangement 27 Mar to 29 Sept for groups of 10+.**

Inspirational small garden next to the Alpine Garden Society office. The garden shows a wide range of alpine plants that are easy to grow in contemporary gardens over a long season. Visitors can see different settings to grow alpines, inc rock and tufa, scree, a dry mediterranean bed, shade and sunny areas, also a dedicated alpine house, and many pots and troughs with alpines and small bulbs. Volunteers will be on hand on open days to provide more information, and planting demonstrations may be available on some dates.

✽ ☕))

National Garden Scheme gardens are identified by their yellow road signs and posters. You can expect a garden of quality, character and interest, a warm welcome and plenty of home-made cakes!

GROUP OPENING

3 ALVECHURCH GARDENS

Alvechurch, B48 7LF. Group Co-ordinator Martin Wright. *3m N of Redditch, 3m NE of Bromsgrove. Alvechurch is on the B4120 close to J2 of M42. Gardens are signed from all roads into the village. Pick up your map when you pay for your ticket at the first garden visited.* **Sat 17, Sun 18 June (1-5). Combined adm £6, chd free. Light refreshments in the village, where there are several local pubs & cafes**

THE ALLOTMENTS
Eileen McHugh.

NEW **55 BIRMINGHAM ROAD**
Roger & Sue Wardle.

NEW **5 CALLOW HILL ROAD**
Mr & Mrs O'Malley.

28 CALLOW HILL ROAD
Martin Wright.

CORNER HOUSE
Janice Wiltshire.

THE MOAT HOUSE
Mike & Tracy Fallon.

RECTORY COTTAGE
Celia Hitch, 0121 445 4824, celia@rectorycottage-alvechurch.co.uk.

4 SNAKE LANE
Jason Turner & Paul Emery.

8 TRANTER AVENUE
Kevin Baker.

2 WHARF COTTAGE
Janette Poole.

WYCHWOOD HOUSE
Chris & Stephanie Miall.

Large village with new development and historically interesting core with buildings spanning medieval to Edwardian and St Laurence Church dedicated in 1239. There is a selection of lovely open gardens ranging from a riverside previous rectory with waterfall to a corner plot gardened for wildlife. There are professionally landscaped terraced gardens and cottage gardens with lots of colour. Also a large moated garden on the old site of the Bishop of Worcester's Summer Palace. Around the gardens there are mature trees, rose beds, shrubberies, herbaceous and fruit and vegetable beds. Some gardens inc sculptures. The village allotments will also be open for viewing. The gardens are walking distance from Alvechurch railway station (on the Cross-City Line 11 miles SW of Birmingham New Street). Refreshments available in local village pubs and cafes.

4 BADGE COURT

Purshull Green Lane, Elmbridge, Droitwich, WR9 0NJ. Stuart & Diana Glendenning, 01299 851216, dianaglendenning1@gmail.com. *5m N of Droitwich Spa. 2½m from J5 M5. Turn off A38 at Wychbold down side of the Swan Inn. Turn R into Berry Ln. Take next L into Cooksey Green Ln. Turn R into Purshull Green Ln. Garden is on L.* **Visits by arrangement May to July for groups of 5 to 30. Visits are available in the afternoon or early evening. Adm £5, chd free. Home-made teas.**
The 2½ acre garden is set against the backdrop of a C16 house (not open). There is something for all gardeners inc mature specimen trees, huge range of clematis and roses, lake with waterfall, stumpery, topiary garden, walled garden, Mediterranean garden, specialist borders, long herbaceous border, large potager vegetable garden and Japanese garden.

5 BIRTSMORTON COURT

Birtsmorton, nr Malvern, WR13 6JS. Mr & Mrs N G K Dawes, rosaliedawes@btinternet.com. *7m E of Ledbury. Off A438 Ledbury/Tewkesbury rd.* **Sun 11 June (2-5.30). Adm £8, chd free. Home-made teas. Visits also by**

Wick Village

arrangement 15 Apr to 15 Sept for groups of 20 to 40.
10 acre garden surrounding beautiful medieval moated manor house (not open). White garden, built and planted in 1997 surrounded on all sides by old topiary. Potager, vegetable garden and working greenhouses, all beautifully maintained. Rare double working moat and waterways inc Westminster Pool laid down in Henry VII's reign to mark the consecration of the knave of Westminster Abbey. Ancient yew tree under which Cardinal Wolsey reputedly slept in the legend of the Shadow of the Ragged Stone. Sorry - no dogs.

6 BRIDGES STONE MILL

Alfrick Pound, WR6 5HR.
Sir Michael & Lady Perry. *6m NW of Malvern. A4103 from Worcester to Bransford r'about, then Suckley Rd for 3m to Alfrick Pound.* **Sun 27 Aug (2-5.30). Adm £7, chd free. Home-made teas.**
Once a cherry orchard adjoining the mainly C19 flour mill, this is now a 2½ acre year-round garden laid out with trees, shrubs, mixed beds and borders. The garden is bounded by a stretch of Leigh Brook (an SSSI), from which the mill's own weir feeds a mill leat and small lake. A rose parterre and a traditional Japanese garden complete the scene. Wheelchair access by car to courtyard.

7 BROCKAMIN

Old Hills, Callow End, Worcester, WR2 4TQ.
Margaret Stone, 01905 830370, stone.brockamin@btinternet.com. *5m S of Worcester. ½m S of Callow End on the B4424, on an unfenced bend, turn R into the car park signed Old Hills. Walk towards the houses keeping R.* **Sun 19 Feb (11-4); Sun 19 Mar, Sun 30 Apr, Sun 4 June, Sun 17 Sept (2-5). Adm £5, chd free. Home-made teas. Visits also by arrangement 18 Feb to 15 Oct for groups of 10+. Donation to Plant Heritage.**
This is a plant specialist's 1½ acre informal working garden, parts of which are used for plant production rather than for show. Situated next to common land. Mixed borders with wide variety of hardy perennials where plants are allowed to self seed. Plant Heritage National Collections of Symphyotrichum (Aster) novae-angliae and some hardy geraniums. Open for snowdrops in Feb, daffodils and pulmonarias in March and April, geraniums in June and asters in September. Seasonal pond/bog garden and kitchen garden. Teas with home-made cakes and unusual plants for sale. An access path reaches a large part of the garden.

8 8 BROOKWOOD DRIVE

Barnt Green, Birmingham, B45 8GG. Mr Mike & Mrs Liz Finlay, 07721 746369, liz@lizfinlay.com. *5m N of Bromsgrove. Please park courteously in local roads or in village. Brookwood Drive is on R towards top of Fiery Hill Rd approx. ¼m from station. Press the Access button to open security gates if closed.* **Sun 21 May, Sun 9 July (10-5). Adm £5, chd free. Home-made teas. Visits also by arrangement 15 Apr to 31 Aug for groups of 6 to 12 on a first come basis, subject to owner availability.**
A multi-themed mature garden with numerous water features and colourful borders. There is a formal white garden, wildlife pond and cutting garden surrounded by large rhododendrons which give a colourful display in late spring. Parking on Brookwood Drive for disabled only.

9 1 CHURCH COTTAGE

Church Road, Defford, WR8 9BJ. John Taylor & Ann Sheppard, 01386 750863, ann98sheppard@btinternet.com. *3m SW of Pershore. A4104 Pershore to Upton rd, turn into Harpley Rd, Defford. Don't go up Bluebell Lane as directed by SatNav, black & white cottage at side of church. Parking in village hall car park.* **Sat 27, Sun 28 May (1-5). Adm £5, chd free. Home-made teas. Visits also by arrangement Feb to Sept for groups of 10 to 30.**
True countryman's ⅓ acre cottage garden. Japanese style feature with 'dragons den'. Specimen trees, rare and unusual plants, water features, perennial garden, vegetable garden, poultry, streamside bog garden. New features in progress. Wheelchair access to most areas, narrow paths may restrict access to some parts.

10 CONDERTON MANOR

Conderton, nr Tewkesbury, GL20 7PR. Mr & Mrs W Carr, 01386 725389, carrs@conderton.com. *5½m NE of Tewkesbury. From M5 - A46 to Beckford - L for Overbury/ Conderton. From Tewkesbury B4079 to Bredon - then follow signs to Overbury. Conderton from B4077 follow A46 directions from Teddington r'about.* **Visits by arrangement Mar to Nov. Light refreshments.**
7 acre garden with flowering cherries and bulbs in spring. Formal terrace with clipped box parterre, huge rose and clematis arches, mixed borders of roses and herbaceous plants, bog bank and quarry garden. Many unusual trees and shrubs make this a garden for all seasons. Visitors are particularly encouraged to visit in spring and autumn when the trees are at their best. This is a garden/ small arboretum of particular interest for tree lovers. The views towards the Cotswolds escarpment are spectacular and it provides a peaceful walk of about an hour. Some gravel paths and steps - no disabled WC.

11 NEW ◆ COTSWOLD GARDEN FLOWERS

Sands Lane, Badsey, Evesham, WR11 7EZ. Mr Bob Brown, 01386 833849, info@cgf.net, www.cgf.net. *Sands Lane is the last road on the L when leaving Badsey for Wickhamford. Garden is nearly ½m down lane.* **For NGS: Mon 10 Apr, Mon 29 May, Sun 30 July, Mon 28 Aug, Sun 1 Oct (11-5). Adm £5, chd free. Light refreshments. For other opening times and information, please phone, email or visit garden website.**
The garden consists of one acre of stockbeds with many thousands of kinds of plants many of which are unusual and rare. On open days visitors can follow or be taken on a guided tour.

12 COWLEIGH LODGE

16 Cowleigh Bank, Malvern, WR14 1QP. Jane & Mic Schuster, 07854 015065, dalyan@hotmail.co.uk. *7m SW from Worcester. From Hereford, turn R after Storridge church, follow rd approx 2½ m, then L onto Cowleigh Bank. From Worcester R at Link Top - Hornyold Rd, R onto St. Peter's Rd, follow yellow signs.* **Sun 25 June (11-5). Adm £6, chd free. Home-made teas. Visits also by arrangement in June for groups of 10+Refreshments £3.00 per person.**

The now mature 1 acre "quirky' garden on the slopes of the Malvern Hills has a formal rose garden, grass beds, bamboo walk, colour themed beds, nature path leading to a wildlife pond, acer bank, chickens. Large vegetable plot and orchard with views overlooking the Severn Valley. Explore the polytunnel and then relax with a cuppa and slice of home-made cake served with a smile. This is the 9th year of opening of this developing and expanding garden - visitors from previous years will be able to see the difference. Lots of added interest with staddle stones, troughs, signs and other interesting artefacts. The garden is definitely now a mature one.

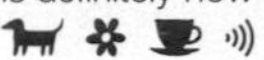

13 COWLEIGH PARK FARM

Cowleigh Road, Malvern, WR13 5HJ. John & Ruth Lucas, 01684 566750, info@cowleighparkfarm.co.uk, www.cowleighparkfarm.co.uk. *On the edge of Malvern. Leaving Malvern on the B4219 the driveway for Cowleigh Park Farm is on the R just before the derestrict speed sign. Coming from the A4103 we are on the L immed after the 30mph sign.* **Sat 13 May, Sat 15 July (1.30-5). Adm £5, chd free. Light refreshments. Visits also by arrangement 21 Apr to 22 Oct.**

The 1½ acre garden at Cowleigh Park Farm surrounds a Grade II listed timber framed former farmhouse (not open). Whilst no longer a farm, the property has views to adjacent orchards and inc lawns, spring fed ponds, a waterfall and stream. The focus in established beds and borders is to be bee and wildlife friendly. The garden contains multiple seating areas and a summerhouse. The whole garden can be viewed from wheelchair accessible places but some parts contain steep grassy slopes that may not be accessible by wheelchair.

14 THE DELL HOUSE

2 Green Lane, Malvern Wells, WR14 4HU. Kevin & Elizabeth Rolph, 01684 564448, kande@dellhousemalvern.uk, www.dellhousemalvern.uk. *What three words app - remember.shrub. robot. 2m S of Gt Malvern Behind former church on corner of Wells Rd & Green Ln. Small car park for those pre-booked. Approach downhill from Wells Road, NOT uphill. Do not use Postcode in SatNav.* **Sun 9, Mon 10 Apr, Sun 11 June, Sun 27 Aug (12-6); Sun 8 Oct (1-5). Adm £5, chd free. Pre-booking essential, please visit www.ngs.org.uk for information & booking. Light refreshments. Visits also by arrangement for groups of up to 20, even at short notice.**

Two acre wooded hillside garden of the 1830s former rectory, now a B&B. Peaceful and natural, the garden contains many magnificent specimen trees inc a Wellingtonia Redwood. Informal in style with meandering bark paths, historic garden buildings, garden railway and a paved terrace with outstanding views. Spectacular tree carvings by Steve Elsby, and other sculptures by various artists. Featured in 'A survey of Historic Parks & Gardens in Worcestershire'. Pre-booking only required for those arriving by car. (Limited parking).

GROUP OPENING

15 ECKINGTON GARDENS

Hilltop, Nafford Road, Eckington, WR10 3DH. Group Coordinator Richard Bateman. *3 gardens - 1 in Upper End, 1 close by in Nafford Rd, 3rd about 1m along Nafford Rd. A4104 Pershore to Upton & Defford, L turn B4080 to Eckington. In centre, by war memorial turn L into New Rd (becomes Nafford Rd).* **Sat 3, Sun 4 June (11-5). Combined adm £7, chd free. Light refreshments at Mantoft.**

HILLTOP FARM
Richard & Margaret Bateman.

MANTOFT
Mr & Mrs M J Tupper, 01386 750819.
Visits also by arrangement May to Sept for groups of 5 to 30. Please liaise with garden owner re refreshments.

NAFFORD HOUSE
George & Joanna Stylianou.

3 very diverse gardens set in/close to lovely village of Eckington. Hilltop – 1 acre garden with sunken garden/pond, rose garden, herbaceous borders, interesting topiary inc cloud pruning and formal hedging to reduce effect of wind and having 'windows' for views over the beautiful Worcestershire countryside. Sculptures made by owner. Vegetable yurt (added 2018) has been successful in allowing pollination and preventing damage to young plants. Mantoft - Wonderful ancient thatched cottage with 1½ acres of magical gardens. Fishpond with ghost koi, Cotswold and red brick walls, large topiary, treehouse with seating, summerhouse and dovecote, pathways, vistas and stone statues, urns and herbaceous borders. Featured in Cotswold Life - should not be missed. Nafford House is 2 acre mature natural garden, wood with walk and slopes to River Avon, formal gardens around the house/ magnificent wisteria. Some wheelchair access issues, particularly at Nafford House.

16 THE FIRS

Brickyard Lane, Drakes Broughton, Pershore, WR10 2AH. Ann & Ken Mein, annmmein@googlemail.com. *From J7 M5 take B4084 towards Pershore. At Drakes Broughton turn L to Stonebow Rd. First R into Walcot Ln then 2nd R into Brickyard Ln. The Firs is 200 yds on L.* **Sun 25 June, Sun 2 July (1-5). Adm £4.50, chd free. Cream teas. Visits also by arrangement 26 June to 6 Aug for groups of 15+.**

2 acre garden with over 200 trees, predominantly Silver Birch. A Rockery and gravel path lead down to the terrace. Informal beds filled with agapanthus, sedums, roses, hydrangeas and grasses; a fruit and vegetable garden and a fernery surround the house. The Barn houses the Pottery with a variety of pots for sale. Lots of seating areas with views towards Bredon Hill. Gentle slopes, grassy paths.

17 THE FOLLY
87 Wells Road, Malvern, WR14 4PB. David & Lesley Robbins, 01684 567253, lesleycmedley@btinternet.com.
1½m S of Great Malvern & 9m S of Worcester. Approx 8m from M5 via J7 or J8. Situated in Malvern Wells on A449, 0.7m N of B4209 & 0.2m S of Malvern Common. Parking off A449 0.25m N on lay-by or side road, or in 0.1m N turn E on Peachfield Rd which runs by Malvern Common. **Sun 30 Apr, Sun 7 May (1.30-5). Adm £5, chd free. Light refreshments, cash only. Visits also by arrangement 17 Apr to 30 June for groups of up to 25.**

A steeply sloping garden on Malvern Hills with views over the Severn Vale. Three levels are accessed by steps, paved/gravel paths and ramps. Landscaped areas for potager and greenhouse, courtyard, formal terrace, pergola, mature cedars and sculptures. Varied planting inc climbers, shrubs, succulents, hostas, ferns, hellebores and spring blooms. There is seating on each level. Gravel and scree gardens, small cottage garden, stumpery and shrubbery linked by winding paths with an intimate atmosphere as views are concealed and revealed.

In 2022, 20,000 people were supported by Perennial caseworkers and prevent teams thanks to our funding

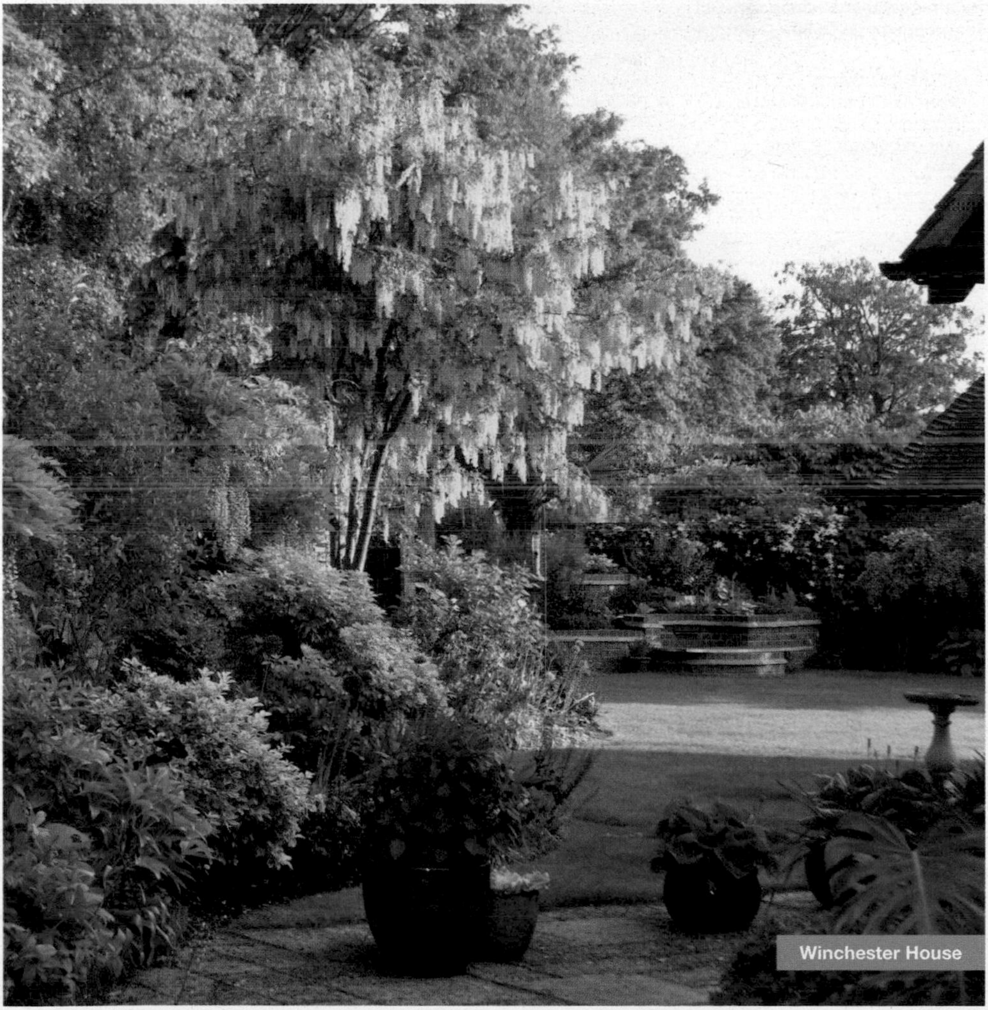
Winchester House

GROUP OPENING

18 HANLEY SWAN NGS GARDENS

Hanley Swan, WR8 0DJ. Group Co-ordinator Brian Skeys, 01684 311297, brimfields@icloud.com, www.brimfields.com. *5m E of Malvern, 3m NW of Upton upon Severn, 9m S of Worcester & M5. From the B4211 towards Great Malvern & Guarlford (Rhydd Rd) turn L to Hanley Swan continue to Orchard Side WR80EA for entrance tickets & maps.* **Sat 10, Sun 11 June (1-5). Combined adm £6, chd free. Light refreshments at 19 Winnington Gardens. Visits also by arrangement 12 June to 1 Oct for groups of up to 30.**

MEADOW BANK
Mrs Lesley Stroud & Mr Dave Horrobin, 01684 310917, djhorrobin@gmail.com.

ORCHARD SIDE
Mrs Gigi Verlander, 01684 310602, gigiverlander@icloud.com.

THE PADDOCKS
Mr & Mrs N Fowler.

SUNDEW
Mr Nick & Mrs Alison Harper.

19 WINNINGTON GARDENS
Brian & Irene Skeys, 01684 311297, brimfields@icloud.com, www.brimfields.com.
Visits also by arrangement 15 May to 1 Oct for groups of up to 30.

YEW TREE COTTAGE
Mr & Mrs Read.

Entrance tickets from Orchard Side, a country garden offering lots of interest of both plants and unusual items dotted around the garden. There is a large koi pond that sits just in front of the peaceful Zen Den where visitors are welcome to sit and contemplate. Sundew sits in a modest plot - since 2016 the owners have added planting, hard landscaping, new borders to a well-stocked garden. The Paddocks is a wildlife garden with ponds, a tadpole nursery, mixed borders and a new greenhouse with cacti and succulents. Meadow Bank is a modern interpretation of a cottage garden, with a hot border, dahlias grown for show and collections of iris and auriculas. Yew Tree Cottage, a C17 black and white cottage (not open) and well, within a cottage garden, with field views at rear. 19 Winnington Gardens is a garden of rooms. Mixed borders enclosed with climbing roses, a small oriental garden, fruit trees, raised herb bed, pelargonium collection, irises, collection of garden vintage tools. 6 gardens different in size and style. Wildlife photos, vintage garden tools on display. Plant sales.Teas.

The Firs

19 HIGHWAYS

Hanley William, Tenbury Wells, WR15 8QT. Mr & Mrs K Baker, 01584 781216, annejbaker@outlook.com. *7m from Tenbury Wells, 9m from Bromyard. From Worcester take the A443 to Tenbury, turn L at the turn marked Eastham. From Tenbury take the B4204. From Bromyard take the B4203, turn L to Tenbury at the Xrds.* **Visits by arrangement 1 May to 1 Sept for groups of 12 to 30. Limited parking, please car share where possible. Adm £6, chd free. Home-made teas.**
A 1 acre garden developed over the last 16 years, with several herbaceous borders, natural stream and wildlife pond. Woodland area with established trees, newly planted arboretum, bog garden and some very impressive gunnera! Vegetable garden, koi pond and pergolas with clematis and climbing roses. Views over the Teme valley. Steps and sloping lawn.

20 HIRAETH

30 Showell Road, Droitwich, WR9 8UY. Sue & John Fletcher, 07752 717243 or 01905 778390, sueandjohn99@yahoo.com. *1m S of Droitwich. On The Ridings estate. Turn off A38 r'about into Addyes Way, 2nd R into Showell Rd, 500yds on R. Follow the yellow signs.* **Sun 7 May, Sun 25 June, Sun 30 July, Sun 13 Aug (2-5). Adm £3.50, chd free. Visits also by arrangement Apr to Sept for groups of up to 30. Please discuss adm & refreshments at time of booking.**
⅓ acre gardens, front and rear, containing many plants, herbaceous, hostas, ferns, acer trees, various other trees, a 300+ year old olive tree, oak sculptures, metal animals inc giraffes, elephant, ducks, birds, reptiles, stone statues and verse plaques. A new patio was laid in 2019 and in late Spring 2022 we have removed our koi pond and waterfall, filled the resultant hole with circa 20 tonnes of soil etc, installed a further gravel pathway and re-planted the area. The garden is an oasis of colours, and was described some years ago as 'a haven on the way to heaven', and should not be missed!

21 ◆ LITTLE MALVERN COURT

Little Malvern, WR14 4JN. Mrs T M Berington, 07856 035599, littlemalverncourt@hotmail.com, www.littlemalverncourt.co.uk. *3m S of Malvern. On A4104 S of junction with A449.* **For NGS: Every Fri 17 Mar to 31 Mar (2-5). Adm £6, chd free. Light refreshments only. For other opening times and information, please phone, email or visit garden website.**
10 acres attached to former Benedictine Priory, magnificent views over Severn Valley. Garden rooms and terrace around house designed and planted in early 1980s; chain of lakes; wide variety of spring bulbs, flowering trees and shrubs. Notable collection of old fashioned roses. Topiary hedge and fine trees. Regional Finalist, The English Garden's The Nation's Favourite Gardens 2019.

22 MADRESFIELD COURT

Madresfield, Malvern, WR13 5AJ. Trustees of Lord Beauchamp's 1963 Settlement, www.madresfieldestate.co.uk. *2m E of Malvern. Entrance ½m S of Madresfield village, just outside Malvern.* **Sat 20 May (2-5). Adm £10, chd £4. Tea.**
Gardens mainly laid out in 1865, based on three avenues of oak, cedar and lombardy poplar, within and around which are specimen trees and flowering shrubs. Meadows within the avenues covered in daffodils, and later, fritillaries, bluebells, cowslips etc. Recent rhododendron plantings. Holly hedge enclosure with 100m walk of peonies and irises, next to a crescent tunnel of pollarded limes. Overall a parkland garden of approx 60 acres.

23 MILLBROOK LODGE

Millham Lane, Alfrick, Worcester, WR6 5HS. Andrea & Doug Bright, andreabright@hotmail.co.uk. *11m from M5 J7. A4103 from Worcester to Bransford r'about, then Suckley Rd for 3m to Alfrick Pound. Turn L at sign for Old Storridge - No Through Road.* **Sun 28 May, Sun 9 July (1-5). Adm £5, chd free. Home-made teas. Visits also by arrangement Apr to Sept for groups of 12 to 30.**
3½ acre garden and woodland developed by current owners over 25 years. Large informal flowerbeds planted for year-round interest, from spring flowering bulbs, camellias and magnolias through to asters and dahlias. Pond with stream and bog garden. Fruit and vegetable garden and gravel garden. Peaceful setting in an area of outstanding natural beauty and near a nature reserve.

24 ◆ MORTON HALL GARDENS

Morton Hall Lane, Holberrow Green, Redditch, B96 6SJ. Mrs A Olivieri. *In centre of Holberrow Green, at a wooden bench around a tree, turn up Morton Hall Ln. Follow NGS signs to gate opp Morton Hall Farm.* **For NGS: Sat 26 Aug (10-5). Adm £10, chd free. Pre-booking essential, please phone 01386 791820, email morton.garden@mhcom.co.uk or visit www.mortonhallgardens.co.uk for information & booking. Light refreshments. Gluten-free & vegetarian options. For other opening times and information, please phone, email or visit garden website.**
Perched atop an escarpment with breathtaking views, hidden behind a tall hedge, lies a Georgian country house with a unique garden of outstanding beauty. A garden for all seasons, it features one of the country's largest fritillary spring meadows, sumptuous herbaceous summer borders, a striking potager, a majestic woodland rockery and an elegant Japanese Stroll Garden with tea house. Tulip Festival on first May bank holiday weekend. Visits by appointment from April to September. ALL visits must be pre-booked via the Morton Hall Garden website. For NGS Open Day, click on the YELLOW BANNER on the Morton Hall Gardens homepage. Pre-booking closes at 9 am on 26.08.23.

Our 2022 donation supports 1,700 Queen's Nurses working in the community in England, Wales, Northern Ireland, the Channel Islands and the Isle of Man

25 NIMROD, 35 ALEXANDRA ROAD

Malvern, WR14 1HE. Margaret & David Cross, 01684 569019, margaret.cross@ifdev.net. *From A449. From Malvern Link, pass train stn, ahead at T-lights, 1st R into Alexandra Rd. From Great Malvern, go through Link Top T-lights (junction with B4503), 1st L.* **Sun 30 Apr, Mon 1 May (10.30-5). Adm £5, chd free. Home-made teas. Visits also by arrangement 30 Apr to 30 Sept for groups of 5 to 20.**

Overlooked by the Malvern Hills, the garden is on two levels with mature horse chestnuts, western red cedars, wildlife pond, small wildflower meadow, shady woodland area, cottage garden, Japanese garden, rockeries and a New Zealand influenced area. An arbour inspired by Geoff Hamilton, also a treehouse and den for accompanied children. Elgar wrote some of the Enigma variations in a bell tent here. With help, the garden can be accessed by wheelchair, there are few steps but gravel paths and gradients. If advised we can offer closer parking.

GROUP OPENING

26 NORTH WORCESTER GARDENS

Northwick, Worcester, WR3 7BZ. *Four town gardens off A449, Ombersley Rd, S of Claines r'about. Look for the Yellow signs from the main road.* **Sat 1, Sun 2 July (11-5). Combined adm £6, chd free.**

5 BECKETT DRIVE
Jacki & Pete Ager, 01905 451108, agers@outlook.com.
Visits also by arrangement 20 May to 6 Aug for groups of 10 to 20.

19 BEVERE CLOSE
Mr Malcolm & Mrs Diane Styles.

10 LUCERNE CLOSE
Mark & Karen Askwith.

27 SHELDON PARK ROAD
Mr Alan & Mrs Helen Kirby.

Four town gardens on the northern edge of Worcester each having its own identity. The diversity of this quartet of gardens would satisfy anyone from a dedicated plantsman to an enthusiastic amateur while also offering plenty of ideas for landscaping, planting and structures. The landscaped garden at 5 Beckett Drive inc borders planted at different levels using palettes of harmonising or contrasting colours complemented by innovative features. 10 Lucerne Close is a cottage style garden literally packed with a wide variety of plants with secluded seating areas. 19 Bevere Drive has a stunning collection of bonsai trees planted in an oriental setting. 27 Sheldon Park Road is a contemporary style garden planned with low maintenance in mind and expertly planted with many different foliage plants. This garden won the Worcester News Best Garden 2022 competition. The New Inn public house is on the route between the gardens. Worcester Garden Centre with café is half a mile away.

27 OAK TREE HOUSE

504 Birmingham Road, Marlbrook, Bromsgrove, B61 0HS.
Di & Dave Morgan, 0121 445 3595, meandi@btinternet.com. *On main A38 midway between M42 J1 & M5 J4. Park in old A38 - R fork 250 yds N of garden or small area in front of Miller & Carter Pub car park 200 yds S or local roads.* **Sat 13, Sun 14 May (1.30-5). Adm £4, chd free. Light refreshments. Visits also by arrangement May to Aug for groups of 10 to 30.**

Plantswoman's cottage garden overflowing with plants, pots and interesting artifacts. Patio with plants and shrubs for spring, small pond and waterfall. Plenty of seating, separate wildlife pond, water features, alpine area, rear open vista. Scented plants, hostas, dahlias and lilies. Conservatory with art by owners. Visitor HS said: "Such a wonderful peaceful oasis". Also: 'Wynn's Patch' - part of next door's garden being maintained on behalf of the owner.

28 3 OAKHAMPTON ROAD

Stourport-On-Severn, DY13 0NR. Sandra & David Traynor, 07970 014295, traynor@clickspeedphotography.co.uk. *Between Astley Cross Inn & Kings Arms. From Stourport take A451 Dunley Rd towards Worcester. 1600 yds turn L into Pearl Ln. 4th R into Red House Rd, past the Kings Arms Pub, & next L to Oakhampton Rd. Extra parking at Kings Arms Pub.* **Sun 4, Sun 25 June, Sun 30 July, Sun 27, Mon 28 Aug (10-6). Adm £5, chd free. Home-made teas. Visits also by arrangement June to Sept for groups of up to 14.**

Beginning in March 2016 our plan was to create a garden with a decidedly tropical feel to inc palms from around the world, with tree ferns, bananas and many other strange and unusual plants from warmer climes that would normally be considered difficult to grow here as well as a pond and small waterfall. Not a large garden but you'll be surprised what can be done with a small space. Home-made cakes using eggs and honey from our neighbours whose bees visit our garden flowers.

29 THE OLD HOUSE

Naunton Beauchamp, Pershore, WR10 2LQ. Mrs Gayle Rowe. *In the centre of Naunton Beauchamp. From A422 in Upton Snodsbury, B4082 1.8m to L turn to village. From A44 in Pinvin B4082 for 2.7m to R into village, drive through village past open garden, parking is another 300m on R in a field.* **Sun 11, Sun 25 June, Sun 17 Sept (1.30-5). Adm £5, chd free. Home-made teas.**

A charming 1 acre informal garden with water features, bridges, garden sculptures, orchard, vegetable garden and lawns set around a 300 year old part timber framed house (not open) with beautiful views out over fields of ancient plough and furrow and old orchards. There is partial wheelchair access to the garden, some is on flat lawns some on paths. Limited disabled parking on drive, book space with owner.

30 OVERBURY COURT

Overbury, GL20 7NP. Sir Bruce Bossom & Penelope Bossom, 01386 725111(office), gardens@overburyenterprises.co.uk. *5m NE of Tewkesbury. Overbury signed off A46. Turn off village road beside the church. Park by the gates & walk up the drive. What3words app - cars. blurs.crunches.* **Sat 1 Apr (11-4). Adm £5, chd free. Open nearby Whitcombe House. Visits also by arrangement Apr to Sept for groups of 10 to 30.**

A 10 acre historic garden, at the centre of a picturesque Cotswold village, nestled amongst Capability Brown inspired parkland. The garden compromises vast formal lawns skirted by a series of rills and ponds

Acorn Children's Hospice

which reflect the ancient plane trees that are dotted throughout the garden. The south side of the house has a formal terrace with mixed borders and yew hedging, overlooking a formal lawn with a reflection pool and yew topiary. Running parallel to the pool is a long mixed border which repeats its colours of silver and gold down to the pool house. Some slopes, while all the garden can be viewed, parts are not accessible to wheelchairs.

31 PEAR TREE COTTAGE
Witton Hill, Wichenford, WR6 6YX. Pamela & Alistair Thompson, 01886 888295, peartree.pam@gmail.com, www.peartreecottage.me. *13m NW of Worcester & 2m NE of Martley. From Martley, take B4197. Turn R into Horn Lane then 2nd L signed Witton Hill. Keep L & Pear Tree Cottage is on R at top of hill.* **Sun 14 May (11-5); Sun 27 Aug (12-9.30). Adm £5, chd £1. Home-made teas. Sun 27 Aug - Wine & Pimms after 6pm. Visits also by arrangement May to Aug. Please discuss refreshments when booking.**
A Grade II listed black and white cottage (not open) south west facing gardens with far reaching views across orchards to Abberley Clock Tower. The ¾ acre garden comprises gently sloping lawns with mixed and woodland borders, shade and plenty of strategically placed seating. The garden exudes a quirky and humorous character with the odd surprise and even includes a Shed of the Year Runner Up. Partial wheelchair access.

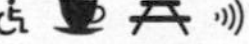

32 RAVELIN
Gilberts End, Hanley Castle, WR8 0AS. Mrs Christine Peer, 01684 310215, cvpeer55@btinternet.com. *From Worcester/Callow End B4424 or from Upton B4211 to Hanley Castle. Then B4209 to Hanley Swan. From Malvern B4209 to Hanley Swan. At pond/Xrds turn to Welland. ½m turn L opp Hall, to Gilberts End.* **Sat 20, Sun 21 May, Sat 16, Sun 17 Sept, Sat 7, Sun 8 Oct (1-5). Adm £5, chd free. Home-made teas. Visits also by arrangement Apr to Oct.**
60 yr old ever evolving ½ acre garden with unusual plants full of colour and texture in all seasons. Interesting to plant lovers gardeners and flower arrangers alike with views overlooking fields and the Malvern Hills. Year-round interest provided by a wide variety of hellebores, hardy geraniums, aconitums, heucheras, Michaelmas daisies, grasses and dahlias and a 55 yr old silver pear tree. Thought to be built on medieval clay pottery works in the royal hunting forest. Garden containing herbaceous and perennial planting with gravel garden, woodland area, pond, summerhouse and plenty of seating areas around the garden. A quiz for children.

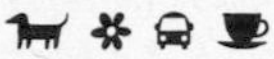

33 REST HARROW

California Lane, Welland, Malvern, WR13 6NQ. Mr Malcolm & Mrs Anne Garner, 01684 310503, anne.restharrow@gmail.com. *4.6m S of Gt Malvern In un-adopted California Lane off B4208 Worcester Rd. From Gt Malvern A449 towards Ledbury, L onto Hanley Rd/B4209 signed Upton. After about 1m R (Blackmore Park Rd/B4209). After 1m R onto B4208. After ⅓m R (California Lane) - garden 300 yds on L.* **Sat 24, Sun 25 June, Sat 2, Sun 3 Sept (1.30-5). Adm £6, chd free. Light refreshments. Visits also by arrangement 12 June to 15 Sept for groups of 10+.**

1½ acres developed over 15 years with 5 acre wildflower meadow, woodland and stunning views of Malvern Hills. Colourful and diverse flower beds, unusual plants, roses, alstroemeria, stocks and shrubs. Potager kitchen garden, fruit trees and rustic trellis made from our own pollarded trees. Sit, relax, enjoy the views or stroll down through the wildflower meadow to the wooded wetland border area. Wheelchair access in garden but not down in field.

34 RHYDD GARDENS

Worcester Road, Hanley Castle, Worcester, WR8 0AB. Bill Bell & Sue Brooks, 01684 311001, NGS@Rhyddgardens.co.uk, www.rhyddgardens.co.uk. *2Km N of Hanley Castle. Gates 200m N of layby on B4211.* **Sat 16, Sun 17 Sept (12-4.30). Adm £5, chd free. Home-made teas. Visits also by arrangement.**

Two walled gardens and a 60 foot greenhouse from the early 1800s set in 6 acres with wonderful views of the entire length of the Malvern ridge. One walled garden is set out with formal paths and borders bounded by box hedging. We have planted fruit trees in espaliers and cordons as they would have been when the garden was first set out and have a nature area with walks and some woodland. Teas and homemade cakes on the lawn, or in the greenhouse if inclement weather. Self-guided tour leaflets available. Wheelchair access to the main walled garden with grass and paving paths. Parking near gates can be arranged in advance.

35 THE RIVER SCHOOL

Oakfield House, Droitwich Road, Worcester, WR3 7ST. Christian Education Trust Worcester, www.riverschool.co.uk. *2.4m N of Worcester City Centre on A38 towards Droitwich. At J6 M5, take A449 signed for Kidderminster, turn off at 1st turning marked for Blackpole. Turn R to Fernhill Heath & at T- junction with A38 turn L. The school is ½m on R.* **Sat 22 Apr, Thur 27 July (10.30-3.30). Adm £5, chd free. Light refreshments in the Lewis Room near garden entrance.**

A former Horticultural College garden being brought back to life. For 35 years after WW2 it was known as Oakfield Teacher Training College for Horticulture. It now features a Forest School, apiary, wildlife pond, and children's vegetable plots within a walled garden. It has many less common trees and shrubs with others coming to light as we develop the Estate.

36 ◆ RIVERSIDE GARDENS AT WEBBS

Wychbold, Droitwich, WR9 0DG. Webbs of Wychbold, 01527 860000, www.webbsdirect.co.uk. *2m N of Droitwich Spa. 1m N of M5 J5 on A38. Follow tourism signs from M5.* **For opening times and information, please phone or visit garden website.**

2½ acres. Themed gardens inc colour spectrum, tropical and dry garden, rose garden, vegetable garden, National Collection of Harvington hellebores, seaside garden, bamboozeleum and self-sufficient garden. The New Wave garden is a natural wildlife area and inc seasonal interest with grasses and perennials. *New in 2022* Extended pathways and a new woodland walk. There are willow wigwams and wooden tepees made for children to play in, a bird hide, the Hobbit House and beehives which produce honey for our own food hall. Our New Wave Garden and new Woodland Walk (2022) has had new paths installed and wheelchair accessible.

NPC

37 ROTHBURY

5 St Peter's Road, North Malvern, WR14 1QS. John Bryson, Philippa Lowe & David, www.facebook.com/RothburyNGS. *7m W of M5 J7 (Worcester). Turn off A449 Worcester to Ledbury Rd at B4503, signed Leigh Sinton. Almost immed take the middle rd (Hornyold Rd). St Peter's Rd is ¼m uphill, 2nd R.* **Sun 21 May (1.30-5); Mon 29 May (11.30-5); Sun 18 June, Sun 23 July (1.30-5). Adm £5, chd free. Home-made teas inc the famous carrot cake & gluten free options.**

Set on slopes of Malvern Hills, ⅓ acre plant-lovers' garden surrounding Arts and Crafts house (not open), created by owners since 1999. Herbaceous borders, rockery, pond, small orchard. Siberian irises in May, roses in June and magnificent Eucryphia glutinosa in July. A series of hand-excavated terraces accessed by sloping paths and steps. Views and seats. Partial wheelchair access. One very low step at entry, one standard step to main lawn and one to WC. Decking slope to top lawn. Dogs on leads.

38 ◆ SPETCHLEY PARK GARDENS

The Estate Office, Spetchley Park, Worcester, WR5 1RS. Mr Henry Berkeley, 01905 345106, enquiries@spetchleygardens.co.uk, www.spetchleyparkestate.co.uk. *2m E of Worcester. On A44, follow brown signs.* **For NGS: Sun 9 Apr, Sun 2 July (10.30-5). Adm £10, chd £5. For other opening times and information, please phone, email or visit garden website.**

Surrounded by glorious countryside lays one of Britain's best kept secrets. Spetchley is a garden for all tastes and ages, containing one of the biggest private collections of plant varieties outside the major botanical gardens and weaving a magical trail for younger visitors. Spetchley is not a formal paradise of neatly manicured lawns or beds but rather a wondrous display of plants, shrubs and trees woven into a garden of many rooms and vistas. Plant sales, gift shop and coffee shop serving homemade treats, and picnics during the open season. Gravel paths, and grassed areas.

39 ◆ STONE HOUSE COTTAGE GARDENS
Church Lane, Stone, DY10 4BG. Louisa Arbuthnott, 07817 921146, louisa@shcn.co.uk, www.shcn.co.uk. *2m SE of Kidderminster. Via A448 towards Bromsgrove, next to church, turn up drive.* **For opening times and information, please phone, email or visit garden website.**
A beautiful and romantic walled garden adorned with unusual brick follies. This acclaimed garden is exuberantly planted and holds one of the largest collections of rare plants in the country. It acts as a shop window for the adjoining nursery. Open Thursday to Sat mid April to late August 10-5. Partial wheelchair access.

40 THE TYNINGS
Church Lane, Stoulton, Worcester, WR7 4RE. John & Leslie Bryant, 01905 840189, johnlesbryant@btinternet.com. *5m S of Worcester; 3m N of Pershore. On the B4084 between M5 J7 & Pershore. The Tynings lies beyond the church at the extreme end of Church Ln. Ample parking.* **Visits by arrangement June to Sept. Adm £6, chd free. Light refreshments inc in adm.**
Acclaimed plantsman's ½ acre garden, generously planted with a large colection of rare trees and shrubs. Features inc specialist collection of lilies, many unusual climbers and rare ferns. The colour continues into late summer with cannas, dahlias, euonymus and tree colour. Surprises around every corner. Lovely views of adjacent Norman Church and surrounding countryside. Plants labelled and plant lists available.

GROUP OPENING

41 NEW THE WALLED GARDEN & NO. 53
Rose Terrace, Worcester, WR5 1BU. Julia & William Scott. *Tickets from The Walled Garden Close to the city centre, ½m from the Cathedral. Via Fort Royal Hill, off London Rd. Park on 1st section of Rose Terrace or surrounding streets & walk 20 yds down track to The Walled Garden* **Sat 15, Wed 19 Apr, Sat 3, Wed 7 June (1-5). Combined adm £5, chd free. Tea at The Walled Garden only.**

NEW 53 FORT ROYAL HILL
Professor Chris Robertson MBE.

THE WALLED GARDEN
William & Julia Scott.

The C19 Walled Kitchen Garden is formal in layout, with relaxed planting. A peaceful historic garden in the city with a focus on chemical free planting especially herbs and their uses. Seating areas around the garden in shade and in the sun. No 53 a very small part-walled garden divided into three 'rooms', including shade tolerant planting. A tranquil, 'secret' garden 5 mins walk from the City centre with containers and bee-friendly planting.

42 WALNUT COTTAGE
Lower End, Bricklehampton, Pershore, WR10 3HL. Mr Richard & Mrs Janet Williams. *2½ m S of Pershore on B4084, then R into Bricklehampton Lane to T-junction, then L. Cottage is on R.* **Sat 24, Sun 25 June (2-5). Adm £6, chd free. Wine.**
1½ acre garden with views of Bredon Hill, designed into rooms, many created with high formal hedging of beech, hornbeam, copper beech and yew. There is a small 'front garden' with circular gravel path, well-stocked original garden area with pond and arches to the side of the house. Magnolia garden with several species and magnificent tree garden with specimens from around the world. Over 200 roses, in colour themed beds, climbing over a pergola or up trees, greenhouse, raised koi carp pond and metal stairway leading to roof-based viewing platform surrounded by roses. The garden continues to evolve with plenty of seating and interesting artefacts.

43 WARNDON COURT
St Nicholas Lane, Worcester, WR4 0SL. Drs Rachel & David Pryke, 01905 611268, rachelgpryke@btinternet.com. *½m from J6 of M5, Worcester N. St Nicholas Ln is off Hastings Drive.* **Sun 19 Feb (12-3). Adm £4, chd free. Light refreshments in St Nicholas Church Barn. Sun 21 May, Sun 18 June (12-4). Adm £6, chd free. Home-made teas in St Nicholas Church Barn. For snowdrop opening we will offer hot chocolate & Welsh cakes. Visits also by arrangement May to Sept for groups of 10 to 30.**
Warndon Court is a 2 acre family garden surrounding a Grade II* listed farmhouse (not open) featuring a circular route with formal rose gardens, terraces, two ponds, pergolas, topiary (inc a scruffy dragon), pretty summerhouse (gallery), a potager and woodland walk along the dry moat and through the secret garden. It has bee-friendly wildlife areas and is home to great-crested newts and slow-worms. Grade I listed St Nicholas Church will also be open to visitors. There will be an exhibition of original oil paintings and display of vintage cars. The gardens around the house can be accessed across the lawn. The potager is accessible but the woodland walk is bumpy with slopes at each end.

44 WESTACRES
Wolverhampton Road, Prestwood, Stourbridge, DY7 5AN. Mrs Joyce Williams, 01384 877496, Koijoy62@yahoo.co.uk. *3m W of Stourbridge. A449 in between Wall Heath (2m) & Kidderminster (6m). Ample parking Prestwood Nurseries (next door).* **Sat 24, Sun 25 June, Sat 26, Sun 27, Mon 28 Aug (11-4). Adm £4, chd free. Light refreshments. Visits also by arrangement July & Aug for groups of 5 to 20.**
¾ acre plant collector's garden with unusual plants and many different varieties of acers, hostas, shrubs. Woodland walk, large fish pool. Covered tea area with home-made cakes. Come and see for yourselves, you won't be disappointed. Described by a visitor in the visitors book as 'A garden which we all wished we could have, at least once in our lifetime'. Garden is flat. Disabled parking.

Our annual donations to Parkinson's UK meant 7,000 patients are currently benefiting from support of a Parkinson's Nurse

45 WHARF HOUSE
Newnham Bridge, Tenbury Wells, WR15 8NY. Gareth Compton & Matthew Bartlett, 01584 781966, gco@no5.com, www.wharfhousegardener.blog. *Off A456 in hamlet of Broombank, between Mamble & Newnham Bridge. Follow signs. Do not rely on SatNav.* **Sat 1, Sun 2 July (10-5). Adm £5, chd free. Home-made teas. Visits also by arrangement May to Sept for groups of 10+.**
2 acre country garden, set around an C18 house and outbuildings (not open). Mixed herbaceous borders with colour theming: white garden, bright garden, spring garden, canal garden, long double borders, intimate courtyards, a scented border, stream with little bridge to an island, vegetable garden. The garden is on several levels, with some uneven paths and only partial wheelchair access.

46 WHITCOMBE HOUSE
Overbury, Tewkesbury, GL20 7NZ. Faith & Anthony Hallett, 01386 725206, faith.hallett1@gmail.com. *9m S of Evesham, 5m NE Tewkesbury. Leave A46 at Beckford to Overbury (2m). Or B4080 from Tewkesbury through Bredon/Kemerton (5m). Or small lane signed Overbury at r'about junction A46, A435 & B4077. Approx 5m from J9 M5.* **Sat 1 Apr (11-4). Light refreshments in the Village Hall (Apr). Open nearby Overbury Court. Sun 25 June, Sun 27 Aug (1.30-4.30). Home-made teas in the Village Hall (Apr). Adm £5, chd free. Visits also by arrangement 1 Apr to 10 Sept for groups of up to 30. Visits welcomed, morning, afternoon & evening. Refreshments available.**
Cotswold stone walled shrub and herbaceous garden flanked by mature Beech, Birch, Acer and Catalpa. Spring fed gravel brook bordered by primula, hosta, hydrangea, lavender and rose. Pastel colours merge with cool white and blue and later hot colours turn to mellow yellow. Nooks, arches, vine, fig, vegetable parterre over bridges with terrace and benches on island for peaceful reflection. Plant store open. C18 Listed Cotswold stone walled House (not open). Lovely village of Overbury with St Faith's Church dating from Norman times. Wheelchair access up gravel path through iron gate at south entrance. Also by double wooden gates at back of house by prior arrangement.

47 ◆ WHITE COTTAGE & NURSERY
Earls Common Road, Stock Green, Inkberrow, B96 6SZ. Mr & Mrs S M Bates, 01386 792414, smandjbates@outlook.com, whitecottage.garden. *2m W of Inkberrow, 2m E of Upton Snodsbury. A422 Worcester to Alcester, turn at sign for Stock Green by Red Hart Pub, 1½m to T- junction, turn L, 500 yds on the L.* **Sun 16, Sun 30 Apr, Sun 28, Mon 29 May, Sun 11 June (11-4.30). Adm £4.50, chd free. Home-made teas.**
2 acre garden with large herbaceous and shrub borders, island beds, stream and bog area. Spring meadow with 1000s of snakes head fritillaries. Formal area with lily pond and circular rose garden. Alpine rockery and new fern area. Large collection of interesting trees inc Nyssa sylvatica, Parrotia persica, and acer 'October Glory' for magnificent autumn colour and many others. Nursery and garden open most Thursdays (10.30-5), please check before visiting. For further information please phone, email or visit garden website and Facebook page.

48 ◆ WHITLENGE GARDENS
Whitlenge Lane, Hartlebury, DY10 4HD. Mr & Mrs K J Southall, 01299 250720, keith.southall@creativelandscapes.co.uk, www.whitlenge.co.uk. *5m S of Kidderminster, on A442. A449 Kidderminster to Worcester L at T-lights, A442 signed Droitwich, over island, ¼m, 1st R into Whitlenge Ln. Follow brown signs.* **For NGS: Sat 1, Sun 2 Apr, Sat 24, Sun 25 June, Sat 2, Sun 3 Sept (10-5). Adm £5, chd £2.50. Light refreshments in the adjacent tea rooms. Full menu from salads to hot meals, pies etc. Always a special of the day. For other opening times and information, please phone, email or visit garden website.**
3 acre show garden of professional garden designer inc large variety of trees, shrubs etc. Features inc a twisted brick pillar pergola, 2½ metre diameter solid oak moon gate set into reclaimed brickwork, and this year a new four turreted, moated castle folly with vertical wall planter set between two water falls, then walk through giant gunnera leaves to the Fairy Garden. There is a full size Standing Stone Circle, a 400 sq metre turf labyrinth and a children's play/pet corner. Extensive plant nursery and Tea room. Wheelchair access but mix of hard paths, gravel paths and lawn.

GROUP OPENING

49 WICK VILLAGE
School Lane, Wick, Pershore, WR10 3PD. *1m E of Pershore on B4084. Sign to Wick is almost opp Pershore Horticultural College. Tickets for gardens at car park next to St Marys. Please park in the official car park - opp the Manor House.* **Sun 11 June (1-5). Combined adm £5, chd free. Light refreshments. Soft drinks only.**

LAMBOURNE HOUSE
Mr & Mrs G Power.

THE OLD FORGE
Sean & Elaine Young.

THE OLD STABLES, YOCK LANE
Mr Tony & Mrs Amalia Knight.

Wick Village gardens offers the visitor a range of extraordinary garden visiting experiences. We can show you gardens with low maintenance, small and large gardens which are planted with an array of plants which are trendy modern day in vogue ornamentals the more unusual enthusiast and specialist plants. As well as ornamental gardens we can show you the keen vegetable growers gardens containing vegetables both everyday and the more unusual. We last opened in 2021 and since then gardens in the village have developed and new gardens are still emerging. Wick is an historic village on the edge of Pershore. Cold drinks will be available. Teas may not be but please check online near the time. Many of the gardens are suitable for wheelchair access.

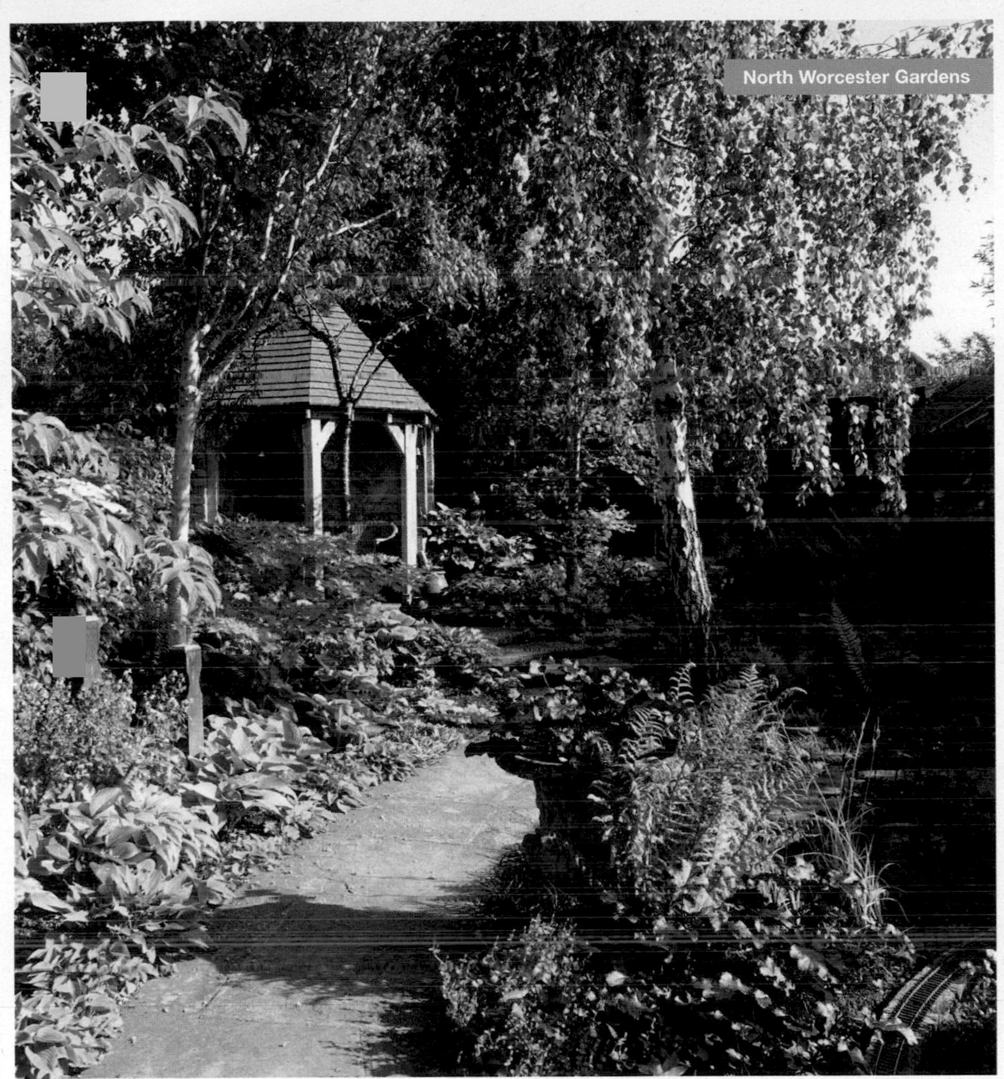

North Worcester Gardens

50 NEW **WINCHESTER HOUSE**
Abbey Manor Park, The Squires, EVESHAM, WR11 4HQ. Dave and Tricia Alesbury. *Near Twyford, edge of Evesham. Top of Greenhill (A4184) turn into The Squires (no through road), fields on R, 1st turn on L, 2nd house in on L. Please park on The Squires on orchard side only. Disabled parking on drive.* **Sat 18, Sun 19 Mar, Sat 22, Sun 23 Apr, Sat 20, Sun 21 May (1-5). Adm £5, chd free. Home-made teas.**
A one acre garden developed over 20 years, divided into two areas by an original Victorian wall, which formed part of Abbey Manor grounds. Formal planting surrounds the house with a landscaped wood beyond the wall. A wide variety of shrubs, perennials, native and ornamental trees are on display designed around two ponds, a stream and a striking tiered water feature. Patio area and path around house is paved. Brick ramp to woodland area, paved path to greenhouse and top patio, rest of woodland area is grass path.

Our 2022 donation to ABF The Soldier's Charity meant 372 soldiers and veterans were helped through horticultural therapy

YORKSHIRE

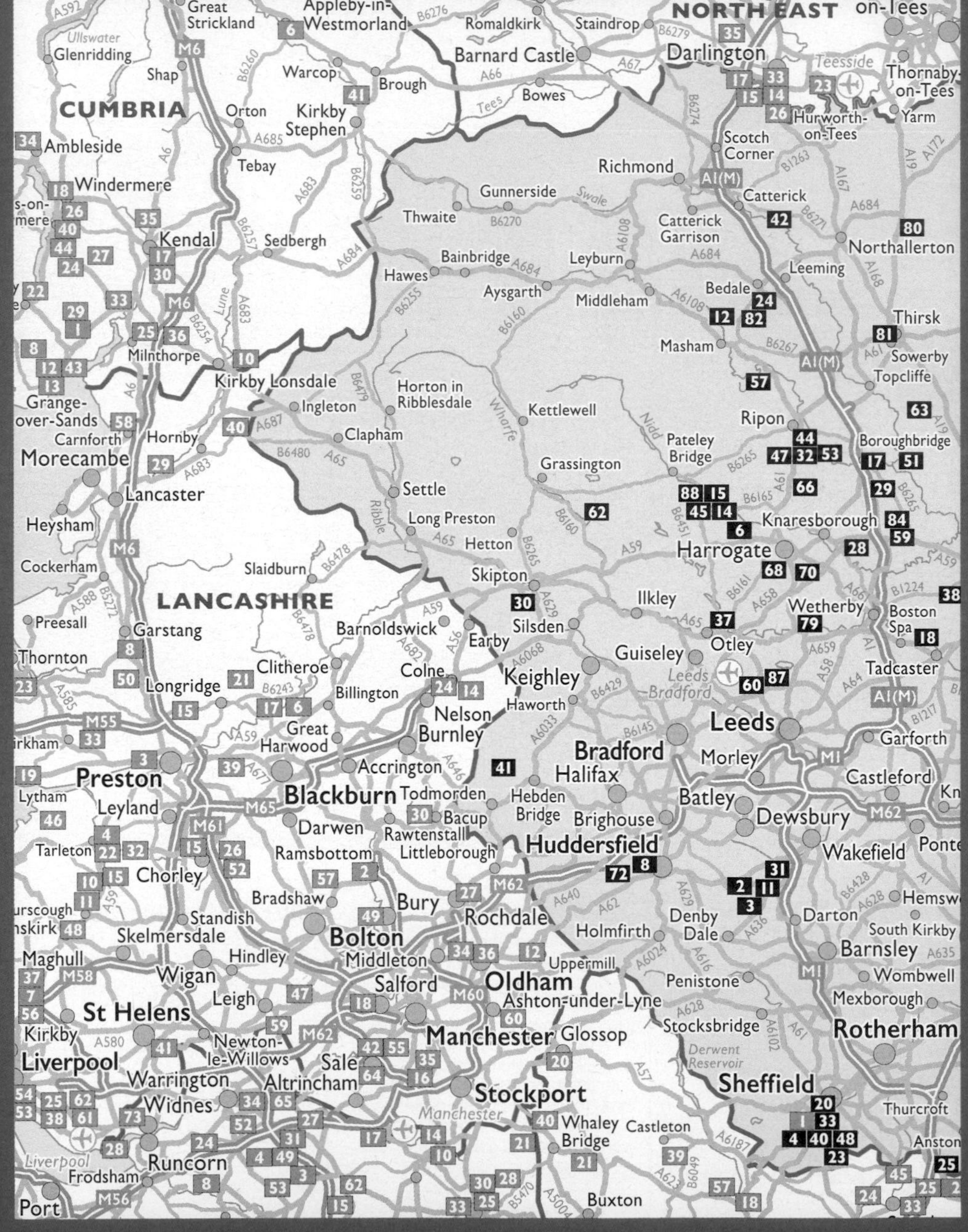

NORTH EAST
CUMBRIA
LANCASHIRE
Darlington
Richmond
Northallerton
Thirsk
Ripon
Harrogate
Knaresborough
Skipton
Settle
Ilkley
Otley
Wetherby
Keighley
Leeds
Bradford
Halifax
Huddersfield
Wakefield
Barnsley
Rotherham
Sheffield
Manchester
Stockport
Oldham
Bolton
Blackburn
Burnley
Preston
Lancaster
Morecambe
Kendal
Liverpool
Wigan
St Helens
Barnard Castle
Appleby-in-Westmorland
Kirkby Stephen
Sedbergh
Hawes
Leyburn
Masham
Pateley Bridge
Grassington
Kettlewell
Horton in Ribblesdale
Kirkby Lonsdale
Ingleton
Clapham
Tadcaster
Boroughbridge
Castleford
Dewsbury
Batley
Brighouse
Hebden Bridge
Todmorden
Holmfirth
Penistone
Glossop
Buxton
Castleton

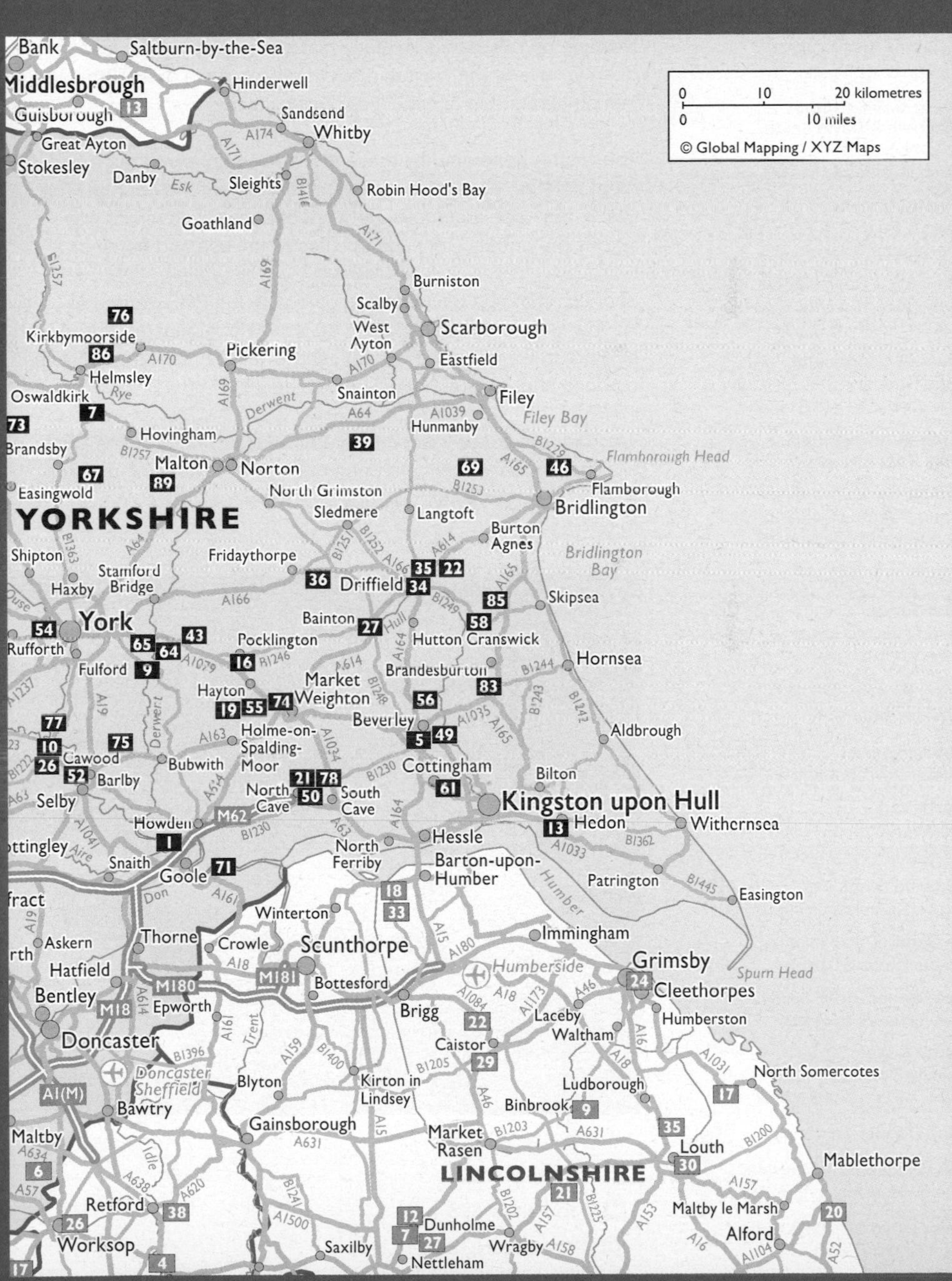

0 10 20 kilometres
0 10 miles
© Global Mapping / XYZ Maps
YORKSHIRE
LINCOLNSHIRE
Middlesbrough
Saltburn-by-the-Sea
Whitby
Robin Hood's Bay
Scarborough
Filey
Filey Bay
Flamborough Head
Flamborough
Bridlington
Bridlington Bay
Pickering
Malton
Norton
Helmsley
Kirkbymoorside
York
Driffield
Beverley
Hornsea
Kingston upon Hull
Hedon
Withernsea
Goole
Selby
Scunthorpe
Grimsby
Cleethorpes
Spurn Head
Doncaster
Gainsborough
Market Rasen
Louth
Mablethorpe
Alford
Retford
Worksop

Volunteers

County Organisers

East Yorks
Helen Marsden 07703 529112
helen.marsden@ngs.org.uk

North Yorks
David Lis 01439 788846
david.lis@ngs.org.uk

South & West Yorks
Elizabeth & David Smith
01484 644320
elizabethanddavid.smith@ngs.org.uk

County Treasurer
Angela Pugh 01423 330456
angela.pugh@ngs.org.uk

Publicity & Social Media
Sally Roberts 01423 871419
sally.roberts@ngs.org.uk

Booklet Coordinator
Jane Cooper 01484 604232
jane.cooper@ngs.org.uk

Booklet Advertising
Sally Roberts (as above)

Group Visits Co-ordinator
Mandy Gordon 01423 331643
mandy.gordon@ngs.org.uk

Assistant County Organisers

East Yorks
Ian & Linda McGowan 01482 896492
ianandlinda.mcgowan@ngs.org.uk

Hazel Rowe 01430 861439
hazel.rowe@ngs.org.uk

Natalie Verow 01759 368444
natalieverow@aol.com

North Yorks
Annabel Alton 07803 907042
annabel.alton@ngs.org.uk

Veronica Brook 01423 340875
veronica.brook@ngs.org.uk

Jo Gaunt 07443 505291
jo.gaunt@ngs.org.uk

Dee Venner 01765 690842
dee.venner@ngs.org.uk

South & West Yorks
Felicity Bowring 01729 823551
felicity.bowring@ngs.org.uk

Jane Cooper (as above)

Jane Hudson 01484 866697
jane.hudson@ngs.org.uk

Peter Lloyd 07958 928698
peter.lloyd@ngs.org.uk

Yorkshire, England's largest county, stretches from the Pennines in the west to the rugged coast and sandy beaches of the east: a rural landscape of moors, dales, vales and rolling wolds.

Nestling on riverbanks lie many historic market towns, and in the deep valleys of the west and south others retain their 19th century industrial heritage of coal, steel and textiles.

The wealth generated by these industries supported the many great estates, houses and gardens throughout the county. From Hull in the east, a complex network of canals weaves its way across the county, connecting cities to the sea and beyond.

The Victorian spa town of Harrogate with the RHS garden at Harlow Carr, or the historic city of York with a minster encircled by Roman walls, are both ideal centres from which to explore the gardens and cultural heritage of the county.

We look forward to welcoming you to our private gardens – you will find that many of them open not only on a specific day, but also by arrangement for groups and individuals - we can help you to get in touch.

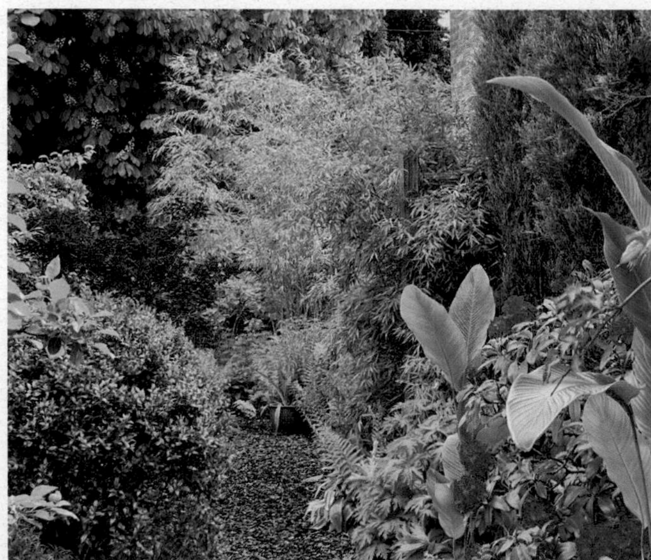

@YorkshireNGS
@YorkshireNGS
@YorkshireNGS

Above: The Poplars

OPENING DATES

All entries subject to change. For latest information check **www.ngs.org.uk**

Map locator numbers are shown to the right of each garden name.

February

Snowdrop Festival

Sunday 26th
Devonshire Mill 16

March

Sunday 26th
Fawley House 21
Holmfield 36

April

Sunday 2nd
Clifton Castle 12
Ellerker House 19
Goldsborough Hall 28

Monday 10th
Holmfield 36

Tuesday 11th
Highfield Cottage 34
NEW Highfield House 35

Wednesday 19th
The Old Vicarage 59

Saturday 22nd
249 Barnsley Road 2

Sunday 23rd
249 Barnsley Road 2
◆ RHS Garden Harlow Carr 68

May

Saturday 6th
The Cottage 13

Sunday 7th
The Cottage 13
Low Hall 45
Scape Lodge 72

Tuesday 9th
Markenfield Hall 47

Sunday 14th
◆ Jackson's Wold 39
NEW The Poplars 64
◆ Stillingfleet Lodge 77
Thirsk Hall 81
Whixley Gardens 84

Wednesday 17th
Primrose Bank Garden and Nursery 65

Sunday 21st
Grafton Gardens 29
Rudding Park 70

Friday 26th
Thimbleby Hall 80

Sunday 28th
Bank Wheel 1
Fernleigh 23
Galehouse Barn 26
NEW New House Farm 52
The Ridings 69

Monday 29th
Pilmoor Cottages 63

Tuesday 30th
Highfield Cottage 34
NEW Highfield House 35

June

Thursday 1st
Land Farm 41

Friday 2nd
◆ Shandy Hall Gardens 73
NEW Thorp Perrow 82

Saturday 3rd
NEW Thorp Perrow 82

Sunday 4th
Firby Hall 24
The Orchard 60
Shiptonthorpe Gardens 74
NEW Wytherstone Gardens 86
◆ The Yorkshire Arboretum 89

Monday 5th
Shiptonthorpe Gardens 74

Thursday 8th
Skipwith Hall 75

Saturday 10th
Linden Lodge 43
Old Sleningford Hall 57

Sunday 11th
NEW Bentley Grange Garden 3
3 Embankment Road 20
NEW Hutton Wandesley Walled Garden 38
Linden Lodge 43
Old Sleningford Hall 57

Tuesday 13th
Markenfield Hall 47

Sunday 18th
Birstwith Hall 6
Clifton Castle 12
Grafton Gardens 29
◆ Jackson's Wold 39
Prospect House 66

Tuesday 20th
NEW Swindon House Farm 79

Wednesday 21st
Whixley Gardens 84

Friday 23rd
◆ Shandy Hall Gardens 73

Sunday 25th
NEW Home Farm 37
NEW Langton Farm 42
The Old Vicarage 58
Prospect House 66
Sleightholmedale Lodge 76
NEW Woodhouse Farm 85
Yorke House & White Rose Cottage 88

Friday 30th
NEW The Minster Secret Garden 49

July

Saturday 1st
NEW The Minster Secret Garden 49

Sunday 2nd
NEW Beverley Gardens 5
Fern House, 5 Wold Road 22
Myton Grange 51

Tuesday 4th
Markenfield Hall 47

Wednesday 5th
Mires Beck Nursery 50

Saturday 8th
12 Brendon Drive 8
Cawood Gardens 10

Sunday 9th
Cawood Gardens 10
Dacre Banks Gardens 15
23 The Paddock 61
The Ridings 69

Saturday 15th
Stonefield Cottage 78

Sunday 16th
The Nursery 54
Stonefield Cottage 78
32 West Street 83

Monday 17th
The Nursery 54

Tuesday 18th
The Nursery 54
Saltmarshe Hall 71

Sunday 23rd
Goldsborough Hall 28

Wednesday 26th
The Grange 30

Saturday 29th
249 Barnsley Road 2

Sunday 30th
249 Barnsley Road 2

August

Friday 4th
Great Cliff Exotic Garden 31

Saturday 5th
Mansion Cottage 46

Sunday 6th
Fern House, 5 Wold Road 22
Greencroft 32
Mansion Cottage 46

Wednesday 9th
The Grange 30

Saturday 12th
Great Cliff Exotic Garden 31

Sunday 13th
Great Cliff Exotic Garden 31
Highfield Cottage 34

Thursday 17th
Duncanne House 17

Langton Farm

© Andrew Lawson

Sunday 27th
Fernleigh 23
138 Greystones Road 33

Monday 28th
Pilmoor Cottages 63

September

Sunday 3rd
138 Greystones Road 33

Wednesday 6th
◆ Parcevall Hall Gardens 62

Sunday 10th
Littlethorpe Manor 44
◆ Stillingfleet Lodge 77

Sunday 17th
NEW Hutton Wandesley Walled Garden 38

By Arrangement

Arrange a personalised garden visit with your club, or group of friends, on a date to suit you. See individual garden entries for full details.

249 Barnsley Road 2
90 Bents Road 4
Birstwith Hall 6
NEW Bramleys 7
Bridge House 9
Cawood Gardens 10
The Circles Garden 11
Cobble Cottage, Whixley Gardens 84
Cow Close Cottage 14
East Wing, Newton Kyme Hall 18
Fawley House 21
Fern House, 5 Wold Road 22
Fernleigh 23
Firvale Perennial Garden 25
Galehouse Barn 26
Glencoe House 27
The Grange 30
Greencroft 32
Holmfield 36
The Jungle Garden 40
NEW Langton Farm 42
Linden Lodge 43
Mansion Cottage 46
115 Millhouses Lane 48
NEW New House Farm 52
NEW 25 New Walk, Beverley Gardens 5
The Nursery 54
The Old Priory 55
The Old Rectory 56
The Old Vicarage 59
The Orchard 60
23 The Paddock 61
Pilmoor Cottages 63
Prospect House 66
Rewela Cottage 67
The Ridings 69
Skipwith Hall 75
Well House, Grafton Gardens 29
Whixley Gardens 84
Yorke House & White Rose Cottage 88

THE GARDENS

1 BANK WHEEL

Ferry Lane, Airmyn, Goole, DN14 8LS. Frank & Angela Meneight. *On the E edge of Airmyn village. Between Howden and Goole. From M62 Jct 36 or 37 follow signs for Airmyn. Ferry Ln is a narrow cul-de-sac close to Boothferry Bridge.* **Sun 28 May (11-4). Adm £4, chd free. Home-made teas.**

1 acre mature country garden surrounded by trees inc formal and wild areas with plenty of seating and interesting statuary. Mixed borders, greenhouse, summerhouse, decorative pond, small orchard with meadow. Woodland area with nature pond, stream, waterfall and bridges, winding path, snowdrops, aconites, bluebells in spring.

2 249 BARNSLEY ROAD

Flockton, Wakefield, WF4 4AL. Nigel & Anne Marie Booth, 01924 848967, nigel.booth1@btopenworld.com. *On A637 Barnsley Rd. M1 J38 or 39 follow signs for Huddersfield. Parking on Manor House Rd (WF4 4AL) & Hardcastle Ln. Please park carefully.* **Sat 22, Sun 23 Apr, Sat 29, Sun 30 July (1-5). Adm £5, chd free. Home-made teas. Visits also by arrangement Apr to Sept for groups of 15+.**

An elevated ⅓ acre, south- facing garden with panoramic views. The garden is packed with an abundance of spring colour, created from 1000s of bulbs, perennials, shrubs and trees. Make a return visit in the summer to view a transformation, with up to 60 hanging baskets and over 150 pots, creating the 'wow factor' summer garden. Massive plant sale with over 100 different varieties of plants on sale. There will be a ukulele group playing live in the garden at the summer opening. Partial wheelchair access. Disabled drop-off point at the bottom of the drive.

3 NEW BENTLEY GRANGE GARDEN

Ash Lane, Emley, Huddersfield, HD8 9QX. Rachael Moorhouse, bentleygrange.co.uk. *5 mins from M1 J38 & Yorkshire Sculpture Park. Off A636 Denby Dale-Wakefield rd at Bentley Grange Feeds.* **Sun 11 June (12-4). Adm £5, chd free. Light refreshments.**

Set within stunning scenery this ¼ acre country garden has sweeping lawns surrounded by colour-themed borders containing a variety of trees, shrubs and plants. 2 ponds, water features, pergolas and scented and herb gardens create interest and divide the space into separate 'rooms' whilst the secret garden has a different feel. Constructed on several levels it has a tranquil white planting scheme. Most of garden can be accessed via lawns, with some sloping level changes.

4 90 BENTS ROAD

Bents Green, Sheffield, S11 9RL. Mrs Hilary Hutson, 01142 258570, h.hutson@paradiseregained.net. *3m SW of Sheffield. From inner ring road in Sheffield nr Waitrose, follow A625. Bents Rd approx 3m on R.* **Visits by arrangement July & Aug for groups of up to 50. Adm £4, chd free. Light refreshments.**

Plantswoman's NE facing garden with many unusual and borderline-hardy species. Patio with alpine troughs for year-round interest and pots of colourful tropical plants in summer. Mixed borders surround a lawn which leads to mature trees underplanted with shade-loving plants at end of garden. Front garden peaks in late summer with hot-coloured blooms. Front garden and patio flat and accessible. Remainder of back garden accessed via 6 steps with handrail, so unsuitable for wheelchairs.

GROUP OPENING

5 NEW BEVERLEY GARDENS

Keldgate, Beverley, HU17 8JA. *0.5m from Minster. From S or W turn R at mini r'about down Keldgate, garden on R.* **Sun 2 July (10.30-4.30). Combined adm £8, chd free. Light refreshments at The Minster.**

NEW **131-133 KELDGATE**
Sue McCallum Gerard McElwee.

NEW **THE MINSTER SECRET GARDEN**
Beverley Minster.
(See separate entry)

NEW **25 NEW WALK**
Mr & Mrs C Green, 07768 684629, hilary.green@hotmail.com.
Visits also by arrangement 1 July to 16 Sept for groups of 6+.

3 new beautiful and unique gardens opening together for the 1st time. 131-133 Kedgate is an informal garden with a mix of young trees, shrubs and herbaceous perennials. Careful attention has been given to encourage wildlife. Minster Secret Gardens is a green woodland oasis in the lee of the ancient Minster and restored to a working garden producing flowers for the Minster. 25 New Walk is a Victorian town house garden full of a variety of interesting plants. Wheelchair access at The Minster Secret Garden.

6 BIRSTWITH HALL

High Birstwith, Harrogate, HG3 2JW. Sir James & Lady Aykroyd, 01423 770250, ladya@birstwithhall.co.uk. *5m NW of Harrogate. Between Hampsthwaite & Birstwith villages, close to A59 Harrogate/Skipton rd.* **Sun 18 June (2-5). Adm £5, chd free. Home-made teas. Visits also by arrangement 29 May to 31 July for groups of 10+.**

Charming and varied 4 acre garden nestling in secluded Yorkshire dale. Formal garden and ornamental orchard, extensive lawns leading to picturesque stream and large pond. Walled garden and Victorian greenhouse.

7 NEW BRAMLEYS

The Terrace, Oswaldkirk, York, YO62 5XZ. Bridget Hannigan, 07825 211154, bridgethannigan@hotmail.com,www.instagram.com/gardenblissyork. *20m N of York, 4m S of Helmsley. Single track rd on bend of B1363 in centre of Oswaldkirk. Parking opp on the Ampleforth rd.* **Visits by arrangement July to Sept for groups of up to 30. Min donation £50. Evenings or weekends. Adm £5, chd free. Tea, coffee & cake available.**

Plantswoman's hillside garden developed from an empty field over the last 6 years. Nestled in The Hag, with stunning views of the Coxwold-Gilling Gap and Howardian Hills beyond. Winding paths lead up through naturalistic planting and wildflower orchard. Seating areas and summerhouse.

8 12 BRENDON DRIVE

Birkby, Huddersfield, HD2 2DF. Jon Caddick. *Off Birkby Road.* **Sat 8 July (11-3.30). Adm £3, chd free. Pre-booking essential, please visit www.ngs.org.uk for information & booking. Light refreshments.**

The inspiration for our garden is to create a family space with an atmosphere of relaxation and sensory experiences. Small but practical, the garden is packed with striking blooms, colour and interest. Meandering paths, a pond and idyllic views are complemented by the sound of water and wildlife attracted to the garden. Architectural plants and wide borders create colour, drama and variety.

9 BRIDGE HOUSE

Main Street, Elvington, York, YO41 4AA. Mrs W C Bundy, 01904 608297, wendy@bundy.co.uk. *6m E from York ring road (A64). On B1128 from York, last house on R before bridge.* **Visits by arrangement 22 May to 1 Sept for groups of 5 to 40. Adm £5, chd free. Light refreshments.**

This 2 acre garden was carved out of the River Derwent's floodplain about 40 years ago. It survives the annual winter flooding of the river which can last up to 2 months and is up to 2-3 feet deep. Formal Rose garden, mixed borders, shrubbery, a large pond with a kitchen garden and orchard. Hostas and ferns thrive here. Various devices in the kitchen garden are used to keep above water level. Slopes to the main garden are relatively steep.

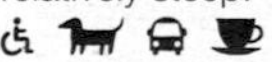

GROUP OPENING

10 CAWOOD GARDENS

Cawood, nr Selby, YO8 3UG. 01757 268571, davidjones051946@gmail.com. *10m S of York on B1222 5m N of Selby & 7m SE of Tadcaster. Village maps given at both gardens.* **Sat 8, Sun 9 July (12-5). Combined adm £6, chd free. Home-made teas. Visits also by arrangement in July for groups of 10 to 30.**

9 ANSON GROVE
Brenda Finnigan, 01757 268888, beeart@ansongrove.co.uk.

21 GREAT CLOSE
David & Judy Jones, 01757 268571, davidjones051946@gmail.com.

2 contrasting gardens in an attractive historic village. Walk across Castle Garth to the remains of Cawood Castle and C11 church with memorial garden on the bank of the River Ouse. 9 Anson Grove is a compact oriental-style garden with tranquil pools, bridge, pagoda and Zen area. Narrow paths and secluded seating areas give different views through bamboo and acers. 21 Great Close is a flower arranger's garden, designed and built by the owners over the past 50 years. Interesting trees and shrubs combine with herbaceous borders, a collection of dierama and a cutting garden. 2 ponds linked by a stream lead to a summerhouse, rose pergola and colourful terrace with views over the garden and surrounding countryside. Arts & crafts items on sale. Partial wheelchair access.

11 THE CIRCLES GARDEN

8 Stocksmoor Road, Midgley, nr Wakefield, WF4 4JQ. Joan Gaunt, mandy.gordon@ngs.org.uk. *Equidistant from Huddersfield, Wakefield & Barnsley, W of M1. Turn off A637 in Midgley at the Black Bull Pub (sharp bend) onto B6117 Stocksmoor Rd. Park on rd.* **Visits by arrangement Apr to Sept for groups of 5 to 20. Adm £4, chd free. Home-made teas.**

An organic and self-sustaining plantswoman's ½ acre garden on gently sloping site overlooking fields, woods and nature reserve opposite. Designed and maintained by owner. Herbaceous, bulb and shrub plantings linked by grass and gravel paths, woodland area with mature trees, meadows, fernery, greenhouse, fruit trees, viewing terrace with pots. About 100 hellebores propagated from owner's own plants. South African plants, hollies, and small bulbs of particular interest.

12 CLIFTON CASTLE

Ripon, HG4 4AB. Lord & Lady Downshire. *2m N of Masham. On rd to Newton-le-Willows & Richmond. Gates on L just before turn to Charlcot.* **Sun 2 Apr, Sun 18 June (2-5). Adm £5, chd free. Home-made teas.**

Impressive gardens and parkland with fine views over lower Wensleydale. Formal walks through the wooded 'pleasure grounds' feature bridges and follies, cascades and abundant wildflowers. The walled kitchen garden is similar to how it was set out in the C19. Recent wildflower meadows have been laid out with modern sculptures. Gravel paths and steep slopes to river.

13 THE COTTAGE

3 Fletcher Gate, Hedon, Hull, HU12 8ET. Mr Ian & Mrs Yvonne Mcfarlane. *6m E of Hull. Take A1033 to Hedon, L at r'about onto Hull Rd which joins Fletcher Gate. Garden on R after the 3rd zebra X-ing. No parking at the property, 2 free car parks 2 mins walk away.* **Sat 6, Sun 7 May (1-4). Adm £5, chd free. Light refreshments.**

Mature ⅓ acre suburban secret garden around Grade II listed cottage. Unusual planting in hidden areas accessed via steps, sloping and grassed paths. Small pond, summerhouse and lots of seating. Small stream originally constructed in early 1900s in a gravel garden. Sitting area in reclaimed palm house. Many farm and garden implements. Some small areas may not be accessible to wheelchairs.

14 COW CLOSE COTTAGE

Stripe Lane, Hartwith, Harrogate, HG3 3EY. William Moore & John Wilson, 01423 779813, cowclose1@btinternet.com. *8m NW of Harrogate. From A61(Harrogate-Ripon) at Ripley take B6165 to Pateley Bridge. 2m beyond Burnt Yates turn R, signed Hartwith/ Brimham Rocks onto Stripe Ln. Parking available.* **Visits by arrangement 15 June to 31 July for groups of 10 to 40. Adm £5, chd free. Tea.**

⅔ acre country garden on sloping site with stream and far reaching views. Large borders with drifts of interesting, well-chosen, later flowering summer perennials and some grasses contrasting with woodland shade and streamside plantings. Gravel path leading to vegetable area. Courtyard area, terrace and seating with views of the garden. Orchard and ha-ha with steps leading to wildflower meadow. The lower part of the garden can be accessed via the orchard.

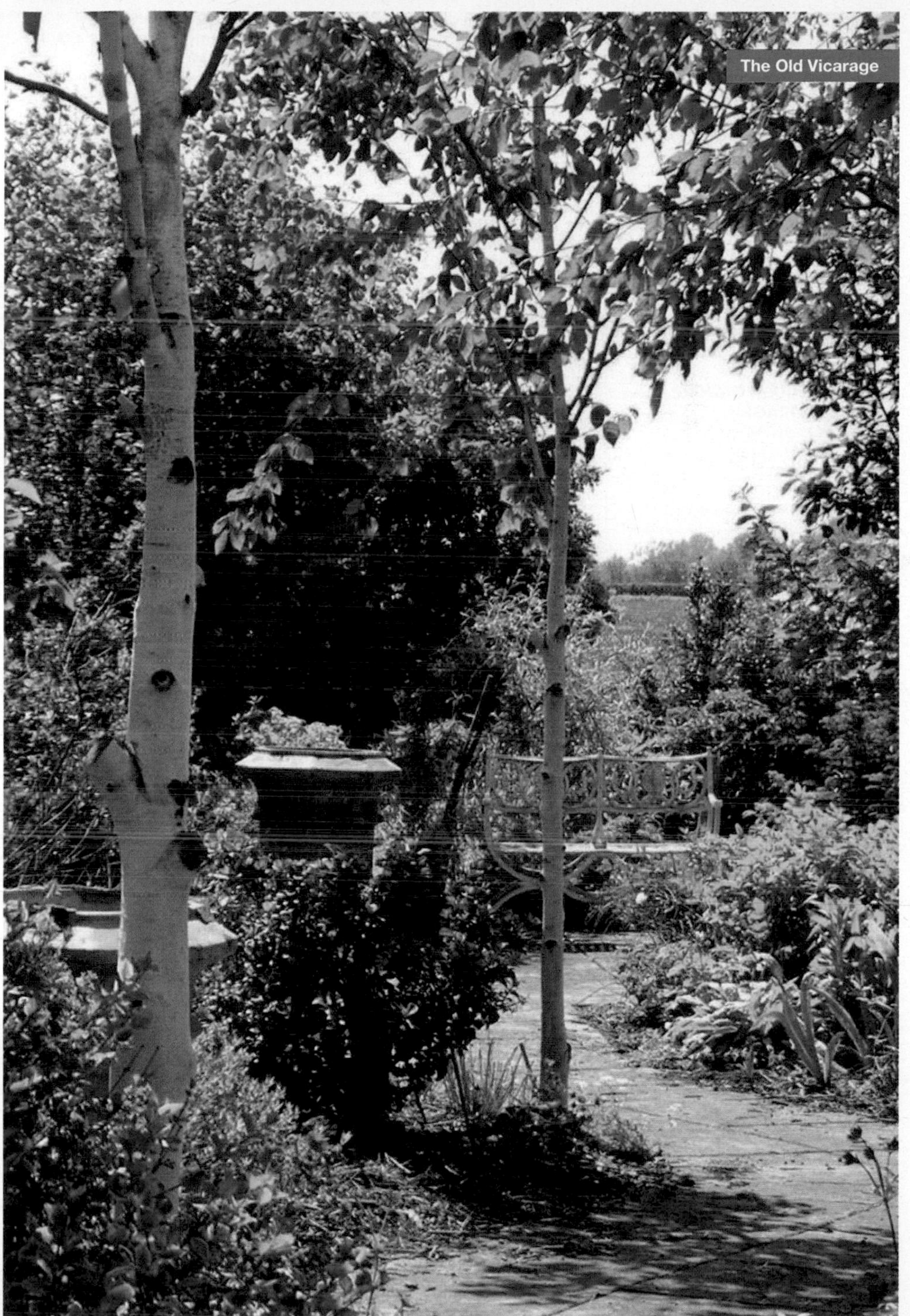

The Old Vicarage

GROUP OPENING

15 DACRE BANKS GARDENS

Nidderdale, HG3 4EW. www.yorkehouse.co.uk. *4m SE Pateley Bridge, 10m NW Harrogate,10m N Otley, on B6451. On-site parking at both gardens.* **Sun 9 July (12-5). Combined adm £8, chd free. Home-made teas at Yorke House and Low Hall. Visitors are welcome to picnic in the orchard at Yorke House.**

LOW HALL
Mrs P A Holliday.
(See separate entry)

YORKE HOUSE & WHITE ROSE COTTAGE
Tony, Pat, Mark & Amy Hutchinson.
(See separate entry)

Dacre Banks Gardens are located in the beautiful countryside of Nidderdale and are designed to take advantage of the scenic Dales landscape. The gardens are linked by an attractive level walk along the valley, but each may be accessed individually by car. Low Hall has a romantic walled garden set on different levels around the historic C17 family home (not open) with herbaceous borders, shrubs, climbing roses, productive vegetable area and a tranquil water garden. Yorke House has extensive colour-themed borders and water features with beautiful waterside plantings together with plentiful seating areas. The newly developed garden at White Rose Cottage is specifically designed for wheelchair users and features a colourful cottage garden, woodland plantings and large collection of hostas.

16 DEVONSHIRE MILL

Canal Lane, Pocklington, York, YO42 1NN. Sue & Chris Bond, 01759 302147, chris.bond.dm@btinternet.com, www.devonshiremill.co.uk. *1m S of Pocklington. Canal Ln, off A1079 on opp side of the rd from the canal towards Pocklington.* **Sun 26 Feb (11-4.30). Adm £5, chd free. Home-made teas.**

Drifts of double snowdrops, hellebores and ferns surround the historic Grade II listed watermill. Explore the 2 acre garden with mill stream, orchards, woodland, herbaceous borders, hen run and greenhouses. The old mill pond is now a vegetable garden with raised beds and polytunnel. Over the past 30 years the owners have developed the garden on organic principles to encourage wildlife.

17 DUNCANNE HOUSE

Roecliffe Lane, Boroughbridge, York, YO51 9LN. Colette & Tom Walker. *Off Roecliffe Ln leaving Boroughbridge, on L along small private drive, just before Boroughbridge Manor Care Home.* **Evening opening Thur 17 Aug (5.30-8). Adm £12, chd free. Pre-booking essential, please visit www.ngs.org.uk for information & booking. Wine and canapes.**

A secluded town garden, which borrows enclosure from surrounding gardens. From the front garden a birch glade leads to theatrical rear garden with sculptural grass bank, lawns, grass paths and 2 vistas terminated by treehouse and rustic tea pavilion. The product of a landscape architect husband and garden tour operator wife team, who have lovingly developed the garden over the last 10 yrs. Wheelchair accessible via grass paths.

18 EAST WING, NEWTON KYME HALL

Croft Lane, Newton Kyme, nr Boston Spa, LS24 9LR. Fiona & Chris Royffe, 07983 272182, plantsbydesign@btinternet.com, www.plantsbydesign.info. *2m from Tadcaster or Boston Spa. Follow directions for Newton Kyme village from A659.* **Visits by arrangement in July for groups of 10 to 25. Home-made teas.**

Contemporary designed garden in dramatic, historic setting with views of Kyme Castle, St Andrews Church and C18 Newton Kyme Hall (not open). Sculptural planting, herb and cutting garden, small meadow. The garden emphasises the broad scene and setting for the East Wing and planted areas create a rich variety of visual interest through the seasons as well as attracting wildlife. Garden design exhibition.

19 ELLERKER HOUSE

Everingham, York, YO42 4JA. Mr & Mrs M Wright, www.ellerkerhouse.weebly.com. *15m SE of York. 5½m from Pocklington. Just out of Everingham towards Harswell on R.* **Sun 2 Apr (10-4). Adm £6, chd free. Home-made teas. Sandwiches, sausage rolls etc served at lunch time.**

5 acre garden with many fine old trees, lawns surrounded by colour themed herbaceous borders planted with old roses, unusual shrubs and herbaceous plants for all year colour. Daffodils, spring bulbs and alpines planted around the lake in a stumpery. 11 acres of woodland, thatched oak hut and several sitting areas. Entry inc Rare Plant Fair with many plant stalls. See website for details. Most of the garden is accessible by wheelchair.

20 3 EMBANKMENT ROAD

Broomhill, Sheffield, S10 1EZ. Charlotte Cummins. *1½m W of city centre. A61 ring rd follow A57 to Manchester. In Broomhill turn R onto Crookes Rd, 1st R on to Crookesmoor Rd. Embankment Rd is 2nd on L.* **Sun 11 June (10.30-4.30). Adm £3.50, chd free. Light refreshments. Good selection of home-made cakes.**

Immaculate, compact, city garden with cottage style planting. Well-stocked borders of interesting, carefully selected herbaceous perennials. Front garden features clipped box, hostas and a lavender bed. Rear garden with steep steps to small, elevated lawn, herbaceous borders and many pots. Collection of over 50 astrantias and more than 100 varieties of hosta. Many homegrown plants for sale.

21 FAWLEY HOUSE

7 Nordham, North Cave, nr Brough, Hull, HU15 2LT. Mr & Mrs T Martin, 07951 745033, louisem200@hotmail.co.uk, www.nordhamcottages.co.uk. *15m W of Hull. Leave M62E at J38. L at '30' & signs: Wetlands & Polo. At L bend, turn R into Nordham. From Beverley, B1230 to N Cave. R after church & over bridge.* **Sun 26 Mar (12-5). Adm £5, chd free. Home-made teas in beamed cottage tea room with log burners, near garden entrance. Visits also by arrangement 13 Feb to 16 June for groups of 20+. Adm inc home-made teas.**

Tiered, 2½ acre formal garden with lawns, mature trees, hedging, gravel pathways. Lavender beds, mixed shrub/herbaceous borders and hot double herbaceous borders. Apple espaliers, pears, soft fruit, produce and herb gardens. Terrace with pergola and vines. Sunken garden with white border. Further woodland area with naturalistic planting and spring bulbs. Quaker well, stream and spring with 3 bridges, ferns and hellebores near mill stream. Snowdrops and aconites early in year. Treasure hunt for children. Self catering accessible accommodation at Nordham Cottages http://nordhamcottages.co.uk/. Wheelchairs welcome on terrace for garden views & teas. Pea gravel paths.

22 FERN HOUSE, 5 WOLD ROAD

Nafferton, Driffield, YO25 4LB. Peter & Jennifer Baker, 01377 255224. *N Nafferton. Follow directions rather than SatNav: From Driffield bypass to Nafferton on A614. Then 1st L & 1st L on Wold Rd. Park on st.* **Sun 2 July, Sun 6 Aug (10-4). Adm £4, chd free. Visits also by arrangement 1 June to 3 Sept for groups of up to 30.**
Designed and developed by owners since 2019, this small garden feels spacious with over 50 varieties of fern amongst many other plants in the borders. Mixed planting is also in quirky containers and up trellises and walls. A pond, a living wall and two greenhouses, one entirely full of ferns. Seats to view different aspects of a constantly evolving garden.

23 FERNLEIGH

9 Meadowhead Avenue, Meadowhead, Sheffield, S8 7RT. Mr & Mrs C Littlewood, 01142 747234, littlewoodchristine@gmail.com. *4m S of Sheffield city centre. From city centre. A61, A6102, B6054 r'about, exit B6054. 1st R Greenhill Ave, 2nd R. From M1 J33, A630 to A6102, then as above.* **Sun 28 May, Sun 27 Aug (11-5). Adm £3.50, chd free. Home-made teas. Visits also by arrangement 8 May to 16 Aug for groups of 10 to 40.**
Plantswoman's ⅓ acre cottage style suburban garden. Large variety of unusual plants set in different areas provide year-round interest. Several seats to view different aspects of garden. Auricula theatre, patio, gazebo and greenhouse. Miniature log cabin with living roof and cobbled area with unusual plants in pots. Sempervivum, alpine displays, collection of epimedium and wildlife hotel. Over 30 peonies end of May. Wide selection of homegrown plants for sale. Animal Search for children.

24 FIRBY HALL

Firby, Bedale, DL8 2PW. Mrs S Page. *½ m along Masham Rd out of Bedale, follow sign to Firby. Hall gates on L after ½ m.* **Sun 4 June (12-4.30). Adm £7.50, chd free. Home-made teas.**
The Hall sits in 4 acres with a walled garden to the north and 2 lakes to the south, 1 of which features a folly. The walled garden and greenhouse were restored in 2019. The main garden continues to undergo renovation: the ha-ha was restored during the 2020 lockdown as were the 110m long herbaceous beds. Ongoing work is focused on the main west facing lawns. Some steps but most of the garden is wheelchair accessible.

25 FIRVALE PERENNIAL GARDEN

Winney Hill, Harthill, nr Worksop, S26 7YN. Don & Dot Witton, 01909 771366, donshardyeuphorbias@btopenworld.com, www.donseuphorbias.webador.co.uk. *12m SE of Sheffield, 6m W of Worksop. M1 J31 A57 to Worksop. Turn R to Harthill. Allotments at S end of village, 26 Casson Drive at N end on Northlands Estate.* **Visits by arrangement Mar to June for groups of up to 50. Adm £3, chd free. Home-made teas at 26 Casson Drive, Harthill S26 7WA.**
Interesting and unusual large allotment with 13 island beds displaying 500+ herbaceous perennials inc the National Collection of hardy euphorbias with over 100 varieties flowering between March and October. Organic vegetable garden. Refreshments, WC and plant sales at 26 Casson Drive, a small garden with mixed borders, shade and seaside garden.

NPC

26 GALEHOUSE BARN

Bishopdyke Road, Cawood, Selby, YO8 3UB. Mr & Mrs P Lloyd and Madge & Paul Taylor, 07768 405642, junelloyd042@gmail.com. *On B1222 1m out of Cawood towards Sherburn in Elmet.* **Sun 28 May (12-5). Combined adm with New House Farm £6, chd free. Home-made teas. Visits also by arrangement 8 Apr to 30 Sept for groups of up to 20.**
The Barn: A plantaholic's informal cottage garden, created in 2015, to encourage birds and insects. Raised beds with tranquil seating area. The Farm: South facing, partly shaded varied herbaceous border. North facing exposed shaded border redeveloped 2017, ongoing for spring and autumn interest. Small experimental 50 shades of white garden. Raised beds for vegetables. Partial wheelchair access.

27 GLENCOE HOUSE

Main Street, Bainton, Driffield, YO25 9NE. Liz Dewsbury, 01377 217592, efdewsbury@gmail.com. *6m SW of Driffield on A614. 10m N of Beverley on B1248 Malton Rd. House on W of A614 in centre of the village. Parking around village or in layby 280m N towards Bainton r'about.* **Visits by arrangement June & July for groups of 5 to 30. Adm £5, chd free. Home-made teas.**
A tranquil 3 acre garden developed over 40+ years inc a cottage and kitchen garden, parkland and wildlife pond. The cottage garden is resplendent with trees, shrubs, roses, clematis and masses of herbaceous perennials. Paths lead to an area of mown grass planted with specimen trees and an orchard. A grove of native and unusual trees and a large wildlife pond. Wheelchair access to paved area in cottage garden but difficult elsewhere.

In 2022, our donations to Carers Trust meant that 456,089 unpaid carers were supported across the UK

28 GOLDSBOROUGH HALL
Church Street, Goldsborough, HG5 8NR. Mr & Mrs M Oglesby, 01423 867321, info@goldsboroughhall.com, www.goldsboroughhall.com. *2m SE of Knaresborough. 3m W of A1M. Off A59 (York-Harrogate). Spring: parking at Hall top car park. Summer: parking E of village in field off Midgley Ln. Disabled parking only at front of Hall.* **Sun 2 Apr (11-4); Sun 23 July (11-5). Adm £5, chd free. Light refreshments. Sandwiches, scones and cakes along with tea & coffee. Donation to St Mary's Church, Goldsborough.**
Historic 12 acre garden and formal landscaped grounds in parkland setting around Grade II*, C17 house, former residence of HRH Princess Mary, daughter of George V and Queen Mary. Gertrude Jekyll inspired 120ft double herbaceous borders, rose garden and woodland walk. Large restored kitchen garden and large glasshouse which produces fruit and vegetables for the Hall's commercial kitchens. Quarter-mile Lime Tree Walk planted by royalty in the 1920s, orchard and flower borders featuring 'Yorkshire Princess' rose, named after Princess Mary. Vegetable garden with fountain, rill and glasshouse. Gravel paths and some steep slopes.

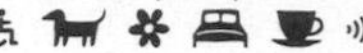

GROUP OPENING

29 GRAFTON GARDENS
Marton Cum Grafton, York, YO51 9QJ. Mrs Glen Garnett. *2½m S of Boroughbridge. Turn off the A168 or B6265 to Marton or Grafton, S of Boroughbridge.* **Sun 21 May, Sun 18 June (1-5). Combined adm £6, chd free.**

PADDOCK HOUSE
Tim & Jill Smith.

WELL HOUSE
Glen Garnett, 01423 322884. **Visits also by arrangement May to July.**

These 2 gardens in adjacent rural villages are connected by a public footpath. Paddock House is on an elevated site with extensive views down a large sloping lawn to a wildlife pond. A plant lover's garden where the house is encircled by a profusion of pots and extensive plant collections combining cottage gardening with Mediterranean and Tropical styles. A curved terrace of Yorkshire stone and steps using gravel and wood sleepers leads to many seating areas and a cutting garden with a small greenhouse. Well House in Grafton nestles under the hillside, with long views to the White Horse. This 1½ acre garden was begun 40 yrs ago and is constantly changing. A traditional English cottage garden with herbaceous borders, climbing roses and ornamental shrubs with a variety of interesting species. Paths meander through the borders to an orchard with chickens. Refreshments at The Punch Bowl pub, a 5 min walk from Well House.

Yorke House

30 THE GRANGE

Carla Beck Lane, Carleton in Craven, Skipton, BD23 3BU. Mr & Mrs R N Wooler, 07740 639135, margaret.wooler@hotmail.com. *1½m SW of Skipton. Turn off A56 (Skipton-Clitheroe) into Carleton. Keep L at Swan Pub, continue to end of village then R into Carla Beck Ln.* **Wed 26 July, Wed 9 Aug (12-4.30). Adm £5, chd free. Home-made teas. Visits also by arrangement July & Aug for groups of 20+. Donation to Sue Ryder Care Manorlands Hospice.**

Over 4 acres of wonderfully varied garden set in the grounds of Victorian house (not open) with mature trees and panoramic views towards The Gateway to the Dales. The garden has been restored and expanded by the owners over the past 3 decades. Bountiful herbaceous borders with many unusual species, rose walk, parterre, mini-meadows and water features. Topiary and large greenhouse. Extensive vegetable and cut flower beds. Oak seating placed throughout the garden invites quiet contemplation - a place to 'lift the spirits'. Gravel paths and steps in some areas.

31 GREAT CLIFF EXOTIC GARDEN

Cliff Drive, Crigglestone, Wakefield, WF4 3EN. Kristofer Swaine, www.greatcliffexoticgarden.co.uk. *1m from J39 M1. From M1 (J39) take A636 towards Denby Dale, past Cedar Court Hotel then L at British Oak pub onto Blacker Ln. Parking on Cliff Rd m'way bridge a 3 min walk from the garden.* **Evening opening Fri 4 Aug (5-8). Sat 12, Sun 13 Aug (10-5). Adm £4, chd free. Pre-booking essential, please visit www.ngs.org.uk for information & booking. Light refreshments. Vegan options available.**

An exotic garden on a long narrow plot. Possibly the largest collection of palm species planted out in Northern England inc a large Chilean wine palm. Colourful and exciting borders with zinnias, cannas, ensete, bananas, tree ferns, agaves, aloes, colcasias and bamboos. Jungle hut, winding paths and a pond that traverses the full width of the garden.

32 GREENCROFT

Pottery Lane, Littlethorpe, Ripon, HG4 3LS. David & Sally Walden, 01765 602487, s-walden@outlook.com. *1½m SE of Ripon town centre. Off A61 Ripon bypass, signs to Littlethorpe, R at church. From Bishop Monkton take Knaresborough Rd towards Ripon then R to Littlethorpe.* **Sun 6 Aug (12-4). Adm £5, chd free. Home-made teas. Visits also by arrangement July & Aug for groups of 20+.**

½ acre country garden with long herbaceous borders packed with colourful late summer perennials, annuals and exotics. Circular garden with views through to large wildlife pond and surrounding countryside. Ornamental features inc gazebo, temple pavilions, formal pool, stone wall with mullions, gate to pergola and a water cascade.

33 138 GREYSTONES ROAD

Sheffield, S11 7BR. Mr Nick Hetherington. *From Sheffield inner Ring Rd, W on A625 (Ecclesall Rd) for 1.8m; R onto Greystones Rd.* **Sun 27 Aug, Sun 3 Sept (10-4). Adm £4, chd free. Light refreshments.**

This typical, small suburban plot has been developed into a lovely garden over 20 yrs. It features many trees and acers, combined with stunning late tender perennials including banana plants, ricinous, cannas, agapanthus and dahlias. Several sculptures created by the owner are displayed amongst the plants. The garden has a tropical feel with splashes of colour, mainly orange and purple.

34 HIGHFIELD COTTAGE

North Street, Driffield, YO25 6AS. Debbie Simpson. *30m E of York. From A614/A166 into Driffield (York Rd, then North St.) Highfield Cottage is white detached house opp park nr Indian takeaway.* **Tue 11 Apr, Tue 30 May (10.30-4.30). Combined adm with Highfield House £9, chd free. Sun 13 Aug (10.30-3). Adm £5, chd free. Home-made teas at Highfield House on 11th Apr & 30th May.**

A ¾ acre suburban garden bordered by mature trees and a stream. Structure provided by numerous yew and box topiary featuring a pergola and sculptures. Lawns with island beds, a small orchard and a fern and hosta area. The garden has been changed since joining the National Garden Scheme to create more distinctive areas and is slowly moving to white planting to provide contrast to the topiary. Refreshments are weather dependent.

35 NEW HIGHFIELD HOUSE

Windmill Hill, Driffield, YO25 5YP. Andrew Lampard, 01377 256231, enquiries@thehighfieldhouse.com, www.thehighfieldhouse.com. *30m E of York. From A614/A166 into Driffield York Rd, follow North St which becomes Windmill Hill. Follow high red brick wall around bend to the entrance.* **Tue 11 Apr, Tue 30 May (10.30-4.30). Combined adm with Highfield Cottage £9, chd free. Light refreshments in the garden. Afternoon teas, lunches & evening meals in licensed restaurant by pre-booking at the hotel.**

Secluded Victorian mansion within traditional grounds. Lawns, herbaceous borders and mature woodland with fine specimens of beech, horse chestnut, yew and pine. Woodland paths through drifts of wild garlic, bluebells and fritillaries. Chalk spring fed stream with pools and weirs, rockery and fernery. New orchards, kitchen garden and glasshouse with cacti, succulents, tillandsias and wisteria. Garden games. Woodland path is uneven, inaccessible by wheelchair, otherwise all main areas of interest can be visited.

Our donation of £2.5 million over five years has helped fund the new Y Bwthyn NGS Macmillan Specialist Palliative Care Unit in Wales. 1,900 inpatients have been supported since its opening

36 HOLMFIELD

Fridaythorpe, YO25 9RZ. Susan & Robert Nichols, 01377 236627, snicholswire@gmail.com. *9m W of Driffield. From York A166 through Fridaythorpe. After 1m turn R by small copse of trees. 1st house on ln.* **Sun 26 Mar, Mon 10 Apr (12-5). Adm £5, chd free. Home-made teas. Visits also by arrangement Apr to June for groups of 10 to 50.**

Informal 2 acre country garden on gentle south facing slope, developed from a field since 1988. Family friendly garden with mixed borders, bespoke octagonal gazebo, 'Hobbit House', sunken trampoline, large lawn, tennis court and hidden paths for hide and seek. Productive fruit cage, vegetable and cut flower area. Collection of phlomis. Display of wire sculptures. Bee friendly planting. Some gravel areas, sloping lawns. Wheelchair access possible with help.

37 NEW HOME FARM

Farnley Park, Farnley, Otley, LS21 2QF. Gilly Cromack. *1 mile N of Otley on B6451 opp Farnley Hall. Parking in Farnley Business Park.* **Sun 25 June (10-4.30). Adm £4, chd free. Home-made teas.**

Formal lawns and mixed borders beside the house lead to romantic and imaginative secluded garden created from woodland by current owner. Rose-filled ancient trees and island beds planted informally inc digitalis and clematis. Pond and stream also encourage habitat rich in birds and wildlife. An enchanting garden created with love and grandchildren in mind.

38 NEW HUTTON WANDESLEY WALLED GARDEN

Hutton Wandesley, York, YO26 7LL. Mrs Sasha York. *For satnavs use YO26 7NA. Follow signs down Hutton St to the garden from the main rd. Bear R to garden.* **Sun 11 June, Sun 17 Sept (1-5). Adm £6, chd free. Home-made teas.**

Hutton Wandesley Walled Garden was relandscaped and planted in 2022. The Walled Garden has been designed and landscaped by the owner, Mrs Sasha York, and extends to 2½ acres. The layout is a quadrant with gardens within gardens, taking inspiration from the original 1872 design. The 4 main areas comprise a perennial meadow, Italian area, cutting garden and the 4th quadrant is laid to lawn. Wheelchair friendly garden.

39 ◆ JACKSON'S WOLD

Sherburn, Malton, YO17 8QJ. Mr & Mrs Richard Cundall, 07966 531995, jacksonswoldgarden@gmail.com, www.jacksonswoldgarden.com. *11m E of Malton, 10m SW of Scarborough. Signs from A64. A64 E to Scarborough. R at T-lights in Sherburn, take the Weaverthorpe rd, after 100m R fork to Helperthorpe & Luttons. 1m to top of hill, turn L at garden sign.* **For NGS: Sun 14 May, Sun 18 June (1-5). Adm £5, chd free. Home-made teas. For other opening times and information, please phone, email or visit garden website.**

Spectacular 2 acre country garden. Many old shrub roses underplanted with unusual perennials in walled garden and woodland paths lead to further shrub and perennial borders. Lime avenue with wildflower meadow. Traditional vegetable garden inc roses and flowers with a Victorian greenhouse. Adjoining nursery. Tours by appointment. Regional Finalist, The English Garden's The Nation's Favourite Gardens 2021.

40 THE JUNGLE GARDEN

124 Dobcroft Road, Millhouses, Sheffield, S7 2LU. Dr Simon & Julie Olpin, 07710 559189, simonolpin@blueyonder.co.uk. *3m SW of city centre. Dobcroft Rd runs between A625 & A621.* **Visits by arrangement 1 July to 15 Oct for groups of up to 20. Informative, guided tours. Adm £5 or £12 incl refreshments. Home-made teas. Donation to Sheffield Children's Hospital.**

Not a traditional garden, but a fascinating mature space of specialist interest using mainly hardy exotics creating a jungle effect. Long (250ft) narrow site, densely planted with mature trees and shrubs inc many trachycarpus, European fan palms, large bamboos, several tree ferns, mature eucalypts and a number of species of mature sheffleras. The planting has a South East Asian theme. A good selection of home-made cakes.

41 LAND FARM

Edge Lane, Colden, Hebden Bridge, HX7 7PJ. Mr J Williams. *8m W of Halifax. At Hebden Bridge (A646) after 2 sets of T-lights take turning circle to Heptonstall & Colden. After 2¾ m in Colden village turn R at Edge Lane 'no through rd'. In ¾ m turn L down lane.* **Thur 1 June (11-5). Adm £5, chd free. Home-made teas.**

An intriguing 6 acre upland garden within a sheltered valley, created by the present owner over the past 50 yrs. In that time the valley has also been planted with 20,000 trees which has encouraged a habitat rich in birds and wildlife. Moss garden and well established meconopsis and cardiocrinum lilies. Vistas around thought-provoking sculptures have been created throughout the garden. Partial wheelchair access, please telephone 01422 842260 for more information.

42 NEW LANGTON FARM

Great Langton, Northallerton, DL7 0TA. Richard & Annabel Fife, 07840561039, annabelfife1661@gmail.com. *5m W of Northallerton. B6271 in Great Langton between Northallerton & Scotch Corner.* **Sun 25 June (2-5). Adm £5, chd free. Home-made teas. Visits also by arrangement May to Aug.**

Garden designer's organic garden created since 2000. Romantic flower garden with mixed borders, roses, poppies, astrantias, delphiniums, white lilies and pebble pool. Formal and informal gravel areas, nuttery. There is also a lovely pear walk to River Swale.

43 LINDEN LODGE

Newbridge Lane, nr Wilberfoss, York, YO41 5RB. Robert Scott & Jarrod Marsden, 07900 003538, rdsjsm@gmail.com. *Equidistant between Wilberfoss, Bolton & Fangfoss. From York on A1079, ignore signs for Wilberfoss, take nxt turn to Bolton village. After 1m at Xrds, turn L onto Newbridge Ln.* **Sat 10, Sun 11 June (12-5). Adm £5, chd free. Light refreshments. Visits also by arrangement 1 May to 15 Sept for groups of 20+.**

A 1 acre garden with 5 acres of developing meadow, trees, pathways, hens and vegetable garden. Owner designed and constructed since

2000. Gravel paths edged with brick or lavender, many borders with unusual mixed herbaceous perennials, shrubs and feature trees. Wildlife pond, summerhouse, nursery, glasshouse and fruit cage. Orchard and woodland area. Formal garden with pond/water feature. Far reaching views towards the Yorkshire Wolds. Plant sales and card/gift wrap stall.

44 LITTLETHORPE MANOR

Littlethorpe Road, Littlethorpe, Ripon, HG4 3LG. Mrs J P Thackray, www.littlethorpemanor.com. *Outskirts of Ripon nr racecourse. Ripon bypass A61. Follow Littlethorpe Rd from Dallamires Ln r'about to stable block with clock tower. Map supplied on application.* **Sun 10 Sept (1.30-5). Adm £7, chd free. Home-made teas in marquee.**

11 acres. Walled garden with herbaceous planting, roses and gazebo. Sunken garden with ornamental plants and herbs. Brick pergola with wisteria, blue and yellow borders. Formal lawn with fountain pool, hornbeam towers, yew hedging. Box-headed hornbeam drive with Aqualens. Large pond with classical pavilion and boardwalk. New contemporary physic garden with rill, raised beds and medicinal plants. Wheelchair access - gravel paths, some steep steps.

45 LOW HALL

Dacre Banks, Nidderdale, HG3 4AA. Mrs P A Holliday. *10m NW of Harrogate. On B6451 between Dacre Banks & Darley.* **Sun 7 May (1-5). Adm £5, chd free. Home-made teas. Opening with Dacre Banks Gardens on Sun 9 July.**

Romantic walled garden set on differing levels designed to complement historic C17 family home (not open). Spring bulbs, rhododendrons and azaleas round tranquil water garden. Asymmetric rose pergola underplanted with auriculas and lithodora links the orchard to the garden. Vegetable garden and conservatory. Extensive herbaceous borders, shrubs and climbing roses give later interest. Bluebell woods, lovely countryside and farmland all around, overlooking the River Nidd. 80% of garden can be seen from a wheelchair but access involves 3 stone steps.

46 MANSION COTTAGE

8 Gillus Lane, Bempton, Bridlington, YO15 1HW. Polly & Chris Myers, 07749 776746, chrismyers0807@gmail.com. *2m NE of Bridlington. From Bridlington take B1255 to Flamborough. 1st L at T-lights - Bempton Ln, turn 1st R into Short Ln then L at end. L fork at church.* **Sat 5, Sun 6 Aug (10-4). Adm £4.50, chd free. Light refreshments. Visits also by arrangement 11 June to 31 Aug for groups of 10 to 30.**

20 years of NGS opening. A truly hidden, private and secret garden with exhuberant, packed, vibrant borders. Visitors' book says 'a veritable oasis', 'the garden is inspirational', with a surprise around every corner. Japanese influenced area, mini hosta walk, 100ft border, summerhouse, and art studio. Vegetable plot, cuttery, late summer hot border, bee and butterfly border, deck and lawns. Produce, plants and home made soaps for sale with delicious sweet and savoury small plates served from the conservatory.

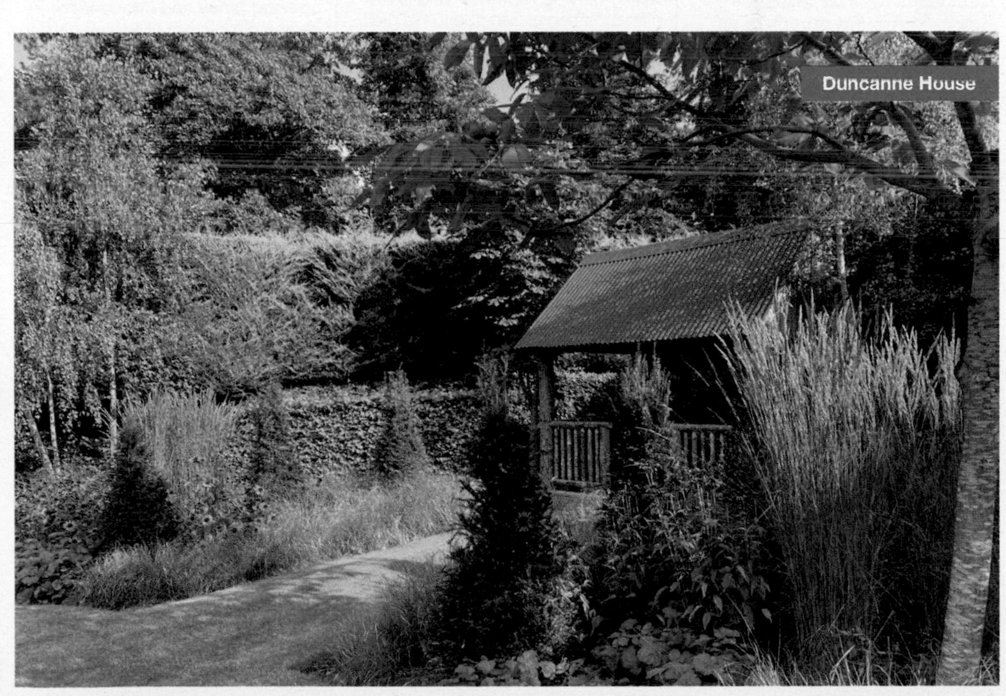
Duncanne House

47 MARKENFIELD HALL

Ripon, HG4 3AD. Lady Deirdre Curteis, www.markenfield.com. *3 miles S of Ripon. A61 between Ripon & Ripley. Turning between 2 low stone gateposts W of main rd. Do not follow SatNav.* **Tue 9 May, Tue 13 June, Tue 4 July (2-3.30). Adm £5, chd free. Pre-booking essential, please visit www.ngs.org.uk for information & booking. Home-made teas.**

The gardens surrounding this moated, medieval manor house are the work of the owner Lady Deirdre Curteis and gardener Giles Gilbey. Mature planting combines with recently designed areas where walls with espaliered apricots and figs frame a mix of hardy perennials. In 2018 the Farmhouse Border was replanted and now blends seamlessly with the main East Border. Inc woodland tour.

48 115 MILLHOUSES LANE

Sheffield, S7 2HD. Sue & Phil Stockdale, 01142 365571, phil.stockdale@gmail.com. *Approx 4m SW of Sheffield City Centre. Follow A625 Castleton/Dore Rd, 4th L after Prince of Wales pub, 2nd L. Alternatively take A621 Baslow Rd. After Tesco garage take 2nd R, then 1st L.* **Visits by arrangement 8 May to 31 Aug for groups of up to 20. Adm £4, chd free. Light refreshments.**

Plantswoman's ⅓ acre south facing level cottage style garden with many choice and unusual perennials and bulbs, providing year-round colour and interest. Large collection of 60+ hostas, roses, peonies, iris and clematis with many tender and exotic plants inc aeoniums, echeverias, bananas and echiums. Seating areas around the garden. Many home-propagated plants for sale. Most of the garden is accessible for wheelchairs.

49 NEW THE MINSTER SECRET GARDEN

20 Minster Moorgate, Beverley, HU17 8HP. Beverley Minster. *100metres W of Minster. Minster is on S side of the town, can't be missed. Minster Moorgate goes W from NW corner. Garden signposted through passage between houses.* **Fri 30 June, Sat 1 July (10.30-4). Adm £4, chd free. Light refreshments in Minster. Opening with Beverley Gardens on Sun 2 July.**

A green woodland oasis, ½ acre, in centre of Beverley cultivated by volunteers with an emphasis on encouraging wildlife, adjoining the Minster stonemasons' yard and in the lee of the ancient Minster. Fruit trees inc medlar and greengage. Vegetable plots surround an old piggery. Herbaceous beds, providing flowers for the Minster. Pond and Victorian brick-built bothy. Garden registered under the Quiet Garden Scheme nearby with wildflower sanctuary garden in churchyard. Some uneven flat paths. No steps.

50 MIRES BECK NURSERY

Low Mill Lane, North Cave, Brough, HU15 2NR. Graham Elliott, www.miresbeck.co.uk. *Between N & S Cave. Do not follow SatNav if entering N Cave from the A63/M62. Take rd to S Cave, turn R 400 yds after leaving the village. From S Cave, turn 1st L after prison.* **Wed 5 July (10-4). Adm £5, chd free. Light refreshments.**

A registered charity that provides horticultural work experience for adults with learning disabilities. The 14 acre site features herbaceous borders, rose garden, vegetable beds, wildflower woodland walk and Hull's official Garden of Sanctuary. We grow over 300 herbaceous perennials, 50 herbs and 100 Yorkshire wildflowers for garden centres and heritage sites in Yorkshire and Lincolnshire. Plants for sale. Tarmac main paths, and compressed gravel side paths.

51 MYTON GRANGE

Myton On Swale, York, YO61 2QU. Nick & Annie Ramsden. *15m N of York. From the N go through Helperby on York Rd. After ½m follow yellow signs towards Myton. From the S leave A19 through Tollerton & Flawith. Turn L at Xrds.* **Sun 2 July (1-5). Adm £7, chd free. Home-made teas.**

This garden, attached to a Victorian farmhouse, once formed part of the Myton Estate. Extending to ¾ acre, the site is adjacent to the River Swale and inc a paved terrace garden, formal parterre, circular garden with mixed shrub and herbaceous border and lawn with topiary borders. There will be a talk about the restored Victorian Stud Farm buildings, and history of the Myton Estate at 3pm.

52 NEW NEW HOUSE FARM

Lordship Lane, Wistow Lordship, Selby, YO8 3RR. Prof Chris Thomas and Dr Helen Billington, helen.billington@gmail.com. *1m SE of Wistow and 2m NE of Selby. From B1223 in Wistow take Lordship Ln out of village for 1m. From B1223 in Selby turn R at mini r'about on Monk Ln, onto Lordship Ln for 2m.* **Sun 28 May (12-5). Combined adm with Galehouse Barn £6, chd free. Refreshments at at Galehouse Barn. Visits also by arrangement June to Sept for groups of 5 to 20.**

2 hectares of wildlife garden, developed since 2000 by professional ecologists, incorporating wildflower meadows, mixed species shelterbelt, newt-filled ponds, courtyard garden, bee-friendly border, vegetable gardens and a banana-filled subtropical glasshouse. Partial wheelchair access.

53 ◆ NEWBY HALL & GARDENS

Ripon, HG4 5AE. Mr R C Compton, 01423 322583, info@newbyhall.com, www.newbyhall.com. *4m SE of Ripon. (HG4 5AJ for SatNav). Follow brown tourist signs from A1(M) or from Ripon town centre.* **For opening times and information, please phone, email or visit garden website.**

40 acres of extensive gardens and woodland laid out in 1920s. Full of rare and beautiful plants. Formal seasonal gardens and stunning double herbaceous borders slope down to River Ure. National Collection of cornus. Newly refurbished rock garden. Miniature railway and adventure gardens for children. Free parking. Licensed restaurant, shop and plant nursery. Wheelchair map available. Disabled parking. Manual and electric wheelchairs available on loan, please call to reserve.

NPC

54 THE NURSERY

15 Knapton Lane, Acomb, York, YO26 5PX. Tony Chalcraft & Jane Thurlow, 01904 781691, janeandtonyatthenursery@hotmail.co.uk. *2½m W of York. From A1237 take B1224 towards Acomb. At r'about turn L (Beckfield Ln.). After 150m turn L.* **Sun 16, Mon 17, Tue 18 July (1-6). Adm £4, chd free. Home-made teas. Visits also by arrangement May to Sept for**

groups of 10+. Apple or Tomato tastings may be possible for groups by arrangement.
A former suburban commercial nursery, now an attractive and productive 1 acre organic, private garden. Over 100 fruit trees, many in trained form. Many different vegetables grown both outside and under cover in 20m greenhouse. Productive areas interspersed with informal ornamental plantings provide colour and habitat for wildlife. The extensive planting of different forms and varieties of fruit trees make this an interesting garden for groups to visit by arrangement at blossom and fruiting times in addition to the summer open days.

55 THE OLD PRIORY

Everingham, YO42 4JD. Dr J D & Mrs H J Marsden, 01430 860222, helen.marsden@ngs.org.uk. *15m SE of York, 6m from Pocklington. 2m S of A1079. On E side of village.* **Visits by arrangement 26 May to 30 June. Home-made teas.**
2 acre rural garden reflecting surrounding countryside. Created in 1990s to enable self-sufficiency in vegetables, meat, most fruit, logs and timber for furniture. Walled vegetable garden with polytunnel. Borders and lawn near house are on dry sandy loam. Garden slopes down to bog. Roughly mown pathway through woodland, along lakeside and lightly grazed pasture.

56 THE OLD RECTORY

Arram Road, Leconfield, Beverley, HU17 7NP. David Baxendale, 01964 502037, davidbax@newbax.co.uk. *Garden entrance on L 80yds along Arram Rd next to double bend sign before the church.* **Visits by arrangement 4 Jan to 31 May for groups of up to 10. Adm £5.**
Approx 3 acres of garden and paddock. The garden is particularly attractive from early spring until mid summer. Notable for aconites, snowdrops, crocuses, daffodils and bluebells. Later hostas, irises, lilies and roses. There is a small wildlife pond with all the usual residents inc grass snakes. Well established trees and shrubs, with new trees planted when required. No refreshments available but visitors are welcome to bring a picnic.

57 OLD SLENINGFORD HALL

Mickley, nr Ripon, HG4 3JD. Jane & Tom Ramsden. *5m NW of Ripon. Off A6108. After N Stainley turn L, follow signs to Mickley. Gates on R after 1½m opp cottage.* **Sat 10, Sun 11 June (12-4). Adm £7.50, chd free. Home-made teas. Donation to other charities.**
A large English country garden and award winning permaculture forest garden. Early C19 house (not open) and garden with original layout. Wonderful mature trees, woodland walk and Victorian fernery, romantic lake with islands, watermill, walled kitchen garden, beautiful long herbaceous border, yew and huge beech hedges. Several plant and other stalls. Picnics around the mill pond very welcome. Of particular interest to anyone interested in permaculture. Reasonable wheelchair access to most parts of garden. Disabled WC at Old Sleningford Farm next to the garden.

58 THE OLD VICARAGE

North Frodingham, Driffield, YO25 8JT. Professor Ann Mortimer. *From Driffield take B1249 E for approx 6m, garden on R opp church. Entrance at T-junc of rd to Emmotland & B1249. From North Frodingham take B1249 W for ½m. Park in farmyard 50yds E opp side B1249.* **Sun 25 June (10.30-4.30). Adm £5, chd free. Home-made teas. Open nearby Woodhouse Farm.**
1½ acre plantsman's garden, owner developed over 25 years. Many themed areas inc rose garden, jungle, desert, fountain, scented, kitchen gardens, glasshouses. Numerous classical statues, unusual trees and shrubs, large and small ponds, orchard, nuttery. Children's interest with 'Jungle Book' dinosaur and wild animal statues. Neo-Jacobean revival house, built 1837, mentioned in Pevsner (not open). The land occupied by the house and garden was historically owned by the family of William Wilberforce.

59 THE OLD VICARAGE

Church Street, Whixley, YO26 8AR. Mr & Mrs Roger Marshall, biddymarshall@btinternet.com. *8m W of York, 8m E of Harrogate, 6m N of Wetherby. off A59 3m E of A1M Junc 47 (Postcode not for SatNav).* **Wed 19 Apr (11-4.30). Adm £5, chd free. Light refreshments. Opening with Whixley Gardens on Sun 14 May, Wed 21 June. Visits also by arrangement May to July.**
The walls, house and various structures within this village garden are festooned with climbers. Mixed borders, old roses, hardy and half-hardy perennials, topiary, bulbs and many hellebores give interest all year. Gravel and old brick paths lead to hidden seating areas creating the atmosphere of a romantic English garden.

60 THE ORCHARD

4A Blackwood Rise, Cookridge, Leeds, LS16 7BG. Carol & Michael Abbott, 01132 676764, michael.john.abbott@hotmail.co.uk. *5m N of Leeds centre, 5 mins from York Gate garden. Off A660 (Leeds-Otley) N of A6120 Ring Rd. Turn L up Otley Old Rd. At top of hill turn L at T-lights (Tinshill Ln). Please park in Tinshill Ln.* **Sun 4 June (11.30-4.30). Adm £4, chd free. Light refreshments. Pop up cafe and cover for inclement weather. Visits also by arrangement 14 May to 30 July for groups of 10+.**
⅓ acre plantswoman's hidden oasis. A wrap around garden of differing levels made by owners using stone found on site, planted for year-round interest. Extensive rockery, unusual fruit tree arbour, oriental style seating area and tea house, linked by grass paths, lawns and steps. Mixed perennials, hostas, ferns, shrubs, bulbs and pots amongst paved and pebbled areas.

The National Garden Scheme searches the length and breadth of England, Wales, Northern Ireland and the Channel Islands for the very best private gardens

61 23 THE PADDOCK

Cottingham, HU16 4RA. Jill & Keith Stubbs, 07932 713281, keith@cottconsult.karoo.co.uk. *Between Beverley & Hull. From Humber Br A164 towards Beverley. R onto Castle Rd, past hospital. From Beverley A164 towards Humber Br, L onto Harland Way 1st r'about. Over L/x-ings and Garden at S of village.* **Sun 9 July (10.30-4.30). Adm £5, chd free. Cream teas. All home baked cakes. Visits also by arrangement 15 May to 31 Aug for groups of 8 to 15.**

A secret garden behind a small mixed frontage. Archway to themed areas within the garden inc Japanese, Mediterranean, fairy and mixed herbaceous areas with patios and lawns. 2 ponds, a rill, ornamental and tree sculptures. Relandscaped areas and new sculptures. Plants for sale. Also selling young artist and blacksmith's garden ornaments and sculptures. Assisted access for disabled through front gate.

62 ◆ PARCEVALL HALL GARDENS

Skyreholme, Skipton, BD23 6DE. Walsingham College, 01756 720311, parcevallhall@btconnect.com, www.parcevallhallgardens.co.uk. *9m N of Skipton. Signs from B6160 Bolton Abbey-Burnsall rd or off B6265 Grassington-Pateley Bridge & at A59 Bolton Abbey r'about.* **For NGS: Wed 6 Sept (10-4). Adm £8, chd free. Tea. For other opening times and information, please phone, email or visit garden website.**

The only garden open daily in the Yorkshire Dales National Park. 24 acres on a sloping south facing hillside in Wharfedale sheltered by mixed woodland. Terrace garden, rose garden, rock garden and ponds. Mixed borders, spring bulbs, tender shrubs and autumn colour.

63 PILMOOR COTTAGES

Pilmoor, nr Helperby, YO61 2QQ. Wendy & Chris Jakeman, 01845 501848, cnjakeman@aol.com. *20m N of York. From A1M J48. From B'bridge follow rd towards Easingwold. From A19 follow signs to Hutton Sessay then Helperby. Garden next to mainline railway.* **Mon 29 May, Mon 28 Aug (11-5). Adm £5, chd free. Light refreshments. Visits also by arrangement 2 Apr to 30 Sept.**

A year-round garden for rail enthusiasts and garden visitors alike. A ride on the 7¼' gauge railway runs through 2 acres of gardens and gives you the opportunity to view the garden from a different perspective. The journey takes you across water, through a little woodland area, past flower filled borders, and through a tunnel behind the rockery and water cascade. 1½ acre wildflower meadow and pond. Clock-golf putting green.

♿ 🐕 ✿ ☕

64 NEW THE POPLARS

Main Street, Newton upon Derwent, York, YO41 4DA. Peter & Christina Young, vimeo.com/user112588780. *9m E of York. From A1079 at Wilberfoss, turn S towards Sutton upon Derwent. After 0.75m, turn R into Newton, then L. Garden on R past pub.* **Sun 14 May (1.30-5.30). Adm £5, chd free. Home-made teas.**

A plant-lover's paradise. Over 100 different trees and shrubs around a Victorian house and barns provide structure and shelter for a succession of flowers through the seasons. Glasshouses full of tender plants that spill out into the gardens in summer. Meadow walk leads to 2 acre arboretum with over 200 woody species around a wildlife pond. Ceramic sculptures.

65 PRIMROSE BANK GARDEN AND NURSERY

Dauby Lane, Kexby, York, YO41 5LH. Sue Goodwill & Terry Marran, www.primrosebank.co.uk. *4m E of York. At J of A64 & A1079 take rd signed to Hull. After 3m, just as entering Kexby, turn R onto Dauby Ln, signed for Elvington. From the E travel on A1079 towards York. Turn L in Kexby.* **Wed 17 May (11-4). Adm £5, chd free.**

2 acres of rare and unusual plants, shrubs and trees. Bulbs, hellebores and flowering shrubs in spring, followed by planting for year-round interest. Extensive eranthis collection. Courtyard garden, mixed borders, summerhouse and pond. Lawns, contemporary rock garden, shade and woodland garden with pond, stumpery, and shepherd's hut. Adjoining award-winning nursery. Poultry and Hebridean sheep. Dogs allowed on a short lead in car park and at designated tables outside the tearoom. CL Caravan Site (CAMC members only) adjoining the nursery. Most areas of the garden are level and are easily accessible for wheelchairs. Accessible WC available.

♿ ✿ 🚌 ☕ 🔊

66 PROSPECT HOUSE

Scarah Lane, Burton Leonard, Harrogate, HG3 3RS. Cathy Kitchingman, 07989 195773, cathyrk@icloud.com, www.abrightprospect.co.uk. *5 m S of Ripon 5½ m from A1, junc 48. Exit A61 signed Burton Leonard. Parking marshals on Station Ln. Drop off only outside Prospect House.* **Sun 18, Sun 25 June (1.30-5). Adm £5, chd free. Pre-booking essential, please visit www.ngs.org.uk for information & booking. Home-made teas. Visits also by arrangement 24 Apr to 29 July for groups of 15 to 40.**

1 acre walled, landscaped garden with ornamental pond, pergola, large oval lawned area, cutting and vegetable beds. Colour-themed herbaceous long border, 'hot' borders, physic bed, woodland area. Also mature hedging, trees and seasonal interest throughout. Additional new planting areas are being established. A renovated outhouse now converted into a potting area used for garden workshops.

♿ ✿ ☕ 🔊

67 REWELA COTTAGE

Skewsby, YO61 4SG. John Plant & Daphne Ellis, 01347 888125, rewelacottage@gmail.com, www.rewelahostas.com. *4m N of Sheriff Hutton, 15m N of York. After Sheriff Hutton, towards Terrington, turn L towards Whenby & Brandsby. Turn R just past Whenby to Skewsby. Turn L into village. 500yds on R.* **Visits by arrangement 15 May to 15 Sept for groups of 10+.Adm inc light refreshments. Adm £9, chd free.**

Situated in a lovely quiet country village, Rewela Cottage was designed from an empty paddock, to be as labour saving as possible using unusual trees and shrubs for year-round interest. Their foliage, bark and berries enhance the well-designed structure of the garden. The garden owner now specialises in growing and selling hostas. Over 300 varieties of hostas for sale. Some gravel paths may make assistance necessary.

Littlethorpe Manor

68 ◆ RHS GARDEN HARLOW CARR

Crag Lane, Harrogate, HG3 1QB. Royal Horticultural Society, 01423 565418, harlowcarr@rhs.org.uk, www.rhs.org.uk/harlowcarr. *1½ m W of Harrogate town centre. On B6162 (Harrogate - Otley).* **For NGS: Sun 23 Apr (9.30-5). Adm £12.95, chd £6.55. Standard adm charge subject to change. For other opening times and information, please phone, email or visit garden website.**

Harlow Carr offers a variety of growing landscapes, from running and still water, to woodland and wildflower meadows. Highlights inc the lavish main borders, bursting with generous prairie-style planting and the lush, moisture-loving plants around Streamside. Betty's Cafe Tearooms, gift shop, plant centre and children's play area inc treehouse and log ness monster. Wheelchairs and mobility scooters available, advanced booking recommended.

69 THE RIDINGS

South Street, Burton Fleming, Driffield, YO25 3PE. Roy & Ruth Allerston, 01262 470489. *11m NE of Driffield. 11m SW of Scarborough. From Driffield B1249, before Foxholes turn R to Burton Fleming. From Scarborough A165 turn R to Burton Fleming.* **Sun 28 May, Sun 9 July (12-4). Adm £4, chd free. Home-made teas. Visits also by arrangement Apr to July. Adm incl teas.**

Secluded cottage garden with colour-themed borders surrounding neat lawns. Grass and paved paths lead to formal and informal areas through rose and clematis covered pergolas and arbours. Box hedging defines well-stocked borders with roses, herbaceous plants and trees. Seating in sun and shade offer vistas and views. Greenhouse and summerhouse. Terrace with water feature. Indoor model railway. Terrace, tea area and main lawn accessible via ramp.

70 RUDDING PARK

Follifoot, Harrogate, HG3 1JH. Mr & Mrs Simon Mackaness, 01423 871350, www.ruddingpark.co.uk. *3m S of Harrogate off the southern bypass A658. Follow brown tourist signs. Use hotel entrance.* **Sun 21 May (1-5). Adm £5, chd free. Light refreshments.**

20 acres of attractive formal gardens, kitchen gardens and lawns around a Grade I Regency House extended and now used as an hotel. Humphry Repton parkland. Formal gardens designed by Jim Russell with extensive rhododendron and azalea planting. Contemporary designs by Matthew Wilson featuring grasses and perennials.

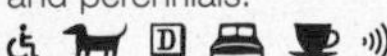

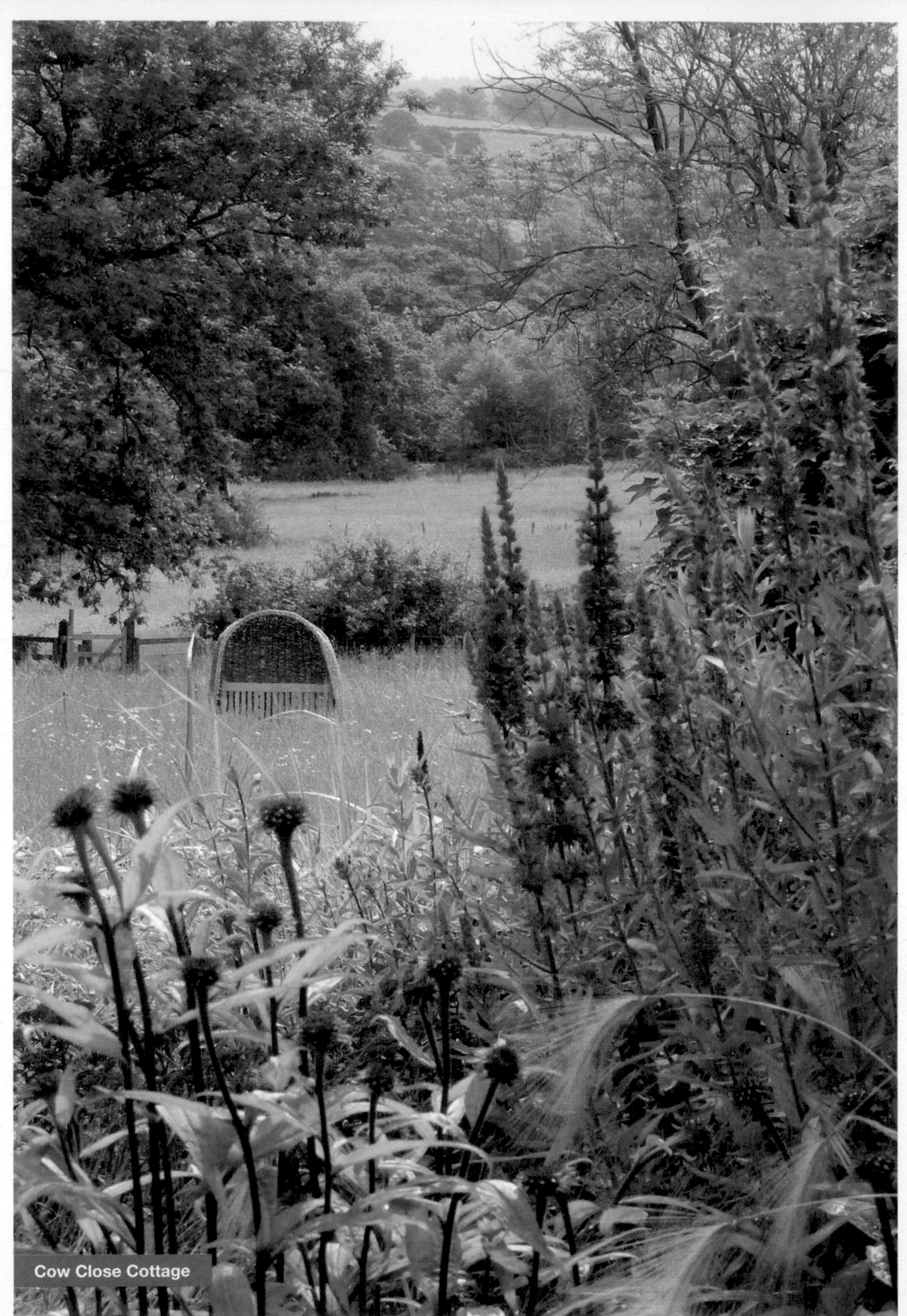

Cow Close Cottage

71 SALTMARSHE HALL

Saltmarshe, Howden, DN14 7RX. Mark Chittenden, 01430 434920, info@saltmarshehall.com, www.saltmarshehall.com. *6m E of Goole. From Howden (M62, J37) follow signs to Howdendyke & Saltmarshe.* **Tue 18 July (10-4). Adm £5, chd free. Light refreshments available outside. Afternoon teas in Hall and picnic hampers available via pre-booking.**

Saltmarshe Hall sits within a 17 acre estate. 5 acres of ornamental gardens, the remainder consisting of parkland and woodland. The gardens consist of herbaceous borders, an ornamental pond, walled garden, small orchard and lime avenue. Enjoy a pre-booked afternoon tea in the River Garden with its beautiful white floral palette. Wheelchair to the house via a ramp. Garden is accessed via gravel paths, narrow walkways and shallow steps.

72 SCAPE LODGE

11 Grand Stand, Scapegoat Hill, Golcar, Huddersfield, HD7 4NQ. Elizabeth & David Smith. *5m W of Huddersfield. From J23 or 24 M62, follow signs to Rochdale. From Outlane village, 1st L. At top of hill, 2nd L. Park at Scapegoat Hill Baptist Church (HD7 4NU) or in village. 5 mins walk to garden. 303/304 bus.* **Sun 7 May (1.30-4.30). Adm £5, chd free. Home-made teas. Donation to Mayor of Kirklees Charity Appeal.**

Stunning ⅓ acre contemporary country garden at 1000ft in the Pennines on a steeply sloping site with far-reaching views. Gravel paths lead between mixed borders on many levels. Colour-themed informal planting sits comfortably in the landscape and gives year-round interest. Steps lead to a terraced kitchen and cutting garden. Gazebo, pond, shade garden, large collection of pots and tender plants.

73 ◆ SHANDY HALL GARDENS

Thirsk Bank, Coxwold, York, YO61 4AD. The Laurence Sterne Trust, 01347 868465, info@laurencesternetrust.org.uk, www.laurencesternetrust.org.uk. *N of York. From A19, 7m from both Easingwold & Thirsk, turn E signed Coxwold. Park on rd.* **For NGS: Evening opening Fri 2, Fri 23 June (6.30-8). Adm £4, chd free. For other opening times and information, please phone, email or visit garden website.**

Home of C18 author Laurence Sterne. 2 walled gardens, 1 acre of unusual perennials interplanted with tulips and old roses in low walled beds. In old quarry another acre of trees, shrubs, bulbs, climbers and wildflowers encouraging wildlife, inc nearly 450 recorded species of moths. Moth trapping demonstration. Partial wheelchair access. Gravel car park, steps down to Wild Garden.

GROUP OPENING

74 SHIPTONTHORPE GARDENS

Shiptonthorpe, York, YO43 3PQ. *2m NW of Market Weighton. Roughly halfway between York and Hull on A1079, 2 m NW of Market Weighton. Parking on Shiptonthorpe playing fields car park off Station Rd.* **Sun 4, Mon 5 June (11-5). Combined adm £5, chd free. Light refreshments at village hall and there is a café in local garden centre.**

LANGDALE END
Mr & Mrs Thompson.

NEW **ORANMORE COTTAGE**
Susan & Paul Kraus.

NEW **SHIPTON COTTAGE**
Royston C Harvey.

WAYSIDE
Susan Sellars.

NEW **YORK HOUSE**
Tracey Baty.

5 contrasting gardens with different gardening styles. Langdale End is an eclectic maze-like garden with a mix of contemporary and cottage garden styles, featuring tropical plants, an Asian inspired garden, water features and a pond. Wayside has interesting planting in a variety of garden areas, with developing vegetable and fruit growing areas with greenhouse. 3 more gardens are opening for the 1st time. Oranmore cottage is a surprising 'hidden garden' divided into 2 distinct areas featuring decorative architectural plants and carefully chosen herbaceous perennials. Shipton cottage is a medium sized country cottage garden encompassing all shades and shapes of greenery be it as shrubs, herbaceous, ferns or trees to create an atmosphere of peacefulness. A step away from traditional flower beds. Finally, there is York House, a wildlife-friendly garden with a bit of everything inc prairie-style and cottage garden planting, an edible garden and the chatter of chickens. Wheelchairs possible in some gardens.

75 SKIPWITH HALL

Skipwith, Selby, YO8 5SQ. Sir Charles and Lady Forbes Adam, 07076 821903, rosalind@escrick.com, www.escrick.com/hall-gardens. *9m S of York, 6m N of Selby. From York A19 Selby, L in Escrick, 4m to Skipwith. From Selby A19 York, R onto A163 to Market Weighton, then L after 2m to Skipwith.* **Thur 8 June (1-4). Adm £6, chd free. Home-made teas. Visits also by arrangement 1 May to 7 July for groups of 10 to 30.**

4 acre walled garden of Queen Anne house (not open). Mixture of historic formal gardens (in part designed by Cecil Pinsent), wildflower walks and lawns. 'No dig' kitchen garden with herb maze, Italian garden, collection of old-fashioned shrub roses and climbers. Orchard with espaliered and fan-trained fruit. Arboretum with woodmeadow. Wheelchair access: gravel paths.

76 SLEIGHTHOLMEDALE LODGE

Fadmoor, YO62 7JG. Patrick & Natasha James. *6m NE of Helmsley. Parking limited in wet weather. Garden is 1st property in Sleightholmedale, 1m from Fadmoor.* **Sun 25 June (1-5.30). Adm £5, chd free. Home-made teas.**

A glorious S facing, 3 acre hillside garden with views over a peaceful valley in the North York Moors. Cultivated for over 100 years, wide herbaceous borders and descending terraces lead down the valley with beautiful, informal planting within the formal structure of walls and paths. The garden features roses, delphiniums and other classic English country garden perennials.

77 ◆ STILLINGFLEET LODGE
Stewart Lane, Stillingfleet, York, YO19 6HP. Mr & Mrs J Cook, 01904 728506, vanessa.cook@stillingfleetlodgenurseries.co.uk, www.stillingfleetlodgenurseries.co.uk. *6m S of York. From A19 York-Selby take B1222 towards Sherburn in Elmet. In village turn opp church.* **For NGS: Sun 14 May, Sun 10 Sept (1-5). Adm £6, chd £1. Home-made teas. For other opening times and information, please phone, email or visit garden website.**
Organic, wildlife garden subdivided into smaller gardens, each based on a colour theme with emphasis on use of foliage plants. Wildflower meadow, natural pond, 55yd double herbaceous borders and modern rill garden. Rare breeds of poultry wander freely in garden. Adjacent nursery. Garden courses run all summer (see website.) Art exhibitions in the cafe. Gravel paths and lawn. Ramp to cafe if needed. No disabled WC.

78 STONEFIELD COTTAGE
27 Nordham, North Cave, Brough, HU15 2LT. Nicola Lyte. *15m W of Hull. M62 E, J38 towards N Cave. Turn L towards N Cave Wetlands, then R at LH bend. Stonefield Cottage is on R, ¼m along Nordham.* **Sat 15, Sun 16 July (10-5). Adm £5, chd free. Light refreshments.**
A hidden and surprising 1 acre garden, with an emphasis throughout on strong, dramatic colours and sweeping vistas. Rose beds, mixed borders, vegetables, a riotous hot bed, boggy woodland and wildlife pond. Jacquemontii underplanted with red hydrangeas. Collections of hellebores, primulas, ferns, astilbes, hostas, heucheras, dahlias, hemerocallis and hydrangeas.

79 NEW SWINDON HOUSE FARM
Spring Lane, Kirkby Overblow, Harrogate, HG3 1HT. Penny Brook, pennybrook65@gmail.com. *1m S of Kirkby Overblow village. From Spring Ln/Swindon Ln junc, the drive is exactly ½m on R.* **Tue 20 June (11-4). Adm £5, chd free. Light refreshments. Home-made cakes.**
An English country garden with mixed perennial borders, pond garden, orchard, roses, mature mixed native hedges and trees. Places to sit and enjoy the rural views. The garden is surrounded by farmland, and a wildflower area was created in the paddock last year. Lots of pots, troughs and a potager with espaliers, raised vegetable and flower beds, thyme path and small camomile lawn. Parking for wheelchair users so that all areas are accessible.
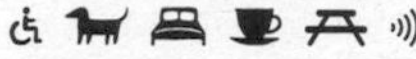

80 THIMBLEBY HALL
Thimbleby, Northallerton, DL6 3PY. Mr & Mrs A Shelley. *From A19, follow signs for Osmotherly then R in the village centre onto South End Rd towards Thimbleby. Proceed through the ford and entrance lodges are on the L.* **Fri 26 May (1-5). Adm £7, chd free. Home-made teas.**
Country estate on the western edge of the North York Moors. A number of formal garden areas inc terraced gardens, herbaceous garden and lake within an extensive parkland setting with sweeping views of the surrounding countryside. The gardens have been substantially extended by the owner with formal planting and a large banked area to the west of the house with spring flowering shrubs. Mature trees feature on the estate and especially impressive is an avenue of redwoods framing views across the lake to the south of the Hall.

81 THIRSK HALL
Kirkgate, Thirsk, YO7 1PL. Willoughby & Daisy Gerrish, www.thirskhall.com. *In the centre of Thirsk. On Kirkgate, next to St Mary's Church.* **Sun 14 May (12-5). Adm £6, chd free. Home-made teas.**
Thirsk Hall is a Grade II* listed townhouse completed by John Carr in 1777 (not open). Behind the house the unexpected 20 acre grounds inc lawns, herbaceous borders, kitchen gardens, walled paddocks and parkland. The present layout and planting are the result of a sensitive restoration, blending formal and informal beds, shrubbery and mature trees. Wheelchair access. Sculpture Park.

82 NEW THORP PERROW
Bedale, DL8 2PR. Sir Henry Ropner. *For satnav use DL8 2PS. From Bedale follow signs towards Masham (B6268), ignore signs for Thorp Perrow along Firby Rd for approx 1 m. L after signs for Snape. Arboretum approx ¼m on L.* **Fri 2, Sat 3 June (10-4). Adm £5, chd free. Light refreshments in the tearoom.**
The 4 acres of Thorp Perrow Gardens are adjacent to the 100 acre Arboretum, laid out in the 1800s. The gardens surround the hall with formal lawns, herbaceous borders and the remnants of the Italian renaissance style topiary all overlooked by the tradition summerhouse. Historical woodland with a natural spring in the centre. The gardens and greenhouses are in the stages of restoration. A new vegetable patch, peach house and a young orchard have been added to grounds. Some steps but most of the garden is wheelchair accessible.

83 32 WEST STREET
Leven, HU17 5LF. Mrs Janet Phillips. *8.5 miles NE of Beverley. Turn off Main St by the Hare & Hounds pub.* **Sun 16 July (10-4). Adm £4, chd free. Light refreshments.**
A quirky garden with themed areas inc tropical jungle planting, fairies and dinosaurs, an oriental style nook and rockery. This family garden was begun 10 yrs ago and has gradually evolved into a wonderfully chaotic mix of themes with some unusual plants and features inc a glass bottle wall. Street parking.

9,500 patients and their families are being supported across four Horatio's gardens thanks to our annual donations

GROUP OPENING

84 WHIXLEY GARDENS
York, YO26 8AR. 01423 330474, biddymarshall@btinternet.com. *8m W of York, 8m E of Harrogate, 6m N of Wetherby. 3m E of A1(M) off A59 York-Harrogate. Signed Whixley.* **Sun 14 May, Wed 21 June (11-4.30). Combined adm £7, chd free. Light refreshments and home-made teas at The Old Vicarage. Visits also by arrangement May to July for groups of 20+.**

COBBLE COTTAGE
John Hawkridge & Barry Atkinson, 01423 331419, cobblecottage@outlook.com. **Visits also by arrangement May to July for groups of 20+.**

THE OLD VICARAGE
Mr & Mrs Roger Marshall.
(See separate entry)

Attractive rural yet accessible village nestling on the edge of the York Plain with beautiful historic church and Queen Anne Hall (not open). The gardens are at opposite ends of the village with good footpaths. A plantsman's and flower arranger's garden at Cobble Cottage, has views to the Hambleton Hills. Close to the church, The Old Vicarage, with a ¾ acre walled flower garden, overlooks the old deer park. The walls, house and various structures are festooned with climbers. Gravel and old brick paths lead to hidden seating areas creating the atmosphere of a romantic English garden. Wheelchair access only to The Old Vicarage.

85 NEW WOODHOUSE FARM
Foston-On-The-Wolds, Driffield, YO25 8BH. Dave and Diana Blanchard. *7 m SE of Driffield. On road between Foston & Beeford. 1m from each. Down long tarmac drive.* **Sun 25 June (11-4.30). Adm £5, chd free. Light refreshments. Open nearby The Old Vicarage.**
1 acre partly walled garden surrounding Grade II listed farmhouse (not open). Created from dereliction 15 years ago. Mainly south facing with extensive views over surrounding countryside. Mixed borders, garden rooms, Mediterranean themed sun patio, courtyard garden, polytunnel. Extensive range of plants, with many unusual species. Mainly paved but some gravel paths.

86 NEW WYTHERSTONE GARDENS
Pockley, York, YO62 7TE. Damien & Martha Byrne Hill. *2m NE of Helmsley. Signed from A170. Turn L into Wytherstone House, before the church in Pockley. Severely restricted parking. Disabled parking on drive & all other parking in field opp.* **Sun 4 June (11-4.30). Adm £5, chd free. Light refreshments in village hall, a short walk from garden & parking.**
A plantsman's garden set in 8 acres of rolling countryside on edge of North York Moors. Divided by beech hedges, creating interlinked areas incl spring ericaceous, terraced, fern, peonia, foliage and bamboo gardens and small arboretum. Plants not thought hardy in north England grow happily in the free draining soil. Plants for sale. Some steep slopes, but most of the garden is accessible for wheelchairs.

87 ◆ YORK GATE
Back Church Lane, Adel, Leeds, LS16 8DW. Perennial, 01132 678240, yorkgate@perennial.org.uk, www.yorkgate.org.uk. *5m N of Leeds. A660 to Adel, turn R at lights before 'Dastaan' restaurant. L on to Church Ln. After church, R onto Back Church Ln. Garden on L.*
For opening times and information, please phone, email or visit garden website.
An internationally acclaimed 1 acre jewel of a garden with 14 individual garden 'rooms'. Many unusual plants and architectural topiary. Recent extension to the garden with new facilities inc a new cafe and gift shop as well as exciting new gardens, plant nursery and heritage exhibition. Owned by Perennial, the charity that looks after horticulturists and their families in times of need. Narrow gravel & cobbled paths make most of the original garden inaccessible to wheelchairs. The new gardens, tea room, shop and WC are accessible.

88 YORKE HOUSE & WHITE ROSE COTTAGE
Dacre Banks, Nidderdale, HG3 4EW. Tony, Pat, Mark & Amy Hutchinson, 01423 780456, pat@yorkehouse.co.uk, www.yorkehouse.co.uk. *4m SE of Pateley Bridge, 10m NW of Harrogate, 10m N of Otley. On B6451 near centre of Dacre Banks. Car park on-site.* **Sun 25 June (11-4.30). Adm £5, chd free. Cream teas. Visitors welcome to use picnic area in orchard. Opening with Dacre Banks Gardens on Sun 9 July. Visits also by arrangement 12 June to 30 July for groups of 10+.**
Award-winning English country garden in the heart of Nidderdale. A series of distinct areas flowing through 2 acres of ornamental garden. Colour-themed borders, natural pond and stream with delightful waterside plantings. Secluded seating areas and attractive views. Adjacent cottage has a recently developed garden designed for wheelchair access. Large collection of hostas. Orchard picnic area. All main features and car parking accessible to wheelchair users.

89 ◆ THE YORKSHIRE ARBORETUM
Castle Howard, York, YO60 7BY. The Castle Howard Arboretum Trust, 01653 648598, marketing@yorkshirearboretum.org, www.yorkshirearboretum.org. *15m NE of York. Off A64. Follow signs to Castle Howard then Yorkshire Arboretum signs at the obelisk r'about.* **For NGS: Sun 4 June (10-4). Adm £10, chd free. Home-made teas in the arboretum cafe. For other opening times and information, please phone, email or visit garden website. Donation to Plant Heritage.**
A glorious 120 acre garden of trees from around the world set in a stunning landscape of parkland, lakes, woodland and meadows. Walks, lakeside trails, guided tours, family activities, courses and events. We welcome visitors of all ages to enjoy the space, serenity and beauty of this sheltered valley. Internationally renowned collection of trees in a beautiful setting, accompanied by a diversity of wildflowers, birds, insects and other wildlife.

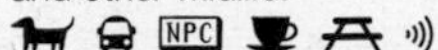

CHANNEL ISLANDS

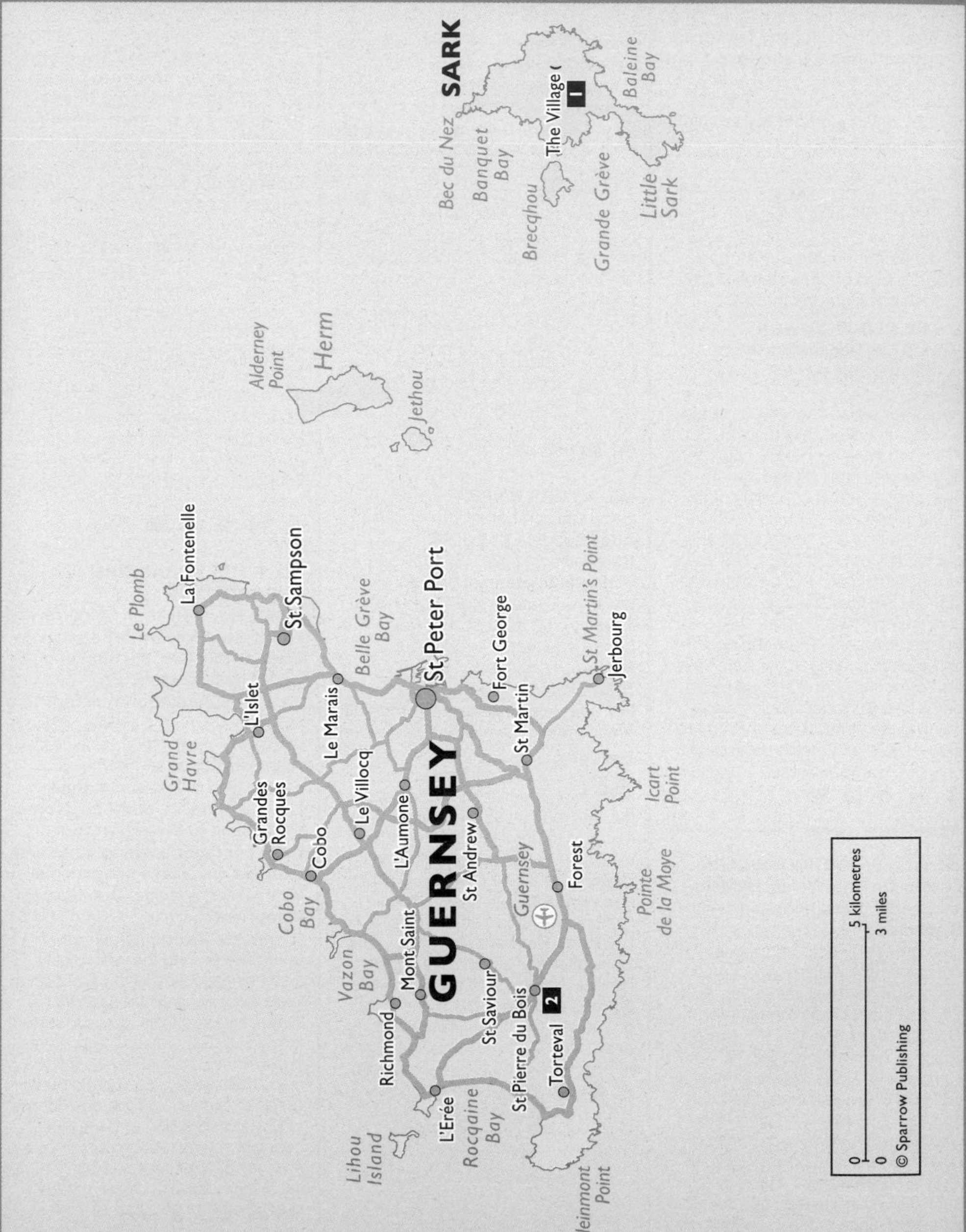

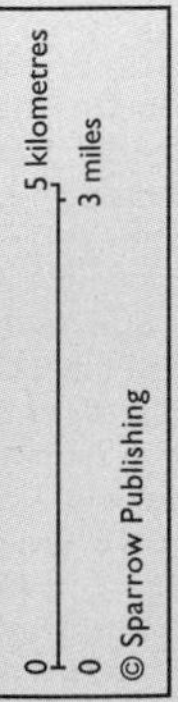

North of Saint-Malo and west of the Cotentin Peninsula, the spectacular beaches and rugged coastline of the Channel Islands have a definite subtropical feel.

The mild winters and endless summer sunshine see a stream of visitors and islanders enjoy the peace and tranquillity offered in Guernsey, along with many scenic cliff top walks, prehistoric remains, woodland valleys, idyllic parish gardens, secluded town gardens and the garden gems of our neighbouring islands of Sark and Herm.

If you'd like to open your garden for the National Garden Scheme or join our wonderful volunteer team, please contact Area Organiser Patricia McDermott for more information.

Volunteers

Area Organiser
Patricia McDermott
patricia.mcdermott@ngs.org.uk

Publicity
Alison Carney
alison.carney@ngs.org.uk

Booklet Coordinator
Ellie Phillips
ellie.phillips@ngs.org.uk

THE GARDENS

1 LE GRAND DIXCART

Sark, Guernsey, GY10 1SD. Helen Magell, 01481 832943, helen@horse.gg. *Next to Stocks Hotel. Take the boat from Guernsey to Sark & follow the signs to Stocks Hotel. Once at the hotel continue up Dixcart Ln towards La Coupee & Le Grand Dixcart will be on your R.* **Sat 17, Sun 18 June (11-3). Adm £4, chd £2. Home-made teas & cakes.**
The gardens surround an old farmhouse and date from 1565. They are formed into five distinct areas over 2 acres and are cultivated for beauty, wildlife and food. There is a large mandala style permaculture area providing many different vegetables and cutting flowers alongside habitats for wildlife, ponds, lawns, herbaceous borders, two glasshouses, woodland, sculptures and an orchard. Wheelchair access around Sark can be difficult but once you are actually at the gardens there are just grassy slopes and paths.

2 NEW LONGFRIE HOUSE

Rue De Longfrie, St. Pierre Du Bois, Guernsey, GY7 9RZ. Nicola Brink & Carl Jensen. *Yellow farmhouse opp Longfrie Inn car park.* **Sat 24 June (10-5). Adm by donation.**
A multigenerational house with a garden that can be enjoyed by all ages. The garden is a haven for friends, family and birds.

OPENING DATES

All entries subject to change. For latest information check **www.ngs.org.uk**

Map locator numbers are shown to the right of each garden name.

June

Saturday 17th
Le Grand Dixcart 1

Sunday 18th
Le Grand Dixcart 1

Saturday 24th
NEW Longfrie House 2

Le Grand Dixcart

NORTHERN IRELAND

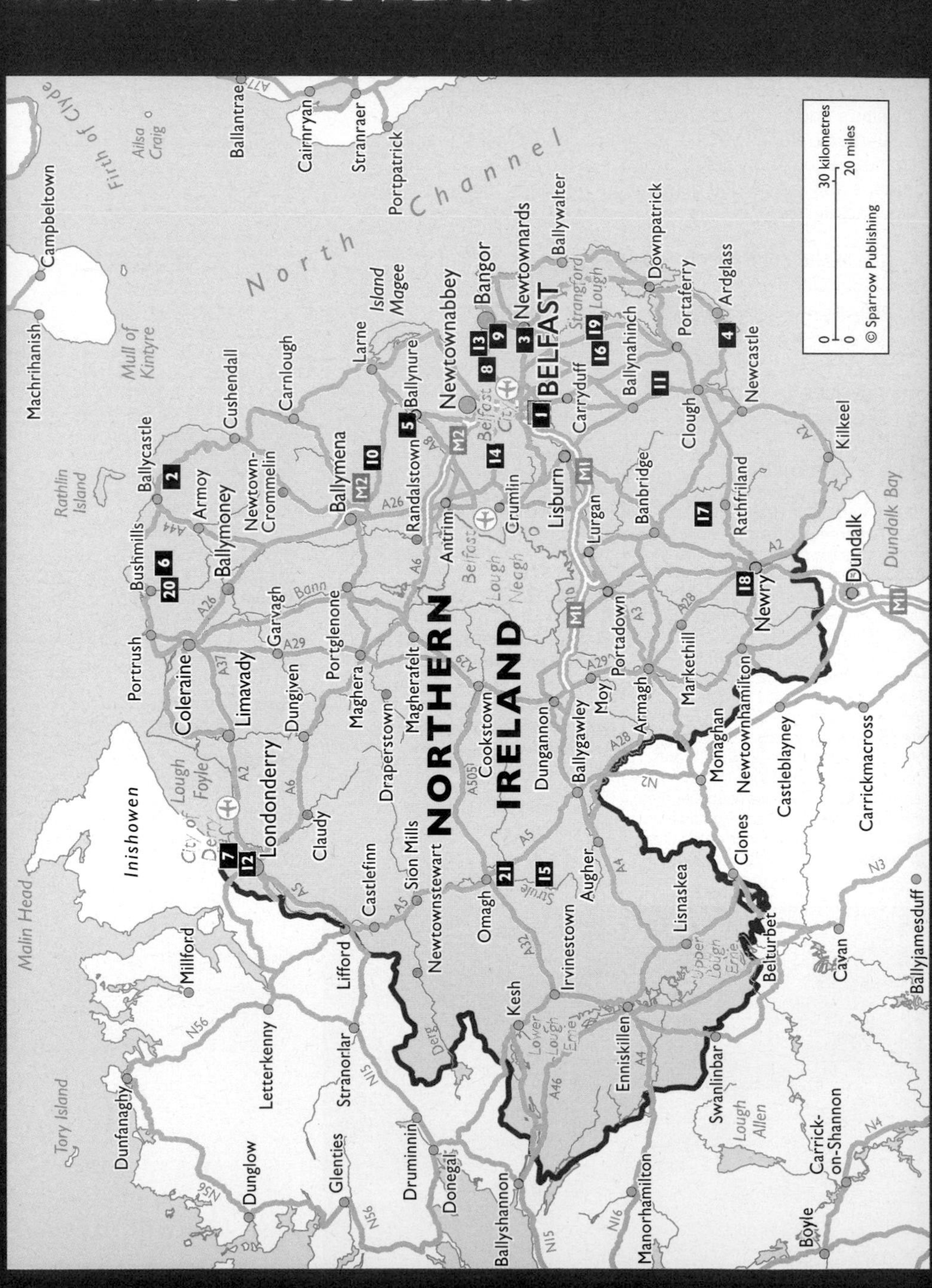

Northern Ireland is the most recent area to join the National Garden Scheme – not just one County but six.

An interesting and varied twenty-one friendly private garden owners will open in this inaugural season. Our programme will start in February for snowdrops running through to late summer.

The region is exposed to the ameliorating effects of the North Atlantic current, a north eastward extension of the Gulf Stream. These mild and humid climatic conditions allow the growth of many tender garden plants and have, in sum, made Northern Ireland a green country in all seasons.

Northern Ireland boasts some of the best public gardens in Europe. The six counties also offer a wealth of large and small private gardens. Please plan to visit as many of the gardens as possible with the assurance of the renowned Northern Ireland warm welcome.

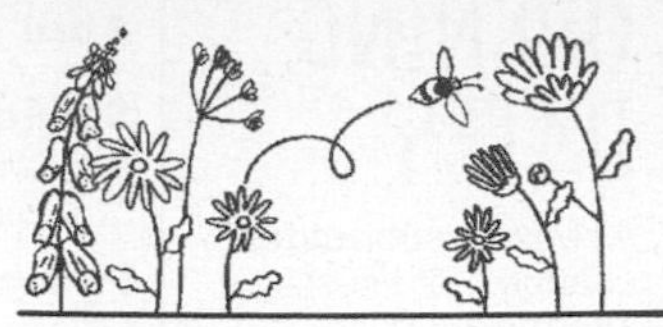

Volunteers

Area Organiser
Trevor Edwards
07860 231115
trevor.edwards@ngs.org.uk

Area Treasurer
Jackie Harte
07977 537842
jackie.harte@ngs.org.uk

Assistant Area Organisers

Trevor Browne
028 9061 3878
trevor.browne@ngs.org.uk

Pat Cameron
07866 706825
pat.cameron@ngs.org.uk

Patricia Clements
07470 474527
patricia.clements@ngs.org.uk

Fionnuala Cook
028 4066 9669
fionnuala.cook@ngs.org.uk

Patricia Corker
07808 926779
pat.corker@ngs.org.uk

Ann Fitzsimons
07706 110367
ann.fitzsimons@ngs.org.uk

Rosslind McGookin
028 2587 8848
rosslind.mcgookin@ngs.org.uk

Sally McGreevy
07871 427025
sally.mcgreevy@ngs.org.uk

Margaret Orr

@ngsnorthernireland
@ngsnorthernireland

Left: The McKelvey Garden

OPENING DATES

All entries subject to change. For latest information check **www.ngs.org.uk**

Map locator numbers are shown to the right of each garden name.

February

Snowdrop Festival

Saturday 11th
Billy Old Rectory Peace Garden 6

Sunday 12th
Billy Old Rectory Peace Garden 6

March

Friday 31st
Old Balloo House & Barn 19

April

Saturday 1st
Old Balloo House & Barn 19

Friday 7th
Old Balloo House & Barn 19

Saturday 8th
Old Balloo House & Barn 19

Saturday 29th
NEW ◆ Brook Hall Estate & Gardens 7

Sunday 30th
NEW ◆ Brook Hall Estate & Gardens 7

May

Saturday 27th
Clandeboye Estate 9

Sunday 28th
Clandeboye Estate 9

June

Saturday 17th
Holly House Garden 14

Sunday 18th
Holly House Garden 14

Saturday 24th
Billy Old Rectory Peace Garden 6

Sunday 25th
Billy Old Rectory Peace Garden 6

July

Saturday 1st
NEW Hampstead Hall 12

Sunday 2nd
NEW Hampstead Hall 12

Saturday 22nd
NEW 20 Circular Road East 8

Sunday 23rd
NEW 20 Circular Road East 8

August

Saturday 5th
Beechmount House Garden 5
Kilcootry Barn 15

Sunday 6th
Beechmount House Garden 5
Kilcootry Barn 15

Saturday 26th
Helen's Bay Organic 13

Sunday 27th
Helen's Bay Organic 13

By Arrangement

Arrange a personalised garden visit with your club, or group of friends, on a date to suit you. See individual garden entries for full details.

Adrian Walsh Belfast Garden 1
13 Ballynagard Road 2
Barley House 3
NEW The Barn Gallery and Rose Gardens 4
Clayburn 10
NEW 61 Crawfordstown Road 11
Linden 16
NEW 67 Lisnacroppan Road, 17
The McKelvey Garden 18
Old Balloo House & Barn 19
10 Riverside Road 20
Tattykeel House Garden 21

Clandeboye Estate

THE GARDENS

1 ADRIAN WALSH BELFAST GARDEN

59 Richmond Park, Stranmillis, Belfast, BT9 5EF. Adrian Walsh, 07808 156856. *Travelling from the r'about at Stranmillis College & going along the Stranmillis Rd towards Malone Rd, Richmond Park is the 2nd exit on the L. No 59 is on the L.* **Visits by arrangement June to Oct for groups of up to 25. Adm £5, chd free.**

Described by Shirley Lanigan, in 'The Open Gardens of Ireland', as 'a paradise garden' and 'a veritable sea of plants', this is an imaginatively designed naturalistic city garden that combines a vibrant mix of annuals, perennials, grasses, shrubs and trees set within a formal layout.

2 13 BALLYNAGARD ROAD

Ballyvoy, Ballycastle, BT54 6PW. Tom & Penny McNeill, 07754 190687, pennymcneill@yahoo.co.uk. *3m outside Ballycastle. Drive through Ballycastle, pass hotel on L, 2nd exit at r'about to Cushendall. From Ballyvoy drive 0.75m, Ballynagard Rd is on R. Drive 0.75m. Arrive at 1st bungalow on R.* **Visits by arrangement June to Aug for groups of up to 8. Adm £5, chd free.**

A challenging ½ acre site 400 ft above sea level with stunning rural and sea views. Designed to meet prevailing environmental conditions and to protect each section from severe winds. Seasonal planting with mature shrubs and rockery on a terraced site create cosy contrasting rooms with seating area. Regret unsuitable for dogs or young children due to steep paths through rockery.

3 BARLEY HOUSE

3 Brooklands Park, Newtownards, BT23 4XY. Miss Gillian Downing, 02891 815263, downingoriginals@gmail.com. *Newtownards, Co. Down. From Dundonald follow A20 to the r'about at Ards Shopping Centre. Take the 1st exit onto Blair Mayne Rd N. Turn L onto Manse Rd. Continue until you reach Brooklands Pk on the L.* **Visits by arrangement May to Aug. Adm £5, chd free.**

A constantly evolving garden with emphasis on roses, trees, shrubs and herbaceous perennials grown to attract birds, bees and butterflies. It has an extensive range of plants and the garden areas have features aimed to enhance and intrigue. It has been artistically planted to give year-round interest. Most of the garden is wheelchair accessible.

♿ ✿ 🚌

4 NEW THE BARN GALLERY AND ROSE GARDENS

49 Rossglass Road South, Killough, Downpatrick, BT30 7RA. Mr Bernard Magennis, 02844 842082 Mob 07874 059030, bernardjjmagennis@hotmail.com, facebook.com/barngalleryandrosegardens. *1.7m W of Killough, from A2, follow St John's Point. Parking opp' Our Lady Star of the Sea Church, Rossglass. The garden is a 2 min stroll along the coast. Limited disabled parking at garden.* **Visits by arrangement June to Sept for groups of up to 15. Adm £6, chd free. Home-made teas by arrangement.**

The rose gardens and rose walk exhibit circa 300 roses of over one hundred varieties. The Walk leads up the sloping hill on the farm to a stunning view point across Dundrum Bay to Newcastle, Kilkeel and the Mournes. Original stone built farmyard, fishpond, waterfall, rockery, herbaceous perennials etc. Artist's studio and gallery.

☕ 🛋 ·))

5 BEECHMOUNT HOUSE GARDEN

85 Ballyalbanagh Road, Ballyclare, BT39 9SP. Mr David Henderson. *N of Ballyeaston nr Ballyclare. From Ballyclare town centre take the Rashee Rd after 1.8m at 5 Corners Guest House fork R onto the Sawmill Rd at junction turn R. Ballyalbanagh Rd is next R.* **Sat 5, Sun 6 Aug (2-5). Adm £5, chd free. Pre-booking essential, please visit www.ngs.org.uk for information & booking.**

A large country garden on all sides of house with herbaceous plants, mixed shrub borders and trees. Lawns and a large elevated rockery offer stunning views of the surrounding Co Antrim countryside.

·))

6 BILLY OLD RECTORY PEACE GARDEN

5 Cabragh Road, Castlecat, Bushmills, BT57 8YH. Mrs Meta Page. *B/money bypass straight at Kilraughts Rd take 2nd R to B66 sign Dervock turn L (B66) sign Bushmills thro Derrykeighan after 2.3m at Castlecat - R sign Billy ½m & L - Haw Rd at Church R -Cabragh Rd.* **Sat 11, Sun 12 Feb (1-4). Combined adm with Benvardin Gardens £8, child free. Sat 24, Sun 25 June (1-5). Adm £5, chd free. Home-made teas.**

A mature 3 acre garden on an historic site. Front of the Georgian rectory is a large lawn with mature trees, an ancient well and a developing woodland garden with a small fernery. Rear has another large lawn with contrasting borders of roses, herbaceous, shrubs and pond area. Also kitchen, greenhouse, herb, vegetable and fruit gardens - a large old orchard and area with wildflowers and annuals. Limited refreshments during Snowdrop Opening.

✿ ☕

7 NEW ◆ BROOK HALL ESTATE & GARDENS

65-67 Culmore Road, Londonderry, BT48 8JE. Mr Gilliland, 028 7135 8968, info@brookhall.co.uk, www.brookhall.co.uk. *3m N of Derry~Londonderry. From the Culmore r'about off the Foyle Bridge, take the exit towards Moville (A2). Continue on the Culmore Road (A2) for 1m to the entrance on the R.* **For NGS: Sat 29, Sun 30 Apr (2-5). Adm £5, chd free. Tea in the Visitors Centre, check on the day for additional refreshment locations. For other opening times and information, please phone, email or visit garden website.**

Brook Hall Estate & Gardens is a C18 demesne home to one of the finest private arboretums in the north west of Ireland with unique collections of conifers, rhododendrons, magnolias and camellias. The C17 walled garden on the edge of the River Foyle, once used to feed the people of the City of Derry, now home to many of the gardens magnolia and camellia collections. Hard surface paths suitable for wheelchairs and prams through the arboretum and gardens. Please note that some paths may be quite steep.

♿ 🚌 ☕ ·))

8 NEW 20 CIRCULAR ROAD EAST

Holywood, BT18 0HA. Mr Daniel & Mrs Anne McCaughan. *A2 Belfast towards Bangor turn L at Cultra Inn. Then turn R at Circular Road East. To travel by train, Cultra railway station is max. 10 min. walk to the garden.* **Sat 22, Sun 23 July (2-5). Adm £5, chd free. Tea.**

There are 115 named varieties of hydrangeas plus others not identified in this one acre garden which has heavy clay soil. It also has an interesting collection of shrubs, trees and perennials. Some lawn has been turned into a meadow, now in its third year. A small pond and two boggy areas offer opportunity for moisture loving plants.

9 CLANDEBOYE ESTATE

Bangor, BT19 1RN. Mrs Karen Kane, www.clandeboye.co.uk. *Main entrance is off the main Belfast to Bangor A2 circa 2m before Bangor.* **Sat 27, Sun 28 May (2-5). Adm £10, chd free. Pre-booking essential, please visit www.ngs.org.uk for information & booking. Light refreshments at the Banqueting Hall.**

A series of intimate walled gardens adjoin the courtyard and house. These inc the delightful bee garden, the Chapel walk and the intimate conservatory garden. Garden features in Shirley Lanigan's 'The Open Gardens of Ireland'.

10 CLAYBURN

30A Ballynulto Road, Glenwherry, Ballymena, BT42 4RJ. Judith & Hugh Jackson, 07539 712991. *What3words app - schematic. composers.remedy. From the A36 Ballymena to Larne Rd, ½ m E of its junction with the B94, turn L onto Ballynulto Rd. The garden is 0.8m on the R.* **Visits by arrangement 15 June to 15 Sept for groups of up to 8. Adm £5, chd £5. Refreshments available by arrangement.**

This prairie-style garden is nestled in a fold of moorlands south of Slemish Mountain. Taking advantage of the surrounding landscape, the mass planting of native trees and layers of colourful herbaceous plants add to the spectacular views from the site: the retained and extended dry stone walls helping to protect the planting in this location.

11 NEW 61 CRAWFORDSTOWN ROAD

Drumaness, Ballynahinch, BT24 8LZ. Mr Liam & Mrs Brigid McDonald, 07470 728803. *A24 Ballynahinch to Newcastle, after 2m turn L to Drumaness. Through village, straight at mini r'about onto Crawfordstown Rd. No 61 on L after 1m. Pass black railings. Garden signed.* **Visits by arrangement 18 June to 29 July for groups of up to 35. Adm £5, chd free. Light refreshments.**

Two acre country garden in which rock is a dominant feature. This has been put to good use creating paths and beds for alpines etc. A recently developed woodland walk contains birch, larch, conifers and oak trees. Roses, hydrangeas, shrubs and perennial plants add to the colour palette. Some parts, but not all of the garden, can be accessed by wheelchair.

12 NEW HAMPSTEAD HALL

40 Culmore Road, Londonderry, BT48 7RS. Mr Liam & Mrs Nora Greene. *Follow the signs for the city centre after crossing the Foyle Bridge. After double set of T-lights take 2nd R into Baronscourt. House is 3rd driveway on R.* **Sat 1, Sun 2 July (2-5). Adm £5, chd free. Light refreshments.**

A fascinating city garden in the suburbs of Derry. It is comprised of a Japanese style garden, an Italian garden and a formal garden in front of an elegant Georgian House. Described and illustrated in Shirley Lanigan's 'The 100 Best Gardens in Ireland'.

13 HELEN'S BAY ORGANIC

Coastguard Avenue, Helen's Bay, BT19 1JY. Mr John McCormick, www.helensbayorganic.com. *Coastguard Ave. off Craigdarragh Rd, Helen's Bay, Co Down. Coming from A2 (Main Belfast-Bangor Road). Drive 200m past the railway bridge on Craigdarragh Rd then turn L Coastguard Ave. Go over 2 speed bumps & enter first farm gate on L.* **Sat 26, Sun 27 Aug (2-5). Adm £5, chd free.**

An urban market garden, established in 1991, producing over 50 varieties of organic vegetables for direct retailing. Also hosts a community garden and allotments which inc an extensive range of fruit and vegetables. An ideal space to get ideas to combine flowers, fruit, food and biodiversity in your garden.

14 HOLLY HOUSE GARDEN

3 Ballyutoag Hill, Nutts Corner, Crumlin, BT29 4UH. Mr Will Hamilton, www.hollyhousegardens.com. *In the hills over looking West Belfast. Belfast > Crumlin Rd, following signs for Crumlin/Int Airport (A52). After 3m Horse Shoe Bend, 4m turn L onto Ballyutoag Hill & the garden is the 1st entrance on the R.* **Sat 17, Sun 18 June (2-5). Adm £5, chd free. Light refreshments.**

Holly House Garden is an area of six acres created out of farmland over the past 20+ yrs. The garden comprises woodlands, herbaceous borders and an iris garden where thousands of spring and summer bulbs have been planted. There is an alpine bed, wildlife ponds and a newly planted shrub and woodland garden. Also, in contrast to the traditional parts there is a modern contemporary garden. Partially wheelchair friendly.

15 KILCOOTRY BARN

Fintona, Omagh, BT78 2JF. Miss Anne Johnston, 07761 918951, annejohnston678@btinternet.com. *Situated 2m from Fintona on the A5 & approx 6m from Omagh on the L. Google map directions are best found searching for Kilcootry Barn rather than postcode* **Sat 5, Sun 6 Aug (10.30-4.30). Adm £6, chd free. Light refreshments provided by Pandora's box.**

A 2½ acre mature cottage garden featuring herbaceous seasonal perennials, water feature, fruit trees, vegetable garden, children's play area, alpacas and ponies in delightful surroundings. Some areas have uneven surfaces.

16 LINDEN

24 Raffrey Road, Killinchy, Newtownards, BT23 6SF. Mr Alan & Mrs Margaret McAteer, 07770 773611. *From Comber r'about take A22 towards Killyleagh, R at Saintfield Rd, after 1.8m L to Raffery Rd. From the Saintfield Xrds take Todds Hill Rd onto Station Rd for 1.6m then L bend. After 1m go R & then L at the staggered junction. Go along road signed Killinchy 4 for*

1.6m, Raffrey Rd is on the R. **Visits by arrangement 15 May to 30 Sept (excl 8 Jul - 31 Aug) for groups of up to 8. Adm £5, chd free.**
A new 2 acre country garden with formal and informal areas. Year-round colour from perennials and grasses plus a large, rose covered pergola, a small parterre, an orchard in perennial flower meadow, soft fruit, vegetables, a greenhouse, woodland planting and experimental growing area. A pleached lime 'hedge' is being developed. Wheelchair access most areas.

17 NEW 67 LISNACROPPAN ROAD,

Rathfriland, Newry, BT34 5NZ. Mrs Helen Harper, 028 4065 1649, helenlharper@yahoo.co.uk. *From A1 take B10 Rathfriland Rd. After 7½m turn R up steep hill onto Lisnacroppan Rd After 2m garden is on R. Park on R after cottage. From Rathfriland take B10 for 2m & turn L.* **Visits by arrangement May to Aug for groups of up to 8. Adm £5.**
Forth Cottage garden is inc in Shirley Lanigan's book "The Open Gardens of Ireland". A cottage garden with three distinct and contrasting areas. Summer emphasis on violas, geraniums, orchids, alpines and drought tolerant plants. There is partial wheelchair access.

18 THE MCKELVEY GARDEN

7 Mount Charles North, Bessbrook, Newry, BT35 7DW. Mr William & Mrs Hilary McKelvey, 02830 838006. *In the village of Bessbrook. Enter village from junction beside Morrow's Garage, 300yds up hill past terraced houses turn L through gate.* **Visits by arrangement 2 May to 31 Aug for groups of up to 50. Adm £5.**
A connoisseurs garden situated in a village setting containing snowdrops, alpine crevice beds, salvias and a large collection of clematis planted among herbaceous and shrub borders. Can be accessed by a side entrance.

19 OLD BALLOO HOUSE & BARN

15 - 17 Comber Road, Killinchy, Newtownards, BT23 6PB. Ms Lesley Simpson & Miss Moira Concannon, 07484 649767, lsimpsonballoo@gmail.com. *Balloo, Killinchy, Co. Down. Between Comber & Killyleagh on the A22. On the Xrds opp Sofaland shop.* **Fri 31 Mar, Sat 1, Fri 7, Sat 8 Apr (11-5). Adm £5, chd free. Visits also by arrangement 31 Mar to 8 Apr.**
Restored and newly created gardens around late Georgian house. This year open earlier in the season for daffodil days. Part of the garden is not accessible to everyone due to the steep incline. Unsuitable for young children and wheelchairs.

20 10 RIVERSIDE ROAD

Bushmills, BT57 8TP. Mrs Pam Traill, 02820 731219, pmtraill@gmail.com. *1½ m from Bushmills on Riverside rd, off the B66. Turn L by Drum Lodge, 100 yds further on R beware ramps.* **Visits by arrangement 11 Feb to 30 June. Adm £5, chd free.**
Rambling cottage garden with unusual shrubs, deep borders, woodland spring walk, rockery and arboretum. Springtime is a sea of snowdrops and aconites and May/June see the perennials in the borders at their best.

21 TATTYKEEL HOUSE GARDEN

115 Doogary Road, Omagh, BT79 0BN. Mr Hugh & Mrs Kathleen Ward, 028 8224 9801, info@tattykeelhouse.com, www.tattykeelhouse.com/gardens. *Tattykeel House is approx 2½m from Omagh on the S side of the A5 Omagh to Ballygawley Rd. There's a sign outside the entrance Tattykeel House & Studio.* **Visits by arrangement May to Aug for groups of 5 to 30. Adm £5, chd free. Refreshments to be arranged at booking.**
A country garden of approximately 1½ acres created over a 30 year period, planted with conifers, shrubs, roses and perennials. There is a sheltered seating area, a Japanese influenced area, interesting features and a collection of well grown climbers on the house. The garden underwent numerous exciting and significant improvements in early 2022

Kilcootry Barn

Garreglllwyd

WALES • CYMRU

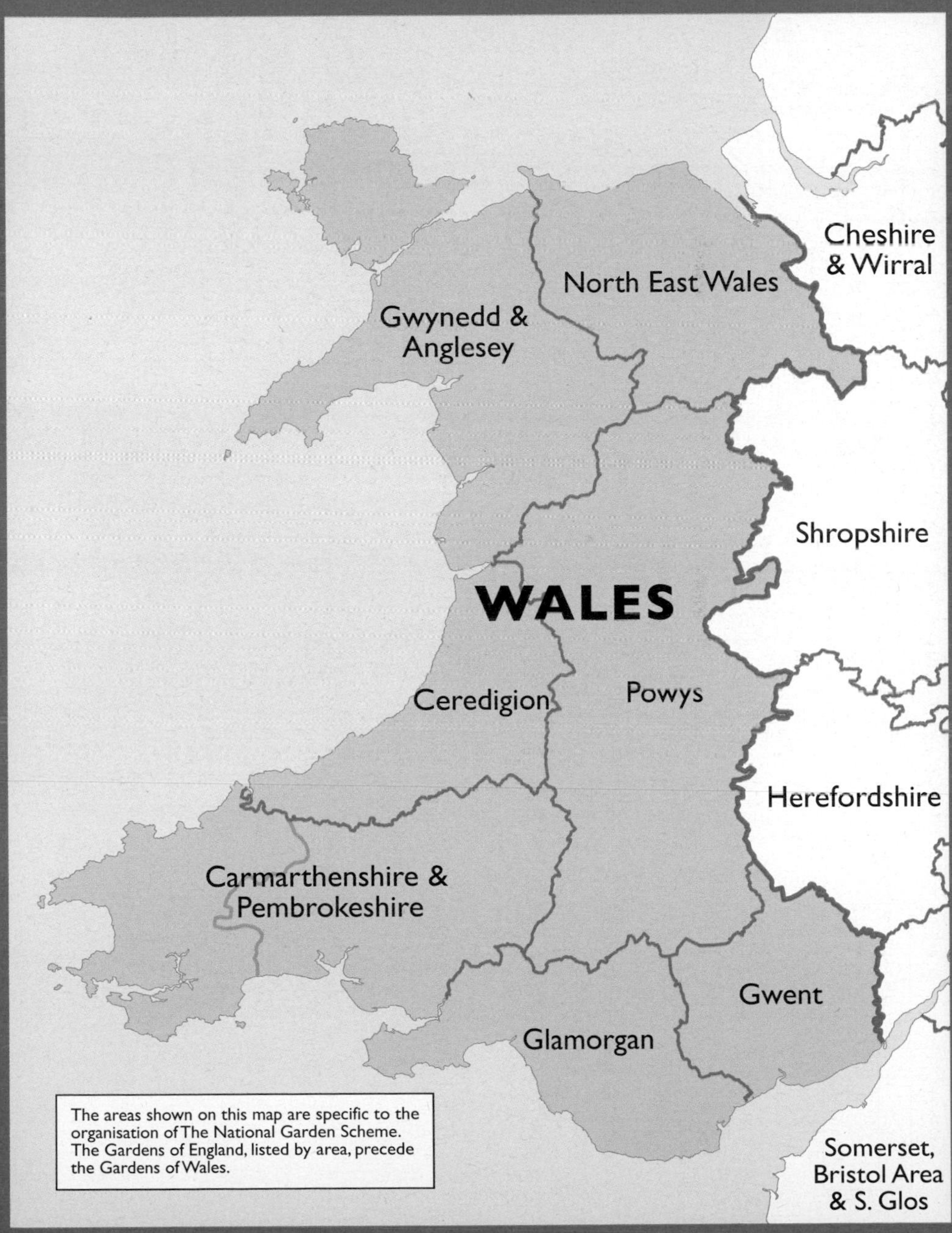
Cheshire & Wirral
North East Wales
Gwynedd & Anglesey
Shropshire
WALES
Ceredigion
Powys
Herefordshire
Carmarthenshire & Pembrokeshire
Gwent
Glamorgan
The areas shown on this map are specific to the organisation of The National Garden Scheme. The Gardens of England, listed by area, precede the Gardens of Wales.
Somerset, Bristol Area & S. Glos

CARMARTHENSHIRE & PEMBROKESHIRE

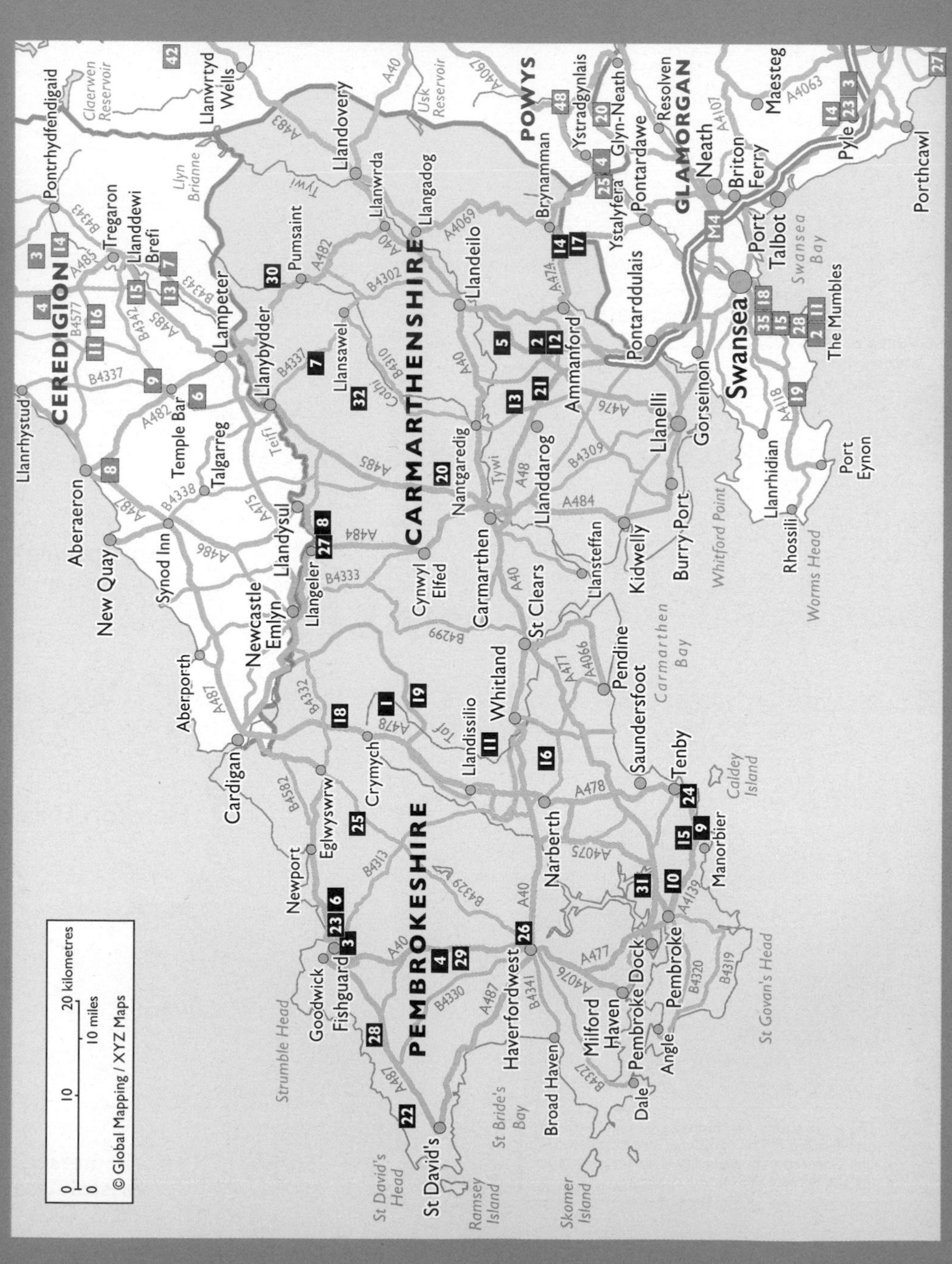

From the rugged Western coast and beaches to the foothills of the Brecon Beacons and Black Mountain, these counties offer gardens as varied as the topography and weather.

We are delighted to welcome several new gardens for 2023. They range from the lovingly designed Shoals Hook Farm in Haverfordwest, to the colour-filled beds at Neuadd y Felin, Ammanford, and the plantsperson's hill-top hideaway at The Retreat, near Tenby. Look for the yellow "new" arrow in the garden entries.

There are gardens for snowdrops (Gelli Uchaf); historic gardens at Norchard and Upton Castle; an upland garden with professional, abundant planting and views at Rhyd y Groes, a young, ambitious formal garden at Pen y Cwm, or why not enjoy an evening opening at the National Botanic Garden of Wales? From small independent nurseries (Cold Comfort Plants, Wolfscastle and The Perennial, St David's) and specialist growers (West Wales Willows) to small gardens crammed with plants and ideas on how to use space (Cefn Amlwg, Moelfryn, and Pont Trecynny), and the larger smorgasbords at Llwyngarreg and Pentresite. Wherever your choice leads you, you will be warmly welcomed – with tea, cake, and plants for sale, to further tempt you. Please do not be put off those gardens which may not have fixed openings – check the "By Arrangement" list of gardens on the Opening Dates page and do not hesitate to make that phone call – there are many surprising and delightful gardens to discover!

O arfordir a thraethau gorllewinol garw i odre'r Bannau Brycheiniog a'r Mynydd Du, mae'r siroedd hyn yn cynnig gerddi mor amrywiol â'r topograffi a'r tywydd.

Rydyn ni'n falch iawn o groesawu sawl gardd newydd ar gyfer 2023. Maent yn amrywio o Fferm Shoals Hook, a ddyluniwyd yn gariadus yn Hwlffordd, i'r gwelyau llawn lliw yn Neuadd y Felin, Rhydaman, a chuddfan pen bryn y garddwr yn The Retreat, ger Dinbych-y-pysgod. Chwiliwch am y saeth "newydd" melyn yn y cofnodion gardd.

Mae gerddi ar gyfer eirlysiau (Gelli Uchaf); gerddi hanesyddol yn Norchard ac Upton Castle; gardd ucheldir gyda phlannu proffesiynol, toreithiog a golygfeydd yn Rhyd y Groes; gardd ffurfiol ifanc, uchelgeisiol ym Mhen y Cwm, neu beth am fwynhau agoriad nos yng Ngardd Fotaneg Genedlaethol Cymru? O feithrinfeydd annibynnol bach (Cold Comfort Plants, Wolfscastle a The Perennial, Tyddewi) a thyfwyr arbenigol (West Wales Willows) i erddi bach sydd yn llawn planhigion a syniadau ar sut i ddefnyddio lle (Cefn Amlwg, Moelfryn, a Phont Trecynny), a'r smorgasbordau mwy o faint yn Llwyngarreg a Phentresite. Ble bynnag mae eich dewis yn eich arwain, bydd croeso cynnes i chi – gyda the, cacennau, a phlanhigion ar werth, i'ch temtio ymhellach. Peidiwch ag oedi cyn ymweld â'r gerddi hynny nad oes ganddynt amseroedd agor sefydlog efallai – gwiriwch y rhestr o gerddi "Trwy Drefniant" ar y dudalen Dyddiadau Agor a pheidiwch ag oedi cyn gwneud yr alwad ffôn honno – mae llawer o erddi annisgwyl a hyfryd i'w darganfod!

Volunteers

County Organisers
Jackie Batty
01437 741115
jackie.batty@ngs.org.uk

County Treasurer
Brian Holness
01437 742048
brian.holness@ngs.org.uk

Assistant County Organisers
Elena Gilliatt
01558 685321
elenamgilliatt@hotmail.com

Gayle Mounsey
07900 432993
gayle.mounsey@gmail.com

Liz and Paul O'Neill
01994 240717
lizpaulfarm@yahoo.co.uk

Social Media
Mary-Ann Nossent
07985077022
maryann.nossent@ngs.org.uk

@CarmsandPembsNGS
carmsandpembsngs

OPENING DATES

All entries subject to change. For latest information check **www.ngs.org.uk**

Map locator numbers are shown to the right of each garden name.

April

Monday 10th
Treffgarne Hall 29

Saturday 15th
Pen-y-Garn 21

Sunday 30th
Treffgarne Hall 29

May

Sunday 14th
◆ Dyffryn Fernant 6

Saturday 20th
◆ National Botanic Garden of Wales 13

Sunday 21st
Cold Comfort Farm 4
Pont Trecynny 23

Saturday 27th
NEW Cefn y Dre 3

Sunday 28th
NEW Cefn y Dre 3

June

Every Tuesday
NEW ◆ The Perennial 22

Saturday 3rd
NEW ◆ The Perennial 22

Sunday 4th
Pen-y-Garn 21
◆ Upton Castle Gardens 31

Tuesday 6th
Cefn Amlwg 2
Moelfryn 12

Saturday 10th
NEW Neuadd y Felin 14

Sunday 11th
NEW Neuadd y Felin 14
Norchard 15
The Retreat 24

Saturday 17th
NEW Pen y Bryn 17
NEW ◆ The Perennial 22
Rhyd-y-Groes 25
Skanda Vale Hospice Garden 27

Sunday 18th
Cold Comfort Farm 4
NEW Pen y Bryn 17
Pont Trecynny 23
Rhyd-y-Groes 25

Wednesday 21st
◆ National Botanic Garden of Wales 13

Friday 23rd
NEW West Wales Willows 32

Saturday 24th
NEW Cefn y Dre 3
◆ The Grange 9
◆ Lamphey Walled Garden 10
NEW West Wales Willows 32

Sunday 25th
NEW Cae Bach 1
NEW Cefn y Dre 3
◆ The Grange 9
◆ Lamphey Walled Garden 10

Wednesday 28th
Cold Comfort Farm 4

July

Tuesday 4th
Cefn Amlwg 2
Moelfryn 12

Sunday 9th
Pentresite 20
Treffgarne Hall 29

Saturday 15th
NEW Neuadd y Felin 14
Skanda Vale Hospice Garden 27

Sunday 16th
NEW Neuadd y Felin 14
Pont Trecynny 23

Saturday 22nd
Rhyd-y-Groes 25

Sunday 23rd
Rhyd-y-Groes 25

Saturday 29th
NEW Cefn y Dre 3

Sunday 30th
NEW Cefn y Dre 3

Pentresite

August

Tuesday 1st
Cefn Amlwg 2
Moelfryn 12

Sunday 20th
NEW Cae Bach 1
◆ Dyffryn Fernant 6

Saturday 26th
◆ National Botanic Garden of Wales 13

September

Every Monday and Saturday from Monday 18th
NEW ◆ The Perennial 22

Sunday 3rd
Pentresite 20

Saturday 9th
◆ The Grange 9

Sunday 10th
◆ The Grange 9

By Arrangement

Arrange a personalised garden visit with your club, or group of friends, on a date to suit you. See individual garden entries for full details.

NEW Cae Bach 1
Cefn Amlwg 2
Cold Comfort Farm 4
Dwynant 5
Gelli Uchaf 7
Glandwr 8
◆ The Grange 9
Llwyngarreg 11
Moelfryn 12
Norchard 15
The Old Rectory 16
NEW Pen y Cwm 18
Pencwm 19
Pentresite 20
Pen-y-Garn 21
Pont Trecynny 23
Rhyd-y-Groes 25
NEW Shoals Hook Farm 26
Stable Cottage 28
Treffgarne Hall 29
Ty'r Maes 30

THE GARDENS

1 NEW CAE BACH

Hermon, Glogue, nr Crymych, SA36 0DS. Liz & Will North, 01239 831663, elizabethmmnorth@gmail.com. *2m SE of Crymych. From Crymych take Hermon rd at Ysgol Bro Preseli, at T jct. in Hermon, turn L. Cae Bach is 300 yds on R. Disabled parking only at house, other parking signposted in village.* **Sun 25 June, Sun 20 Aug (11-5). Adm £4, chd free. Home-made teas. Visits also by arrangement 3 June to 27 Aug for groups of 5 to 30. Teas on request when booking.**

A 1½ acre newly established garden designed to attract wildlife, particularly bees and butterflies. Inc herbaceous borders, ponds, grasses, Japanese garden with rill, rose garden (50 varieties), greenhouse, exotic area, conifer/heather area, wildflower meadow area, vegetable and fruit trees and fruit cage. Most areas wheelchair accessible.

2 CEFN AMLWG

Llandeilo Road, Gorslas, Llanelli, SA14 7LU. Ann & Collin Wearing, 01269 842754, collin.wearing@yahoo.co.uk. *Approx 2m from Cross Hands r'about heading to Llandeilo, From Llandeilo towards Cross Hands on A476 go through Carmel village into Castell y Rhingyll. 1st & 2nd properties on R.* **Tue 6 June, Tue 4 July, Tue 1 Aug (10-4). Combined adm with Moelfryn £6, chd free. Light refreshments. Visits also by arrangement 14 May to 12 Aug. Teas on request when booking.**

The garden provides a mix of neat and tidy manicured lawn and ornamental beds with colour, planted to attract bees and butterflies. A large fish pond, raised vegetable beds, fruit cage and seating areas. A summerhouse allows you to sit and soak in the delights, the lawn leads your eyes to explore the beautifully laid out garden. Cleverly designed structures add interest. See separate entry for description of Moelfryn.

3 NEW CEFN Y DRE

Fishguard, SA65 9QS. Gaye Williams & Geoff Stickler. *2m E of Scleddau on A40 and 1.5m S of Fishguard. Turn opp Gate Inn on A40 in Scleddau. After 1½ m turn R down No Through Road. Garden at end of road. From A487 in Fishguard take Llanychaer Rd for 100 yds then up hill to No Through Road as above.* **Sat 27, Sun 28 May, Sat 24, Sun 25 June, Sat 29, Sun 30 July (11-4.30). Adm £4, chd free. Pre-booking essential, please phone 01348 875663, email cefnydre@me.com or visit www.cefnydre.co.uk for information & booking. Teas.**

One of Pembrokeshire's Historic Homes, with open lawned area, and extensive, varied border planting under mature trees. Herb bed, large kitchen garden, polytunnel, small greenhouse and propagation area.

4 COLD COMFORT FARM

Wolfscastle, Haverfordwest, SA62 5PA. Judy & Paul Rumbelow, 07809 560409, judy.rumbelow@gmail.com. *7m N of Haverfordwest. Signed off A40 at Wolfscastle. Turn towards Hayscastle at Wolfe Inn. 1m along lane on L.* **Sun 21 May, Sun 18 June (11-5), open nearby Pont Trecynny. Wed 28 June (11-5). Adm £4, chd free. Home-made teas. Visits also by arrangement May to July for groups of up to 40.**

2 acres of developing wildflower meadow alongside wrap around farm garden and small plant nursery. Perennial beds, showcasing nursery stock, rockery, gravel garden, raised beds, greenhouses, polytunnel and compost. Welcoming garden under development on sloping plot, full of planting ideas. Outstanding views of the Preseli Mountains. Find us on facebook at ' Cold Comfort Plants'

5 DWYNANT

Golden Grove, SA32 8LT. Mrs Sian Griffiths, 01558 668727, sian.41@btinternet.com, www.airbnb.co.uk/rooms/31081717. *14m E of Carmarthen, 3m from Llandeilo. Take B4300 Llandeilo to Carmarthen. Take L turn to Gelli Aur, pass church & vicarage then 1st R onto Old Coach Rd. Dwynant is approx ¼ m on R.* **Visits by arrangement Apr to Sept for groups of 2+. Adm £4, chd free. Home-made teas on request when booking.**

A ¾ acre garden set on a steep slope designed to sit comfortably within a verdant countryside environment with beautiful scenery and tranquil woodland setting. A spring garden with lily pond, selection of plants and shrubs inc azaleas, rhododendrons, rambling roses set amongst a carpet of bluebells. Seating in appropriate areas to enjoy the panoramic view and flowers. (For accommodation see garden website above).

6 ◆ DYFFRYN FERNANT

Llanychaer, Fishguard, SA65 9SP. Christina Shand & David Allum, 01348 811282, christina@dyffrynfernant.co.uk, www.dyffrynfernant.co.uk. *3m E of Fishguard, then ½ m inland. A487 E, 2m from Fishguard turn R towards Llanychaer. Follow lane for ½ m, entrance on L. Look for yellow garden signs.* **For NGS: Sun 14 May, Sun 20 Aug (11-5). Adm £9, chd free. Correct cash at gate or pre-purchase on garden's own website. For other opening times and information, please phone, email or visit garden website.**

A modern, 6 acre garden which, over two decades, has grown out of the ancient landscape. Distinctly different areas inc a lush bog garden, exotic courtyard, and a field of ornamental grasses combine to create a garden that unfolds as you journey though it. Abundant sitting places and a garden library invite visitors to take their time. RHS Partner Garden. Scented azaleas and peonies in May. Dahlias, ornamental grasses and exotics in August.

Our 2022 donation enabled 7,000 people affected by cancer to benefit from the garden at Maggie's in Oxford

7 GELLI UCHAF

Rhydcymerau, Llandeilo, SA19 7PY. Julian & Fiona Wormald, 01558 685119, thegardenimpressionists@gmail.com, www.thegardenimpressionists.com. *See garden website for detailed directions. 5m SE of Llanybydder. 1m NW of Rhydcymerau. From the B4337 in Rhydcymerau take minor road opp bungalows. After 300yds turn R up track. Limited parking so essential to phone or email first.* **Visits by arrangement Feb to Oct for groups of up to 20. Home-made teas on request when booking. Adm £5, chd free.**

Complementing a C17 Longhouse and 11 acre smallholding this 1½ acre garden is mainly organic. Trees and shrubs are underplanted with hundreds of thousands of snowdrops, crocus, cyclamen, daffodils, with woodland shrubs, clematis, rambling roses, hydrangeas and autumn flowering perennials. Extensive views, shepherd's hut and seats to enjoy them. Year-round flower interest with naturalistic plantings. 6 acres of wildflower meadows, two ponds and stream.

8 GLANDWR

Pentrecwrt, Llandysul, SA44 5DA. Jo Hicks, 01559 363729, leehicks@btinternet.com. *15m N of Carmarthen, 2m S of Llandysul, 7m E of Newcastle Emlyn. On A486. At Pentrecwrt village, take minor rd opp Black Horse pub. After bridge keep L for ¼m. Glandwr is on R.* **Visits by arrangement June to Aug. Teas on request when booking. Just a phone call! Perfect for small groups. Adm £3.50, chd free.**

Delightful 1 acre enclosed mature garden, with a natural stream. Inc a rockery, various flower beds (some colour themed), many clematis and shrubs. Do spend time in the woodland with interesting trees, shade loving plants and ground cover, surprise paths, and secluded places to hear the bird song.

9 ◆ THE GRANGE

Manorbier, Tenby, SA70 7TY. Joan Stace, 01834 871311, www.grangegardensmanorbier.com. *4m W of Tenby. From Tenby, take the A4239 for Pembroke. ½m after Lydstep, The Grange is on L, at the Xrds. Parking limited.* **For NGS: Sat 24, Sun 25 June (11-4.30), open nearby Lamphey Walled Garden. Sat 9, Sun 10 Sept (11-4.30). Adm £4, chd free. Home-made teas. Visits also by arrangement 15 June to 17 Sept. Teas on request when booking. For other opening times and information, please phone or visit garden website.**

5 acre country garden of two parts. Older established garden around Grade II listed house (not open) has colourful herbaceous borders, rose garden, outdoor chess set and swimming pool (private). Newer garden was reclaimed from a heavy clay bog field which was drained and now consists of a lake with islands, ponds and dry river bed, newly-planted with wide variety of grasses as a 'prairie' garden.

10 ◆ LAMPHEY WALLED GARDEN

Lamphey, Pembroke, SA71 5PD. Mr Simon Richards, 07503 976766, ullapoolsi@hotmail.co.uk, www.facebook.com/lowerlampheypark. *½m from Lamphey village off the Ridgeway at Lower Lamphey Park. On A477 turn S at Milton towards Lamphey 1m. At T-jct, turn R to Lamphey, after 1m take R up track for walled garden. Parking behind cottages.* **For NGS: Sat 24, Sun 25 June (10-4). Adm £5, chd free. Cream teas. Open nearby The Grange. For other opening times and information, please phone, email or visit garden website. Donation to The Bumblebee Conservation Trust.**

Built in late 1700's, 1 acre walled garden renovated with extensive plantings from 2015. Several hundred species, a plantsperson's delight with a nature first feel. 50 Salvia species, rare perennials and unusual shrubs with old roses and a lavender avenue. Espalier apples and pears plus a Heritage vegetable garden. Bumblebee heaven. Sloping grass paths allow wheelchair access to majority of garden.

11 LLWYNGARREG

Llanfallteg, Whitland, SA34 0XH. Paul & Liz O'Neill, 01994 240717, lizpaulfarm@yahoo.co.uk, www.llwyngarreg.co.uk. *19m W of Carmarthen. A40 W from Carmarthen, turn R at Llandewi Velfrey, 2½m to Llanfallteg. Go through village, garden ½m further on: 2nd farm on R. Disabled car park in bottom yard on R.* **Visits by arrangement Feb to Nov. Open most days, must ring first. Adm £6, chd free. Teas on request when booking.**

Llwyngarreg is always changing, delighting plant lovers with its many rarities inc species primulas, many huge bamboos with roscoeas, hedychiums and salvias extending the season through to riotous autumn colour. Trees and rhododendrons are underplanted with perennials. The exotic sunken garden and gravel gardens continue to mature. Springs form a series of linked ponds across the main garden, providing colourful bog gardens. Spot the subtle mobiles and fun constructions hidden around the 4 acre garden. Wildlife ponds, fruit and vegetables, composting, numerous living willow structures, mobiles, swing, chickens, goldfish. Partial wheelchair access.

12 MOELFRYN

Llandeilo Road, Gorslas, Llanelli, SA14 7LU. Elaine & Graeme Halls. *Approx 2m from Cross Hands r'about heading to Llandeilo. From Llandeilo towards Cross Hands on A476 go through Carmel village into Castell y Rhingyll. 1st & 2nd properties on R.* **Tue 6 June, Tue 4 July, Tue 1 Aug (10-4). Combined adm with Cefn Amlwg £6, chd free. Light refreshments. Visits also by arrangement 14 May to 12 Aug. Please book & arrange teas with with Cefn Amwlg on 01269 842754.**

Organic and environmentally friendly our garden has a herb labyrinth, cottage garden borders, vegetable beds and tunnels, fruit cage, an orchard, small wildflower meadow area, shrubs, small wooded/wild patch, hens, hedgehog house, bug hotels, bird boxes and bat boxes, small pond and bog garden, Interesting arches, compost bins, water butts and a collection of 40 varieties of mints. Home-made organic ice-cream. Mint tasting quiz can be booked in advance. Hen keeping and holding advice can be given. See separate entry for description of Cefn Amlwg.

13 ◆ NATIONAL BOTANIC GARDEN OF WALES

Middleton Hall, Llanarthne, SA32 8HN. 01558 667149, www.botanicgarden.wales. *Between Cross Hands and Carmarthen. From Carmarthen take A48 eastbound, after 8m take slip rd signed B4310 to Nantgaredig. Follow brown*

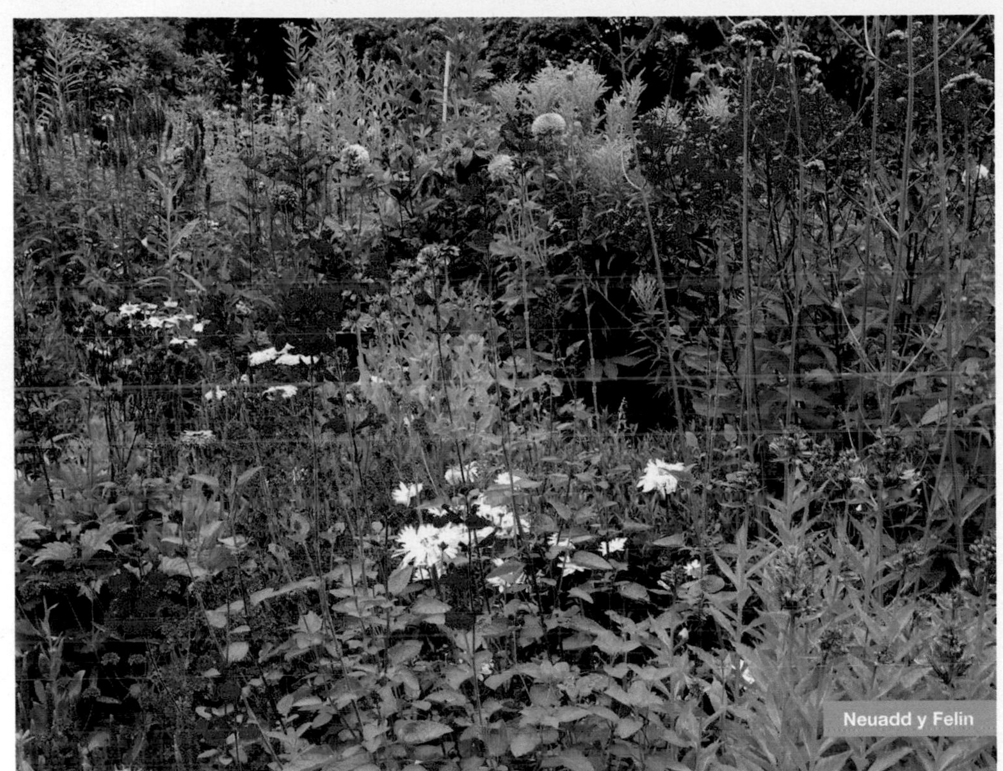
Neuadd y Felin

signs. **For NGS: Evening opening Sat 20 May, Wed 21 June, Sat 26 Aug (6-8). Adm £15, chd £7.50. Pre-booking essential, please visit www.ngs.org.uk for information & booking. Wine & nibbles & talk-tour. For other opening times and information, please phone or visit garden website.**
The National Botanic Garden of Wales is spread across 568 acres of picturesque landscape. At 23 yrs old the site encompasses the beautifully laid out garden and its living collections covering approximately 70 acres, and the wider estate, inc the Regency Restoration, Waun Las National Nature Reserve and working, sustainable farm. The Garden includes the Great Glasshouse, an iconic Fosters + Partners creation dedicated to mediterranean flora; Tropical House; Walled Garden with systematic beds; wild woodlands and diverse grasslands. Popular with locals and far-travelled visitors, this is a botanic garden for the C21. Limited buggy service on evening openings.

14 NEW NEUADD Y FELIN

Neuadd Road, Garnant, Ammanford, SA18 1UF. Terry & Sara Knight. *Garnant. on A474, 5m W of Pontardawe, 5m E of Ammanford. Access to garden entrance is best from Nant Gwinau Road, turn R as you enter Garnant on A474 from Pontardawe. From Ammanford direction turn L just before you leave Garnant on A474.* **Sat 10, Sun 11 June, Sat 15, Sun 16 July (10.30-4.30). Adm £4.50, chd free. Home-made teas.**
A mature 2 acre smallholding inc different areas of interest: large garden with sloping perennial beds; rose garden; large dahlia display; mature shrubs and shaded planted areas;40 foot white wisteria pergola; wildlife pond and fish pond; meadow with willow walk; and riverside walk with seated area. Located on the site of an old C14 watermill.

15 NORCHARD

The Ridgeway, Manorbier, Tenby, SA70 8LD. Ms H Davies, 07790 040278. *4m W of Tenby. From Tenby, take A4139 for Pembroke. ½ m after Lydstep, take R at Xrds. Proceed down lane for ¾ m. Norchard on R.* **Sun 11 June (12-5). Adm £6, chd free. Home-made teas. Open nearby The Retreat. Visits also by arrangement 19 Mar to 2 July for groups of 20+.**
Historic gardens at medieval hall house nestled in tranquil location sheltered by ancient oak woodland. Strong structure with formal and informal areas. Early walled gardens, restored Elizabethan parterre, ornamental kitchen garden, subtropical planting and orangery. 1½ acre orchard of old (many local) varieties. Young arboretum and meadow looking towards grist mill and millpond with moorhens and ducks. Extensive collections of roses, daffodils and tulips. Partial wheelchair access. Access to lower level of the kitchen garden via steps only.

16 THE OLD RECTORY

Lampeter Velfrey, Narbeth, SA67 8UH. Jane & Stephen Fletcher, 01834 831444, jane_e_fletcher@hotmail.com. *3m E of Narbeth. Next to church in Lampeter Velfrey. Parking in church car park.* **Visits by arrangement Mar to Sept. Individuals welcome. Adm £4, chd free. Light refreshments.**

Behind a formal front garden the woodland garden is a green hollow of historic charm; wildlife welcoming, with ancient and newly planted trees and shrubs. Spring brings snowdrops, daffodils and bluebells to the woodlands. Later rhododendrons, hydrangeas, acers, roses, daisies and geraniums bring pops of colour to the green backdrop. Approx 2 acres inc veg, ponds and unique garden buildings.

17 NEW PEN Y BRYN

Gwaun Cae Gurwen, Ammanford, SA18 1DY. Helen & Alex Brook. *8m East of Ammanford off the A474. At railway Xing in GCG drop down into Quarry Place (very small lane, opposite CK Foodstore). Go over waterfalls, turn L, follow road round to R & go under low bridge, property on L.* **Sat 17, Sun 18 June (11-5). Adm £4, chd free. Pre-booking essential, please phone 07551 393030 or email hjh63@hotmail.co.uk for information & booking. Tea.**

A garden of two halves. Upper formal gardens offering herbaceous borders, a pond and no-dig fruit and veg beds. The woodland is full of ferns and other woodland planting, a pond with marginal planting and several winding paths.

18 NEW PEN Y CWM

Blaenffos, Boncath, SA37 0HY. Adrian & Sarah Brown, 01239 842050. *6m S of Cardigan. On A478, from S, enter Blaenffos village, through 'the trees' garden approx. 200m on R. From N, pass the village shop, garden approx. 400m on L.* **Visits by arrangement May to Aug for individuals & groups of up to 10. Home-made teas on request when booking. Adm £5, chd free.**

The garden at Pen Y Cwm has been developed since 2016 from paddocks previously used for Alpacas. Within a formal layout there are herbaceous borders with hedges of yew, beech and hornbeam. There is a young orchard and a collection of interesting trees scattered across two acres. Many seats allow visitors to relax and enjoy the garden and views of the Preselis. Wheelchair access to most areas over grass paths, some gentle slopes.

19 PENCWM

Hebron, Whitland, SA34 0JP. Lorna Brown, 07967 274830, lornambrown@hotmail.com. *10m N of Whitland. From A40 take St Clears exit & head N to Llangynin then on to Blaenwaun. Through village, after speed limit signs take 1st L. Over Xrds, 1¼m then 2nd lane on L marked Pencwm.* **Visits by arrangement Apr to Oct Home-made teas on request when booking. Adm £4, chd free.**

A secluded garden of about 1 acre set among large native trees, designed for year-round interest and for benefit of wildlife. A wide variety of exotic specimen trees and shrubs inc magnolias, rhododendrons, hydrangeas, bamboos and acers. Drifts of bluebells and other spring bulbs and good autumn colour. Inc boggy area and pond with appropriate planting. Wellies recommended at most times.

Cefn y Dre

20 PENTRESITE
Rhydargaeau Road, Carmarthen, SA32 7AJ. Gayle & Ron Mounsey, 07900 432993, gayle.mounsey@gmail.com. *4m N of Carmarthen. Take A485 heading N out of Carmarthen, once out of village of Peniel take 1st R to Horeb & cont for 1m. Turn R at NGS sign, 2nd house down lane.* **Sun 9 July (11-5); Sun 3 Sept (11-4). Adm £5, chd free. Home-made teas. Visits also by arrangement June to Sept. Teas on request when booking. Entrance driveway not suitable for coaches.**
Approx. 2 acre garden developed over the last 16 yrs with extensive lawns, colour filled herbaceous and mixed borders, on several levels. A bog garden and magnificent views of the surrounding countryside. There is now a new area planted with trees and herbaceous plants. This garden is south facing and catches the south westerly winds from the sea.

21 PEN-Y-GARN
Foelgastell, Cefneithin, Carmarthen, SA14 7EU. Mary-Ann Nossent & Mike Wood, 07985 077022, maryann.nossent@ngs.org.uk, www.instagram.com/pen_y_garn. *10m SE Carmarthen, A48 N from Cross Hands. Take 1st L to Foelgastell R at T-junction 300metres sharp L, 300metres 1st gateway on L. A48 S Bot. Gdns turning r'about R to Porthythyd 1st L before T junction. 1m 1st R & 300 metres on L.* **Sat 15 Apr, Sun 4 June (11-4.30). Adm £4, chd free. Home-made teas. Visits also by arrangement June to Aug for groups of up to 20. Limited parking. Teas for groups on request when booking.**
1½ acres set in former limestone quarry sympathetically developed to sit within the landscape, with five distinct areas cultivated herbaceous borders and wild no mow lawns. A shady area with woodland planting and wild ponds; no dig kitchen garden; terraced borders with shrubs and herbaceous planting; lawns and pond. The garden is on several levels with slopes and steps. Steep in places.

22 NEW ◆ THE PERENNIAL
Llanrhian Road, St David's, SA62 6DB. Ms Pippa Allen, 07398 804148, perennialgardensandcafe@gmail.com, www.perennialstdavids.co.uk. *3m E of St David's on Llanrhian Rd. between St David's & Fishguard.* **For NGS: Sat 3 June (11-4). Every Tue 6 June to 27 June (10.30-5). Sat 17 June (11-4). Every Mon and Sat 18 Sept to 30 Sept (10.30-5). Adm £4, chd free. Light refreshments. For other opening times and information, please phone, email or visit garden website.**
An independent, nursery, café and farm shop with a formal garden at the centre featuring plants from the nursery. Polytunnels for salad crops, veg and cut flowers; developing small woodland; pond area; new orchard; compost and propagating areas; coastal/countryside views. Focus on sustainability, "no dig" and recycling principles; peat and pesticide free; member of Pollinators Assurance Scheme. We want to inspire visitors to grow plants for food and for pleasure, whatever their space.

23 PONT TRECYNNY
Garn Gelli Hill, Fishguard, SA65 9SR. Wendy Kinver *1½m N of Fishguard. Driving up the hill from Fishguard to Dinas turn R ½way up the rd & follow signs. SatNav will take you to a lay-by opp the garden. Parking v limited.* **Sun 21 May, Sun 18 June (10-1), open nearby Cold Comfort Farm. Sun 16 July (10-1). Adm £5, chd free. Pre-booking essential, please phone 01348 873040 or email wendykinver@icloud.com for information & booking. Visits also by arrangement 20 May to 20 Aug for groups of 10 to 30.**
A diverse garden of 3½ acres. Meander through the meadow planted with native trees, pass the pond and over a bridge which takes you along a path, through an arboretum, orchard and gravel garden and into the formal garden full of cloud trees, exotic plants and pots,which then leads you to the stream and vegetable garden.

24 THE RETREAT
Retreat Road, Penally, Tenby, SA70 7PL. Richard & Clare Rhys Davies. *1m SW of Tenby. Take A4139 to Pembroke. Turn R. after Kiln Park into Penally Village. Park in village & walk up Retreat Rd. (steep) or footpath. Strictly no parking or turning in Retreat Rd please.* **Sun 11 June (12.30-5). Adm £5, chd free. Open nearby Norchard.**
Tucked away on a hill, up a private, single track road (steep access), backing onto meadows and bluebell woods, with stunning coastal views. An eclectic, plant person's garden of herbaceous perennials punctuated by semi-tropicals, all billowing colourfully down the hillside from full sun into shadier areas. Hardy geraniums are a favourite with around 50 varieties. Wildlife friendly. This hillside garden is a plant person's delight but is too steep for access by wheelchairs or unassisted visitors with mobility restrictions.

25 RHYD-Y-GROES
Brynberian, Crymych, SA41 3TT. Jennifer & Kevin Matthews, 01239 891363, rhydygroesgardeners@gmail.com. *12m SW of Cardigan. 5m W of Crymych, 16m NE of Haverfordwest, on B4329, ¾m downhill from cattlegrid (from Haverfordwest) & 1m uphill from signpost to Brynberian (from Cardigan).* **Sat 17, Sun 18 June, Sat 22, Sun 23 July (1-5). Adm £5, chd £1.50. Light refreshments. Visits also by arrangement June to Aug for groups of 12+. Min charge £60 irrespective of numbers on the day.**
Upland garden with extensive views over boggy moor and barren hillside. A 4 acre oasis of abundant, diverse planting to suit varied growing conditions. Mixed borders, shrubbery, boggy corner, and prairie planting. Wildflower meadow that is better in June. Woodland and shade areas featuring many hydrangeas best in July. Ample seating. Beautiful, mollusc-proof planting demonstrates what can be achieved without using slug killer, pesticides and fungicides, resulting in a garden that is full of wildlife.

26 NEW SHOALS HOOK FARM
Shoals Hook Lane, Haverfordwest, SA61 2XN. Karen & Robert Hordley, 07712 268899, shoalshook@icloud.com. *What3words app - clips.november.samplers. From A40 take B4329 to H/west town centre (Prendergast). Before fork with Fishguard Rd, turn L into Back Ln, turn L to Shoals Hook Ln, con 1m, property on L.* **Visits by arrangement 1 Mar to 12 Oct excl 1 May- 15 Jun. Mon-Thu only. Teas on request at booking. Adm £5, chd free.**
Approx 6 acre garden on our smallholding. Designed and made by ourselves over 26 yrs. Inc flower borders, veg garden, fruit trees, woodland walk, lake garden and specimen trees. Planted for the different seasons and to encourage wildlife. Many spring bulbs at the start of the year, ending with a late summer prairie style grass garden area.

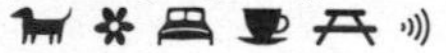

27 SKANDA VALE HOSPICE GARDEN
Saron, Llandysul, Carmarthen, SA44 5DY. brotherfrancis@skandavale.org, www.skandavalehospice.org. *On A484, in village of Saron, 13m N of Carmarthen. Between Carmarthen & Cardigan.* **Sat 17 June, Sat 15 July (11-4.30). Adm £4, chd free. Home-made teas.**
A tranquil garden of approx 1acre, built for therapy, relaxation and fun. Maintained by volunteers, the lawns and glades link garden buildings with willow spiral, wildlife pond, colourful borders, sculptures and stained glass. Planting schemes of blue and gold at entrance inspire calm and confidence, leading to brighter red, orange and white. Hospice facilities open for visitors to view. A garden for quiet contemplation, wheelchair friendly, with lots of small interesting features. Craft stall. For accommodation please visit: www.skanda-hafan.com.

28 STABLE COTTAGE
Rhoslanog, Mathry, Haverfordwest, SA62 5HG. Mr Michael & Mrs Jane Bayliss, 01348 837712, michaelandjane1954@michaelandjane.plus.com. *Between Fishguard & St David's. Head W on A487 turn R at Square & Compass sign. ½m, at hairpin take track L. Stable Cottage on L with block paved drive.* **Visits by arrangement May to Aug for groups of 2 to 20. Limited parking, please car share where possible. Adm £3.50, chd free. Refreshments by prior arrangement.**
Garden extends to approx ⅓ of an acre. It is divided into several smaller garden types, with a seaside garden, small orchard and wildlife area, scented garden, small vegetable/kitchen garden, and two Japanese areas - a stroll garden and courtyard area.

29 TREFFGARNE HALL
Treffgarne, Haverfordwest, SA62 5PJ. Martin & Jackie Batty, 01437 741115, jmv.batty@gmail.com. *7m N of Haverfordwest, signed off A40. Go up through village & follow rd round sharply to L, Hall ¼m further on L.* **Mon 10, Sun 30 Apr, Sun 9 July (1-5). Adm £5, chd free. Home-made teas. Visits also by arrangement Mar to Sept. Teas for groups on request when booking.**
Stunning hilltop location with panoramic views: handsome Grade II listed Georgian house (not open) provides formal backdrop to garden of 4 acres with wide lawns and themed beds. A walled garden, with double rill and pergolas, planted with a multitude of borderline hardy exotics. Also large scale sculptures, summer border, meadow patch, gravel garden, heather bed and stumpery. Planted for year-round interest. The planting schemes seek to challenge the boundaries of what can be grown in Pembrokeshire.

30 TY'R MAES
Ffarmers, Llanwrda, SA19 8JP. John & Helen Brooks, 01558 650541, johnhelen140@gmail.com. *7m SE of Lampeter. 8m NW of Llanwrda. 1½m N of Pumsaint on A482, opp turn to Ffarmers. Please ignore SatNav.* **Visits by arrangement Mar to Oct. Individuals welcome. Adm £5, chd free. Home-made teas on request when booking.**
4 acre garden with splendid views. Herbaceous and shrub beds – formal design, exuberantly informal planting, full of cottage garden favourites and many unusual plants. Interesting arboretum with over 200 types of tree; wildlife and lily ponds; pergola, gazebos, post and rope arcade covered in climbers. Gloriously colourful from early spring till late autumn. Wheelchair note: Some gravel paths.

31 ◆ UPTON CASTLE GARDENS
Cosheston, Pembroke Dock, SA72 4SE. Prue & Stephen Barlow, 01646 689996, info@uptoncastle.com, www.uptoncastle.com. *4m E of Pembroke Dock. 2m N of A477 between Carew & Pembroke Dock. Follow brown signs to Upton Castle Gardens through Cosheston.* **For NGS: Sun 4 June (10-4.30). Adm £6, chd free. Home-made teas. For other opening times and information, please phone, email or visit garden website.**
Privately-owned, listed, historic gardens and arboretum of 35 acres surround C13 castle (not open). Terraces of herbaceous borders and formal rose garden with over 150 roses provide constant summer interest. Walled productive kitchen garden, chapel garden with millennial yew. Rare magnolias, rhododendrons, camellias, and new hydrangea beds. Woodland with rare and champion trees. RHS Partner Garden. Walk on the Wild Side: Woodland walks funded by C.C.W. and Welsh Assembly Government. Medieval chapel as featured on Time Team. Partial wheelchair access.

32 NEW WEST WALES WILLOWS
The Mill, Gwernogle, Carmarthen, SA32 7SA. Justine & Alan Burgess, www.facebook.com/WestWalesWIllows. *3m from Brechfa. On A40 at Nantgaredig, take B4310 to Brechfa. ½m after Brechfa, turn L towards Gwernogle. After 2½m pass Chapel on R, go up steep hill & bear R. Take 1st R turn & The Mill is on the R.* **Fri 23, Sat 24 June (10-4). Adm £4.50, chd free. Home-made teas.**
Located in a scenic valley, halfway up a mountain, the National Plant Collection of Salix/Willow is set in a sloping field with grass pathways. Approx 260 varieties are on show, inc varieties for pollinators, autumn colour, bio fuel, hedging and windbreaks. The field also has sample living willow structures on display inc domes, tepee, long arch/tunnel, hedges and a new arbour.

NPC

West Wales Willows

CEREDIGION

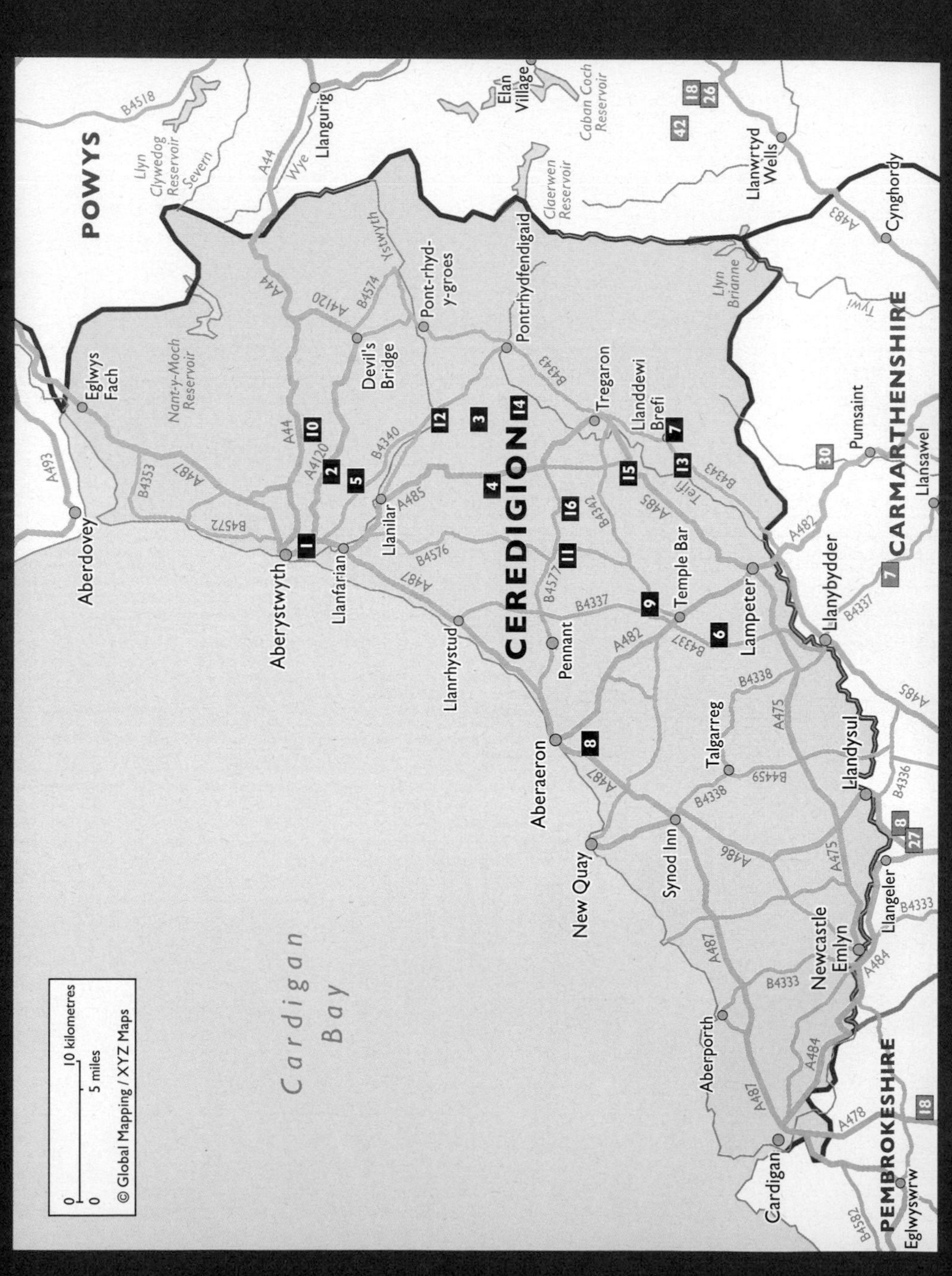
POWYS
CEREDIGION
CARMARTHENSHIRE
PEMBROKESHIRE
Cardigan Bay
Aberdovey
Eglwys Fach
Aberystwyth
Llanfarian
Llanilar
Devil's Bridge
Pont-rhyd-y-groes
Pontrhydfendigaid
Llangurig
Elan Village
Caban Coch Reservoir
Claerwen Reservoir
Llyn Clywedog Reservoir
Nant-y-Moch Reservoir
Llyn Brianne
Llanwrtyd Wells
Cynghordy
Tregaron
Llanddewi Brefi
Pumsaint
Llansawel
Llanrhystud
Pennant
Temple Bar
Lampeter
Llanybydder
Aberaeron
Talgarreg
Llandysul
New Quay
Synod Inn
Llangeler
Newcastle Emlyn
Aberporth
Cardigan
Eglwyswrw
Severn
Wye
Ystwyth
Teifi
Tywi
A44
A493
A487
A4120
A485
A482
A475
A486
A484
A478
A483
B4518
B4574
B4340
B4353
B4572
B4576
B4577
B4337
B4342
B4343
B4338
B4459
B4336
B4333
B4582
0 5 miles
0 10 kilometres
© Global Mapping / XYZ Maps

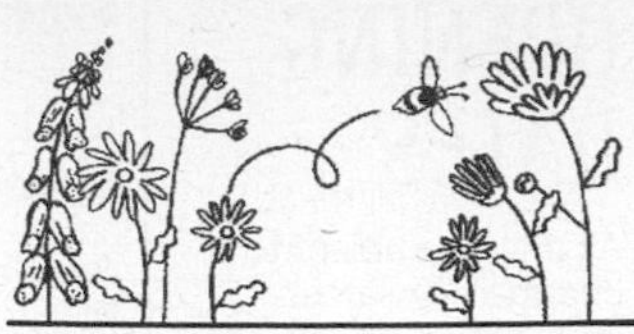

Ceredigion is essentially a rural county, the second most sparsely populated in Wales, devoid of any large commercial area.

Much of the land is elevated, particularly towards the east of the county. There are steep- sided wooded valleys, fast flowing rivers and streams, acres of moorland and a dramatic coastline with some lovely sandy beaches. From everywhere in the county there are breathtaking views of the Cambrian Mountains and, from many places, glimpses of the stunning Cardigan Bay. The gardens in Ceredigion reflect this natural beauty and sit comfortably in the rugged landscape. They range from fairly large formal gardens to equally interesting cottage gardens and vegetable plots. Many gardens put an emphasis on naturalistic planting with woodland and water features to encourage biodiversity. Some of the gardens have open days whilst private visits by arrangement are offered by some – please see individual garden details.

Whichever gardens you plan to visit you can be sure of a warm welcome from the garden owners and the opportunity to enjoy delicious homemade refreshments to complete a perfect day out! We look forward to meeting you soon!

Sir wledig yw Ceredigion yn ei hanfod, yr ail fwyaf gwasgaredig ei phoblogaeth yng Nghymru, heb unrhyw ardal fasnachol fawr.

Mae llawer o'r tir yn uchel, yn enwedig tua dwyrain y sir. Mae yma ddyffrynnoedd coediog ag ochrau serth, afonydd a nentydd cyflym, erwau o rostir ac arfordir dramatig gyda rhai traethau tywodlyd hyfryd. O bob man yn y sir ceir golygfeydd godidog o Fynyddoedd Cambria o sawl man, a chipolwg o Fae godidog Ceredigion. Mae gerddi Ceredigion yn adlewyrchu'r harddwch naturiol hwn ac yn eistedd yn gyfforddus yn y dirwedd arw. Maent yn amrywio o erddi ffurfiol gweddol fawr i erddi bythynnod a lleiniau llysiau yr un mor ddiddorol. Mae llawer o erddi yn rhoi pwyslais ar blannu naturiol gyda nodweddion coetir a dŵr i annog bioamrywiaeth. Mae gan rai o'r gerddi ddiwrnodau agored tra bod ymweliadau preifat trwy drefniant yn cael eu cynnig gan rai – gweler manylion gerddi unigol.

Pa bynnag erddi yr ydych yn bwriadu ymweld â nhw gallwch fod yn sicr o groeso cynnes gan berchnogion yr ardd a'r cyfle i fwynhau lluniaeth cartref blasus i gwblhau diwrnod allan perffaith!

Edrychwn ymlaen at gwrdd â chi yn fuan!

Volunteers

County Organiser
Stuart Bradley
07872 451821
stuart.bradley@ngs.org.uk

County Treasurer
Elaine Grande
01974 261196
e.grande@zoho.com

Booklet Co-ordinator
Shelagh Yeomans
07796 285003
shelaghyeo@hotmail.com

Publicity & Social Media
Caroline Palmer
01970 615403
caroline.palmer@ngs.org.uk

Samantha Wynne-Rhydderch
07899 911483
samantha.wynne-rhydderch
@ngs.org.uk

Assistant County Organisers
Gay Acres
01974 251559
gayacres@aol.com

Joanna Kennaugh
07872 451821
joanna.kennaugh@ngs.org.uk

@Ceredigion Gardens

OPENING DATES

All entries subject to change. For latest information check **www.ngs.org.uk**

Map locator numbers are shown to the right of each garden name.

May

Saturday 13th
Bryngwyn 2

Sunday 14th
Bwlch y Geuffordd 5

Sunday 21st
Penybont 12

Sunday 28th
Yr Efail 15

June

Sunday 4th
Rhos Villa 13

Sunday 11th
Ffynnon Las 8

Saturday 17th
Bryngwyn 2

Sunday 18th
Llanllyr 9

Monday 19th
Bryngwyn 2

Sunday 25th
Ysgoldy'r Cwrt 16

July

Sunday 2nd
Aberystwyth Allotments 1

Saturday 8th
NEW Brynmeheryn 3

Sunday 9th
◆ Cae Hir Gardens 6

Sunday 23rd
Yr Efail 15

August

Sunday 6th
Penybont 12

Monday 28th
Bryngwyn 2

Wednesday 30th
Bryngwyn 2

By Arrangement

Arrange a personalised garden visit with your club, or group of friends, on a date to suit you. See individual garden entries for full details.

Bryngwyn 2
NEW Brynmeheryn 3
Bwlch y Geuffordd Gardens 4
NEW Delfryn 7
Ffynnon Las 8
Llanllyr 9
Melindwr Valley Bees, Tynyffordd Isaf 10
NEW Pencnwc 11
Penybont 12
Tanffordd 14
Yr Efail 15
Ysgoldy'r Cwrt 16

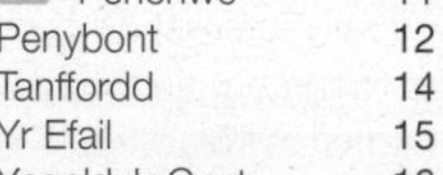

Bwlch y Geuffordd Gardens

THE GARDENS

1 ABERYSTWYTH ALLOTMENTS

5th Avenue, Penparcau, Aberystwyth, SY23 1QT. Aberystwyth Town Council. *On S side of R.Rheidol on Aberystwyth by-pass. From N or E, take A4120 between Llanbadarn & Penparcau. Cross bridge then take 1st R into Minyddol. Allotments ¼m on R.* **Sun 2 July (1-5). Adm £4, chd free. Home-made teas.**

There are 48 plots in total on two sites just a few yards from each other. The allotments are situated in a lovely setting alongside River Rheidol close to Aberystwyth. Wide variety of produce grown, vegetables, soft fruit, top fruit, flowers, herbs and a newly created wildlife pond. Car parking available. For more information contact Brian Heath 01970 617112. Sample tastings from allotment produce. Grass and gravel paths.

2 BRYNGWYN

Capel Seion, Aberystwyth, SY23 4EE. Mr Terry & Mrs Sue Reeves, 01970 880760, sueterr02@btinternet.com. *5m E of Aberystwyth on A4120. On the A4120 between the villages of Capel Seion & Pant y Crug. Parking for up to 15 cars available on site.* **Sat 13 May, Sat 17, Mon 19 June, Mon 28, Wed 30 Aug (1.30-5.30). Adm £4, chd free. Home-made teas. Visits also by arrangement 16 May to 30 June.**

Bluebells in May. Traditional wildflower-rich hay meadows managed for wildlife. As a result of conservation grazing, hedgerow renovation and tree planting, habitat has been restored such that numbers and diversity of wildflowers and wildlife have increased. Explore along mown paths. Small orchard containing Welsh heritage apples and pears, option to gather wildflower seeds in August. Wildflower seed sales.

3 NEW BRYNMEHERYN

Tynygraig, Ystrad Meurig, SY25 6AH. Sharron Johnson, brynmeherynforestry@gmail.com. *Look for yellow NGS signs from Tynygraig. Continue on route past Swyddffynnon Xrds, turn L down sign no through road to end over small bridge From lledrod, look for NGS signs on R.* **Sat 8 July (12-4.30). Adm £5, chd £2.50. Light refreshments. Visits also by arrangement 15 July to 29 July.**

We have 5 acres of gardens separated into different themes. There are three large ponds; the first is purely for wildlife; the second is the tropical area, moss and fernery; the third is a large ornate pond. There is a sunken garden with the most beautiful azaleas in the spring, a vegetable area and bordered back garden, a large grass labyrinth which looks out over the Cambrian mountains.

4 BWLCH Y GEUFFORDD GARDENS

Bronant, Aberystwyth, SY23 4JD. Mr & Mrs J Acres, 01974 251559, gayacres@aol.com, bwlch-y-geufforddd-gardens.myfreesites.net. *12m SE of Aberystwyth, 6m NW of Tregaron off A485. Take turning opp village school in Bronant for 1½m then L up ½m uneven track.* **Visits by arrangement. Cash only please. Max bus size 25 seats. Adm £5, chd £2.50. Tea.**

1000ft high, 3 acre, constantly evolving wildlife and water garden. An adventure garden for children. There are a number of themed gardens, inc Mediterranean, cottage garden, woodland, oriental, memorial and jungle. Plenty of seating. Unique garden sculptures and buildings, inc a cave, temple, gazebo, jungle hut, treehouse and willow den. Musical instruments. Children's adventure garden, musical instruments, pond dipping, dragon hunt, beautiful lake, temple and labyrinth, treehouse, sculptures, wildlife rich, particularly insects and birds. Regret, no dogs allowed.

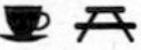

5 BWLCH Y GEUFFORDD

New Cross, Aberystwyth, SY23 4LY. Manuel & Elaine Grande. *5m SE of Aberystwyth. Off A487 from Aberystwyth, take B4340 to New Cross. Garden on R at bottom of small dip. Parking in lay-bys opp house.* **Sun 14 May (10.30-4.30). Adm £4, chd free. Home-made teas.**

Landscaped hillside 1½ acre garden, fine views of Cambrian mountains. Embraces its natural features with different levels, ponds, mixed borders merging into carefully managed informal areas. Banks of rhododendrons, azaleas, bluebells in spring. Full of unusual shade and damp-loving plants, flowering shrubs, mature trees, clematis and climbing roses scrambling up the walls of the old stone buildings. Partial wheelchair access only to lower levels around house. Some steps and steep paths further up the hillside.

6 ◆ CAE HIR GARDENS

Cribyn, Lampeter, SA48 7NG. Julie & Stuart Akkermans, 01570 471116, caehirgardens@gmail.com, caehirgardens.com. *5m W of Lampeter. Take A482 from Lampeter towards Aberaeron. After 5m turn S on B4337. Garden on N side of village of Cribyn.* **For NGS: Sun 9 July (10-5). Adm £6, chd £2. Cream teas. For other opening times and information, please phone, email or visit garden website.**

A Welsh Garden with a Dutch History, Cae Hir is a true family garden of unassuming beauty, made tenable by its innovative mix of ordinary garden plants and wild flowers growing in swathes of perceived abandonment. At Cae Hir the natural meets the formal and riotous planting meets structure and form. A garden not just for plant lovers, but also for design enthusiasts. 5 acres of fully landscaped gardens. Tearoom serving a selection of homemade cakes and scones and a 'Soup of the Day'. Limited wheelchair access.

7 NEW DELFRYN

Llanddewi Brefi, Tregaron, SY25 6SB. Tim Brock, tim.brock108@gmail.com. *Take the youth hostel road out of Llanddewi Brefi & after ¼ m take the 1st R. Follow the narrow lane & take the next sharp L. The garden is further up this lane on the R.* **Visits by arrangement June to Sept for groups of up to 10. Adm £4.50. Tea.**

Unique fern garden with the Plant Heritage national collection of hart's-tongue (asplenium scolopendrium) cultivars. The rock garden is situated on a hillside and the ferns are growing in individual stone circles. There are over a hundred different fern cultivars to see. Guided tour of the fern garden.

NPC

8 FFYNNON LAS

Ffosyffin, Aberaeron, SA46 0HB. Liz Roberts, 01545 571687, lizhomerent@hotmail.co.uk. *A short distance off the A487. 1m S of Aberaeron. Turn off A487 opp The Forge Garage in Ffosyffin, 300m up the road take the L turn at the T junction.* **Sun 11 June (12-4). Adm £5, chd free. Home-made teas. Visits also by arrangement 29 Apr to 17 June for groups of up to 25. Donation to Pancreatic Cancer Research.**

A 2 acre garden that has been in the making for over 15 yrs, Ffynnonlas is a beautiful area that delivers on many different aspects of gardening. There are large lawns, several beds of mature shrubs and flowers. A small lake and two smaller ponds that are separated by a Monet style bridge with lilies. There is a wildflower meadow as a work in progress that has spectacular wild orchids in spring. A rockery with water cascade, vegetable garden with raised beds. Grass paths, level ground.

9 LLANLLYR

Talsarn, Lampeter, SA48 8QB. Mrs Loveday Gee & Mr Patrick Gee, 01570 470900, lgllanllyr@aol.com. *6m NW of Lampeter. On B4337 to Llanrhystud. From Lampeter, entrance to garden on L, just before village of Talsarn.* **Sun 18 June (2-6). Adm £5, chd free. Home-made teas. Visits also by arrangement Apr to Sept.**

Large early C19 garden on site of medieval nunnery, renovated and replanted since 1989. Large pool, bog garden, formal water garden, rose and shrub borders, gravel gardens, rose arbour, allegorical labyrinth and mount, all exhibiting fine plantsmanship. Year-round appeal, interesting and unusual plants. Spectacular rose garden planted with fragrant old fashioned shrub and climbing roses. Specialist Plant Fair by Ceredigion Growers Association. Garden mostly flat.

10 MELINDWR VALLEY BEES, TYNYFFORDD ISAF

Capel Bangor, Aberystwyth, SY23 3NW. Vicky Lines, 01970 880534, vickysweetland@googlemail.com. *Capel Bangor. From Aberystwth take L turn off A44 at E end of Capel Bangor. Follow this road round to R until you see sign with Melindwr Valley Bees.* **Visits by arrangement 31 May to 7 Aug. Individuals & groups up to 10 welcome. Adm £4, chd free. Cream teas £4.50.**

A bee farm dedicated to wildlife with permaculture and forest garden ethos. Fruit, vegetables, culinary and medicinal herbs and bee friendly plants within our wildlife zones. Beehives are situated in dedicated areas and can be observed from a distance. Ornamental and wildflower areas in a cottage garden theme. Formal and wildlife ponds. Grass paths, not suitable for wheelchairs. Vintage tractors and other machinery. Subtropical greenhouse with carnivorous plants, orchids, exotic fruits.

11 NEW PENCNWC

Penuwch, Tregaron, SY25 6RE. Ms Andrea Sutton, 01974 821413, andreasutton@uclmail.net. *7m W of Tregaron on B4577. on the B4577, in the corner where the B4576 turns off in direction of Bwlchllan. This turning is opp cottage called Bear's Hill. Driveway to Pencnwc is 100 yds from the Tregaron Road on R.* **Visits by arrangement for groups of up to 5. Adm £4, chd free.**

7½ acre landscape garden 900ft up with view of Irish Sea between long bare hill and treed hill. Stone and slate house is in the centre. Land reaches out in a star shape around, all sloping down to marshland, now holding a large lake with boathouse. Springs are collected into hillside pools that run to the lake. Woodland, wildlife, especially birds. Walks, bridges, dry stone walls and castles, seats.

12 PENYBONT

Llanafan, Aberystwyth, SY23 4BJ. Norman & Brenda Jones, 01974 261737, tobrenorm@gmail.com. *9m SE of Aberystwyth. Ystwyth Valley. 9m SE of Aberystwyth. B4340 via Trawscoed towards Pontrhydfendigaid. Over stone bridge. ¼m up hill. Lane on R by row of ex Forestry houses.* **Sun 21 May, Sun 6 Aug (11-5). Adm £4, chd free. Home-made teas. Visits also by arrangement May to Aug for groups of up to 30. We welcome your Club or WI for an afternoon or evening.**

Our 10th NGS year. Carefully designed hillside garden backed by Ystwyth Forest and overlooking the valley. An acre of heavenly beauty and continuity of interest from spring bulbs May rhododendrons and azaeleas, roses from June. July brings a Mediterranean feel with large swathes of lavender, grapevines, olives followed by lots of hydrangeas in bright blue, and whites turning to pink. Original design, maturing fast. Seats with stunning views over the borrowed landscape, forest, valley, hills. Kites and buzzards. Sloping ground, gravel paths. Partial wheelchair access.

13 RHOS VILLA

Llanddewi Brefi, Tregaron, SY25 6PA. Andrew & Sam Buchanan. *From Lampeter take the A485 towards Tregaron. After Llangybi turn R opp junction for Olmarch. The property can be found on the R after 1½m.* **Sun 4 June (11-4.30). Adm £4, chd free. Home-made teas.**

A ¾ acre garden creatively utilising local materials. Secret pathways meander through sun and shade, dry and damp. A variety of perennials and shrubs are inter-planted to create interest throughout the year. A productive vegetable and fruit garden with semi-formal structure contrasts the looser planting through the rest of the garden. Interesting and beautifully crafted paths and walls constructed from local materials.

14 TANFFORDD

Swyddffynnon, Ystrad Meurig, SY25 6AW. Jo Kennaugh & Stuart Bradley, 07872 451821, stustart53@outlook.com. *Swyddffynnon. 1m W of Ystrad Meurig. 5m E of Tregaron. Tanffordd is ¼m from Swyddffynnon village on the road to Tregaron. Please do not follow SatNav.* **Visits by arrangement May to Sept for individuals & groups of up to 20. Please discuss refreshments at booking. Adm £4, chd free.**

Wildlife garden, rich in biodiversity, set in a 5 acre smallholding with large pool. There is a small woodland area and various beds of shrubs and herbaceous planting containing some more unusual plants. Vegetables, polytunnel and poultry enclosures with chickens and ducks. Wander along to meet the friendly donkeys and ponies who share a buttercup filled field in late spring and early summer. Large natural pond with an island. Some unusual shrubs and trees.

15 YR EFAIL

Llanio Road, Tregaron, SY25 6PU. Mrs Shelagh Yeomans, 07796 285003, shelagh.yeomans@ngs.org.uk. *3m SW of Tregaron. On B4578 between Llanio & Stag's Head.* **Sun 28 May, Sun 23 July (10-4.30). Adm £4, chd free. Home-made teas. Visits also by arrangement for groups of up to 30.**

Savour one of the many quiet spaces to sit and reflect amongst the informal gardens of relaxed perennial planting, inc a wildlife pond, shaded areas, bog and gravel gardens. Be inspired by the large productive vegetable plots, three polytunnels and greenhouse. Wander along grass paths through the maturing, mostly native woodland. Enjoy home-made teas incorporating homegrown fruit and vegetables. Changing seasonal bilingual quiz sheet. Seasonal vegetables, flowers and plants for sale. Find us on facebook at 'Yr Efail' Gravel and grass paths accessible to wheelchairs with pneumatic wheels.

16 YSGOLDY'R CWRT

Llangeitho, Tregaron, SY25 6QJ. Mrs Brenda Woodley, 01974 821542. *1½m N of Llangeitho. From Llangeitho, turn L at school signed Penuwch. Garden 1½m on R. From Cross Inn take B4577 past Penuwch Inn, R after brown sculptures in field. Garden ¾m on L.* **Sun 25 June (11-5). Adm £4, chd free. Home-made teas. Visits also by arrangement Apr to Oct.**

One acre hillside garden, with four natural ponds which are a magnet for wildlife plus a fish pond. Areas of wildflower meadow, rockery, bog, dry and woodland gardens. Established rose walk. Rare trees, large herbaceous beds with ornamental grasses. Azalea and acer collection in shade bed, bounded by a mountain stream, with two natural cascades and magnificent views. Large Iris ensata and Iris laevigata collections in a variety of colours.

GLAMORGAN

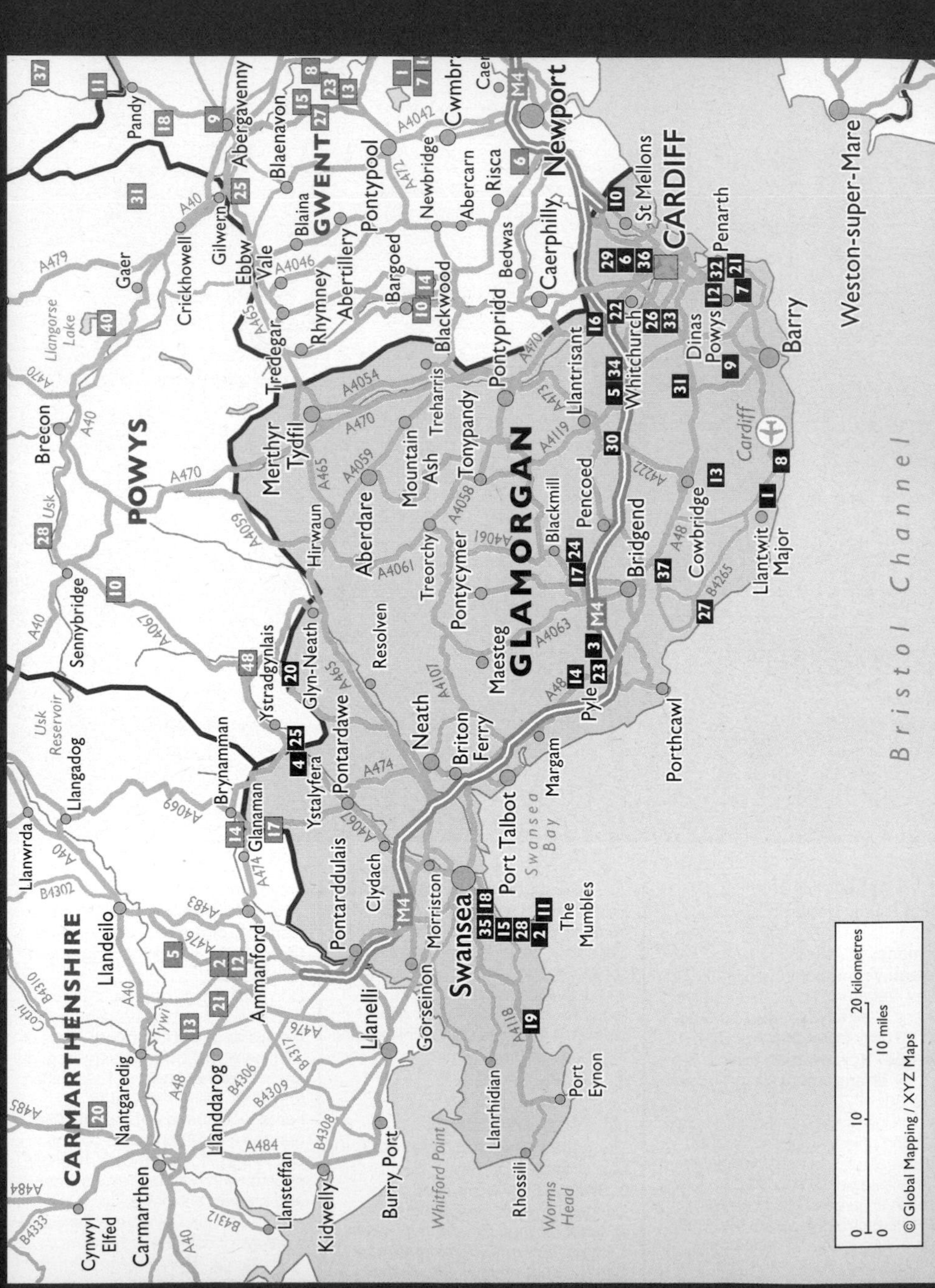

POWYS
GWENT
GLAMORGAN
CARMARTHENSHIRE
CARDIFF
Newport
Swansea
Bristol Channel
Swansea Bay
Brecon
Sennybridge
Usk Reservoir
Llangadog
Llanwrda
Llandeilo
Ammanford
Brynamman
Glanaman
Ystradgynlais
Ystalyfera
Pontardawe
Glyn-Neath
Resolven
Hirwaun
Aberdare
Merthyr Tydfil
Mountain Ash
Treharris
Tonypandy
Treorchy
Pontycymer
Maesteg
Neath
Briton Ferry
Port Talbot
Margam
Pyle
Porthcawl
Bridgend
Pencoed
Blackmill
Llantrisant
Pontypridd
Caerphilly
Whitchurch
St Mellons
Penarth
Dinas Powys
Barry
Cowbridge
Llantwit Major
Weston-super-Mare
Pandy
Abergavenny
Crickhowell
Gilwern
Gaer
Llangorse Lake
Ebbw Vale
Tredegar
Rhymney
Blaina
Blaenavon
Abertillery
Bargoed
Blackwood
Pontypool
Newbridge
Abercarn
Risca
Bedwas
Cwmbran
Llanelli
Gorseinon
Pontarddulais
Clydach
Morriston
The Mumbles
Llanrhidian
Port Eynon
Rhossili
Worms Head
Whitford Point
Burry Port
Kidwelly
Llansteffan
Carmarthen
Cynwyl Elfed
Nantgaredig
Llanddarog
0 10 20 kilometres
0 10 miles
© Global Mapping / XYZ Maps

Glamorgan is a large county stretching from the Brecon Beacons in the north to the Bristol Chanel in the south, and from the city of Cardiff in the east to the Gower Peninsula in the west.

We are delighted to be opening our amazing gardens once again. Our garden owners cannot wait to share their passion with you and are eager to welcome you through their garden gates.

One of our beneficiaries, Horatio's is opening for the first time at Horatio's Garden Wales Llandough Hospital, Cardiff as well as 3 Community Gardens; V21 Community Garden Cardiff, Victoria Community Garden Penarth and St. Peter's Community Hall & Garden Cardiff.

Look out for gardens that are opening By Arrangement, you can then choose when you would like to visit. A quick email or phone call will transport you to a garden opening just for you.

Very many thanks also for your support, enjoy your garden visits and do visit our website for the most up to date information, plus "Pop Up" openings, our Beneficiaries and thousands of wonderful garden images www.ngs.org.uk

Mae Morgannwg yn ymestyn o Fannau Brycheiniog yn y gogledd i Fôr Hafren yn y de, ac o ddinas Caerdydd yn y dwyrain i Benrhyn Gŵyr yn y gorllewin.

Hyfrydwch o'r mwya' fydd cael agor ein gerddi unwaith eto. Mae'r perchnogion fel bob amser yn awyddus i rannu eu brwdfrydedd gyda chi ac i'ch croesawu i'w gerddi.

Mae un o'n buddiolwyr, Horatio yn agor am y tro cynta - Gardd Horatio Cymru yn Ysbyty Llandochau Caerdydd.

Mae tair gardd gymunedol yn agor hefyd - Gerddi Cymunedol V21 Caerdydd, Gerddi Cymunedol Fictoria ym Mhenarth, a Neuadd a Gerddi Cymunedol St Peter yng Nghaerdydd.

Peidied anghofio'r gerddi hynny sy'n agor drwy drefniant. Cewch ddewis pryd hoffech ymweld. Bydd galwad ffôn neu e-bost yn ddigon i sicrhau ymweliad personol i chi.

Diolchwn i chi hefyd am eich cefnogaeth. Mwynhewch eich ymweliadau â'n gerddi, ac ymwelwch â'n gwefan am y wybodaeth ddiweddara gan gynnwys gerddi 'pop up', gwybodaeth am ein Buddiolwyr a rhai miloedd o luniau gerddi. www.ngs.org.uk

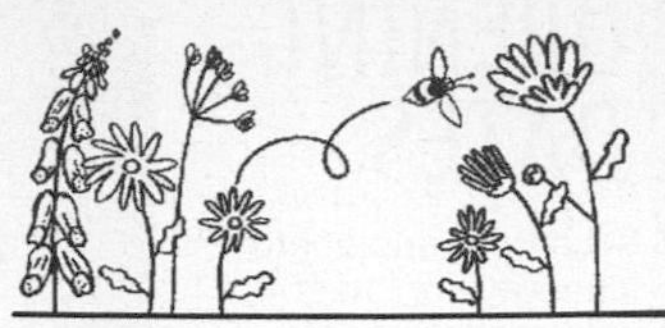

Volunteers

County Organiser
Rosamund Davies
01656 880048
rosamund.davies@ngs.org.uk

County Treasurer
Steven Thomas 01446 772339
steven.thomas@ngs.org.uk

Publicity
Rhian James 07802 438299
rhian.james@ngs.org.uk

Social Media – Instagram & Twitter
Rhian Rees 01446 774817
rhian.rees@ngs.org.uk

Social Media – Facebook
Tony Leyshon 07896 799378
anthony.leyshon@icloud.com

Booklet Co-ordinator
Lesley Sherwood 02920 890055
lesley.sherwood@ngs.org.uk

Talks Co-ordinator
Frances Bowyer
02920 892264
frances.bowyer@ngs.org.uk

Health and Gardens Co-ordinator
Miranda Workman 02920 766225
miranda.parsons@talktalk.net

Community Gardens Co-ordinator
Cheryl Bass 07969 499967
cheryl.bass@ngs.org.uk

Assistant County Organisers
Dr. Isabel Graham
isabelgraham84@gmail.com

Sol Blytt Jordens 01792 391676
sol.blyttjordens@ngs.org.uk

Ceri Macfarlane 01792 404906
ceri@mikegravenor.plus.com

Sarah Boorman 07969 499967
sarah.boorman@ngs.org.uk

@GlamorganNGS
@GlamorganNGS
@ngsglamorgan

OPENING DATES

All entries subject to change. For latest information check **www.ngs.org.uk**

Map locator numbers are shown to the right of each garden name.

February

Snowdrop Festival

Sunday 19th
Slade 27

April

Saturday 15th
Slade 27

Sunday 16th
Gileston Manor 8
Slade 27

Sunday 30th
Hen Felin & Swallow Barns 9

May

Sunday 7th
9 Willowbrook Gardens 35

Monday 8th
9 Willowbrook Gardens 35

Saturday 20th
110 Heritage Park 10
Maggie's Swansea 18

June

Friday 2nd
100 Pendwyallt Road 22

Saturday 3rd
17 Maes y Draenog 16
100 Pendwyallt Road 22
NEW 9 Railway Terrace 24

Sunday 4th
17 Maes y Draenog 16
Y Bwthyn 37

Horatio's Garden Cardiff

Saturday 10th
Dinas Powys 7
Uplands 31

Sunday 11th
The Cedars 2
Dinas Powys 7
NEW 11 Windermere Avenue 36

Saturday 17th
NEW Boverton House 1
Penarth Gardens 21

Sunday 18th
Creigiau Village Gardens 5
22 Dan-y-Coed Road 6
NEW Maes Canol 15
Penarth Gardens 21
28 Slade Gardens 28

Friday 23rd
NEW V21 Community Garden 33

Saturday 24th
NEW Boverton House 1
4 Clyngwyn Road 4
Rhos y Bedw 25

Sunday 25th
Cefn Cribwr Garden Club 3
4 Clyngwyn Road 4
Hen Felin & Swallow Barns 9
Rhos y Bedw 25

July

Saturday 1st
NEW Horatio's Garden Cardiff 12
St Peter's Community Hall & Garden 26

Sunday 2nd
4 Hillcrest 11
2 Llys Castell 14
St Peter's Community Hall & Garden 26
12 Uplands Crescent 32

Friday 7th
NEW Mulberry Hill 19

Sunday 9th
Gileston Manor 8
NEW Mulberry Hill 19
12 Uplands Crescent 32

Saturday 15th
38 South Rise 29

Sunday 16th
Maes-y-Wertha Farm 17
38 South Rise 29

Saturday 29th
Pwyllygarth Farm 23
Swn y Coed 30

Sunday 30th
Pwyllygarth Farm 23
Swn y Coed 30

August

Sunday 13th
12 Uplands Crescent 32

Sunday 20th
12 Uplands Crescent 32
Waunwyllt 34

February 2024

Sunday 11th
Slade 27

By Arrangement

Arrange a personalised garden visit with your club, or group of friends, on a date to suit you. See individual garden entries for full details.

22 Dan-y-Coed Road 6
Gileston Manor 8
Llandough Castle 13
17 Maes y Draenog 16
Nant Melyn Farm 20
12 Uplands Crescent 32
9 Willowbrook Gardens 35

THE GARDENS

1 NEW BOVERTON HOUSE

Boverton, Llantwit Major, CF61 1UH. Mr John Wainwright. *Boverton at the E end Llantwit Major. Garden entrance is approx 50m E of the Boverton Post Office, located on Boverton Rd where it bends round heading N to the B4265 main airport road. Please follow yellow NGS signs.* **Sat 17, Sat 24 June (10-5). Adm £4, chd free.**

A walled garden next to the River Hoddnant, recently brought back to use. Formerly an orchard and kitchen garden, the garden has a unique character and has been designed to create a beautiful garden space and encourage wildlife, whilst retaining its unique historic character. Wheelchair access throughout the garden. However, access from the public road into the garden does involve a raised kerb which may require assistance.

2 THE CEDARS

20A Slade Road, Newton, Swansea, SA3 4UF. Mr Ian & Mrs Madelene Scott. *Take the A4067 to Oystermouth, turn R at White Rose pub cont along Newton Rd keeping R at fork to T-junction, turn R & 1st R into Slade Rd, follow yellow signs.* **Sun 11 June (2-5). Adm £5, chd free. Home-made teas.**

South facing garden on a sloping site consisting of rooms subdivided by large shrubs and trees. Small kitchen garden with greenhouse. Ornamental pond. Herbaceous perennials, number of fruit trees. It is a garden which affords all year interest with azalea, camellias, rhododendrons, magnolias and hydrangeas.

GROUP OPENING

3 CEFN CRIBWR GARDEN CLUB

Cefn Cribwr, Bridgend, CF32 0AP. www.cefncribwrgardeningclub.com. *5m W of Bridgend on B4281.* **Sun 25 June (11-5). Combined adm £6, chd free.**

6 BEDFORD ROAD
Carole & John Mason.

13 BEDFORD ROAD
Mr John Loveluck.

2 BRYN TERRACE
Alan & Tracy Birch.

CEFN CRIBWR GARDEN CLUB ALLOTMENTS
Cefn Cribwr Garden Club.

CEFN METHODIST CHURCH
Cefn Cribwr Methodist Church.

77 CEFN ROAD
Peter & Veronica Davies & Mr Fai Lee.

25 EAST AVENUE
Mr & Mrs D Colbridge.

15 GREEN MEADOW
Tom & Helen.

HILL TOP
Mr & Mrs W G Hodges.

6 TAI THORN
Mr Kevin Burnell.

3 TY-ISAF ROAD
Mr Ryland & Mrs Claire Downs.

Cefn Cribwr is an old mining village atop a ridge with views of Swansea to the west, Somerset to the south and home to Bedford Park and the Cefn Cribwr Iron Works. The village hall is at the centre with teas, cakes and plants for sale. The allotments are to be found behind the hall. Children, art and relaxation are just some of the themes to be found in the gardens besides the flower beds and vegetables. There are also water features, fish ponds, wildlife ponds, summerhouses and hens adding to the diverse mix in the village. Themed colour borders, roses, greenhouses, recycling, composting and much more. The chapel grounds are peaceful with a meandering woodland trail. Visitors may travel between gardens courtesy of the Glamorgan Iron Horse Vintage Society.

4 4 CLYNGWYN ROAD

Ystalyfera, Swansea, SA9 2AE. Paul Steer, www.artinacorner.blogspot.com. *Follow NGS signs from Rhos y Bedw.* **Sat 24, Sun 25 June (1-5). Adm £3, chd free. Light refreshments. Open nearby Rhos y Bedw.**

The Coal Tip Garden is a small personal space created in order to help us relax. Its main character is enclosure and a sense of rest. It is not a flowery garden but is formed out of shrubs and trees - forming a tapestry of hedging with arches and niches being cut out in order to place seats and sculpture and to produce a visual rhythm. A key philosophy is using native perennials. An intimate space designed for rest and contemplation.

GROUP OPENING

5 CREIGIAU VILLAGE GARDENS

Maes Y Nant, Creigiau, CF15 9EJ. *W of Cardiff (J34 M4). From M4 J34 follow A4119 to T-lights, turn R by Castell Mynach Pub, pass through Groes Faen & turn L to Creigiau. Follow NGS signs.* **Sun 18 June (11-5). Combined adm £7, chd free. Home-made teas at 28 Maes y Nant.**

28 MAES Y NANT
Mike & Lesley Sherwood.

26 PARC Y FRO
Mr Daniel Cleary.

WAUNWYLLT
John Hughes & Richard Shaw.
(See separate entry)

Creigiau Village Gardens - 3 vibrant and innovative gardens. Each quite different, they combine some of the best characteristics of design and planting for modern town gardens as well as cottage gardens. Each has its own forte; Waunwyllt's ½ acre is divided into garden rooms; 28 Maes y Nant, cottage garden planting reigns; 26 Parc y Fro is newly planted with wildlife at its heart. Enjoy a warm welcome, home-made teas and plant sales.

Our 2022 donation to Hospice UK meant that 911 NHS and Social Care staff accessed individual counselling support

6 22 DAN-Y-COED ROAD

Cyncoed, Cardiff, CF23 6NA. Alan & Miranda Workman, 029 2076 6225, miranda.parsons@talktalk.net. *Dan y Coed Rd leads off Cyncoed Rd at the top & Rhydypenau Rd at the bottom. No 22 is at the bottom of Dan y Coed Rd. There is street parking & level access to the R of the property.* **Sun 18 June (2-6). Adm £5, chd free. Home-made teas. Visits also by arrangement Apr to Sept for groups of 6 to 20.**
A medium sized, much loved garden. Owners share a passion for plants and structure, each year the lawn gets smaller to allow for the acquisition of new features. Hostas, ferns, acers and other trees form the central woodland theme as a backdrop is provided by the Nant Fawr woods. Year-round interest has been created for the owner's and visitor's greater pleasure. There is a wildlife pond, many climbing plants and a greenhouse with cacti, succulents and pelargoniums.

GROUP OPENING

7 DINAS POWYS

Dinas Powys, CF64 4TL. *Approx 6m SW of Cardiff. Exit M4 at J33, follow A4232 to Leckwith, onto B4267 & follow to Merry Harrier T-lights. Turn R & enter Dinas Powys. Follow yellow NGS signs.* **Sat 10, Sun 11 June (11-5). Combined adm £5, chd free. Donation to Dinas Powys Voluntary Concern & Dinas Powys Community Library.**

21 CARDIFF ROAD
Rob & Pam Creed.

23 CARDIFF ROAD
Eoghan Conway & David Manfield.

30 MILLBROOK ROAD
Mr & Mrs R Golding.

32 MILLBROOK ROAD
Mrs G Marsh.

NIGHTINGALE COMMUNITY GARDENS
Keith Hatton.

THE POUND
Helen & David Parsons.

5 SIR IVOR PLACE
Ceri Coles.

WEST CLIFF
Alan & Jackie Blakoe.

There are many gardens to visit in this small friendly village, all with something different to offer. Displays of vegetables and fruit in the community garden, gardens with exuberant planting providing yearlong interest, wildlife ponds, chickens and a mini vineyard. There will also be plants for sale and refreshments in the different gardens. There are many restful and beautiful areas to sit and relax. Good wheelchair access at the community gardens. Partial access elsewhere.

8 GILESTON MANOR

Gileston, Vale of Glamorgan, CF62 4HX. Lorraine & Joshua, 01446 754029, enquiries@gilestonmanor.co.uk, www.gilestonmanor.co.uk. *From Cardiff airport take B4265 to Llantwit Major. After 3m turn L at petrol station. Go under bridge, turn R follow yellow NGS signs. No turning at Cenotaph.* **Sun 16 Apr, Sun 9 July (11-4). Adm £6, chd free. Home-made teas. Visits also by arrangement Mar to Oct for groups of 5+.**
Surrounded by beautiful Welsh countryside with breathtaking views across the jurassic coast, you will find our historic 9 acre estate. Gileston Manor, in the Vale of Glamorgan, has been loved and lived in since 1320 and recently, lovingly renovated and restored by the current owners. There is so much to see and explore and will definitely delight any keen gardener - a perfect blend of old and new. The garden bar will also be open for drinks. From the parking area we have flat gravel pathways to access the Walled Garden and The Rookery.

GROUP OPENING

9 HEN FELIN & SWALLOW BARNS

Vale of Glamorgan, Dyffryn, CF5 6SU. 07961 542452, janet.evans54@me.com. *Dyffryn village. 3m from Culverhouse. Cross r'about take the A48 up the hill to st Nicholas, L at T-lights, past Dyffryn. House, R at T-junction & follow the road to big yellow sign.* **Sun 30 Apr (11-5); Sun 25 June (10-6.30). Combined adm £5, chd free. Home-made teas in Hen Felin.**
Dyffryn hamlet is a hidden gem in the Vale of Glamorgan; despite the lack of a pub there is a fantastic community spirit. Yr Hen Felin: Beautiful cottage garden with stunning borders, breathtaking wildflower meadows, oak tree with surrounding bench, 200 year old pig sty, wishing well, secret garden with steps to river, lovingly tended vegetable garden and chickens. Mill stream running through garden with adjacent wildflowers. Swallow Barns: Cross the bridge over the river and pass under the weeping ash to enter Swallow Barns garden, a mature garden with packed herbaceous borders - formal and informal, orchard, hens, herb garden, lavender patio, willow arches leading to woodland walk and deep secluded pond. We welcome all visitors - old and new. Unprotected river access, children must be supervised in both gardens. Home-made jams, Pimm's at the pond, cheese and wine, artists' stall, face painting, white elephant stall, jewellery stall and plant stall. Wheelchair access possible to most of the two gardens.

10 110 HERITAGE PARK

St Mellons, Cardiff, CF3 0DS. Sarah Boorman. *Leave A48 at St Mellons junction, take 2nd exit at r'about. Turn R to Willowdene Way & R to Willowbrook Drive. Heritage Park is 1st R. Park outside the cul-de-sac.* **Sat 20 May (1-5). Adm £4, chd free. Home-made teas. Gluten free options available.**
An unexpected gem within a modern housing estate. Evergreen shrubs, herbaceous borders, box topiary and terracotta pots make for a mix between modern cottage garden and Italian style. With numerous seating areas and a few quirky surprises this small garden is described by neighbours as a calm oasis. The garden would be able to be viewed from the patio area.

11 4 HILLCREST

Langland, Swansea, SA3 4PW. Mr Gareth & Mrs Penny Cross. *Approach from L, turn at Langland corner on Langland Rd Mumbles; take first L. off Higher Lane; take 3rd L. up Worcester Rd & bungalow faces you at top.* **Sun 2 July (1.30-5). Adm £5, chd free. Home-made teas.**
Redesigned and developed over the last 9 years, our hilltop garden backs

27 Victoria Road

onto woodland overlooking Swansea bay and attracts many birds. We have planted large herbaceous borders with mainly perennial plants/shrubs to provide interest and colour through the seasons. A wildlife pond and fruit/vegetable garden add to interest.

12 NEW HORATIO'S GARDEN CARDIFF

Penlan Road, Llandough, Penarth, CF64 2XX. Owen Griffiths, www.horatiosgarden.org.uk. *University Hospital Llandough. Past main entrance to the University Hospital opp the main car park & by the spinal ward. Parking is free for 4 hrs.* **Sat 1 July (12-4). Adm £5, chd free. Light refreshments.**

Sarah Price's design was inspired by the Welsh landscape. The planting inc valerian, Welsh poppy, field maple, and crab apples. Circular openings within the perimeter fence frame the countryside and the sea in the distance. The fence is covered with climbers inc clematis, grapes, and roses. It provides patients, visitors and staff with a beautiful place to spend time in the garden year-round. Fully wheelchair accessible.

13 LLANDOUGH CASTLE

Llandough, Cowbridge, CF71 7LR. Mrs Rhian Rees, 07802 438299, rhianedrees@gmail.com. *1½m outside Cowbridge. At the T- lights in Cowbridge turn onto the St Athan Rd. Continue then turn R to Llandough. Drive into the village follow the car park signs. The car park is a 5 min walk from the gardens.* **Visits by arrangement 6 Feb to 15 Sept for groups of 10 to 30. Adm £6.50, chd free. Home-made teas.**

Set within castle grounds and with a backdrop of an ancient monument, the 3½ acres of garden inc a potager with a hint of the Mediterranean, formal lawns and herbaceous beds, a wildlife pond with waterfall and a woodland garden with stumpery and sculpture. Over 90,000 spring bulbs have been planted inc snowdrops, narcissi and tulips. Potager, wildlife pond, formal borders, stumpery.

14 2 LLYS CASTELL

Coed Hirwaun, Port Talbot, SA13 2UX. Mair & John Jones. *Coed Hirwaun is situated on the A48, between Margam Country Park & Pyle, please follow yellow signs.* **Sun 2 July (11-5). Adm £5, chd free. Light refreshments.**

Approx. ⅓ acre featuring raised beds planted with a mix of cottage style plants. There are vegetable beds and soft fruit beds along side a small ornamental fish pond. The garden is split into two parts by a stream, the second part is a woodland containing mature trees, accessed by a bridge or pathway. New this year is a gabion wall with raised herbaceous planting leading to a wildlife pond.

15 NEW MAES CANOL
Mayals Green, Mayals, Swansea, SA3 5JR. Dewi & Cathy Evans. *Garden entrance is close to Fairwood Rd & the house is accessed by a 100m long driveway which runs parallel with Fairwood Rd.* **Sun 18 June (1-5). Adm £5, chd free. Home-made teas.**
Maes Canol has a large garden surrounded by traditional houses in a suburb of Swansea. The garden is mature and planted with trees and shrubs. There is a courtyard garden and terrace close to the house. A pond provides a focal point in the garden at a distance from the house.

16 17 MAES Y DRAENOG
Tongwynlais, Cardiff, CF15 7JL. Mr Derek Price, 07717 462295, derekprice7@btinternet.com. *N of M4, J32. From S: M4 , J32, take A4054 into village. R at Lewis Arms pub, up Mill Rd. 2nd R into Catherine Drive, park in signed area (no parking in Maes y Draenog). Follow signs to garden.* **Sat 3, Sun 4 June (12-5). Adm £5, chd free. Home-made teas. Visits also by arrangement 3 June to 30 June for groups of up to 16.**
A hidden garden, in the shadow of Castell Coch, fed by a mountain stream having a wooden footbridge to naturalised areas and woodland area. With wonderful late spring flower displays, against a woodland backdrop. Developed over many years with a wide variety of lavender beds and herbaceous borders, both front and rear late spring and summer displays, summerhouse, patios, greenhouse and vegetable area. A good variety of plants in different borders around house. Rear of house is set against woodland and fields, while front areas have mature roses and herbaceous plants, borders.

17 MAES-Y-WERTHA FARM
Bryncethin, CF32 9YJ. Stella & Tony Leyshon. *3m N of Bridgend. Follow sign for Bryncethin, turn R at Masons Arms. Follow sign for Heol-y-Cyw garden about 1m outside Bryncethin on R.* **Sun 16 July (12-7). Adm £5, chd free. Home-made teas. Pimms. Live music all day.**
A 3 acre hidden gem outside Bridgend. Entering the garden you find a small Japanese garden fed by a stream, this leads you to informal mixed beds and enclosed herbaceous borders. Ponds and rill are fed by a natural spring. A meadow with large lawns under new planting gives wonderful vistas over surrounding countryside. Mural in the summerhouse by contemporary artist Daniel Llewelyn Hall. His work is represented in the Royal Collection and House of Lords. Fresh hand made sandwiches available.

18 MAGGIE'S SWANSEA
Singleton Hospital, Sketty Lane, Sketty, Swansea, SA2 8QL. Maggie's Swansea, www.maggies.org/swansea. *On the grounds of Singleton Hospital. Next to the Genetic Building & close to the chemotherapy day unit at the back of the main hospital. Please follow orange signs.* **Sat 20 May (10-12). Adm £3, chd free. Tea, coffee, squash, cakes & biscuits.**
Maggie's Swansea's gardens wrap around the building, and overlook into Swansea Bay. The garden, designed by Kim Wilkie, attracts wildlife, heightening the natural and tranquil feel, and there is also a fully functional allotment. The Centre sits among a small wooded area, and the wings of the design also help to shelter the outside seating areas, meaning that visitors can enjoy sitting out for as much of the year as possible. Decking all the way around the building.

19 NEW MULBERRY HILL
Penmaen, Gower, Swansea, SA3 2HQ. Mrs Sian Burgess, 01792 371344, sianandpeter@btinternet.com. *10m W of Swansea. A4118 to Penmaen. Turn at church. Car Park after cattle grid can be used for visiting garden(300 yds). Limited parking at Mulberry Hill.* **Fri 7, Sun 9 July (1-5). Adm £6, chd free. Home-made teas.**
Elements of a Victorian garden with some original features. Hidden behind a high wall is a garden of "rooms" with sea views: a yew tunnel and flower beds surround a fish pond; a terrace with borders leads down to a lawn and herbaceous borders and eventually to a lower garden with pond, bog garden and meadow area. Mature trees, hedges and shrubs throughout.

20 NANT MELYN FARM
Seven Sisters, Neath, SA10 9BW. Mr Craig Pearce, cgpearce@hotmail.co.uk. *From J43 on M4 take A465 towards Neath. Exit for Seven Sisters at r'about take 3rd exit, 6m for Seven Sisters you come to Pantyffordd sign, turn L under low bridge.* **Visits by arrangement May to Aug for groups of up to 20. Adm £4.50, chd free. Tea.**
A spacious interesting garden with many features which inc a stunning natural waterfall, a meandering woodland stream covered by a canopy of entwined trees, a picturesque Japanese garden, beautiful lawned areas with winding pathways.

GROUP OPENING

21 PENARTH GARDENS
Penarth, CF64 3HY. *Penarth. Follow yellow signs to Lower Penarth.* **Sat 17, Sun 18 June (10-5). Combined adm £6, chd free. Home-made teas at 27 Victoria Road, Penarth. Vegan & gluten free options.**

NEW 6 MELIDEN ROAD
Suzanne Hughes.

NEW 5 ROSEBERRY PLACE
Cathryn Mayo.

NEW SEAFIELD HOUSE
Anja Ernest.

NEW VICTORIA COMMUNITY GARDEN
Haydn Mayo.

NEW 27 VICTORIA ROAD
Emma Berry.

During the Victorian Era, Penarth was known as the 'Garden by the Sea'. Come and visit a selection of traditional and contemporary town gardens all within walking distance of each other. There is also a chance to visit an exciting community garden in the grounds of All Saints Church.

22 100 PENDWYALLT ROAD
Whitchurch, Cardiff, CF14 7EH. *Near J32 of M4. Opposite side of rd to The Village Hotel. Road runs parallel to main rd to Whitchurch village. Enter cul-de-sac opposite side of road to Whitworth Sq. Go uphill in the cul-de-sac past 3 blocks*

Clyngwyn Road

of flats. **Fri 2, Sat 3 June (12-5). Adm £5, chd free.**
6 year old large garden divided into three areas all attracting wildlife. Dense planting of interesting trees, shrubs, grasses and bamboos. Absence of lawns. Long pergola. Huge ponds. A place to lose yourself. Tarmac and coarse gravel allow access and good views of the garden.

23 PWYLLYGARTH FARM

Garth Street, Kenfig Hill, Bridgend, CF33 6EU. Cheryl Bass. *In Kenfig Hill. Farmhouse directly off main road in Kenfig Hill. At the Spar follow yellow signs.* **Sat 29, Sun 30 July (10-5). Adm £5, chd free. Light refreshments.**
Pwllygarth Farm, built by Welsh heiress Emily Charlotte Talbot in 1904 as a working farm. The front and side gardens are a mix of rose gardens, sedum planted walls, raised rockery beds, small woodland garden and a landscaped paved area. The relaxing rear garden offers a Japanese garden, butterfly rockery, summerhouse, reading corner and a wildlife pond incorporating grasses and creative planting. To the rear of the property is a car park area that will feature some classic cars. The garden is accessible via a wheelchair and there are no steps that split the garden areas.

24 NEW 9 RAILWAY TERRACE

Heol Laethog, Bryncethin, Bridgend, CF32 9JE. Ms Sue Deary. *Approx 1m from Bryncethin & just under a mile from Pencoed. Along the common road over the railway bridge, follow road along by the children's park over a cattle grid then follow yellow signage.* **Sat 3 June (10.30-4.30). Adm £3.50, chd free. Home-made teas.**
An old miner's cottage circa 1900 on Coity common with a sloping small south facing garden that has been terraced and has two different levels divided by some steps. There are a number of fruit trees inc apple, pear and fig trees, also magnolias and a palm tree. There are herbaceous borders planted with roses and many other plants. There is a seated area on the top terrace and a rose covered arbour on the lower level and a small water feature with uninterrupted views of the common. The main ethos of the garden is to try and keep it as a cottage garden inc reusing and repurposing various items inc old water tanks for planting and planting vegetables and flowers side by side.

25 RHOS Y BEDW

4 Pen y Wern Rd, Ystalyfera, Swansea, SA9 2NH. Robert & Helen Davies. *13m N of Swansea. M4 J45 take A4067. Follow signs for Dan yr Ogof caves across 5 r'abouts. After T-lights follow yellow NGS signs. Parking above house on road off to R.* **Sat 24, Sun 25 June (1-5). Adm £3.50, chd free. Home-made teas. Gluten free options.**
A haven of peace and tranquillity with spectacular views, this glorious compact garden with its amazing array of planting areas is constantly evolving. Our diverse planting areas inc cottage, bank and bog gardens also an array of roses and a knot garden are sure to provide inspiration. A garden with something different around every corner to be savoured slowly, relax and enjoy. Range of craft items made by the garden owners will be on sale with a donation to the NGS.

110 Heritage Park

26 ST PETER'S COMMUNITY HALL & GARDEN

211 St Fagans Road, Fairwater, Cardiff, CF5 3DW. St Peter's Church, www.stpeterscommunitygarden.org.uk. *On the St Fagans Rd opp Gorse Place. Next door to Church. A48 to Culverhousecross r'bout take A48 Cowbridge Rd West to Ely r'bout 1st L. At T-lights go L B4488 to Fairwater Green, follow yellow NGS signs.* **Sat 1, Sun 2 July (11-4). Adm £4.50, chd free. Home-made teas.**

Secret garden in city suburb. Unusual combination of flower beds, vegetable beds and nature reserve, all created by volunteers. Features inc two natural ponds, wild flowers, fruit bushes and herb border. Welsh heritage apple trees, seasonal vegetable beds, polytunnel and greenhouses. Quiet Garden for tranquil contemplation. Rainbow garden for children's activities. Two beehives. Sunny patio and community hall for all day refreshments and to browse the plants, crafts and preserves for sale. Buy a raffle ticket for our "Honey Draw" and enjoy our city oasis. Disabled WC available. Most of the Garden has wheelchair access.

27 SLADE

Southerndown, CF32 0RP. Rosamund & Peter Davies, 01656 880048, rosamund.davies@ngs.org.uk, www.sladeholidaycottages.co.uk. *5m S of Bridgend. M4 J35 Follow A473 to Bridgend. Take B4265 to St. Brides Major. Turn R in St. Brides Major for Southerndown, then follow yellow NGS signs.* **Sun 19 Feb (2-5), Sat 15, Sun 16 Apr (11-5). Adm £6, chd free. Home-made teas. 2024: Sun 11 Feb.**

Hidden away Slade garden is an unexpected jewel to discover next to the sea with views overlooking the Bristol Channel. The garden tumbles down a valley protected by a belt of woodland. In front of the house are delightful formal areas a rose and clematis pergola and herbaceous borders. From terraced lawns great sweeps of grass stretch down the hill enlivened by spring bulbs and fritillaries. Heritage Coast wardens will give guided tours of adjacent Dunraven Gardens with slide shows every hour from 2pm (Apr opening only). Partial wheelchair access.

28 28 SLADE GARDENS

West Cross, Swansea, SA3 5QP. Pete & Helen Shetherline. *At the end of Slade Gardens facing Oystermouth Cemetery. Wheelchair drop off only in Slade Gdns. Parking in cemetery car park at top of lane from Newton Rd indicated with white sign. Approx 150 metres level walk on tarmac. Additional parking on Bellevue Rd.* **Sun 18 June (1.30-5). Adm £5, chd free. Home-made teas.**

This is a renovation of the old garden of the original Lodge to Oystermouth Cemetery. It is tucked under the steep limestone woodland at the eastern end of the cemetery valley and faces south overlooking the Victorian part of the cemetery. It is planted in cottage garden style with perennials and selected shrubs on the side of the hill separated from a lawn by a limestone wall. The garden can be enjoyed from the terrace of the house which is reached without climbing steps and where tea and cakes can be enjoyed. The view over the garden and 'hidden Valley' of the cemetery is a lovely surprise even for those who know Mumbles well. The scented roses should be in full bloom.

29 38 SOUTH RISE

South Rise, Llanishen, Cardiff, CF14 0RH. Dr Khalida Hasan. *N of Cardiff, from Llanishen Village Station Rd past Train Stn go R down The Rise or further down onto S Rise directly. Following yellow signs.* **Sat 15, Sun 16 July (11-5). Adm £5.50, chd free. Light refreshments. Cakes & South Asian savouries (e.g. samosa, chick pea chaat, pakora) available.**

A relatively new garden backing on to Llanishen Reservoir gradually establishing with something of interest and colour all year-round. Herbaceous borders, vegetables and fruit plants surround central lawn. Wildlife friendly; variety of climbers and exotics. In front shrubs and herbaceous borders to a lawn. Stepping stones leading to children's play area and vegetable plot also at the back. Wheelchair access to rear from the side of the house.

30 SWN Y COED

Tyla Garw, Pontyclun, CF72 9HD. Mair & Owen Hopkin. *N of Pontyclun. M4 J34, A4119, at r'about 1st exit A473, through lights, l at r'about,2nd l r'about, over Xing immed R past pub. 1st house on R after forestry entrance.* **Sat 29, Sun 30 July (11-4). Adm £5, chd free. Light refreshments inc gluten & dairy free options.**

This family friendly garden started from a blank canvas 5 years ago, initially laid to lawn. Raised vegetable beds were installed and a 75m natural hedge planted along the side boundary to encourage wildlife. This was supplemented with fruit trees, flower and herb borders to attract insects. The lawn provides a clearing to the surrounding forestry attracting a variety of birds. Greenhouse. Disabled parking on drive. Ramped access to rear patio and WC. Path adjacent to herb border.

31 UPLANDS

Gwern-y-Steeple, Peterston Super Ely, CF5 6LG. David Richmond. *From the A48 between St Nicholas & Bonvilston take the Peterston Super Ely turning & follow yellow NGS arrow, park at small green near Gwern-y-Steeple sign.* **Sat 10 June (12-5). Adm £4, chd free.**

Uplands has a rustic heart within a cottage garden design. The front is vegetables and currant bushes. The back was planted in 2019, on once all grass. There are four main herbaceous borders, young fruit trees, ornamental grasses, ferns, shrubs and roses, with many places to relax and enjoy the garden. Medium sized greenhouse with citrus and fig tree. Several sculptures located around the garden.

32 12 UPLANDS CRESCENT

Llandough, Penarth, CF64 2PR. Mr Dean Mears, 07910 638682, dean.mears@ntlworld.com. *Head for Llandough Hospital & a small turning head with a Weeping Willow clearly visible in the front garden. Follow NGS signs.* **Sun 2, Sun 9 July, Sun 13, Sun 20 Aug (2-4.30). Adm £4, chd free. Visits also by arrangement May to Sept for groups of up to 6.**

An exotic medium size garden full of variety, unusual and big plants, an unusual fern collection and a pond. Banana plants and Gunnera fill the corners and a Foxglove tree. Tree Ferns, Phormium, Tetrapanax, Cordyline and Giant Reed add height to the garden. An area of potted tropical plants, Brugmansia, Canna, Alocasia, Strelitzia and palms add further interest.

33 NEW V21 COMMUNITY GARDEN

Sbectrwm, Bwlch road, Cardiff, CF5 3EF. Mr Roy Bailey, www.V21.org.uk. *From the A48 turn R onto St Fagans Rd. Continue onto Norbury Rd leading to Finchley rd & Bwlch Rd.* **Fri 23 June (10-4). Adm £4.50, chd free. Home-made teas at the on-site café.**

The V21 Community Garden is developed and maintained by students with additional needs. The garden is designed to be an accessible learning environment. The garden has a newly developed herbaceous flower border, a wildlife pond, and shade garden and areas for growing vegetables. There is also a new polytunnel to extend the growing season. Wheelchairs can access via main paths.

34 WAUNWYLLT

Pant y Gored, Creigiau, Cardiff, CF15 9NF. John Hughes & Richard Shaw. *Heol Pant y Gored, Creigiau.* **Sun 20 Aug (11-5). Adm £5, chd free. Home-made teas. Opening with Creigiau Village Gardens on Sun 18 June.**

Waunwyllt is a garden of approx ½ an acre, divided into several 'garden rooms'. Winding paths lead the visitor through a tranquil garden set against a woodland backdrop with many secluded seating areas along the way and something of interest around every corner. Colour coordinated borders and strong design contribute to an attractive and stylish garden.

35 9 WILLOWBROOK GARDENS

Mayals, Swansea, SA3 5EB. Gislinde Macphereson, ngs@willowgardens.idps.co.uk. *Nr Clyne Gardens. Go along Mumbles Rd to Blackpill. Turn R at Texaco garage up Mayals Rd. At the top of Clyne Park, turn R into Westport Ave. Willowbrook Gardens is 2nd turning on the L as you go up Westport Ave.* **Sun 7, Mon 8 May (12.30-6). Adm £5, chd free. Light refreshments. Visits also by arrangement 15 Mar to 30 Sept for groups of up to 30.**

Informal ½ acre mature garden on acid soil, designed to give natural effect with balance of form and colour between various areas linked by lawns; unusual trees suited to small suburban garden, especially conifers and maples; rock and water garden. Sculptures, ponds and waterfall.

36 NEW 11 WINDERMERE AVENUE

Roath Park, Cardiff, CF23 5PQ. Kathy and Glyn Jones. *2½m from Cardiff city centre & 75m from Roath Park attractions. Windermere Avenue is off Lake Road West.* **Sun 11 June (12-5). Adm £5, chd free. Home-made teas.**

Mature, medium size suburban garden with working greenhouse. Stepping stone pond, pergola, summerhouse and veg patch, that feels like it's set in the countryside. Organically and mindfully gardened where the hand of the gardener is lightly applied, serendipitous self seeding is encouraged and wildlife welcomed.

37 Y BWTHYN

Y Bwthyn, Corntown, Bridgend, CF35 5BB. Mrs Joyce Pegg. *Approx 4m from J35 of M4. On the B4524 which is the main road through Corntown. If travelling W from the Cardiff direction its on the R just past the meadows turning.* **Sun 4 June (10.30-4). Adm £5, chd free. Light refreshments.**

Over 30yrs of hard labour, some guesswork and considerable good luck has resulted in a delightful garden. The area at the front of the house is a mixture of hot colour combinations whilst at the rear of this modest sized garden the themes are of a more traditional cottage garden style which inc colour themed borders as well as soft fruit, herbs and vegetables. There is a shallow step at the entrance to the side gate of the property.

GWENT

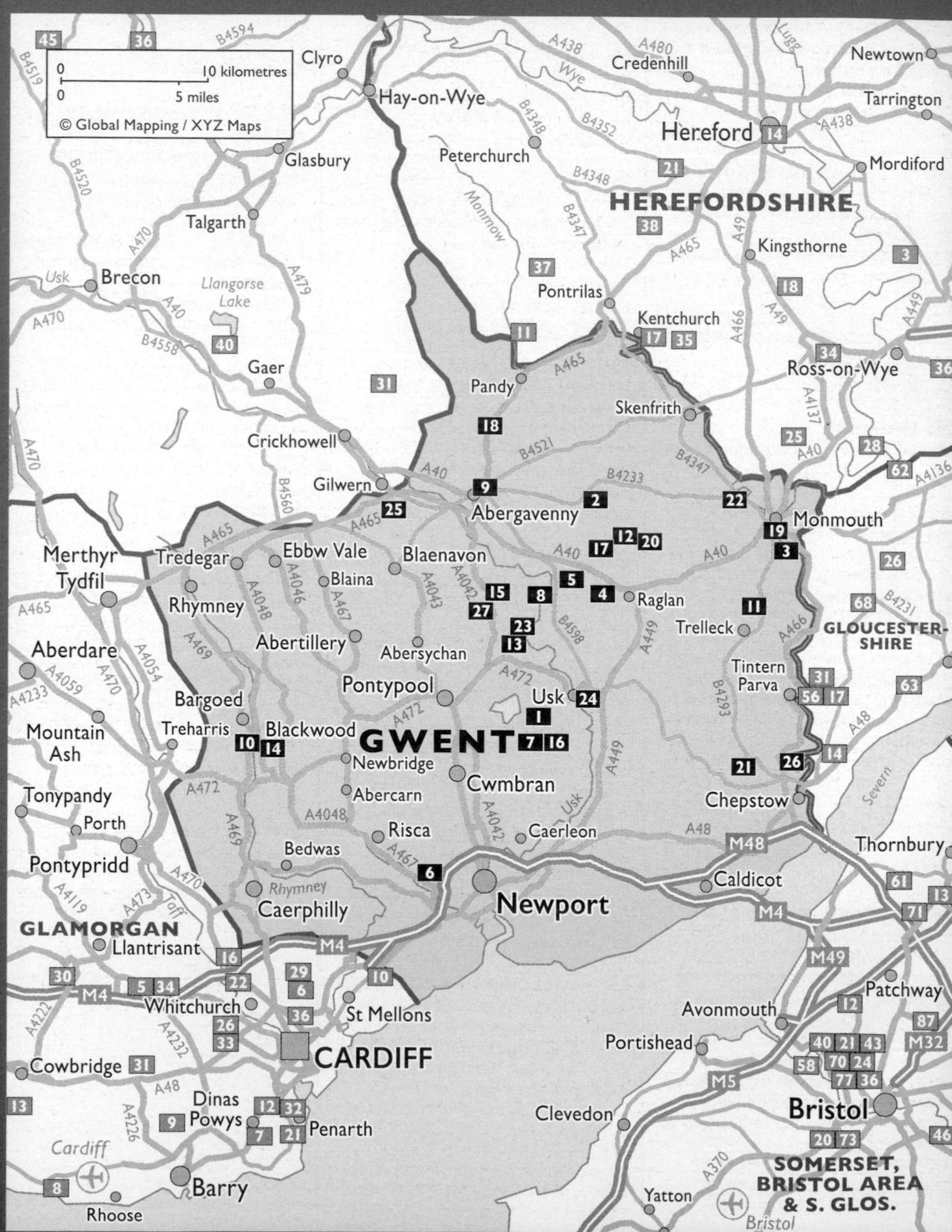

0 10 kilometres
0 5 miles
© Global Mapping / XYZ Maps
GWENT
HEREFORDSHIRE
GLOUCESTER-SHIRE
GLAMORGAN
SOMERSET, BRISTOL AREA & S. GLOS.
Clyro
Hay-on-Wye
Glasbury
Talgarth
Brecon
Llangorse Lake
Gaer
Crickhowell
Gilwern
Credenhill
Hereford
Peterchurch
Pontrilas
Kentchurch
Pandy
Skenfrith
Abergavenny
Monmouth
Newtown
Tarrington
Mordiford
Kingsthorne
Ross-on-Wye
Merthyr Tydfil
Tredegar
Ebbw Vale
Blaina
Blaenavon
Rhymney
Raglan
Trelleck
Aberdare
Abertillery
Abersychan
Pontypool
Usk
Tintern Parva
Bargoed
Treharris
Blackwood
Mountain Ash
Newbridge
Cwmbran
Abercarn
Tonypandy
Porth
Pontypridd
Risca
Bedwas
Caerleon
Chepstow
Caerphilly
Newport
Caldicot
Thornbury
Llantrisant
Whitchurch
St Mellons
CARDIFF
Cowbridge
Dinas Powys
Penarth
Barry
Rhoose
Cardiff
Avonmouth
Patchway
Portishead
Clevedon
Bristol
Yatton
Severn
Wye
Usk
Monmow
Lugg
Rhymney
Taff
M4
M48
M49
M5
M32

Welcome to the ancient county of Gwent. The area known as Gwent covers Monmouthshire, Blaenau Gwent, Torfaen, Newport and Caerphilly. It is a county of valleys and hills, of castles and farms, of winding country lanes and hidden gems.

It reaches from the Bristol Channel in the south to the Brecon Beacons national park in the north, the English border to the east and the industrial valleys to the west. The variety in the landscape has led to a wonderful range of gardens whose owners kindly welcome visitors to their gardens in aid of the NGS.

The gardens open for visitors range from small jewels of town gardens to gracious estates, from manicured lawns to hillside gardens that blend into the landscape and offer fabulous views. There is truly something for everyone, of all ages, to enjoy. A number of our gardens open by arrangement for small groups. This can be a wonderful opportunity to learn more about the garden and how it has been developed. Please check the handbook for more information.

Visiting a garden is such a wonderful way to support charities and even better when a cuppa and cake is involved! We look forward to seeing you in 2023 and thank you for your support.

Croeso i sir hynafol Gwent. Mae'r ardal a elwir yn Gwent yn cynnwys Sir Fynwy, Blaenau Gwent, Torfaen, Casnewydd a Chaerffili. Mae'n sir o ddyffrynnoedd a bryniau, cestyll a ffermydd, lonydd gwledig troellog a thrysorau cudd.

Mae'n ymestyn o Fôr Hafren yn y de i barc cenedlaethol Bannau Brycheiniog yn y gogledd, y ffin â Lloegr i'r dwyrain a'r cymoedd diwydiannol i'r gorllewin. Mae'r amrywiaeth yn y dirwedd wedi arwain at amrywiaeth wych o erddi y mae eu perchnogion yn croesawu ymwelwyr i'w gerddi er budd y Cynllun Gerddi Cenedlaethol.

Mae'r gerddi sydd ar agor i ymwelwyr yn amrywio o drysorau bach o erddi tref i ystadau grasol, o lawntiau twt i erddi ar ochr bryniau sy'n ymdoddi i'r dirwedd ac yn cynnig golygfeydd gwych. Mae yna wirioneddol rhywbeth i bawb, o bob oed, ei fwynhau. Mae nifer o'n gerddi ar agor drwy drefniant ar gyfer grwpiau bach. Gall hwn fod yn gyfle gwych i ddysgu mwy am yr ardd a sut y cafodd ei datblygu. Gwiriwch y llawlyfr am ragor o wybodaeth.

Mae ymweld â gardd yn ffordd mor wych o gefnogi elusennau a hyd yn oed yn well pan fydd paned a chacen yn rhan ohono! Edrychwn ymlaen at eich gweld yn 2023 a diolch i chi am eich cefnogaeth.

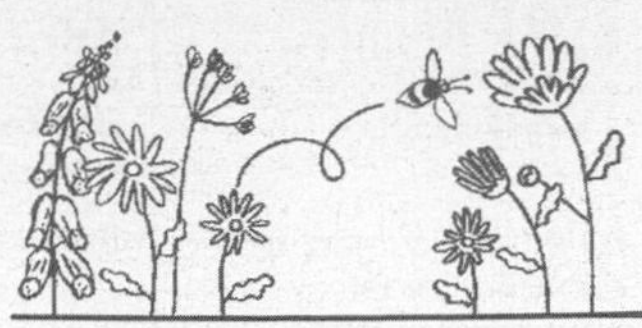

Volunteers

County Organiser
Debbie Field
01873 832752 / 07885 195304
wenalltisaf@gmail.com

County Treasurer
David Warren
01873 880031
david.warren@ngs.org.uk

Publicity
Penny Reeves
01873 880355
penny.reeves@ngs.org.uk

Social Media
Roger Lloyd 01873 880030
droger.lloyd@btinternet.com

Assistant County Organiser
Cherry Taylor
07803 853681
cherry.taylor@dynamicmarkets.co.uk

Veronica Ruth
07967 157806 / 01873 859757
veronica.ruth@ngs.org.uk

Jenny Lloyd
01873 880030 / 07850 949209
jenny.lloyd@ngs.org.uk

Booklet Co-Ordinator
Veronica Ruth (as above)

@gwentngs
@GwentNGS
@gwentngs

OPENING DATES

All entries subject to change. For latest information check **www.ngs.org.uk**

Map locator numbers are shown to the right of each garden name.

February

Snowdrop Festival

Sunday 12th
Bryngwyn Manor 4

March

Sunday 26th
Llanover 15

April

Saturday 1st
Bryngwyn Manor 4

Sunday 9th
The Old Vicarage 20

Sunday 23rd
High Glanau Manor 11

Saturday 29th
Glebe House 8

Sunday 30th
Glebe House 8

May

Sunday 7th
Park House 21

Sunday 21st
April House 1
Monmouth Gardens 19

Saturday 27th
Hillcrest 14

Sunday 28th
Baileau 2
Hillcrest 14

Monday 29th
Hillcrest 14

June

Saturday 3rd
Longhouse Farm 17

Sunday 4th
Longhouse Farm 17
◆ Wyndcliffe Court 26

Sunday 11th
High House 12
Rockfield Park 22
Wenallt Isaf 25

Friday 16th
Trengrove House 23

Saturday 17th
NEW Cwm Farm 7
NEW Long Owl Barn 16

Sunday 18th
Highfield Farm 13
Trengrove House 23

Saturday 24th
Bryngwyn Manor 4
Usk Open Gardens 24

Sunday 25th
Baileau 2
Bryngwyn Manor 4
Mione 18
Usk Open Gardens 24

Friday 30th
Trengrove House 23

July

Sunday 2nd
NEW The Growing Space Garden 9
Mione 18
Trengrove House 23

Saturday 8th
14 Gwerthonor Lane 10

Sunday 9th
14 Gwerthonor Lane 10
Mione 18

Sunday 16th
Clytha Park 5
Highfield Farm 13

Sunday 23rd
Hillcrest 14

August

Sunday 6th
Hillcrest 14

Sunday 13th
Highfield Farm 13

Sunday 27th
Wenallt Isaf 25

September

Sunday 3rd
Hillcrest 14

Sunday 10th
Highfield Farm 13

By Arrangement

Arrange a personalised garden visit with your club, or group of friends, on a date to suit you. See individual garden entries for full details.

April House 1
Birch Tree Well 3
Glebe House 8
NEW The Growing Space Garden 9
Highfield Farm 13
Hillcrest 14
Llanover 15
Longhouse Farm 17
Wenallt Isaf 25
Y Bwthyn 27

Long Owl Barn

THE GARDENS

1 APRIL HOUSE

Coed y Paen, Usk, NP15 1PT. Charlotte Fleming, charlotte.fleming00@gmail.com. *2m W of Usk. From Usk bridge go S towards Caerleon. Take 1st R towards Coed y Paen & go uphill for 1.9m. Garden is on L. SatNav gets you here.* **Sun 21 May (11-5). Adm £5, chd free. Pre-booking essential, please visit www.ngs.org.uk for information & booking. Home-made teas. Visits also by arrangement 1 Apr to 1 Sept for groups of 5 to 12.**

Long ½ acre site with glorious views over the Usk Valley and Wentwood forest. Garden developed over past 5 yrs from bramble thicket. Low maintenance shrub and prairie-style and herbaceous borders. Fruit and vegetables.

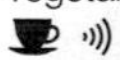

2 BAILEAU

Abergavenny, NP7 8TA. Sue Wilson & Seb Gwyther. *Nr Llantilio Crossenny. Between Llantilio Crossenny & Treadam. What3words app - dislikes.piglet.views.* **Sun 28 May, Sun 25 June (11-6). Adm £5, chd free. Light refreshments.**

A mature cottage style garden around an ancient farmhouse (not open). Packed with fruit and vegetables, the garden inc a rose walk, a crab apple walk, herbaceous borders, a circular ornamental vegetable garden and an old orchard. Activities for children, dogs welcome. Plenty of spots for a picnic. Views to Blorenge and Sugarloaf mountains. Baked goods, light lunches, drinks and treats for sale, making creative use of abundant produce from the garden.

3 BIRCH TREE WELL

Upper Ferry Road, Penallt, Monmouth, NP25 4AN. Jill Bourchier, 01600 775327, gillian.bourchier@btinternet.com. *4m SW of Monmouth. Approx 1m from Monmouth on B4293, turn L for Penallt & Trelleck. After 2m turn L to Penallt. On entering village turn L at Xrds & follow yellow signs. Approach lanes are single track & steep.* **Visits by arrangement 20 Apr to 30 Sept for groups of up to 25. Unsuitable for coaches. Adm £4, chd free. Home-made teas.**

Situated in the heart of the Lower Wye Valley, amongst the ancient habitat of woodland, rocks and streams these three acres can be viewed from a lookout tower. The garden features bluebells, specialist hydrangeas as well as unusual plants and trees attracting butterflies, bees and insects. An addition this year is the newly planted Reflective Garden with a brookside walk featuring many ferns. Children are very welcome (under supervision).

4 BRYNGWYN MANOR

Bryngwyn, Raglan, NP15 2JH. Peter & Louise Maunder. *2m W of Raglan. Take B4598 (old Abergavenny - Raglan Rd signed Clytha). Turn S (between the two garden centres) at Croes Bychan (opposite village hall). House ¼m up lane on L. What3words app - goat. blubber.composts.* **Sun 12 Feb (11-3); Sat 1 Apr (11-4). Adm £3, chd free. Light refreshments. Sat 24, Sun 25 June (11-5). Adm £6, chd free. Cream teas.**

3 acres. Winter snowdrops, daffodil walk, mature trees, walled parterre garden, mixed borders, lawns, ponds and shrubbery. Family friendly afternoon out, with children's activities, loads of space to run about, and scrumptious teas. A series of practical garden workshops are being held at Bryngwyn with RHS gardener Dean Peckett – for details please see ngs.org.uk/product-category/tickets/gardening-workshops. Each session costs £40 and inc refreshments. All profits donated to the NGS. All areas of the garden can be accessed without using the steps, however, ground is uneven and mainly grass paths.

5 CLYTHA PARK

Abergavenny, NP7 9BW. Jack & Susannah Tenison. *Between Abergavenny (5m) & Raglan (3m). Entrance by main gate on the old Raglan-Abergavenny A40.* **Sun 16 July (2-5). Adm £6, chd free. Home-made teas.**

Large C18/19 garden around lake with wide lawns and specimen trees, original layout by John Davenport, with C19 arboretum, and H. Avray Tipping influence. Visit the 1790 walled garden and the newly restored greenhouses. Gravel and grass paths.

7 NEW CWM FARM

Coedypaen, Pontypool, NP4 0TB. Lee & Louisa Morgan. *2m from Usk towards the village of Coed y Paen. Main parking at Cwm Farm. Walking distance to Long Owl Barn, where parking is limited.* **Sat 17 June (11-5). Combined adm with Long Owl Barn £8, chd free. Home-made teas.**

Nestled in the Usk Valley, the garden of this C16 smallholding has been created over the last 8 yrs. Set in 2 acres, perennial borders and wildflower areas are contrasted against structural elements of topiary and hedging. A large pond and stream is surrounded by a wild meadow with mown paths and a Victorian style greenhouse stands within a vegetable garden. Most of the garden is suitable for a wheelchair.

8 GLEBE HOUSE

Llanvair Kilgeddin, Abergavenny, NP7 9BE. Mr & Mrs Murray Kerr, 01873 840422, joanna@amknet.com. *Midway between Abergavenny (5m) & Usk (5m) on B4598.* **Sat 29, Sun 30 Apr (2-6). Adm £6, chd free. Home-made teas. Visits also by arrangement Apr to June. Lane not suitable for coaches.**

Borders bursting with spring colour inc tulips, narcissi and camassias. South facing terrace with wisteria and honeysuckle, decorative vegetable garden and orchard underplanted with succession of bulbs. Some topiary and formal hedging in 1½ acre garden set in AONB in Usk valley. Old rectory of St Mary's, Llanfair Kilgeddin which will also be open to view famous Victorian scraffito murals. Some gravel and gently sloping lawns.

In 2022 the National Garden Scheme donated £3.11 million to our nursing and health beneficiaries

9 NEW THE GROWING SPACE GARDEN
Hereford Road, Mardy, Abergavenny, NP7 6HU.
Jim Quinn, 07732 117011. *The rear of Mardy Park Resource Centre. 1.4m N of centre of Abergavenny, on the Hereford Rd opp the Crown & Sceptre public house.* **Sun 2 July (10.30-4). Adm £4, chd free. Visits also by arrangement 2 July to 16 July.**
The gardens are laid out around disabled access paths, herbaceous borders, a prairie-style border, and fruit and veg beds. There is also a tropical border and large raised beds. We have 2 polytunnels and a small craft /carpentry workshop. The gardens are tended by a hard working team of volunteers of mixed ability and ages from 18 to 96. Adjoining parkland is perfect for a picnic. Good access, via tarmac paths with a hand rail; however the garden is on a slight slope.

10 14 GWERTHONOR LANE
Gilfach, Bargoed, CF81 8JT.
Suzanne & Philip George. *8m N of Caerphilly. A469 to Bargoed, through the T-lights next to sch then L filter lane at next T-lights onto Cardiff Rd. First L into Gwerthonor Rd, 4th R into Gwerthonor Lane.* **Sat 8, Sun 9 July (11-6). Adm £4, chd free. Light refreshments.**
The garden has a beautiful panoramic view of the Rhymney Valley. A real plantswoman's garden with over 800 varieties of perennials, annuals, bulbs, shrubs and trees. There are numerous rare, unusual and tropical plants combined with traditional and well loved favourites (many available for sale). A pond with a small waterfall adds to the tranquil feel of the garden.

11 HIGH GLANAU MANOR
Lydart, Monmouth, NP25 4AD. Mr & Mrs Hilary Gerrish, 01600 860005, helenagerrish@gmail.com, www.highglanaugardens.com.
4m SW of Monmouth. Situated on B4293 between Monmouth & Chepstow. Turn R into private rd, ¼m after Craig-y-Dorth turn on B4293. **Sun 23 Apr (2-5.30). Adm £6, chd free. Home-made teas.**
Listed Arts and Crafts garden laid out by H Avray Tipping 100yrs ago. Original features inc impressive stone terraces with far reaching views over the Vale of Usk to Blorenge, Skirrid, Sugar Loaf and Brecon Beacons. Pergola, herbaceous borders, Edwardian glasshouse, rhododendrons, azaleas, tulips, orchard with wildflowers. Originally open for the NGS in 1927. Garden guidebook by owner, Helena Gerrish, available to purchase. Gardens lovers cottage to rent.

12 HIGH HOUSE
Penrhos, NP15 2DJ. Mr & Mrs R Cleeve. *4m N of Raglan. From r'about on A40 at Raglan take exit to Clytha. After 50yds turn R at Llantilio Crossenny. Follow NGS signs, 10mins through lanes.* **Sun 11 June (2-6). Adm £5, chd free. Home-made teas.**
3 acres of spacious lawns and trees surrounding C16 house (not open) in a beautiful, hidden part of Monmouthshire. South facing terrace and extensive bed of old roses. Swathes of grass with tulips, camassias, wildflowers and far reaching views. Espaliered cherries, pears and scented evergreens in courtyard. Large extended pond, orchard with chickens and ducks, large vegetable garden. Partial wheelchair access, some shallow steps, sloping lawn, gravel courtyard.

13 HIGHFIELD FARM
Penperlleni, Goytre, NP4 0AA. Dr Roger & Mrs Jenny Lloyd, 01873 880030, jenny.plants@btinternet.com, www.instagram.com/davidrogerlloyd. *4m W of Usk, 6m S of Abergavenny. Turn off the A4042 at Goytre Arms, over railway bridge, bear L. Garden ½m on R. From Usk off B4598, turn L after Chain Bridge, then L at Xrds. Garden 1m on L. What3words app - unheated. lifetimes.chickens.* **Sun 18 June, Sun 16 July, Sun 13 Aug, Sun 10 Sept (11-4). Adm £6, chd free. Home-made teas. Visits also by arrangement June to Sept for groups of up to 50.**
This is a garden defined by its plants. There are over 1400 cultivars, with many rarities, densely planted over 3 acres to generate an exuberant display across the seasons. It provides an intimate, immersive experience with this diverse array of herbaceous, shrubs and trees. ' 'Madness garden' opened in 2022. Huge sale of plants from the garden. Live music. Art sale. Access to almost all garden without steps.

14 HILLCREST
Waunborfa Road, Cefn Fforest, Blackwood, NP12 3LB.
Mr M O'Leary, 01443 837029, olearymichael18@gmail.com. *3m W of Newbridge. B4254/A469 at T/L head to B'wood. At X take lane ahead,1st on L at top of hill. A4048/B4251 cross Chartist Bridge, 2nd exit on next 2 r'abouts. End of road turn L & immed R onto Waunborfa. 400m on R.* **Sat 27, Sun 28, Mon 29 May (11-6); Sun 23 July, Sun 6 Aug, Sun 3 Sept (11-5). Adm £5, chd free. Light refreshments. Visits also by arrangement Apr to Sept.**
A cascade of secluded gardens of distinct character over 1½ acres. Magnificent, unusual trees, interesting shrubs, perennials and annuals. Choices at every turn, visitors are well rewarded as hidden delights and surprises are revealed. Well placed seats encourage a relaxed pace to fully appreciate the garden's treasures. Tulips in April, glorious blooms of the Chilean Firebushes, Handkerchief Tree and cornuses in May and many trees in their autumnal splendour in October. Lowest parts of garden not accessible to wheelchairs.

15 LLANOVER
Abergavenny, NP7 9EF. Mr & Mrs M R Murray, 07753 423635, elizabeth@llanover.com, www.llanovergarden.co.uk. *4m S of Abergavenny, 15m N of Newport, 20m SW Hereford. On A4042 between Abergavenny & Goytre. At Llanover, turn into the driveway with tall black gates opp the pull-in bus stop.* **Sun 26 Mar (2-5). Adm £7, chd free. Home-made teas. Visits also by arrangement 3 Jan to 15 Dec for groups of up to 5. Please email to arrange visit in advance.**
Benjamin Waddington, the direct ancestor of the current owners, purchased the house and land in 1792. Subsequently he created a series of ponds, cascades and rills which form the backbone of the 15 acre garden as the stream winds its way from its source in the Black Mountains to the River Usk. There are herbaceous borders, a drive lined with narcissi, spring bulbs, wild flowers, a water garden, champion

trees and two arboreta. The house (not open) is the birthplace of Augusta Waddington, Lady Llanover, C19 patriot, supporter of the Welsh language and traditions. Gwerinyr Gwent will be performing Welsh folk dances during the afternoon in traditional Welsh costume. Gravel and grass paths and lawns. No disabled WC.

16 NEW LONG OWL BARN

Coedypaen, Pontypool, NP4 0TB. Mike & Tina Booth. *2m W of Usk, 1m from Llandegveth reservoir. Take turning to Coed-y-paen from Llanbadoc or Llangybi then follow yellow signs. From Cwmbran Crematorium, take Tre-Herbert lane towards Llangybi, then L turn to Coed-y-Paen. Main parking at Cwm Farm.* **Sat 17 June (12-5). Combined adm with Cwm Farm £8, chd free. Home-made teas.**

In a rural setting with extensive views, there are meandering paths, colourful borders and wildlife areas with plenty to explore in a little over 1 acre. Ponds, a vegetable plot and greenhouse.

17 LONGHOUSE FARM

Penrhos, Raglan, NP15 2DE. Mr & Mrs M H C Anderson, 01600 780389, m.anderson666@btinternet.com. *Midway between Monmouth & Abergavenny. 4m from Raglan. Off Old Raglan/Abergavenny rd signed Clytha. At Bryngwyn/Great Oak Xrds turn towards Great Oak - follow yellow NGS signs from red phone box down narrow lane.* **Sat 3, Sun 4 June (2-6). Adm £5, chd free. Home-made teas. Visits also by arrangement 5 June to 1 Oct. Please discuss refreshments at booking.**

The garden has matured over 25 years with a few recent tweaks. The woodland walk and its series of ponds and stream will continue to develop. The productive vegetable garden has an additional fruit cage and potting shed. The avenue of malus trees seen from the south-facing terrace are a special feature supported by borders planted with year-round colourful plants.

18 MIONE

Old Hereford Road, Llanvihangel Crucorney, Abergavenny, NP7 7LB. Yvonne & John O'Neil. *5m N of Abergavenny. From Abergavenny take A465 to Hereford. After 4.8m turn L - signed Pantygelli. Mione is ½m on L.* **Sun 25 June, Sun 2, Sun 9 July (10.30-5). Adm £5, chd free. Home-made teas.**

Beautiful garden with a wide variety of established plants, many rare and unusual. Pergola with climbing roses and clematis. Wildlife pond with many newts, insects and frogs. Numerous containers with diverse range of planting. Several seating areas, each with a different atmosphere. Enjoy our new benches in a secret hideaway under the pergola. Lovely home-made cakes, biscuits and scones to be enjoyed sitting in the garden or pretty summerhouse.

National Garden Scheme gardens are identified by their yellow road signs and posters. You can expect a garden of quality, character and interest, a warm welcome and plenty of home-made cakes!

Cwn Farm

High Glanau Manor

GROUP OPENING

19 MONMOUTH GARDENS

Worcester Street, Monmouth, NP25 3DF. *Central Monmouth. Park at Glendower St (NP25 3DF). Tickets/maps at car park & each garden.* **Sun 21 May (12-5). Combined adm £10, chd free. Home-made teas at North Parade House throughout the afternoon.**

NEW **MANSARD HOUSE**
NP25 3HW.
Richard & Liz Wells.

THE NELSON GARDEN
NP25 3ER.
www.nelsongarden.org.uk.

NORTH PARADE HOUSE
NP25 3PB.
Tim Haynes & Lisa O'Neill.

NEW **ST JOHNS**
NP25 3DG.
Simon & Hilary Hargreaves.

4 very different town gardens open under the banner of Monmouth Gardens for the first time. The Nelson Garden dates back to Roman times and when Lord Nelson visited Monmouth in 1802 he took tea there with Lady Hamilton. The memorial pavilion (1840) marks this visit. St Johns, Glendower St opens for the first time this year. This charming walled garden has views over Chippenham Fields and is under restoration. It has a sunken central lawn and deep herbaceous borders. There are several mature trees of note. Entrance to both these gardens is via Chippenham Fields. Mansard House, Vine Acre lies approximately ½ m from the town up the Hereford Road. Spectacular views of the Skirrid and the Black Mountains from this contemporary garden which features an architectural planting scheme and striking sculpure. North Parade House, Hereford Road is a surprisingly large and secluded walled garden with mature specimen trees, herbaceous borders and a kitchen garden. Plant sale at the Nelson Garden.

20 THE OLD VICARAGE

Penrhos, Raglan, Usk, NP15 2LE. Mrs Georgina Herrmann. *3m N of Raglan. From A449 take Raglan exit, join A40 & move immed into R lane & turn R across dual carriageway. Follow yellow NGS signs.* **Sun 9 Apr (2-5.30). Adm £5, chd free. Light refreshments.**

Designed by Welsh architect, George Mair, this unusual Victorian old vicarage (1867, not open) is set in rolling countryside and is surrounded by a traditional garden, Lovely all year but especially in spring with a variety of trees, shrubs and spring bulbs, a parterre with a charming gazebo, kitchen garden, wildlife areas and ponds make this a garden, always changing, well worth a visit.

21 PARK HOUSE
School Lane, Itton, Chepstow, NP16 6BZ. Professor Bruce & Dr Cynthia Matthews. *Itton Monmouthshire. From M48 take A466 Tintern. At 2nd r'about turn L D4290 After blue sign Itton turn R Park House is at end of lane. Parking 200m before house. From Devauden B4293 1st L in Itton.* **Sun 7 May (10-5). Adm £5, chd free.**
Approx 1 acre garden with large vegetable areas and many mature trees, rhododendrons, azaleas, camellias in a woodland setting. Bordering on Chepstow Park Wood. Magnificent views over open country.

22 ROCKFIELD PARK
Rockfield, Monmouth, NP25 5QB. Mark & Melanie Molyneux. *On arriving in Rockfield village from Monmouth, turn R by phone box. After approx 400yds, church on L. Entrance to Rockfield Park on R, opp church, via private bridge over river.* **Sun 11 June (10.30-4). Adm £7, chd free. Home-made teas on the terrace.**
Rockfield Park dates from C17 and is situated in the heart of the Monmouthshire countryside on the banks of the River Monnow. The extensive grounds comprise formal gardens, meadows and orchard, complemented by riverside and woodland walks. Possible to picnic on riverside walks. Main part of gardens can be accessed by wheelchair but not steep garden leading down to river.

23 TRENGROVE HOUSE
Nantyderry, Abergavenny, NP7 9DP. Guin Vaughan & Chris Jofeh. *Approx 4m N of Usk, 6m SE of Abergavenny. From Abergavenny, L off the A4042 after Llanover, signed to Nantyderry. From Usk, B4598 N to Chainbridge, then immed L after bridge.* **Fri 16, Sun 18, Fri 30 June, Sun 2 July (11-5.30). Adm £6, chd free. Home-made teas.**
Garden designer's own 'single-handed' 2½ acre country garden developed over 20yrs along 'right plant, right place' lines. Informal borders planted for a range of conditions with interesting and some unusual shrubs, trees, perennials and grasses. I acre meadow, managed to encourage only naturally occurring species with many wildflowers and grasses. Tranquil and atmospheric.

GROUP OPENING

24 USK OPEN GARDENS
Maryport Street, Usk, NP15 1BH. *Main car park postcode is NP15 1AD. From M4 J24 take A449 8m N to Usk exit. Signposts to free parking around town. Blue badge parking in main car parks. Map of gardens provided with ticket.* **Sat 24, Sun 25 June (10-5). Combined adm £7.50, chd free. Home-made teas. Donation to local charities.**
Usk's floral public displays are a wonderful backdrop to the gardens. Around ten private gardens opening. Gardeners' Market with interesting plants. Lovely day out for all the family with lots of places to eat and drink inc places to picnic. Various cafes, pubs and restaurants available for refreshments, plus volunteer groups offering teas and cakes. Wheelchair accessible gardens and gardens allowing well-behaved dogs on leads noted on passport/map available from the ticket desks at the free car park at Usk Memorial Hall (NP15 1AD). Some gardens/areas of gardens are partially wheelchair accessible. Accessibility noted on ticket/map.

25 WENALLT ISAF
Twyn Wenallt, Gilwern, Abergavenny, NP7 0HP. Tim & Debbie Field, 01873 832753, wenalltisaf@gmail.com. *3m W of Abergavenny. Between Abergavenny & Brynmawr. Leave the A465 at Gilwern & follow yellow NGS signs through the village. Do not follow SatNav. What3Words app - sensibly. gone.connects.* **Sun 11 June, Sun 27 Aug (2-5.30). Adm £6, chd free. Home-made teas inc gluten free & lactose free cakes. Visits also by arrangement 15 May to 30 Sept. Not suitable for coaches.**
An everchanging garden of nearly 3 acres designed in sympathy with its surroundings and the challenges of being 650ft up on a north facing hillside. Far reaching views of the magnificent Black Mountains, mature trees, rhododendrons, viburnum, hydrangeas, borders, vegetable garden, small polytunnel, orchard, chickens, bees. Child friendly with plenty of space to run about.

26 ◆ WYNDCLIFFE COURT
St Arvans, NP16 6EY. Mr & Mrs Anthony Clay, 07710 138972, sarah@wyndcliffecourt.com, www.wyndcliffecourt.com. *3m N of Chepstow. Off A466, turn at Wyndcliffe signpost coming from the Chepstow direction.* **For NGS: Sun 4 June (2-6). Adm £10, chd free. Pre-booking essential, please visit www.ngs.org.uk for information & booking. Home-made teas in the house. Refreshments inc in adm For other opening times and information, please phone, email or visit garden website.**
Exceptional and unaltered garden designed by H. Avray Tipping and Sir Eric Francis in 1922. Arts and Crafts 'Italianate' style. Stone summerhouse, terracing and steps with lily pond. Yew hedging and topiary, sunken garden, rose garden, bowling green and woodland. Walled garden with fruit trees and new platinum jubilee planting. Rose garden completely replanted by Sarah Price in 2017.

27 Y BWTHYN
Pencroesoped, Llanover, Abergavenny, NP7 9EL. Jacqui & David Warren, warrens.ybwthyn@gmail.com. *6m S of Abergavenny, just outside Llanover. Detailed directions provided ahead of visit.* **Visits by arrangement 17 July to 25 Aug for groups of 10 to 30. Home-made teas.**
Beautifully varied 1½ acre garden. Sweeping herbaceous, mixed and prairie-style borders, gravel garden, lawns, pond and bog garden, ornamental kitchen garden and greenhouse, wildflower banks, and a magnificent veteran oak. Lovely views of the surrounding hills and countryside, and fascinating house history and art exhibition too. Most garden routes are step-free. Some steep grass slopes and narrow gravel paths. Sorry, no wheelchair access to the WC.

GWYNEDD & ANGLESEY
GWYNEDD A MÔN

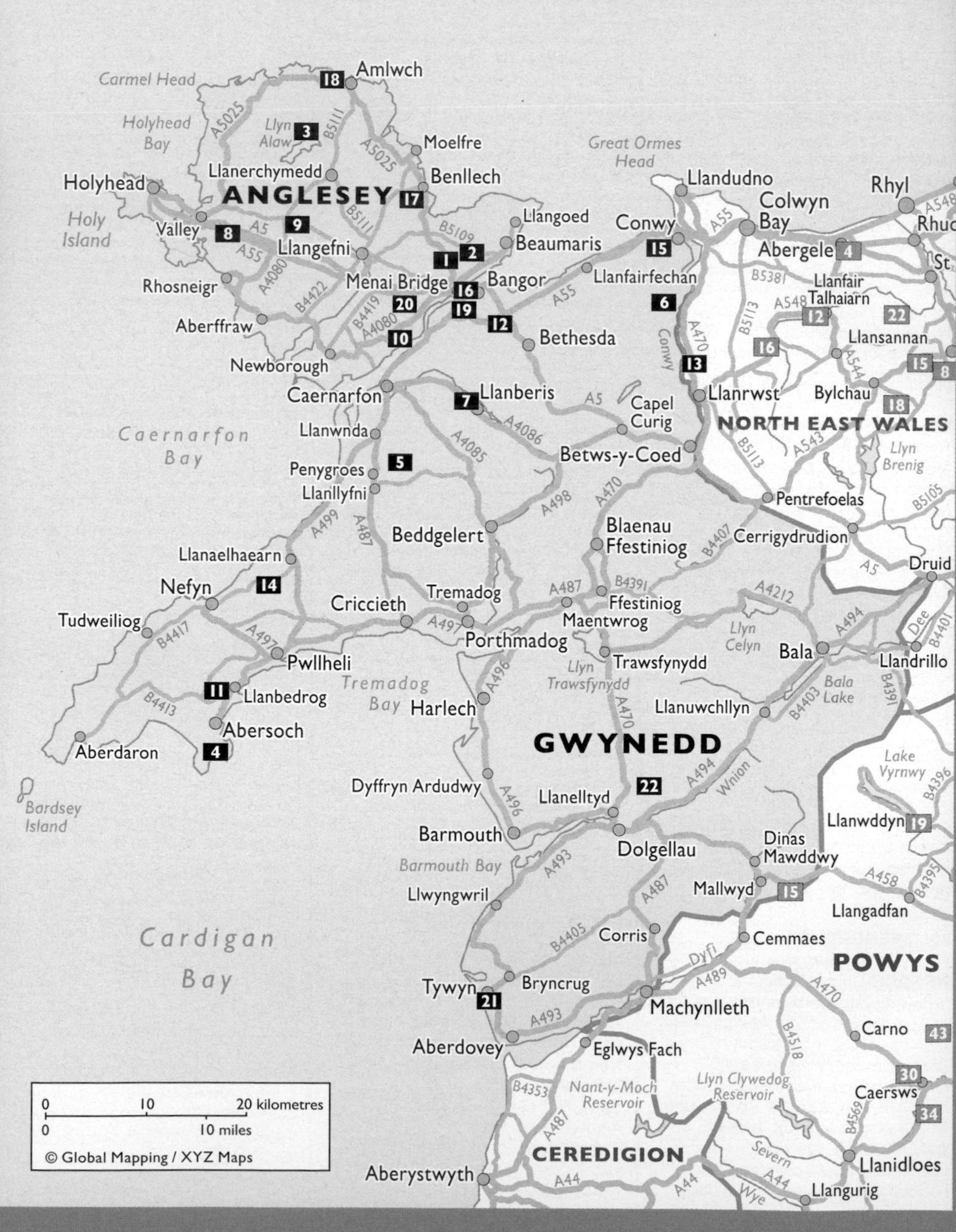

Rich in history and with two contrasting Areas of Outstanding Natural Beauty (AONB) a visit to Gwynedd and Anglesey offers the chance to discover both the wonderful mountains of Snowdonia National Park and the fantastic maritime seascape of the Isle of Anglesey. Together they form some of the most impressive landscapes in the UK.

This land offers garden lovers a wonderful diversity of gardens, open to visitors under the National Garden Scheme. Beautiful cottage gardens high in the mountains; gardens with amazing views of surrounding landscapes; gardens benefiting from the gentler climate created by the Gulf Stream warming the coastline and gardens celebrating sustainability, nurturing wildlife and packed with ideas to try in your own garden.

As we gardeners sense how global warming is affecting the way our seasons are changing, a visit gives you a special opportunity to talk to garden owners who are deeply immersed and connected with their plots. They can tell you about how their plants are coping and how they are adapting the way they garden over time. People who open their gardens for the NGS love to share their knowledge and experience – with tea and cake, it's a perfect day out.

Yn gyfoethog mewn hanes a gyda dwy Ardal o Harddwch Naturiol Eithriadol (AHNE) cyferbyniol, mae ymweliad â Gwynedd ac Ynys Môn yn cynnig cyfle i ddarganfod mynyddoedd hyfryd Parc Cenedlaethol Eryri a morlun morwrol gwych Ynys Môn. Gyda'i gilydd maent yn ffurfio rhai o'r tirweddau mwyaf trawiadol yn y DU.

Mae'r tir hwn yn cynnig amrywiaeth hyfryd o erddi i'r rheini sy'n caru gerddi, sy'n agored i ymwelwyr o dan y Cynllun Garddio Cenedlaethol. Gerddi bwthyn hardd yn uchel yn y mynyddoedd; gerddi gyda golygfeydd anhygoel o'r tirweddau cyfagos; gerddi sy'n elwa o'r hinsawdd dynerach a grëwyd gan Lif y Gwlff yn cynhesu'r arfordir a gerddi sy'n dathlu cynaliadwyedd, yn meithrin bywyd gwyllt ac yn llawn syniadau i roi cynnig arnynt yn eich gardd eich hun.

Wrth i ni arddwyr synhwyro sut mae cynhesu byd-eang yn effeithio ar y ffordd y mae ein tymhorau'n newid, mae ymweliad yn rhoi cyfle arbennig i chi siarad â pherchnogion gerddi sydd wedi ymgolli'n ddwfn ac yn gysylltiedig â'u lleiniau. Gallant ddweud wrthych am sut mae eu planhigion yn ymdopi a sut maent yn addasu'r ffordd y maent yn garddio dros amser. Mae pobl sy'n agor eu gerddi ar gyfer y Cynllun Garddio Cenedlaethol wrth eu bodd yn rhannu eu gwybodaeth a'u profiad - gyda the a chacen, mae'n ddiwrnod allan perffaith.

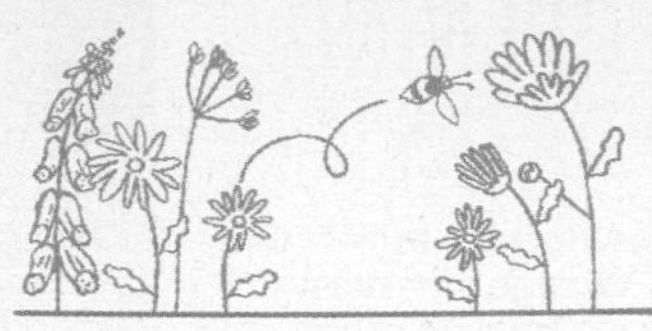

Volunteers

County Organiser
Kay Laurie
07971 083361
kay.laurie@ngs.org.uk

County Treasurer
Vacancy

Publicity
Rebecca Andrews
07901 878009
rebecca.andrews@ngs.org.uk

Assistant County Organisers
Hazel Bond
07378 844295
hazelcaenewydd@gmail.com

Janet Jones
01758 740296
janetcoron@hotmail.co.uk

Delia Lanceley
01286 650517
delia@lanceley.com

Heather Broughton

@gwyneddandangleseyngs

OPENING DATES

All entries subject to change. For latest information check **www.ngs.org.uk**

Map locator numbers are shown to the right of each garden name.

Treborth Botanic Garden, Bangor University

April

Saturday 22nd
Llanidan Hall 10

Wednesday 26th
◆ Plas Cadnant Hidden Gardens 16

Sunday 30th
Cae Newydd 3
Gilfach 6

May

Sunday 7th
Maenan Hall 13
NEW Trefnant Bach 20

Saturday 13th
Mynydd Heulog 14

Sunday 14th
◆ Gardd y Coleg 5
Gilfach 6
Mynydd Heulog 14
Ty Capel Ffrwd 22

Saturday 20th
Glan Llyn 7
Gwaelod Mawr 8

Sunday 21st
Gwaelod Mawr 8
Llys-y-Gwynt 12

Sunday 28th
Cae Newydd 3

June

Sunday 4th
◆ Pensychnant 15
Sunningdale 18

Sunday 11th
Ty Cadfan Sant 21

Saturday 17th
Crowrach Isaf 4

Sunday 18th
Crowrach Isaf 4

Saturday 24th
Gwyndy Bach 9
Llanidan Hall 10
NEW Llwydiarth 11

Sunday 25th
NEW Llwydiarth 11

July

Saturday 1st
Sunningdale 18

Saturday 8th
Llanidan Hall 10

Sunday 9th
Arcady 1
◆ Pensychnant 15
Ty Cadfan Sant 21

Saturday 22nd
NEW Rhoslan 17

Sunday 30th
Maenan Hall 13

August

Sunday 6th
Gilfach 6

September

Saturday 9th
NEW Trefnant Bach 20

Sunday 10th
NEW Bodlwyfan 2

Saturday 16th
Treborth Botanic Garden, Bangor University 19

Sunday 17th
◆ Gardd y Coleg 5

Sunday 24th
Llys-y-Gwynt 12

By Arrangement

Arrange a personalised garden visit with your club, or group of friends, on a date to suit you. See individual garden entries for full details.

Crowrach Isaf 4
Gilfach 6
Llys-y-Gwynt 12
Mynydd Heulog 14
Ty Cadfan Sant 21

In 2022, 20,000 people were supported by Perennial caseworkers and prevent teams thanks to our funding

THE GARDENS

1 ARCADY

Llansadwrn, Menai Bridge, LL59 5SE. James Weisters & Liz Mangham. *3m N of Menai Bridge. A5025 Amlwch/Benllech exit from Britannia Bridge. Approx 3m turn R to Llansadwrn (1m after Pentraeth Motors). 1m Arcady on L. Parking in adjoining field.* **Sun 9 July (11-4). Adm £5, chd free. Light refreshments.**

Surrounded by farmland and with views of Snowdon range. Approx 1 acre of quirky, structured garden originally laid out by the artist Ed Povey - pagoda and meditation garden, laburnum walk, small orchard, lawn, ponds and hidden places. Gravel paths with steps. Loose stones and uneven surfaces. Level gravel drive to lawn with view of main garden but steps and uneven surfaces prevents access to other areas.

2 NEW BODLWYFAN

Llansadwrn, Menai Bridge, LL59 5SN. Mrs Andrea Bristow. *3m N of Menai Bridge. Menai Bridge towards Pentraeth B5025, after 2m R signed Llansadwrn.* **Sun 10 Sept (11-4.30). Adm £4, chd free. Home-made teas served under oak framed Be' ti'n Galw.**

Trace Bodlwyfan's story from Victorian, the 1960s to today with its dry gravel garden, beautiful landscaped areas and wild garden full of insects. In 2017 elements by RHS Chelsea silver-gilt garden designer were incorporated. Apple orchard and rambling pumpkins; Eupatorium and fallen apples attract butterflies. Experience sitting on the living sofa. Level access to decked seating area with view of garden. Garden is flat though paths are not paved.

3 CAE NEWYDD

Rhosgoch, Anglesey, LL66 0BG. Hazel & Nigel Bond. *3m SW of Amlwch. A5025 from Benllech to Amlwch, follow signs for leisure centre & Lastra Farm. Follow yellow NGS signs (approx 3m), car park on L.* **Sun 30 Apr, Sun 28 May (11-4). Adm £5, chd free. Light refreshments. April opening d-i-y teas.**

A mature country garden of 2½ acres which blends seamlessly into the open landscape with stunning views of Snowdonia and Llyn Alaw. Variety of shrubs, trees and herbaceous areas, large wildlife pond, polytunnel, greenhouses. Collections of fuchsia, pelargonium, cacti and succulents. An emphasis on gardening for wildlife throughout the garden, lots of seating to enjoy this peaceful space. Hay meadow with beehives, visitors are welcome to bring a picnic. Garden area closest to house suitable for wheelchairs.

4 CROWRACH ISAF

Bwlchtocyn, LL53 7DY. Margaret & Graham Cook, 01758 712860, crowrach_isaf@hotmail.com. *1½m SW of Abersoch. Follow road through Abersoch & Sarn Bach, L at sign for Bwlchtocyn for ½m until junction & no-through road sign. Turn R, parking 50 metres on R.* **Sat 17, Sun 18 June (1.30-4.30). Adm £5, chd free. Cream teas. Visits also by arrangement 1 June to 2 Sept for groups of 15+.**

2 acre plot developed from 2000, inc island beds, windbreak hedges, vegetable garden, wildflower area and wide range of geraniums, unusual shrubs and herbaceous perennials. Views over Cardigan Bay and Snowdonia. Grass and gravel paths, some gentle slopes. Parking at garden for disabled visitors.

5 ◆ GARDD Y COLEG

Carmel, LL54 7RL. Pwyllgor Pentref Carmel Village Committee. *Garden at Carmel village centre. Parking on site.* **For NGS: Sun 14 May, Sun 17 Sept (12-4.30). Adm £3, chd free. Light refreshments.**

Approx ½ acre featuring raised beds planted with ornamental and native plants mulched with local slate. Benches and picnic area, wide pathways suitable for wheelchairs. Spectacular views. Garden created and maintained by volunteers. Development of the garden is ongoing. Ramped access.

6 GILFACH

Rowen, Conwy, LL32 8TS. James & Isoline Greenhalgh, 01492 650216, isolinegreenhalgh@btinternet.com. *4m S of Conwy. At Xrds 100yds E of Rowen S towards Llanrwst, past Rowen Sch on L, turn up 2nd drive on L.* **Sun 30 Apr, Sun 14 May, Sun 6 Aug (2-5.30). Adm £4, chd free. Home-made teas. Visits also by arrangement Apr to Aug.**

1 acre country garden on south facing slope with magnificent views of the River Conwy and mountains; set in 35 acres of farm and woodland. Collection of mature shrubs is added to yearly; woodland garden, herbaceous border and small pool. Spectacular panoramic view of the Conwy Valley and the mountain range of the Carneddau. Classic cars. Large coaches can park at bottom of steep drive.

7 GLAN LLYN

Llanberis, Caernarfon, LL55 4EL. Mr Bob Stevens. *On A4086, ½m from Llanberis village. Next door to the Galty Glyn Hotel (Pizza & Pint Restaurant) opp DMM factory.* **Sat 20 May (10-6). Adm £4, chd free. Light refreshments.**

A 3 acre woodland edge garden inc 2 acres of woodland, wildlife ponds, stream, wildflower area, raised sphagnum bog garden, two green roofs, three glasshouses for cacti and succulents, Australasian and South African beds, sand bed, many unusual trees, shrubs and herbaceous perennials. The garden is on fairly steep sloping ground, no wheelchair access to the woodland.

8 GWAELOD MAWR

Caergeiliog, Anglesey, LL65 3YL. Tricia Coates. *6m E of Holyhead. ½m E of Caergeiliog. From A55 J4. r'about 2nd exit signed Caergeiliog. 300yds, Gwaelod Mawr is 1st house on L.* **Sat 20, Sun 21 May (11-4). Adm £5, chd free. Light refreshments.**

2 acre garden created by owner over 20 yrs with lake, large rock outcrops and palm tree area. Spanish style patio and laburnum arch lead to sunken garden and wooden bridge over lily pond with fountain and waterfall. Peaceful Chinese orientated garden offering contemplation. Separate koi carp pond. Abundant seating throughout. Mainly flat, with gravel and stone paths, no wheelchair access to sunken lily pond area.

9 GWYNDY BACH
Llandrygarn, Tynlon, Anglesey, LL65 3AJ. Keith & Rosa Andrew. *5m W of Llangefni. From Llangefni take B5109 towards Bodedern, cottage exactly 5m out on L. Postcode good for SatNav.* **Sat 24 June (1-4.30). Adm £4, chd free. Home-made teas.**
¾ acre artist's garden, set amidst rugged Anglesey landscape. Romantically planted in informal intimate rooms with interesting rare plants and shrubs, box and yew topiary, old roses and Japanese garden with large koi pond (deep water, children must be supervised). Gravel entrance to garden.

10 LLANIDAN HALL
Brynsiencyn, LL61 6HJ. Head Gardener. *5m E of Llanfair Pwll. From Llanfair PG follow A4080 towards Brynsiencyn for 4m. After Hooton's farm shop on R take next L, follow lane to gardens.* **Sat 22 Apr, Sat 24 June, Sat 8 July (10-4). Adm £5, chd free. Light refreshments (weather dependant) are served outside. Donation to CAFOD.**
Walled garden of 1¾ acres. Physic and herb gardens, ornamental vegetable garden, herbaceous borders, water features and many varieties of old roses. Sheep, rabbits and hens to see. Children must be kept under supervision. Well behaved dogs on leads welcome. Llanidan Church will be open for viewing. Hard gravel paths, gentle slopes.

11 NEW LLWYDIARTH
Mynytho, Pwllheli, LL53 7RW. David & Anne Mitchell. *Mynytho, Gwynedd. Garden on main road through Mynytho village. 5½m from Pwllheli 3½m from Abersoch.* **Sat 24, Sun 25 June (12-4). Adm £4, chd free. Home-made teas.**
Just 2 years ago this garden was an overgrown field, but now with hedges, trees, shrubs, grasses, perennials and roses planted, it is already developing character and interest. A large wildlife pond, filled with aquatic plants and surrounded with tons of sandstone, dug from the field itself, now makes a stunning rockery. There is a lush green Mediterranean style back yard.

12 LLYS-Y-GWYNT
Pentir Road, Llandygai, Bangor, LL57 4BG. Jennifer Rickards & John Evans, 01248 353863, mjrickards@gmail.com. *3m S of Bangor. 300yds from Llandygai r'about at J11, A5 & A55, just off A4244. Follow signs for services (Gwasanaethau). Turn off at No Through Rd sign, 50yds beyond. Do not use SatNav.* **Sun 21 May, Sun 24 Sept (11-4). Adm £5, chd free. Home-made teas. Visits also by arrangement Feb to Nov. By arrangement adm inc tea, coffee, biscuits.**
Interesting, harmonious and very varied 2 acre garden inc magnificent views of Snowdonia. An exposed site inc Bronze Age burial cairn. Winding paths and varied levels planted to create shelter, year-round interest, microclimates and varied rooms. Ponds, waterfall, bridge and other features use local materials and craftspeople. Wildlife encouraged, well organised compost. Good family garden.

13 MAENAN HALL
Maenan, Llanrwst, LL26 0UL. The Hon Mr & Mrs Christopher Mclaren. *2m N of Llanrwst. On E side of A470, ¼m S of Maenan Abbey Hotel.* **Sun 7 May, Sun 30 July (10.30-5). Adm £5, chd free. Light refreshments. Donation to Ogwen Valley Mountain Rescue Organisation.**
A superbly beautiful 4 hectares on the slopes of the Conwy Valley, with dramatic views of Snowdonia, set amongst mature hardwoods. Both the upper part, with sweeping lawns, ornamental ponds and retaining walls, and the bluebell carpeted woodland dell contain copious specimen shrubs and trees, many originating at Bodnant. Magnolias, rhododendrons, camellias, pieris, cherries and hydrangeas, amongst many others, make a breathtaking display. Upper part of garden accessible but with fairly steep slopes.

15 ◆ PENSYCHNANT
Sychnant Pass, Conwy, LL32 8BJ. Pensychnant Foundation; Warden Julian Thompson, 01492 592595, jpt.pensychnant@btinternet.com, www.pensychnant.co.uk. *2½m W of Conwy at top of Sychnant Pass. From Conwy: L at Lancaster Sq. into Upper Gate St; after 2½m, Pensychnant's drive signed on R. From Penmaenmawr: fork R by shops, up Sychnant Pass; after walls at top of Pass, U turn L into drive.* **For NGS: Sun 4 June, Sun 9 July (11-5). Adm £4, chd free. Light refreshments & FairTrade cakes. For other opening times and information, please phone, email or visit garden website.**
Wildlife Garden. Diverse herbaceous cottage garden borders surrounded by mature shrubs, banks of rhododendrons, ancient and Victorian woodlands. 12 acre woodland walks with views of Conwy Mountain and Sychnant. Woodland birds. Picnic tables, archaelogical trail on mountain. A peaceful little gem. Large Victorian Arts and Crafts house (open) with art exhibition. Partial wheelchair access, please phone for advice.

Our annual donations to Parkinson's UK meant 7,000 patients are currently benefiting from support of a Parkinson's Nurse

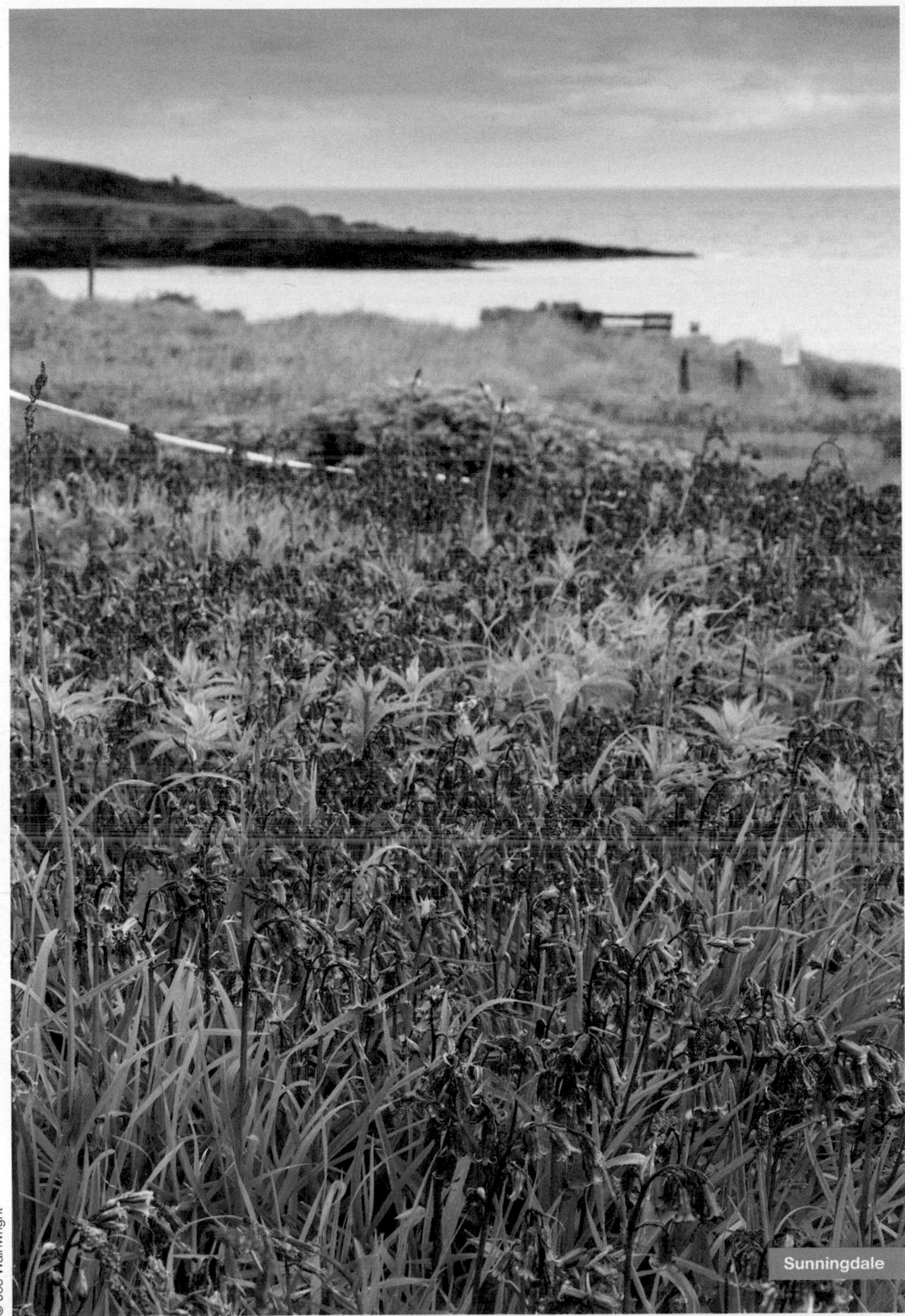

Sunningdale

16 ◆ PLAS CADNANT HIDDEN GARDENS

Cadnant Road, Menai Bridge, LL59 5NH. Mr Anthony Tavernor, 01248 717174, plascadnantgardens@gmail.com, www.plascadnantgardens.co.uk. *½m E of Menai Bridge. Take A545 & leave Menai Bridge heading for Beaumaris, then follow brown tourist information signs. SatNav not always reliable.* **For NGS: Wed 26 Apr (12-5). Adm £9, chd free. Light refreshments in traditional Tea Room. For other opening times and information, please phone, email or visit garden website. Donation to Wales Air Ambulance.**

Early C19 picturesque garden undergoing restoration since 1996. Valley gardens with waterfalls, large ornamental walled garden, woodland and early pit house. Also Alpheus water feature and Ceunant (Ravine) which gives visitors a more interesting walk featuring unusual moisture loving Alpines. Restored area following flood damage. Guidebook available. Visitor centre open. Partial wheelchair access to parts of gardens. Some steps, gravel paths, slopes. Access statement available. Accessible Tea Room and WC.

17 NEW RHOSLAN

Garreglwyd, Benllech, LL74 8RG. Mr Jim Beverley. *Enter Benllech driving from Bangor, take 1st L. First L turning off A5025 is shortly after petrol station. Driving into Garreglwyd garden is positioned immed in front of you as the crescent road sweeps R.* **Sat 22 July (10-4). Adm £4, chd free. Light refreshments.**

The gardens around a bungalow Rhoslan is the home of a retired Head Gardener. With a range of small shrubs, trees and colourful annuals. Rear garden 12 by 14 paces with hedged borders filled with bulbs, herbaceous plants and some shrubs. Raised beds to grow a selection of herbs and salads. A garden for reflection and meditation, in evolution to embrace the owner's art and crafts. A variety of ideas for small gardens. Composting system used in garden to recycle vegetation and weeds, to provide garden soil enrichment on a sandy soil, for mulching and potting compost. Look out on Website for a possible additional open day in May.

18 SUNNINGDALE

Bull Bay Road, Bull Bay, Amlwch, LL68 9SD. Mike & Gill Cross. *1½m NW of Amlwch. On A5025 through Amlwch towards Cemaes. No parking at house.* **Sun 4 June, Sat 1 July (11-4). Adm £5, chd free. Home-made teas.**

A seaside garden. Headland has cliffs, steps, wild flowers and seating, spectacular views and sheer drops! Front garden has raised pond and planting to cope with hostile weather. The relatively sheltered rear garden has lots of different rooms, no large lawn here! There is a wealth of plants, paths, seats, pots and a raised bed vegetable area. Lots of new features have been added. Access to Headland and a completely different atmosphere in the rear garden.

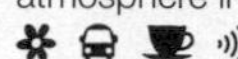

Bodlwyfan

19 TREBORTH BOTANIC GARDEN, BANGOR UNIVERSITY

Treborth, Bangor, LL57 2RQ. Natalie Chivers, treborth.bangor.ac.uk. *On the outskirts of Bangor towards Anglesey. Approach Menai Bridge either from Upper Bangor on A5 or leave A55 J9 & travel towards Bangor for 2m. At Antelope Inn r'bout turn L just before entering Menai Bridge.* **Sat 16 Sept (10-1). Adm £4, chd free. Home-made teas inc vegan & gluten free options.**

Owned by Bangor University and used as a resource for teaching, research, public education and enjoyment. Treborth comprises planted borders, species rich natural grassland, ponds, arboretum, Chinese garden, ancient woodland, and a rocky shoreline habitat. Six glasshouses provide specialised environments for tropical, temperate, orchid and carnivorous plant collections. Partnered with National Botanic Garden of Wales to champion Welsh horticulture, protect wildlife and extol the virtues of growing plants for food, fun, health and wellbeing. Glasshouse and garden Q&As. Wheelchair access to some glasshouses and part of the garden. Woodland path is surfaced but most of the borders only accessed over grass.

20 NEW TREFNANT BACH

Anglesey Bees, Llanddaniel Fab, Gaerwen, LL60 6ET. Dafydd & Dawn Jones, 07816 188573, dafydd@angleseybees.co.uk, www.angleseybees.co.uk. *0.4m from the centre of Llanddaniel Fab. A5 from LlanfairPG, 1½km W, turn L for Llanddaniel. Village centre, by bus shelter follow signs down farm track. Parking on site. Alternatively park in village & 5 min walk.* **Sun 7 May, Sat 9 Sept (11-5). Adm £5, chd free. Home-made teas.**

Beekeepers' magical 7½ acre wildlife garden with beautiful large spring-fed pond with islands and stream; honeybee friendly woodland and woodland walks; grazed pastures; small garden with herbaceous plants; vegetable raised beds, orchard, soft fruit; fun summerhouse; information boards; woodland treasure hunt; suitable picnic areas. Honey-bee and wildlife friendly woodland planted Feb 2019 with Woodland Trust support. Apiary and Anglesey Bees centre for beekeeping training and experiences. Local honey for sale. Tree-friendly Shropshire sheep. Very gentle slopes, short-mowed grassed or hard paths. Refreshment area accessible with hard surface. Regret WC unsuitable for wheelchair access.

21 TY CADFAN SANT

National Street, Tywyn, LL36 9DD. Mrs Katie Pearce, 07816 604851, Katie@tycadfansant.co.uk. *A493 going S & W. L into one way, garden ahead. Bear R, parking 2nd L. A493 going N, 1st R in 30mph zone, L at bottom by garden, parking 2nd L.* **Sun 11 June, Sun 9 July (9.30-4.30). Adm £5, chd free. Cream teas. Visits also by arrangement 28 Apr to 15 Oct. Refreshments by prior arrangement.**

Large eco friendly garden. In the front, shrubbery, mixed flower beds and roses surround a mature copper beech. Up six steps the largely productive back garden has an apiary in the orchard, fruit, vegetables, flowers and a polytunnel, plenty of seating. Special diets catered for. Seasonal produce as available, honey, plants and crafts. Partial wheelchair access due to steps to rear garden.

22 TY CAPEL FFRWD

Llanfachreth, Dolgellau, LL40 2NR. Revs Mary & George Bolt. *4m NE of Dolgellau, 18m SW of Bala. From Dolgellau 4m up hill to Llanfachreth. Turn L at War Memorial. Follow lane ½m to chapel on R. Park & walk down lane past chapel to cottage.* **Sun 14 May (11.30-5). Adm £4, chd free. Cream teas.**

True cottage garden in Welsh mountains. Azaleas, rhododendrons, acers; large collection of aquilegia. Many different hostas give added strength to spring bulbs and corms. Stream flowing through the garden, 10ft waterfall and on through a small woodland bluebell carpet. For summer visitors there is a continuous show of colour with herbaceous plants, roses, clematis and lilies, inc Cardiocrinum giganteum. Harp will be played in the garden.

Llwydiarth

NORTH EAST WALES

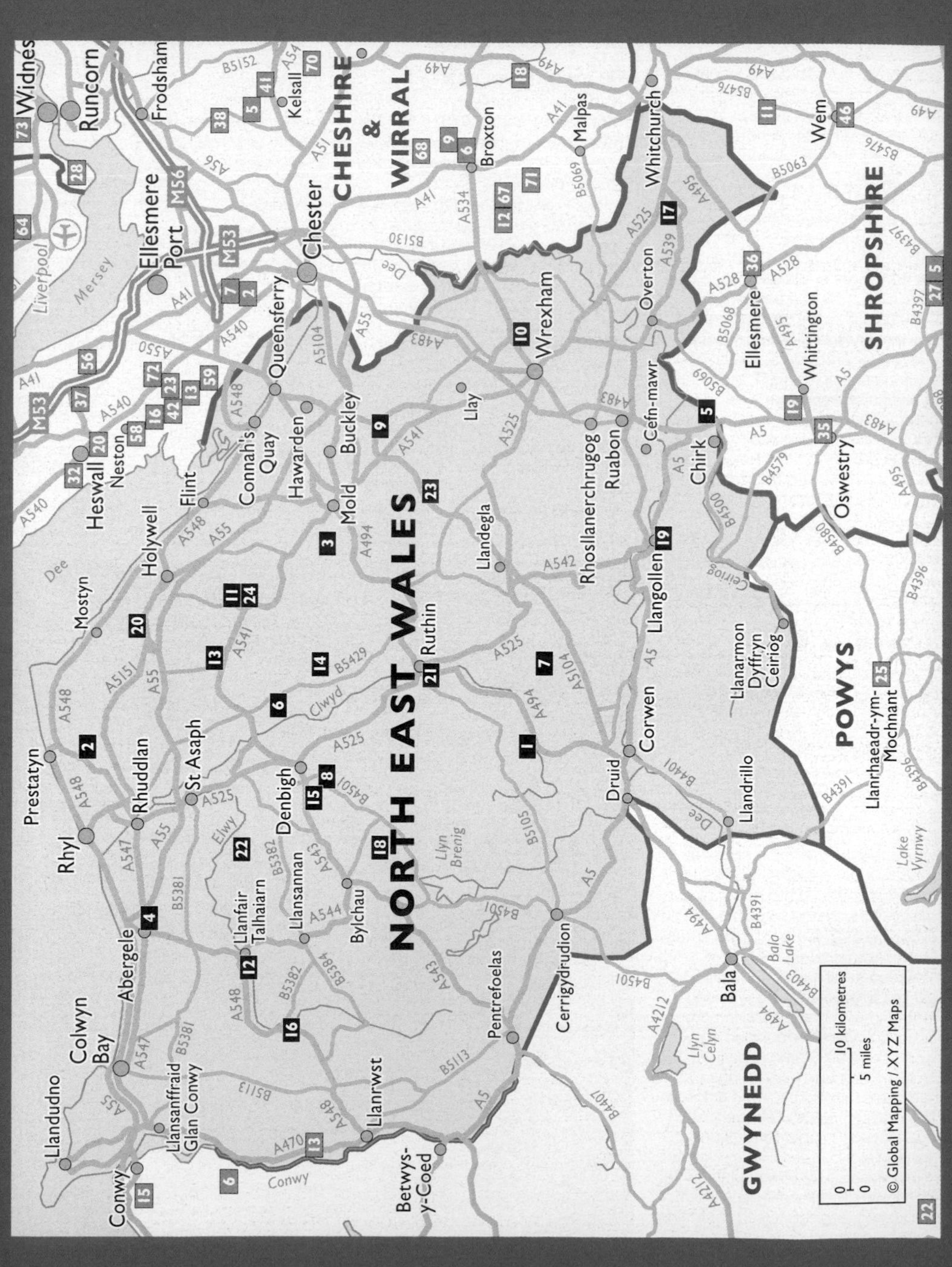

With its diversity of countryside from magnificent hills, seaside vistas and rolling farmland, North East Wales offers a wide range of gardening experiences.

Our gardens have a wealth of designs and come in all shapes and sizes. Visitors will have something to see from snowdrops in February to the colourful days of autumn.

Come and enjoy the beauty and variety of our gardens and enjoy a cup of tea and slice of cake. Our garden owners await your visit.

Gyda'i amrywiaeth o gefn gwlad o fryniau godidog, golygfeydd glan môr a thir ffermio bryniog, mae Gogledd Ddwyrain Cymru yn cynnig ystod eang o brofiadau garddio.

Mae gan ein gerddi gyfooth o ddyluniadau ac maent ym mhob lliw a llun. Bydd gan ymwelwyr rywbeth i'w weld o eirlysiau ym mis Chwefror hyd at ddyddiau lliwgar yr hydref.

Dewch i fwynhau harddwch ac amrywiaeth ein gerddi a mwynhau paned o de a thafell o gacen. Mae perchnogion ein gerddi yn aros i chi ymweld.

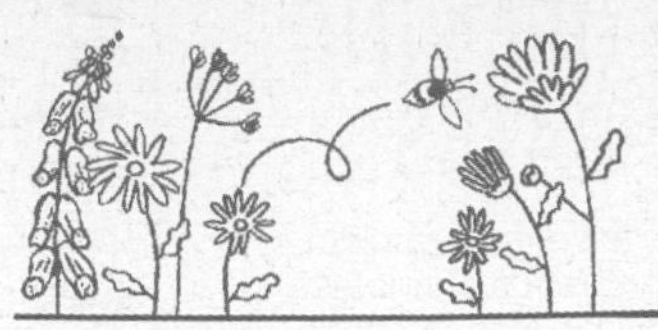

Volunteers

County Organiser
Jane Moore
07769 046317
jane.moore@ngs.org.uk

County Treasurer
Iris Dobbie
01745 886730
iris.dobbie@ngs.org.uk

Booklet Co-ordinator
Position vacant
Please email hello@ngs.org.uk for details

Assistant County Organisers
Fiona Bell
07813 087797
bell_fab@hotmail.com

Kate Bunning
01978 262855
kate.bunning@ngs.org.uk

Lesley Callister
01824 705444
lesley.callister@ngs.org.uk

Anne Lewis
01352 757044
anne.lewis@ngs.org.uk

Pat Pearson
01745 813613
pat.pearson@ngs.org.uk

Helen Robertson
01978 790666
helen.robertson@ngs.org.uk

Left: White Croft

OPENING DATES

All entries subject to change. For latest information check **www.ngs.org.uk**

Map locator numbers are shown to the right of each garden name.

February

Snowdrop Festival

Wednesday 8th
Clwydfryn 6

Wednesday 15th
Aberclwyd Manor 1

March

Wednesday 1st
Aberclwyd Manor 1

Wednesday 15th
Aberclwyd Manor 1

Wednesday 29th
Aberclwyd Manor 1

April

Wednesday 12th
Aberclwyd Manor 1

Sunday 16th
NEW The Homestead 17

Wednesday 26th
Aberclwyd Manor 1

May

Monday 1st
Dibleys Nurseries 7

Wednesday 10th
Aberclwyd Manor 1

Saturday 13th
Plas Nantglyn 18

Sunday 14th
Plas Nantglyn 18

Wednesday 24th
Aberclwyd Manor 1

Saturday 27th
Hafodunos Hall 16

Monday 29th
Garthewin 12

June

Sunday 4th
Brynkinalt Hall 5

Wednesday 7th
Aberclwyd Manor 1

Saturday 10th
33 Bryn Twr and Lynton 4
Scott House 21

Sunday 11th
33 Bryn Twr and Lynton 4
Scott House 21

Saturday 17th
NEW Glasfryn Hall 13

Sunday 18th
NEW Glasfryn Hall 13
Gwaenynog 15

Wednesday 21st
Aberclwyd Manor 1

Sunday 25th
NEW 17 Ffordd Walwen 11
White Croft 24

July

Sunday 2nd
◆ Plas Newydd 19

Wednesday 5th
Aberclwyd Manor 1

Wednesday 19th
Aberclwyd Manor 1

Sunday 23rd
Dove Cottage 9

August

Wednesday 2nd
Aberclwyd Manor 1

Wednesday 16th
Aberclwyd Manor 1

Wednesday 30th
Aberclwyd Manor 1

September

Sunday 10th
◆ Erlas Victorian Walled Garden 10

Wednesday 13th
Aberclwyd Manor 1

Thursday 14th
Brynkinalt Hall 5

Wednesday 27th
Aberclwyd Manor 1

By Arrangement

Arrange a personalised garden visit with your club, or group of friends, on a date to suit you. See individual garden entries for full details.

Aberclwyd Manor 1
5 Birch Grove 2
Bryn Bellan 3
Dolhyfryd 8
Garthewin 12
Glyn Arthur 14
NEW The Homestead 17
Saith Ffynnon Farm 20
Tal-y-Bryn Farm 22
Tyn Rhos 23

Gwaenynog

THE GARDENS

1 ABERCLWYD MANOR

Derwen, Corwen, LL21 9SF. Mr & Mrs G Sparvoli, 01824 750431, irene662010@live.com. *7m from Ruthin. Travelling on A494 from Ruthin to Corwen. At Bryn SM Service Station turn R, follow sign to Derwen. Aberclwyd gates on l before Derwen. Do not follow SatNav directions.* **Weds 15 Feb, 1, 15, 29 Mar, 12, 26 Apr, 10, 24 May, 7, 21 June, 5, 19 July, 2, 16, 30 Aug, 13, 27 Sept (11-4). Adm £5, chd free. Cream teas. Visits also by arrangement 22 Feb to 30 Sept for groups of 10 to 25.**

4 acre garden on a sloping hillside overlooking the Upper Clwyd Valley. The garden has many mature trees underplanted with snowdrops, fritillaries and cyclamen. An Italianate garden of box hedging lies below the house and shrubs, ponds, perennials, roses and an orchard are also to be enjoyed within this cleverly structured area. Mass of cyclamen in Sept. Abundance of spring flowers. Snowdrops in February. Mostly flat with some steps and slopes.

2 5 BIRCH GROVE

Woodland Park, Prestatyn, LL19 9RH. Mrs Iris Dobbie, 01745 886730, iris.dobbie@ngs.org.uk. *A547 from Rhuddlan turn up The Avenue, Woodland Park after railway bridge 1st R into Calthorpe Dr, 1st L Birch Gr. 10 mins walk from town centre - at top of High St, turn R & L onto The Avenue.* **Visits by arrangement 8 May to 11 Sept for groups of 5 to 20. Adm £4, chd free.**

The gardens consist of a variety of borders inc woodland, grass, herbaceous, alpine, shrub, drought, tropical and a simulated bog garden with a small pond. The more formal front lawned garden has borders of mixed colourful planting and box balls. A small greenhouse is fully used for propagation. Gravel drive with few steps. Plenty of parking and not far from the beach, Offa's Dyke, town centre. A variety of borders and planting.

3 BRYN BELLAN

Bryn Road, Gwernaffield, CH7 5DE. Gabrielle Armstrong & Trevor Ruddle, 01352 741806, gabrielle@indigoawnings.co.uk. *2m W of Mold. Leave A541 at Mold on Gwernaffield Rd (Dreflan), ½m after Mold derestriction signs turn R to Rhydymwyn & Llynypandy. Bryn Bellan is 300yds along on R. Parking in courtyard.* **Visits by arrangement for groups of 10 to 40. Please discuss refreshments when booking.**

A tranquil, elegant garden, perfect for morning coffee, afternoon tea or evening glass of wine with nibbles. A partly walled upper garden with circular sunken lawn featuring a Sequoia and white and green themed mixed borders. The lower garden has an ornamental cutting garden, two perennial borders, an orchard and bijou potting shed. spring bulbs, iris, peonies, roses, hydrangeas and cyclamen. A very photogenic garden, pre shoot visits and shoots by appointment. Some gravel paths.

4 33 BRYN TWR AND LYNTON

Lynton, Highfield Park, Abergele, LL22 7AU. Mr & Mrs Colin Knowlson & Bryn Roberts & Emma Knowlson-Roberts. *From A55 heading W take slip road into Abergele town centre. Turn L at 2nd set of T-lights signed Llanfair TH, 3rd road on L. For SatNav use LL22 8DD.* **Sat 10, Sun 11 June (1-5). Adm £5, chd free. Light refreshments.**

Bryn Twr is a family garden with chickens, shrubs, roses, pots and lawn. Lynton completely different, intense cottage style garden: vegetables, trees, shrub, roses, ornamental grasses and lots of pots. Garage with interesting fire engine; classic cars and memorabilia; greenhouse over large water capture system that was part of an old swimming pool.

5 BRYNKINALT HALL

Brynkinalt, Chirk, Wrexham, LL14 5NS. Iain & Kate Hill-Trevor, www.brynkinalt.co.uk. *6m N of Oswestry, 10m S of Wrexham. Come off A5/A483 & take B5070 into Chirk village. Turn into Trevor Rd (beside St Mary's Church). Continue past houses on R. Turn R on bend into Estate Gates. N.B. Do not use postcode with SatNav.* **Sun 4 June, Thur 14 Sept (10.30-4.30). Adm £6, chd free. Home-made teas.**

5 acre ornamental woodland shrubbery, overgrown until recently, now cleared and replanted, rhododendron walk, historic ponds, well, grottos, ha-ha and battlements, new stumpery, ancient redwoods and yews. Also 2 acre garden beside Grade II* house (see website for opening), with modern rose and formal beds, deep herbaceous borders, pond with shrub/mixed beds, pleached limes and hedge patterns. Home of the first Duke of Wellington's grandmother and Sir John Trevor, Speaker of House of Commons. Stunning rhododendrons and formal West Garden. Major film location for Lady Chatterley's Lover.

6 CLWYDFRYN

Bodfari, LL16 4HU. Keith & Susan Watson. *5m outside Denbigh. ½ way between Bodfari & Llandyrnog on B5429. Yellow signs at bottom of lane.* **Wed 8 Feb (11-3). Adm £5, chd free. Home-made teas.**

¾ acre garden developed from a field, which has had many changes over the past 30 years. Extensive collection of snowdrops. Orchard with various fruit trees now well established. Vegetable and colourful garden planting below the orchard. Alpine house with sand plunge beds for alpines and bulbs. Greenhouse with many interesting succulents. Garden access to a paved area at back of house for wheelchair users.

7 DIBLEYS NURSERIES

Cefn Rhydd, Cricor, Llanelidan, Ruthin, LL15 2LG. Rex Dibley, www.dibleys.com. *7m S of Ruthin. Follow brown signs off A525 nr Llysfasi College.* **Mon 1 May (10-5). Adm £6, chd £1. Light refreshments. Donation to Plant Heritage.**

8 acre woodland garden with a wide selection of rare and unusual trees set in beautiful countryside. In late spring there is a lovely display of rhododendrons, magnolias, cherries and camellias. Much of our grassland promotes native wildflowers. Our ¾ acre glasshouses are open showing a display of streptocarpus and other house plants. Houseplant shop. Partial wheelchair access to glasshouses, uneven ground and steep paths in arboretum and elsewhere.

8 DOLHYFRYD

Lawnt, Denbigh, LL16 4SU. Captain & Mrs Michael Cunningham, 01745 814805, virginia@dolhyfryd.com. *1m SW of Denbigh. On B4501 to Nantglyn, from Denbigh - 1m from town centre.* **Visits by arrangement 30 Jan to 1 Nov for groups of 10+. Light refreshments.**

Established garden set in small valley of River Ystrad. Acres of crocuses in late Feb/early Mar. Paths through wildflower meadows and woodland of magnificent trees, shade loving plants and azaleas; mixed borders; walled kitchen garden. Many woodland and riverside birds, inc dippers, kingfishers, grey wagtails. Many species of butterfly encouraged by new planting. Much winter interest, exceptional display of crocuses. Gravel paths, some steep slopes.

9 DOVE COTTAGE

Rhos Road, Penyffordd, Chester, CH4 0JR. Chris & Denise Wallis, 01244 547539, dovecottage@supanet.com. *6m SW of Chester. Leave A55 at J35 take A550 to Wrexham. Drive 2m, turn R onto A5104. From A541 Wrexham/Mold Rd in Pontblyddyn take A5104 to Chester. Garden opp train stn.* **Sun 23 July (2-5). Adm £5, chd free. Home-made teas.**

Approx 1½ acre garden, shrubs and herbaceous plants set informally around lawns. Established vegetable area, two ponds (one wildlife), summerhouse. Raised board walk through planted woodland area. Gravel paths.

10 ◆ ERLAS VICTORIAN WALLED GARDEN

Bryn Estyn Road, Wrexham, LL13 9TY. Erlas Victorian Walled Garden, 01978 265058, info@erlas.org, www.erlas.org. *From A483 follow signs for the Wrexham Ind Est. From A5156 follow signs to Wrexham on A534 (Holt Rd). At 2nd r'about on Holt Rd take 1st L on to Brynestyn Rd,*

Dibley's Nurseries

for ½m. **For NGS: Sun 10 Sept (10.30-3.30). Adm £4, chd free. Light refreshments. For other opening times and information, please phone, email or visit garden website.**

Home of the Erlas Victorian Walled Garden charity, this is a place of work, solace and inspiration for adults with a range of abilities. The Walled Garden has many delights inc a centuries-old mulberry tree, fruit, vegetables, herbs and an orchard with a mixture of apple and pear varieties. Our ecology area is a haven for flora and fauna. We have wheelchair access throughout the garden, however it is on a slope, but the gradient is not too extreme.

11 NEW 17 FFORDD WALWEN

Lixwm, Holywell, CH8 8LW. Mr & Mrs Bridge. *Parking on field opposite Whitecroft.* **Sun 25 June (1-5). Combined adm with White Croft £5, chd free. Tea at White Croft.**

A brand new garden, which has been designed, made and planted in 2021-2022 by the owners. The south facing front garden inc trees, ponds, pergola and perennial planting, grasses and roses. The rear smaller garden has a 'courtyard feel' and has raised beds. The garden is supportive of wildlife. A good opportunity to view a newly planned garden.

12 GARTHEWIN

Llanfairtalhaiarn, LL22 8YR. Mr Michael Grime, 01745 720288, michaelgrime12@btinternet.com. *6m S of Abergele & A55. From Abergele take A548 to Llanfair TH & Llanrwst. Entrance to Garthewin 300yds W of Llanfair TH on A548 to Llanrwst. SatNav misleading.* **Mon 29 May (2-6). Adm £6, chd free. Home-made teas. Visits also by arrangement 1 Apr to 1 Nov.**

Valley garden with ponds and woodland areas. Much of the 8 acres have been reclaimed and redesigned providing a younger garden with a great variety of azaleas, rhododendrons and young trees, all within a framework of mature shrubs and trees. Teas in old theatre. Chapel open.

13 NEW GLASFRYN HALL

South Street, Caerwys, Mold, CH7 5AF. Mrs Lise Roberts, orielglasfryn.com. *In Caerwys, Flintshire. Follow the signs for Caerwys either from the A55 North Wales expressway or the A541 Mold - Denbigh Rd. Off the junction on South St & Pen Y Cefn Rd follow signs for Oriel Glasfryn Gallery.* **Sat 17 June (10-5); Sun 18 June (10-4). Adm £3.50, chd free. Teas, coffees & cake available.**

A Victorian Villa set in a beautiful Victorian garden with an interesting variety of mature trees and stunning views of the Clwydian hills - croquet lawn, Victorian tea house and pond, walled parterre garden. Sculptures and an art gallery showcasing Welsh Art add a modern and distinctive character to the garden. Teas and coffees available in the grounds from a horsebox cafe. Art gallery and sculpture for sale within garden.

14 GLYN ARTHUR

Llandyrnog, LL16 4NB. Mr & Mrs P Rowley Williams, 01824 790511, rw@glynarthur.co.uk. *5m E of Denbigh. At r'about off A543 head to Llandyrnog; over next r'about pass Kimmel Arms pub. Turn L at Xrds & R at next by bus shelter. L at Xrds by White Cott, up hill ½m to grass triangle turn L Glyn Arthur.* **Visits by arrangement Mar to Sept for groups of 5 to 30. Adm £5, chd free. Home-made teas. Donation to Llanychan church.**

Situated in steep secluded valley in the Clwydian Hills with spectacular views towards Snowdonia. Azaleas and rhododendrons; herbaceous borders; recent landscaping alterations; vegetable and fruit garden; pond and bluebell walk.

15 GWAENYNOG

Denbigh, LL16 5NU. Major & Mrs Tom Smith. *1m W of Denbigh. On A543, Lodge on L, ¼m drive.* **Sun 18 June (2-5). Adm £5, chd free. Home-made teas.**

2 acres inc the restored walled garden where Beatrix Potter wrote and illustrated the Tale of the Flopsy Bunnies. Also a small exhibition of some of her work. Long Herbaceous borders and island beds, some recently replanted, espalier fruit trees, rose pergola and vegetable area. C16 house (not open) visited by Dr Samuel Johnson during his Tour of Wales. Grass paths.

16 HAFODUNOS HALL

Llangernyw, Abergele, Conwy, LL22 8TY. Dr Richard Wood. *1m W of Llangernyw. ½way between Abergele & Llanrwst on A548. Signed from opp Old Stag Public House. Parking available on site.* **Sat 27 May (12-5). Adm £5, chd free. Home-made teas in Victorian conservatory.**

Historic garden undergoing restoration after 30 years of neglect surrounds a Sir George Gilbert Scott Grade I listed Hall, derelict after arson attack. Unique setting. ½m tree lined drive, formal terraces, woodland walks with ancient redwoods, laurels, yews, lake, streams, waterfalls and a gorge. Wonderful rhododendrons. Uneven paths, steep steps. Children must have adult supervision. Most areas around the hall accessible to wheelchairs by gravel pathways. Some gardens are set on slopes.

17 NEW THE HOMESTEAD

Horsemans Green, Whitchurch, SY13 3DY. Mr Elwyn & Mrs Jenny Hughes, 01948 830433, jdh53@hotmail.co.uk. *From Wrexham travel 9m on A525 towards Whitchurch & turn R into Horseman's Green. From Whitchurch travel 5m on the A525 & turn L into Horseman's Green.* **Sun 16 Apr (2-5). Adm £5, chd free. Home-made teas. Visits also by arrangement.**

For 20 yrs we opened The Old Rectory, Llanfihangel Glyn Myfyr, for the National Garden Scheme. We have moved and look forward to inviting visitors to our new garden which has an emphasis on low maintenance, but without losing any of our passion for plants. It has taken 3 yrs to reimagine this new garden. We have planted an abundance of acers, magnolias, cornus species, perennials, bulbs.

In 2022, our donations to Carers Trust meant that 456,089 unpaid carers were supported across the UK

18 PLAS NANTGLYN

Nantglyn, Denbigh, LL16 5PW. Janette & Richard Welch. *5m S of Denbigh. From Denbigh follow signs for Nantglyn in SW direction; nr phone box in Nantglyn straight over Xrds & bear R at fork, house 300yds on L. Map Reference SJ 003613.* **Sat 13, Sun 14 May (2-5.30). Adm £5, chd free. Home-made teas. Donation to Hope House and Ty Gobaith Children's Hospices.**

Large old established gardens with a fine collection of rhododendrons which should be at their best in May. Formal yew hedging and topiary enclose a rose garden and small vegetable plot beyond. Herbaceous borders. Fine trees inc some tall beeches along the drive and a cedar of Lebanon which is just reaching maturity. Pond, summerhouse, terrace and splendid views.

19 ◆ PLAS NEWYDD

Llangollen, LL20 8AW. Denbighshire County Council, 01978 862834, plasnewydd@denbighshire.gov.uk, www.denbighshire.gov.uk. *Follow brown sign from A5 in Llangollen.* **For NGS: Sun 2 July (10-4). Adm by donation. For other opening times and information, please phone, email or visit garden website.**

Plas Newydd was the home of the famous Ladies of Llangollen (1780-1814) and they transformed it into a Gothic fantasy of projecting stained glass and elaborately carved oak. The peaceful and tranquil garden has a formal parterre and rose garden, a dell area and a streamside walk along the Cyflymen

20 SAITH FFYNNON FARM

Downing Road, Whitford, Holywell, CH8 9EN. Mrs Jan Miller-Klein, 01352 711198, Jan@7wells.org, www.7wells.co.uk. *1m outside Whitford village. From Holywell follow signs for Pennant Park Golf Course. Turn R just past the Halfway House. Take 2nd lane on R (signed for Downing & Trout Farm) & Saith Ffynnon Farm is the 1st house on R.* **Visits by arrangement 15 May to 1 Oct for groups of up to 8. Adm £5, chd £2. Home-made teas by special arrangement in advance. Donation to Plant Heritage.**

Wildlife garden 1 acre and re-wilded meadows of 8 acres, inc ponds, woods, wildflower meadows, butterfly and bee gardens, Medieval herbal, natural dye garden and the Plant Heritage National Collection of Eupatorium. Garden at it's best early Sep - early Oct. Guided walks with the owner. Hard surface from gate to patio and two sections of the garden. No wheelchair access on the damp meadows.

NPC

21 SCOTT HOUSE

Corwen Road, Ruthin, LL15 2NP. Scott House Residents. *A494 Corwen Road. Park in town, then walk from town square onto Castle St, garden 300m beyond Ruthin Castle Hotel, entrance on R. Disabled parking at garden* **Sat 10, Sun 11 June (11-4). Adm £5, chd free. Home-made teas.**

Redesigned 2½ acre shared garden surrounding a fine 1930s Arts and Crafts house former nurses home to Ruthin Castle hospital. Breathtaking open views of the Vale of Clwyd. Avenue of Limes and a magnificent Cedar of Lebanon stands sentinel over a newly planted arboretum. Long herbaceous borders planted for summer long interest. New highly scented rose garden and recently constructed vegetable plot. Partial wheelchair access due to steps and terraces.

22 TAL-Y-BRYN FARM

Llannefydd, Denbigh, LL16 5DR. Mr & Mrs Gareth Roberts, 01745 540208, falmai@villagedairy.co.uk, www.villagedairy.co.uk. *3m W of Henllan. From Henllan take rd signed Llannefydd. After 2½m turn R signed Llaeth y Llan. Garden ½m on L.* **Visits by arrangement. Refreshments to be discussed on booking. Adm by donation.**

Medium sized working farmhouse cottage garden. Ancient farm machinery. Incorporating ancient privy festooned with honeysuckle, clematis and roses. Terraced arches, sunken garden pool and bog garden, fountains and old water pumps. Herb wheels, shrubs and other interesting features. Lovely views of the Clwydian range. Water feature, new rose tunnel, vegetable tunnel and small garden summerhouse.

23 TYN RHOS

Ffordd Y Rhos, Treuddyn, Mold, CH7 4NJ. Karen & Robert Waight, 01352 770638, robert.waight@btinternet.com. *4m S of Mold. On A541 at Pontblyddyn, turn onto A5104 toward Corwen. After 2.3m turn R onto Ffordd-y-Llan, signed for Treuddyn. After 0.4m turn L at Xrds onto Ffordd-y-Rhos. After 0.2m take farm track on R.* **Visits by arrangement 17 Apr to 31 Oct. Adm £4, chd free. Home-made teas. Gluten free options available on request.**

1 acre garden, 2 acre wood (Waight's Wood), orchard, vegetable plot. Main garden planning by Jenny Hendy, laid to herbaceous planting, enhanced during summer with homegrown annuals. Path winds past the natural pond, stream and under planting to the British woodland containing native bulbs, leading to paddocks and hives. A graveled yard and small steps (2 inch) need to be negotiated. For the intrepid wheelchair user the woodland walk is a possibility.

24 WHITE CROFT

19 Ffordd Walwen, Lixwm, Holywell, CH8 8LW. Mr John & Mrs Mary Jones. *From A541 turn onto the B5121 for Lixwm, From A55 come off at J32 or J32A. Drive S, follow Lixwm signs. Car park available.* **Sun 25 June (1-5). Combined adm with 17 Ffordd Walwen £5, chd free. Light refreshments.**

Wrap around garden inc cottage style planting, dahlia bed, rock garden, vegetable plot, fruit trees. greenhouse, stream and fish pond, pergola, balcony views over Clwydian Range and Japanese style garden. The garden has been established over the last 11 yrs. It was originally just lawn and conifer trees. A double garage containing a Westfield Kit car and a sports car.

In 2022 we donated £450,000 to Marie Curie which equates to 23,871 hours of community nursing care

Plas Nantglyn

POWYS

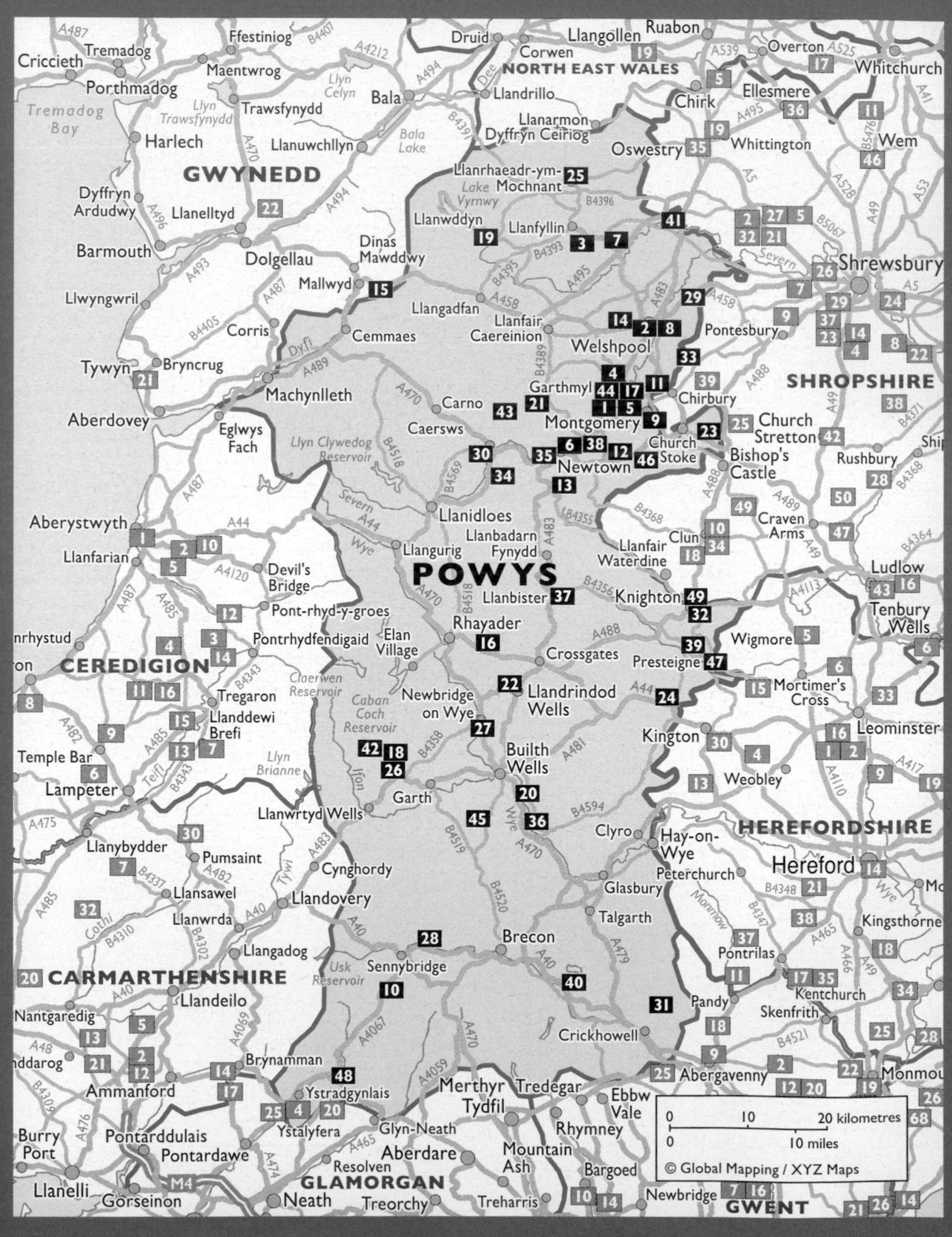
POWYS
GWYNEDD
NORTH EAST WALES
SHROPSHIRE
CEREDIGION
HEREFORDSHIRE
CARMARTHENSHIRE
GLAMORGAN
GWENT
Welshpool
Montgomery
Newtown
Llanidloes
Rhayader
Llandrindod Wells
Builth Wells
Brecon
Crickhowell
Hay-on-Wye
Knighton
Presteigne
Machynlleth
Llanfyllin
Llanwrtyd Wells
Sennybridge
Ystradgynlais
Shrewsbury
Hereford
Aberystwyth
Abergavenny
Merthyr Tydfil
0 10 20 kilometres
0 10 miles
© Global Mapping / XYZ Maps

A three hour drive through Powys takes you through the spectacular and unspoilt landscape of Mid Wales, from the Berwyn Hills in the north to south of the Brecon Beacons.

Through the valleys and over the hills, beside rippling rivers and wooded ravines, you will see a lot of sheep, pretty market towns, half-timbered buildings and houses of stone hewn from the land.

The stunning landscape is home to many of the beautiful National Garden Scheme gardens of Powys. Gardens nestling in valleys, gardens high in the hills, wild-life gardens, riverside gardens, walled gardens, grand gardens, cottage gardens, gardens in picturesque villages, gardens in towns and even village openings where you can enjoy a feast of gardens in one day - they can all be found in Powys! If you have something to celebrate or just fancy a private visit with your club or friends why not arrange a personalised visit to one of our gardens that open by arrangement?

Here in Powys all is the spectacular, the unusual, the peaceful and the enchanting all opened by generous and welcoming garden owners. We can't wait to share them with you!

Mae taith tair awr drwy Bowys yn eich tywys drwy dirwedd ysblennydd a digyffwrdd Canolbarth Cymru, o Fryniau'r Berwyn yn y gogledd i'r de o Fannau Brycheiniog.

Trwy'r dyffrynnoedd a thros y bryniau, wrth ymyl afonydd crychdonnol a cheunentydd coediog, fe welwch lawer o ddefaid, trefi marchnad tlws, adeiladau hanner pren a thai o gerrig wedi'u naddu o'r tir.

Mae'r dirwedd drawiadol yn gartref i lawer o erddi hardd y Cynllun Gerddi Cenedlaethol ym Mhowys. Gerddi sy'n swatio mewn cymoedd, gerddi yn uchel yn y bryniau, gerddi bywyd gwyllt, gerddi ar lan yr afon, gerddi muriog, gerddi crand, gerddi mewn pentrefi prydferth, gerddi mewn trefi a hyd yn oed agoriadau pentref lle gallwch fwynhau gwledd o erddi mewn un diwrnod - gellir dod o hyd iddynt i gyd ym Mhowys! Os oes gennych rywbeth i'w ddathlu neu ddim ond awydd ymweliad preifat gyda'ch clwb neu ffrindiau, beth am drefnu ymweliad wedi'i bersonoli ag un o'n gerddi sy'n agor trwy drefniant?

Yma ym Mhowys mae'r ysblennydd, yr anarferol, y heddychlon a'r hudolus i gyd wedi'i agor gan berchnogion gerddi hael a chroesawgar. Ni allwn aros i'w rhannu gyda chi!.

Volunteers

North Powys County Organiser
Susan Paynton 01686 650531
susan.paynton@ngs.org.uk

County Treasurer
Jude Boutle 07702 061623
jude.boutle@ngs.org.uk

Publicity
Sue McKillop 07753 289701
sue.mckillop@ngs.org.uk

Talks
Helen Anthony 07986 061051
helen.anthony@ngs.org.uk

Social Media
Roo Nicholls 07989788640
roo.nicholls@ngs.org.uk

Nikki Trow 07958 958382
nikki.trow@ngs.org.uk

Booklet Co-ordinator
Position vacant

Assistant County Organisers
Ann Thompson 07979 645489
ann.thompson@ngs.org.uk

Simon Cain 07958 915115
simon.cain@ngs.org.uk

Kate Nicoll 07960 656814
Kate.nicoll@ngs.org.uk

Chris Smith 07775 656299
chris.smith@ngs.org.uk

Elizabeth Fairhead 07824 398840
elizabeth.fairhead@ngs.org.uk

South Powys County Organiser
Christine Carrow 01591 620461
christine.carrow@ngs.org.uk

County Treasurer
Steve Carrow 01591 620461
steve.carrow@ngs.org.uk

Publicity Officer
Gail Jones 07974 103692
gail.jones@ngs.org.uk

@powysngs
@PowysNGS
@powysngs

OPENING DATES

All entries subject to change. For latest information check **www.ngs.org.uk**

Map locator numbers are shown to the right of each garden name.

Extended openings are shown at the beginning of each month

March

Sunday 26th
Bryngwyn Hall 7

April

Sunday 30th
NEW Garregllwyd 16

May

Monday 1st
NEW Garregllwyd 16

Saturday 6th
◆ Gregynog Hall & Garden 21

Sunday 7th
Cwm Farm 11
◆ Gregynog Hall & Garden 21
Pontsioni House 36

Saturday 13th
◆ Dingle Nurseries & Garden 14

Sunday 14th
◆ Dingle Nurseries & Garden 14

Saturday 20th
Rock Mill 38

Sunday 21st
NEW Llanrhaeadr Ym Mochnant Gardens 25
Llwyn Madoc 26
Rock Mill 38

Saturday 27th
Bachie Uchaf 3
NEW Hill Cottage 22

Sunday 28th
Bachie Uchaf 3
Glanwye 20
NEW Hill Cottage 22
Llysdinam 27

June

Saturday 3rd
The Rock House 37

Sunday 4th
The Neuadd 31
Pontsioni House 36
The Rock House 37
Tranquility Haven 39

Thursday 8th
White Hopton Farm 46

Saturday 10th
Bron Hafren 5

Sunday 11th
Bron Hafren 5
Hurdley Hall 23

Saturday 17th
NEW 14 Adelaide Drive 2
Tremynfa 41

Sunday 18th
NEW 14 Adelaide Drive 2
Tremynfa 41
Tyn y Cwm 42

Saturday 24th
The Hymns 24

Sunday 25th
The Hymns 24
Tyn-y-Graig 43

July

Every Friday from Friday 21st
◆ Welsh Lavender 45

Saturday 1st
Lower Wernfigin Barns 28

Sunday 2nd
Berriew Village Gardens 4
Lower Wernfigin Barns 28
Treberfydd House 40

Sunday 9th
Hurdley Hall 23

Wednesday 12th
Tranquility Haven 39

Saturday 15th
1 Glanrafon 19
NEW The Meadows 30

Sunday 16th
Cwm-Weeg 13
1 Glanrafon 19
NEW The Meadows 30

Sunday 23rd
Crai Gardens 10
Vaynor Park 44

Saturday 29th
Bryn Teg 6
Ponthafren 35

Sunday 30th
Bryn Teg 6
Ponthafren 35

August

Every Friday to Friday 11th
◆ Welsh Lavender 45

Saturday 5th
Clawdd-Y-Dre 9

Sunday 6th
Clawdd-Y-Dre 9

Saturday 12th
1 Church Bank 8

Sunday 13th
1 Church Bank 8

Wednesday 16th
Tranquility Haven 39

Saturday 19th
NEW Willowbrook 47

Sunday 20th
NEW Willowbrook 47

September

Sunday 10th
Garthmyl Hall 17

Wednesday 13th
Tranquility Haven 39

October

Saturday 7th
◆ Dingle Nurseries & Garden 14

Sunday 8th
◆ Dingle Nurseries & Garden 14

Saturday 14th
◆ Gregynog Hall & Garden 21

Sunday 15th
◆ Gregynog Hall & Garden 21

November

Sunday 5th
Tranquility Haven 39

December

Sunday 10th
Tranquility Haven 39

By Arrangement

Arrange a personalised garden visit with your club, or group of friends, on a date to suit you. See individual garden entries for full details.

Abernant 1
1 Church Bank 8
NEW Cwm Sidwell 12
NEW Dugoed Bach 15
Gilwern Barn 18
The Hymns 24
Llysdinam 27
Maesfron Hall and Gardens 29
NEW The Meadows 30
The Neuadd 31
No 2 The Old Coach House 32
NEW Pen-y-Bank 33
Plas Dinam 34
Rock Mill 38
Tranquility Haven 39
Tremynfa 41
Tyn y Cwm 42
NEW Willowbrook 47
NEW 57 & 58 Ynyswen 48
1 Ystrad House 49

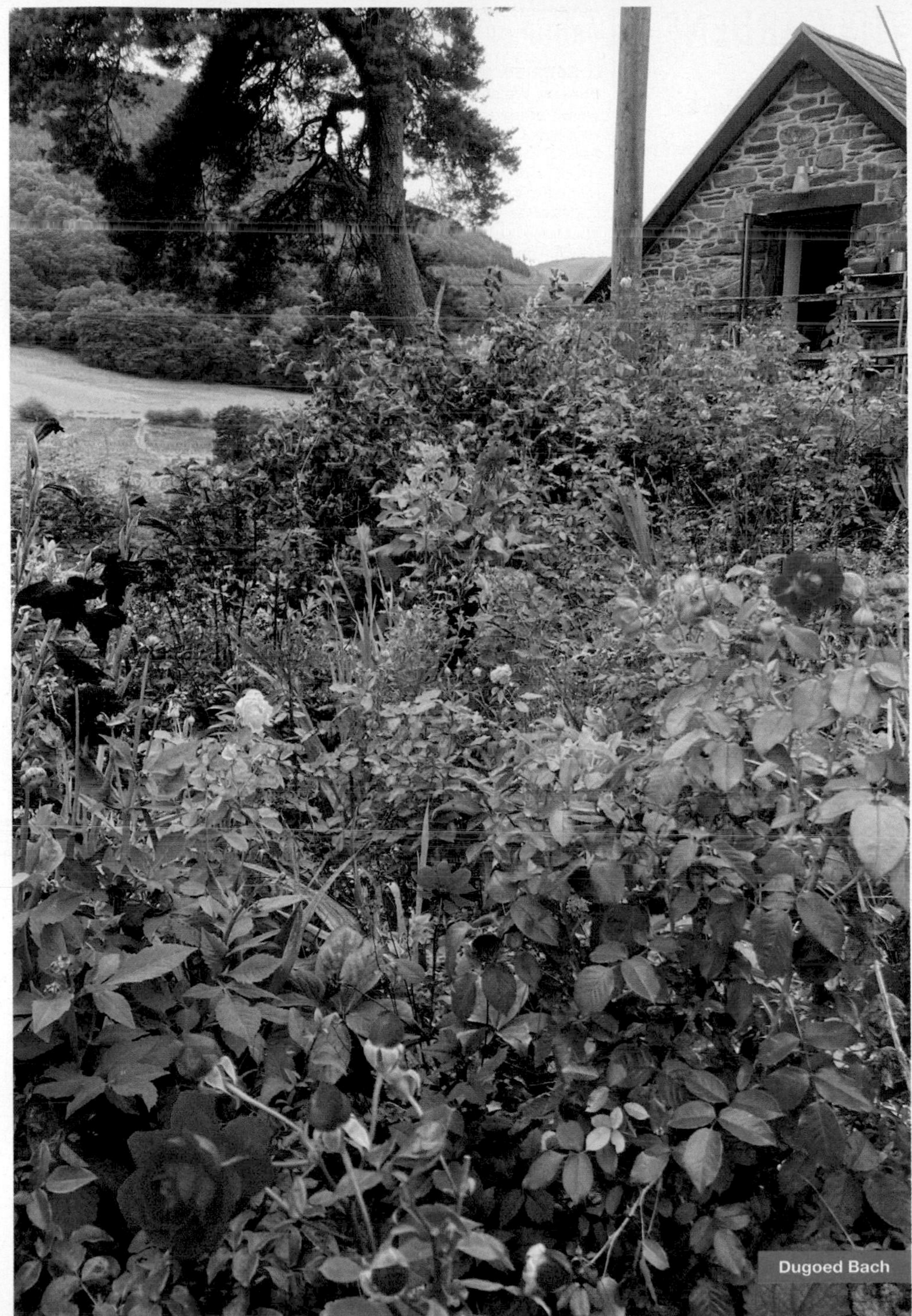

Dugoed Bach

THE GARDENS

1 ABERNANT

Garthmyl, SY15 6RZ. Mrs B M Gleave, 01686 640494. *1½m S of Garthmyl. On A483 midway between Welshpool & Newtown (both 8m). Approached over steep humpback bridge with wooden statue of workman. Straight ahead to house & parking.* **Visits by arrangement Apr to July for groups of up to 12. Adm £4, chd free.**
3 acre cottage garden with stunning cherry blossom orchard. Knot garden, formal rose garden, rockery, pond, shrubs, ornamental trees, archaic sundials, fossilised wood and stone heads. Additional woodland of 9 acres borrowed views of the Severn Valley. 85 cherry trees in blossom in late April. Roses in late June.

2 NEW 14 ADELAIDE DRIVE

Red Bank, Welshpool, SY21 7RQ. Jackie Lindsay. *5 min drive from centre Welshpool. Follow one way system A458 NW turn L up Red Bank after church, straight across mini r'about & 1st R Adelaide Drive. No.14 is 7th bungalow on R.* **Sat 17, Sun 18 June (2-5). Adm £4, chd free. Home-made teas.**
¼ acre gently sloping edge of town garden wrapped around bungalow. Bursting with colour. Created from scratch since beginning 2019. Front garden mainly roses and shrubs while the back garden has some young trees, a large herbaceous bed, a small vegetable garden and an old orchard.

3 BACHIE UCHAF

Bachie Road, Llanfyllin, SY22 5NF. Glyn & Glenys Lloyd. *S of Llanfyllin. Going towards Welshpool on A490 turn R onto Bachie Rd after Llanfyllin primary sch. Keep straight for 0.8m. Take drive R uphill at cottage on L.* **Sat 27, Sun 28 May (12-5). Adm £5.50, chd free. Home-made teas.**
Inspiring, colourful hillside country garden. Gravel paths meander around extensive planting and over streams cascading down into ponds. Specimen trees, shrubs, rhododendrons and azaleas and vegetable garden. Enjoy the wonderful views from one of the many seats; your senses will be rewarded.

GROUP OPENING

4 BERRIEW VILLAGE GARDENS

Berriew, Welshpool, SY21 8BA. www.berriew.com. *A483 5m S Welshpool. From Welshpool take A483 S for approx 4m. Turn R onto B4390. Cont through village & turn R to car park at sch. Map & adm tickets at car park. All within walking distance of car park.* **Sun 2 July (12-5). Combined adm £6, chd free. Home-made teas at The Old School.**

3 CHURCH TERRACE
Mrs Jane Hancock.

NEW **9 MAES BEUNO**
Jenny Linsdell.

NEW **16 MAES BEUNO**
John Edwards.

NEW **26 MAES BEUNO**
Helen & Kristian Hickson-Booth.

THE OLD COURT HOUSE
Michael Davis & Andrew Logan.

RHIEW HOUSE
Mr Richard & Mrs Fiona Noyce.

The picturesque village of Berriew is on the Montgomeryshire Canal with the R Rhiew flowing through its heart. Black and white cottages, church, two pubs, shops, William O'Brien artist/blacksmith forge and Andrew Logan Museum of Sculpture. Seven very different gardens: The Old Court House idyllic situation on banks of R Rhiew terrace with metal staircase by Berriew sculptor/metalworker William O'Brien, raised beds for soft fruits and vegetables: Rhiew House combines a degree of order and colour with a sense of wildness: 3 Church Terrace ¼ acre plantsman's garden with river below and diverse range of rare and unusual plants to create interest throughout and three densely planted small gardens in Maes Beuno (first time opening) bursting with flowers and vegetables give inspiration for what can be achieved in small spaces. Partial wheelchair access.

5 BRON HAFREN

Garthmyl, Montgomery, SY15 6RT. Rod & Debbie Kent. *5m S Welshpool. A483 opp Nags Head turn on B4835 towards Montgomery. After 600 metres L through double gates diagonally opp Kings Nursery & park in field.* **Sat 10, Sun 11 June (1.30-5). Adm £5, chd free. Home-made teas.**
1½ acre mature garden on banks R Severn. Riverside walk with view of ornate Grade II listed bridge. Surrounding the Victorian house are lawns, orchard, mixed borders, shrubbery, raised veg beds, spinney and large redwood. Interesting mature specimen trees, outbuildings inc original Ty Bach. Uneven surface in paddock parking area. Wheelchair access to refreshments over gravel.

6 BRYN TEG

Bryn Lane, Newtown, SY16 2DP. Novlet Childs. *N side of Newtown. Take Llanfair Caereinion Rd towards Bettws Cedewain & turn L before hospital. Up hill on L.* **Sat 29 July (10-5); Sun 30 July (1-5). Adm £4, chd free. Home-made teas at Ponthafren, Long Bridge Street, Newtown. Open nearby Ponthafren.**
Amazing exotic secret Caribbean garden in centre of Newtown planted to remind me of my childhood in Jamaica. An exciting walk through the jungle. High above the head are banana leaves and colourful climbers. A winding path takes you on a journey through a moon gate into another land. Sounds of water fill the air; explosion of colourful intermingling flowers and plants. A huge number of plants on many levels. All shapes, sizes and colours mixed together as found in tropical jungles. Feel transported to another continent. Even in the rain you will get that jungle experience. Manual wheelchair access to half of the garden. Not suitable for mobility scooters.

7 BRYNGWYN HALL

Bwlch-y-Cibau, Llanfyllin, SY22 5LJ. Auriol Marchioness of Linlithgow, www.bryngwyn.com. *3m SE Llanfyllin. From Llanfyllin take A490 towards Welshpool for 3m turn L up drive just before Bwlch-y-Cibau.* **Sun 26 Mar (11-4). Adm £7.50, chd free. Home-made teas.**
Stunning grade II* listed 9 acre garden with 60 acres parkland

design inspired by William Emes. Prunus subhirtella 'Autumnalis', varieties of hamamelis, mahonia, early flowering daphnes, corylopsis and chimonanthus. Woodland garden carpeted with snowdrops then stunning show of thousands of daffodils, camassias and fritillaries in the long grass down to serpentine lake. Unusual trees, shrubs and unique Poison Garden.

8 1 CHURCH BANK

Welshpool, SY21 7DR. Mel & Heather Parkes, 01938 559112, melandheather@live.co.uk. *Centre of Welshpool. Church Bank leads onto Salop Rd from Church St. Follow one way system, use main car park then short walk. Follow yellow NGS signs.* **Sat 12, Sun 13 Aug (12-5). Adm £3.50, chd free. Home-made teas. Visits also by arrangement Apr to Aug for groups of 5 to 30.**

An intimate jewel in the town with many interesting plants and unusual features where the sounds of water fill the air. Densely planted garden with a Gothic arch and zig zag path leading to a shell grotto, bonsai garden and fernery. You can also explore the ground floor of this C17 barrel maker's cottage and walk into a large garden room which houses a museum of tools and motor memorabilia.

9 CLAWDD-Y-DRE

Lions Bank, Montgomery, SY15 6PT. Nadine & Daniel Roach and Geoff Ferguson. *From centre of Montgomery (parking) proceed up hill along Church Bank past St Nicholas Church. Turn R into Lions Bank past Spider Cottage.* **Sat 5, Sun 6 Aug (2-5). Adm £5, chd free. Home-made teas.**

Sitting on the old town walls, this well established ¾ acre garden features mature trees, yew hedging and landscaped rockeries. Gravel pathways lead to a manicured lawn and a productive orchard with both fruit and nut trees. Displaying a wide range of shrubs and herbaceous perennials the garden also boasts a delightful gazebo with stunning panoramic views across the valley to the Shropshire Hills.

GROUP OPENING

10 CRAI GARDENS

Crai, LD3 8YP. *13m SW of Brecon. Turn W off A4067 signed 'Crai'. Village hall is 50yds straight ahead. Park here for adm, info about gardens, map, teas.* **Sun 23 July (2-5). Combined adm £6, chd free. Home-made teas in Village Hall served until 4.30.**

Set against the backdrop of Fan Gyhirych and Fan Brycheiniog, at 1000 ft above sea level the Crai valley is a hidden gem, off the beaten track between Brecon and Swansea. Those in the know have long enjoyed visiting our serene valley, with its easy access to the hills and its fabulous views. In difficult climatic conditions - fierce wind and heavy rain are frequent visitors - the Crai Gardens reflect a true passion for gardening. The gardens come in a variety of size, purpose and design, and inc a range of shrubs, perennials and annuals; vegetables and fruits; raised beds; water features; patio containers and window boxes; manicured lawns. Not to forget the chickens and ducks. And to complete your Sunday afternoon, come and enjoy the renowned hospitality of the Crai ladies by sampling their delicious home-made cakes in the village hall. It is too far between the gardens to expect to walk and there is not a lot of parking so bring your bicycle if you can.

11 CWM FARM

Forden, Welshpool, SY21 8NB. Mr Michael & Mrs Gemma Hughes. *3m S Welshpool. from A483 turn L A490 for 2.4m then ahead on B4388 for 1.1m.* **Sun 7 May (12-5). Adm £5, chd free. Home-made teas inc gluten & dairy free options.**

Charming C19 farmhouse down wooded cwm with impressive crag. 5 acre garden in stunning location with panoramic view of Corndon and Roundton Hills. Bluebells, herbaceous beds, Welsh apple orchard, pond, stream, wildflower meadow, wide lawns and paths to wander and places to sit. Engaging mix of wild and cultivated. Gemma's studio will be open. Wheelchair drop off point by farmhouse. Highest view point from top meadow has steep steps so not accessible.

12 NEW CWM SIDWELL

Kerry, Newtown, SY16 4NA. Jan Rogers, 07772 414783, janetrogers62@hotmail.co.uk. *6m E Newtown. From Newtown take A489. After 4m turn L on B4368. After 1.4m turn R signed Goetre Follow yellow arrows.* **Visits by arrangement 3 June to 31 Aug for groups of up to 20. Parking for up to 5 cars. Adm £4, chd free. Light refreshments.**

Cottage garden with lovely views brought back from the wilderness over the last couple of years. Roses, sweet peas, herbaceous perennials, vegetable garden and cut flower garden all grown from seed. Pockets of colour whichever way you look. Pond and chickens. Flower and vegetable seedlings for sale.

13 CWM-WEEG

Dolfor, Newtown, SY16 4AT. Dr W Schaefer & Mr K D George, 01686 628992, wolfgang@cwmweeg.co.uk, www.cwmweeg.co.uk. *4½m SE of Newtown. Off Bypass, take A489 E from Newtown for 1½m, turn R towards Dolfor. After 2m turn L down asphalted farm track, signed at entrance. Do not follow SatNav. Also signed from Dolfor village on NGS days.* **Sun 16 July (2-5). Adm £6, chd free. Home-made teas & cream teas.**

2½ acre garden set within 24 acres of wildflower meadows and bluebell woodland with stream centred around C15 farmhouse (open by prior arrangement). Formal garden in English landscape tradition with vistas, grottos, sculptures, stumpery, lawns and extensive borders terraced with stone walls. Translates older garden vocabulary into an innovative C21 concept. Extensive under cover seating area in the large Garden Pavilion. Partial wheelchair access. For further information please see garden website.

The National Garden Scheme searches the length and breadth of England, Wales, Northern Ireland and the Channel Islands for the very best private gardens

Rhiew House, Berriew Gardens

© Simon Morgan

14 ◆ DINGLE NURSERIES & GARDEN

Welshpool, SY21 9JD. Mr & Mrs D Hamer, 01938 555145, info@dinglenurseriesandgarden.co.uk, www.dinglenurseryandgarden.co.uk. *2m NW of Welshpool. Take A490 towards Llanfyllin & Guilsfield. After 1m turn L at sign for Dingle Nurseries & Garden. Follow signs & enter the Garden from adjacent plant centre.* **For NGS: Sat 13, Sun 14 May, Sat 7, Sun 8 Oct (9-5). Adm £3.50, chd free. Tea & coffee. For other opening times and information, please phone, email or visit garden website.**

4½ acre internationally acclaimed RHS partner garden on south facing site, sloping down to lakes. Huge variety of rare and unusual trees, ornamental shrubs and herbaceous plants give year-round interest. Set in the hills of mid Wales this beautiful well known garden attracts visitors from Britain and abroad. Plant collector's paradise. Open all yr except 24 Dec - 2 Jan.

15 NEW DUGOED BACH

Mallwyd, Machynlleth, SY20 9HR. Katie & Elaine Weston, 07745 529597, katiewestonuk@gmail.com. *14m N of Machynlleth, 25m W of Welshpool. Directly on A458 1m E of Mallwyd r'about. What3words app - generally. nitrate.drain.* **Visits by arrangement 21 June to 29 Sept for groups of 5 to 15. Adm £5, chd free. Home-made teas.**

A rural ¾ acre cottage garden set in Snowdonia National Park with magnificent views. Large variety of plants, and planting themes on a sloping site, with challenging conditions. Orchard, wildflower meadow and ponds provide habitats for wildlife and our own bees. There are collections of irises, roses, dahlias and many others planted to provide colour and interest throughout the year.

16 NEW GARREGLLWYD

Nantmel, Rhayader, LD6 5PE. Stephanie Morgan. *5m E of Rhayader. From Rhayader take A44 E for 3½m to Nantmel. Turn L by Dolau Chapel signed Abbeycwmhir. Follow lane for 1½m then turn R by red warning triangle signed Garregllwyd. Follow track for 0.4 m.* **Sun 30 Apr, Mon 1 May (2-5). Adm £5, chd free. Home-made teas.**

3 acre landscaped garden at 1000' with stunning panoramic views of mid Wales. Large ponds with abundance of wildlife, unusual specimen trees, daffodils, bluebells, rhododendrons and raised vegetable beds/greenhouses growing seasonal veg. Designed to cope with exposed altitude with minimal maintenance and to encourage wildlife. Variety of seating areas to enjoy the ever changing weather conditions.

17 GARTHMYL HALL

Garthmyl, Montgomery, SY15 6RS. Julia Pugh, 07716 763567, hello@garthmylhall.co.uk, www.garthmylhall.co.uk. *On A483 midway between Welshpool & Newtown (both 8m). Turn R 200yds S of Nag's Head Pub.* **Sun 10 Sept (12-5). Adm £5, chd free. Home-made teas.**
Grade II listed Georgian manor house (not open) surrounded by 5 acres of grounds. 100 metre herbaceous borders, newly restored 1 acre walled garden with gazebo, circular flowerbeds, lavender beds, wildflower meadow, pond, two fire pits and gravel paths. Fountain, three magnificent Cedar of Lebanon and giant redwood. Partial wheelchair access. Accessible WC.

18 GILWERN BARN

Beulah, LD5 4YG. Mrs Penelope Bourdillon, 01591 620203, pbourdillon@gmail.com, www.gilwerngarden.co.uk. *2m N of Beulah. Turn L exactly ¾m from Beulah on B4358. Follow rd for over 1m. Past Cefnhafdref Farm. L at T junction. In 600 yds fork L through stone gateposts.* **Visits by arrangement May to Aug. Adm £5, chd free.**
Terraced garden on very challenging site situated on steep rocky hillside in the beautiful and secluded Cammarch Valley. Roses, herbaceous and shrub borders. Fine walling and gate posts making use of stone and slate found in the garden. Man-made waterfall and rill. The situation, on a steep rocky slope, is fairly dramatic and makes it interesting. The waterfall is quite a feature. Late flowering roses. The peonies should be worth seeing in late May. A new rill, recently planted.

19 1 GLANRAFON

Llanwddyn, Oswestry, SY10 0LU. Margaret Herbert. *24m W of Oswestry. From Llanfyllin, follow brown signs to Lake Vyrnwy. Through village of Llanwddyn & turn L across dam then L past Artisans Cafe & park in public car park by playground. Follow yellow signs to garden.* **Sat 15, Sun 16 July (2-5). Adm £4.50, chd free.**
1 acre secluded garden under development. As a keen plantsperson, I have been establishing this steeply sloping, densely planted wildlife garden with hundreds of varieties of shrubs and perennials over the past 5 years. Very productive raised vegetable beds, fruit cages and polytunnels, large pond. Engaging mix of wild and cultivated. Large greenhouse with ornamental and edible crops. Fruit trees, bushes and vegetables, many of which are unusual varieties. Yard with chickens and ducks. Wheelchairs can access the upper & lower levels via access road & separate gates. Car parking strictly for disabled badge holders at sculpture park.

20 GLANWYE

Builth Wells, LD2 3YP. Mr & Mrs H Kidston. *2m SE Builth Wells. From Builth Wells on A470, after 2m R at Lodge Gate. From Llyswen on A470, after 6m L at Lodge Gate. Will be signposted. Suggest not using SatNav as unreliable.* **Sun 28 May (2-5). Adm £5, chd free. Home-made teas.**
Large Victorian garden, spectacular rhododendrons, azaleas. Herbaceous borders, extensive yew hedges, lawns, long woodland walk with bluebells and other woodland flowers. Magnificent views of upper Wye Valley.

21 ◆ GREGYNOG HALL & GARDEN

Tregynon, Newtown, SY16 3PL. Gregynog Trust, 01686 650224, enquiries@gregynog.org, www.gregynog.org. *5m N of Newtown. From Newtown A483 turn L B4389 for Llanfair Caereinion (car parking charge applies).* **For NGS: Sat 6, Sun 7 May, Sat 14, Sun 15 Oct (10-4). Adm by donation. Light refreshments in Courtyard Cafe. For other opening times and information, please phone, email or visit garden website.**
Gardens are Grade I Listed due to association with the C18 landscape architect William Emes. Set within 750 acres, a designated National Nature Reserve with SSSI. Parkland with small lake and traces of a water garden. A mass display of rhododendrons, azaleas and unique yew hedge surround the sunken lawns. Unusual trees, woodland walks and arboretum. Good autumn colour. Some gravel paths.

22 NEW HILL COTTAGE

North Avenue, Llandrindod Wells, LD1 6BY. Mr Neville & Mrs Sylvia Clare. *At r'about junction of Cadwallader Way & North Avenue.* **Sat 27, Sun 28 May (1-5). Adm £4.50, chd free. Cream teas.**
A peaceful and secluded garden in Llandrindod Wells principally inspired by oriental influences. Paths wind through shady plantings surrounding a lawn, across an attractive carp pond via an oriental style bridge. The garden features rhododendrons, azaleas, camellias, sculptural ferns. There is an authentic Nordic cabin, a hidden Alice In Wonderland themed area and vegetable plot.

23 HURDLEY HALL

Hurdley, Churchstoke, SY15 6DY. Simon Cain & Simon Quin. *2m from Churchstoke. Take turning for Hurdley off A489, 1m E of Churchstoke. Garden is a further 1m up the lane.* **Sun 11 June, Sun 9 July (11-5). Adm £6, chd free. Home-made teas.**
Beautiful traditional 2 acre garden in stunning location; herbaceous and mixed borders leading to 18 acres of Coronation Meadows, woodland and orchard. Featured in Country Life in 2022 and Regional Finalist in The English Garden's Favourite Garden Competition. Visit in June for meadow flowers and roses and July for lavender and herbaceous border.

24 THE HYMNS

Walton, Presteigne, LD8 2RA. E Passey, 07958 762362, thehymns@hotmail.com, www.thehymns.co.uk. *5m W of Kington. Take A44 W, then 1st R for Kinnerton. After approx 1m, at the top of small hill, turn L (W).* **Sat 24, Sun 25 June (11-5). Adm £6, chd free. Light refreshments. Visits also by arrangement 1 Apr to 23 Sept.**
In a beautiful setting in the heart of the Radnor valley, the garden is part of a restored C16 farmstead, with long views to the hills, and The Radnor Forest. It is a traditional garden reclaimed from the wild, using locally grown plants and seeds, and with a herb patio, wildflower meadows and a short woodland walk. It is designed for all the senses: sight, sound and smell.

GROUP OPENING

25 NEW LLANRHAEADR YM MOCHNANT GARDENS

Market Street, Llanrhaeadr Ym Mochnant, Oswestry, SY10 0JN. *The village is 12m W of Oswestry on the B4580. The open gardens are all walking distance from the Village Hall Car Park in Back Chapel Lane, or the village car park in Park Street.* **Sun 21 May (12-5). Combined adm £6, chd free. Home-made teas in Llanrhaeadr ym Mochnant Public Hall, Back Chapel Street.**

NEW **BRYNARFON, WATERFALL ROAD**
Mrs Olly Williams.

NEW **NO 2 MAES DERW**
Juliet & Gerard Moony.

NEW **NO 5 MAES DREW**
Mrs Susan Morris.

PLAS YN LLAN
Kate & Fergus Nicoll,
01691 780747,
kate.fergus.nicoll@gmail.com.

NEW **ROCK HOUSE**
Mrs Pauline Kirby.

NEW **TRE TYLLUAN**
Mr Peter Carr.

The lively village of Llanrhaeadr ym Mochnant is set on the edge of the Berwyn Mountains with local shops and beauty spots such as the famous waterfall, Pistyll Rhaeadr. We have a selection of small village gardens offering a range of styles from cottage, courtyard, terrace and prairie style plantings. Innovative use of small spaces, often steeply sloping down to the beautiful river Rhaeadr. All of the gardens attract a multitude of wildlife: birds, bees, bats and butterflies. Some have trained fruit trees and productive vegetable plots, as well as pots of bulbs for late spring interest. Local coffee shop Gegin Fach will be open for light lunches. The Wynnstay Arms offers a carvery lunch which would need to be booked in advance. Many show the challenges of working in steeply sloping gardens.

26 LLWYN MADOC

Beulah, Llanwrtyd Wells, LD5 4TT. Patrick & Miranda Bourdillon, 01591 620564, miranda.bourdillon@gmail.com. *8m W of Builth Wells. On A483 at Beulah take rd towards Abergwesyn for 1m. Drive on R. Parking on field below drive - follow signs.* **Sun 21 May (2-5.30). Adm £6, chd free. Cream teas.**
Terraced garden in attractive wooded valley overlooking lake; yew hedges; rose garden with pergola; azaleas and rhododendrons.

27 LLYSDINAM

Newbridge-on-Wye, LD1 6NB. Sir John & Lady Venables-Llewelyn & Llysdinam Charitable Trust, 07748 492025, llysdinamgardens@gmail.com, llysdinamgardens.org. *5m SW of Llandrindod Wells. Turn W off A470 at Newbridge-on-Wye; turn R immed after crossing R Wye; entrance up hill.* **Sun 28 May (2-5). Adm £5, chd free. Cream teas. Visits also by arrangement.**
Llysdinam Gardens are among the loveliest in mid Wales, especially noted for a magnificent display of rhododendrons and azaleas in May. Covering some 6 acres, they command sweeping views down the Wye Valley. Successive family members have developed the gardens over the last 150 years to inc woodland with specimen trees, large herbaceous and shrub borders and a water garden, all of which provide varied, colourful planting throughout the year. The Victorian walled kitchen garden and extensive greenhouses grow a wide variety of vegetables, hothouse fruit and exotic plants. Open by arrangement in March to view the daffodils. Limited spaces for booked groups of between 4-10 for guided walk, tea and cake at £12pp Gravel paths.

28 LOWER WERNFIGIN BARNS

Trallong, Brecon, LD3 8HW. Mark Collins & Alan Loze. *Trallong, Nr Sennybridge, Brecon. A40 W from Brecon, 2nd signpost to Trallong (R). Down, over river, up to Trallong Common then follow yellow signs. A40 E from Sennybridge take 1st sign to Trallong then the same.* **Sat 1, Sun 2 July (11-5.30). Adm £6, chd free. Home-made teas.**
A terraced south facing garden with dense planting of herbaceous perennials and shrubs. Rear garden planted with hydrangeas, azaleas and rhododendrons. A newly planted area of woodland and field walk. A small orchard, vegetable and fruit garden, all managed, like the flower garden without chemicals. In front of the house is a large courtyard with formal rill, fishpond and seating areas.

29 MAESFRON HALL AND GARDENS

Trewern, Welshpool, SY21 8EA. Dr & Mrs TD Owen, 01938 570600, maesfron@aol.com, www.maesfron.co.uk. *4m E of Welshpool. On N side of A458 Welshpool to Shrewsbury Rd.* **Visits by arrangement Apr to Sept for groups of 10+. Adm inc guided tour of the gardens & historic buildings. Adm £5, chd free. Home-made teas available by prior arrangement.**
Largely intact Georgian estate of 6 acres. House (partly open) built in Italian villa style set in south facing gardens on lower slopes of Moel-y-Golfa with panoramic views of The Long Mountain. Terraces, walled kitchen garden, chapel, restored Victorian conservatories, tower, shell grotto and hanging gardens below tower. Explore ground floor, wine cellar, old kitchen and servants quarters. Parkland walks with wide variety of trees. Refreshments served in reception rooms or on terrace. Some gravel, steps and slopes.

30 NEW THE MEADOWS

Carno Road, Caersws, SY17 5JA. Pete & Lesley Benton, 07929 038936, bentlesl@aol.com. *7m W of Newtown. From Newtown A489 for 6m & turn R on A470 thru Caersws & car park on R. Garden on L 200m along. Disabled & drop off point at garden.* **Sat 15, Sun 16 July (2-5). Adm £4, chd free. Home-made teas. Visits also by arrangement 1 July to 29 July for groups of up to 20. Optional Garden Treasure Hunt for group.**
South facing level garden on village edge, originally The Manse. An eclectic mix of garden areas inc small bog garden with home-made waterfall, shrubs, flowers, vegetables + over100 containers with colourful annuals, veg and perennials. A number of seating areas surrounded by flowers and shrubs to take a while and relax. Quirky and upcycled items around the garden. Disabled drop off area outside house.

Plas Yn Llan

31 THE NEUADD
Llanbedr, Crickhowell, NP8 1SP. Robin & Philippa Herbert, 01873 812164, philippahherbert@gmail.com. *1m NE of Crickhowell. Leave Crickhowell by Llanbedr Rd. At junction with Great Oak Rd bear L, cont up hill for approx 1m, garden on L. Ample parking.* **Sun 4 June (2-6). Adm £6.50, chd free. Home-made teas in the courtyard. Visits also by arrangement May to Sept for groups of up to 12.**
Robin and Philippa Herbert have worked on the restoration of the garden at The Neuadd since 1999 and have planted many unusual trees and shrubs in the dramatic setting of the Brecon Beacons National Park. One of the major features is the walled garden, which has both traditional and decorative planting of fruit, vegetables and flowers. There is also a woodland walk with ponds and streams and a formal garden with flowering terraces. Spectacular views, water feature, rare trees and shrubs and a plant stall and teas. The owner uses a wheelchair and most of the garden is accessible, but some steep paths.

32 NO 2 THE OLD COACH HOUSE
Church Road, Knighton, LD7 1ED. Mr Richard & Mrs Jenny Vaughan, 01547 520246. *Out of Knighton on the A488 road to Clun, take 1st L into Church Rd, enter by courtyard of 1 Ystrad House.* **Visits by arrangement June to Aug. Adm £5, chd free. Home-made teas at 1 Ystrad House.**
The garden at the Old Coach House is a quiet haven of rooms as you move along a winding path edged by the top borders of shrubs, colourful perennials and circular lawns. An archway of climbers leads to the lower lawn, flanked by beech hedging, apple trees, a greenhouse, fruit bushes, vegetable patch, shrubs, hostas and ferns, arriving at a private quiet riverside woodland glade. The garden can be accessed through the side gate for wheelchair users, where the lawns are generally level.

33 NEW PEN-Y-BANK
Marton, Welshpool, SY21 8JY. Robin & Penny Kenward, 01938 580216. *6m NE Montgomery. B4386 N side of Marton turn L on 'S' bend to Marton Hill & Trelystan. Follow this for1m; do not take 1st R but cont uphill. Entrance is on R.* **Visits by arrangement 5 June to 28 July for groups of 10 to 20. Weekdays only. Adm £6, chd free. Home-made teas.**
A 5 acre, elevated, hillside setting comprising an herbaceous garden, an area of ornamental trees, a kitchen garden and wildflower meadow. The herbaceous area is very colourful mid-spring to mid-summer. Several cornus, acer and betula varieties also parrotia, sorbus, quercus and platanus etc make up the trees; the meadow contains common spotted orchids. The garden also has several rough areas. Beautiful views. Please mention any food allergies or intolerances when booking.

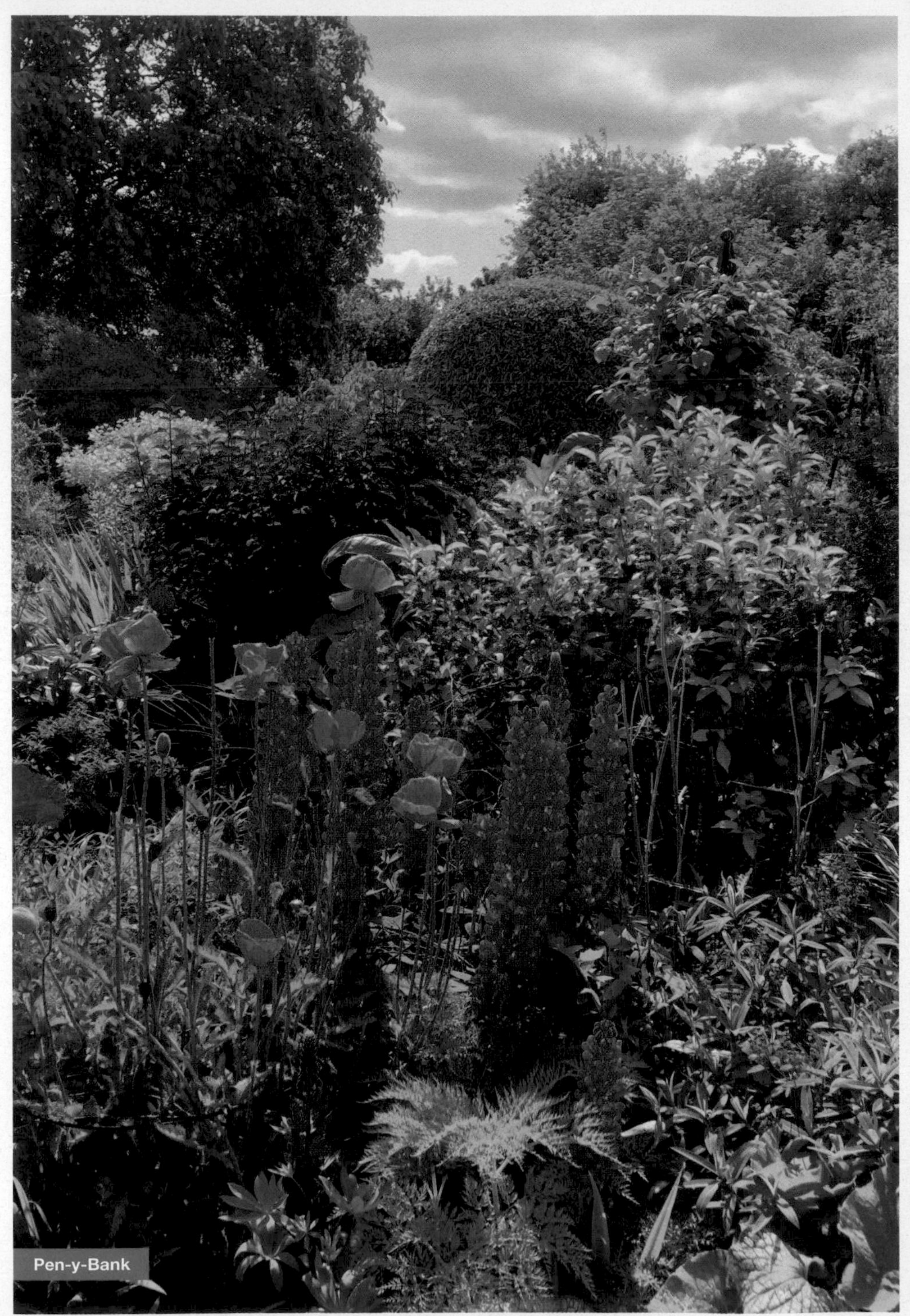

Pen-y-Bank

34 PLAS DINAM

Llandinam, Newtown, SY17 5DQ. Eldrydd Lamp, 07415 503554, eldrydd@plasdinam.co.uk, www.plasdinamcountryhouse.co.uk. *7½m SW Newtown. on A470.* **Visits by arrangement 3 Apr to 1 Nov for groups of 10+. Please discuss refreshments at booking. Adm £5, chd free.**

12 acres of parkland, gardens, lawns and woodland set at the foot of glorious rolling hills with spectacular views across the Severn Valley. A host of daffodils followed by one of the best wildflower meadows in Montgomeryshire with 36 species of flowers and grasses inc hundreds of wild orchids; Glorious autumn colour with parrotias, liriodendrons, cotinus etc. Millennium wood. From 1884 until recently the home of Lord Davies and his family (house not open).

35 PONTHAFREN

Long Bridge Street, Newtown, SY16 2DY. Ponthafren Association, www.ponthafren.org.uk. *Park in main car park in town centre, 5 mins walk. Turn L out of car park, turn L over bridge, garden on L. Limited disabled parking, please phone for details.* **Sat 29 July (10-5); Sun 30 July (1-5). Adm by donation. Home-made teas. Open nearby Bryn Teg.**

Ponthafren is a registered charity that provides a caring community to promote positive mental health and well-being for all. Open door policy so everyone is welcome. Interesting community garden on banks of River Severn run and maintained totally by volunteers: sensory garden with long grasses, herbs, scented plants and shrubs, quirky objects. Productive vegetable plot. Lots of plants for sale. Covered seating areas positioned around the garden to enjoy the views. Partial wheelchair access.

36 PONTSIONI HOUSE

Aberedw, Builth Wells, LD2 3SQ. Mr & Mrs Jonathan Reeves. *5m SE of Builth Wells. On B4567 between Erwood Bridge & Aberedw on Radnorshire side of R Wye.* **Sun 7 May, Sun 4 June (2.30-6). Adm £5, chd free. Home-made teas.**

With a background of old ruins and steep rocky woodland, this Wye Valley garden with herbaceous, shrub borders, terraces and natural rockery merge with lawns. Small walled vegetable and fruit garden. Walks through wildflower meadow along a mile of old railway line with bluebell woods and walks up to the Aberedw Rocks. Dogs welcome along old railway line. Spectacular rocky and woody situation. Extensive bluebells.

37 THE ROCK HOUSE

Llanbister, LD1 6TN. Jude Boutle & Sue Cox. *10m N of Llandrindod Wells. Off B4356 just above Llanbister village.* **Sat 3, Sun 4 June (12-5). Adm £5, chd free. Home-made teas inc gluten free & vegan options.**

About an acre of informal mature hillside garden, 1000ft up with views over the Radnorshire Hills. Wildlife ponds, bluebell meadow and a laburnum arch. It is a bit of a battle with nature this high up but we feel after around 30 years of making the garden it has finally settled into the space. The Cwtch - an epic summerhouse/garden retreat.

38 ROCK MILL

Abermule, Montgomery, SY15 6NN. Rufus & Cherry Fairweather, 01686 630664, fairweathers66@btinternet.com. *1m S of Abermule on B4368 towards Kerry. Best approached from Abermule village on B4368 as there is an angled entrance into field for parking.* **Sat 20, Sun 21 May (2-5). Adm £6, chd free. Home-made teas. Visits also by arrangement in May for groups of 10 to 30.**

A river runs through this magical 3 acre award-winning garden in a beautiful wooded valley. Colourful borders and shrubberies, specimen trees, terraces, woodland walks, bridges, extensive lawns, fishponds, orchard, herb and vegetable beds, beehives, roundhouse, remnants of industrial past (corn mill and railway line) offer much to explore. Child friendly activities (supervision required) inc croquet, animal treasure hunt, wilderness trails, cockleshell tunnel. Sensible shoes and a sense of fun/ adventure recommended. Much to explore.

39 TRANQUILITY HAVEN

7 Lords Land, Whitton, Knighton, LD7 1NJ. Val Brown, 01547 560070, valerie.brown1502@gmail.com. *approx 3m from Knighton & 5m Presteigne. From Knighton take B4355 after approx. 2m turn R on B4357 to Whitton. Car park on L by yellow NGS signs.* **Sun 4 June (2-5). Evening opening Wed 12 July, Wed 16 Aug, Wed 13 Sept (6-8). Sun 5 Nov (2-4). Evening opening Sun 10 Dec (6-8). Adm £4.50, chd free. Light refreshments available for visits by arrangement. Visits also by arrangement 1 Apr to 10 Dec for groups of up to 30.**

Amazing Japanese inspired garden with borrowed views to Offa's Dyke. Winding paths pass small pools and lead to Japanese bridges over natural stream with dippers and kingfishers. Sounds of water fill the air. Enjoy peace and tranquillity from one of the seats or the Japanese Tea House. Dense oriental planting with Cornus kousa satomi, acers, azaleas, unusual bamboos and wonderful cloud pruning.

40 TREBERFYDD HOUSE

Llangasty, Bwlch, Brecon, LD3 7PX. Sally Raikes, www.treberfydd.com. *6m E of Brecon. From Abergavenny on A40, turn R in Bwlch on B5460. Take 1st turning L towards Pennorth & cont 2m along lane. From Brecon, turn L off A40 in Llanhamlach towards Pennorth. Go through Pennorth, 1m on.* **Sun 2 July (1-5.30). Adm £6, chd free. Home-made teas.**

Grade I listed Victorian Gothic house with 10 acres of grounds designed by W A Nesfield. Magnificent Cedar of Lebanon, avenue of mature Beech, towering Atlantic Cedars, Victorian rockery, herbaceous border and manicured lawns ideal for a picnic. Wonderful views of the Black Mountains. Plants available from Commercial Nursery in grounds - Walled Garden Treberfydd. House tours £3.00. Easy wheelchair access to areas around the house, but herbaceous border only accessible via steps.

41 TREMYNFA

Carreghofa Lane, Llanymynech, SY22 6LA. Jon & Gillian Fynes, 01691 839471, gillianfynes@btinternet.com. *Edge of Llanymynech village. From N leave Oswestry on A483 to Welshpool. In Llanymynech turn R at Xrds (car wash on corner). Take 2nd R then follow yellow NGS signs. 300yds park signed field, limited disabled parking nr garden.* **Sat 17, Sun 18 June (1-5). Adm £5, chd free. Cream teas inc gluten free option. Visits also by arrangement 12 June to 23 June for groups of 10 to 35. Afternoon or evening groups welcome.**

South facing 1 acre garden developed over 15 yrs. Old railway cottage set in herbaceous and raised borders, patio with many pots of colourful and unusual plants. Garden slopes to productive fruit and vegetable area, ponds, spinney, unusual trees, wild areas and peat bog. Patio and seats to enjoy extensive views inc Llanymynech Rocks. Pet ducks on site, Montgomery canal close by. 100s of homegrown plants and home-made jams for sale.

42 TYN Y CWM

Beulah, Llanwrtyd Wells, LD5 4TS. Steve & Christine Carrow, 01591 620461, steve.carrow@ngs.org.uk. *10m W of Builth Wells. On A483 at Beulah take rd towards Abergwesyn for 2m. Drive drops down to L.* **Sun 18 June (2-5.30). Adm £6, chd free. Home-made teas. Visits also by arrangement May to Aug for groups of up to 20.**

Garden started 21 years ago, lower garden has spring/woodland area, raised beds mixed with vegetables, fruit trees, fruit and flowers. Perennial borders, summerhouse gravel paths through rose and clematis pergola. Grass and dahlia beds. Upper garden, partly sloped, inc bog and water gardens. Perennial beds with unusual slate steps. Beautiful views. Property bounded by small river. Children's quiz.

43 TYN-Y-GRAIG

Bwlch y Ffridd, Newtown, SY16 3JB. Simon & Georgina Newson. *4m N Caersws. B4568 from Newtown Turn R after Aberhafesp. At fork bear L past community centre. At xrds turn R. At fork bear L. After cattle grid take first R onto rough track.* **Sun 25 June (11-5). Adm £4.50, chd free. Home-made teas.**

A garden at 300m with views of surrounding hills, a variety of borders and a hectare of meadow managed for biodiversity. Garden planted to achieve different moods inc a herb area, bright, white and pastel borders, rose garden, farmyard borders, fernery, fruit trees and vegetable beds. Meadow with woodland and large pond. Open studio. Partial wheelchair access. Grass paths, some steps, most can be avoided. Disabled visitors may be dropped off at house.

44 VAYNOR PARK

Berriew, Welshpool, SY21 8QE. Mr & Mrs William Corbett-Winder. *5m S Welshpool. Leave Berriew going over bridge & straight up the hill on the Bettws rd. Entrance to Vaynor Park is on R ¼m from speed derestriction sign.* **Sun 23 July (1-5). Adm £6, chd free. Home-made teas.**

Beautiful C17 house (not open)with 5 acre garden. Spectacular long herbaceous borders at their peak with cardoons, crambe, geraniums, etc, box edged rose parterre, banks of hydrangeas; topiary yew birds, box buttresses and spires bring formality. Courtyard with lime green hydrangea paniculata and Annabelle. Woodland garden. Orangery. Spectacular views. Home to the Corbett-Winder family since 1720.

45 ◆ WELSH LAVENDER

Cefnperfedd Uchaf, Maesmynis, Builth Wells, LD2 3HU. Nancy Durham & Bill Newton-Smith, 01982 552467, farmers@welshlavender.com, www.welshlavender.com. *Approx 4½m S of Builth Wells & 13m N from Brecon Cathedral off B4520. The farm is 1⅓m from turn signed Farmers' Welsh Lavender.* **For NGS: Every Fri 21 July to 11 Aug (10-5). Adm £6, chd free. Pre-booking essential, please visit www.ngs.org.uk for information & booking. Light refreshments. For other opening times and information, please phone, email or visit garden website.**

Jeni Arnold's stylish wild planting of the steep bank above the ever popular wild swimming pond is a riot of colour best seen in June. Fields of blue lavender peak from mid July to mid August. Walk in the lavender fields, learn how the distillation process works, and visit the farm shop to try body creams and balms made with lavender oil distilled on the farm. Swim in the pond before enjoying coffee, tea and light refreshments. Partial wheelchair access. Large paved area adjacent to teas and shop area easy to negotiate.

46 WHITE HOPTON FARM

Wern Lane, Sarn, Newtown, SY16 4EN. Claire Austin, www.claireaustin-hardyplants.co.uk. *From Newtown A489, E towards Churchstoke for 7m & turn R in Sarn follow yellow NGS signs.* **Thur 8 June (10-4). Adm £7, chd free. Home-made teas.**

Horticulturist and author Claire Austin's private 1½ acre plant collector's garden designed along the lines of a cottage garden with hundreds of different perennials. Front garden mainly full of May blooming perennials, back garden, which is split into various areas, has a June and July garden, a small woodland walk, a Victorian fountain and mixed rose borders. Fabulous views over the Kerry Vale. Stunning Peony and Iris fields. Wide range of Claire Austin's Hardy Plants for sale at the nursery. Toilet facilities on site.

NPC

47 NEW WILLOWBROOK

Knighton Road, Presteigne, LD8 2ET. Fiona Collins & David Beech, 07867 385694, fm.collins@hotmail.co.uk. *Outskirts of Presteigne on B4355 towards Knighton, turning L within the 40 mile zone, just before the bridge.* **Sat 19, Sun 20 Aug (12-5). Adm £5, chd free. Home-made teas. Visits also by arrangement 1 Aug to 20 Sept for groups of up to 25.**

Begun in 2013, yew hedges create garden rooms and vistas: pergola cottage garden planted with climbing roses and wisteria; formal 'Italian' style pond garden; double west facing 'hot' borders. Rose border, rill, box parterre, sunken garden. Yew and box topiary. Kitchen garden with raised vegetable beds, fruit trees, soft fruit. Greenhouses, cold frames, polytunnel. Level grass & gravel paths.

48 NEW **57 & 58 YNYSWEN**

Penycae, Swansea, SA9 1YT. Rebecca Buck & Wendy Mannion, 07913 743457. *M4 exit 45, 15m on A4067 towards Brecon. L after Ynyswen rd sign. L at T junction, Blue house on R, no. 57. 13m from A40/A4067 junction, Sennybridge.* **Visits by arrangement June to Aug for groups of up to 10. Adm £5, chd free. Home-made teas.**

Two identically sized gardens with vastly different results, one a wildlife and sculpture garden with hardy exotics, fruit and perennials. The other recently transformed from plain grass and old sheds to smart family garden filled with plants and vegetables grown from seed by newly inspired owner.

49 **1 YSTRAD HOUSE**

1 Church Road, Knighton, LD7 1EB. John & Margaret Davis, 01547 528154, jamdavis@ystradhouse.plus.com. *At junction of Church Rd & Station Rd. Take the turning opp Knighton Hotel (A488 Clun) travel 225yds along Station Rd. Yellow House, red front door, at junction with Church Rd. Enter by side gate in Station Rd.* **Visits by arrangement June to Aug. Adm £5, chd free. Home-made teas inc gluten free cakes.**

A town garden behind a Regency Villa of earlier origins. A narrow entrance door opens revealing unexpected calm and timelessness. A small walled garden with box hedging and greenhouse lead to broad lawns, wide borders with soft colour schemes and mature trees. More intimate features: pots, urns and pools add interest and surprise. The formal areas merge with wooded glades leading to a riverside walk. Croquet on request. Lawns and gravelled paths mostly flat, except access to riverside walk.

9,500 patients and their families are being supported across four Horatio's gardens thanks to our annual donations

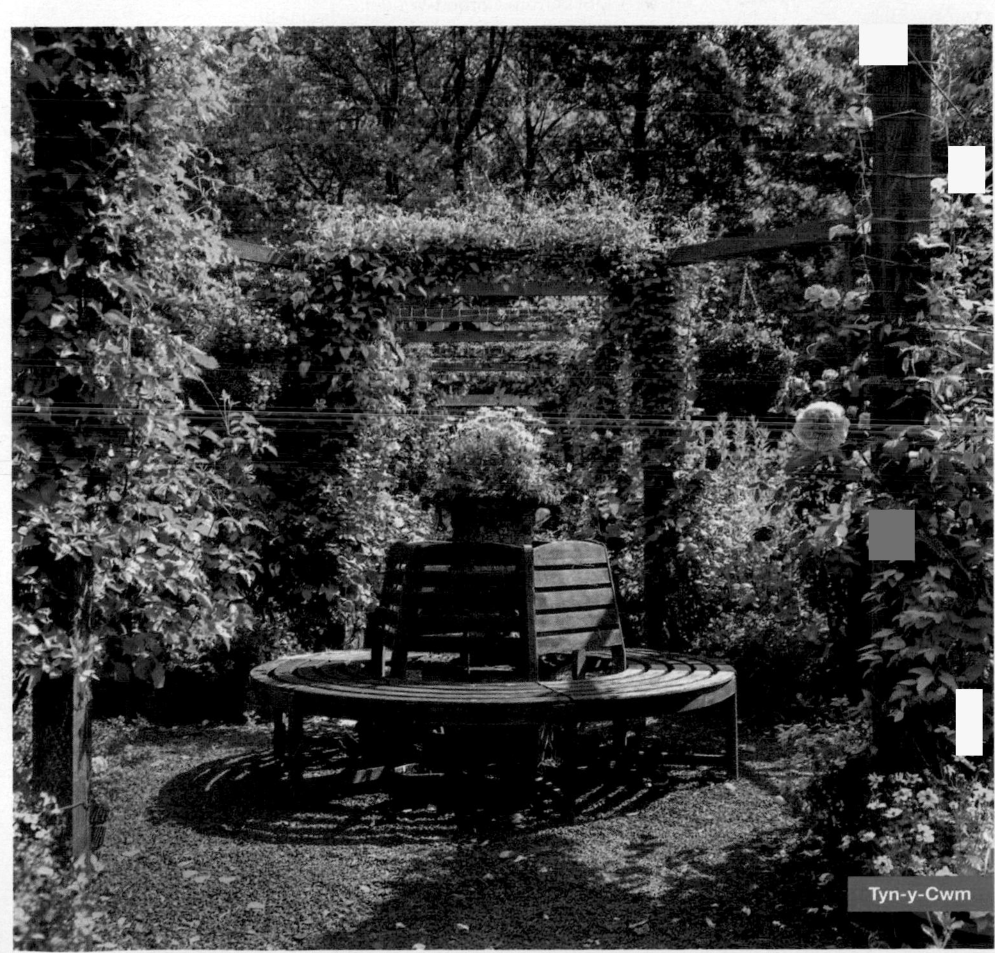

Tyn-y-Cwm

Early Openings 2024

Plan your garden visiting well ahead – put these dates in your diary!

Gardens across the country open early – before the next year's guide is published – with glorious displays of colour including hellebores, aconites, snowdrops and carpets of spring bulbs.

Cheshire & Wirral

Sun 25 February
(12.30–4)
Bucklow Farm

Devon

Fri 2, Fri 9, Sat 17 February
(2–4.30)
Higher Cherubeer

Essex

Sun 28 January
(10–4)
Green Island

Glamorgan

Sun 11 February
(11–5)
Slade

Gloucestershire

Sun 28 January,
Sun 11 February
(11–4)
Home Farm

Sun 11, Sun 18 February
(11–4)
Trench Hill

Sat 17, Sun 18 February
(11–3)
Cotswold Farm

Hampshire

Sun 11, Mon 12, Sun 18, Mon 19 February
(2–5)
Little Court

Herefordshire

Sun 25 February
(11–4)
The Picton Garden

Kent

By arrangement in February
The Old Rectory, Fawkham

Sat 3, Sun 4 February
(11–3)
Knowle Hill Farm

Sat 10, Sun 18 February
(12–4)
Copton Ash

Northamptonshire

Sun 25 February
(11–3)
67-69 High Street

Somerset, Bristol & South Gloucestershire

By arrangement in February
1 Birch Drive

Staffordshire, Birmingham & West Midlands

Sun 18 February
(12–4)
5 East View Cottages

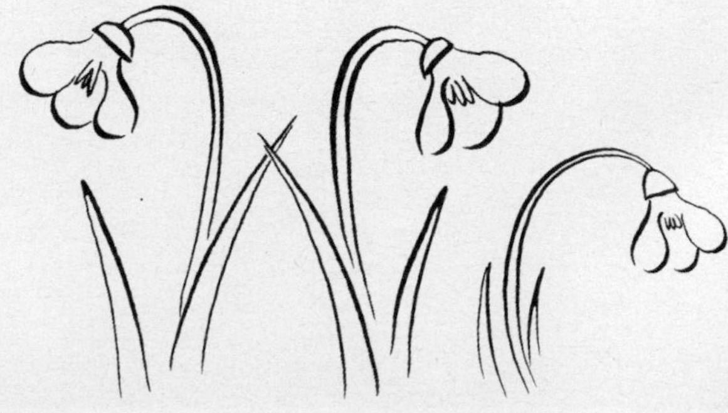

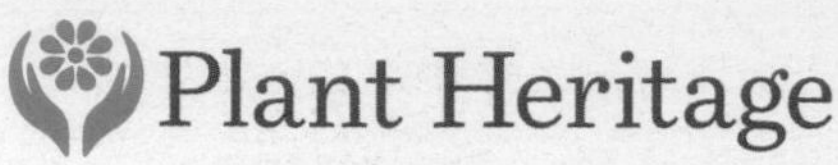

Seventy gardens that open for the National Garden Scheme are holders of a Plant Heritage National Plant Collection although this may not always be noted in the garden description. These gardens carry the NPC symbol.

www.plantheritage.org.uk, info@plantheritage.org.uk, 01483 447540

Abies spp.
The Yorkshire Arboretum, Yorkshire

Acer (excl. palmatum cvs.)
Blagdon, North East

Alnus
Stone Lane Gardens, Devon

Araliaceae (excl. Hedera)
Meon Orchard, Hampshire

Asplenium scolopendrium cvs.
Delfryn, Ceredigion

Aster & related genera (autumn flowering)
The Picton Garden, Herefordshire

Astilbe
Holehird Gardens, Cumbria
Marwood Hill Garden, Devon

Astrantia
Norwell Nurseries, Norwell Gardens, Nottinghamshire

Betula
Ness Botanic Gardens, Cheshire
Stone Lane Gardens, Devon

Camellia (autumn and winter flowering)
Green Island, Essex

Camellia japonica
Antony Woodland Garden & Woodland Walk, Cornwall

Camellias & Rhododendrons introduced to Heligan pre-1920
The Lost Gardens of Heligan, Cornwall

Carpinus
Sir Harold Hillier Gardens, Hampshire

Carpinus betulus cvs.
West Lodge Park, London

Catalpa
West Lodge Park, London

Ceanothus
Eccleston Square, London

Cercidiphyllum
Sir Harold Hillier Gardens, Hampshire
Hodnet Hall Gardens, Shropshire

Chrysanthemum (Hardy)
Norwell Nurseries, Norwell Gardens, Nottinghamshire
Hill Close Gardens, Warwickshire

Clematis viticella
Longstock Park Water Garden, Hampshire

Clematis viticella cvs.
Roseland House, Cornwall

Codonopsis & related genera
Woodlands, Lincolnshire

Colchicum
East Ruston Old Vicarage, Norfolk

Convallaria
Kingston Lacy, Dorset

Cornus
Sir Harold Hillier Gardens, Hampshire

Cornus (excl. C. florida cvs.)
Newby Hall & Gardens, Yorkshire

Corokia
33 Wood Vale, London

Corylus
Sir Harold Hillier Gardens, Hampshire

Cotoneaster
Sir Harold Hillier Gardens, Hampshire

Cyclamen (excl. persicum cvs.)
Higher Cherubeer, Devon

Daboecia
Holehird Gardens, Cumbria

Daffodil Dispersed Noel Burr Cultivars
Mill Hall Farm, Sussex

Dierama spp.
Yew Tree Cottage, Staffordshire

Dionaea
Beaufort, Shropshire

Dispersed Collection of Noel Burr Narcissi
Bourne Botanicals, Sussex

Erica & Calluna - Sussex heather cvs
Nymans, Sussex

Erythronium
Greencombe Gardens, Somerset

Eucalyptus
Meon Orchard, Hampshire

Eucalyptus spp.
The World Garden at Lullingstone Castle, Kent

Eucryphia
Whitstone Farm, Devon

Euonymus (deciduous)
The Place for Plants, East Bergholt Place Garden, Suffolk

Eupatorium
Saith Ffynnon Farm, North East Wales

Euphorbia (hardy)
Firvale Perennial Garden, Yorkshire

Fraxinus
The Lovell Quinta Arboretum, Cheshire

Galanthus
127 Stoke Road, Bedfordshire

Gaultheria (incl. Pernettya)
Greencombe Gardens, Somerset

Geranium phaeum cvs. & primary hybrids
The New Barn, Leicestershire

Geum cvs.
1 Brickwall Cottages, Kent

Hamamelis
Sir Harold Hillier Gardens, Hampshire

Hamamelis cvs.
Green Island, Essex

Heliotropium
Hampton Court Palace, London

Helleborus (Harvington hybrids)
Riverside Gardens at Webbs, Worcestershire

Hilliers (Plants raised by)
Sir Harold Hillier Gardens, Hampshire

Hoheria
Abbotsbury Gardens, Dorset

Hosta (European and Asiatic)
Hanging Hosta Garden, Hampshire

Hypericum sect. Androsaemum & Ascyreia spp.
Holme for Gardens, Dorset

Hypericum spp. & cvs.
Sir Harold Hillier Gardens, Hampshire

Iris ensata
Marwood Hill Garden, Devon

Juglans
Upton Wold, Gloucestershire

Lapageria rosea (& named cvs.)
Roseland House, Cornwall

Lewisia
'John's Garden' at Ashwood Nurseries, Staffordshire

Ligustrum
Sir Harold Hillier Gardens, Hampshire

Lithocarpus
Sir Harold Hillier Gardens, Hampshire

Malus (ornamental)
Barnards Farm, Essex

Meconopsis (large perennial spp. & hybrids)
Holehird Gardens, Cumbria

Metasequoia
Sir Harold Hillier Gardens, Hampshire

Narcissus (Engleheart cvs) dispersed
2 Holmwood Cottages, Suffolk

Paeonia (hybrid herbaceous)
White Hopton Farm, Powys

Pennisetum spp. & cvs. (hardy)
Knoll Gardens, Dorset

Penstemon
Kingston Maurward Gardens and Animal Park, Dorset

Persicaria virginiana
Sweetbriar, Kent

Photinia
Sir Harold Hillier Gardens, Hampshire

Picea spp.
The Yorkshire Arboretum, Yorkshire

Pinus
Sir Harold Hillier Gardens, Hampshire

Pinus spp.
The Lovell Quinta Arboretum, Cheshire

Plectranthus
Sweetbriar, Kent

Podocarpus & related Podocarpaceae (incl. Graham Hutchins introductions)
Meon Orchard, Hampshire

Polystichum
Holehird Gardens, Cumbria
Greencombe Gardens, Somerset

Primula auricula (Border)
12 Meres Road, Staffordshire

Pseudopanax
Sweetbriar, Kent

Pterocarya
Upton Wold, Gloucestershire

Quercus
Chevithorne Barton, Devon
Sir Harold Hillier Gardens, Hampshire

Quercus (Sir Bernard Lovell collection)
The Lovell Quinta Arboretum, Cheshire

Rhododendron (Ghent Azaleas)
Sheffield Park and Garden, Sussex

Rhododendron (Kurume Azalea Wilson 50)
Trewidden Garden, Cornwall

Rhus
The Place for Plants, East Bergholt Place Garden, Suffolk

Rosa (Hybrid Musk intro by Pemberton & Bentall 1912-1939)
Dutton Hall, Lancashire

Rosa (rambling)
Moor Wood, Gloucestershire

Salix
West Wales Willows, Carmarthernshire & Pembrokeshire

Sambucus
Cotswold Garden Flowers, Worcestershire

Sarracenia
Beaufort, Shropshire

Saxifraga sect. Ligulatae: spp. & cvs.
Waterperry Gardens, Oxfordshire

Saxifraga sect. Porphyrion subsect. Porophyllum
Waterperry Gardens, Oxfordshire

Sorbus (British endemic spp.)
Blagdon, North East

Stewartia (Asian spp.)
High Beeches Woodland and Water Garden, Sussex

Streptocarpus
Dibleys Nurseries, North East Wales

Symphyotrichum (Aster) novae-angliae
Brockamin, Worcestershire

Taxodium spp. & cvs.
West Lodge Park, London

Toxicodendron
The Place for Plants, East Bergholt Place Garden, Suffolk

Tulbaghia
Marwood Hill Garden, Devon

Tulipa (historic tulips)
Blackland House, Wiltshire

Vaccinium
Greencombe Gardens, Somerset

Yucca
Spring View, Burwell Village Gardens, Cambridgeshire
Renishaw Hall & Gardens, Derbyshire

Society of Garden Designers

The D symbol at the end of a garden description indicates that the garden has been designed by a Fellow, Member, Pre-Registered Member, Student or Friend of the Society of Garden Designers.

Fellow of the Society of Garden Designers (FSGD) is awarded to Members for exceptional contributions to the Society or to the profession

Rosemary Alexander FSGD
Sarah Massey FSGD
Dan Pearson FSGD
Nigel Philips FSGD
Anne-Marie Powell FSGD
David Stevens FSGD
Tom Stuart Smith FSGD
Joe Swift FSGD
Robin Templar Williams FSGD
Julie Toll FSGD
Cleve West FSGD

Member of the Society of Garden Designers (MSGD) is awarded after passing adjudication

Susan Ashton MSGD
Fi Boyle MSGD
Barry Chambers MSGD (retired)
Peter Eustance MSGD
Mark Fenton MSGD
Jill Fenwick MSGD
Phil Hirst MSGD
Nic Howard MSGD
Arabella Lennox-Boyd MSGD
Robert Myers MSGD
Joe Perkins MSGD
Emma Plunket MSGD
Debbie Roberts MSGD
Libby Russell MSGD
Charles Rutherfoord MSGD
James Scott MSGD
Ian Smith MSGD
Sue Townsend MSGD
Matthew Wilson MSGD

Pre-Registered Member is a member working towards gaining Registered Membership

Jackie Cahoon
Kristina Clode
Claire Merriman
Matthew Stephens
Helen Thomas
Rebecca Winship

Student

Will Jennings

Friend

Ian Kitson
Caz Renshaw

Acknowledgements

Each year the National Garden Scheme receives fantastic support from the community of garden photographers who donate and make available images of gardens: sincere thanks to them all. Our thanks also to our wonderful garden owners who kindly submit images of their gardens.

Unless otherwise stated, photographs are kindly reproduced by permission of the garden owner.

The 2023 Production Team: Vicky Flynn, Louise Grainger, Vince Hagan, Sarah Hosker, Kay Palmer, Christina Plowman, Helena Pretorius, George Plumptre, Gail Sherling-Brown, Catherine Swan, Georgina Waters, Anna Wili.

CONSTABLE
First published in Great Britain in 2023 by Constable

A CIP catalogue record for this book is available from the British Library.

ISBN: 978-1-4087-1932-9

Designed by Level Partnership
Maps by Mary Spence © Global Mapping and XYZ Maps
Typeset in Helvetica Neue
Printed and bound in Italy by Rotolito Lombarda S.p.A.

Constable
An imprint of Little, Brown Book Group
Carmelite House, 50 Victoria Embankment,
London EC4Y 0DZ

An Hachette UK Company
www.hachette.co.uk www.littlebrown.co.uk

If you require this information in alternative formats, please telephone 01483 211535 or email hello@ngs.org.uk

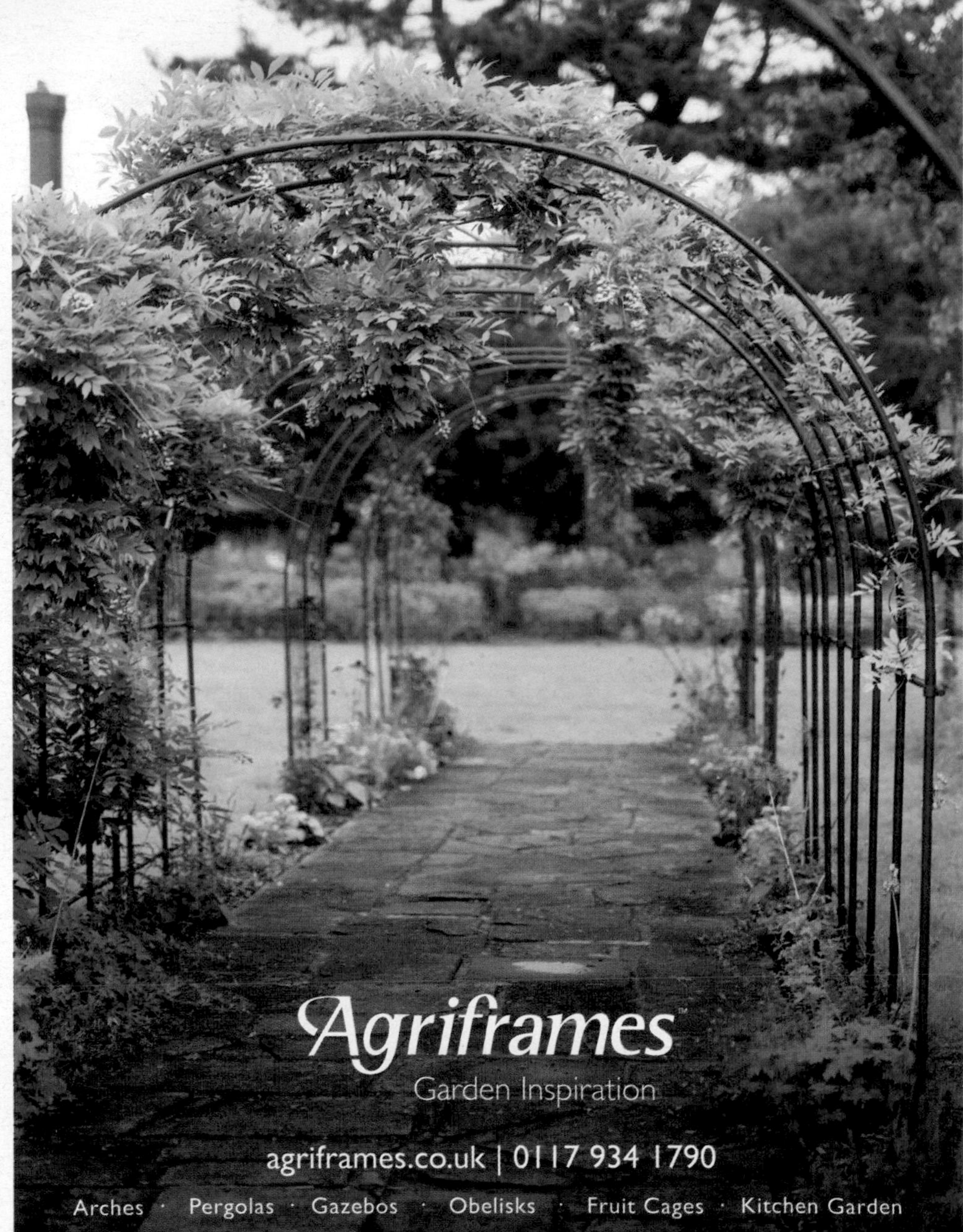
Agriframes
Garden Inspiration
agriframes.co.uk | 0117 934 1790
Arches · Pergolas · Gazebos · Obelisks · Fruit Cages · Kitchen Garden